Lecture Notes in Artificial Intelligence 2903

Edited by J. G. Carbonell and J. Siekmann

Subseries of Lecture Notes in Computer Science

Springer
Berlin
Heidelberg
New York
Hong Kong
London
Milan
Paris
Tokyo

Tamás D. Gedeon   Lance Chun Che Fung (Eds.)

AI 2003: Advances in Artificial Intelligence

16th Australian Conference on AI
Perth, Australia, December 3-5, 2003
Proceedings

Springer

Series Editors

Jaime G. Carbonell, Carnegie Mellon University, Pittsburgh, PA, USA
Jörg Siekmann, University of Saarland, Saarbrücken, Germany

Volume Editors

Tamás D. Gedeon
Lance Chun Che Fung
Murdoch University, School of Information Technology
South Street, Murdoch, Australia
E-mail: l.fung@murdoch.edu.au

Tamás D. Gedeon now at:
Australian National University, Department of Computer Science
Canberra, Australia
E-mail: tom.gedeon@anu.edu.au

Cataloging-in-Publication Data applied for

A catalog record for this book is available from the Library of Congress.

Bibliographic information published by Die Deutsche Bibliothek
Die Deutsche Bibliothek lists this publication in the Deutsche Nationalbibliografie; detailed bibliographic data is available in the Internet at <http://dnb.ddb.de>.

CR Subject Classification (1998): I.2, F.1, F.4.1, H.3, H.2.8, H.4

ISSN 0302-9743
ISBN 3-540-20646-9 Springer-Verlag Berlin Heidelberg New York

Springer-Verlag is a part of Springer Science+Business Media

springeronline.com

Typesetting: Camera-ready by author, data conversion by DA-TeX Gerd Blumenstein
Printed on acid-free paper SPIN: 10974582 06/3142 5 4 3 2 1 0

Preface

The 16th Australian Conference on Artificial Intelligence (AI 2003) was held in Perth, Western Australia for the third time (1990 and 1997 were the previous dates). The remoteness of Perth resulted in a different approach to organizing the conference, which all four of the publicly funded universities in Western Australia combined resources to organize and run. It was held on the campus of the University of Western Australia, which has a somewhat traditional campus feel with on-campus accommodation and an environment conducive to academic discussion and debate.

The conference was organized under the auspices of the National Committee on Artificial Intelligence and Expert Systems of the Australian Computer Society (ACS), the main body representing information technology in Australia. The goal of the conference series is to promote the exchange of ideas and collaboration across the different subdisciplines of artificial intelligence by serving as a forum for the somewhat spatially separated researchers and practitioners in Australia and the surrounding region.

Given the current climate in the world, it was marvelous to see so many papers submitted, especially as there are other conferences covering similar ground in the region.

The conference could not have been a success without the contributions of the sponsors. First, the National Committee on Artificial Intelligence and Expert Systems, which provided substantial seed funding. Second, the four universities that each contributed to the funds. And, third, the IEEE WA section, which contributed the student prize.

We would like to take this opportunity to thank all the members of the organizing committee, especially for travelling across Perth to meet each other regularly. We would also like to thank all the reviewers who unselfishly reviewed many papers. This is a time-consuming process that is a vital part of the research community's activities, and a conference cannot function without this peer-review process. We would also like to thank all the student volunteers and others we haven't mentioned for their contributions.

December 2003 Geoff West and Svetha Venkatesh

Introduction

AI 2003 was the 16th in the series of annual conferences on artificial intelligence held in Australia. This conference is the major forum for the presentation of artificial intelligence research in Australia, and has traditionally attracted strong international participation, a trend continued this year.

High quality papers make a conference a success. This volume is based on the proceedings of AI 2003. Full-length versions of all submitted papers were reviewed by an international program committee, with each paper receiving at least two independent reviews. From the 179 submissions received, a total of 68 papers were accepted for oral presentation, and a further 19 for poster presentation. This is an acceptance rate of 38% for oral presentation and 50% overall. The full papers of all presentations are included in this volume.

In addition to the refereed papers, the conference featured a program of tutorials, and plenary talks by four invited speakers covering a number of the subdisciplines of artificial intelligence. The speakers were: Kotagiri Ramamohanarao (University of Melbourne), Mohan S. Kankanhalli (National University of Singapore), Claude Sammut (University of New South Wales), and B. John Oommen (Carleton University, Canada).

We would like to thank the members of the Program Committee and the panel of reviewers who produced some 400 paper reviews under time constraints. Without your help, this conference could not have succeeded. In addition, we are also grateful to Springer-Verlag for its support and collaborations.

December 2003 Tom Gedeon and Lance Fung

Organization

General Co-chairs

Svetha Venkatesh	Curtin University of Technology
Geoff A.W. West	Curtin University of Technology

Program Co-chairs

Tamás D. Gedeon	Murdoch University
Lance C.C. Fung	Murdoch University

Organizing Committee

Svetha Venkatesh	Curtin University of Technology
Geoff A.W. West	Curtin University of Technology
Raj P. Gopalan	Curtin University of Technology
Tamás D. Gedeon	Murdoch University
Lance C.C. Fung	Murdoch University
Graham Mann	Murdoch University
Cara MacNish	University of Western Australia
Luigi Barone	University of Western Australia
Phil Hingston	Edith Cowan University

Program Committee

Abdul Sattar
Achim Hoffman
Adrian Pearce
Alan Blair
Bob Dale
Bob McKay
Cara MacNish
Chengqi Zhang
Frederic Maire
Geoff Webb
Geoff West
Graham Mann
Graham Williams
Hussein Abbass
Ingrid Zukerman
Irwin King
John Debenham
John Slaney
Jong-Hwan Kim
Kevin Wong Kok Wai
Kit Po Wong
Leila Alem
Leon Sterling
Liya Ding
Luigi Barone
Mark Reynolds
Maurice Pagnucco
Mehmet Orgun
Michael Brooks
Norman Foo
Ong Sing Goh
Paul Darwen
Phil Hingston
Raj Gopalan

Ruhul Sarker
S.V.N. Vishwanathan
Shamim Khan
Shlomo Geva
Svetha Venkatesh
Sylvie Thiebaux
Toby Walsh
Van Le
Yue Xu
Yuefeng Li
Yusuf Pisan

Other Reviewers

Alan Aizhong Lin
Andrew Czarn
Anthony Dick
Anthony Senyard
Baikunth Nath
Barry O'Sullivan
C.P. Tsang
Chris Beck
Daniel Pooley
David Albrecht
Diarmuid Pigott
Domonkos Tikk
Doug Aberdeen
Guido Governatori
Hongxing He
Jie Chen
Jing Ye
Jitian Xiao
John Bastian
Lifang Gu
Mark Hennessy
Mihai Lazaraescu
Ong Yew Soon
Owen Lamont
Pushkar Piggott
Raj Gururajan
Rex Kwok
Rhys Hill
Roberto Togneri
Rodney Topor
Seow Kiam Tian
Sieteng Soh
Susannah Soon
Thomas Juan
Thomas Meyer
Tim French
Wanquan Liu
Wei Lui
Willie Guo
Xudong Luo
Yan Zheng Wei
Yannick Pencolé
Zahia Guessoum

Table of Contents

Knowledge Discovery and Data Mining I

Knowledge Discovery and Data Mining II

Expert Systems

Neural Networks Applications

Belief Revision and Theorem Proving

Reasoning and Logic

Machine Learning I

AI Applications

Neural Networks

Intelligent Agents

Computer Vision

AI & Medical Applications

Machine Learning II

Machine Learning and Language

Artificial Intelligence I

AI & Business

Soft Computing

Language Understanding

Theory

Artificial Intelligence II

Discovery of Emerging Patterns and Their Use in Classification

Kotagiri Ramamohanarao and James Bailey

Department of Computer Science and Software Engineering
University of Melbourne, Australia
{rao,jbailey}@cs.mu.oz.au

Abstract. Emerging patterns are sets of items whose frequency changes significantly from one dataset to another. They are useful as a means of discovering distinctions inherently present amongst a collection datasets and have been shown to be a powerful method for constructing accurate classifiers. In this paper, we present different varieties of emerging patterns, discuss efficient techniques for their discovery and explain how they can be used in classification.

1 Introduction

Discovery of powerful distinguishing features between datasets is an important objective in data mining. An important class of patterns that can represent strong contrasts is known as *emerging patterns*. An emerging pattern is a set of items whose frequency changes significantly from one dataset to another. Emerging patterns have been successfully used in constructing highly accurate classifiers [19, 10, 20]. In particular, for predicting the likelihood of diseases such as leukaemia [21] and discovering patterns in gene expression data [22].

As an example, consider the following database defined on six attributes: (with all possible attribute values given in parentheses) Attr1 (a, b, c), Attr2 (d, e, f), Attr3 (g, h, i), Attr4 (j, k, l), Attr5 (m, n, o) and Attr6 (p, q, r).

The minimal emerging patterns in this dataset are sets of items which appear in one class and not in the other (by minimal, we mean that no proper subset of the items in the pattern should also be an emerging pattern). Inspecting the table, we see that patterns appearing in Class 1 include $\{a, i\}$, $\{o, p\}$, and $\{b, g\}$. Patterns appearing in Class 2 include $\{f\}$ and $\{k\}$. These patterns represent very strong changes (zero frequency in one dataset and non-zero frequency in

Table 1. Sample Database

Class 1 Instances	Class 2 Instances
$\{a, d, i, l, o, p\}$	$\{c, d, i, l, o, r\}$
$\{b, d, g, l, o, r\}$	$\{a, f, g, k, o, q\}$
$\{c, d, h, l, o, r\}$	$\{b, d, h, j, m, p\}$

T.D. Gedeon and L.C.C. Fung (Eds.): AI 2003, LNAI 2903, pp. 1–12, 2003.

the other) and they are given a special name to indicate this property - *jumping emerging patterns.* Other useful, but less restrictive emerging patterns can also be defined. e.g. *Constrained emerging patterns* are minimal sets of items which occur $\geq \alpha$ times in one class and $\leq \beta$ times in the other. The set of items $\{d, o\}$ is such a pattern appearing in Class 1, for $\alpha = 3$ and $\beta = 1$.

In this paper, we discuss the principal types of emerging patterns, and focus on two principal issues

- *Methods for Efficient Pattern Discovery:* This is a challenge, especially for high volume, high dimensionality datasets, since in the worst case, the number of patterns present in the data can be exponential.
- *Classification Methods*: Constructing emerging pattern classifiers requires a number of different decisions to be made. What type of patterns should be mined, which patterns should be used in classification and the how are patterns weighted in the scoring function.

An outline of the rest of the paper is as follows. In section 2, we give some necessary background and terminology. Section 3 focuses on jumping emerging patterns and discusses their relationship to hypergraph transversals. Section 4 looks at more general kinds of emerging patterns. Section 5 examines algorithms for efficient pattern mining and section 6 discusses how to use emerging patterns in classification. In Section 7, we discuss related work and in section 8 provide a summary and outline directions for future research.

2 Terminology

To help formally define emerging patterns, some preliminary definitions are required. Assume a dataset D, defined upon some set of attributes $\{A_1, A_2, ..., A_n\}$, with each attribute A_i in turn defined by some number of values, its domain $domain(A_i)$. By aggregating all such domain values across all attributes, we obtain all the items $I = \{domain(A_1) \cup domain(A_2) \cup ... \cup domain(A_n)\}$, in our system. The dataset consists of a number of transactions, where each transaction, T, is some collection of the items in the system e.g. $T = \{v_1, v_2, ..., v_n\}$, $v_i \epsilon\ domain(A_i)$. An *itemset* is some subset of a transaction. The support of an itemset S in D is the number of transactions S occurs in - i.e. $(count_D(S))$. The relative support of S in D is the support of S divided by the number of transactions in D. $(count_D(S)/|D|)$. Assume two data sets D_p (the positive dataset) and D_n (the negative dataset). The growth rate of an itemset i in favour of D_p is defined as $\rho = \frac{supportD_p(i)}{supportD_n(i)}$. An itemset M satisfying a constraint $C = (support(M) \geq \alpha)$ is maximal if no proper superset of M satisfies C. An itemset N satisfying a constraint $C' = (support(N) \leq \beta)$ is minimal if no proper subset of N satisfies C'. An emerging pattern e is minimal if the itemset e is minimal.

An Emerging Pattern is an itemset whose support in one set of data differs from its support in another. Thus a ρ Emerging Pattern, favouring a class of data D_p, is one in which the growth rate of an itemset (in favour of D_p) is $\geq \rho$.

This growth rate may be finite or infinite. In the case that the growth rate ρ is infinite, we call the pattern a *jumping emerging pattern* (JEP) (i.e. it is present in one and absent in the another). It is this kind of pattern we first focus on.

3 Jumping Emerging Patterns and Hypergraphs

Jumping emerging patterns (JEPs) are particularly interesting due to their ability to represent strong contrasts. Finding all JEPs in D_p, with respect to D_n, requires discovery of all minimal itemsets which occur in at least one transaction of D_p and not in any transaction of D_n. It turns out that JEPs are closely connected to a problem in the theory of hypergraphs, a topic which itself has many applications in discrete mathematics and computer science.

One of the central questions in hypergraph theory is the problem of how to compute the minimal transversals of a given hypergraph. Example application areas related to this question range from minimal diagnosis and propositional circumscription [27], to learning boolean formulae [18], boolean switching theory [11] and discovering functional dependencies in databases [24].

The hypergraph minimal transversal problem is particularly significant from a data mining perspective. Indeed, the algorithmic complexity of mining maximal frequent itemsets and minimal infrequent itemsets is closely linked to the complexity of computing minimal hypergraph transversals [6, 18]. Although there has been considerable research in the data mining community with regard to efficient mining of maximal frequent itemsets (e.g. [4, 7, 17]), the companion problem of data mining of minimal infrequent sets is less well studied.

Infrequent itemsets are itemsets whose support is less than α in the dataset being mined (where α is some threshold ≥ 1). Minimal infrequent itemsets are infrequent itemsets such that no proper subset is infrequent. They are closely connected to JEPs. Consider the following example in table 2:

Suppose we are interested in finding jumping emerging patterns between Class A and Class B. Then transactions in Class A contain the following JEPs:

Transaction 1 $\{b\}$
Transaction 2 $\{ce, cg, cj\}$
Transaction 3 $\{cj, fj, hj\}$
Transaction 4 $\{af, ah, fj, hj\}$

Table 2. Sample Dataset

Trans_id_A	Class A	Trans_id_B	Class B
1	b,d,g,j	1	a,d,g,j
2	c,e,g,j	2	a,d,g,i
3	c,f,h,j	3	c,f,h,i
4	a,f,h,j	4	a,e,g,j

The relationship to hypergraphs is natural. A hypergraph is defined by a set of vertices $V = \{v_1, v_2, ..., v_n\}$ and a set of edges E, where each edge is some subset of V. A *transversal* of a hypergraph is any set of vertices that contains at least one element of every edge. A *minimal transversal* is a transversal such that no proper subset is also a transversal

Each transaction t in Class A can be thought of as inducing a hypergraph with respect to all the transactions in Class B. The vertex set corresponds to the elements of t. Hypergraph edges are individually defined by subtracting a transaction in the negative dataset from the vertex set. So, for transaction $3 = \{c, f, h, j\}$ in class A, we have a hypergraph with vertex set $\{c, f, h, j\}$ and edges:

Hyperedge 1 $\{c, f, h\}$
Hyperedge 2 $\{c, f, h, j\}$
Hyperedge 3 $\{j\}$
Hypedge 4 $\{c, f, h\}$

The three minimal transversals of this hypergraph are $\{cj, fj, hj\}$, which correspond precisely to the JEPs listed above for transaction 3 in class A. They also correspond to minimal infrequent itemsets within class B, using threshold of $\alpha = 1$. Mining of JEPs in class A can thus be achieved by computing the minimal transversals for each possible choice of t in A.

There are several algorithms which can be used for computing minimal transversals. With respect to the data mining context, most of these techniques do not perform efficiently in practice. In particular, there is a performance gap on high dimensional datasets, that have a huge number of relatively short patterns. In section 5, we discuss some techniques to address this issue.

4 The Emerging Pattern Landscape

An emerging pattern (EP) has been defined as an itemset that has some specified growth rate ρ.

If we further specify thresholds α and β, we can also define another type of pattern, called a *constrained emerging pattern* (CEP). This is a minimal itemset i satisfying the following two conditions: i) $supportD_p(i) \geq \alpha$ instances, ii) $supportD_n(i) \leq \beta$ instances. Clearly, by setting $\beta = 0$, CEPs reduce to JEPs. By having a no-zero value of β, however, greater robustness to noise can be achieved, with (hopefully) little sacrifice in discriminating power

An advantage of JEPs was that they represent very sharp contrasts between D_p and D_n. Their disadvantage is that low quality datasets may contain an abundance of low support JEPs and few high support JEPs. This is because the requirement that a JEP never occur within D_n is often too strict. Strong inherent features of D_p sometimes do appear within D_n, due to the presence of noise or recording errors. Such features do not qualify as JEPs and hence would

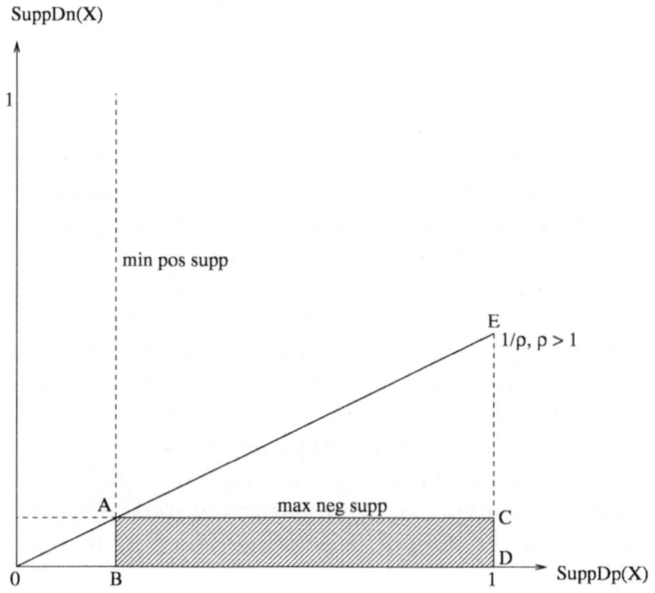

Fig. 1. Space of Emerging Patterns

not be found in JEP discovery, a potentially serious problem when using JEPs as the basis of a classifier.

Unlike JEPs, EPs may occur within D_n, provided the growth rate threshold ρ is satisfied. Hence, they are potentially more stable and resistant to noise. The disadvantage, though, is that the contrasts may no longer be as 'sharp'. e.g. An EP with $\rho = 5$ could have $supportD_p = 100$ and $supportD_n = 20$, values which arguably do not discriminate as sharply between D_p and D_n, as a JEP of $supportD_p = 50$. Indeed, in practice, our experience is that JEP-based classifiers are superior to EP based classifiers for exactly this reason.

Figure 1 illustrates the support plane for EPs, JEPs and CEPs. JEPs occupy the x-axis from O to D. EPs occupy the trapezoid ABDE, while CEPs occupy the shaded rectangle ABDC. CEPs can thus be viewed as occupying an intermediate space between the entire set of EPs and those JEPs that lie on the BD axis

Other varieties of emerging patterns can also be defined, by incorporation of further constraints. One example is "interesting emerging patterns" [14]. These are similar to constrained emerging patterns, but additionally need to satisfy an interestingness constraint, specified using a chi-square statistical measure.

5 Efficient Pattern Mining

We now briefly describe techniques for mining i) JEPs, ii) CEPs and iii) General EPs. In general this is a very challenging problem, since if the dimensionality of the datasets is n, then the search space is 2^n. For large n, it is prohibitive to examine this exhaustively and more sophisticated methods are required.

As discussed in section 3, mining JEPs with respect to a single transaction from D_p, corresponds to the problem of enumerating the minimal transversals of a hypergraph. We have developed a partitioning algorithm [3], which performs this task faster than any other algorithms we are aware of. The idea is to recursively partition the input hypergraph into a set of smaller, manageable hypergraphs, which can then have their minimal transverals enumerated by an optimisation of the well-known method of Berge [5].

Of course JEPs must be mined from every transaction in D_p, not just a single one. One method is to just iterate through every transaction in D_p and solve the corresponding hypergraph problem to obtain the complete result. It is possible to improve on this, however, by use of a tree structure to store the transactions in D_p and D_n [1]. This allows transactions in both D_p and D_n to overlap, thus reducing the number of separate hypergraph problems which need to be solved.

CEP mining can be accomplished by an extension of JEP mining. Step i) is to represent both the positive and the negative datasets being mined in terms of their border descriptions. Step ii) is to mine the CEPs by operating on the relevant borders.

Step 1 constructs two borders. One representing the positive dataset, such that only those (maximal) itemsets with support $\geq \alpha$ are present and the other border representing the negative dataset with maximal itemsets having support $\geq \beta$. Finding these borders can be achieved by utilising a maximal frequent itemset generator (e.g. any of [4, 7, 17]). Once the borders are computed, the method for mining JEPs can then be applied to gain the desired patterns in step ii). i.e. Finding all minimal itemsets which occur in the positive border and are absent from the negative border. Observe that each pattern output from this process satisfies the original user specified support constraints.

Efficient mining of general EPs (defined using just a nonzero growth rate ρ) appears difficult. Work in [28] employed a set-enumeration tree [28] and optimisations similar to the MaxMiner algorithm [4]. Another possibility is to mine sets of CEPs for different values of α and β and then conduct a post-pruning step to check whether the growth rate ρ is satisfied. This is an incomplete method, since some EPs may not be discovered. Another incomplete EP mining method is presented in [12], where a set enumeration tree is built and patterns are only mined to a certain depth in the tree.

5.1 Incremental Maintenance of JEPs

Efficient algorithms for mining JEPs should aim to avoid having to completely re-do computation, if small changes are made to either the positive dataset D_p or the negative dataset D_n. Assuming that the changes to D_p and D_n are small compared to the volume of the old data, it is likely that many of the JEPs will remain valid. Incremental maintenance techniques aim to avoid expensive re-computation by efficiently making use of the previous set of EPs. Work in [20] gives efficient incremental maintenance rules for i) Inserting new instances into D_p, ii) Inserting new instances into D_n, iii) Deleting instances from D_p and iv) Deleting instances from D_n. Cases where an item is deleted or new item is inserted are also examined.

6 Classification with Emerging Patterns

We now discuss how a set of emerging patterns may be used in classification. We do not present any experimental results, but work in [2, 12, 14, 13, 19, 22, 10, 20] attests to the high accuracy of classifiers which can be built.

First assume that all continuous attributes from the datasets have been discretised. Our method of choice is the entropy technique from [15].

6.1 Support Aggregation

Suppose we have two dataset classes D_1 and D_2 and assume that the (minimal) EPs have been mined for each of them. Let the set of EPs for D_1 be E_1 and the set of EPs for D_2 be E_2. Now given a test instance t, for which we need to assign a class label, we generate a new set E_1', which contains all patterns from E_1 that are a subset of t. Similarly generate the set E_2'. A score is now calculated for each of the classes. The score for D_1 is the sum of the individual relative supports for each EP in E_1'. Similarly for D_2. The test instance is assigned a label corresponding to the class with the higher score.

If there are more than two classes, then the procedure is similar. If the classes are $D_1, D_2, \ldots D_n$, then the EPs for class D_1 are found by comparing D_1 against the (negative) dataset $D_2 \cup D_3 \cup \ldots D_n$. The EPs for class D_3 are found by comparing D_3 with respect to the (negative) dataset $D_1 \cup D_2 \cup D_4 \cup D_5 \ldots \cup D_n$ etc.

Observe that in the simple aggregation method, the impact of each individual EP is equal to its relative support. Classifiers that use simple aggregation were discussed in [19, 10].

6.2 Bayesian Scoring

Although easy to implement, using simple aggregation to compute a class score is not based on any solid statistical foundation.

A way of attacking this problem is to use a scoring function based on Bayes theorem, which says that the chosen class should be the one which maximises

$$P(C_i|T) = P(T, C_i)/P(T) = P(C_i)P(T|C_i)/P(T)$$

where C_i is the class label, $T = \{a_1 a_2 \ldots a_n\}$ is the test case, $P(Y|X)$ denotes the conditional probability of Y given X and probabilities are estimated from the training sample. Since classification focuses on prediction of a single class, the denominator $P(T)$ can be ignored, since it will not affect the relative ordering of classes. Since it is very difficult in practice to calculate the probability $P(T, C_i)$, one must use approximations.

The approximation used here incorporates the use of a set of emerging patterns to derive a product approximation of $P(T, C_i)$, using the chain rule of probability $P(X_1, X_2, \ldots, X_n) = P(X_1)P(X_2|X_1)\ldots P(X_n|X_1, \ldots, X_{n-1})$. The product approximation of the probability of an n-item itemset t for a class C_i, contains a sequence of at most n subsets of T, such that each itemset contains at least one item not covered in the previous itemsets. e.g. Consider a test instance $T = \{a_1a_2a_3a_4a_5\}$ and a set of emerging patterns contained in the test instance $B = \{\{a_1, a_2, a_3\}, \{a_2, a_5\}, \{a_3, a_4\}, \{a_1, a_4, a_5\}\}$. The approximation using this set B of EPs is $P(C_i)P(a_1a_2a_3|C_i)P(a_5|a_2C_i)P(a_4|a_3C_i)$. The product approximation is created by adding one EP at a time until no more EPs can be added (either all the items of the remaining EPs are already covered or no more EPs are available). The probabilities $P(a_1a_2a_3|C_i)$, $P(a_5|a_2C_i)$ and $P(a_4|a_3C_i)$ are then computed from the dataset.

Of course the order in which EPs are added will affect the resulting approximation. Adding in order of support is advantageous, since greater support is likely to mean greater reliability of the resulting factor in the approximation.

A Bayesian approach to emerging patterns is described in [13], which in turn is based on the large Bayes classifier from [25].

6.3 Pairwise Classification

Although JEPs and CEPs have good classification performance on two class problems, their performance for multi-class problems is less impressive [2]. The crucial observation here is that EP-based classifiers seem inherently designed for dealing with two-classes, because multi-class problems get collapsed into two class ones, as earlier described.

To overcome this, it is possible to employ a pairwise technique. Suppose there are n different classes. The pairwise mechanism sees an $n(n-1)/2$ number of mining operations (one for each pair), replacing the n number that would normally be performed. For each pair, a winner is determined using a scoring function (e.g. simple aggregation). Following this process, each class has a tally of wins in its favour. The class with the highest number of wins is assigned to the test case. Discretisation is performed for each pair of classes, rather than once at the beginning.

Employing this pairwise strategy greatly improves classification accuracy in such multi-class scenarios. This appears to be because success of each class D_p in generating patterns is related to the negative dataset (D_n). If these sets are similar, a small number of patterns will result. If the negative set size far exceeds the positive set size, few patterns will be generated. Having more balanced pairs of D_p and D_n means that a greater number of EPs can be generated.

6.4 Lazy Classification

The classifiers so far discussed have been eager. i.e. They do a once only computation of all the EPs and then use them if they are contained within the given test instance t. In contrast, it is also possible to compute EPs in a lazy fashion, based on knowledge of the test. This results in better classifier accuracy, since discretisation can now be targeted towards the test instance [20].

Assume the sets $D_p = \{M_1, \ldots, M_m\}$ and $D_n = \{N_1, N_2, \ldots, N_n\}$ and assume continuous valued attributes have *not* been discretised. Consider a fixed test instance t. The following steps are used

- Take the intersection of each training instance with t, namely $t \cap M_1, \ldots, t \cap M_m$ and $t \cap N_1, \ldots, t \cap N_n$. This operation is equivalent to removal of irrelevant training values.
- Select the maximal itemsets from $\{t \cap M_1, \ldots, t \cap M_m\}$ and similarly from $\{t \cap N_1, \ldots, t \cap N_m\}$. Denote the former collection of maximal itemsets as max_{R_m} and the latter as max_{R_n}.
- Find all JEPs in max_{R_m} and sum their relative supports to get a score for Class 1. Similarly find all JEPs in max_{R_n} and sum their relative supports to get a score for Class 2.

Intersection is performed as follows. Assume $attr_A$ is a continuous valued attribute having domain $[0, 1]$ (if not, it can be scaled into this range). Given a training instance s, $t \cap s$ will contain the $attr_a$ value of t, if the $attr_A$ value of s is in the neighbourhood $[x_1 - \alpha, x_1 + \alpha]$, where x_1 is the attribute value for t. The parameter α is called the neighbourhood factor, which can be used to adjust the length of the neighbourhood. Experiments have shown 0.12 to be a suitable value in practice.

7 Related Work

The concept of emerging patterns was first introduced in [8]. Jumping emerging patterns appeared first appeared in [19, 9] and constrained emerging patterns in [2].

Classification has a long history and their exist a multitude of machine learning algorithms for it. In this paper, we have only focused on emerging pattern methods for classification.

The CBA classifier [23] uses association rules for classification that can be interpreted as being a kind an emerging pattern. Rules must satisfy a minimum threshold (typically 1% relative support) and a minimum confidence (typically 50%). Translated into our framework, this would mean mining all emerging patterns having minimum relative support in D_p of 1% and maximum relative support of β in D_n, where $\beta \leq \alpha$. This constraint on the β value is much more permissive than what is used in the CEP classifier in [2], which uses the low value of $\beta = 1$.

Pairwise classification was discussed in [16], where it was applied to a number of classifiers and was seen to result in increased accuracies on some datasets. It was also observed that the gains achieved were roughly proportional to the number of classes. This is in line with our reported results in [2].

Emerging patterns are similar to version spaces [26]. Given a training set of positive instances and a set of negative instances, a version space is the set of all generalisations that each match (or are contained in) every positive instance and no negative instance in the set. Therefore the consistency restrictions with the training data are quite different for JEP spaces.

8 Summary and Future Work

In this paper, we have examined a number of kinds of emerging patterns and described techniques for using them in classification and methods for efficiently mining them. Future research directions include i) Methods for discovery of the top k EPs, ranked by support. ii) Further developing an understanding of the foundations of EPs and their relationship to other structures in logic and machine learning.

References

[1] J. Bailey, T. Manoukian, and K. Ramamohanarao. Fast Algorithms For Mining Emerging Patterns. In *Proceedings of the Sixth European Conference on Principles of Data Mining and Knowledge Discovery*, pages 39–50, 2002. 6

[2] J. Bailey, T. Manoukian, and K. Ramamohanarao. Classification using constrained emerging patterns. In *Proceedings of the 4th International Conference on Web Age Information Management*, 2003. 7, 8, 9

[3] J. Bailey, T. Manoukian, and K. Ramamohanarao. A fast algorithm for computing hypergraph transversals and its use in mining emerging patterns. In *Proceedings of the 3rd IEEE International Conference on Data Mining*, 2003. 6

[4] R. J. Bayardo. Efficiently Mining Long Patterns from Databases. In *Proceedings of the International Conference on Management of Data*, pages 85–93, 1998. 3, 6

[5] C. Berge. *Hypergraphs, North Holland Mathematical Library*, volume 45. Elsevier Science Publishers B.V (North-Holland), 1989. 6

[6] E. Boros, V. Gurvich, L. Khachiyan, and K. Makino. On the complexity of generating maximal frequent and minimal infrequent sets. In *STACS 2002, 19th Annual Symposium on Theoretical Aspects of Computer Science, Antibes - Juan les Pins, France, March 14-16, 2002, Proceedings*, pages 133–141, 2002. 3

[7] Doug Burdick, Manuel Calimlim, and Johannes Gehrke. MAFIA: A Maximal Frequent Itemset Algorithm for Transactional Databases. In *Proceedings of the 17th International Conference on Data Engineering*, pages 443–452, 2001. 3, 6

[8] G. Dong and J. Li. Efficient Mining of Emerging Patterns: Discovering Trends and Differnces. In *Proceedings of the Fifth International Conference on Knowledge Discovery and Data Mining*, pages 43–52, 1999. 9

[9] G. Dong, J. Li, and X. Zhang. Discovering Jumping Emerging Patterns and Experiments on Real Datasets. In *Proceedings of the Ninth International Database Conference on Heterogeneous and Internet Databases*, pages 15–17, 1999. 9

[10] G. Dong, X. Zhang, L. Wong, and J. Li. CAEP: Classification by Aggregating Emerging Patterns. In *Proceedings of the Second International Conference on Discovery Science*, pages 30–42, 1999. 1, 7

[11] T. Eiter and G. Gottlob. Identifying the minimal transversals of a hypergraph and related problems. *Siam Journal of Computing*, 24(6):1278–1304, 1995. 3

[12] H. Fan and K. Ramamohanarao. An Efficient Single-Scan Algorithm for Mining Essential Jumping Emerging Patterns. In *Proceedings of the Sixth Pacific Asia Conference on Knowledge Discovery in Databases*, pages 456–462, 2002.
[13] H. Fan and K. Ramamohanarao. A bayesian approach to use emerging patterns for classification. In *Database Technologies 2003, Proceedings of the 14th Australasian Database Conference, ADC 2003, Adelaide, South Australia*, volume 17. Australian Computer Society, 2003.
[14] H. Fan and K. Ramamohanarao. Efficiently mining interesting emerging patterns. In *Proceedings of the 4th International Conference on Web Age Information Management*, 2003.
[15] U. M. Fayyad and K. B. Irani. Multi-Interval Discretization of Continuous-Valued Attributes for Classification Learning. In *Proceedings of the Thirteenth International Joint Conference on Artificial Intelligence*, pages 1022–1027, 1993.
[16] J. Furnkranz. Pairwise Classification as an Ensemble Technique. In *Proceedings of the Thirteenth European Conference on Machine Learning*, pages 97–110, 2002.
[17] K. Gouda and M. J. Zaki. Efficiently Mining Maximal Frequent Itemsets. In *Proceedings of the First International Conference on Data Mining*, pages 163–170, 2001.
[18] Dimitrios Gunopulos, Roni Khardon, Heikki Mannila, and Hannu Toivonen. Data mining, hypergraph transversals, and machine learning. In *Proceedings of the Sixteenth ACM SIGACT-SIGMOD-SIGART Symposium on Principles of Database Systems, May 12-14, 1997, Tucson, Arizona*, pages 209–216. ACM Press, 1997.
[19] J. Li, G. Dong, and K. Ramamohanarao. Making use of the most Expressive Jumping Emerging Patterns for Classification. In *Proceedings of the Fourth Pacific Asia Conference on Knowledge Discovery in Databases*, pages 220–232, 2000.
[20] J. Li, G.Dong, and K. Ramamohanarao. DeEPs: Instance-Based Classification by Emerging Patterns. In *Proceedings of the Fourth European Conference on Principles of Data Mining and Knowledge Discovery*, pages 191–200, 2000.
[21] J. Li, H. Liu, J. R. Downing, A. Yeoh, and L. Wong. Simple rules underlying gene expression profiles of more than six subtypes of acute lymphoblastic leukemia (ALL) patients. *Bioinformatics*, 19:71–78, 2003.
[22] J. Li and L. Wong. Emerging Patterns and Gene Expression Data. In *Proceedings of the Twelfth Workshop on Genome Informatics*, pages 3–13, 2001.
[23] B. Liu, W.Hsu, and Y. Ma. Integrating Classification and Association Rule Mining. In *Proceedings of the Fourth International Conference on Knowledge Discovery and Data Mining*, pages 80–86, 1998.
[24] H. Mannila and R. Kari-Jouko. Algorithms for inferring functional dependencies from relations. *Data and Knowledge Engineering*, 12(1):83–99, 1994.
[25] Dimitris Meretakis and Beat Wüthrich. Extending naïve bayes classifiers using long itemsets. In *Proceedings of the fifth ACM SIGKDD international conference on Knowledge discovery and data mining*, pages 165–174. ACM Press, 1999.
[26] T. M. Mitchell. Generalization as Search. *Artficial Intelligence*, 18(2):203–226, 1982.
[27] R. Reiter. A theory of diagnosis from first principles. *Artificial Intelligence*, 32(1):57–95, 1987.
[28] X. Zhang, G.Dong, and K. Ramamohanarao. Exploring Constraints to Efficiently Mine Emerging Patterns from Large High-Dimensional Datasets. In *Proceedings of the Sixth International Conference on Knowledge Discovery and Data Mining*, pages 310–314, 2000.

Robot Soccer: Science or Just Fun and Games?*

Claude Sammut

School of Computer Science and Engineering
University of New South Wales, Sydney, Australia
claude@cse.unsw.edu.au

Abstract. RoboCup is an international robot soccer competition that has been running since 1997. A frequent criticism of any such competition is that all of a team's development is narrowly focussed on a very specific task and consequently, little contribution is made to science. In this paper, we describe our involvement in the RoboCup four-legged robot league and show how the competition has, indeed, resulted in contributions beyond robot soccer. We describe innovations in vision, localisation and locomotion. In particular, we discuss the stimulus that RoboCup has given to system integration and cooperative behaviour between multiple agents.

1 Introduction

The RoboCup competition was created to provide an incentive for researchers to work, not only on specific problems in robotics, but to build integrated systems to solve a very complex but well-defined problem. However, a frequent criticism of any such competition is that all of a team's efforts are narrowly focussed on shortcuts for winning games and consequently, little contribution is made to science. In this paper, we suggest that the answer to the question: "is robot soccer science or fun and games", is "it's both".

The benefits of the competition are evident in the progress from one year to the next. The competitive nature of the research has forced participants to evaluate very critically their methods in a realistic task as opposed to experiments in a controlled laboratory. It has also forced researchers to face practical hardware and real time constraints. In particular, there are very few domains in which such strong emphasis is placed on multi-agent cooperation. Despite the serious nature of the scientific endeavour, the "fun" aspect of RoboCup is important as a way of encouraging bright students into robotics and AI.

In this paper, we will describe the involvement of the rUNSWift team in the RoboCup four-legged robot league and the innovations that have resulted in vision, localisation, locomotion and cooperative multi-agent behaviour. Hopefully, this will answer the question of whether there is science in RoboCup. The work described has been the accumulated effort of UNSW teams from 1999 to 2003

* The work described here is the collective effort of successive UNSW RoboCup teams since 1999. All the team members are listed in the acknowledgements.

T.D. Gedeon and L.C.C. Fung (Eds.): AI 2003, LNAI 2903, pp. 12–23, 2003.

and, in 2003, jointly with NICTA. The participation of all team members is gratefully acknowledged. They have been finalists in all five years of the official RoboCup competition[1] and three times world champions, including 2003.

AI has often suffered from either attempting to tackle problems that were overly ambitious or else using toy problems that have been simplified to the extent that the real problems have been removed. Although robot soccer is a highly constrained domain, it is, nevertheless, sufficiently complex that it preserves most of the central problems of robotics. It seems to have struck a good balance between complexity and attainability. Furthermore, the rules of the competition adapt, each year, introducing more challenges as teams come to terms with previously unsolved problems.

For a mobile robot to operate effectively it must be capable of at least the following:

– perceive its environment through its sensors;
– use its sensory inputs to identify objects in the environment;
– use its sensors to locate the robot relative to other objects in the environment;
– plan a set of actions in order to achieve some goal;
– execute the actions, monitoring the progress of the robot with respect to its goal;
– cooperate with other agents that have a the same goal and thwart the efforts of opponent agents.

Robot soccer requires a solution to all of these problems. Most importantly, they must be integrated into a complete system that must operate in real-time in a highly dynamic environment. So the solutions must be practical and robust. In the following sections, we first describe the robot soccer competition and then discuss how we have approached each of the above problems.

2 The Four-legged Robot League

The four-legged robot league is one of several leagues that currently form the RoboCup competition. In some leagues, the teams are required to design and build the robot hardware themselves. In the case of the four-legged robot league, all teams use the same robotic platform, manufactured by Sony. Since all teams use the same hardware, the competition is based on a combination of a team's creativity in programming and also its discipline in making it work reliably.

The robot used in the four-legged league Sony's AIBO. Although designed for the entertainment market, this robot is quite sophisticated. The model ERS-210A robot, used in 2003, has an on board MIPS R5200 384MHz processor, 176x144 colour CMOS camera, accelerometers, contact sensors, speaker and stereo microphones. Each of the four legs has three degrees of freedom, as does

[1] A demonstration competition run in 1998, won by CMU. The principal team member, Will Uther, is now with NICTA.

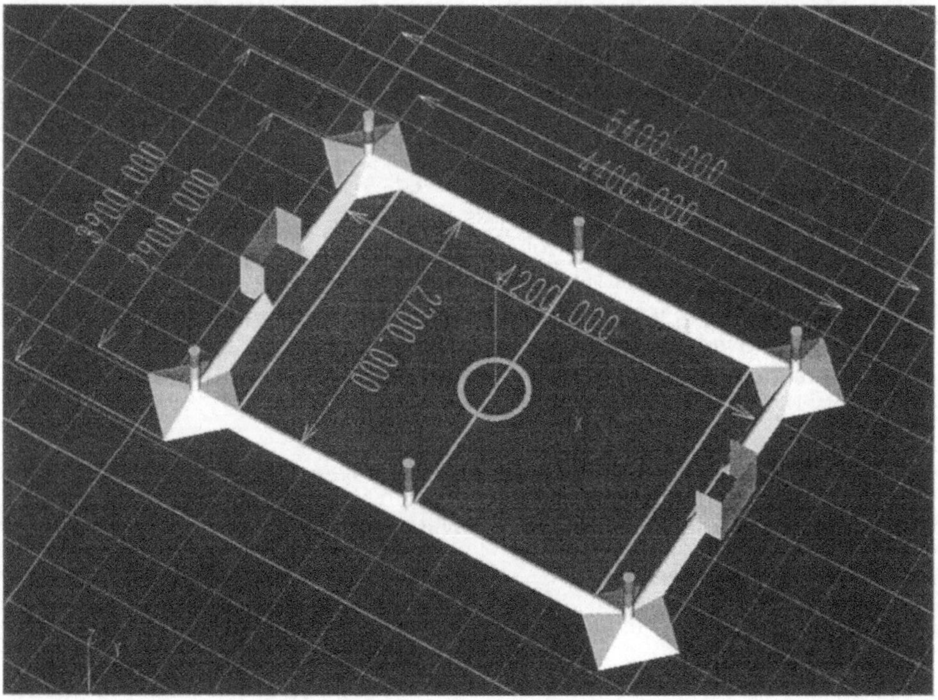

Fig. 1. The RoboCup playing field

the head. Programs are written off-board in C++ and loaded onto a memory stick that is inserted into the robot. All of Sony's ERS robots run a proprietary operating system called Aperios, which provides an interface for the user-level programs to the sensors and actuators.

Four robots make a team, with one usually being designated the goalie. The field for in the four-legged robot league is shown in Figure 1. It has an angled white border designed to keep the ball within the field. The game ball is coloured orange and the two goals are coloured yellow and blue. The field also contains six coloured landmark poles to aid the robots in localising themselves in the field. In front of each goal there is the penalty region that only the goalie is allowed to defend.

Games are 10 minutes per half. The robots from each team wear "uniforms" that are red of blue stickers. The blue team defends the blue goal and the red team defends the yellow goal. Teams sides at half-time and also swap colours. Because colour recognition can be quite difficult, as we well see in the next section, a team may play better as one colour or the other, so to even out this effect, team colours are swapped.

3 Vision

3.1 Colour Classification

A colour camera is the robot's primary sensor. The ERS-210A, has a 176x144 resolution camera mounted in the "snout" of the robot. This delivers 25 frames per second. The information in each pixel is in YUV format, where Y is the brightness and U and V are the colour components. Since all the objects on the field are colour coded, the aim of the first stage of the vision system is to classify each pixel into the eight colours on the field: orange for the ball, blue and yellow for the goals and beacons, pink and green for the beacons, light green for the field carpet, dark red and blue for the robot uniforms.

Despite the large amount of activity in digital camera technology, the recognition of colours remains a difficult problem and is an area in which RoboCup has made a significant contribution. The main difficulty in colour recognition is that the characteristics of the light source have a major influence on how a camera detects colour. Therefore, we need a vision system that can be easily adapted to new surroundings. One way of doing this is by training the vision system to recognise colours, given samples of images from a particular environment.

The first step is to take sample images of different objects at different locations on the field. Then for each image, every pixel is classified by "colouring in" the image by hand, with the help of a custom painting tool. All these pixels form the set of training data for a learning algorithm. Each pixel can be visualised as a point in a 3 dimensional space, as shown in Figure 2(a).

Figure 2(b) shows the regions grown by Quinlan's C4.5 decision tree learning program [3]. C4.5's input is a file of examples where each example consists of a set of attribute values followed by a class value. In this case, the attributes are the Y, U and V values of a pixel and the class value is the colour manually

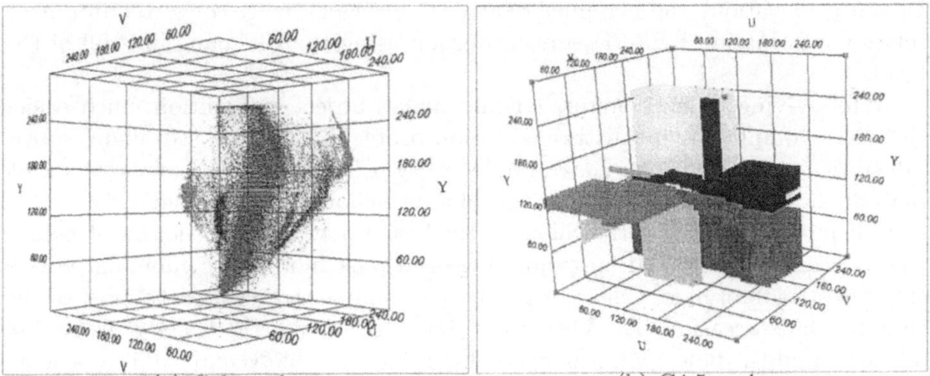

(a) Colour data (b) C4.5 regions

Fig. 2. Using C4.5 to learn to recognise colours

assigned to that pixel. C4.5 turns the training data into a set of classification rules that should be able to determine the colour of a new, unclassified pixel.

Although colour classification with C4.5 generated decision trees has proved to be quite reliable, the disadvantage of this approach is that it requires quite a lot of manual labour to collect and label the sample images. This means that it takes time to recalibrate if conditions change, for example, of crowds around the field cast shadows, etc. Therefore, it is desirable to further automate colour calibration. Recently, Cameron and Barnes [1] have developed a method for using the constraints of the field to suggest the colour of region, without requiring human assistance. Unsupervised learning, using clustering techniques may also help to automate colour recognition.

3.2 Object Recognition

Once colour classification is completed, the object recognition module takes over to identify the objects in the image. The possible objects are: the goals, the beacons, the ball and the blue and red robots. A blob formation algorithm links neighbouring pixels of the same colour in the image to form blobs. Based on the blobs, we identify the objects, along with their distance, heading and elevation relative to the camera and the bas of the neck of the robot.

Objects are identified in the order: beacons first, then goals, the ball and finally the robots. Since the colour uniquely determines the identity of an object, once we have found the bounding box around each colour blob, we have enough information to identify the object and compute its parameters. Because we know the actual size of the object and the bounding box determines the apparent size, we can calculate the distance from the snout of the robot (where the camera is mounted) to the object. We also calculate heading and elevation relative to the nose of the robot and the bounding box's centroid. However to create a world model, needed for strategy and planning, measurements must be relative to a fixed point. The neck of the robot is chosen for this purpose. Distance, elevations and headings relative to the camera are converted into neck relative information by a 3D transformation using the tilt, pan, and roll of the head.

While giving objects unique colours makes object recognition much easier than in a completely unconstrained vision problem, there are still many subtle difficulties to be overcome. For example, every beacon is a combination of a pink blob directly above or below a green, blue or yellow blob. One side of the field has the pink on the top of the colour in the beacons, while the other has it below. The beacons are detected by examining each pink blob and combining it with the closest blob of blue, yellow or green. This simple strategy works most of the time but fails occasionally. When the robot can just see the lower pink part of a beacon and the blue goal, it may combine these two blobs and call it a beacon. A simple check to overcome this problem is to ensure that the bounding boxes of the two blobs are of similar size and the two centroids are not too far apart.

Over the years of that we have been competing, the vision system has accumulated a large library of such situation-specific heuristics. Knowledge-based image understanding seems to be a requirement for a reliable vision system. However, like all knowledge acquisition tasks, we have the problem of building a consistent and reliable set of heuristics. It is very easy to add new rules to the knowledge base that interfere with previous rules. Discovering such interactions is difficult because they might occur in only a few frames, remembering that the camera images are processed at 25 frames per second. If a heuristic triggers incorrectly at a crucial moment, for example, while attempting to grab the ball for a kick, its effects could be serious. For this reason, most teams create debugging tools for monitoring image processing. Since each robot has on onboard wireless LAN card, it is possible to transmit debugging information to another computer. We currently have a semi-automated system for capturing the firing of heuristics and checking them against the desired result. An obvious improvement will be to apply machine learning.

To achieve reasonable performance, most "real-world" applications of robotics require a lot of task-specific engineering. Thus, any method that can help in automating the acquisition of domain knowledge is of interest. The RoboCup environment, once again, provides a challenging, but bounded, test bed for such work.

3.3 Line Detection

RoboCup imposes a very strict discipline on the implementor. Being a fast moving game, it is important that the robots are able to process incoming sensor data and make decisions as quickly as possible. It is usually preferable to compromise on the amount of processing done so that the system can keep up with the frame rate of the camera. For this reason, we try to find the simplest image processing operations that yield enough information to be able to play effectively. Of course, how much processing we can afford changes as hardware improves. However, the real-time constraints force us to develop very efficient algorithms, which often requires some radical rethinking of how something should be done.

A recent addition to the robot's repertoire, thanks to a faster CPU, is the detection of lines such as the field boundaries, centre and goal lines. In the early stages of development, we attempted to use well-known techniques such as the Hough Transform. However, this provided to be both time consuming and not well-suited to the task for which it was intended, namely, helping to localise the robot on the field. After much experimentation, another knowledge-based method was used. Knowing the lines that ought to be visible on the soccer field, the system collects edge-points and attempts to map them onto the lines that should be in the robot's field of view, assuming it has a rough idea of where it is. There are usually several candidate lines. One is chosen according the which mapping has the smallest error.

One might argue that such domain-specific methods are of little general use and therefore, the scientific merit robot soccer is diminished. On the contrary, we believe that to achieve adequate performance in practical tasks, domain specific knowledge is essential. Our challenge is to understand this kind of knowledge and develop ways of learning domain-specific methods.

4 Localisation

A common problem in mobile robots of all kinds is localisation, that is, maintaining an accurate estimate of its position. The most complex form of the problem requires the robot to build a map of an unknown environment while simultaneously locating itself within the map. In robot soccer, this is simpler because we have exact measurements of the field. Even given an accurate map, it is still not straightforward to localise since all sensory data are inherently noisy. Another important variant of the localisation problem occurs when there are several agent that can share information. In principal, sharing information should improve localisation, but since no robot knows its exact location and all sensors are noisy, how can the diverse streams of information be merged? This is the problem of *distributed data fusion.*

Information from the robot's vision system, as well as odometry from the locomotion module, are combined to create a world model. Since the dimensions of the field are known in advance, the world model is constructed by placing the mobile objects on a map of the field. As objects are observed to move, their positions in the map are updated. However, since observations can be inaccurate, the update must be done cautiously. We use a method based on the Kalman filter [2], which maintains as estimate of the error in the current position.

One of the innovations in 2003 was to combine information from objects recognised by their colour with the new line detections algorithms described earlier. There are situations in which localisation based on edges is more accurate than by using landmarks. The ball often strays to the field boundaries or into the corners. When a robot approaches either area, it very often loses sight of the landmarks and soon loses localisation. It gets lost. In such cases, we use line detection as a backup. Being somewhat more expensive to execute, line detection is used sparingly. While our method of line detection was the most accurate to appear in the 2003 competition, the German Team developed a method that was faster and therefore could be used more frequently. Arguably, they made the better trade-off between speed and accuracy.

Figure 3 shows a snapshot of the world model being maintained by one of the blue robots during a game. The ellipses around the objects indicate the size of the error in x and y while the shaded sector indicates the error on heading of the robot. As long as the robot has sight of at least two coloured landmarks, its position estimate will be quite accurate. Since distance is estimated by the size that an object appears in an image, distance to landmark is usually less accurate than the robot's angle, relative to the landmark, hence the elliptical shape of the error estimate.

During a game, each robot on the field receives information from every other robot. This means that a robot must reconcile the world models of all three other robots and its own. Each robot maintains an additional, *wireless model* that is a representation of the world as given by a robot's team mates. It is kept separate from the robot's own world model. We associate with each team mate a variance which represents our confidence in the accuracy of the information from that robot. The variance of an object in the wireless model is the sum of

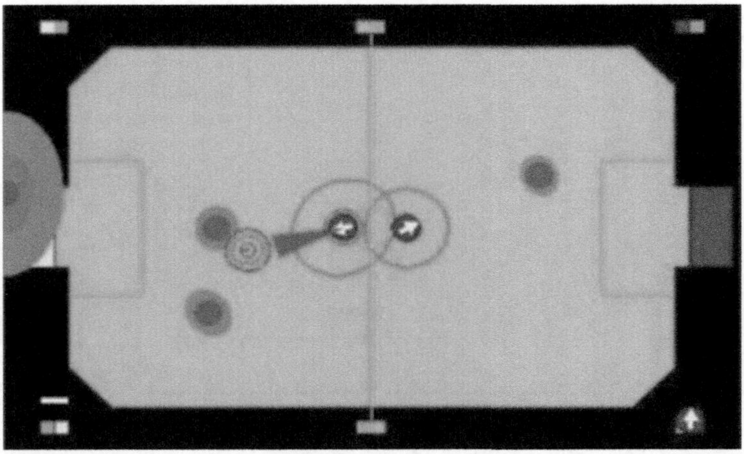

Fig. 3. The world model

the team mate's variance, the object's variance, as believed by the team mate, and a small variance for the latency over the wireless network.

The general problem of distributed data fusion remains a major challenge for multi-agent systems. Among the problems to be solved is in combining several world models how do we know that if they refer to the same object. For example, Figure 4, shows two world models on the left. One robot is aware of two red robots and one blue robot. The other is aware of three red robots and one blue robot. Our present solution is a heuristic. If all the supposed robots are placed in one world model, where the variances of two robots overlap, they are combined into one.

Once the information is merged, the best (lowest variance) four robots of each colour are stored in the wireless model. The same process is also applied to combine information for balls. In this case, only the best ball is selected. In the wireless model, three extra robots are also stored. These describe the location of the team mates according to their own belief.

The solutions to the distributed data fusion problem being developed for robot soccer are relevant to other applications. Another competition in RoboCup is concerned with rescue. Teams in the rescue league construct robots that are intended to search through collapsed buildings and similar dangerous environments, searching for survivors and delivering aid. Because of the complexity of the environment, this presents many problems that are avoided when moving about a soccer field. However, any real rescue operation will involve multiple robots, possibly of different construction. They will have to share information to coordinate their efforts and they will have to decide on the identity of objects just as described above. The techniques being tested in the soccer competition for this problem, and others, will carry across to the more socially relevant tasks such as rescue.

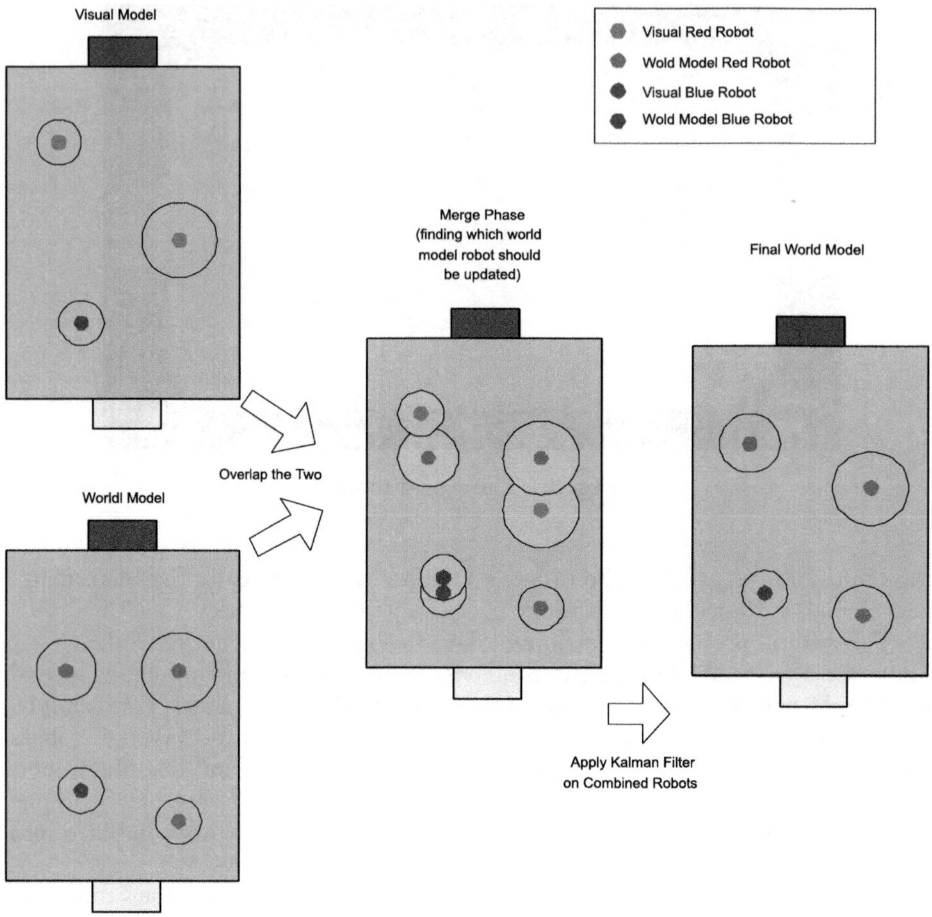

Fig. 4. Updating Multiple Robot Locations

5 Locomotion

In RoboCup, speed is crucial. The team that is able to reach the ball first, usually wins. Speed results from a combination of fast and reliable vision, localisation, decision making and, obviously, locomotion.

Although Sony provides a library of walking routines with the Aibo, these are not designed for playing soccer and lack the speed and manoeuvrability needed to play effectively. UNSW's 2000 team [4] wanted to be able to drive the robots as if controlled by a three degree of freedom joystick. That is, the robot should be able to travel in any direction, regardless of which way it was facing. To achieve this, they devised a parameterised walk. The paw of each leg described a rectangle as it moved. A particular kind of walk or turn could be specified by giving the dimensions and orientations of the rectangles. The joint

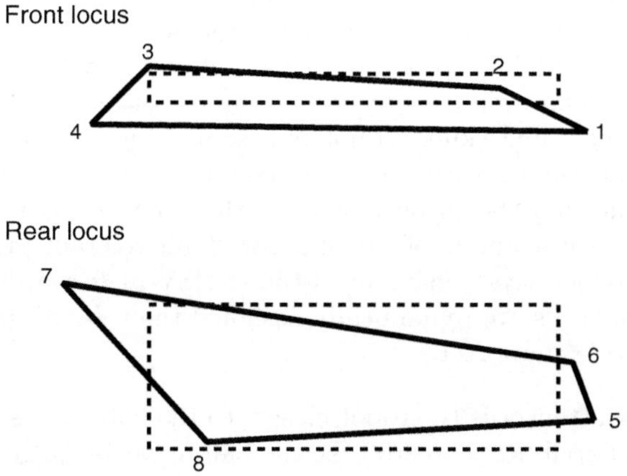

Fig. 5. The robot's paw describes a quadrilateral. The robot's walk can easily be changed by adjusting the dimensions and orientation of the rectangle

angles are automatically computed by inverse kinematics. This innovation was such a success that the team never had less than double digit scores in its games.

Each year, teams publish reports on their efforts, usually accompanied by source code. In subsequent years, almost all of the teams adopted the "UNSW walk". As a result of this sharing of ideas, the standard of the league has steadily risen, putting pressure on the leading teams to make further leaps forward. To improve the speed of the robot, the team re-examined the parameterised walk. Perhaps there were trajectories other than a rectangular path that would allow the robot to walk faster or more smoothly? Min Sub Kim and Will Uther employed an optimisation algorithm to try to discover a better walk. During the 2003 competition, a robot could be seen pacing back forth across the field, collecting data for the optimisation. At the end of the process the walk was 10% faster than any other teams and also smoother, making the camera shake less. The trajectories found by the optimisation are shown in Figure 5.

The process began with an initial rectangular trajectory. During the process of learning to walk faster, the corner points were shifted to obtained a more efficient walk. Previously, front and back legs had followed the same path, but through learning, it was discovered that having different trajectories improved the walk.

6 Game Play Behaviours

Developing game playing behaviours involves lengthy and painstaking work testing and tuning different skills for approaching the ball, kicking, handling special cases such as when the robot is against the field boundary, etc. These skills are described in detail in reports available online[2]. In this section we discuss the importance of developing strategies that are robust.

No matter how good the robot's skills are, it will inevitably fumble the ball. Opponent robots (end even friendly robots) are constantly attacking the ball so even if the ball handling skills do not fail, something unexpected will happen because of the interaction with other robots. One of the most important lessons that students entering the project learn is that they must program for when things go wrong. An example of this kind of "fault tolerant programming" can be seen in the cooperative behaviour of the rUNSWift forwards.

The three forwards are dynamically assigned the roles of: primary attacker, supporter and wide supporter.

- The primary attacker is the robot closest to the ball that is heading towards the opponent goal. Its job is to take the ball towards the goal and if it is not being challenged by opponents, it may shoot for goal.
- The supporter is the next closest robot. It follows the ball, but backs away from the primary attacker so as not to interfere with it.
- The wide supporter stays on the side of the field opposite to where the ball is. When the ball is nearer its own half, the wide supporter adopts a defensive role. It is prepared for the ball popping into the clear so that it can move the ball up field. When the ball is close to the opponent goal, it will move in, ready for a cross from its team mates.

As the ball moves around, these roles change dynamically. For example, if the primary attacker loses the ball behind it, the supporter will move in and become the primary attacker, while the other two robots reassign themselves depending on their proximity to the ball.

The effect of these roles, as observed during the 2003 competition, was to create a wave of attacks. The first robot, as primary attacker would move the ball towards the opponent goal. If it lost the ball, a second robot would take over, moving the ball closer. If it also lost the ball, the third robot was nearby to finish off the goal. Thus, by providing constant backup to team mates, the rUNSWift team mounted an attack that almost always ended in a score.

A potential weakness of the rUNSWift strategy was that in being very aggressive, it could leave the goalie on its own to defend. However, no opponents could take advantage of this until the final against the University of Pennsylvania team. Their robots were programmed with a very powerful kick that could take the ball from an opponent and move the ball far down field. They also had another player positioned to attack when the ball came its way. The final settled into a pattern of rUNSWift attacks and UPenn counterattacks.

Fortunately, we had anticipated this as a possibility. The wide supporter was programmed so that if the ball got behind the rUNSWift attackers, it would execute a fast "run" back and circle behind the ball to block the opposition's attack. Nevertheless, the game was very close one, ending 4-3 in rUNSWIft's favour.

---

[2] http://www.cse.unsw.edu.au/~robocup

7 Conclusion

As the average quality of the teams improves each year, we are seeing more close games that may be won or lost because of relatively small differences. We are also, seeing a shift from gaining an advantage from better low-level skills to trying to gain the upper hand through cooperation and strategy. As these trends continue, we learn more about how to create robust behaviour, how to anticipate potential weaknesses and how to prepare for the unexpected. All of these are excellent training for applications of robotics in a wide variety of tasks outside of competitions.

The RoboCup four-legged robot league has been a source of many interesting research problems for robotics and artificial intelligence. In succeeding years, the rules of the game are changed to introduce new challenges and gradually move the game closer to human soccer. The official aim of the RoboCup federation is, by 2050, to have a humanoid robot soccer team that can take on, and beat, the human world champions. The unofficial, and much important aim, is that as a side effect of the grand challenge, the competition will advance robotics and AI to benefit science and society in many practical ways.

Acknowledgements

The results described here are due to the efforts of successive UNSW RoboCup teams: John Dalgliesh and Mike Lawther (1999); Bernhard Hengst, Son Bao Pham and Darren Ibbotson (2000); Spencer Chen, Martin Siu, Tom Vogelgesang and Tak Fai Yik (2001); James Brooks, Albert Chang, Min Sub Kim, Benjamin Leung, Canh Hao Nguyen, Andres Olave, Timothy Tam, Nick von Huben, David Wang and James Wong (2002); Ricky Chen, Eric Chung, Ross Edwards, Eileen Mack, Raymond Sheh, Nicodemus Sutanto, Terry Tam, Alex Tang and Nathan Wong (2003). Special thanks to team co-supervisors Bernhard Hengst, Will Uther (both with NICTA), Tatjana Zrimec and Brad Hall. The financial support of the UNSW School of Computer Science and Engineering, the Faculty and Engineering and National ICT Australia is gratefully acknowledged.

References

[1] Cameron, D. and Barnes, N. (2003) Knowledge-Based Autonomous Dynamic Colour Calibration. In *Proceedings of the Robocup Symposium*, Padua: Springer. 16

[2] Kalman, R,E. (1960) A New Approach to Linear Filtering and Prediction Problems, *Transactions of the ASME–Journal of Basic Engineering*, **82**(D), pp 35-45. 18

[3] Quinlan., J. R. (1993). *C4.5: Programs for Machine Learning.* Morgan Kaufmann, San Mateo, CA. 15

[4] UNSW United Team Report.
http://www.cse.unsw.edu.au/~robocup/2002site/2000PDF.zip 20

On How to Learn from a Stochastic Teacher or a Stochastic Compulsive Liar of Unknown Identity

B. John Oommen[1], Govindachari Raghunath[2], and Benjamin Kuipers[3]

[1] Fellow of the IEEE. School of Computer Science
Carleton University, Ottawa, ON, K1S 5B6, Canada
oommen@scs.carleton.ca

[2] oommen@scs.carleton.ca

[3] Fellow of the AAAI and IEEE. Department of Computer Sciences
University of Texas, Austin, Texas 78712, USA
kuipers@cs.utexas.edu

Abstract. We consider the problem of a learning mechanism (robot, or algorithm) that learns a parameter while interacting with either a *stochastic teacher* or a *stochastic compulsive liar*. The problem is modeled as follows: the learning mechanism is trying to locate an unknown point on a real interval by interacting with a stochastic environment through a series of guesses. For each guess the environment (teacher) essentially informs the mechanism, possibly erroneously, which way it should move to reach the point. Thus, there is a non-zero probability that the feedback from the environment is erroneous. When the probability of correct response is $p > 0.5$, the environment is said to be *Informative*, and we have the case of learning from a stochastic teacher. When this probability is $p < 0.5$ the environment is deemed *Deceptive*, and is called a *stochastic compulsive liar*.

This paper describes a novel learning strategy by which the unknown parameter can be learned in both environments. To the best of our knowledge, our results are the first reported results which are applicable to the latter scenario. Another main contribution of this paper is that the proposed scheme is shown to operate equally well even when the learning mechanism is unaware whether the environment is Informative or Deceptive. The learning strategy proposed herein, called CPL–ATS, partitions the search interval into three equi-sized sub-intervals, evaluates the location of the unknown point with respect to these sub-intervals using fast-converging ϵ-optimal L_{RI} learning automata, and prunes the search space in each iteration by eliminating at least one partition. The CPL-ATS algorithm is shown to be provably converging to the unknown point to an arbitrary degree of accuracy with probability as close to unity as desired. Comprehensive experimental results confirm the fast and accurate convergence of the search for a wide range of values for the environment's feedback accuracy parameter p. The above algorithm can be used to learn parameters for non-linear optimization techniques.

Keywords: Learning Automata, Statistical Learning, Parameter Estimation, Stochastic Optimization, Pattern Recognition.

T.D. Gedeon and L.C.C. Fung (Eds.): AI 2003, LNAI 2903, pp. 24–40, 2003.

1 Introduction

Consider the problem of a robot (algorithm, learning mechanism) moving along the real line attempting to locate a particular point $\lambda^\star$. To assist the mechanism, we assume that it can communicate with an Environment ("Oracle") which guides it with information regarding the direction in which it should go. If the Environment is deterministic the problem is the "Deterministic Point Location Problem" which has been studied rather thoroughly [1]. In its pioneering version [1] the problem was presented in the setting that the Environment could charge the robot a cost which was proportional to the distance it was from the point sought for. The question of having multiple communicating robots locate a point on the line has also been studied [1, 2]. In the stochastic version of this problem, we consider the scenario when the learning mechanism attempts to locate a point in an interval with stochastic (i.e., possibly erroneous) instead of deterministic responses from the environment. Thus when it should really be moving to the "right" it may be advised to move to the "left" and vice versa.

Apart from the problem being of importance in its own right, the stochastic point location problem also has potential applications in solving optimization problems. In many optimization solutions – for example in image processing, pattern recognition and neural computing [5, 9, 11, 12, 14, 16, 19], the algorithm works its way from its current solution to the optimal solution based on information that it currently has. A crucial question is one of determining the *parameter* which the optimization algorithm should use. In many cases the parameter of the scheme is related to the second derivative of the criterion function, which results in a technique analogous to a Newton's root solving scheme. The disadvantages of the latter are well known – if the starting point of the algorithm is not well chosen, the scheme can diverge. Additionally, if the second derivative is small, the scheme is ill-defined. Finally, such a scheme requires the additional computation involved in evaluating the (matrix of) second derivatives [14, 16, 19]. To tackle this problem we suggest that our strategy to solve the stochastic point location problem can be invoked to learn the best parameter to be used in any algorithm.

The results that are claimed here are stated in the body of the paper. Sketches of their proofs are included in the Appendix. The more detailed proofs are found in the unabridged version of the paper [15].

2 The Stochastic Point Location Problem

The goal of the learning mechanism is to determine the optimal value of some parameter $\lambda^\star \in [0, 1)$. Although the mechanism does not know the value of $\lambda^\star$, we assume that it has responses from an intelligent environment E which is capable of informing it whether the current estimate $\hat{\lambda}$ is too small or too big. To render the problem both meaningful and distinct from its deterministic version, we would like to emphasize that the response from this environment is assumed "faulty". Thus, E may tell us to increase $\hat{\lambda}$ when it should be decreased, and vice versa with a non-zero probability $1-p$. Note that the quantity p reflects on the "effectiveness" of the environment, E . Thus, whenever $\hat{\lambda} < \lambda^\star$, the environment correctly suggests that we increase $\hat{\lambda}$ with probability p. It could as well have incorrectly recommended that we decrease $\hat{\lambda}$ with probability $1-p$. Similarly, whenever $\hat{\lambda} > \lambda^\star$, the environment tells us to decrease $\hat{\lambda}$ with probability p, and to increase it with probability $(1-p)$.

We further distinguish between two types of environments – *Informative* and *Deceptive*. An environment is said to be "Informative" if the probability p of its giving a correct feedback is greater than 0.5. If $p < 0.5$, the environment is said to be "Deceptive". Thus a Deceptive environment is more likely to give erroneous feedback than an Informative environment. This together with the fact that the learning mechanism is not aware of the nature of the environment complicates the learning process and its convergence.

3 Related Work

Oommen [9] proposed and analyzed an algorithm that operates by discretizing the search space while interacting with an *Informative* environment. This algorithm takes advantage of the limited precision available in practical implementations to restrict the probability of choosing an action to only finitely many values from the interval $[0, 1)$. Its main drawback is that the steps are always very conservative. If the step size is increased the scheme converges faster, but the accuracy is correspondingly decreased. Bentley and Yao [3] solved the deterministic point location problem of searching in an unbounded space by examining points $f(i)$ and $f(i+1)$ at two successive iterations between which the unknown point lies, and doing a binary search between these points. Although it may appear that similar binary search can be applied in the stochastic point location problem, the faulty nature of the feedback from the environment may affect the certainty of convergence of the search, and hence a more sophisticated search strategy is called for. Thus, whereas in Bentley's and Yao's algorithm we could confidently discard regions of the search space, we have to now resort to stochastic methods, and work so that we minimize the probability of rejecting an interval of interest. This is even more significant and sensitive when the learning mechanism is unaware whether the environment is Informative or Deceptive. A novel strategy combining learning automata and pruning was used in [10] to search for the parameter in the continuous space when interacting with an Informative environment. In this paper we extend the results of [10] further to operate also in the continuous space, but to work equally well for both *Informative* and *Deceptive* environments. To the best of our knowledge this is the first reported result for learning in a Deceptive environment.

A completely different approach to locating a point will be to partition the search space into intervals and choose the mid-point of each interval repeatedly as an estimate for the unknown point, and use a majority voting scheme on the feedbacks obtained to eliminate one interval. Applying Chernoff bounds would then allow us to precisely compute the number of steps sufficient for ensuring correct pruning with a certain level of confidence [13]. The difference between such an "estimation" approach and a learning automaton approach such as ours is that in the former case, a fixed number of sampling steps has to be done *irrespective* of how strong the feedback is, before determining the pruning. In contrast, in a learning automata approach, when one of the actions is superior, the reinforcement mechanism ensures that this action is sampled more frequently than the other. Thus the better an action is with respect to another, the faster it's optimality is determined. Another drawback with the simple majority voting scheme is that in each epoch it can prune at most one partition. On the other hand, our scheme allows more than two thirds of the search space to be pruned in a single epoch. A third and most compelling drawback of the former scheme in our problem setting is that it will not be feasible to detect a point in a deceptive environment where the probability of correct feedback $p < 0.5$.

4 Continuous Point Location with Adaptive Tertiary Search

The solution presented in this paper is based on the Continuous Point Location with Adaptive Tertiary Search (CPL–ATS) strategy introduced in [10]. The basic idea behind both the solutions is to systematically explore a current interval for the parameter. This exploration is a series of guesses, each one more accurate than the previous one. Initially, we guess the mid-point of the given interval to be our estimate. We then partition the given interval into disjoint sub-intervals, eliminating at least one of the subintervals from further search and recursively searching the remaining interval until the search interval is at least as small as the required resolution of estimation. Crucial to the CPL–ATS scheme is the construction of partitions and the elimination process.

The given search interval is divided into three partitions. Each of these three partitions is independently explored using an ϵ-optimal fast converging two-action learning automaton where the two actions are those of selecting a point from the left and right half of the partition under consideration. The elimination process then utilizes the the ϵ-optimality property of the underlying automata and the monotonicity of the intervals to systematically eliminate at least one of the partitions. The resultant is a new pruned interval consisting of at most two contiguous partitions from the original three. At each stage the mid-point of the remaining interval is taken to be the estimate of λ^*. We shall assume that the individual learning automaton used is the well-known L_{RI} scheme with parameter θ, although any other ϵ-optimal scheme can be used just as effectively.

5 Notations and Definitions

Let $\Delta = [\sigma, \gamma)$ s.t. $\sigma \leq \lambda^* < \gamma$ be the current search interval containing λ^* whose left and right (smaller and greater) boundaries on the real line are σ and γ respectively. We partition Δ into three equi-sized disjoint partitions Δ^j, $j \in \{1, 2, 3\}$, such that, $\Delta^j = [\sigma^j, \gamma^j)$. To formally describe the relative locations of intervals we define an interval relational operator $\prec$ such that, $\Delta^j \prec \Delta^k$ iff $\gamma^j < \sigma^k$. Since points on the real interval are monotonically increasing, it is easy to see that, $\Delta^1 \prec \Delta^2 \prec \Delta^3$. For every partition Δ^j , we define L^j and R^j as its *left* half and *right* half respectively as:

$$L^j = \{x \,|\, \sigma^j \leq x < mid(\Delta^j)\}, \text{and} R^j = \{x \,|\, mid(\Delta^j) \leq x < \gamma^j\},$$

where $mid(\Delta^j)$ is the mid point of the partition Δ^j. A point $x \in L^j$ will be denoted by x_L^j, and a point $x \in R^j$ by x_R^j.

To relate the various intervals to $\lambda^\star$ we introduce the following relational operators.

$\lambda^\star \textcircled{<} \Delta^j$ iff $\lambda^\star < \sigma^j$.	i.e., $\lambda^\star$ is to the left of the interval Δ^j.
$\lambda^\star \textcircled{>} \Delta^j$ iff $\lambda^\star > \gamma^j$.	i.e., $\lambda^\star$ is to the right of the interval Δ^j.
$\lambda^\star \textcircled{=} \Delta^j$ iff $\sigma^j \leq \lambda^\star < \gamma^j$.	i.e., $\lambda^\star$ is contained in the interval Δ^j.
$\lambda^\star \textcircled{\leq} \Delta^j$ iff $\lambda^\star \textcircled{<} \Delta^j$ or $\lambda^\star \textcircled{=} \Delta^j$	i.e., $\lambda^\star$ is either to the left of or inside Δ^j.
$\lambda^\star \textcircled{\geq} \Delta^j$ iff $\lambda^\star \textcircled{>} \Delta^j$ or $\lambda^\star \textcircled{=} \Delta^j$	i.e., $\lambda^\star$ is either to the right of or inside Δ^j.

These operators can be shown to satisfy the usual laws of transitivity.

6 Construction of the Learning Automata

In the CPL–ATS strategy, with each partition Δ^j we associate a 2-action L_{RI} automaton $\mathcal{A}^j$ $(\Sigma^j, \Pi^j, \Gamma^j, \Upsilon^j, \Omega^j)$ where, Σ^j is the set of actions, Π^j is the set of action probabilities, Γ^j is the set of feedback inputs from the environment, Υ^j is the set of action probability updating rules and Ω^j is the set of possible decision outputs of the automaton at the end of each epoch. The environment E is characterized by the probability of correct response p which is later mapped to the penalty probabilities c_k^j for the two actions of the automaton $\mathcal{A}^j$. The overall search strategy CPL–ATS, in addition uses a decision table Λ to prune the search interval by comparing the output decisions Ω^j for the three partitions. Thus $\mathcal{A}^j$, $j \in \{1, 3\}$, together with E and Λ completely define the CPL–ATS strategy.

1. *The set of actions of the automaton*(Σ^j)
   The two actions of the automaton are $\alpha_{k=0,1}^j$, where, α_0^j corresponds to selecting the *left* half L^j of the partition Δ^j, and α_1^j corresponds to selecting the *right* half R^j.
2. *The action probabilities*(Π^j)
   $P_k^j(n)$ represent the probabilities of selecting action $\alpha_{k=0,1}^j$ at step n. Initially, $P_k^j(0) = 0.5$, for $k = 0, 1$.
3. *The feedback inputs from the environment to each automaton* (Γ^j)
   It is important to recognize a subtle, but crucial point in the construction of the learning automata in CPL–ATS. From the automaton's point of view, the two actions are those of selecting either the left or the right half from

its partition. However, from the environment's point of view, the automaton presents a current guess for the true value of $\lambda^\star$, and it gives a feedback based on the relative position (or direction) of the guessed value with respect to $\lambda^\star$. Thus there is a need to map the intervals to a point value and the feedback on the point value to the feedback on the choice of the intervals. Let the automaton select either the left or right half of the partition and then pick a point randomly (using a continuous uniform probability distribution) from this sub-interval which is presented as the current estimate for $\lambda^\star$. Then the feedbacks $\beta(n)$ at step n are defined by the conditional probabilities,

$$\begin{aligned} Pr[\beta(n) = 0 \,|\, x_L^j \in L^j \text{ and } x_L^j \geq \lambda^\star] &= p \\ Pr[\beta(n) = 1 \,|\, x_L^j \in L^j \text{ and } x_L^j < \lambda^\star] &= p \\ Pr[\beta(n) = 0 \,|\, x_R^j \in R^j \text{ and } x_R^j < \lambda^\star] &= p \\ Pr[\beta(n) = 1 \,|\, x_R^j \in R^j \text{ and } x_R^j \geq \lambda^\star] &= p \end{aligned} \tag{1}$$

Note that, the condition $x_L^j \in L^j$ indicates that the action α_0^j was selected, and the condition $x_R^j \in R^j$ indicates the other action α_1^j was selected.

4. *The action probability updating rules* (Υ^j)
First of all, since we are using the L_{RI} scheme, we ignore all the penalty responses. Upon reward, we obey the following updating rule :
If $\alpha_{k=0,1}^j$ was rewarded then,

$$\begin{aligned} &P_{1-k}^j[n+1] \leftarrow \theta . P_{1-k}^j[n] \\ &P_k^j[n+1] \leftarrow 1 - P_{1-k}^j[n+1] \end{aligned}$$

where $0 \ll \theta < 1$ is the L_{RI} reward parameter.

5. *The decision outputs at each epoch* (Ω^j)
From the action probabilities we infer the decision Ω^j of the L_{RI} automaton $\mathcal{A}^j$, after a fixed number N_∞, of iterations. Typically, N_∞ is chosen so as to be reasonably sure that the automaton has converged. Ω^j indicates that the automaton has inferred whether $\lambda^\star$ is to the *left*, *right* or *inside* the partition. The set of values that Ω^j can take and the preconditions are given by:

$$\Omega^j = \begin{cases} \mathit{Left} \text{ if} & P_0^j(N_\infty) \geq 1 - \epsilon. \\ \mathit{Right} \text{ if} & P_1^j(N_\infty) \geq 1 - \epsilon. \\ \mathit{Inside} \text{ Otherwise.} & \end{cases}$$

6. *The decision table for pruning the search space* (Λ)
Once the individual automata for the three partitions have made a decision regarding where they reckon $\lambda^\star$ to be, the CPL–ATS reduces the size of the search interval by eliminating at least one of the three partitions. The new pruned search interval Δ^{new} for the subsequent learning phase (epoch) is generated according to the pruning decision table Λ shown in Table 1.

Table 1. Decision table(Λ) to prune the search space based on the automata outputs Ω^j

Ω^1	Ω^2	Ω^3	**New Sub-interval** Δ^{new}
$Left$	$Left$	$Left$	Δ^1
$Inside$	$Left$	$Left$	Δ^1
$Right$	$Left$	$Left$	$\Delta^1 \cup \Delta^2$
$Right$	$Inside$	$Left$	Δ^2
$Right$	$Right$	$Left$	$\Delta^2 \cup \Delta^3$
$Right$	$Right$	$Inside$	Δ^3
$Right$	$Right$	$Right$	Δ^3

Theorem 1. *If the automaton* $\mathcal{A}^j = (\Sigma^j, \Pi^j, \Gamma^j, \Upsilon^j, \Omega^j)$ *interacts with the environment E and gets feedbacks obeying Equation (1), then the effective penalty probabilities* $c^j_{k=0,1}$ *for the two actions* $\alpha^j_{k=0,1}$ *are given by:*

$$c^j_0 = (1-p) + (2p-1).Pr(x^j_L < \lambda^\star \mid \alpha^j_0 \text{ was chosen}) \quad (2)$$

$$c^j_1 = p - (2p-1).Pr(x^j_R < \lambda^\star \mid \alpha^j_1 \text{ was chosen}) \quad (3)$$

¿From the theory of learning automata [6, 8], we know that the for any 2-action L_{RI} scheme, if $\exists k \in \{0,1\}$ such that $c^j_k < c^j_{1-k}$, then the action α^j_k is optimal and for this action $P^j_k(N_\infty) \to 1$.

By the construction of the automaton, once the left or right sub-interval is chosen, a point is picked from this interval using a uniform probability distribution. Therefore, the cumulative probability distributions for picking a guess value in a selected interval are given by:

$$Pr(x^j_L < x \mid x^j_L \in L^j) = \begin{cases} 0 & \text{if } x < \sigma^j \\ \frac{x-\sigma^j}{mid(\Delta^j)-\sigma^j} & \text{if } \sigma^j \le x < mid(\Delta^j) \\ 1 & \text{if } x \ge mid(\Delta^j). \end{cases} \quad (4)$$

$$Pr(x^j_R < x \mid x^j_R \in R^j) = \begin{cases} 0 & \text{if } x < mid(\Delta^j) \\ \frac{x-mid(\Delta^j)}{\gamma^j-mid(\Delta^j)} & \text{if } mid(\Delta^j) \le x < \gamma^j \\ 1 & \text{if } x \ge \gamma^j \end{cases} \quad (5)$$

By substituting $\lambda^\star$ for x, in the above equations, we get $Pr(x^j_L < \lambda^\star \mid x^j_L \in L^j)$ and $Pr(x^j_R < \lambda^\star \mid x^j_R \in R^j)$ in Equations (2) and (3) respectively.

7 Properties of CPL-ATS in an Informative Environment

We shall now state some results about how CPL-ATS behaves in an Informative Environment.

Lemmas 1 and 2 essentially use the ϵ-optimality property of L_{RI} automata to prove that they produce the correct decision output for each partition under an

Informative environment. Lemma 3 uses the monotonicity of the real interval to establish some restrictions on the combination of decision outputs for adjacent partitions. These are then used in Theorem 2 to prove that the decision table in Table 1 is complete for the Informative environment. Theorem 3 establishes that after elimination of one or more partitions, the remaining interval still contains $\lambda^\star$, thereby assuring convergence. It should be borne in mind that these are still probabilistic results, although the probability is shown to be potentially as close to unity as we want, provided that we choose the parameters for the L_{RI} automata appropriately.

Lemma 1. *For an Informative environment E, given the L_{RI} scheme with a parameter θ which is arbitrarily close to unity, the following is true:*

$$\begin{array}{l} \textit{If } (\lambda^\star \circledless \Delta^j), \textit{ then } Pr(\Omega^j = Left) \rightarrow 1. \\ \textit{If } (\lambda^\star \circledgtr \Delta^j), \textit{ then } Pr(\Omega^j = Right) \rightarrow 1. \\ \textit{If } (\lambda^\star \circleddash \Delta^j), \textit{ then } Pr(\Omega^j = \{Left,\ Inside \text{ or } Right\}) \rightarrow 1. \end{array}$$

Lemma 2. *For an Informative environment E, given the L_{RI} scheme with a parameter θ which is arbitrarily close to unity, the following is true:*

$$\begin{array}{l} \textit{If } (\Omega^j = Left) \quad \textit{then } Pr(\lambda^\star \circledless \Delta^j) \rightarrow 1 \\ \textit{If } (\Omega^j = Right) \ \textit{then } Pr(\lambda^\star \circledgtr \Delta^j) \rightarrow 1 \\ \textit{If } (\Omega^j = Inside) \textit{ then } Pr(\lambda^\star \circleddash \Delta^j) \rightarrow 1 \end{array}$$

Lemma 3. *In an Informative environment E, if the CPL-ATS learning mechanism uses the same L_{RI} scheme at all levels of the recursion, and the parameter θ is arbitrarily close to unity, the following is true:*

$$\begin{array}{l} \textit{If } (\Omega^j = Left) \quad \textit{then } Pr(\Omega^{j+1} = Left) \rightarrow 1 \\ \textit{If } (\Omega^j = Inside) \textit{ then } Pr(\Omega^{j+1} = Left) \rightarrow 1 \\ \textit{If } (\Omega^j = Right) \ \textit{then } Pr(\Omega^{j+1} = \{Left,\ Right \text{ or } Inside\}) \rightarrow 1. \end{array}$$

Theorem 2. *If the environment is Informative and if the partitions use the same L_{RI} scheme with parameters θ as close to unity as needed, then the decision table given in Table 1 is complete.*

A consequence of this theorem is that any entry not shown in the decision table is said to be *inconsistent* in the sense that for an Informative environment and appropriate θ, the probability of occurrence of this entry is arbitrarily close to zero.

Theorem 3. *If the algorithm uses the same L_{RI} scheme at all levels of the recursion with a parameter θ arbitrarily close to unity and N_∞ sufficiently large, then for an Informative environment, the unknown $\lambda^\star$ is always contained in the new search-interval Δ^{new} resulting from the application of the decision rules of Table 1.*

8 Properties of CPL-ATS in a Deceptive Environment

Let E be an environment of a 2-action learning automaton. Let $c_{k\in\{0,1\}}$ be the penalty probabilities of the two actions $\alpha_{k\in\{0,1\}}$ of any automaton $\mathcal{A}^j$ that operates in this environment. Then another environment E* with penalty probabilities $c'_{k\in\{0,1\}}$, is said to be the *dual* of E if and only if under E*, $c'_k = 1 - c_k$.

The following lemma is a natural corollary of the above definition.

Lemma 4. *If α^j_k is the ϵ-optimal action for an L_{RI} automaton $\mathcal{A}^j$ under a given environment E, then α^j_{1-k} is the ϵ-optimal action under its dual environment E^*, and vice versa.*

Lemma 5. *Let E be an environment with the probability of correct feedback p and coresponding penalty probabilities $c^j_{k\in\{0,1\}}$ respectively. If E* is a new environment constructed such that its probability of correct feedback $p' = 1 - p$, then the penalty probabilities $c'^j_{k\in\{0,1\}}$ for its two actions $\alpha^j_{k\in\{0,1\}}$ are $c'^j_k = 1 - c^j_k$.*

The above lemma follows easily by substituting $1 - p$ in place of p in Equations (2) and (3) respectively. Thus we arrive at a dual of a given environment merely by complementing its parameter p.

Let E be the given Deceptive environment. By definition then, we have its probability of correct response $p < 0.5$. We now construct a dual E* of this environment with a corresponding probability of correct response $p' = 1 - p$. Then this dual environment is Informative since, $p' > 0.5$. Thus if the learning autmaton can somehow determine whether a given environment is Deceptive, then Lemma 4 and 5 assure us that by interchanging the actions (or equivalently the penalties and rewards), the automaton will still be able to converge to the optimal action with as high a probability as we want.

The following results are quite straightforward.

Theorem 4. *Given a Deceptive environment E,*

$$\begin{array}{l} \textit{If } (\lambda^\star \olessthan \Delta^j), \textit{ then } Pr(\Omega^j = Right) \rightarrow 1 \\ \textit{If } (\lambda^\star \ogreaterthan \Delta^j), \textit{ then } Pr(\Omega^j = Left) \rightarrow 1 \\ \textit{If } (\lambda^\star \circledequal \Delta^j), \textit{ then } Pr(\Omega^j = \{Left,\ Inside\ or\ Right\}) \rightarrow 1. \end{array}$$

Theorem 5. *Suppose it is given that $\lambda^\star \ogreaterthan \Delta^1$, $\lambda^\star \circledequal \Delta^2$ and $\lambda^\star \olessthan \Delta^3$. Then under a Deceptive environment, the decision output vector $\boldsymbol{\Omega}$ for the three automata $\mathcal{A}^j, j \in \{1,2,3\}$, will be inconsistent with the decision table of Table 1. Conversely, if for the given environment and $\lambda^\star$ as above, the decision output vector $\boldsymbol{\Omega}$ of the automata is inconsistent with the decision table for large N_∞ and $\theta \rightarrow 1$, then the environment is Deceptive.*

Theorem 5 suggests a simple mechanism for determining whether or not an environment is Deceptive.

Theorem 6. *Let $\mathcal{I} = [0,1)$ be the original search interval in which $\lambda^\star$ is to be found. Let $\mathcal{I}' = [-1,2)$ be the initial search interval used by CPL–ATS. Then, CPL–ATS always determines whether or not an environment is Deceptive after a single epoch.*

Theorem 6 gives us a simple and practical mechanism for detecting Deceptive environments. Thus, we start with an initial search interval $\mathcal{I}' = [-1, 2)$, equipartition it and run the three L_{RI} automata $\mathcal{A}^j, j = 1, 2, 3.$ for one epoch. At the end of the epoch, we use Table 1 to check if the decision vector $\boldsymbol{\Omega}$ has an entry. By appealing to Theorem 6, we conclude that the environment is Informative or Deceptive accordingly. If the environment was found to be Deceptive, we simply flip the probability update rules. i.e., we essentially treat every reward as a penalty and vice versa. Lemma 4 guarantees that we will then converge to the optimal action. If instead, the environment was found to be Informative we simply proceed with the search.

Note that the expanded interval $\mathcal{I}'$ is needed only for the first epoch, to detect the environment's nature. Once this is detected, we use the original interval $\mathcal{I}$ to search for $\lambda^\star$. It is assumed that the fact that we use the expanded intervals $[-1, 0)$ and $[1, 2)$ does not affect the responses given by the environment.

9 Experimental Results

The parameter learning mechanism, CPL–ATS, described in this paper was experimentally evaluated to verify the validity of our analytic results and to examine its rate of convergence. To verify the power of the scheme and to study its effectiveness for various conditions, simulation experiments were conducted for various values of θ , the reward factor of the L_{RI} automata, and for various values of p, the probability of the environment correctly providing the feed back. In all the experiments it was assumed that $\lambda^\star \in [0, 1)$. In each case, a single screening epoch was run using the expanded interval $[-1, 2)$ to detect whether or not the environment is Informative. After that, the given original search interval $[0, 1)$ was used as the starting point. Each epoch consisted of 250 iterations (N_∞) of the three L_{RI} automata. At the end of each epoch the decision table Table 1 was consulted to prune the current search interval and the algorithm was recursively evoked. The recursion was terminated when the width of the interval was less than twice the desired accuracy.

The results of our experiments are truly conclusive and confirm the power of the CPL-ATS scheme. Although several experiments were conducted using various $\lambda^\star$ and parameter values, we report for brevity sake, only one set of results, namely, those for $\lambda^\star = 0.9123$. For this value, several independent replications with different random number streams were performed to minimize the variance of the reported results. The reported results are averaged over the replications.

The mean asymptotic value of the estimate for $\lambda^\star$ are shown in Table 2 for various values of p and θ. The final estimates agree with the true value 0.9123 of $\lambda^\star$to the first three decimal places for *all* values of p shown in the table. Figure 1 shows the convergence of the algorithm for $p = 0.1, 0.35$ and 0.8. This figure plots the running estimate at the end of each epoch. The values shown are actually the averaged over 50 replications for each set of parameters.

We first note that the convergence of the algorithm is almost identical for $p = 0.1$ and $p = 0.8$ even though the former represents a highly unreliable

Table 2. Asymptotic value of $\hat{E}[\lambda(N_\infty)]$ as it varies with p and θ. In all the cases, $\lambda^\star = 0.9123$, $N_\infty = 250$ and $\epsilon = 0.005$. The values shown are averaged over 50 independent experiments

p	$\theta = 0.8$	$\theta = 0.85$	$\theta = 0.9$
0.10	0.912298	0.912273	0.912194
0.15	0.912312	0.912298	0.912222
0.20	0.912193	0.912299	0.912236
0.80	0.912317	0.912284	0.912234
0.85	0.912299	0.912275	0.912202
0.90	0.912302	0.912275	0.912202

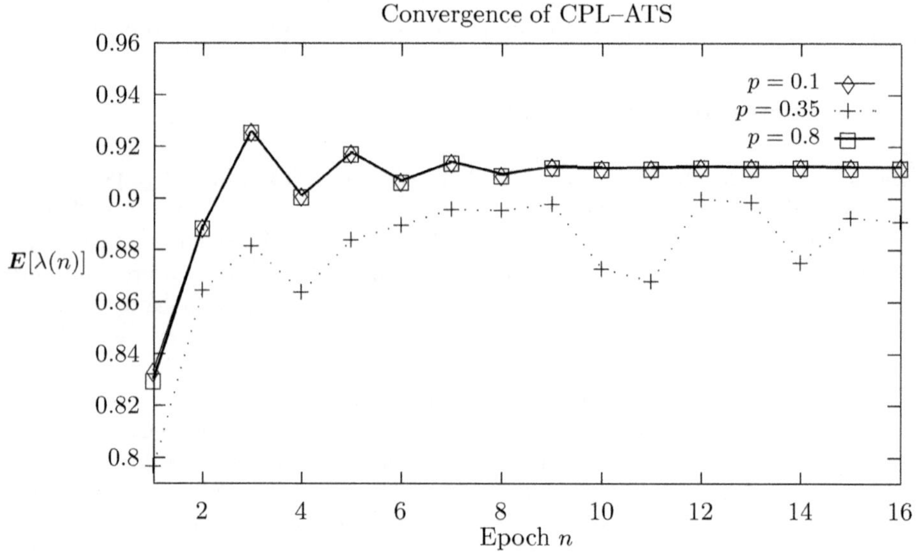

Fig. 1. Convergence of estimate $\hat{E}[\lambda(n)]$ for $\theta = 0.8$. The value shown are averaged over 50 replications. The unknown parameter $\lambda^\star = 0.9123$

environment whereas the latter is a highly reliable environment. This is not surprising because, after having detected in epoch 0 that the environment is not Informative, the CPL–ATS strategy switches the reward and penalty feedbacks, effectively behaving as $p' = 1-p$. Thus, even in the very first epoch a value of 0.83 is achieved which is within 9 percent error of the true value of $\lambda^\star = 0.9123$. In two more epochs, for all the four p values shown, the estimate is 0.925926 which is within 1.5 percent error. Note however, that for $p = 0.35$ the convergence is relatively sluggish. It took 25 epochs for it to reach the terminal value of 0.906453 which is within 0.64 percent of the true $\lambda^\star$. Thus, as p is increasingly closer to 0.5, θ must be set close to unity and N_∞ (the number of iterations per epoch) has to be increased to achieve convergence.[1] Thus, a p value very close to 0.5 represents a highly inconsistent feedback wherein half the time the environment directs the learning mechanism in one direction and the rest of the time in another.

10 Conclusions

In this paper we have considered the problem of a learning mechanism locating a point on a line when it is interacting with a random stationary environment which essentially informs it, possibly erroneously, which way it should move to reach the point being sought. The first reported paper to solve this problem [9] presented a solution which operated in a discretized space. Later in [10], we presented a new scheme by which the point can be learned using a combination of various learning principles. The heart of this strategy involved performing a controlled random walk on the underlying space using the L_{RI} updating rule and then intelligently pruning the space using an adaptive tertiary search. The overall learning scheme is shown to be ϵ-optimal. We have also shown that the results of [10] can also be generalized to the cases when the environment is a stochastic compulsive liar (i.e., is Deceptive). Although the problem is solved in its generality, its application in non-linear optimization has also been suggested. We are currently investigating how our scheme can be utilized to catalyze the convergence of various optimization problems including those involving neural networks as described in [9, 11, 19].

Acknowledgements

The first author is grateful to Professor Riccardo Baeza-Yates for introducing him to this very interesting problem in Santiago, Chile. The first author was partially supported by the Natural Sciences and Engineering Research Council of Canada. The work of the second author was supported by the Canadian Commonwealth Scholarships and Fellowships Program.

References

[1] Baeza-Yates, R. A., Culberson, J. C. and Rawlins, G. J. E., "Searching with uncertainty", *Proc. Scandinavian Workshop on Algorithms and Theory*, SWAT 88, *LNCS 318*, 1988, pp. 176-189. 25

[2] Baeza-Yates, R. A. and Schott, R., "Parallel Searching in the Plane", *Proc. 1992 International Conference of the Chilean Computer Society, IC-SCCC 92*, 1992, pp. 269-279. 25

[3] Bentley, J. L. and Yao, A. C-C., "An Almost Optimal Algorithm for Unbounded Searching", *Information Processing Letters*, Aug. 1976, pp.82-87. 26

[1] For the parameter values that we experimented, we found that the algorithm consistently converges for p in the ranges $(0, 0.4)$ and $(0.6, 1.0)$.

[4] Karlin, S., and Taylor, H. M., *A First Course on Stochastic Processes* , 2nd. ed.,Academic Press , 1974.

[5] Kashyap, R. L., and Oommen, B. J., "Scale Preserving Smoothing of Polygons", *IEEE Trans. on Pat. Anal. and Mach. Intel.*, Nov. 1983, pp. 667-671. 25

[6] Lakshmivarahan, S., *Learning Algorithms Theory and Applications*, Springer-Verlag, 1981. 30

[7] Lancôt, J. K. and Oommen, B. J., "Discretized Estimator Learning Automata", IEEE *Trans. on Syst. Man and Cybern.*, Vol. SMC-22, November/December 1992, pp. 1473- 1483.

[8] Narendra, K. S., and Thathachar, M. A. L., *Learning Automata*, Prentice-Hall, 1989. 30

[9] Oommen, B. J., "Stochastic Searching on the Line and its Applications to Parameter Learning in Nonlinear Optimization", *IEEE Transactions on Systems, Man and Cybernetics*, Vol.SMC-27, 1997, pp.733-739. 25, 26, 35

[10] Oommen, B. J. and Raghunath, G., "Automata Learning and Intelligent Tertiary Searching for Stochastic Point Location", *IEEE Transactions on Systems, Man and Cybernetics*, Vol. SMC-28B, 1998, pp. 947-954. 26, 27, 35

[11] Pao, Y.-H., *Adaptive Pattern Recognition and Neural Networks*, Addison Wesley, Reading, Mass., 1989. 25, 35

[12] Pavlidis, T., *Structural Pattern Recognition*, Springer-Verlag, New York, 1977. 25

[13] Pelc, A., "Searching with Known Error Probability", Theoretical Computer Science, 63: pp. 185-202. 27

[14] Press, W. H., Flannery, B. P., Teukolsky, S. A., and Vetterling, W. T., *Numerical Recipes : The Art of Scientific Computing*, Cambridge University Press, Cambridge, 1986. 25

[15] Oommen, B. J., Raghunath, G., and Kuipers, B., "Parameter Learning from Stochastic Teachers and Stochastic Compulsive Liars", Technical Report in preparation, School of Computer Science, Carleton University, Ottawa. The report can be obtained by contacting the first author. 25, 36

[16] Rao, S. S., *Optimization :Theory and Applications*, John Wiley, New Delhi 2nd. Ed., 1984. 25

[17] Rich, E. and Knight, K. *Artificial Intelligence.* McGraw Hill Inc., 1991.

[18] Tsetlin, M. L., *Automaton Theory and the Modelling of Biological Systems*, New York and London, Academic, 1973.

[19] Wasserman, P. D., *Neural Computing : Theory and Practice*, Van Nostrand Reinhold, New York, 1989. 25, 35

Appendix

In the following we provide only sketches of proof for major lemmas and theorems due to limitations of space. The complete proofs for all the lemmas and theorems stated in this paper can be found in the unabridged version which is also available as a technical report [15].

Theorem 1 *If the automaton* $\mathcal{A}^j = (\Sigma^j, \Pi^j, \Gamma^j, \Upsilon^j, \Omega^j)$ *interacts with the environment E gets feedbacks obeying equation 1, then the effective penalty probabilities* $c^j_{k=0,1}$ *for the two actions* $\alpha^j_{k=0,1}$ *are given by:*

$$c^j_0 = (1-p) + (2p-1).Pr(x^j_L < \lambda^\star \mid \alpha^j_0 \text{ was chosen})$$
$$c^j_1 = p - (2p-1).Pr(x^j_R < \lambda^\star \mid \alpha^j_1 \text{ was chosen})$$

Proof: By definition of the penalty probability we have,

$$\begin{aligned}
c_0^j &= Pr(\beta(n) = 1 \,|\, \text{sub-interval } L^j \text{ is chosen at step } n) \\
&= Pr(\beta(n) = 1 \,|\, x_L^j \in L^j, x_L^j < \lambda^\star).Pr(x_L^j < \lambda^\star \,|\, x_L^j \in L^j) + \\
&\quad Pr(\beta(n) = 1 \,|\, x_L^j \in L^j, x_L^j \geq \lambda^\star).Pr(x_L^j \geq \lambda^\star \,|\, x_L^j \in L^j) \\
&= p.Pr(x_L^j < \lambda^\star \,|\, x_L^j \in L^j) + (1-p).(1 - Pr(x_L^j < \lambda^\star \,|\, x_L^j \in L^j)) \text{ (by eq. 1)} \\
&= (1-p) + (2p-1).Pr(x_L^j < \lambda^\star \,|\, x_L^j \in L^j) \\
&= (1-p) + (2p-1).Pr(x_L^j < \lambda^\star \,|\, \alpha_0^j \text{ was chosen}).
\end{aligned}$$

In the same line of reasoning for the action α_1^j we can derive c_1^j. □

Lemma 1 *For an Informative environment E, given the L_{RI} scheme with a parameter θ which is arbitrarily close to unity, the following is true:*

If $(\lambda^\star \circledless \Delta^j)$, then $Pr(\Omega^j = Left) \to 1$.
If $(\lambda^\star \circledgtr \Delta^j)$, then $Pr(\Omega^j = Right) \to 1$.
If $(\lambda^\star \ominus \Delta^j)$, then $Pr(\Omega^j = \{Left,\ Inside\ or\ Right\}) \to 1$.

Proof: Consider first the case $\lambda^\star \circledless \Delta^j$. From Equation (4) and (5), we get $Pr(x_L^j < \lambda^\star) = Pr(x_R^j < \lambda^\star) = 0$. Substituting these in equations 2 and 3 we get the values $c_0^j = 1 - p$ and $c_1^j = p$. Since for an informative environment $p > 0.5$, we immediately have $c_0^j < c_1^j$. Learning automata theory then assures that for any ϵ-optimal scheme such as the L_{RI} scheme used here for each $\mathcal{A}^j$, α_0^j is the optimal action and hence $P_0^j[N_\infty] \to 1$. Similar arguments for $\lambda^\star \circledgtr \Delta^j$ lead to the conclusion $P_1^j(N_\infty) \to 1$.

Now consider the third case when $\lambda^\star \ominus \Delta^j$. By the definition of $\ominus$ we have, $\sigma^j \leq \lambda^\star < \gamma^j$. Consider first the possibility, $\sigma^j \leq \lambda^\star < mid(\Delta^j)$. In this case, by Equations (4), (5) we get

$$c_0^j = 1 - p + (2p-1).\frac{\lambda^\star - \sigma^j}{mid(\Delta^j) - \sigma^j}$$
$$c_1^j = p$$

Since $0.5 < p \leq 1$ in the above expressions for c_0^j and c_1^j, we can see that $c_0^j < c_1^j$, and hence α_0^j is the optimal action and for an ϵ-optimal scheme $P_0^j[N_\infty] \to 1$.

Similarly when $mid(\Delta^j) < \lambda^\star < \sigma^j$, we get,

$$c_0^j = p$$
$$c_1^j = p - (2p-1).\frac{\lambda^\star - mid(\Delta^j)}{\gamma^j - mid(\Delta^j)} \ .$$

Since $p > 0.5$, it follows that $(2p-1) > 0$ and $c_0^j > c_1^j$, and therefore, α_1^j is the optimal action, and from the ϵ-optimality of the L_{RI} scheme, we get $P_1^j[N_\infty] \to 1$.

When $\lambda^\star = mid(\Delta^j)$, $c_0^j = c_1^j$ and so there is no optimal action and hence $\epsilon < P_0^j[N_\infty], P_1^j[N_\infty] < 1 - \epsilon$.

The lemma immediately follows from the decision output rule (Ω^j) in the construction of the automaton $\mathcal{A}^j$ based on the above values for $P_k^j[N_\infty]$. □

Lemma 2 *For an Informative environment E, given the L_{RI} scheme with a parameter θ which is arbitrarily close to unity, the following is true:*

$$\begin{array}{ll} \text{If } (\Omega^j = Left) & \text{then } Pr(\lambda^\star \olessthan \Delta^j) \to 1 \\ \text{If } (\Omega^j = Right) & \text{then } Pr(\lambda^\star \ogreaterthan \Delta^j) \to 1 \\ \text{If } (\Omega^j = Inside) & \text{then } Pr(\lambda^\star \circledequal \Delta^j) \to 1 \end{array}$$

Proof: By the first hypothesis $\Omega^j = Left$, we have $Pr(\Omega^j = Left) = 1$, $Pr(\Omega^j = Right) = 0$, and $Pr(\Omega^j = Inside) = 0$. Therefore,

$$Pr(\lambda^\star \olessthan \Delta^j) = Pr(\lambda^\star \olessthan \Delta^j \mid \Omega^j = Left). \tag{6}$$

But by Baye's rule for conditional probabilities,

$$Pr(\lambda^\star \olessthan \Delta^j \mid \Omega^j = Left) = \tag{7}$$

$$\frac{Pr(\Omega^j = Left \mid \lambda^\star \olessthan \Delta^j).Pr(\lambda^\star \olessthan \Delta^j)}{Pr(\Omega^j = Left \mid \lambda^\star \olessthan \Delta^j).Pr(\lambda^\star \olessthan \Delta^j) + Pr(\Omega^j = Left \mid \lambda^\star \ogreaterthan \Delta^j).Pr(\lambda^\star \ogreaterthan \Delta^j)}$$

By Lemma 1, the product in the second term of the denominator tends to zero giving us

$$Pr(\lambda^\star \olessthan \Delta^j \mid \Omega^j = Left) \to 1.$$

When this is substituted in Equation (6), it leads to : $Pr(\lambda^\star \olessthan \Delta^j) \to 1$. By similar arguments (which we omit here in the interest of brevity), the lemma can be shown to hold for the other two hypotheses. □

Theorem 2 *If the environment is Informative and if the partitions use the same L_{RI} scheme with parameters θ as close to unity as needed, then the decision table given in Table 1 is complete.*

Proof: In any decision table with three input variables and three possible values for each of these variables, we expect a total of 27 potential entries. However, Lemma 3 imposes constraints on the output value combinations of the automata. A straightforward enumeration of these possibilities will show that the only possible combinations of outputs for the three L_{RI} automata are the seven entries shown in the decision table. Consequently Table 1 is complete. □

Theorem 3 *If the algorithm uses the same L_{RI} scheme at all levels of the recursion with a parameter θ arbitrarily close to unity and N_∞ sufficiently large, then for an Informative environment, the unknown $\lambda^\star$ is always contained in the new search-interval Δ^{new} resulting from the application of the decision rules of Table 1.*

Proof: Consider the first row of Table 1 where we see that $\Omega^1 = Left$, $\Omega^2 = Left$ and $\Omega^3 = Left$. Appealing to Lemma 2 for each of the automata outputs, we get $Pr(\lambda^\star \circledleq \Delta^1) \rightarrow 1$, $Pr(\lambda^\star \circledleq \Delta^2) \rightarrow 1$ and $Pr(\lambda^\star \circledleq \Delta^3) \rightarrow 1$. When we consider the fact that $\Delta^1 \prec \Delta^2 \prec \Delta^3$, the above three reduce to the equivalent predicate, $Pr(\lambda^\star \circledleq \Delta^1) \rightarrow 1$. By the definition of $\circledleq$ we have the two possibilities – $\lambda^\star \olessthan \Delta^1$ and $\lambda^\star \circledequal \Delta^1$. But, since $\lambda^\star \circledequal \Delta$ and $\Delta = \Delta^1 \cup \Delta^2 \cup \Delta^3$, we can rule out $\lambda^\star \olessthan \Delta^1$. Therefore, $Pr(\lambda^\star \olessthan \Delta^1) \rightarrow 1$. Thus, the partition Δ^1 that remains after pruning still contains $\lambda^\star$ with as high a probability as we want. The same arguments can be repeated for each of the other entries, and are omitted in the interest of brevity. □

Theorem 4 Given a Deceptive environment E,

$$\begin{array}{l} \text{If } (\lambda^\star \olessthan \Delta^j), \text{ then } Pr(\Omega^j = Right) \rightarrow 1 \\ \text{If } (\lambda^\star \ogreaterthan \Delta^j), \text{ then } Pr(\Omega^j = Left) \rightarrow 1 \\ \text{If } (\lambda^\star \circledequal \Delta^j), \text{ then } Pr(\Omega^j = \{Left,\ Inside \text{ or } Right\}) \rightarrow 1. \end{array}$$

Proof: Since E is a Deceptive environment, the probability of it giving correct responses is $p < 0.5$. Now construct a dual E* for this environment such that the probability of its correct feedback $(p' = 1-p) > 0.5$. Clearly, E* is an Informative environment, and hence we can apply Lemma 1 to it. Consequently, if we are given that $\lambda^\star \olessthan \Delta^j$, by Lemma 1, we know that for E* , $Pr(\Omega^j = Left) \rightarrow 1$. Lemma 4 however tells us that if α^j_k was the optimal action for E, then α^j_{1-k} is the optimal option for E* and vice versa. Therefore, we conclude that for E, $Pr(\Omega^j = Right) \rightarrow 1$. Similar arguments hold good for the other two hypotheses. □

Theorem 5 *Suppose $\lambda^\star$ is such that $\lambda^\star \ogreaterthan \Delta^1$, $\lambda^\star \circledequal \Delta^2$ and $\lambda^\star \olessthan \Delta^3$. Then under a Deceptive environment, the decision output vector $\boldsymbol{\Omega}$ for the three automata $\mathcal{A}^j, j = 1, 2, 3$, will be inconsistent with the decision table of Table 1. Conversely, if for the given environment the decision output vector $\boldsymbol{\Omega}$ of the automata is inconsistent with the decision table for large N_∞ and $\theta \rightarrow 1$, then the environment is Deceptive.*

Proof: Applying Theorem 4 to each of $\lambda^\star \ogreaterthan \Delta^1$, $\lambda^\star \circledequal \Delta^2$ and $\lambda^\star \olessthan \Delta^3$, we have,

$$\begin{array}{l} Pr(\Omega^1 = Left) \rightarrow 1 \\ Pr(\Omega^2 = \{Left,\ Inside \text{ or } Right\}) \rightarrow 1 \\ Pr(\Omega^3 = Right) \rightarrow 1 \end{array}$$

By inspecting the decision table (Table 1), we see that there are no entries for this output vector $\boldsymbol{\Omega} = [Left, \{Left,\ Inside \text{ or } Right\}, Right]$ and hence the entry is inconsistent with Table 1. The converse is proved by contradiction by alluding to the completeness result of Theorem 2 for an Informative environment. Therefore, whenever the decision output vector $\boldsymbol{\Omega}$ is inconsistent with Table 1, we can safely conclude that the environment is Deceptive. □

Theorem 6 *Let $\mathcal{I} = [0, 1)$ be the original search interval in which $\lambda^\star$ is to be found. Let $\mathcal{I}' = [-1, 2)$ be the initial search interval used by CPL–ATS. Then,*

CPL–ATS always determines whether or not an environment is Deceptive after a single epoch.

Proof: First of all, we can see that $\lambda^\star \in \mathcal{I}'$ because, $\mathcal{I} \subset \mathcal{I}'$. When we divide $\mathcal{I}'$ into three equi-partitions we get, $\Delta^1 = [-1, 0)$, $\Delta^2 = [0, 1)$ and $\Delta^3 = [1, 2)$. Since $\lambda^\star \in \mathcal{I} = \Delta^2$, we have, $\lambda^\star \ogreaterthan \Delta^1$, $\lambda^\star \circledequal \Delta^2$, $\lambda^\star \olessthan \Delta^3$, which is the pre-condition for Theorem 5. Hence, by appealing to Theorem 5 we see that if the environment was Deceptive, we would get an inconsistent decision vector. If not, by Theorem 2 we would get a consistent decision vector. Thus, after *one* epoch we conclude decisively about the nature of the environment. □

Multimedia Analysis and Synthesis

Mohan S. Kankanhalli

School of Computing
National University of Singapore
mohan@comp.nus.edu.sg
http://www.comp.nus.edu.sg/~mohan

Abstract. We describe novel approaches to multimedia analysis and synthesis problems. We first present the experiential sampling technique which has the ability to focus on the analysis task by making use of the contextual information. Sensor samples are used to gather information about the current environment and attention samples are used to represent the current state of attention. In our framework, the task-attended samples are inferred from context and maintained by a sampling based dynamical system. The multimedia analysis task can then focus on the attention samples only. Moreover, past experiences and the current environment can be used to adaptively correct and tune the attention model. This experiential sampling based analysis method appears to be a promising technique for general multimedia analysis problems. We then present the multimedia synthesis technique based on analogies. This method aims to synthesize new media objects based on some existing objects on which appropriate transformation can be applied using analogical reasoning. This multimedia synthesis technique appears particularly useful in the area of computational media aesthetics.

1 Introduction

1.1 Multimedia Analysis

Multimedia processing often deals with live spatio-temporal data which have the following attributes:

- They possess a tremendous amount of redundancy
- The data is dynamic with temporal variations
- It does not exist in isolation – it exists in its context with other data. For instance, visual data comes along with audio, music, text, etc.

However, many current multimedia analysis approaches do not fully consider the above attributes which leads to inefficiency and lack of adaptability. The inefficiency arises from the inability to filter out the relevant aspects of the data and thus considerable resources are expended on superfluous computations on redundant data. Hence

T.D. Gedeon and L.C.C. Fung (Eds.): AI 2003, LNAI 2903, pp. 41-52, 2003.

speed-accuracy tradeoffs cannot properly be exploited. If the ambient experiential context is ignored, the approaches cannot adapt to the changing environment. Thus, the processing cannot adapt itself to the task at hand.

On the other hand, we have solid evidence that humans are superb at dealing with large volumes of disparate data using their sensors. Especially the human visual system is quite successful in understanding the surrounding environment at appropriate accuracy quite efficiently. We argue that like in the case of human perception, multimedia analysis should be placed in the context of its experiential environment. It should have the following characteristics: 1. The ability to "focus" (have attention), i.e., to selectively process the data that it observes or gathers based on the context. 2. Experiential exploration of the data. Past analysis should help improve the future data assimilation. In return, these two attributes would help the analysis to deal with the redundancy and diversity of the spatial-temporal data which is particularly important for real time applications.

In order to achieve this, we describe a novel technique called experiential sampling, i.e., sampling multimedia data according to the context. As shown in Figure 1, by sensing the contextual information in the experiential environment, a sampling based dynamic attention model is built to maintain the focus towards the interest of the current analysis task. Only the relevant samples survive for performing of the final task. These samples precisely capture the most important data. What is interesting is the past samples influence future sampling via feedback. This mechanism ensures that the analysis task benefits from past experience.

As an illustrative example of an analysis task, the face detection problem in videos is described. The experiential sampling technique can be utilized for many other applications involving multimedia analysis such as object detection, object recognition, object tracking (face recognition or traffic sign recognition), context aware video streaming and surveillance.

1.2 Multimedia Synthesis

A well-produced video always makes a strong impression on the viewer. However due to the limitations of the camera, the ambient conditions, or the skill of the videographer, sometime the quality of captured videos falls short of expectation, such as those from a war, medical check-up, surveillance or home videos. On the other hand, we have vast amount of superbly-captured videos available on the web and digital libraries. We describe the novel approach of video analogies which provides a powerful ability to improve the quality of videos by utilizing computable video features. We denote the mechanism of video analogies as A:B::A':B'. B is a target video whose quality we wish to improve and A is a high-quality source video having similar content. During the learning phase, we find the correlation between this pair. We obtain the correspondence by using feature learning. Then for the target video B, we utilize this correspondence to transfer some desired trait of A (which is A') in B in order to synthesize a new video B'. Thus the new video B' will obtain the desired features A' of the source video A while retaining the merits of the target video B. We demonstrate the power of the analogies technique by describing two applications – video rhythm adjustment and audio-video mixing. We describe the details of the technique in each case and provide experimental results to establish the efficacy of the proposed technique.

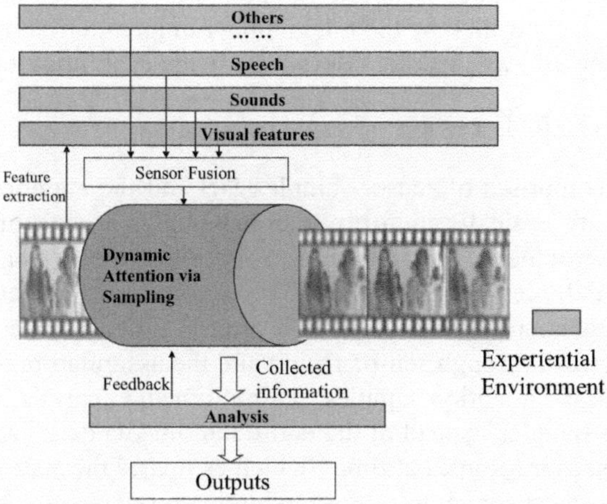

Fig. 1. Experiential sampling for multimedia analysis

2 Experiential Sampling

2.1 Preliminaries

Context in multimedia analysis is any information that needs to be specified to characterize the current state of the multimedia system. It includes the current environment, a priori knowledge of the system domain, current goals and the past states. The current goal and prior knowledge provide a top-down approach to analysis. It also determines which features of the visual scene and other accompanying data type should be used to represent the environment. The past states encapsulate the experiences till the current state. More importantly, when we consider the experiential environment, the analysis task needs to systematically integrate the top-down and bottom-up approaches.

2.2 Sensor Sampling

Studies on human visual system show that the role of experiences used in top-down visual perception increases in importance and can become indispensable when the viewing conditions deteriorate or when a fast response is desired. In addition, humans get information about the objects of interest from different sources of different modalities. Therefore, when we analyze one particular data type (say spatio-temporal visual data) in multimedia, we cannot constrain our analysis to this data type only. Sensing other accompanying data like audio, speech, music, and text can help us find out where is the important data. Therefore, it is imperative to develop a sampling framework which can sense and fuse all environmental context data for the purpose of

multimedia analysis. The current environment is first sensed by uniform random sensor samples and based on the context, we compute the attention samples to discard the irrelevant data. Spatially, higher attended samples will be given more weight and temporally, attention is controlled by the total number of attention samples.
In the sampling framework, we represent the environment e_t at time t as:

$$e_t = \{S(t), A(t)\} \tag{1}$$

The environment e_t comprises of sensor samples $S(t)$ and the attention samples $A(t)$. The sensor samples are basically uniform random samples at any time t which constantly sense the environment. The attention samples are the dynamically changing samples which essentially represent the data of interest at time t. The attention samples are actually derived dynamically and adaptively at each time instance from the sensor samples in our framework through sensor fusion and the assimilation of the past experience. Once we have the attention samples, the multimedia analysis task at hand can work only with these samples instead of the entire multimedia data. In our framework, $S(t)$ is a set of $N_S(t)$ sensor samples at time t which estimates the state of the multimedia environment. These sensor samples are randomly and uniformly generated. Since the number of the sensor samples do not change with time, N_S denotes the number of sensor samples at any point in time. $S(t)$ is then defined as:

$$S(t) = \left\{s(t); \Pi^S(t)\right\} \tag{2}$$

where s(t) depends on the type of multimedia data.

2.3 Attention Sampling

Our sampling based dynamic attention model systematically integrates the top-down and bottom-up approaches to infer attention from the environment based on the context. Thus, the number of attention samples dynamically evolves so the number will be increased when more attention is required and vice-versa. Moreover feedback from the final analysis task is used to tune the attention model with time. The attention in a scene can be represented by a multi-modal probability density function. Any assumptions about the form of this distribution would be limiting. However, not making any assumption about this distribution leads to intractability of computation. Therefore, we adopt a sample-based method to represent the visual attention. For example, in the one dimensional case, the visual attention is maintained by N samples $a(t)$=[$s^1(t)$,…,$s^N(t)$] and their weights =[$\pi^1(t)$,…, $\pi^N(t)$] as shown in Figure 2. It provides a flexible representation with minimal assumptions. The number of samples employed can be adjusted to achieve a balance between the accuracy of the approximation and the computation load. Moreover, it is easy to incorporate within a dynamical system which can model the temporal continuity of visual attention.

We represent the dynamically varying $N_A(t)$ number of attention samples $A(t)$ using:

$$A(t) = \left\{a(t); \Pi^A(t)\right\} \tag{3}$$

where $a(t)$ depends on the type of multimedia data. Consider the traffic monitoring application shown in Figure 3, Figure 3 (a) has more motion activity and hence needs more attention samples to represent this motion attention. As shown in Figure 3 (b), 567 attention samples (marked as yellow points) are required to represent this motion attention using our method. In contrast, Figure 3 (c) has less motion and needs fewer attention samples. As shown in Figure 3 (d), no attention samples are needed.

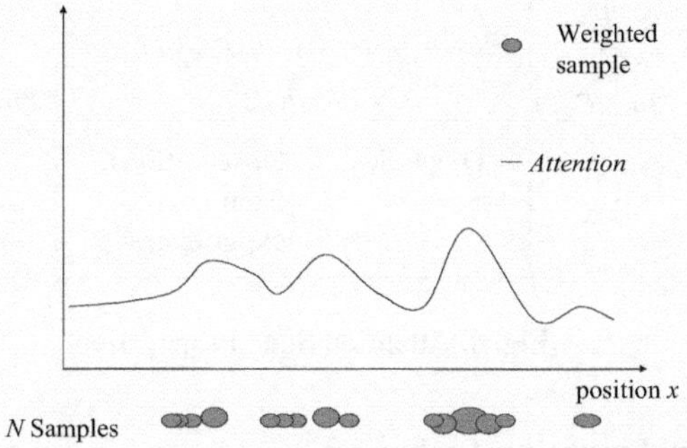

Fig. 2. The multi-modal attention can be represented by N samples $a(t)=[s_1(t),\ldots,s_N(t)]$ and their weights $=[\pi_1(t),\ldots,\pi_N(t)]$

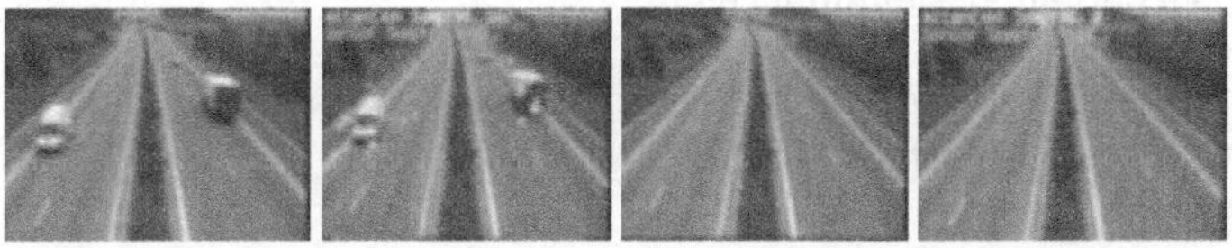

(a) frame 37 (b) frame 37 (c) frame 479 (d) frame 479

Fig. 3. Temporal motion attention. (a) more motion activity (b) 567 attention samples are employed to represent this motion attention. (c) need less attention at this time (d) No attention samples are needed at this time

Dynamical Evolution of Attention

Attention is inferred from the observed experiences coming from the experiential environments. That is, we try to estimate the probability density of the attention (which is the state variable of the system) at time t using $P(a_t|E_t)$. Note that E_t consists of all the observed experiences until time t which is $E_t=\{e_1,\ldots,e_t\}$, at is the "attention" in the scene and $a(t)$ is the sampled representation of at. Attention has temporal continuity which can be modeled by a first-order Markov process state-space model. The new state depends only on the immediately preceding state, independent of the earlier history. This still allows quite general dynamics, including stochastic difference equations of arbitrary order. Therefore,

$$P(a_t \mid a_{t-1},\ldots,a_0) = P(a_t \mid a_{t-1}) \tag{4}$$

Based on the above state space model, the a posteriori density $P(a_t|E_t)$ can iteratively be obtained by knowing the observations (likelihood) $P(e_t|a_t)$, the temporal continuity (dynamics) $P(a_t|a_t\text{-}1)$ and the previous state density $P(a_{t-1}|E_{t-1})$ as shown in Figure 4.

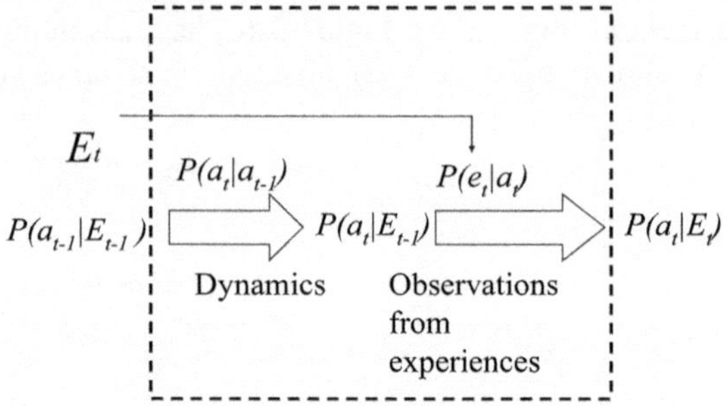

Fig. 4. Attention State Propagation

2.4 Experiential Sampling Technique

In this section, we only provide the outline of the experiential sampling algorithm. The interested readers may consult [3] for details. Briefly, the ES is as follows:

Algorithm: Experiential Sampling (ES):

1. Initialization: t=0
2. {SS(t)} ← Uniform Sampling
3. Asat(t) ← sum of {SS(t)}
4. Ns(t) ←Asat(t)
5. if Ns(t) =< 0; t=t+1; goto step 2
6. {AS(t)} ←Importance Resampling from {SS(t)}
7. *for* each AS, perform the analysis task
8. t=t+1; goto step 2.

As a general analysis framework, the experiential sampling technique can be used for a variety of multimedia analysis tasks, especially real-time applications like traffic monitoring and surveillance. As a test example, we have applied this framework for the face detection problem in videos. We use the adaboost face detector as the multimedia analysis task. Face detection is only performed on the attention samples to achieve robust real time processing. Most importantly, past face detection results serve as context to adaptively correct the attention samples and the skin color model in the attention inference stage. This allows the face detector to cope with a variety of changing visual environments

We will now describe the results of the experiential sampling technique when applied to the face detection problem. Sensor samples are employed to obtain the context. The face attention is maintained by the attention samples. The face detector is executed only on the attention samples which indicate the most probable face data. As shown in Figure 5, N_S number of sensor samples is set to 200.The number and spatial distribution of attention samples can dynamically change according to the face attention. In Figure 5(a), there is no motion in the frame, so N_A, the number of attention samples is zero. No face detection is performed. In Figure 5(b), when a chair enters, it alerts the motion sensor and attention is aroused. N_A increases to 414. Face detection is performed on the 414 attention samples. But the face detector verifies that there is no face there. In Fig 5(c) as the chair stops, there is no motion and so the attention samples vanish. In Figure 5(d)-(h) attention samples come on with the face until the face vanishes. Note the number of attention samples is different in each case.

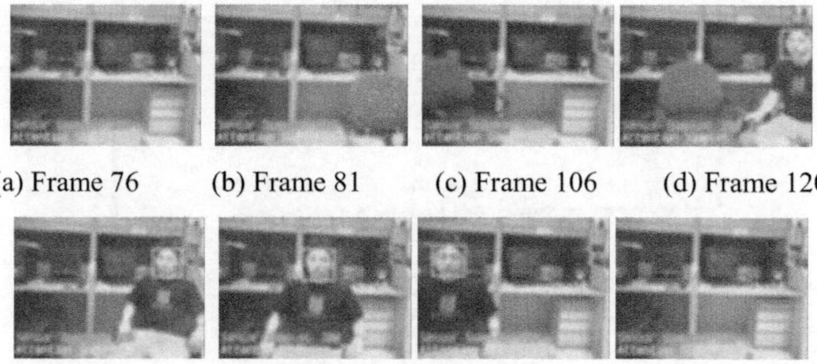

(a) Frame 76 (b) Frame 81 (c) Frame 106 (d) Frame 120

(e) Frame 147 (f) Frame 152 (g) Frame 158 (h)Frame 268

Fig. 5. Face detection sequence 1. (a) static frame N_A=0 (b)A chair moves N_A =414 (c)the chair stopped. N_A =0 (d) a person comes. N_A =791 (e) a person. N_A =791 (f) one person. N_A =791 (g) one person. N_A =791 (h) static frame. N_A =2

3 Synthesis Using Analogies

3.1 Preliminaries

We will now describe a method to synthesize multimedia objects using analogical reasoning. The idea is to transfer the desired characteristics from a multimedia object into the given object. We now provide a general description of our scheme with respect to digital video. Given a target video and a designated source video with similar content, we match one common feature of the video pair. Utilizing this feature correspondence, we transfer some other feature of the source video to the target video. For instance, we can build texture correspondence between the video pair and then transfer the color feature of the source video to the target video. We would like to point out here is that the compared videos should be similar in terms of video content at the shot level so that a proper analogy is set up.

The framework of video analogies involves two phases as shown in Figure 6: learning and transfer. The inputs are a source video and a target video. At the learning stage, the corresponding computable features in the source video **A** and the target video **B** need to be extracted and compared for similarity (**A:B**). During the transfer stage, we establish a new function to transfer the desired features (**A'**) from the source video to the target video to create the new video **B'** by employing the analogy **A:B::A':B'**. The source video is assumed to be a high-quality clip which we wish to emulate. The target video possibly has some artifacts or shortcomings which we wish to overcome by mimicking the source video. We have observed that our framework of video analogies can potentially be utilized for several problems:

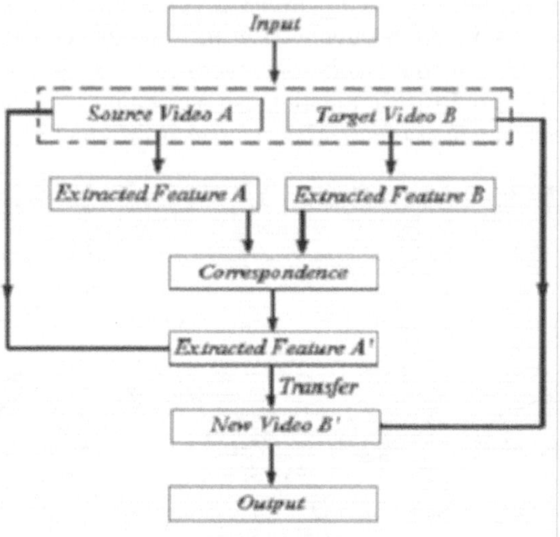

Fig. 6. Flowchart for video analogies

- *Color transfer*: We can transfer suitable colors from a source video to a target video (X-ray or infrared video). This can generate a more vivid video.
- *Texture transfer*: By transferring the specific texture to some areas, we can overlap and patch areas on video frame such as annoying video logos.
- *Motion trajectory transfer*: By transferring the desired motion to the target video, we can actually remove or reduce shakes caused by camera vibration. We can also borrow shakes from a source video to add excitement to a target video.
- *Music matching*: Automatic music matching for atmosphere enhancement is a crucial step in video editing. From the source videos, we can track features in order to add similar music to the target video.
- *Aesthetic styles transfer*: Artistic styles in a professional movie embody the experience and knowledge of the directors. Most of these artistic features are at the semantic level; however some of them are apparent at the low level itself such as color, lighting, motion trajectory and rhythm. Such stylistic aspects can be advantageously transferred.

Although analogies based digital video handling has many applications, here we describe our work on two problems: video rhythm adjustment and audio-video mixing.

3.2 Video Rhythm Adjustment

Video rhythm refers to the duration and frequency of segmented events; it is subject to the pace of the events and the relationships between these events. The overall rhythm is usually determined by the transitions between video components such as shots, scenes, and sequences. Usual home videos often do not possess a proper aesthetic rhythm. In order to emphasis special scenes, actions, characters and atmosphere in a video, video rhythm adjustment is extremely important. In this section, we discuss video rhythm adjustment based on video analogies. Our motivation is that video rhythm adjustment can be used to emphasize the content and significantly enhance the atmosphere.

In order to perform video rhythm adjustment, we work at the video clip level. Given a source video V_1 and a target video V_2, we segment them into several clips based on the events which are subject to our needs and obtain the clip lengths (total frame number). Suppose V_1 has n clips: $S_1, S_2,\ldots, S_n$, V_2 has m clips: $s_1, s_2,\ldots, s_m$, we can find the clip proportion of $V_1(S_1: S_2:\ldots: S_n)$. From this ratio, we can determine the relative duration or frame numbers of the video clips. We use it to modify the target video by keeping the ratio invariant. Namely:

$$C_i = C_{i-1} \cdot S_i/S_{i-1}, (m>i> 2) \tag{5}$$

where C_i is the new length of video clips, C_0 can be fixed, for instance, $C_0= S_0$.

In order to automatically detect the rhythm changes in a video shot, we use the video motion feature. We subtract two adjacent frames in the shot. i.e. $\Delta_i = |F_j-F_{j-1}|$, $F_j\in V_i, j<S_i$. If the motion in a clip is large, the difference will be significant, or else the distinction is minor. Thus we can find these minor differences by calculating the density of minor differences in this portion. i.e. $\Omega(\Delta_j) = \sum_{T>\Delta i}/ \sum_{j<Si}$. If a density is the highest one among all segmentations of this clip, we think the rhythm in this portion is slow. In practice, we need to add frames or drop frames from the clips of target video according to equation (6):

$$d_i = C_i - s_i \tag{6}$$

If d_i >0, then we need add d_i frames in the i-th clip. New frames can be created by frame interpolation or replication. If d_i <0, then we need drop d_i frames from the i-th clip. We add or drop video frames according to equation (7).

$$C_{ij} = [j * C_i / s_i], j =1,2,\ldots, s_i \tag{7}$$

The procedure to drop frames from a clip is rather easy, however the procedure to add frames for a clip is difficult, requiring interpolation and frame synthesis. In our implementation, we just replicate the adjacent frames to expand the sequence which is sufficient in many cases.

Figure 7 depicts two groups for video rhythm adjustment. Given a video clip from a movie (a), we map the proportion of clip length to the target video (b), we get an exciting video clip (c). Group 1 is a video about Michael Jackson's dance performance to which we transfer the rhythm of a piece of clip of the movie "Hero". This transfer emphasizes the dancing skills of Michael Jackson. Group 2 is the video about creaming of Bill Gates, to which we transfer the rhythm of one clip of the movie of Jackie Chan (Flying Motorcycle) and obtain a new vivid clip that accentuates the creaming effect.

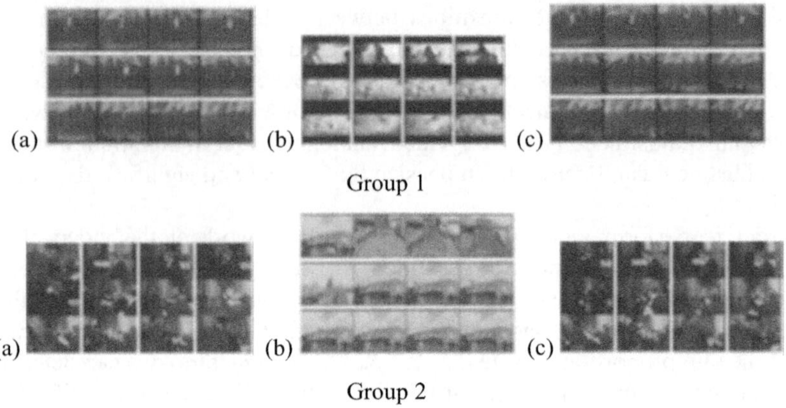

Fig.7. Video rhythm transfer

3.3 Audio-Video Mixing

There have been efforts in the recent years to make home videos look more pleasing to viewers by mixing it with appropriate music. Most of the existing software enables the user to add music of his preference. It assumes that the user has enough knowledge about the aesthetic mixing principles. We use the analogies technique to add audio to video by synthesizing appropriate music based on the video content. We have developed a system that takes in music examples selected by the user and generates new music by applying the aesthetic rules of audio-video mapping. We have **A** (an example) and **B** (the output pitch profile based on the given video). We need to transfer certain aspects of **A** (melody, rhythm, etc.) to **B**. For transferring melodies, we derive an underlying **A'** and pose the problem as **A':A::B:B'**. Given the hue profile of the video, we arrive at the melody profile for video that closely corresponds to the example melody track. The derived melody profile follows the contours of the example.

The music examples in our experiments are melodies mainly selected from western classical instrumental music. Figure 8 gives the pitch contour of a melody. The profile in solid line indicates the original pitch and the profile in dotted line is the Haar wavelet approximation of the music example. The pitch contour of the video as shown in Figure 9 is derived from the hue of every frame of the video. The sequence comparison method gives us the notes that are 'similar' to the music example. The velocity of the note is also computed from the brightness of video and assigned to every note. The pitch, volume so generated are re-assembled in the midi format and then converted to midi music. The matched contour is shown in Figure 10. A user survey done on the analogy results suggests that the music generated is acceptable though some parts of music may seem repetitive or may not be musically pleasing. However, this analogies based method in general is preferred over the rule based generation of chord music with which we had experimented earlier.

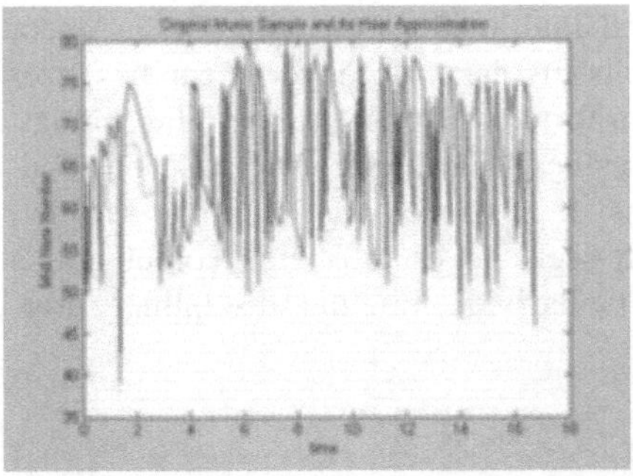

Fig. 8. Actual Pitch Contour of a Bach melody

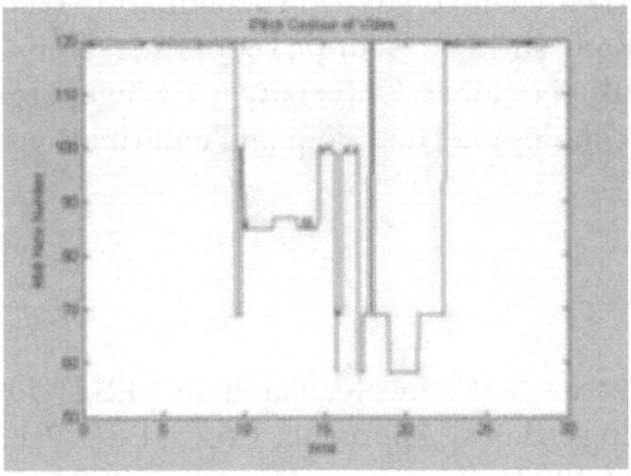

Fig.9. Desired Pitch Contour of a Video Clip Obtained from Aesthetic Mapping

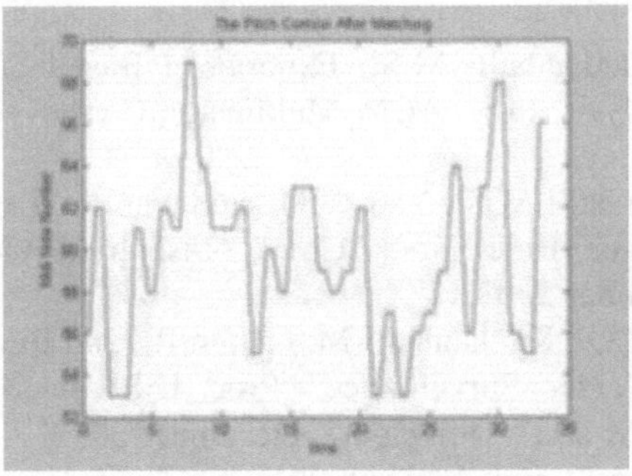

Fig.10. Pitch Contour of the Synthesized Music for the Video Clip

4 Conclusion

We have described a novel sampling based framework for multimedia analysis called experiential sampling. Based on this framework, we can utilize the context of the experiential environment for efficient and adaptive computations. This technique can be extended for general multimedia processing when operating with multiple data streams with possibly missing data. Moreover, it can be incorporated into a dynamical feedback control system for continuous systems. The analogy synthesis technique is a powerful multimedia synthesis technique which can be used for many consumer entertainment applications. It is particularly useful in the new area of computational media aesthetics which seeks to establish the computational foundations of media aesthetics. This is a particularly exciting interdisciplinary area of research.

Acknowledgement

The work described in the paper has been done in collaboration with international colleagues: Jun Wang (Delft University of Technology), Wei-Qi Yan (National University of Singapore), Ramesh Jain (GeorgiaTech), S H Srinivasan (Satyam Computer Services), Meera Nayak (National University of Singapore) and Marcel Reinders (Delft University of Technology). Their help and contributions have been invaluable.

References

[1] Nayak M., Srinivasan S.H., and Kankanhalli M.S., Music Synthesis for Home Videos: An Analogy Based Approach, Proc. IEEE Pacific-Rim Conference on Multimedia (PCM 2003), Singapore, December 2003.

[2] Mulhem P., Kankanhalli M.S., Hassan H., and Yi J., Pivot Vector Space Approach for Audio-Video Mixing, IEEE Multimedia, Vol. 10, No. 2, (2003) 28-40.

[3] Wang J. and Kankanhalli M.S., Experience Based Sampling Technique for Multimedia Analysis, Proc. ACM Multimedia Conference 2003, Berkeley, November 2003.

[4] Wang J., Kankanhalli M.S., Yan W.Q., and Jain R.. Experiential Sampling for Video Surveillance, Proc. First ACM Workshop on Video Surveillance 2003, Berkeley, November 2003.

[5] Wang J., Yan W.Q., Kankanhalli M.S., Jain R., and Reinders M.J.T., Adaptive Monitoring for Video Surveillance, Proc. IEEE Pacific-Rim Conference On Multimedia (PCM 2003), Singapore, December 2003.

[6] Yan W.Q., Wang J., and Kankanhalli M.S., Video Analogies. [manuscript under preparation]

Modelling Message Handling System

Insu Song and Pushkar Piggott

School of Information Technology
Faculty of Engineering and Information Technology
Gold Coast Campus, Griffith University
PMB 50 Gold Coast Mail Centre Q 9726
Fax: 07 5552 8066, Phone: 07 5552 8639
insu.song@kopo.com, p.piggott@griffith.edu.au
http://www.gu.edu.au

Abstract. This paper introduces a new approach to designing a Message Handling Assistant (MA). It provides a generic model of an MA and an intention extraction function for text messages using speech act theory and the belief-desire-intention (BDI) model of rational agency. The model characterizes the desired behaviors of an MA and the relationships between the MA, its user, and other human agents with whom the user interacts through message exchange. The intention extraction function formulates intentions from performatives identified by a probabilistic dialogue act classifier using constraints for felicitous human communication. This paper also proposes a semantic communication framework which integrates key agent and Internet technologies for composing and exchanging semantic messages.

Keywords: Intelligent agents, Belief revision and update, Knowledge discovery and data mining, Knowledge acquisition, Ontology, Applications.

1 Introduction

The recent proliferation of agent technology [1] has produced many proposals to reduce workload through learning and emergent behaviors (e.g., Maes [2]). Commonly, a personal assistant agent learns by observing the user's interaction with other applications and agents.

The information contained in email and other messages the user sends and receives is often closely related to their daily schedule and activities. Messages also often explicitly express both the user's and the correspondent's intentions. Despite this, the unstructured nature of such messages makes it hard to capture useful modelling information from them or to process them automatically. Thus this rich source of user information is largely untapped, and most message management is left to the user.

Several email handling assistants [2, 3, 4] have been proposed, but most approaches are ad-hoc and only address very limited services. In particular,

T.D. Gedeon and L.C.C. Fung (Eds.): AI 2003, LNAI 2903, pp. 53–64, 2003.

little effort has been made to develop a message handling system based on the popular human communication models of *speech act* and *rational agent theory*.

In this paper, we define and characterize a generic model of a Personal Assistant (PA) and a Message Handling Assistant (MA) based on speech act theory [5, 6, 7, 8] and the BDI model of rational agency [9, 10, 11]. This simple model demonstrates why and how identifying the intentions of messages plays a crucial role in user modelling, and in designing PAs and MAs. In addition, we develop a simple function that extracts intentions from text messages using speech act theory and probabilistic dialogue act classifiers. Finally, as a more grand vision toward fully mechanized message processing and composing, we propose a semantic communication framework (SCF) that combines key AI technologies: multiagent, agent communication language, ontology, and semantic Web. We believe that the exploration presented in this paper provides a sound model for a message handling systems that can provide more sophisticated services than are currently available.

This paper is organized as follows: Section 2 defines and characterizes a generic MA. Section 3 develops a model of an intention extraction function. In Section 4, we propose the semantic communication framework (SCF). In Section 5, we discuss how our approach differs from others. The paper finishes with concluding remarks in Section 6.

2 Modelling a Message Handling Assistant

This section defines and characterizes a generic model of a message handling assistant (MA). It uses the abstraction mechanism provided by the BDI model and speech act theory to express the desired behaviors of an MA and the relationships between an MA, its user, and other human agents with whom the user interacts through message exchange.

This enables us to clearly understand the characteristics of an MA, how to model a user, and how to design an MA that satisfies certain requirements. In particular, we are interested in the role of the intentions contained in the messages that the user exchanges with other human agents. The result of this analysis provides justification for why and how identifying the intentions of messages plays a crucial role in user modelling and the design of an MA.

2.1 Belief-Desire-Intention (BDI) Model of Rational Agency

We adapt the BDI framework described by Wooldridge [9], which is based on Rao and Georgeff's belief-desire-intention (BDI) framework [10] of rational agency. Their theory is based on the intentional logic which was developed by Cohen and Levesque [11] in 1990 to build their speech act theory [6, 7]. Speech act theory is a popular human communication model and it is the basis of agent communication languages (ACLs) [12]. The best-known implementation of the BDI model is the Procedural Reasoning System (PRS) and its descendants [13, p.140]: JAM, dMARS, AgentSpeck(L).

In PRS, the agent's mental states are represented in data structures that correspond to its belief, desire, and intention. The agent's beliefs are a model of its environment, usually represented in first-order logic as predicates. The agent updates its beliefs when it senses changes in its environment. Based on these beliefs and its intentions (committed goals), it considers possible options (desires). The agent may drop intentions or adopt new ones based on its beliefs, options, and its intentions. In PRS, desires and intentions are given as a plan library which consists of a body and various conditions. The body of a plan is the actual plan that specifies a series of actions or sub-goals to be achieved. The conditions are used to choose some plans as options and others as committed plans.

The availability of practical implementations of the BDI model, and its compatibility with speech act theory, make it a viable choice for designing a message handling system that is based on speech act theory and an agent communication language.

Agent Environment: We make the following assumptions about the agent and its environment to isolate various implementation issues:

1. The environment of the agent consists only of objects. An agent is an object with its own thread of control, mental status [14]. Users are also agents.
2. Messages are the only sensor input and action output of agents and objects. Therefore the only way an agent can get input is through messages from other objects, and the only way an agent can perform actions is by sending messages to other objects.

Furthermore, we are only interested in the messages that are related to assisting users. In addition, we only consider those messages that carry senders' intentions similarly to agent communication language (ACL). In this model, human communication messages (for example, email, news, instant messaging, and mobile phone SMS) are also viewed as messages exchanged between human agents that carry senders' intentions. This view removes many unnecessary complexities in modelling agents while retaining the expressiveness to model most agents of interest.

2.2 Personal Assistant (PA)

We now define the characteristics that capture our intuition about the roles and behaviors of a generic PA. In 1994, Maes [2] described a PA as:

> "personal assistant who is collaborating with the user in the same work environment [...] gradually more effective as it learns the user's interests, habits and preferences."

In short, the purpose of a PA is to assist its user. What it is supposed to assist with defines its type. For example, a message handling assistant assists in its user's message processing. A PA also need to know how it is supposed to assist. Intuitively, knowing the following is crucial to learning how to assist:

1. What the user is doing.
2. What the user is going to do.
3. What the user might need later and what the user might want to do later.

In terms of the BDI theory, the above items can be described as the followings. (1) is what the user has just intended. This can be acquired by capturing intentions of the messages sent by the user to other objects or to the assistant. (2) describes the user's current intentions. (3) fits into the user's desires. The user's desires can be built from the user's recurring intentions. This intuition tells us that intentions convey most of the information required to know how to assist the user.

Furthermore, through the BDI theory, an agent's intentions can be used to deduce other mental states of the agent. For instance, the BDI model of rational agency defines various properties and assumptions about the interrelationships between the beliefs, desires, and intentions of agents that might be appropriate for rational agents. Some examples are shown below:

$$(Int\ user\ \varphi) \Rightarrow (Bel\ user\ \varphi) \tag{1}$$

$$(Int\ user\ \varphi) \Rightarrow (Des\ user\ \varphi) \tag{2}$$

Formula (1) is called intention-belief consistency [15]. It says that an agent believes what it intends. Although not all assumptions are appropriate in every situations [9, p.101], they can be used to deduce other mental states from the intentions captured.

How a PA should view its user's intentions can be captured by the following two assumptions:

$$(Int\ user\ \varphi) \Rightarrow (Bel\ pa\ (Int\ user\ \varphi)) \tag{3}$$

$$(Int\ user\ \varphi) \Rightarrow (Bel\ pa\ \varphi) \tag{4}$$

Formula (3) says that if *user* intends φ, then *pa* should believe that *user* intends φ. (4) makes the somewhat stronger assumption that if *user* intends φ, then *pa* should believe φ.

Finally, we define a PA to be a structure written as:

$$PA \equiv \langle m^s, m^r, m^{rel}, F, B_0, I_0, D_0 \rangle,$$

where m^s, m^r, m^{rel} are the sets of messages that a PA can send, can accept, and must pursue to observe, respectively. The three sets are defined by what a PA is intended to assist with and what are relevant to it. B_0, I_0, and D_0 are the initial beliefs, intentions, and desires, respectively. F is a set of functions that includes a belief revision function, an option generator, etc.

2.3 Message Handling Assistant

A Message Handling Assistant (MA) is a type of a PA with more specific descriptions of what it is supposed to do. Intuitively, we can say that an MA is a PA that processes incoming messages on behalf of its user. The set of messages that an MA is interested includes communication messages that its user exchanges with other human agents.

To process incoming messages on behalf of its user, an MA must observe incoming messages to find out what are the intentions of the message. An MA should view intentions of other agents differently from its user's intention. The following two assumptions capture how an MA should view other agent's intentions on its user:

$$(Int\ x\ (Int\ user\ p)) \Rightarrow (Bel\ ma\ (Int\ x\ p)) \tag{5}$$

$$(Int\ x\ (Bel\ user\ p)) \Rightarrow (Bel\ ma\ (Bel\ x\ p)) \tag{6}$$

Formula (5) says that if agent x intends *user* to intend p, *ma* should believe that x intends p. (6) says that if x intends *user* to believe p, *ma* should believe that x believes p. (5) and (6) are compatible with the request speech act and assertion speech act defined in [6], respectively.

We give a simple scenario to demonstrate how a message can be handled if the intention of the message can be identified. Let's suppose an MA has just received a message, containing the following two sentences, from x for its user.

```
I am a customer interested in your products.
Please send me a product catalogue.
```

The intention of the first sentence is that the sender wants the user believe that the sender is a customer. Intention of the second sentence is that the sender wants a catalogue being sent from the user's computer. These intentions can be written as:

1: $(Int\ x\ (Bel\ user\ customer(x)))$
2: $(Int\ x\ (Int\ user\ send(catalogue, x))$

If the MA has a belief that the user intends to send a product catalogue to anyone who asks, the MA can intend to send a catalogue to x. This rule can be instantiated for x as:

$$(Bel\ ma\ ((Int\ x\ send(catalogue, x)) \Rightarrow (Int\ user\ send(catalogue, x))))$$
$$\Rightarrow (Int\ ma\ send(catalogue, x)),$$

where x is the sender of the message. The MA learns this kind of rules by capturing the user's intentions using formula (3) and (4), and the sender's intentions using formula (5) and (6).

2.4 Related Works

Recently, various email assistant systems have been developed to provide automatic email classification and processing. MailCat [3] relies on text classifiers to classify incoming emails into a predefined set of classes. Bergman et al. attempt to integrate various agents to build email and personal assistant to provide more sophisticated services like vacant notification using user's schedule information [4]. However, very little efforts have been made to consider emails as human communication and process them based on speech acts and the sender's intentions.

3 Extracting Intentions From Messages

In this section, we propose a model of an intention extraction function, $Iext$, that maps human communication messages (e.g, email) to a set of intentions. The development of the model takes two steps. First, we use speech act theories developed by Singh [5], Cohen and Levesque [6, 7, 8] to model a function that maps performatives to intentions. We then discuss the use of dialog act classifiers for extracting performatives from messages.

3.1 The Intention Extraction Function: $Iext$

The intention extraction function, $Iext$, defines a relationship between messages and sender's intentions. Since all sentences in a message could be put in the form of performatives by using appropriate performative verbs [5], a sentence can be written as a pair $\langle a, p \rangle$, where a is a performative and p is a proposition. A message can contain several sentences. Thus, a message, m, can be represented as an ordered set of $\langle a_i, p_i \rangle$, i.e., $m = (\langle a_1, p_1 \rangle, \ldots, \langle a_n, p_n \rangle)$, where n is the number of sentences in the message. Assuming messages are independent, we can define intention extractor $Iext$ as a function that maps a message to a set of intentions:

$$Iext : M \rightarrow \wp(I), \tag{7}$$

where M is the set of all messages and I is the set of all intentions. If we assume naively sentence independence, $Iext$ can be defined as:

$$Iext : S \rightarrow I, \tag{8}$$

where S is the set of all sentences.

3.2 Intentions from Speech Acts

In this section, we discuss how speech act theory can be used to deduce the message sender's intentions. Then, we propose a simple algorithm for the intention extraction function $Iext(m)$. Speech act theory is an attempt to build a logical theory that would explain dialogue phenomena in terms of the participants' mental states [7]. The theory models how a speaker's mental states lead to communicative actions on what assumptions the speaker makes about the hearers.

To clarify terms used in speech act theory, let us consider an example of speech act: "Please shut the door." This sentence can be rewritten as "I request you to shut the door." Here, the performative verb used is "request" and the proposition is "shut the door". We can classify the speech act of the sentence as *directive* illocutionary force. In this paper, we adapt the classification made by Singh [5].

Using the formal speech act theory developed by Cohen and Levesque [6, 7, 8], we can deduce speakers' intentions from their speech acts. For example, upon receiving a sentence containing a request speech act and a proposition *prop*, we can deduce the following mental states of the speaker using the definition of the request action in [6]:

$$(Bel\ speaker\ \neg p \wedge \neg q),$$
$$(Int\ speaker\ p),$$
$$(Des\ speaker\ \diamond q),$$

where q is $\exists e(Done\ hearer\ e; prop?)$, and p is a mutual belief between *speaker* and *hearer* about certain conditions for the speech act to be appropriate. The intuition is that when certain conditions meet (e.g., *hearer* is cooperative) *hearer* will do an action e to bring about *prop*.

With their definitions, we could try to model a complete model of message handling. In fact, various agent communication models have been developed (e.g. KQML, FIPA-ACL) based on their speech act theory. However, human communication messages do not have clear definitions of the sender's intended communication act. Using knowledge-based natural language parser is not only too complex, but also error prone [16].

Here we propose a simple method that can acquire a sender's intention by adapting normative constraints for felicitous communication proposed by Singh [5]. First, we introduce some notions to define an illocutionary force extractor function, $Fext$. Let $A = F \times Prop$ be the set of all speech acts, where F is the set of all illocutionary forces and $Prop$ is the set of all propositions. If $a \in A$ is a speech act, it is defined as $a = \langle i, p \rangle$, where i is an illocutionary force and p is a proposition. Then, the illocutionary force extractor function, $Fext$, is defined as:

$$Fext : S \rightarrow A, \tag{9}$$

where S is the set of all sentences. This function is modelled in section 3.3 using dialog act classifiers.

Once we have the type of the speech act identified for a sentence, we can reason what the sender's intention is for saying the sentence. That is, if the illocutionary force of a sentence is directive (e.g., I request you p), the intention of the sender is to make the user intend p. If it is assertive (e.g, I inform you p), the intention is to make the user believe p. If it is commissive (e.g., I promise you p), the intention is to make the user believe that the sender is intending p. If it is permissive, the intention is to make the user believe that the user can intend p. If it is prohibitive, the intention is to make the user believe that the

```
Function Iext(message): Returns a set of intentions, I.
    Set I := { };
    Set x := Sender(message);
    For every sentence s in message;
        < i,p >← Fext(s);
        If i is directive, add (Int x (Int user p)) to I;
        If i is assertive, add (Int x (Bel user p)) to I;
        If i is declarative, add (Int x (Bel user p)) to I;
        If i is commissive, add (Int x (Bel user (Int x p))) to I;
        If i is permissive, add (Int x (Bel user can(user, p))) to I;
        If i is prohibitive, add (Int x (Bel user ¬can(user, p))) to I;
    End For;
    Return I;
```

Fig. 1. Intention extraction function: Iext

user cannot intend p. From this, we can formulate an algorithm for the intention extraction function, $Iext$, shown in (Fig. 1). Formula (1), (2), (5), and (6) can be used to deduce the sender's mental states from the intentions extracted.

Once an MA acquires the sender's intentions and if the MA is required to satisfy the intentions, it needs know if the intention can be satisfied or not. Singh [5] proposed a formal semantics for speech acts that provides semantics what it means by a speech act is satisfied in term of possible world semantics. For example, "shut the door" is *whole-heartedly* satisfied if the hearer knows how to shut the door, intends to shut the door, and the door is shut by hearer in some future after the speaker utters the speech act. In regard to an MA, a speech act of a message is *whole-heartedly* satisfied if the MA knows how to bring about the goal, intends to perform it, and performs it in some future time.

3.3 Dialog Act Classifiers

Traditionally syntactic and semantic processing approaches have been used for dialogue act classification, but they are error prone [16] and requires intensive human effort in defining linguistic structures and developing grammars [17]. Several statistical approaches have been proposed to overcome these problems.

Garner et al. [18] developed a probabilistic dialog act classifier. Their classifier takes an utterance of a speaker and classifies it into one of pre-defined dialog acts. Their classifier is simply a Bayesian:

$$D = \arg\max_{D'} P(W|D')P(D') \quad (10)$$

where D is the most probable dialogue act describing the illocution of a string of certain words W. Their classifier showed recognition rate of 50% using a very simple unigram model. They proposed to use a more involved N-gram Markov model to improve recognition rate.

Reithinger et al. [16] developed a dialog act classifier used in VERBMOBIL. VERBMOBIL is a translation system for face-to-face human dialog translation.

They claim that its recognition rate for their predefined 43 dialog acts is 75% with English dialogues. In [19], both the dialog act and the concepts of a sentence are extracted by Bayesian classifiers. It reports that the recognition rate of 26 dialog acts is 58% on travel arrangement conversations whereas the conventional grammar-based approach achieves the rate of 50%.

Another advantage of probabilistic approaches over traditional syntactic and semantic processing is that no expensive and error-prone deep processing is involved. Using these classifiers, we can define our illocutionary force extractor function, *Fext*:

```
Function Fext(sentence): Returns a speech act <i, p>.
    (D,c) <- DialogueAct(sentence);
    p <- Proposition(sentence,D,c);
    i <- Illocution(D);
    Return <i, p>;
```

DialogAct(sentence) is a function that returns the performative of the sentence and a concept. It can be implemented using the probabilistic dialog act classifier [19]. Proposition(sentence,D,c) parses the sentence syntactically for arguments and returns a proposition. Illocution(D) is an indexing function that simply maps performative verbs into predefined illocutionary forces.

4 Semantic Communication Framework

Much of the difficulties in messages processing is not only from the complexity of the system itself, but also from the difficulties in interpreting human written messages. Despite the increasing number of communication messages that users have to handle, machine inaccessible messages virtually leave all messages to be handled manually.

The same problem has been already recognized in Web technologies. The simplicity of Web technology despite its successes has already caused bottlenecks that hinder searching, extracting, maintaining, and generating information [20]. In 1999, Tim Berners-Lee envisioned a Semantic Web after observing the problems in Web technology. He proposed a semantic representation of data accompanied with ontologies. Already much efforts have been made toward such vision and several representation standards of the Semantic Web have been developed: XML, XMLS, RDF, and RDFS. Furthermore, ontology languages such as OIL, DAML, and DAML + OIL. DAML + OIL have been proposed as the basis of a semantic working group of the W3C [20].

On the other hand, agent communication languages (e.g., KQML, FIPA-ACL) have been developed based on speech act theory. Speech act theory helped defining the type of message by using the concept of the illocutionary force, which constraints the semantics of the communication act itself [12, p.87].

Similarly, those approaches can be applied to structuring human communication messages to simplify system design. That is, if the sender's intentions are

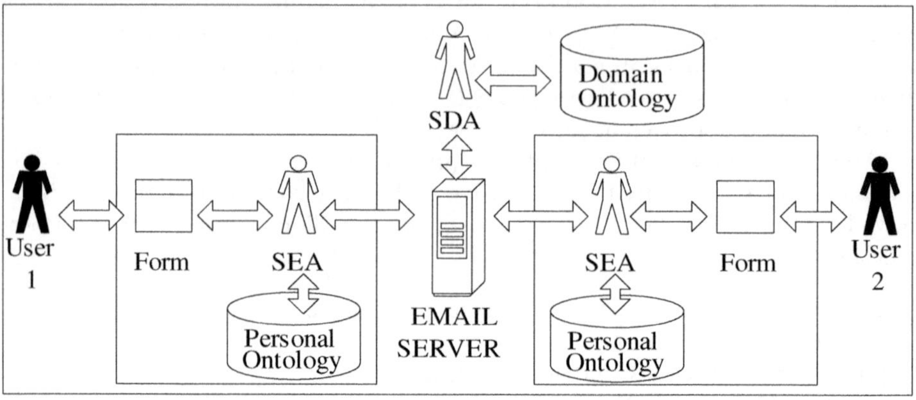

Fig. 2. The overall structure of Semantic Communication Framework (SCF). This shows Smart Email Assistants (SEAs) and Smart Domain Assistant (SDA) providing a control structure for the ontology exchange service

clearly defined and the contents of messages are machine accessible, message processing can be much simplified.

However, forcing users to structure their messages will simply not work. Fortunately, recent developments on ontology tools for semantic Web shows some hope of making this approach feasible. That is, if there are tools available that can provide free semantic mark-ups for composing messages, users can structure their messages without much additional effort. This issue in Web page authoring has been discussed by Hendler [21].

From this discussion, it seems clear that combining the semantic Web technology and an agent communication language might provide a solution for full automatic message processing. The semantic Web provides both layout and semantics data. Standards developed for agent communication languages can be used to provide the overall control structure.

Therefore, we propose Semantic Communication Framework (SCF). In this framework, MAs facilitate exchange of ontologies and Web forms between users. Messages are also structured as semantic Web forms with constructs borrowed from agent communication languages. Furthermore, MAs help users compose messages with ontologies and Web forms that have been retrieved from other users. Unlike Web servers, MAs do not run scripts in Web form, and they must work asynchronously, on-line/off-line, and have access to local resources. The main purpose of the framework is to vision how ontology, semantic Web technology, and agent technology can be integrated to provide automatic message progressing and composing. The major components of the framework is shown in (Fig. 2). In the figure, the email assistants retrieve the recipients' ontologies and forms in order to assist their users in composing semantic messages which define clearly what the senders' intentions are.

5 Discussion

Our approach departs from current ad-hoc and application specific approaches by developing a message handling system based on sound formal theories of human communication and rational behavior. Our simple but general model captures an intuitive description of the desired characteristics of a message handling system.

Interpreting a theory of rational agency on a problem to develop a generic agent model that can be reused called agent-based or agent-oriented approach [22]. In this paper, we apply an agency theory to a software agent description, and also use it to characterize the relationships between a software agent and human agents. In addition, we use speech act theory to identify appropriate relationships between an MA, the user, and the other human agents with whom the user interacts.

The proposed model of an intention extraction function defines how the intentions of messages can be extracted without using complex knowledge-based approaches. This function provides a new message classification framework and vital information on what needs to be done with each message.

Furthermore, our brief introduction of a semantic communication framework identifies fundamental problems in automatic message handling and key technologies that can be combined to provide a feasible solution for future automatic message composing and handling systems.

6 Conclusion

We have developed a simple model of a generic message handling assistant from an intuitive description of their characteristics and desired behaviors. The simple model enabled us to see how intentions can be used in user modelling and how the intentions of both the user and the sender can be used in processing messages.

From this motivation, we proposed a model of an intention extraction function that extracts intentions contained in messages. The function first converts each sentence of a message to a tuple of (*performative, proposition*). The tuples are then converted to intentions using the normative constraints for felicitous communication proposed by Singh [5].

We further proposed a semantic communication framework which integrates key agent and Internet technologies for composing and exchanging semantic messages.

References

[1] Russell, S., Norvig, P.: Artificial Intelligence a Modern Approach. AI. Prentice_Hall (2003) 53

[2] Maes, P.: Agents that reduce work and information overload. Communications of the ACM **37** (1994) 30–40 53, 55

[3] Segal, R., Kephart, J. O.: Mailcat: An intelligent assistant for organizing e-mail. In: International Conference on Autonomous Agents. (1999) 276–282 53, 57

[4] Bergman, R., Griss, M., Staelin, C.: A personal email assistant. Technical Report HPL-2002-236, Hewlett Packard Laboratories (2002) 53, 58
[5] Singh, M. P.: A semantics for speech acts. In: Readings in Agents. Morgan Kaufmann, San Francisco, CA, USA (1997) 458–470 (Reprinted from *Annals of Mathematics and Artificial Intelligence, 1993*). 54, 58, 59, 60, 63
[6] Cohen, P. R., Levesque, H. J.: Performatives in a rationally based speech act theory. In: Meeting of the Association for Computational Linguistics. (1990) 79–88 54, 57, 58, 59
[7] Cohen, P. R., Levesque, H. J.: Rational interaction as the basis for communication. In: Intentions in Communication. MIT Press, Cambridge, Massachusetts (1990) 221–255 54, 58, 59
[8] Cohen, P. R., Levesque, H. J.: Communicative actions for artificial agents. In: Proc. the First International Conference on Multi-Agent Systems (ICMAS'95), San Francisco, CA, USA, The MIT Press: Cambridge, MA, USA (1995) 65–72 54, 58, 59
[9] Wooldridge, M.: Reasoning about Rational Agents. The MIT Press: Cambridge, MA, USA (2000) 54, 56
[10] Rao, A. S., Georgeff, M. P.: Decision procedures for BDI logics. Journal of Logic and Computation **8** (1998) 293–343 54
[11] Cohen, P. R., Levesque, H. J.: Intention is choice with commitment. Artificial Intelligence, AI **42** (1990) 213–261 54
[12] Huhns, M. N., Stephens, L. M.: Multiagent systems and societies of agents. In: Multiagent Systems: A Modern Approach to Distributed Artificial Intelligence. The MIT Press, Cambridge, MA, USA (1999) 79–120 54, 61
[13] van der Hoek, W., Wooldridge, M.: Towards a logic of rational agency. Logic Journal of the IGPL **11** (2003) 135–159 54
[14] d'Inverno, M., Luck, M.: Understanding Agent Systems. Springer Series on Agent Technology. Springer-Verlag, Berlin, Germany (2001) 55
[15] Bratman, M. E.: Intentions, Plans, and Practical Reason. Harvard University Press, Cambridge, Massachusetts, U. S. A. (1987) 56
[16] Reithinger, N., Klesen, M.: Dialogue act classification using language models. In: Proc. Eurospeech '97, Rhodes, Greece (1997) 2235–2238 59, 60
[17] Wang, Y. Y., Waibel, A.: Statistical analysis of dialogue structure. In: Proc. Eurospeech '97, Rhodes, Greece (1997) 2703–2706 60
[18] Garner, P., Browning, S., Moore, R., Russell, M.: A theory of word frequencies and its application to dialogue move recognition. In: Proc. ICSLP '96. Volume 3., Philadelphia, PA (1996) 1880–1883 60
[19] Toshiaki, F., Koll, D., Waibel, A.: Probabilistic dialogue act extraction for concept based multilingual translation systems. In: Proc. ICSLP '98. Volume 6., Sydney, Australia (1998) 2771–2774 61
[20] Dieter Fensel, M. A. M.: The semantic web: a brain for humankind. Intelligent Systems, IEEE [see also IEEE Expert] **16** (2001) 24–25 61
[21] Hendler, J.: Agents and the semantic web. IEEE Intelligent Systems Journal **16** (2001) 62
[22] Petrie, C. J.: Agent-based software engineering. In: Proc. 5th International Conference on the Practical Application of Intelligent Agents and Multi-Agent Technology (PAAM 2000), Manchester, UK, The Practical Application (2000) 63

A New Approach for Concept-Based Web Search

Seung Yeol Yoo and Achim Hoffmann

School of Computer Science
and Engineering University of New South Wales Sydney 2052
NSW, Australia
{syy,achim}@cse.unsw.edu.au

Abstract. Searching for information on the Internet remains a difficult task, despite considerable progress in search engines, such as Google. One difficulty for many users is to formulate a suitable search query. In this paper we propose a new interactive query refinement process that helps the user to articulate their information needs by supporting the word sense disambiguation of search terms as well as by dynamically generating potentially new relationships among several search terms, based on analysing the retrieved documents.
The main functionality of our system, called WebConceptualizer, presented in this paper support the following: 1) the user's awareness of the different word senses of query terms. 2) to visualise a network of concepts surrounding the query terms in a graph structure that allows the user to operate at the conceptual level rather than the term level when articulating their information need. 3) the identification of documents whose content reflects a particular word sense rather than just the query words as such.
Our initial experiments with the implemented prototype shows a good accuracy in recognising the correct word sense of the main topic of the document. The user interface has not been systematically evaluated as yet. However, the usability seems rather good based on a verbal reports from a few test users.

Keywords: Knowledge Acquisition, Interactive Query Refinement for Concept-Based Web Search, Ontology.

1 Introduction

Searching for information on the World-Wide Web remains a difficult task, despite considerable progress in search engines, such as Google. Support for users in formulating suitable queries are rather limited in current search engines. Another weakness of current search engines concerns the appropriate retrieval of documents which cover a number of topics, where a single topic is only discussed in a particular section of that document. It would be desirable to segment the document and to allow the retrieval of the various segments each covering a single topic.

Ideally, a user iteratively refines their search queries by evaluating preliminary search results and by modifying the query resulting in a more precise

T.D. Gedeon and L.C.C. Fung (Eds.): AI 2003, LNAI 2903, pp. 65–76, 2003.

description of the user's information need. In many cases users find it difficult to precisely articulate their information need. An iterative process that makes the user aware of different meanings of the search terms in different contexts promises substantial potential for improving the current state of web search engines. Furthermore, providing the user with possible additional search terms that would identify a particular word sense of a previously entered search term is also desirable.

Existing approaches to Web search engines that attempt to satisfy a user's information need can be distinguished by the following aspects:

1. What are the main measures for their approaches to web search?
2. To what extent is the context in which search terms occur in a document taken into account.
3. How is the user-intended word sense of a search term determined, if at all?

In this paper we present our work towards addressing both of the above problems. Our work goes beyond current approaches in the latter two of the above mentioned aspects.

This paper is organised as follows: The following section 2 presents current approaches to web search and analyses them with respect to the above criteria. In section 3 our approach of conceptual search is presented. Finally, we present first experimental results in section 4. Section 5 contains the conclusions that can be drawn so far as well as it outlines future work.

2 Current Approaches to Web Search Engines

In the following we review the main ideas underlying current web search engines. We will discuss the following three types of search engines and their respective user interfaces in turn: Term-based, Taxonomy-based, and Concept-based approaches.

2.1 Term-Based Approaches

The Term-Based Approach (e.g., as used in Google [1]) considers a vector model of documents and queries. A vector of term frequencies in both, the document as well as the query, is used as the basis to compute similarities between a query and the various documents available. It can also be used to compute similarities between different documents. This is a well-known approach from the long-established field of information retrieval. In the case of handling html-documents, the term vector may also include various meta-information represented in the html-code.

Term-based approaches tend to show the following shortcomings:

- Short and generic query terms have shortcomings to represent the user's information needs, since the user's information needs are based not on words but rather on topics being addressed in the documents.

- Boolean queries are not very useful, as it turned out to be too complicated for most people. The complication is caused by the difficulty of expressing the user's information need, which is concept-based rather then term-based. I.e. to exhaustively express a certain concept it is often necessary to formulate a very complex boolean, term-based, query as many terms may have a relevant relationship with a single concept. Techniques, such as latent semantic indexing [2] are targeted to address this problem. However, the success of such techniques is limited. An interactive and incremental query refinement process which increasingly captures the concepts relevant to the user's information need is very desirable.

The term-based approaches generate usually long lists of documents upon a web search query. The performance of such an approach is usually measured by the two quantities of recall (the fraction of relevant documents actually presented in the list to the user) and precision (the fraction of relevant documents in the list of retrieved documents).

2.2 Taxonomy-Based Approaches

The goal of taxonomy-based approaches is to resolve the information-overload problem, caused by a usually long list of retrieved documents in a term-based approach, by providing a set of document clusters (or categories) and organising them in a hierarchical structure. According to which topical theme these clusters are formed is either determined by a term taxonomy provided by human experts or is dynamically determined on the basis of the retrieved documents. Advantages and disadvantages of each way of forming topical themes is discussed below using the examples of NorthernLight and Grouper.

One of the early taxonomy-based approaches are web catalogues, such as Yahoo [3], which consist of a huge human-classified catalogue of documents which can be browsed through by following a pre-defined hierarchical structure. The assignment of documents to the appropriate category is accurate only in the context the human classifier assumed. Not only the number of documents on the web grows rapidly they also change their content. Furthermore, new relationships between terms emerge over time which require an re-categorisation of already manually categorised documents. E.g. the concept of travel agent may now also include an automatic web agent that can book a trip - not only the traditional human travel agent.

An alternative approach to manage the ongoing growth of Web is to take the help of Web users, such as the Open Directory Project (ODP) [4], where everyone is invited to contribute by providing new categories and classifying documents.

Both systems allow to query for related categories. It allows a user to start from somewhere relatively close to the appropriate category in the taxonomy and end up at the appropriate category after some search.

However, they still have following limitations:

- The manual update, in Yahoo, is not fast enough for World-Wide Web and its dynamic nature. On the other hand, in the ODP, the fact that many users contribute to it and that there will be many different views of what is a sensible grouping as well as different understandings of involved concepts, the ODP does not represent a universally usable catalogue.
  Even within a single organisation it is often difficult to find agreement on the boundary of certain topics and terms among different users, not be to a user outside of the organisation. For example, someone considers "Agent" as "a representative (person) who acts on behalf of other persons or organisations." On the other hand, someone will include "Soft agent" in their understanding of "Agent". Thus, conceptual abstractions will only be appropriate in a given context or set of contexts.
- In ODP, it is not easy to identify relevant categories as query result because the categories have static category descriptions. Ideally, each category would extract a dynamic category description from the meta-data of included documents. For example, when the user queries for "Agent" coming from the "Data Mining" category, a new category should be created and connected to the related pre-existing categories.
- The category structure is static. If some parts (i.e. meanings in certain contexts) of a category change, the connections among categories may need to be changed as well. For example, initially "Agent" has no connection with "Mobile Computing". But when "Mobile Agent" related documents are added to the "Agent" category, a new connection between "Agent" and "Mobile Computing" should be created.

Another approach consists of automatically creating a hierarchical view of a ranked list, such as in the Scatter/Gather System [5], WiseNut [6], Vivisimo [7], NorthernLight [8], or Grouper [9, 10]. Their goals are: 1) to create a hierarchical view automatically for each query, 2) to assign only relevant documents for a query into each category at run time, and 3) to provide a user interface which allows iterative and hierarchical refinement of the search process.

- Scatter/Gather System [5]: It allows a user to select one or more categories and to re-cluster them in an iterative and hierarchical manner. A category can be subdivided into more fine-grained categories in order to be manageable for browsing. However, because the re-clustered categories do not consider the terms' dependencies, they show poor semantics and the re-clustered sub-categories do not reflect their relationships with the other categories. For example, let us assume a user starts his query with "AI" and "Database" categories, and they share some documents about "Data Mining". When he re-clusters the "AI" category and obtains, among others, a "Data Mining" sub-category, the "Data Mining" sub-category should be connected with both, the "AI" category as well as the "Database" category.
- NorthernLight [8]: It performs a semi-post-retrieval clustering, based on a static, manually constructed hierarchy of topics, which again has the problem of untimely updates. It indexes documents by classifying each document

against multiple topic tags from a 20,000 terms' subject hierarchy manually developed by librarians. To cluster query-matched documents, it tries to extract the most important topic tags among the query-matched documents. The extracted topic tags become the result categories. The importance of topic tags is assessed based on the occurrence frequencies of terms in the documents.
However, the NorthernLight approach suffers from the following weaknesses:

1. The sub-categories which NorthernLight uses are, at least for technical or scientific queries, often not well chosen, only a fraction of those categories are often adequate.
2. the manually constructed hierarchical topic-tag structure does not consider all the relevant semantic connections between categories. As a consequence semantically linked categories cannot be reached by following the offered sub-categories.
3. documents which cannot easily be assigned to a single specific category are usually assigned to a very general category. This makes the retrieval of those documents difficult as there are too many documents associated with those broader categories.
4. the query results are too much dependent on the order of query term combinations. For example, a query of "Intelligent Agents" returns 4,402 items. In contrast to that contains the result for the initial query "Agent" a sub-category "Intelligent Agents" which has only 486 items.

This phenomenon is due to the fact that documents in the hierarchical categorisation are mostly only categorised into a single category, hence leaving those documents out which fall into the competing sub-categories at the same level. See also Figure 1. This is common problem of taxonomy-based approaches.

– Grouper [9, 10]: It executes a query-focused post-retrieval clustering, relying on automatic detection of phrases from retrieved documents to group search results. Grouper's principal idea is the "User-Cluster Hypothesis": "Users have a mental model of the topics and subtopics of the documents presented in the result set: similar documents will tend to belong to the same mental category in the users' model. Thus the automatic detection of clusters of similar documents can help the user in browsing the result set." To identify suitable clusters, it is generally assumed that "documents on the same topic will share common phrases." Grouper II [10] generates interactively a hierarchical structure by providing dynamic indices for non-merged phrase clusters. The major problems of this approach lie in the fact that the resulting groupings are much less than the one ideally being formed. Technical reasons for that seem to include the fact that the dynamic indices do not support any contextual information to assist the user in their selection. Furthermore, just matching phrases is not enough to discover the semantic similarity among documents as it is not clear how important a particular phrase for the content of a document really is.

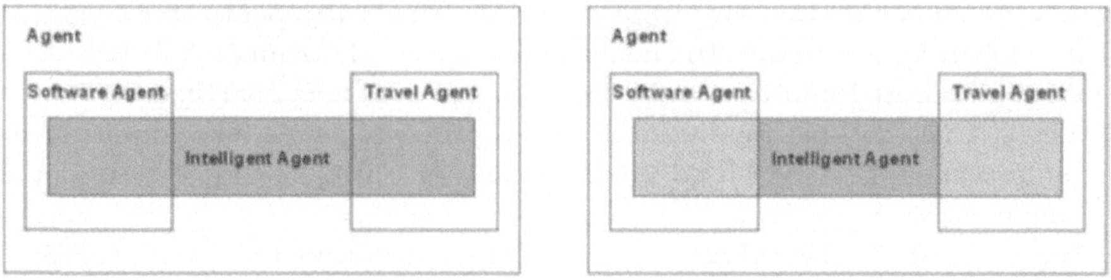

Fig. 1. The problem of using hierarchical categories and phrase matching. In the left hand image the entire grey area is retrieved upon the initial query of " Intelligent Agent". In the right hand image the retrieved documents are shown only in the grey area in the middle. The shaded area on the left and right is categorised into "Software Agent" and "Travel Agent" and hence not found under the sub-category of "Intelligent Agent"

3 Our Concept-Based Approach

More modern approaches in Information Retrieval have started to consider term dependencies to achieve better retrieval results. They represent such term dependencies as a vector model. For example, a context vector for a term t can be generated by using the terms occurring close to term t in a text, see e.g. [11, 2], [12, 13], or [14].

3.1 Our Approach

Our approach goes beyond the work discussed above. We support an interactive and incremental querying process: The objective is to have a user-system interaction where the system reflects the content of actually available documents and guides the user to refine their search query to an extent that results in far superior quality of the eventually found documents. This is somewhat similar to NorthernLight or Grouper but it is not limited to terms. It rather operates on the conceptual level.

The rationale behind it is that even experienced users are unable to formulate sufficiently precise queries as they are not aware of all the documents and the different ways search terms are used in different contexts in documents.

This is envisaged to be achievable by allowing the user to browse through a comprehensive network of terms and their different word senses in which it is possible to refine the articulation of the user's information needs resulting in a manageable number of retrieved documents.

Interactive Conceptual Query Refinement Our approach aims at allowing an interactive session with the user where the user articulates their information needs in form of increasingly detailed queries. This is supported by providing the user with information concerning the various possible meanings of query terms as well as information about the associated documents. The user has access to the related documents to allow them to verify what meaning of a term they are really interested in.

The main functions we provide in the current implementation are the following:

1. A starting point for the conceptual search by disambiguating query terms. In our approach the user is guided towards expressing their information needs in concepts and their relationships, rather than in often ambiguous terms. To start off this process for the initial query terms all different word senses listed in WordNet [15, 16] are used to build an initial concept graph. To provide a visual representation of the concepts and their relationships (it is originally textually represented), we employ a graphical depiction of the mathematical lattice based on Formal Concept Analysis [17, 18]. Each node in the lattice represents a concept. The visual interface allows the user to modify the displayed concept structure by using operators, such as concept deletion, concept combination, or concept separation.
2. The query terms are interpreted at the conceptual level (Synsets in WordNet) and incorporated into the conceptual model of the user's information need.

Formal Concept Analysis (FCA) is used to construct a context-based and data-centred model of the concepts surrounding the different word senses of the query terms. The lattice-based conceptual structure can provide a meaningful and comprehensible browsing structure. The mathematic background of Formal Concept Analysis is provided in detail in [17, 18].

4 Experiments

The following sample session shows how a user can use our WebConceptualizer to view and browse the different word senses of the query terms and supplementary information.

4.1 Sample Search Session

Firstly, when the user inputs query terms, the terms are passed to the text preprocessing module to filter out stop-words, to obtain the stems of key-words, and to get the syntactic category (possible parts of speech) of the stems.

For instance, the input term "Java" has three senses (land, coffee, and programming language) and, each sense (Synset on WordNet) has its own description. Each description consists of a list of synonymous words, definitions, as well as pointers that describe the relations (e.g., hypernym/hyponym) between a sense and other senses. The semantic relations among the concepts are represented in a lattice structure as in the window in the upper left-hand corner of Figure 2. Each node, in the lattice structure, is a Formal Concept keeping its own contextual information. A path, from "Top" node to "Bottom" indicates a sequence of increasingly specialised concepts, each being a subconcept of

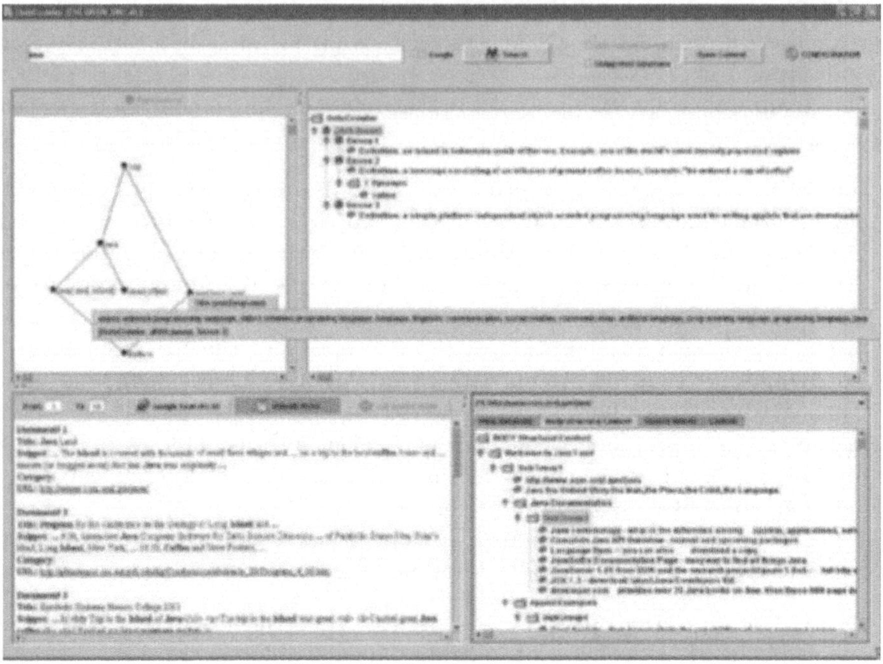

Fig. 2. Our concept-based user interface

the concept associated to the preceding node above it. For example, the "Java" node is the generalised concept for both "Java (island)" and "Java (coffee)". Conversely, "Java (coffee)" is the specialisation of concept "Java" in Figure 2. This concept graph can be changed by adding or deleting a sense of a term, i.e. adding or deleting a node in the graph. The various paths going through a node are automatically generated (in case of node addition) or rerouted/deleted in case of a node deletion.

Finally, the constructed graph, reflecting the user's information need at the conceptual level (with selected word senses of search terms), can be stored in a semi-structured RDF format at the user's request. For example, the description of concepts and their relationships over the term "Java" with respect to "programming language" can be stored as a concept structure for the user about "Java". The stored concept structure can be reused for later searches. As it does not depend on temporary data such as traditional bookmarks or user profiles but rather on the conceptual description of the user's interests, it can be re-used in a more general way. For example, a bookmark can retrieve only an html-document via a fixed URL, but conceptual descriptions can be reused for the identification of conceptual similarity for new html-documents.

To illustrate this, let us assume that another interest of the user concerns "Education" related documents, and he wants to restrict the meaning of "Java" to the sense of "programming language". As the user already had predefined and stored knowledge through a previous search session about his interest in "Java", he can reuse that knowledge. In WordNet, "Java" and "Education" do not have any lexical connection, which is a limitation of WordNet but should not impede the capabilities of our tool.

However, not all relevant relationships can realistically be predefined. For example, "Java" (as programming language) may be an educational course, but "Java" (as coffee) cannot be (at least much less likely). Therefore, a good retrieval system should allow the user to make word-sense specific decisions. In our WebConceptualizer, the user can instantiate the connections between two concepts by using the mentioned concept operations.

Our current implementation offers the following features and information (see Figure 2):

1. The different word senses of the query terms (obtained from WordNet) in the upper right sub-window in Figure 2.
2. The concept graph allowing manipulation by deleting/adding conceptual nodes (word senses) as well as combining different concepts by logical operators. (See upper left sub-window in Figure 2.)
3. "Html Summary" which summarises the header part of the document, and extracts keywords used as hyper-link labels or highlighted words (e.g., head titles, words in boldface). See the bottom right sub-window in Figure 2.
4. "Content Structure" which analyses the syntactic structure of html tags, such as a header title and its sub-header titles which have a semantic relationship within the document. An example is shown in the bottom right sub-window in Figure 2.
5. "Query Terms" which provide a set of candidate keywords extracted from the documents based on word frequency. By using those keywords in the next query refinement the articulation of the user's information need can be made more precise. Those additional keywords can also expand the conceptual graph in the upper left sub-window.
6. The system also provides a measure of conceptual matching between a node in the conceptual graph and the retrieved documents. This is discussed in more detail in Subsection 4.2.

4.2 Experimental Results on Topic Recognition

For each of the 10 sample documents about "Java" the list below show the manually determined main topics (determining the different word senses of "Java"): For example, H1's main topics are "Programming Language" and "Coffee" and H6's main topics are "Coffee" and "Island".

Programming Language: **H1**, **H2**, **H3**, **H4**, **H5**, **H9**
Coffee: **H1**, **H3**, **H6**, **H8**, H9, **H10**
Island: H1, H2, **H3**, **H6**, **H8**, **H9**, H10

Table 1. Similarity Measure of the various documents and the shown word senses of "Java"

	Program(%)	Coffee(%)	Island(%)
H1	***77.62***	***48.26***	*47.1*
H2	***4.62***	11.63	*88.37*
H3	***33.33***	***33.9***	***59.32***
H4	***95.65***	48.89	51.11
H5	***93.75***	44.8	46.4
H6	22.22	***40.74***	***59.26***
H7	ASP	ASP	ASP
H8	18.18	***27.08***	***64.58***
H9	***90.79***	*44.81*	***44.81***
H10	47.62	***85.83***	*14.17*

Table 2. Conceptual Similarity Measure of document

	Broadcast(%)	Course(%)	Software(%)
H1	***55.17***	37.14	53.57
H2	***60.0***	*30.51*	*60.0*
H3	***78.35***	***19.53***	56.76
H4	37.5	***43.37***	53.12
H5	*52.0*	***48.0***	38.46
H6	18.4	***54.55***	*44.86*
H7	50.93	39.83	***58.41***
H8	*36.54*	18.09	***81.72***
H9	44.44	24.6	***73.5***
H10	*45.83*	31.4	***66.28***

For each document their conceptual similarity for those three topics are calculated as shown in Table 1. The numbers show the calculated degree of match between the various word senses and the main topic(s) being discussed in the document. The italic numbers indicate that the corresponding topic is discussed in the document (as determined manually) while the boldfaced italic numbers indicate the manually determined main topic or main topics of the documents.

In Table 2 we see results for the three word senses of "program" (in broadcasting, in educational courses and in computer programming).

In our current implementation, the entire document is used for the conceptual similarity calculation by using the information on the different word senses in WordNet. Even though, the similarities are calculated on the basis of the entire document, main topics from those html-documents in Table 1 were recognised with a precision of 88%. For the 10 documents in Table 2 main topics were recognised in 80% of the cases. While this is not a systematic or rigorous evaluation it provides some evidence that a reasonably accurate recognition of the user word sense in t document topic can be recognised successfully.

The main topics in H1, H3, H4, H5, H6, H8, H9 and H10 were successfully identified. H7 is a document which was generated by ASP, and we are currently unable to process such documents. Hence, the only failure to identify the main topic correctly occurred in H2. In case of H2, the terms "coffee" and "island" are frequently used together with the term "Java", however they are domain jargon in the "programming" domain. For example, in "Java coffee", the combination of "coffee" cannot affect the original word sense of "programming language" of "Java".

We found it rather easy for the user to browse the conceptual hierarchy as shown in the graph to refine the search query to reflect the information need more precisely.

5 Conclusions and Future Work

In this paper we discussed the limitations of previous approaches to more intelligent web search and browsing facilities (term-based and taxonomy-based approaches). The novelty of our approach lies in a new user-interaction model for a concept-based web search. Our approach allows to exploit conceptual dependencies in the process of querying for and the automatic categorisation of documents. Documents are selected on the basis of conceptual similarity rather than term-based similarity. The user is supported in specifying the specific sense of an entered search term which in turn is evaluated on the basis of the associated terms indicative for the chosen word sense of the term t rather than merely the occurrence the involved term t as other approaches do.

In this paper we present the ideas of an advanced user interaction model for web search. The necessary techniques to automatically produce highly accurate categorisations and to produce semantically appropriate related terms etc. are still under further development. We believe that the presented ideas as well as the current prototype implementation of WebConceptualizer are inspiring and stimulate constructive discussions.

Future work will investigate the applicability of statistical approaches to identifying new terms indicative for a certain concept from a collection of documents rather than from a static source, such as WordNet. A successful technique will allow to keep up to date with the latest developments of term usage as well as the development of new concepts being discussed in web documents. Furthermore, it will also ensure a more complete coverage of concepts than a source, such as WordNet, is currently able to be.

Another issue of interest for our further research is the observation that the scope of the context of an individual term occurrence is not necessarily the entire document: Even though the entire document is what is being classified, it does not mean that the frequencies of terms over the entire document should be used to classify the document, since the document may describe several topics. By extracting only the relevant document segments (i.e. paragraph, sentence, words), one would expect a more accurate classification of documents. For example, traditional methods, such as Inverse Document Frequency and Deviation Measure of Terms, can be applied to the context-focused partitions for retrieval performance.

References

[1] Google (2003) http://www.google.com/. 66
[2] Deerwester, S., Dumais, S.T., Furnas, G.W., Landauer, T.K., Harshman, R.: Indexing by latent semantic analysis. JASIS **41** (1990) 391–407 67, 70
[3] Yahoo (2003) http://www.yahoo.com/. 67
[4] Open directory project (2003) http://dmoz.org/. 67
[5] Scatter/gather (2000) http://www.sims.berkeley.edu/~hearst/sg-overview.html. 68
[6] Wisenut (2003) http://www.wisenut.com/. 68
[7] Vivisimo (2003) http://vivisimo.com/. 68
[8] Northern light (2002) http://www.northernlight.com/. 68
[9] Zamir, O., Etzioni, O.: Grouper: A dynamic clustering interface to web search results. Computer Networks **31** (1999) 1361–1374 68, 69
[10] Grouper (2000) http://www.cs.washington.edu/research/projects/WebWare1/www/metacrawler/. 68, 69
[11] Bollmann-Sdorra, P., Raghavan, V.V.: On the necessity of term dependence in a query space for weighted retrieval. JASIS **49** (1998) 1161–1168 70
[12] Billhardt, H., Borrajo, D., Maojo, V.: A context vector model for information retrieval. JASIST **53** (2002) 236–249 70
[13] Schütze, H.: Dimensions of meaning. In: Proceedings of Supercomputing '92, IEEE (1992) 787–796 70
[14] Rungsawang, A.: DSIR: the First TREC-7 Attempt. In E. Voorhees, D.H., ed.: Proceedings of the Seventh Text REtrieval Conference (TREC 1998), Department of Commerce, National Institute of Standards and Technology (1998) 366–372 70
[15] Wordnet (2003) http://www.cogsci.princeton.edu/ wn/. 71
[16] Fellbaum, C., ed.: WordNet - An electronic lexical database. MIT PRESS, CAMBRIDGE, MA (1998) 71
[17] Wille, R.: Restructuring lattice theory: An approach based on hierarchies of concepts. In: Ordered Sets. Volume 83 of Series C., NATO Advanced Study Institute (1982) 445–470 71
[18] Wille, R., Ganter, B.: Formal Concept Analysis: mathematical foundations. Springer Verlag (1999) 71

Representing the Spatial Relations in the Semantic Web Ontologies

Hyunjang Kong[1], Kwanho Jung[1], Junho Choi[1], Wonpil Kim[1],
Pankoo Kim[2], and Jongan Park[3]

[1] Dept. of Computer Science, Chosun University
Gwangju 501-759, Korea
{kisofire,khjung,spica,kwpil}@mina.chosun.ac.kr
[2] Corresponding author, Dept. of CSE, Chosun University
pkkim@chosun.ac.kr
[3] Dept. of EICE, Chosun University
japark@chosun.ac.kr

Abstract. The core of the study into the semantic web is ontologies. Ontologies have been actively studied in many areas for a long time. There has been a great deal of effort reported in building ontologies. In the study of a semantic web, the ontologies are constructed using the RDF Schemas and OWL Capabilities. The RDF Schemas and OWL Capabilities can deal with the many relationships that occur when ontologies are constructed, but it is not inadequate for managing spatial relations. This paper defines the new axioms for representing the spatial relations based on the Description Logic as a web ontology regarding nations was built. The ontology was then represented using the OWL.

1 Introduction

As the web advances, users want a more semantic information search. For a semantic search, ontology is very important in the study of the semantic web. Many studies regarding ontologies involved developing methods for building ontologies as well as the various markup languages for expressing the web ontologies. Although there are many methods reported for assembling web ontologies over the last five years, they are still inadequate for constructing the perfect web ontologies.

In the study of ontologies, careful consideration should be paid to the accurate classification of the specific item and the adequate way for representing the relations that occur among concepts when the domain ontologies are constructed. When the relationships among the concepts are built, various other relationships occur. Recently, there was an attempt to build web ontologies using the OWL Capabilities based on Description Logic[1,2].

This paper emphasizes the importance of spatial relationships for constructing the web ontologies, and suggests a method for representing the spatial relationships.

T.D. Gedeon and L.C.C. Fung (Eds.): AI 2003, LNAI 2903, pp. 77-87, 2003.

This paper is organized as follows. Section 2 explains the related works on the Semantic Web, the Markup Languages, Description Logic and Spatial Description Logic[3][4][5]. Section 3 shows a representation of the spatial relationships, and expresses the spatial relationships using the OWL language. Section 4 illustrates the propriety of our suggestion. Finally, conclusion and future works are presented in section 5.

2 Related Works

2.1 Semantic Web and Ontologies

The current web aims to expand the quantity but the semantic web, known as a trust web, means an expansion of quality. The semantic web appears correct to be called an evolution of the web than calling it a reformation of the web. Tem Berners-Lee who proposes the semantic web said that *The Semantic Web is an extension of the current web in which information is given a well-defined meaning, better enabling computers and people to work in cooperation*[3]. The Semantic Web will enable intelligent services such as information brokers, search agents, information filters etc. Such intelligent services on the knowledgeable web should surpass the currently available versions of these services, which are limited in their functionality, and only work as stand-alone services that do not interoperate.

Two things constituting the semantic web are ontology, which represents the semantic constitution, and markup language, which represents well-defined information. Humans and machines should communicate with each other in order to realize the semantic web to process and interpret information. The languages for representing information are XML for representing information structures and RDF, DAML, and OWL for representing the information meaning, have been developed and standardized in various ways using W3C[1, 2, 3, 15].

2.2 Description Logic

Description Logics are considered to be a structured fragment of predicate logic. The basic types of a concept language are the concept, roles, features, and the individual constants. A concept is a description gathering the common properties among a collection of individuals. Inter-relationships between these individuals are represented either by means of the roles or the features. The individual constants denote the single individuals. According to the syntax rules in table 1, concepts (denoted by the letters C and D) are built out of atomic concepts (denoted by the letter A), roles (denoted by the letter R), and features (denoted by the letter P), and individual constants (denoted by the letter a). The roles are built from the atomic roles (denoted by the letter T) and features [8].

Description Logics can support a definition language for role expression of the concepts and represents complex concepts and relationships from simple concepts and relationships. The concept is the same as the class and individual relation. Description Logics are used for database schema to represent structure information. In addition, Description Logics are fundamental to RDF/RDFS/OIL/DAML/OWL. Table 1 shows the constructor, syntax and semantics in using Description Logics.

Table 1. Constructor, Syntax and Semantics of Description Logic

Constructor	Syntax	Semantics
Concept	A	$A^I \subseteq \Delta^I$
Role name	R	$R^I \subseteq \Delta^I \times \Delta^I$
Conjunction	$C \cap D$	$C^I \cap D^I$
Value restriction	$\forall R.C$	$\{X \in \Delta^I \mid \forall y.(x,y) \in R^I \Rightarrow y \in C^I\}$
Existential quantification	$\exists R$	$\{X \in \Delta^I \mid \forall y.(x,y) \in R^I\}$
Top	T	Δ^I
Bottom	$\perp$	Ø
Negation(C)	$\neg A \neg C$	$\Delta^I \setminus C^I$
Disjunction(U)	$C \cup D$	$C^I \cup D^I$
Existential restriction(E)	$\exists R.C$	$\{X \in \Delta^I \mid \exists y.(x,y) \in R^I \wedge y \in C^I\}$
Inverse role	R	$\{(y,x) \mid (x,y) \in R^I\}$

2.3 The Description Logic $\mathcal{ALC}(\mathcal{D})$

$\mathcal{ALC}(\mathcal{D})$ is an extension of the well-known Description Logic $\mathcal{ALC}$ by the so-called concrete domains. First, a concrete domain is specified [5].

Concrete Domain

A concrete domain $\mathcal{D}$ = (**dom**($\mathcal{D}$), **pred**($\mathcal{D}$)) consists of

- a set **dom**($\mathcal{D}$)(the domain), and
- a set of predicate symbols **pred**($\mathcal{D}$).

Each predicate symbol $p \in$ **pred**($\mathcal{D}$) is associated with an arity n and n ary relation $P^{\mathcal{D}} \subseteq \mathrm{dom}(\mathcal{D})^n$

Definition of $\mathcal{ALC}(\mathcal{D})$

Let N_C, N_R, and N_F be disjoint sets of *concept*, *role*, and *feature names*. The set of *$\mathcal{ALC}(\mathcal{D})$-concepts* is the smallest set such that

1. every concept name is a concept and
2. if C, D are concepts, R is a role or a feature name, $P \in$ **pred**($\mathcal{D}$) is an n-ary predicate name, and $u_1,\ldots,u_n$ are feature chains, then $(C \sqcup D)$, $(C \sqcap D)$, $(\neg C)$,$(\forall R.C)$,$(\exists R.C)$, and $P(u_1,\ldots,u_n)$ are concepts.

Concepts of the form $P(u_1,\ldots,u_n)$ are called *predicate restrictions*, and concepts of the form $(\forall R.C)$(resp.$(\exists R.C)$) are called *universal* (resp. *existential*) *value restrictions*. In order to fix the exact meaning of these concepts, their semantics are defined in the usual model-theoretic way.

Interpretation

An interpretation $I = (\Delta^I, \cdot^I)$ consists of a set Δ^I and an interpretation function $\cdot^I$. The sets $\Delta^{\mathcal{D}}$ and Δ^I must be disjoint.

The interpretation function maps each concept name, C, to a subset, C^I of Δ^I, each role name, R, to a subset R^I of $\Delta^I \times \Delta^I$, and each feature name f to a partial function f^I from Δ^I to $\Delta^D \cup \Delta^I$, where $f^I(a) = x$ will be written as $(a,x) f^I$. If $u=f_1 \ldots f_n$ is a feature chain, then u^I denotes the composition, $f^I_1 \circ \cdots \circ f^I_n$, of the partial functions, $f^I_1 \cdots f^I_n$. The semantics of the concept terms in $\mathcal{ALC}(\mathcal{D})$ can then be extended as follows :

$(C\ D)^I = C^I \cap D^I$
$(C\ D)^I = C^I \cup D^I$
$(\neg C)^I = \Delta^I \setminus C^I$
$(\exists R.C)I = \{a \in \Delta^I \mid \exists b \in \Delta^I: (a,b) \in R^I \wedge b \in C^I\}$
$(\forall R.C)^I = \{a \in \Delta^I \mid \forall b: (a,b) \in R^I \rightarrow b \in C^I\}$
$(\exists u_1, \cdots, u_n.P)^I = \{a \in \Delta^I \mid \exists x_1, \cdots, x_n \in \Delta^{\mathcal{D}} : (a, x_1) \in u^I{}_1 \wedge \cdots \wedge$
$(a, x_n) \in u^I{}_n \wedge (x_1, \cdots, x_n) \in P^{\mathcal{D}}$

2.4 The Description Logic $\mathcal{ALC}$(Drcc8)

The *Region Connection calculus* RCC-8[6][7][8] is a language for qualitative spatial representation and reasoning where the spatial regions are regular subsets of a topological space. The regions themselves do not need to be internally connected i.e. a region may consist of different disconnected pieces.

As the concrete domain in $\mathcal{ALC}(\mathcal{D}_{recc8})$, $\Delta^{\mathcal{D}}{}_{rcc8}$ is the set of all the non-empty regular closed subsets of the topological space R^2. $\Phi^{\mathcal{D}}{}_{rcc8}$ is obtained by imposing a union, intersection, composition and converse operations over the set of the elementary binary relationships between the regions i.e.(PO, NTPP, TPP, EQ, TPP^{-1}, $NTPP^{-1}$, EC, DC) where the intended meaning of the elements are respectively Proper Overlap, Non Tangential Proper Part, Tangential Proper Part, External Connection, and DisConnected. Fig. 1 shows some elementary relationships between two regions, X and Y.

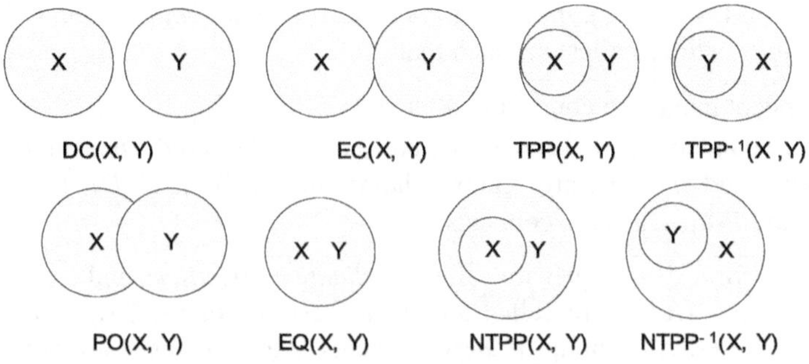

Fig. 1 Examples for the eight base relationships of RCC-8

3 Defining the New Axioms for Representing the Spatial Relationships Based on Description Logic

Many relationships occur when web ontologies are constructed. The spatial relationships of the variety relation also take place quite often. The main idea of this paper is to handle the spatial relationships that occur when web ontologies are constructed. The existing methods for representing the relationships do not have the capability to manage the spatial relationships.

In order to deduce the relationships between the spatial regions, $\mathcal{ALCRP(D)}$ was developed by extending the DL $\mathcal{ALC(D)}$. It provides a foundation to support the spatio-terminological reasoning with DLs[9][10].

This paper defines the new axioms for representing the spatial relationships based on the $\mathcal{ALCRP(D)}$ and uses them construct web ontologies. There are many spatial relationships. Fig.2 shows the hierarchy of the spatial relationships.

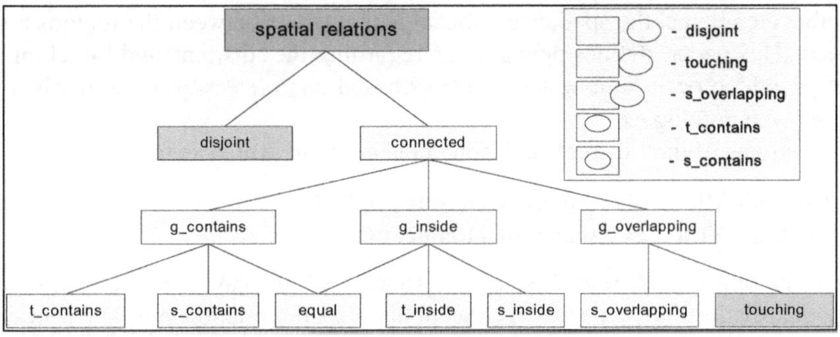

Fig. 2. The hierarchy of the spatial relationships

However, it is too difficult to define all the spatial relationships. Therefore, two spatial relationships, which are 'disjoint' and 'touching', are basically defined.

This section introduces the concept that the $\mathcal{ALCRP(D)}$ is more suitable for describing the spatial relationships. Using $\mathcal{ALCRP(D)}$s role-forming predicate-based operator, a set of complex roles can be defined based on the mentioned $\mathcal{RCC}$-**8** $\mathbf{S_2}$ predicates. Subsequently, 'disjoint' and 'touching' could be defined as follows:

$$\mathbf{Disjoint} \doteq \exists(\mathbf{has\_area})(\mathbf{has\_area}).\mathbf{dc}$$
$$\mathbf{Touching} \doteq \exists(\mathbf{has\_area})(\mathbf{has\_area}).\mathbf{ec}$$

where has_area is the feature to relate abstract individuals. Fig. 3 illustrates the spatial domain and abstract domain regarding the 'disjoint' and 'touching' relationships.

The 'dc' and 'ec' predicates of **RCC8** can be represented as follows:

$$\mathbf{C(X_1, X_2)} \quad \exists x \; x \in X_1 \cap X_2$$
$$\mathbf{DC(X_1, X_2)} \quad \neg\exists x \; x \in X_1 \cap X_2$$
$$\mathbf{EC(X_1, X_2)} \quad C(X_1, X_2) \wedge \neg\exists x \; x \in \|X_1 \cap \|X_2$$

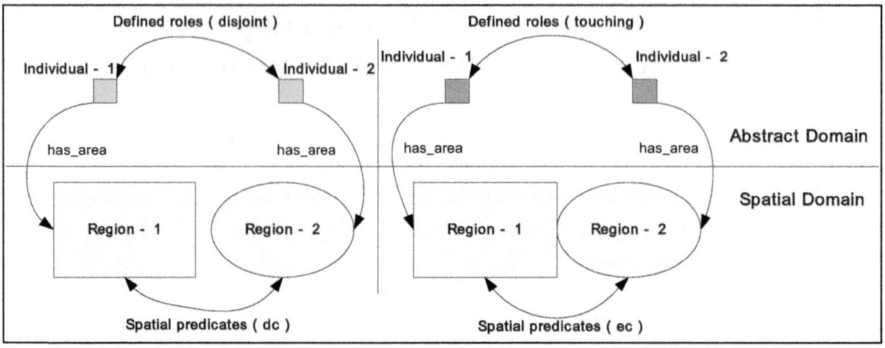

Fig. 3. The spatial and abstract domain regarding 'disjoint' and 'touching'

The RCC language contains only one primitive predicate C(X, Y), which represents 'region X, is connected with region Y'.

In above contents, the spatial relationships that occur between the regions are represented. This paper, defines new axioms regarding the 'disjoint' and 'touching' relationships, which occur among the concepts based on $\mathcal{ALCRP(D)}$ are used when constructing web ontologies.

New axioms of the 'disjoint' and 'touching' are represented as follows:

s_disjointWith $\doteq$ **∃(concept 1)(concept 2).dc**
s_touchingWith $\doteq$ **∃(concept 1)(concept 2).ec**

The new axioms – 's_disjointWith' and 's_touchingWith' make it possible to represent the spatial relationships when each concept has a spatial relationship. The necessity of the new axioms, which are defined above, are proven using an example that is the steps of building a web ontology for nations.

These are the steps of building a web ontology for nations.

Step 1: Collect all the concepts regarding the nations.
(ex) Asia, Africa, Europe, America, Korea, China, Japan, London, etc…

Step 2: Classify the collected concepts.

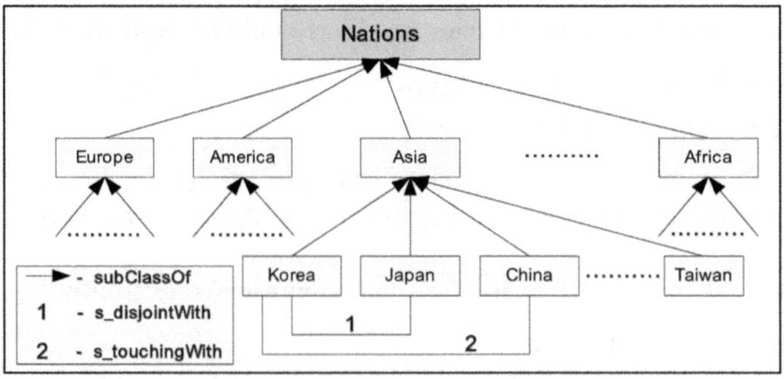

Fig 4. The classification of the concepts regarding nations

Step 3: Define the relationship between concepts using the existing representation method.

- Most concepts that are represented in above Fig. 3 are possible to define use of 'subClassOf' relationship.

Step 4: Find the new relationships that cannot be defined using the existing method only.

- The relationships – 'Korea↔China', 'Korea→Japan', etc, need the new relationships for defining the concepts.
- In this step, the newly occurred relationships can be represented using the new axioms that is defined in the 's_disjointWith' and 's_touchingWith'.
- According to the Fig. 5, the relationship among Korea, China, and Japan can be defined.

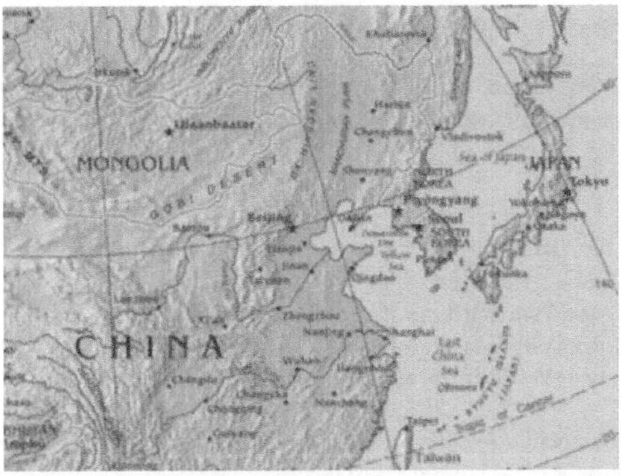

Fig. 5. A map of Asia

Step 5: Build a complete web ontology on the nations using the new axioms.

In above example, the occurrence of spatial relationships is inevitable and the plan for representing the spatial relationships is essential for constructing more perfect web ontologies. Therefore, ways of representing the spatial relationships should be undertaken in the future.

4 The Representation of the Spatial Relationships Using New Axioms

The purpose of this section is initially to add the 's_disjointWith' and 's_touchingWith' to the OWL axioms and build the web ontology on the nations using the OWL language.

Table 2. An example that is nations.owl

```
<?xml version="1.0" encoding="UTF-8"?>
<rdf:RDF xmlns:rdf="http://www.w3.org/1999/02/22-rdf-syntax-ns#"
      xmlns:rdfs="http://www.w3.org/2000/01/rdf-schema#"
       xmlns:owl="http://www.w3.org/2002/07/owl#"

  <owl:Class rdf:ID="Nations">
     <rdfs:subClassOf rdf:resource="http://www.w3.org/2002/07/owl#Thing"/>
  </owl:Class>

  <owl:Class rdf:ID="Europe">
     <rdfs:subClassOf rdf:resource="#Nations"/>
  </owl:Class>
  <owl:Class rdf:ID="America">
     <rdfs:subClassOf rdf:resource="# Nations "/>
  </owl:Class>
  <owl:Class rdf:ID="Asia">
     <rdfs:subClassOf rdf:resource="# Nations "/>
  </owl:Class>
  <owl:Class rdf:ID="Africa">
     <rdfs:subClassOf rdf:resource="# Nations "/>
  .........

</owl:Class>
  <owl:Class rdf:ID="Korea">
     <rdfs:subClassOf rdf:resource="#Asia"/>
     <owl:s_disjointWith rdf:resource="#Japan"/>
     <owl:s_disjointWith rdf:resource="#Taiwan"/>
     <owl:s_touchingWith rdf:resource="#China"/>
</owl:Class>
<owl:Class rdf:ID="Japan">
     <rdfs:subClassOf rdf:resource="#Asia"/>
     <owl:s_disjointWith rdf:resource="#Korea"/>
     <owl:s_disjointWith rdf:resource="#China"/>
</owl:Class>
<owl:Class rdf:ID="China">
     <rdfs:subClassOf rdf:resource="#Asia"/>
     <owl:s_disjointWith rdf:resource="#Japan"/>
     <owl:s_touchingWith rdf:resource="#Korea"/>
</owl:Class>
<owl:Class rdf:ID="Taiwan">
     <rdfs:subClassOf rdf:resource="#Asia"/>
  ..........

</rdf:RDF>
```

The above nations.owl routine attempts to express the nations ontology using the OWL language. Most of the concepts, properties and relationships of the ontology on nations can be represent easily using the OWL Capabilities. The main focus is to represent the spatial relationships in this example. The new axioms, 's_disjointWith' and 's_touchingWith', are defined and applied. The 's_disjointWith' and 's_touchingWith' will be very useful for building web ontologies.

There is a great deal of multimedia data in the web. The retrieval and representation of multimedia data will be a good issue as the semantic web progresses. Therefore, this study regarding representation of the spatial relationships will be a core part for building the multimedia data ontology.

5 Conclusions and Future Works

This study attempted to represent the spatial relationships, and defined new axioms, 's_disjointWith' and 's_touchingWith'. These axioms were applied to construct the web ontologies. The result of this study shows that if new axioms are used for building web ontologies, the representation of the spatial relationships will easier and more flexible. It can be expected that a study on extending the representation of the spatial relationships will be a remarkable part for constructing the perfect web ontology.

More research on the representation of the spatial relationships will be needed in the future. A future study will aim to define more complex spatial relationships and make many more spatial axioms.

References

[1] Peter F. Patel-Schneider, Patrick Hayes, Ian Horrocks, "OWL Web Ontology Language Semantics and Abstract Syntax, W3C Working Draft 31 March 2003", http://www. w3.org/TR/2003/WD-owl-semantics-20030331.

[2] D. Brickley, R. Guha (eds.), "Resource Description Framework (RDF) Schema Specification, W3C Candidate Recommendation 27 March 2000, http://www. w3.org/TR/2000/CR-rdf-schema-20000327.

[3] T. Berners-Lee, J. Hendler, and O. Lassila, "The Semantic Web", Scientific Am., vol.284, no.5, May 2001, pp.34-43.

[4] F.Wolter and M.Zakharyaschev, "Modal description Logics: Modalizang roles", *Fundamenta Informaticae*, 39:411-438, 1999.

[5] V. Haarslev, C.Lutz, and R. Moller, "A description logic with concrete domains and a role-forming predicate operator," Journal of Logic and Computation 9(S), pp. 351-384, 1999

[6] Yongjuan Zou, "Integrating concrete Domains into Description Logic: A Survey", April 23, 2003.

[7] V. Haarslev, C.Lutz, and R. Moller, "Foundations of spatioterminological reasoning with description logics," Proceedings of the Sixth International Conference on Principles of Knowledge Representation and Reasoning(KR '98), pp. 112-123, June 1998

[8] V. Haarslev, R. Moller, and M. Wessel, "Visual spatial query languages: A semantics using description logic," Diagrammatic Representation and Reasoning, 2000

[9] Cohn, Z. Cui, and D. Randell, "A spatial logic based on regions and connection," Proc. Third International Conference on Principles of Knowledge Representation and Reasoning(KR '92), 1992

[10] Nebel and J. Renz, "On the complexity of qualitative spatial reasoning:A maximal tractable fragment of the region connection calculus," Artificial Intelligence, 1992
[11] Guarino, N, and Giaretta, P., "Ontologies and Knowledge bases: towards a terminological clarification", In N. Mars, Ed. Toward Very Large Knowledge Bases: Knowledge Building and Knowledge Sharing 1995, PP. 25-32.
[12] Ankolekar, F. Huch and K. Sycara, "Concurrent Semantics for the Web Services Specification Language Daml-S", In Proc. of the Coordination 2002 conf., 2002.
[13] W. Kim, H. Kong, K. Oh, Y. Moon and P. Kim, "Concept Based Image Retrieval Using the Domain Ontology", Computational Science and Its Applicatons(ICCSA 2003), PP 401-410, 2003.
[14] Chandrasekaran, J. Josephson, and R. Benjamins, "What Are Ontologies, and Why do we Need Them?", IEEE Intelligent Systems, 14,1:20-26, 1999.
[15] F. vanHarmelen, P.F. Patel-Schneider, and I. Horrocks, "A Modal-Theoretic Semantics for DAML+OIL", http://www.daml.org, March 2001.
[16] Stephan Tobies. The complexity of reasoning with cardinality restrictions and nominals in expressive description logics. *J. of Artificial Intelligence Research*, 12:199-217, May 2000.
[17] M. Morurovic, F. Wolter, and M. Zakharyaschev. Modalized description logic – how much? Technical report, Computer Science Department, University of Leipzig, Germany, 2000.
[18] E. Franconi, G. De Giacomo, R.M.MacGregor,W. Nutt, C. A.Welty, and F. Sebastiani, editors. Collected Papers from the International Description Logics Workshop (DL'98). CEUR, May 1998.
[19] Horrocks. Optimising Tableaux Decision Procedures for Description Logics. PhD thesis, University of Manchester, 1997.
[20] Horrocks and G. Gough. Description logics with transitive roles. In M.-C. Rousset, R. Brachmann, F. Donini, E. Franconi, I. Horrocks, and A. Levy, editors, Proceedings of the InternationalWorkshop on Description Logics, pages 25–28, Gif sur Yvette, France, 1997. Universit´e Paris-Sud.
[21] Franz Baader and Ralf K¨usters. Matching in description logics with existential restrictions. In Proc. of the 1999 Description Logic Workshop (DL'99).
[22] Franz Baader and Paliath Narendran. Unification of concepts terms in description logics. J. of Symbolic Computation, 31(3):277–305, 2001.
[23] Diego Calvanese. Finite model reasoning in description logics. In Luigia C. Aiello, John Doyle, and Stuart C. Shapiro, editors, Proc. of the 5th Int. Conf. on the Principles of Knowledge Representation and Reasoning (KR'96), pages 292–303. Morgan Kaufmann, Los Altos, 1996.
[24] Horrocks and Patel-Schneider, 1998d] Ian Horrocks and Peter F. Patel-Schneider. Optimising propositional modal satisfiability for description logic subsumption. In Proc. of the 4th Int. Conf. on Artificial Intelligence and Symbolic Computation (AISC'98), 1998.

[25] Ulrich Hustadt and Renate A. Schmidt. Issues of decidability for description logics in the framework of resolution. In R. Caferra and G. Salzer, editors, Automated Deduction in Classical and Non-Classical Logics, volume 1761 of Lecture Notes in Artificial Intelligence, pages 191–205. Springer, 2000.
[26] Daphne Koller, Alon Levy, and Avi Pfeffer. P-CLASSIC: A tractable probabilistic description logic. In Proc. of the 14th Nat. Conf. on Artificial Intelligence(AAAI'97), pages 390–397. AAAI Press/The MIT Press, 1997.

Inductive Construction of Ontologies from Formal Concept Analysis

Michael Bain

School of Computer Science and Engineering
University of New South Wales
Sydney NSW 2052, Australia
mike@cse.unsw.edu.au

Abstract. We present an approach to representation and structuring of theories and ontologies based on a formalism of propositional logic programs. Formal concept analysis is adopted to identify structure in theories. This structure, in the form of conjunctive concepts and the relations between them, is used for representation change in theories based on feature construction and iterative program transformation. Ontologies are represented as sets of propositional definite clauses containing named concepts having a superclass–subclass relationship derived from a concept lattice built using formal concept analysis. Logic programming methods are used to incrementally construct and revise such ontologies. An information compression measure is used to guide the operation of structuring theories and ontologies. The clauses defining the ontology are proved to preserve the relationships which hold between formal concepts in the concept lattice. This framework enables inheritance inference not possible from the structured theories alone. Experimental results are presented from an application to a sample of descriptions of computer science academics' research interests and a reconstruction experiment on randomly generated theories with added noise.

1 Introduction

At one extreme, general ontology projects seem to have some distance to go before attaining their goals, for example capturing most of everyday human knowledge. At the other, it is questionable how useful very narrowly-focused domain- or application-specific ontology projects will be. Although they may be sufficiently small to be built by domain experts and knowledge engineers, this can be a time-consuming process. Certainly they are likely to be limited in interoperability, i.e. usage in combination with other ontologies. Since this capability is usually cited as a goal of ontology construction [12] it seems likely that such efforts can be only partly successful. Therefore some middle ground between maximum coverage, which appears to be unattainable, and maximum specificity, which appears too limited to meet the requirements for an ontology, would seem to be a desirable objective. This suggests that methods of automating ontology construction may be useful and may be a suitable area for the application of machine learning.

T.D. Gedeon and L.C.C. Fung (Eds.): AI 2003, LNAI 2903, pp. 88–99, 2003.

1.1 Background

Formal Concept Analysis (FCA) is one approach to identifying structure present in domains. Introduced by Wille (see [3] for an overview), FCA is based on a complete lattice of all *formal concepts* in a domain. A concept in this formalism is an ordered pair of sets, one a set of attributes or descriptors of the concept, the other a set of object indices denoting all instances of the concept in the domain. The set of descriptors of a concept is the maximal set common to all the instances of the concept. These concepts form a partial order from which a *concept lattice* is constructed.

FCA has proved to be useful in exploring the conceptual structure of a domain, for example in a graphical representation or via a browser interface. A drawback of FCA, however, is that it does not allow for the introduction of concept *names* as new intermediate-level terms. This is necessary for re-use of conceptual structures, so that they may be referred to by name, e.g. to be used in incremental learning or theory revision.

Inverse resolution is a technique from Inductive Logic Programming (ILP) [8]. It comprises operators based on inverting the resolution rule of deductive inference. A key aspect of inverse resolution is the introduction of *theoretical* terms, additional to the *observational* terms given a priori for the domain. These theoretical terms are combined with observational terms to enable the construction of hierarchical concept definitions.

In [1] inverse resolution and FCA were combined in a method to identify potentially interesting and useful concepts in a concept lattice and revise the underlying formal context and the lattice it generates to invent new descriptors and extract their definitions. The approach was developed using FCA and inverse resolution operators for both a theory and its lattice.

Results with a system called Conduce which is an implementation of this method showed that it could be applied to both unsupervised and supervised learning problems. In unsupervised learning the task is to recover implicit classes in the data and organise them into structured relationships. For classification (supervised learning) the system can find new features which, when added to the attribute set of a decision-tree learner, can improve the predictive accuracy of the induced classifier.

1.2 Motivation

We propose augmenting a knowledge-base by the use of a hierarchically structured taxonomy or ontology which contains concept names and the relations between them. One advantage could be to simplify querying such a knowledge base. If queries may contain intermediate-level terms they can be much more compact than would otherwise be the case. For example, instead of searching for things with "beak, tail, two legs and wings" we could query on "bird".

Although Conduce [1] can take a "flat" theory and structure it, the new concept names introduced in this process stand in an *inverse* relationship to each other, in terms of the call-graph dependencies, to the "natural" taxonomical hierarchy dependencies, in terms of sub-class or super-class dependencies.

We use the following example to illustrate the problem. Initially we have a flat theory, i.e. there is no clause with a descriptor in its body identical to the descriptor in the head of some other clause.

sparrow :- beak, tail, legs(2), wings, homeothermic, brown.
eagle :- beak, tail, legs(2), wings, homeothermic, golden.
gorilla :- legs(2), homeothermic, hairy, vegetarian.
human :- legs(2), homeothermic, hairless, omnivore.

Suppose the concept with intent "legs(2), homeothermic" is selected. The theory is revised to contain a new clause defining the new descriptor "bipedal_homeotherm" in terms of the selected concept. A further revision is applied with respect to the selected concept with intent "bipedal_homeotherm, beak, tail, wings". The theory after the sequence of two revisions using the inter-construction operator (defined in [8]) looks as follows.

sparrow :- bird, brown.
eagle :- bird, golden.
gorilla :- bipedal_homeotherm, hairy, vegetarian.
human :- bipedal_homeotherm, hairless, omnivore.
bipedal_homeotherm :- legs(2), homeothermic.
bird :- bipedal_homeotherm, beak, tail, wings.

Evidently this revision has produced some of the results we required. It is structured and compressed, and it contains new terms (here with user-supplied names). However, some of the key conceptual structure, apparent in the associated concept lattice which is not shown here, is missing. Specifically, by inspection we know that the concept named "bird" is a sub-class of the concept named "bipedal_homeotherm". But if we were told that "sparrow" was true, we could not conclude that "bird" was also true. Similarly, knowing that "bird" was true, we could not conclude that "bipedal_homeotherm" was true. Suppose we were to construct an associated ontology with respect to the theory, comprising in this case the following clauses.

bipedal_homeotherm :- bird.
bird :- sparrow.
bird :- eagle.

Now the desired inferences are possible. It is the automatic construction of such ontology clauses that we investigate in this paper.

2 Formal Concept Analysis

Detailed coverage of Formal Concept Analysis (FCA) is in [3]. The following basic definitions are included to make the paper more self-contained.

Definition 1. (Formal context) *A formal context is a triple* $\langle \mathcal{G}, \mathcal{M}, \mathcal{I} \rangle$. $\mathcal{G}$ *is a set of objects,* $\mathcal{M}$ *is a set of descriptors, and* $\mathcal{I}$ *is a binary relation such that* $\mathcal{I} \subseteq \mathcal{G} \times \mathcal{M}$.

The notation $\langle y, x \rangle \in \mathcal{I}$ or alternatively $y\mathcal{I}x$ is used to express the fact that an object $y \in \mathcal{G}$ has an attribute or descriptor $x \in \mathcal{M}$.

Definition 2. (Formal concept) *A formal concept is a pair of sets* $\langle Y, X \rangle$, *where* $Y \subseteq \mathcal{G}$ *and* $X \subseteq \mathcal{M}$. *Each pair must be complete with respect to* $\mathcal{I}$, *which means that* $Y' = X$ *and* $X' = Y$, *where* $Y' = \{x \in \mathcal{M} | \forall y \in Y, y\mathcal{I}x\}$ *and* $X' = \{y \in \mathcal{G} | \forall x \in X, y\mathcal{I}x\}$.

The set of descriptors of a formal concept is called its intent, while the set of objects of a formal concept is called its extent. In this paper, since we tend to focus on intent of formal concepts, we will more often write $\langle X, Y \rangle$ for a formal concept. For a set of descriptors $X \subseteq \mathcal{M}$, X is the intent of a formal concept if and only if $X'' = X$. A dual condition holds for the extent of a formal concept. This means that any formal concept can be uniquely identified by either its intent or its extent alone. Intuitively, the intent corresponds to a kind of maximally specific description of all the objects in the extent.

The correspondence between intent and extent of complete concepts is a Galois connection between the power set $\mathcal{P}(\mathcal{M})$ of the set of descriptors and the power set $\mathcal{P}(\mathcal{G})$ of the set of objects. The Galois lattice $\mathcal{L}$ for the binary relation is the set of all complete pairs of intents and extents, with the following partial order.

Definition 3. (Concept order) *Given two concepts* $N_1 = \langle X_1, Y_1 \rangle$ *and* $N_2 = \langle X_2, Y_2 \rangle$, $N_1 \leq N_2 \leftrightarrow X_1 \supseteq X_2$. *The dual nature of the Galois connection means we have the equivalent relationship* $N_1 \leq N_2 \leftrightarrow Y_1 \subseteq Y_2$.

For a concept N, $I(N)$ denotes its intent and $E(N)$ denotes its extent.

3 Ontology Definitions

An ontology is defined with respect to a theory. The set of terms permitted to appear in a theory is referred to as the *vocabulary* of the theory. In this paper we consider only a propositional vocabulary for both theories and ontologies, Since the methods we use are derived from ILP and logic programming, theories and ontologies are represented as sets of definite clauses.

Definition 4. (Vocabulary) *An ontology is defined with respect to a vocabulary of terms partioned into a set Ob of observational terms and a set Th of theoretical terms. In terms of formal concept analysis the vocabulary corresponds to the set of descriptors,* $Ob \cup Th = \mathcal{M}$.

In practical applications we would expect the observational terms to be given. The theoretical terms are machine-generated symbols which can be substituted later, either by user-supplied names or algorithmically generated labels designed to be humanly comprehensible.

Definition 5. (Base-level heads) *A base-level head is the head h of a clause in a theory such that $h \in Ob$.*

Some functions are used to simplify the definitions and algorithm description. For convenience we assume that clauses are unique, that each has associated with it a unique index, and each is represented as a triple $\langle K, H, B \rangle$.

Definition 6. (Clause index) *The function $id(C)$ returns the index K of the clause C.*

Definition 7. (Clause head) *The function $hd(C)$ returns the head H of the clause C.*

Definition 8. (Clause body) *The function $bd(C)$ returns the body B of the indexed clause C.*

Definition 9. (Clause for index) *The function $cl(K)$ returns the clause C with index K.*

We can treat ontology clauses as theory clauses, i.e. as indexed definite clauses. For convenience, the index can be omitted when writing the clause whenever it is not necessary in the context.

Definition 10. (Ontology clause) *An ontology clause C is a definite clause with exactly two literals. The head is a single descriptor $hd(C) \in Th$. The body is a set containing a single descriptor b. If b is in the set of base-level heads then C is referred to as an isa clause. If $b \in Th$ then C is referred to as an ako clause.*

The *isa* clauses represent membership of a class by an object, such as the fact that sparrows and eagles are birds. The *ako* clauses represent superclass-subclass subsumption, such as birds being a kind of "bipedal_homeotherm" [10].

Definition 11. (Ontology) *An ontology is a set of ontology clauses.*

We assume the standard semantics developed for definite clause logic programs [7]. This allows straightforward inference of inheritance, query-answering at the level of intermediate terms, and hierarchical construction of the ontology.

4 An Algorithm for Ontology Construction

The method of structuring named concepts into an ontology is based on the idea of "folding" from program transformation. This has been widely studied in Logic Programming [4]. In fact we use an extension of the typical method of folding in Logic Programming called "disjunctive folding" [11].

Algorithm 1 (Ontocon)

Input: *theory* $\mathcal{T}_{t-1}$, *concept* n_{t-1},
ontology $\mathcal{O}_{t-1}$, *new theory clause* $w \leftarrow A$
Output: *ontology* $\mathcal{O}_t$

Begin
/ construct new ontology clauses */*
$Cs = \emptyset$
For each clause $C \in \mathcal{T}_{t-1}$ *s.t.* $id(C) \in E(n_{t-1})$ *Do*
    $Cs = Cs \cup \{w \leftarrow hd(C)\}$;
EndFor
/ repeated folding of ontology clauses */*
$OCs = \mathcal{O}_{t-1} \cup Cs$;
Repeat
    potential-folds $= \{\langle folded, folder \rangle\}$ *where*
        folded, folder $\subseteq OCs$ *such that*
        folded defines a single predicate and
        each 2-tuple denotes a disjunctive fold; (see text for details)
    If potential-folds $= \emptyset$ *Then*
        $\mathcal{O}_t = \mathcal{O}_{t-1}$;
        Halt;
    Else
        select maximum cardinality fold from potential-folds;
        $OCs' =$ *apply-disjunctive-folding(fold, OCs);*
        $OCs = OCs'$;
    Endif
End

Ontocon comprises two stages. First, following application of updates to theory $\mathcal{T}_{t-1}$ and its associated concept lattice $\mathcal{L}_{t-1}$ by Conduce using the inter-construction operator, there is a single clause defining the new predicate, w. Each of the revised clauses in $\mathcal{T}_{t-1}$ is used to construct a set of new *base-level* ontology clauses. This is referred to as *expansion* since it constitutes an expansion of the existing ontology [2].

Expansion introduces a new set of ontology clauses. Owing to the restricted syntax of ontology clauses, the set of clauses defining a single theoretical term w can be represented as a single *disjunctive* clause $w \leftarrow B$, where $B = b_1 \vee b_2 \vee \ldots \vee b_m$. Each literal in the body is taken from the body of one of the corresponding set of ontology clauses. For example, we might have "bird :- sparrow ; eagle.", in Prolog syntax. Note that this is simply syntactic sugar for the actual set of clauses defining a predicate in the ontology.

However, following expansion we do not have any structure in the ontology to relate the new named concept to previously added concepts, although such relationships were present in the concept lattice. Therefore it is necessary to relate new concept definitions to existing concept definitions in the ontology.

This can be done by incorporating the new concepts in such a way that the pattern of invocation between clauses reflects the subsumption relationship in the concept lattice of the concepts they represent.

The method is based on a disjunctive folding operator, as follows:

$p \leftarrow A \vee B$ $q \leftarrow A$	$p \leftarrow q \vee B$ $q \leftarrow A$

In the clauses upper-case letters represent disjunctions of literals (descriptors) while lower-case letters represent single literals. The operator is shown as a set of preconditions (to the left of the vertical line) and postconditions (to the right). Read left to right, the changes to operand clauses for the disjunctive folding operator are evident. The upper clause on the left is the "unfolded" clause and on the right is the "folded" clause. The lower clause is the "folder" clause (which is unchanged). Note that this single-step disjunctive folding is a compressive, structuring revision of the ontology. The second stage of the Ontocon algorithm, referred to as *structuring*, repeatedly applies disjunctive folding to assimilate the new concept into the ontology.

4.1 An Example

We illustrate the method with the example from Section 1.2. On the first iteration, with the introduction of the new descriptor "bipedal_homeotherm" to the theory we get the following ontology clauses:

```
bipedal_homeotherm :- sparrow.
bipedal_homeotherm :- eagle.
bipedal_homeotherm :- gorilla.
bipedal_homeotherm :- human.
```

On the second iteration, the new descriptor "bird" is introduced to the theory and we have the following new ontology clauses:

```
bird :- sparrow.
bird :- eagle.
```

However, the ontology can now be simplified, by applying disjunctive folding to replace the clauses :

```
bipedal_homeotherm :- sparrow.
bipedal_homeotherm :- eagle.
```

with the single clause: bipedal_homeotherm :- bird.

4.2 Ontocon Preserves Conceptual Structure

The following definitions and results apply to Algorithm 1.

Definition 12. (Expansion) *For a given formal concept N with name w expansion introduces a new set of ontology clauses, or a single disjunctive ontology clause $C = w \leftarrow B$, where $b \in B = hd(cl(y))$ for all $y \in E(N)$. We notate this $C = expand(N)$.*

We define the folding operator described above.

Definition 13. (Folding) *Let C be a (disjunctive) ontology clause (the unfolded clause) and D be an ontology clause (the folder clause) such that every literal in the body of D occurs in the body of C. Then C' (the folded clause) is obtained by replacing the set of literals in the body of C which also occur in the body of D by the head of D.*

Unfolding is the inverse operation.

Definition 14. (Unfolding) *Let C be a (disjunctive) ontology clause (the folded clause) and D be an ontology clause (the folder clause) such that the head q of D is in the body of C. Then C' (the unfolded clause) is obtained by replacing q in the body of C by the body of D.*

Definition 15. (Structuring) *Let C be a (disjunctive) ontology clause defining a predicate p which appears in the head of C. Then by repeated application of folding C' is an ontology clause defining p such that the no further folding is possible between C' and some other clause $D \neq C'$ in the ontology.*

Definition 16. (Elimination) *Let C be a (disjunctive) ontology clause defining a predicate p which appears in the head of C. Then by repeated application of unfolding C' is an ontology clause defining p such that the body of C' contains only elements of Ob. We notate this $C' = elim(C)$. Elimination is undefined otherwise.*

Elimination is the process of replacing all occurrences of theoretical terms by observations terms. For example, elimination applied to "animal :- bird ; fish." might give "animal :- sparrow ; eagle ; shark.". See [6] for a discussion of theoretical term elimination.

Definition 17. (Ontological size) *Let C be a (disjunctive) ontology clause. The "size" of C in the ontology is the cardinality of the body of C after elimination. We notate this $size(C) = |bd(elim(C))|$.*

Definition 18. (Ontological subsumption) *Let C, D be (disjunctive) ontology clauses. C subsumes D in the ontology, written $C \sqsupseteq D$, iff $size(C) \geq size(D)$.*

Lemma 2. (Elimination equal to expansion) *For any (disjunctive) ontology clause C' defining a predicate p introduced and assimilated into an ontology by Algorithm 4 for a given formal concept N $elim(C') = expand(N)$.*

Proof. *Expansion in Algorithm 4 to give a clause C is followed by structuring which comprises a sequence of folding steps to give C'. In each step the clause defining p is either the folder clause or the unfolded clause which is transformed to the folded clause. If the former, the clause is unchanged. If the latter, the step is reversible, i.e. to obtain the unfolded clause from the folded clause. By applying the reverse step in each such case, elimination produces C.*

Lemma 3. (Structuring preserves ontological subsumption) *Let C, D be (disjunctive) ontology clauses in an ontology constructed by Algorithm 4 from concepts N, M, respectively. in concept lattice $\mathcal{L}$. If $N \geq M$ in $\mathcal{L}$ then $C \sqsupseteq D$.*
Proof. *Assume the opposite, that $\neg(C \sqsupseteq D)$. Then $\neg(size(C) \geq size(D))$, by Definition 18. But by Definition 12 and Lemma 2 $size(C) = |E(N)|$ and $size(D) = |E(M)|$. By Definition 3 for concept lattices $N \geq M$ implies $E(N) \supseteq E(M)$. This contradicts the assumption.* □

Theorem 1. (Ontocon preserves conceptual structure) *Algorithm 4 (Ontocon) preserves conceptual structure in terms of concept ordering.*
Proof. *At each iteration Ontocon applies expansion and structuring. By Lemma 3 this preserves concept ordering. No other operations are performed on the ontology by Ontocon. Therefore the algorithm preserves concept ordering.*

5 Results and Discussion

As a first test of this approach we applied Ontocon to a data set of research interests collected from researchers at the School of Computer Science and Engineering, University of New South Wales. This dataset was collected by Mihye Kim and Paul Compton [1] as part of a research project [5].

This domain contains, for each of the 81 individuals surveyed in the study, a list of their research interests. A concept lattice of 460 nodes (i.e. formal concepts) and 1195 edges (i.e. superclass-subclass relationships) was generated from this initial theory by Conduce. Following repeated application of interconstruction by Conduce to the theory and associated concept lattice a structured theory was generated. The associated concept lattice was reduced to 315 nodes and 764 edges. This theory contained a total of 37 new named concepts found by the system and defined in terms of the concept intents.

These concepts form the denotation of the new terms in the ontology generated by Ontocon. Of these, 20 concepts were refined into structured ontology clauses, i.e. ako clauses. At the terminals of this hierarchy 50 isa clauses represented researchers' interests. Due to multiple interests, 44 individuals were represented in these structured ontology clauses. Inheritance between concepts can be thought of as "ako" links, whereas inheritance between concepts and individuals can be thought of as "isa" links [10]. The actual rendering of links into something resembling a natural language, for example a subset of English, is always a domain-specific issue. In the domain of research interests of individuals

[1] See `http://pokey.cse.unsw.edu.au/servlets/RI/`

Table 1. Each entry is the mean (standard deviation) of 100 trials

	Ontology concepts only	Extra "noise" concepts
Clauses	9.66 (6.69)	9.62 (6.86)
Nodes	16.48 (10.83)	35.90 (19.55)
Edges	24.14 (17.51)	74.89 (44.06)
SDC	0.0 (0.0)	1.05 (0.85)

in a university we could conceivably treat the meanings of "ako" and "isa" as follows. "X isa Y" could be read as "X interested-in Y". "X ako Y" could be read as "X extra-interest-of Y".

Inheritance is inferred then according to the following schema.

If X interested-in Y
and Y extra-interest-of Z
then X interested-in Z

So if we know that the individual "Researcher 35" is interested in "Learning and Philosophy" and we know that "Learning and Philosophy" are additional interests of those interested in "Case Based Reasoning and Knowledge Based Systems" we can conclude that Researcher 35 has an interest in those subjects too.

5.1 Reconstructing Ontologies

A quantitative reconstruction experiment was undertaken to evaluate how well the method could recover randomly generated ontologies. First we generate an ontology with certain key parameters (depth, branching factor, etc.) randomly assigned. Then this ontology is "flattened" using unfolding which produces a theory. Ontocon is applied to this theory and the recovered ontology is compared to the original using a quantitative measure of structural similarity.

The structural dissimilarity coefficient (SDC) was proposed in [13]. It measures the average graphical distance between terminal nodes when two hierarchies (here ontologies) are compared. A value of zero indicates identical structures. In Table 1 it is evident that ontologies are reconstructed exactly. With added spurious "noise" concepts, seen in the increased number of nodes (concepts) and edges in the concept lattice, recovery is still good. An SDC of 1-2 was found in cases of very similar structures (by expert evaluation) in [13].

5.2 Ontologies

Answering the question "what *is* an ontology ?" is not straightforward. A detailed definition comes from Noy and McGuinness [9]. This is developed into a series of instructions for constructing ontologies. There are four main steps.

1. defining classes in the ontology
2. arranging the classes in a taxonomic (subclass-superclass) hierarchy
3. defining slots and describing allowed values for these slots
4. filling in the values of slots for instances

It can be argued that our methods Ontocon and Conduce implement each of these steps, as follows.

1. defining structured clauses in terms of formal concepts using the inter-construction operator [Conduce]
2. constructing ontology clauses [Ontocon]
3. each named concept is a "slot", and the clause defining that named concept describes the values it can take [Conduce]
4. values of slots are filled by the examples in the domain

We claim therefore that Ontocon and Conduce can reasonably be said to implement ontology construction.

6 Conclusions and Future Directions

We have presented a method for theory structuring and ontology learning with a proof of a key desirable property and results from its application. The approach can identify implicit structure in a domain. It is oriented to achieving compression, structuring and hence improved comprehensibility of a theory in the form of a knowledge base and an associated ontology in the form of a taxonomical hierarchy. This allows simplification of theory itself and expands the set of queries applicable to the knowledge base to allow intermediate-level concepts not in the original vocabulary.

Using a logic program representation enables the use of standard techniques such as predicate invention and disjunctive folding. We rely on semantics from Logic Programming to set up ontological inference. Formal Concept Analysis provides a rigorous framework for the definition of concepts and their ordering. This enables a straightforward proof that our methods preserve the important properties of subclass-superclass hierarchical ordering from the concept lattice to the ontology clauses. Results on the research interests domain and in the reconstruction experiments show the discovery of some reasonable structure.

This work is preliminary. For example, a method of using first-order concept descriptions and relations between them is almost certainly required for many real-world ontology construction tasks. A method of using first-order concept descriptions and relations between them should be investigated as a priority. The approach also needs to be tested on more data sets.

Acknowledgements

I thank Mihye Kim and Paul Compton for useful discussions on these topics and for providing the research interests data set. Research supported by the Australian Research Council.

References

[1] M. Bain. Structured Features from Concept Lattices for Unsupervised Learning and Classification. In B. McKay and J. Slaney, editors, *AI 2002: Proc. of the 15th Australian Joint Conference on Artificial Intelligence*, LNAI 2557, pages 557–568, Berlin, 2002. Springer. 89

[2] N. Foo. Ontology Revision. In G. Ellis, R. Levinson, W. Rich, and J. Sowa, editors, *Proc. Third Intl. Conf. on Conceptual Structures*, LNAI 954, pages 1–14, Berlin, 1995. Springer. 93

[3] B. Ganter and R. Wille. *Formal Concept Analysis: Mathematical Foundations.* Springer, Berlin, 1999. 89, 90

[4] C. J. Hogger. *Essentials of Logic Programming.* Clarendon Press, Oxford, 1990. 92

[5] M. Kim and P. Compton. Incremental Development of Domain-Specific Document Retrieval Systems. In S. Handschuh, R. Dieng-Kuntz, and S. Staab, editors, *Proceedings of the First International Conference on Knowledge Capture*, pages 69–77, Calgary, 2001. University of Calgary. 96

[6] R. Kwok. *Formalising Aspects of Theoretical Term Usage.* PhD thesis, University of Sydney, Sydney, Australia, 1998. 95

[7] J. W. Lloyd. *Logic Programming, 2nd Edition.* Springer-Verlag, Berlin, 1987. 92

[8] S. Muggleton. Inverse Entailment and Progol. *New Generation Computing*, 13:245–286, 1995. 89, 90

[9] N. Fridman Noy and D. L. McGuinness. Ontology Development 101: A Guide to Creating Your First Ontology. Technical Report KSL-01-05, Knowledge Systems Laboratory, Stanford University, CA, 2001. 97

[10] T. Richards. *Clausal Form Logic: An Introduction to the Logic of Computer Reasoning.* Addison-Wesley, Wokingham, England, 1989. 92, 96

[11] A. Roychoudhury, K. N. Kumar, C. R. Ramakrishnan, and I. V. Ramakrishnan. Beyond Tamaki-Sato Style Unfold/Fold Transformations for Normal Logic Programs. *International Journal on Foundations of Computer Science*, 13(3):387–403, 2002. 92

[12] S. Schultze-Kremer. Ontologies for Molecular Biology. *Pacific Symposium on Biocomputing*, 3:693–704, 1998. 88

[13] B. Zupan. *Machine learning based on function decomposition.* PhD thesis, University of Ljubljana, Slovenia, 1997. 97

Dynamic Variable Filtering for Hard Random 3-SAT Problems

Anbulagan, John Thornton, and Abdul Sattar

Knowledge Representation and Reasoning Unit (KRRU)
School of Information Technology
Griffith University, Gold Coast Campus
PMB 50 Gold Coast Mail Centre, QLD 9726, Australia
{a.anbulagan,j.thornton,a.sattar}@griffith.edu.au
Telephone: (07) 5552 8730, Fax: (07) 5552 8066

Abstract. The DPL (Davis-Putnam-Logemann-Loveland) procedure is one of the most effective methods for solving SAT problems. It is well known that its efficiency depends on the choice of the branching rule. One of the main improvements of this decision procedure has been the development of better branching variable selection through the use of unit propagation look-ahead (UPLA) heuristics (e.g., [12]). UPLA heuristics perform one of the following actions during two propagations of a free variable at each search tree node: detecting a contradiction earlier, simplifying the formula, or weighing the branching variable candidates. UPLA heuristics can be considered as polynomial time reasoning techniques. In this paper, we propose a new branching rule which does more reasoning to narrow the search tree of hard random 3-SAT problems. We term this reasoning technique the *Dynamic Variable Filtering (DVF) heuristic*. In our empirical study we develop four different variants of the DPL procedure: two (*ssc*34 and *ssc*355) based on this new heuristic and another two (*Satz*215-0 and *Satz*215*sT*) based on static variable filtering heuristics. *ssc*355 additionally integrates the Neighborhood Variable Ordering (NVO) heuristic into *ssc*34. We then compare the best known versions of the state-of-the-art *Satz* DPL procedure (*Satz*215), with each of our four procedures. Our results suggest that improved branching rules can further enhance the performance of DPL procedures. On hard random 3-SAT problems, our best procedure (*ssc*355) outperforms *Satz*215 with an order of magnitude in terms of the number of branching nodes in the search tree. While the run-times for dynamic variable filtering are still uncompetitive with *Satz*215, we have yet to explore the benefits that can be gained from avoiding redundant propagations and we still can improve the performance of the NVO heuristic. A further interesting property of dynamic filtering is that all backtracking can be eliminated during the DPL unit rule process. This property can also be explored in our future work for improving DPL procedure efficiency.

Keywords: Constraints, Problem Solving, Search.

T.D. Gedeon and L.C.C. Fung (Eds.): AI 2003, LNAI 2903, pp. 100–111, 2003.

1 Introduction

The satisfiability (SAT) problem is central in mathematical logic, artificial intelligence and other fields of computer science and engineering. In conjunctive normal form (CNF), a SAT problem can be represented as a propositional formula $\mathcal{F}$ on a set of Boolean variables $\{x_1, x_2, ..., x_n\}$. A literal l is then a variable x_i or its negated form $\bar{x_i}$, and a clause c_i is a logical *or* of some literals such as $x_1 \vee x_2 \vee \bar{x_3}$. A propositional formula $\mathcal{F}$ consists of a logical *and* of several clauses, such as $c_1 \wedge c_2 \wedge \ldots \wedge c_m$, and is often simply written as a set $\{c_1, c_2, ..., c_m\}$ of clauses.

Given $\mathcal{F}$, the SAT problem involves testing whether all the clauses in $\mathcal{F}$ can be satisfied by some consistent assignment of truth values $\{true, false\}$ to the variables. If this is the case, $\mathcal{F}$ is satisfiable; otherwise it is unsatisfiable. The SAT problem, as the original of all NP-Complete problems [3], is fundamentally important to the study of computational aspects of decision procedures. When each clause in $\mathcal{F}$ contains exactly k literals, the restricted SAT problem is called k-SAT. 3-SAT is the smallest NP-Complete sub-problem of SAT.

One of the best known and most widely used algorithms to solve SAT problems is the DPL (Davis-Putnam-Logemann-Loveland) procedure [5]. Many SAT solvers such as Posit [7], Tableau [4], *Satz* [12], *Satz*214 [14] and *cnfs* [6] are based on this procedure. DPL essentially enumerates all possible solutions to a given SAT problem by setting up a binary search tree and proceeding until it either finds a satisfying truth assignment or concludes that no such assignment exists. It is well known that the search tree size of a SAT problem is generally an exponential function of the problem size, and that the branching variable selected by a branching rule at a node is crucial for determining the size of the sub-tree rooted at that node. A wrong choice may cause an exponential increase of the sub-tree size. Hence, the actual performance of a DPL procedure depends significantly on the effectiveness of the branching rule used.

Much of the research on DPL has focussed on finding clever branching rules to select the branching variable that most effectively reduces the search space. In this paper, we too propose a new branching rule based on a *dynamic variable filtering heuristic* as a polynomial time reasoning technique aimed at significantly narrowing the search tree for solving hard random 3-SAT problems. The key idea underlying this new branching rule is to further detect failed literals that would remain undiscovered using a unit propagation look-ahead (UPLA) branching rule, before choosing a branching variable. In other words, we perform more reasoning in the open space between the UPLA heuristic and the MOMS (Maximum Occurrences in clause of Minimum Size) heuristic in the actual DPL branching rule. To test this idea, we use *Satz*215 (the best version of the *Satz* DPL procedure) where we simply replace its branching rule by a new branching rule. The new rule allows filtering of free variables, and at the same time reduces the sub-problem size at each node until the filtering process is saturated.

We develop two DPL procedures that use a dynamic variable filtering heuristic, and two other DPL procedures that use a static filtering heuristic. We then analyse these four DPL procedures with respect to *Satz*215, using a large sample of hard random 3-SAT problems.

An empirical study of the proposed variants of DPL indicates significant performance improvements can be obtained, with the two dynamic filtering heuristics consistently outperforming *Satz*215 in terms of mean search tree size when solving hard random 3-SAT problems. In some tests, the best dynamic filtering procedure was able to reduce the search tree size by an order of magnitude over *Satz*215, and for the 300 variable problems (1000 problems solved), the best dynamic filtering procedure produced a mean search tree size of 2679 nodes, in contrast to *Satz*215's mean size of 6634 nodes. Not surprisingly, given the additional reasoning involved in the branching rule, the new heuristics proved less competitive in terms of run-times, with *Satz*215 running approximately twice as fast on the largest (300 variable) problems.

Li and Gérard [15] discussed the hardness of proving an unsatisfiable hard random 3-SAT problem with 700 variables, and empirically calculated the approximately optimal number of branching nodes. They conjectured that obtaining this optimal sized tree was not possible using a branching heuristic. However, our results indicate that a dynamic variable filtering heuristic can achieve an optimal number of branching nodes. Therefore our work shows, in principle, that optimal branching can be achieved. The second major issue is whether optimal branching can be achieved efficiently. At present our results show *Satz*215 still performs significantly better than the dynamic filtering heuristics in terms of run-time. However we have not explored the potential gains that could result from avoiding redundant UPLA propagations. This we leave to future work.

The paper is organized as follows: In the next section we present the *Satz* DPL procedure, one of the best known SAT procedures for solving hard random 3-SAT problems and structured SAT problems. In the subsequent section, we present our new dynamic filtering branching rules which perform additional reasoning to find a best branching variable. In section 4, we present a comparison of results for our new DPL procedures with *Satz*215 on a set of hard random 3-SAT problems. Finally, we conclude the paper with some remarks on current and future research.

2 The *Satz* Solver

Satz [12, 13] is one of the best DPL procedures developed for solving both hard random SAT problems and structured SAT problems. Its powerful branching rule allows it to select the best branching variable for generating more and stronger constraints. It is a well-structured program and allows integration of new ideas easily. In 1999, Li further improved *Satz* to produce *Satz*214 [14], and in 2001, Daniel Le Berre suggested the further detection of implied literals within *Satz*214, resulting in the latest and best version of *Satz*, *Satz*215 [11].

Definition: Let $PROP$ be a binary predicate such that $PROP(x, i)$ is true iff x is a variable that occurs both positively and negatively in binary clauses and occurs in at least i binary clauses in $\mathcal{F}$, and let $\mathcal{T}$ be an integer, then $PROP_z(x)$ is defined to be the first of the three predicates $PROP(x,4)$, $PROP(x,3)$, *true* (in this order) whose denotational semantics contains more than $\mathcal{T}$ variables.

In figure 1, we present *Satz*'s branching rule which integrates the UPLA heuristic. The unary predicate $PROP_z$ is used to restrict the number of free variables executed by the heuristic and the parameter $\mathcal{T}$ is empirically fixed to 10. This UPLA heuristic plays a crucial role in the *Satz-family* and is used to reach dead-ends earlier with the aim of minimising the length of the current path in the search tree. We distinguish the UPLA heuristic from the conventional unit propagation procedure (UP) that is usually used in DPL as follows: UP is executed to reduce the size of a sub-problem possessing unit clauses *after* a branching variable selected, while UPLA is integrated in the branching rule and is executed at each search tree node.

Given a variable x_i, the UPLA heuristic examines x_i by adding the two unit clauses possessing x_i and $\bar{x}_i$ to $\mathcal{F}$ and independently making two unit propagations. These propagations result in a number of newly produced binary clauses, which are then used to weigh the variable x_i. This is calculated in figure 1, using the function $diff(\mathcal{F}_1, \mathcal{F}_2)$ which returns the number of new binary clauses in $\mathcal{F}_1$ that were not in $\mathcal{F}_2$. Let $w(x_i)$ be the number of new binary clauses produced by setting the variable to *true*, and $w(\bar{x_i})$ be the number of new binary clauses produced by setting the variable to *false*. *Satz* then uses a MOMS heuristic to branch on the variable x_i such that $w(\bar{x_i}) * w(x_i) * 1024 + w(\bar{x_i}) + w(x_i)$ is the highest. The branching variable selected follows the two-sided Jeroslow-Wang (J-W) Rule [8] designed to balance the search tree.

The UPLA heuristic also allows the earlier detection of the so-called failed literals in $\mathcal{F}$. These are literals l where $w(l)$ counts an empty clause. For such variables, *Satz* immediately tries to satisfy $\bar{l}$. When there is a contradiction during the second unit propagation, *Satz* will directly perform backtracking, else the size of the sub-problem is reduced which allows the selection of a set of best branching variable candidates at each node in search tree.

So, during two propagations of a free variable through the UPLA heuristic, three circumstances can occur:

- The free variable selected becomes a candidate for branching variable.
- Only one contradiction found during two unit propagations, meaning the size of formula $\mathcal{F}$ will be reduced during the other successful unit propagation process.
- Two contradictions are found during two unit propagations causing the search to backtrack to an earlier instantiation.

3 Using Variable Filtering to Narrow the Search Tree

The branching rules used in *Satz* are powered by the UPLA heuristic. The main objective of using UPLA in *Satz* is to detect contradictions earlier or to find a set of best branching variable candidates. In reality, *Satz*'s UPLA heuristic

```
𝓑 := ∅;
For each free variable x_i, do
Begin
   let 𝓕' and 𝓕'' be two copies of 𝓕
   𝓕' := UPLA(𝓕' ∪ {x_i}); 𝓕'' := UPLA(𝓕'' ∪ {x̄_i});
   If both 𝓕' and 𝓕'' contain an empty clause then backtrack();
   else if 𝓕' contains an empty clause then x_i := false; 𝓕 := 𝓕'';
   else if 𝓕'' contains an empty clause then x_i := true; 𝓕 := 𝓕';
   else
      𝓑 := 𝓑 ∪ {x_i};
      w(x_i) := diff(𝓕', 𝓕) and w(x̄_i) := diff(𝓕'', 𝓕);
End;

For each variable x_i ∈ 𝓑, do 𝓜(x_i) := w(x̄_i) * w(x_i) * 1024 + w(x̄_i) + w(x_i);
Branch on the free variable x_i such that 𝓜(x_i) is the highest.
```

Fig. 1. The *Satz* Branching Rule

performs a series of variable filtering processes at each node as a static variable filtering agency. We therefore term *Satz*'s UPLA heuristic a Static Variable Filtering (SVF) heuristic, because it will only perform between one to three filtering processes at each node (depending on the evaluation of $PROP_z(x)$). During the filtering process, some variables are assigned the value *true* or *false* through a forced unit propagation when a contradiction occurs during another unit propagation. Note that UPLA examines a free variable by performing two unit propagations. This process will automatically reduce the size of sub-problem and collects the (almost) best branching variable candidates at each node of the search tree. In the empirical studies presented in [12, 13] the *Satz* solver using UPLA was shown to outperform a range of other UP heuristic based DPL procedures. It was concluded that the superior performance of *Satz* was due to its greater use of variable filtering processes.

Our work is based on the insight that the size of a sub-problem during the variable filtering process can be further reduced in the *Satz-family* of DPL procedures. Here, we propose a new heuristic called the *Dynamic Variable Filtering (DVF) heuristic* that further filters the free variables and at the same time reduces the sub-problem size at each node until the filtering process is saturated. We illustrate the new branching rule powered by the DVF heuristic in figure 2.

We expect this new heuristic to perform better than the UPLA heuristics in terms of reducing the search tree size. To verify this, we carried out an empirical study and modified the branching rule of the DPL procedure *Satz*215[1] for our purpose. Four new DPL procedures based on the variable filtering heuristic are proposed. Two of them (*Satz*215-0 and *Satz*215*sT*) are based on the SVF heuristic, and the other two solvers (*ssc*34 and *ssc*355) are based on the DVF heuristic.

[1] available from http://www.laria.u-picardie.fr/~cli/EnglishPage.html

```
Do
   𝓕_init := 𝓕; 𝓑 := ∅;
   For each free variable x_i, do
   Begin
      let 𝓕' and 𝓕'' be two copies of 𝓕
      𝓕' := UPLA(𝓕' ∪ {x_i}); 𝓕'' := UPLA(𝓕'' ∪ {x̄_i});
      If both 𝓕' and 𝓕'' contain an empty clause then backtrack();
      else if 𝓕' contains an empty clause then x_i := false; 𝓕 := 𝓕'';
      else if 𝓕'' contains an empty clause then x_i := true; 𝓕 := 𝓕';
      else
         𝓑 := 𝓑 ∪ {x_i};
         w(x_i) := diff(𝓕', 𝓕) and w(x̄_i) := diff(𝓕'', 𝓕);
   End;
Until (𝓕 = 𝓕_init);

For each variable x_i ∈ 𝓑, do 𝓜(x_i) := w(x̄_i) * w(x_i) * 1024 + w(x̄_i) + w(x_i);
Branch on the free variable x_i such that 𝓜(x_i) is the highest.
```

Fig. 2. The Dynamic Variable Filtering Branching Rule

***Satz*215-0.** This is same as the *Satz*215 DPL procedure, except we examine all free variables using the SVF heuristic without any restriction to the free variables. This process is executed only once at each node.

***Satz*215*sT*.** This is same as the *Satz*215 DPL procedure, except we refuse to use the T parameter in the branching rule. This DPL procedure performs at most three filtering processes for each variable at each node.

***ssc*34.** This is same as the *Satz*215 DPL procedure, except we replace the branching rule used in *Satz*215 with the DVF heuristic based branching rule. It performs the variable filtering process until the sub-problem cannot be further reduced at each node before a branching variable selected. In fact, *ssc*34 examines the free variables many times using the UPLA heuristic at each node. One might think that this saturation process is very costly, but it is not the case.

***ssc*355.** Since *ssc*34 examines all free variables many times using the UPLA heuristic at each node, we attempt to limit the number of free variables examined by only exploring the neighborhood variables of the current assigned variable. For this purpose, we create the *ssc*355 DPL procedure by integrating a simple Neighbourhood Variable Ordering (NVO) heuristic in *ssc*34. Bessière et. al. [2] proposed a formulation of a dynamic variable ordering heuristic in the CSP domain that takes into account the properties of the neighborhood of the variable. The main objective of our simple NVO heuristic in *ssc*355 is to restrict the number of variables examined by UPLA in the DVF heuristic.

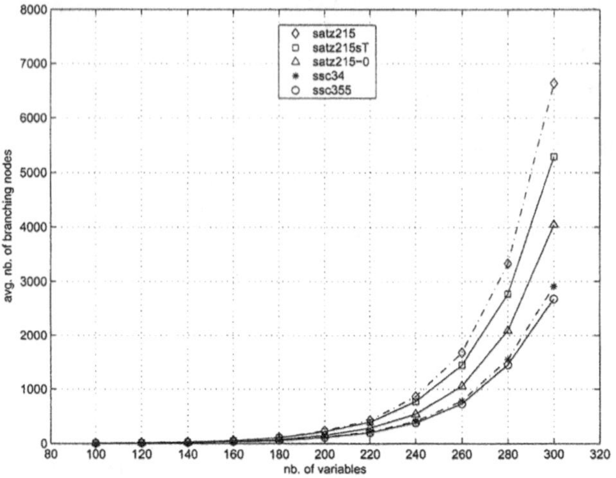

Fig. 3. Mean search tree size of each DPL procedure as a function of n for hard random 3-SAT problems at the ratio m/n=4.25

4 Comparative Experimental Results

We compare the four DPL procedures we introduced in the previous section with *Satz*215 on a large sample of hard random 3-SAT problems generated by using the method of Mitchell et. al. [16]. Given a set $\mathcal{V}$ of n Boolean variables $\{x_1, x_2, ..., x_n\}$, we randomly generate m clauses of length 3. Each clause is produced by randomly choosing 3 variables from $\mathcal{V}$ and negating each with probability 0.5. Empirically, when the ratio m/n is near 4.25 for a 3-SAT problem $\mathcal{F}$, $\mathcal{F}$ is unsatisfiable with a probability 0.5 and is the most difficult to solve. We vary n from 100 variables to 300 variables incrementing by 20, at ratio $(m/n) = 4.25$, with 1000 problems solved at each point by all the five DPL procedures.

Figure 3 shows the performance of five DPL procedures on all problem instances, where the mean search tree size is computed from the number of branching nodes reported by each procedure. It illustrates that *ssc*355's mean search tree size is the smallest, e.g., for 300 variables, *ssc*355's search tree consists of 2679 nodes, in contrast to *Satz*215's search tree of 6634 nodes.

Figure 4 also illustrates that *ssc*355's search tree size is smaller than the other DPL procedures on the *unsatisfiable* problem instances. For example, on the hard random unsatisfiable 3-SAT problems, with 300 variables, *ssc*355's mean search tree size consists of 4405 nodes, in contrast to *Satz*215's mean search tree size of 10328 nodes. Also, the mean search tree size of *ssc*355 achieves the approximate optimal search tree size proposed by Li and Gérard in [15].

Three of the DPL procedures (*Satz*215, *Satz*215-0, and *Satz*215*sT*) integrate the SVF heuristic. The comparative results presented in figures 3 and 4 show that the mean search tree sizes of *Satz*215-0 and *Satz*215*sT* are smaller than

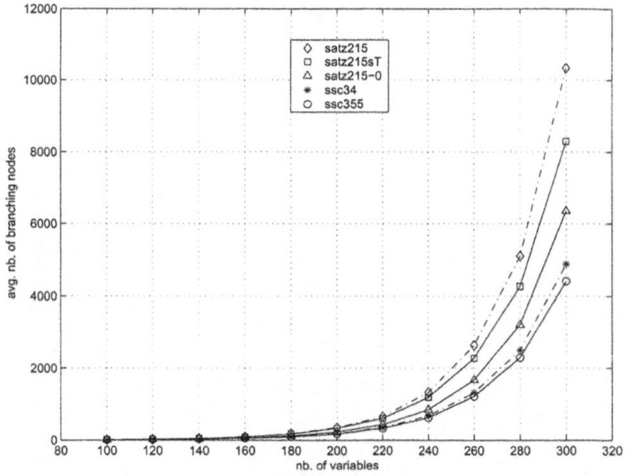

Fig. 4. Mean search tree size of each DPL procedure as a function of n for hard random unsatisfiable 3-SAT problems at the ratio m/n=4.25

$Satz215$. This means that more reasoning at each node has reduced the search tree size.

Overall, in terms of search tree size, the DVF heuristics used in $ssc34$ and $ssc355$ outperform the other SVF heuristic-based DPL procedures. This better performance is explained because the DVF sub-problem at each node is explored more fully to find a contradiction or to simply reduce the sub-problem size through a unit propagation before selecting a best branching variable.

The integration of a simple NVO heuristic on top of DVF in $ssc355$ yields a promising result compared to $ssc34$ (which only uses the DVF heuristic). In terms of search tree size, the gain of $ssc355$ over $ssc34$ increases with the size of the input problem.

Finally, figure 5 illustrates mean run-time of each DPL procedure on a Dual Xeon PC with a 2.4 GHz CPU. At present our results show that $Satz215$ still performs significantly better than the dynamic filtering heuristics in terms of run-time. However we have not explored the potential gains that could result from avoiding redundant UPLA propagations. A further improvement of the NVO heuristic also can ameliorate promisingly the run-time of $ssc355$ DPL procedure. This we leave to future work.

5 Related and Future Work

In addition to real world benchmark problems, hard random 3-SAT problems are used for testing or comparing DPL procedures. Since hard random 3-SAT problems are difficult to solve in an acceptable time when the number of variables is greater than 500, a challenge to prove that a hard random 3-SAT problem with 700 variables is unsatisfiable, was put forward by Selman et. al. in IJCAI'97 [18].

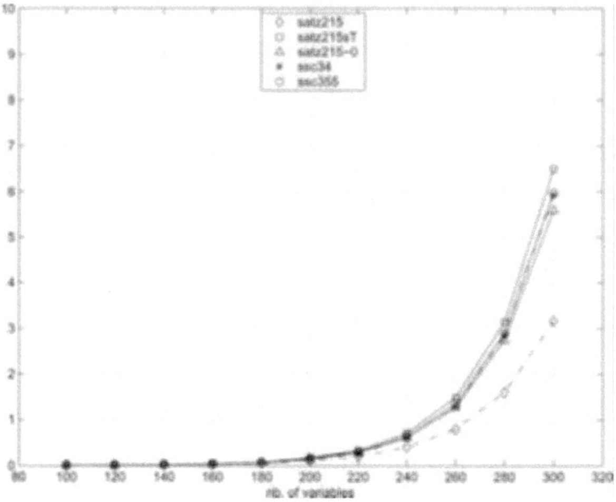

Fig. 5. Mean run-time of each DPL procedure as a function of n for hard random 3-SAT problems at the ratio m/n=4.25

To answer the challenge, Li and Gérard [15] have studied the limit of branching rules in DPL procedures for solving hard random unsatisfiable 3-SAT problems. After running an empirical study for more than five months, they suggested in their paper that the current state-of-the-art DPL procedures are already close to optimal for solving hard random unsatisfiable 3-SAT problems and that the resolution of the challenge problem with 700 variables cannot be proved by a branching rule based DPL procedure.

A final answer to the challenge was realised by Dubois and Dequen [6] with their *cnfs* solver. This solver integrates a backbone search heuristic (introduced by Monasson et. al. [17] in 1999) into a DPL-based branching rule. Dubois and Dequen's implementation allows the DPL procedure to solve hard random unsatisfiable 3-SAT problems with 700 variables (12 problems solved) in a mean run-time of 26 days using an AMD Athlon PC running at 1 GHz under a Linux operating system. When solving hard random unsatisfiable 3-SAT problems with 300 variables (456 problems solved), *cnfs*'s mean search tree size consists of 12739 nodes, in contrast to *satz*214's mean search tree size of 18480 nodes [6]. Although the *cnfs* solver can solve hard random unsatisfiable 3-SAT problems with up to 700 variables, on the other hand its means search tree size is still far from the approximate optimal mean search tree size calculated by Li and Gérard [15].

After reviewing the results presented in [6, 15], we decided the first step in our research would be to reduce the mean search size rather than looking for a more efficient solver. As the result, we have developed the Dynamic Variable Filtering (DVF) heuristic, for reinforcing the branching rule of a DPL procedure. In effect, we further explore the open space between the UPLA heuristic process and the MOMS heuristic process in *Satz*215. The results of our empirical study show that DVF heuristic can achieve an optimal number of branching nodes as presented in [15].

In the last stages of finishing this paper, we obtained the new more powerful version of *cnfs*, the *kcnfs*[2] solver. We ran *kcnfs* on the same hard random unsatisfiable 3-SAT problems with 300 variables used in our earlier study. As a result, *kcnfs*'s mean search tree size came of 7418 nodes (compared to *ssc*355's mean search tree size of 4405 nodes) although *kcnfs* performs three times better than *ssc*355 in terms of run-time.

To further investigate the effectiveness of our DVF heuristics, we also compared our techniques with the *2clseq* solver [1] and *OKsolver* [10]. *2clseq* integrates a UPLA heuristic, binary clause reasoning, using intelligent backtracking and a pruneback technique. Hence it performs more reasoning in each node of the search tree and proved very competitive with the other solvers during 2002 SAT competition for solving real-world problems. The second procedure, *OKsolver*, integrates an adaptive density-based heuristics in its branching rule. In 2002 SAT competition, *OKsolver* won both categories for the random benchmarks (only satisfiable and all problems).

We compared *ssc*355 with *2clseq* and *OKsolver* on our hard random 3-SAT problems with 300 variables. The preliminary results of this comparison show that *ssc*355 still performs significantly better than both solvers in terms of run-time and search tree size. When solving the hard random satisfiable 3-SAT problem *v*300*c*1275*g*4, the run-time of *2clseq* and *OKsolver* are 34.78 and 2.25 seconds with search tree sizes consist of 683 and 2612 nodes respectively. In contrast, *ssc*355 solves the problem in 0.02 seconds with a search tree size of 48 nodes. When running the hard random unsatisfiable 3-SAT problem *v*350*c*1488*g*255, *2clseq* had no result after 6000 seconds, and *OKsolver* solved the problem in 438 seconds with a search tree size of 275,159 nodes. Again *ssc*355 strongly outperformed these techniques, solving the problem in 190 seconds with a search tree size of 65,784 nodes.

The work presented in this paper is a first attempt at building an efficient SAT solver than can approach an optimal branching rule. In principle, our results show DVF can obtain optimal results - hence, if the efficiency issues can be addressed, DVF could prove to be a better heuristic than backbone search. In our future work, we envisage at least three further improvements of our current approach. Firstly, it is clear that savings can be made by avoiding redundant unit propagation searches for variables that remain unchanged between iterations of UPLA. The kind of benefits that we hope to obtain have already been exploited in arc-consistency algorithms. Secondly, further improvements of the NVO heuristic appear promising, as our first implementation is fairly simplistic. Finally, we are also looking at integrating a backjumping technique into DVF, exploiting the feature that backtracking only occurs during the UPLA process in DVF, compared to *Satz*215, where backtracking can also occur during the UP process.

[2] available from http://www.laria.u-picardie.fr/~dequen/sat/

6 Conclusion

In this paper, we have proposed a new heuristic, Dynamic Variables Filtering (DVF), for improving the performance of DPL-based procedures. In terms of search tree size, our preliminary results already show that DVF heuristic outperforms some of the best solvers in the area, while still remaining reasonable efficient in terms of search time.

Our evidence further suggests that the branching rules integrated in DPL procedures can obtain predicted optimal performance by using a DVF heuristic. Finally, we anticipate that the efficiency of DVF can be improved by eliminating the redundant unit propagation, improving the NVO heuristic and integrating a backjumping technique.

References

[1] Bacchus, F. *Enhancing Davis Putnam with Extended Binary Clause Reasoning.* In Proceedings of AAAI Conference, 2002, USA, pp. 613-619. 109

[2] Bessière, C., Chmeiss, A., and Sais, L. *Neighborhood-based Variable Ordering Heuristics for the Constraint Satisfaction Problem.* In Proceedings of Seventh International Conference on Principles and Practice of Constraint Programming, 2001, Paphos, Cyprus, pp. 565-569. 105

[3] Cook, S. A. *The Complexity of Theorem Proving Procedures.* In Proceedings of 3rd ACM Symp. on Theory of Computing, Ohio, 1971, pp. 151-158. 101

[4] Crawford, J. M., and Auton, L. D. Experimental Results on the Crossover Point in Random 3SAT. *Artificial Intelligence Journal*, 1996, Vol. 81, no. 1-2. 101

[5] Davis, M., Logemann, G. and Loveland, D. *A Machine Program for Theorem Proving.* Communication of ACM 5 (1962), pp. 394-397. 101

[6] Dubois, O., and Dequen, G. *A Backbone-search Heuristic for Efficient Solving of Hard 3-SAT Formulae.* In Proceedings of 17th International Joint Conference on Artificial Intelligence, 2001, Seattle, Washington, USA. 101, 108

[7] Freeman, J. W. *Improvements to Propositional Satisfiability Search Algorithms.* Ph.D. Dissertation, Department of Computer and Information Science, University of Pennsylvania, Philadelphia, PA, 1995. 101

[8] Hooker, J. N., Vinay, V. *Branching Rules for Satisfiability.* Journal of Automated Reasoning, 15:359-383, 1995. 103

[9] Jeroslow, R., Wang, J. *Solving Propositional Satisfiability Problems.* Annals of Mathematics and AI 1 (1990), pp. 167-187.

[10] Kullmann, O. *Investigating the behaviour of a SAT solver on random formulas.* Submitted to Annals of Mathematics and AI, 2002. 109

[11] Le Berre, D. *Exploiting the Real Power of Unit Propagation Lookahead.* In Proceedings of Workshop on the Theory and Applications of Satisfiability Testing, 2001, Boston University, MA, USA. 102

[12] Li, C. M., and Anbulagan. *Heuristics Based on Unit Propagation for Satisfiability Problems.* In Proceedings of 15th International Joint Conference on Artificial Intelligence, 1997, Nagoya, Aichi, Japan, pp. 366-371. 100, 101, 102, 104

[13] Li, C. M., and Anbulagan. *Look-Ahead Versus Look-Back for Satisfiability Problems.* In Proceedings of Third International Conference on Principles and Practice of Constraint Programming, 1997, Schloss Hagenberg, Austria, pp. 341-355. 102, 104

[14] Li, C. M. *A Constraint-based Approach to Narrow Search Trees for Satisfiability.* Information Processing Letters, 71, 1999, pp. 75-80. 101, 102

[15] Li, C. M., and Gérard, S. *On the Limit of Branching Rules for Hard Random Unsatisfiable 3-SAT.* In Proceedings of 14th European Conference on Artificial Intelligence, 2000, Berlin, Germany. 102, 106, 108, 109

[16] Mitchell, D., Selman, B., and Levesque, H. *Hard and Easy Distributions of SAT Problems.* In Proceedings of 10th AAAI Conference, 1992, San Jose, CA, pp. 459-465. 106

[17] Monasson, R., Zecchina, R., Kirkpatrick, S., Selman, B., and Troyansky, L. *Determining Computational Complexity for Characteristic 'Phase Transitions'.* Nature, 400, 1999, pp. 133-137. 108

[18] Selman, B., Kautz, H., and McAllester, D. *Ten Challenges in Propositional Reasoning and Search.* In Proceedings of 15th International Joint Conference on Artificial Intelligence, 1997, Nagoya, Aichi, Japan. 107

A Proposal of an Efficient Crossover Using Fitness Prediction and Its Application

Atsuko Mutoh, Tsuyoshi Nakamura, Shohei Kato, and Hidenori Itoh

Nagoya Institute of Technology
Gokiso-cho, Showa-ku, Nagoya 466-8555, Japan
{atsuko,tnaka,shohey,itoh}@ics.nitech.ac.jp

Abstract. Genetic algorithm (GA) is an effective method of solving combinatorial optimization problems. Generally speaking most of search algorithms require a large execution time in order to calculate some evaluation value. Crossover is very important in GA because discovering a good solution efficiently requires that the good characteristics of the parent individuals be recombined. The Multiple Crossover Per Couple (MCPC) is a method that permits a variable number of children for each mating pair, and MCPC generates a huge search space. Thus this method requires a huge amount of execution time to find a good solution. This paper proposes a novel approach to reduce time needed for fitness evaluation by "prenatal diagnosis" using fitness prediction. In the experiments based on actual problems, the proposed method found an optimum solution 50% faster than the conventional method did. The experimental results from standard test functions show that the proposed method using the Distributed Genetic Algorithm is applicable to other problems as well.

1 Introduction

Genetic algorithm (GA) is stochastic search algorithm based on the mechanics of natural selection and natural genetics. Although GA is an effective method of solving combinatorial optimization problems, it also requires much execution time because of the need to calculate the fitness for many search points. One way to reduce the execution time is to accelerate evolution so that a good solution can be found at an earlier generation. When GA is applied to actual problems, the most time consuming aspect is the fitness evaluation. Thus, the best way to reduce the time would be to find a more efficient method for fitness calculations, and then to accelerate evolution. While there have been some studies on acceleration methods, there have been very few studies on reducing the calculation time for fitness evaluation.

Crossover is one of the most important genetic operations in GA, and is used to perform direct information exchange between individuals in a population. Therefore, the performance of the GA depends on the crossover operator implemented. In the past we have researched more efficient methods of crossover [1].

T.D. Gedeon and L.C.C. Fung (Eds.): AI 2003, LNAI 2903, pp. 112–124, 2003.

In order to discover a good solution efficiently, it is very important to recombine good traits of the parent individuals. The Multiple Crossover Per Couple (MCPC) is a method that permits a variable number of children for each mating pair [2] and often used in combination with other acceleration methods [3]. Although MCPC generates a larger search space, this method leads to an increase in execution time in each generation. It is not easy to find the proper parameter (the number of crossovers) for each problem, yet it is an issue that must be addressed with every problem. To overcome this issue, self adaptations of parameters for MCPC was proposed, yet this raises a separate concern of how to find a fixed number of adaptations for each parameter [4]. Using the Best Combinatorial Crossover (BCX) method [5], a modified version of MCPC was proposed, but this method only works efficiently for the one-point crossover problem.

Recent developments in gene-analysis technology has made it possible to check for hereditary diseases via "prenatal diagnosis" and has been of great interest to us. We have proposed that removing unborn children with bad diagnoses will allow for a large decrease in execution time of GA [6]. The fitness of the unborn child is predicted using methods such as Artificial Neural Network (ANN). As a related work with ours, Grefenstette [7] had proposed and developed a general predictive model for the population fitness based on correlations between the fitness of the parents and the fitness of the children, while our proposed method is built based on each gene expression of the individuals. We predict the unborn child's fitness based on the gene sequences of individuals in previous generations, using a previously proposed method [8] [9]. The purpose of these methods was to reduce the burden of human Interactive Genetic Algorithm (IGA) operators, and not to reduce fitness evaluation time. This paper proposes a novel crossover method to reduce execution time for GA, in such a way that the "prenatal diagnosis" [6] is applied to the MCPC [2].

Section 2 revisits and discusses the MCPC method proposed to generate better children, and proposes a new crossover method using fitness prediction. The proposed method will reduce the execution time while exploiting the characteristic of the conventional method. Section 3 gives an experiment which compares the proposed method with the conventional method. In the experiment, we use a parameter optimization problem for Switched Reluctance Motor (SRM) control as an actual problem where the calculation cost of fitness evaluation is high. Section 4 gives an experiment using five standard test functions. Section 5 applies the proposed method to the Distributed Genetic Algorithm (DGA) based on experimental results in Section 4. Section 6 presents our conclusion.

2 Proposed Method

In this section a brief overview of Multiple Crossover Per Couple (MCPC) and Multiple Crossover Per Couple with Prediction (MCPCP) is given.

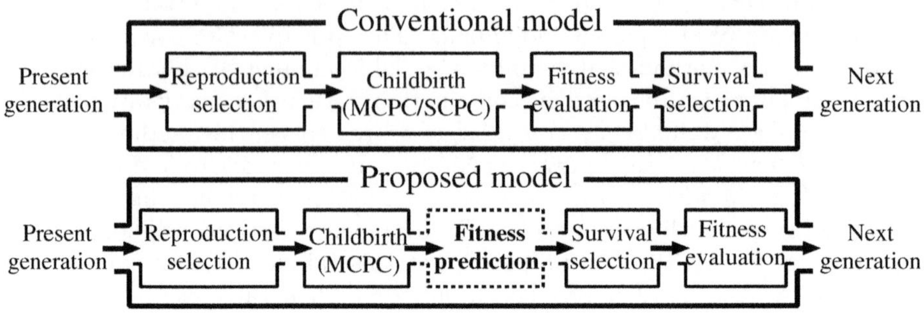

Fig. 1. Generation-Alternation model. (Upper:Conventional, Lower:Proposed)

2.1 Multiple Crossover Per Couple (MCPC)

In the conventional generation-alternation model illustrated in Fig. 1, children are born using genetic operations after the reproduction selection stage. Survival selection is then performed based on children's fitness calculation. Either of two methods of genetic operations can be used. One is the Single Crossover Per Couple (SCPC) method where children are born with only one genetic crossover, and the other is the Multiple Crossover Per Couple (MCPC) method where child are born with two or more crossovers. Although MCPC performs well on various optimization problems [2] it is not without its drawbacks. The MCPC method is able to find a good solution at an early generation for some situations; however, for others the increase in calculation time for each generation causes the total calculation time to greatly increase. In addition, its performance depends on the number of crossovers and the kind of problems.

2.2 Multiple Crossover Per Couple with Prediction (MCPCP)

To overcome the weaknesses in MCPC noted above, we propose a method using MCPC with Prediction (MCPCP). In the proposed generation-alternation model in Fig. 1, the fitness evaluation after childbirth is replaced with fitness prediction. The fitness evaluation is performed only on individuals chosen to survive after the initial fitness prediction. As a result, MCPCP reduces execution time compared with MCPC. In this paper, the model which uses SCPC and MCPC will be referred to as the conventional method, and the model which uses MCPCP will be referred to as the proposed method.

In particular, MCPCP obeys the algorithm outlined in Fig. 2. First, Parent1 and Parent2 are chosen as a mating pair randomly without replacement from a parent population. Then, C crossovers are performed. The number of crossovers C is decided in advance. Children generated by C crossovers are added to the child_pool. Then, $2C$ individuals in the child_pool have their fitness predicted, and Parent1 and Parent2 are replaced by the two individuals with the highest predicted value. Finally, Parent1 and Parent2 are evaluated for their fit-

```
procedure Multiple Crossover Per Couple with Prediction
begin
 i = 0;
 Initialize child_pool;
 Choose Parent1 and Parent2 randomly without replacement;
 while i < C do
 begin
    Parent1 and Parent2 are crossed;
    Children are added to child_pool;
    i = i + 1;
 end
 Fitness prediction for all individuals in child_pool;
 Replace Parent1 and Parent2 with best two individuals;
 Fitness evaluation for Parent1 and Parent2;
 Parent1 and Parent2 become part of the next generation;
end
```

Fig. 2. Outline of proposed algorithm

ness, and they become part of the next generation. As a result, MCPCP has a reduced calculation time compared with MCPC under problems in which fitness evaluation time is large, since the number of fitness evaluation per generation is small (only two).

2.3 Method of Fitness Prediction

The prediction scheme should be much simpler than the actual evaluation. On the other hand, the performance of searching for the solution depends on its accuracy, which leads to trade-offs. Prediction method should be carefully chosen for each application.

Two previously proposed prediction methods [9] involve using an Artificial Neural Network (ANN) and using the distance measure between individuals in past and current generations. The latter method obtained better results in a shorter calculation time than the former. Since reduction of calculation cost is important for our proposed model, we used the latter method. This method calculates weights W_i by using the similarity function $S(m_{new}, m_i)$ between one individual offspring m_{new} and n individuals $m_i (i = 1, \ldots, n)$ from previous generations in the GA search space (Eq. (2)). It is necessary to set the similarity function $S(m_{new}, m_i)$ for each problem; in this case we use the reciprocal of the Euclidean distances $D(m_{new}, m_i)$ (Eq. (1)) , which had gained good experimental results in [9]. The average of fitness values P_i of the individuals in the past generation(s), weighted by the reciprocals of the distance, is defined as the predicted fitness value P_{new} of the present individual (Eq. (3)).

$$S(m_{new}, m_i) = \frac{1}{D(m_{new}, m_i)}. \tag{1}$$

$$W_i = \frac{S(m_{new}, m_i)}{\sum_{j=1}^{n} S(m_{new}, m_j)}. \tag{2}$$

$$P_{new} = \sum_{i=1}^{n} W_i P_i. \tag{3}$$

In the past [9], we have found that it is more efficient to use individuals $m_i (i = 1, \ldots, n)$ from only one generation back to predict the new fitness, since the calculation cost increase with the number of candidates n. Thus for our proposed model, we choose to look one generation back to predict offspring fitness.

2.4 Target Problems for the Proposed Method

Since calculating the time for fitness evaluation is the most time intensive part, in this paper we approximate the fitness evaluation time as the total calculation time for the GA. Let T_e and T_p be, respectively, the average time for fitness evaluation and fitness prediction per individual. Allowing C crossovers with m individuals per generation leads to the following calculation times.

$$Conventional: \quad T_{mcpc} \simeq m \cdot C \cdot T_e. \tag{4}$$

$$Proposed: \quad T_{mcpcp} \simeq m \cdot (T_e + C \cdot T_p). \tag{5}$$

Assuming no error in fitness prediction we would like to fulfill the requirement that $T_{mcpc} > T_{mcpcp}$, or Eq. (6).

$$T_e > \frac{C}{C-1} T_p. \tag{6}$$

However, since prediction error is not necessarily zero, we would expect good results to occur in problems with $T_p \simeq 0$, or Eq. (7).

$$T_e >> T_p. \tag{7}$$

3 A Real-World Experiment

GA is of interest in the field of mechatronics and power electronics, because it reduces the laborious trial and error used to determine various design parameters [10]. Switched Reluctance Motor (SRM) is determined by the geometry of structural design. It has been studied by exploring the parameter for SRM optimal control such that the output performance is maximized [11]. Previous work achieved a large reduction of execution time using the GA compared with other search methods. However, execution time for evaluating the fitness of each individual is huge. In this section, we apply and evaluate the proposed method of the GA as it relates to this problem.

Table 1. Range and spacing of parameters

	Range	Spacing
PWM pattern $T^*[\mu sec]$	0.00~42.5	0.18
turn-on angle $\theta_0[deg]$	22.50~28.10	0.09
commutation angle $\theta_c[deg]$	37.50~43.10	0.09

Table 2. Parameters and operators

Parameter	Value
Chromosome Length	24
Population Size	50
Selection method	Tournament
Crossover Rate	1.0
Crossover Method	Uniform
Mutation Rate	0.03
Terminal Time (sec)	10,000

3.1 A Parameter Optimization Problem for SRM Control

We will first briefly describe the problem of discovering the best parameter for SRM control under operating condition. Specifically, an individual has a 24-bit chromosome consisting of 3 parameters -Pulse Width Modulation (PWM) pattern T^*, turn-on angle θ_0, and commutation angle θ_c- which are the voltage control parameters for SRM. We explored parameters for which the square ratio of *torque/ampere* is maximized and speed-ripple factor is minimized. This problem has an epistasis and many local minima, although it has a unimodal large-scale trend. The range and spacing of each control parameters is shown in Table 1.

3.2 Fitness Evaluation

We will now describe a method of fitness evaluation for this problem. First, we calculate instantaneous current waveform $i(\theta)$ and instantaneous torque waveform $\tau(\theta)$ using the parameters T^*, θ_0, and θ_c. Eq. (8) gives the fitness based on $i(\theta)$ and $\tau(\theta)$. We define a good individual as one with a high fitness score.

$$F(T^*, i(\theta), \tau(\theta)) = \begin{cases} -T^* & ;\text{if } i(\theta) > 30A \\ -\left|\frac{\tau^* - \tau_{ave}}{\tau^*}\right| & ;\text{else if} \left|\frac{\tau^* - \tau_{ave}}{\tau^*}\right| > 0.02 \\ \frac{\tau_{ave}/I^2}{(\tau_{ave}/I^2)_{max}} + k_{rip}\frac{(\omega_{rip})_{min}}{\omega_{rip}} & ;\text{else,} \end{cases} \tag{8}$$

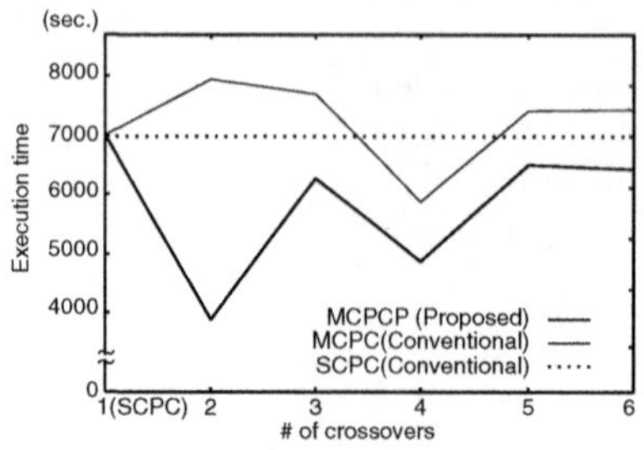

Fig. 3. The average execution time needed to discover an optimal solution for each number of crossovers

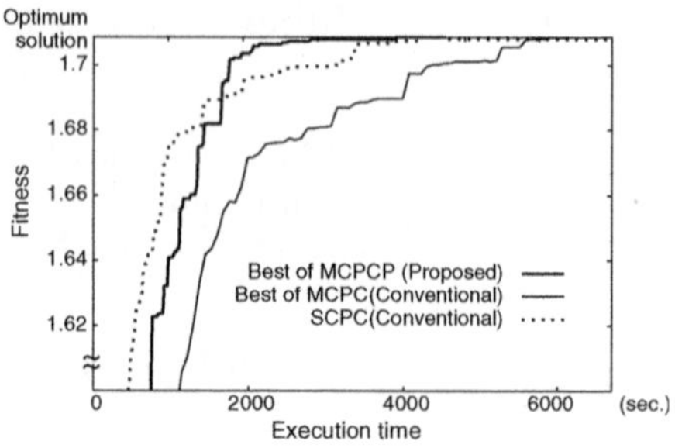

Fig. 4. Fitness transition at each execution time

where

$$
\begin{aligned}
&\tau^* = 14.7[N \cdot m] \times 0.8\\
&k_{rip} = 1.0\\
&\omega_{rip} = \frac{\omega_{max} - \omega_{min}}{\omega^*} \times 100\\
&\omega^* = 100 \times \frac{2\pi}{60},\\
&\omega = \frac{1}{J} \int (\tau_{ave} - \tau(\theta)) d\theta\\
&J = 1.258 \times 10^{-2} [kg \cdot m^2]\\
&I = \sqrt{\int i(\theta)^2 d\theta}.
\end{aligned}
$$

To be sure this is a suitable problem to apply the proposed method to, we found that the average calculation time for fitness evaluation T_e was 0.35 sec. The average calculation time for fitness prediction T_p was 0.1×10^{-4} sec. This satisfies Eq. (7). Therefore, we expect positive results from the proposed method.

3.3 Results and Discussion

The experimental conditions are detailed in Table 2. The experiment of 20 runs was performed in such a way that run 1 performed only one crossover per couple (SCPC) , $i(2 \leq i \leq 6)$ executed i crossovers per couple (MCPC), and $i(2 \leq i \leq 6)$ executed i crossovers per couple with prediction (MCPCP). The number of crossovers in the experiment is based on [2].

The average execution time needed to discover an optimal solution for each number of crossovers for is shown in Fig. 3. Esquivel et al. note a weakness of MCPC, that it will give a wrong response depending on the number of crossovers [2]. MCPC reaches an optimal solution faster than SCPC only when it is applied with four crossover. In other words, the number of crossovers for which MCPC works is limited. On the other hand, MCPCP reaches an optimal solution faster than SCPC and MCPC for any number of crossovers, although for it also the calculation time depends on the number of crossovers.

Fig. 4 shows the transition of the fitness value at each time on this problem for SCPC, the best value of MCPC and MCPCP. In this experiment, we observe that SCPC performs well at the beginning of the search but not in the later stage. On the other hand, although MCPC and MCPCP do not perform well when the number of choices in the beginning of the search is increased, that is, when the variety in the population is large, we observe that MCPC and MCPCP do not fall into local minima at the later stage and provide the optimal solution faster than SCPC. Additionally, the MCPCP method we proposed provides the optimal solution faster than MCPC, thus reducing the calculation time of fitness evaluation with fitness prediction.

4 An Experiment with Standard Test Functions

In this section, the performance of the proposed method is examined by optimizing five standard test functions[1]. The proposed method attempts to reduce execution time by performing fitness prediction instead of fitness evaluation. Thus, problems in which a calculation time for fitness evaluation is far larger than that for fitness prediction, as shown in Eq. (7), are the targets for the proposed method. This problem is not one of those target problems, since the calculation time for fitness evaluation is very small. However, in this paper we examine the performance of the proposed method by focusing on the number of fitness evaluations, since the time required to find the solution approaches the calculation time for fitness evaluation in target problems which meet Eq. (7).

4.1 Test Functions

The optimization problems used here are the minimization of Schwefel's function 1.2 (f1), Rosenbrock's function (f2), Griewank's function (f3), Schwefel's function (f4), and Rastrigin's function (f5) with 10 design variables, as shown in Eq. (9)∼ (13). The optimum solution is zero, which is the minimization of the function. Characteristics of each function are shown in Table 3. Schwefel's function 1.2 and Rosenbrock's function are unimodal functions, but they have a strong epistasis among their variables. Griewank's function has very small but numerous minima around the global minimum, although it has a unimodal shape on a large scale. Schwefel's function has many local minima, and it has a global minimum at one of the four corners in a two dimensional case. Rastrigin function also has many local minima, but it has no epistasis among its variables. For these functions, one design variable is represented by 10 bits, 10 design variables means the chromosome is 100 bits long.

$$f_1(x_1, \cdots, x_n) = \sum_{i=1}^{n}(\sum_{j=1}^{i} x_j)^2. \qquad (-64 < x_i \leq 64) \quad (9)$$

[1] We used a binary-coded GA in this paper, although it is possible that real-coded GA is more useful than binary-coded GA for this problem.

Table 3. Characteristics of test functions

function name	modality	epistasis
Schwefel 1.2 (f1)	unimodal	high
Rosenbrock (f2)	unimodal	high
Griewank (f3)	multimodal	medium
Schwefel (f4)	multimodal	nothing
Rastrigin (f5)	multimodal	nothing

Table 4. Parameters and operators

Parameter	Value
Chromosome Length	100
Population Size	100
Selection method	Roulette
Crossover Rate	1.0
Crossover Method	1-Point
Mutation Rate	0.01
Terminal # of Evaluations	5,000,000

$$f_2(x_1, \cdots, x_n) = \sum_{i=1}^{n-1} [100(x_{i+1} - x_i^2)^2 + (1 - x_i)^2]. \qquad (-2.048 < x_i \le 2.048) \quad (10)$$

$$f_3(x_1, \cdots, x_n) = 1 + \sum_{i=1}^{n} \frac{x_i^2}{4000} - \prod_{i=1}^{n} (cos(\frac{x_i}{\sqrt{i}})). \qquad (-512 < x_i \le 512) \quad (11)$$

$$f_4(x_1, \cdots, x_n) = \sum_{i=1}^{n} -x_i sin(\sqrt{|x_i|}). \qquad (-512 < x_i \le 512) \quad (12)$$

$$f_5(x_1, \cdots, x_n) = 10n + \sum_{i=1}^{n} (x_i^2 - 10cos(2\pi x_i)). \qquad (-5.12 < x_i \le 5.12) \quad (13)$$

4.2 Results and Discussion

The experimental condition is shown in Table 4. The experiment of 20 runs is performed on each function, in such a way that run 1 executed only one crossover per couple (SCPC), $i(2 \le i \le 6)$ executed i crossovers per couple (MCPC), and $i(2 \le i \le 6)$ executed i crossovers per couple with prediction (MCPCP). The number of crossovers in the experiment is based on [2]. The results of the experiment, the average number of evaluations where the optimum is discovered for each number of crossovers, are shown in Table 5. When the optimum is not discovered, the fitness values at the terminal number of evaluations (=5,000,000) for each number of crossovers. In this table, the bold type indicates that the proposed method performs well compared with the conventional method.

In this experiment, we observe that the MCPCP method we proposed provided or approaches the optimal solution faster than the conventional method according to tests f1, f2, and f5. On the other hand, MCPCP under test f3 does not perform well on average. MCPCP under test f4 does not provide the optimal solution for some numbers of crossovers until the terminal generation. It is difficult to predict exactly in multimodal functions having many local minima, because the fitness prediction uses all individuals in the previous generation. It is supposed that the reason why the proposed method does not perform well in

Table 5. The average number of evaluations (#) where the optimum is discovered ($\times 10^3$) or the average fitness at the terminal

	Best value of SCPC&MCPC (Conventional)	MCPCP (Proposed) for each number of crossovers				
		2	3	4	5	6
f1	#164.60	**#138.00**	**#143.60**	**#108.00**	**#131.90**	**#132.40**
f2	0.61	**0.50**	**0.53**	**0.56**	**0.58**	**0.54**
f3	0.05	0.07	0.05	0.06	0.05	0.06
f4	#4,716.70	**#2,695.80**	**#4,247.70**	5.92	5.92	**#4,475.80**
f5	#125.30	**#96.10**	**#91.50**	**#86.50**	**#91.30**	**#91.30**

multimodal functions appears to be that it settles on a local minima because of prediction error. The proposed method performs well on parts of f4 and f5 in spite of their multimodal shapes because the shape is a big-valley structure[2].

From these results, it seems that the proposed method performs well because it had fewer errors in fitness prediction of unimodal functions. However, we need to apply the proposed method to other problems having varied search spaces. Therefore, we apply the proposed method to the Distributed Genetic Algorithm (DGA) [12], since it is effective in solving multimodal functions. The next section gives the method and a re-experiment using the DGA.

5 An Application to the Distributed Genetic Algorithm

In this section we show a re-experiment using the DGA for two functions f3 and f4 which had bad results in previous section. Fitness prediction is performed per subpopulation when we apply the proposed method to the DGA.

5.1 A Distributed Genetic Algorithm (DGA)

Fig. 5 shows the overview of the DGA [12]. In the DGA, a large population is divided into smaller subpopulations, and a traditional GA is executed on each subpopulation separately. Some individuals are selected from each subpopulation and migrated to different subpopulations periodically. The variety of the individuals in the whole population is increased by converging on local minima at each subpopulation. This large variety improves the effectiveness of the search in DGA compared with conventional GA.

Additional coding is not needed to apply the proposed method to the DGA. However, past individuals $m_i (i = 1, \ldots, n)$ used for prediction should be in the same subpopulation as the individual m_{new} that we want to predict. As a result, we can make full use of the characteristics of the DGA executing on each subpopulation separately, and we expect that individuals will converge on local solution early at each subpopulation.

[2] A function with big-valley structure has many local minima, but a unimodal shape on a large scale.

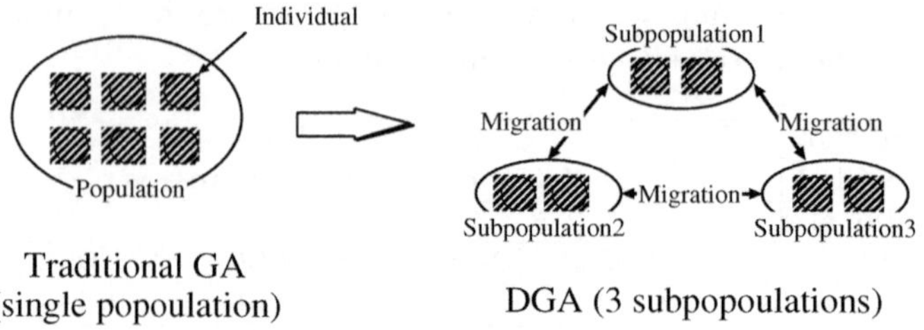

Fig. 5. Overview of the DGA

Table 6. The average number of evaluations (#) where the optimum is discovered ($\times 10^3$) or the average fitness at the terminal

	Best value of SCPC&MCPC (Conventional)	MCPCP (Proposed) for each number of crossovers				
		2	3	4	5	6
f3	0.0049	**0.0010**	**0.0016**	**0.0016**	**0.0012**	**0.0016**
f4	#52.8	**#40.0**	**#42.4**	**#48.0**	**#44.8**	**#46.8**

5.2 Results and Discussion

We used much the same experiment as the previous section. In the experiment, we examined the prediction about the two functions f3 and f4 which had bad results in previous section. The parameters for the DGA based on [13] are follows: the population size is 400; the number of subpopulations is 40; the migration interval and the migration rate of f3 are 7 and 0.1 respectively; the migration interval and the migration rate of f4 are 5 and 0.5 respectively. Table 6 shows the results of the experiments of f3 and f4. The result of f3 is the average fitness at the terminal number of evaluations (=5,000,000) for each number of crossovers. The result of f4 is the average number of evaluations where the optimum is discovered at each number of crossovers.

In this additional experiment, we observed that the proposed method obtains the optimal solution with fewer numbers of evaluation and the average fitness at the terminal is higher than the conventional method at all number of crossovers. Thus, the proposed method works efficiently with multimodal shaped problems when applied to the DGA.

6 Conclusion

We have proposed a method to reduce execution time using the fitness prediction as a modified version of MCPC that increases search points. The proposed method is intended for problems in which fitness evaluation time is large. In this paper, the proposed method has been applied and examined with a parameter optimization problem for SRM control. In this experiment, the proposed method could obtain an optimum solution faster than SCPC and MCPC, the conventional method. We also examined this method with the standard test function to check applicability to other common problems. We observed that the proposed method performed well in experiments with these test functions except for multimodal functions having many local minima. In this case, prediction error increases because we need to use all individuals in the previous generations for prediction, and we guess that the incorrect prediction causes the solution to fall into the local minima. On the other hand, we had good results using fitness prediction for each subpopulation by using DGA about two functions having bad experimental results. Thus, the proposed method with the DGA can be applicable to various problems.

When GA is applied to various engineering applications, the search space is unknown, and many applications need large amounts of high-fitness individuals because of spread of the search space. However, the increase of the prediction error may cause the probability of selecting precisely best two individuals per crossover to decrease. This means that the applicable number of crossovers depends on the prediction error. Thus, it is important to increase the prediction accuracy to improve the search performance by increasing the number of crossovers. It has been reported the effect of the prediction accuracy depends on the generation and the number of past generations needed to predict [14]. In the future, we would like to improve the effectiveness of the proposed method. Therefore, it is important to improve the prediction accuracy by modifying the proposed method. Thus, it has need to consider other research or examine other method of fitness prediction.

References

[1] A. Mutoh, S. Oono, K. Moriwaki, T. Nakamura, N. Inuzuka and H. Itoh, 1999, *An Evolutionary Method Using Crossover in a Food Chain Simulation*, Proc. 5th European Conference on Artificial Life, LNAI 1674, pp. 89-93, Springer. 112

[2] S. Esquivel, A, Leiva, and R, Gallard, 1997, *Multiple Crossover Per Couple in Genetic Algorithms*, IEEE International Conference on Evolutionary Computation, pp. 103-106. 113, 114, 118, 120

[3] I. Ono, Y. Nagata, and S. Kobayashi, 1998, *A Genetic Algorithm Taking Account of Characteristics Preservation for Job Shop Scheduling Problems*, Intelligent Autonomous Systems, pp. 711-178, IOS Press. 113

[4] Esquivel S., Leiva A. and Gallard R., 1998, *Self Adaptation of Parameters for MCPC in Genetic Algorithms*, Proc. 4th Congreso Argentino de Ciencias de la Computacion del Comahue, pp. 419-425. 113

[5] M. Miki, T. Hiroyasu, J. Yoshida, and I. Ohmukai, 2002, *New Crossover Scheme for Parallel Distributed Genetic Algorithms*, Proc. the IASTED International Conference Parallel and Distributed Computing and Systems, pp.145-150. 113

[6] A. Mutoh, T. Nakamura, and H. Itoh, 2001, *An Evolutionary Method of an Extended Binary Decision Diagram using Fitness Prediction*, Proc. The 2001 International Conference on Parallel and Distributed Processing Techniques and Applications, Vol.3, pp.1265-1270. 113

[7] J. J. Grefenstette, 1995, *Predictive Models Using Fitness Distributions of Genetic Operators*, Foundations of Genetic Algorithms 3, pp. 139-161, Morgan Kaufmann Publishers. 113

[8] B. Johanson, and R. Poli, 1998, *GP-Music: An Interactive Genetic Programming System for Music Generation with Automated Fitness Raters*, Genetic Programming 1998: Proc. 3rd Annual Conference, pp.181-186, Morgan Kaufmann Publishers. 113

[9] M. Ohsaki, and H. Takagi, 1998, *Improvement of Presenting Interface by Predicting the Evaluation Order to Reduce the Burden of Human Interactive EC Operators*, Proc. IEEE International Conference on System, Man, and Cybernetics, pp.1284-1289. 113, 115, 116

[10] K. S. Tang, K. F. Man, and D. W. Gu, 1996, *Structured Genetic Algorithm for Robust H_∞ Control Systems Design*, IEEE Trans. on Industrial Electronics, Vol.43, No.5, pp.575-582. 116

[11] S. Fubuki, T. Kosaka, and N. Matsui, 2001, *Non-linear Drive Characteristics Analysis and GA-based Search for Optimum Control Parameters of SRM Drives*, The Papers of Technical Meeting on Rotating Machinery, IEE Japan, RM01-51, pp.13-18. (in Japanese) 116

[12] R. Tanase, 1989, *Distributed genetic algorithms*, Proc. 3rd International Conference on Genetic Algorithms, pp. 434-439. 121

[13] T. Hiroyasu, M. Miki, and J. Kamiura, 2002, *A presumption of parameter settings for Distributed Genetic Algorithms by using Design Of Experiments*, IPSJ Transactions on Mathematical Modeling and Its Applications, Vol.43, No.SIG10, pp.199-217. (in Japanese) 122

[14] H. Ochiai, S. Saegusa, M. Ohsaki, K. Hayashibe, 2002, *Evaluation and Formulation of the Performance of a Method to Predict Fitness Values*, The papers of 12nd Tokai fuzzy workshop, pp.14.1-14.4. (in Japanese) 123

A New Hybrid Genetic Algorithm for the Robust Graph Coloring Problem

Ying Kong[1], Fan Wang[2], Andrew Lim[2], and Songshan Guo[1]

[1] Dept. of Computer Science, Sun Yat-sen University
Guangzhou, 510275, China

[2] Dept. of Industrial Engineering and Engineering Management
Hong Kong University of Science and Technology, Clear Water Bay
Hong Kong, China

Abstract. The RGCP (Robust Graph Coloring problem) is a new variant of the traditional graph coloring problem. It has numerous practical applications in real world like timetabling and crew scheduling. The traditional graph coloring problem focuses on minimizing the number of colors used in the graph. RGCP focuses on the robustness of the coloring so that the coloring is able to handle uncertainty that often occurs in the real world. By that, we mean that given a fixed number of colors we would like to color the graph so that adjacent vertices are assigned different colors with the consideration of the possible appearance of the missing edges. In this paper, we present a new hybrid genetic algorithm (GA), which embeds two kinds of local search algorithms - enumerative search and random search within the GA framework. In addition, we integrate a partition based encoding scheme with a specialized crossover operator into our GA method. We also propose an adaptive scheme that alternates between two local search methods to increase performance. Experimental results show that our new algorithm outperforms the best published results in terms of the quality of solutions and the computing time needed.

1 Introduction

The graph coloring problem is a well-known NP-complete problem, which has numerous applications in real world [7]. It has many applications in timetabling and scheduling. The goal of the coloring problem is to use the minimal number of colors to color the vertices of a given graph, with the constraint that all pairs of adjacent vertices must receive the different colors. The graph coloring problem is an important problem. DIMACS, in its second challenge, has put the graph coloring problem as one of its challenge problems with many benchmark data [6]. Unfortunately, the graph coloring problem is NP-complete [4], as a result, many heuristics algorithms have been proposed [1, 6, 5] to solve it.

The RGCP (robust graph coloring problem) is a very useful extension of the classical graph coloring problem. The problem was first introduced by [8]. RGCP focuses on building robust coloring for a given graph by a fixed number

T.D. Gedeon and L.C.C. Fung (Eds.): AI 2003, LNAI 2903, pp. 125–136, 2003.

Table 1. An instant of one-day timetable of flight route

ID	Route	Departure Time	Return Time
1	HKG-PEK-HKG	09:00	16:00
2	HKG-XMN-HKG	09:00	14:00
3	HKG-PEK-HKG	15:00	22:00
4	HKG-KMG-HKG	16:00	23:00
5	HKG-PVG-HKG	08:00	15:00

of colors, taking into consideration the possibility of appearing and penalizing those coloring where both vertices of an missing edge having the same color. The penalty function depends on the application domain. In [8], a GA based method was presented to solve RGCP. The experiments in that paper showed that the algorithm had acceptable results and running time, as compared to approached based on binary (0/1) programming.

In this paper, we present a new hybrid GA based approach to solve RGCP. A new encoding method has also been devised which has the ability to reach all solution configurations. This is an improvement over the previous method will may not reach certain solution. A specialized crossover operator is created and integrated into the hybrid GA search framework with two kinds of local search operators. The experimental results indicate that our new approach to be superior in quality and also in computing time.

In Section 2, we shall present the problem statement with an analysis of the time complexity of RGCP. In Section 3, the binary programming model of RGCP is presented together with the previous GA approach. We present our new encoding method for search space and our crossover operator in Section 4. The complete hybrid GA is introduced in Section 5. Computation results are given in Section 6. Finally, we present the conclusion in Section 7.

2 Problem Statement and Time Complexity

2.1 Crew Assignment Problem – An Application

Before we formalize the statement of RGCP, we would like to introduce - robust crew assignment for a regional airline. Suppose you are in charge of four crew teams for five round-trip domestic flight routes from Hong Kong to other mainland China cites, illustrated in Table 1. In Table 1, columns 1 to 4 are identity number (ID), the flight route, the time of departure from and to the home base airport (Hong Kong International Airport - HKG) within one day respectively.

The constraints for crew assignment are (i) each flight route should has one and only one crew team; (ii) each crew team can serve at most one flight route at any time; (iii) each crew team should has at least 60 minutes or more ground time in HKG to transfer from one flight route to another, a crew member may be asked to take one more flight route due to insufficient personnel. Considering the

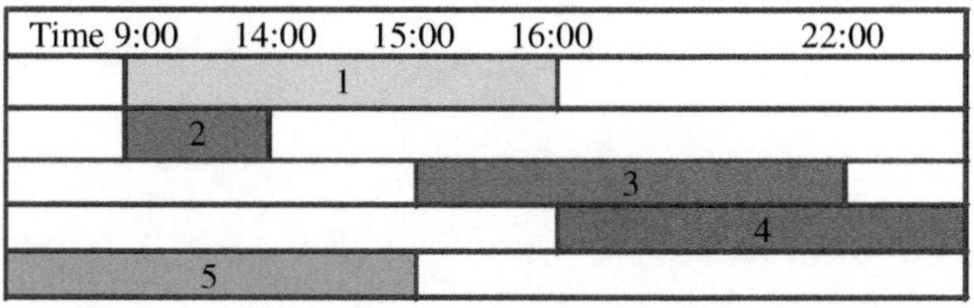

Fig. 1. overlap based on the timetable of Table 1

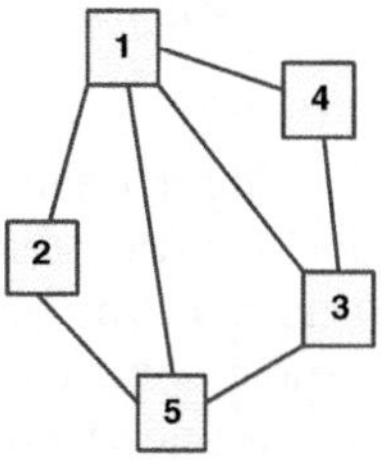

Fig. 2. Graph model responding the timetable in Table 1

above constraints for crew assignment, we know the most important relationship is the overlap of flying duration. For example, the overlap relationship based on timetable of Table 1 is shown in Figure 1. We can use a graph to model the overlap relationships where each vertex represents one flight route. An edge exists between two vertices if a crew member can work on the two flight routes represented by the two vertices. This means that the end time of the earlier route must be earlier than the start time of the later route by 60 minutes. For instance, based on the overlap relationship in Figure 1, the corresponding graph model is showed as Figure 2.

We can model the crew assignment problem as a graph coloring problem where each vertex represents a flight route and each color represents one crew team. For the graph in Figure 2, it is easy to see that the minimum number of colors needed is 3. This means that only 3 teams of crew can handle all 5 flight routes without violating any constraint. One possible optimal coloring for the graph in Figure 2 is (Solution 1): $[C(1) = 1, C(2) = C(3) = 2, C(4) = C(5) = 3]$

This means that we assign crew team 1 to route HKG-PEK-HKG; schedule crew team 2 to take route HKG-XMN-HKG and route HKG-PEK-HKG (the team has exactly 60 minutes for ground transfer); schedule crew team 3 to handle route HKG-PVG-HKG and HKG-KMG-HKG.

Next, let us consider flight delays which occur very often. For example, due to the bad weather of Xiamen (XMN), the flight route No.2 has been delayed to return to HKG at 16:00 instead of the original schedule at 14:00. This change results in the change of the overlap relationship between flight routes No.2, No.3, and No.4. The modified graph model is shown in Figure 3. In Figure 3, three colors is not enough. However, if the previous coloring solution for the graph of

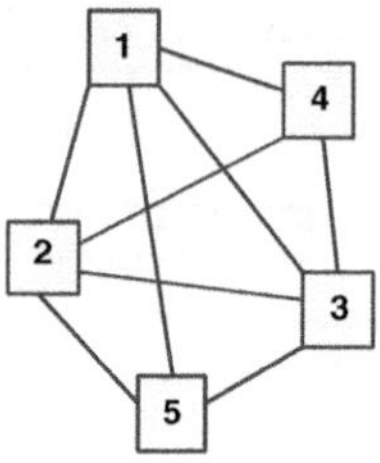

Fig. 3. Graph model if flight route No.2 delays to return at 16:00

Figure 2 is (Solution 2): $[C(1) = 1, C(2) = 2, C(3) = 3, C(4) = C(5) = 4]$, it can obviously handle this uncertain flight delay without any crew assignment rescheduling. In fact, in many situations, we prefer to get a robust result rather than the other with the minimum number of colorings. The aim of robust crew assignment problem is to seek the most robust assignment. Here, the definition of 'Robustness' takes into account the uncertainty of the flight delays which is a norm in day to day operations.

One simple way to model the uncertainty is to estimate the probabilities of the presence of the missing edges. In the robust crew assignment problem, we have the following equation, considering the overlap rate between two flights route No.i and No.j,

$$h_{ij} \equiv \frac{T_{tr}}{\max\{T_i^{dep}, T_j^{dep}\} - \min\{T_i^{ret}, T_j^{ret}\} + \alpha} \tag{1}$$

where T_i^{dep} and T_i^{ret} represent the departure and return time for flight route No.i, T_{tr} means the minimal ground transfer time, and α is a constant. From the above equation, we can obtain the following uncertainty matrix H which presents all uncertain changes on the fixed timetable of Table 1 (c is set to 10)

$$H = \begin{bmatrix} * & - & - & - & - \\ & * & 0.86 & 0.46 & - \\ & & * & - & - \\ & & & * & 0.86 \\ & & & & * \end{bmatrix} \tag{2}$$

In addition, we formalize the objective function for a given graph topology $G = (V, E)$ with given uncertain information matrix H,

$$\min R(G) \equiv -log \sum_{(i,j)\in\overline{E}, C(i)=C(j)} h_{ij} \tag{3}$$

For the solution 1, the value of robustness is $R = -log(p(2,3) + p(4,5)) = -log(0.86{+}0.86) = 0.236$. On the other hand, for the solution 2, $R = -log(p(4,5)) = -log(0.86) = 0.066$. From the above calculation results, we say that Solution 2 is more robust than Solution 1.

2.2 The General Problem

The Robust Graph Coloring Problem (RGCP) can be defined as follows:

Given the graph $G = (V, E)$ with $|V| = n$, let $c > 0$ be an integer number and the penalty set $P = \{p_{ij}, (i,j) \in \overline{E}\}$, The objective function of RGCP is to find

$$\min R(G) \equiv -log \sum_{(i,j)\in\overline{E},C(i)=C(j)} p_{ij} \tag{4}$$

where C is a coloring mapping i.e. $C : V \rightarrow 1, 2, \cdots, c$ satisfying $C(i) \neq C(j), \forall(i,j) \in E$. The RGCP can be characterized by (n, c, P). Depending on various application domains, the penalty set may have different definitions.

2.3 Time Complexity

As a special instance of RGCP when all penalties are zero, i.e. $p_{ij} = 0, \forall(i,j) \in \overline{E}$ and $R(C) = 0$, the RGCP problem becomes the classical decision minimal coloring problem which has been proven to be a NP-complete problem [Garey and Johnson, 1979]. Hence, RGCP is a NP-complete problem.

3 Binary Programming and GA

We introduce the binary programming model to solve the RGCP exactly. Let the decision variable $x_{jk} = 1$ if $C(j) = k$, otherwise $x_{jk} = 0$, where $j(1 \leq j \leq n)$ is the index of vertex, $k(1 \leq k \leq c)$ is the index of color and C represents the coloring mapping. We define the auxiliary variables

$$y_{ij} = \begin{cases} 1 \ \exists k, x_{ik} = x_{jk} = 1 \\ 0 \ otherwise \end{cases} \quad \forall(i,j) \in \overline{E} \tag{5}$$

Then the RGCP can be stated as

$$\min R(G) \sum_{(i,j)\in\overline{E}} p_{ij}y_{ij} \tag{6}$$

subject to the following constraints

$$\begin{array}{l} \sum_{k=1}^{c} x_{ik} = 1, \forall i \in \{1, 2, \cdots, n\} \\ x_{ik} + x_{jk} \leq 1, \forall(i,j) \in E, \forall k \in \{1, 2, \cdots, c\} \\ x_{ik} + x_{jk} - 1 \leq y_{ij}, \forall(i,j) \in \overline{E}, \forall k \in \{1, 2, \cdots, c\} \end{array} \tag{7}$$

As the RGCP is NP-complete, the binary programming method can only solve very small instances optimally in acceptable computing time.

We briefly describe the GA method presented in [8] to to solve the RGCP.

There are several key components in any classical GA algorithm. These include search space encoding, fitness function, operators for population (selection, crossover and mutation) and the parameters setting such as the maximum number of iterations, the population size and so on. In that paper, the search space encoding is based on the sequence of vertices. Once the sequence is defined, the solution is constructed greedily as follows: for any vertex, the least feasible color was computed taking into account the colors of adjacent vertices that were colored previously. This encoding method cannot guarantee the feasibility of coloring from every sequence, i.e. eventually more than c colors may be needed and a large penalty P_1 is added to the fitness function to penalize any invalid coloring during GA search process.

The selection operator is the basic Monte-Carlo method that gives each individual in the population a probability of selection. The probability is the rate of the value of its fitness function over the sum of all individuals. The crossover is the standard single point crossover method which randomly selects a crossover point and exchange the two sequences at the crossover point. The numbering may be readjusted to be unique based on the relative order of the "crossovered" results.

There are several drawbacks for the above method. Firstly, the encoding method is not sufficient enough to cover the whole space of coloring. In other words, the optimal solution may not be reachable. For example, the sequence (1, 2, 3, 4, 5) can be used to construct Solution 1 for the graph in Figure 2. However, no sequence can construct the solution 2 successfully for the same graph by the above greedy construction algorithm. The reason is that the above greedy construction algorithm focuses on the minimal number of colors only in each search step, but not on the robustness of the coloring. In addition, the crossover method does not handle the objective function (robustness) effectively. Finally, the classical GA method is a little unrefined as an efficient search method.

4 Encoding and Crossover Operator

We use an encoding method based on the partition approach, where a set of vertices belonging to a group will be assigned the same color. In other words, an individual s in the population can be presented as $s = \{V_1, V_2, \cdots, V_c\}$, where $V_i = \{j | C(j) = i, 1 \leq j \leq n\}, 1 \leq i \leq c$. For instance, the Solution 1 for the graph in Figure 2 can be presented as {{1}, {2,3},{4,5},{}} and the Solution 2 can be presented as {{1},{2},{3},{4,5}}. It is obvious that the above partition based encoding can represent any coloring. Compare with the order and color based approach, our search space is c^n instead of $n!$, since the order of the vertices assigned with the same color will be ignored in the search. In practice, we always have $c \ll n$, hence $c^n \ll n!$.

One efficient crossover operator for the individuals encoded based on the partition approach is the Greedy Partition Crossover (GPX) mentioned in [3]. The description of GPX is presented below:

Algorithm 1 The GPX crossover operator

Input: individual s^1 and s^2 as parents
$s^1 = \{V_1^1, V_2^1, \cdots, V_c^1\}$
$s^2 = \{V_1^2, V_2^2, \cdots, V_c^2\}$
Output: offspring $s = \{V_1, V_2, \cdots, V_c\}$
for all $i(1 \leq i \leq c)$ **do**
  if i is odd **then**
    A=1;
  else
    A=2;
  end if
  $V_i = V_j^A$;
  remove all vertices of V_i from both s^1 and s^2;
end for
Assign randomly the vertices of $V/(V_1 \cup V_2 \cup \cdots \cup V_c)$;

		s^1 – parent 1				s^2 – parent 2				s – offspring			
		V_1^1	V_2^1	V_3^1	V_4^1	V_1^2	V_2^2	V_3^2	V_4^2	V_1	V_2	V_3	V_4
Input		1,9	2,3	4,5	6,7,8	1,6	2,9	3,8	4,5,7				
i=1	*A*=1	1,9	2,3	4,5		1	2,9	3	4,5	6,7,8			
i=2	*A*=2	1	3	4,5		1		3	4,5	6,7,8	2,9		
i=3	*A*=1	1	3			1		3		6,7,8	2,9	4,5	
i=4	*A*=2		3					3		6,7,8	2,9	4,5	1
Random Assign										6,7,8	2,9	4,5,3	1
Output										6,7,8	2,9	4,5,3	1

Fig. 4. The GPX crossover operator: An example

Figure 4 illustrates how the offspring are produced by GPX, where the two parents are $s^1 = \{\{1,9\},\{2,3\},\{4,5\},\{6,7,8\}\}$ and $s^2 = \{\{1,6\},\{2,9\},\{3,8\},\{4,5,7\}\}$ when c is 4.

5 The Hybrid GA

5.1 Algorithm Description

We use the hybrid GA to solve RGCP using the new encoding scheme and crossover operator introduced in Section 4. The general framework of our algorithm is given in Figure 5.

First, the algorithm generates an initial population from random colorings. It will then perform Selection, Crossover, Local Search and Update steps iteratively for T_{max} iterations. The details of each step in each generation is described below:

- Selection: The population m is divided into $m/2$ pairs randomly. The crossover operation is performed on each pair with a probability of p_c

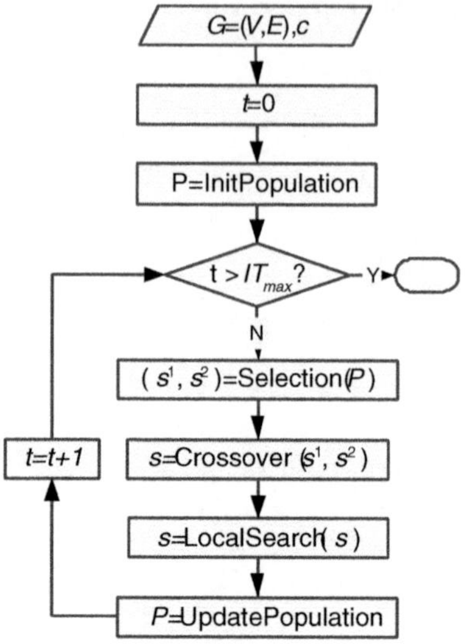

Fig. 5. The framework of the hybrid GA

- Crossover: a offspring is created by CPX
- Local Search: A local search algorithm is applied to improve the offspring to get a new individual using a fixed number of iterations
- Population Update: The improved individual is inserted into the population by replacing the parent individual with the worst value of fitness function

We replace mutation operator with a local search procedure to focus the search process.

5.2 Selection Operator

We set the deterministic sampling as the selection operator in our GA. The deterministic sampling is given by the following equation:

$$N_i = M \frac{F(individual_m)}{\sum_{m=1}^{M} F(individual_m)} \tag{8}$$

M is the size of population and the fitness function $F(G) = \frac{1}{R(G)}$ if the coloring is valid, otherwise $F(G) = \frac{1}{R(G)+P_1}$. In deterministic sampling, N_i is first determined for each individual. And then, the integer value of the expected occurrence is copied directly into the population. Secondly, the fractional portion of the N_i is used as a sort order for the individuals in the population, i.e., the individuals with the greatest fractional component head a list. Finally, the remaining individuals are selected from the top of the list until the population is full.

5.3 Two Local Search Operators

In general, the objective of the local search in our hybrid GA is to maximize fitness function subjected to the neighborhood search space of changing only one vertex. We designed two kinds of local search operators in our algorithm, one is known as enumerative search and the other is called random search.

- *Enumerative search*
  Step 1: For each vertex, the algorithm first calculates the reduction of the fitness function if the color for that vertex is removed. The vertex with minimum reduction to the fitness function is marked as v'.
  Step 2: We try all possible colors on the vertex v'. The color with has maximum increase of the fitness function is denoted as c'. We assign color c' to vertex v'. The above two steps are repeated L_e times to obtain the final result. In Step 1, we iterate all vertices and in Step 2, we iterate through all colors, therefore the computing time is $O(L_e * (n + c))$. If we were to try all possible coloring switches naively, the computing time is $O(L_e * (n * c)$.
- *Random search*
  Step 1: We randomly choose a vertex v' in the given graph and calculate the reduction of the fitness function when the existing color for v' is removed.
  Step 2: Try all possible colors on vertex v'. Let color c' be the color with the maximum increase of the fitness function. Assign color c' to vertex v'. The above two steps will be iterated L_r times. Since each iteration enumerates all colors in Step 2, the time of computing is $O(L_r * (1 + c))$.

The performance comparison of the two local search operators is presented in Section 6. Since they have different performance according to the size of input graph, we make our algorithm adaptive and more efficient by integrating both local search methods into a hybrid GA that selects either method based on a threshold such as n or c/n.

6 Experimental Results

First, we design a set of experiments to compare the performances between the two local search operators - enumerative search and random search on various input data. There are a total of 16 input sizes, n, where n ranges from 10 to 1000. For each size, we randomly generate 50 instances where the missing edge penalties are generated with the uniform distribution in the interval [0,1]. The graph density is fixed to be 0.5. We have the same parameters setting ($m = 20, IT_{max} = 50, p_c = 0.6, P_1 = 10^5$) as previous GA method [8]. We try $L_e = 10$ and $L_e = 20$ for the enumerative search and $L_r = 50$ for the random search. For our experiments, we use a Pentium III personal computer with a 500MHz CPU and 256M RAM. The previous paper proposed in [8] uses the same hardware configuration.

The experimental results are shown in Table 2. For each row identified by ID from 1 to 16, the graph size and the number of colors allowed, the results

Table 2. Enumerative Local Search v.s. Random Local Search (the optimal solution marked *)

ID	n	c	Enumerative Search ($L_e = 10$)		Enumerative Search ($L_e = 20$)		Random Search ($L_r = 50$)	
			R	CPU Time	R	CPU Time	R	CPU Time
1	10	4	3.04	0.58	3.03	0.67	*2.96	1.26
2	10	5	1.52	0.5	1.51	0.8	*1.48	1.09
3	15	5	7.18	0.49	7.02	1.03	*6.33	1.19
4	15	6	4.07	0.64	4.1	0.86	*3.73	1.14
5	20	5	15.58	0.67	14.92	0.66	*13.12	0.72
6	20	8	4.17	0.68	4.06	1.06	*3.78	1.52
7	20	10	1.68	0.92	1.71	1.04	*1.48	1.43
8	50	18	11	1.46	10.9	2.32	*8.38	2.41
9	100	35	25.28	2.8	23.34	3.07	*15.45	3.24
10	100	50	4.18	2.92	3.74	4.29	*2.40	3.94
11	250	70	103.08	9.24	98.73	11.41	*80.71	7.64
12	250	90	47.65	10.6	37.62	13.49	*36.54	9.44
13	500	150	193.85	37.66	*139.29	49.11	156.2	31.1
14	500	200	113.23	47.71	*59.72	68.23	81.1	48.2
15	500	250	92.46	73.47	*36.05	100.14	109.99	120.09
16	1000	400	306.19	264.49	*220.36	372.03	267.84	351.37

for both two local search operators including the average minimum value R of objective function and the corresponding average CPU time in seconds.

From the Table 2, the enumerative search of $L_e = 10$ has the fastest running time. However, the its solution is the worst among the three methods. For example, when L_e increases to 20, the solution obtained for $(n, c) = (500, 200)$ improves from 113.23 to 59.72. However, its running time is also increased accordingly. When we compare random search with $L_r = 50$ and enumerative search with $L_e = 20$, the random search is superior in terms of the run time except in the test set ID.15 where the ratio c/n is too big. When $n < 500$, the random search based method outperforms the enumerative search. However, when $n \geq 500$, the enumerative search is better than the random search.

Secondly, we designed another set of experiments to compare the performance between our new hybrid GA and GA proposed in [8] for medium and large graphs. The experimental results are given in Table 3. We have 9 sizes of input graph (row), where the number of vertices n ranges from 50 to 2000. For each size, we randomly generated 50 instances for the graph where the penalty for the each missing edge is 1. This has the same setting as previous published GA[8]. In each row, we provide the number of vertices n, the fixed number of colors c, the average minimum value of the objective function R with the average CPU running time in seconds by our hybrid GA using random local search of $L_r = 50$. In addition, the results by published GA in [8] are also listed for comparison.

From Table 3, it is clear that our new hybrid GA outperforms the published GA significantly, not only obtaining better quality of solutions, but also reducing

Table 3. Hybrid GA v.s. Published GA (the optimal solution marked *)

ID	n	c	Hybrid GA (without Alternating) R	CPU Time	(with Alternating) R	CPU Time	Published GA R	CPU Time
1	50	18	*46.00	2.35	*46.00	2.35	46	1.54
2	100	35	*95.00	4.22	*95.00	4.22	97	5.5
3	250	70	*330.00	12.26	*330.00	12.26	334	54.32
4	250	80	*270.00	12.79	*270.00	12.79	280	46.47
5	250	90	*230.00	14.22	*230.00	14.22	238	39.38
6	500	200	400.04	56.56	*400.00	54.82	411	166.14
7	1000	300	1202.28	262.89	*1202.20	227.52	1235	646.42
8	2000	500	3106	977.49	*3006	1006.33	N/A	N/A
9	2000	700	2006	1276.31	*1948	1175.60	N/A	N/A

the computing by almost 75%. Furthermore, our hybrid GA can solve inputs with large sizes. For instance, it can obtain results for $n = 2000$ within 20 minutes using a 500MHz PC.

Finally, in Table 3, we also show the results for the adaptive alternating local search. The threshold for alternating between the two local search methods is fixed to be $n = 500$, where the algorithm will select the random local search if $n < 500$, otherwise it will select the enumerative local search. From Table 3, the alternating method improves the performance for test cases with large sizes. For instance, in ID.9 when $(n, c) = (2000, 700)$, it brings 3% relative improvement in solution with the same amount of running time.

7 Conclusion

In this paper, we studied RGCP and presented a new hybrid GA approach to solve the problem. Computational results indicated that our hybrid GA to be superior to the previous method, not only in solution quality but also in computational efficiency.

References

[1] Brelaz, D. New Methods to Color Vertices of a Graph, Communications of ACM, 22(1979)251-256 125
[2] Davis, L. Handbook of Genetic Algorithms. Van Nostrand Reinhold. New York, NY(1991)
[3] Galinier, P. and Hao, J.-K. Hybrid Evolutionary Algorithm for Graph Coloring, Journal of Combinatorial Optimization, 3(1999)379-397. 130
[4] Garey M. R. and Johnson D. S. Computer and Intractability, Freeman: San Francisco(1979) 125
[5] Johnson, D. S. Aragon, C. R., McGeoch, L. A. and Schevon, C. Optimization by Simulated Annealing: an Experimental Evaluation; part II, Graph Coloring and Number Partitioning, Operations Research, 39(3)(1991)378-406. 125

[6] Johnson, D. S. and Trick, M. A. (Eds.) Proceedings of the 2nd DIMACS Implementation Challenge, DIMACS Series in Discrete Mathematics and Theoretical Computer Science, American Mathematical Society, 26(1996). 125

[7] Pardalos, P. M., Mavridou, T. and Xue, J. The Graph Coloring Problems: A Bibliographic Survey, Handbook of Combinatorial Optimization, Du, D.-Z. and Pardlos, P. (EDS.), Kluwer Academic Publishers, 2(1996)331-395. 125

[8] Yanez, J. and Ramirez J. The Robust Coloring Problem", European Journal of Operational Research, 148(3)(2003)546-558. 125, 126, 129, 133, 134

Estimating Problem Metrics for SAT Clause Weighting Local Search

Wayne Pullan, Liang Zhao, and John Thornton

School of Information Technology, Griffith University
Gold Coast, Qld., 4215, Australia
Phone: (07)55529002, Fax: (07)55528002
{w.pullan,j.thornton}@griffith.edu.au

Abstract. Considerable progress has recently been made in using clause weighting algorithms to solve SAT benchmark problems. While these algorithms have outperformed earlier stochastic techniques on many larger problems, this improvement has generally required extra, problem specific, parameters which have to be fine tuned to problem domains to obtain optimal run-time performance. In a previous paper, the use of parameters, specifically in relation to the DLM clause weighting algorithm, was examined to identify underlying features in clause weighting that could be used to eliminate or predict workable parameter settings. A simplified clause weighting algorithm (Maxage), based on DLM, was proposed that reduced the parameters to a single parameter. Also, in a previous paper, the structure of SAT problems was investigated and a measure developed which allowed the classification of SAT problems into random, loosely structured or compactly structured. This paper extends this work by investigating the behaviour of Maxage with regard to the structural characteristics of SAT problems. The underlying motivation for this study is the development of an adaptive, parameterless clause weighting algorithm.

Keywords: Constraints, Search.

1 Introduction

The propositional satisfiability (SAT) problem is fundamental in solving many practical problems in mathematical logic, inference, machine learning, constraint satisfaction, and VLSI engineering. Theoretically, the SAT problem is the *core* of a large family of computationally intractable NP-complete problems. Several such NP-complete problems have been identified as central to a variety of areas in computing theory and engineering. Therefore, methods to solve the satisfiability problem play an important role in the development of computing theory and systems.

In this paper, as with most other work on SAT algorithms, we only consider propositional formulae in conjunctive normal form. That is, formulae of the form $F = \bigwedge_i \bigvee_j l_{ij}$ where each l_{ij} is a propositional variable or its negation. The l_{ij} are

T.D. Gedeon and L.C.C. Fung (Eds.): AI 2003, LNAI 2903, pp. 137–149, 2003.

termed *literals* while the disjunctions $\bigvee_j l_{ij}$ are the *clauses* of F. The goal of all SAT algorithms is to find an assignment of the truth values to the propositional variables in F that results in no unsatisfied (*false*) clauses.

Algorithms for solving SAT problems can be divided into two categories: *complete* and *incomplete*. Complete SAT algorithms perform a systematic traversal of the search space and will always find a solution if one exists. Incomplete SAT algorithms are stochastic algorithms in that they may find a solution but, if they fail, it cannot be concluded that no solution exists.

Some of the best known incomplete SAT algorithms are local search algorithms that, while they differ in detail, all basically implement a local search strategy which starts with an initial random assignment of truth values to all propositional variables. In each subsequent search step, the algorithm selects a variable using some heuristic and negates the truth value of that variable (i.e. *true* to *false* or *false* to *true*). Variable negations are typically performed with the goal of minimising an objective function based on the currently unsatisfied clauses.

A version of local search, that has recently become very effective in SAT problem solving, modifies the objective function by associating a weight with every clause of the given formula. These algorithms then aim to minimise the total weighted objective function rather than the number of unsatisfied clauses. By appropriate manipulation of clause weights, these clause weighting algorithms are able to escape from local minima and other attractive non-solution areas in the underlying search space. The major, inherent disadvantage in these algorithms is that additional, problem dependent, parameters are required to control the clause weights.

In 1993, clause weighting local search algorithms for SAT were simultaneously proposed in [5] and [8]. Various enhancements to clause weighting followed in the mid-90s, particularly Jeremy Frank's work on multiplicative weighting and weight decay [1]. Early clause weighting algorithms avoided plateau search by adding weight to all unsatisfied clauses as soon as a plateau was encountered [5]. However, it was not until the development of DLM [9] that these insights were translated into significant performance improvements. The main differences between Discrete Lagrangian Multiplier (DLM) and earlier clause weighting techniques are in the use of a tabu list [2] to guide the search over plateau areas, and a weight reduction heuristic that periodically reduces clause weights. The Smoothed Descent and Flood (SDF) algorithm [6] uses multiplicative weighting and a continuous renormalisation of relative weights after each increase. While SDF produced some improvement over DLM in terms of the number of variable negations required on smaller sized problems, there is a significant run-time overhead in maintaining SDF's real valued weights. SDF subsequently evolved into the Exponentiated Sub-Gradient (ESG) method [7] which has been improved on by the Scaling and Probabilistic Smoothing (SAPS) [3] method.

Another recent algorithm is Maxage [10], which is based on DLM and has comparable performance but the problem dependent parameters have been reduced from 14 to just a single parameter, the DECREASE parameter, which controls the frequency with which clause weights are decayed. The major objective of this paper is to determine if an estimate of the optimal value of the Maxage DECREASE parameter can be determined from characteristics of SAT problems. The underlying motivation for this study is the development of an adaptive, parameterless clause weighting algorithm, based on Maxage, that determines an initial estimate for DECREASE and then, using measurable characteristics of the search process, adaptively modifies DECREASE to produce an effective search.

This paper is structured as follows. Section 2 presents a more detailed discussion of clause weighting, particularly with reference to Maxage, while Section 3 contains a description of SAT problem characteristics. An analysis of four benchmark SAT problems, with regard to these characteristics, is presented in Section 4 and Section 5 presents an analysis of how these characteristics influence the Maxage DECREASE parameter. Finally, Section 6 contains a conclusion and suggestions for future research.

2 Local Search

At a high level, a local search algorithm can be viewed as a mechanism for traversing a highly multi-dimensional hyper-surface with the objective of locating a global minima. While features of the hyper-surface may be extremely complicated, the perception of the hyper-surface by the local search algorithm is very limited in that it only knows the hyper-surface immediately adjacent to its current position and has no memory or knowledge of the hyper-surface in any other location. In fact, its knowledge is limited to, for a single step in any given direction, the rate at which the hyper-surface is changing. To efficiently traverse the hyper-surface the local search algorithm must avoid:

- **Search Cycles** which are basically some configuration of closed paths on the hyper-surface. They arise because of the presence of local minima or some combination of other hyper-surface features.
- **Unguided Travel** which occurs when the hyper-surface is locally flat (plateau) and there is no basic, under-lying guidance for the search.

Search cycles that have short path lengths can be successfully handled by mechanisms such as Tabu lists [2], which prevent re-selection of a variable before some number of other variables have been modified. However, search cycles with longer path lengths or a variety of paths are much more difficult to detect and escape from. Tabu lists can also have a beneficial effect when traversing a hyper-surface plateau as they tend to provide an underlying direction for the search.

2.1 Non-clause Weighting Local Search Algorithms

The hyper-surface traversed by non-clause weighting local search algorithms for SAT problems is generally that formed by evaluating the number of false clauses for each assignment of variables identified during the search. That is, the local search is performing the global optimisation problem (for a SAT problem with n variables $(x_1, \ldots, x_n)$ and m clauses $(c_1, \ldots, c_m)$):

$$min \ \ f(x_1, \ldots, x_n) = \sum_{i=1}^{m} b_i \tag{1}$$

where $b_i = 0$ if clause c_i is *true* and $b_i = 1$ if clause c_i is *false*. At each step in the search, these algorithms evaluate the effect of negating each variable in terms of the reduction in the number of false clauses and will generally select the variable which causes the largest decrease in f. The fundamental differences in these algorithms are typically in the tie-breaking rules, if more than one variable gives the best decrease, and how they handle search cycles and plateaus on the hyper-surface (random moves, random restarts and Tabu lists).

2.2 Clause Weighting Local Search Algorithms

Clause weighting local search algorithms traverse the weighted false clause hyper-surface formed by the weighted cost of a problem solution. That is, a clause weighting local search is addressing the global optimisation problem:

$$min \ \ g(x_1, \ldots, x_n) = \sum_{i=1}^{m} w_i b_i, \quad w_i \geq 1 \tag{2}$$

where w_i is the weight associated with clause c_i. At each step in the search, these algorithms evaluate the effect of negating each variable in terms of the reduction in the weighted cost of the false clauses and will generally select that variable which causes the largest decrease in g. Clause weights act to deform the weighted false clause hyper-surface (S_g) from the false clause hyper-surface (S_f) and are typically incremented whenever a local minimum or extended plateau is found by the algorithm. This action tends to remove local minima [5] and plateaus from S_g. To prevent S_g from becoming too deformed and "rugged" (and thus losing any natural underlying guidance from S_f), a clause weight reduction mechanism is normally incorporated into the search.

Clause weighting local search algorithms tend to focus on satisfying the more difficult clauses of a SAT problem as the weights for clauses that are difficult to satisfy will, on average, be higher than those for clauses that are easier to satisfy. This will make satisfying these clauses more attractive to the algorithm and is analogous to the common problem solving heuristic of attacking the most difficult parts of a problem first, and then addressing the less constrained resources until a solution is found. It is also important to note that all global minima, corresponding to $f = g = 0$, will be in the same positions on S_g as they are on S_f. If this were not the case, clause weighting local search algorithms would be at a considerable disadvantage to non-clause weighting local search algorithms in that they would be trying to locate global minima that are moving on S_g as clause weights change (as is the case when the global minima are such that $f > 0$).

There are some inherent disadvantages in clause weighting algorithms, namely that additional, problem dependent parameters are required to control the clause weighting. These include the amount by which to increase or decrease the clause weights and at what point reweighting should occur. Also, the possibility of clause weighting cycles exists where clause weights are repetitively increased and decreased causing a search cycle in the sequence of variables to be negated.

2.3 Clause Weighting in Maxage

As shown in Fig. 1, Maxage [10] has only one parameter, the DECREASE clause weighting parameter, which specifies how many clause weight increases must occur before a clause weight reduction is performed. Unfortunately, as with the parameters for DLM, the Maxage DECREASE parameter is problem dependent and must be pre-determined for each particular class of problem. This typically requires performing a large number of experiments, where DECREASE is systematically varied, to identify the optimal value of DECREASE for that class of problem. Another issue is that these pre-determined values are a compromise in that they are the optimal, on average, over the entire search. It is reasonable to assume that the actual optimal value for DECREASE, at any point in the search, depends on the nature of S_f at the current location of the search and this varies, with most problems, during the search. When DECREASE is very close to one, clause weighting has little effect as clause weights are rapidly returned to their default value. That is, g tends to be very close to f and there is relatively little influence on the search from clause weighting. When DECREASE is large, clause weights tend to become large as the downward re-weighting of clauses is performed less frequently. This tends to make the difference between f and g more significant and g becomes a more "rugged" function than f, in the sense that there is greater potential for the change in g to be larger than that for f at each step of the search.

2.4 Determination of Optimal DECREASE for Maxage

The benchmark problems that will be used in this study are four of the problem domains for which existing Maxage parameters have already been developed, namely random 3-SAT, parity learning, graph colouring and blocks world planning. Using the SATLIB[1] and DIMACS[2] benchmark libraries, we selected *f2000* for random 3-SAT, *par16-1-c* for parity learning, *bw_large.c* for blocks world planning and *g125.17* for the graph colouring problem. These problems provide a range of SAT problem types, from random to structured. In addition, they provide the largest range for the Maxage DECREASE parameter and all have reasonable computational requirements (which ensures that clause weighting becomes a relevant factor in the search). The failure rate of Maxage, for a maximum

[1] http://www.intellektik.informatik.tu-darmstadt.de/SATLIB/
[2] ftp://dimacs.rutgers.edu/pub/challenge/satisfiability/benchmarks/cnf/

procedure *MAX-AGE*
begin
  Generate a random starting point
  Initialise counters and clause weights to zero
  while solution not found and $flips <$ maxFlips **do**
    $B \leftarrow$ set of best weighted cost single flip moves
    if no improving $x \in B$ **then**
      if oldest $x \in B$ has $age(x) \geq maxAge$ **then**
        $B \leftarrow x$
        $maxAge \leftarrow maxAge + 1$
      else if $random(p) \leq P$ **then**
        $B \leftarrow \emptyset$
      end if
    end if
    if $B \neq \emptyset$ **then**
      Randomly pick and flip $x \in B$
      $age(x) \leftarrow ++flips$
    else
      Increase weight on all false clauses
      if $++increases$ % DECREASE $= 0$ **then**
        Decrease weight on all weighted clauses
    end if
  end while
end

Fig. 1. The MAX-AGE Algorithm

of 1,000,000 steps, as DECREASE varies in the range 1 . . . 100 is shown in Fig. 2 for each of the four benchmark problems. From the data presented in the graphs, we see that the optimal DECREASE is large (46) for *par16-1-c*, is smaller (10) for the randomly generated *f2000*, is six[3] for *bw_large.c* and is smallest (four) for *g125.17*. Clearly some aspects of the structure of SAT problems have an impact on the optimal value for DECREASE and this question is investigated in the next section.

3 SAT Problem Characteristics

The overall goal of this section is, from a static analysis of a SAT problem, to identify characteristics of S_f that are relevant to the optimal setting for the DECREASE parameter of Maxage. Using statistical characteristics from the uniformly randomly generated SAT problem *f2000* as a basis, we categorise other SAT problems by comparing their statistical properties with those of randomly

[3] The computational requirements when DECREASE is large for *bw_large.c* are much greater than when DECREASE is six.

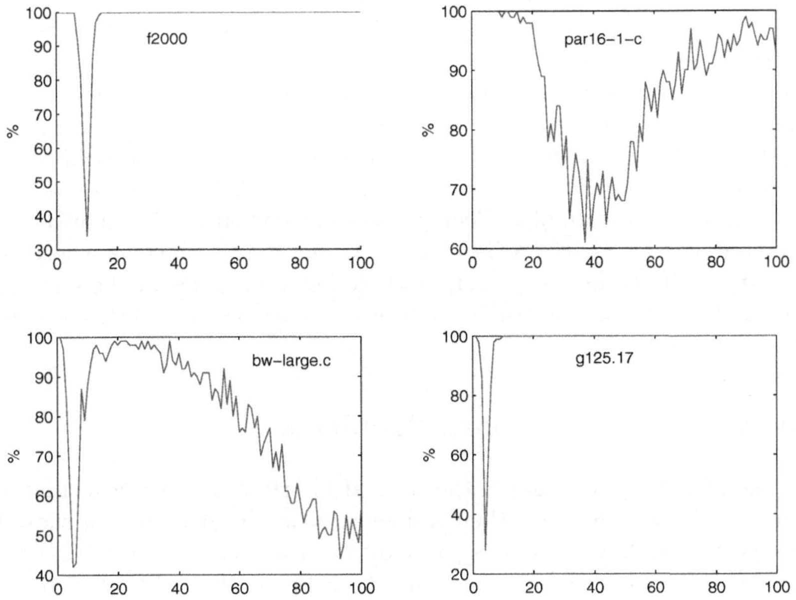

Fig. 2. The failure rate for Maxage as DECREASE varies from 1 ... 100 . For each value of DECREASE, Maxage was executed from 100 random starting points with a maximum of 1,000,000 steps allowed. The vertical axis records the number of times for which Maxage failed to find a global minima

generated SAT problems. Random SAT problems are constructed by first uniformly randomly selecting the variables for each clause and then, with probability 0.5, negating each variable. This generation technique ensures that the distributions of most problem characteristics will closely approximate the normal distribution. In particular, the distribution of u_i, the number of times each is variable is used, the distribution of the number of unique variables that each variable shares a clause with (or neighbourhood density)[4], and the presence of dependent variables can all be used to classify SAT problems as random or structured. A fourth measure, the distribution of positive and negated literals for each variable can also be used as a test of randomness. In addition, this measure also provides information as to the nature of S_f. If we define u_i^d as the absolute difference between the number of positive occurrences (u_i^+) and the number of negated occurrences (u_i^-) of x_i in clauses then, the distributions of u_i^+ and u_i^- are indicators of the structure of the S_f being traversed by the search. They place a range on the maximum value of f_i, the change in f caused by negating x_i. Clearly, $-u_i^+ \leq f_i^+ \leq u_i^-$, where f_i^+ denotes the change in the number of *false* clauses when x_i goes from *false* to *true*. Correspondingly, $-u_i^- \leq f_i^- \leq u_i^+$, where f_i^- denotes the change in the number of *false* clauses when x_i goes from *true* to *false*. Clearly, if u_i^+ and u_i^- are small, both f_i^+ and f_i^- will also be small. Of interest also is, how are f_i^+ and f_i^- distributed within these ranges. Our hypothesis is that if u_i^d is close to zero, then this will bias both f_i^+ and f_i^- towards the zero point of their ranges. The rationale for this is that a small u_i^d states that there are potentially as many clauses which will go from *true* to *false* as will go from *false* to *true* when x_i is negated. Conversely, if u_i^d is not close to zero, then this will bias both f_i^+ and f_i^- towards one of the extremes of their ranges.

Supporting the argument that the distribution of u_i^d can be used as an estimator of the structure of S_f is the degenerate 1-SAT case where all clauses only contain a single variable. Clearly, the distribution of u_i^d directly reflects the characteristics of S_f. For k-SAT ($k > 1$), there are situations where negating a variable will not falsify a clause (where both variables in the clause were *true*) so, in these cases, the distribution of u_i^d is a worst case estimator of the characteristics of the S_f.

4 Structure of Benchmark Problems

As expected, *f2000* approximated the normal distribution with a median of 12 while the distributions for the other problems showed that there is a wide range in the usage of variables in clauses. In addition, the counts of independent and dependent variables for the four benchmark problems showed that both *par16-1-c* and *bw_large.c* are classified as non-random problems. With regards to the distributions of variable signs and with reference to Fig. 3, the following points can be made for each of the benchmark problems:

- **f2000**. For the randomly generated problem *f2000*, the distributions of u_i^+ and u_i^- have medians around 6.4 and there are no outliers. From the distribution of u_i^d it can be seen that for approximately 12% of variables, $u_i^d = 0$, the median is 0.045 and the variance is 13.2. From these observations it is reasonable to conclude that, S_f for *f2000* will have a relatively small proportion of plateaus and the other areas will contain relatively small features. Accordingly, in this study, we classify S_f for *f2000* as "choppy" and expect f_i to be relatively small as each variable is negated.
- **par16-1-c**. With regard to u_i^+ and u_i^-, their distributions have medians ± 5.7 and it can be seen that there are just a few large outliers. From the distribution of u_i^d it can be seen that for approximately 80% of variables, $u_i^d = 0$ and the variance is 0.17. Both these factors place a small range on f_i^+ and f_i^- resulting in a relatively smooth S_f for *par16-1-c*. That is, S_f for *par16-1-c* contains a higher proportion of plateaus than S_f for *f2000* and also contains some areas with approximately the same features as *f2000*. In this study, we classify S_f for *par16-1-c* as "flat" and expect f_i to be predominantly zero, but not infrequently small, as each variable is negated.
- **bw_large.c** For the *bw_large.c* problem, the distribution for u_i^- has a median of -31.0 while that for u_i^+ is 6.0. This gives a large range for both f_i^+ and f_i^-. From the distribution of u_i^d it can be seen that it is has a median of -25.0 with a variance of 62.0. This implies that f_i^+ and f_i^- will be towards an extreme of their ranges. From these observations it is reasonable to conclude that, S_f for *bw_large.c* will have virtually no plateaus and mainly consists of relatively

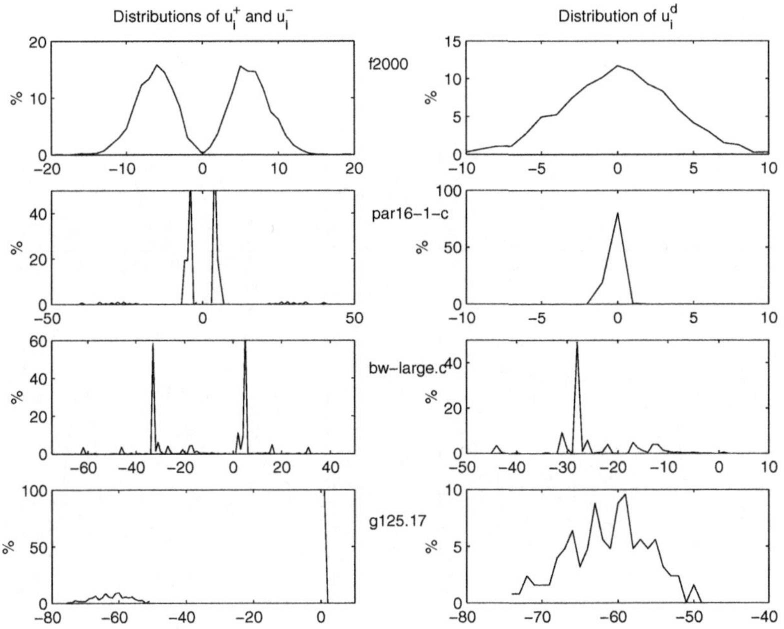

Fig. 3. Distributions of u_i^+, u_i^- and u_i^d for each of the four benchmark problems. The vertical axis is, in each plot, the percentage frequency for which the horizontal axis value appeared. For the distributions of u_i^+ and u_i^-, u_i^+ is plotted as positive values and u_i^- as negative values

large features. Accordingly, in this study, we classify S_f for *bw_large.c* as "moderate" and expect f_i to be relatively large as each variable is negated.

– **g125.17** For the *g125.17* problem, the distribution for u_i^- has a median of -62 while that for u_i^+ is one. This gives a large range for both f_i^+ and f_i^-. From the distribution of u_i^d it can be seen that it is has a median of -61 with a variance of 27. This implies that f_i^+ and f_i^- will be towards an extreme of their ranges. From these observations it is reasonable to conclude that, S_f for *g125.17* will have virtually no plateaus and mainly consists of large features. Accordingly, in this study, we classify S_f for *g125.17* as "rugged" and expect f_i to be large as each variable is negated.

To confirm the observations detailed above, random sampling of S_f was performed for the benchmark problems and the results are presented in Fig. 4. This random sampling was performed by first generating uniformly random assignments for variables and then evaluating the change in f as each variable is negated. This process was repeated 1,000,000 times for each problem. As can be seen, the results support the arguments presented above in that S_f for *f2000* contains a moderately small proportion of plateaus (22% of variable negations resulted in $f_i = 0$) and $|f_i| < 6$ in all other areas, S_f for *par16-1-c* contains

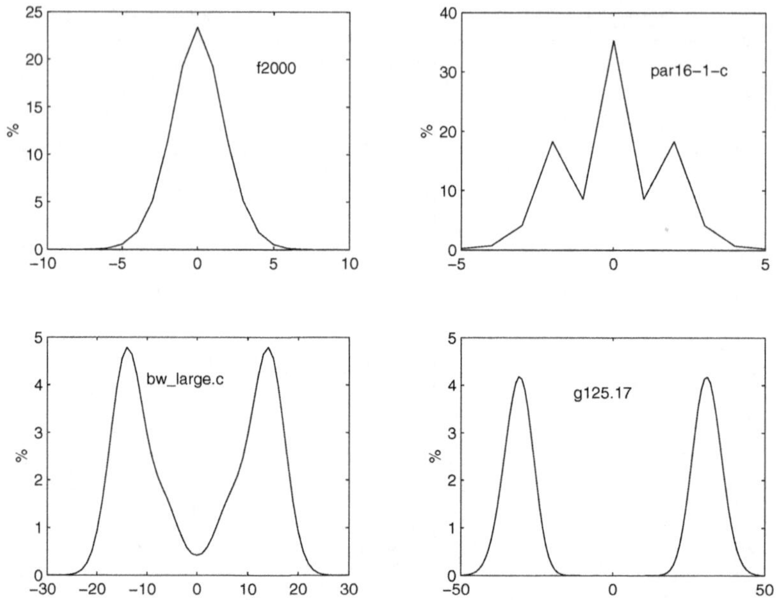

Fig. 4. Distributions of sampled f_i for each of the four benchmark problems. The vertical axis is, in each plot, the percentage frequency for which the horizontal axis value appeared

a larger proportion of plateaus (37% of variable negations resulted in $f_i = 0$) and $|f_i| < 5$ in all other areas. For *bw_large.c*, S_f has virtually no plateaus and consists of moderately large features where $|f_i| \leq 25$, while for *g125.17*, S_f has no plateaus and consists only of large features where $15 \leq |f_i| \leq 50$.

5 SAT Problem Characteristics and Maxage

The basic rationale for the analysis of SAT problems was to investigate characteristics of the benchmark problems to identify if there is any relationship between the optimal value for DECREASE and measurable properties of a SAT problem. If this can be shown to be the case, then some form of estimate for the optimal value of DECREASE could be programmatically determined at the start of a Maxage search by an analysis of the SAT problem.

From Figs. 2 and 3 and the discussion presented above, our hypothesis is that a larger value of DECREASE (i.e. more clause weighting) will be optimal for problems where S_f is relatively smooth and a smaller value optimal when S_f is relatively rugged.

Intuitively, as traversing a predominantly flat S_f, takes some number of variable assignments, clause weighting needs to be more aggressive (higher DECREASE) to prevent unguided travel. Conversely, if there is a small proportion

Table 1. Estimated DECREASE compared with the experimentally determined optimal DECREASE values for benchmark and other SATLIB SAT problems

Problem	Estimated DECREASE	Optimal DECREASE	Problem	Estimated DECREASE	Optimal DECREASE
f2000	10	10	par16-1-c	46	46
bw_large.c	6	6	g125.17	4	4
aim-200-6_0-yes1-1	56	≥ 10	ais10	6	> 50
BMS_k3_n100_m429_140	15	≥ 15	flat200-1	7	10
CBS_k3_n100_m449_b90_889	11	10	ii32a1	7	> 50
RTI_k3_n100_m429_140	14	$8 \ldots 15$	logistics.b	7	> 50
uf100-0953	9	10	sw100-1	7	$3 \ldots 7$

or no flat areas on S_f, the main role for clause weighting is escaping from search cycles, in particular local minima, which is a relatively localised phenomena of S_f and takes relatively fewer variable assignments. Accordingly clause weighting needs to be less aggressive (lower DECREASE) so that S_g stays close to S_f and the inherent features of S_f are used as much as possible during the search.

The measures u_i^+, u_I^- and u_i^d ranked S_f for the benchmark problems as (smoothest to most rugged) *par16-1-c*, *f2000*, *bw_large.c* and *g125.17*. This is in accordance with the ranking of their optimal DECREASE value (46, 10, 6, 4). This suggests the following algorithm for setting the DECREASE parameter: If all four tests classify the SAT problem as random, then the optimal value for DECREASE is 10 otherwise, if $u_i^d, i = 1 \ldots n$ approximates a normal distribution with a median of zero, DECREASE is a linear function of the frequency of the median, otherwise DECREASE is a linear function of the median of the distribution.

The results obtained, using this algorithm, are compared to experimentally determined optimal values for DECREASE in Table 1 for the benchmark plus other representative SAT problem classes from the SATLIB library. For these problems, the estimated DECREASE was within the range of the optimal DECREASE with a tolerance of ± 1 in 6 of the 10 cases. In the remaining cases, DECREASE was slightly underestimated for *flat200-1* at 7 (optimal value 10) and significantly underestimated for *ais10*, *ii32a1* and *logistics.b*. However, these last three problems represented special cases, having approximately linear relationships between runtime and DECREASE, such that increasing values of DECREASE produced better run-times. The implication of such a relationship is that these problems do not require a weight decrease scheme. A closer analysis showed that these problems had unusual distributions of neighbourhood densities (i.e. the connections between variables that share clauses, see Section 3). This implies that a further analysis of neighbourhood density may prove useful in further distinguishing problem structure. Overall, the results show the distribution of u_i^d provides the basis for a usable estimate of DECREASE. Our intention is to use this estimate as the starting point in an adaptive version of Maxage, which will modify DECREASE based on measures such as search mobility and coverage. Our initial work in this area (not reported here) has already shown the feasibility of this approach.

6 Conclusion

The aim of this study was to move towards developing an adaptive clause weighting algorithm, based on Maxage, that required no external, problem dependent parameter(s). This algorithm will analyse the SAT problem to determine an initial value for the DECREASE parameter and then adaptively modify this value during the search using run-time characteristics that determine how effectively the search is proceeding. This paper proposes a method for estimating the initial value for DECREASE.

We consider this study a step towards developing more intelligent, adaptive constraint solving technologies. Future research will include:

- the identification of structure sub-types within SAT problems to refine the initial setting of the DECREASE parameter.
- an investigation of why, for some SAT problems, there is a wide range of optimal DECREASE values whereas for other problems it is unique.
- the development of adaptive control mechanisms so that the DECREASE parameter can be adaptively adjusted during the search.
- an investigation to determine if clause weighting is able to identify clauses which are "globally" difficult to satisfy or if it is a localised activity which simply gives the search adequate mobility and the global minima is arrived at without any use of "global" knowledge.
- an investigation to determine what makes clause weighting algorithms such effective global optimisation algorithms. Is it in the ability to escape search cycles and the guided travel across plateaus or is it in the initial focus on satisfying the more "difficult" clauses and then addressing the "easier" clauses?
- incorporating a parameterless clause weighting algorithm as the local search within a hybrid genetic algorithm which will open the way to pool based, parallel search algorithms for SAT problems.

References

[1] J. Frank. Learning short term weights for GSAT. In *Proceedings of the Fourteenth National Conference on Artificial Intelligence (AAAI-97)*, pages 384–389, 1997. 138

[2] F. Glover. Tabu search: Part 1. *ORSA Journal on Computing*, 1(3):190–206, 1989. 138, 139

[3] F. Hutter, D. A. D. Tompkins, and H. H. Hoos. Scaling and probabilistic smoothing: Efficient dynamic local search for SAT. In *LNCS 2470: Proceedings of Constraint Programming, 2002*, pages 233–248, 2002. 138

[4] O. Kravchuk, W. Pullan, J. Thornton, and A. Sattar. An investigation of variable relationships in 3-SAT problems. In *AI 2002: Advances in Artificial Intelligence*, pages 579–590, 2002. 143

[5] P. Morris. The Breakout method for escaping local minima. In *Proceedings of the Eleventh National Conference on Artificial Intelligence (AAAI-93)*, pages 40–45, 1993. 138, 140

[6] D. Schuurmans and F. Southey. Local search characteristics of incomplete SAT procedures. In *Proceedings of the Seventeenth National Conference on Artificial Intelligence (AAAI-00)*, pages 297–302, 2000. 138

[7] D. Schuurmans, F. Southey, and R. C. Holte. The exponentiated subgradient algorithm for heuristic boolean programming. In *Proceedings of the Seventeenth International Joint Conference on Artificial Intelligence (IJCAI-01)*, pages 334–341, 2001. 138

[8] B. Selman and H. Kautz. Domain-independent extensions to GSAT: Solving large structured satisfiability problems. In *Proceedings of the Thirteenth International Joint Conference on Artificial Intelligence (IJCAI-93)*, pages 290–295, 1993. 138

[9] Y. Shang and B. Wah. A discrete Lagrangian-based global search method for solving satisfiability problems. *J. Global Optimization*, 12:61–99, 1998. 138

[10] J. Thornton, W. Pullan, and J. Terry. Towards fewer parameters for SAT clause weighting algorithms. In *AI 2002: Advances in Artificial Intelligence*, pages 569–578, 2002. 138, 141

Information Extraction via Path Merging

Robert Dale[1], Cecile Paris[2], and Marc Tilbrook[1]

[1] Centre for Language Technology, Macquarie University, Sydney
www.clt.mq.edu.au
[2] Intelligent Interactive Technology Group, CSIRO, Sydney
www.cmis.csiro.au/iit

Abstract. In this paper, we describe a new approach to information extraction that neatly integrates top-down hypothesis driven information with bottom-up data driven information. The aim of the KELP project is to combine a variety of natural language processing techniques so that we can extract useful elements of information from a collection of documents and then re-present this information in a manner that is tailored to the needs of a specific user. Our focus here is on how we can build richly structured data objects by extracting information from web pages; as an example, we describe our methods in the context of extracting information from web pages that describe laptop computers. Our approach, which we call **path-merging**, involves using relatively simple techniques for identifying what are normally referred to as named entities, then allowing more sophisticated and intelligent techniques to combine these elements of information: effectively, we view the text as providing a collection of jigsaw-piece-like elements of information which then have to be combined to produce a representation of the useful content of the document. A principle goal of this work is the separation of different components of the information extraction task so as to increase portability.

Keywords: Natural language understanding, natural language generation

1 Introduction

Information Extraction (IE [4, 2, 7]; the process of identifying a pre-specified set of key data elements from a free-text data source) is widely recognised as one of the more successful spin-off technologies to come from the field of natural language processing. The DARPA-funded Message Understanding Conferences (see, for example, [6]) resulted in a number of systems that could extract from texts, with reasonable results, specific information about complex events such as terrorist incidents or corporate takeovers. In each case, the task is manageable because (a) some other means has determined that the document being analysed falls within the target domain, and (b) the key information required is typically only a very small subset of the content of the document. A major component task is **named entity recognition** [1, 8], whereby people, places and organizations

T.D. Gedeon and L.C.C. Fung (Eds.): AI 2003, LNAI 2903, pp. 150–160, 2003.

San Salvador, 19 Apr 89 (ACAN-EFE)
Salvadoran President-elect Alfredo Cristiani condemned the terrorist killing of Attorney General Roberto Garcia Alvarado and accused the Farabundo Marti National Liberation Front (FMLN) of the crime.
...
Garcia Alvarado, 56, was killed when a bomb placed by urban guerrillas on his vehicle exploded as it came to a halt at an intersection in downtown San Salvador.
...
Vice President-elect Francisco Merino said that when the attorney general's car stopped at a light on a street in downtown San Salvador, an individual placed a bomb on the roof of the armored vehicle.

Fig. 1. A typical text used in information extraction

are located and tracked in texts; other processing can then take the results of this process to build higher order data structures, establishing, for example, who did what to who and when.

The domain-dependency that is typical of IE systems means that the cost of porting a system to a new domain can be high. As a result, in recent years there has been a move towards IE based on statistical techniques and machine learning, and these have been quite successful for the lower level tasks of named entity identification (see, for example, [3]). However, the larger scale knowledge integration tasks involved in IE are often better served by more traditional, knowledge-based approaches. In this paper we present a framework for information extraction that focuses on these higher-level processes. Particularly in cases where the knowledge to be extracted is complex in structure, statistical approaches can only solve part of the problem; informational elements, once extracted, have somehow to be built into a larger whole. We describe a new mechanism that relies on a hand-constructed knowledge template, along with a general inference mechanism we call **path merging**, to reconcile and combine the informational elements found in a text. The template provides top-down hypotheses as to the information we might find in a text; the named entities identified in the text provide bottom-up data that is merged with these hypotheses.

2 Background

A key aspect of any information extraction task is that only a small portion of the information available in a text is important. Figure 1 shows fragments of an input text in the terrorist incident domain; Table 1 shows the data structure that might be constructed as a result of analysing this document:

The general approach taken in these systems is that we first identify the named entities and then work out the relations between them. Even for relatively flat output structures such as the one shown here, domain-specific hand-crafted rules of considerable complexity are required to build the templates.

Table 1. The extracted results from the text in Figure 1

Incident: Date	19 Apr 89
Incident: Location	El Salvador: San Salvador (CITY)
Incident: Type	Bombing
Perpetrator: Individual ID	urban guerrillas
Perpetrator: Organization ID	FMLN
Perpetrator: Confidence	Suspected or Accused by Authorities: FMLN
Physical Target: Description	vehicle
Physical Target: Effect	Some Damage: vehicle
Human Target: Name	Roberto Garcia Alvarado
Human Target: Description	attorney general: Roberto Garcia Alvarado
Human Target: Effect	Death: Roberto Garcia Alvarado

The goal of the KELP[1] project is to develop technology that can extract information from a collection of web pages that describe similar things, and then collate and re-present this information in such a way as for a user to make it easy to compare those things. Suppose you are interested in purchasing a new cell phone: you could check out a consumer magazine, or visit a web site that presents comparisons of the different available models, but those sources are typically not up-to-date. You might visit the manufacturers' web pages to obtain information from the original sources, but this is a painful, slow process, and comparison is hindered by the different terminology each vendor uses; further, all these sources provide information for an 'average' visitor, rather than tailoring what they present to your particular needs and interests.

In KELP, we aim to build technology that can mine the information from these source web pages, and then, using techniques we have discussed elsewhere (see, for example, [5, 9]), re-present it in a form that is tailored to needs and interests captured in a specific user profile.

Our initial experiments have been in the context of web pages that describe laptop computers. Based on an analysis of around 120 web pages describing laptop computers, we developed a template that captures the range of information we might want to extract regarding a laptop computer, and then set about developing techniques to extract the relevant data from these web pages. Part of a typical filled template, or, as we call these structures within KELP, a **knowledge object** or KO, is shown in Figure 2. We have elided this structure to make it fit the space available; it should be obvious that the quantity and complexity of the data to be extracted is significant.

In our initial experiments, we focussed our attention on the information encoded in tables on web pages: for complex products such as laptop computers, this is a typical means of information provision, although it should not be forgotten that there is also a considerable amount of useful content presented in

[1] KELP stands for *Knowledge Extraction and Linguistic Presentation*, emphasising that the project is concerned both with natural language analysis and natural language generation techniques.

```
<?xml version="1.0" ?>
  <laptop_info>
    <laptop_id>
      <manufacturer>NEC</manufacturer>
      <series>Versa</series>
      <model>Premium PIII 1GHz</model>
    </laptop_id>
  <components>
    <cpu>
      <cpu_type>Pentium III</cpu_type>
      <cpu_speed>
        <number>1</number>
        <unit>GHz</unit>
      </cpu_speed>
    </cpu>
    <memory>
      <installed_memory>
        <number>128</number>
        <unit>MB</unit>
      </installed_memory>
      <maximum_memory>
        <number>512</number>
        <unit>MB</unit>
      </maximum_memory>
    </memory>
    <storage>
      <hard_drive>
        <number>20</number>
        <unit>GB</unit>
      </hard_drive>
      <removable>
        <cd_drive>24x</cd_drive>
      </removable>
    </storage>
  ...
</laptop_info>
```

Fig. 2. A portion of a filled knowledge object

free-form text. We started out by giving our information extraction module clues as to the locations of information-bearing tables, so that, typically, the information extractor had to first work out whether the cells of the table contained object names (for example, *Versa Premium*), attributes (*installed memory*), or values (*256Mb*), and then combine these components to produce the resulting KO. This approach worked reasonably well, largely because of the predictable layout of information within tables. However, it turns out that tables are used for many general-purpose formatting tasks on the web, and so identifying the information-bearing table in the first place is far from trivial.

3 Our Approach

In part to overcome some of the problems in locating the important tabular information, and also to enable the processing of free-form text, we decided to explore techniques that were capable of using information regardless of where in a document it is found.

Our starting point is a definition of a KO, as shown in the previous section; defined in our current system via an XML DTD, this is a hierarchical structure that specifies the nature of the data to be extracted from the source documents.

The simple key to our approach is to recognize that fragments of the text can be correlated with different parts of the KO structure. For example, we know that the manufacturer will be a company name; and we know that the hard disk capacity will be measured in Mb or Gb. Thus, we can assign type information to the leaf nodes in this structure. At the same time, words in the text can serve as indicators of particular attributes: so, for example, if a sentence contains the phrase *removable storage*, we can use this to hypothesise the presence of a particular attribute in the text. We think of each such text fragment as a piece of evidence; faced with evidence from the text of a collection of attributes and values, the goal is then to combine this information to populate a KO.

The architectural model we use thus consists of the following components.

- An **annotation resource file** provides a correlation between between arbitrarily complex textual patterns and the knowledge object constituents for which these patterns provide evidence.
- A **text scanner** processes each input document, searching for the patterns specified in the annotation resource file. Each time a match is found, this generates a hypothesis, in the form of a **path fragment** associated with a piece of text. The consequence of processing a document is thus a collection of path fragments that capture the evidence found in the document.
- The **path combiner** then takes this set of path fragments and attempts to put these together to build a complete knowledge object. Path fragments may contribute to multiple hypotheses about the object being constructed, so the combiner uses the target knowledge object template as a source of constraints on what is possible.
- In most situations, this will not produce a single KO as a result; there will still be ambiguities resulting from textual fragments providing evidence for more than one KO constituent. At this point, we resort to a collection of **inference strategies** to resolve the ambiguities.

Of course, if we had full broad-coverage natural language processing available, then this would achieve the same result: we could just parse the entire text, carry out semantic analysis, and build a representation of the text. However, such an approach is not feasible given the current state of NLP technology, so our aim here is to build on the simpler pattern-matching approaches found in the IE literature, but to augment this with the scope for more sophisticated higher-level processing. The architectural breakdown we adopt stratifies the knowledge used in a way that supports easy maintenance and portability: the annotation resource file is a relatively simple declarative knowledge source developed anew for each domain; the text scanner and path combiner are generic components that do not embody any domain-specific knowledge; and higher-level knowledge of the domain is factored out into the KO template and the inference strategies.

Section 3.1 below explains in more detail the nature and role of paths and path equations; and Section 3.2 shows how path-merging is carried out. Section 3.3 explains how arbitrary inference can be incorporated into this process.

3.1 Paths

We can view a KO as a graph structure (and for present purposes a tree) into which extracted information can be placed. The arcs between nodes in this tree are labelled with the same names as the element tags in the XML DTD; a path is a sequence of arcs in the tree. Each attribute corresponds to path from the root of tree, and each value is a data element that resides at the end of that path. Paths may share common initial subsequences: this means that we use hierarchy in the tree to cluster related pieces of information into subtrees.

We use the notation A:B:C to notate **path fragments**; if a path is from the root of the tree, the path contains an initial ':', as in :A:B:C. If the path has a value at its end, then we use a terminal ':' in the path to indicate this, as in A:B:C:. A **path equation** indicates the value at the end of a path: for example

:laptop:iodevices:keyboard:numkeys: = 131

Each piece of evidence we find in a text can be annotated with a path fragment: this fragment can be of various kinds depending upon how strong the evidence is.

Complete Paths. A path fragment can be **complete**, in which case there is no doubt that a particular value is the value of a particular attribute. So, for example, if we find the string *US keyboard* in the text, we might take this to be conclusive evidence for the following path equation:

:laptop:iodevices:keyboard:type: = US

Initial Path Fragments. A path fragment can be **initial**: this means that we don't have a complete path but we do have some initial sequence of arcs in a path. So, for example, if we find the string *on-board memory*, this might correspond to the following annotation:

:laptop:memory:on-board

We don't know at this point what aspect of the on-board memory is being described; it might be size or speed, for example.

Medial Path Fragments. A path fragment can be **medial**: this means that we don't have a complete path but we do have the some sequence of arcs in the middle of a path. So, for example, if we find a string like *memory capacity (maximum)*, we do not know if this corresponds to main memory, graphics memory, or perhaps some other kind of memory. This corresponds to the following path fragment:

memory:maximum:bytesize

Final Path Fragments. Finally, a path fragment can be **final**. This means we have some sequence of arcs at the end of a path. So, a string like *2Gb* corresponds to the following pair of path fragments:

bytesize:unit: = Gb
bytesize:number: = 2

To operationalise these notions, the annotation resource file correlates arbitrarily complex textual patterns with path fragments. These patterns are then used by the text scanner to generate hypotheses about the information in the text, expressed in terms of the path fragments; the complete analysis of a text thus results in a set of textual clues and their corresponding hypothesised path fragments.

3.2 Path Merging

A KO consists of set of paths, and a fully instantiated KO has a value for each path that makes up the KO. We can characterise an instantiated KO by a set of path equations:

:laptop:model:manufacturer: = Dell
:laptop:model:series: = Inspiron
...
:laptop:iodevices:mouse:type: = optical
:laptop:iodevices:mouse:numbuttons: = 3
...

From a text, we derive a set of path fragments. From the examples in Section 3.1 above, we would have the following set of path fragments:

:laptop:iodevices:keyboard:type: = US
:laptop:memory:on-board
memory:maximum:bytesize
bytesize:unit: = Gb
bytesize:number: = 2

Our goal is therefore to take this collection of path fragments and to derive from them a set of path equations that define an instantiated KO.

Formally there are many combinations of path fragments that we could entertain.[2] A number of cases need to be considered.

First, we do not have to do anything to path fragments which are complete; note, however, that we do not want to just forget about these, since their presence may rule out some other possible combinations (in the example above, if we are tempted to merge two path fragments that would give us a different keyboard type, then we have a good reason for not doing this).

Second, two path fragments may share some arcs: so, for example, in the above, a possible combination results in

memory:maximum:bytesize:unit: = Gb

This is a possible combination but not a necessary one: since arc labels are not unique, it's possible that this particular bytesize:unit fragment does not belong with the memory:maximum:bytesize fragment.

Third, from a formal point of view, any pair of paths can be combined, with the exception that an initial path can only appear at the front of a combined path, and a terminal path can only appear at the end of a combined path. In such cases, where there is no overlap, there is essentially missing material in the middle. We indicate missing material using '...'. So, we might have a combined path that looks something like the following':

:laptop:memory:...:bytesize:unit: = Gb

This is a case where we have some possible evidence for a memory size but we don't know if it is the standard on-board memory, the expanded-to-maximum memory size, or some other aspect of memory size. [3] Of course, not all formally possible combined paths are actually possible. The KO definition provides a way of ruling out impossible path combinations. Our technique for filtering the paths is as follows.

First, we take the set of paths that constitute a KO, called the KO **path set**; this will look something like the following:

:laptop:model:manufacturer:
:laptop:model:series:
...
:laptop:iodevices:mouse:type:
:laptop:iodevices:mouse:numbuttons:
...

[2] The ideas discussed here have been strongly influenced by work in graph-based unification [10], although we do not currently make use of a unification engine in our approach.

[3] Note that the presence of medial paths makes the situation more difficult than it would otherwise be; rather than just pairs of fragments being combined, in principle any number of medial path fragments can be placed between pairs of initial and final paths.

Then, we take the set of path fragments derived from the document. We first separate out the medial paths and the complete paths, leaving the initial paths and final paths. We then produce all possible initial × final combinations, resulting in what we call the IF **path set**. Each element of the IF path set has the form

:A:B:C:...:X:Y:Z:

We then compare each element of the IF path set against the KO path set. Any match of an IF path against the KO path set constitutes an **instantiation**: it provides a possible substitution for the '...' part. Note that any IF path may result in multiple instantiations. If an IF path results in no instantiations, then it is not completable given the KO definition, and can be discarded. The remaining IF paths make up the **filtered** IF **path set**; and each element of this set may correspond to multiple instantiations. We notate instantiations as follows:

:A:B:C:[P:Q:R]:X:Y:Z:

This indicates that P:Q:R is a possible completion derived from the KO path set. In effect, material between square brackets is hypothesized on the basis of top-down knowledge; we have not extracted direct evidence for it from the text.

Next we take the medial paths and see if they support any of these instantiations by matching them against the instantiations. Again, a medial path may support multiple instantiations. Note that a medial path may actually overlap with the initial or final path in an instantiation.

In the abstracted example used here, suppose we have the following medial fragments available: Q, P:Q, P:Q:R, C:P, and R:X. When a medial path matches one of our instantiations, we notate this by moving the square brackets, echoing the idea that the square brackets indicate material for which we have no direct evidence, and each medial path adds some evidence. The effect of this process is shown by means of the **instantiation equations** in Figure 3.

The results here correspond to a **filtered** IMF **path set**: a set of paths built from the initial, medial and final path fragments in the text, and filtered using the KO path set. Note that any given initial, final or medial path fragment may figure in more than one element of the filtered IMF path set, which is to say that it can contribute evidence to more than one instantiation.

:A:B:C:[P:Q:R]:X:Y:Z: + Q = :A:B:C:[P]:Q:[R]:X:Y:Z:
:A:B:C:[P:Q:R]:X:Y:Z: + P:Q = :A:B:C:P:Q:[R]:X:Y:Z:
:A:B:C:[P:Q:R]:X:Y:Z: + P:Q:R = :A:B:C:P:Q:R:X:Y:Z:
:A:B:C:[P:Q:R]:X:Y:Z: + C:P = :A:B:C:P:[Q:R]:X:Y:Z:
:A:B:C:[P:Q:R]:X:Y:Z: + R:X = :A:B:C:[P:Q]:R:X:Y:Z:

Fig. 3. Instantiation Equations

For our present purposes, we make the assumption that each path fragment derived from the text should only actually contribute to one instantiated path.[4] We want to now reduce the set of instantiations in the filtered IMF path set so that we end up with a set where each I, M or F element only plays a role once. Each path in the filtered IMF path set can be thought of as being a combination of its contributing I, M and F fragments; some combinations are more likely than others. We therefore need to assess each combination, and where there is a competing demand for a piece of evidence, determine which of the alternative uses of that evidence is most likely.

3.3 Adding Reasoning

So far we have constructed a set of hypotheses regarding the information contained in a text using a combination of bottom-up knowledge from the textual source itself, and top-down knowledge from the KO. As we have seen above, this does not necessarily result in a completely instantiated KO, and so we may need to add to this process heuristics that allow us to choose between competing hypotheses.

In our present work, we have only begun to explore how this might be approached. Note that the architectural separation we have adopted allows the incorporation at this stage of arbitrary intelligence to the process, thus focussing the knowledge-based processing in one place; here, we describe the incorporation of a simple heuristic based on a distance metric.

Ultimately, where we have competing instantiations, we can assign probabilities to each. One simple probability measure is based on the distance between the initial path fragment (which generally corresponds to the attribute being considered) and the final path fragment (which corresponds to the value being assigned to that attribute): we can assign scores so that those instantiations that have closer I and F evidence will score higher. Then, we select from this set a smaller set of paths that (a) uses each piece of evidence only once and (b) takes account of the scores. The result is a set of hypotheses that populate the KO using all the data located in the source document, with each piece of evidence being used once.

This is, of course, a very simple technique, but one that seems to work well on the basis of our initial experiments, at least for information extracted from free-form text. Far more elaborate heuristics could be incorporated in the same way; a key future target for us here is to incorporate heuristics that take account of table layout information in order to determine the appropriate correlation of table row and column headers.

---

[4] This is an important assumption to make processing more straightforward, but note that it may not actually hold: in an expression like *the main processor and the co-processor both have 512Mb of dedicated RAM*, the *512Mb of dedicated RAM* actually contributes to two instantiated paths.

4 Conclusions

We have presented an approach to information extraction that separates out the different knowledge sources and types of knowledge required. The approach makes use of both top-down and bottom-up knowledge sources, and neatly partitions domain-specific and domain-independent processing: although we have not yet attempted this, we believe the mechanisms described here should be easily portable to a new domain by constructing a knowledge object template for that domain, and an appropriate annotation resource file. The text scanner and path combining modules are domain independent, as are the current inference strategies; as the work develops, we would expect the inference strategies to break down into those which are domain-dependent and those which are domain-independent.

Acknowledgements

We acknowledge the generous support of CSIRO in funding the work presented here, and the helpful comments of numerous colleagues in the Centre for Language Technology.

References

[1] A Borthwick et al. Exploiting diverse knowledge sources via maximum entropy in named entity recognition. In *Proceedings of the Sixth Workshop on Very Large Corpora*, pages 152–160, 1998. 150

[2] D Appelt, J Hobbs, J Bear, D Israel, M Kameyana, and M Tyson. Fastus: a finite-state processor for information extraction from real-world text. 1993. 150

[3] D M Bikel, S Miller, R Schwartz, and R Weischedel. Nymble: a high-performance learning name-finder. In *Proceedings of the Fifth Conference on Applied Natural Language Processing*, pages 194–201. Morgan Kaufmann Publishers, 1997. 151

[4] J. Cowie and W. Lehnert. Information extraction. *Communications of the ACM*, 39(1):80–91, 1996. 150

[5] R Dale, S J Green, M Milosavljevic, C Paris, C Verspoor, and S Williams. Using natural language generation techniques to produce virtual documents. In *Proceedings of the Third Australian Document Computing Symposium (ADCS'98)*, Sydney, Australia, August 21 1998. 152

[6] Defense Advanced Research Projects Agency. *Proceedings of the Sixth Message Understanding Conference (MUC-6)*. Morgan Kaufmann, 1995. 150

[7] P Jackson and I Moulinier. *Natural Language Processing for Online Applications: Text Retrieval, Extraction and Categorization*. John Benjamins, Amsterdam, 2002. 150

[8] Andrei Mikheev, Claire Grover, and Marc Moens. XML tools and architecture for named entity recognition. *Markup Languages*, 1(3):89–113, 1999. 150

[9] E Reiter and R Dale. *Building Natural Language Generation Systems*. Cambridge University Press, 2000. 152

[10] S. Shieber. *An Introduction to Unification-Based Approaches to Grammar*. CSLI Lecture Notes. Chicago University Press, Chicago, 1986. 157

Natural Language Agreement Description for Reversible Grammars

Stefan Diaconescu

SOFTWIN
Str. Fabrica de glucoza, Nr. 5, Sect.2, Bucharest, Romania
sdiaconescu@softwin.ro

Abstract. The paper presents a very general method to describe the agreement in the natural languages. The method can be used in automatic translation. It allows the analysis of a text and then the generation of the text in target language so it can be embeded in the reversible grammars. It allows also in the case of analysis the treatment of the agreement errors. The method is based on a set of simple expressions having logical values. We define a tetravalent logic to work with these expressions and that can be implemented in an automatic treatment. The agreement expressions are transformed in a normal form that can be used in both text analysis and text generation.

1 Introduction

It is already well known the agreement definition given by Steele: "The term agreement commonly refers to some systematic covariance between a semantic or formal property of one element and a formal property of another [20]. But the attitude concerning the agreement is quite contradictory: starting from a lack of interest [13] in the case of the very low inflected languages (like English) until a special interest [9][18][9][10][14] in the case of high inflected languages [16].

The agreement description we will present is a declarative one.The declarative grammars (like different types of constraint grammars [15][12]) are suitable to be used as reversible grammars [17]. The parsing (syntactic analysis) and the generation can be realized using the same reversible grammar [21].

The syntactic analysis allows to determine a deep structure of a surface text. This deep structure contains syntactic elements (terminals that have associated some categories with some values) and relations between these elements. In the course of the analysis, the agreement rules must indicate how the combination of the syntactic categories and values must be associated with the syntactic elements (considering the relations that exist between these syntactic elements).

The generation allows obtaining a surface text using the deep structure. The deep structure can be obtained during a process of machine translation or by the inference process based on a data base or knowledge base. During the generation, the agreement rules must force some values of the syntactic categories associated to some elements in order to agree with others elements.

T.D. Gedeon and L.C.C. Fung (Eds.): AI 2003, LNAI 2903, pp. 161–172, 2003.

There are are three types of using reversible grammars [5]: 1 - the parsing and the generation are done by the same interpreter using one grammar; 2 - the parsing and the generation are done by different interpreters using the same grammar; 3 - the parsing and the generation are done by two different programs that use two different compiling of the same grammar.

We consider that the agrement description we will present (section 2) can be used for all three modes. This agreement description that can be used by human is handled with a special tetravalent logic (section 3) and is transformed in a normal form (section 4) that is better for automatic treatment. The agreement (under its normal form) is used in the syntactic analysis and text generation so it is particularly suitable for reversible grammars (section 5).

2 A Way to Represent the Grammatical Agreement

We will consider that the term agreement will refer not only to the relation between words but also to the relation between syntactic notions (non terminals) that are used in grammar descriptions. According to the above definition (section 1) the agreement is a sort of matching between different elements that appear in a text.

We will consider that two parts are implied in such a matching: a controller C and a target T and we note the agreement by C <- T. In our notation, the arrow look to the controller (and not to the target, like in [10][9]). The controller can be simple C = c (formed by one element) or multiple C = c_1, c_2, c_3 ... (formed by two or many elements). If the controller is a multiple one, we will consider that two consecutive elements will have an *agreement value* that can substitute in the text the two elements from the agreement point of view. We can have for example the following sequence of transformations:

c_1, c_2, c_3 <- T

$c_{1,2}$, c_3 <- T

$c_{1,2,3}$ <- T

We consider the agreement as an asymmetric relation. The target has a subordinate position because its features must agree with the controller. For example, if a subject is a multiple controller, than the predicate must have the plural number. (Observation: this is the reason for us to consider here the predicate as subordinate though usually the subject is considered as a subordinate).

Let us suppose that we have a description of the syntax of a natural language. We are not interested here what is the method used for this description but we suppose that it contains at least three elements: non-terminals, terminals (i. e. elements that belong to the natural language we describe) and rules that indicate how the non-terminals and terminals (we will note NT&T the terminals and non-terminal) are sequenced in the syntax description. Each NT&T have associated some syntactic categories with their values. Among the rules that are used in the syntactic description there are the agreement rules, very important especially for the languages with a strong inflection and quite free word order.

These agreement rules contain the relations that must exist between the syntactic categories of the NT&T. The number of this agreement rules can be very large. We will consider that somewhere in the grammar description there is a section that contains syntactic rules. A syntactic rule contains many NT&T. Each NT&T have associated a label and a list of syntactic categories, each category having one or many values. The NT&T labels will be used by the agreement rules description. We can not give here the complete syntax of the agreement description. We will present only the main features of this syntax and few examples in order to give the flavor of this description. The general form of the agreement description is as follows:

Agreement if ({condition}) true ({expression}) false ({expression}) not applicable ({expression}) undefined ({expression}).

The {condition} is a logical expression that uses the logical operators "and", "or" and negation "~". The logical operands are {simple expressions} that have two forms:

{operand} + {operand} <- {operand}

{operand} <- operand}

The first form is used for multiple controllers and the second for simple controllers. An {operand} is a label of an NT&T and a set of categories with their values. These categories and values can be described using AVT (Attribute Value Trees - see [11]). The {expression} can contain others "if"s or some {action list}. This action list contains error messages and indicates how the analysis will be continued after an error was found.

A condition can have one of four values (in a tetravalent logic as we can see in section 3): TRUE, FALSE, NOTAPPLICABLE, UNDEFINED. Therefore, after "if({condition expression})" a list of many alternatives will follow (and not only two as in the case of "if" in a bivalent logic). If some of the four alternatives are identical, we can use "else".

Example 1: An agreement expression of the form if ($Label_1$ (person = I) -> $Label_2$ (person = I)) true (OK) else (Message = "Person agreement error", OK) will be read: "If the NT&T with the label $Label_1$ from the syntactic description has the person I and the NT&T with the label $Label_2$ from the syntactic description has the person I then continue the analysis, otherwise give the error message "Person agreement error" and continue the analysis as it was not an error.

Example 2: Let us have two non terminals that are found in a grammar rule:

$Label_1$: {$non\text{-}terminal_1$}(a = av1, av2)(b = bv_1, bv_2, bv_3)(c = cv_1, cv_2, cv_3)

$Label_2$: {non $terminal_2$}(d = dv_1, dv_2)(e = ev_1, ev_2, ev_3)(f = fv_1, fv_2, fv_3)

Let us have a simple expression of the form:

$Label_1$(a = av_1, av_2)(b = bv_2,bv_3) -> $Label_2$(e = ev_1, ev_2)(f = fv_2)

During the syntactic analysis, {$non\text{-}terminal_1$} and {$non\text{-}terminal_2$} will have some lexical or syntactic categories with some values. The operand $Label_1$(a = av_1, av_2)(b = bv_2, bv_3) will subsume the non terminal {$non\text{-}terminal_1$} when the {$non\text{-}terminal_1$} will have associated (by the syntactic analysis): the category a with the value av_1 or av_2 and the category b with the value bv_1 or bv$_3$. The operand Label$_2$(e = ev$_1$, ev$_2$)(f = vf$_2$) will subsume the non terminal {non-terminal$_2$} when {non-terminal$_2$} will have associated (by the syntactic analysis): the category e with the value ev$_1$ or ev$_2$ and the category f with the value fv$_2$.

Example 3: We will refer to the general agreement rules (in Romanian language) for the number (sg = singular, pl = plural), gender (m = masculine, f = feminine, n = neuter), (p = animate, l = inanimate) and person (I, II, III) that must be observed by a predicate with verb to the passive voice when we have two subjects (i.e. a multiple subject).

Let us have some non terminals:

Label$_1$: {complex nominal group} [nominal group type = conjunctive] (position against the subject = left) (position against the predicate = left) (animation = p, l) (gender = m, f, n) (number = sg, pl) (person = I, II, III)

Label$_2$: {complex nominal group} [nominal group type = conjunctive, disjunctive] (position against the subject = right) (position against the predicate = left) (animation = p, l) (gender = m, f, n) (number = sg, pl) (person = I, II, III)

Label$_3$: {verbal group} [predicate type = verbal, nominal] [voice = active, reflexive, passive] (gender = m, f, n) (person = I, II, III) (number = sg, pl)

Here there are few rules for persons and numbers:

Agreement if (

/*1,1*/ Label$_1$(animation = p) (gender = m) (number = sg, pl) (person = II) + Label$_2$(animation = p) (gender = m, f) (number = sg, pl) (person = III) <- Label$_3$(gender = m) (number = plural) (person = II) or

/*1,2*/ Label$_1$(animation = p) (gender = m, f) (number = sg, pl) (person = II) + Label$_2$(animation = p) (gender = m, f) (number = sg, pl) (person = III) <- Label$_3$(gender = m) (number = plural) (person = II) or

/*1,3*/ Label$_1$(animation = p) (gender = m) (number = sg, pl) (person = II) + Label$_2$(animation = l) (gender = m, f, n) (number = sg, pl) (person = III) <- Label$_3$(gender = m) (number = plural) (person = II) or

/*2,1*/ Label$_1$(animation = p) (gender = m) (number = sg, pl) (person = III) + Label$_2$(animation = p) (gender = m, f) (number = sg, pl) (person = II) <- Label$_3$(gender = m) (number = plural) (person = II) or

/*2,2*/ Label$_1$(animation = p) (person = III) (gender = m, f) (number = sg, pl) (person = III) + Label$_2$(animation = p) (gender = m, f) (number = sg, pl) (person = II) <- Label$_3$(gender = m) (number = plural) (person = II) or

/*2,4*/ Label$_1$(animation = l) (person = III) (gender = m, f, n) (number = sg, pl) (person = III) + Label$_2$(animation = p) (gender = m) (number = sg, pl) (person = II) <- Label3(gender = m) (number = plural) (person = II) or

/*3,1*/ Label$_1$(animation = p) (gender = m) (number = sg, pl) (person = II) + Label$_2$(animation = p) (gender = m, f) (number = sg, pl) (person = II) <- Label$_3$(gender = m) (number = plural) (person = II) or

/*3,2*/ Label$_1$(animation = p) (gender = m, f) (number = sg, pl) (person = II) + Label$_2$(animation = p) (gender = m) (number = sg, pl) (person = II) <- Label$_3$(gender = m) (number = plural) (person = II)) true (Ok)

/* Continue for others situations*/

Using the above language we made a complete description of the accord in the Romanian language. We found that the Romanian language has 1350 agreement situations between a controller (simple or multiple) and a target. The situation's number can be even larger if we will try to capture also the error situations (that are quite frequent, and now we can see why!). Because it is a formal description, it can be used in automatic treatments.

else (/*..*)

3 A Tetravalent Logic

In order to work with agreement rules we will define a more formal tetravalent logic

a) Truth values. We will use the following four truth values: i - "TRUE" noted with "1"; ii - "FALSE" noted with "0"; iii - "NOTAPPLICABLE" noted with "#"; iv - "UNDEFINED" noted with "*".

The variables that will take values in the this tetravalent logic are simple expressions described above. A simple expression contains operands. We will say that an operand subsumes or do not subsumes an NT&T obtained during the syntactic analysis. By "subsumption" we mean the followings:

Let us have in the agreement description an operand A_1 of the form:

$\text{Label}_1(Y_1 = y_{1,1}, y_{1,2},...) (Y_2 = y_{2,1}, y_{2,2}, ...) (Y_3 = y_{3,1}, y_{3,2}, ...) ...$

Let us have an interpretation A_2 of the NT&T corresponding to the label Label_1 obtained by the syntactic analysis:

$(X_1 = x_{1,1}, x_{1,2},...) (X_2 = x_{2,1}, x_{2,2}, ...) (X_3 = x_{3,1}, x_{3,2}, ...) ...$

The operand A_1 will subsume the NT&T A_2 if:

- A_1 and A_2 have not common lexical/syntactic categories or

- If A_1 and A_2 have some common lexical/syntactic categories (for example $Y_1 = X_1$, $Y_2 = X_2$, $X_3 = Y_3$) and the lexical/syntactic category values taken from A_1 are found among the corresponding lexical/syntactic category values taken from A_2. In the above example (where $Y_1 = X_1$, $Y_2 = X_2$, $Y_3 = Y_3$):

$y_{1,1}, y_{1,2}, ...$ must be found among $x_{1,1}, x_{1,2}, ...$

$y_{2,1}, y_{2,2}, ...$ must be found among $x_{2,1}, x_{2,2}, ...$

$y_{3,1}, y_{3,2}, ...$ must be found among $x_{3,1}, x_{3,2}, ...$

...

We will say that an operand A_1 do not subsume an NT&T if A_1 and the NT&T have at least one common category and, for at least one common category, at least one category value from A_1 are not found among the corresponding category value from NT&T.

A simple expression is TRUE if all its operands have the property "subsumption" of some NT&T.

A simple expression is FALSE if the operands representing the controller subsume the corresponding NT&T and the operands representing the targets do not subsume the corresponding NT&T.

A simple expression is NOTAPPLICABLE if at least one of the operands representing the controller does not subsume the corresponding NT&T.

A simple expression can not be UNDEFINED. An undefined value can appear by logical operations with simple expressions as we will see below. A logical variable or a logical expression is UNDEFINED if we do not know if the corresponding value is TRUE or FALSE.

The subsumption property can be defined in a more complex (and more general) way using the AVT (Attribute Value Tree). In this case, the subsumption property must be replaced by the unification, i.e. the operand subsumes the corresponding NT&T if the operand is unifiable with the corresponding NT&T [11].

b) Basic unary logical operations. Let us have a logical variable x that can have one of the four truth values: 1, 0, #, *. We will define the following basic unary logical operations:

- "true" or "identity" noted by the variable itself x or by ^{1}x;
- "false" or "negation" noted by x or ^{0}x;
- "not applicable" noted by $^\#$x;
- "undefined" noted by *x.

The definitions are done in the Table 1. We can use also parentheses: (x) or x, ˜(x) or ˜x. (Observation: In a binary logic we can define $2^2 = 4$ unary operations. In a tetravalent logic we can define $4^4 = 256$ unary operations.)

c) Basic binary logical operations. Let us have two tetravalent logical variables x and y. We will define the basic binary logical operations "and" (noted by ".") and "or" (noted by "+") according to Table 2. (Observation: In a binary logic we can define $2^{2*2} = 16$ binary operations. In a tetravalent logic we can define $4^{4*4} = 4\ 294\ 967\ 296$ binary operations.)

d) Other properties. We can very easy verify that the usual logical properties are true: De Morgan Relations, the distributiveness of "and" against "or", the distributiveness of "or" against "and", the logical implication, the double logical implication. Using the double logical implication a <-> b = ˜a . ˜b + a . b, the fact that c is true, false, not applicable or undefined can be expressed as follows:

c <-> 1, c <-> 0, c <-> #, c <-> *

In the next section we will see how the agreement expressions can be transformed in tetravalent logical expressions.

Table 1. The unary operations in the tetravalent logic

x	"true" x, ^{1}x	"false" ˜x, ^{0}x	"not app." $^\#$x	"undefined" *x
1	1	0	*	#
0	0	1	#	*
#	#	*	1	0
*	*	#	0	1

Table 2. The definition of "and" and "or" in the tetravalent logic

x	y	x . y	x + y
1	1	1	1
1	0	0	1
1	#	#	1
1	*	*	1
0	1	0	1
0	0	0	0
0	#	0	#
0	*	0	*
#	1	#	1
#	0	0	#
#	#	#	#
#	*	0	1
*	1	*	1
*	0	0	*
*	#	0	1
*	*	*	*

4 Normalizing the Agreement Rules

The agreement rules (section 2) are written manually. We will present now a transformation of these rules to a tetravalent logical form (section 3) that can be used both in the automatic analysis process and automatic generation process. The transformation itself can be done automatically too.

Each rule from an agreement rule list can be represented intuitively as follows:

if ({condition}) true({a new rule beginning with "if" or an action list}) false({a new rule beginning with "if" or an action list}) not applicable({a new rule beginning with "if" or an action list}) undefined({a new rule beginning with "if" or an action list})

We can write more compact:

if (condition) true (a) false (b) not applicable (c) undefined (d)

By definition this expression is equivalent with a logical expression of the form of a conjunction of implications. These implications have the form:

- for true(a): (c <-> 1) -> a
- for false(b): (c <-> 0) -> b
- for not applicable(c): (c <-> #) -> c
- for undefined (d): (c <-> *) -> d

These forms can be also noted:

if(condition <-> 1) then (a)
if(condition <-> 0) then (b)
if(condition <-> #) then (c)
if(condition <-> *) then (d)

One or a group of clauses true(...), false(...), not applicable(...), undefined(...) can appear as being replaced by a single clause:

then({a new rule beginning with "if" or an action list})

To normalize an agreement rule list means to obtain a new agreement list under the form of a conjunction of rules of the form :

if({conjunctive condition})then({action list})

By {conjunctive condition} we mean a logical expression in conjunctive form (a conjunction of disjunctions).

The steps of such a transformation are as follows:

1. Transforming "else if". The form "...else if ({condition expression})..." will be replaced with "...else (if ({condition expression})...)" i.e. we insert an open parenthesis between "else" and "if" and we insert a closed parenthesis at the end of the expression.

2. Transforming "else". The transformation rule results immediately from the following example. Let us have a rule containing an "else":

if ({condition}) false(x_1) not applicable(x_2) else(x_3)

We will replace "else(x_3)" as follows:

if ({condition}) true(x_3) false(x_1) not applicable(x_2) undefined(x_3)

We replaced "else(x_3)" with the missing clauses ("true", "undefined") from the four possible clauses ("true", "false", "not applicable", "undefined"). The new clauses will have between the parentheses the content of the parentheses of "else".

After this transformation, all the expressions will have the same form: an "if" followed by one or many clauses from the four possible and the "else" will disappear".

3. Transforming an "if" that appeared in the parentheses of a "true", "false", "not applicable" or "undefined". After the above transformations there is not a clause "else". All the expression is now a sort of tree that has as nodes: "if({condition})", "true(...)", "false(...)", "not applicable(...)", "undefined(...)", "{action list}". An "if" node has at most four sons (at most one of each "true" , "false", "not applicable" or "undefined" sons) and at least one of the four types "true", "false", "not applicable" and "undefined" sons. Each node of the type "true", "false", "not applicable" and "undefined" has as sons a node of type "if" or a node of type action list. We will bring this tree to the form of a conjunction of expressions of the form:

if({condition}) then ({action list})

We can see that the root of the tree is an "if" node and the leaves are action list nodes. All the paths i that link the root with the leaves are of the form:

node($c_{i,1}$), $xx_{i,1}$ (...), nod($c_{i,2}$), $xx_{i,2}$(...), ...nod($c_{i,j}$), $xx_{i,j}$(...), ... nod($c_{i,ni}$), $xx_{i,ni}$(...), a_i

where $xx_{i,j}$ (...) are nodes of the type "true", "false" or "not applicable", "undefined" and a_i is an "action list" node. We will apply the following procedure:

a) We write all the paths of the above form.

b) For each path:

- We build a condition of the form:
$C_i = (c_{i,1}$ <-> $x_{i,1})$. $(c_{i,2}$ <-> $x_{i,2})$ . ... $(c_{i,j}$ <-> $x_{i,j})$ . ... $(c_{i,n}$ <-> $x_{i,n})$
- We build the expression:
if(C_i) then (a_i)
- Finally we put in a conjunction all the built expression:
if(C_1) then (a_1) . if(C_2) then (a_2) . .... . if(C_i) then (a_i) . .... . if(C_n) then(an)

Example 1: Let us have the expression:
if(c_1) true(if(c_2) true(a_1)) false (if(c_3) true(a_2))
After the transformation we obtain:
if((c_1 <-> 1) . (c_2 <-> 1)) then (a_1) . if((c_1 <-> 0) . (c_3 <-> 1)) then (a_2)

Example 2: Let us have the expression:
if(c_1) true(if(c_2) true(if(c_3) true(a_1)) false (if(c_4) true(a_2))) false (if(c_5) not applicable (a_3) undefined(a_4)) undefined (if(c_6) true (if(c_7) true(if(c_8) true(a_5))) undefined (if (c_9) true (a_6)))

The tree will become:
if((c_1 <-> 1) . (c_2 <-> 1) . (c_3 <-> 1)) then(a_1) . if((c_1 <-> 1) . (c_2 <-> 0) . (c_4 <-> 1)) then(a_2) . if((c_1 <-> 0) . (c_5 <-> #)) then(a_3) . if((c_1 <-> 0) . (c_5 <-> ∗)) then(a_4) . if((c_1 <-> ∗) . (c_6 <-> 1) . (c_7 <-> 1) . (c_8 <-> 1)) then(a_5) . if((c_1 <-> ∗) . (c_6 <-> ∗) . (c_9 <-> 1)) then(a_6)

The validity of these transformations can be demonstrated to the general case but here we will demonstrate only a particular case in the following example.

Example 3: Let us have the expression:
if(c) true (if(c_1) true(a_1)) false (if(c_2) true(a_2)) not applicable (if(c_3) true(a_3)) undefined (if(c_4) true(a_4))

The expression will become:
if((c <-> 1) . (c_1 <-> 1)) then(a_1) . if((c <-> 0) . (c_2 <-> 1)) then(a_2) . if((c <-> #) . (c_3 <-> 1)) then(a_3) . if((c <-> ∗) . (c_4 <-> 1)) then(a_4)

It is very easy to demonstrate that the two expression are equivalent.

4. Normalizing the conditions. The obtained condition expressions are logical expression that contains:

- logical operators ”.” and ”+” (the priority of the operator ”.” is greater than the priority of the operator ”+”);
- the negation ”~” (the priority of ”~” is greater tan the priority of ”.”);
- operands that are simple expressions;
- parentheses that change the priority of the operations;
- the logical constants: 1, 0, #, ∗ .

We do not enter here in the structure of the simple expressions. We will consider that each simple expression is noted by a letter: a, b, c, etc. To normalize the conditions means to put the conditions in the disjunctive form. There different simple algorithms to do this transformation. Such an algorithm is the following:

a) We apply the negation that is found before the parentheses using De Morgan relations until there are no more negations before the parentheses.

b) We open the parenthesis using the distributiveness of ”.” against ”+”.

5. The decomposition of the conjunctive forms. After the above transformations the agreement rules became conjunctions of expressions of the form:

if({condition}) then ({action list})

where {condition} is in the disjunctive form:

if((a1 and a2 ....) or (b1 and b2 and ...) or (c1 and c2 and ...)...) then (action list)

These forms can be rewritten as follows:

if(a1 and a2 and ...) then (action list) and if(b1 and b2 and ...) then (action list) and if(c1 and c2 and ...) then (action list) and ....

This equivalence can be demonstrated for the general case but we give here only a demonstration in a particular case, for illustration. Let us have the expression:

if((a and b) or (c and d))then(action list)

We rewrite the expression using only logical symbols and noting action list by A:

(a . b + c . d) -> A = ˜(a . b + c . d) + A = ˜(a . b) . ˜(c . d) + A = (˜(a . b) + A) . (˜(c . d) + A) = ((a . b) -> A) . ((c . d) -> A))

We pass to the initial notation and we obtain:

if(a and b) then (action list) and if(c and d) then (action list)

This normalized form will simplify not only the agreement check during the syntactic analysis but also especially the generation process where the controllers will force the attributes of the targets.

5 The Agreement Evaluation

As we showed in the section 3, the agreement between different NT&T is described under the form of an agreement rule list:

if({expression$_1$}) then ({action-list$_1$}) and if({expression$_2$}) then ({action-list$_2$}) and if({expression$_3$}) then ({action-list$_3$}) ... if({expression$_n$}) then ({action-list$_n$})

Each {expression$_i$} is a logical expression using as operands simple expressions.

a) During the syntactic analysis, the agreement check will be realized by the "execution" of this agreement rules. The execution consists in the following steps:

1. We evaluate all the simple expressions that appears in the expression i obtaining the corresponding truth values (see the section 3 point (a)).

2. We evaluate the truth value of each expression {expression i} using the above calculated values for the operands.

3. If the truth value of the expression i is "TRUE" (i.e. it is not "FALSE", "NOTAPPLICABLE" or "UNDEFINED") then the actions from the {action-list$_i$} will be executed. An action list can contain the actions OK, KO and error message. Executing an error message means to put this error message in a list of messages that will be eventually showed to the user. The execution of an

OK means in fact no operation. The execution of a KO will mean to note the apparition of an error.

4. When all the n expressions were evaluated:

- If no KO appeared, then the treatment is positively finished.
- If at least a KO is appeared, then treatment is negatively finished.

The syntactic analysis is accordingly continued.

b) During the text generation, the treatment concerns only the simple expressions and consists of:

1. We search a matching between the controller part of the simple expressions and the corresponding elements from the deep structure (this depends on how the deep structure is represented). Let us have the simple expression operand$_1$ -> operand$_2$ + operand$_3$ or operand$_1$ -> operand$_2$. We note {controller} the controller part and {target} the target part of this expressions. We consider that in the deep structure we have the corresponding elements {controller'}and {target'}. To have a matching between {controller} and {controller'} means that {controller} subsumes {controller'} (see the section 4).

2. When we find such a matching, the corresponding {target'} is forced to have such an attribute/value combinations that {target} can subsume {target'}.

We can see that the same representation of the agreement can be used in the automatic treatment of the analysis and of the generation.

6 Conclusions

The method of agreement description we presented permits a compact and systematic representation of natural language reversible grammars. The theoretical basis and the agreement expression handling do not imply excessively complicated problems of implementation in a computing system. The presented elements were embedded in a more general language GRAALAN (Grammar Abstract Language). Using this method and also others features of GRAALAN like the possibility to use macros) a complete description of the agreement in Romanian language was implemented.

References

[1] Baker, M.: Complex Predicates and Agreement in Polysynthetic Languages, in A. Alsina, J. Bresnan, and P. Sells (eds.) Complex Predicates, CSLI, Stanford, (2002), 249-290.

[2] Baker, M.: On Zero Agreement and Polysynthesis, Rutgers University (2002)

[3] Barlow M., Agreement as a Discourse Phenomenon. In: Folia Lingusitica XXXIII/2, (1999) 187-210

[4] Barlow, M. and Ferguson, C.: Agreement in Natural Language: Approaches, Theories, Descriptions, CSLI Publication (1988)

[5] Block, Hans Ulrich: Compiling Trace & Unification Grammar for Parsing and Generation 162

[6] Chung, S. : The Design of Agreement, University of Chicago Press (2000)

[7] Corbett, G..: Morphology and Agreement. In: Zwicky , A. and Spencer, A. (eds) A Handbook of Morphology, Blackwell Oxford, (1998) 191-205
[8] Corbett, G.: Constraints on Agreement. In: Caron, B. (ed.) Proceedings of the 16th International Congress of Linguists. Pergamon. Oxford. CD publication, (1998)
[9] Corbett, G.: Agreement: Terms and boundaries. In: Griffin, W. (ed.): The Role of Agreement in Natural Language Proceedings of the Texas Linguistic Society Conference, Austin, Texas (2001) 161, 162
[10] Corbett, G.: Agreement: Canonical instances and the extent of the phenomenon. In: DeCesaris, J., Booij G., Ralli A. and Scalise, S. (eds.) Proceedings of the Third Mediterranean Morphology Meeting, Barcelona (2001) 161, 162
[11] Diaconescu, S.: Natural Language Understanding using Generative Dependency Grammar, in Proceedings of the twenty-second Annual International Conference of the British Computer Society's Specialist Group on Artificial Intelligence (SGES), (ES2002), Liverpool, UK, (2002) 163, 166
[12] Gertjan van Noord: Reversibility in Natural Language Processing, dissertation, 1993 161
[13] Hudson, R: Subject-verb agreement in English, UCL (1998) 161
[14] Katho, Andreas: Agreement and the Syntx Morphology Interface in HPSG, UC Berkley (1997) 161
[15] Neumann, G. and Gertjan van Noord: Reversiility and Self-Monitoring in Natural Language Generation 161
[16] Osenova, P. : On Subject-Verb Agreement in Bulgarian (An HPSG-based account)In: Proc. of the fourth Formal Description of Slavic Languages Conference, Potsdam, Germany, (2001) 161
[17] Isabelle,Pierre: Towards Reversible MT Systems, in MT Summit II, 1989 161
[18] Pollard, C. J., Sag, I. A.: Head-Driven Phrase Structure Grammar, University of Chicago Press and Stanford CSLI Publications (1994) 161
[19] Reiter, E. and Dale R.: Building Natural Language Systems, Cambridge University Press, (2000)
[20] Steele, Susan: Word order variation: a typological study. In Greenberg, J. H., Ferguson C. A. and Moravsik, E. A. (edc.), Universals of Human Language: IV: Syntax, Stanford University Press, (1978), 585-623 161
[21] Strzalkowski, Tomek (ed): Reversible Grammar in Natural Language Processing, USA Kluwer Academic Publishers, Boston Hardbound, 1993 161

Token Identification Using HMM and PPM Models

Yingying Wen[1,2], Ian H. Witten[2], and Dianhui Wang[3]

[1] School of Computer Science and Software Engineering
Monash University, Clayton, Victoria 3800, Australia
ywen@csse.monash.edu.au
[2] Department of Computer Science
The University of Waikato, Hamilton, New Zealand
{yingying,ihw}@cs.waikato.ac.nz
[3] Department of Computer Science and Computer Engineering
La Trobe University, Victoria 3086, Australia
dhwang@cs.latrobe.edu.au

Abstract. Hidden markov models (HMMs) and prediction by partial matching models (PPM) have been successfully used in language processing tasks including learning-based token identification. Most of the existing systems are domain- and language-dependent. The power of retargetability and applicability of these systems is limited. This paper investigates the effect of the combination of HMMs and PPM on token identification. We implement a system that bridges the two well known methods through words new to the identification model. The system is fully domain- and language-independent. No changes of code are necessary when applying to other domains or languages. The only required input of the system is an annotated corpus. The system has been tested on two corpora and achieved an overall F-measure of 69.02% for TCC, and 76.59% for BIB. Although the performance is not as good as that obtained from a system with language-dependent components, our proposed system has power to deal with large scope of domain- and language-independent problem. Identification of date has the best result, 73% and 92% of correct tokens are identified for two corpora respectively. The system also performs reasonably well on people's name with correct tokens of 68% for TCC, and 76% for BIB.

1 Related Research

Token identification task is to automatically identify the boundaries of a variety of phrases of interest in raw text and mark them up with associated labels. The systems reported in the Message Understanding Conference are limited to the following tokens: person, organization, location, date, time, money and percent. For us, however, the token identification task has no restriction—tokens are defined by a system designer and could encompass any type of information that is of interest.

T.D. Gedeon and L.C.C. Fung (Eds.): AI 2003, LNAI 2903, pp. 173–185, 2003.

Some learning algorithms have been reported such as decision trees, maximum entropy models and hidden markov models.

Sekine [1] and Bennett *et al.* [2] both implemented their token identification systems using decision trees. Their decision trees are based on almost identical features, such as part-of-speech, character type information and special dictionaries. While the two systems are similar, there are significant differences between them. Another system using decision trees is proposed by Baluja *et al.* [3]. Like the systems described by both Sekine and Bennett *et al.*, they utilized a part-of-speech tagger, dictionary lookups, and word-level features, such as all-uppercase, initial-caps, single-character, and punctuation features.

Borthwick *et al.* [4] described a token identification system based on a maximum entropy framework. The system used a variety of knowledge sources, such as orthographic, lexical, section and dictionary features, to make tagging decisions. For any particular class label, there are four states: *label_start*, *label_continue*, *label_end* and *label_unique*. The first three states are for the case that more than one consecutive words are identified as the same class. The fourth is for the case that only one word is identified in a particular class. In addition, there is a special label—*other*, which indicates that the word is not part of a class. For example, the phrase "Jenny Bentley lives in Hamilton" is marked as "person_start, person_end, other, other, location_unique". One label is assigned to every word in the text. This approach is essentially the same as that described by Sekine [1]. Borthwick *et al.* employed Viterbi's [5] search algorithm to find the highest probability legal path. For example, label_end can only be assigned to a word that follows a word with either label_start or label_continue. The system is a purely statistical one, and contains no hand-generated patterns.

Another system for token identification that uses a maximum entropy model is reported by Mikheev *et al.* [6]. The model uses contextual features of tokens, for example the position of tokens in a sentence, whether they appear in lowercase in general, whether they were used in lowercase somewhere else in the same document and so on. This system makes decisions using the answers provided by the Maximum Entropy model.

IdentiFinder [7] is a well-known system. It uses a variant of a hidden Markov model to identify tokens like names, dates and numerical quantities. Each state of the HMM corresponds to a token class. There is a conditional state for "not a token class". Each individual word is assumed to be either part of some predetermined class or not part of any class. According to the definition of the task, one of the class labels or the label that represent "none of the classes" is assigned to every word. IdentiFinder uses word features, which are language-dependent, such as capitalization, numeric symbols and special characters, because they give good evidence for identifying tokens.

This paper considers of the effect of the combination of HMMs and PPM on token identification. The system bridges the two well known methods through words new to the identification model. The main characteristics of the proposed system is that it is fully domain- and language-independent.

Two corpora, The Computists' Weekly—formerly known as The Computists' Communique (TCC).[1] and The Collection of Computer Science Bibliographies (BIB).[2], are employed to evaluate the techniques presented in this paper.

The rest of the paper is organized as follows. The next section describes the algorithms of the models. Section 3 demonstrates how the models are used in token identification. Section 4 evaluates our proposed system. We conclude the paper in the last section.

2 HMMs and PPM

The idea of the system is to identify tokens from word-level to character-level when encounter unknown tokens. It is achieved by bridging HMM and PPM models. Due to limitation of space, we briefly describe the algorithms of the models in this section. Please refer to [8] and [9, 10] for more details.

2.1 HMMs

A hidden Markov model is a finite-state automaton with stochastic state transitions and symbol emissions [8]. It is a particular model based on a sequence of events, and consists of a set of states and a set of output symbols. The automaton generates a sequence of symbols by starting from the initial state, transitioning to a new state, emitting an output symbol, transitioning to another state, emitting another symbol, and so on, until the final state is reached and the last symbol is emitted.

For each member of the set of states, $S = \{S_1, S_2, ..., S_N\}$, there are two probability distributions. One governs the outgoing state transitions, which indicates how likely another state is to follow; the other governs the emission of symbols in the observation vocabulary $V = \{V_1, V_2, ..., V_M\}$, which indicates how likely a symbol is to be generated in the particular state. N and M are the number of states and number of symbols respectively.

We assume that time is discrete, and the model transitions between states at each time unit. In the case of a first-order Markov model, which is used in the undertaken research, the probability of moving from state S_i to state S_j is stored in the *state transition matrix*, $A = \{a_{ij}\}$, where:

$$a_{ij} = \texttt{Pr}[q_t = S_j | q_{t-1} = S_i], \qquad 1 \leq i, j \leq N. \qquad (1)$$

In this and future equations, t refers to the time instant, q_t is the variable that records the state assignment to the tth symbol, and S_j, the jth member of the set of possible states, is the assigned value. In other words, the probability of being in the current state is determined by the previous state.

When the HMM moves between states, it emits an output symbol after each transition. Exactly which output symbol is emitted depends on the *output symbol*

[1] http://www.computists.com

[2] http://liinwww.ira.uka.de/bibliography/index.html

distribution B, which defines the probability of emitting a particular symbol in a particular state. For first-order HMM, B is a two dimensional matrix defined as $B = \{b_j(k)\}$, where:

$$b_j(k) = \mathtt{Pr}[o_t = V_k | q_t = S_j], \qquad 1 \le j \le N, \quad 1 \le k \le M. \tag{2}$$

Here, o_t is the variable that records the tth symbol emission, and V_k, the kth member of the observation vocabulary, is the emitted symbol.

To complete the model we need an initial probability distribution $\pi = \{\pi_i\}$ over states, where:

$$\pi_i = \mathtt{Pr}[q_1 = S_i], \qquad 1 \le i \le N. \tag{3}$$

Let us assume that an HMM model has been constructed for a particular kind of sequence, and we are presented with a new example of such a sequence, $O = o_1, o_2, ..., o_T$. The problem of finding the most likely state sequence $Q = q_1, q_2, ...,$
q_T that produces the given symbol sequence is called *decoding*. There are several possible ways of solving this problem. We have used *Viterbi* algorithm [5, 11], which is to recover the state sequence that has the highest probability of having produced the given observation sequence.

2.2 PPM Models

Models that take a few immediately preceding symbols into account to make a prediction are called *finite-context* models of order m, where m is the number of preceding symbols used [9]. The PPM technique uses finite-context models of characters [10]. It is a so-called character-level model. It uses the last few characters in the input string to predict the upcoming one. By considering such a context, each character can be better predicted. The prediction is done by using the counts of occurrences of each context. The probabilities associated with each character that has followed the context are used to predict the probability for the upcoming character.

PPM uses fixed-order context models with different values of m, up to a pre-determined maximum. The maximum number is a given constant, which is call the *order* of the model. The bigger the order, the more information is considered. But increasing the order does not guarantee better compression, because the contexts become rarer as the order grows.

Several orders are blended together in PPM to obtain a good probability estimate for the current character. The prediction starts with a given maximum order m and checks the occurrence of the order m context. If the order m context has occurred with the upcoming character following it, the corresponding counts are used to predict the probability. If the context has not been seen in the past, the model then uses the order $m - 1$ context.

Consider the case where the context has occurred but never followed by the upcoming character. This is called the *zero-frequency* situation [12]—the character will be predicted using a zero count. In this case, a special transmission

Table 1. PPM after processing the string *tobeornottobe*

Order 2				
Prediction			c	p
be	→	o	1	1/2
	→	*esc*	1	1/2
eo	→	r	1	1/2
	→	*esc*	1	1/2
no	→	t	1	1/2
	→	*esc*	1	1/2
ob	→	e	2	3/4
	→	*esc*	1	1/4
or	→	n	1	1/2
	→	*esc*	1	1/2
ot	→	t	1	1/2
	→	*esc*	1	1/2
rn	→	o	1	1/2
	→	*esc*	1	1/2
to	→	b	2	3/4
	→	*esc*	1	1/4
tt	→	o	1	1/2
	→	*esc*	1	1/2

Order 1				
Prediction			c	p
b	→	e	2	3/4
	→	*esc*	1	1/4
e	→	o	1	1/2
	→	*esc*	1	1/2
n	→	o	1	1/2
	→	*esc*	1	1/2
o	→	b	2	3/8
	→	r	1	1/8
	→	t	1	1/8
	→	*esc*	3	3/8
r	→	n	1	1/2
	→	*esc*	1	1/2
t	→	o	2	1/2
	→	t	1	1/6
	→	*esc*	2	1/3

Order 0			
Prediction		c	p
→	b	2	3/26
→	e	2	3/26
→	n	1	1/26
→	o	4	7/26
→	r	1	1/26
→	t	3	5/26
→	*esc*	6	3/13

Order −1			
→	A	1	1/|A|

called *escape* is used to drop the model down one order, and the order $m-1$ model is used to make the prediction.

Another possible situation is that the character has never occurred in the past—an unknown character. Then even order 0 cannot be used. This is another instance of the zero-frequency problem. The model then escapes down to a bottom-level model, order −1, that predicts all characters equally.

To illustrate the PPM modeling technique, Table 1[13] shows the four models with order 2, 1, 0 and −1 after the string *tobeornottobe* has been processed.

In Table 1, c represents the occurrence, *esc* and p are probabilities for escape and symbol, respectively. They are determined by the following equations. $|A|$ is the size of the alphabet.

$$esc = \frac{t}{2n}, \tag{4}$$

$$p = \frac{2c-1}{2n}, \tag{5}$$

where t is the distinct number of characters that have followed a particular context, n is the number of times a context has appeared.

The model in Table 1 is used as follows. Suppose the character following *tobeornottobe* is *o*. Since the order-2 context is *be*, and the upcoming symbol has already been seen once in this context, the order-2 model is used and the probability is 1/2. If the next character, instead of *o*, were *t*, this has not been seen in the current order. Consequently an order-2 escape probability of 1/2 is used and the context is truncated to the order-1 context *e*. Again it has not been seen in this context, so an order-1 escape probability of 1/2 is used and the context is truncated once more to the null context, corresponding to order 0. Finally the character t is predicted with a probability of 5/26. Thus the prediction of t is done in three steps, using order 2 to order 0 context respectively, with

Table 2. Effect of context and current character with order-2 PPM

		Context			
		tobeornottobe		*nottobeorto*	
		Probability	Model used	Probability	Model used
Upcoming character	*o*	½	Order 2	½×½×5/6	Order 0
	t	½×½×5/26	Order 0	½×1/6	Order 1

a probability of $1/2 \times 1/2 \times 5/26$. If the upcoming character had been x instead of t, a final level of escape to order -1 would have occurred with a probability of $3/13$, and x would be predicted with a probability of $1/256$ (assuming that the alphabet $|\mathrm{A}| = 256$).

The probabilities predicted by PPM are based on the occurrences of the prior context and the characters that have followed each context every time the context has occurred in the training text. Table 2 shows how the previous context being processed and current character affect the result in terms of the order of model and probabilities by using the same prior contexts to predict different characters and vice versa.

3 Token Identification

3.1 HMM Based Approach

For the token identification application, the observation sequence is a sequence of words in text. The symbols emitted in each state are words, and the HMM is a word-level model.

In the system, each sequence corresponds to a sentence in text, and each state corresponds to a type of token that the program will identify and mark up. Example token class include people's names, geographical locations, monetary amounts and e-mail addresses. Each type of token will be marked in the text by a unique tag. N, the number of states in the model, is the number of different token classes, and is determined by the training data. Because the system uses a word-level HMM model, M, the size of the output vocabulary, is the number of different words that appear in the training data.

The matrix A in HMM gives the probability that the current word belongs to a particular token type given that the previous word belongs to a particular token type as well. We also call it the *contextual probability*. Distribution B is the probability of the same words being seen in a particular token class. It is token-dependent: different token classes have different probabilities for a certain word. B is also called the *lexical probability*. The initial distribution π is the probability that each type of token starts a sentence.

For instance, in the following sentence, a fragment in annotated version of The Computists' Weekly, 1998,

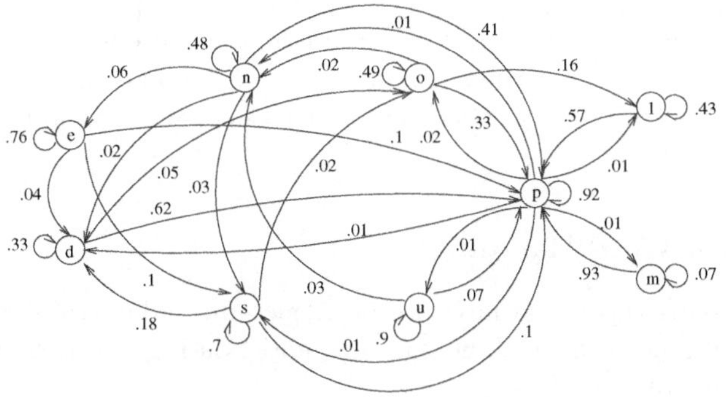

Fig. 1. A HMM obtained from the training data

<o>Polytechnic University</o> in <l>Brooklyn</l> will get <m>$190M</m> from the <n>Othmer</n> estate, about four times the school's previous endowment.

The sequence of four words "Polytechnic", "University", "in" and "Brooklyn" contributes to the four-element class sequence <o><o><p><l>. It contributes the probabilities of transitioning from organization (o) to organization, state o to o; organization to plain text (p), o to p; and plain text to location (l), p to l. Thus probabilities are given by the elements of matrix A. The words themselves will be counted as the appearances in the corresponding token class to make up the elements of matrix B, for example, words "Polytechnic" and "University" labeled as organization would increase the counts for their occurrences in this class. "Polytechnic" as part of organization would also increase the probability of token <o> starting a sentence.

The model is trained on the training set of the corpus. The system processes the training data in two passes. The first pass counts the number of token classes, N, the number of different words, M, and the vocabularies for each token type. The second pass counts the number of events and calculates the A and B matrices.

For illustration, Figure 1 is an example of an HMM obtained from part of the training data. It is annotated with transition probabilities and token labels: email address (e), dates (d), people's name (n), sources (s), organizations (o), URLs (u), locations (l), money (m) and plain text (p). The figure shows that it is possible for words in the plain text (p) class to follow words in any other token classes, and these words can also be followed by words in any other class except email (e). It is reasonable that words in all token classes can be surrounded by plain text. The location class (l), monetary class (m) and URL (u) class have no direct relationship between each other. They never appear one after another and always have tokens in other classes between them. This is understandable from the grammatical point of view. Probabilities from plain text to some other token class, such as date, source and location, are very low. This is not because the events are rare but because they are overwhelmed by plain text words, which makes the denominator bigger and results in smaller numbers.

The figure indicates that there are no tokens in fax and phone classes in this particular set of training data, because classes, along with vocabularies, depend on the training data.

3.2 Unknown Word Handling

Unknown words are those that have not been encountered in training data. There are two kinds of unknown word: neologisms, and existing words that happen not to appear in the training data.

One of the main goals of token identification is to choose the correct label in cases where a word can have more than one label assignment. Additionally, a system must deal with words that have not been encountered in the training data, and so are not found in the lexicon.

The lexicon for the HMM is built during training, so the model contains all words. If an unknown word is encountered during decoding, there is no entry in the model. The emission probability in the state transition matrix B is unknown. To ensure that the process continues and works in a proper way, some policy must be adopted to estimate the probability that the current state will emit the given unknown word.

A PPM model is then constructed for each token class, using all tokens in a class in the training data. Whenever a word that has not been encountered in training is seen, and is therefore not contained in the lexicon, the value of $b_j(k)$ is assigned the probability that is predicted by an appropriate PPM model. The more words in a class that occur in the training data, the more likely it is that tokens in the same class can be identified in new text.

4 Performance Evaluation

4.1 Measurement Metric

Three standard measures, *recall*, *precision* and *F-measure* [14, 15], along with *error-rate* are used to evaluate the accuracy of the token identification system. They are calculated by using the corresponding manually marked-up fragment in the training corpus as the gold standard. For easy reference, let us call this gold standard *hand mark-up*. To define them, the following terms are used:

N	Number of tokens occurring in the standard text;
c	Number of tokens correctly marked up by the system;
e	Number of tokens incorrectly marked up by the system;
$n = c + e$	Number of tokens marked up by the system.

The measures take into account two aspects of the mark-up: the label itself, and the boundary where the label is inserted. A token is considered to be correctly marked up when both label and boundaries are correct. For example

The board has been begging and bribing <n>Steve Jobs</n> to stay on, but he hasn't accepted yet.

"Steve Jobs" is correctly marked as a person's name and it contributes one count to c.

Recall and precision are widely used to assess the quality of an information retrieval system in terms of how many of the relevant documents are retrieved (recall) and how many of the retrieved documents are relevant (precision). In the token identification task, recall is the proportion of the correct tokens which are actually identified by the system, while precision is the proportion of tokens identified by the system which are correct. They are written as:

$$\text{Recall} = \frac{c}{N}, \tag{6}$$

$$\text{Precision} = \frac{c}{n}. \tag{7}$$

The two measures do not always provide an adequate evaluation because there are some extreme situations where one of them is very small while the other is large. For example, if the system identifies few tokens compared to the number of N and they are all correct, recall will be very small whereas precision is 100%. It is better to have a measure that yields a high score only when recall and precision are balanced. A widely used measure is the F-measure [14, 15]:

$$\text{F-measure} = \frac{(\beta^2 + 1) \times \text{Precision} \times \text{Recall}}{\beta^2 \times \text{Precision} + \text{Recall}} \tag{8}$$

where values of β between 0 and ∞ give varying weights to recall and precision. In this paper, $\beta = 1$, gives equal importance to recall and precision, therefore, F-measure is the harmonic mean of recall and precision:

$$\text{F-measure} = \frac{2 \times \text{Precision} \times \text{Recall}}{\text{Precision} + \text{Recall}}. \tag{9}$$

The measure of error-rate is used just for easy analysis of the result. It is defined as

$$\text{Error-rate} = \frac{e}{N}. \tag{10}$$

This is normally used on its own as an overall indicator of the quality of identification. However, it can give misleading results—an extreme condition is where the system only identifies a single word, leading to a very small error-rate of $1/N$ despite the fact that all tokens but one remain unidentified.

If the system marks up the same number of tokens as the hand mark-up, recall and precision both become equal to one minus the error-rate. A perfect identification system will have an error-rate of zero, and recall and precision of 100%.

Table 3. Result for five TCC test files

File	Recall (%)	Precision (%)	F-measure (%)	Error-rate (%)
test1	64.86	72.37	68.41	24 76
test2	65.67	74.58	69.84	22.39
test3	68.72	78.34	73.21	18.99
test4	60.96	74.79	67.17	20.55
test5	66.26	66.67	66.46	33.13
Average	65.29	73.35	69.02	23.96

4.2 TCC Corpus

The first series of evaluations used the TCC corpus. The corpus was first split into sentences because the system processes input sentence by sentence. The available TCC corpus contains 38 issues of the newsletter. 25 issues, which are randomly selected, are used as the training set.

To see how the system works, 5 of the remaining issues are used as the test set with all the tags removed. PPM models are used for the unknown words.

Table 3 shows the results for each of the five test files in terms of recall, precision, F-measure and error-rate. It also gives the average values of these measures.

In the table, the value of precision is always higher than that of recall. This indicates that the system marks less tokens than it should do. Consequently, some tokens are either missed out—left unmarked, or merged into another token by the system.

Figure 2 depicts the proportion of correctly marked tokens (dark gray) and errors (light gray) over the corresponding numbers of tokens in the hand mark-up for each class and overall.

The figure shows that organization class has a very poor result, with only about 20% of correct tokens identified. This is probably due to the lower per-

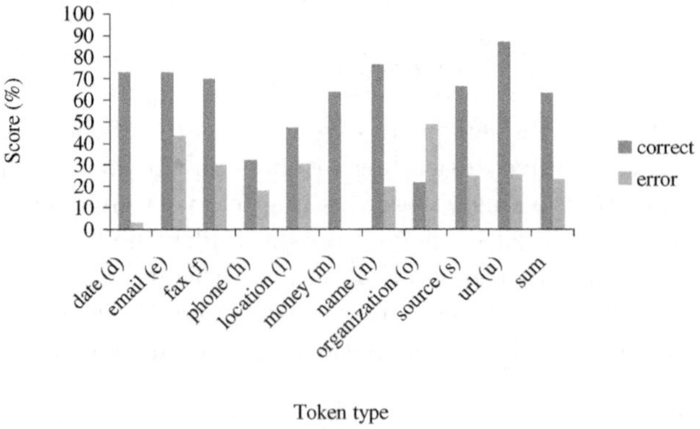

Fig. 2. Detailed result for five TCC test files

centage of words in this class in the training data. As mentioned before, the more words in a class that occur in the training data, the more likely it is that tokens in the same class can be identified in new text. The proportion of words in classes such as fax number and sums of money are also small, however, they have special so-called indicators, for example "fax" and "$". This increases the performance for these classes. However, a name of an organization is more likely to comprise diverse words, such as:

<o>Bell Labs</o>
<o>John Benjamins</o>
<o>Holland Academic Graphics</o>

The percentage of correct tokens in date, email and fax classes are about the same. However, the system is more successful on date class because the error rate is very low. This is understandable because date has much less ambiguity than email address and fax number. Most tokens in date class are distinguishable in both words and format. Errors happen in special cases such as "30-<d>40 years</d>" and word stemming. But they are rare. On the other hand, email addresses can be mixed up with URLs due to the definition of a word. Some sources such as "<s>comp.ai.genetic</s>" also increase the error in email address and URL classes. There is no distinction between fax number and phone number except the word "fax/Fax", which is a strong indicator to distinguish fax number from any other classes.

It is interesting to notice that a token which is a money amount is either marked up correctly or left unmarked. Due to the format of the token in this particular class, there is no marking ambiguities.

Generally speaking, name of a person is relatively harder to identify than other classes. The figure shows that it gets the second best result for both the correct ratio and error ratio. As we have noticed, there are some names that occur repeatedly. For example, "Ken", "Bill Park" and "Brandon" have occurred several times in the entire corpus and none of them has been found to be identified incorrectly. However, most of the names are not in this special case.

4.3 Bibliographies

The text in bibliographies is more structured than that in the TCC newsletter, which the model takes advantage of. A bibliography entry contains name, title and date. Most of them have page number(s), and some provide organization, publisher, location and source. In the experiments, the BIB corpus of 2400 references were selected randomly from the bibliography collection. The model is trained on 2200 entries, and the experiments are done by running the system on 100 of the remaining references with labels removed.

Figure 3 shows the proportion of correctly marked tokens (dark gray) and errors (light gray). As we can see, tokens in date class have the best identification result. From the figure, not a single token is marked up as an organization whether correct or not. It is because that there are only 4 organizations in the

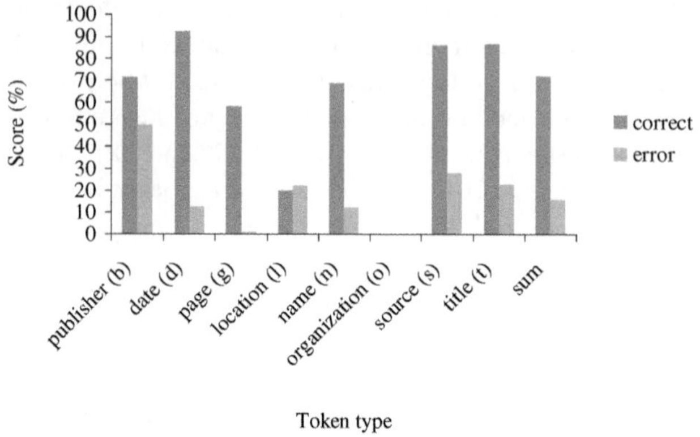

Fig. 3. Detailed result for 100 bibliographic entries

test data. 3 of them are marked up incorrectly in the hand mark-up and one is left unmarked.

Table 4 shows the average value of recall, precision, F-measure and error-rate for 100 bibliographic entries. It also shows the average result of 5 TCC issues for comparison. Overall, the system was successfully applied on the new text. The figure indicates that the performance is 7.57% better in F-measure than that of TCC.

5 Conclusion

The token identification system described in this paper combines two well-known language models through unknown words for developing a domain- and language-independent system. The system has been tested on two data sets without any changes in the code. The results show that 69.02% of F-measure for TCC and 76.59% for BIB are achieved. Although the performance is not as good as that obtained from a system which includes language-dependent components, our system has power to deal with large scope of domain- and language-independent problem. The performance evaluation is made by accuracy, however, the scope of the system's applicability should be taken into account. There is always a trade-off between this two issues. The more specific the application, the higher accuracy a system provides. This research emphasizes retargetability and generality. It is understandable that the system is degraded because of the lack of language

Table 4. Average result for 100 bibliographic entries and 5 TCC issues

Corpus	**Recall** (%)	**Precision** (%)	**F-measure** (%)	**Error-rate** (%)
BIB	72.02	81.78	76.59	16.05
TCC	65.29	73.35	69.02	23.96

dependence. Experiment results demonstrate that the presented techniques in this paper perform reasonably well on date and people's name for TCC and BIB.

References

[1] Sekine, S.: NYU: Description of the Japanese NE system used for MET-2. In: Proceedings of MUC-7 1998. (1998) 174

[2] Bennett, S.W., Aone, C., Lovell, C.: Learning to tag multilingual texts through observation. In: Proceedings of the Second Conference on Empirical Methods in Natural Language Processing, Providence, Rhode Island (1997) 109–116 174

[3] Baluja, S., Mittal, V.O., Sukthankar, R.: Applying machine learning for high performance named-entity extraction. In: Proceedings of the Conference of the Pacific Association for Computational Linguistics, Waterloo, CA (1999) 365–378 174

[4] Borthwick, A., Sterling, J., Agichtein, E., Grishman, R.: Exploiting diverse knowledge sources via maximum entropy in named entity recognition. In: Proceedings of the Sixth Workshop on Very Large Corpora, Montreal, Canada (1998) 174

[5] Viterbi, A.J.: Error bounds for convolutional codes and an asymptotically optimum decoding algorithm. IEEE Transactions on Information Theroy **IT-13** (1967) 260–269 174, 176

[6] Mikheev, A., Moens, M., Grover, C.: Named entity recognition without gazetteers. In: Proceedings of EACL, Bergen, Norway. (1999) 174

[7] Bikel, D.M., Schwartz, R., Weischedel, R.M.: An algorithm that learns what's in a name. Machine Learning Journal **34** (1999) 211–231 174

[8] Rabiner, L.R.: A tutorial on hidden Markov models and selected applications in speech recognition. Proceedings of the IEEE **77** (1989) 257–286 175

[9] Witten, I.H., Moffat, A., Bell, T.C.: Managing Gigabytes-Compressing and Indexing Documents and Images. 2 edn. ISBN 1-55860-570-3. Morgan Kaufmann Publishers, San Francisco, California (1999) 175, 176

[10] Cleary, J.G., Witten, I.H.: Data compression using adaptive coding and partial string matching. IEEE Trans on Communications **32** (1984) 396–402 175, 176

[11] Forney, J.G.D.: The Viterbi algorithm. Proceedings of the IEEE **61** (1973) 268–278 176

[12] Witten, I.H., Bell, T.C.: The zero-frequency problem: Estimating the probabilities of novel events in adaptive text compression. IEEE Transactions on Information Theory **37** (1991) 1085–1093 176

[13] Teahan, W.J., Wen, Y., McNab, R., Witten, I.H.: A compression-based algorithm for Chinese word segmentation. Computational Linguistics **26** (2000) 375–393 177

[14] Van Rijsbergen, C.J.: Information Retrieval. Second edn. Butterworths, London (1979) 180, 181

[15] Lewis, D.D.: Evaluating and optimizing autonomous text classification systems. In: Proceedings of the Eighteenth Annual International ACM Special Interest Group on Information Retrieval. (1995) 246– 254 180, 181

Korean Compound Noun Term Analysis Based on a Chart Parsing Technique

Kyongho Min[1], William H. Wilson[2], and Yoo-Jin Moon[3]

[1] School of Computer and Information Sciences
Auckland University of Technology
Private Bag 92006, Auckland, 1020 New Zealand
kyongho.min@aut.ac.nz
[2]School of Computer Science and Engineering, UNSW
Sydney, 2052 Australia
billw@cse.unsw.edu.au
[3]Department of MIS, Hankook University of Foreign Studies
Yongin, Kyonggi, 449-791 Korea
yjmoon@hufs.ac.kr

Abstract. Unlike compound noun terms in English and French, where words are separated by white space, Korean compound noun terms are not separated by white space. In addition, some compound noun terms in the real world result from a spacing error. Thus the analysis of compound noun terms is a difficult task in Korean NLP. Systems based on probabilistic and statistical information extracted from a corpus have shown good performance on Korean compound noun analysis. However, if the domain of the actual system is expanded beyond that of the training system, then the performance on the compound noun analysis would not be consistent. In this paper, we will describe the analysis of Korean compound noun terms based on a longest substring algorithm and an agenda-based chart parsing technique, with a simple heuristic method to resolve the analyses' ambiguities. The system successfully analysed 95.6% of the testing data (6024 compound noun terms) which ranged from 2 to 11 syllables. The average ambiguities ranged from 1 to 33 for each compound noun term.

1 Introduction

Search engines can be improved by two search-related methods: effective indexing schemes for document contents and efficient ranking systems to pinpoint precise and necessary information for users rather than showing massive amounts of retrieved documents. However, the processing of compound noun terms influences the performance of information retrieval systems.

Unlike compound noun terms in English and French, where words are separated by white space, Korean compound noun terms are not separated by white space. In addi-

T.D. Gedeon and L.C.C. Fung (Eds.): AI 2003, LNAI 2903, pp. 186-195, 2003.

addition, some compound noun terms in the real world result from a spacing error. Thus the recognition of a compound noun term as a key word in IR, or as a class feature in document classification, has been studied to test the performance of systems in English or French [1]. However, analysis of monolithic compound noun terms in Korean has been performed to test its effect on IR systems, machine translation systems, and document classification systems [7], [9], [12]. For example, the two interpretations of "black coffee mug" would be $_{np}$($_{adj}$ black ($_{noun}$ coffee ($_{noun}$ mug))) – ("검은" ("black") • "커피 잔" ("coffee mug")) - or $_{np}$(($_{adj}$ "black" ($_{noun}$ "coffee")) ($_{noun}$ "mug"))) – "블랙커피•잔" (("black coffee") • "mug"). These different results greatly affect the performance of NLP-based application systems.

Past research has studied Korean compound noun terms based on a statistical approach using min-max frequencies of terms from a corpus [11], a N-gram approach using a corpus [8], a probabilistic approach using frequencies of terms from a corpus and their similarities to the corpus content [10], a combination of multiple approaches such as rule-based analysis, statistical information, and heuristic rules [12], and a statistical approach based on mutual information extracted from a corpus [7]. Most systems used a tagged corpus to extract compound noun formation rules or statistical information to analyse testing data on compound noun terms. The use of a corpus requires a large amount of reliable data to get effective compound noun analysis. In addition, the domain coverage of the corpus used for training the system and extracting rules will greatly affect the performance of the analysis system. Thus the performance of the compound noun analysis would not be consistent, if the domain of the actual system is expanded beyond that of the training system.

However, compound noun term analysis based on linguistic information and syntactic rules can give consistent performance regardless of different domains of the terms. This method does not depend on the size of data used for extracting rules and statistical information but greatly depends on the size of the lexicon used [4]. Thus, in this paper, a compound noun analysis system based on linguistic information and a syntactic rule (NP → N/PN+{N}) is implemented (PN signifies Proper Noun). Some systems employ heuristic methods to reduce the ambiguities of a compound noun term, such as two/three noun terms for a compound noun term [11], or default patterns [12]. However, in this paper, all possible candidate analyses were produced and then the best analysis is chosen using a heuristic method.

The system in this paper employs a longest substring algorithm to search for all legal noun terms in a compound noun term and a left-to-right chart parsing technique to interpret the syntactic structure of the compound noun term. The compound noun analysis system produces ambiguous syntactic interpretations that require resolution of structural ambiguities. A heuristic and simple ranking system is applied to resolve the structural ambiguities. The next section will describe the method of compound noun term analysis, and the experimental results will be described in section 3. Section 4 will describe some problems with the approaches that this paper has taken, further improvement, and the conclusion of this paper.

2 Compound Noun Term Analysis

This section describes a method of analysis of Korean compound noun terms in three phases: morphological analysis based on HAM (Hangul Analysis Module [5]) system, analysis of compound noun terms, and the choice of the best analysis among alternatives.

2.1 Morphological Analysis of Words

Korean is a highly inflected and free-ordered language, requiring very complex computational processing for successful Korean NLP. In addition, Korean morphological analysis is very difficult; first, a single syllable (a morpheme/word) in a word can have different meaning when used by itself. For example, "bank" – a financial institution - is"은행- */eun-hang/*" in Korean with two syllables ("은- */eun/*", "행-*/hang/*"). If the word is treated as a compound noun by a morphological analysis system, then it would be a "silver row" in English ("silver" for"은- */eun/*" and "row" for"행- */hang/*"). Second, Korean has no capital letters to enable one to distinguish a proper noun from other nouns (common, mass, abstract, etc.) Third, Korean is highly inflectional and derivational. In the case of verb inflections, there are different forms of inflections mixed together (imperative, question, past, passive, etc.)

For the purpose of efficient and effective Korean morphological analysis, HAM [5] is employed. HAM reads a Korean text file and returns multiple morphological analyses of each word in the text. The major shortcoming of the system is the wrong analysis of Korean expressions for foreign words. If the Korean expression of a foreign word includes some case markers (i.e. subjective ("은", "는"),objective ("을", "를")), affixes, or inflectional morphemes, then the foreign term is analysed as a noun with a case marker, because the dictionary used cannot cover all possible foreign terms. Thus Korean morphological analysis depends greatly on the size of the lexicon the system uses.

HAM analyses each Korean word unit (e.g. a word or a phrase in English) into a stem (noun, verb, adjective) and its inflectional and derivational unit. From the results of HAM analysis, if the stem is identified as a compound noun term, then the compound noun analysis module is invoked to analyse the term to collect class features for the Korean document classification system [9]. If its syntactic category is noun from HAM analysis, and the stem is not found in the lexicon (which is used, not in the HAM system, but in a compound noun term analysis system), then the stem is identified as a compound noun term.

However, in real world text, the term may be the Korean representation of a foreign word, a proper noun (i.e. name), or a relatively unknown word that is not in the lexicon, if its size is small. Thus Korean compound noun analysis would greatly depend on the size of the lexicon. Unsurprisingly, as the size of the lexicon increases, the ambiguities of compound noun term analysis, and spurious analyses would also increase [4]. In this paper, a supplementary user lexicon is added to cover words such as foreign words ('소핑몰" – "shopping mall"), names ("현대" – "Hyundai"), newly-coined words, and derivational nouns, etc. The supplementary user lexicon has 7750 entries that are noun terms (3375 entries) and proper noun terms (4375 entries).

2.2 Compound Noun Analysis

The Korean morphological analyser HAM focuses on the analysis of inflectional and derivational morphemes. HAM cannot analyse or interpret compound noun terms. In this section, the compound noun analysis system operating on the output of HAM's morphological analysis will be described. If a term is identified as a compound noun term, then first, possible noun terms for the compound noun term are extracted by a longest substring algorithm [3], and a left-to-right and agenda-based chart parsing algorithm [2] is then applied to get the best analysis of the compound noun term.

As mentioned, Korean is highly ambiguous in word formation because each single syllable is meaningful when used by itself. For example, some compound noun formations in Korean are as follows: NP → {N}, NP → Prefix + {N}, NP → {N} + suffix, NP → {N} + DN, etc. (a word belonging to DN (dependent noun) is not used by itself in a phrase/sentence and depends on a specific word belonging to the Noun). This paper focuses on the analysis of NP → {N} and NP → PN + {N}.

```
Initialise variables
        StartIndex = 0; EndIndex = 2; n = Length of the input (Korean string)
Loop (until StartIndex == n)
   Loop (until EndIndex == n)
        Get a Substring(x) from StartIndex to EndIndex
        Search the Substring(x) in the dictionary
        If (Substring(x) is in the dictionary and its POS is noun)
          Push the Substring to Found_Strings (see Fig.3) with its starting
          position, i.e. as (StartIndex: Substring(x))
        Else
          Discard the Substring
        End If
        Increase EndIndex by 2 (i.e. because Korean is a 2-byte language)
   End loop
   Increase StartIndex by 2
End loop
```

Fig. 1. Longest substring algorithm

A compound noun term is analysed in two steps. First, a longest substring algorithm searches for all possible longest substrings, that are legal lexical entries whose POS = NOUN, from an identified compound noun term. The substrings are extracted from the leftmost position (the first starting index of a substring, i.e. string[0]) to the end of the string by moving the starting position by a syllable to the right (i.e. 2 bytes for Korean) until the starting index is equal to the length of the string (Fig. 1).

For example, a Korean compound noun term, "정보검색" ("information retrieval" in English) would produce the extracted substrings "정" (emotion), "정보"(information), "보" (a beam or a reservoir), "보검" (a treasured sword), "검" (sword), "검색" (retrieval), "색" (colour) in order. If the term "정보검" as a proper noun is in the lexicon, then the longest substring algorithm extracts this term too. Fig. 2 tabulates the results for "정보•검색" (information • retrieval).

Using positional indices: $_0$ 정 $_1$ 보 $_2$ 검 $_3$ 색 $_4$, legal substrings found are
"정" from 0 to 1, - emotion – with a penalty score 6
"정보" from 0 to 2, - information – with a penalty score 4
*"정보검" from 0 to 3 is not a legal substring.
"보" from 1 to 2, - a beam or a reservoir – with a penalty score 4
"보검" from 1 to 3, - a treasured sword – with a penalty score 2
*"보검색" from 2 to 4 is not a legal substring
"검" from 2 to 3, - a sword – with a penalty score 2
"검색" form 2 to 4, - retrieval – with a penalty score 0
"색" form 3 to 4, - colour – with a penalty score 0

Fig. 2. Example of a longest substring algorithm of "정보검색"

Set Found_Strings (i.e. noun terms) to contain all substrings extracted by the longest substring algorithm.
Set Agenda_List to contain the strings from the Found_Strings whose starting position is 0 only.
Loop (until no Agenda_item at the starting position only).
Agenda_item (e.g. "A") = remove the longest substring from the Agenda _List
Compute penalty score of this item (see penalty score part).
Retrieve all substrings (e.g. "B" and "C") after the position of the agenda_item.
Combine all substrings with the agenda item (for example, "A" + "B" and "A" + "C").
If the combined string covers the whole compound noun string,
Then add the string and penalty score onto a list of resulting compound noun term.
Else add the string and penalty score onto the Agenda_List.
End If
End Loop
Select the best analysis with the least penalty score.

Fig. 3. Analysis of a compound noun term using a chart parsing technique

After retrieving all possible legal substrings from the identified compound noun term, an agenda-based left-to-right chart parsing technique with the found legal substrings is applied to get the best analysis of the compound noun term. The chart parsing technique tries to find all possible parse trees covering an input sentence [3]. Thus a strategy to resolve ambiguities resulting from the chart parsing algorithm is required. We applied a heuristic penalty scheme to choose the best analysis from ambiguous analyses of the compound noun term. Fig. 3 shows an agenda-based chart parsing algorithm to analyse compound noun terms.

The chart parsing algorithm stores legal substrings in a place called Found_Strings in Fig. 3. The legal substrings are found by the longest substring algorithm [3] using a structure similar to a stack of inactive arcs in a syntactic chart parser. From the found substrings (i.e. Found_Strings), agenda items are selected and they invoke the formation of a longer substring to cover the whole compound noun term. Only a substring whose starting position is 0 (e.g. leftmost substrings only) becomes an agenda item.

Fig. 4 shows an example of Korean compound noun analysis with states of Found_Strings, Agenda_List, and analyses found for a compound noun term.

Found_Strings = { (starting at position 0: "정", "정보"),
(starting at position 1: "보", "보검"),
(starting at position 2: "검", "검색") ,
(starting at position 3: "색") }

Agenda_List = {(starting at position 0: "정", "정보")}

Loop until no Agenda item starting at position 0.
Pop an agenda item with the longest substring at position 0 ("정보").
Move to position 2 because the length of the agenda item is 2.
Retrieve substrings at position 2 from Found_Strings (e.g. (starting at position 2: "검", "검색")) and combine them to the popped agenda item,
thus two substrings are produced e.g. "정보" + "검", "정보" + "검색".
If the combined substring covers the whole string (e.g. compound noun),
then the string is the result of analysis of the compound noun term.
else the string is added to the agenda list, "정보"+"검" is added to the
Agenda_List (e.g. {(starting at position 0: "정", "정보"+"검")}).
End Loop;

Fig. 4. Example of the agenda-based chart parsing technique with **"정보검색"** ("information retrieval" in English)

2.3 Selection of the Best Analysis

The best analysis from the alternatives is chosen by a simple heuristic method, called a penalty score scheme. This scheme prefers a smaller number of terms in a compound noun analysis. The penalty score of each substring is computed while performing the longest substring algorithm. The penalty score of an extracted substring is the length of syllables (i.e. 2 bytes for a Korean syllable) left behind after extracting the substring. For example, the penalty score of substring **"정보" ("information")** extracted from compound noun term "$_0$정 $_1$보 $_2$검 $_3$색 $_4$" ("information retrieval") starting at position 0 is 4 because two syllables **"검색"** ("retrieval") are left behind. The penalty score of substring **"검"** ("sword") starting at position 2 is 2 because a single syllable **"색"** ("colour") is left behind. The penalty score of substring **"검색"** ("retrieval") starting at position 2 is 0. By a left-to-right chart parsing technique, the penalty score of analysis **"정보•검색"** is 4 (i.e. the sum of each term's penalty score) while the penalty score for the analysis **"정보•검•색"** is 6. Thus the analysis with the smaller penalty score (i.e. **"정보•검색"**) is chosen as the best analysis.

If there are analyses with the same penalty scores, then the analysis with the smallest number of noun terms is chosen as the best.

3 Experimental Results

The system was implemented and tested in Perl5.0 under the Linux OS. The text data (i.e. 800 articles across eight topics) was collected from an online Korean daily newspaper. Among the data, 6024 compound noun terms were identified. The collected compound noun terms are 2 to 11 syllables. Four syllables terms form 70.9% of the total, and two syllables terms form 0.2% of the total (Table 2). The domains of the collected compound noun terms are: economy, politics, international news, information technology, domestic news, culture and life, sports, and entertainment.

Table 1. Distribution of Noun/Proper Noun (PN) Terms in Lexicons

Syllables	Basic Lexicon	User Lexicon (Noun)	User Lexicon (PN-name)	Total	Portion
	No. Of terms	No. of terms	No. of terms	No. of terms	Percentage
1	648	28	131	807	0.9%
2	41623	484	774	42881	46.8%
3	30419	2154	2060	34633	37.8%
4	8851	522	783	10156	11.1%
5	1700	138	376	2214	2.4%
6	486	41	131	658	0.7%
>6	166	8	120	294	0.3%
Total	83893	3375	4375	91643	100.0%

The system used two lexicons: a basic lexicon and a supplementary user lexicon. The basic lexicon has 122881 entries and a supplementary lexicon has 7750 entries (3375 entries for noun terms and 4375 entries for proper noun terms). The supplementary lexicon covers foreign words, jargon, acronyms, names (e.g. place, person, institution), newly-coined words, and derivational words that occurred in the collected data. Table 1 shows the distribution of syllables belonging to Noun or Proper Noun in the lexicons used in this paper. Two and three syllable words in the lexicon are 84.6% of the total Noun/Proper Noun terms in the lexicon.

Table 2. Results of Compound Noun Terms Analysis

No of Syllables	Test Data	Portion	Success	Failure	Success Rate	Failure Rate
2	14	0.2%	14	0	100.00%	0.00%
3	133	2.2%	122	11	91.73%	8.27%
4	4274	70.9%	4072	202	95.27%	4.73%
5	935	15.5%	920	15	98.40%	1.60%
6	505	8.4%	471	34	93.27%	6.73%
7	103	1.7%	100	3	97.09%	2.91%
8	39	0.6%	37	2	94.87%	5.13%
> 8	21	0.3%	20	1	95.24%	4.76%
Total	6024	100.0%	5756	268	95.55%	4.45%

3.1 Result of Compound Noun Analysis

Different lengths of compound noun terms (i.e. 2 syllables to 11 syllables) were tested. The average number of analyses ranges from 1 to 33; see Fig. 5. The minimum number of analyses is 1 and the maximum number of analyses is 130 - this resulted from the compound noun term 영어•독서•지도자•양성•과정 (English • Reading • Teacher • Training • Course). Fig. 5 shows the relationships between the number of syllables in a compound noun term and the number of its analyses. As the number of syllables increases, the number of analyses increases sharply.

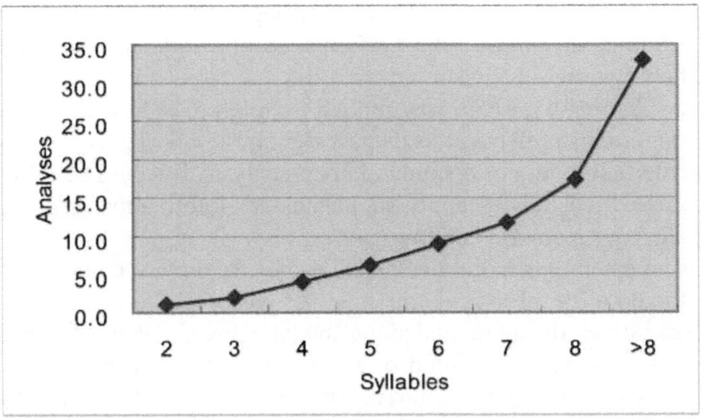

Fig. 5. Relationship between Syllables and Average Number of Analyses

The result of compound noun term analysis shows 95.55% success rate on average across the length of syllables. In the case of 3 syllables, the success rate is the worst. In the case of 2 syllables, the success rate is 100% because there is only one analysis for two-syllable compound noun terms. The system failed to correctly analyse a compound noun term with three syllables ($S_1S_2S_3$) because the penalty scheme allocated less score to the rightmost longest substring in the compound noun term. Thus the correct analysis of compound noun term $S_1S_2S_3$ could be $S_1 \bullet S_2S_3$. However, the system suggested $S_1S_2 \bullet S_3$ as the best analysis of the three syllables term (e.g. "광신호" ("photonic/light signal") → "광신" ("fanaticism") • "호" ("arc"), "광" ("photonic/light") • "광신" ("signal"), "광" ("photonic/light") • "신" ("footwear/god") • "호" ("arc")). This issue increased the failure rate of analysis of 3-syllable compound noun terms.

A system applying multiple methods to compound noun analysis showed a 95.6% success rate [12] and another system based on a min-max algorithm by using statistical information obtained from a corpus showed 97.3% of success rate of compound noun segmentation [11]. Compared to both systems, the result of our system showed quite a good performance without using a large tagged corpus.

4 Further Improvement and Conclusion

The critical drawback of the rule-based system (e.g. morphological/syntactic rules) is to know the optimal size of the lexicon. There are a lot of relatively unknown words produced from foreign words, newly-coined words, names, and jargon. It is possible to add the whole words into a lexicon. However, if the size of the lexicon is larger, then the compound noun analysis produces more ambiguities that make the ambiguity resolution more difficult. Thus the system requires optimization according to the size of the lexicon. In addition, a system for recognising relatively unknown words automatically is required to improve the performance of compound noun analysis and recognition of entity terms.

The longest substring algorithm was applied from left to right with its penalty score scheme. This algorithm preferred the longest rightmost subtring as the best analysis, and increased the failure rate of compound noun analysis. It would be a further development task to find out how to apply supplementary information (e.g. statistical information) to improve resolution of ambiguities.

Korean has no capital letters to distinguish names from other noun terms. Thus it is not effective to add all legal lexical entries in the real world to a lexicon because the number of alternatives of compound noun analysis greatly depends on the size of lexicon a system uses. Thus a method of processing compound noun term with relatively unknown words would be required. The unknown word identification process will reduce the size of lexicon. In addition, a process for bilingual term analysis would be helpful for the identification of compound noun terms like "ASEM•회의" ("ASEM • Meeting").

A Korean compound noun term analysis method based on morphological and syntactic rules employing the longest substring algorithm and a left-to-right agenda-based chart parsing algorithm has been described. Unsurprisingly, the method produced many ambiguities as the length of the compound noun term increased. However, a simple heuristic method to select the best analysis among many alternatives showed a 95.6% success rate on compound noun terms collected from an online newspaper, across eight topics.

Acknowledgement

We greatly appreciate the permission of Prof. Kang, S. from Kookmin University, to use the Linux version of his Korean Morphology Analysis, HAM. His application contributes to the implementation of document classification system in this paper.

References

[1] Basili, R., Moschitti, A., and Pazienza, M.: NLP-driven IR: Evaluating Performance over a Text Classification Task. Proceedings of IJCAI'01. Seattle Washington (2001)

[2] Earley, J.: An Efficient Context-Free Parsing Algorithm. CACM 13(2) (1970) 94-102

[3] Hirshberg, D.S.: Algorithms for the Longest Common Subsequence Problem. The Journal of ACM. 24(4) (1977) 664 – 675

[4] Kando, N., Kageura, K., Yoshoka, M., and Oyama, K.: Phrase Processing Methods for Japanese Text Retrieval. SIGIR Forum, 32(2) (1998) 23-28

[5] Kang, S.: Korean Morphological Analysis Program for Linux OS, http://nlp.kookmin.ac.kr. (2001)

[6] Kim, J., Kwak, B., Lee, S., Lee, G., and Lee, J.: A Corpus-Based Learning Method of Compound Noun Indexing Rules for Korean. Information Retrieval 4(2) (2001) 115-132

[7] Kwak, B., Kim, J., Lee, G., and Seo, J.: Corpus-based Learning of Compound Noun Indexing. Proceedings of ACL2000 Workshop on Recent Advances in NLP and IR. Hong Kong (2000) 57-66

[8] Lee, J. and Ahn, J.: Using n-Grams for Korean Text Retrieval. Proceedings of SIGIR'96. Zurich Switzerland (1996) 216-224

[9] Min, K., Wilson, W.H., and Moon, Y.: Preferred Document Classification for a Highly Inflectional/Derivational Language. In: McKay, B., Slaney, J. (eds.): AI2002: Advances in Artificial Intelligence. Lecture Notes in Artificial Intelligence, Vol. 2557. Springer-Verlag, Berlin Heidelberg (2002) 12-23

[10] Park, H., Han, Y., Lee, K., and Choi, K.: A Probabilistic Approach to Compound noun Indexing in Korean Texts. Proceeding of COLING'96. Copenhagen Denmark (1996) 514-518

[11] Yoon, J.: Compound Noun Segmentation Based on Lexical Data Extracted from Corpus. Proceedings of 6th ANLP. Seattle (2000) 196-203

[12] Yun, B., Kwak, Y., and Rim, H.: Resolving Ambiguous Segmentation of Korean Compound Nouns Using Statistics and Rules. Computational Intelligence, 15(2) (1999) 101-113

A Language Modeling Approach to Search Distributed Text Databases

Hui Yang and Minjie Zhang

School of Information Technology and Computer Science,
University of Wollongong
Wollongong, 2500, Australia
{hy92,minjie}@uow.edu.au

Abstract: As the number and diversity of distributed information sources on the Internet exponentially increase, it is difficult for the user to know which databases are appropriate to search. Given database language models that describe the content of each database, database selection services can provide assistance in locate relevant databases of the user's information need. In this paper, we propose a database selection approach based on statistical language modeling. The basic idea behind the approach is that, for the databases that are categorized into a topic hierarchy, individual language models are estimated at different search stages, and then the databases are ranked by the similarity to the query according to the estimated language model. Two-stage smoothed language models are presented to circumvent the inaccuracy due to word sparseness. Experimental results demonstrate such a language modeling approach is competitive with current state-of-the-art database selection approaches.

1 Introduction

The rapid proliferation of online text databases on the Internet has made it difficult for a user to determine which databases to search for desired information. To reduce network traffic as well as the cost of searching useless databases, the idea of sending the user query only to potentially useful databases has become more and more attractive to both users and information science researchers.

Database classification, the techniques to partition multiple, distributed web databases based on their subject contents into a structured hierarchy of topics [1, 9], provide a useful and efficient way to organize and manage a vase number of web databases. At the same time, it is also helpful to make the task of database selection simplified.

In this paper, we present a novel database selection approach based on database classification. The language model explored in this paper is, in practice, a two-stage database language model which is the combination of a class-based language model and a term-based language model. For text databases that have been categorized into a

T.D. Gedeon and L.C.C. Fung (Eds.): AI 2003, LNAI 2903, pp. 196-207, 2003.

hierarchical structure of topics, the task of database selection is conceptually decomposed into two distinct steps:

(1) First, with a class-based language model, the search only focuses on databases in confined domains, e.g., a certain specific subject area that a user is interested in.
(2) Second, selection algorithm computes the likelihood of individual chosen databases to the query using a term-based language model, and further selects the best databases for the query.

Obviously, the two-stage language model approach explicitly captures the different influences of the category-specific search stage and the term-specific search stage on the optimal settings of selection parameters. Our experimental results reported here have demonstrated that the exact performance of this two-stage language model is significantly better than other traditional selection methods such as traditional well-known $df \times icf$ method.

This work is an extension of our previous research [9] which introduces a clustering method for hierarchically categorizing multiple, distributed web databases. Our study has the following contributions:

- First, we acquire database models independently according to different search stages and intentionally keep these models separate.
- Second, we consider general ways of combining different models of a database by introducing suitable parameters, which can be further individualized and optimized for each database.

We believe that such a simple and effective language model as a solid baseline method opens up the door for further improving the database selection performance.

The paper is organized as follows: in Section 2 we discuss the basic theory of statistical language modeling approach. Section 3 lays out the two-stage database language model and develops the formulas for a simple realization of it. Experimental methodology and a series of experiments to evaluate our language model are presented in Section 4 and 5. Finally, conclusion and the contribution of this work are summarized in Section 6.

2 The Statistical Language Modeling Approach

Due to the relative simplicity and effectiveness of statistical methods that have been applied successfully in a wide variety of speech and language recognition areas, statistical language model has recently attracted significant attention as a new alterative to traditional retrieval models [7, 6]. Firstly, let us take a brief look at the standard statistical language model.

In automatic speech recognition, language models are capable of assigning a probability $P(W)$ to a word string W ($W = w_1 w_2 \cdots w_n$). The probability of the word string W occurring in English text can be characterized by a set of conditional probabilities $P(w_i \mid w_1^{i-1})$, which can be expressed as a product

$$P(W) = \prod_{i=1}^{n} P(w_i \mid w_1^{i-1}) \tag{1}$$

where $P(w_i \mid w_1^{i-1})$ is the probability that w_i will be occurred when all of the previous words w_1^{i-1} ($w_1^{i-1} = w_1 w_2 \cdots w_{i-1}$) are occurred.

The task of a statistical language model is to make it tractable to estimate the probabilities $P(w_i \mid w_1^{i-1})$. At present, the dominant technology in language modeling used is the unigram model, which makes a strong assumption that each word occurs independently, and consequently, the probability of a word string becomes the product of the probabilities of the individual words. Therefore, Equation 1 can be simplified as

$$P(W) = \prod_{i=1}^{n} P(w_i) \tag{2}$$

For a vocabulary of size *V*, the unigram model is a multinomial, i.e., multinomial distributions over words in *V*, which has *V-1* independent parameters.

The basic idea of the statistical language modeling approach for database selection is to view each database *S* as a language sample and to estimate the conditional probability $P(Q \mid S)$, i.e., the probability of generating a query $Q = \{q_1, q_2, \cdots, q_N\}$ given an observation of a database *S*:

$$P(Q \mid S) = \prod_i P(q_i \mid S) \tag{3}$$

The databases are ranked by the probability $P(Q \mid S)$.

Applying Bayes' rule of probability, the posterior probability of the database *S* to the query *Q* can be written as

$$P(S \mid Q) = \frac{P(Q \mid S) P(S)}{P(Q)} \tag{4}$$

The *P(Q)* term represents the probability that query *Q* is generated from a document-independent language model. Since *P(Q)* is constant for all databases given a specific query *Q*, it does not affect the ranking of the databases and can be ignored for the purpose of ranking databases. Therefore, Equation 4 can be rewritten as

$$P(S \mid Q) \propto P(Q \mid S) P(S) \tag{5}$$

As to the *P(S)* term, it is a query-independent term, which measures the quality of the databases according to the user's general preference and information needs. In this work, since a user query probably involves in a variety of topics and we cannot beforehand predict its content, the prior term *P(S)* is assumed to be uniform over all databases, and so does it not affect database ranking. As a result, in practice, only the $P(Q \mid S)$ term really has final influence on determining the relevance of the database to the query.

Just as in the use of language model for speech recognition, one of the most obvious practical problems in applying statistical language modeling to database selection is the problem of sparse data, that is, it is possible to assign a probability of zero to a database that is missing one or more of the query terms. The data sparseness problem can be avoided by data smoothing methods which assign non-zero probability to the missing query terms so as to improve the accuracy and reliability of the models. In the following section, a brief overview of the smoothed two-stage database language models for database selection is described as follows.

3 Two-Stage Database Language Models for Database Selection

3.1 Hierarchical Structure of Topics for Text Databases

Firstly, to facilitate the construction of a two-stage database language model, we first describe a structured hierarchy of topics shown as Figure 1.

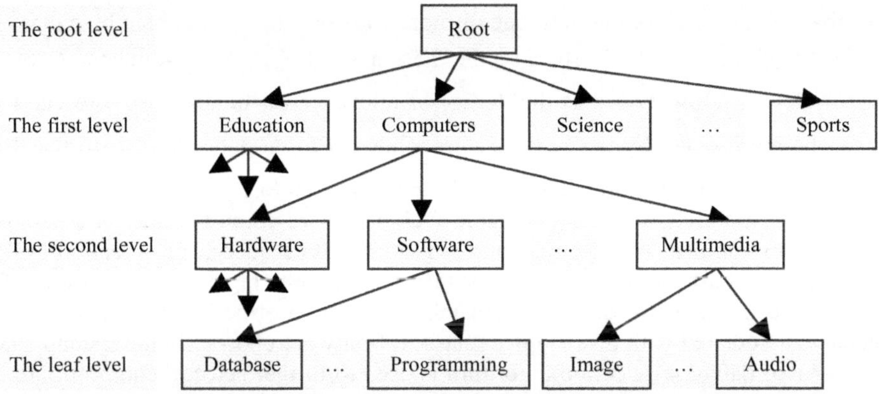

Fig. 1. A Small Fraction of the Topic Hierarchical Structure

Formally, we can give some formal definitions for a hierarchical structure of topics as follows:

Definition 1. A structured hierarchy of topics is a rooted directed tree which contains a number of topics (classes), namely, $c_1, c_2, \cdots, c_K$, organized into multiple levels and each node corresponds to a topic.

In general, in such a topic hierarchy, topics are ordered from general broad topics (high level) to specific and narrow ones (low level). There exist parent-child relationships between two adjacent layers. Each parent class has a set of child classes and these child classes together cover different aspects of the parent class.

Definition 2. In each parent-child pair, the weight of the connection between parent class c_k^p and child class c_k is denoted by $w_{c_k^p c_k}$ which represents the degree of their association.

Definition 3. For each topic (class) c_k in the topic hierarchy, it is represented by a feature space F^{c_k} which is denoted as $F^{c_k} = \{f_1, f_2, \cdots, f_M\}$, where f_i ($1 \leq i \leq M$) is a distinct feature vector in the feature space F^{c_k}.

The features are the words that are strongly associated with one specific category (class). The features have enough discrimination power to distinguish this class from a set of classes in the hierarchical classification scheme.

Definition 4. Cluster c^p is a parent class that consists of a number of child classes, $c^p = \{c_1, c_2, \cdots, c_T\}$, where T is the number of the child classes. F^{c^p} is a feature space of the cluster c^p, which is described as $F^{c^p} = \{f_1, f_2, \cdots, f_S\}$, where F^{c^p} is the feature space set of all the child classes, namely, $F^{c^p} = F^{c_1} \cup F^{c_2} \cup \cdots \cup F^{c_T}$.

Definition 5. With a hierarchical classification scheme with categories, $c_1, c_2, \cdots, c_K$, a database S can be classified into one or more classes, which is denoted by a class set C^S, $C^S = \{c_1^S, c_2^S, \cdots, c_K^S\}$. Each class in the class set C^S is a 2-dimension vector $\{c_i^S, t_i^S\}$ ($1 \leq i \leq K$), where t_i^S is the degree of relevance of the database S to class c_i. t_i^S can be represented by the posterior probability $P(c_i \mid S)$ of class c_i to the database S, whose value is normalized into the range from 0 to 1.

The topic hierarchy and its associated web databases could be used in a database selection environment as follows: once a user submits a query, the selection system selects the databases to search in two stages: in the first stage, the system identifies one or more topics that best matches/match the user's information need, and then the databases associated with these topics (classes) firstly are chosen. In the second stage, a term-based database selection algorithm is used to further select the appropriate databases from those chosen databases based on the degree of relevance of the databases to the query.

3.2 The Smoothed Two-Stage Language Models

The generic language model for the two-stage database selection procedure can be refined by a concrete class-based model $P(Q \mid C)$ for generating the specific subject classes that best match the query and a concrete term-based model $P(Q \mid S)$ for generating the most likely similar databases to the query. Different specifications lead to different selection formulas. Now, we present the generic two-stage models with data smoothing methods which consists of two individual models, a smoothed class-based language model and a smoothed term-based model.

3.2.1 The Smoothed Class-Based Language Model

Given a user query Q, it may be possible to make reasonable predictions of determining the appropriate topic(s) that the user is interested in by using the feature vectors in the feature space. Let $F^{c_k} = \{f_1, f_2, \cdots, f_M\}$ denotes a feature space of class c_k, $Q = \{q_1, q_2, \cdots, q_N\}$ denotes a user query and $F^C = \{f_1, f_2, \cdots, f_{|V|}\}$ denotes the feature vocabulary of all the classes in the topic hierarchy. For a certain topic class c_k, the likelihood function of class c_k to a query is described as follows,

$$P(Q \mid c_k) = \prod_i P(q_i \mid c_k) = \prod_i P(q_i \mid F^{c_k}) \tag{6}$$

where $P(q_i \mid c_k)$ is the probability of query term q_i in the feature space of class c_k.

Due to computation cost, the feature space F^{c_k} is actually a much small feature set of fixed size and content. This makes it difficult to distinguish the effects of the missing query terms in the feature space F^{c_k}. To avoid the sparse data problem, we describe a smoothing technique called the Jelinek-Mercer method (also called linear interpolation) [3] that combines the probability estimates of several language models within a unified model that suffers less from the data sparseness problem. The probability estimates of this linearly interpolated language model will become more reliable and more accurate.

In order to capture the common and non-discriminating words in a query, we assume that a query is generated by sampling words from a three-component mixture of language models with one component being $P(q \mid c_k)$, and the two others being the parent-class language model $P(q \mid c_k^p)$ and the class-corpus language model $P(q \mid C)$. That is,

$$P_{JK}(Q \mid c_k) = \prod_i (\lambda_1 P(q_i \mid c_k) + \lambda_2 P(q_i \mid c_k^p) + \lambda_3 P(q_i \mid C)) \tag{7}$$

where c_k^p is the parent class of the class c_k, and C is the class corpus of the topic hierarchy; λ_1, λ_2 and λ_3 are coefficients, which control the influence of each model, $0 \le \lambda_1, \lambda_2, \lambda_3 \le 1$ and $\lambda_1 + \lambda_2 + \lambda_3 = 1$.

Obviously, the class-based language model is effectively smoothed in two steps.

(1) First, it is smoothed with the correlation information about parent-child pairs. Since the feature space of the parent class c_k^p is the feature space set of all the child classes (recall Definition 4), the size of the feature space $F^{c_k^p}$ of the parent class c_k^p is far larger than that of the feature space F^{c_k}. Therefore, such a parent-class model $P(q \mid c_k^p)$ can help us differentiate the contribution of different missing query terms in the feature space F^{c_k}. In consideration of the associative degree between parent class and child classes, there exists the following relationship between λ_1 and λ_2:

$$\lambda_2 = w_{c_k^p c_k} \lambda_1 \tag{8}$$

where $w_{c_k^p c_k}$ is the weight of the connection between parent class c_k^p and child class c_k (recall Definition 2). Combining this equation with the class model, we can rewrite Equation 7 as follows,

$$P_{JK}(Q \mid c_k) = \lambda_1 P(Q \mid c_k) + \lambda_1 w_{c_k^P c_k} P(Q \mid c_k^P) + [1-(1+w_{c_k^P c_k})\lambda_1]P(Q \mid C) \quad (9)$$

$$= \prod_i (\lambda_1 P(q_i \mid c_k) + \lambda_1 w_{c_k^P c_k} P(q_i \mid c_k^P) + [1-(1+w_{c_k^P c_k})\lambda_1]P(q_i \mid C)) \quad (10)$$

(2) Second, in order to avoid the probability that some query terms may still be missing in the feature space of the parent class, this class-based language model is further interpolated with a class-corpus model. The class-corpus model $P(q \mid C)$ will give such query term that which did not appear in the above two feature spaces F^{c_k} and $F^{c_k^P}$, a much smaller probability, since the feature space F^C is the feature vocabulary of all the classes in the topic hierarchy. Note that the coefficient λ_1 can be optimized by EM (Estimate Maximization) algorithm [2]. .

Assume that there are u topics $(c_1, c_2, \cdots, c_u)$ selected by the class-based language model, the likelihood probability of the databas associated with the chosen topics to the query Q can be calculated as

$$P(Q \mid S) = \sum_{k=1}^{u} P_{JK}(Q \mid c_k) P(c_k \mid S) \quad (11)$$

where $P(c_k \mid S)$ is the posterior probability of class c_k for the database S (recall Definition 5). The database will be ordered by the likelihood probabilities, and only the top K databases in the ranking list will be chosen as preliminary relevant databases to the query.

3.2.2 The Smoothed Term-Based Language Model

Once a number of relevant databases are chosen from the category-specific search stage, the next step is to further refine the search range for search efficiency and effectiveness. A straightforward method to estimate the probabilities of individual query terms is the Maximum Likelihood Estimation (MLE) [5]. Our term-based language model is a unigram language model based on the MLE. Given a user query Q, the probability $P(Q \mid S)$ of a database S to the query can be expressed as

$$P_{MLE}(Q \mid S) = \prod_i P_{MLE}(q_i \mid S) = \prod_i \frac{C(q_i \mid S)}{\sum_{t_j} C(t_j \mid S)} \quad (12)$$

where $C(q_i \mid S)$ is the frequency occurrence of query term q_i in database S, and $\sum_{t_j} C(t_j \mid S)$ is the sum of the times of all term occurrence in database S. The MLE assigns the probability of the observed data in the database as high as possible and assigns zero probability to the missing data.

For a database with a large number of documents, there are enough data for $P_{MLE}(q_i \mid S)$ to be robustly estimated. However, the MLE still probably assign zero probability to the query terms that do not appear in the database. To compensate this sparse data problem, one approach is to model the term distributions using Dirichlet distributions [4]. With the Dirichlet distribution with parameters $\alpha \boldsymbol{m}$ ($\boldsymbol{m}=\{m_1, m_2, \cdots, m_n\}$), the $P_{MLE}(Q \mid S)$ can be rewritten as

$$P_{DP}(Q \mid S) = \prod_i Dirichlet(P_{MLE}(q_i \mid S)) = \prod_i \frac{C(q_i \mid S) + \alpha m_i}{\sum_{t_j} C(t_j \mid S) + \alpha} \tag{13}$$

where α is a positive scalar; m_i is the posterior probability of query term q_i, which is denoted as $P(q_i \mid C')$. The probability $P(q_i \mid C')$ can be estimated based on a collection language model $P(t \mid C')$. The collection language model $P(t \mid C')$ is constructed with a large amount of training data set C', which contains the average probability of the term through a geometric distribution. The model $P(t \mid C')$ is used as a reference model to smooth the probability distribution of query terms. For a term t_i, $P(t_i \mid C')$ is normalized, i.e., $\sum_{t_i} P(t_i \mid C') = 1$.

4 Experimental Design

4.1 The Data Sets

In order to demonstrate the effectiveness of our modeling techniques, we conduct a number of experiments to evaluate the selection performance of two-stage database language models on the Reuter 21578 data set (Http://www.research.att.com/~lewis/~reuters21578. html). The Reuters 21578 data set contains 21,450 Reuters news articles from 1987 with 135 categories. Due to the documents with multiple labels, we distinguish some labels that tend to subsume other labels as the high level topics and therefore construct a 3-layer hierarchical structure.

4.2 Evaluation Metrics

At the category-specific search stage, we draw the 11 point average precision-recall curve to compare the selection performance between the JK model and non-interpolated model without the interpolation of parent class and class corpus. The 11 point average precision-recall (P / R) curve plots the average precision values at each of the 11 recall points 0, 0.1, 0.2, …, 1.0, where *precision P* is the ration of the number of relevant classes to the total number of classes selected, and *recall R* is the ration of the number of relevant classes to the total number of relevant classes in the class set.

At the term-specific search stage, we use the mean-squared root error metric to compare the effectiveness of variations of two types of language models (i.e., the MLE model, the DP model) to the basic language model which is the well-known vector-space model (VS) with the Cosine function and the famous $df \times icf$ formula [8]. Note that the names of all the models are abbreviated with the two initial letters (e.g., JK for Jelinek-Mercer smoothing).

The mean-squared root error of the collection ranking for a single query is calculated as:

$$Error = \frac{1}{|C|} \cdot \sqrt{\sum_{i \in C} (O_i - R_i)^2} \quad (14)$$

where: (1) O_i is the position of database S_i in the optimal relevance-based ranking O_Q given a query Q. The optimal ranking O_Q is produced based on the following two criteria:

- The number of relevant topics in the databases. If database S_i has more classes than database S_j, then S_i is ranked ahead of S_j. That is, Rank (S_i, S_j) = {S_i, S_j}.
- The number of relevant documents in the databases. If database S_i has more documents associated with relevant classes than database S_j, then S_i is ranked ahead of S_j. That is, Rank (S_i, S_j) = {S_i, S_j}.

(2) R_i is the position of database S_i in the selection ranking based on the probability $P(Q|S)$ obtained from the language model. The database with the largest value of $P(Q|S)$ is ranked 1, the database with second largest value is ranked 2, and so on; (3) C is the set of collections being ranked.

The mean-squared root error metric has the advantage that it is easy to understand (an optimal result is 0), and it does not require labeling a database 'relevant' or 'not relevant' for a particular query.

5 Experimental Results

In this section, we present experimental results to test the robustness of using the smoothed two-stage language model for database selection.

5.1 The Effects of Various Values of Smoothing Parameters in Individual Language Models

To examine the effect of the smoothing parameter in each specific smoothing method on selection effectiveness, we vary the value of the smoothing parameters in a wide range in order to observe all possible difference in smoothing. In each run, we set the smoothing parameter the same value to report the average selection performance at each parameter value. Then, we compare selection performance by plotting the average precision or mean-squared root error against the variation in these values.

Jelinek-Mercer (JK) Smoothing: In Figure 2, we compare the precision of the JK model at different settings of the smoothing parameter λ_1 (Recall Equation 10). The plots in Figure 2 show that the average precisions for different values of λ_1 over Reuters 21578 data set. It is easily noted that the choice of the appropriate smoothing parameter λ_1 can have a large impact on selection performance. When λ_1 varies the range between 0.6~0.8, the results are statistically significant improvement in precision. We can see that the optimal value of λ_1 is a little high, which suggests that although the parent-class model and the class-corpus model can help smooth the word sparseness, the basic class model $P(q \mid c_k)$, in practice, plays a major role in class selection.

Dirichlet Prior (DP) Smoothing: To see the detail change, we experiment with a wide range of value of the DP smoothing parameter α (recall Equation 13). It is clear that, from the results in Figure 3, the optimal value of α seems to be set a relative large value, about between 1000~2000.

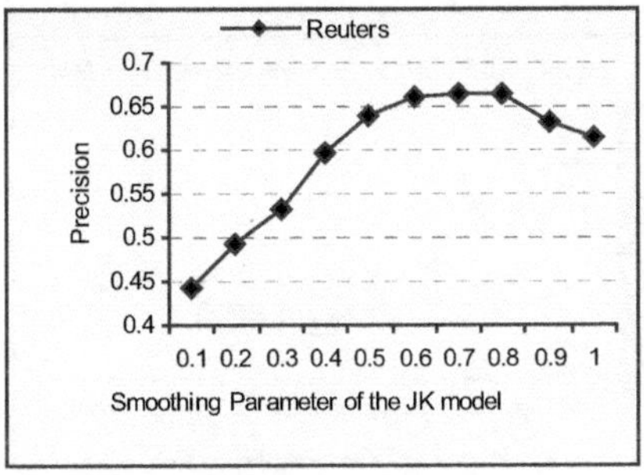

Fig. 2. The Effect of Different Value of the JK coefficient λ_1 on Selection Performance

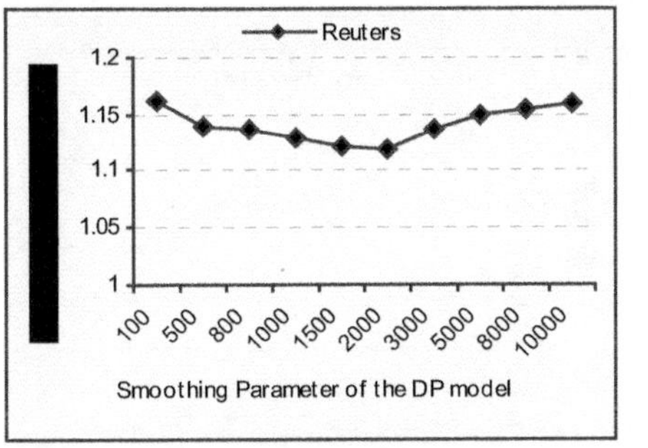

Fig. 3. The Effect of Different Values of the DP parameter α on Selection Performance

5.2 The JK Model vs The Non-interpolated Model

The class-based language model obtained using the JK smoothing method is expected to perform better than the simple non-interpolated model. In order to prove it, we compare the precision-recall chart of the JK model with that of the non-interpolated model. The 11 point precision-recall curves are shown in Figure 4. Note that the figure for the JK model uses the best choice of the smoothing parameter λ_1 (λ_1=0.7).

Figure 4 clearly shows that on the 11 point precision-recall chart, the JK language modeling approach achieved better precision at all levels of recall, and most of them statistically significant, by 7.54% on average respectively. It means that the linear interpolated approach is an attractive way in which the parent-class model and the class-corpus model help effectively smooth the word sparseness and therefore improve the selection effectiveness.

5.3 The Effects of Different Term-Based Language Models on Selection Performance

In addition to the vector-space (VS) model whose mean-squared root error we consider to be the evaluation baseline for comparison, two different language models, the MLE model and the DP model are used in our experiments. The selection performances of searching a set of distributed databases are shown in Figure 5.

Compared with the behavior of the VS model, the DP model improves selection performance significantly across the Reuters 21578 data set, but the MLE model performs a little worse. This is understandable that some query terms missed at the database lead to a bad probability estimate for the query.

Note that the results of the DP model are still very competitive when compared with the performances of two other models, the MLE model and the VS model. As we see, in all results, the performance of the DP model smoothed with the optimal parameter value α (α=1000) is statistically better than the best performance of the VS model. It appears that the DP model is actually a quite robust language model in database selection.

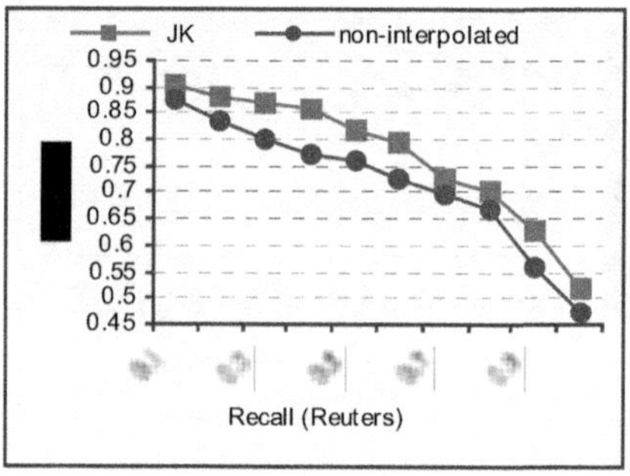

Fig. 4. Selection Performance of the JK and the Non-interpolated Model

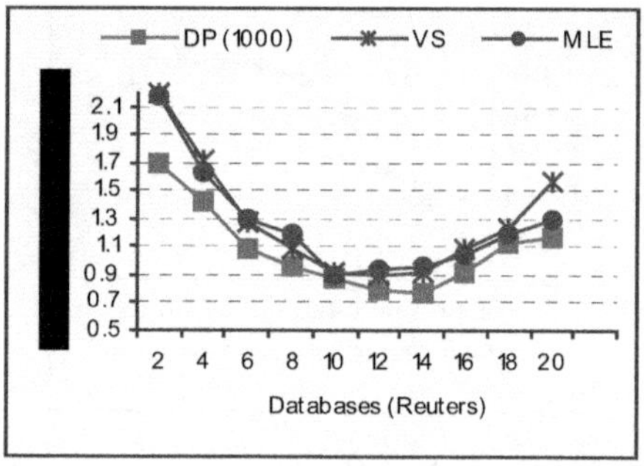

Fig. 5. The effects of different term-based Model language models on selection performance

6 Conclusion and Future Works

We have presented a novel way of looking at the problem of database selection based on a two-stage language model that is rooted on the solid foundation of statistical nature language processing. For the databases categorized into a topic hierarchy, instead of searching them in an ad hoc manner, we intentionally keep the database language models separated according to individual natures of different search stages. It makes our model easy to be understood and be expanded. We also conducted a number of experiments to prove the effectiveness and robustness of our language model. The experimental results have demonstrated that this approach holds great promise of improving search performance. Due to the simplicity of our model, we believe that our model can be easily extended to incorporate with any new language-based techniques under a general, well-ground framework.

We plan to investigate the following matters in the future work. First, due to time constraint, our current language models do not incorporate many familiar ideas into our selection system such as relevant feedback and semantic-based information retrieval. It is possible that additional knowledge added to the models will further improve our selection system. For example, using relevant feedback techniques, a user can provide more meaningful words which facilitate the formation of better queries. Second, as mentioned previously, we simply take all the words occurring in the databases independent. Such unigram models completely ignore word phrases that are likely to be beneficial to probability estimate of the query. To compensate the shortcoming, we will consider bigram and possibly trigram models in our general language model for database selection. Finally, we only employ simple smoothing methods to our current work. We are planning to explore other more elaborate smoothing methods to extend our models. It is also possible to further improve selection results.

Acknowledgement

This research was supported by a university grant from Wollongong University under contract SDRG03-227381010.

References

[1] Chen, H., Schuffels, C., and Orwig, R.: Internet Categorization and search: a Self-Organizing Approach. Journal of Visual Communication and Image Representation, Vol. 7 (1), (1996) 88-102.

[2] Dempser, A. P., Laird, N. M. and Rubin, D. B.: Maximum likelihood from incomplete data via the EM algorithm. Int. J. Journal of the Royal Statistical Society. Vol. 39(B). (1977) 1-38.

[3] Jelinek, F. and Mercer, R: Interpolated estimation of Marvok source parameters from sparse data. Patter Recognition in Practices. North Holland, Amsterdam (1980) 381- 402.

[4] MacKey, D. and Peto, L.: A Hierarchical Dirichlet Language Model. Int. J. National Language Engineering, Vol. 1(3) (1995) 289-307.

[5] Mood, A. M. and Graybill, F. A.: Introduction to The Theory of Statistics. Second Edition, McGraw-Hill (1963).

[6] Miller, D. J., Leek, T., and Schwartz, R. M.: A Hidden Markov Model Information Retrieval System. Proceedings of the 22th annual international ACM SIGIR conference on Research and development in information retrieval. Berkeley, California, United States (1999) 214 – 221.

[7] Ponte, J. M. and Croft, W. B.: A language modeling approach to information retrieval. Proceedings of the 21th annual international ACM SIGIR conference on Research and development in information retrieval. Melbourne, Australia (1998)214 – 221.

[8] Salton, G. and Mcgill, M.: Introduction of Modern Information Retrieval. McGrag-Hill, New York (1983).

[9] Yang, H., and Zhang, M.: Hierarchical Classification for Multiple, Distributed Web Databases. Proceedings of the18th International Conference on Computers and Their Applications, Honolulu, Hawaii, US, (2003) 155-160.

Combining Multiple Host-Based Detectors Using Decision Tree

Sang-Jun Han and Sung-Bae Cho

Dept. of Computer Science, Yonsei University
134 Shinchon-dong, Sudaemoon-ku, Seoul 120-749, Korea
{sjhan,sbcho}@cs.yonsei.ac.kr

Abstract. As the information technology grows interests in the intrusion detection system (IDS), which detects unauthorized usage, misuse by a local user and modification of important data, have been raised. In the field of anomaly-based IDS several artificial intelligence techniques are used to model normal behavior. However, there is no perfect detection method so that most of IDSs can detect the limited types of intrusion and suffers from its false alarms. Combining multiple detectors can be a good solution for this problem of conventional anomaly detectors. This paper proposes a detection method that combines multiple detectors using a machine learning technique called decision tree. We use conventional measures for intrusion detection and modeling methods appropriate to each measure. System calls, resource usage and file access events are used to measure user's behavior and hidden Markov model, statistical method and rule-base method are used to model these measures which are combined with decision tree. Experimental results with real data clearly demonstrate the effectiveness of the proposed method that has significantly low false-positive error rate against various types of intrusion.

Keyword: anomaly detection, decision tree, combining detectors

1 Introduction

Today's high reliance on computers raises concerns about the computer security. As the damage from security incidents increases, several tools and equipments become essential to computer systems. Intrusion detection is to find attacks exploiting illegal uses or misuses. An IDS is software to detect attacks exploiting illegal uses or misuses and modification of important data by analyzing system calls, system logs, activation time, and network packets of each operating system [1]. Generally, intrusion detection techniques can be divided into two groups according to the type of data they use: misuse detection and anomaly detection [2].

Misuse detection has the advantage that known attacks can be detected reliably with low false-positive error and economically. The shortcoming is that it cannot detect unknown attacks. Anomaly detection is better than misuse detection in terms of detecting novel attacks and its low false-negative error rate.

T.D. Gedeon and L.C.C. Fung (Eds.): AI 2003, LNAI 2903, pp. 208–220, 2003.

However, it suffers from high false-positive error rate because unseen normal behaviors are considered as attacks. Another drawback of anomaly detection technique is that the detectable types of intrusion are limited according to the measures and modeling methods used.

In this paper, to overcome drawbacks of the conventional anomaly detection techniques, we propose an anomaly-based detection technique that combines multiple detectors. First of all, we develop four appropriate detection methods that use system call events, resource usage of process, file access events as the measure of normal behavior with appropriate modeling methods. Next, we combine these detectors using decision tree. The proposed detection method is expected better performance because it can model normal behaviors from various perspectives.

The rest of this paper is organized as follows. In Section 2, we give a brief overview of the related works. The overall design and detailed description of the proposed methods are presented in Section 3. Experimental results are shown in Section 4.

2 Related Works

Various techniques are used for anomaly-based intrusion detection. Expert system, statistics, artificial neural network and hidden Markov model (HMM) are widely used for modeling normal behaviors. The representative studies on anomaly detection are summarized in Table 1.

Statistics is the most widely used technique, which defines normal behavior by collecting data relating to the behavior of legitimate users over a period of time [3]. The representative IDS based on statistics is NIDES (Next-generation Intrusion Detection Expert Systems), which measures the similarity between a subject's long-term behavior and short term behavior for intrusion detection [8]. The detection rate is high because it can use various types of audit data

Table 1. The representative studies on intrusion detection

ES: Expert System, NN: Neural Network, ST: Statistics, HMM: Hidden Markov Model

Organization	Name	Period	ES	NN	ST	HMM
AT&T	ComputerWatch [4]	1987-1990	X			
UCDavis	NSM [5]	1989-1995			X	
	GrIDS [6]	1995-	X			
SRI International	IDES [7]	1983-1992			X	
	NIDES [8]	1992-1995			X	
	EMERALD [9]	1996-			X	
CS Telecom	Hyperview [10]	1990-1995	X	X		
Univ. of New Mexico	C. Wrannder et. al [11]	1995			X	X
Yonsei Univ.	Park and Cho [12]	1999-				X

and detect intrusion based on the previous experimental data. The shortcoming is that it is not sensitive to some behavior and detectable types of intrusion are limited.

Hyperview of CS Telecom uses neural network for intrusion detection [10]. It consists of 2 modules: neural network and expert system. The neural network in Hyperview uses temporal sequence of audit data as inputs and 60 types of audit data. It has the advantage of easy representation of nonlinear relationship between input and output. The defects of neural networks are that its computational load is very heavy and it is difficult to interpret the relationship between inputs and outputs.

An HMM is useful technique because it is good for modeling system call sequences. The representative study is the technique proposed by C. Warender of New Mexico University [11]. It uses system call audit trails to measure normal behaviors. While HMM produces better performance in modeling system call events than other methods, it requires very long time for modeling normal behaviors. The solution for this problem might be to improve the performance of computer system or to reduce the training data. The technique that reduces audit trails by filtering audit trails from abstracted information around the change of privilege flows can save the computational costs significantly while maintaining good performance [12].

3 Proposed Method

Among the intrusion types that frequently occur, buffer overflow, S/W security error, configuration error and denial of service attacks are prevalent on host. According to the hacking trends in May, June, and July 2002 reported by CERTCC the majority of attacks occurred is buffer overflow. Recently massive access to internet raises the issue of denial of service attack. This paper focuses on the two intrusion types to develop sophisticated detection method.

In this paper, we use common measures for host-based detection system: system call event, file system information and resource usage of process. Though there are several methods for modeling each measure, we select modeling methods appropriate to each measure considering the relationship between the characteristics of intrusion traces and the power of modeling methods. However, because each intrusion type leaves anomalies at individual measure, there are undetectable intrusion types in each measure and modeling method. To overcome this drawback, it is necessary to remedy shortcomings of each method through combining multiple detectors. We construct decision tree that combines multiple measure models and detect intrusions with it . The general architecture of the proposed method can be seen in Fig 1.

3.1 HMM with System Call Events

Sun Microsystem's Basic Security Module (BSM) which is auditing facility for Solaris provides an adequate representation of the behavior of the program be-

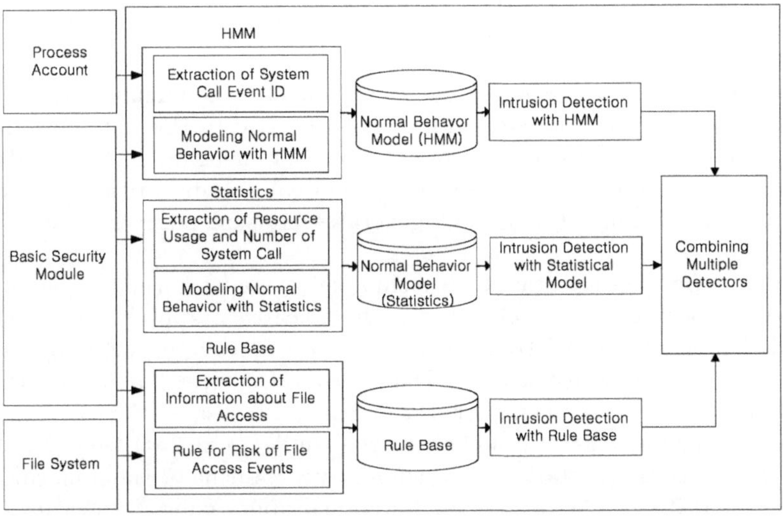

Fig. 1. Overall structure of the proposed method

cause any privileged activities that might be generated by a program are captured by BSM. Usually audit trail from BSM consists of several measures. The victim process of a certain attack generates system call events that are significantly different from events generated at normal situation. Thus, we can detect intrusions effectively by building the model of system call events from normal situation and noting significant deviation from the model. In this paper, for modeling system call event sequences, we use HMM that is widely used for speech recognition because it is very useful for modeling sequence information. HMM can be successfully applied to modeling system call event sequences [13].

Intrusion detection with HMM consists of two phases: normal behavior modeling and anomaly detection. The first phase is normal behavior modeling, which is determining HMM parameters to maximize the probability $Pr(O \mid \lambda)$ with which input sequence is generated out of given normal behavior model. Because no analytic solution is known for it, an iterative method called Baum-Welch reestimation is used. Anomaly detection, the second phase, matches current behavior against the normal behavior model, and calculates the probability with which it is generated out of the model. Forward-backward procedure is used for this purpose [14]. The probability is used to decide whether normal or not with a threshold.

3.2 Statistics with Resource Usage and System Call Events

Most of Unix-based operating systems serve Process Account (PACCT) as audit data. It provides the resource usage of processes: CPU time, memory usage, I/O usage, etc. Denial of service attack sharply raises the resource usage of victim process. This type of attack can be detected by noting processes that show unusual resource usage compared with normal behavior using PACCT audit data.

PACCT audit data is useful to detect DoS attacks. Unfortunately, we cannot detect attack type that targets resource not recorded to PACCT. For example, attack that consumes process table leaves no anomalies in PACCT. However, it unusually generates large amount of system call events. In this case, we can detect that by noting process that generates unusual number of system call events.

In this paper, we use statistical technique to model normal resource usage and the number of system call events. This statistical approach is a modified method that has used in NIDES. For each audit record generated by user, we generate a single test statistic value denoted by T that summarizes the degree of abnormality in the user's behavior in the near past. Large values of T indicate abnormal behavior, and values close to zero indicate normal behavior. In the case of PACCT, the T statistic is a summary judgment of the abnormality of measures in PACCT. We denote the individual abnormality of measures by S, each of which measures the degree of abnormality of behavior with respect to specific features such as CPU time, memory usage and I/O usage. T statistic has been set equal to the weighted sum of the S statistics as follows:

$$T = a_1 s_1 + a_2 s_2 + \ldots + a_n s_n \tag{1}$$

where s_n is the S score of each measure and a_n is the weight to each measure.

Each S statistics is derived from a corresponding statistic called Q. In fact, each S statistic is a normalizing transformation of the Q statistic so that the degree of abnormality for different types of features can be added on a comparable basis. The value of Q corresponding to the current audit record represents the number of audit records that are arrived in the recent past. In order to transform Q to S, we have built a normal historical profile of all previous values of Q and compare the current value of Q with this normal profile to determine if the current value is anomalous.

Small value of Q indicates a recent past that is similar to historical behavior, and large values of Q indicates a recent past that is not similar to historical behavior. Given that k is an index of appropriate audit records, t_k is the time that elapses between the kth and the most recent audit records, r is the decay rate and D_k is the change that occurs between the $(k+1)$st and kth appropriate audit records, Q is defined as follows [15]:

$$Q \sum_{k \geq 1} D_k \times 2^{-rt_k} \tag{2}$$

3.3 Rule-Base with File Access Events

Generally, attacks tend to access files of which the security level is high. For example, it executes file which has a SETUID privilege to acquire root privilege or attempts to read files owned by root to destroy the computer system or obtain a critical data. Owing to this tendency of abnormal behavior, it is adequately suspicious behavior for ordinary user to access files which have higher security level.

The best-known security policy related with file access is the Bell-LaPadula (BLP) model which has been formulated by Bell and LaPadula [16]. In an access to some information, there are three primary elements: subject, object and access attribute. The subject corresponds to user or program, the object corresponds to files and the access attribute corresponds to the kind of access model: read, write, execute, etc. In order to determine if a specific access mode is allowed, the BLP model compares the clearance of a subject with the classification of the object and determination is made as to whether the subject is authorized for the specific access mode [17]. The BLP model includes several rules referenced frequently. One of them is no read up rule, which allows access if the current level of the subject dominates the level of the object. It prevents user from accessing information for which they are not allowed to access.

In this paper, we audit file access and evaluate the risk of access event by analyzing the content of that. The security levels of subject and object are divided into 4 categories: root, administrator, ordinary user and guests. The risk of access event is evaluated to one of 21 levels according to the difference of security levels between subject and object. The access event of which the difference between two levels is bigger has higher risk. If the level of object is lower than that of subject the risk is not increased due to the security level difference. If the access event contains file that allows more operations, it has higher risk. For example, when the file mode is 755 the access event is riskier than that of 744.

3.4 Combining Multiple Detectors

In this paper, we have proposed an intrusion detection technique that combines multiple detectors in order to increase detectable attack types and reduce false-positive error rate. We use decision tree to combine multiple detectors. Decision tree is one of the most widely used methods for inductive inference. Constructing decision tree is to build up tree data structure that predicts the class of a case in terms of the attribute values of the case. Each node of tree specifies a test of some attribute of the case and each branch corresponds to one of the possible values for this attribute. The tree-construction algorithm infers trees by growing them from the root downward and selecting the next best attribute for each new branch added to the tree.

We have used Quinlan's C4.5 which is a well-known decision tree generation algorithm [18]. C4.5 uses an information-theoretical approach based on the energy entropy. It builds the decision tree using a divide-and-conquer approach: select an attribute, divide the training set into subsets characterized by the possible values of the attribute, and follow the same partitioning procedure recursively with each subset until no subset contains objects from more than one class [19]. The single class subsets correspond them to the leaves. The entropy-based criterion that has been used for the selection of the attribute is called the gain ratio criterion.

Let T be the set of classes of cases associated at the node. At first, the frequency of cases in T which belongs to some class C_j is computed. If there is a class C_j whose frequency is 1 (i.e. all cases belongs to cases) or higher than some value a leaf node is made with associated class C_j. If there is no class that has high frequency the information gain of each attribute is calculated. The information gain is a measure of the effectiveness of an attribute in classifying the data [20]. The attribute which has maximum information gain is selected. If the selected attribute is continuous, the tree branches into binary nodes with a certain threshold. If the selected attribute is discrete, the tree branches for each possible value of that attribute. Let T' one of the sets that are produced by the test on the selected attribute and the same operation is applied recursively on each non-empty T'. If T' is empty, the child node is a leaf. The overall process of tree-construction algorithm is given as follows:

```
[Step 1] compute class frequency of T
[Step 2] if one or few cases
             return a leaf
         else create a decision node N
[Step 3] for each attribute A
             compute the best attribute A
[Step 4] if A is continuous
             find Threshold
[Step 5] for each T' in the splitting T
             if T is Empty
                 child of N is a leaf
             else
                 child of N = form a tree
```

An audit data set that contains labeled attacks mixed with normal behavior is collected for training decision tree. Each of the four detectors evaluates this data set and the evaluation result constitutes the decision tree training data that conforms to data format of the decision tree generator. Each record of training data consists of four continuous real value (evaluation results) and labeled as two classes: attack and normal. The decision tree generator constructs decision

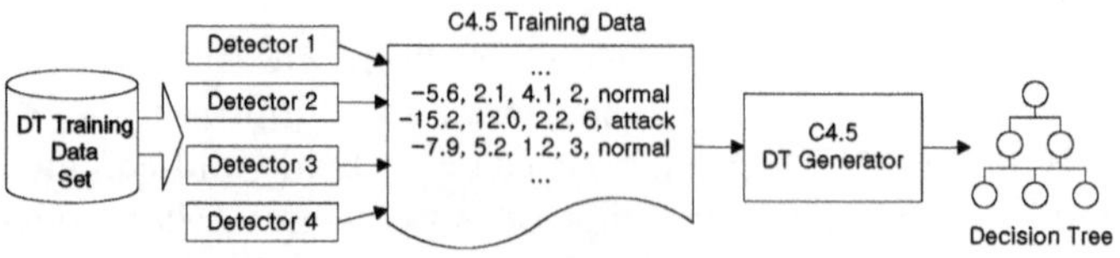

Fig. 2. Constructing the decision tree

tree and generated tree is used to combine multiple detectors. Overall process of constucuting decision tree is shown in Fig. 2.

The decision tree is constructed with decision tree training data set as shown in Fig. 3. 8 rules are generated by decision tree generator since each leaf node corresponds to one rule. The rules that are more operative (frequently used to classify items) are given as follows:

```
(1) IF ((HMM/System Call <= -11.7) AND (Rule-base/File Access > 11))
      THEN ATTACK
(2) IF ((HMM/System Call > -11.7) AND (Statistics/Resource Usage <= 6.7))
          AND Statistics/System Call <= 10)
      THEN NORMAL
(3) IF ((HMM/System Call > -11.7) AND (Statistics/Resource Usage <= 6.7)
      AND (Statistics/System Call > 10) AND (Rule Base/File Access <= 4))
      THEN ATTACK
(4) IF ((HMM/System Call > -11.7) AND (Statistics/Resource Usage > 6.7)
          AND (Rule Base/File Access <= 1))
      THEN ATTACK
```

Rule 1 classifies the process that produces suspicious system call sequences and accesses risky files as attack. This rule detects most of buffer overflow attacks because it generates abnormal system call event sequences and accesses files that have SETUID previledge to obtain root privilege. Rule 2 discriminates most of normal behaviors because it classifies the processes that generate normal system call sequences and show ordinary resource usage and system call as normal behavior. Rule 3 and 4 detect denial of service type attacks. Rule 3 classifies as attack processes that show ordinary resource usage and system call sequences but use large amount of system call such as process table consuming attack. Rule 4 determines processes that shows unusual resource usage such as disk and memeory consumption as attack. Both of rules that detect the denial of services attacks include the 'smaller than' condition with the risk of file access because this type of attack does not require to access the risky files.

4 Experiments

We have collected normal behaviors from six graduate students for 2 weeks using Solaris 7 operating system. They have mainly used text editor, compiler and programs of their own writing. Total 13 megabytes (160,448 records) of BSM audit data and 840 kilobytes of PACCT audit data have been collected from 16,470 commands. We also collect audit data that contain labeled attacks for testing in the same operating system. It contains 9 cases of u2r buffer overflow intrusion and 4 cases of denial of services. Audit data for constructing decision tree have also been collected. It contains 28 cases of labeled attack mixed with normal behavior. The attacks used in our experiments are as shown in Table 2.

In this paper, the experiments have been conducted in three phases. At first, we have built behavior profile of normal data and tested the performance of each detector. Next, we construct decision tree with decision tree training algorithm to

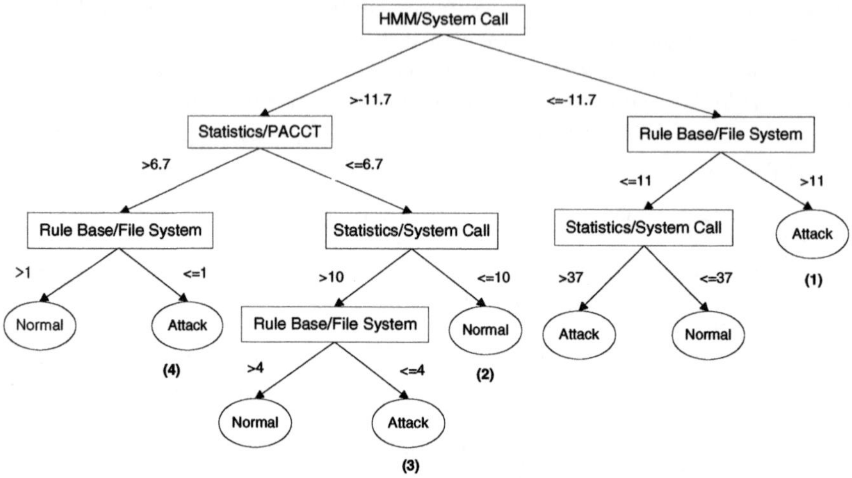

Fig. 3. The constructed decision tree

Table 2. The attacks used in our experiments

Attack Type	Attack Name
Buffer Overflow	kcms_configure buffer overflow vulnerability lpset -r buffer overflow vulnerability xlock heap buffer overflow vulnerability
Denial of Service	Consumption of Disk Consumption of Process Table Consumption of Memory

combine multiple detectors. The performance of combined method is evaluated with test data set. The overview of the experiment is shown in Fig. 4.

4.1 Performance Metrics

The important metrics in evaluating the performance of IDS are detection rate and false-positive error rate. We calculate the detection rate as the rate of detected intrusions. Each attack counts as one detection if at least one process which is used in attack process is evaluated as intrusion. The false-positive error rate is calculated as the rate of mis-classified processes.

In the experiments, we visualize the performance of detection method through ROC (Receiver Operating Characteristics) curve. An ROC curve depicts the changes in attack detection rate and false-positive error rate as modifiable parameter is changed. In this paper, evaluation threshold is used as a modifiable parameter. A desirable intrusion detection system must show a high detection

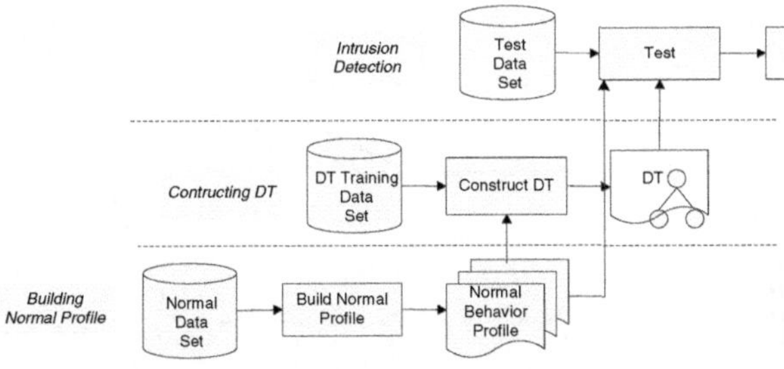

Fig. 4. Overview of the experiment

rate at low false-positive error rate. In ROC curve, top-left curve is more desirable.

We use discriminability to measure numerically the performance of a detection method and efficiency to compare integration methods with others. The discriminability is a measure of the average intensity difference perceived by an observer between samples including the signal and samples not including a signal. It has higher value at high detection rate and low false-positive error rate [21]. Generally, it has been noted as d' and defined as follows:

$$d' = z(H) - z(F) \tag{3}$$

where z is the inverse of the normal distribution function, H is detection rate and F is false-positive error rate.

4.2 Experimental Results

At first, we have conducted experiments without the integration of detection methods in order to identify the characteristics of each method and find optimal parameters of each method. The experimental result of individual detection methods does not show good performance owing to the characteristics of measure and modeling methods. We have compared the detection methods with the best parameters: The number of states is 3 and the length of input sequence is 8 in HMM and the weight ratio to resource usage is 2:1:2 (CPU time: memory usage: I/O usage). The false-positive error rate has been sharply raised after a certain degree of detection rate as shown in Fig. 5.

The experiment with the proposed method using the constructed decision tree is conducted using the test data set with the parameters of each method that show the best performance in the preliminary experiments. We have compared false-positive error rate and discriminability of each method at 100% detection rate as shown in Table 3. The result shows the combined detection method dramatically reduces the false-positive error rate.

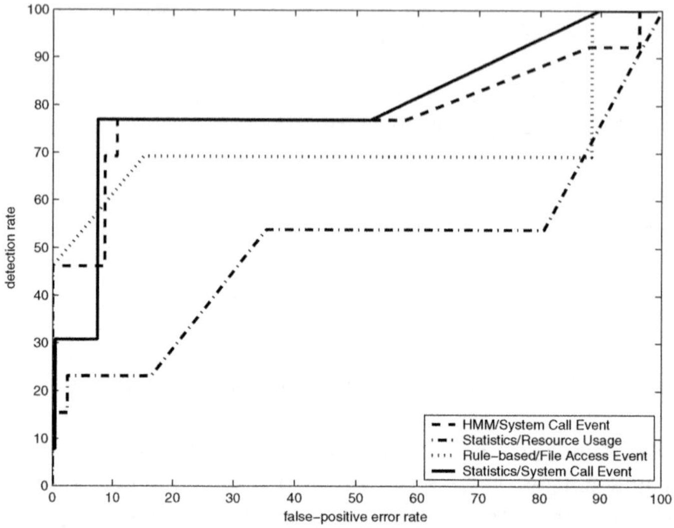

Fig. 5. ROC for individual detection methods with best parameters

Table 3. A comparison of each detection method

Method	Detection rate	F-P error rate	d'
HMM/System Call Event	100%	99.177%	1.866
Statistics/Resource Usage	100%	80.658%	3.400
Statistics/System Call Event	100%	99.177%	1.866
Rule base/File Access Event	100%	88.477%	3.066
Combining with Decision Tree	100%	0.412%	5.733

5 Conclusions and Future Work

In this paper, we have presented effective modeling methods for three audit data and proposed a novel intrusion detection technique that integrates the detection methods. The proposed method uses multiple measure and modeling methods and combines the results of individual detectors using decision tree to overcome the drawbacks of the conventional anomaly detection techniques.

We evaluate the performance of each detection method and compare the results with proposed method. The proposed method shows 0.412% false-positive error rate at 100% detection rate whereas the other methods show more than 80% false-positive error rate at the same detection rate. It indicates that we can overcome the drawbacks by integrating several methods.

Decision tree has the advantage that the user can intuitively and visually understand rules which are generated by decision tree that combines multiple detectors autonomously. In the future, it is needed to perform experiments with larger data set such as the data set of DARPA Intrusion Detection System

Evaluation program for more accurate evaluation of the proposed method. In addition, it is also needed to consider the computaional cost for using multiple models.

Acknowledgements

This research was supported by University IT Research Center Project.

References

1. H. S. Vaccaro and G. E. Liepins, "Detection of anomalous computer session activity," *In Proceedings of IEEE Symposium on Research in Security and Privacy*, pp. 280-289, 1989. 208
2. S. Axelsson, "Intrusion detection systems: A survey and taxonomy," *Technical Report 99-15*, Department of Computer Engineering, Chalmers University, March 2000. 208
3. E. Biermann, E. Cloete and L. M. Venter, "A comparison of intrusion detection systems," *Computers & Security*, vol 20, no. 8, pp. 676-683, December 2001. 209
4. C. Dowel and P. Ramstedt, "The computer watch data reduction tool," *In Proceedings of the 13th National Computer Security Conference*, pp. 99-108, Washington DC, USA, October 1990. 209
5. T. Heberlein, G. Dias, K. Levitt, B. Mukherjee, J. Wood, and D. Wolber, "A network security monitor," *In Proceedings of the 1990 IEEE Symposium on Research in Security and Privacy*, pp. 296-304, Los Alamitos, CA, USA, 1990. 209
6. S. Staniford-Chen, S. Cheung, R. Crawford, M. Dilger, J. Frank, J. Hoagland, K Levitt, C. Wee, R. Yip, and D. Zerkle, "GrIDS-A graph based intrusion detection system for large networks," *In Proceedings of the 19th National Information Systems Security Conference*, vol. 1, pp. 361-370, October, 1996. 209
7. T. F. Lunt, A. Tamaru, F. Gilham, R. Jagannathan, C. Jalali, and P. G. Neuman, "A real-time intrusion-detection expert system (IDES)," *Technical Report Project 6784*, CSL, SRI International, Computer Science Laboratory, SRI International, February 1992. 209
8. D. Anderson, T. F. Lunt, H. Javits, A. Tamaru and A. Valdes, "Detecting unusual program behavior using the statistical components of NIDES," *NIDES Technical Report*, SRI International, May 1995. 209
9. P. A. Porras and P. G. Neumann, "EMERALD: Event monitoring enabling responses to anomalous live disturbances," *In Proceedings of the 20th National Information Systems Security Conference*, pp. 353-365, Baltimore, Maryland, USA, October 1997. 209
10. H. Debar, M. Becker and D. Siboni, "A neural network component for an intrusion detection system," *In Proceedings of 1992 IEEE Computer Society Symposium on Research in Security and Privacy*1, pp. 240-250, Oakland, CA, May 1992. 209, 210
11. C. Warrender, S. Forrest and B. Pearlmutter, "Detecting intrusion using calls: Alternative data models," *In Proceedings of IEEE Symposium on Security and Privacy*, pp. 133-145, May 1999. 209, 210
12. S.-B. Cho and H.-J. Park, "Efficient anomaly detection by modeling privilege flows with hidden Markov model," *Computers & Security*, vol. 22, no. 1, pp. 45-55, 2003. 209, 210

13. S.-B. Cho, "Incorporating soft computing techniques into a probabilistic intrusion detection system," *IEEE Transactions on Systems, Man and Cybernetics-Part C:Applications and Reviews*, vol. 32, no. 2, pp. 154-160, May 2002. 211
14. L. R. Rabiner and B. H. Juang, "An introduction to hidden Markov models," *IEEE ASSP Magazine*, pp. 4-16, January 1986. 211
15. H. S. Javitz and A. Valdes, "The SRI IDES statistical anomaly detector," *NIDES Technical Report*, SRI International, 1991. 212
16. Bell, D. E. and LaPadula, L. J., "Secure computer systems: Unified exposition and multics interpretation," *Mitre Technical Report ESD-TR-75-306*, Mitre Corporation, March 1976. 213
17. A. G. Amoroso, *Fundamentals of Computer Security Technology*, PTR Prentice Hall, New Jersy, 1994. 213
18. J. R. Quinlan, *C4.5: Programs for Machin Learning*, Morgan Kaufmann, 1993. 213
19. S. Ruggieri, "Efficient C4.5," *IEEE Transactions on Knowledge and Data Engineering*, vol. 14, no. 2, March 2002. 214
20. T. Mitchell, *Machine Learning*, McGraw-Hill, 1997. 214
21. N. A. Macmillan and C. D. Creelman, *Detection Theory: A User's Guide*, Cambridge University Press, Cambridge, 1991. 217

Association Rule Discovery with Unbalanced Class Distributions

Lifang Gu[1], Jiuyong Li[2], Hongxing He[1], Graham Williams[1], Simon Hawkins[1], and Chris Kelman[3]

[1] CSIRO Mathematical and Information Sciences
GPO Box 664, Canberra, ACT 2601, Australia
{lifang.gu,hongxing.he,graham.williams,simon.hawkins}@csiro.au
[2] Department of Mathematics and Computing
The University of Southern Queensland
jiuyong@usq.edu.au
[3] Commonwealth Department of Health and Ageing
christopher.kelman@health.gov.au

Abstract. There are many methods for finding association rules in very large data. However it is well known that most general association rule discovery methods find too many rules, many of which are uninteresting rules. Furthermore, the performances of many such algorithms deteriorate when the minimum support is low. They fail to find many interesting rules even when support is low, particularly in the case of significantly unbalanced classes. In this paper we present an algorithm which finds association rules based on a set of new interestingness criteria. The algorithm is applied to a real-world health data set and successfully identifies groups of patients with high risk of adverse reaction to certain drugs. A statistically guided method of selecting appropriate features has also been developed. Initial results have shown that the proposed algorithm can find interesting patterns from data sets with unbalanced class distributions without performance loss.

Keywords: knowledge discovery and data mining, association rules, record linkage, administrative data, adverse drug reaction

1 Introduction

The aim of association rule mining is to detect interesting associations between items in a database. It was initially proposed in the context of market basket analysis in transaction databases, and has been extended to solve many other problems such as the classification problem. Association rules for the purpose of classification are often referred to as *predictive association rules*. Usually, predictive association rules are based on relational databases and the consequences of rules are a pre-specified column, called the class attribute.

One of the problems with conventional algorithms for mining predictive association rules is that the number of association rules found is too large to handle,

T.D. Gedeon and L.C.C. Fung (Eds.): AI 2003, LNAI 2903, pp. 221–232, 2003.

even after smart pruning. A new algorithm, which mines only the optimal class association rule set, has been developed by one of the authors [9] to solve this problem.

A second problem with general predictive association rule algorithms is that many interesting association rules are missed even if the minimum support is set very low. This is particularly a problem when a dataset has very unbalanced class distributions, as is typically the case in many real world datasets.

This paper addresses the problem of finding interesting predictive association rules in datasets with unbalanced class distributions. We propose two new interestingness measures for the optimal association rule algorithm developed earlier and use them to find all interesting association rules in a health dataset containing classes which are very small compared to the population.

In collaboration with the Commonwealth Department of Health and Ageing and Queensland Health, CSIRO Data Mining has created a unique cleaned and linked administrative health dataset bringing together State hospital morbidity data and Commonwealth Medicare Benefits Scheme (MBS) and Pharmaceutical Benefits Scheme (PBS) data. The Queensland Linked Data Set (QLDS) links de-identified, administrative, unit-level data, allowing de-identified patients to be tracked through episodes of care as evidenced by their MBS, PBS and Hospital records [16]. The availability of population-based administrative health data set, such as QLDS, offers a unique opportunity to detect common and rare adverse reactions early. This also presents challenges in developing new methods, which detect adverse drug reactions directly from such administrative data since conventional methods for detecting adverse drug reactions work only on data from spontaneous reporting systems, or carefully designed case-control studies [2, 5, 11].

This paper presents an association algorithm which uses QLDS to identify groups with high risk of adverse drug reaction. Section 2 describes an algorithm for mining interesting class association rule sets. Section 3 presents a method of feature selection based on statistical analysis. In Section 4 we give a brief description of QLDS and the features selected. Results from mining the optimal class association rule set are then presented in Section 5. Section 6 concludes the paper with a summary of contributions and future work.

2 Interesting Association Rule Discovery

Association rule mining finds interesting patterns and associations among patterns. It is a major research topic in data mining since rules are easy to interpret and understand. However, general association rule discovery methods usually generate too many rules including a lot of uninteresting rules. In addition, most algorithms are inefficient when the minimum support is low. In this section, we propose two new interestingness criteria which overcome problems of existing approaches and can be used in finding interesting patterns in data sets with very unbalanced class distributions.

2.1 Interestingness Criterion Selection

We first introduce the notation used in this paper.
A set of attributes is $\{A_1, A_2, \ldots, A_m, C\}$. Each attribute can take a value from a discrete set of values. For example, $A_i \in \{a_{i_1}, a_{i_2}, \ldots, a_{i_p}\}$. C is a special attribute which contains a set of categories, called classes. A record is $T \in A_1 \times A_2 \times \ldots \times A_m \times C$, and a set of records from a data set is denoted by $D = \{T_1, T_2, \ldots, T_n\}$. A pattern is a subset of a record, $P \subset D$. The support of a pattern P is the ratio of the number of records containing P to the number of all records in D, denoted by $\sup(P)$. A rule is in the form of $P \Rightarrow c$. The support of the rule is $\sup(P \cup c)$ [1]. The confidence of the rule is defined as $\mathrm{conf}(P \Rightarrow c) = \sup(Pc)/\sup(P)$.

If a data set has very unbalanced class distributions many existing interestingness criteria are unable to capture interesting information from the data set. In the following we will use a two-class example to demonstrate this. Assume that $\sup(c_1) = 0.99$ and $\sup(c_2) = 0.01$. Note that $\neg c_1 = c_2$ since there are only two classes.

One of the mostly used interestingness criteria is *support*. A minimum support of 0.01 is too small for class c_1, but too large for class c_2. Another such criterion is *confidence*, which can be written for class c_2 as $\mathrm{conf}(P \Rightarrow c_2) = \frac{\sup(Pc_2)}{\sup(Pc_1)+\sup(Pc_2)}$. We can see that any noise in class c_1 will have a significant impact on class c_2, e.g. causing $\sup(Pc_1) \sim \sup(Pc_2)$. As a result, we can hardly find any high confidence rules in the smaller class, c_2. Similarly, *gain* [6] suffers the same problem as confidence.

Other alternatives can be similarly evaluated. For example, *conviction* [3], defined as $\mathrm{conviction}(P \Rightarrow c) = \frac{\sup(P)\sup(\neg c)}{\sup(P\neg c)}$, measures deviations from the independence by considering outside negation. It has been used for finding interesting rules in census data. A rule is interesting when conviction is far greater than 1. The conviction for class c_2 in our example is written as $\mathrm{conviction}(P \Rightarrow c_2) = \frac{\sup(P)\sup(c_1)}{\sup(Pc_1)}$. Since $\sup(c_1) \approx 1$, we have $\sup(Pc_1) \approx \sup(P)$. As a result, $\mathrm{conviction}(P \Rightarrow c_2) \approx 1$. This means we will not find any interesting rules in the small class using the *conviction* metric.

On the other hand, we can prove that *lift* [15], *interest* [3], *strength* [4]) or the *p-s metric* [12] all favour rules in small classes. Here we use lift to show this. The metric *lift* is defined by $\mathrm{lift}(P \Rightarrow c) = \frac{\sup(Pc)}{\sup(P)\sup(c)}$. A rule is interesting if its lift is far greater than 1. For large class c_1 we have $\mathrm{lift}(P \Rightarrow c_1) = \frac{\sup(Pc_1)}{\sup(P)\sup(c_1)} \approx 1$ since $\sup(c_1) \approx 1$. As a result, we can hardly find any high lift rules from the large class using the *lift* metric.

A statistical metric is fair for rules from both large and small classes. One such statistical metric is the *odds-ratio*, which is defined as $\text{Odds-Ratio}(P \Rightarrow c) = \frac{\sup(Pc)\sup(\neg P\neg c)}{\sup((\neg P)c)\sup(P\neg c)}$. However, odds-ratio does not capture the interesting rules we are looking for. We show this with the following examples.

---

[1] For convenience, we abbreviate $P \cup c$ as Pc in the rest of the paper.

Notation

	P	$\neg P$
c_1	$\text{sup}(Pc_1)$	$\text{sup}(\neg Pc_1)$
c_2	$\text{sup}(Pc_2)$	$\text{sup}(\neg Pc_2)$

Example 1

	P	$\neg P$
c_1	0.297	0.693
c_2	0.006	0.004

Example 2

	P	$\neg P$
c_1	0.792	0.198
c_2	0.0095	0.0005

In Example 1, the probability of pattern P occurring in class c_2 is 0.6, which is twice as that in class c_1. In Example 2, the probability of pattern P occurring in class c_2 is 0.95, which is only 1.19 times of that in class c_1. Hence rule $P \Rightarrow c_2$ is more interesting in Example 1 than in Example 2. However, odds-ratio$(A \Rightarrow c_2)$ is 3.5 in Example 1 and 4.75 in Example 2 and this leads to miss the interesting rule in Example 1.

To have an interestingness criterion that is fair for all classes regardless of their distributions, we propose the following two metrics.

- Local support, defined in Equation 1:

$$\text{lsup}(P \Rightarrow c) = \text{sup}(Pc)/\text{sup}(c) \tag{1}$$

  When we use local support, the minimum support value will vary according to class distributions. For example, a local support value of 0.1 means 0.099 support in class c_1 and 0.001 in class c_2.
- Exclusiveness, defined in Equation 2:

$$\text{excl}(P \Rightarrow c_i) = \frac{\text{lsup}(P \Rightarrow c_i)}{\sum_j^{|C|} \text{lsup}(P \Rightarrow c_j)} \tag{2}$$

  Such a metric is normalised ($[0, 1]$) and fair for all classes. If pattern P occurs only in class c_i, the exclusiveness will reach one, which is the maximum value.

We now discuss the practical meaning of the exclusiveness metric. From the formula, it can be seen that it is a normalised lift, i.e.,

$$excl(P \Rightarrow c_i) = \frac{\text{lift}(P \Rightarrow c_i)}{\sum_j^{|C|} \text{lift}(P \Rightarrow c_j)} \tag{3}$$

The term $\text{lift}(P \Rightarrow c_i)$ is the ratio of the probability of P occurring in class c_i to the probability of P occurring in data set D. Hence the higher the lift, the more strongly P is associated with class c_i. However as discussed above, $\text{lift}(P \Rightarrow c_i) \approx 1$ when $\text{sup}(c_i) \approx 1$. As a result, it is difficult to get a uniform minimum lift cutoff for all classes. From Equation 2, it can be seen that $|C| \times \text{excl}(P \Rightarrow c_i)$ is the ratio of the lift of P in Class c_i to the average lift P in all classes. Therefore, it is possible to set a uniform exclusiveness cutoff value for all classes regardless of their distributions. More importantly, the metric exclusiveness reveals extra information that lift fails to identify. For example, if we have three classes c_1, c_2 and c_3, and $\text{sup}(c_1) = 0.98$, $\text{sup}(c_2) = 0.01$ and $\text{sup}(c_3) = 0.01$, it is possible that we have a pattern P such that both $\text{lift}(P \Rightarrow c_2)$ and $\text{lift}(P \Rightarrow c_3)$ are very high. However, exclusiveness will not be high for either class since P is not exclusive to any single class.

The above two metrics (local support and exclusiveness) and lift are used in our application for identifying groups of high risk to adverse drug reactions.

2.2 Efficient Interesting Rule Discovery

There are many association rule discovery algorithms such as the classic Apriori [1]. Other algorithms are faster, including [7, 14, 17], but gain speed at the expense of memory. In many real world applications an association rule discovery method fails because it runs out of memory, and hence Apriori remains very competitive. In this application, we do not use any association rule discovery algorithm since it is not necessary to generate *all* association rules for interesting rule discovery.

In the following, we first briefly review definitions of a general algorithm for discovering interesting rule sets and informative rule sets [8, 10], and then argue that the algorithm fits well with our application.

Given two rules $A \Rightarrow c$ and $AB \Rightarrow c$, we call the latter more specific than the former or the former more general than the latter. Based on this concept, we have the following definition.

Definition 1. *A rule set is an informative rule set if it satisfies the following two conditions:* 1) *it contains all rules that satisfy the minimum support requirement; and* 2) *it excludes all more specific rules with a confidence no greater than that of any of its more general rules.*

A more specific rule covers a subset of records that are covered by one of its more general rules. In other words, a more specific rule has more conditions but explains less cases than any of its more general rules. Hence we only need a more specific rule when it is more interesting than all of its more general rules. Formally, $P \Rightarrow c$ is interesting only if for all $P' \subset P$ Interestingness$(P \Rightarrow c) > Interestingness(P' \Rightarrow c)$. Here Interestingness stands for an interestingness metric. In the current application we use local support, lift and exclusiveness as our metrics. Local support is used to vary the minimum support among unbalanced classes to avoid generating too many rules in one class and too few rules in another class. Lift is used to consider rules in a single small class, and exclusiveness is used for comparing interesting rules among classes. We prove in the following that using the informative rule set does not miss any interesting rules instead of the association rule set.

Lemma 1. *All rules excluded by the informative rule set are uninteresting by the lift and exclusiveness metrics.*

Proof. In this proof we use $AB \Rightarrow c$ to stand for a more specific rule of rule $A \Rightarrow c$. $AB \Rightarrow c$ is excluded from the informative rule set because we have $\mathrm{conf}(AB \Rightarrow c) \leq \mathrm{conf}(A \Rightarrow c)$.

We first prove that the lemma holds for lift. We have

$$\mathrm{lift}(A \Rightarrow c) = \frac{\mathrm{sup}(Ac)}{\mathrm{sup}(A)\mathrm{sup}(c)} = \frac{\mathrm{conf}(A \Rightarrow c)}{\mathrm{sup}(c)} \geq \frac{\mathrm{conf}(AB \Rightarrow c)}{\mathrm{sup}(c)} = \mathrm{lift}(AB \Rightarrow c)$$

As a result, rule $AB \Rightarrow c$ is uninteresting according to the lift criterion.

For the exclusiveness, we first consider a two-class case, c and $\neg c$. We have $\mathrm{conf}(A \Rightarrow \neg c) = 1 - \mathrm{conf}(A \Rightarrow c)$. Since $\mathrm{conf}(AB \Rightarrow c) \leq \mathrm{conf}(A \Rightarrow c)$, we have conf$(AB \Rightarrow \neg c) \geq$ conf$(A \Rightarrow \neg c)$. Further we have lift$(AB \Rightarrow c) \leq$ lift$(A \Rightarrow c)$, lift$(AB \Rightarrow \neg c) \geq$ lift$(A \Rightarrow \neg c)$ and therefore

$$\begin{aligned}\text{excl}(A \Rightarrow c) &= \frac{\text{lift}(A \Rightarrow c)}{\text{lift}(A \Rightarrow c) + \text{lift}(A \Rightarrow \neg c)} \geq \frac{\text{lift}(A \Rightarrow c)}{\text{lift}(A \Rightarrow c) + \text{lift}(AB \Rightarrow \neg c)} \\ &\geq \frac{\text{lift}(AB \Rightarrow c)}{\text{lift}(AB \Rightarrow c) + \text{lift}(AB \Rightarrow \neg c)} = \text{excl}(A \Rightarrow c)\end{aligned}$$

Hence, $AB \Rightarrow c$ is uninteresting according to the exclusiveness metric.

For more than two classes, we do not provide a direct proof. Instead we have the following rational analysis. Let c_i be any class. $\sum_j^{|C|} \text{conf}(A \Rightarrow c_j) = \sum_j^{|C|} \text{conf}(AB \Rightarrow c_j) = 1$, When conf$(AB \Rightarrow c_i) \leq$ conf$(A \Rightarrow c_i)$, we must have at least one class c_j such that conf$(AB \Rightarrow c_j) \geq$ conf$(A \Rightarrow c_j)$. Further, we have lift$(AB \Rightarrow c_i) \leq$ lift$(A \Rightarrow c_i)$ and lift$(AB \Rightarrow c_j) \geq$ lift$(A \Rightarrow c_j)$. As a result, pattern AB is more interesting in class c_j than in class c_i, and rule $AB \Rightarrow c_i$ is uninteresting by exclusiveness.

Thus we can use the informative rule set instead of the association rule set for interesting rule discovery.

The advantages of using the informative rule set are listed as following. Firstly, the informative rule set is much smaller than the association rule set when the minimum support is small. Secondly, the informative rule set can be generated more efficiently than an association rule set. Thirdly, this efficiency improvement does not require additional memory, and actually the informative (IR) rule set generation algorithm [8, 10] uses less memory than Apriori.

The IR algorithm was initially implemented on transactional data sets where there are no pre-specified classes. Optimal Class Association Rule set generator (OCAR)[9] is a variant of the IR algorithm on relational data sets for classification. Since a relational data set is far denser than a transactional data set, OCAR is significantly more efficient than Apriori.

In our application we used a modified OCAR, which uses the exclusiveness as the interestingness metric instead of the estimated accuracy as used in the original OCAR.

Rule discovery requires appropriate features to find significant rules. Feature selection is therefore discussed in the next section.

3 Feature Selection Method

We use statistical methods such as bivariate analysis and logistic regression to identify the most discriminating features associated with patient classes.

3.1 Bivariate Analysis

Assume that the dependent variable represents a yes/no binary outcome, e.g., having the disease or not having the disease, an $m \times 2$ frequency table (Table 1) can be used to illustrate the calculation of the χ^2 value for a categorical

Table 1. The $m \times 2$ frequency table. Here the row total, R_i, and the column total, C_j are the sums of all cells in a row and column respectively, i.e., $R_i = n_{i1} + n_{i2}$ and $C_j = \sum_{i=1}^{m} n_{ij}$

Independent Variable level	Dependent Variable Yes	No	Row Total
1	n_{11}	n_{12}	R_1
2	n_{21}	n_{22}	R_2
3	n_{31}	n_{32}	R_3
$\vdots$	$\vdots$	$\vdots$	$\vdots$
m	n_{m1}	n_{m2}	R_m
Column Total	C_1	C_2	

independent variable with m levels. Specifically, we have

$$\chi^2 = \sum_{i=1}^{m} [\frac{(n_{i1} - E_{i1})^2}{E_{i1}} + \frac{(n_{i2} - E_{i2})^2}{E_{i2}}] \quad (4)$$

where E_{ij} is the expected count in cell ij, which is equal to $C_j R_i / N$, $i = 1, \cdots, m$ and $j = 1, 2$. The term, N, refers to the total number of counts.

The calculated χ^2 value for each independent variable can be compared with a critical (or cut-off) χ^2 value for $m-1$ degrees of freedom at a required p value. The value p denotes the degree of confidence with which to test the association. For example, a p value of 0.01 indicates that the association will be tested at the 99% confidence. If the calculated χ^2 value is larger than the cut-off value, it can be concluded that the independent variable is associated with the dependent variable in the study population. Otherwise, the independent variable is not related to the dependent variable and will not be selected as a feature.

Similarly if the independent variable is continuous, the t value can be used to test for correlation between the dependent and independent variables. Since our class association rule discovery algorithm only takes categorical variables, calculation details of the t value are skipped.

3.2 Logistic Regression

An alternative multivariate statistical method is logistic regression. A logistic regression model can be written as the following equation:

$$\ln(\frac{p}{1-p}) = \alpha + \beta_1 x_1 + \beta_2 x_2 + \cdots + \beta_n x_n \quad (5)$$

where p is the probability, at which one of the binary outcomes occurs (e.g., the probability of having the disease), α is the intercept, and β_i is the coefficient of the independent variable x_i. The coefficient, β_i, can be transformed into a more meaningful measure, the odds ratio (OR), by the following formula:

$$OR_i = e^{\beta_i} \quad (6)$$

Odds ratios can be used to infer the direction and magnitude of the effects of the independent variables on the probability of the outcome. An odds ratio greater than 1 indicates that the probability of the outcome (e.g., having the disease) will increase when a continuous independent variable increases a unit in its value or when a specific group of a categorical independent variable is compared to its reference group. For example, if the dependent variable is the *probability of having migraine*, the independent variable is *gender*, and *male* is used as the reference group, an odds ratio of 2.0 then implies that females are 2.0 times more likely to have migraine than males. As a result, only independent variables with odd ratio values much larger or smaller than 1 will be selected as important features.

4 Data

The Queensland Linked Data Set [16] has been made available under an agreement between Queensland Health and the Commonwealth Department of Health and Ageing (DoHA). The data set links de-identified patient level hospital separation data (1 July 1995 to 30 June 1999), Medicare Benefits Scheme (MBS) data, and Pharmaceutical Benefits Scheme (PBS) data (1 January 1995 to 31 December 1999) in Queensland.

Each record in the hospital data corresponds to one inpatient episode. Each record in MBS corresponds to one MBS service for one patient. Similarly, each record in PBS corresponds to one prescription service for one patient. As a result, each patient may have more than one hospital, or MBS or PBS record.

4.1 Selection of Study Population

PBS data in QLDS contain mostly prescription claims for concessional or repatriate cardholders. There are a total of 733,335 individuals in PBS and 683,358 of them appear as always concessional or repatriate during our five year study period. This represents 93% of all individuals in PBS. Since the drug usage history of these 683,358 individuals is covered more completely in PBS, a subset of them is chosen as our study population, i.e, those who take a particular type of drug.

The adverse drug reaction to be investigated in this study is angioedema associated with the use of ACE inhibitors [13]. This is a known adverse reaction and the aim of this investigation is to confirm its existence from administrative health data using the proposed algorithm. Drugs are identified in PBS using the WHO codes, which are based on the Anatomical and Therapeutic Classification (ATC) system. Adverse events are identified in hospital data using principal diagnosis codes. Table 2 shows the number of ACE inhibitor users split into two classes, those having and not having angioedema. It can be seen that the distribution of the two classes is very unbalanced and Class 1 is only 0.088% of the whole study population. However, this is the class we are interested in characterising. Section 2 has already described how to find rules to characterise such an under-represented class.

Table 2. Study population split into two classes

	Angioedema		
	Yes (Class 1)	No (Class 0)	Total
Counts	116	131,184	132,00
Percentage	0.088	99.912	100

4.2 Feature Selection

The aim of the feature selection process is to select those variables which are strongly associated with the dependent variable, based on statistical analysis described in Section 3. For our particular study the dependent variable is the probability of having angioedema for ACE inhibitor users.

From the hospital data we initially extract variables, such as age, gender, indigenous status, postcode, the total number of bed days, and 8 hospital flags. The last two variables are used to measure the health status of each patient. From the PBS data, 15 variables (the total number of scripts of ACE inhibitors and 14 ATC level-1 drug flags) are initially extracted. The variable "total number of scripts" can be used to indicate how long an individual has been taking ACE inhibitors. The ATC level-1 drug flags are used to investigate adverse reactions caused by possible interactions between ACE inhibitors and other drugs.

Using the feature selection method described in Section 3, the 15 most discriminating features are selected. These features include age, gender, hospital flags, and flags for exposure to other drugs. The optimal class association rule discovery algorithm then run on the extracted data.

5 Results

The interesting association rule discovery algorithm described in Section 2 is applied to the data set with 132,000 records. There are several input parameters to the algorithm. Two of them are the minimum local support and the maximum length (number of variables used in each rule). In our tests we set the minimum local support to 0.05 and the maximum length to 6. The rules (thousands) are sorted by their value of interestingness in descending order. Only the top few rules with highest interestingness values are described here for verbosity. The top three rules and their characteristics are shown in Table 3.

Rule 1
- Gender = Female
- Age $\geq$ 60
- Took genito urinary system and sex hormone drugs = Yes
- Took Antineoplastic and immunimodulating agent drugs = Yes
- Took musculo-skeletal system drugs = Yes

Rule 2
- Gender = Female
- Had circulatory disease = Yes
- Took systemic hormonal preparation drugs = Yes
- Took musculo-skeletal system drugs = Yes
- Took various other drugs = Yes

Rule 3
- Gender = Female
- Had circulatory disease = Yes
- Had respiratory disease = Yes
- Took systemic hormonal preparation drugs = Yes
- Took various other drugs = Yes

For example, the group identified by Rule 1 has a lift value of 5.14 for Class 1. This indicates that individuals identified by this rule are 5.14 times more likely to have angioedema than the average ACE inhibitor users.

Figure 1 provides graphic presentation of Rule 1 in terms of a probability tree. According to the probability tree, female ACE inhibitor users are 1.12 times more likely to have angioedema than the average ACE inhibitor users. For female ACE inhibitor users aged 60 years or older, the likelihood increases to 1.14. As the tree goes further down, i.e., females aged 60 years or older who have also taken genito urinary system and sex hormone drugs, the likelihood increases further to 1.61. If individuals who have all the characteristics mentioned above and have also used antineoplastic and immunimodulating agent drugs, the likelihood goes up to 4.21 times of the average ACE inhibitor users. Finally, in addition to all the above mentioned characteristics, the factor of exposure to musculo-skeletal system drugs, raises the risk factor of having angioedema up to 5.14 times of the average ACE inhibitor users.

6 Discussion and Conclusions

We have developed an algorithm for mining optimal class association rule sets and have applied the algorithm to a very unbalanced data set to identify groups with high risks of having adverse drug reactions. A feature selection method based on statistical analysis has also been developed to select appropriate features for our data mining algorithm.

Results from testing the association of angioedema with usage of ACE inhibitors have identified groups that are, on average, 4 to 5 times more likely to have an adverse reaction (angioedema) than the average ACE inhibitor users. We note that while the data mining approach has successfully identified key areas in the data worthy of exploration and explanation, conclusions relating to the suitability of ACE inhibitor usage for particular populations need to be further investigated and confirmed by medical specialists, or existing medical studies.

Table 3. Characteristics of groups identified by Rules 1 to 3

	Rule 1	Rule 2	Rule 3
Number of patients in the group	1,549	1,629	1,999
Percentage of this group	1.17	1.23	1.51
Number of patients in Class 0	1,542	1,622	1,991
Number of patients in Class 1	7	7	8
Local support in Class 1	6.03%	6.03%	6.90%
Interestingness	0.838	0.831	0.820
Lift of Class 1	5.14	4.89	4.55

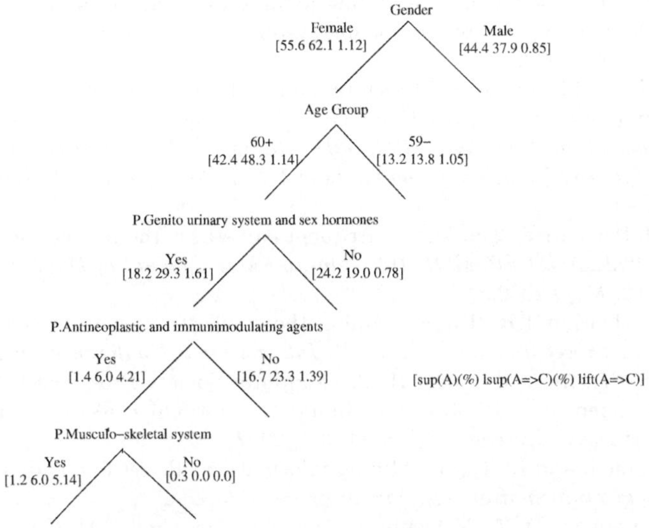

Fig. 1. Probability tree of Rule 1

This study has focused on analysing a known adverse reaction to confirm the approach. The more challenging task of identifying unknown adverse reactions from administrative health data like QLDS is ongoing.

Acknowledgements

The authors wish to acknowledge the Commonwealth Department of Health and Ageing and Queensland Health for providing data suitable for linking. These data are particularly useful for the development of methodologies appropriate for the study of health outcomes in large populations.

References

[1] R. Agrawal and R. Srikant. Fast algorithms for mining association rules in large databases. In *Proceedings of the Twentieth International Conference on Very Large Databases*, pages 487–499, Santiago, Chile, 1994. 225

[2] A. Bate, M. Lindquist, I.R. Edwards, S. Olsson, R.Orre, A. Landner, and R.M. De Freitas. A bayesian neural network method for adverse drug reaction signal generation. *European Journal of Clinical Pharmacology*, 54:315–321, 1998. 222

[3] S. Brin, R. Motwani, J. D. Ullman, and Shalom Tsur. Dynamic itemset counting and implication rules for market basket data. In *Proceedings, ACM SIGMOD International Conference on Management of Data: SIGMOD 1997: May 13–15, 1997, Tucson, Arizona, USA*, volume 26(2), pages 255–264, NY, USA, 1997. ACM Press. 223

[4] V. Dhar and A. Tuzhilin. Abstract-driven pattern discovery in databases. *IEEE Transactions on Knowledge and Data Engineering*, 5(6), 1993. 223

[5] W. DuMouchel. Bayesian data mining in large frequency tables, with an application to the fda spontaneous reporting system. *American Statistical Association*, 53(3):177–190, 1999. 222

[6] T. Fukuda, Y. Morimoto, S. Morishita, and T. Tokuyama. Data mining using two-dimensional optimized association rules: scheme, algorithms, and visualization. In *Proceedings of the 1996 ACM SIGMOD International Conference on Management of Data, Montreal, Quebec, Canada, June 4–6, 1996*, pages 13–23, New York, 1996. ACM Press. 223

[7] J. Han, J. Pei, and Y. Yin. Mining frequent patterns without candidate generation. In *Proc. 2000 ACM-SIGMOD Int. Conf. on Management of Data (SIGMOD'00)*, pages 1–12, May, 2000. 225

[8] J. Li, H. Shen, and R. Topor. Mining the smallest association rule set for prediction. In *Proceedings of 2001 IEEE International Conference on Data Mining (ICDM 2001)*, pages 361–368. IEEE Computer Society Press, 2001. 225, 226

[9] J. Li, H. Shen, and R. Topor. Mining the optimal class assciation rule set. *Knowledge-based Systems*, 15(7):399–405, 2002. 222, 226

[10] J. Li, H. Shen, and R. Topor. Mining informative rule set for prediction. *Journal of intelligent information systems*, in press. 225, 226

[11] J.P. Ottervanger, H. A. Valkenburg, D.E. Grobbee, and B.H. Ch. Stricker. Differences in perceived and presented adverse drug reactions in general practice. *Journal of Clinical Epidemiology*, 51(9):795–799, 1998. 222

[12] G. Piatetsky-Shapiro. Discovery, analysis and presentation of strong rules. In G. Piatetsky-Shapiro, editor, *Knowledge Discovery in Databases*, pages 229–248. AAAI Press / The MIT Press, Menlo Park, California, 1991. 223

[13] M. Reid, B. Euerle, and M. Bollinger. Angioedema, 2002. 228

[14] P. Shenoy, J. R. Haritsa, S. Sudarshan, G. Bhalotia, M. Bawa, and D. Shah. Turbo-charging vertical mining of large databases. In *Proceedings of the ACM SIGMOD International Conference on Management of Data (SIGMOD-99)*, ACM SIGMOD Record 29(2), pages 22–33, Dallas, Texas, 1999. ACM Press. 225

[15] G. I. Webb. Efficient search for association rules. In *Proceedinmgs of the 6th ACM SIGKDD International Conference on Knowledge Discovery and Data Mining (KDD-00)*, pages 99–107, N. Y., 2000. ACM Press. 223

[16] Graham Williams, Deanne Vickers, Rohan Baxter, Simon Hawkins, Chris Kelman, Richard Solon, Hongxing He, and Lifang Gu. Queensland linked data set. Technical Report CMIS 02/21, CSIRO Mathematical and Information Sciences, Canberra, 2002. Report on the development, structure and content of the Queensland Linked Data Set, in collaboration with the Commonwealth Department of Health and Ageing and Queensland Health. 222, 228

[17] M. J. Zaki, S. Parthasarathy, M. Ogihara, and W. Li. New algorithms for fast discovery of association rules. In *Proceedings of the Third International Conference on Knowledge Discovery and Data Mining (KDD-97)*, page 283. AAAI Press, 1997. 225

Efficiently Mining Frequent Patterns from Dense Datasets Using a Cluster of Computers

Yudho Giri Sucahyo[1], Raj P. Gopalan[1], and Amit Rudra[2]

[1] Department of Computing, Curtin University of Technology
{sucahyoy,raj}@computing.edu.au
[2] School of Information Systems, Curtin University of Technology
Kent St, Bentley, Western Australia 6102
rudraa@cbs.curtin.edu.au

Abstract. Efficient mining of frequent patterns from large databases has been an active area of research since it is the most expensive step in association rules mining. In this paper, we present an algorithm for finding complete frequent patterns from very large dense datasets in a cluster environment. The data needs to be distributed to the nodes of the cluster only once and the mining can be performed in parallel many times with different parameter settings for minimum support. The algorithm is based on a master-slave scheme where a coordinator controls the data parallel programs running on a number of nodes of the cluster. The parallel program was executed on a cluster of Alpha SMPs. The performance of the algorithm was studied on small and large dense datasets. We report the results of the experiments that show both speed up and scale up of our algorithm along with our conclusions and pointers for further work.

1 Introduction

Discovering Association Rules from large databases has been a topic of considerable research interest for nearly a decade. Most of the research effort has focused on the computationally expensive step of mining frequent patterns that satisfy support constraints specified by users. Though the problem is intractable in the worst case, efficient mining algorithms have been developed for many practical data sets like typical market basket data. However, several datasets of current interest such as DNA sequences, proteins and sales transactions with large time windows contain dense data with long patterns and they significantly slow down the current algorithms. The amount of processing time for dense data also increases almost exponentially as the support level of frequent patterns is reduced. Performing the mining tasks in parallel appears to be a logical and inexpensive way of extending the range of data that can be mined efficiently.

Zaki [1] has presented an excellent survey of parallel mining of association rules, though there has been relatively less work in this area so far. Parallel association

T.D. Gedeon and L.C.C. Fung (Eds.): AI 2003, LNAI 2903, pp. 233-244, 2003.

mining algorithms have been developed for distributed memory, shared memory and hierarchical systems. Most of them are in fact extensions of sequential algorithms. For example, Count Distribution is based on Apriori, ParEclat on Eclat, and APM on DIC. Recent progress in parallel computing using clusters of PCs and workstations provides an alternative to the more expensive conventional parallel processors [2]. Besides the lower cost, a cluster may be chosen for other advantages such as performance, development tools, communication bandwidth, integration (clusters are easier to integrate into existing networks), price/performance and scaling (clusters can be easily enlarged). Research on parallel data mining using clusters, while not new, is quite limited. The previous work on the use of clusters for association rule mining has studied mainly the scale up and speed up of Apriori based algorithms [3].

Han et al, proposed the pattern-growth approach for mining frequent patterns in [4] as a better alternative to Apriori like algorithms. We improved the pattern-growth approach by using a compact data representation [5]. Our algorithm is significantly more efficient than FP-Growth [4], and scales up for very large datasets. It is an interesting challenge to develop a parallel pattern growth algorithm using our compact data representation on the shared nothing nodes of a cluster of workstations.

In this paper, we present such an algorithm for mining frequent patterns. We have developed a data distribution scheme that supports flexible interactive mining while minimizing the data transfer. The transaction database is partitioned by projections suitable for parallel mining described later. Each projection is stored on the local disk of a node in the cluster. The projections are distributed in a round-robin fashion to balance the load on the nodes during mining. However, the size of individual projections will vary depending on the data characteristics of a given transaction database. The total number of projections corresponds to the number of different items present in the database. As the number of nodes is usually much smaller than the number of projections, several projections are stored at each node. The data is distributed only once, but mining can be carried out as many times as required using different minimum support levels and other relevant parameters. The projections at each node of the cluster are mined independently. The mining algorithm uses the compact prefix tree representation originally described in [5] for storing data in memory. The details of data distribution and the mining algorithm are given in later sections.

The performance of our algorithm was studied in two stages. First, the algorithm that executes at each node was compared against other significant sequential algorithms including OpportuneProject (OP) [6], FP-Growth [4] and the best available implementation of Apriori [7], using a number of widely used dense test datasets. Then the performance of parallel execution of our algorithm was studied on the same data sets at lower support levels. This approach is justified since mining is a highly computation intensive process at low support levels on dense datasets. The scalability of the algorithm was further tested using very large dense datasets generated with the synthetic data generator [8]. These datasets had characteristics similar to Connect-4, which is a widely used small dense dataset. The performance results indicate that our algorithm is faster than other significant algorithms on dense datasets and is scalable to very large dense datasets.

The structure of the rest of this paper is as follows: In Section 2, we define relevant terms used in association rules mining, discuss the density measure and present the compact transaction tree. In Section 3, we present the data distribution scheme and the parallel algorithm. The experimental results on various datasets are presented in Section 4. Section 5 contains conclusions and pointers for further work.

2 Preliminaries

In this section, we define the terms used for describing association rule mining, discuss a simple density measure for datasets and describe the compact transaction tree used for grouping transactions with common subsets of items.

2.1 Definition of Terms

We give the basic terms needed for describing association rules using the formalism of [9]. Let $I=\{i_1,i_2,\ldots,i_n\}$ be a set of items, and D be a set of transactions, where a transaction T is a subset of I ($T \subseteq I$). Each transaction is identified by a *TID*. An association rule is an expression of the form $X \Rightarrow Y$, where $X \subset I$, $Y \subset I$ and $X \cap Y = \varnothing$. Note that each of X and Y is a set of one or more items and the quantity of each item is not considered. X is referred to as the *body* of the rule and Y as the *head*. An example of association rule is the statement that 80% of transactions that purchase A also purchase B and 10% of all transactions contain both of them. Here, 10% is the *support* of the itemset {A, B} and 80% is the *confidence* of the rule $A \Rightarrow B$. An itemset is called a *large itemset* or *frequent itemset* if its *support* ≥ *support threshold* specified by the user, otherwise the itemset is *small* or *not frequent*. An association with the *confidence* ≥ *confidence threshold* is considered as a valid association rule.

2.2 A Simple Density Measure for Datasets

As mentioned in [9], the transactions in the database could be represented by a binary table as shown in Fig. 1. Counting the support for an item is the same as counting the number of 1s for that item in all the transactions. If a dataset contains more 1s than 0s, it can be considered as a dense dataset, otherwise, it is a sparse dataset. In this paper, we use the percentage of 1s in the total of 1s and 0s as the density measure. Given two datasets, we can determine if one of them is relatively denser compared to the other from the ratio of 1s to the total of 1s and 0s in each.

Tid	Items
1	1 2 3 5
2	2 3 4 5
3	1 2 4 5
4	3 4 5
5	3 4 5
6	1 2 3 4 5
7	1 2 3 5
8	1 2 4 5

⇨

1	2	3	4	5
1	1	1	0	1
0	1	1	1	1
1	1	0	1	1
0	0	1	1	1
0	0	1	1	1
1	1	1	1	1
1	1	1	0	1
1	1	0	1	1

Fig. 1. The transaction database

2.3 Compact Transaction Tree

A Compact Transaction Tree was introduced in [5] to group transactions that contain either identical sets of items or some subsets of items in common. The compact tree has only about half the number of nodes of prefix trees used by various well-known algorithms such as FP-Growth.

Fig. 2a shows a complete prefix tree with 4 items assuming the transaction count to be one at every node. The corresponding compact tree is in Fig. 2b. A complete prefix tree has many identical subtrees in it. In Fig. 2a, we can see three identical subtrees *st1, st2,* and *st3*. In the compact tree, we store information of identical subtrees together. Given a set of n items, a prefix tree would have a maximum of 2^n nodes, while the corresponding compact tree will contain a maximum of 2^{n-1} nodes, which is half the maximum for a full tree. In practice, the number of nodes for a transaction database will need to be far less than the maximum for the problem of frequent item set discovery to be tractable.

The nodes along the leftmost path of the compact tree have entries of all itemsets corresponding to the paths ending at each of the nodes. Other nodes only have itemsets that are not present in the leftmost path to avoid duplicates. For example, node 4 in the leftmost path has itemsets 1234, 234, 34, and 4 (we omit the set notation for simplicity). Node 4 in the path 134 does not have itemsets 34 and 4 since it has been registered at node 4 in the leftmost path.

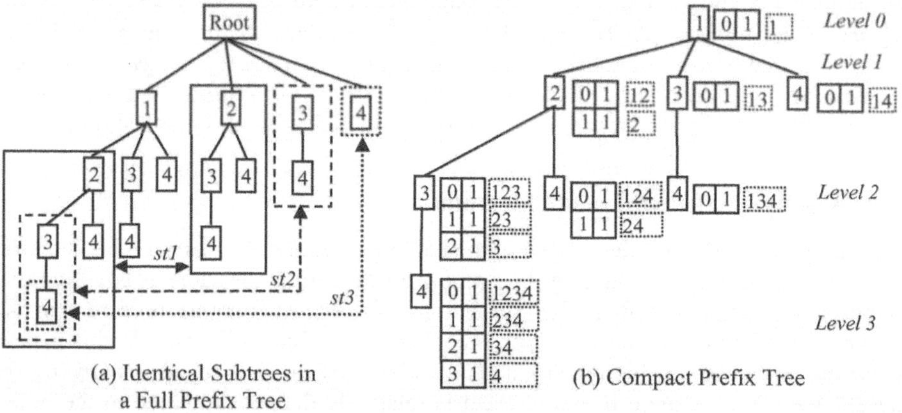

(a) Identical Subtrees in a Full Prefix Tree

(b) Compact Prefix Tree

Fig. 2. Compact Transaction Tree

Attached to each node, an array represents the count of transactions for different item subsets. The array index represents the level of the node containing the starting item of the subset in the tree and the value in each cell is the transaction count. The doted rectangles in Fig. 2b show the itemsets corresponding to the nodes in the compact tree only for illustration (they are not part of the data structure). Similarly, the index entries are implicit and only the transaction count entries need to be explicitly stored.

3 Mining Large Dense Datasets on a Cluster of Computers

Our algorithm for mining large dense datasets in a cluster environment consists of two major components: (1) a partitioning scheme for distributing the large dataset to different nodes and (2) an efficient main memory based procedure for mining the partition at each node. In this section, we first describe the partitioning scheme. It is followed by the data structures and algorithm for in-memory mining at each node. Then the cluster algorithm based on the master-slave scheme of a coordinator and a set of participant nodes is described.

3.1 Partitioning the Transaction Database

The entire database is first partitioned and distributed to nodes of the cluster. If there are a total of N distinct items in the database, N-1 projections will be created. These projections are allocated to m nodes of the cluster in a round-robin fashion. For each item i in a transaction, we copy all items that occur after i (in lexicographic order) to projection i. Therefore, a given transaction could be present in many projections. For example, if we have items 1 2 3 5 in a transaction, we put a transaction 1 2 3 5 in the projection for item 1, put a transaction 2 3 5 in the projection for item 2, and put a transaction 3 5 in the projection for item 3.

Tid	Items
1	1 2 3 5
2	2 3 4 5
3	1 2 4 5
4	3 4 5
5	3 4 5
6	1 2 3 4 5
7	1 2 3 5
8	1 2 4 5

Tid	Items
1	1 2 3 5
2	1 2 4 5
3	1 2 3 4 5
4	1 2 3 5
5	1 2 4 5

Tid	Items
1	2 3 5
2	2 3 4 5
3	2 4 5
4	2 3 4 5
5	2 3 5
6	2 4 5

Tid	Items
1	3 5
2	3 4 5
3	3 4 5
4	3 4 5
5	3 4 5
6	3 5

Tid	Items
1	4 5
2	4 5
3	4 5
4	4 5
5	4 5
6	4 5

```
/* Input: DB  Output: ItemTable */
Procedure ParseDB
For each transaction in DB
  For each item in trans
    If item in ItemTable
      Increment count of item
    Else
      Insert item with count = 1
    End If
  End For
End For
Sort ItemTable by frequency
  ascending and assign index number
  for each item
Write ItemTable to frequency file
For each transaction in DB
  Map each item in transaction
    according to its index number
  Sort transaction based on index
  For each index number in trans
    Write suffix of the transaction
      starting from current index
      to projection[index]
  End For
End For
```

Fig. 3. Parallel Projection Example and Algorithm

Fig. 3 shows the contents of each projection for the sample database following the parallel projection step and the algorithm. The frequency of all items in the database is kept in a frequency file that will be replicated at all the nodes. The frequency file is used to construct the compact tree along with the associated Itemtable for each projection (see Section 3.2).

The main advantage of this scheme is that the entire database is partitioned into parallel projections and distributed to the nodes only once. The projections can be subsequently mined many times with different parameter settings of support and confidence as desired by the user. However, the total size of the projections will be larger than the size of the original database.

3.2 Mining Frequent Itemsets from a Projection

The database partition at a node may consist of several projections since the number of nodes in the cluster is usually much smaller than the number of items. Mining a partition involves the mining of all the projections of that partition. Each projection can be mined independently. Therefore, if a node has multiple processors as in the case of Alpha SMPs, each processor can mine a different projection in parallel. Mining a projection consists of three steps: (1) Constructing a compact tree for the projection; (2) Mapping the compact tree to a special data structure named Item-TransLink (ITL) and (3) Mining the frequent itemsets from ITL. These steps are described in detail in the following subsections.

3.2.1 Building a Compact Transaction Tree for a Projection

The transactions of each projection are represented in a compact tree as discussed in Section 2.3. As transactions containing common items are grouped together using the compact tree, the cost of traversing the transactions is reduced in the mining phase.

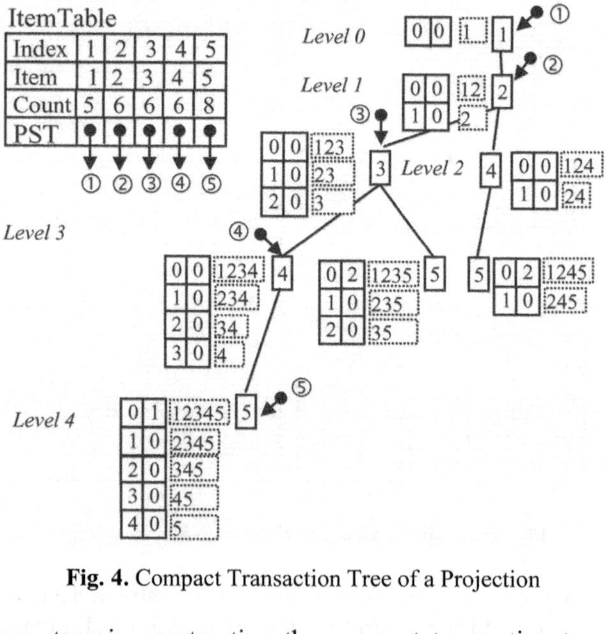

Fig. 4. Compact Transaction Tree of a Projection

There are two steps in constructing the compact transaction tree: (1) Read the frequency file (mentioned in Section 3.1) to identify individual frequent (1-freq) items since only 1-freq items will be needed to mine frequent patterns. (2) Using the output of the first step, read only 1-freq items from transactions in the projection, and then insert the transactions into the compact transaction tree.

In the first step, the 1-freq items are entered in the ItemTable. After step 2, each node of the tree will contain a 1-freq item and a set of counts indicating the number of transactions that contain subsets of items in the path from the root as mentioned in Section 2.3.

Consider the first projection of the sample database in Fig. 3 and let the minimum support be 5 transactions. The ItemTable and compact tree for the projection are shown in Fig. 4. The ItemTable contains a pointer to the subtree (PST) of each item. Each node of the tree has additional entries to keep the count of transactions represented by the node. For example, the entry (0,0) at the leaf node of the leftmost branch of the tree represents itemset 12345 that occurs once in this projection. The first 0 indicates the starting level in the tree which makes 1 its first item. The second number is the count of the transaction. Similarly entry (0,2) at the leaf node of the rightmost branch of the tree means there are two transactions of 1245. In the implementation of the tree, the counts at each node are stored in an array and the level of an entry is the array index that is not stored explicitly.

3.2.2 Mapping the Compact Tree to Item-TransLink (ITL)

Performance of mining can be improved by reducing the number of columns traversed in the conceptual binary representation of transactions, for counting the support of itemsets. We use group intersection in our algorithm to compute the support of itemsets. To map groups in the tree to an array representation for faster group intersections, we had previously developed a data structure called Item-Trans Link (ITL) that has features of both vertical and horizontal data layouts (see Fig. 5a).

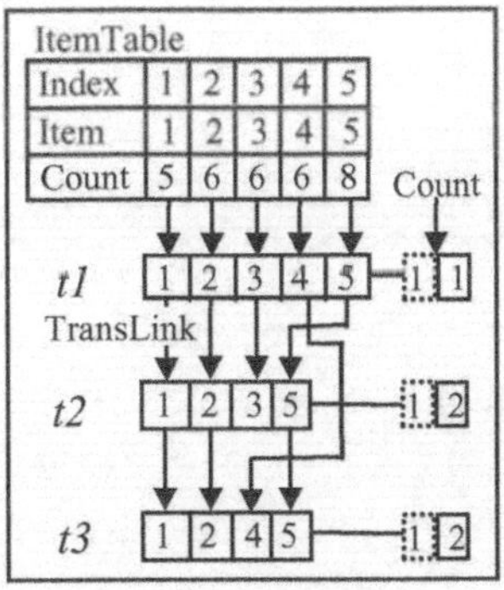

(a) The Item-TransLink (ITL) Data Structure

Prefix	TempList (count)	Patterns (count)
1	3 (3), 4 (3), 2 (5), 5 (5)	1 (5), 1 2 (5), 1 5 (5)
1 2	3 (3), 4 (3), 5(5)	1 2 5 (5)

(b) Mining Frequent Patterns Recursively (support of the pattern shown in brackets)

Fig. 5. The Item-TransLink (ITL) Data Structure and Mining Frequent Patterns

The main features of ITL are described below.

1. Item identifiers are mapped to a range of integers and transaction identifiers are ignored as the items of each transaction are linked together.
2. ITL consists of ItemTable and the transactions of the current projection linked to it (TransLink).
3. ItemTable contains all frequent items of the database. Each item is stored with its support count and a link to the first occurrence of that item in TransLink.

4. A row in TransLink represents a group of transactions along with a count of occurrences of each transaction in the current projection. The items of a row are arranged in sorted order and each item is linked to the next row where it occurs.

Fig. 5a shows the result of mapping the compact tree of Fig. 4 to ITL. For example, transaction group *t2* in the TransLink represents the transaction 1235 that occurs twice. In large datasets, the count of each transaction group is normally much higher, and a number of subgroups will be present too. In the implementation of ITL, the count entries at each row are stored in an array and so the array index is not stored explicitly. The doted rectangles that show the index at each row, is only for illustration and not part of the data structure.

```
Procedure Par-ITL
Read the frequency file
If there is any frequent item
   ConstructTree
   ConstructITL
   MineFI
End If

/* Input: projection  */
/* Output: compact trans tree */
Procedure ConstructTree
For each transaction in the projection
   For each frequent item in transaction
     Insert the item into the tree
   End For
End For

/* Input: trans tree    Output: ITL */
Procedure ConstructITL
For each path in the tree traversed by
  depth first search
  Map the path as an entry in TransLink
  Attach the count entries of the path
  Establish links in TransLink to
    previous occurrences of each item
End For
```

```
/* Input: ITL Output: Freq Patterns */
Procedure MineFI
Initialize Freq_Patterns
For each frequent item x ∈ ItemTable
  Freq_Patterns := Freq_Patterns ∪ x
  Prepare and fill tempList for x
  Sort tempList by count ascending
  For each frequent item y ∈ tempList
    Freq_Patterns := Freq_Patterns ∪ xy
    For each freq item z ∈ tempList
      after y
         RecMine(xy, z)
    EndFor
  End For
End For

Procedure RecMine(prefix, testItem)
tlp:= group-list of prefix
tli:= group-list of testItem
tl_current:= Intersect(tlp,tli)
If size(tl_current) ≥ min_sup
  new_prefix := prefix ∪ testItem
  Freq_Patterns := Freq_Patterns ∪
    new_prefix
  For each frequent item z ∈ tempList
    after testItem
    RecMine(new_prefix, z)
  End For
End If
```

Fig. 6. Algorithm to Mine Frequent Patterns in a Projection

3.2.3 Mining Frequent Patterns in a Projection

Each item in the ItemTable is used as a starting point to mine all longer frequent patterns for which it is a prefix. Fig. 6 shows the algorithm for mining individual projections.

For example, starting with item 1, the vertical link is followed to get all other items that occur together with item 1. These items with their support counts and transaction-count lists are entered in a table named TempList as in Fig. 5b (transaction lists are not shown in the figure). For prefix 1, items {2, 5} are frequent (count ≥ 5). Therefore, 1 2, and 1 5 are frequent item sets. After generating the 2-frequent-item sets for prefix 1, the associated group lists of items in the TempList, are used to recursively generate frequent sets of 3 items, 4 items and so on by intersecting the transaction-count lists. For example, the transaction-count list of 1 2 is intersected with that of 1 5 to generate the frequent itemset 1 2 5. At the end of recursive calls with item 1, all frequent patterns that contain item 1 would be generated: 1 (5), 1 2 (5), 1 5 (5), 1 2 5 (5).

3.3 Parallel Mining in a Cluster Environment

We now describe the framework for mining frequent patterns using the multiple processors of a cluster of Alpha 21264C computers. It is based on master-slave architecture using Message Passing Interface (MPI) [10] calls for distributing the tasks to various nodes. The coordinator node (master) assigns mining tasks to the participant nodes (slaves). The parallel program running on each participant node performs the following tasks:

Perform disk I/O: It opens and reads the dataset projections allocated to it (in the earlier step described in Section 3.1). These projections are already present on the local disk of the node.

Mine dataset projections for frequent itemsets: It mines the allocated projections for frequent itemsets as described in Section 3.2. As the projections can be mined independently, the mining activity at each node of the cluster is not dependent on other nodes. Once a mining task has been allocated to a node by the coordinator, there is no need for further communication during the execution.

Create datasets of frequent itemsets: The frequent itemsets mined at each node are written to the local disk. They are copied to another specified directory as a post-step.

Fig. 7 outlines the algorithm used for parallel execution of tasks at all nodes.

```
MPI_Initialize
Find my rank
If coordinator
   Start timing work
   Distribute work to participants
Else if participant
   Mine allocated dataset projections
End If
Time end of work
Finalize work
```

Fig. 7. Scheme for Distributing the Work among Nodes

4 Performance Study

In this section, we report the performance of our algorithm, which was studied in two stages. First, a sequential version of the algorithm was compared against other well known efficient algorithms including FP-Growth, OpportuneProject (OP), Eclat, and the fastest available implementation of Apriori on small dense datasets. Then the performance of parallel execution of our algorithm was studied on the same data sets at lower support levels. We adopted this approach since mining is a highly computation intensive process at low support levels on dense data sets and therefore the performance gain from parallel processing should be apparent on these data sets. The scalability of the algorithm was further tested using very large dense datasets generated with the synthetic data generator [8].

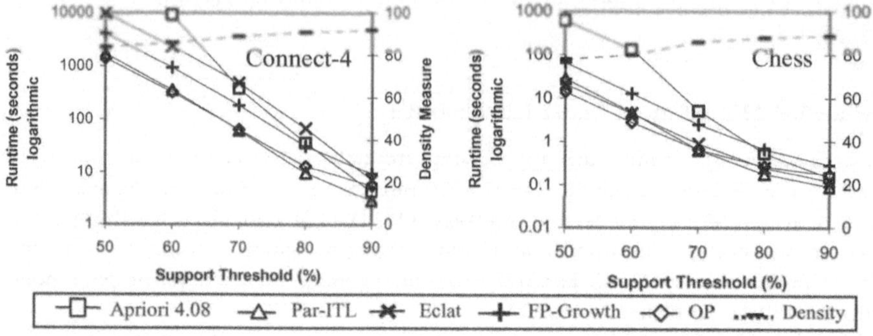

Fig. 8. Comparison of Par-ITL with Sequential Algorithms on Connect-4 and Chess

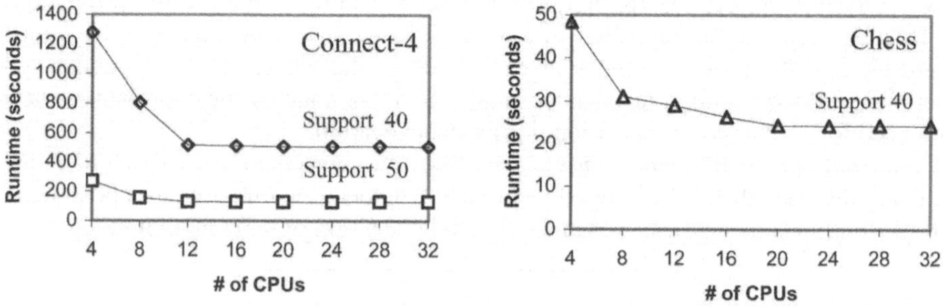

Fig. 9. Parallel Execution of Par-ITL on Connect-4 and Chess

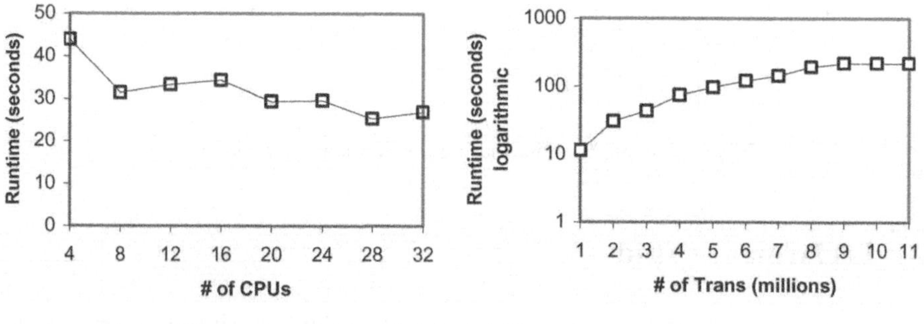

(a) 3 millions transactions (b) Scalability: Support 60, Density 67%

Fig. 10. Parallel Execution on Large Datasets

The sequential programs in the performance study were all written in C++ and compared on an 866 MHz Pentium III PC, 512 MB RAM, 30 GB HD running MS Windows 2000. The clusters used for testing the parallel program had 4 to 32 processors of Alpha 21264C, running Unix64 each with hard disk capacity of 36 GB, connected by a Elan3 fast link interconnect. However, due to resource restrictions imposed by the facility, we performed the tests with a maximum of 32 processors only. For parallel mining, we used C++ as well as MPI routines to perform message passing between the processors running under True Unix64.

The small dense datasets used to test the performance were Chess (3196 trans, 75 items, average transactions 37) and Connect-4 (67557 trans, 129 items, average transaction 43) from Irvine Machine Learning Database Repository [11]. To show the scalability of the algorithm, we generated large dense datasets ranging from 1-11 million transactions using the synthetic data generator [8]. The characteristics of those datasets are similar to Connect-4 with 129 distinct items and average length of transactions 43. The density of the dataset was 67% at support 60.

As shown in Fig. 8, the sequential version of our program performs better than all other sequential algorithms on Connect-4 (support 70-90) and Chess (support 70-90). At lower support levels, its performance is better than all others except OP. In general, our algorithm performs best at a density of 50% or higher. As the support threshold gets lower, the relative performance of OP improves.

Fig. 9 shows the speed up achieved on Connect-4 and Chess datasets by the different runs of our parallel mining program using various configurations of processors. For Connect-4 dataset, we also see that the speed up is consistent for the support levels of 40 and 50 used in the performance study. We observe that a significant speed up was achieved when the configuration was changed from 4 processors to 8 and then to 12 processors. However, beyond 12 processors no further speed up is achieved on Connect-4. The time taken for mining on a processor includes the fixed time of setting up the task and the variable time of mining the given set of projections. By using additional processors, the variable time for mining is shared among more processors, but the fixed time required remains unchanged. So adding more processors beyond a certain number does not reduce the run time.

Fig. 10a shows the performance of the parallel algorithm on a medium sized dataset (3 million transactions, 129 items) and Fig. 10b shows the scalability of our algorithm to datasets of 1-11 million transactions. These results indicate that our algorithm is scalable to very large dense datasets.

5 Conclusions

In this paper, we have presented an algorithm for mining complete frequent patterns from very large dense datasets using a cluster of computers. The algorithm implements a parallel projection approach for partitioning the transaction data. We compared the sequential version of the algorithm against other well-known algorithms. Our algorithm outperformed other algorithms such as OP, FP-Growth, Eclat and Apriori on small dense datasets for most support levels. We further showed the speed up and scale up of our algorithm on small and large dense data sets on clusters of varying number of computers.

Significant speed up was achieved by parallel mining on a cluster of multiprocessors. However, for the test datasets we used, the speed up curve flattened beyond a certain number of processors, though the minimum number of processors that gave the maximum speed up varies for each dataset and support level. We plan to study this characteristic further to determine the factors that affect the speed up of mining algorithms on clusters.

The present implementation assumes that each projection will fit into the available memory when compressed using the compact tree. Very large projections will need to be partitioned further into smaller sub-projections. This is a necessary extension of the program for dealing with huge datasets. The parallel projection scheme we used is flexible but requires a large amount of disk space at each node, particularly for dense datasets. Alternative schemes that significantly reduce disk space requirements are being investigated.

Acknowledgements

This work was supported by a grant of machine time on the Australian Partnership for Advanced Computing (APAC) national facilities [12]. We especially thank APAC for the pro-active technical support they provide. We also thank Christian Borgelt for providing Apriori and Eclat, Jian Pei for the executable code of FP-Growth and Junqiang Liu for providing us the OpportuneProject program.

References

[1] Zaki, M.J., Parallel and distributed association mining: A survey. IEEE Concurrency (Special Issue on Data Mining). (Oct/Dec 1999) 14-25

[2] Baker, M., Buyya, R., Cluster Computing: The Commodity Supercomputing. Software-Practice and Experience. **1**(1) (Jan 1999) 1-26

[3] Jin, R., Agrawal, G., An Efficient Association Mining Implementation of Cluster of SMPs. Proc. of workshop on Parallel and Distributed Data Mining (PDDM) (2001)

[4] Han, J., Pei, J., Yin, Y., Mao, R.: Mining Frequent Patterns without Candidate Generation: A Frequent-pattern Tree Approach. to appear in Data Mining and Knowledge Discovery: An International Journal, Kluwer Academic Publishers (2003)

[5] Gopalan, R.P., Sucahyo, Y.G.: Improving the Efficiency of Frequent Pattern Mining by Compact Data Structure Design. Proc. of 4th Int. Conf. on Intelligent Data Engineering and Automated Learning (IDEAL), Hong Kong (2003)

[6] Liu, J., Pan, Y., Wang, K., Han, J.: Mining Frequent Item Sets by Opportunistic Projection. Proceedings of ACM SIGKDD, Edmonton, Alberta, Canada (2002)

[7] http://fuzzy.cs.uni-magdeburg.de/~borgelt/

[8] http://www.almaden.ibm.com/cs/quest/syndata.html

[9] Agrawal, R., Imielinski, T., Swami, A.: Mining Association Rules between Sets of Items in Large Databases. Proc. of ACM SIGMOD, Washington DC (1993)

[10] Gropp, W., Lusk, E. Skjellum, A. Using MPI: Portable Parallel Programming with the Message-Passing Interface 2nd ed. MIT Press, Cambridge, MA (1999)

[11] http://www.ics.uci.edu/~mlearn/MLRepository.html

[12] APAC–Australian Partnership for Advanced Computing, http://nf.apac.edu.au/ (June 2003)

Weighted MCRDR: Deriving Information about Relationships between Classifications in MCRDR

Richard Dazeley and Byeong-Ho Kang

School of Computing, University of Tasmania, Hobart, Tasmania 7001, Australia[1].
Smart Internet Technology Cooperative Research Centre
Bay 8, Suite 9/G12 Australian Technology Park Eveleigh NSW 1430[1].
{rdazeley,bhkang}@utas.edu.au

Abstract. Multiple Classification Ripple Down Rules (MCRDR) is a knowledge acquisition technique that produces representations, or knowledge maps, of a human expert's knowledge of a particular domain. However, work on gaining an understanding of the knowledge acquired at a deeper meta-level or using the knowledge to derive new information is still in its infancy. This paper will introduce a technique called Weighted MCRDR (WM), which looks at deriving and learning information about the relationships between multiple classifications within MCRDR by calculating a meaningful rating for the task at hand. This is not intended to reduce the knowledge acquisition effort for the expert. Rather, it is attempting to use the knowledge received in the MCRDR knowledge map to derive additional information that can allow improvements in functionality of MCRDR in many problem domains. Preliminary testing shows that there exists a strong potential for WM to quickly and effectively learn meaningful weightings.

1 Introduction

Multiple Classification Ripple Down Rules (MCRDR) [1, 2] is an incremental Knowledge Acquisition (KA) methodology which allows the expert to perform both the KA process and the maintenance of a Knowledge Based System (KBS) over time [3]. MCRDR allows for multiple independent classifications and is an extension of its predecessor Ripple Down Rules (RDR), which only allowed for single classifications of each case presented. The underlying concept behind these approaches is to use the expert's knowledge within the context it is provided [4]. Thus, if the expert disagrees with a particular conclusion made by the system, knowledge can be added to improve future results.

While MCRDR works very effectively in a number of domains, there is implicit information contained within the structure itself that is not being extracted or used in

---

[1] Collaborative research project between both institutions

T.D. Gedeon and L.C.C. Fung (Eds.): AI 2003, LNAI 2903, pp. 245-255, 2003.

its existing form. For example, when multiple classifications for a case occur, it would be useful to know the relationship between those classes and how that relationship changes the meaning or importance of the case as a whole when compared to cases with differing class configurations. Potentially, such information could reveal important details that may not have been consciously realised by the user or that could have taken significant rule creation for the user to have been able to capture within the standard MCRDR structure.

This paper is going to present one means for addressing this issue through an extension to MCRDR, referred to as Weighted MCRDR (WM). WM provides the ability to extract and learn information about the interrelationships between the various classifications found in MCRDR, as well as derive a meaningful value for a given case through either direct or indirect means. This extension could provide the standard MCRDR algorithm with the ability to provide more functionality and usability to existing domains, as well as opening up additional application possibilities. Preliminary testing shows that there exists a strong potential for WM to quickly and effectively learn meaningful weightings.

2 Multiple Classification Ripple Down Rules (MCRDR)

MCRDR uses a collection of rules in an n-ary tree and evaluates each case by first evaluating the root and then moving down level by level. This continues until either a leaf node is reached or until none of the child rules evaluate to *true*. Due to the fact that any, or all, of a node's children have the potential to fire, the possibility exists for a number of conclusions or classifications to be reached by the system for each case presented [5]. The system then lists the collection of classifications found and the paths they followed. The path is given as the justification for the conclusion.

KA is achieved in the system through the expert inserting new rules when a misclassification has occurred. The new rule must allow for the incorrectly classified case to be distinguished from the existing stored cases that could reach the new rule [6]. This is accomplished by the user identifying key differences between the current case and one of the earlier cornerstone cases already stored. This is continued for all past cases that could reach the new rule in the tree structure, until there is a composite rule created that uniquely identifies the current case from all of the previous cases. The idea here is that the user will select differences that are representative of the new class they are trying to create. Stopper rules, which cancel a particular evaluation path, are added in the same way [6].

Previously, simulation studies using RDR illustrated, despite the random nature of rule creation and insertion, that the knowledge bases created are as compact and accurate as those produced by induction [7, 8]. Furthermore, Kang [1] showed that MCRDR produces somewhat better knowledge bases, even when used in single classification domains. It is also arguable that a multiple classification system, such as MCRDR, provides an approach to dealing with a number of problem types, such as configuration and scheduling [5, 9]. RDR and MCRDR, as well as variations to these methodologies, have been used in a number of application areas, such as the PEIRS pathology system [10], resource allocation [11] and document management [12].

3 Weighted MCRDR (WM)

However, MCRDR has a lack of cohesion between the individual classifications generated by a single case, due to each classification being uniquely derived with no consideration for what other classification paths may have also been followed. Therefore, in certain problem domains, a particular case's multiple classifications may all be individually correct but still not capture its essence or accurately represent the level of importance the case has to the expert.

For example, in an email classification system that sorts emails into various categories of varying importance to the user, such as that developed by Deards [13], there exists a class for work related email and one for advertising spam. The spam category catches advertising emails that are of little use to the user, while the work category holds material that requires the expert's immediate attention, before all the other emails received. However, if the system receives an email selling something directly relevant to the user's work then it should classify it as being both spam and work related. The problem with this classification is that neither category individually describes the case adequately, yet both are correct: while it is spam the user may wish to read it; and, then again, it is not as high a priority as the usual work material and should not be brought to the users' attention immediately.

To handle this in traditional MCRDR, the expert would need to create a new set of rules so that such mails could be categorised separately. This, however, can become a tedious and never ending task for emails that only appear infrequently. It also requires the user to be aware of the required classification, which can be difficult as the user generally has little or no knowledge of the MCRDR structure. Finally, it makes little sense for the user to create a new category when the existing classifications are already correct from the user's point of view. The intention of WM is to try to capture these relationships between various classifications that may exist, either consciously or otherwise. If we can identify a set of relative values for the various relationships, this information could be used to improve the functionality available to the user. For instance, in the above email example we could list the emails from each classification in their order of importance, using the relationship's value as a gauge.

3.1 WM Implementation

The most straight forward approach to calculating a rating based on the relationships in an MCRDR classification set is to include a weight at each rule node and simply find the weighted-sum of all the terminal rules found, thereby, giving a value for the case. However, the method used for acquiring each rule's value must take into consideration all the other concluding rules. In addition, the method would need to be able to change the value to cater for other possible relationships that are not evident when a new rule or class is initially created. Finally, it would be useful if such a system was able to provide a number of varying results in applications where dissimilar tasks may need to be rated differently.

1. **Pre-process Case**
   Initialise Case c
   $c \leftarrow$ Identify all useful data elements.

2. **Classification**
   Initialize *list l* to store classifications
   Loop
      If child's rule evaluates Case c to *true*
         $l \leftarrow$ goto step 2 (generate all classifications in child's branch).
   Until no more children
   If no children evaluated to true then
      $l \leftarrow$ Add this nodes classification.
   Return l.

3. **Rate Case**
   $\bar{i} \leftarrow$ Generate input vector from l.
   NN $\leftarrow \bar{i}$
   v $\leftarrow$ *NN output value.*

4. **Return RM evaluation**
   Return list l of classifications for case c and
   Value v of case c.

Fig. 1. Algorithm for WM

Using the notion of finding a weighted-sum leads directly to the use of a single layer perceptron neural network, which is fundamentally a weighted-sum that has been thresholded, using the sigmoid function (equation 1) to lie between a lower and upper bound. Thus, each rule in the MCRDR tree would be given an associated input node in the neural network. Using such a network also has the added advantage of providing a means for each terminating rule to learn the optimal weight across all of the various relationships, by using the *generalized-Δ-rule* [14]. In addition, any number of outputs can be used (where each output is effectively a new perceptron network) for each different task that such values may be needed.

$$f(net) = 1 / \left(1 + e^{-k\ net}\right) \tag{1}$$

Thus, the full WM algorithm, given in pseudo code in Figure 1 and shown diagrammatically in Figure 2, consists of two primary components. Firstly, a document or case is pre-processed to identify all of the usable data elements, such as stemmed words or a patient's pulse. The data components are then presented to the standard MCRDR engine, which classifies them according to the rules previously provided by the user. Secondly, for each rule identified an associated input neuron in the neural network will fire. The network finally produces a number of outputs, $\bar{v}$, for the case presented. The system, therefore, essentially provides two separate outputs; the case's classifications and the associated set of values generated from those classifications.

For example, in figure 2, the document {a b b a c f i} has been pre-processed to a set of unique tokens {a, b, c, f, i}. It is then presented to the MCRDR component of the WM system, which ripples the case down the rule tree finding three classifications *Z*, *Y*, and *U* from the terminating rules 1, 5, and 8. In this example, which is using the RA method (discussed below), the terminating rules then cause the three associated neurons to fire and feed forward through the neural network producing a single value of 0.126 to be outputted. Thus, this document has been allocated a set of classifications that can be used to store the document appropriately, plus a single rating measuring its overall level of importance to the user.

3.2 Neuron Association Methods

There are three possible methods for the association of neurons to the MCRDR structure: the Class Association method (CA), the Rule Association method (RA) and the Rule Path Association method (RPA). The different methods arise from the possibility of many paths through the tree that result in the same class as the conclusion. The *class association* method, where each unique *class* has an associated neuron, can reduce the number of neurons in the network and potentially produce faster, but possibly less general, learning. The *rule association* method, where each *rule* has an associated neuron and only the terminating rule's neuron fires, allows for different results to be found for the same class depending on which path was used to generate that class as the conclusion. Therefore, it is more capable of finding variations in meaning and importance within a class than expected by the user that created the rules. The *rule path association* method, where all the *rules* in the *path* followed, including the terminating rule, cause their associated neuron to fire, would be expected to behave similarly but may find some more subtle results learnt through the paths rules, as well as being able to learn meaning hidden within the paths themselves.

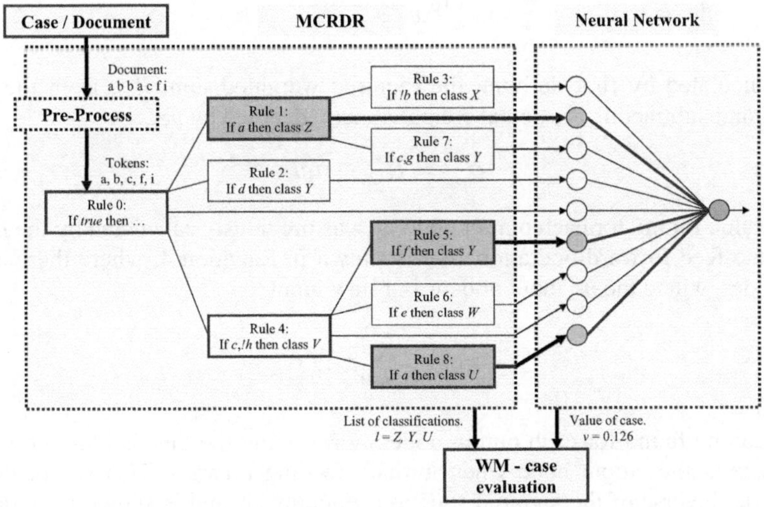

Fig. 2. WM illustrated diagrammatically

3.3 Adding New Input Neurons

However, a perceptron network does not provide a standard way for calculating the initial weight for a rule node when it is first added to the MCRDR tree. This can be resolved, though by first adding new input nodes to the network each time a rule is added and secondly, calculating the initial weight for the new network node for each of the allocated outputs. The simplest approach is to give the new input connections created a random start up weight in the same fashion used when initialising the network. However, we already have the system's *correct rating* for the case that is causing the new rule to be created, which WM should use to foster faster learning. For this purpose the system's *correct rating* is assumed to have been determined for the case either through directly querying the user or through indirectly deriving it from observed user behaviour.

The approach taken in this implementation of WM captures this important information by directly calculating the required weight that provides us with the correctly weighted-sum for the given case without the need for a training period. This provides the system with an immediate understanding of the new case's general value. It can also then be used as the foundation for the new node's relationships with other terminating rules to be learnt.

3.4 The Single-Shot Δ-Rule in WM

In order to calculate the weight needed for the new input connection, w_{no}, the system must first calculate the error in the weighted-sum, δ_{ws}, and divide this by the input at the newly created input node, x_n, which is always 1 in this implementation, where there are $n>0$ input nodes and $o>0$ output nodes.

$$w_{no} = \frac{\delta_{ws}}{x_n} \tag{2}$$

δ_{ws} is calculated by first deriving the required weighted-sum, R_{ws}, from the known error, δ, and subtracting the actual weighted-sum, denoted by *net*.

$$\delta_{ws} = R_{ws} - net \tag{3}$$

The value for *net* for each output node, *o*, was previously calculated by the network during the feed forward operation, and is shown in Equation 4, where there are $n>1$ input nodes, where the nth input node is our new input.

$$net_o = \sum_{i=0}^{n-1} x_i w_{io} \tag{4}$$

R_{ws}, can be found for each output node, by reversing the thresholding process that took place at the output node when initially feeding forward. This is calculated by finding the inverse of the sigmoid function, Equation 1, and is shown in Equation 5, where *f(net)* is the original thresholded value that was outputted from that neuron.

$$R_{ws_o} = \frac{\log\left[\frac{f(net)_o + \delta_o}{1-\left(f(net)_o + \delta_o\right)}\right]}{k} \tag{5}$$

Thus, the full *single-shot Δ-rule*, used for each of the new input's connections to all of the output nodes is given in equation 6. Due to the use of the sigmoid function, it is clear that at no time can the system try and set the value of the output to be outside the range $0 > (f(net) + \delta) > 1$ as this will cause an error.

$$w_{no} = \frac{\left(\log\left[\frac{f(net)_o + \delta_o}{1-\left(f(net)_o + \delta_o\right)}\right]\Big/k\right) - \left(\sum_{i=0}^{n-1} x_i w_{io}\right)}{x_n} \tag{6}$$

4 Testing with a Simulated Expert

The problem with testing a system, such as WM, is the use of expert knowledge that cannot be easily gathered without the system first being applied in a real world application. This is a similar problem that has been encountered with the testing of any of the RDR methodologies [2, 8, 15]. Thus, these systems built their KB incrementally through the use of a simulated expert. The simulated expert essentially provided a rule trace for each case run through another KBS with a higher level of expertise in the domain than the RDR system being trained [2, 5]. WM, however, has the additional problem of needing to learn from results returned by the system, derived either directly or indirectly from the expert.

Thus, in order to perform preliminary testing of the rating component of the system, while still being able to create a KB in the MCRDR tree, a simulated expert was also developed for WM, with the ability to both classify cases, as well as form an opinion as to a case's overall importance. Basically, the simulated expert randomly generates weights, representing the level that each possible attribute in the environment contributes to each possible class, which is used to define rules for the MCRDR tree. Likewise, an additional weight is generated for each possible attribute, indicating what level of importance the case has overall to the simulated expert, allowing the simulated expert to form preferences for particular cases.

The environment then created many sets of documents consisting of randomly generated collections of attributes and passed each set to the WM system for classification and rating. The order allocated by the WM system for each set was then compared against the simulated expert's expected ordering. In testing the system, assumptions were made that the expert's interest in a case could be accurately measured and that the behaviour was constant, without noise or concept drift.

5 Results and Discussion

One useful application for a system such as WM is the rating of documents so they can be appropriately ordered according to the user's preference. The ability of WM to accomplish this is illustrated in the example cross sections shown in Figure 3. The first graph shows the order in which WM places the documents prior to any learning. The second graph, after only 10 document sets, clearly illustrates that it has been able to correctly order many of the documents.

To show how the system performs over time the difference was calculated between each documents' place in the WM ordering minus the place it should have been placed according to the simulated expert. These differences were then averaged over all the documents presented in that group and subtracted from the total number of documents in the set. Figure 4, shows how the system performed over the first 50 document sets. It can be seen in this example that the assignment of immediate weights to a position and the ongoing trainability of the network allows the system to immediately start to provide a reasonably accurate solution.

a)
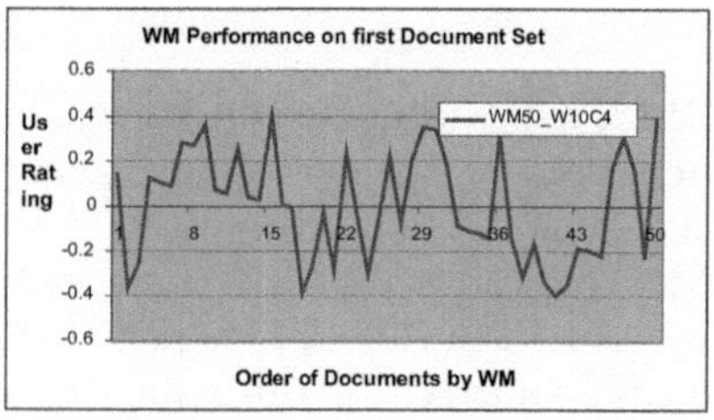

b)
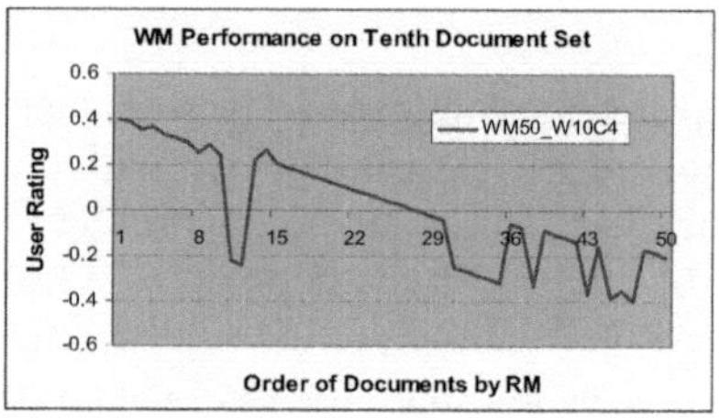

Fig. 3. Ability of WM to order cases according to user preference. a) Shows WM's performance prior to any training. b) Shows WM's performance after 10 document sets. Both graphs have the highest rated cases on the left and the lowest on the right according to RM

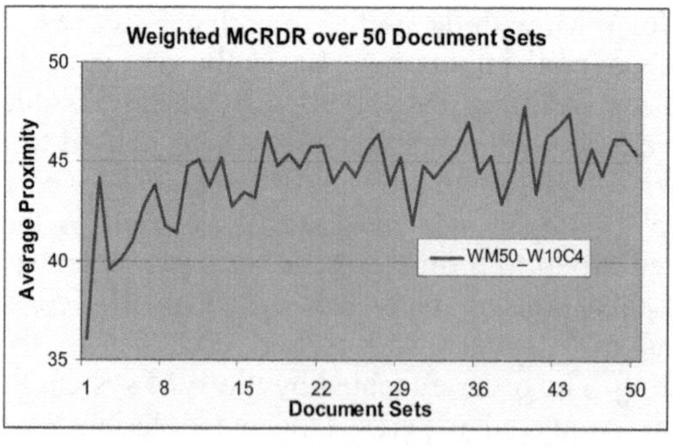

Fig. 4. Proximity of WM to simulated users ordering. The average proximity is the absolute difference between the place allocated by WM and the place given by the simulated expert, averaged over all the cases in each set and subtracted from the total number of documents in the set. Each document set contains 50 cases

While these results are only preliminary, they do show that WM is capable of learning ratings that allow us to order documents into the users preferred order quickly. Additionally, due to the network learning appropriate ratings it shows that it has been able to identify patterns in the structure of the MCRDR tree and the way a case has been classified within that structure. Finally, it can be seen that a system that uses MCRDR to organise data items for long term storage and later retrieval, could easily include extra functionality by extracting this meta-knowledge from MCRDR.

6 Conclusion and Future Work

The system described in this paper was developed to provide extra functionality to the MCRDR system by finding values for representing the relationships between classifications generated by the knowledge base. It is designed to learn simple linear relationships quickly, so that the system can respond to new information without the need for a number of training examples. To facilitate this system the *single-step-Δ-rule* was developed for the input connection's initial weight, which provides the ability of the system to step immediately to the indicated value required by the system.

The system has undergone preliminary testing with a simulated expert using a randomly generated data set. These tests were done primarily to show that the system was able to learn quickly and to be used for parameter tuning purposes. Clearly a more rigorous testing regime needs to be used in order to fully justify the algorithm's ability to learn.

This paper was an introduction to one area where MCRDR can be used as a basis for deriving further information about a user's knowledge and understanding of a domain. It still holds the possibility of being refined further such as:

- Providing a means to learn non-linear relationships.
- Through the addition of a per-neuron reducing gain to allow the system to better cope with noise when creating new input nodes.
- Building in a windowing technique to allow the system to deal effectively with concept drift.
- Incorporating Temporal Difference Learning through the addition of an eligibility trace [16]. This would allow the system to predict the likelihood of future documents, linked to by the one being analyzed, being of importance.
- There exists the possibility of using the error information that is propagated back through the network to identify which rules in MCRDR are likely to be incorrect and only asking the user about those particular cases, thereby, reducing the workload of knowledge acquisition for the user.

Acknowledgements

The authors would like to thank Dr Ray Williams and Mr Robert Ollington for their helpful discussions.

References

[1] B. H. Kang, Validating Knowledge Acquisition: Multiple Classification Ripple Down Rules. 1996, University of New South Wales: Sydney.

[2] B. H. Kang, P. Compton and P. Preston. Multiple Classification Ripple Down Rules: Evaluation and Possibilities. in *The 9th Knowledge Acquisition for Knowledge Based Systems Workshop*. 1995. Department of Computer Science, University of Calgary, Banff, Canada: SRDG Publications.

[3] R. Martínez-Béjar, Ibáñez-Cruz, F., Le-Gia, T, Cao, T., and Compton, P., FMR: an incremental knowledge acquisition system for fuzzy domains., *Lecture Notes in Artificial Intelligence.*, (1999). **1621**: p. 349-364.

[4] P. Compton and R. Jansen, A philosophical basis for knowledge acquisition., *Knowledge Acquisition*, (1990). **2**: p. 241-257.

[5] B. H. Kang, P. Preston and P. Compton. Simulated Expert Evaluation of Multiple Classification Ripple Down Rules. in *Eleventh Workshop on Knowledge Acquisition, Modeling and Management*. 1998. Voyager Inn, Banff, Alberta, Canada.

[6] P. Preston, P. Compton, G. Edwards and B. H. Kang. An Implementation of Multiple Classification Ripple Down Rules. in *Tenth Knowledge Acquisition for Knowledge-Based Systems Workshop*. 1996. Department of Computer Science, University of Calgary, Calgary, Canada UNSW, Banff, Canada: SRDG Publications.

[7] P. Compton, P. Preston and B. H. Kang. The Use of Simulated Experts in Evaluating Knowledge Acquisition. in *The 9th Knowledge Acquisition for Knowledge Based Systems Workshop*. 1995. Department of Computer Science, University of Calgary, Calgary, Canada: SRDG Publications.

[8] P. Compton, P. Preston, B. H. Kang and T. Yip, Local Patching Produces Compact Knowledge Bases, in A Future for Knowledge Acquisition, W. V. d. Velde, Editor. 1994, Springer-Verlag: Berlin, Germany. p. 104-117.

[9] P. Compton, B. H. Kang, P. Preston and M. Mulholland. Knowledge Acquisition Without Analysis. in *Knowledge Acquisition for Knowledge Based Systems*. 1993. Berlin: Springer Verlag.

[10] G. Edwards, P. Compton, R. Malor, A. Srinivasan and L. Lazarus, PEIRS: a pathologist maintained expert system for the interpretation of chemical pathology reports., *Pathology*, (1993). **25**: p. 27-34.

[11] D. Richards and P. Compton. Knowledge Acquisition First, Modelling Later. in *Proceedings of the Tenth European Knowledge Acquisition Workshop*. 1997.

[12] B. H. Kang, K. Yoshida, H. Motoda and P. Compton, A help desk system with intelligent interface, *Applied Artificial Intelligence*, (1997). **11**(7-8): p. 611-631.

[13] E. A. Deards, MCRDR Applied to Email Classification, in *Department of Computer Science*. 2001, University of Tasmania: Hobart, Australia.

[14] R. Beale and T. Jackson, *Neural Computing: An Introduction*. 1992, Bristol, Great Britain: IOP Publishing Ltd.

[15] Y. Mansuri, J. G. Kim, P. Compton and C. Sammut. A comparison of a manual knowledge acquisition method and an inductive learning method. in *Australian workshop on knowledge acquisition for knowledge based systems.* 1991. Pokolbin, Australia.
[16] R. Sutton and A. Barto, *Reinforcement Learning: An Introduction.* 1998, Cambridge, Massachusetts: A Bradford Book, The MIT Press.

Fuzzy Cognitive Map Learning Based on Nonlinear Hebbian Rule

Elpiniki Papageorgiou[1], Chrysostomos Stylios[2], and Peter Groumpos[1]

[1] Laboratory for Automation and Robotics
Department of Electrical and Computer Engineering
University of Patras, Rion 26500, Greece
Tel. +30 2610 997293, Fax. +30 26120 997309
{epapageo,groumpos}@ee.upatras.gr
[2] Computer Science Department, University of Ioannina
P.O. Box 1186, Ioannina, Greece
Tel. +30651298818
stylios@cs.uoi.gr

Abstract. Fuzzy Cognitive Map (FCM) is a soft computing technique for modeling systems. It combines synergistically the theories of neural networks and fuzzy logic. The methodology of developing FCMs is easily adaptable but relies on human experience and knowledge, and thus FCMs exhibit weaknesses and dependence on human experts. The critical dependence on the expert's opinion and knowledge, and the potential convergence to undesired steady states are deficiencies of FCMs. In order to overcome these deficiencies and improve the efficiency and robustness of FCM a possible solution is the utilization of learning methods. This research work proposes the utilization of the unsupervised Hebbian algorithm to nonlinear units for training FCMs. Using the proposed learning procedure, the FCM modifies its fuzzy causal web as causal patterns change and as experts update their causal knowledge.

Keywords: Unsupervised learning, Nonlinear Hebbian learning, fuzzy cognitive maps, neural networks, Hebbian rule.

1 Introduction

FCMs were proposed by Kosko to represent the causal relationship between concepts and analyze inference patterns [1,2]. FCMs represent knowledge in a symbolic manner and relate states, variables, events, outputs and inputs in a cause and effect approach. Comparing FCMs to either expert system or neural networks, they have several desirable qualities such as: FCM is relatively easy to represent structured knowledge, and the inference is computed by numeric matrix operation. FCMs are

T.D. Gedeon and L.C.C. Fung (Eds.): AI 2003, LNAI 2903, pp. 256-268, 2003.

appropriate to explicit the knowledge, which has been accumulated for years observing the operation and behavior of a complex system. Fuzzy Cognitive Maps have already been applied in many scientific areas, such as medicine, manufacturing, decision support systems, political science [3,4,5,6,7,8].

Till today, very few research efforts have been made to investigate and propose a learning algorithm suitable for FCMs [9,10]. This research work proposes a learning procedure based on the nonlinear Hebbian learning rule to improve the FCM structure. The introduction of FCM training eliminates the deficiencies in the usage of FCM and enhances the dynamical behavior and flexibility of the FCM model. A criterion function similar to that of the Hebbian rule for linear units is used and the proposed learning algorithm is used to train the FCM model of a process chemical control problem proving the efficiency of the algorithm.

The extension of the Hebbian learning rule suggesting nonlinear units drew the attention of research community [11,12,13] and it was applied in many problems [12,14]. There were showed that the use of units with nonlinear activation functions, which employ Hebbian learning, might lead to robust principal component analysis and also a nonlinear Hebbian learning rule by minimizing a given criterion function was proposed [15]. Furthermore, for a better understanding of the implementation of nonlinear units by exploring the statistical characteristics of the criterion function, i.e. how the operation of the nonlinear activation is being optimized and interpreted using a probability integral transformation you can refer to [16,17].

The outline of this paper follows. Section 2 describes the FCM modeling methodology, how a FCM is developed and how it models a system. Section 3 discusses nonlinear Hebbian learning algorithm. Section 4 introduces the Hebbian Learning Algorithm to nonlinear units for FCM and it presents the mathematical justification of the algorithm for FCMs and a methodology to implement this algorithm. In section 5, the proposed algorithm is implemented to train the FCM model of a process control problem and section 6 concludes the paper and discusses the usefulness of the new training methodology for FCMs.

2 Fuzzy Cognitive Maps Background and Description

Fuzzy cognitive maps have their roots in graph theory. Euler formulated the first graph theory in 1736 [18] and later on the directed graphs (digraphs) was used for studying structures of empirical world [19]. Signed digraphs were used to represent the assertions of information [20] and the term "cognitive map" described the graphed causal relationships among variables. The term "fuzzy cognitive map" was used for first time by Kosko [1] to describe a cognitive map model with two significant characteristics: (a) Causal relationships between nodes are fuzzified and (b) The system has dynamical involving feedback, where the effect of change in one node affects other nodes, which in turn can affect the node initiating the change.

The FCM structure is similar to a recurrent artificial neural network, where concepts are represented by neurons and causal relationships by weighted links connecting the neurons.

Concepts reflect attributes, characteristics, qualities and senses of the system. Interconnections among concepts of FCM signify the cause and effect relationship that a concept has on the others. These weighted interconnections represent the direction and degree with which concepts influence the value of the interconnected concepts. Figure 1 illustrates the graphical representation of a Fuzzy Cognitive Map.

The interconnection strength between two nodes C_j and C_i is w_{ji}, with w_{ji} taking on any value in the range -1 to 1.

There are three possible types of causal relationships among concepts:

- $w_{ji} > 0$ which indicates positive causality between concepts C_j and C_i. That is, the increase (decrease) in the value of C_j leads to the increase (decrease) on the value of C_i.
- $w_{ji} < 0$ which indicates negative causality between concepts C_j and C_i. That is, the increase (decrease) in the value of C_j leads to the decrease (increase) on the value of C_i.
- $w_{ji} = 0$ which indicates no relationship between C_j and C_i.

The directional influences are presented as all-or-none relationships, so the FCMs provide qualitative as well as quantitative information about these relationships [9]. Generally, the value of each concept is calculated, computing the influence of other concepts to the specific concept, [5], by applying the following calculation rule:

$$A_i^{(k+1)} = f(A_i^{(k)} + \sum_{\substack{j \neq i \\ j=1}}^{N} A_j^{(k)} \cdot w_{ji}) \tag{1}$$

where $A_i^{(k+1)}$ is the value of concept C_i at time $k+1$, $A_j^{(k)}$ is the value of concept C_j at time k, w_{ji} is the weight of the interconnection between concept C_j and concept C_i and f is the sigmoid threshold function.

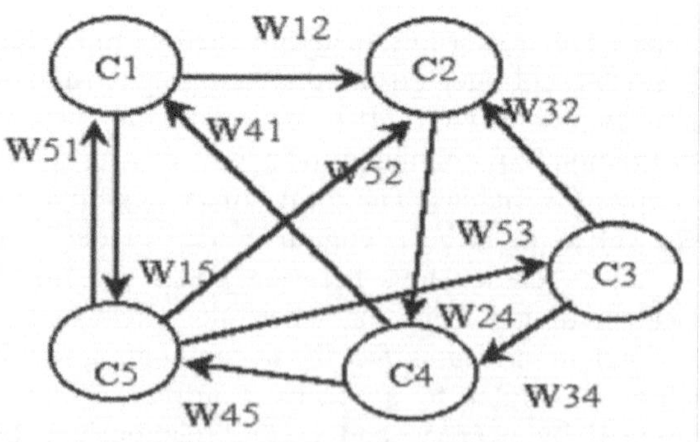

Fig.1. A simple Fuzzy Cognitive Map

The methodology for developing FCMs is based on a group of experts who are asked to define concepts and describe relationships among concepts and use IF-THEN rules to justify their cause and effect suggestions among concepts and infer a linguistic weight for each interconnection [8]. Every expert describes each interconnection with a fuzzy rule; the inference of the rule is a linguistic variable, which describes the relationship between the two concepts according to everyone expert and determines the grade of causality between the two concepts.

Then the inferred fuzzy weights that experts suggested, are aggregated and an overall linguistic weight is produced, which with the defuzzification method of Center of Area (CoA) [21,22], is transformed to a numerical weight w_{ji}, belonging to the interval [-1, 1] and representing the overall suggestion of experts. Thus an initial matrix $\mathbf{w}^{initial} = [w_{ji}]$, *i,j*=1,...,N, with $w_{ii} = 0$, *i*=1,..,N, is obtained.

The most significant weaknesses of the FCMs are their dependence on the expert's opinion and the uncontrollable convergence to undesired states. Learning procedures is a mean to increase the efficiency and robustness of FCMs, by modifying the FCM weight matrix.

3 Nonlinear Extension of Hebbian Rule

A weight-learning rule requires the definition and calculation of a criterion function (error function) and examining when the criterion function reaches a minimum error that corresponds to a set of weights of NN. When the error is zero or conveniently small then a steady state for the NN is reached; the weights of NN that correspond to steady-state define the learning process and the NN model [23].

According to the well-known Hebb's learning law, during the training session, the neural network receives as input many different excitations, or input patterns, and it arbitrarily organizes the patterns into categories. Hebb suggested the biological synaptic efficacies change in proportion to the correlation between the firing of the pre-and post-synaptic neuron [22,24]. Given random pre-synaptic input patterns **x**, weight vector **w**, and output $y = \mathbf{w}^T \mathbf{x}$, the criterion function J maximized by Hebb's rule may be written as:

$$J = E\{y^2\} \tag{2}$$

An additional constraint such as $\|\mathbf{w}\| = 1$ is necessary to stabilize the learning rule derived from eq. (2). A stochastic approximation solution based on (1) leads to the single neuron Hebbian rule [11]:

$$\Delta w_{ji} = \eta_k y_i (x_j - w_{ji} y_i) \tag{3}$$

where η_k is the learning rate at iteration k.

An extension of (3) to nonlinear units proposed with various possible nonlinear activation functions and their criterion functions [12]. Considering the case where the output of the linear unit is transformed using a nonlinear activation function (usually a sigmoid type function) with the criterion function in eq. (2). The following optimization problem is solving adjusting the nonlinear units' weights adaptively:

$$\begin{aligned} &\text{maximize} \quad J = E\{z^2\} \\ &\text{subject to:} \quad \|\mathbf{w}\| = 1 \end{aligned} \tag{4}$$

where $z = f(y)$, and f is a sigmoid function.

It is difficult to determine a closed form solution of eq. (4); therefore we employ a stochastic approximation approach, which leads to the following nonlinear Hebbian learning rule (the derivation is given in the [17])

$$\Delta w_{ji} = \eta_k z \frac{dz}{dy}(x_j - w_{ji} y_i) \tag{5}$$

The learning rule in eq. (5) is one of the three possible cases of nonlinear Hebbian rules discussed in Oja et al. [12,13]. If z is a linear function of y, say $z = y$, then the learning rule is reduced to the simple Hebbian rule for a linear unit.

For the case of the Hebbian learning rule in a linear unit, the objective function in eq. (2) has an obvious interpretation, i.e., projection of the input patterns in the direction of maximum variance, or maximum output value of neuron. On the other hand the same criterion function applied to nonlinear units, may lead to results radically different from those produced by linear units. Note that both linear and nonlinear learning rules are seeking a set of weight parameters such that the outputs of the unit have the largest variance. The nonlinear unit constraints the output to remain within a bounded range, e.g., $z = 1/(1+\exp(-y))$ limits the output within [0,1]. The restriction of the nonlinear unit outputs significantly distinguishes nonlinear units from their linear counterparts and affects the mechanism of variance maximization.

4 Learning in FCMs

FCM learning involves updating the strengths of causal links so that FCM converge in a desired region. An advanced learning strategy is to modify FCMs by fine-tuning its initial causal link or edge strengths through training algorithms similar to that in artificial neural networks.

There are just a few FCM learning algorithms [9,10], which are based on artificial neural networks training. Kosko proposed the Differential Hebbian, a form of unsupervised learning, but without mathematical formulation and implementation in real problems [9]. There is no guarantee that DHL will encode the sequence into the FCM and till today no concrete procedures exist for applying DHL in FCMs. Another algorithm is named Adaptive Random for FCMs based on the definition of Random Neural Networks [10]. Recently, a different approach has been proposed for FCM learning based on evolutionary computation [25], where evolution strategies have been used for the computation of the desired system's configuration. This technique is exactly the same used in neural networks training; it doesn't take into consideration the initial structure and experts' knowledge for the FCM model, but uses data sets determining input and output concepts in order to define the cause-effect relationships satisfied the fitness function. The calculated weights appear large deviations from the actual FCM weights and in real problems they have not the accepted physical magnitude. So, a formal methodology and general learning algorithm suitable for FCM learning and for practical applications is still needed.

4.1 The Proposed Approach Based on Nonlinear Hebbian Learning (NHL)

The proposed learning algorithm is based on the premise that all the concepts in FCM model are triggering at each iteration step and change their values. During this triggering process the weight w_{ji} of the causal interconnection of the related concepts is updated and the modified weight $w_{ji}^{(k)}$ is derived for iteration step k.

The value $A_i^{(k+1)}$ of concept C_i, at iteration $k+1$, is calculated, computing the influence of interconnected concepts with values A_j to the specific concept C_i due to modified weights $w_{ji}^{(k)}$ at iteration k, through the following equation:

$$A_i^{(k+1)} = f(A_i^{(k)} + \sum_{\substack{j \neq i \\ j=1}}^{N} A_j^{(k)} \cdot w_{ji}^{(k)}) \tag{6}$$

Furthermore, some of concepts are defined as output concepts (OCs). These concepts stand for the factors and characteristics of the system that interest us, and we want to estimate their values, which represent the final state of the system. The distinction of FCM concepts as inputs or outputs is determined by the group of experts for each specific problem. Any of the concepts of the FCM model may be inputs or outputs. However, experts select the output concepts and they consider the rest as initial stimulators of the system. The learning algorithm that extracts hidden and valuable knowledge of experts can increase the effectiveness of FCMs and their implementation in real problems.

Taking the advantage of the general nonlinear Hebbian-type learning rule for NNs, [12], we introduce the mathematical formalism incorporating this learning rule for FCMs, a learning rate parameter and the determination of input and output concepts. This algorithm relates the values of concepts and values of weights in the FCM model.

The proposed rule has the general mathematical form:

$$\Delta w_{ji} = \eta_k A_j (A_i - A_j w_{ji}) \tag{7}$$

where the coefficient η_k is a very small positive scalar factor called learning parameter. The coefficient has determined using experimental trial and error method that optimizes the final solution.

This simple rule states that if $A_i^{(k)}$ is the value of concept C_i at iteration k, and A_j is the value of the triggering concept C_j which triggers the concept C_i, the corresponding weight w_{ji} from concept C_j towards the concept C_i is increased proportional to their product multiplied with the learning rate parameter minus the weight decay at iteration step k.

The training weight algorithm takes the following form:

$$w_{ji}^{(k)} = w_{ji}^{(k-1)} + \eta_k A_j (A_i^{(k)} - A_j w_{ji}^{(k-1)}) \tag{8}$$

At every simulation step the value of each concept of FCM is updated, using the equation (6) where the value of weight $w_{ji}^{(k)}$ is calculated with equation (8).

Also, we introduce two criteria functions for the proposed algorithm. One criterion is the maximization of the objective function J, which has been defined by Hebb's rule in equation (4). The objective function J has been proposed for the NHL, examining the desired values of output concepts (OCs), which are the values of the activation concepts we are interested about. The J is defined as:

$$J = \sum_{i=1}^{l} (OC_i)^2 \tag{9}$$

where l is the number of OCs.

The second criterion is the minimization of the variation of two subsequent values of OCs, represented in equation:

$$\left| OC_j^{(k+1)} - OC_j^{(k)} \right| < e \tag{10}$$

These criteria determine when the iterative process of the learning algorithm terminates. The term e is a tolerance level keeping the variation of values of OC(s) as low as possible and it is proposed equal to $e = 0.001$.

Through this process and when the termination conditions are met, the final weight matrix $\mathbf{W}^{updated}$ is derived.

4.2 The Proposed Learning Algorithmic Process

The schematic representation of the proposed NHL algorithm is given in Figure 2. Considering an *n*-node FCM-model, the execution phase of the proposed algorithm is consisted of the following steps:

Step 1: Read input state $\mathbf{A}^0$ and initial weight matrix $\mathbf{w}^0$

Step 2: Repeat for each iteration step k

2.1: Calculate A_i according to equation (6)

2.2: Update $w_{ji}^{(k)}$ according to equation (8)

2.3: Calculate the two criterion functions

Step 3: Until the termination conditions are met.

Step 4: Return the final weights $w^{updated}$ and concept values in convergence region.

All the FCM concepts are triggering at the same iteration step and their values are updated due to this triggering process.

5 Implementation to a Process Control Problem

In this section the proposed Nonlinear Hebbian Learning (NHL) is implemented to modify the FCM model of a simple process control problem [8].

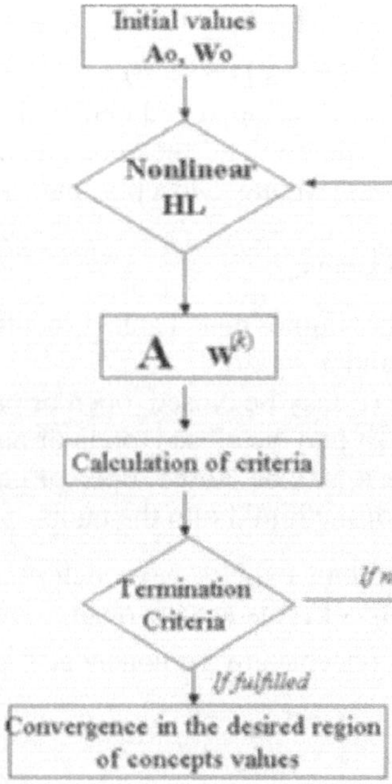

Fig. 2. Schematic representation of NHL training algorithm

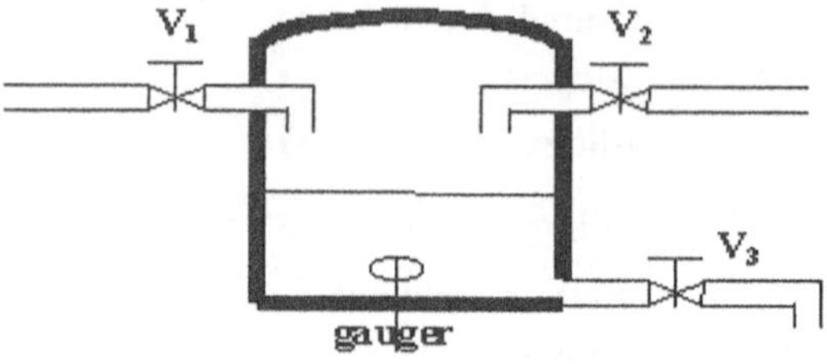

Fig. 3. The illustration for simple process example

At this process problem there is one tank and three valves that influence the amount of liquid in the tank; figure 3 shows an illustration of the system. Valve 1 and valve 2 empty two different kinds of liquid into tank1 and during the mixing of the two liquids a chemical reaction takes place into the tank. A sensor measures the specific gravity of the produced liquid into tank. When value of specific gravity is in the range between (G_{max}) and (G_{min}), this means that the desired liquid has been produced in tank. Moreover, there is a limit on the height of liquid in tank, which cannot exceed an upper limit (H_{max}) and a lower limit (H_{min}). So the control target is to keep these variables in the following range of values:

$$0.74 \leq G \leq 0.80$$

$$0.68 \leq H \leq 0.70$$

A FCM model for this system is developed and depicted on figure 4. Three experts constructed the FCM, they jointly determined the concepts of the FCM and then each expert drawn the interconnections among concepts and assigned fuzzy weight for each interconnection [5].

The FCM is consisted of five concepts:

Concept 1 – the amount of the liquid that Tank 1 contains is depended on the operational state of Valves 1, 2 and 3;
Concept 2 – the state of Valve 1 (it may be closed, open or partially opened);
Concept 3 - the state of Valve 2 (it may be closed, open or partially opened);
Concept 4 - the state of Valve 3 (it may be closed, open or partially opened);
Concept 5 –the specific gravity of the liquid into the tank.

For this specific control problem, experts have determined the initial values of concepts and weights, and which concepts are the desired output concepts (OCs). For this problem the desired output concepts are the concepts C_1 and C_5.

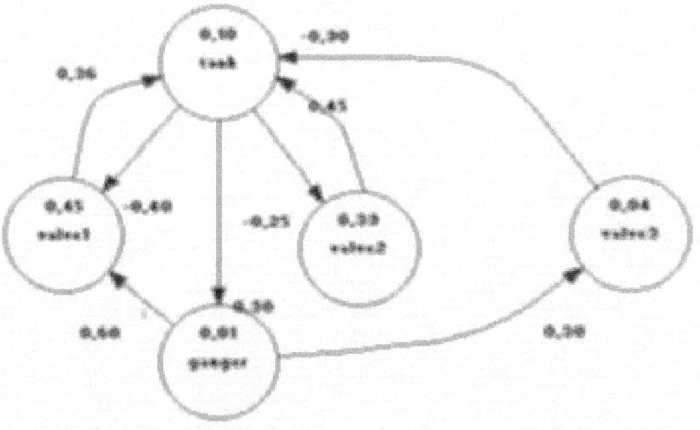

Fig. 4. The FCM model of the chemical process

Experts suggested the appropriate initial weights of the FCM model that are shown in the following weight matrix:

$$\mathbf{W}^{initial} = \begin{bmatrix} 0 & -0.4 & -0.25 & 0 & 0.3 \\ 0.36 & 0 & 0 & 0 & 0 \\ 0.45 & 0 & 0 & 0 & 0 \\ -0.9 & 0 & 0 & 0 & 0 \\ 0 & 0.6 & 0 & 0.3 & 0 \end{bmatrix}$$

Now the NHL algorithm can be applied to train the FCM and modify the weights. The training process starts by applying the initial values of concepts $\mathbf{A}^{0}_{first}$=[0.4 0.708 0.612 0.717 0.3] and weights $\mathbf{W}^{initial}$. The suggested value of learning rate η, at equation (8), after trial and error experiments, is determined as 0.01. Notably, at each iteration step, all weights are updated using the equation (6).

The algorithmic procedure continues, until the synchronously satisfaction of the two termination criteria are met. The result of training the FCM is a new connection weight matrix $\mathbf{W}$ that maximizes the objective function J and satisfy synchronously the second criterion. This algorithm iteratively updates the connection weights based on the equation (8), and equation (6) calculates the activation values of concepts based on the described asynchronous updating mode.

The convergent state is reached, after 16 recursive cycles, and it is $\mathbf{A}$=[0.692 0.738 0.676 0.747 0.743]. The updated weight matrix after 16 cycles, is:

$$\mathbf{W}^{updated} = \begin{bmatrix} 0 & -0.207 & -0.112 & 0.064 & 0.264 \\ 0.298 & 0 & 0.061 & 0.069 & 0.067 \\ 0.356 & 0.062 & 0 & 0.063 & 0.061 \\ -0.516 & 0.070 & 0.063 & 0 & 0.068 \\ 0.064 & 0.468 & 0.060 & 0.268 & 0 \end{bmatrix}$$

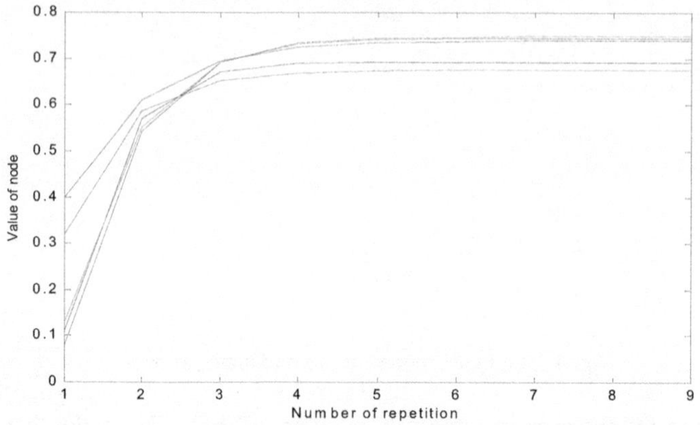

Fig. 5. Variation values of concepts for 9 simulation steps

These new values for weights describe new relationships among the concepts of FCM. It is noticeable that the initial zero weights no more exist, and new interconnections with new weights have been assigned, only diagonal values remain equal to zero. This means that all concepts affect the related concepts, and the weighed arcs show the degree of this relation.

5.1 Evaluation of the Modified Weight Matrix

Let's make now a testing using a random initial vector $\mathbf{A}^{0}_{random}$, and with the previously derived weight matrix, $\mathbf{w}^{updated}$ as initial. The randomly produced initial values are

$\mathbf{A}^{0}_{random}$=[0.1 0.45 0.37 0.04 0.01]. Applying the NHL algorithm, it stops after 9 simulation steps and the derived concept vector is A_{random}:

$\mathbf{A}_{random}$=[0.693 0.738 0.676 0.747 0.744], as shown in Fig. 5. This new state vector has the same values as the previous state vector $\mathbf{A}$, driving the FCM in the same convergence-desired region.

Also, we have tested this FCM model for 1000 test cases with random initial values of concepts, using the weight matrix $\mathbf{w}^{updated}$ and we end up at the same result for concepts values. So, if we use the FCM model of the process control problem with the modified values of weights for any initial values of concepts this model is driving in the desired region of concept values.

The NHL rule, which represented with equation (8), is suitable for updating the weights of FCMs with the same manner as in neural networks. Here the output concept is the one that have been fired by its interconnected concepts and it is initially defined for each specific problem. This output concept is the learning rate signal in the unsupervised learning procedure and due to the feedback process fires sequentially the other concepts and updates their cause-effect interconnections of FCM through the previous described Hebbian-type learning rule.

6 Conclusions and Future Directions

In this paper, the unsupervised learning method (NHL) based on nonlinear Hebbian-type learning rule is introduced to adapt the cause-effect relationships among concepts of FCM and to eliminate the deficiencies that appear in operation of FCMs. In this way, we take into account the dynamic characteristics of the learning process and the environment.

The unsupervised Hebbian rule is introduced to train the FCM and accompanied with the good knowledge of a given system or process can contribute towards the establishment of FCM as a robust technique. Also a formal methodology suitable for practical application has been developed and desired convergence regions for FCM process control have been produced.

In future work, further developments of the learning algorithm will be examined and the implementation of this learning approach on more complex problems will be investigated.

References

[1] Kosko, B.: Fuzzy Cognitive Maps, Int. J. Man-Machine Studies, Vol. 24 (1986), 65-75.

[2] Kosko, B.: Neural Networks and Fuzzy Systems, Prentice-Hall, New Jersey, (1992).

[3] Papageorgiou, E., Stylios, C.D., Groumpos, P.P.: Decision Making in external beam radiation therapy based on FCMs, Proceedings of the 1st International IEEE Symposium on Intelligent Systems, IS 2002, Varna, Boulgaria, 10-12 September, (2002), CD-ROM

[4] Papageorgiou, E.I., Stylios, C.D., Groumpos, P.P.: An Integrated Two-Level Hierarchical Decision Making System based on Fuzzy Cognitive Maps, IEEE Transactions on Biomedical Engineering, vol. 50, no. 12, (December 2003).

[5] Stylios, C.D., Groumpos, P.P.: Fuzzy Cognitive Maps in Modelling Supervisory Control Systems, Journal of Intelligent & Fuzzy Systems, vol. 8, (2000), 83-98.

[6] Khan, S., Chong, A., and Gedeon, T.A.: Methodology for Developing Adaptive Fuzzy Cognitive Maps for Decision Support, *Proc. 3rd Australia-Japan Evolutionary Systems,* Canberra 23-25 November, (1999), 93-100.

[7] Papageorgiou, E.I., Stylios, C.D., Groumpos, P.P.: Activation Hebbian Learning Rule for Fuzzy Cognitive Map Learning, *Proc. of 15th IFAC World Congress of Inter. Federation of Automatic Control,* Barcelona, Spain, July 21-26,CD-ROM, 2002.

[8] Stylios, C.D., Groumpos, P.P., Georgopoulos, V.C.: An Fuzzy Cognitive Maps Approach to Process Control Systems, *Journal of Advanced Computational Intelligence*, vol. 3, No. 5, 1999, pp. 409-417.

[9] Kosko, B.: Fuzzy Engineering, Prentice-Hall, New Jersey, (1997).

[10] Aguilar, J.: Adaptive Random Fuzzy Cognitive Maps. In: Garijio, F.J., Riquelme, J.C., Toro, M. (eds.): IBERAMIA 2002, Lecture Notes in Artificial Intelligence 2527, Springer-Verlag Berlin Heidelberg (2002), 402-410.

[11] Oja, E.: A simplified neuron model as a principal component analyzer. Journal of Mathematical Biology, Vol. 16, (1982), 267-273

[12] Oja, E., Ogawa, H., Wangviwattana, J.: Learning in nonlinear constrained Hebbian networks. In. T. Kohonen et al.(eds.): Artificial Neural Networks (1991), Amsterdam: North-Holland, 385-390.

[13] Oja, E.: Neural networks, principal components and subspaces. International Journal of Neural Systems, Vol. 1, (1989), 61-68

[14] Lee, W.C., Oslhausen, B.A.: A nonlinear Hebbian network that learns to detect disparity in random-dot stereograms. Neural Computation, Vol. 8, (1996), 545-566.

[15] Oja, E., Karhunen, J.: Nonlinear PCA, algorithm and applications. Report A18, Finland: Helsinski University of Technology (1993).

[16] Hassoun, M.: *Fundamentals of Artificial Neural Networks,* MIT Press, Bradford Book, Massachusetts, (1995).

[17] Sudjianto, A., Hassoun, M.: Statistical basis of nonlinear hebbian learning and application to clustering. Neural Netwoks, Vol. 8, No. 5, (1995), 707-715.

[18] Biggs, N.L., Llooyd, E.K., Wilson, R.J.: Graph Theory 1736-1936. Clarendon Press, Oxford, (1976).

[19] Harary, F., Norman, R.J., Cartwright, D.: Structural models: an introduction to the theory of directed graphs. John Wiley & Sons, New York, (1965), pp.201.

[20] Axelrod, R.: Structure of Decision, the cognitive maps of political elites. Princeton University Prass, Princeton, New Jersey, (1976), pp. 404

[21] Lin, C.T., Lee, C.S. G: *Neural Fuzzy Systems: A Neuro-Fuzzy Synergism to Intelligent Systems*, Prentice Hall, Upper Saddle River, N.J, (1996).

[22] Jang, J. S. R., Sun, C. T., Mizutani, E.: *Neuro-Fuzzy & Soft Computing.* Prentice-Hall, Upper Saddle River, (1997).

[23] Haykin, S.: Neural Networks: A Comprehensive Foundation (2nd ed.), Macmillan, (1994).

[24] Zurada, J.: Introduction to Artificial Neural Systems. West Publishing Company, (1992)

[25] Koulouriotis, D.E., Diakoulakis, I.E., Emiris, D.M.: Learning Fuzzy Cognitive Maps using evolution strategies: A novel schema for modeling a simulating high-level behavior. *Pro. 2001 IEEE Congress on Evolutionary Computation,* Seoul, Korea, 2001, vol. 1, pp. 364-371.

MML Inference of Decision Graphs with Multi-way Joins and Dynamic Attributes

Peter J. Tan and David L. Dowe

School of Computer Science and Software Engineering, Monash University
Clayton, Vic 3800, Australia
ptan@bruce.csse.monash.edu.au

Abstract. A decision tree is a comprehensible representation that has been widely used in many supervised machine learning domains. But decision trees have two notable problems - those of replication and fragmentation. One way of solving these problems is to introduce the notion of decision graphs - a generalization of the decision tree - which addresses the above problems by allowing for disjunctions, or joins. While various decision graph systems are available, all of these systems impose some forms of restriction on the proposed representations, often leading to either a new redundancy or the original redundancy not being removed. Tan and Dowe (2002) introduced an unrestricted representation called the decision graph with multi-way joins, which has improved representative power and is able to use training data with improved efficiency. In this paper, we resolve the problem of encoding internal repeated structures by introducing dynamic attributes in decision graphs. A refined search heuristic to infer these decision graphs with dynamic attributes using the Minimum Message Length (MML) principle (see Wallace and Boulton (1968), Wallace and Freeman (1987) and Wallace and Dowe (1999)) is also introduced. On both real-world and artificial data, and in terms of both "right"/"wrong" classification accuracy and logarithm of probability "bit-costing" predictive accuracy (for binary and multinomial target attributes), our enhanced multi-way join decision graph program with dynamic attributes improves our Tan and Dowe (2002) multi-way join decision graph program, which in turn significantly outperforms both C4.5 and C5.0. The resultant graphs from the new decision graph scheme are also more concise than both those from C4.5 and from C5.0. We also comment on logarithm of probability as a means of scoring (probabilistic) predictions.

1 Introduction

In spite of the success of decision tree systems in ("right"/"wrong") supervised classification learning, the search for a confirmed improvement of decision trees has remained a continuing topic in the machine learning literature. Two well-known problems from which the decision tree representation suffers have provided incentives for such efforts. The first one is the replication problem, which leads to the duplication of subtrees from disjunctive concepts. The effect of the

T.D. Gedeon and L.C.C. Fung (Eds.): AI 2003, LNAI 2903, pp. 269–281, 2003.

replication problem is that many decision tree learning algorithms require an unnecessarily large amount of data to learn disjunctive functions. The second problem is the fragmentation problem, which occurs when the data contains attributes with more than 2 values. Both of the problems increase the size of decision trees and reduce the number of instances in the individual nodes. Several decision graph representations have been introduced to resolve these problems. Decision graphs can be viewed as generalizations of decision trees, and both have decision nodes and leaves. The feature that distinguishes decision graphs from decision trees is that decision graphs may also contain joins (or disjunctions), which are represented by two (or more) nodes having a common child. This representation specifies that two subsets have some common properties, and hence can be considered as one subset. Tan and Dowe recently presented a general decision graph representation called decision graphs with multi-way joins [20], which was more expressive than previous decision graph representations [14, 12, 13, 8, 6, 11, 7]. We also introduced an efficient MML coding scheme for the new decision graph representation. However, the decision graph with multi-way joins [20] is not able to make efficient use of subtrees with internal repeated structures, as has been innovatively done for decision trees in [21]. In this paper, we refine our recent representation [20] by introducing dynamic attributes in decision graphs to solve this problem. We also point out some drawbacks in the search heuristic which led to premature joins in our multi-way join decision graphs [20] and resolve it by proposing a new search heuristic for growing decision graphs. We further advocate (in section 5.1) the merits of logarithm of probability - for binomial [4, 5, 3, 2, 9, 20], multinomial [3, 20] and other [2] distributions - as opposed to other approaches (see e.g., [16, 15]) to scoring probabilistic predictions.

2 Related Works

2.1 Minimum Message Length (MML) and MML Inference

The Minimum Message Length (MML) principle [22, 23, 26, 24] provides a guide for inferring the model of best fit given a set of data. MML inferences involve assigning a code length to each candidate model and searching for the model with minimum two-part message length (code length of the model plus the code length of the data given the model) [23, 26, 24].

MML and the subsequent Minimum Description Length (MDL) principle [19, 8] (see also [24] for a survey) are widely used for model selection in various machine learning problems. In practice, MML and MDL work very well on inference of decision trees. Among efforts that have been put into the development of tree-based classification techniques in recent years, Quinlan and Rivest [18] proposed a method for inferring decision trees using MDL. Wallace and Patrick subsequently [27] presented a refined coding scheme for decision trees using MML in which they identified and corrected some errors in Quinlan and Rivest's derivation of the message length, including pertaining to the issue of probabilistic prediction (cf. section 5.1). Wallace and Patrick also introduced a "Look Ahead" heuristic of arbitrarily many ply for selecting the test attribute at a node. We re-use the Wallace and Patrick decision tree coding [27] as part of the coding scheme for our new decision graph program. For further details of the implementation, please see [20].

2.2 Decision Graphs Currently in the Literature

As we mentioned in section 1, it is important to resolve the replication and fragmentation problems of decision trees. Many attempts have been made to extend decision trees to decision graphs or graph-like systems. A binary decision graph scheme using MML was introduced by Oliver and Wallace [14, 12, 13]. Other schemes include a generalized decision tree system using MDL [19] proposed by Mehta et al. [8], the HOODG (Hilling-climbing Oblivious read-Once Decision Graphs) system proposed by Kohavi [6], and a decision graph representation called the branch program proposed by Mansour and McAllester [7]. For a more detailed discussion on these systems, see [20, Section 1].

The decision graph system proposed by Tan and Dowe [20] allows multi-way joins. For a similar scheme, see the [10, Appendix]. Directed acyclic graphs were used in the Tan and Dowe system [20] as in both the Oliver and Wallace system [14, 12, 13] and the Kohavi system [6]. The main idea behind the coding of the decision graphs with multi-way joins is to decompose a decision graph into a sequence of decision trees and joining patterns [20]. In this way, encoding a decision graph is equivalent to encoding a sequence of decision trees and joining patterns in order. An efficient coding scheme for decision graphs can be achieved by re-using some of the well-proved Wallace-Patrick decision tree coding scheme [27] and devising an efficient coding of the joining patterns [20].

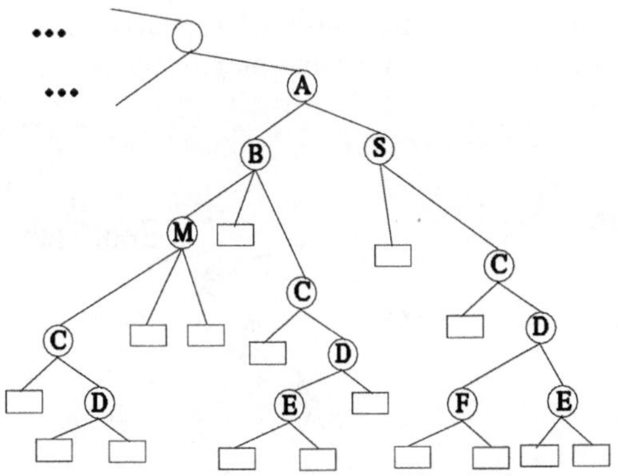

Fig. 1. A decision tree with internal repeated structures (involving C and D)

3 Internal Repeated Structures and Linked Decision Forests

As discussed in the previous sections, decision graphs are able to represent some duplicated sub-concepts efficiently by uniting these subtrees into one tree. However, in many tree learning problems, these subtrees are not entirely identical but rather share repeated internal structures. For example, the tree in Figure 1 contains three subtrees with internal repeated structures (involving C and D).

The repeated internal structure problem was first brought forward and studied by Uther and Veloso [21]. Their solution to this problem was to introduce a new representation called the decision linked forest [21], in which the decision tree is turned into a sequence of attribute trees and a root tree. The attribute trees were formed by abstracting the topological structures of the repeated internal structures in the original trees. The attribute trees were then treated as new attributes in the root tree. The new coding scheme only encodes the repeated sub-concepts once by forming attribute trees as new attributes available to decision trees in the linked decision forest. The scheme can be explained by Figure 2, which shows how linked decision forests provide a more efficient solution to resolve the internal repeated structure problems in Figure 1.

3.1 Decision Graphs with Dynamic Attributes

Uther and Veloso's novel use of linked decision forests [21] eliminated inefficient coding of the internal repeated structures in a decision tree. However, the root tree of a linked decision forest, where the inference process occurs, is still a decision tree, so the fragmentation problem of decision trees (which is solved by decision graphs) remains unresolved in the linked decision forest. Uther and Veloso [21] claimed that none of the existing decision graph programs was able to resolve the problem of encoding internal repeated structures. We have addressed this issue by introducing dynamic attributes in our decision graph with

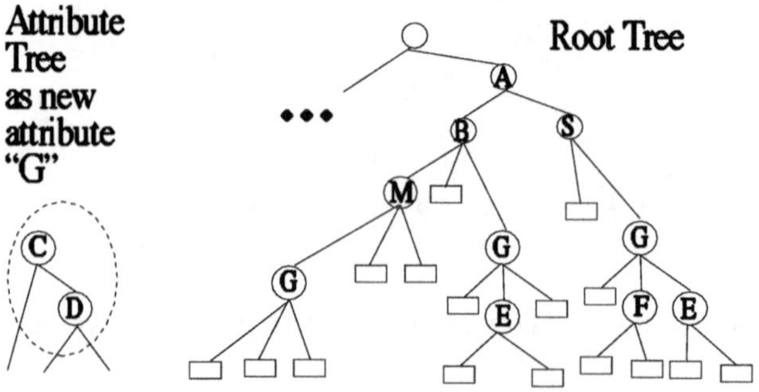

Fig. 2. A linked decision forest [21] with an attribute tree (c.f. Fig. 1)

multi-way joins, which essentially generalises both decision graphs with multi-way joins [20] and Uther-Veloso linked decision forests [21] - as is explained in detail below.

Whenever there is an M-way join operation in a decision graph, a new attribute is created and is made available for every node in the subtrees under the node resulting from the M-way join operation. From the node, back traces are performed along the M joining routes until they reach the root of the decision graph. Then the root and the M routes define a new attribute with arity M. The purpose of this attribute is to separate the data into M categories corresponding to the way by which the data arrived at the M-way join. When there are several subtrees in a decision graph with internal repeated structures, they are joined to form one subtree consisting of the internal structure. Then, the leaf nodes where there are differences among corresponding leaves in the original subtrees are split on the new attribute. As such, the decision graphs are able to join subtrees with internal repeated structure (immediately before each one would split on this structure) so that the repeated structure is only encoded once. (Ideally, the new dynamic attribute is not immediately split on.) Once created, the new attribute becomes common knowledge to both sender and receiver and thus there is no transmitting cost on its description. We contend that this scheme provides a more efficient and elegant solution to this problem than linked decision forests and certainly decision graphs described in much of the literature. (We have opted for this scheme rather than for linked decision graph forests.) A solution to resolve the internal repeated structure problems in Figure 1 by our new decision graphs with dynamic attributes is shown in Figure 3.

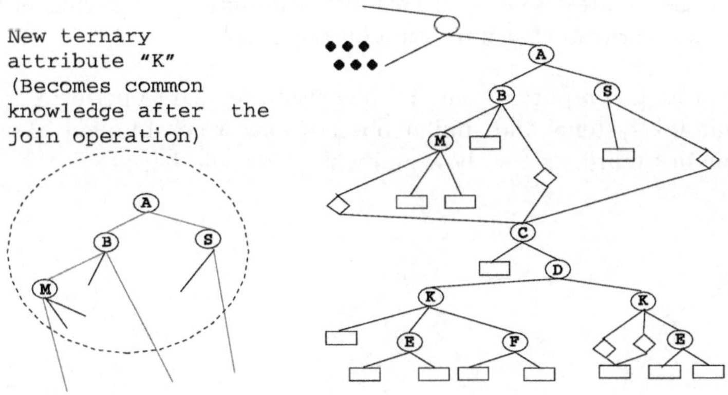

Fig. 3. A decision graph with an dynamic attribute (c.f. Fig. 1 and Fig. 2)

4 Growing a Decision Graph

While growing a decision tree, the order in which the leaf nodes are expanded is irrelevant to the resultant tree since splitting a leaf node would have no effect on the following actions taken on other leaves. However, the order in which the leaf nodes are expanded or joined is often crucial while growing a decision graph. Such a significant difference between decision tree inference and decision graph inference makes the algorithm used in the former inadequate for the latter. In this section we investigate the MML decision graph growing algorithm implemented for Oliver and Wallace's binary join decision graph program [12, 13, 14], and explain a drawback in their search heuristic which makes their program unable to infer the optimal graph in some circumstances. In section 4.2, we will present our new algorithm for inferring decision graphs with multi-way joins in detail.

4.1 Oliver and Wallace's MML Decision Graph Generation Algorithm

Oliver and Wallace's algorithm extends a decision graph by iteratively performing the following procedures until no further improvement can be achieved.

1. For each Leaf, L, determine the attribute A on which it should be split. Record, but do not perform, the alteration (Split L on A) along with its saving in message length.
2. For each pair of leaves, L1 and L2, perform a tentative join. Record, but do not perform, the alteration (Join L1 and L2) along with its saving in message length.
3. Choose the alteration (whether from step 1 - a Split, or from step 2 - a Join) that has the greatest saving. If this alteration creates a saving in message length, then perform that alteration on the graph.

Oliveira et al. [11] reported that this algorithm tended to perform premature joins on complex systems and similar observations were obtained in our tests. The example in Figure 4 shows how and why this could happen.

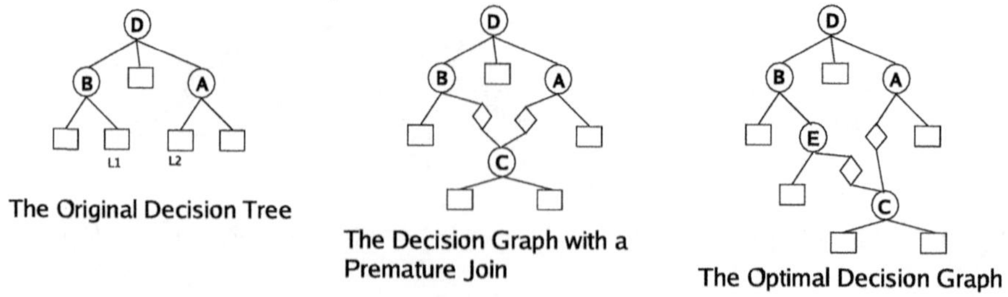

Fig. 4. An example illustrating how the premature joins are generated

Suppose it is decided to grow the decision tree shown on the left of Figure 4. Thus, for leaf L1, splitting on attribute C will save S_1 bits in message length while splitting on attribute E will yield S_1' bits saving in message length. If $S_1' > S_1$, then according to the algorithm above, the alteration that splits L1 on attribute E is recorded. For leaf L2, the same is done for the alteration that splits L2 on attribute C with S_2 bits saving in message length. When performing a tentative join, the same is done again for the alteration that joins L1 and L2 with S_j bits saving in message length. When estimating the saving from a tentative join, a lookahead search whose aim is to look for the subtree with the minimum message length is conducted on the node resulting from the join. Since expanding the node resulting from joining leaf L1 and leaf L2 would be viewed as merging expanded L1 with expanded leaf L2, so the saving is $S_j = S_1 + S_2 - S_g + S_t$, where S_g is the cost in message length to transmit the join, and S_t is the cost to transmit the topological structure of one of the subtrees. In the case when $S_j > S_1'$, the resultant graph would be the graph shown in the middle of Figure 4 instead of the optimal one shown on the right of Figure 4. The Oliver and Wallace algorithm has a bias toward joins because it compares the sum of the savings from expanding the two leaf nodes with the saving from expanding just one leaf node. This shows why Oliver and Wallace's decision graph growing algorithm produces premature joins in some circumstances.

4.2 The New MML Decision Graph Growing Algorithm

If we implement the above algorithm in our decision graph inference scheme, the fact that we allow multi-way joins could only increase such bias. So we propose the following algorithm to eliminate the premature joins. To grow a decision graph, we begin with a graph having one node, with the root being a leaf. We grow the graph by performing the following procedures iteratively until no further improvement can be achieved.

1. For each leaf L, perform tentative splits on each available attribute in the leaf, and determine the attribute A that will lead to the shortest message length when L is split on A. Record, but do not perform, the alteration (Split L on A) along with its rate of communication saving - the communication saving divided by the number of data items in the leaf.
2. For each leaf L, perform tentative joins with other leaves. Record, but do not perform, the alterations (join L_i and L_j; ...; join $L_i, L_j, \ldots, L_k$; etc.) along with its rate of communication savings - the communication saving divided by the number of data items in the join.
3. Sort the alterations from step 1 and step 2 by their communication savings. Choose the alteration (whether from step 1 or from step 2) that has greatest rate of saving.

When splitting on any continuous-valued attributes, we implement a simple single cut-point search algorithm, in which the information gained from the cut is the objective function. Then the cost in message length to state the cut-point is log(the number of values of the attribute in this node - 1). In each iteration, we manipulate the data (i.e., split a leaf or join leaves) so that the greatest rate of saving in message length can be achieved. Thus, it is guaranteed that in the later iterations we will generate a decision graph better or not worse than the possible optimal graph expected in the current iteration. Of course, the algorithm with this search heuristic is only locally optimal.

5 Experiments

One artificially generated data set and eight real-world data sets were used in our tests. The only artificial data set is the XD6 data set [18, 27, 14, 20], which consists of 10 (9 input, 1 output) binary attributes. It was generated according to the boolean function of attributes 1 to 9:

$(A1 \wedge A2 \wedge A3) \vee (A4 \wedge A5 \wedge A6) \vee (A7 \wedge A8 \wedge A9)$

with 10% noise added to the target attribute. The other eight real-world data sets were downloaded from the UCI machine learning repository [1] and have been widely tested in other decision tree or decision graph systems [14, 6, 17]. In order to rigorously examine the proposed algorithms, 10 10-fold cross-validations were performed on each of the nine data sets. This amounted to 10x10=100 tests for one single data set. Each pair of training/test data from these tests was fed into four different decision tree and graph algorithms: the well-known decision tree classification programs C4.5, C5 [17], the decision graph with multi-way joins [20] and our new decision graphs with multi-way joins and dynamic attributes.

The experimental results are presented in Tables 1, 2, 3 and 4. In table 1, the run time recorded the execution time of one test by the algorithms on a PIII 1G Linux Redhat7.3 PC. In table 2, “Error Rate” describes the rate of “right”/“wrong” classification errors. In table 3, “pr costing” describes Good’s (binomial) probabilistic costing [4, 5], or logarithmic ‘bit costing’ [2, 3, 20, 5, 4, 9]. In table 4, we compare the size of resultant decision trees and graphs by recording the number of leaf nodes in them. For the data sets on which 10 10-fold cross-validations were performed, the rate of classification errors and probabilistic costings are presented as mean ± standard deviation, $\mu \pm \sigma$.

Table 1. Summary of Data Sets

Data-set Name	size	Discrete Attributes	Continuous Attributes	Number of Classes	Run Time (C4.5, C5)	Run Time dGraph[20]	Run Time dG (dyn atts)
abalone	4177	1	7	29	1.35s	1062s	1132s
car	1728	6	0	4	0.03s	2.07s	2.42s
cmc	1473	8	2	2	0.06s	11.0s	13.7s
credit	690	9	6	2	0.04s	29.9s	38.9s
led	500	7	0	10	0.01s	4.50s	5.90s
scale	625	4	0	3	0.01s	1.10s	1.70s
tic-tac-toe	958	9	0	3	0.01s	3152s	4031s
vote	435	16	0	2	0.00s	0.01s	0.01s
XD6	500	9	0	2	0.01s	2.55s	3.22s

5.1 Comparing and Scoring Probabilistic Predictions

Decision trees and graphs are often used as classifiers in many machine learning problems. In the case in which the target attribute is multinomial, each leaf node in a tree or graph is given a class label corresponding to the class with the highest inferred probability for this node. However, the multinomial distribution in each leaf node can also be interpreted as a probabilistic prediction model. In this way, the decision trees and decision graphs are not only classifiers, but they can also provide a probabilistic prediction model. Provost and Domingos [16] showed that with some modifications, tree inductions programs can produce very high quality probability estimation trees (PETs). Perlich, Provost and Simonoff [15] also observed that for large data sets, tree induction often produces probability-based rankings that are superior to those generated by logistic regression. Thus, in addition to the conventional classification accuracy, a metric called probabilistic costing [5, 4, 2, 3, 20, 9] was implemented in our tests for comparisons of probabilistic predictions with C4.5 and C5. It is defined as $-\sum_{i=1}^{n}\log(p_i)$, where n is the total number of test data and p_i is the predicted probability of the true class associated with the corresponding data item [5, 4, 2, 3]. The reader can interpret the metric as the optimal coding length for the test data given the resultant tree and graph models. This metric can be used to approximate (within a constant) the Kullback-Leibler distance between the true (test) model and the inferred model. Its relation to log-likelihood via $-\log(\prod_{i=1}^{n} p_i) = -\sum_{i=1}^{n}\log(p_i)$, its relation to Kullback-Leibler distance and its corresponding general applicability to a wide range of probability distributions (recall section 1) [5, 4, 2, 3, 20, 9] strongly recommend this log(prob) bit costing as a statistically-based general alternative to metrics such as ROC and AUC (Area Under Curve) [16, 15].

Logarithm of probability (bit) scoring, Pr_cost, enables us to compare probabilistic prediction accuracy of inferences from an identical training data set by various decision tree and graph algorithms. The lower the value of the Pr_cost, the more consistent the predicted probabilistic model is with the true model.

Table 2. Test Results ('right'/'wrong' Error rates) %

Data-set Name	C4.5	C5	dGraph with M-way joins [20]	dGraph with dynamic atts
abalone	78.9 ± 1.9	79.0 ±1.8	74.3 ±2.1	74.3 ± 2.1
car	7.8 ± 2.2	7.8±2.1	8.5 ±2.8	6.7 ± 2.8
cmc	48.2 ±3.6	48.5 ± 3.6	48.4 ±3.6	48.2 ±3.6
credit	14.4 ±3.6	14.5 ± 3.6	14.2±4.3	14.2 ±4.3
led	30.0 ±5.2	30.0 ± 5.2	30.0±5.8	30.0 ±5.8
scale	35.6 ±4.9	35.4 ± 3.3	22.0±5.3	22.0 ±5.3
tic-tac-toe	14.4 ±3.4	14.0 ± 3.5	11.9±4.8	10.7 ±4.9
vote	5.0 ± 3.1	5.0 ± 3.1	4.4±3.3	4.4 ± 3.3
XD6	14.1 ±4.9	14.2 ± 5.0	9.2±4.0	9.2 ±4.0

Table 3. Test Results ($-\log(Prob)$ Costing) bits

Data-set Name	C4.5 (+0.5)	C5 (+0.5)	dGraph with M-way joins [20] (+0.5)	dGraph with dyn atts (+0.5)
abalone	1810.3 ± 26.3	1814.5±25.9	1269.6±32.0	1269.8±33.4
car	60.0 ± 9.9	61.2±9.8	49.8 ±10.5	40.7 ±12.0
cmc	221.4±13.3	222.1 ±13.4	202.4±9.9	202.2 ±9.9
credit	38.3 ±8.1	38.4 ± 8.0	35.2±7.5	35.2 ±7.5
led	79.3 ±9.8	79.3±9.7	76.2±10.0	76.2 ±10.0
scale	82.9 ±6.8	80.1±6.6	55.0±10.0	55.0 ±10.0
tic-tac-toe	46.4 ±8.4	45.4 ±7.1	44.7±13.5	41.1 ±14.7
vote	9.8 ± 6.2	9.8 ±6.1	8.6±5.5	8.6 ± 5.5
XD6	28.5 ±7.8	28.4 ±7.1	22.4±6.7	22.4 ±6.7

Given an array of occurrences of events of an m-state multinomial distribution $(c_1, c_2, \ldots, c_m)$, the probability of a certain event j can be estimated by (either) $\hat{p_j} = \frac{c_j+0.5}{(\sum c_i)+m/2}$ [22, p187 (4), p194 (28), p186 (2)][26][25, p75][20] or $\hat{p_j} = \frac{c_j+1}{(\sum c_i)+m}$ [22, p187 (3), p189 (30)][20], the latter being known as the Laplace estimate and also corresponding (with uniform prior) to both the posterior mean and the minimum expected Kullback-Leibler distance estimator [24]. In our experiments, the first (+0.5) was used in MML multinomial message length calculations, $\hat{p_j}$ estimations and calculations of the log(prob) bit costing.

5.2 Discussions of Above Test Results

Tables 2 and 3 clearly show the decision graph with dynamic attributes to always be either outright first or (sometimes) equal first. When testing on the data sets with disjunctions (like abalone, scale, tic-tac-toe and XD6), decision graph with dynamic attributes has a much lower error rate. On other data sets, it returns results not worse than those from C4.5 and C5. Results from the decision graph with dynamic attributes are either identical or marginally better than those from the decision graphs with multi-way joins [20]. In the cases that the decision graph with dynamic attributes performs better, generated dynamic attributes are found in the resultant graphs. This proves the importance of the dynamic attributes and also shows that decision graphs with dynamic attributes are a superset of decision graphs with multi-way joins [20]. In table 3, the Pr_cost from both kinds of decision graph are clearly lower than those from both C4.5 and C5. This suggests that the decision graphs inferred by MML are resistant to over-fitting. As such, the decision graphs not only produce excellent "right"/"wrong" predictions, but also provide inferred probabilistic models that are clearly more consistent with those inferred from the test data (cf. also [20, Tables 1 to 3]).

From table 4, we find that the resultant multi-way join decision graphs [20] are similar in size to the decision graphs with dynamic attributes. The sizes of

Table 4. Size of Resultant Tree or Graph (Number of leaf nodes)

Data-set Name	C4.5	C5	dGraph with M-Way joins [20]	dGraph with dynamic atts
abalone	2103	1062	315	313
car	170	122	36	38
cmc	230	132	10	10
credit	34	15	7	7
led	42	22	22	22
scale	38	34	21	21
tic-tac-toe	132	88	37	35
vote	16	8	5	5
XD6	52	25	5	5

the resultant graphs tend to be substantially smaller than the sizes of both the C4.5 and C5 trees. From table 1, both kinds of MML decision graph take longer to infer than C4.5 and C5 trees. Tables 3, 2, and 4 suggest it is well worth the wait. Nonetheless, we do intend to trim the decision graph searches.

6 Conclusion and Discussion

In this paper, we have refined the decision graph with multi-way joins [20] representation by introducing dynamic attributes in the decision graphs. Using the Minimum Message Length principle, an improved coding scheme for inferring the new decision graphs has been devised to address some of the inefficiencies in previous decision tree and decision graph coding schemes. Our experimental results demonstrated that our refined coding scheme compares favourably with other decision tree inference schemes, namely both C4.5 and C5. This favourable comparison holds true both for 'right'/'wrong' prediction accuracy and especially for I.J. Good's logarithm of probability bit costing (recall section 5.1 and table 3), as well as for both artificially generated and real-world data.

In future work, we hope to both speed up the decision graph searches and compare with more programs, such as, e.g., Yin and Han's CPAR [28].

References

[1] C.L. Blake and C.J. Merz. UCI repository of machine learning databases, 1998. `http://www.ics.uci.edu/~mlearn/MLRepository.html`. 276

[2] D.L. Dowe, G.E. Farr, A.J. Hurst, and K.L. Lentin. Information-theoretic football tipping. In N. de Mestre, editor, *Third Australian Conference on Mathematics and Computers in Sport*, pages 233–241. Bond University, Qld, Australia, 1996. `http://www.csse.monash.edu.au/~footy`. 270, 276, 277

[3] D.L. Dowe and N. Krusel. A decision tree model of bushfire activity. In *(Technical report 93/190) Dept Computer Science, Monash University, Clayton, Vic. 3800, Australia*, 1993. 270, 276, 277

[4] I.J. Good. Rational Decisions. *Journal of the Royal Statistical Society. Series B*, 14:107–114, 1952. 270, 276, 277

[5] I.J. Good. Corroboration, Explanation, Evolving Probability, Simplicity, and a Sharpened Razor. *British Journal of Philosophy of Science*, 19:123–143, 1968. 270, 276, 277

[6] Ron Kohavi. Bottom-up induction of oblivious read-once decision graphs: Strengths and limitations. In *National Conference on Artificial Intelligence*, pages 613–618, 1994. 270, 271, 276

[7] Yishay Mansour and David McAllester. Boosting using branching programs. In *Proc. 13th Annual Conference on Comput. Learning Theory (CoLT)*, pages 220–224. Morgan Kaufmann, San Francisco, 2000. 270, 271

[8] Manish Mehta, Jorma Rissanen, and Rakesh Agrawal. MDL-based Decision Tree Pruning. In *The First International Conference on Knowledge Discovery & Data Mining*, pages 216–221. AAAI Press, 1995. 270, 271

[9] S.L. Needham and D.L. Dowe. Message length as an effective Ockham's razor in decision tree induction. In *Proc. 8th International Workshop on Artificial Intelligence and Statistics (AI+STATS 2001)*, pages 253–260, Key West, Florida, U.S.A., Jan. 2001. 270, 276, 277

[10] Julian R. Neil. *MML discovery of Causal Models.* PhD thesis, Monash University, Clayton 3800, Australia, Computer Science and Software Engineering, 2001. 271

[11] Arlindo L. Oliveira and Alberto L. Sangiovanni-Vincentelli. Using the minimum description length principle to infer reduced ordered decision graphs. *Machine Learning*, 25(1):23–50, 1996. 270, 274

[12] J.J. Oliver. Decision Graphs - An Extension of Decision Trees. In *Proc. 4th International Workshop on Artif. Intelligence and Statistics*, pages 343–350, 1993. 270, 271, 274

[13] J.J. Oliver, D.L. Dowe, and C.S. Wallace. Inferring Decision Graphs Using the Minimum Message Length Principle. In *Proceedings of the 5th Joint Conference on Artificial Intelligence*, pages 361–367. World Scientific, Singapore, 1992. 270, 271, 274

[14] J.J. Oliver and C.S. Wallace. Inferring Decision Graphs. In *Workshop 8 International Joint Conference on AI (IJCAI)*, Sydney, Australia, August 1991. 270, 271, 274, 276

[15] C. Perlich, F. Provost, and J.S. Simonoff. Tree induction versus logistic regression: a learning-curve analysis. *Journal of Machine Learning Research*, 4:211–255, 2003. 270, 277

[16] Foster Provost and Pedro Domingos. Tree induction for probability-based ranking. *Machine Learning*, 52:199–215, Sept. 2003. 270, 277

[17] J.R. Quinlan. *C4.5 : Programs for Machine Learning.* Morgan Kaufmann,San Mateo,CA, 1992. The latest version of C5 is available from `http://www.rulequest.com`. 276

[18] J.R. Quinlan and R. Rivest. Inferring Decision Trees Using the Minimum Description Length Principle. *Information and Computation*, 80:227–248, 1989. 270, 276

[19] J.J. Rissanen. Modeling by shortest data description. *Automatica*, 14:465–471, 1978. 270, 271

[20] P.J. Tan and D.L. Dowe. MML inference of decision graphs with multi-way joins. In *Proc. 15th Australian Joint Conf. on AI, LNAI 2557 (Springer)*, pages 131–142, Canberra, Australia, 2-6 Dec. 2002. 270, 271, 273, 276, 277, 278, 279

[21] W.T.B. Uther and M.M. Veloso. The Lumberjack Algorithm for Learning Linked Decision Tree. In *Proc. 6th Pacific Rim International Conf. on Artificial Intel-*

ligence (PRICAI'2000), LNAI 1886 (Springer), pages 156–166, 2000. 270, 272, 273

[22] C.S. Wallace and D.M. Boulton. An Information Measure for Classification. *Computer Journal*, 11:185–194, 1968. 270, 278

[23] C.S. Wallace and D.M. Boulton. An Invariant Bayes Method for Point Estimation. *Classification Society Bull.*, 3:11–34, 1975. 270

[24] C.S. Wallace and D.L. Dowe. Minimum Message Length and Kolmogorov Complexity. *Computer Journal, Special Issue - Kolmogorov Complexity*, 42(4):270–283, 1999. 270, 278

[25] C.S. Wallace and D.L. Dowe. MML Clustering of multi-state, Poisson, von Mises circular and Gaussian distributions. *Statistics and Computing*, 10(1):73–83, Jan 2000. 278

[26] C.S. Wallace and P.R. Freeman. Estimation and Inference by Compact Coding. *Journal of the Royal Statistical Society. Series B*, 49(3):240–265, 1987. 270, 278

[27] C.S Wallace and J.D. Patrick. Coding Decision Trees. *Machine Learning*, 11:7–22, 1993. 270, 271, 276

[28] Xiaoxin Yin and Jiawei Han. CPAR: Classification based on predictive association rules. In *SIAM International Conference on Data Mining*. San Francisco, CA, USA, May 2003. 279

Selection of Parameters in Building Fuzzy Decision Trees

Xizhao Wang[1], Minghua Zhao[1], and Dianhui Wang[2]

[1] Machine Learning Center, School of Mathematics and Computer Science
HeBei University, Baoding, China
Phone/Fax: +86-312-5079638
wangxz@mail.hbu.edu.cn

[2] Department of Computer Science and Computer Engineering
La Trobe University, Melbourne, VIC 3083, Australia
Phone: +61-3-94793034, Fax: +61-3-94793060
csdhwang@ieee.org

Abstract. Compared with conventional decision tree techniques, fuzzy decision tree (FDT) inductive learning algorithms are more powerful and practical to handle with ambiguities in classification problems. The resultant rules from FDTs can be used in decision-making with similar nature of our human beings by inference mechanism. A parameter, namely significant level (SL) α, plays an important role in the entire process of building FDTs. It greatly affects the computation of fuzzy entropy and classification results of FDTs, however this important parameter value is usually estimated based on users by domain knowledge, personal experience and requirements. As a result of this, it will be hard to build a high performance FDT without an optimal SL option in practice. This paper aims at developing a method to determine an optimal SL value through analyzing the relationship between the fuzzy entropy and α. The main contribution of this work is to provide some guidelines for selecting the SL parameter α in order to build FDTs with better classification performance. Six data sets from the UCI Machine Learning database are employed in the study. Experimental results and discussions are given.

1 Introduction

Inductive learning [1], an important branch in machine learning field, extracts generic rules from a collection of data. Decision tree based inductive learning algorithm ID3 proposed by Quinlan in 1986 is well known and representative. Decision tree, as an effective method for knowledge representation, has some advantages in descriptions, effectiveness and efficiency. Many algorithms in building DTs are based on assumptions that attribute-values and classification-values are exactly known. This constrains application scopes and is not a suitable manner to meet the request for automated acquisition of uncertain knowledge. In order to overcome this shortcoming,

T.D. Gedeon and L.C.C. Fung (Eds.): AI 2003, LNAI 2903, pp. 282-292, 2003.

M.Umanol and C.Z.Janikow proposed fuzzy-ID3 algorithm [3,4] as a fuzzy version of traditional ID3 algorithm [2]. The significance of use of fuzzy logic in decision tree techniques is that it helps to handle ambiguity in classification problems with a parallel manner to our human's thought and sense.

In building FDTs, each expanded attribute will not classify data in crisp way as the original decision trees perform, attribute values will be allowed to have some overlaps. The entire process of building FDT is based on a significance level α [5]. This parameter plays an important role and controls the degree of such overlaps to some extent, which, however, is usually assigned by domain experts based on their working experience and knowledge base. Therefore, it is difficult to reach the best classification performance. Through analyzing the relationship between α and fuzzy entropy, this paper discusses analytical properties of the fuzzy entropy function associated with the parameter α, and shows that the change trend of fuzzy entropy function is problem dependent. The main contribution of this paper is to develop an experimental method for obtaining the optimal value of α so that it will result in the best FDT in terms of classification performance.

2 FDT Inductive Learning and Definition of the Parameter α

Optimization in DT inductive learning was proven to be a NP problem [6]. Recently, researchers working in this field focus on developing some good heuristic algorithms, where the key is to properly select the attributes in building the DT [7-9]. The concept of entropy is derived from information theory proposed by Shannon [10]. Quinlan's ID3 algorithm was based on entropy to select expanded attribute. The fuzzy-ID3 algorithm was constructed based on fuzzy entropy, a fuzzy extension of the concept of entropy in FDT induction learning. The following sections give some details on algorithm descriptions and relevant parameters.

2.1 FDT Heuristic Algorithm ID3 Based on Fuzzy Entropy

We view the nodes in FDT as fuzzy subset defined on universe E, which are usually obtained by ∩ operator with different linguistic values. Let D be a considered node.

Definition 1. We use i attributes of A_i to partition node D, let $Range(A_i) = \{T_{i1}, T_{i2}, \cdots, T_{ik_i}\}(1 \le i \le m)$, for each $j(1 \le j \le k_i)$, T_{ij} is a fuzzy vector defined on E. A partition is shown as below:

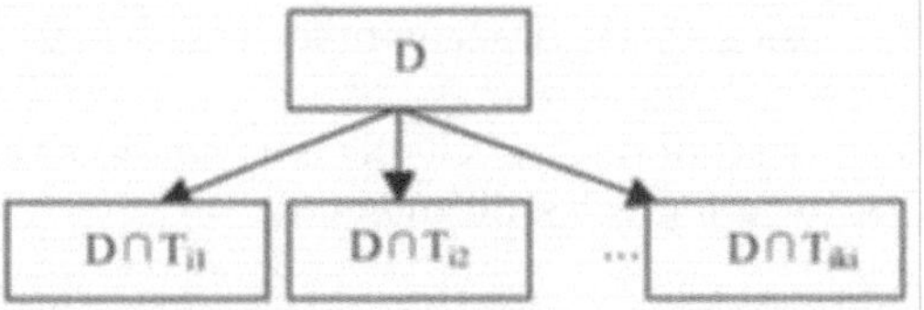

Fig. 1. A fuzzy partition

The fuzzy set $\{T_{i1}, T_{i2}, \cdots, T_{ik_i}\}$ defines a partition of D. Then, partition fuzzy entropy involved by A_i can be calculated as below. For the node D, a crisp partition about class {P, N} forms a fuzzy partition of D, that is, $\{P \cap D, N \cap D\}$. The fuzzy entropy of this partition is defined as:

$$E(D \cap T_{ij}) = -\frac{a}{a+b}\log_2\frac{a}{a+b} - \frac{b}{a+b}\log_2\frac{b}{a+b} \quad j = 1,2,\cdots,k_i, \tag{1}$$

where $a = M(P \cap D \cap T_{ij}), b = M(N \cap D \cap T_{ij})$.

Let $m^* = \Sigma_j M(D \cap T_{ij})$. Then, partition fuzzy entropy associated with A_i in node D is measured by:

$$FE(D, A_i) = \sum_{i=1}^{k_i} \frac{M(D \cap T_{ij})}{m^*} E(D \cap T_{ij}) \tag{2}$$
,

where M is the summation operator on all membership of the cases in considered fuzzy set, and

$$f_P(D) = M(D \cap P)/M(D),\ f_N(D) = M(D \cap N)/M(D) \tag{3}$$

denote the relative frequency of cases in node D belonged to P class and N class, respectively. Using these notations, we can describe the fuzzy ID3 algorithm.

Given a leaf criterion $\delta (0<\delta<1)$, consider node D:

Step 1: calculate $f_P(D)$ and $f_N(D)$, if they are larger than the δ, then sign the node as leaf, turn *step 4.*

Step 2: calculate each fuzzy entropy $FE(D, A_j)$ of the attribute, which are not used in father node, select the attribute possessing the minimum as expanded attribute on this node, signed as A_i.

Step 3: generate child node based on the branch of attribute Ai in node D, sign D as expanded node.

Step 4: judge if there are other non-leaf nodes that are not expanded:
  If yes, consider one of them, turn *step 1*,
  Else, stop, export the result, end.

The process of FDT inductive learning can be summarized below:

Step 1: fuzzify the training data using fuzzy cluster techniques based on iterative self-organization algorithm.

Step 2: build up FDT using fuzzy ID3 algorithm based fuzzy entropy mentioned above.

Step 3: translate FDT to a set of rules. Each rule is a path from root to leaf.

Step 4: apply the rules to perform classification task.

2.2 The Introduction of Parameter α

Definition 2. Given a fuzzy set E with membership $E(x)$, select a significance level α, 0< α <1, the fuzzy subset E_α is defined as

$$E_\alpha = \begin{cases} E(x) & \text{if } E(x) \geq \alpha \\ 0 & \text{if } E(x) < \alpha \end{cases} \tag{4}$$

The fuzzy subset E_α is a filter of E, from the view of evidence theory, E is evidence, and while E_α is strong evidence. The significance level α provides a filter to reduce the ambiguity (overlapping) in partitioning. The higher the α takes, the lower the ambiguity will be. However, a high α may lead to the omission of some objects that have no evidence exceed α therefore they do not belong to any subset of the partition. In the process of building FDT, we use fuzzy-ID3 algorithm based on fuzzy entropy described by Definition 1. Before building the FDT, the original data is intercepted by α to reduce the fuzziness of partition.

3 The Effect of α in FDT Inductive Learning

3.1 The Importance of Parameter α

In the process of expanding decision tree by ID3 algorithm, intercepted data will join the calculation of fuzzy entropy and further influence the selection of expanded attribute and the size of generated FDT, and finally the result in terms of rule complexity and classification accuracy. The following flow chart of building the FDT will show the importance of parameter α in FDT:

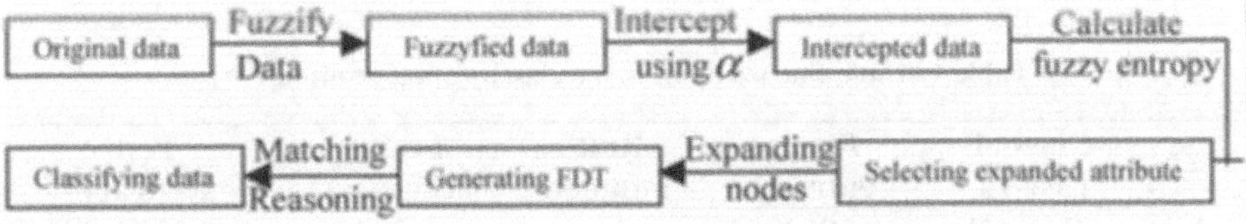

Fig. 2. The process of the influence of α to FDT

The α plays a key role for selection of expanded attribute, generation of decision tree and final classification result. So the relationship between α and fuzzy entropy is needed to explore, which will establish the base for further study on the parameter α.

The dataset *iris* in UCI-machine-learning database is employed as an example. The same cases will belong to different classes with different α. By attribute A4 of *iris*, Table 1 shows the change of class that case 16 belongs to □class of case 16□and the influence of α to the classification result of FDT with fuzzy entropy, rule number and classification accuracy for *iris* database. Here the true level β =0.9.

As α increases, the fuzzy entropy of A4 decreases gradually in the root node, so the information gain of A4 increases, while rule number and the overlapping that each class cover cases decrease, and classification accuracy becomes better. Obviously, different α will lead to different classification performance. Because the significant influence of α to FDT performance, it is essential to optimize the parameter α for obtaining the better results.

3.2 The Analytical Relationship between α and Fuzzy Entropy

From the above discussion, the higher the α takes, the less the classification fuzziness will be. Also, the crisper the classification performs, the fewer the fuzzy entropy of extended attribute is. This leads to stable classification accuracy at the higher level and the stable leaf number. Now, we extend Definition 1 on fuzzy entropy to more than two classes case.

Definition 3. Let U be a set of examples considered, $U = \{u_i | i = 1,2,\cdots,N\}$, there are k classes $C_1, C_2,\cdots,C_k$. Each attribute of each example is defined as a linguistic value, and each linguistic value, which is the attribute-value of examples, is a fuzzy set defined on U. For simplicity, let A be an attribute-value, $A = (a_1, a_2 \cdots, a_N)$, $a_i(i = 1, 2, \cdots, N)$ denotes the degree which the attribute of i example belongs to its attribute-value A. Then the fuzzy entropy of nodes is defined as

$$E(D) = -\sum_{j=1}^{K} \frac{\sum_{i=1}^{N}(C_{ji}\Lambda a_i)}{\sum_{j=1}^{K}\sum_{i=1}^{N}(C_{ji}\Lambda a_i)} \log \frac{\sum_{i=1}^{N}(C_{ji}\Lambda a_i)}{\sum_{j=1}^{K}\sum_{i=1}^{N}(C_{ji}\Lambda a_i)} \qquad (5)$$

Table 1. The influence of α to the classification result of FDT

α	**Class of case 16**	**Fuzzy entropy**	**Rule number**	**Train accuracy**	**Test accuracy**
0.1	C1 or C2	0.391275	11	0.920026	0.972589
0.2	C1 or C2	0.340673	10	0.933118	0.976523
0.3	C1 or C2	0.226829	10	0.945236	0.977778
0.4	C1	0.157356	9	0.962377	1
0.5	C1	0.121731	3	0.962377	1

Observe the changing trend of fuzzy entropy function with the increase of α from (5) for a simple case, a crisp partition of class{P,N}, (5) becomes

$$E(D) = -\frac{\sum_{i=1}^{N}(P_i \Lambda a_i)}{\sum_{i=1}^{N} a_i} \log \frac{\sum_{i=1}^{N}(P_i \Lambda a_i)}{\sum_{i=1}^{N} a_i} - \frac{\sum_{i=1}^{N}(N_i \Lambda a_i)}{\sum_{i=1}^{N} a_i} \log \frac{\sum_{i=1}^{N}(N_i \Lambda a_i)}{\sum_{i=1}^{N} a_i} \quad (6)$$

The left hand side of (6) is a function with $3N$ dimension, which becomes $3N+1$ dimension while adding the α.

Let attribute I be 3 attribute-value such as *attribute* $1 = \{A, B, C\}$ in which $A = \{a_1, a_2 \cdots, a_N\}$, $a_i \in [0,1]\,(i = 1,2,\cdots,N)$ denotes the degree which the attribute of i example belongs to its attribute-value A, consider 2-class problem such as (0,1), then (6) becomes (7) below:

$$E = f(x) = -x \log x - (1-x)\log(1-x)\,, x \in [0,1]\,. \quad (7)$$

Variable x is also a function of α in (7), the form is

$$x = g(\alpha) = \frac{a_1 \wedge 1 + a_2 \wedge 0 + \ldots + a_N \wedge 1}{a_1 + a_2 + \ldots + a_N}\,, \quad (8)$$

where a_i is the value of attribute-value A. $E = f(x)$ will increase as x increases in the interval $x \in [0,0.5]$, and will decrease as x increases in the interval $x \in [0.5,1]$. Function $x = g(\alpha)$ is a step-function; the piece number of the ladder is N, which is the number of attribute-value. The monotone character of function (8) is discussed using two instances as follows:

With the increase of α from 0 to 1, if it causes some item of numerator and denominator of (8) decrease at the same time, then x decrease;

If the increase of α cannot influence numerator of (6), for example, α increase at a_2 in (8), and any nonzero item of numerator does not become 0, while the corresponding item of denominator of (8) become 0, then x increase.

Here α does not appear in (8) explicitly, while α can intercept a_i using the method like (4).

4 Parameter Optimization

Firstly, we analyze the change trend of fuzzy entropy with the increase of α. From (8), only if α takes N value as $a_i \in [0,1]\,(i = 1,2,\cdots,N)$, then the fuzzy entropy E in (7) will be influenced. So, the relationship between α and fuzzy entropy involves two functions: $E = f(x)\; x \in [0,1]$, $x = g(\alpha)\; \alpha \in [0,1]$. The fuzzy entropy E becomes E' while taking the value of α as $a_1, a_2, \cdots, a_N$ below:

Step1: based on some idiographic attribute-value $A=\{a_1,a_2,\cdots,a_N\}$ and (8), calculate the value of x and E;

Step2: for (8), if $x\in[0,0.5]$, with the increase of α, there are two probabilities:

A□Some item of numerator and denominator of (8) both become 0, then x decrease to $x^{'}$, then $x=g(\alpha)$ is a step-down function which leads to $x^{'}\in[0,0.5]$, because $E=f(x)$ is a step-up function as $x\in[0,0.5]$, $E=f(g(\alpha))$ is also a step-down function, so we have $E^{'}<E$.

B□The numerator of (8) keeps no change, one item of denominator become 0, then x increases to $x^{'}$, in this case $x=g(\alpha)$ is a step-up function. There are two in stances:

(a) If $x^{'}<0.5$, and that $E=f(x)$ is a step-up function at [0,0.5], then $E=f(g(\alpha))$ is a step-up function, $E^{'}>E$;

(b) If $x^{'}>0.5$ and $x^{'}-0.5>x-0.5$, based on the character of $E=f(x)$,
$E=f(g(\alpha))$ is a step-down function, $E^{'}<E$;
if $x^{'}-0.5<x-0.5$, based on the character of $E=f(x)$,
$E=f(g(\alpha))$ is a step-up function, $E^{'}>E$;

***Step3*:** for (8), if $x\in[0.5,1]$, with the increase of α, there have two probability:

A□The numerator of (8) don't change, one item of denominator become 0, then x increase to $x^{'}$, then $x=g(\alpha)$ is a step-up function which leads to $x^{'}\in[0.5,1]$, because $E=f(x)$ is step-down function when $x^{'}\in[0.5,1]$, $E=f(g(\alpha))$ is step-down function, $E^{'}<E$;

B□Some item of numerator and denominator of (8) both become 0, then x decrease to $x^{'}$, then $x=g(\alpha)$ is a step-down function, here are 2 instances as below:

(a) If $x^{'}>0.5$, and that $E=f(x)$ is a step-down function at [0.5,1], then
$E=f(g(\alpha))$ is a step-up function, $E^{'}>E$;

(b) If $x^{'}<0.5$, and $x^{'}-0.5>x-0.5$, based on the character of $E=f(x)$, $E=f(g(\alpha))$ is step-down function, $E^{'}<E$;
if $x^{'}-0.5<x-0.5$, based on the character of
$E=f(x)$, $E=f(g(\alpha))$ is a step-up function, $E^{'}>E$;

In this way, we can know when using some attribute-value as partition node, the change trend of fuzzy entropy of that node will increase along with the increase of α.

Secondly, we give a guideline for obtaining the optimal parameter. A small database car_type is employed to study the change trend of fuzzy entropy.

Table 2. Car_type Dataset

No	Hight(H)	Weight(W)	Length(L)	class
1	2.2	2.8	4	P
2	3.2	4.4	16	N
3	3.8	10	6	P
4	3.0	18	7	N
5	3.0	25	6	N
6	3.8	5.1	6	P
7	3.0	17	14	N
8	3.4	19	13	N

In the algorithm described above, each attribute-value obtain a minimum fuzzy entropy (min_entropy) and the corresponding α (optimal α), at the same time, each attribute obtain a minimum fuzzy entropy (min_entropy_sum) and the corresponding α (optimal_sum α). The obtained results are showed in Table 3.

Table 3. The minimum fuzzy entropy of each attribute-value and its corresponding α

Database name	Attr	Attr-Value	Min_ entropy	Optimal α	Min_entropy _sum	Optimal _sum α
	A_1	A_{11}	0.000000	0.23	0.539762	0.23
		A_{12}	0.625712	0.49		
		A_{21}	0.595286	0.40		
Car_type	A_2	A_{22}	0.000000	0.40	0.287884	0.40
		A_{23}	0.000000	0.40		
	A_3	A_{31}	0.666574	0.12	0.419186	0.12
		A_{32}	0.000000	0.49		

From Table 3, we can see that fuzzy entropy of attribute A_2 is the smallest for car_type dataset, i.e., 0.287884. For minimum fuzzy entropy of the 3 attribute-value of A_2, the fuzzy entropy of A_{22} takes the smallest value, the corresponding parameter α is α=0.40, which can chose as the optimal value of α.

This experiment demonstrates that in the process of building fuzzy decision tree the selected extended attribute at root-node is also A_2 at the moment. Based on this rule, the method of obtaining optimal α is proposed as follows: as a new data comes, the optimal attribute of root-node is firstly tested, then, the attribute with minimum fuzzy entropy is detected. As α takes N values as $a_i \in [0,1]\ (i = 1,2,\cdots,N)$, for each attribute-value of this optimal attribute, we can obtain the value E' of N fuzzy entropy using the above algorithm, each attribute-value select minimum E' and its corresponding parameter α, in which we take the biggest α as the optimal α of database for classification practice. While applying the above algorithm for different databases, it is observed that the change trend of fuzzy entropy varies in different ways. As a result of this, the optimal value of α will be problem dependent.

5 Experimental Results and Discussions

In this study, several datasets downloaded from UCI Machine Learning database are used. Table 4 below gives some statistics of these data sets.

Firstly, let α=0.5. It was found that the attribute take the minimum fuzzy entropy at root node for each database, details are given in Table 5-1 below.

Secondly, let α take N values as $a_i \in [0,1]\,(i=1,2,\cdots,N)$. Each attribute of these databases gains a set of fuzzy entropy, from which we chose the attributes corresponding to the minimum fuzzy entropy. The smallest one in these minimum fuzzy entropies is denoted by min_attr_entropy, its corresponding attribute and parameter are denoted by min_attr and optimal_cut_sum, respectively. Furthermore, for the value of this attribute (attr_value), we denote its minimum fuzzy entropy as min_attr_value_entropy, and its corresponding parameter α as optimal_cut.

From Table 5-1 and Table 5-2, the attribute having minimum fuzzy entropy is consistent with the selected attribute at root node. Attribute A4 of *iris* is the selected expanded attribute at root node, the optimal α which the 3 attribute-value of A4 corresponding to is 0.27, 0.4, 0.46 respectively, so the biggest value 0.46 is selected which is consistent with the optimal α that attribute A4 corresponds to. From Table 5-2, the optimal parameters α for these databases are: pima: 0.48, liver: 0.46, ecoli: 0.49, wine: 0.49, rice: 0.44, housing: 0.49 respectively.

Now, we study the effect of α on classification results. Constraining $0<\alpha<0.5$, we observe the change trend of classification rate for training set, test set and rule number for the databases. The results are depicted below.

It is observed in Figure 3 that the optimal parameter α obtained by the proposed algorithm makes the classification performance best in terms of recognition rates for training set and test set, and rule number as well. For example, as α=0.46 for liver, and α=0.44 for rice, the classification rates reach the highest points, and the rule numbers become the lowest.

Table 4. Statistics of 6 databases

Database name	Attribute number	The number of attribute-value
Pima	8	2
liver	6	3
ecoli	7	2
iris	4	3
wine	13	2
Rice	5	3

Table 5-1. The optimal attributes at root node

Database name	Pima	Liver	Ecoli	Iris	Wine	Rice	Housing
The optimal attribute	A2	A5	A6	A4	A7	A3	A13

Table 5-2. The attribute of minimum fuzzy entropy and the corresponding α

Database name	min_attr	min_attr_entropy	optimal_cut_sum	Attr_value	min_attr_value_entropy	Optimal_cut
Pima	A2	0.541514	0.48	A21	0.457574	0.47
				A22	0.666760	0.48
Liver	A5	0.660609	0.46	A51	0.692567	0.00
				A52	0.589134	0.46
				A53	0.543974	0.46
Ecoli	A6	0.903120	0.49	A61	0.897752	0.49
				A62	0.912927	0.49
Iris	A4	0.121731	0.46	A41	0.000000	0.27
				A42	0.248067	0.46
				A43	0.130226	0.40
Wine	A7	0.629253	0.49	A71	0.561929	0.47
				A72	0.672104	0.49
Rice	A4	0.390087	0.44	A41	0.192071	0.43
				A42	0.688028	0.23
				A43	0.199210	0.44
Housing	A13	0.729959	0.49	A131	0.809863	0.46
				A132	0.744533	0.38
				A133	0.340317	0.49

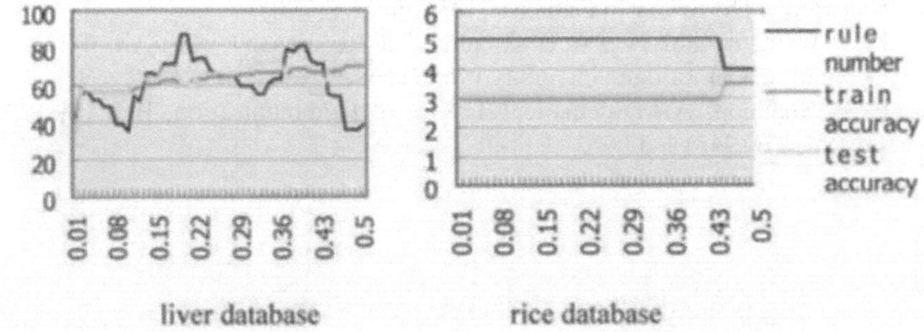

Fig. 3. The value of α vs. classification rates and rule number

6 Conclusion

The significance level (SL) greatly affects the computation of fuzzy entropy and FDT's performance in terms of rule number and classification accuracy. This paper develops a method to determine an optimal SL value through analyzing the relationship between the fuzzy entropy and α. The main contribution of this work is to provide some guidelines for selecting the SL parameter α in order to build FDTs with better classification performance.

Instead of taking one optimal value of α in the whole tree, this paper proposes to use different values of α in different nodes. In this way, a fuzzy decision tree may correspond to many local optimal α. This idea will further improve the power and quality of FDT, but the computational burden will be increased.

References

[1] R.S.Michalski, J.G.Carbonell and T.M.Mitchell(Eds.). Machine Learning: An Artificial Intelligence Approach. Vol. I,Tioga, Palo Alto, CA, (1983)98-139

[2] Quinlan J.R. Induction of Decision Trees. Machine Learning, 1(1986) 81-106

[3] M.Umano, H.Okamolo, I.Hatono, H.Tamura, F.Kawachi, S.Umezu and J.Kinoshita. "Fuzzy Decision Trees by Fuzzy ID3 Algorithm and Its Application to Diagnosis System", Proceedings of Third IEEE International Conference on Fuzzy Systems, 3(1994) 2113-2118

[4] C.Z.Janikow, "Fuzzy Processing in Decision Trees," Proceeding of the International Symposium on Artificial Intelligence, (1993)360-370

[5] Y.Yuan and M.J.Shaw, Induction of fuzzy decision trees, Fuzzy Sets Syst., 69(1995)125-139

[6] J. R. Hong, A new learning algorithm for decision tree induction (in Chinese), Journal of Computer, 18(1995) 470-474

[7] Breiman L., Friedman J. H., Olshen R.A. and Stone C.J. Classification and Regression Trees. Wadworth International Group (1984)

[8] Smyth P. and Goodman R.M.. Rule Induction Using Information Theory. Knowledge Discovery in Database, MIT Press, (1990)

[9] Press W.H., Teukolsky S.A. et al. Numerical Recipes in C: The Art of Scientific Computing. Cambridge University Press, (1988)

[10] C. E. Shannon, A Mathematical Theory of Communication, Bell. Syst. Teh. Journal, 27(1948) 1-65

Tool Condition Monitoring in Drilling Using Artificial Neural Networks

Vishy Karri and Tossapol Kiatcharoenpol

School of Engineering, University of Tasmania
PO. 252-65, Hobart, Tasmania, 7001, Australia
Vishy.Karri@utas.edu.au
TKO@postoffice.utas.edu.au

Abstract. A reliable and sensitive technique for monitoring tool condition in drilling is essential help for practising engineers. It is commonly known that the unattended use of a drill bit until it reaches the ultimate failure can potentially damage to machine tool and work-piece resulting in considerable down time and productivity loss. Thus there is a need for such tools to save high maintenance costs in case of the catastrophic failure. A system in drilling that can estimate tool life in terms of the number of hole to failure as condition monitoring techniques in the form of a digital display is significantly beneficial. In this paper, a tailor-made novel feed forward network is proposed to predict tool life in terms of the number of holes to failure. These involved the development of predictive model, test rig design and a digital display to assist engineers with on-line tool life. To entitle the network to cater for various cutting conditions, a knowledge base as training and testing data have to be generated on the experimental data in a comprehensive working range of drilling. Consequently, the experiments were performed in thirty-two cutting conditions based on the combination of three basic cutting parameters, which are feed rate, spindle speed and drill diameter. The neural networks were trained and the architecture of networks was appropriately selected by benchmarking the Root Mean Square error (RMS). The results of the novel network, Optimisation layer by layer (OLL), have shown the ability to accurately predict the number of holes to failure with a 100% success rate at both training and testing stages. To highlight OLL predictive capability, a brief comparison with Backpropagation Neural Network (BPNN) is carried out.

1 Introduction

It has been recognized that the tool condition is an important performance feature in machining and the tool condition monitoring is usually considered as the estimation of tool life in practical work. Tool life may be defined as the useful life of a tool from the start of a cut to some agreed end point identified by a failure criterion. Tool-life is

T.D. Gedeon and L.C.C. Fung (Eds.): AI 2003, LNAI 2903, pp. 293-301, 2003.

commonly expressed as the actual cutting time to failure but other measures such as the length of work-piece cut, the volume of metal removed and the number of components produced to failure can be also used [1]. From a maintenance point of view, it is very expensive if cutting tool breakage occurs during machining operation due to end of tool life. Therefore it is very important to accurately estimate tool life as a tool condition monitoring technique under varying cutting conditions. To establish a knowledge base for tool life prediction catering for various cutting conditions, a number of experiments must be performed which are very costly in aspects of time and material. Thus, in recent years, an artificial neural network approach has been efficiently used to establish the tool life predictive model based on a finite set of data. Many works have proved the performance of neural networks to estimate tool life in major machining operations including turning [2, 3] and drilling [4].

In this work, the neural network approach was proposed to estimate tool life in drilling operations. The number of holes to failure is used as a predicted feature because of its two practical advantages. Firstly, the job scheduling can be effectively done by acquiring tool life information in terms of the maximum number of components or holes that can be produced at any cutting conditions. Secondly as assistance to engineers for controlling, the number of holes to failure can be real-time displayed on a monitor at the shop floor level. Furthermore, operators can easily follow the visual display, which informs them how many holes they can produce in a particular cutting condition that a drill bit has been using.

Optimisation Layer by Layer [5] has been proposed to predict the number of holes to failure at various cutting conditions consisting of drill diameter, feed rate and spindle speed. Furthermore, the performance benchmarking of OLL and Backpropagation Neural Network (BPNN) was studied in this application. OLL is known not only for being fast in learning, but also for being superior in accurate modeling. It has shown some results better than those of traditional neural networks in drilling performance predictions such performance as cutting forces [6, 7], and hole oversize [8] and internal surface roughness [9]. A brief introduction of neural networks is initially discussed, followed by the development of neural network. Then the investigation of results is carried out.

2 A Tailor-Made Neural Network for Cutting Tool Condition Monitoring

It is important to note that while the objective of each neural network is to predict the number of holes to failure, the architecture and algorithms used by each network to achieve this are significantly different.

The basic architecture of OLL, as shown in Figure 1, consists of an input layer, one or more hidden layers and an output layer. All input nodes are connected to all hidden neurons through weighted connections, W_{ji}, and all hidden neurons are connected to all output neurons through weighted connections, V_{kj}.

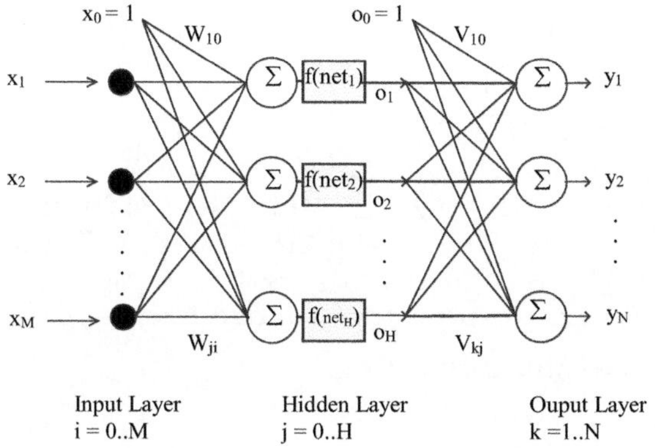

Fig. 1. Basic structure of OLL with one hidden layer

The basic ideas of the OLL learning algorithm are that the weights in each layer are modified dependent on each other, but separately from all other layers and the optimisation of the hidden layer is reduced to a linear problem [5]. The algorithm of one hidden layer is summarised as following:

Training Algorithm of OLL with one hidden layer (Fig. 1)

Step 1 Initialize weights

- Set all weights (W_{ji} ,V_{kj}) to small random values
- Set weight factor μ=0.0001,

Step 2 Optimization of output-hidden layer weights
- The gradient of cost function with respect to V is calculated to derive the optimal weight V for all training patterns. Thus,

$$V_{jk} = A^{-1}.b$$

where the A and b matrix are given by:

$A_{(j,j1)}$ = matrix $[a_{j,j1}]$; $a_{jj1} = \Sigma^{p}[o_j\, o_{j1}]$: j,j1=0..H
$b_{(j,k)}$ = matrix $[b_{j,k}]$; $b_{jk} = \Sigma^{p}[t_k\, o_j]$: k = 1..N
p = number of training patterns
t_k = target output of output neuron k

Step 3 Optimization of the input-hidden layer weights
- Transform non-linear part into linear problem. Then the linearized weights in each output layer node can be calculated as follow :

$$Vlin_{kj} = \Sigma^{j}\ [f'(net_j)\ V_{kj}\ ]$$

where f'(net) = derivative of the sigmoidal function
- Calculate weight correction term (ΔW_{opt}) for all training patterns.

$\Delta W_{opt} = Au^{-1}.bu$

where Au = matrix $[a_{(ij,hm)}]$;

bu = vector$[b_{(j,i)}]$

$a_{(ji,hm)}$: for $(j \neq h) = \Sigma^p \Sigma^k [(Vlin_{kj} x_i)(Vlin_{kh} x_m)]$

: for $(j = h) = \Sigma^p \Sigma^k (Vlin_{kj} x_i)(Vlin_{kh} x_m) + \mu/H^* \, abs(V_{kj}) f''(net_j) \, x_i \, x_m$

$b_{ji} = \Sigma^p \Sigma^k [(t_k - y_k) \, Vlin_{kj} \, x_i]$

H = number of neurons in hidden layer

- Calculate weight test (W_{test})

$W_{test} = W_{old} + \Delta W_{opt}$

Step 4 Update of the input-hidden layer weights

- Base on W_{test}, the new RMS error is calculated

If (New RMS > RMS) then go back to Step 3 and increase μ ($\mu = \mu*1.2$)

Else update weights

$W_{new} = W_{test} \; (=W_{old} + \Delta W_{opt})$

and decrease μ ($\mu = \mu*0.9$) for next iteration

Step 5 Do step 2 - 4 until test stop condition is true. (acceptable RMS or end of number of iterations)

3 Development of Knowledge Base and a Neural Network Model

For the cutting tool condition monitoring in drilling, four inputs were used. The inputs were the drill diameter (mm), the spindle speed (rev/min), the feed rate (mm/min) and the cutting time that the tool has been used (sec) and the output was the number of holes to failure. The architecture for number of hole to failure is shown in Figure 2.

To train the neural network based model, a knowledge base of the inputs with a corresponding tool life in terms of the number of hole to failure must be generated. A set of drilling experiments was carried out on the universal CNC-controlled milling machine. The general HSS twist drill bits were used to drill 1045 steel work-piece with 31.75 mm thickness. The cutting conditions were chosen from the combinations of diameter, feed rate and spindle speed. The experimental cutting conditions are illustrated in the Table 1.

In any particular cutting condition, a drill bit was used for drilling work-pieces until the drill bit reached the end of tool life. Then the number of holes produced were recorded and also listed in the Table 1.

An average flank wear equal to 0.3 mm is the criterion to define the effective tool life for HSS tools, according to the recommendation of International Standard organization (ISO). By using a digital camera and associated software, pictures of both cutting edges have been taken and enlarged to measure the wear size. The actual scale of wear land can be calculated by comparing with the diameter of drill bits. Figure 3 illustrates a typical measurement carried out in this investigation.

Table 1. Experimental cutting conditions and corresponding tool life

No.	Diameter (mm)	Feed (mm/min)	Speed (rpm)	Tool life (No of holes produced)
1	7	80	1000	7
2	7	80	1100	8
3	7	80	1200	7
4	7	80	1300	7
5	7	100	1000	7
6	7	100	1100	7
7	7	100	1200	5
8	7	100	1300	8
9	7	120	1000	3
10	7	120	1100	6
11	7	120	1200	4
12	7	120	1300	5
13	7	140	1000	3
14	7	140	1100	4
15	7	140	1200	4
16	7	140	1300	3
17	9	120	900	6
18	9	120	1000	10
19	9	120	1100	11
20	9	120	1200	10
21	9	140	900	6
22	9	140	1000	8
23	9	140	1100	8
24	9	140	1200	11
25	9	160	900	4
26	9	160	1000	6
27	9	160	1100	6
28	9	160	1200	9
29	9	180	900	4
30	9	180	1000	6
31	9	180	1100	6
32	9	180	1200	7

A simple formula for calculating wear land is given by

$$A = \frac{x}{y} \times D \qquad (1)$$

A = Actual size of wear (mm)
x = Dimension of wear in the picture
y = Dimension of diameter in the picture
D = Actual size of diameter (mm)

The four places (see fig. 3.) of flank wear were averaged to represent the average flank wear.

Based on the maximum number of holes produced at each cutting conditions, 238 data patterns for training and testing a neural network were generated. 214 data patterns were used as training data to build the model and the remaining 24 data patterns were used as extensive testing data to validate the model.

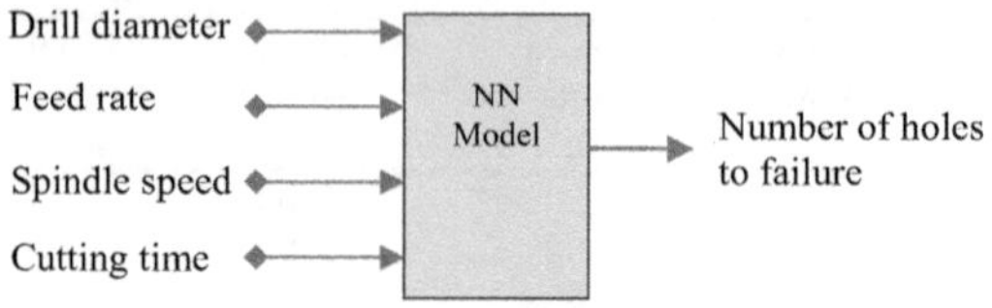

Fig. 2. Neural network model for predicting the number of holes to failure

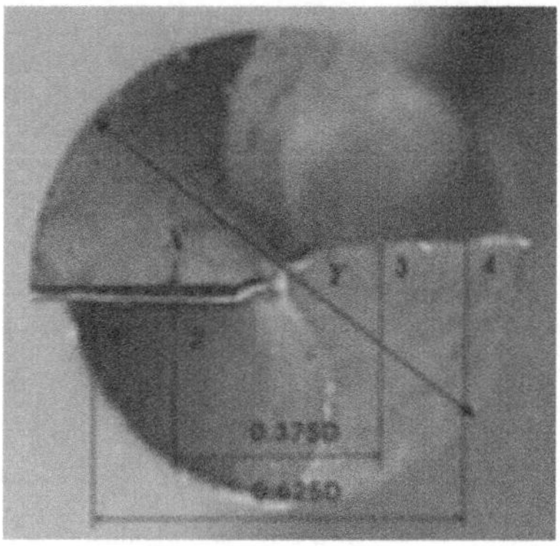

Fig. 3. Picture of flank wear and measurement places

The optimisation of the neural networks is done by systematically training and testing the networks with varied parameter settings. As a benchmark, the root mean square (RMS) error of the test data set is recorded and compared. The RMS error easily combines the results from all the test data patterns into one number without any cancellation of negative and positive errors. The formula of RMS [10-12] is given by

$$\text{RMS error} = \sqrt{\frac{\sum (\text{target} - \text{output})^2}{\text{no. of data}}} \tag{2}$$

In OLL, the number of neurons in the hidden layer has to be optimized. Therefore OLL networks with one hidden layer containing two to ten neurons were tested. The results have shown that the trend of RMS errors decreases when the number of hidden neurons increases as illustrated in the figure 4.

Although the lower RMS error can be found by increasing the number of hidden neurons, the additional hidden neuron numbers increase a network complexity leading to tremendous computing time for gaining small marginal accuracy. Hence, the OLL with ten hidden neurons giving the lowest RMS in the study range were chosen and the RMS error for such a configuration was 0.010. For BPNN, since it is a well-established neural network, the detailed algorithm and architecture are not discussed in this work, but appropriate references are provided in [10-12]. The lowest RMS error of 0.083 was obtained when using the neural network with four hidden layer neurons.

4 Results and Discussion

In this section while the predictive capability of OLL is shown, a parallel comparison with BPNN is carried out. Since the unit of the number of holes to failure is an integer in drilling operation on the shop floor, all outputs of networks must be rounded to the nearest integer. The direct comparison between the actual outputs from the experiment and rounded outputs from prediction of testing data is presented in Figure 5.

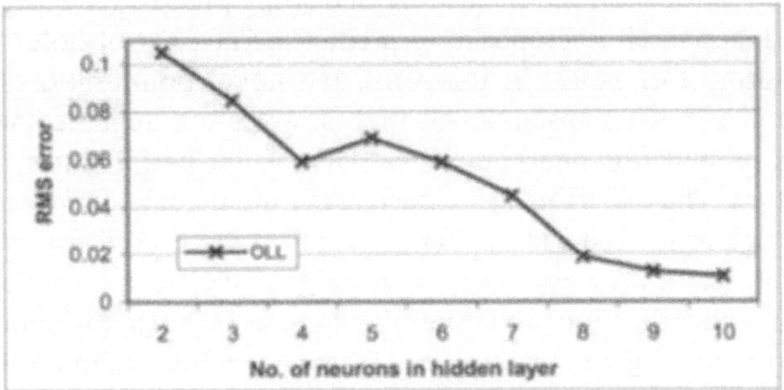

Fig. 4. RMS errors of OLL with number of hidden neurons

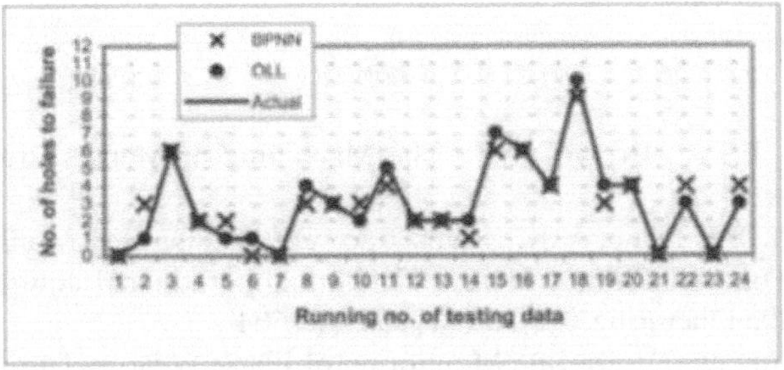

Fig. 5. Actual and predicted number of holes to failure by neural networks

Table 2. The accuracy of neural networks at training and testing stages

Methods	% Success rate	
	Training	Testing
OLL	100%	100%
BPNN	47%	50%

Based on the number of correct estimations, the percentages of accuracy or success rates of prediction are calculated and summarized in Table 2. It can be seen that OLL can predict with a 100% success rate in both stages while BPNN can predict with a 47% success rate in the training stage and a 50% success rate at the testing stage.

OLL is clearly showing the better predictive capability than the established BPNN. The OLL structure has dealt with the non-linearity of the problem more successfully than BPNN. The predictive capability reassures practising engineers for a reliable intelligent model for cutting tool condition monitoring, which reduces expensive maintenance cost.

5 Conclusion

The advantages of tool condition monitoring in drilling process have been highlighted. A neural network approach is proposed to predict the number of holes to failure as a condition monitoring technique. In this work the novel optimisation layer by layer network (OLL) has been chosen as a first architecture together with the well-established back propagation neural network (BPNN) for performance prediction as a second architecture. The experiments at various cutting parameters were carried out to create a knowledge base, 214 data patterns of which were used for training and the remaining 24 data patterns were used for testing of the neural networks. OLL has shown a superior capability for predicting the number of holes to failure with 100 % correctness at both training and testing data. With such a high quantitative accuracy, it is no doubt in the use of the neural network model to monitor tool condition by real-time displaying the number of hole to failure on the shop floor.

References

[1] Armarego, E.J.A., Brown, R.H.: The Machining of Metals. Prentice-Hall Inc. (1969)

[2] Choudhury, S.K., Jain, V.K., Rama Rao, C.V.V.: On-line Monitoring of Tool Wear in Turning Using a Neural Network. International Journal of Machine Tools & Manufacturing, Vol. 39 (1999) 489-504

[3] Liu, Q., Altintas, Y.: On-line Monitoring of Flank Wear in Turning with Multi-layered Feed-forward Neural Network. International Journal of Machine Tools & Manufacturing, Vol. 39 (1999) 1945-1959

[4] Lee, B.Y., Liu, H.S., Tarng, Y.S.: Abductive Network for Predicting Tool Life in Drilling. IEEE Transactions on Industry Application, Vol. 35(1) (1999) 190-195

[5] Ergezinger, S. and Thomsen, E.: An Accelerated Learning Algorithm for Multilayer Perceptrons: Optimisation Layer by Layer. IEEE Transactions on neural networks, Vol. 6(1) (1995) 31-42

[6] Karri, V., Kiatcharoenpol, T.: Prediction Of Thrust And Torque In Drilling Using Conventional And Feedforward Neural Networks. Australia Journal of Intelligent Information Processing System, Vol. **7**(1/2) (2002) 33-38

[7] Moore, T.J., Performance Estimation in Drilling Using Artificial Neural Networks.
Master Thesis, University of Tasmania, (2000)

[8] Karri, V., Kiatcharoenpol, T.: Radial Force and Hole Oversize Prediction in Drilling Using Traditional and Neural Networks. International Conference of Manufacturing Automation, Hong Kong (2002)

[9] Karri, V. and Kiatcharoenpol, T.: Prediction of Internal Surface Roughness in Drilling Using Three Feedforward Neural Networks - A Comparison. 9th International Conference On Neural Information Processing, Singapore (2002)

[10] Caudill, M., Butler, C.: Naturally Intelligent Systems. MIT Press, Cambridge, (1990)

[11] Caudill, M., Butler, C.: Understanding Neural Networks - Computer Explorations. Vol. 1 Basic Networks. MIT Press, Cambridge, (1992)

[12] Rumelhart, D.E., Mccellan, J.L.: Parallel Distributed Processing: Explorations in the Microstructure of Cognition. Vol. 1. MIT Press, Cambridge, (1988)

Software Verification of Redundancy in Neuro-Evolutionary Robotics

Jason Teo[1] and Hussein A. Abbass[2]

[1] School of Engineering and Information Technology, Universiti Malaysia Sabah
Kota Kinabalu, Sabah, Malaysia
j.teo@ums.edu.my

[2] Artificial Life and Adaptive Robotics (A.L.A.R.) Lab
School of Information Technology and Electrical Engineering
University of New South Wales @ Australian Defence Force Academy
Canberra, ACT, Australia
h.abbass@adfa.edu.au

Abstract. Evolutionary methods are now commonly used to automatically generate autonomous controllers for physical robots as well as for virtually embodied organisms. Although it is generally accepted that some amount of redundancy may result from using an evolutionary approach, few studies have focused on empirically testing the actual amount of redundancy that is present in controllers generated using artificial evolutionary systems. Network redundancy in certain application domains such as defence, space, and safeguarding, is unacceptable as it puts the reliability of the system at great risk. Thus, our aim in this paper is to test and compare the redundancies of *artificial neural network* (ANN) controllers that are evolved for a quadrupedal robot using four different evolutionary methodologies. Our results showed that the least amount of redundancy was generated using a self-adaptive Pareto *evolutionary multi-objective optimization* (EMO) algorithm compared to the more commonly used single-objective *evolutionary algorithm* (EA) and weighted sum EMO algorithm. Finally, self-adaptation was found to be highly beneficial in reducing redundancy when compared against a hand-tuned Pareto EMO algorithm.

1 Introduction

The automatic synthesis of autonomous controllers through artificial evolution has become a key area of research in robotics [12, 13]. This concept known as evolutionary robotics emphasizes the utilization of artificial evolution as the primary mechanism for driving the self-organization process in automatically generating controllers for both wheeled [4, 5], abstract [8, 10] and legged robots [9, 15]. There are a number of theoretical as well as technological considerations that needs to be addressed in order to conduct such artificial evolution. Theoretical considerations include the genetic encodings of the robot's controller and/or morphology, the kinds of algorithms for driving the evolutionary process and the types of

T.D. Gedeon and L.C.C. Fung (Eds.): AI 2003, LNAI 2903, pp. 302–314, 2003.

controllers suitable to act as the robot's controllers. Technological considerations include the robotic platforms suitable for evolutionary design approaches and techniques for automatically generating robots with variable controllers as well as morphologies. The emphasis of most studies have been on the role of genetic encodings (eg. [7]) and the role of fitness functions (eg. [5]). A very recent investigation explored how morphological complexity itself affects the emergence of more complex behavior in virtual legged creatures [2]. However, considerably little has been said about the role of controllers that are generated through such artificial evolutionary means. In particular, few studies have specifically focused on the empirical determination of the amount of redundancy that is present in the neural networks generated through such methods.

Hence, the aim of this paper is to conduct an empirical comparison of the amount of redundancy present in locomotion controllers that are automatically generated using four different artificial evolutionary approaches. In our experiments, we will attempt to maximize the robot's locomotion distance as well as to minimize the number of hidden nodes in its *artificial neural network* (ANN) controller where *evolutionary multi-objective optimization* (EMO) is used. The capacity of an ANN is determined by its Vapnik-Chervonenkis (VC) dimension, which in turn is determined by the number of free parameters in the network such as the connection weights [6]. One way to control the number of free parameters in the network is by controlling the number of hidden units present in the ANN. Thus, implementing a suitably-sized hidden layer within the ANN architecture is paramount for the following reasons:

1. Firstly, finding the ANN controller with the minimum network size will reduce the amount of computation that needs to be carried out by the robot's controller, thereby further enhancing its efficiency during operation.
2. Secondly, to be able to use the controller as some type of complexity measure [16], we need to ensure that the amount of redundancy in the network is minimized as far as possible in order to avoid false indications given by large redundant networks.
3. Thirdly, although redundancy may be beneficial for life-long learning, evolving networks with unseen redundancy should be avoided in order to reduce the risk of unpredictable behavior. Redundancy can be later added manually, with its corresponding effects analyzed by the designer. Hence, minimizing the number of redundant hidden units can reduce the amount of "*surprise*" [14] arising from evolved solutions.

2 The Artificial Evolutionary System

Morphology of the Quadrupedal Robot The robot is a basic quadruped with 4 short legs. Each leg consists of an upper limb connected to a lower limb via a hinge (one degree-of-freedom) joint and is in turn connected to the torso via another hinge joint. Hinge joints are allowed to rotate between 0 to 1.57 radians. Each hinge joint is actuated by a motor that generates a torque producing rotation of the connected body parts about that hinge joint. The mass

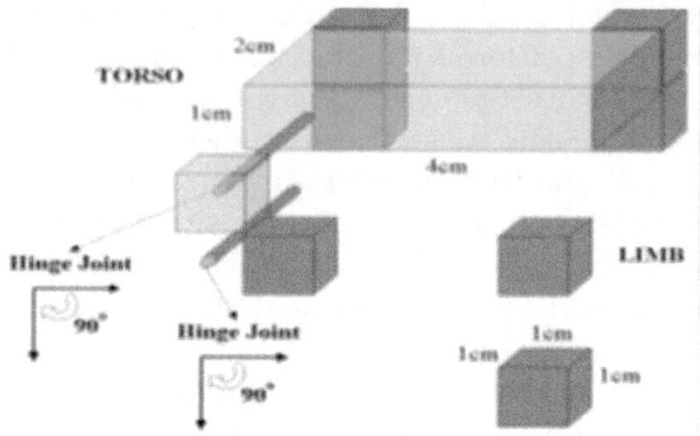

Fig. 1. Geometric description of the robot

of the torso is 1g and each of the limbs is 0.5g. The torso has dimensions of 4 x 1 x 2cm and each of the limbs has dimensions of 1 x 1 x 1cm. The robot has 12 sensors and 8 actuators. The 12 sensors consist of 8 joint angle sensors corresponding to each of the hinge joints and 4 touch sensors corresponding to each of the 4 lower limbs of each leg. The joint angle sensors return continuous values in radians whereas the touch sensors return discrete values, 0 if no contact and 1 if contact is made. The 8 actuators represent the motors that control each of the 8 articulated joints of the robot. These motors are controlled via outputs generated from the ANN controller which is then used to set the desired velocity of rotation of the connected body parts about that joint. The Vortex physics engine [3] was employed to generate the physically realistic artificial creature and its simulation environment. Vortex is a commercial-off-the-shelf (COTS) simulation toolkit which consists of a set of C++ routines for robust rigid-body dynamics, collision detection, contact creation, and collision response.

Genotype Representation Our chromosome is a class that contains the weight matrix Ω and one vector ρ; where ω_{jo} refers to the weights connecting the hidden-output layers and ω_{ij} refers to the weights connecting the hidden-output layers. $\rho_h \in \rho$ is a binary value that works as a switch to turn a hidden unit on or off and is used to indicate if hidden unit $h = 1, \ldots, H$ exists in the network or not. The sum, $\sum_{h=0}^{H} \rho_h$, represents the actual number of hidden units in a network. The use of ρ allows a hidden node to evolve even if it is not active during certain periods of the evolutionary optimization process. The chromosome also includes the crossover rate δ and mutation rate η for the self-adaptive algorithm. A direct encoding method was chosen as an easy-to-implement and simple-to-understand encoding scheme. Other encoding methods are possible.

Controller Architecture The ANN architecture used in this study is a fully-connected feed-forward network with recurrent connections on the hidden units as well as direct input-output connections. Recurrent connections were included to allow the robot's controller to learn time-dependent dynamics of the system.

Direct input-output connections were also included in the controller's architecture to allow for direct sensor-motor mappings to evolve that do not require hidden layer transformations. An input-bias is incorporated in the calculation of the activation of the hidden as well as output layers.

Fitness Functions The fitness f_1 of each locomotion controller represented in the genotype g is defined to be simply

$$f_1 = \Uparrow d(g) \tag{1}$$

where d refers to the horizontal Euclidean distance (measured in cms) achieved by the simulated robot as controlled by the ANN at the end of the evaluation period of 500 timesteps as given by

$$d = \sqrt{(a_0 - a_{499})^2 + (b_0 - b_{499})^2} \tag{2}$$

subject to the following constraints

$$a_0, b_0 = \text{initial coordinates as set by the user.} \tag{3}$$

In our case, $a_0 = 0$ and $b_0 = 0$, and

$$(a_{t+1}, b_{t+1}) = (a_t, b_t) + \nu(A_t, T_t, \hat{Y}_{ot}, M) \tag{4}$$

where the function ν takes as input the values returned by the robot's joint angle sensors A and touch sensors T, the controller's previous outputs $\hat{Y}_o$ and the creature's morphology M, and

$$\hat{Y}_{ot} = \sigma_o(\sum_j \omega_{jo}\sigma_j(\sum_i (\omega_{ij}A_t + \omega_{ij}T_t))) \tag{5}$$

where the controller's outputs $\hat{Y}_{ot}$ at timestep t is calculated using a sigmoidal transfer function σ_o, and w_{jo} refers to the weights connecting the hidden-output layers and w_{ij} refers to the weights connecting the hidden-output layers. There is no strict fitness threshold that separates viable from non-viable neuro-controllers since a large variety of locomotion behaviors are generated, including both static and dynamic gaits, which subsequently return a range of different fitness values. However, neuro-controllers with $f < 6$ generally do not produce sustained locomotion in the robot and are thus considered to be low quality solutions.

In experiments involving multi-objective evolutionary optimization of the locomotion controller, a second fitness f_2 is defined to be

$$f_2 = \Downarrow \sum_{h=0}^{H} \rho_h \tag{6}$$

where the number of active hidden units used in the ANN controller is counted by summing the binary vector ρ_h, which is part of the genotype g. Therefore, the first objective is to maximize the horizontal locomotion achieved by the simulated robot and where a second objective is involved, to minimize the use of nodes in the hidden layer of the ANN controller, which will in turn determine the VC-dimension of the controller as explained in Section 1. Unless stated explicitly, the fitness of a controller always refers to the first fitness function f_1 which measures the locomotion distance.

3 Evolving the Robot Controllers

We now present the setup and results obtained with each of the four different evolutionary methodologies, beginning with our proposed self-adaptive Pareto EMO algorithm which is based on the SPANN algorithm [1]. This is followed by a weighted sum EMO algorithm and a single-objective EA, both of which are self-adaptive as well, and finally by a hand-tuned Pareto EMO algorithm. In evolving the locomotion controllers for the quadrupedal robot, the evolutionary parameters were set as follows in all experiments: 1000 generations, 30 individuals, maximum of 15 hidden units, 500 timesteps and 10 repeated runs for each setup.

SPANN: A Self-Adaptive Pareto EMO Algorithm SPANN is a Pareto EMO algorithm which optimizes the f_1 and f_2 fitness functions as distinctly separate objectives. It uses a *differential evolution* (DE) crossover operator with a Gaussian step and implements elitism through breeding children only from the current Pareto set. The crossover and mutation rates are also self-adapted in SPANN. The implementation details for SPANN can be found in [1]. The version of SPANN used here has a modified repair function for the crossover and mutation rates (addition/subtraction of a random number between [0,1] instead of truncation) as well as using purely evolutionary learning instead of a hybrid of an EA and back-propagation. In contrast to the other algorithms that follow, SPANN does not require any hand-tuning of parameters.

A Weighted Sum EMO Algorithm (WS-EMO) For this second set of experiments, we used a single objective that combined the two objectives f_1 and f_2 using a weighted sum rather than a true Pareto approach as in SPANN. The weighting of the individual objectives was done in a relative manner using a parameter denoted by γ. The two objectives were unified as follows:

$$f(overall) = 100.0 - [(\gamma \times f_1') + ((1 - \gamma) \times f_2)] \tag{7}$$

10 different values were used for γ ranging from 10% to 100% in increments of 10%. An approach similar to the $(\mu + \lambda)$ evolutionary strategy is used where the 15 best individuals of the population are carried over to the next generation without any modification to the genotype at all. This is to allow a setup similar to SPANN, where the upper bound on the number of Pareto solutions is simply 1 + 15, the maximum number of hidden units allowed. The crossover and mutation

operators function as in SPANN and the rates for these genetic operators are also self-adaptive.

A Single-Objective EA (SO-EA) In the third set of experiments, we used a conventional EA which optimizes only one objective. The only objective being optimized in the following evolutionary runs is the locomotion distance achieved by the robot's ANN controller while the size of the hidden layer is kept fixed. Hence, only the f_1 fitness function is used to evaluate the genotypes. Apart from the change of optimizing two objectives to one, the single-objective algorithm is otherwise similar to the SPANN algorithm in all other respects. The crossover and mutation rates are self-adaptive and are identical to their counterparts in SPANN except that crossover and mutation now excludes any changes to the number of hidden units used in the ANN controller since this component is fixed in the single-objective EA. As in the weighted sum EMO algorithm discussed in Section 3, the $(\mu + \lambda)$ strategy is used in this single-objective EA where the 15 best individuals of the population are carried over to the next generation without any modification to the genotype at all. Sixteen separate sets of evolutionary runs were conducted corresponding to each one of the different number of nodes used in the hidden layer ranging from 0–15 (the range allowed in the multi-objective runs).

A Hand-Tuned EMO Algorithm (HT-EMO) In the last set of experiments, we used an EMO algorithm with user-defined crossover and mutation rates rather than self-adapting parameters in the SPANN algorithm. Apart from the non-self-adapting crossover and mutation rates, the hand-tuned EMO algorithm is otherwise similar to the SPANN algorithm in all other respects. 3 different crossover rates (**c**) and mutation rates (**m**) were used: 10%, 50% and 90% for both rates giving a total of 9 different combinations. As with SPANN, the fitness of each genotype in these experiments was evaluated according to both the f_1 and f_2 objective functions, which measures the locomotion distance achieved and number of hidden units used by the controller respectively.

4 Methodology for Testing Redundancy

We now compare the amount of redundancy present in the overall best evolved controllers in terms of the hidden units as well as weight synapses obtained from SPANN against the weighted sum, single-objective and hand-tuned methodologies. For this set of experiments, we selected the overall best controller evolved for locomotion distance obtained from SPANN, the weighted sum EMO algorithm, the single-objective EA and the hand-tuned EMO algorithm as representative ANN architectures optimized using the four different evolutionary methodologies. These controllers, which are used in the lesioning experiments that test for redundancy in the ANN architecture, are listed in Table 1. In our analysis, redundancy is considered to be present when a controller can allow deletion of:

Table 1. Best controllers in terms of locomotion found by SPANN, WS-EMO, SO-EA and HT-EMO used in the lesioning experiments

Algorithm	Locomotion Distance	No. of Hidden Units
SPANN	17.6994	4
WS-EMO	21.8228	7
SO-EA	22.4069	14
HT-EMO	19.5051	9

(1) entire hidden units, or (2) individual weight synapses, without loss of fitness of more than 1 unit of locomotion distance compared to the intact controller originally evolved. The first comparison involved deletion of entire nodes in the hidden layer and can be regarded as macro-lesioning of the controller. This was done in a fashion similar to [11] where all possible combinations of hidden units used in the ANN controller were systematically deleted. The lesioned controller is then re-evaluated and the new fitness achieved recorded. For example, if the best evolved controller had 4 hidden units, then all possible combinations of networks with 1 node lesioned are first evaluated, followed by all combinations of networks with 2 nodes lesioned and so forth terminating when the next level of hidden unit removal causes the fitness evaluations of the lesioned controllers to fall below the redundancy threshold. The second comparison involved the deletion of individual weight synapses which can be regarded as micro-lesioning of the controller. The test for weight synapse redundancy was carried out in a greedy fashion due to the very large numbers of possible weight synapse combinations: first find the least loss of locomotion capability with 1 weight synapse lesioned, then proceed to find the next least loss of locomotion capability with another weight synapse lesioned by keeping the weight synapse found in the preceding step lesioned, and so forth terminating when the next level of weight synapse removal causes the fitness evaluations of the lesioned controllers to fall below the redundancy threshold. This second redundancy test is less drastic compared to the first test since the deletion of a single hidden node would cause entire sets of weight synapses to be also deleted in a single step. As such, the second redundancy test of lesioning only at the weight synapse level allows for a finer investigation into the controller's evolved architecture.

Results: Hidden Unit Redundancy None of the overall best controllers evolved for locomotion distance using SPANN, weighted sum and single-objective methodologies showed any redundancy in terms of hidden units present in the ANN controller. All possible combinations of controllers with a single hidden node removed from the optimized architecture using these algorithms resulted in locomotion fitness below the redundancy threshold as shown in Table 2. However, the overall best controller evolved using the hand-tuned EMO algorithm did have one redundant hidden unit (the eighth node) which could be lesioned without causing the controller's capabilities to fall below the fitness threshold.

Table 2. Comparison of locomotion distance fitness of overall best controller evolved using SPANN, WS-EMO, SO-EA and HT-EMO with 1 hidden node lesioned

Algorithm	Redundancy Fitness Threshold	Best Lesioned Fitness	Worst Lesioned Fitness	Average Lesioned Fitness ± Standard Deviation
SPANN	16.6994	12.8242	8.6991	10.9156 ± 1.9204
WS-EMO	20.8228	18.6352	8.5114	14.4979 ± 3.8776
SO-EA	21.4069	17.5729	2.1421	12.9751 ± 4.8180
HT-EMO	18.5051	*18.6472*	1.6899	10.8691 ± 7.0303

This phenomenon together with the results from further levels of hidden unit lesioning of the overall best controller evolved using the hand-tuned EMO algorithm are discussed in the next paragraph. Surprisingly, the lesioning of a single hidden unit appeared to also have the most detrimental effect on the best controller evolved using the hand-tuned EMO algorithm in terms of the average loss of locomotion fitness compared to all other algorithms. Furthermore, the lesioning of a particular hidden node in the best controller evolved using the hand-tuned EMO algorithm also produced the worst 1-node lesioned controller, which only achieved a minuscule locomotion distance of just over 1.6 units. The removal of entire hidden nodes from the optimized ANN controllers seemed to result in large scale loss of locomotion capability suggesting that macro-lesioning of these evolved architectures is too drastic due to removal of not only a single hidden node but an entire set of weight synapses connecting to and originating from the lesioned hidden node. If the redundancy test were to be concluded at this coarse level, then the results would have indicated that no redundancy was present at all in the evolved controllers obtained using SPANN, the weighted sum EMO algorithm and the single-objective EA. However, a redundancy test at the finer weight synapse level, which is presented in the next section, showed otherwise. Before proceeding with the analysis at the weight synapse level, we first discuss the results obtained from the lesioning of different combinations of 2 hidden nodes from the overall best controller evolved using the hand-tuned EMO algorithm since lesioning of 1 hidden node did show redundancy in this particular controller. None of the controllers which had 2 hidden nodes removed could

Table 3. Locomotion distance fitness of overall best controller evolved using HT-EMO with 2 hidden nodes lesioned

Algorithm	Redundancy Fitness Threshold	Best Lesioned Fitness	Worst Lesioned Fitness	Average Lesioned Fitness ± Standard Deviation
HT-EMO	18.5051	18.1569	1.3326	9.4544 ± 5.6139

Table 4. Number of redundant synapses in the best evolved controllers from SPANN, WS-EMO, SO-EA and HT-EMO

	SPANN	WS-EMO	SO-EA	HT-EMO
No. of Redundant Synapses	1	3	2	281

produce locomotion fitness above the redundancy threshold as shown in Table 3. Compared to the overall best controller evolved using the self-adaptive SPANN algorithm, the hand-tuned EMO algorithm evolved a controller with more redundancy at the hidden unit level. Hence, the self-adaptive crossover and mutation rates appeared to have benefited the Pareto evolutionary optimization process in that SPANN was able to find a more compact network with less redundancy compared to the hand-tuned algorithm, which had pre-determined and fixed crossover and mutation rates during the optimization process.

Results: Weight Synapse Redundancy A summary comparing the number of redundant weight synapses present in the best evolved controller obtained from SPANN against those obtained using the hand-tuned, weighted sum and single-objective algorithms is given in Table 4. For SPANN, there was only one particular weight synapse that could be lesioned without causing the controller's performance to fall below the fitness threshold. No other lesioning of a single synapse could produce a fitness above the redundancy threshold. Additionally, no controller with 2 synapses lesioned could produce controllers that maintained their performance above the fitness threshold (Fig. 2). This meant that the best evolved controller from SPANN only had a redundancy of one weight synapse and furthermore this occurred only with a specific weight synapse, which was

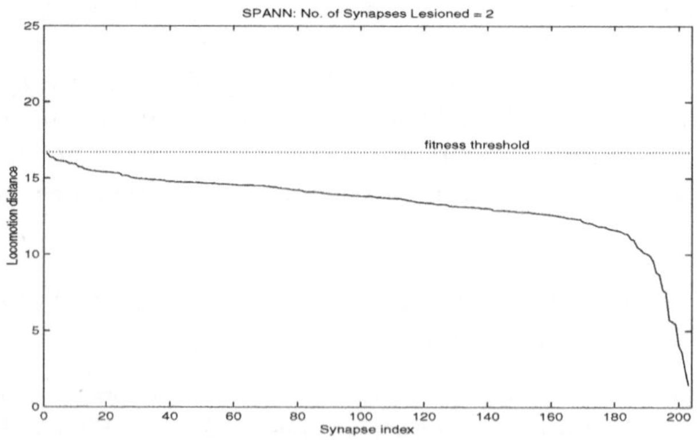

Fig. 2. Performance of best evolved SPANN controller with lesioning of 2 weight synapses

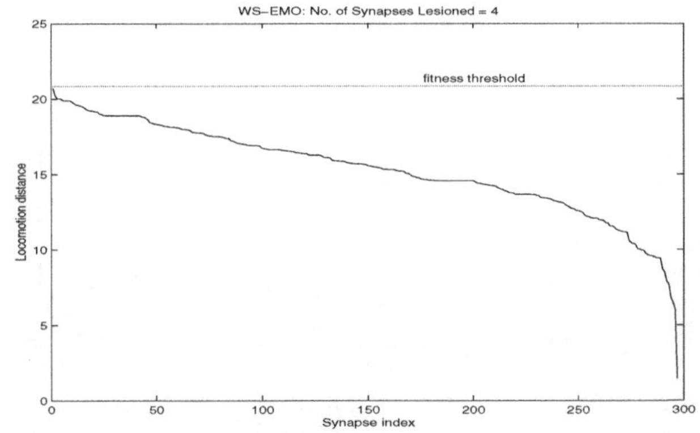

Fig. 3. Performance of best evolved WS-EMO controller with lesioning of 4 synapses

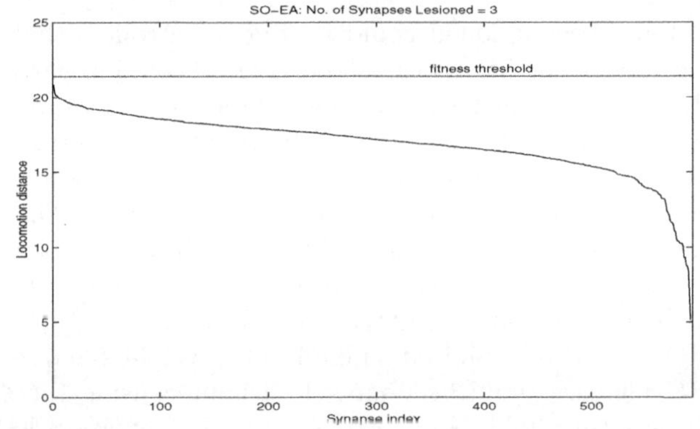

Fig. 4. Performance of best evolved SO-EA controller with lesioning of 3 synapses

the connection between the input from the joint sensor that measures the angle between the torso and the upper back left limb and hidden layer's first node. For the weighted sum EMO, up to three synapses could be lesioned without causing the controller to fall below the fitness threshold. No controller with 4 synapses lesioned could produce controllers that maintained their performance above the fitness threshold (Fig. 3). In terms of the number of different weight synapses that could be removed and still produced controllers that performed above the threshold, 6 controllers were found when 1 synapse was lesioned, followed by 2 controllers when 2 synapses were lesioned and finally by only 1 controller when 3 synapses were lesioned. Hence there was more synaptic redundancy present in the overall best controller evolved using the weighted sum method compared to SPANN. For the single-objective EA, up to 2 synapses could be lesioned without causing the evolved controller to fall below the fitness threshold. No

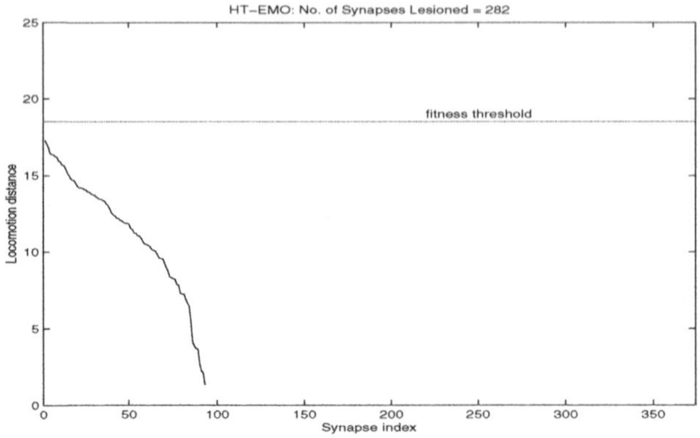

Fig. 5. Performance of best evolved HT-EMO controller with lesioning of 282 synapses

controller with 3 synapses lesioned could produce controllers that maintained their performance above the fitness threshold (Fig. 4). Only a particular weight synapse could be removed when 1 synapse was lesioned which produced a controller that performed above the threshold. Following this lesioning, only another specific weight synapse could be removed at the 2-synapse lesioning level that resulted in a controller that performed above the threshold. Thus, there was also more synaptic redundancy present in the overall best controller evolved using the single-objective EA compared to SPANN but less synaptic redundancy compared to the hand-tuned and weighted sum method. The controller from the hand-tuned EMO algorithm exhibited a high level of weight synapse redundancy where up to 281 synapses could be removed without causing the controller to fall below the fitness threshold. No controller with 282 synapses lesioned could maintain its performance above the fitness threshold (Fig. 5). In line with results obtained from macro-lesioning at the hidden unit level, a much higher level of weight synapse redundancy should be expected in this controller compared to the overall best controllers evolved using the other algorithms. This is the case since the analysis from the previous section showed that an entire hidden node could be removed without causing the controller to fall below the fitness threshold, which correspondingly means that the entire set of synapses connected to and originating from this particular hidden unit were redundant. Thus, at least 30 weight synapses that are connected to this hidden unit can be removed without causing the lesioned controller to fall below the fitness threshold. The number of different weight synapses that could be removed varied considerably at different levels of lesioning. For example, only approximately 25 different synapses could be removed at the 100-synapse level without causing the controller's locomotion capabilities to fall below the redundancy threshold. On the other hand, approximately 75 different synapses could be removed at the 200-synapse level without causing the lesioned controller to fall below the fitness threshold. The fluctuations observed with regards to the number of different synapses that could be removed at various levels of weight synapse lesioning are most probably due to the greedy nature in which the synapse lesioning takes place combined with the complex dynamics that takes place within the evolved ANN. A weight synapse lesioned presently will provide the best performance at the current level but the effects of this lesioning may at certain stages become highly sensitive to further lesioning and vice versa due to the combinatorial effects that occur between different weight synapses and hidden nodes in the network, which cannot be ascertained by this one-step lookahead algorithm.

5 Conclusion & Future Work

The experiments showed that there was more redundancy present in the best controllers evolved using the hand-tuned, weighted sum and single-objective methodologies compared to the self-adaptive Pareto EMO approach. Furthermore, the use of self-adaptation appeared to have a very significant impact in terms of reducing redundancy when comparing the self-adaptive against the hand-tuned algorithms. However, self-adaptation produced the least best locomotion distance. Although a loss of four units of locomotion distance is not a drastic loss given the total distance achieved by the creature, it is important in some applications to decide on this trade-off between performance and redundancy in advance. A possible solution is to generate good controllers with redundancy then manually find and remove the redundancy. However, such a solution is unacceptable in certain applications such as evolvable hardware. Imagine a robot being sent to Mars and the hardware is required to evolve a controller for a certain task. The previous method of manually searching for redundancy would cause significant additional computational and memory cost on the actual hardware which will be highly undesirable.

For future work, it would be interesting to compare the redundancies of evolved controllers using direct versus indirect encoding schemes such as those involving developmental mechanisms.

References

[1] Hussein A. Abbass. Speeding up back-propagation using multiobjective evolutionary algorithms. *Neural Computation* (to appear), 2003. 306

[2] Josh C. Bongard and Rolf Pfeifer. A method for isolating morphological effects on evolved behavior. *7th International Conference on the Simulation of Adaptive Behavior*, pp. 305–311. MIT Press, Cambridge, MA, 2002. 303

[3] CM Labs. Vortex [online]. `http://www.cm-labs.com`, 2002. 304

[4] Dario Floreano and Joseba Urzelai. Evolution and learning in autonomous mobile robots. In *Bio-Inspired Computing Machines*, pp. 317–364. PPEUR, Lausanne, Switzerland, 1998. 302

[5] Dario Floreano and Joseba Urzelai. Evolutionary robotics: The next generation. *7th International Symposium on Evolutionary Robotics*, pp. 231–266. AAI Books, Ontario, 2000. 302, 303

[6] Simon Haykin. *Neural Networks: A Comprehensive Foundation.* Prentice-Hall, Upper Saddle River, NJ, 2nd edition, 1999. 303
[7] Gregory S. Hornby and Jordan B. Pollack. Creating high-level components with a generative representation for body-brain evolution. *Artificial Life*, 8(3):223–246, 2002. 303
[8] Phil Husbands, Inman Harvey, Nick Jakobi, Adrian Thompson, and Dave Cliff. Evolutionary robotics. In *Handbook of Evolutionary Computation*, pp. 1–11. IOP Publishing, Bristol, Philadelphia, 1997. 302
[9] Nick Jakobi. Running across the reality gap: Octopod locomotion evolved in a minimal simulation. *1st European Workshop on Evolutionary Robotics*, pp. 39–58. Springer-Verlag, Berlin, 1998. 302
[10] Hod Lipson and Jordan B. Pollack. Automatic design and manufacture of robotic lifeforms. *Nature*, 406:974–978, 2000. 302
[11] Orazio Migliono and Richard Walker. Genetic redundancy in evolving populations of simulated robots. *Artificial Life*, 8(3):265–277, 2002. 308
[12] Stefano Nolfi and Dario Floreano. *Evolutionary Robotics.* MIT Press, Cambridge, MA, 2000. 302
[13] Jordan B. Pollack, Hod Lipson, Sevan G. Ficici, Pablo Funes, and Gregory S. Hornby. Evolutionary techniques in physical robotics. In *Creative Evolutionary Systems*, pp. 511–523. Morgan Kaufmann, San Francisco, 2002. 302
[14] Edmund M. A. Ronald and Moshe Sipper. Surprise versus unsurprise: Implications of emergence in robotics. *Robotics and Autonomous Systems*, 37:19–24, 2001. 303
[15] Jason Teo and Hussein A. Abbass. Coordination and synchronization of locomotion in a virtual robot. *9th International Conference on Neural Information Processing*, vol. 4, pp. 1931–1935, Singapore, 2002. 302
[16] Jason Teo, Minh Ha Nguyen, and Hussein A. Abbass. Multi-objectivity as a tool for constructing hierarchical complexity. *2003 Genetic and Evolutionary Computation Conference*, LNCS 2723, pp. 483–494. Springer-Verlag, Berlin, 2003. 303

A Firearm Identification System Based on Neural Network

Jun Kong*, D.G. Li and A. C. Watson

School of Computer and Information Science Edith Cowan University
2 Bradford Street, Mount Lawley 6050, Perth, Western Australia
{j.kong,d.li@ecu.edu.au}

Abstract. In this paper, a Firearm Identification system based on Self-Organizing Feature Map (SOFM) neural network is proposed. We focus on the cartridge case identification of rim-firing mechanism. Experiments show that the model proposed has high performance and robustness by integrating the SOFM neural network and the decision-making strategy. This model will also make a significant contribution towards the further processing, such as the more efficient and precise identification of cartridge cases by combination with more characteristics on cartridge cases images.

1 Introduction

The analysis of marks on bullet casings and projectiles provides a precise tool for identifying the firearm from which a bullet is discharged [1,2]. Characteristic markings on the cartridge and projectile of a bullet are produced when a gun is fired. Over thirty different features within these marks can be distinguished, which in combination produce a "fingerprint" for identification of a firearm [3]. This forensic technique is the vital element for legal evidence, in cases where the use of firearms is involve. Given a means of automatically analyzing features within such a firearm fingerprint, it will be possible to identify not only the type and model of a firearm, but also each individual weapon as effectively as human fingerprint identification can be achieved.

Due to the skill required and intensive nature of ballistics identification, law enforcement agencies around the word have expressed considerable interest in the application of ballistics imaging identification systems to both greatly reduce the time for a positive identification and to introduce reliability (or repeatability) to the process. Several ballistics identification systems are available either in a commercial form or in a beta-test state. A Canadian company, Walsh Automation, has developed a commercial system called "Bulletproof", which can acquire and store images of projectiles and cartridge cases, and automatically search the image database for particular striations on projectiles but not impressed markings or striations on

* Jun Kong is also at Department of Computer Science, Northeast Normal University, China.

T.D. Gedeon and L.C.C. Fung (Eds.): AI 2003, LNAI 2903, pp. 315-326, 2003.

cartridge cases. This inherent limitation of the system with respect to cartridge cases of the system has prohibited its use. The Edith Cowan University of Australia, in conjunction with the Western Australia Police, has developed a prototype database called FIREBALL [4]. It has the capability of storing and retrieving images of cartridge cases heads, and of interactively obtaining position metrics for the firing-pin impression, ejector mark, and extractor mark. The limitation of the system is that the position and shape of the impression images must be traced manually by users. This will decrease the efficiency of this system.

The papers on the automatic identification of cartridge cases are hardly to be found. Le-Ping Xin [5] proposed a cartridge cases based identification system for firearm authentication. His work was focused on the cartridge cases of center-firing mechanism. And he also provided a decision strategy from which the high recognition rate would be achieved interactively. Chenyuan Kou et al. [6] described a neural network based model for the identification of the chambering marks on cartridge cases. But no experiment results were given in their paper.

In this paper, a system for identifying the firing pin marks of cartridge cases images automatically using the Self-Organizing Feature Map (SOFM) neural network is proposed. We mainly focus on the consideration of rim-firing pin mark identification. The system will also make a significant contribution towards the efficient and precise identification of ballistics specimens in the further processing, such as the locating and coding of ejector marks, extractor marks and chambering marks. In Section 2, the SOFM neural network and the methods of image processing in our study is described briefly. The capturing and preprocessing of cartridge cases images are presented in Section 3. The model based on SOFM for identification of cartridge cases images is proposed in Section 4. Section 5 gives a numeric experiment. Finally, Section 6 is a short conclusion.

2 SOFM and Image Processing

2.1 SOFM Neural Network

We pick the Self-Organizing Feature Map (SOFM) neural network as our classifying model in our identification system. The SOFM has been applied to the study of complex problems such as speech recognition, combinatorial optimization, control, pattern recognition and modeling of the structure of the visual cortex [7,8,9] and [10]. The SOFM we used is a kind of un-supervised neural network models, it in effect represents the result of a vector quantization algorithm that places a number of reference or codebook vectors into a high-dimension input data space to approximate defined between the reference vectors, the relative values of the latter are made to depend on ate to its data set in an ordered fashion. When local-order relations are each other as if there neighboring values would lies along an "elastic surface". By means of the self-organizing algorithm, this "surface" becomes defined as a kind of nonlinear regression of the reference vectors through the data points [11]. We employ the standard Kohonen's SOFM algorithm summarized in Table 1, the topology of SOFM is shown in Fig. 1.

2.2 Image Processing and Feature Extraction

Contrast Enhancement. One of the general functions in image preprocessing is the contrast enhancement transformation [12], and function is expressed in Equation (1). Low-contrast images can result from poor lighting conditions, lack of dynamic rang in the imaging sensor, or even wrong setting of a lens aperture during image acquisition. The idea behind contrast enhancement is to increase the dynamic range of the gray levels in the image being processed. The image shown in Fig. 2.b is transformed by contrast enhancement. The locations of points (x_1, y_1) and (x_2, y_2) control the shapes of the transformation function. If $x_1 = y_1$ and $x_2 = y_2$, the transformation is a linear function that produces no changes in gray level. If $x_1 = x_2$, $y_1 = 0$ and $y_2 = 255$, the transformation becomes a **threshold function**.

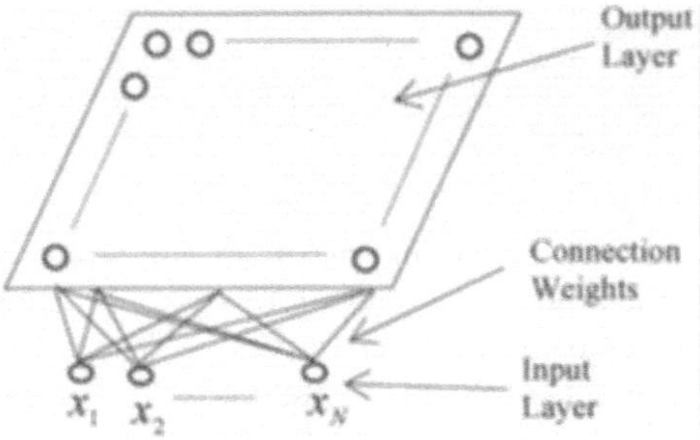

Fig. 1. The topology of SOFM

Table 1. The unsupervised SOFM algorithm

SOFM1:	Initialize the weights for the given size map. Initialize the learning rate parameter, neighborhood size and set the number of unsupervised learning iterations.
SOFM2:	Present the input feature vector $x = [x_{1,} x_2, \cdots, x_n, \cdots x_N]$ in the training data set, where x_n is the n th element in the feature vector.
SOFM3:	Determine the winner node c such that $\lVert x - w_c \rVert = \min_i \{\lVert x - w_i \rVert\}$.
SOFM4:	Update the weights, w_i's, within the neighborhood of node c, $N_c(t)$, using the standard update rule: $w_i(t+1) = w_i(t) + \alpha(t)[x_n - w_i(t)]$, where $i \in N_c(t)$.
SOFM5:	Update learning rate, $\alpha(t)$, and neighborhood size, $N_c(t)$. $\alpha(t+1) = \alpha(0)\{1 - t/K\}$; $\lvert N_c(t+1) \rvert = \lvert N_c(0)\{1 - t/K\} \rvert$, where K is a constant and is usually set to be equal to the total number of iterations in the self-organizing phase.
SOFM6:	Repeat 2-5 for the specified number of unsupervised learning iterations.

Polar Transaction. Polar transformation is also a useful tool in stage of image preprocessing. In our study, the polar transformation can bring us some advantages:

1. In the test phase (see Section 4), we only move the detecting windows over the testing images in direction of horizontal and vertical rather than rotating the testing images or detecting windows. This will decrease the numerical error and increase the efficiency.
2. Under the Polar Systems, we can get more information about the testing images. Some images that have similar shapes may be different in shapes and be distinguished in Polar Systems.

$$f(x) = \begin{cases} \dfrac{y_1}{x_1} x, & x < x_1 \\ \dfrac{y_2 - y_1}{x_2 - x_1}(x - x_1) + y_1, & x_1 \le x \le x_2 \\ \dfrac{255 - y_2}{255 - x_2}(x - x_2) + y_2, & x > x_2 \end{cases} \tag{1}$$

Feature Extraction. Feature extracting plays an important role in identification system. In the real application, the speed of feature extracting techniques is also crucial factor. So we pick up the first derivatives [12] as the images features. First derivatives in image processing are implemented using the magnitude of the gradient. For a function $f(x, y)$, the gradient of f at coordinates (x, y) is defined as follows:

$$\nabla f = [G_x, G_y]^T = \left[\frac{\partial f}{\partial x}, \frac{\partial f}{\partial y}\right]^T. \tag{2}$$

The magnitude of this vector is given by

$$\nabla f = mag(\nabla f) = [G^2{}_x + G^2{}_y]^{1/2} = \left[\left(\frac{\partial f}{\partial x}\right)^2 + \left(\frac{\partial f}{\partial y}\right)^2\right]^{1/2}. \tag{3}$$

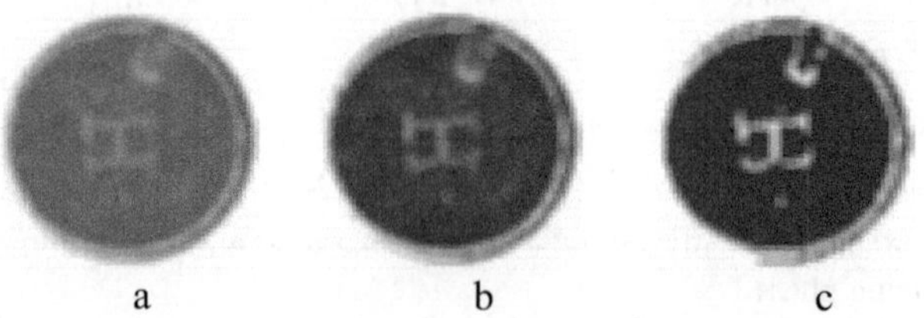

Fig. 2. Low-contrast image a. Result of contrast enhancement b. Result of threshold c

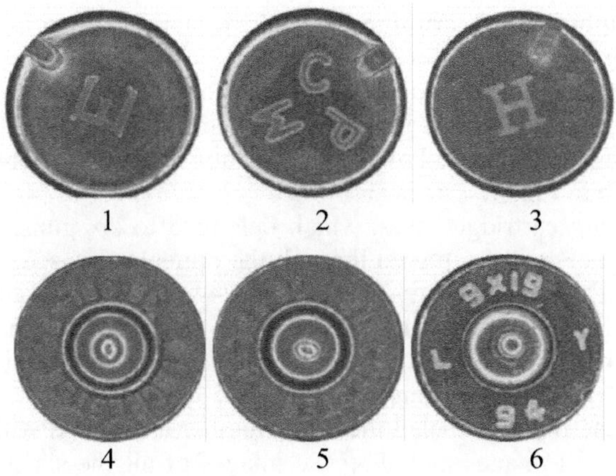

Fig. 3. Rim-firing mechanism 1, 2 and 3. and Center-firing mechanism 4, 5 and 6

For a digital image $f(i,j)$, where (i,j) is the coordinate of a pixel, and $f(i,j)$ is the gray value (for gray scale image), the gradient of f at coordinate (i,j), we can use:

$$G[f(i,j)] = \{[f(i,j) - f(i+1,j)]^2 + [f(i,j) - f(i,j+1)]^2\}^{1/2}, \qquad (4)$$

where the positions of points in Formula (4) are shown as:

$f(i,j)$	$f(i,j+1)$
$f(i+1,j)$	$f(i+1,j+1)$

We set a threshold Δ, $\Delta = 128$ in our study, for $G[f(i,j)]$ as follows:

$$g(i,j) = \begin{cases} 0, & G[f(i,j)] < \Delta \\ G[f(i,j)], & G[f(i,j)] \geq \Delta \end{cases}. \qquad (5)$$

3 Cartridge Cases Images in Our Study

There are typically three types of weapons encountered in ballistics: rifles, handguns, and shotguns. The firing mechanism of the weapon will be generally of two types: the firing pin will be either rim-firing mechanism or a center-firing mechanism, as shown in Fig. 3.

In the real application, all the images of cartridge cases are obtained through the optical microscope. So some information such as the depth of the impression will be dismissed. Other factors such as the lighting conditions, the material of cartridge cases, and the stamp letters of manufacturer can bring strong noise into the cartridge cases images or damage the shapes of the cartridge cases images. All these would bring many difficulties to feature extraction and identification. The lighting conditions for the images capturing of cartridge cases are crucially importance. In order to produce high contrast of striation (firing-pin mark) on the cartridge, the illuminator must be installed at an angle of greater than 45 degree from normal to the plane of the head of the cartridge [1]. It is vary important to choose a proper tool for capturing the cartridge cases images. The Vide Pro 32 image capture broad is adopted in our image capturing, as shown in Fig. 4.

The 150 rim-firing cartridge cases, which belonged to six guns, provided by the Western Australia Police are captured through the optical microscope, one image for each, formed 150 BMP files in gray scale size by 244×240 pixels, and classified into six types by shape of firing pin marks. They are: 1. U-shaped pin mark. 2. Axe-head pin mark. 3. Rectangular (Short) pin mark. 4. Rectangular (Long) pin mark. 5. Square pin mark. 6. Slant pin mark. Examples of the six types are shown in Fig. 5 (The numbers below these figures labeled the class number associated with each cartridge cases). We choose 50 images including the images of all the six guns randomly to form the set C_0 and form the testing set T for the rest images. Then, the images of set C_0 are processed through the following steps (shown in Fig. 6), contrast enhancement, polar transformation, feature extraction, and threshold transform.

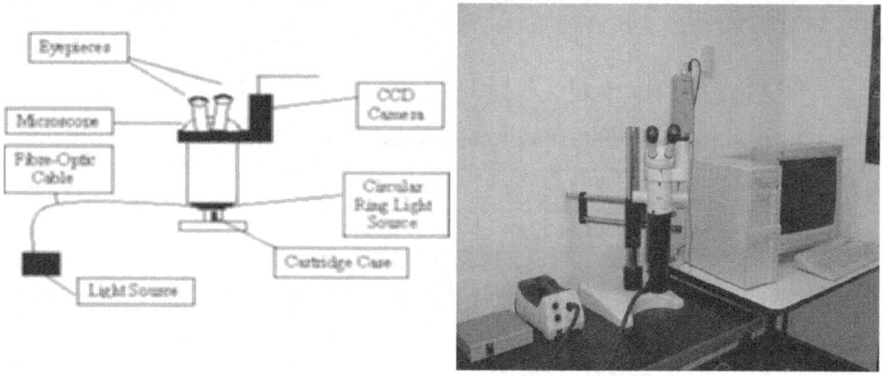

Fig. 4. Vide Pro 32 Image Capture Broad

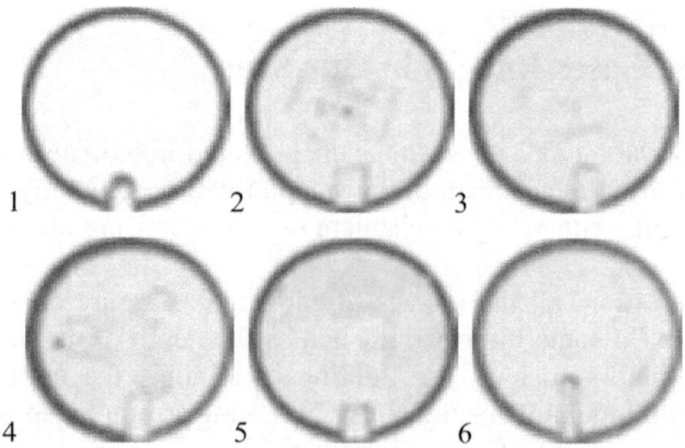

Fig. 5. Six type of cartridge cases images longed to six guns

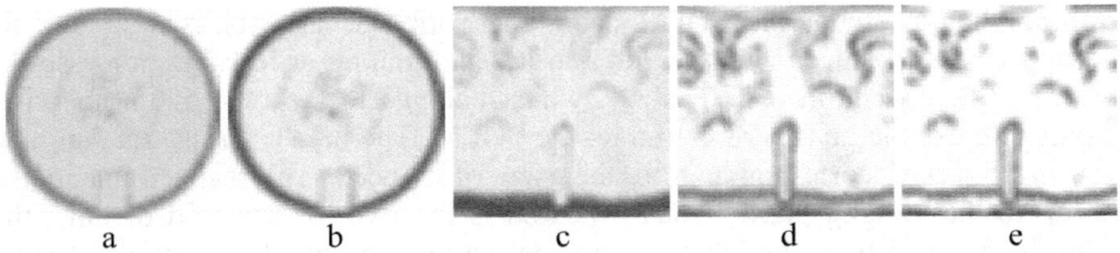

Fig. 6. The original image a, the contrast stretching b, the polar transformation c, the gradient d, the threshold e

Table 2. The size (in pixel) of six type windows

Type 1	20×96	Type 2	20×96
Type 3	20×120	Type 4	24×116
Type 5	20×120	Type 6	24×168

Having been finished the above transformations for the images of every type, we need a "window" operation:

First, windows, size by $n_i \times m_i$ pixels, are used to copy the sub-images---the firing pin marks of the cartridge cases images processed before, where i stands for the label of the class to which the firing pin marks belong. The sizes of six type windows associated with six type firing pin marks are as follows in Table 2. Second, the images (the firing pin marks) within these six type windows are copied into windows with size normalized by 48×196 pixels to meet the need of having unified input units of SOFM. The process is shown in Fig. 7. In addition, in order to make our model have some robustness to slight variations in the testing cartridge cases images, we process part of images obtained as mentioned above in the manners: a. Shifted up to two pixels by the direction left, right, up, and down. b. Scaled by factor 0.95 and 0.90. All the images we obtained through these processing above, with the number of 350, are combined into a training set C for the model based on SOFM, which will be discussed in the following section.

4 SOFM Identification Model

In this section, a firearm identification model based on cartridge cases images using SOFM is proposed. The structure of the model, the training, testing, and decision-making rule is given in details in following parts, respectively.

Identification Model. The system proposed is comprised of three stages as shown in Fig. 8, the preprocessing stage mentioned in Section 3, the classification stage based on SOFM neural network and the decision-making stage.

Training. In our study, the SOFM neural network acts as a classier among the training (or testing) specimens presented to it. The training or learning processing is as same as that mentioned in Table 1, which belongs to the type of unsupervised learning (we use the images of C to train the SOFM. The number of neurons in input layer is 48×196, corresponding to the size of windows normalized mentioned before). In the training phase, when a neuron is inactive for a period of time, that neuron is removed from the lattice. A neuron may be considered inactive if it is not chosen frequently as the winner over a finite time interval. After being trained, the positions of neurons in the output layer of SOFM stand for the classes to which the training specimens (or the testing specimens) belong.

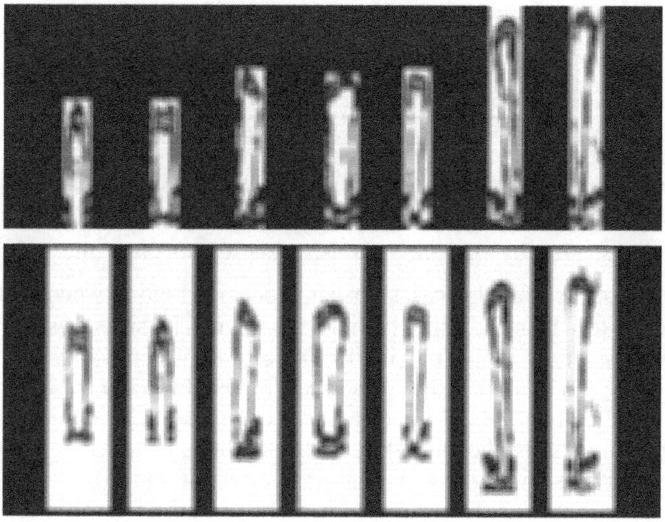

Fig. 7. Some specimens within windows with size normalization. The first row shows the six firing pin marks within six type windows. The second row shows the firing pin marks within windows with size normalization

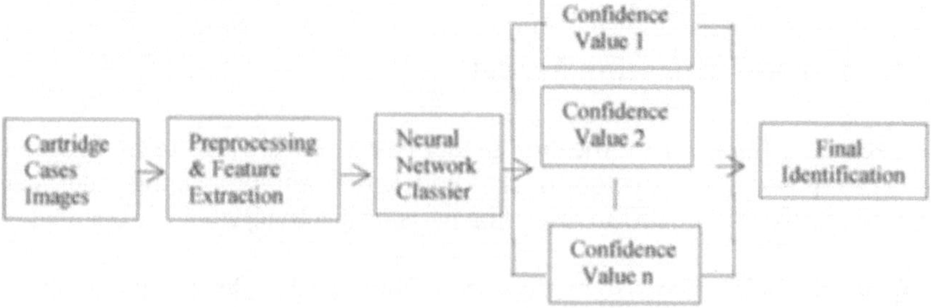

Fig. 8. The proposed identification system

Testing. The Testing Process Is as Follows:

Step 1. Select a testing cartridge case image from the testing set T, and present this testing pattern to the first stage of identification system--the preprocessing stage.

Step 2. Select a type of window from all types in turn, and move this window over the testing pattern processed in Step1 at every location by every pixel horizontally and vertically, pick up the sub-images.

Step 3. Present all the sub-images to the SOFM in turn, and calculate the confidence values with Formula (6) for each sub-image. Return Step2 until all type windows are used up.

Step 4. Present these confidence values to the decision-making stage, and calculate the finnal result for the testing cartridge case image by Formula (7) and (8).

Decision-Making Strategy. For the reasons of noise, lighting conditions, and the trademarks on the head of cartridge cases images, the following situation could generally be encountered in the testing phase:

a. For a testing cartridge case image, when a type of detecting window is used over the image, more than one sub-image under this type window is classified to include a firing pin mark.

b. For a particular testing cartridge case image, when all types of windows are used over the pattern, more than one sub-image under the different windows is classified to include a type of firing pin mark.

We use a final decision-making mechanism in decision-making stage in our study to solve these problems mentioned above and improve the performance and accuracy, defining a Confidence Function $D(i,j)$ for the testing pattern i to the jth class which measures the ratio between the testing pattern distance to the weight vectors and the average distance of training patterns to the weight vectors, as follows:

$$D(i,j) = D(j) / D(i,j) , \tag{6}$$

where $D(j)$ is the average distant when all the training patterns, which belong to the jth class, are tested using the jth type window, and $D(i,j)$ is the distant resulted when the ith testing pattern is tested using the jth type window. Defining a decision-making rule as follows:

$i \in$ Class K, if

$$D(i,k) = \min_j \{D(i,j) > \Delta_j\}, \quad j = 1,\ 2, \cdots n , \tag{7}$$

where Δ_j $j = 1,\ 2, \cdots n$, is an appropriate threshold selected for the jth class by experiments. In generally, the unbalanced distribution of training patterns which we get in real word, results the unbalance trainings in the neural network for each class. Hence, Δ_j for every class should not be unique. Defining a rejection rule, Testing pattern i is rejected by all classes, if

$$D(i,j) < \Delta_j \ , \quad j = 1,\ 2,\cdots n \ , \tag{8}$$

where Δ_j $j = 1,\ 2,\cdots n$, is same as in Formula (7).

5 Experimental Results

In our study, we use the following experimental parameters for the training of SOFM shown in Table 3.

Table 3. Experimental parameters for SOFM and results over training set C

Input Layer	Output Layer	$\eta(0)$	$\Lambda_i(0)$
48×196	9×9	0.6	7
Training pattern	Right rate	Error rate	Rejection rate
350	100%	0%	0%

We have the experiment results over testing set T as follows:

Table 4. Experiment results

Testing pattern	Right rate
100	97.00%
Rejection rate	Error rate
3.00%	0%

Analysis of Experiment Results. From the results of Table 4, we can see: 1. Identification model proposed in our study can make the combination of locating and identifying of firing pin marks of cartridge cases into one stage. 2. It shows that the model proposed has high performance and robustness for the testing patterns in aspects as follows: a. Having high accuracy in location and identification of firing pin marks. b. Having robustness to the noise patterns, to the damaged and deformed patterns. c. Having some robustness to the scaled patterns.

We also see that there still are rejections for some patterns, and we found that the rejection is caused mainly by the following reasons: the high noise on the cartridge images; the letters of trademark on the cartridge images; the similitude of one pattern with others in some location.

Further Work. In order to improve our model to achieve higher performance, we will do some further researching in following aspects: 1. To improve the quality of images capturing and preprocessing. 2. To generate some higher levels classifier to distinguish the patterns those be similar and like to be confused. 3. To extract some fine and robust features with more complex techniques to represent the patterns (training or testing). 4. To add some complex decision making mechanism. 5. To integrate multiple classifier combination using different features sets.

6 Conclusion

In this paper, we mainly focus on the consideration of rim-firing pin mark identification. This study is investigating a system for identifying the firing pin marks of cartridge cases images automatically using a neural network model. The identification model in our study can make the combination of location and identification of firing pin mark of cartridge case images into one stage. It shows that the model proposed has high performance and robustness for real testing patterns. The efficiency of this system will also make a significant contribution towards the efficient and precise identification of ballistics specimens in the further processing, such as the locating and coding of ejector marks and extractor marks.

Reference

[1] C.L. Smith, and J.M. Cross, (1995): Optical Imaging Techniques for Ballistics Specimens to Identify Firearms. Proceedings of the 29th Annual 1995 International Carnahan Conference on Security Technology, pp. 275-289, Oct. 1995, England.

[2] R. Saferstein (ED), (1988) Forensic Science Handbook: Volume 2. Englewood Cliffs: Prentice Hall, 1988.

[3] G.Burrard, (1951): Identification of Firearms and Forensic Ballistics. London: Herbert Jenkins,1951.

[4] C.L. Smith, J.M. Cross, and G.J. Variyan, (1995): FIREBALL: An Interactive Database for the Forensic Ballistic Identification of Firearms. Research Report, Australian Institute of Security and Applied Technology, Edith Cowan University, Western Australia, 1995.

[5] Le Ping Xin, (2000): A Cartridge Identification System for Firearm Authentication, Signal Processing Proceedings, 2000. WCCC_ICSP 2000. 5th International Conference on Volume: 2, P1405-1408.

[6] Chenyuan Kou, Cheng-Tan Tung and H. C. FU, (1994): FISOFM: Firearms Identification based on SOFM Model of Neural Network, Security Technology, 1994. Proceedings. Institute of Electrical and Electronics Engineers 28th Annual 1994 International Carnahan Conference on , 12-14 Oct. Pages: 120-125.

[7] T. Kohonen, (1990): Self-organizing Maps, Springer, Berlin, 1995, The self-organising Maps, Proc. IEEE 78 (9) (1990) 1464-1480.

[8] S.B. Cho: Pattern Recognition with Neural Networks Combined by Genetic Algorithm, Fuzzy Set and System 103(1999) 339-347.

[9] T.M. Ha, H. Bunke: Off-line Handwritten Numeral Recognition by Perturbation Method, IEEE Trans. Pattern Anal. Mach. Intell. 19(5) (1997) 535-539.

[10] P.N. Suganthan: Structure Adaptive Multilayer Overlapped SOMs with Partial Supervision for Handprinted Digit Classification, Proceedings of International Joint Conference on Neural Networks, WCCI'98, Alaska, May 1998.

[11] T. Kohonen, Tutorial Notes, (1993) International Symposium on Artificial Neural Networks, pp. 9-15, Dec.20-22, 1993.
[12] Rafael C. Gonzalez, Richard E. Woods: Digital Image Processing, Second Edition, Beijing: Publishing House of Electronics Industry, 2002, 7, 519-566.

Predicting the Australian Stock Market Index Using Neural Networks Exploiting Dynamical Swings and Intermarket Influences

Heping Pan, Chandima Tilakaratne, and John Yearwood

School of Information Technology & Mathematical Sciences,
University of Ballarat
Mt Helen, Victoria 3350, Australia
h.pan@ballarat.edu.au

Abstract. This paper presents a computational approach for predicting the Australian stock market index - AORD using multi-layer feed-forward neural networks from the time series data of AORD and various interrelated markets. This effort aims to discover an optimal neural network or a set of adaptive neural networks for this prediction purpose, which can exploit or model various dynamical swings and inter-market influences discovered from professional technical analysis and quantitative analysis. Four dimensions for optimality on data selection are considered: the optimal inputs from the target market (AORD) itself, the optimal set of interrelated markets, the optimal inputs from the optimal interrelated markets, and the optimal outputs. Two traditional dimensions of the neural network architecture are also considered: the optimal number of hidden layers, and the optimal number of hidden neurons for each hidden layer. Three important results were obtained: A 6-day cycle was discovered in the Australian stock market; the time signature used as additional inputs provides useful information; and a minimal neural network using 6 daily returns of AORD and 1 daily returns of SP500 plus the day of the week as inputs exhibits up to 80% directional prediction correctness.

1 Introduction

Predicting financial markets has been one of the biggest challenges to the AI community since about two decades ago. The objective of this prediction research has been largely beyond the capability of traditional AI because AI has mainly focused on developing intelligent systems which are supposed to emulate human intelligence. However, the majority of human traders cannot win consistently on the financial markets. In other words, human intelligence for predicting financial markets may well be inappropriate. Therefore, developing AI systems for this kind of prediction is not simply a matter of re-engineering human expert knowledge, but rather an iterative process of

T.D. Gedeon and L.C.C. Fung (Eds.): AI 2003, LNAI 2903, pp. 327-338, 2003.

knowledge discovery and system improvement through data mining, knowledge engineering, theoretical and data-driven modeling, as well as trial and error experimentation.

Multi-layer feed-forward neural networks, also known as multi-layer Perceptrons, are sub-symbolic connectionist models of AI, which have been proven both in theory and in practical applications to be able to capture general nonlinear mappings from an input space to an output space. The capacity of neural networks to represent mappings was investigated by Cybenko [3, 4] and Hornik et al. [6]. It is now known that a single nonlinear hidden layer is sufficient to approximate any continuous function, and two nonlinear hidden layers are enough to represent any function.

Using neural networks to predict financial markets has been an active research area since the 1990s [1-2, 5, 7-9, 11-26]. Most of these published works are targeted at US stock markets and other international financial markets. Very little is known or done on predicting the Australian stock market. From our empirical knowledge of technical and quantitative analysis and our first-hand observations in the international stock markets, it appears clearly to us that every stock market is different, and has its unique "personality" and unique position in the international economic systems. While the empirical knowledge gathered from predicting US and other international markets is definitely helpful, developing neural networks particularly devoted to predicting the Australian stock market requires highly specialized empirical knowledge about this market and its dynamic relationships to the US stock markets and other international financial markets.

There are quite a few non-trivial problems involved in developing neural networks for predicting the Australian stock market. Two traditional problems are the optimal number of hidden layers and the optimal number of hidden neurons for each hidden layer. Theoretically sound and practically sufficient solutions for these two problems can be found. However, the most critical problem in this research is the selection of the optimal input vector of features to be extracted from the time series data of the market. In our case, the primary source of data is the Australian stock market index, either the Australian all ordinary index (AORD), or the ASX/S&P 200 index. The secondary sources of data are US stock market indices and other international financial market indices. From the perspective of a trading system development, the second most important problem is the optimal prediction horizon, or the time frame. That is how far into the future the neural networks can predict with the highest reliability.

The remainder of this paper is organized as follows: Section 2 defines the problem of this prediction research and formalizes the major sub-problems; Section 3 describes the neural network approach to the prediction problem; Sections 4 to 5 present our solutions to the problems of optimal input and output selection; Section 6 points out further possibilities for research; Section 7 concludes the paper.

2 The Stock Index Prediction Problem

Let $X(t)$ be the target index at the current time t, note that $X(t)$ is a vector of five components:

$$X(t) = (X.O(t), X.H(t), X.L(t), X.C(t), X.V(t)) \tag{1}$$

where O, H, L, C, V denote respectively the open, high, low, close index level and the traded volume for the trading period of time *t*. In general, we take each single day as the standard resolution of time, so we use daily charts as the standard data. Let $Y_k(t)$ be the index of another market which is considered to be interrelated to the target market $X(t)$, where $k = 1,2,\cdots,K$, and K denotes the total number of interrelated markets selected. Similarly, $Y_k(t)$ is a vector:

$$Y_k(t) = (Y_k.O(t), Y_k.H(t), Y_k.L(t), Y_k.C(t), Y_k.V(t)) \tag{2}$$

We shall call $Y_k(t)$ an inter-market of $X(t)$, and

$$Y(t) = \{Y_k(t) \mid k = 1,2,\cdots K\} \tag{3}$$

the inter-markets – the complete set of selected inter-markets of $X(t)$. We assume the availability of historical time series data, usually called charts in technical analysis, $DX(t)$ of the target market and $DY(t)$ of the inter-markets, defined as

$$DX(t) = \{X(t) \mid t = t - N + 1, t - N + 2, \cdots, t - 2, t - 1, t\} \tag{4}$$

$$DY(t) = \{DY_k(t) \mid k = 1,2,.\cdots K\} \tag{5}$$

where

$$DY_k(t) = \{Y_k(t) \mid t = t - N + 1, t - N + 2, \cdots, t - 2, t - 1, t\} \tag{6}$$

and N is the total number of trading days for the standard time resolution (or minutes for intra-day charts). The time t starts from $N-1$ trading days back into the past, taking today as t. So the current time is just after today's market close for the target market and all the inter-markets.

The problem of prediction is to use the given historical chart data $DX(t)$ and $DY(t)$ of the last N trading days to predict the index of the target market T days into the future

$$(DX(t), DY(t)) \mapsto X(t+T) \tag{7}$$

where T can range from 1 to 10 for short-term prediction, or to 22 for monthly prediction, or to 256 for yearly prediction, and so on. Chaos theory tells us that the precision and reliability of prediction decays exponentially in time.

3 A Neural Network Approach for Stock Index Prediction

The approach of using neural networks for predicting the stock market index starts with defining a mapping M from a n-dimensional input space $\{x(x_1, x_2, \cdots, x_n)\}$ to an m-dimensional output space $\{z(z_1, z_2, \cdots, z_m)\}$:

$$M : x(x_1, x_2, \cdots, x_n) \mapsto z(z_1, z_2, \cdots, z_m) \quad (8)$$

Here we assume we have defined n real-valued features from the available data-set $DX(t)$ and $DY(t)$ at the current time t, and we want to predict m real values which may correspond to one or more target index levels in the future.

There are only two sensible architectures of neural network to consider for the stock index prediction problem: if we assume the stock index prediction problem can be modeled as a nonlinear continuous function (mapping), we should use a three-layer Perceptron with one hidden nonlinear layer, and this is the basic architecture; however, in general, this mapping should be assumed to be just any function which may contain discontinuities, so the general architecture should be a four-layer Perceptron with two hidden nonlinear layers. In both cases, the output layer should be linear because the outputs can be positive or negative real values. Pan and Foerstner [10] proposed an MDL-principled approach for determining the bounds of the number of hidden neurons.

The three-layer Perceptron with one hidden nonlinear layer of h hidden neurons can be represented as follows:

$$z_j = \sum_{k=1}^{h} w_{kj} y_k + b_j , \quad \text{for } j = 1,2,\cdots,m \quad (9)$$

$$y_k = \phi(\sum_{i=1}^{n} w_{ik} x_i + a_k), \text{ for } k = 1,2,\cdots,h \quad (10)$$

where y_k is the output of the k-th hidden neuron, a_k, b_j are the bias for the k-th hidden neuron and the j-th output neuron respectively, $\phi(\cdot)$ is the nonlinear transfer function for the hidden neurons, which generally takes the form of sigmoid function

$$\phi(x) = \frac{1}{1 + e^{-x}} \quad (11)$$

For the four-layer Perceptron with two hidden nonlinear layers, we have similar formulas as (10) for each of the hidden layers, but the two hidden layers may have different numbers of hidden neurons.

4 Inspiration and Data Mining for Optimal Input Selection

The vector of input features includes two sub-vectors: those features extracted from the target market index and those extracted from the inter-market indices. The basic candidates for the features from the target market index are the relative returns of the closing index for the short term and geometrically spanning over the intermediate term. The relative return of the time interval τ is defined as

$$r_\tau(t) = \frac{X.C(t) - X.C(t-\tau)}{X.C(t-\tau)} \tag{12}$$

where $r_1(t)$ refers to daily returns, $r_5(t)$ to weekly returns, $r_{22}(t)$ to monthly returns, and so on. Our empirical knowledge from technical analysis suggests the following candidates:

$$(r_1(t), r_1(t-1), r_1(t-2), r_1(t-3), r_1(t-4), r_1(t-5)) \tag{13}$$

$$(r_5(t-6), r_5(t-11), r_5(t-16), r_5(t-21), r_5(t-26), r_5(t-31), \cdots)) \tag{14}$$

A more general approach for spanning over the past through a geometrical scale space is to use Fibonacci ratios, such as

$$(r_2(t-6), r_3(t-8), r_5(t-11), r_8(t-16), r_{13}(t-24), r_{21}(t-37), \cdots) \tag{15}$$

It is known that the relative return series of (13) and (14) do not have a long-term memory, so they may only capture short-term market cycles and candlestick chart patterns. One of the most significant weaknesses of using relative return series is the inability of representing important index support/resistance levels spanning over intermediate to longer terms. To overcome this weakness, we should consider the second form of relative return:

$$q_\tau(t) = \frac{X.C(t) - X.C(t-\tau)}{X.C(t)} \tag{16}$$

Then a vector similar to (13) can be defined as

$$(r_1(t), r_2(t), r_3(t), r_4(t), r_5(t), r_6(t))$$

Similarly features of geometrical time span can be defined.

According to our knowledge of inter-market technical analysis and our observations of the international stock markets and other financial markets, we have selected the following international markets as the inter-markets of the target market:

$$Y_1 = \text{US S\&P 500 Index} \tag{17}$$

$$Y_2 = \text{US Dow Jones Industrial Average Index} \tag{18}$$

$$Y_3 = \text{US NASDAQ 100 Index} \tag{19}$$

$$Y_4 = \text{Gold \& Silver Mining Index} \tag{20}$$

$$Y_5 = \text{AMEX Oil Index} \tag{21}$$

The space of possible inter-markets for the Australian stock market is not limited to these, however, these inter-markets of (17)-(21) provide the most influential parts of the global economical system affecting the Australian stock market. Candidates of the additional input features selected from these inter-markets are the relative returns in the form of either (12) or (16). If the prediction horizon is just the next day, the relative returns of the inter-markets at the latest time t are sufficient; if the horizon is longer, then at least a short-term time series of the relative returns of one of the inter-markets should be used, together with the relative return time series of the target market such as (13) or (15).

Figures 1 & 2 show the autocorrelation and partial autocorrelation of the target market X = Australian All Ordinary Index (AORD) relative return as defined by (12). From the autocorrelation and partial autocorrelation figures, we can clearly see that there is a 6-day cycle in the Australian stock market. This is surprising as we usually assume there is a 5-day cycle, known as the Day of the Week.

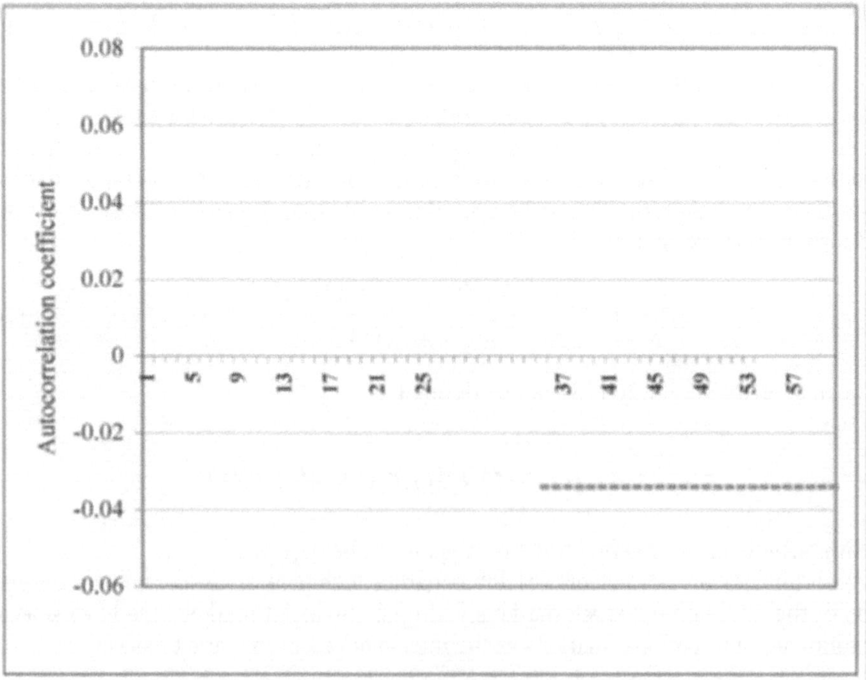

Fig. 1. Autocorrelation function of X

Figures 3 & 5 show the cross correlation between X and each Y_k, $k = 1,2,3$ as defined by (17)-(19). The correlations between $X(t)$ and $Y_k(t-1)$, for $k = 1,2,3$, are extremely high, and those between $X(t)$ and $Y_k(t)$ are also very significant. Significant correlations between $Y_1(t)$ and each of $Y_4(t), Y_5(t)$ are also found as expected according to our empirical knowledge. Therefore, $Y_4(t), Y_5(t)$ should also be included in the inter-markets of $X(t)$.

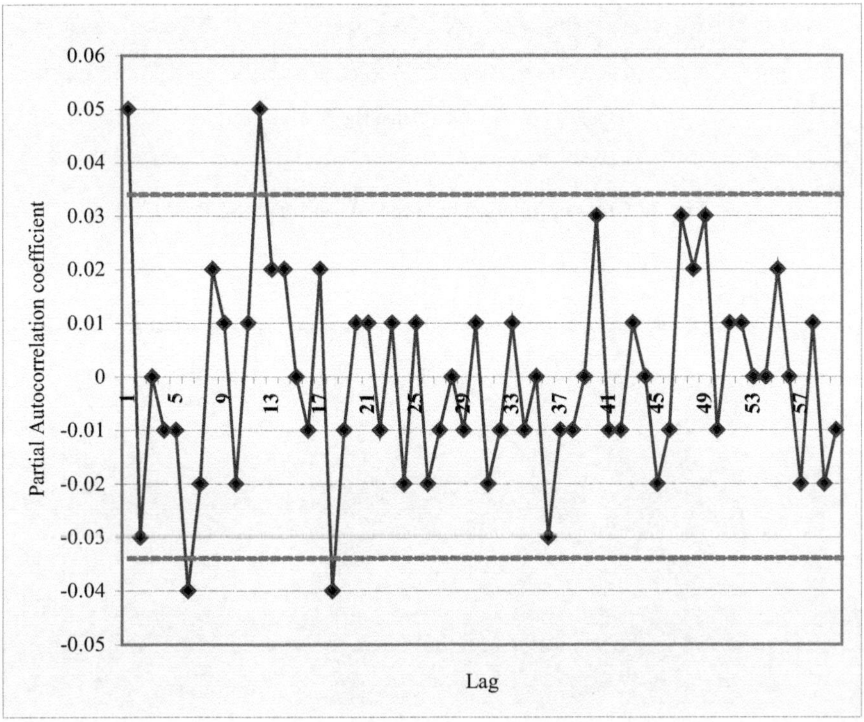

Fig. 2. Partial autocorrelation coefficients of X

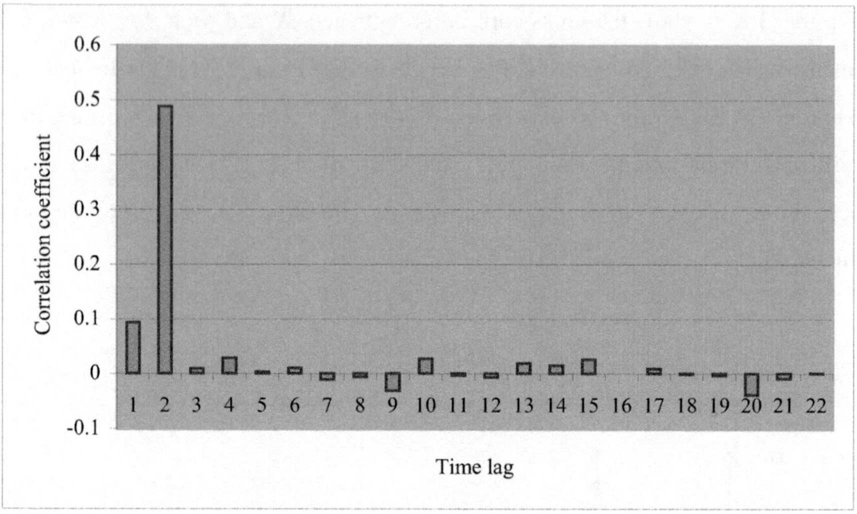

Fig. 3. Cross correlation between X and Y_1 = S&P 500

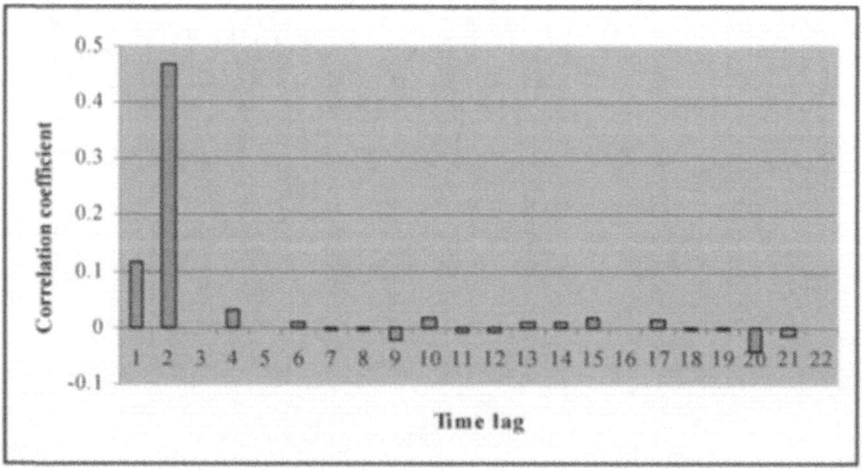

Fig. 4. Cross correlation between X and Y_2 = Dow Jones

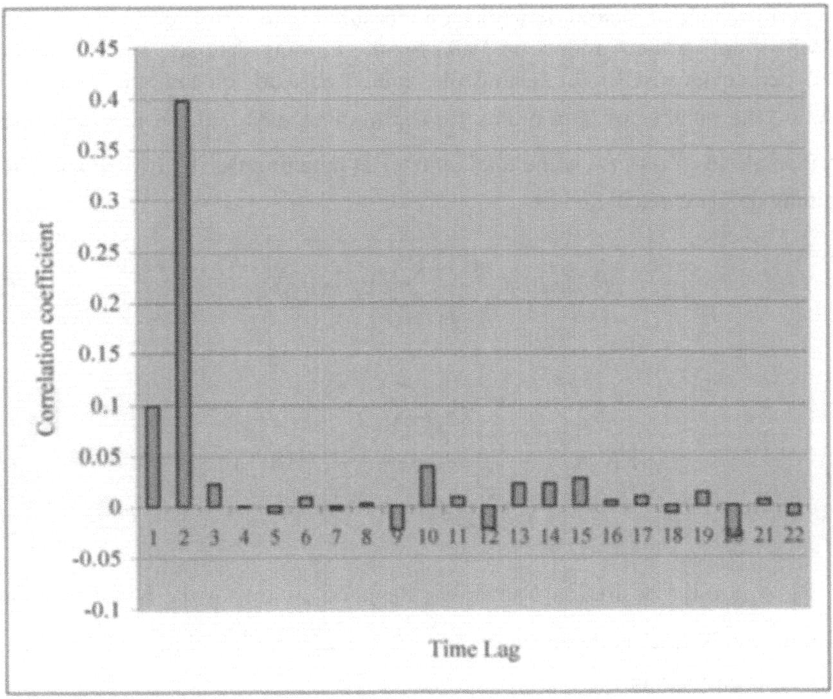

Fig. 5. Cross correlation between X and Y_3 = NASDAQ

5 Inspiration and Experimentation for Optimal Output Selection

With a series of short-term daily and weekly returns as input vector, we expect to be able to only predict 1 to 10 days into the future. Therefore, we have limited our output candidates to be $\{X(t+1), X(t+2), \cdots, X(t+10)\}$. According to the result of our statistical data mining (Figures. 1& 2), the Australian stock index, AORD, shows clearly cyclic behaviour with period 6 days. Therefore, we decided to use the last 6 daily returns of AORD in the input vector. Among the inter-markets (17)-(21) considered, Y_1 - (US S&P 500 Index) shows the highest correlation to the target market. So it was also included in the input vector. In addition, the day of the week as a time signature was also included as an additional input in our experiments. The number of hidden layers are limited to 1 or 2, and the number of hidden neurons for each hidden layer ranges from 2 to 50 in our experiments.

The historical index time series data of $DX(t)$ and $DY(t)$ between January 1990 and May 2003 were used. The whole data set is divided into two data sets: 20% of the data are randomly selected and put into the test data set, and the remaining 80% into the training data set.

The performance of neural networks are measured in terms of root mean square error (RMSE) and variance reduction (VR) for the training data set, and the sign correctness percentage (SCP) for testing the trained network on the test data set. Let N_1, N_2 be the number of data points for the training data and the test data respectively, then $N_1 + N_2 = N$ is the total number of data points. The RMSE and VR for the training data are calculated as

$$RMSE = \sqrt{\frac{1}{N_1}\sum_{k=1}^{N_1}(z_k - o_k)^2} \tag{22}$$

$$VR = \left(1 - \frac{\sum_{k=1}^{N_1}(z_k - o_k)^2}{\sum_{k=1}^{N_1}(z_k - \bar{z})^2}\right) \cdot 100\% \tag{23}$$

where z_k, o_k are the desired and actually calculated output for the k-th data point in the training data set, $\bar{z}$ is the mean of $\{z_k \mid k = 1,2,\cdots,N_1\}$. The SCP for the testing data set is defined as

$$SCP = \frac{|\{sign(z_k) = sign(o_k) \mid k = 1,2,\cdots,N_2\}|}{N_2} \tag{24}$$

where $|\{\}|$ denotes the number of elements in the given set.

A number of neural networks were trained. The best result is achieved on predicting the next day market direction and magnitude $X(t+1)$ with the selected inputs. However, it must be pointed out that prediction of different time frames requires different input selections, which may lead to significantly different network architectures. Table 1 shows the results of using different inputs with 1 hidden layer of 2 hidden neurons for predicting $X(t+1)$:

Table 1. Training and testing results with different inputs

Input Vector	RMSE	VR	SCP
$X(t-k)$, $k = 0,1,\cdots 5$, only	0.57	50.80%	65%
With $Y_1(t)$ (S&P 500) added	0.54	56.85%	76%
With the Day of the Week added	0.53	57.26%	80%

According to the SCP scores, the US S&P 500 index and the day of the week have added 11% and 4 % additional information to the prediction based on the Australian stock market index alone.

6 Further Possibilities

So far we have only investigated very basic aspects of the problem. The results obtained are already very useful and the trained neural networks have been used routinely to generate daily predictions on the Australian stock index in our real-money stock and index futures trading. However, it should be pointed out that developing a specific neural network or a set of neural networks for this prediction purpose is a never-ending evolutionary process. Further possibilities for improving the correctness, accuracy and reliability of the prediction include selection of technical indicators, in particular, those based on wavelet and fractal analysis, prediction of marginal probability distribution, ensembles or committees of neural networks, profit-driven genetic training algorithms, finer representation of time signatures such as using complex numbers for the day of the week, the day or week of the month, the day or week or month of the year in the input vector, and so on.

7 Conclusions

A minimal neural network has been developed for predicting the Australian stock market index AORD, which takes the last 6 daily returns of the target market, the last daily return of US S&P 500 index, and the day of the week as the input vector. It has only one hidden layer of 2 hidden neurons. It has achieved correctness in directional prediction of 80%. In additional to this development, a 6-day cycle has been discovered in the Australian stock market for the first time. These results shed strong light on further research and development directions.

References

[1] Azoff E.M. (1994): Neural Network Time Series Forecasting of Financial Markets. Wiley.

[2] Baestaens D.E., van den Berg W.M. and Vaudrey H. (1995): Money market headline news flashes, effective news and the DEM/USD swap rate: An intraday analysis in operational time. *Proc. 3rd Forecasting Financial Markets Conference*, London.

[3] Cybenko G. (1988): Continuous valued neural networks with two hidden layers are sufficient. Technical report, Department of Computer Science, Tufts University, Medford, Massachusetts, USA.

[4] Cybenko (1989): Approximation by superpositions of a signoidal function. *Mathematics of Controls, Signals and Systems*, 2: 303-314.

[5] Gately (1996): Neural Networks for Financial Forecasting. Wiley.

[6] Hornik K., Sinchcombe M. and White H. (1989): Multilayer feedforward networks are universal approximators. *Neural Networks*, 2: 259-366.

[7] Murphy J. (1999): Technical Analysis of the Financial Markets: A Comprehensive Guide to Trading Methods and Applications. Prentice Hall Press.

[8] Pan H.P. (2003a): A joint review of technical and quantitative analysis of the financial markets towards a unified science of intelligent finance. Proc. *2003 Hawaii International Conference on Statistics and Related Fields*, June 5-9, Hawaii, USA.

[9] Pan H.P. (2003b): Swingtum – A computational theory of fractal dynamic swings and physical cycles of stock market in a quantum price-time space. Ibid.

[10] Pan H.P. and Foerstner W. (1992): An MDL-principled evolutionary mechanism to automatic architecturing of pattern recognition neural network. *IEEE Proc. 11th International Conference on Pattern Recognition*, the Hague.

[11] Qi M. (1999): Nonlinear predictability of stock returns using financial and economic variables. *Journal of Business and Economic Statistics*, 17: 419-429.

[12] Shih Y.L. (1991): Neural nets in technical analysis. *Technical Analysis of Stocks and Commodities*, 9(2):62-68.

[13] Fishman M.B., Barr D.S. and Loick W.J. (1991): Using neural nets in market analysis. *Technical Analysis of Stocks and Commodities*, 9(4): 18-22.

[14] Katz J.O. (1992): Developing neural network forecasters for trading. *Technical Analysis of Stocks and Commodities*, 10(4).

[15] Kean J. (1992): Using neural nets for intermarket analysis. *Technical Analysis of Stocks and Commodities*, 10(11).

[16] Refenes A.-P., Abu-Mostafa Y., Moody J. and Weigend A. (eds) (1996): *Neural Networks in Financial Engineering.* Proc. 3rd International Conference on Neural Networks in the Capital Markets. World Scientific.

[17] Rogers R. and Vemuri V. (1994): *Artificial Neural Networks Forecasting Time Series.* IEEE Computer Society Press, Los Alamitos, CA.

[18] Ruggerio M. (1994): Training neural nets for intermarket analysis. *Futures*, Sep: 56-58.

[19] Stein (1991): Neural networks: from the chalkboard to the trading room. *Futures*, May: 26-30.

[20] Swales G.S. and Yoon Y. (1992): Applying artificial neural networks to investment analysis. *Financial Analysts Journal*, 48(5).

[21] Utans J. and Moody J.E. (1991): Selecting neural network architectures via the prediction risk: application to corporate bond rating prediction. *Proc. 1st Int. Conference on AI Applications on Wall Street*, IEEE Computer Society Press.

[22] Virili F. and Reisleben B. (2000): Nonstationarity and data preprocessing for neural network predictions of an economic time series. *Proc. Int. Joint Conference on Neural Networks 2000*, Como, Vol. 5, pp. 129-136.

[23] Ward S. and Sherald M. (1995): The neural network financial wizards. *Technical Analysis of Stocks and Commodities*, Dec: 50-55.

[24] White H. (1988): Economic predictions using neural networks: the case of IBM daily stock returns. *Proc. IEEE International Conference on Neural Networks*, Vol. 2, pp. 451-458.

[25] Wong F.S. (1992): Fuzzy neural systems for stock selection. *Financial Analysts Journal*, 48: 47-52.

[26] Yao J.T. and Tan C.L. (2001): Guidelines for financial forecasting with neural networks. *Proc. International Conference on Neural Information Processing*, Shanghai, China, pp. 757-761.

A Tableaux System for Deontic Interpreted Systems

Guido Governatori[1], Alessio Lomuscio[2], and Marek J. Sergot[3]

[1] School of Information Technology and Electrical Engineering, The University of Queensland Brisbane, Australia
guido@itee.uq.edu.au

[2] Department of Computer Science, King's College London, London, UK
alessio@dcs.kcl.ac.uk

[3] Department of Computing, Imperial College, London, UK
mjs@doc.ic.ac.uk

Abstract. We develop a labelled tableaux system for the modal logic $KD45_n^{i-j}$ extended with epistemic notions. This logic characterises a particular type of interpreted systems used to represent and reason about states of correct and incorrect functioning behaviour of the agents in a system, and of the system as a whole. The resulting tableaux system provides a simple decision procedure for the logic. We discuss these issues and we illustrate them with the help of simple examples

1 Introduction

One of the main areas of interest in the use formal methods in Software Engineering involve the use of tools based on mathematical logic for the specification and verification of computing systems. This is true in general but it specially applies to the area of multi-agent systems. Here, multi-modal logics are normally used to *specify* the behaviour of a multi-agent systems. Several formalisms have been designed for this task, most importantly logics for knowledge [5], logics for Belief-Desires-Intentions [15], deontic logics [14], etc. The usual approach in this line of work is to suggest a logic, in terms of its syntax and axiomatisation, to show that it captures the intuitive properties of the concept under investigation, and to show metalogical properties of the logic system, such as its completeness and decidability.

This is adequate for the task of specifying distributed systems but methodologies need to be developed for *verifying* that a system complies with a given specification. Several methods are employed to perform this task, traditionally theorem provers, and more recently model checkers.

It has been argued elsewhere [13, 12] that a formalism for specifying properties of a system could make use of a deontic component. That can be used for example to distinguish in a precise and unambiguous way among properties that *should* hold in a system, properties that *may* hold in a system, properties that simply hold in a system. While deontic concepts are useful on their own, especially when paired to temporal operators, they become even more of importance in multi-agent systems when paired with informational properties such as their knowledge, beliefs, intentions, etc. The formalism of *deontic interpreted systems* was designed to make a step in that direction.

T.D. Gedeon and L.C.C. Fung (Eds.): AI 2003, LNAI 2903, pp. 339–351, 2003.

In deontic interpreted systems a semantics based on interpreted systems [5] is given to interpret a multi-modal language consisting of a family of operators $\{O_i\}$, representing correct functioning behaviour of agent i, a family of operators $\{K_i\}$ representing the knowledge of agent i, and a family of operators $\{\widehat{K}_i^j\}$ representing the knowledge agent i has under the assumption of correctness of agent j. It was argued in [13, 12] that this set of operators could be useful to represent a number of key interesting scenarios, including communication and security examples.

A complete axiomatisation for deontic interpreted systems limited to the fragment of $\{O_i, K_i\}$, was also shown. This comprises the logics $S5_n$ for the modalities K_i for knowledge and the logic $KD45_n^{i-j}$ for the modalities O_i for correct functioning behaviour. While a Hilbert style axiomatisation is a theoretically valuable result, proving properties of particular examples by means of this is notoriously awkward. Automated technologies, such as the ones based on theorem provers or model checkers are called for. In this paper we define and investigate a tableaux system for the full logic above.

The remaining of this paper is organised as follows. In Section 2 we define the language of the logic, the semantics, and present its axiomatisation. In Section 3 we define a tableaux system for it. In Section 4 we present an example, the bit transmission problem, and we prove properties about it by means of the tableaux strategy. In Section 5 we wrap up and point to further work.

2 Deontic Interpreted Systems

We present here the main definitions for the notation we are going to use in this paper, as from [5, 13]. Due to space consideration we are forced to assume working knowledge with some of the technical machinery presented there.

Interpreted Systems Consider n agents in a system and n non-empty sets $L_1, \ldots, L_n$ of local states, one for every agent of the system, and a set of states for the environment L_E. Elements of L_i will be denoted by $l_1, l_1', l_2, l_2', \ldots$. Elements of L_E will be denoted by $l_E, l_E', \ldots$.

A *system of global states for n agents S* is a non-empty subset of a Cartesian product $L_1 \times \cdots \times L_n \times L_E$. When $g = (l_1, \ldots, l_n, l_E)$ is a global state of a system S, $l_i(g)$ denotes the local state of agent i in global state g. $l_E(g)$ denotes the local state of the environment in global state g. An *interpreted* system of global states is a pair $IS = (S, h)$ where S is a system of global states and $h : S \to 2^P$ is an interpretation function for a set of propositional variables P. Systems of global states can be used to interpret epistemic modalities K_i, one for each agent.

$$(IS, g) \models K_i \varphi \text{ iff } \forall g' : l_i(g) = l_i(g') \Rightarrow (IS, g') \models \varphi.$$

Alternatively one can consider generated models $(S, \sim_1, \ldots, \sim_n, h)$ of the standard form, where the equivalence relations $\sim_i$ are defined on equivalence of local states, and then interpret modalities in the standard modal tradition (e.g. [3, 10]). The resulting logic for modalities K_i is $S5_n$; this models agents with complete introspection capabilities and veridical knowledge.

Deontic Interpreted Systems The notion of interpreted systems can be extended to incorporate the idea of correct functioning behaviour of some or all of the components [13].

Given n agents and $n+1$ non-empty sets $G_E, G_1, \ldots, G_n$, a *deontic system of global states* is any system of global states defined on $L_E \supseteq G_E, \ldots, L_n \supseteq G_n$. G_E is called the *set of green states for the environment*, and for any agent i, G_i is called *the set of green states for agent i*. The complement of G_E with respect to L_E (respectively G_i with respect to L_i) is called the *set of red states for the environment (respectively for agent i)*.

The terms 'green' and 'red' are chosen as neutral terms, to avoid overloading them with unintended readings and connotations. The term 'green' can be read as 'legal', 'acceptable', 'desirable', 'correct', depending on the context of a given application.

Deontic systems of global states are used to interpret modalities such as the following

$$(IS, g) \models O_i \varphi \text{ iff } \forall g' : l_i(g') \in G_i \Rightarrow (IS, g') \models \varphi.$$

$O_i \varphi$ is used to represent that φ holds in all (global) states in which agent i is functioning correctly. Again, one can consider generated models $(S, \sim_1, \ldots, \sim_n, R_1^O, \ldots, R_n^O, h)$, where the equivalence relations are defined as above and the relations R_i^O are defined by $g R_i^O g'$ if $l_i(g') \in G_i$, with a standard modal logic interpretation for the operators O_i.

Knowledge can be modelled on deontic interpreted systems in the same way as on interpreted systems, and one can study various combinations of the modalities such as $K_i O_j$, $O_j K_i$, and others. Another concept of particular interest is knowledge that an agent i has *on the assumption that the system (the environment, agent j, group of agents X) is functioning correctly*. We employ the (doubly relativised) modal operator $\widehat{K}_i^j$ for this notion, interpreted as follows:

$$(IS, g) \models \widehat{K}_i^j \varphi \text{ iff } \forall g' : l_i(g) = l_i(g') \text{ and } l_j(g') \in G_j \Rightarrow (IS, g') \models \varphi.$$

An Axiomatisation of Deontic Interpreted Systems The multi-modal language defined by O_i, K_i is axiomatised by the logics $S5_n$ union $KD45_n^{i-j}$ where there are defined as follows:

The component $S5_n$ is defined by the smallest normal multi-modal logic (i.e., closed under the necessitation rule for K_i) satisfying the axioms T, 4, and 5 for each modal operator K_i. Semantically $S5_n$ is determined by the class of Kripke frames $(W, \sim_1, \ldots, \sim_n)$ where each $\sim_i$ is an equivalence relation.

The component $KD45_n^{i-j}$ is defined by the smallest normal multi-modal logic (i.e., closed under the necessitation rule for O_i) satisfying the axioms D, 4, 5 and $\neg O_i \neg \varphi \rightarrow O_j \neg O_i \neg \varphi$. for each pair of modal operators O_i, O_j. Semantically $KD45_n^{i-j}$ is determined by the class of serial, transitive and i-j Euclidean Kripke frames $(W, R_1^O, \ldots, R_n^O)$ where a frame is i-j Euclidean iff for all $w', w'', w''' \in W$ an for all i, j such that $1 \leq i, j \leq n$, we have that $w R_i^O w'$ and $w R_j^O w''$ implies $w R_i^O w''$.

For the operator $\widehat{K}_i^j$, determined semantically by $\sim_i \cap R^{O_j}$, we do not have a complete axiomatisation. In this paper we provide a sound and complete tableuax system for it.

3 Tableaux for Deontic Interpreted Systems

In [1, 9, 2] a tableau-like proof system, called KEM, has been presented, and it has been proven to be able to cope with a wide variety of logics accepting possible world semantics. KEM is based on D'Agostino and Mondadori's [4] classical proof system KE, a combination of tableau and natural deduction inference rules which allows for a restricted ("analytic") use of the cut rule. The key feature of KEM, besides its being based neither on resolution nor on standard sequent/tableau inference techniques, is that it generates models and checks them using a label scheme for bookkeeping states in interpreted systems. In [7, 8, 9] it has been shown how this formalism can be extended to handle various systems of multi-modal logic with interaction axioms. The mechanism KEM uses in manipulating labels is close to the possible world semantic constructions. In the following section we show how to adapt it to deal with deontic interpreted systems.

Label Formalism KEM uses *Labelled Formulas* (*L*-formulas for short), where an *L*-formula is an expression of the form $A : t$, where A is a wff of the logic, and t is a label. In the case of deontic interpreted systems we have a type of labels corresponding to various modalities for each agent; the set of atomic labels for i (Φ^i) is defined as follows:

$$\Phi^i = \Phi^i_O \cup \Phi^i_K \cup \Phi^{ij}$$

Each set of atomic labels for the modalities is partitioned into the (non-empty) sets of variables and constants.

$$\Phi^i_O = V^i_O \cup C^i_O; \qquad \Phi^i_K = V^i_K \cup C^i_K; \qquad \Phi^{ij} = V^{ij} \cup C^{ij} \text{ for any } j$$

where $V^i_O = \{O^i_1, O^i_2, \dots\}$, $C^i_O = \{o^i_1, o^i_2, \dots\}$, $V^i_K = \{K^i_1, K^i_2, \dots\}$, $C^i_K = \{k^i_1, k^i_2, \dots\}$, $V^{ij} = \{IJ_1, IJ_2, \dots\}$, and $C^{ij} = \{ij_1, ij_2, \dots\}$. Finally we add a sets of auxiliary unindexed atomic labels $\Phi^A = V^A \cup C^A$ – here $V^A = \{W_1, W_2, \dots\}$ and $C^A = \{w_1, w_2, \dots\}$–, that will be used in unifications and proofs. With Φ_C and Φ_V we denote, respectively, the set of constants and the set of variables.

The set of labels $\Im$ is then defined inductively as follows: a label is either (i) an element of the set Φ_C, or (ii) an element of the set Φ_V, or (iii) a path term (s', s) where (iiia) $s' \in \Phi_C \cup \Phi_V$ and (iiib) $s \in \Phi_C$ or $s = (t', t)$ where (t', t) is a label. From now on we shall use $t, s, r, \dots$ to denote arbitrary labels.

As an intuitive explanation, we may think of a label $t \in \Phi_C$ as denoting a world (a *given* one), and a label $t \in \Phi_V$ as denoting a set of worlds (*any* world) in some Kripke model. A label $s = (t', t)$ may be viewed as representing a path from t to a (set of) world(s) t' accessible from t (i.e., from the world(s) denoted by t).

For any label $t = (s', s)$ we shall call s' the *head* of t, s the *body* of t, and denote them by $h(t)$ and $b(t)$ respectively. Notice that these notions are recursive (they correspond to projection functions): if $b(t)$ denotes the body of t, then $b(b(t))$ will denote the body of $b(t)$, and so on. We call each of $b(t)$, $b(b(t))$, etc., a *segment* of t. The length of a label t, $\ell(t)$, is the number of world-symbols in it, i.e., $\ell(t) = n \Leftrightarrow t \in \Im_n$. $s^n(t)$ will denote the segment of t of length n and we shall use $h^n(t)$ as an abbreviation for $h(s^n(t))$. Notice that $h(t) = h^{\ell(t)}(t)$.

For any label $t, \ell(t) > n$, we define the *counter-segment-n* of t, as follows (for $0 < n < k < \ell(t)$):

$$c^n(t) = h(t) \times (\cdots \times (h^k(t) \times (\cdots \times (h^{n+1}(t), w_0))))$$

where w_0 is a dummy label, i.e., a label not appearing in t (the context in which such a notion occurs will tell us what w_0 stands for). The counter-segment-n defines what remains of a given label after having identified the segment of length n with a 'dummy' label w_0. The appropriate dummy label will be specified in the applications where such a notion is used. However, it can be viewed also as an independent atomic label.

So far we have provided definitions about the structure of the labels without regard of the elements they are made of. The following definitions will be concerned with the type of world symbols occurring in a label.

Let t be a label and t' an atomic label, in what follows we shall use $(t';t)$ as a notation for the label (t',t) if $t' \neq h(t)$, or for t otherwise.

We say that a label t is *i-preferred* iff $h(t) \in \Phi^i$, and a label t is *i-pure* iff each segment of t of length $n > 1$ is i-preferred, and we shall use $\mathfrak{I}^i$ to denote the set of i-pure labels. A label is *i-compatible* iff each segment of t of length $n > 1$ is either i-preferred or ij-preferred (for any j). A label t is *ij-ground* iff every label of type Φ^{ij} is a constant.

Label Unifications In the course of proofs labels are manipulated in a way closely related to the semantic of the logics under analysis. Labels are compared and matched using a specialised logic dependent unification mechanism. The notion that two labels t and s unify means that the intersection of their denotations is not empty and that we can "move" to such a set of worlds, i.e., to the result of their unification.

According to the semantics each modality is evaluated using an appropriate binary relation on the model and the model results from the combination of the relations. Similarly we provide an unification for each modality, the unification characterising it in the KEM formalism, then we combine them into a single unification for the whole logic. Every unification is built from a basic unification defined in terms of a substitution $\rho : \mathfrak{I}_1 \mapsto \mathfrak{I}$ such that:

$$\rho : \mathbf{1}_{\Phi_C},\ V^i_O \mapsto \Phi^i_O \text{ for any } j,\ V^i_K \mapsto \Phi^i_K \cup C^A \text{ for any } j,\ V^{ij} \mapsto \Phi^{ij},\ V^C \mapsto \mathfrak{I}.$$

The above substitution is appropriate to characterise the logic without interaction among the modal operators. To capture them we have to introduce two specialized substitutions based on it.

$$\begin{aligned} \rho^O &: \rho \cup V^i_O \mapsto \Phi^i_O \cup \Phi^{ji} \text{ for any } j \\ \rho^K &: \rho \cup V^i_K \mapsto \Phi^i_K \cup \Phi^{ij} \cup C^A \text{ for any } j \end{aligned}$$

Accordingly we have that two atomic ("world") labels t and s σ-unify iff there is a substitution ρ such that $\rho(t) = \rho(s)$, with the constraint that a label in V^{ij} cannot unify with another variable. We shall use $[s,t]\rho$ both to indicate that there is a substitution ρ for s and t, and the result of the substitution. The notion of σ-unification (or label unification) is extended to the case of composite labels (path labels) as follows:

$$[i,j]\sigma = k \text{ iff } \exists \rho : h(k) = \rho(h(i)) = \rho(h(j)) \text{ and } b(k) = [b(i), b(j)]\sigma.$$

Clearly σ is symmetric, i.e., $[i,j]\sigma$ iff $[j,i]\sigma$. Moreover this definition offers a flexible and powerful mechanism: it allows for an independent computation of the elements of the result of the unification, and variables can be freely renamed without affecting the result of a unification. Notice that a label W_i σ-unifies with every label. The intuition here is that W_i denotes the set of world in a Kripke model.

We are now ready to introduce the unifications corresponding to the modal operators at hand. The first unification is that for O_i.

$$[s,t]\sigma^O = ([h(s),h(t)]\rho^O,[h^1(s),h^1(t)]\sigma) \text{ iff } \min\{\ell(s),\ell(t)\} \geq 2 \text{ and } s,t \text{ are } ij\text{-ground}$$

Here we notice that the main structure is the structure for a $KD45$ modal operator ($\min\{l(s),l(t)\} \geq 2$) [8, 1]. However here we have that O_i is defined globally over the green states of an interpreted systems, so we can ignore the intermediate steps with the proviso that there are no variables of type IJ. Intuitively we can think of a variable of type IJ as the intersection of the worlds accessible from a given world using R^O_j and $\sim_i$; but in general such intersection can be empty, hence the proviso about the ij-groundness of the labels; moreover the restriction to ρ^O prevents unwanted unifications of labels in V^j_O and in V^i_K. According to the above definition we have that the labels $t = (O^j_1,(K^m_1,w_1))$ and $s = (o^j_1,w_1)$ σ^O-unify. In the same way t σ^O-unifies with $(ij_1,(k^n_1,w_1))$, but not with $(o^j_1,(IJ_1,w_1))$.

The following is the unification for K_i

$$[s,t]\sigma^K = ([h(s),h(t)]\rho^K;[h^1(s),h^1(t)]\sigma) \text{ iff } s \text{ and } t \text{ are } ij\text{-ground, and } ij\text{-compatible}$$

This is the condition for a unification corresponding to an equivalence relation [1, 9]. The important point here are that all the atomic symbols should be compatible. A label such as ij_n denotes a world in the intersection of the world accessible from a given world by R^O_j and $\sim_i$. But this means that it is also one of the world accessible from $\sim_i$.

Let us consider the label $t = (K^i_2,(ij_1,(K^i_1,w_1)))$. The result of the σ^K-unification of t and w_1 is w_1; similarly the unification of t and $s = (ij_2,w_1)$ is s. Notice that the label t does not σ^K-unify with (O^j_1,w_1).

$$\begin{aligned}[s,t]\sigma^* &= ([h(s),h(t)]\sigma,[h^1(s),h^1(t)]\sigma)\\ &\quad \text{iff } s,t \text{ are } ij\text{-compatible, and either } h^2(s) \text{ or } h^2(t) \text{ is } ij\text{-restricted}\end{aligned}$$

This unification is mainly designed for labels of type ij, and it corresponds to the unification for a $K45$ modal operator [8, 1]. A label (IJ_1,w_1) is intended to denote the equivalence class of type IJ associated to w_1. Since $\widehat{K}^j_i$ is not serial the equivalence class associated to a given world may be empty. However if we have that one of the labels in position 2 is a constant of type ij, then we are guaranteed that the equivalence class is not empty, and we can use the unification conditions for equivalence classes. Accordingly the labels $(IJ_1,(K^i_1,w_1))$ and (ij_1,w_1) σ^*-unify, and so do $(IJ_1,(ij_1,w_1))$ and $(IJ_2,(ij_2,w_1))$.

The above three unifications cover occurrences of sequences of compatible labels (relations). However we have to cover occurrences of interleaved labels. To this end we are going to define a recursive unification combining the unifications for the various operators. For convenience we introduce a unification corresponding to their simple combination. Hence $[s,t]\sigma^{\mathrm{DIS}}$ iff either $[s,t]\sigma$ or $[s,t]\sigma^{*}$ or $[s,t]\sigma^{O}$ or $[s,t]\sigma^{K}$. At this point the (recursive) unification for the logic DIS is defined as follows.

$$[s,t]\sigma_{\mathrm{DIS}} = \begin{cases} [s,t]\sigma^{\mathrm{DIS}} \\ [c^{n}(s),c^{m}(t)]\sigma^{\mathrm{DIS}} \end{cases}$$

where $w_0 = [s^{n}(s),s^{m}(t)]\sigma_{\mathrm{DIS}}$.

As we have seen the labels $t = (K^{i}_{2},(ij_1,(K^{i}_{1},w_1)))$ and $s = (O^{j}_{1},w_1)$ neither σ^{O}-unify, nor σ^{K}-unify. However $[t,s]\sigma_{\mathrm{DIS}} = (ij_1,w_1)$, and so the labels σ_{DIS}-unify. We can decompose the unification as follows: $[c^{3}(t),c^{2}(s)]\sigma^{K}$, where $c^{3}(t) = (K^{i}_{2},w_0)$, $c^{2}(s) = w_0$, and $w_0 = [s^{3}(t),s]\sigma_{\mathrm{DIS}}$. $s^{3}(t) = (ij_1,(K^{i}_{1},w_1))$.

Let us consider the following set of labels $\{t = (O^{j}_{1},w_1), s = (K^{i}_{1},w_1), r = (ij_1,w_1)\}$. Intuitively t,s,r denote, respectively, the set of worlds accessible from w_1 by the relation R^{O}_{j}, the set of worlds accessible from w_1 by the relation $\sim_i$, and a world in $R^{O}_{j} \cap \sim_i$. In general labels such as t and s should not unify. The intersection of their denotations may be empty, but in cases like the present one $h(s)$ and $h(t)$ can be mapped to a common label (i.e., ij_1) but with different substitution, and this is not permitted in σ_{DIS}. So we have to introduce a label-unification that takes care of context in which labels occur.

Let $\mathscr{L}$ be a set of labels (i.e., the labels occurring in a KEM-proof). Then $[s,t]\sigma^{\mathscr{L}}_{\mathrm{DIS}}$ iff

1. $[s,t]\sigma_{\mathrm{DIS}}$ or
2. $\exists k \in \mathscr{L}$, $\exists n,m \in Nat$ such that
   - $[s^{n}(s),k]\sigma^{\mathscr{L}}_{\mathrm{DIS}} = [s^{m}(t),k]\sigma^{\mathscr{L}}_{\mathrm{DIS}}$ and
   - $[c^{n}(s),c^{m}(t)]\sigma^{\mathscr{L}}_{\mathrm{DIS}}$ where $w_0 = [s^{n}(s),k]\sigma^{\mathscr{L}}_{\mathrm{DIS}}$

It is easy to verify that that the labels s and t described in the previous paragraph now $\sigma^{\mathscr{L}}_{\mathrm{DIS}}$ unify in the presence of the label r.

Inference Rules For the presentation of the inference rules of KEM we shall assume familiarity with Smullyan-Fitting α, β, ν, π unifying notation [6].

$$\frac{\alpha : t}{\begin{array}{c}\alpha_1 : t\\ \alpha_2 : t\end{array}} \quad \frac{A \wedge B : t}{\begin{array}{c}A : t\\ B : t\end{array}} \quad \frac{\neg(A \vee B) : t}{\begin{array}{c}\neg A : t\\ \neg B : t\end{array}} \quad \frac{\neg(A \to B) : t}{\begin{array}{c}A : t\\ \neg B : t\end{array}} \qquad (\alpha)$$

The α-rules are just the familiar linear branch-expansion rules of the tableau method. For the β-rules (formulas behaving disjunctively) we exemplify only the rules for implication.

$$\frac{\begin{array}{c}\beta : t\\ \beta^{c}_{i} : s\end{array} (i = 1,2)}{\beta_{3-i} : [t,s]\sigma^{\mathscr{L}}_{\mathrm{DIS}}} \quad \frac{\begin{array}{c}A \to B : t\\ A : s\end{array}}{B : [t,s]\sigma^{\mathscr{L}}_{\mathrm{DIS}}} \quad \frac{\begin{array}{c}A \to B : t\\ \neg B : s\end{array}}{\neg A : [t,s]\sigma^{\mathscr{L}}_{\mathrm{DIS}}} \qquad (\beta)$$

The β-rules are nothing but natural inference patterns such as Modus Ponens, Modus Tollens and Disjunctive syllogism generalised to the modal case. In order to apply such rules it is required that the labels of the premises unify and the label of the conclusion is the result of their unification.

$$\frac{\nu : t}{\nu_0 : (X_n, t)} \quad \frac{O_i A : t}{A : (O^i_n, t)} \quad \frac{K_i A : t}{A : (K^i_n, t)} \quad \frac{\widehat{K}^j_i A : t}{A : (IJ_n, t)} \tag{ν}$$

where O^i_n, K^i_n, and IJ_n are new labels.

$$\frac{\pi : t}{\pi_0 : (x_n, t)} \quad \frac{\neg O_i A : t}{A : (o^i_n, t)} \quad \frac{\neg K_i A : t}{A : (k^i_n, t)} \quad \frac{\neg \widehat{K}^j_i A : t}{A : (ij_n, t)} \tag{π}$$

where o^i_n, k^i_n, and ij_n are new labels. ν- and π- rules allow us to expand labels according to the intended semantics, where, with "new" we mean that the label does not occur previously in the tree.

$$\frac{}{A : t \quad | \quad \neg A : t} \tag{PB}$$

The "Principle of Bivalence" represents the semantic counterpart of the cut rule of the sequent calculus (intuitive meaning: a formula A is either true or false in any given world). PB is a zero-premise inference rule, so in its unrestricted version can be applied whenever we like. However, we impose a restriction on its application. Then PB can be only applied w.r.t. immediate sub-formulas of unanalysed β-formulas, that is β formulas for which we have no immediate sub-formulas with the appropriate labels in the branch (tree).

$$\begin{array}{c} A : t \\ \dfrac{\neg A : s}{\times} [\text{ if } [t, s]\sigma^{\mathscr{L}}_{\text{DIS}}] \end{array} \tag{PNC}$$

The rule PNC (*Principle of Non-Contradiction*) states that two labelled formulas are σ_L-complementary when the two formulas are complementary and their labels σ_L-unify.

Theorem 1. $\vdash_{\text{KEM}} A \iff IS \models A$

We sketch only the proof. The main idea is to define a Kripke model where the possible worlds are the labels ($\mathscr{L}$) occurring in a KEM-proof for A, where the accessibility relations are defined as follows: (i) $t \sim_i s$ iff $[(K^i_0, t), s]\sigma^{\mathscr{L}}_{\text{DIS}}$; (ii) $tR^O_i s$ iff $[(O^i_0, t), s]\sigma^{\mathscr{L}}_{\text{DIS}}$; and (iii) $(t, s) \in \sim_i \cap R^O_j$ iff $[(IJ_0, t), s]\sigma^{\mathscr{L}}_{\text{DIS}}$. For (i) and (ii) it is immediate to verify that the frame induced from the above construction is a frame for DIS. The result for (iii) depends on the definition of the substitution ρ where labels of type V^j_O and V^i_K can be mapped to labels in C^{ij}. Hence $t \sim_i (ij_n, t)$ and $tR^O_j(ij_n, t)$.

Proof Search Let $\Gamma = \{X_1, \ldots, X_m\}$ be a set of formulas. Then $\mathscr{T}$ is a KEM-*tree for* Γ if there exists a finite sequence $(\mathscr{T}_1, \mathscr{T}_2, \ldots, \mathscr{T}_n)$ such that (i) $\mathscr{T}_1$ is a 1-branch tree consisting of $\{X_1 : t_1, \ldots, X_m : t_m\}$; (ii) $\mathscr{T}_n = \mathscr{T}$, and (iii) for each $i < n$, $\mathscr{T}_{i+1}$ results from $\mathscr{T}_i$ by an application of a rule of KEM. A branch τ of a KEM-tree $\mathscr{T}$ of L-formulas is said to be σ_{DIS}-*closed* if it ends with an application of *PNC*, open otherwise. As usual with tableau methods, a set Γ of formulas is checked for consistency by constructing a KEM-tree for Γ. It is worth noting that each KEM-tree is a (class of) Hintikka's model(s) where the labels denote worlds (i.e., Hintikka's modal sets), and the unifications behave according to the conditions placed on the appropriate accessibility relations. Moreover we say that a formula A is a KEM-*consequence of a set of formulas* $\Gamma = \{X_1, \ldots, X_n\}$ ($\Gamma \vdash_{\text{KEM}} A$) if a KEM-tree for $\{X_1 : t, \ldots, X_n : t, \neg A : s\}$ is closed, where $s \in C^A$, and $t \in V^A$. The intuition behind this definition is that A is a consequence of Γ when we take Γ as a set of global assumptions [6], i.e., true in every world in a Kripke model.

We now describe a systematic procedure for KEM. First we define the following notions.

Given a branch τ of a KEM-tree, we shall call an L-formula $X : t$ *E-analysed in* τ if either (i) X is of type α and both $\alpha_1 : t$ and $\alpha_2 : t$ occur in τ; or (ii) X is of type β and one of the following conditions is satisfied: (a) if $\beta_1^C : s$ occurs in τ and $[t,s]\sigma_{\text{DIS}}^{\mathcal{L}}$, then also $\beta_2 : [t,s]\sigma_{\text{DIS}}^{\mathcal{L}}$ occurs in τ, (b) if $\beta_2^C : s$ occurs in τ and $[t,s]\sigma_{\text{DIS}}^{\mathcal{L}}$, then also $\beta_1 : [t,s]\sigma_{\text{DIS}}^{\mathcal{L}}$ occurs in τ; or (iii) X is of type ν and $\nu_0 : (m,t)$ occurs in τ for some $m \in \Phi_V$, of the appropriate type, not previously occurring in τ, or (iv) X is of type π and $\pi_0 : (m,t)$ occurs in τ for some $m \in \Phi_C$, of the appropriate type, not previously occurring in τ.

A branch τ of a KEM-tree is *E-completed* if every L-formula in it is E-analysed and it contains no complementary formulas which are not $\sigma_{\text{DIS}}^{\mathcal{L}}$-complementary. We shall say a branch τ of a KEM-tree *completed* if it is E-completed and all the L-formulas of type β in it either are analysed or cannot be analysed. We shall call a KEM-tree *completed* if every branch is completed.

The following procedure starts from the 1-branch, 1-node tree consisting of $\{X_1 : t, \ldots, X_m : s\}$ and applies the inference rules until the resulting KEM-tree is either closed or completed. At each stage of proof search (i) we choose an open non completed branch τ. If τ is not E-completed, then (ii) we apply the 1-premise rules until τ becomes E-completed. If the resulting branch τ' is neither closed nor completed, then (iii) we apply the 2-premise rules until τ becomes E-completed. If the resulting branch τ' is neither closed nor completed, then (iv) we choose an *LS*-formula of type β which is not yet analysed in the branch and apply *PB* so that the resulting *LS*-formulas are $\beta_1 : t'$ and $\beta_1^C : ti'$ (or, equivalently $\beta_2 : ti'$ and $\beta_2^C : t'$), where $t = t'$ if t is restricted (and already occurring when $h(t) \in \Phi_C$), otherwise t' is obtained from t by instantiating $h(t)$ to a constant not occurring in t; (v) ("Modal *PB*") if the branch is not E-completed nor closed, because of complementary formulas which are not $\sigma_{\text{DIS}}^{\mathcal{L}}$-complementary, then we have to see whether a restricted label unifying with both the labels of the complementary formulas occurs previously in the branch; if such a label exists, or can be built using already existing labels and the unification rules, then the branch is closed, (vi) we repeat the procedure in each branch generated by *PB*.

The above procedure is based on on a (deterministic) procedure working for *canonical* KEM-trees. A KEM-tree is said to be canonical if it is generated by applying the rules of KEM in the following fixed order: first the α-, ν- and π-rule, then the β-rule and *PNC*, and finally *PB*. Two interesting properties of canonical KEM-trees are (i) that a canonical KEM-tree always terminates, since for each formula there are a finite number of subformulas and the number of labels which can occur in the KEM-tree for a formula A (of DIS) is limited by the number of modal operators belonging to A, and (ii) that for each closed KEM-tree a closed canonical KEM-tree exists. Proofs of termination and completeness for canonical KEM-trees have been given in [8].

4 The Bit Transmission Problem

The bit-transmission problem [5] involves two agents, a *sender* S, and a *receiver* R, communicating over a faulty communication channel. The channel may drop messages but will not flip the value of a bit being sent. S wants to communicate some information—the value of a bit for the sake of the example—to R. We would like to design a protocol that accomplishes this objective while minimising the use of the communication channel.

One protocol for achieving this is as follows. S immediately starts sending the bit to R, and continues to do so until it receives an acknowledgement from R. R does nothing until it receives the bit; from then on it sends acknowledgements of receipt to S. S stops sending the bit to R when it receives an acknowledgement. Note that R will continue sending acknowledgements even after S has received its acknowledgement. Intuitively S will know for sure that the bit has been received by R when it gets an acknowledgement from R. R, on the other hand, will never be able to know whether its acknowledgement has been received since S does not answer the acknowledgement.

We assume *fairness* ([5], p.164) for the communication channel: every message that is repeatedly sent in the run is eventually delivered.

What we would like to do is to check mechanically that the protocol above guarantees that when sender receives the acknowledgement it then knows (in the information-theoretic sense defined in Section 2) that the receiver knows the value of the bit. In order to do this, first we model the scenario in the interpreted systems paradigm.

An interesting scenario arises when we assume that the agents may not behave as they are supposed to. For example, the receiver may not send an acknowledgement message when it receives a bit ([12]). We deal with this case by considering a new protocol which extends the original one.

Bit Transmission Problem — No Violations First of all we give an axiomatisation of the bit transmission problem (BTP). For a detailed discussion of the BTP in the framework of Deontic Interpreted Systems see [13, 12].

- Sender
  (S1) $\mathbf{recack} \rightarrow K_S\,\mathbf{recack}$
  (S2) $(\mathbf{bit} = n) \rightarrow K_S\,(\mathbf{bit} = n)$, for $n = 1,2$
- Receiver
  (R1) $\mathbf{recbit} \wedge (\mathbf{bit} = n) \rightarrow K_R\,(\mathbf{bit} = n)$, for $n = 1,2$
- Communication
  (C1) $\mathbf{recack} \rightarrow \mathbf{recbit}$

We can derive the following key property

$$\mathbf{recack} \wedge (\mathbf{bit} = n) \rightarrow K_S\,K_R\,(\mathbf{bit} = n) \qquad \text{for } n = 1,2$$

So, if an acknowledgement is received by the sender S, then S is sure that receiver R knows the value of the bit: although the communication channel is potentially faulty, if messages do manage to travel back and forth between the sender and receiver the protocol is strong enough to eliminate any uncertainty in the communication. Let us examine the KEM-proof for this property

1.	$\mathbf{recack} \rightarrow K_S\, \mathbf{recack} : W_1$	
2.	$(\mathbf{bit} = n) \rightarrow K_S\, (\mathbf{bit} = n) : W_1$	
3.	$\mathbf{recbit} \wedge (\mathbf{bit} = n) \rightarrow K_R\, (\mathbf{bit} = n) : W_1$	
4.	$\mathbf{recack} \rightarrow \mathbf{recbit} : W_1$	
5.	$\neg(\mathbf{recack} \wedge (\mathbf{bit} = n) \rightarrow K_S K_R\, (\mathbf{bit} = n)) : w_1$	
6.	$\mathbf{recack} : w_1$	5α
7.	$\mathbf{bit} = n : w_1$	5α
8.	$\neg K_S K_R\, (\mathbf{bit} = n)$	5α
9.	$\neg K_R\, (\mathbf{bit} = n) : (s_1, w_1)$	8π
10.	$K_S\, \mathbf{recack} : w_1$	$1,6\beta$
11.	$K_S\, (\mathbf{bit} = n) : w_1$	$2,7\beta$
12.	$\mathbf{recack} : (S_1, w_1)$	10ν
13.	$\mathbf{bit} = n : (S_2, w_1)$	11ν
14.	$\neg(\mathbf{recbit} \wedge (\mathbf{bit} = n)) : (s_1, w_1)$	$3,9\beta$
15.	$\neg\mathbf{recbit} : (s_1, w_1)$	$13,14\beta$
16.	$\mathbf{recbit} : (S_1, w_1)$	$4,12\beta$
17.	$\times$	$15,16 PNC$

Bit Transmission Problem — Violation by the Receiver Now we admit the possibility that the receiver, in violation of the protocol, may send acknowledgements without having received the bit. In this version, the axiom $(C1)$ does not hold. It is replaced by

$$O_R(\mathbf{recack} \rightarrow \mathbf{recbit}) \qquad (C1*)$$

which represents what holds when R is working correctly according to the protocol. All other parts of the formalisation are unchanged.

A particular form of knowledge still holds. Intuitively if S makes the assumption of R's correct functioning behaviour, then, upon receipt of an acknowledgement, it would make sense for S to assume that R does know the value of the bit. To model this intuition we use the operator $\widehat{K}_i^j$ "knowledge under the assumption of correct behaviour".

We now derive, given $(C1*)$ instead of $(C1)$

$$\mathbf{recack} \wedge (\mathbf{bit} = n) \rightarrow \widehat{K}_S^R K_R\, (\mathbf{bit} = n) \qquad \text{for } n = 1,2$$

We give a KEM-proof for it.

1.	$\mathbf{recack} \rightarrow K_S\, \mathbf{recack} : W_1$	
2.	$(\mathbf{bit} = n) \rightarrow K_S\, (\mathbf{bit} = n) : W_1$	
3.	$\mathbf{recbit} \wedge (\mathbf{bit} = n) \rightarrow K_R\, (\mathbf{bit} = n) : W_1$	
4.	$\mathbf{recack} \rightarrow \mathbf{recbit} : W_1$	
5.	$\neg(\mathbf{recack} \wedge (\mathbf{bit} = n) \rightarrow \widehat{K}_S^R K_R\, (\mathbf{bit} = n)) : w_1$	
6.	$\mathbf{recack} : w_1$	5α
7.	$\mathbf{bit} = n : w_1$	5α
8.	$\neg \widehat{K}_S^R K_R\, (\mathbf{bit} = n)$	5α
9.	$\mathbf{recack} \rightarrow \mathbf{recbit} : (O_1^R, w_1)$	4ν
10.	$K_S\, \mathbf{recack} : w_1$	$1,6\beta$

11.	$K_S(\mathbf{bit} = n) : w_1$	$2,7\beta$
13.	$\neg K_R(\mathbf{bit} = n) : (sr_1, w_1)$	8π
14.	$\neg(\mathbf{recbit} \wedge (\mathbf{bit} = n)) : (sr_1, w_1)$	$3,13\beta$
15.	$\mathbf{recack} : (S_1, w_1)$	10ν
16.	$\mathbf{bit} = n : (S_2, w_1)$	11ν
17.	$\neg\mathbf{recbit} : (sr_1, w_1)$	$14,16\beta$
18.	$\mathbf{recbit} : (sr_1, w_1)$	$9,15\beta$
19.	$\times$	$17,18PNC$

The only step that deserves some attention is step 18. This step is the consequence of a β-rule on 9 and 15. The labels of the relevant formulas are (O_1^R, w_1) and (S_1, w_1). Normally such labels do not unify, and the β-rule would not be applicable. However, thanks to the presence of (sr_1, w_1) in the tree, the labels of 9 and 15 do $\sigma_{\text{DIS}}^{\mathcal{L}}$-unify.

5 Conclusions

In this paper we presented a tableaux-based system for proving properties of a system whose properties can be expressed in an epistemic-deontic language. The tableaux system was used to prove properties about variations of the bit transmission problem, a widely explored protocol to reason about information exchange in communication protocols. The results obtained confirmed results obtained by model checking techniques presented elsewhere [11]. Further work involves an implementation of this method so that experimental results based on larger scenarios can be evaluated.

References

[1] A. Artosi, P. Benassi, G. Governatori, and A. Rotolo. Shakespearian modal logic: A labelled treatment of modal identity. In M. Kracht, M. de Rijke, H. Wansing, and M. Zakharyaschev, editors, *Advances in Modal Logic. Volume 1*, pages 1–21. CSLI Publications, Stanford, 1998. 342, 344

[2] A. Artosi, G. Governatori, and A. Rotolo. Labelled tableaux for non-monotonic reasoning: Cumulative consequence relations. *Journal of Logic and Computation*, 12(6):1027–1060, December 2002. 342

[3] B. Chellas. *Modal Logic: An Introduction.* Cambridge University Press, Cambridge, 1980. 340

[4] M. D'Agostino and M. Mondadori. The taming of the cut. *Journal of Logic and Computation*, 4:285–319, 1994. 342

[5] R. Fagin, J. Y. Halpern, Y. Moses, and M. Y. Vardi. *Reasoning about Knowledge.* MIT Press, Cambridge, 1995. 339, 340, 348

[6] M. Fitting. *Proof Methods for Modal and Intuitionistic Logics.* Reidel, Dordrecht, 1983. 345, 347

[7] D. M. Gabbay and G. Governatori. Fibred modal tableaux. In D. Basin, M. D'Agostino, D. Gabbay, S. Matthews, and L. Viganó, editors, *Labelled Deduction*, volume 17 of *Applied Logic Series*, pages 163–194. Kluwer, Dordrecht, 2000. 342

[8] G. Governatori. Labelled tableaux for multi-modal logics. In P. Baumgartner, R. Hähnle, and J. Posegga, editors, *Theorem Proving with Analytic Tableaux and Related Methods*, volume 918 of *LNAI*, pages 79–94, Berlin, 1995. Springer-Verlag. 342, 344, 347

[9] G. Governatori. *Un modello formale per il ragionamento giuridico*. PhD thesis, CIRFID, University of Bologna, Bologna, 1997. 342, 344

[10] G. E. Hughes and M. J. Cresswell. *A New Introduction to Modal Logic*. Routledge, New York, 1996. 340

[11] A. Lomuscio, F. Raimondi, and M. Sergot. Towards model checking interpreted systems. In *Proceedings of Mochart — First International Workshop on Model Checking and Artificial Intelligence*, 2002. 350

[12] A. Lomuscio and M. Sergot. Violation, error recovery, and enforcement in the bit transmission problem. In *Proceedings of DEON'02*, London, May 2002. 339, 340, 348

[13] A. Lomuscio and M. Sergot. Deontic interpreted systems. *Studia Logica*, 75, 2003. 339, 340, 341, 348

[14] J.-J. C. Meyer and W. Hoek. *Epistemic Logic for AI and Computer Science*, volume 41 of *Cambridge Tracts in Theoretical Computer Science*. Cambridge University Press, 1995. 339

[15] A. S. Rao and M. P. Georgeff. Decision procedures for BDI logics. *Journal of Logic and Computation*, 8(3):293–343, June 1998. 339

Decidability of Propositionally Quantified Logics of Knowledge

Tim French

Murdoch University and the University of Western Australia
Perth, W.A., Australia
tim@csse.uwa.edu.au

Abstract. Logics of knowledge have important applications for reasoning about security protocols and multi-agent systems. We extend the semantics for the logic of necessity with local propositional quantification $\mathcal{L}_{(\Box,\exists,\exists_1,\ldots\exists_k)}$ introduced in [4] to allow reasoning about knowledge in more general (non-hierarchical) systems. We show that these new semantics preserve the properties of knowledge in a multi-agent system, give a significant and useful increase in expressivity and most importantly, have a decidable satisfiability problem. The new semantics interpret propositional (local and non-local) quantification with respect to bisimulations, and the satisfiability problem is shown to be solvable via an embedding into the temporal logic, QCTL.

1 Introduction

Logics of knowledge [5] have important applications for reasoning about security protocols and multi-agent systems. Such a logic allows you to formalize what facts (represented by propositions) are known by which agents or whether one agent knows if another agent knows (or considers possible) some fact. Formalizations of knowledge also allow us to to represent such notions as common knowledge (every agent knows, and every agent knows every agent knows, and so on), and distributed knowledge (what a group of agents could infer if they shared their knowledge). Propositional quantification in modal logics [6, 9] has often been considered as a way of increasing expressivity.

Recently several extensions have been investigated which allow quantification of propositions. This allows us to reason about an agent's knowledge independent of individual propositions. Being able to quantify over propositions (so that they represent arbitrary facts) allows us to examine and verify protocols independent of the context. However the resulting increase in expressive power extends much further and, in the most general case, the language has been shown to be highly undecidable. This problem has been addressed in several ways, including by weakening the properties of knowledge and by enforcing various semantic structures on the knowledge of agents.

In this paper we present a third approach. Rather than restricting any semantic properties of the system, we generalize the notion of propositional quantification to apply to bisimulations of the model. We show that interpretations of

T.D. Gedeon and L.C.C. Fung (Eds.): AI 2003, LNAI 2903, pp. 352–363, 2003.

knowledge, common knowledge and distributed knowledge are unaffected by this change. Furthermore the resulting language will be shown to be decidable. This is significant as it allows us to express a wide range of second order properties which are of practical value (for example "agent 1 knows strictly more than agent 2 and 3 combined"), whilst avoiding any undecidable second-order properties. To the author's knowledge, this combination of expressivity and decidability is not present in any previous languages.

This paper will be organized as follows. We will first introduce the basic logic of knowledge, its syntax and semantics. We will then generalize this to a logic of local propositions [3]. The new notion of quantification will be introduced and we will show that interpretations of knowledge are invariant under this change. We will discuss several formalizations of these semantics and show they are equivalent. We will also explore the expressive power of the language. The next part of the paper will present a proof of decidability for the language. This proof will require us to examine the temporal logic QCTL (since the satisfiability problem is reduced to the satisfiability problem for QCTL). This language is shown to be decidable in [8]. We will prove that this reduction is both sound and correct.

2 The Language

The language $\mathrm{L}_{\mathrm{k}}^{\mathrm{CD}}$, for some natural number k, can express properties of knowledge for k agents and is built from the following abstract syntax:

$$\alpha := x \in \mathcal{V} \mid \alpha_1 \wedge \alpha_2 \mid \neg\alpha \mid K_i\alpha \mid C_G\alpha \mid D_G\alpha$$

where i is the label of an agent and ranges from 1 to k, and $G \subseteq \{1, ...k\}$ represents a set of agents. The interpretations of these formulas are as follows:

- $x \in \mathcal{V}$ are the propositional variables. These are used to represent properties of the system.
- $\wedge$ and $\neg$ are the standard boolean operators.
- $K_i\alpha$ represents the statement, "agent i knows α is true".
- $C_G\alpha$ represents the statement, "it is common knowledge to all agents in G that α is true"
- $D_G\alpha$ represents the statement, "if all the agents in G were to combine their knowledge, they would be able to deduce α".

For details on common semantics for logics of knowledge see [5] We assume that each agent has a set of possible states, and each of these states correspond to a partition of the set of possible worlds. An agent, i, cannot distinguish two worlds if the agent's local states in those two worlds are the same (the worlds are i-local).

We suppose that each agent, i, has a set of local states $\mathcal{L}_i$. Rather than using a distinct label for each world, we let each world be a tuple of agent's local states, along with the set of propositions that are true at that world. We say

$w = (w_0, w_1, ..., w_k)$ is a world where $w_0 \subseteq \mathcal{V}$ and for each i, $w_i \in \mathcal{L}_i$. The model M, w is simply a set of worlds, with one world specified as "this world", (i.e. $M \subseteq \wp(\mathcal{V}) \times \mathcal{L}_1 \times ... \times \mathcal{L}_k$) and we define the satisfiability of a formula at a given world, w, as follows:

$$M, w \models_K x \iff x \in w_0 \quad (1)$$
$$M, w \models_K \neg\alpha \iff M, w \not\models_K \alpha \quad (2)$$
$$M, w \models_K \alpha \wedge \beta \iff M, w \models_K \alpha \text{ and } M, w \models_K \beta \quad (3)$$
$$M, w \models_K K_i\alpha \iff M, w' \models_K \alpha \text{ for all } w' \text{ where } w_i = w_i' \quad (4)$$
$$M, w \models_K C_G\alpha \iff M, w' \models_K \alpha \text{ for all } w' \text{ where } w \cong_G w' \quad (5)$$
$$M, w \models_K D_G\alpha \iff M, w' \models_K \alpha \text{ for all } w' \text{ where } \forall i \in G, w_i = w_i', \quad (6)$$

where $w \cong_G w'$ is the smallest relation recursively defined by

$$w \cong_G w' \iff w = w' \text{ or } \exists u \in M,\ \exists i \in G \text{ such that } w_i = u_i \text{ and } u \cong_G w'.$$

We say α is a validity of $\mathrm{L}_\mathrm{k}^\mathrm{CD}$ if $M, w \models \alpha$ for all models M, w, and α is satisfiable if $\neg\alpha$ is not a validity. We will examine the semantic interpretation for these operators in more detail in the following section.

3 Semantics for Propositional Quantification

We are now ready to introduce propositional quantification into the language of local propositions. We could simply add propositional quantification to $\mathrm{L}_\mathrm{k}^\mathrm{CD}$, but this would unnecessarily complicate the language. We use the notion of local quantifiers to express the properties of knowledge. These were introduced in [3].

The logic of quantified local propositions, (QL_k[1]) is as follows:

$$\alpha \ := \ x \in \mathcal{V} \mid \alpha_1 \wedge \alpha_2 \mid \neg\alpha \mid \exists x\alpha \mid \exists_i x\alpha \mid \Box\alpha.$$

The new operators are existential propositional quantification ($\exists$), existential i-local propositional quantification and necessity. The formula $\exists x\alpha$ states " there is an interpretation of the variable x that makes α true"; the formula $\exists_i x\alpha$ states "there is some interpretation of the variable x that is consistent with agent i's local state, and which makes α true"; and the formula $\Box\alpha$ states "in all worlds α is true". We use $\exists\{x_1, ..., x_n\}\alpha$ as an abbreviation for $\exists x_1 ... \exists x_n\alpha$

This language appears to be very different from the language given above, however we will show that every formula of $\mathrm{L}_\mathrm{k}^\mathrm{CD}$ is expressible in QL_k. To give the formal semantic interpretations of the operators $\Box$, $\exists$ and $\exists_i$ we will require the following definition.

Definition 1. *Given some model M, w we say a model M', w' is an X-*variant *of M if there is some relation $B \subseteq M \times M'$ such that*

[1] While the language is identical to $\mathcal{L}_{(\Box, \exists, \exists_1, ... \exists_k)}$ of [3], we will use QL_k so as to distinguish the semantics.

1. $\forall u \in M\ \exists u' \in M'$ *with* $(u, u') \in B$, *and* $\forall u' \in M'\ \exists u \in M$ *with* $(u, u') \in B$,
2. $(w, w') \in B$, *and*
3. *for all* $(u, u') \in B$
   (a) $u_0 \backslash X = u'_0 \backslash X$,
   (b) $\forall i \in I\ \forall v \in M,\ u_i = v_i \Longrightarrow \exists v' \in M'$ *with* $(v, v') \in B$ *and* $u'_i = v'_i$,
   (c) $\forall i \in I\ \forall v' \in M',\ u'_i = v'_i \Longrightarrow \exists v \in M$ *with* $(v, v') \in B$ *and* $u_i = v_i$.

Given $i \in I$ *we say* M', w' *is an* i-local X-variant of M, w *if* M', w' *is an* X-*variant of* M, w *and for all* $x \in X$, *for all* $u' \in M'$, $x \in u'_0$ *if and only if for all* $v' \in M'$ *with* $v'_i = u'_i$ *we have* $x \in v'_0$. *If* M', w' *is a* $\emptyset$-*variant of* M, w *we say that* M', w' *is* bisimilar *to* M, w.

Essentially, we allow quantification to not only apply to the given model, but also to models that are bisimilar to the given model. Since the semantic description of states forces all possibility relations in a model to be equivalence relations, we can assume the basic axioms of knowledge are preserved. However we should note that we are using a restricted set of bisimulations, since every model is bisimilar to a tree. We are now able to give the semantic interpretations for the new operators.

$$M, w \models_L \Box\alpha \Longleftrightarrow M, w' \models_L \alpha \text{ for all } w' \in M$$
$$M, w \models_L \exists_i x\alpha \Longleftrightarrow M', w' \models_L \alpha \text{ where } M \text{ is an } i\text{–local } \{\text{x}\}\text{–variant of } M, w$$
$$M, w \models_L \exists x\alpha \Longleftrightarrow M', w' \models_L \alpha \text{ where } M \text{ is an } \{x\}\text{-variant of } M, w$$

Alternative semantic interpretations have been offered for logics of knowledge with propositional quantification. In [3], two alternatives based on local propositions were considered. The first, referred to as the *strong* semantics considered propositional quantification over a fixed set of worlds. That is

$$M, w \models_S \exists x\alpha \Longleftrightarrow M', w' \models_S \alpha \tag{7}$$

where for all $w \in M$ there is some $w' \in M'$ such that $w_0\backslash\{x\} = w'_0\backslash\{x\}$ and $w_i = w'_i$ for $i > 0$. These semantics were shown to be too strong, being expressively equivalent to full second order logic.

In the same paper, a set of *weak* semantics were also considered. We will not go into the formal definitions here, but the basic idea was to restrict the interpretation of atoms over sets of worlds. This interpretation was shown to be unable to fully express the knowledge operators required by logics of knowledge.

Finally in [4], a restricted version of the strong semantics were considered where the knowledge relations formed linear hierarchy. That is, for any pair of agents, one knew strictly more than the other. This language was shown to be decidable, and in the same paper a complete axiomatization was given.

The main advantage of the semantics presented here is that we are now able to reason about non-linear hierarchies of knowledge, without losing decidability or any of the necessary expressiveness.

4 Definability

The logic QL_k combines the power of logics of knowledge with propositional quantification. We will first show that our intuitions of propositional quantification are preserved. For example, if α is a validity, $\forall x\alpha$ should also be a validity, (this follows trivially from the definition), and if the atom, x, does not appear in α then $\alpha \to \forall x\alpha$ should be a validity. This follows form the following lemma.

Lemma 1. *Suppose that (M, w) is a model, and (N, v) is a model bisimilar to (M, v). The for all formulas α,*

$$M, w \models_L \alpha \Longleftrightarrow N, v \models_L \alpha. \tag{8}$$

Proof. This can be shown by induction over the complexity of formulas. By definition 1, we know $w_0 = v_0$, so the propositional case is trivial. Likewise the inductive steps for $\neg$ and $\wedge$ are trivial. The inductive steps for the quantifiers $\exists$ and $\exists_i$ follow directly from the fact that bisimilarity is an equivalence relation. This leaves the $\Box$ operator.

By the induction hypothesis we suppose for all bisimilar models $M, w \models_L \alpha$ if and only if $N, t \models_L \alpha$. The result follows from the first condition of definition 1, which requires the relation to be defined for every element of M and every element of N. If there were some $u \in M$ such that $M, u \not\models_L \alpha$, then by the induction hypothesis there must be some $v \in N$ where $N, v \not\models_L \alpha$, and vice-versa. Therefore $M, w \models_L \Box\alpha$ if and only if $N, t \models_L \Box\alpha$ and the proof is complete.

It follows that the quintessential properties of propositional quantification are retained in QL_k. We will now show that the standard knowledge operators can be expressed in QL_k, and retain their semantic interpretation. As was shown in [3], the formulas $K_i(\alpha)$, $C_G(\alpha)$, and $D_G(\alpha)$ can be expressed in terms of these local quantifier and the $\Box$ operator. Let $x, x_1, ..., x_k$ be variables that do not appear in α.

$$K_i(\alpha) = \exists_i x(x \wedge \Box(x \to \alpha)) \tag{9}$$

$$C_G(\alpha) = \exists x(x \wedge \bigwedge_{i\in G} \exists_i y \, \Box(y \leftrightarrow x) \wedge \Box(x \to \alpha)) \tag{10}$$

$$D_G(\alpha) = \exists x[\exists_i x_i]_{i\in G} \left[(x \wedge \bigwedge_{i\in G} x_i) \wedge \Box((\bigwedge_{i\in G} x_i) \to x) \wedge \Box(x \to \alpha)\right] \tag{11}$$

We will now show that as with the semantics discussed in [3], the interpretation of knowledge is retained.

Lemma 2. *For every model M, w,*

$$M, w \models_K K_i\alpha \Longleftrightarrow M, w \models_L K_i\alpha \tag{12}$$

$$M, w \models_K C_G\alpha \Longleftrightarrow M, w \models_L C_G\alpha \tag{13}$$

$$M, w \models_K D_G\alpha \Longleftrightarrow M, w \models_L D_G\alpha \tag{14}$$

Proof. Suppose that $M, w \models_K K_i\alpha$. Then for all worlds $v \in M$, where v is i-local to w, $M, v \models_K \alpha$. Let x be true at exactly these worlds, v. Then x is i-local so $M, w \models \exists_i x(x \wedge \Box(x \rightarrow \alpha))$.

Now suppose that $M, w \models_L K_i(\alpha)$. Then there is some i-local x-variant, (N, t), of (M, s) such that $N, t \models_L x \wedge \Box(x \rightarrow \alpha)$. Therefore for every world $u \in N$ that is i-local to t, we must have $x \in u_0$ and hence, $N, u \models \alpha$. From Lemma 1, it follows that for all $v \in M$, i-local to w, we must have $M, v \models \alpha$ (since we assume that x is not a variable of α. Therefore $M, v \models K_i\alpha$, and we have shown the equivalence (12) holds.

We will omit the proofs of the following remaining two equivalences, as they are similar.

We will now briefly consider the expressive power of our logic. From the above proof, anything that is expressible in $\mathrm{L}_\mathrm{k}^{\mathrm{CD}}$ is expressible in QL_k. For examples of the expressive power of $\mathrm{L}_\mathrm{k}^{\mathrm{CD}}$, see [5]. One of the significant advantages of QL_k is the power to reason about arbitrary hierarchies of knowledge. In the standard logics of knowledge we can express concepts, such as "If agent i knows p is true, then agent j knows p is true", (i.e. $K_i p \rightarrow K_j p$). In QL_k we can express such properties independent of the proposition p, so the hierarchy applies to all formulas expressible in the language, (i.e. $\forall x(K_i x \rightarrow K_j x)$). Furthermore, we can reason about non-linear hierarchies of knowledge. The formula

$$\exists x(K_1 x \wedge \neg D_{2,3} x) \wedge \exists x(K_2 x \wedge \neg D_{1,3} x) \wedge \exists x(K_3 x \wedge \neg D_{1,2} x) \qquad (15)$$

expresses the property that there are three agents, and if any two agents were to combine their knowledge, they would still not know strictly more than the third agent. Such a property could be important for ensuring the security of shared network resources, or the fairness of three player games.

5 Decidability

We can now present the proof of decidability for the language. We will do this via a translation into the temporal logic QCTL [1, 2]. This is because the decidability of QCTL is a very complicated result [8, 7], and we would like to avoid repeating it. The proof of decidability for QCTL involves a new kind of tree automata (amorphous Street tree automata) which act on ω-trees. (In fact, they act on the set of bisimulations of an ω-tree). There is a deterministic translation from a formula α, of QCTL to an amorphous automata which accepts exactly the models of α. Therefore the emptiness of the automata is equivalent to the unsatisfiability of α. Since the emptiness of an amorphous automata can be determined, this solves the satisfiability for QCTL.

The decidability process described above is particularly appropriate for QCTL since trees are a natural model for branching temporal logics. However the models of QL_k involve multiple equivalence relations interacting in an arbitrary manner, so it is not immediately apparent how we could define such a translation. The answer comes from the expressive power of QCTL. Every model of QL_k can be "untangled" along the relations to give an ω-tree. We will find that QCTL has the expressive power to interpret formulas on this tree with respect to the original structure. That is every relation can be syntactically "entangled" back into an equivalence relation. This way we simplify the semantic specification of a formula, but increase the complexity of the syntax to preserve validities in the language.

We will now briefly describe the logic QCTL. The syntax is given as

$$\alpha ::= x \mid \neg\alpha \mid \alpha_1 \vee \alpha_2 \mid \mathrm{AX}\alpha \mid \mathrm{AG}\alpha \mid \exists x\alpha \tag{16}$$

The formula $\mathrm{AX}\alpha$, states that alpha is true at the next moment of time for all possible futures, $\mathrm{AG}\alpha$ states that α is true for all moments of time from here on, regardless of future. The abbreviations $\wedge, \rightarrow, \leftrightarrow$ are defined as usual, and we define $\mathrm{EX}\alpha$ to be $\neg\mathrm{EX}\neg\alpha$, $\mathrm{EG}\alpha$ to be $\neg\mathrm{AF}\neg\alpha$ and $\mathrm{EF}\alpha$ to be $\neg\mathrm{AG}\neg\alpha$. To give the semantics for CTL* we define *$\mathcal{V}$-labeled Kripke frames*:

Definition 2. *A Kripke frame is a tuple (S, R) where*

1. *S is a nonempty set of states, or moments.*
2. *$R \subseteq S^2$ is a total binary relation.*

A $\mathcal{V}$-labeled Kripke frame is a Kripke frame with a valuation $\pi : S \longrightarrow \wp(\mathcal{V})$.

Let $T = (S, R, \pi)$ be a $\mathcal{V}$-labeled Kripke frame, and let R^* be the reflexive, transitive closure of R. We interpret a formula α of QCTL with respect to a $\mathcal{V}$-labeled Kripke frame T and a state $s \in S$. We write $T, s \models \alpha$ where:

$$\begin{aligned}
T, s \models_T x &\iff x \in \pi(s)\\
T, s \models_T \neg\alpha &\iff T, s \not\models_T \alpha\\
T, s \models_T \alpha \vee \beta &\iff T, s \models_T \alpha \text{ or } T, s \models_T \beta\\
T, s \models_T \mathrm{AX}\alpha &\iff \text{for all } t \text{ where } (s,t) \in R,\ T, t \models_T \alpha\\
T, s \models_T \mathrm{AG}\alpha &\iff \text{for all } t \text{ where } (s,t) \in R^*,\ T, t \models_T \alpha\\
T, s \models_T \exists x\alpha &\iff \text{ there is some } x\text{-variant } T', s' \text{ of } T, s \text{ where } T', s' \models_T \alpha.
\end{aligned}$$

For the semantic interpretation of $\exists x\alpha$, we require the following definition:

Definition 3. *Given $X \subseteq \mathcal{V}$, $(T, s_0) = (S, R, \pi, s_0)$ is an X-variant of $(T', s') = (S', R', \pi', s'_0)$ if there exists some relation $B \subseteq S \times S'$ with $(s_0, s'_0) \in B$ and for all $(s, s') \in B$:*

1. *$\pi(s)\backslash X = \pi'(s')\backslash X$.*
2. *For all $t \in S$ such that $(s,t) \in R$, there exists $t' \in S'$ with $(s', t') \in R'$ such that $(t, t') \in B$.*
3. *For all $t' \in S'$ such that $(s', t') \in R'$, there exists $t \in S$ with $(s,t) \in R$ such that $(t, t') \in B$.*

We say α is a validity if for all models, (T, s), we have $T, s \models_T \alpha$, and α is satisfiable if $\neg\alpha$ is not a validity. It is important to note that the truth of formulas of QCTL with respect to a given model is invariant under bisimulation. As all QCTL models are bisimilar to tree, from now one we will consider all models of QCTL to be trees where the relation R acts as the parent-child relation.

The definitions of x-variant are quite similar for QCTL and QL_k. They both use the basic idea of a bisimulations [10, 11] to access the power of propositional quantification whilst abstracting out the non-modal properties of the model. For more details on QCTL see [7].

Our approach will involving translating the satisfiability problem for a formula of QL_k into the satisfiability problem for a formula of QCTL. We must first describe the translation, and secondly show that it preserves satisfiability and unsatisfiability. Let $\mathcal{M}_L$ be the set of models for QL_k and let $\mathcal{M}_T$ be the set of models for QCTL.

Definition 4. *Given some formula α, we define translation $\phi : \mathcal{M}_L \longrightarrow \mathcal{M}_T$ as follows: Let $I_\alpha = \{x_1, ...x_k\}$ be a set of propositions not appearing in α. Given $(M, s) \in \mathcal{M}_L$ Let $(M, s)^\phi$ be the model (S, R, π, s) where*

- $S = \{sw \mid w \in M*\} \subseteq M^*$.
- $\pi(s) = s_0$
- *For all* $tu \in S$ *where* $t \in S$ *and* $u \in M$, *for all* $v \in M$, $\pi(tuv) = (v_0 \backslash I_\alpha) \cup \{x_i \mid u_i = v_i\}$.
- $R = \{(t, tu) \mid t \in S, u \in M\}$.

The QCTL model $(M, s)^\phi$ is similar to the model (M, s), except where (M, s) has k different relations, $(M, s)^\phi$ only has the one. To compensate for this each node is labeled with the propositions I_α, which indicate which local states a node shares with its parent.

We will now define a syntactic translation $*$ on QL_k, such that α is satisfiable for QL_k if and only if α^* is satisfiable for QCTL. Essentially what we are doing is untangling the semantics of QL_k to be represented by a tree, and entangling the syntax of QL_k to retain the necessary properties of knowledge.

Definition 5. *We define the translation $* : QL_k \longrightarrow QCTL$ in several stages. Let α be some formula of QL_k and we define $\mathcal{V}_\alpha = \{y_\beta | \beta \subseteq \alpha\}$ where for each sub-formula $\beta \subseteq \alpha$, y_β is some variable not appearing in either α or I_α. For each sub-formula β of α we define β' recursively as follows:*

$$
\begin{aligned}
x' &= AG(y_x \leftrightarrow (x)) \\
(\neg\gamma)' &= AG(y_\beta \leftrightarrow \neg y_\gamma) \wedge \gamma' \\
(\gamma_1 \wedge \gamma_2)' &= AG(y_\beta \leftrightarrow (y_{\gamma_1} \wedge y_{\gamma_2})) \wedge \gamma_1' \wedge \gamma_2' \\
(\Box\gamma)' &= AG(y_\gamma) \leftrightarrow AG(y_\beta) \wedge EF(y_\beta) \rightarrow AG(y_\gamma) \wedge \gamma' \\
(\exists x\gamma)' &= \exists x(AG(y_\beta \leftrightarrow y_\gamma) \wedge \gamma') \\
(\exists_i x\gamma)' &= \exists x(\gamma' \wedge AG((x \rightarrow AX(x_i \rightarrow x)) \wedge (EX(x_i \wedge x) \rightarrow x) \wedge (y_\beta \rightarrow y_\gamma)))
\end{aligned}
$$

Finally let $\alpha^* = \exists\mathcal{V}_\alpha(\alpha' \wedge y_\alpha)$.

This definition implements the semantic interpretations for each operator. The states at which any sub-formula, β, is true is marked with the proposition y_β and any sub-formula referring to β is then interpreted with respect to that proposition. This allows us to implement the semantic interpretation for QL_k. Particularly, the operators of QCTL are naturally anti-symmetric, irreflexive, and transitive, so the concept of an equivalence relation is enforced in the translation of $\exists_i x\beta$, which considers all states in the model which can be reached by passing only through states with x_i in their label.

Lemma 3. $M, s \models_L \alpha \Longrightarrow (M, s)^\phi \models_T \alpha^*$.

Proof. We first must show that for all $\beta \subseteq \alpha$, the formula β' is satisfiable for QCTL. This can be shown by induction. This is trivial in the case that $\beta \in \mathcal{V}$, so suppose $\beta = \mathcal{O}\gamma$, where $\mathcal{O}$ is some operator, and γ' is satisfiable. Then β' simply dictates that the interpretation of y_β is defined with respect to the semantic interpretation of $\mathcal{O}$ and y_γ. Since propositional quantification is interpreted with respect to bisimulations for QCTL, it follows that α' is not only satisfiable, but also $\exists\mathcal{V}_\alpha\alpha'$ is a validity.

Let $(M, s)^\phi = (S, R, \pi, s)$. We must show that if $(M, s)^\phi \models_L \alpha'$, then for all $\beta \subseteq \alpha$, $M, u \models_L \beta$ if and only if $y_\beta \in \pi(wu)$ (where $w \in M^*$, $u \in M$). Again, this can be shown by a simple induction over the complexity of α where the base case ($\beta \in \mathcal{V}$) is trivial, as are the cases for $\neg$, $\wedge$ and $\exists x$.

The leaves the operators $\Box$ and $\exists_i$, so suppose that for all models, M, s, for all $u \in M$, $M, t \models_L \beta$ then $y_\beta \in \pi(wu)$. In the case of $\Box\beta$, $M, s \models_L \Box\beta$ if and only if β is true at every world in M. By the induction hypothesis it follows that $y_\beta \in \pi(u)$ for all $u \in M$. As the definition of $(\Box\beta)'$ defines the interpretation of $y_{\Box\beta}$ with respect to the entire tree it follows that if $M, u \models_L \Box\beta$ then $y_{\Box\beta} \in \pi(wu)$.

In the case of $\exists_i x\beta$, $M, s \models_L \exists_i x\beta$ if and only if there is i-local x-variant, (N, t), of (M, s) such that $N, t \models_L \beta$. We state without proof (though it is easy to show) that if (N, t) is an i-local x-variant of (M, s), then $(N, t)^\phi$ is an x-variant of $(M, s)^\phi$. From the induction hypothesis $(N, t)^\phi \models_T y_\beta$. The definition of ϕ will ensure that all worlds which share an i-local state will be mapped to a subtree, upon which x_i is always true. The definition of $(\exists_i x\beta)'$ will ensure that the interpretation of x will not vary on this subtree. Therefore $(N, t)^\phi$ is an x-variant of $(M, s)^\phi$ which satisfies

$$(\gamma' \wedge \mathrm{AG}((x \to \mathrm{AX}(x_i \to x)) \wedge (\mathrm{EX}(x_i \wedge x) \to x) \wedge (y_\beta \to y_\gamma))), \qquad (17)$$

and it follows from the definition of an x-variant that for all $u \in M$, if $M, u \models_L \exists_i x\beta$, then $y_{\exists_i x\beta} \in \pi(wu)$.

This completes the induction, so it follows that if $M, s \models_L \alpha$, then $(M, s)^\phi \models_T \alpha'$ implies that $y_\alpha \in \pi(s)$ completing the proof.

We now know that if α is satisfiable, then α^* is also satisfiable. To prove that the satisfiability problem for QL_k is decidable, we must also show that if α^* are satisfiable in QCTL, then α is satisfiable in QL_k. To do this we will show that from any QCTL model that satisfies α^* we can generate a QL_k model that satisfies α. We first define a map from $\mathcal{M}_T$ to $\mathcal{M}_L$.

Definition 6. *We define the translation $\kappa^\alpha : \mathcal{M}_T \rightarrow \mathcal{M}_L$ as follows. We suppose that there are a set of atoms I_α not appearing in α, that agree with the set of atoms used in the translation $*$, and $T, s = (S, R, \pi, s)$ is a model of QCTL.*

We define a function $\kappa : S \rightarrow \wp(\mathcal{V}) \times S^k$ recursively by

- $\kappa(s) = (\pi(s), s, ..., s)$
- *If $\kappa(t) = (a, \tau_1, ..., \tau_k)$, and $(t, u) \in R$, then $\kappa(u) = (\pi(u), \sigma_1, ..., \sigma_k)$ where for all $i \in I$,*
  - *if $x_i \in \pi(u)$ then $\sigma_i = \tau_i$*
  - *if $x_i \notin \pi(u)$ then $\sigma_i = u$.*

We let $\kappa(S)$ be the range of κ, and define $\kappa^\alpha(T, s) = (\kappa(S), \kappa(s))$.

This translation essentially builds up a QL_k model by taking the reflexive, symmetric, transitive closure of the relations represented by the atoms in I_α.

Lemma 4. $T, s \models_T \alpha^* \Longrightarrow \kappa^\alpha(T, s) \models_L \alpha$.

Proof. Suppose that $(T, s) \models_T \alpha^*$ and $\mathcal{V}_\alpha = \{y_\beta | \beta \subseteq \alpha\}$. Then there is a $\mathcal{V}_\alpha$-variant, T', s' of T, s such that $T', s' \models_T \alpha' \wedge y_\alpha$. We state the following fact without proof:

> If (M, s) and (N, t) are bisimilar models of QCTL, then $\kappa(M, s)$ is bisimilar to $\kappa(N, t)$ with respect to QL_k.

This is easy to see from the construction of κ. Since the atoms of $\mathcal{V}_\alpha$ do not appear in α, it is enough to show that

$$T, s \models_T \alpha' \wedge y_\alpha \Longrightarrow \kappa(T, s) \models_L \alpha. \tag{18}$$

This is shown by induction over the complexity of α. Specifically, we will show for all models (T, s), for all $t \in S$, if $T, s \models_T \alpha'$ and $y_\beta \in \pi(t)$, then $\kappa(S), \kappa(t) \models_L \beta$. This is trivial for $\beta \in \mathcal{V}$ (the base case), and the inductive steps for $\neg$, $\wedge$ and $\exists x$ are easy to show (noting that $\alpha' \rightarrow \beta'$ is a validity for all $\beta \subseteq \alpha$).

So suppose that for all models,(T, s), for all $t \in S$, if $T, s \models_T \alpha'$ and $y_\beta \in \pi(t)$, then $\kappa(S), \kappa(t) \models_L \beta$. For the necessity operator, suppose that $y_{\Box\beta} \in \pi(t)$ for some t. By the definition of β', we see that $y_{\Box\beta} \in \pi(t)$ for all $t \in S$. By the induction hypothesis we must have $\kappa(S), \kappa(t) \models_L \beta$.
For the local quantifier, $\exists_i$, suppose that $y_{\exists_i x\beta} \in \pi(t)$ for some t. Therefore there is some x-variant, (T', s'), of (T, s) such that $y_\beta \in \pi(t')$, where $(t, t') \in B$, for some bisimulation, B, defined by the x-variant. By the definition of $\exists_i x\beta)'$, we see that the interpretation of x cannot change from one node to its successor if x_i is true at it's successor. From the definition of κ, we see that x will be i-local in $\kappa(M', s')$. Applying the induction hypothesis we have $\kappa(S'), \kappa(t') \models_L \beta$, and by applying the fact stated above, $\kappa(S), \kappa(t) \models_L \exists_i x\beta$.

This completes the induction and (18) follows.

Theorem 1. *The satisfiability problem for QL_k is decidable.*

Proof. This follows trivially from the above lemmas. Given any formula α, satisfiable in QL_k, we can compute the formula α^*, and decide demonstrate the satisfiability in QCTL, via Lemma 3 and the decision procedure for QCTL. If α is not satisfiable in QL_k by Lemma 4 it is enough to show that α^* is not satisfiable in QCTL.

6 Conclusion

In this paper we have presented an effective interpretation for logics of knowledge with propositional quantification. We have briefly looked at the expressive power of such a logic, and shown how we can express a wide range of useful second-order properties. Most importantly, we have achieved this gain in expressivity without sacrificing the decidability of the logic. The decidability of QL_k was shown via a translation to QCTL. From this translation we can see that the decision process presented is infeasible for automated reasoning. We also note that the language QCTL has a polynomial translation into the language QL_k, and the satisfiability problem for QCTL is known to be non-elementary [8], i.e. the decision procedure is optimal. This does not mean that the logic is of no use. Its expressive power, and the fact that it is decidable, suggest that it is valuable for formalizing the specification of systems. Automating reasoning could be effectively applied to sub-languages, and axiomatizations could be found to assist reasoning about systems and verification of protocols.

One of the main avenues for future work is to find an axiomatization of QL_k working from the foundation laid in [4]. The other area for future work is applying the above decision procedure to a more general class of logics. The process of semantic untangling and syntactic entangling appears to have potential for generating a large class of propositionally quantified multi-modal logics. Furthermore, interpreting propositional quantification with respect to bisimulations will maintain the natural abstractions of the modal logics.

References

[1] E. Clarke and E. Emerson. Synthesis of synchronization skeletons for branching time temporal logic. In *Proc. IBM Workshop an Logic of Programs, Yorktown Heights*, NY, pages 52-71. Springer, Berlin, 1981. 357

[2] E. Emerson and A. Sistla. Deciding full branching time logic. *Information and Control*, 61:175 - 201, 1984. 357

[3] K. Englehardt, R. van der Meyden, and Y. Moses. Knowledge and the logic of local propositions. In *Theoretical Aspects of Rationality and Knowledge, Proceedings of the Seventh Conference*, pages 29-41, 1998. 353, 354, 355, 356

[4] K. Englehardt, R. van der Meyden, and K. Su. Modal logics with a linear hierarchy of local propositional quantifiers. In *Proceedings of AiML 2002*, to appear. 352, 355, 362

[5] R. Fagin, J. Halpern, Y. Moses, and M. Vardi. *Reasoning about knowledge.* MIT Press, 1995. 352, 353, 357
[6] K. Fine. Propositional quantifiers in modal logic. *Theoria*, 36:336-346, 1970. 352
[7] T. French. Decidability of quantified propositional branching time logics. In *Proceedings of the 14th Australian Joint Conference an Artificial Intelligence*, pages 165-176, 2001. 357, 359
[8] T. French. *The Decidability of Modal Logics with Bisimulation Quantifiers.* PhD thesis, Murdoch University, in preparation. 353, 357, 362
[9] D. Kaplan. S5 with quantifiable propositional variables. *Journal of Symbolic Logic*, 35:355, 1970. 352
[10] R. Milner. A calculus of communicating systems. *Lecture Notes in Computer Science*, 92, 1980. 359
[11] David Park. Concurrency and automata an infinite sequences. *Lecture Notes in Computer Science*, 104:167-183, 1981. 359

Some Logics of Belief and Disbelief

Samir Chopra[1], Johannes Heidema[2], and Thomas Meyer[3]

[1] Department of Computer Science
Brooklyn College of CUNY, Brooklyn, New York
schopra@sci.brooklyn.cuny.edu
[2] Department of Mathematics, Applied Mathematics and Astronomy
University of South Africa, Pretoria, South Africa
heidej@unisa.ac.za
[3] National ICT Australia
School of Computer Science and Engineering
University of New South Wales, Sydney, Australia
tmeyer@cse.unsw.edu.au

Abstract. The introduction of explicit notions of rejection, or disbelief, in logics for knowledge representation can be justified in a number of ways. Motivations range from the need for versions of negation weaker than classical negation, to the explicit recording of classic belief contraction operations in the area of belief change, and the additional levels of expressivity obtained from an extended version of belief change which includes disbelief contraction. In this paper we present four logics of disbelief which address some or all of these intuitions. Soundness and completeness results are supplied and the logics are compared with respect to applicability and utility.

1 Introduction

The introduction of explicit notions of *belief rejection* or *disbelief* into logics for knowledge representation can be justified in a number of ways. One is found in the research area of *belief change* [1, 5]. Classical belief change recognises two basic operations on the beliefs of an agent: *revision*, where the aim is to incorporate a new belief into an agent's belief state while still maintaining consistency, and *contraction*, where an agent needs to discard one of its beliefs while trying to retain as many of its remaining beliefs as possible. The argument for introducing explicit disbeliefs into the object language of choice is that belief contraction should be seen as a belief change operation on par with revision, and not just as an intermediate step for performing revision. The following example, based on an example in [3], but originally found in [6], illustrates the point.[1]

Example 1. Consider the situation where an agent revises its current (empty) belief base with $p \to q$, then contracts with q, and then revises with p. In a framework which caters for the explicit representation of beliefs only, the principle of

[1] Observe that this is an example of *base change* [4, 10, 13].

T.D. Gedeon and L.C.C. Fung (Eds.): AI 2003, LNAI 2903, pp. 364–376, 2003.

informational economy dictates that there is only one result for this sequence of base change operations: the set $\{p \to q, p\}$. Since the representational framework does not allow for the explicit recording of the contraction of q, the choice of permissible outcomes is skewed in favour of revision operations. Contrast this with a situation in which disbelieving a sentence ϕ can be represented in the object language as a sentence of the form $\overline{\phi}$. In this case the sets $\{p \to q, \overline{q}\}$ and $\{\overline{q}, p\}$ will be acceptable outcomes as well.

In the example above, classical belief contraction is equated with disbelief *revision*. But the extended version of belief change in which it is also possible to perform contraction with *disbeliefs* has no equivalent in classical belief change. The contraction of disbeliefs provides for a particularly interesting class of operations if disbeliefs are thought of as *guards* against the introduction of certain beliefs, in the spirit of default logic [12]. Disbelief *contraction* corresponds to the removal of such a guard, which might then trigger the addition of previously suppressed beliefs into the current belief set of an agent. An adequate modelling of this type of extended (dis)belief change requires some level of interaction between beliefs and disbeliefs, something which has not been dealt with adequately in the literature. It also presupposes a consequence relation which handles disbeliefs appropriately, since it allows for contraction with disbeliefs which are *consequences* of previously asserted disbeliefs.

Another motivation for the introduction of disbeliefs is that it allows us to reason about rejected beliefs and their consequences [14, 7, 8]: in many situations it makes more sense to adopt a "negative stance", to reason about how to reject new beliefs on the basis of rejecting old ones. Under these circumstances the focus is on the provision of an appropriate consequence relation with regard to disbeliefs. Yet another motivation for an explicit representation of disbeliefs is that it can be viewed as a weaker version of classical negation. The value of such an additional version of negation is that it enables us to resolve well-known paradoxes of belief such as the lottery paradox.

Example 2. Given a finite number of lottery tickets issued in a particular week, I express doubt about the possibility of any individual ticket winning the lottery, although I do believe that exactly one of these tickets will win the lottery.

The lottery paradox would be resolved by replacing classical negation (I believe the claim that ticket number 113 will not win the lottery) with the much weaker notion of disbelief (I disbelieve the claim that ticket number 113 will win the lottery). This should not be seen as an *ad hoc* response to a particular paradox of belief. To the contrary, such a weaker version of negation is prevalent in many practical situations. Here is another example.

Example 3. I have two very good friends, Agnetha and Björn, who are the only suspects in the murder of my friends Annifrid and Benny. I refuse to believe that Agnetha was able to commit the murder (represented as the disbelief $\overline{a}$), and similarly for Björn (represented as the disbelief $\overline{b}$). Yet, the evidence compels me to believe that one of them committed the crime (represented as $a \vee b$).

The challenge is to provide a formal definition of disbelief so that sentences of the form $\overline{\phi}$ can be regarded as a legitimate way of negating ϕ, while at the same time ensuring the consistency of sets like $\{\overline{\phi}, \overline{\psi}, \phi \vee \psi\}$. In this paper we present four logics of belief and disbelief. Two of these, **WBD** and **GBD** have been used implicitly in work on disbelief, but not in the form of a *logic* of belief and disbelief. The first, **WBD**, has a very weak notion of consequence with regard to disbeliefs. It has been used for the explicit recording of belief contraction operations in [3, 6]. It provides an extremely weak link between beliefs and disbeliefs. The relation between beliefs and disbeliefs is made only via a notion of consistency, and is done for the sole purpose of providing an accurate model of classical belief contraction in terms of disbeliefs. The second logic, **GBD**, was presented in [7, 8], albeit in a different form, and is a formalisation of the idea that it is useful to reason about rejected beliefs and their consequences. The notion of consequence, with regard to disbeliefs, associated with **GBD** is stronger than that of **WBD**. We argue in section 4 that it is *too* strong. On the other hand, the connection between beliefs and disbeliefs in **GBD** is just as weak as in **WBD**. We argue that it needs to be strengthened.

The third logic, **BD**, is designed with the express intention of obtaining a weaker version of classical negation. Its version of consequence with respect to disbeliefs seems to be pitched at just the right level. It is weaker than the version provided by **GBD**, but stronger than that of **WBD**. And unlike **WBD** and **GBD** there is a strong and intuitively plausible connection between beliefs and disbeliefs. Indeed, one of the consequences of the belief in a sentence $\neg\phi$ is that ϕ will be disbelieved (that is, $\overline{\phi}$ will hold), as befits a view of disbelief as a weaker version of classical negation. It is our contention that **BD** is the most useful of the logics introduced in this paper.

BN, the fourth and final logic to be discussed, takes matters a step further by obtaining new beliefs from existing disbeliefs. In the process of doing so, disbelief collapses into classical negation - disbelieving $\overline{\phi}$ becomes equivalent to believing $\neg\phi$ - and **BN** becomes equivalent to classical logic in its level of expressivity.

1.1 Formal Preliminaries

We assume a classical propositional logic **PL** with a language $L_{\mathbf{PL}}$ generated from a (possibly countably infinite) set of propositional atoms, together with the usual propositional connectives, and with $\perp$ and $\top$ as canonical representatives of the set of contradictions and tautologies, respectively. V is the set of classical valuations of **PL**. For $\phi \in L_{\mathbf{PL}}$, $M(\phi)$ is the set of models of ϕ. Classical deduction for **PL** is denoted by $\vdash_{\mathbf{PL}}$. For any set X we denote the powerset of X (the set of all subsets of X) by $\mathcal{P}X$. Given a language L and a consequence relation $\vdash_X$ from $\mathcal{P}L$ to L we let $C_X(\Gamma) = \{\alpha \mid \Gamma \vdash_X \alpha\}$ be the consequence operation associated with $\vdash_X$. $C_{\mathbf{PL}}$ denotes classical consequence for propositional logic and satisfies the properties of Inclusion: $X \subseteq C(X)$, Idempotency: $C(X) = C(C(X))$, Monotonicity: $X \subseteq Y$ implies $C(X) \subseteq C(Y)$, and

Compactness: if $\phi \in C(X)$ then $\phi \in C(Y)$ for some finite subset Y of X.[2] An operation C from $\mathcal{PL}$ to $\mathcal{PL}$ is a *Tarskian consequence operation* iff it satisfies Inclusion, Idempotency and Monotonicity.

2 A Language for Beliefs and Disbeliefs

The language L on which we base all of the logics to be presented is a simple extension of $L_{\mathbf{PL}}$. For each potential belief ϕ, expressed as a sentence of $L_{\mathbf{PL}}$, there is a corresponding potential disbelief $\overline{\phi}$.

Definition 1. *$L = L_B \cup L_D$, where $L_B = L_{\mathbf{PL}}$ is the set of potential beliefs and $L_D = \{\overline{\phi} \mid \phi \in L_{\mathbf{PL}}\}$, the set of potential disbeliefs. An* information set *Γ is any subset of L. The* beliefs *in Γ are defined as $\Gamma_B = L_B \cap \Gamma$. The* disbeliefs *in Γ are defined as $\Gamma_D = L_D \cap \Gamma$.*

We use ϕ and ψ to denote potential beliefs and α and β to denote arbitrary sentences of L (potential beliefs and disbeliefs). The language L is fairly restrictive in the sense that $^-$ is not viewed as a propositional connective. Thus, for example, we cannot construct sentences of the form $p \vee \overline{q}$. The reason for this is twofold. Firstly, in many (but not all) motivations for the introduction of disbeliefs, such a level of expressivity is simply not necessary. Secondly, this paper should be seen as a first step towards a description of logics of belief and disbelief. Our intention is to treat $^-$ as a full-blown 'propositional' connective in future research.

3 The Logic WBD

Our first logic is relatively weak in terms of the consequences of explicit disbeliefs. The motivation for the introduction of **WBD** is primarily for the explicit expression of classical belief contraction as in [6, 3], albeit implicitly. A proof theory for **WBD** is obtained from the following three inference rules.

(B) If $\Gamma_B \vdash_{\mathbf{PL}} \phi$ then $\Gamma \vdash \phi$
(D⊥) If $\phi \vdash_{\mathbf{PL}} \bot$ then $\Gamma \vdash \overline{\phi}$
(WD) If $\overline{\psi} \in \Gamma_D$ and $\phi \vdash_{\mathbf{PL}} \psi$ then $\Gamma \vdash \overline{\phi}$

(B) is a supraclassicality requirement. (D⊥) requires that contradictions always be disbelieved, analogous to the case of tautologies always being believed. (WD) requires that disbelieving a sentence $\overline{\psi}$ leads to a disbelief in all those sentences classically stronger than ψ.

Definition 2. *In* **WBD**, *α can be deduced from an information set Γ, written as $\Gamma \vdash_{\mathbf{WBD}} \alpha$, iff (Γ, α) is in the smallest binary relation from $\mathcal{PL}$ to L closed under (B), (D⊥), and (WD).*

[2] Instead of taking **PL** as our "basic logic" we can work with any logic whose associated consequence relation satisfies these four properties.

Observe that beliefs and disbeliefs are completely decoupled in **WBD**. In particular, **WBD** satisfies the following two properties which show that disbeliefs and beliefs are only derivable from the set of disbeliefs and beliefs respectively:

(B$\not\to$D) $\forall \Omega \subseteq L_B$ and $\phi \in L_B$, $\Gamma \vdash \overline{\phi}$ iff $\Gamma_D \cup \Omega \vdash \overline{\phi}$
(D$\not\to$B) $\forall \Delta \subseteq L_D$ and $\phi \in L_B$, $\Gamma \vdash \phi$ iff $\Gamma_B \cup \Delta \vdash \phi$

Proposition 1. $\vdash_{\mathbf{WBD}}$ *satisfies (B$\not\to$D) and (D$\not\to$B).*

The only way in which beliefs and disbeliefs are related in **WBD** is by way of the notion of **WBD**-*inconsistency.*

Definition 3. Γ *is* WB-inconsistent *iff* $\Gamma_B \vdash_{\mathbf{WBD}} \bot$, *and is* WD-inconsistent *iff* $\Gamma_D \vdash_{\mathbf{WBD}} \overline{\top}$. Γ *is* **WBD**-inconsistent *iff* $\Gamma \vdash_{\mathbf{WBD}} \phi$ *and* $\Gamma \vdash_{\mathbf{WBD}} \overline{\phi}$ *for some* $\phi \in L_B$.

Note that **WBD** has a well-behaved Tarskian consequence operation:

Proposition 2. $C_{\mathbf{WBD}}$ *satisfies Inclusion, Idempotency, Monotonicity, Compactness.*

We now turn to a semantics for **WBD**. A useful intuition is to think of the beliefs in Γ as the information an agent has acquired on its own, and each disbelief $\overline{\phi}$ as information it has obtained from a different source, informing it that ϕ does *not* hold. The agent has more faith in its own capabilities than in those of its sources, and information obtained from its sources is therefore seen as less reliable. The information obtained from a specific source is independent of the beliefs of the agent and the information obtained from other sources. A model for **WBD** consists of a set of valuations, corresponding to the worlds that the agent regards as possible, together with a set of sets of valuations, with each element of this set corresponding to the worlds that a particular source of the agent regards as *possible.* A potential belief is satisfied in a model if it is true in all the worlds that the agent regards as possible, and a potential disbelief $\overline{\phi}$ is satisfied in a model if at least one of the sources regards ϕ as impossible. Satisfaction is denoted by $\Vdash$.

Definition 4. *A* **WBD**-model $\mathcal{M}$ *is an ordered pair* $(M, \mathcal{N})$ *where* $M \subseteq V$ *and* $\emptyset \subset \mathcal{N} \subseteq \mathcal{P}V$. *For* $\phi \in L_B$, $\mathcal{M} \Vdash \phi$ *iff* $M \subseteq M(\phi)$. *For* $\overline{\phi} \in L_D$, $\mathcal{M} \Vdash \overline{\phi}$ *iff* $\exists N \in \mathcal{N}$ *such that* $N \subseteq M(\neg\phi)$.

In a **WBD**-model $\mathcal{M} = (M, \mathcal{N})$, M represents the models of the beliefs of the agent, and each $N \in \mathcal{N}$ represents the models of a sentence that the source associated with N holds to be possible. We require an agent to have at least one source of information. That is, we require that $\mathcal{N} \neq \emptyset$. Entailment for **WBD** (denoted by $\models_{\mathbf{WBD}}$) is then defined in the normal model-theoretic fashion.

Definition 5. $\mathcal{M} \Vdash \Gamma$ *iff* $\mathcal{M} \Vdash \alpha$ $\forall \alpha \in \Gamma$. $\Gamma \models_{\mathbf{WBD}} \alpha$ *iff* $\mathcal{M} \Vdash \alpha$ *for every* **WBD**-*model* $\mathcal{M}$ *s.t.* $\mathcal{M} \Vdash \Gamma$.

It turns out that the logic **WBD** is sound and complete.

Theorem 1. $\Gamma \vdash_{\mathbf{WBD}} \alpha$ *iff* $\Gamma \models_{\mathbf{WBD}} \alpha$, *for all* $\alpha \in L$.

Based on the semantics of **WBD** we can also define appropriate notions of (un)satisfiability which, via Theorem 1, can be shown to coincide with inconsistency.

The logic **WBD** is, essentially, the logic on which the work in [6, 3] is based. Both argue for an explicit expression of disbeliefs in order to maintain a record of belief contraction. Although **WBD** seems adequate for this purpose, two criticisms can be levelled at it: the extreme weakness of the notion of consequence associated with disbeliefs, and the complete decoupling of beliefs and disbeliefs. These weaknesses become apparent in scenarios where it is appropriate to perform various belief change operations on beliefs as well as disbeliefs. So, while contracting with a belief might correspond to revising with a disbelief, contracting with a *disbelief* has no equivalent in classical belief change. And such an operation is useful if disbeliefs are thought of as *guards* against the introduction of certain information, in the spirit of default logic [12]. The removal of such a guard might then trigger the addition of beliefs into an agent's current belief set. An adequate modelling of this type of extended (dis)belief change requires some level of interaction between beliefs and disbeliefs. It also presupposes a consequence relation which handles disbeliefs appropriately, since it allows for contraction with disbeliefs which are *consequences* of previously asserted disbeliefs. In section 4 we consider an attempt to rectify the weakness of consequences derived from disbeliefs. In section 5 we address this issue, and also consider the the provision of a link between beliefs and disbeliefs.

4 The Logic GBD

One way to address the weakness of the consequences to be derived from disbeliefs is to regard consequence, with respect to disbeliefs, as the exact dual of consequence with respect to beliefs. This is the approach in [8] where, interestingly enough, the reason for defining such a logic is to define disbelief change.

Let $\overline{\Gamma}_D = \{\neg\phi \mid \overline{\phi} \in \Gamma_D\}$ and consider the following rule:

(GD) If $\overline{\Gamma}_D \vdash_{\mathbf{PL}} \neg\phi$ then $\Gamma \vdash \overline{\phi}$

(GD) requires that the consequences of disbeliefs be the exact dual of classical consequence. The proof theory for the logic **GBD** is then obtained from **WBD** by replacing (WD) with (GD).

Definition 6. *In* **GBD** α *can be deduced from an information set* Γ*, written as* $\Gamma \vdash_{\mathbf{GBD}} \alpha$*, iff* (Γ, α) *is in the smallest binary relation from* $\mathcal{PL}$ *to* L *closed under (B), (D$\bot$), and (GD).*

GBD also has a complete decoupling of beliefs and disbeliefs, as the following proposition shows.

Proposition 3. $\vdash_{\mathbf{GBD}}$ *satisfies (B$\not\to$D) and (D$\not\to$B).*

Like **WBD**, **GBD** has a well-behaved Tarskian consequence operation.

Proposition 4. $C_{\mathbf{GBD}}$ *satisfies Inclusion, Idempotency, Monotonicity and Compactness.*

The logic **GBD** satisfies the following property, which was proposed in [7].

(Rej) If $\Gamma \vdash \overline{\psi}$ and $\Gamma \vdash \overline{\neg(\phi \to \psi)}$ then $\Gamma \vdash \overline{\phi}$

Proposition 5. **GBD** *satisfies (Rej).*

(Rej) can be thought of as a version of the inference rule Modus Tollens. In fact, if disbelief is replaced with classical negation, (Rej) coincides exactly with Modus Tollens.

The three versions of inconsistency for **GBD** are defined as for **WBD**.

Definition 7. *Γ is* GB-inconsistent *iff $\Gamma_B \vdash_{\mathbf{GBD}} \bot$. Γ is* GD-inconsistent *iff $\Gamma_D \vdash_{\mathbf{GBD}} \overline{\top}$. Γ is* **GBD**-inconsistent *iff $\Gamma \vdash_{\mathbf{GBD}} \phi$ and $\Gamma \vdash_{\mathbf{GBD}} \overline{\phi}$ for some $\phi \in L_B$.*

Observe that beliefs and disbeliefs are related only by **GBD**-inconsistency.

A semantics for **GBD** is obtained by considering only those **WBD**-models in which $\mathcal{N}$ has a single element. Intuitively, all disbeliefs are obtained from a single source. This ensures that disbeliefs can be combined to obtain new disbeliefs, which allows for a stronger notion of consequence regarding disbeliefs. **GBD**-entailment is defined as for **WBD**.

Definition 8. *A* **GBD**-model *is an ordered pair $\mathcal{M} = (M, N)$ where $M, N \subseteq V$. For $\phi \in L_B$, $\mathcal{M} \Vdash \phi$ iff $M \subseteq M(\phi)$. For $\overline{\phi} \in L_D$, $\mathcal{M} \Vdash \overline{\phi}$ iff $N \subseteq M(\neg\phi)$. $\mathcal{M} \Vdash \Gamma$ iff $\mathcal{M} \Vdash \alpha$ $\forall \alpha \in \Gamma$. $\Gamma \vDash_{\mathbf{GBD}} \alpha$ iff $\mathcal{M} \Vdash \alpha$ for every* **GBD**-*model $\mathcal{M}$ s.t. $\mathcal{M} \Vdash \Gamma$.*

The logic **GBD** is sound and complete.

Theorem 2. *$\Gamma \vdash_{\mathbf{GBD}} \alpha$ iff $\Gamma \vDash_{\mathbf{GBD}} \alpha$, for all $\alpha \in L$.*

The crucial difference between **WBD** and **GBD** is that **GBD** combines disbeliefs to obtain new disbeliefs, which allows for a much stronger notion of consequence with respect to disbeliefs. In particular, **GBD** satisfies the following property:

(D$\vee$) If $\Gamma \vdash \overline{\phi}$ and $\Gamma \vdash \overline{\psi}$ then $\Gamma \vdash \overline{\phi \vee \psi}$

Proposition 6. **GBD** *satisfies (D$\vee$).*

Such a property is undesirable for a notion of disbelief. Since disbelief is *not* intended to be equivalent to classical negation, it is reasonable to require that it be possible to express notions not expressible in classical logic. One of these is *agnosticism*, in which an agent refuses to commit to a potential belief or its negation. Formally, this amounts to both $\overline{\phi}$ and $\overline{\neg\phi}$ being consequences of a consistent information set Γ. But if the consequence relation satisfies (D$\vee$), it means that the agent is forced to accept $\overline{\phi \vee \neg\phi}$, and therefore $\overline{\top}$ as well. Disbelieving a tautology amounts to GD-inconsistency. It is the analog of classical inconsistency (believing the negation of the tautology), but with classical negation replaced by disbelief. If agnosticism in the sense described above leads to inconsistency with respect to disbeliefs, it is an indication that the consequence relation for **GBD** is too strong to account for a proper treatment of disbeliefs. We now consider a logic which seems to be pitched at the right level in this regard.

5 The Logic BD

In this section we introduce a logic in which there is interaction between beliefs and disbeliefs. Disbelief is seen as a weaker notion of classical negation. As a result, believing the negation of a sentence also results in disbelieving that sentence, although the converse relationship does not hold. Interestingly enough, this coupling of beliefs and disbeliefs comes about as a result of the introduction of the following inference rule, which looks like a simple strengthening of the consequence relation with respect to disbeliefs.

(D) If $\Gamma \vdash \overline{\psi}$ and $\Gamma_B \cup \{\phi\} \vdash_{\mathbf{PL}} \psi$ then $\Gamma \vdash \overline{\phi}$

(D) requires that disbelieving a sentence ψ also leads to a disbelief in those sentences classically stronger than ψ, *but with respect to* Γ_B. Observe that (WD) is the special case of (D) where $\Gamma_B = \emptyset$. The difference between the rules (GD), (WD) and (D) is perhaps best brought out through an example:

Example 4. Consider our earlier example of the murder mystery. Then (WD) commits me to the following: If I refuse to believe that Agnetha killed Annifrid and Benny, then I also refuse to believe that Agnetha and Björn killed them (a similar commitment is enforced by (D) and (GD)). However, in addition, (D) commits me to: If I believe that Agnetha is a Swede, I believe that if you killed Annifrid and Benny then you are a murderer, and I refuse to believe that a Swede can be a murderer, then I also refuse to believe that Agnetha could have killed Annifrid and Benny. This argument cannot be made with either (WD) or (GD). To further illustrate the difference, consider what (GD) requires us to infer in the lottery example. Consider for the time being, a version with just two tickets. I believe that exactly one of t_1 or t_2 will win the lottery; I refuse to believe that t_1 wins the lottery and similarly for t_2. Then (GD) compels me to refuse to believe that exactly one of the two will win. Therefore I end with a contradiction. In contrast, (D) and (WD) do not require such a commitment.

From the example above it should be clear that both (D) and (GD) are stronger than (WD). Furthermore, as the second part of the example makes clear, since (D) is stronger than (WD), some conclusions obtained from the former will not be obtainable from the latter, but others will. The example provided is one of those obtainable from (D) but not (WD). Since (D) and (GD) are incomparable in strength, they will have some conclusions that coincide (as above) but there will also be conclusions obtained from (D) which are not obtainable from (GD) and vice-versa. The example provided for (D) is an instance of a conclusion obtainable from (D) but not from (GD).

Definition 9. *In* **BD**, α *can be deduced from an information set* Γ, *written as* $\Gamma \vdash_{\mathbf{BD}} \alpha$, *iff* (Γ, α) *is in the smallest binary relation from* $\mathcal{PL}$ *to* L *closed under (B), (D*$\bot$*), and (D).*

In the logic **BD**, like **WBD** and **GBD**, the introduction or removal of disbeliefs has no bearing on beliefs, as the following result shows.

Proposition 7. $\vdash_{\mathbf{BD}}$ *satisfies (D*$\not\to$*B).*

However, the manner in which disbeliefs are obtained from beliefs in **BD** can be expressed by the following property which links up a belief in $\neg\phi$ to a disbelief in ϕ.

(B→D) If $\Gamma \vdash \neg\phi$ then $\Gamma \vdash \overline{\phi}$

Proposition 8. $\vdash_{\mathbf{BD}}$ *satisfies (B→D) and does not satisfy (B$\not\to$D).*

Observe that neither **WBD** nor **GBD** satisfies $(B \to D)$.

BD has a well-behaved Tarskian consequence operation:

Proposition 9. $C_{\mathbf{BD}}$ *satisfies Inclusion, Idempotency, Monotonicity and Compactness.*

The different notions of inconsistency for **BD** are defined in the same way as for **WBD** and **GBD**.

Definition 10. *An information set* Γ *is* B-inconsistent *iff* $\Gamma \vdash_{\mathbf{BD}} \bot$, Γ *is* D-inconsistent *iff* $\Gamma \vdash_{\mathbf{BD}} \overline{\top}$, *and* **BD**-inconsistent *iff* $\Gamma \vdash_{\mathbf{BD}} \phi$ *and* $\Gamma \vdash_{\mathbf{BD}} \overline{\phi}$ *for some* $\phi \in L_B$.

Because there is interaction between beliefs and disbeliefs in **BD**, there is also a connection between the different notions of inconsistency for **BD**. In particular, we have the following results.

Proposition 10. *If* Γ *is B-inconsistent then it is also* **BD***-inconsistent, but the converse does not hold.* Γ *is* **BD***-inconsistent iff it is D-inconsistent.*

In the logic **BD**, then, **BD**-inconsistency collapses into D-inconsistency. Given the intuition of disbelief as a weaker version of classical negation, this is a particularly desirable state of affairs. In classical logic, asserting both ϕ and $\neg\phi$ is tantamount to the assertion that $\neg\top$ is the case, while in **BD**, asserting both ϕ and $\overline{\phi}$ amounts to the assertion that $\overline{\top}$ is the case.

Observe that **BD**-inconsistency (or D-inconsistency) amounts to disbelieving the tautology, which leads to a disbelief in every sentence in L_B. So, while B-inconsistency, like classical consistency, leads to the acceptance of *every* sentence in the language, **BD**-inconsistency leads only to the acceptance of every *disbelief* in the language. **BD**-inconsistency can thus be seen as a weaker version of classical inconsistency. We regard this as a particularly attractive feature of the logic **BD**. The logics **WBD** and **GBD** also have similar features, but the connection between classical inconsistency and the weaker version of inconsistency, based on disbeliefs, is not as intuitively appealing.

The semantics for **BD** is obtained by considering only those **WBD**-models for which every $N \in \mathcal{N}$ is a subset of M. That is, the worlds that the sources of an agent may regard as possible have to be worlds that the agent itself regards as possible. **BD**-entailment is defined as for **WBD**-entailment and **GBD**-entailment.

Definition 11. *A* **BD**-model *is a tuple* $(M, \mathcal{N})$ *where* $M \subseteq V$ *and* $\emptyset \subset \mathcal{N} \subseteq \mathcal{P}M$. *For a* **BD***-model* $\mathcal{M} = (M, \mathcal{N})$ *and* $\phi \in L_B$, $\mathcal{M} \Vdash \phi$ *iff* $M \subseteq M(\phi)$. *For* $\overline{\phi} \in L_D$, $\mathcal{M} \Vdash \overline{\phi}$ *iff* $\exists N \in \mathcal{N}$ *s.t.* $N \subseteq M(\neg\phi)$. $\mathcal{M} \Vdash \Gamma$ *iff* $\mathcal{M} \Vdash \alpha$ $\forall \alpha \in \Gamma$. $\Gamma \vDash_{\mathbf{BD}} \alpha$ *iff* $\mathcal{M} \Vdash \alpha$ *for every* **BD***-model* $\mathcal{M}$ *s.t.* $\mathcal{M} \Vdash \Gamma$.

The logic **BD** is sound and complete.

Theorem 3. *$\Gamma \vdash_{\mathbf{BD}} \alpha$ iff $\Gamma \models_{\mathbf{BD}} \alpha$, for all $\alpha \in L$.*

Based on the semantics of **BD** we can define appropriate notions of satisfiability and unsatisfiability which can be shown to coincide with inconsistency.

We conclude this section by pointing out that **BD** is able to handle examples such as the lottery paradox and its variants (cf. examples 2 and 3) in a manner that is intuitively satisfactory. It can be shown that any information set of the form $\Gamma = \{\overline{\phi_1}, \ldots, \overline{\phi_n}, \bigvee_{i=1}^{i\leq n} \phi_i\}$ (with not all ϕ_i inconsistent) is neither B-inconsistent, **BD**-inconsistent, nor D-inconsistent. **BD** thus allows us, for example, to disbelieve the fact that any particular ticket will win the lottery, while still believing that exactly one of the tickets will win the lottery, without collapsing into *any* kind of inconsistency.

6 The Logic BN

We have seen that **BD** allows for the generation of disbeliefs from beliefs. An interesting question is whether it makes sense to do the opposite; that is, to generate beliefs from disbeliefs. We consider two ways of doing so. For the first one, note that the generation of disbeliefs from beliefs in **BD** is achieved by the inference rule (D). On the basis of this, we consider the possibility of generating beliefs from disbeliefs in a similar fashion.

(B′) If $\Gamma \vdash \psi$ and $\Gamma \cup \{\overline{\phi}\} \vdash \overline{\psi}$ then $\Gamma \vdash \phi$

(B′) asserts that if I currently believe ψ, and if the addition of ϕ as a disbelief leads me to disbelieve ψ, then ϕ should be one of my current beliefs. It is analogous to (D), but with the roles of beliefs and disbeliefs reversed. It turns out, however, that (B′) is a derived rule of the logic **BD**.

Proposition 11. *$\vdash_{\mathbf{BD}}$ satisfies the property (B′).*

A more direct way to obtain new beliefs from current disbeliefs is to consider the converse of the property (B→D).

(D→B) If $\Gamma \vdash \overline{\phi}$ then $\Gamma \vdash \neg\phi$.

Observe that **WBD**, **GBD** and **BD** do not satisfy (D→B). A proof theory for the logic we call **BN** is then obtained by adding (D→B) to the inference rules of **BD**.

Definition 12. *For $\alpha \in L$ and $\Gamma \subseteq L$, α can be deduced from Γ in* **BN***, written as $\Gamma \vdash_{\mathbf{BN}} \alpha$, iff (Γ, α) is in the smallest binary relation from $\mathcal{P}L$ to L closed under (B), (WD), (D⊥) and (D→B).*

The introduction of (D→B) does indeed give us a logic that is stronger than **BD**. It turns out, however, that disbelief now collapses into classical negation

Theorem 4. *For every $\phi \in L_B$, $\Gamma \vdash_{\mathbf{BN}} \overline{\phi}$ iff $\Gamma \vdash_{\mathbf{BN}} \neg\phi$.*

BN is therefore exactly as expressive as **PL**.

7 Conclusion and Future Research

Of the four logics of belief and disbelief presented, the logic **BD** is the most deserving of such a label. It covers all the motivations for the explicit introduction of disbeliefs. Contraction with a belief ϕ can be represented explicitly in **BD** as revision with the disbelief $\overline{\phi}$, as in **WBD** and **GBD**. This ability has a useful side-effect. In classical belief change, the result of the pathological case in which a belief set is contracted with a tautology, is taken to be the belief set itself, primarily because it is unclear what else the result could be: it is the only case in which a contraction does not succeed. But in **BD** (as in **WBD** and **GBD**), contraction by ϕ corresponds to a revision by $\overline{\phi}$. So contraction by a tautology is equivalent to revising with the sentence $\overline{\top}$, a disbelief in the tautology. It can be verified that disbelief in the tautology results in *disbelieving* all (propositional) sentences with no effect on the current *beliefs*. The explicit introduction of disbeliefs thus enables us to devise a more intuitive result.

Like **WBD** and **GBD**, **BD** provides a well-behaved notion of consequence with respect to disbeliefs. In **BD**, consequence with respect to disbeliefs is stronger than in **WBD** but weaker than in **GBD**. In particular, (D) is satisfied in **BD** but not in **WBD**, and **BD** does not satisfy (D$\vee$), a property satisfied by **GBD**. By not satisfying (D$\vee$), consequence for **BD** is weak enough to allow for the expression of *agnosticism* (disbelieving both ϕ and $\neg\phi$) without collapsing into inconsistency. In this respect it is superior to the logic **GBD**. There is a link between beliefs and disbeliefs in **BD** lacking in both **WBD** and **GBD**, which ensures that **BD** is a suitable base logic for a version of belief change in which it is possible to contract with disbeliefs. The link between beliefs and disbeliefs ensures that disbelief, as defined in **BD**, is an appropriate weaker version of classical negation, as is apparent from the fact that **BD** satisfies (B$\rightarrow$D) and (D$\not\rightarrow$B) and by its intuitive resolution of the lottery paradox (and its variants).

If disbelief is to be viewed as a weaker version of negation, it is necessary to conduct a proper investigation into the connection between disbelief, classical negation, and the other propositional connectives. In order to do so, it is necessary to treat $^{-}$ as full-blown propositional connective, in which sentences such as $p \vee \overline{q}$, $\neg\overline{p}$, and $\overline{\overline{p}}$ have a well-defined meaning. Comparisons with other logical operators that express the rejection of beliefs—in particular, Nelson's strong negation [9]—will be useful in assessing the properties of our notion of disbelief. In addition, there seems to be a connection with paraconsistent logics. Our work on **BD**, in particular, can be seen as an attempt to combine consequence relations for belief and disbelief and a comparison with the work of [2] will be appropriate.

It has been suggested that there is an obvious connection between **BD** and the epistemic logic **KD45** [11]. The statement $\Box\phi$ in **KD45** corresponds to the assertion that ϕ is believed, while $\Diamond\neg\phi$ corresponds to the assertion that $\neg\phi$ is possible, and hence that ϕ is disbelieved. Observe, though, that statements in **BD** are object-level assertions, or assertions from a first person perspective ("The sky is blue"), while the corresponding statements in epistemic logics are meta-level assertions, or assertions from a third person perspective ("The agent believes the sky is blue"). Also, note that $\Diamond\neg\phi$ is equivalent to $\neg\Box\phi$, which suggests that a disbelief in ϕ is the same as a belief in $\neg\phi$, which is contrary to the intuition for disbeliefs developed in this paper.

In fact, the differences run slightly deeper and are easily illustrated. To compare **BD** with **KD45** we need to restrict **KD45** to sentences of the form $\Box\phi$ and $\Diamond\phi$. Then **KD45** satisfies (B), (D$\perp$) and (D). So it is at least as strong as **BD**. It does not satisfy ($B \not\rightarrow D$), and satisfies ($D \not\rightarrow B$). It does not satisfy (Rej). It supports agnosticism (I disbelieve ϕ as well as $\neg\phi$) but does not satisfy ($D\vee$). In these respects it has the same behaviour as **BD**. But it is stronger than **BD** as can be seen by looking at what happens when the tautology is disbelieved. In **KD45** (indeed, in any normal modal logic i.e., one containing the axiom schema **K**), if $\Diamond(\neg\top)$ can be deduced from Γ, then so can $\Box\phi$ and $\Diamond\phi$ for every propositional sentence ϕ. That is, disbelieving $\top$ in **KD45** leads to a belief as well as a disbelief in every propositional sentence. In contrast disbelieving the tautology in **BD** leads to a disbelief in every propositional sentence, but our beliefs remain unaffected. This is a particularly desirable property—as pointed out in the discussion following Proposition 8. We plan to develop a restricted version of **KD45** as defined above as a separate logic of belief and disbelief. The normal **KD45** characterization requires a more expressive language than we possess at the moment. We will provide a formal proof to the effect that the semantics of the logic **BD** is not adequately captured by a class of Kripke models. Applications of the logics that we have developed to belief revision is a non-trivial task that needs separate study.

Acknowledgements

The first author would like to thank the University of South Africa in Pretoria for the invitation and financial support to visit its Departments of Mathematics and Computer Science in November 2001, when this paper was conceived. The authors would also like to thank Aditya Ghose for some insightful comments and suggestions.

References

[1] Carlos E. Alchourrón, Peter Gärdenfors, and David Makinson. On the logic of theory change: Partial meet functions for contraction and revision. *Journal of Symbolic Logic*, 50:510–530, 1985. 364

[2] C. Caleiro, W. A. Carnielli, Coniglio M. E., Sernadas A., and Sernadas C. Fibring non-truth-functional logics: Completeness preservation. *Journal of Logic Language and Information*, 12(2):183–211, 2003. 374

[3] Samir Chopra, Aditya Ghose, and Thomas Meyer. Non-prioritized ranked belief change. *Journal of Philosophical Logic*, 32(3):417–443, 2003. 364, 366, 367, 369

[4] André Fuhrmann. Theory contraction through base contraction. *Journal of Philosophical Logic*, 20:175–203, 1991. 364

[5] Peter Gärdenfors. *Knowledge in Flux : Modeling the Dynamics of Epistemic States*. The MIT Press, Cambridge, Massachusetts, 1988. 364

[6] Aditya Ghose and Randy Goebel. Belief states as default theories: Studies in non-prioritised belief change. In Henri Prade, editor, *ECAI 98. 13th European Conference on Artificial Intelligence*, pages 8–12, New York, 1998. John Wiley & Sons, Ltd. 364, 366, 367, 369

[7] Anna Gomolińska. On the logic of acceptance and rejection. *Studia Logica*, 60:233–251, 1998. 365, 366, 370

[8] Anna Gomolińska and David Pearce. Disbelief Change. In *Electronic essays on the occasion of the fiftieth birthday of Peter Gärdenfors*, 2001. 365, 366, 369

[9] Yuri Guirevich. Intuitionistic logic with strong negation. *Studia Logica*, 36:49–59, 1977. 374

[10] Sven-Ove Hansson. Changes of disjunctively closed bases. *Journal of Logic, Language and Information*, 2(4):255–284, 1993. 364

[11] G. E. Hughes and M. J. Cresswell. *An introduction to Modal Logic.* Methuen, 1972. 374

[12] Raymond Reiter. A logic for default reasoning. *Artificial Intelligence*, 13:81–132, 1980. 365, 369

[13] Hans Rott. *Change, Choice and Inference.* Oxford University Press, 1996. 364

[14] J. Slupecki, G. Bryll, and U. Wybraniec-Skardowska. Theory of rejected propositions II. *Studia Logica*, 30:97–145, 1972. 365

Axiomatic Analysis of Negotiation Protocols

Dongmo Zhang[1] and Norman Foo[2]

[1] School of Computing and Information Technology
University of Western Sydney, Australia
dongmo@cit.uws.edu.au
[2] School of Computer Science and Engineering
The University of New South Wales, Australia
norman@cse.uws.edu.au

Abstract. This paper presents an axiomatic approach to negotiation protocol analysis. We consider a negotiation procedure as multiple stages of mutual belief revision. A set of postulates in AGM-style of belief revision are proposed to specify rational behavior of negotiation. An explicit construction of negotiation function is given in which negotiation process is viewed as the interaction of two iterated revision operations. As a result the proposed axiomatic system is proved to be consistent. Finally, we examine our approach with an instantiation of Rosenschein and Zlotkin's Monotonic Concession Protocol of Negotiation.

1 Introduction

Negotiation has been investigated from many perspectives, including economics, applied mathematics, psychology, sociology and computer science [18, 3, 20, 11, 5, 17]. Significant advances have been made in both quantitative and qualitative analysis of negotiating processes. Quantitative approaches, especially those which are inspired by game-theory, dominate much of the existing work. Sometimes, numeric utility functions can be used as analytic bases for decision making because they may provide accurate evaluations of situations. However, such numeric evaluations are often unreliable, or even simply unavailable. In real life negotiation, logical reasoning often dominates the process, with numeric analysis playing an auxiliary role in the decision making. Despite this, there has not been much work on logic-based approaches to negotiation[21, 11, 17]. This paper attempts to alleviate this deficiency by introducing a logical framework to capture notions of rational behavior of negotiation. We give an axiomatic analysis of negotiating process based on belief revision theory.

Negotiation is a process of consensus-seeking among two or more parties. The parties in negotiation first verbalize contradictory demands or offers. These demands or offers change with the progress of the negotiation through mutual persuasion or argumentation till an mutual acceptable agreement has been reached. If we consider the demands(or offers) of parties as their beliefs on the matter in question, the change of the demands of each party reflects the change of its beliefs during the progress of negotiation. The parties who are convinced to accept part of the other parties' demands would perform a belief revision. New

T.D. Gedeon and L.C.C. Fung (Eds.): AI 2003, LNAI 2903, pp. 377–389, 2003.

belief states of participants represent their revised demands which are normally closer to each other and apt to reach an agreement. We term such kind of belief revision *mutual belief revision*.

Different from the AGM belief revision in which an agent is principally concerned with minimizing loss of its beliefs [1], each agent in mutual belief revision not only tries to keep as many of its own beliefs as possible but also intends to learn from the other agents as much as possible. Negotiation behaves in a similar manner. A negotiator normally tries her best to keep all her demands and is also ready to accept selectively some demands from his opponents in order to reduce conflicts, avoid failure and maximize gains from negotiation. The logical framework we introduce attempts to model this *combination of cooperation and competition*. As a starting point of the investigation, we restrict ourselves to the case of two agents. We propose a set of postulates to capture rational behavior of mutual belief revision and negotiation. These postulates are mostly inspired by the AGM framework of belief revision and part of Darwiche and Pearl's iterated belief revision.

2 Postulates for Mutual Belief Revision and Negotiation

We assume an agent to have a deductively closed set of beliefs taken from some underlying propositional language $\mathcal{L}$[1]. The language is that of classical propositional logic with an associated consequence operation Cn. Thus a set K of sentences is a *belief set* when $K = Cn(K)$. If F and G are two sets of sentences, $F + G$ denotes $Cn(F \cup G)$.

In this section we propose a set of axioms to specify properties of mutual belief revision and negotiation between two agents. The idea is the following. Suppose that K_1 and K_2 are the current belief states of two agents. During the mutual belief revision, each agent accepts part of beliefs from the other agent and revises her belief states to preserve consistency. As a result, both agents' belief states will be revised and the resulting belief states, denoted by $\mathrm{N}_1(\mathrm{K}_1,\mathrm{K}_2)$ and $\mathrm{N}_2(\mathrm{K}_1,\mathrm{K}_2)$, normally get closer each other. The following picture depicts the changes of belief states in mutual belief revision.

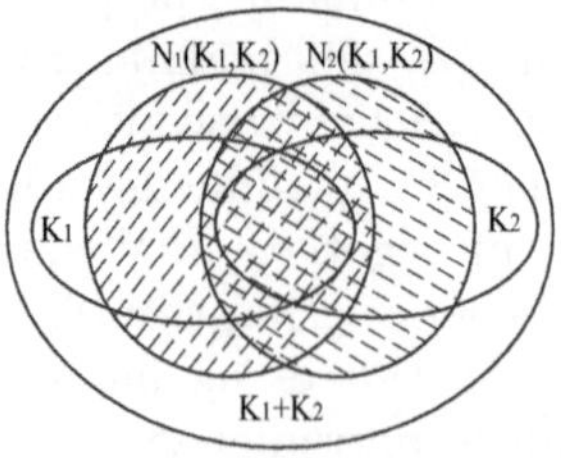

In negotiation setting, revision of belief states of an agent reflects change of its demands or offers in each negotiation round. Therefore if X and Y are the initial demands/offers of two agents, respectively, $\mathrm{N}_1(\mathrm{X},\mathrm{Y})$ and $\mathrm{N}_2(\mathrm{X},\mathrm{Y})$ are their

[1] Different from most literature, we do not assume the language is finite. This allows us to extend the current framework to the first-order case in the future.

revised demands/offers after the round of negotiation. And a possible agreement reached in the negotiation should be just $N_1(X,Y) \cap N_2(X,Y)$ or a subset of it.

Formally, a mutual revision or negotiation function is a two-input and two-output function $N(X,Y) = (N_1(X,Y), N_2(X,Y))$, where X and Y represent the initial belief sets or demands of each agent and $N_1(X,Y)$ and $N_2(X,Y)$ the revised belief sets or demands of each agent respectively. Note that X and Y here are not required to be logically closed.

Definition 1 A function $N : 2^{\mathcal{L}} \times 2^{\mathcal{L}} \to 2^{\mathcal{L}} \times 2^{\mathcal{L}}$ is a *mutual belief revision* or *negotiation function* if it satisfies the following postulates:

(N1) $N_1(X,Y) = Cn(N_1(X,Y))$; $N_2(X,Y) = Cn(N_2(X,Y))$.
(N2) $N_1(X,Y) \subseteq X + Y$; $N_2(X,Y) \subseteq X + Y$.
(N3) $N_1(X,Y)$ is inconsistent iff X or Y is inconsistent;
$N_2(X,Y)$ is inconsistent iff X or Y is inconsistent.
(N4) If $X \cup Y$ is consistent, $N_1(X,Y) = N_2(X,Y) = X + Y$.
(N5) If $Cn(X) = Cn(Y)$, then
$N(X,Z) = N(Y,Z)$ and $N(Z,X) = N(Z,Y)$.
(N6) $Cn(Y) \cap N_1(X,Y) \subseteq N_2(X,Y)$;
$Cn(X) \cap N_2(X,Y) \subseteq N_1(X,Y)$.
(N7) If $F \subseteq Cn(Y) \cap N_1(X,Y)$, then $N(N_1(X,F),Y) = N(X,Y)$.
If $F \subseteq Cn(X) \cap N_2(X,Y)$, then $N(X, N_2(F,Y)) = N(X,Y)$.

Intuitively, (N1) says that the resulting belief state of each agent is logically closed as assume in AGM theory. (N2) states that no third party's information is introduced if there is no conflict between two agents. (N3) says that mutual belief revision can only happen between rational agents. (N4) says that each agent will accept all the beliefs of the other agent if no conflict arises. This is the cooperative aspect of mutual belief revision that an agent is happy to accommodate the other party whenever possible. (N5) assumes that mutual belief revision is syntax-independent, i.e. logically equivalent descriptions of belief states should lead to the same results.

Similar interpretation of these postulates can also be given in terms of negotiation. If we localize the belief state of an agent on the matters of a negotiation, its belief set represents its demands in the negotiation, which is called the *demand set* of the agent. (N1) then states that each negotiator should be aware of that she is responsible to undertake all the items and their consequences of her demands once they are included in an agreement. (N2) assumes that if no conflicts between the demands of two agents, amendments of demands may only be done within both side's initial demand sets[2]. (N3) means that no negotiation can proceed from inconsistent demands. (N4) and (N5) are similar.

It is easy to see that postulates (N1)-(N5) are counterparts of AGM postulates for belief revision. Notice that we have not a counterpart of *success postulate*. It is unreasonable to assume that one side of negotiation would accept all

[2] Note that if there are conflicts between the demands of two agents, the amended demand sets could be anything since $X + Y$ will be inconsistent.

the demands of the other side. In fact, we can view a mutual revision function as two non-prioritized belief revision operators [7].

Postulates (N6) and (N7) deviate slightly in style from AGM. We call (N6) the *principle of no recantation.* It states that once the demands of an agent are accepted by the other agent, it will not allowed for the agent to withdraw them [3]. We consider this as a rational restriction in negotiation protocol. In belief change setting this reflects the principle of information economy. Once an agent knows that her beliefs are accepted by other agents, she will more value them. We remark that (N6) does not imply the following condition:

$$K_1 \cap K_2 \subseteq N_1(K_1, K_2) \cap N_2(K_1, K_2)$$

which says that common items of initial demands must be included in the last agreement of negotiation. This condition is not generally acceptable because sometimes if both sides decide to give up a common item (in order to keep some more beneficial demands), it would not be in the last agreement.

(N7) deals with multiple stages of negotiation or iterated mutual belief revision. It is a typical strategy in negotiation that a negotiator poses its demands in stages. At each stage, the negotiator may reveal only part of her demands, hiding something tough or alternatives, and tries to persuade the other side to change her mind. (N7) says that if one can expect that some of her demands will be definitely accepted by the other side, it is unnecessary to pose them in stages. Therefore it would be useless to put weak demands first.

3 Iterated Belief Revision

It has been shown in the last section that there exists close relationship between AGM belief revision and mutual belief revision. We will show that the AGM revision function is a special case of mutual belief revision. Moreover, mutual belief revision implies an iterated revision mechanism. One of the postulates for iterated belief revision proposed by Darwiche and Pearl is required by mutual belief revision.

To show these, we first recall some basic facts of iterated belief revision in single agent environments.

To best suit the context of mutual belief revision, instead of using the original AGM framework, we shall exploit the multiple version of the AGM theory[16, 22], which allows us to revise a belief set by another belief set. Formally, for any belief set K and a set F of sentences, $K \otimes F$ stands for the new belief set that results when K is revised by the new information F. The operation is required to satisfy the following postulates:

($\otimes$1) $K \otimes F = Cn(K \otimes F)$.
($\otimes$2) $F \subseteq K \otimes F$.

[3] Note that $Cn(X) \cap N_2(X,Y)$ represents the demands of agent 1 which have been accepted by agent 2 after negotiation.

($\otimes$3) K$\otimes$F$\subseteq$ K+F.
($\otimes$4) If F$\cup$K is consistent, K+F $\subseteq$ K$\otimes$ F.
($\otimes$5) K$\otimes$ F is inconsistent if and only if F is inconsistent.
($\otimes$6) If Cn(F_1) = Cn(F_2), K$\otimes F_1$ = K$\otimes F_2$.
($\otimes$7) K$\otimes$($F_1 \cup F_2$) $\subseteq$ (K$\otimes F_1$)+F_2.
($\otimes$8) If $F_2\cup$(K$\otimes F_1$) is consistent, (K$\otimes F_1$)+F_2 $\subseteq$ K$\otimes$($F_1\cup F_2$).

The associated contraction function $\ominus$ can be defined by the following identity:

($\ominus Def$) $K \ominus F = (K \otimes F) \cap K$

A negotiation process normally consists of several stages of mutual belief revision. To simulate such a process, an iterated mechanism of belief revision is required[4]. The following assumption has been accepted by several different iterated belief revision formalisms:

($\otimes$IBR) $(K \otimes F_1) \otimes (F_1 \cup F_2) = K \otimes (F_1 \cup F_2)$

It is easy to see that ($\otimes$IBR) is the multiple version of the postulate ($C1$) in [4]. The following theorem is an easy generalization of Lehmann's consistency result presented in [8].

Theorem 1 *($\otimes$1)-($\otimes$6) are consistent with* ($\otimes$IBR).

From now on, we will call a revision function an *iterated belief revision* if it satisfy the postulates ($\otimes$1)-($\otimes$8) and ($\otimes$IBR). Note that such an iterated revision function is different from Darwiche and Pearl's one since we use the exact AGM postulates. This should be fine because ($\otimes$IBR) goes well with AGM postulates. However, how to adjust our formulism by using the exact form of Darwiche and Pearl's formalism should be a promising research topic for the future.

4 Master-Slave Revision: A Special Case of Mutual Belief Revision

Before we provide a general construction of negotiation function, let's consider a special case of mutual belief revision in which one agent unconditionally accepts the other agent's beliefs while the other stands still. Formally, a mutual revision function N is a *master-slave* revision if it satisfies

(M-S) $X \subseteq N_2(X, Y)$.

The condition says that the second agent (slave) always accepts beliefs of the first agents (master) with no reservation. By the following observation, the first agent can always keep its beliefs.

Lemma 1 *If N is a master-slave revision, then for any X and Y, $X \subseteq N_1(X, Y)$.*

[4] This is similar to some other settings, say belief fusion [14].

The following is the representation theorem for master-slave revision. It is also a proof of the consistency of mutual belief revision postulates.

Theorem 2 *Let N be a master-slave revision. Define a revision function $\otimes$ as follows: for any consistent belief set K and a set F of sentences,*

$K \otimes F \stackrel{def}{=} N_2(F, K)$

Then $\otimes$ satisfies the postulates ($\otimes$1)-($\otimes$8) and ($\otimes$IBR).

Conversely, if $\otimes$ is an iterated belief revision function which satisfies ($\otimes$1)-($\otimes$8) and ($\otimes$IBR), then the mutual belief revision function defined as follows satisfies (N1)-(N7) and (M-S):

$N(X, Y) \stackrel{def}{=} (Cn(Y) \otimes X, Cn(Y) \otimes X).$

This representation theorem shows that (N1)-(N7) plus (M-S) fully specifies a master-slave revision function. Since an iterated revision function always exists, this theorem also shows the consistency of the postulates (N1)-(N7).

In the next section, we will present a general construction of negotiation function which will allow more "balanced" negotiation.

5 Construction of Negotiation Operator

In this section we provide a general construction of negotiation function by using two independent iterated revision operators to simulate the process of belief change by two autonomous agents.

Negotiation Set Let X and Y be two sets of sentences, representing the initial demands or offers of two agents, respectively. We will call the pair (X,Y) a *negotiation encounter*. The naive way to define a negotiation function is to define it as two revision operation that each agent revises its demands by the other agent's demands:

$N(X, Y) \stackrel{def}{=} (Cn(X) \otimes_1 Y,\ Cn(Y) \otimes_2 X)$

Unfortunately, this definition is obviously inappropriate because it implies that one agent accepts the whole demand set of the other without care about her own demands. Typically, one agent accepts only part of demands from her opponent during a stage of negotiation. To model such a synthesis of cooperation and competition in negotiation, we introduce the following concepts.

Definition 2 Given two iterated revision functions $\otimes_1$ and $\otimes_2$. A *deal* over an encounter (X,Y) is a pair (Ψ_1, Ψ_2) such that $\Psi_1 \subseteq Cn(X)$, $\Psi_2 \subseteq Cn(Y)$ and satisfies the fix-point condition:

$$Cn(\Psi_1 \cup \Psi_2) = (Cn(X) \otimes_1 \Psi_2) \cap (Cn(Y) \otimes_2 \Psi_1) \quad (1)$$

Intuitively, a deal is an agreement between two agents in which agent 1 accepts part of demands, Ψ_2, from agent 2 whereas agent 2 accepts part of demands, Φ_1, from agent 1.

A deal δ=(Ψ_1, Ψ_2) *dominates* a deal δ'=(Ψ'_1, Ψ'_2) if they satisfies the following two conditions:

1. $Cn(X) \otimes_1 \Psi'_2 = Cn(X) \otimes_1 \Psi_2$ and $Cn(Y) \otimes_2 \Psi'_1 = Cn(Y) \otimes_2 \Psi_1$; And
2. either $\Psi'_1 \subset \Psi_1$ and $\Psi'_2 \subseteq \Psi_2$ or $\Psi'_1 \subseteq \Psi_1$ and $\Psi'_2 \subset \Psi_2$.

In the other words, δ dominates δ' if at least one agent agrees to accept more demands from the other without sacrificing any agent's profits.

A deal δ is called *pareto optimal* if there does not exist a deal which dominates δ.

A deal δ=(Ψ_1, Ψ_2) over (X,Y) is called *rational* if it satisfies the following conditions:

$Cn(X) \cap (Cn(Y) \otimes_2 \Psi_1) \subseteq Cn(X) \otimes_1 \Psi_2$
$Cn(Y) \cap (Cn(X) \otimes_1 \Psi_2) \subseteq Cn(Y) \otimes_2 \Psi_1$

which means that if the demands of one agent have been accepted by the other, they will be kept in her amended demand set.

Definition 3 A deal is a *negotiable alternative* over (X,Y) if it is pareto optimal and rational. The set of all the negotiable alternatives is called the *negotiation set* of (X,Y), denoted by $\mathcal{NS}$(X,Y).

The following lemma lists some properties of negotiation sets.

Lemma 2 *Let (X, Y) be any encounter.*

1. *For any $(\Psi_1, \Psi_2) \in \mathcal{NS}(X, Y)$, Ψ_1 and Ψ_2 are logically closed.*
2. *If $X \cup Y$ is consistent, $\mathcal{NS}(X, Y) = \{(Cn(X), Cn(Y))\}$.*
3. *If $X \cup Y$ is inconsistent, $(Cn(X) \cap Cn(Y), Cn(X) \cap Cn(Y)) \in \mathcal{NS}(X, Y)$.*
4. *For any $(\Psi_1, \Psi_2) \in \mathcal{NS}(X, Y)$,*
   $\Psi_1 = Cn(X) \cap (Cn(Y) \otimes_2 \Psi_1)$ and $\Psi_2 = Cn(Y) \cap (Cn(X) \otimes_1 \Psi_2)$.

The lemma shows that if $X \cup Y$ is consistent then $(Cn(X), Cn(Y))$ is the unique deal in the negotiation set of the encounter (X, Y). If it is inconsistent, the negotiation set could have more than one elements but it always includes a deal $(Cn(X) \cap Cn(Y), Cn(X) \cap Cn(Y))$, which we call it the *conflict deal* of the encounter.

Example 1 Consider an encounter (X, Y) where $X = \{p, q\}$ and $Y = \{\neg p, \neg q\}$. Let $\otimes_1$ and $\otimes_2$ are any iterated revision functions. Then the conflict deal will be $(Cn(\{p \leftrightarrow q\}), Cn(\{p \leftrightarrow q\}))$, which is in the negotiation set. Two other extreme cases, $(Cn(X), Cn(\{p \leftrightarrow q\}))$ and $(Cn(\{p \leftrightarrow q\}), Cn(Y))$, are also negotiable alternatives. If each agent is going to take some offers from the other, then more "balanced" negotiation could happen. However the result will heavily depend on the evaluation on their demands and counter-demands. Suppose that for agent 1, p is more entrenched than q and for agent 2 $\neg q$ is more entrenched than $\neg p$. Then $(Cn(\{p\}), Cn(\{\neg q\}))$ will be a negotiable alternative whereas $(Cn(\{q\}), Cn(\{\neg p\}))$ is not. □

5.1 Selection Function

Negotiation protocol is applied to regulate negotiations. It rules out irrational negotiation behavior but makes no decisions for negotiation participants. The outcomes of a negotiation mainly depend on the strategies of negotiators and their evaluation on negotiation alternatives. Therefore, a concrete negotiation is a decision-making procedure with which each agent chooses a deal from negotiation set. If both agents choose the same deal, then an agreement is reached; otherwise, the conflict deal will be the result of the negotiation. Anyhow, a negotiation process is nothing but a selection mechanism which chooses a deal from negotiation set. Let γ be a selection function which selects an element from a nonempty set. We will abbreviate $\gamma(\mathcal{NS}(X,Y))$ to $\gamma(X,Y)$. As usual, $\gamma_i(X,Y)$ means the ith component of $\gamma(X,Y)$. Now we define negotiation function as follows:

Definition 4 *Let $\otimes_1$ and $\otimes_2$ be two iterated revision functions and γ a selection function. Define a negotiation function N as follows: for any encounter (X,Y),*
1. if both X and Y are consistent, $N(X,Y) = (Cn(X) \otimes_1 \Phi_2, Cn(Y) \otimes_2 \Phi_1)$ where $(\Phi_1, \Phi_2) = \gamma(X,Y)$; otherwise,
2. $N(X,Y) = (\mathcal{L}, \mathcal{L})$.

It is easy to see that given an encounter (X,Y), if the selection function γ selects the conflict deal to the encounter, then $N(X,Y) = (X,Y)$. The following example shows another extreme case.

Example 2 Let $\otimes$ be an iterated revision operator and $\ominus$ is the associated contraction operator. For any encounter (X,Y), let $\gamma(X,Y) = (Cn(X), Cn(Y) \ominus X)$. Then γ is a selection function that defines a master-slave mutual belief revision operator where $\otimes_1 = \otimes_2 = \otimes$. □

It is easy to verify that any negotiation function N satisfies (N1)-(N6). However (N7) does not necessarily hold. In fact, (N7) requires kinds of uniformity in the selection mechanism over different negotiation situation. Arbitrary selection functions can not embody the rationality of negotiation behavior.

Definition 5 A selection function γ is *downwards compatible* if for any $F_1 \subseteq \gamma_1(X,Y)$ and $F_2 \subseteq \gamma_2(X,Y)$,
$\gamma(Cn(X) \otimes_1 F_2, Cn(Y) \otimes_2 F_1) = (\gamma_1(X,Y) + F_2, \gamma_2(X,Y) + F_1)$.

Example 3 Let γ be a selection function such that $\gamma(X,Y) = (Cn(X) \cap Cn(Y), Cn(X) \cap Cn(Y))$. Then γ is downwards compatible. In other wards, always standing still is a kind of uniform negotiation behavior.

Now we come to the main result of the paper.

Theorem 3 *If γ is a downwards compatible selection function, the negotiation function defined by Definition 4 satisfies (N1)-(N7).*

Since a downwards compatible selection function always exists, this theorem shows again the consistency of the negotiation postulates. We remark that downward compatibility is a sufficient condition for postulate (N7) but not necessary. In fact, we have shown that any master-slave negotiation function satisfies postulate (N7) but its associated selection function is not necessarily downwards compatible. How to find a sufficient and necessary condition for (N7) is open for the future research.

6 Monotonic Concession Protocol

Unlike most game-theory based approaches to negotiation, the purpose of this research is not to design particular negotiation strategies for individual agents. We aim to provide a formal language and an analysis tool to describe and evaluate negotiation protocols. In this section, we will show an example of such an analysis by using Rosenschein and Zlotkin's *Monotonic Concession Protocol* [20].

Monotonic Concession Protocol deals with negotiations between two agents. The agents start by simultaneously proposing one deal from the space of possible deals. An agreement is reached if both agents chose the same deal or one agent offered a deal that exceeded what the other one asked for. If no agreement has been reached, the protocol continues to another round. Each agent has two choices: to stand still or to concede. An agent is not allowed to offer the other agent less than her did in the previous round. If neither agent concedes at some step, then the negotiation ends with the conflict deal. Otherwise the negotiation continues (see [RosenscheinandZlotkin1994] page 40).

The feature of the protocol is that agents cannot backtrack, nor can they both simultaneously stand still in the negotiation more than once. As a result, the negotiation process is monotonic and ensures convergence to a deal.

To make the protocol more specific, let's assume that the initial demands of the agents are X and Y, respectively. At the first round, each agent chooses a deal from the negotiation set $\mathcal{NS}$(X,Y), say $\delta' = (\Psi_1', \Psi_2')$ and $\delta'' = (\Psi_1'', \Psi_2'')$. If $\delta' = \delta''$, then an agreement is reached. Otherwise, each agent makes a decision whether she is going to stand still or to concede. If an agent chose to concede, say agent 1, she should amend her demand set so that the new demand set includes Ψ_2' as well as Ψ_1'', and then chooses another deal from the new negotiation set that extends Ψ_2'.

To such a protocol, if both agents apply iterated belief revision when revise their demands set and they make the same decision on whether to stand-still or to concede for any given pair of demand sets, then we have the following

Theorem 4 *Any negotiation under the protocol satisfies postulates (N1)-(N6).*

Unfortunately, (N7) does not automatically hold. To satisfy (N7), we need more specification on the belief revision and decision-making behavior. We leave this for a longer version of the paper.

7 Related Work

There have been several streams of research which related to this work. One of them is the work on arbitration, belief merging and knowledge fusion [9, 10]. All these researches deal with conflicts between agents. In fact, for any given negotiation function N, we can define an "arbitration" operator $\triangle$ as: for any belief set K_1 and K_2,

$K_1 \triangle K_2 = N_1(K_1, K_2) \cap N_2(K_1, K_2)$

Then we can easily verify that $\triangle$ satisfies most of Revesz's postulates and Liberatore and Schaerf's postulates for arbitration operation[19, 9]. However the differences between negotiation and arbitration are obvious. For instance, the following postulate does not necessarily hold in negotiation setting:

If K_1 is consistent, so is $K_1 \cup (K_1 \triangle K_2)$.

Another stream of work is that of non-prioritized belief revision[7, 2]. Particularly, Booth in [2] showed a way to model non-prioritized belief revision via negotiation process. However, his work aimed to provide a formalism for non-prioritized revision rather than a formal framework for negotiation.

There have been several attempts to consider belief revision in the setting of multi-agent systems [10, 13, 15]. In [15] mutual belief revision is referred to the process of belief revision by which an agent in synchronous multi-agent systems revises its beliefs about other agents' beliefs. Instead of axiomatic analysis, a Kripkean semantical theory was presented for modelling such kind of mutual belief revision.

In terms of logical approach to negotiation, there are numbers of researches on argumentation-based negotiation[21, 11, 17]. Although the goal is similar, the formalisms and emphases of the work are distinct from ours.

8 Conclusion

This paper presented a formal framework for negotiation protocol analysis. A set of AGM-style postulates was presented to model rational negotiation behavior. Its consistency was proved through an explicit construction of negotiation function in which negotiation process was modelled by two related iterated belief revision operations. This model also captured the characteristic of negotiation that negotiation is an organic synthesis of cooperation and competition between agents.

Many problems remain to explore. A representation theorem for the postulates is to be given. To this end, a necessary and sufficient condition for postulate (N7) is needed. We emphasize that the overall purpose of this research is not to present a logical analysis for a particular negotiation protocol but to provide a formal language for describing rational negotiation behavior. The postulates presented in this paper are less than complete. We bleief that more characteristics of negotiation process can be investigated in this framework. For instance, the following property of negotiation can be described in our language, which states sort of terminability of negotiation process:

$N(N_1(X,Y), N_2(X,Y)) = N(X,Y)$

It can be proved that it is consistent but independent with the existing postulates.

Appendix: Proofs of Theorems

Proof of Theorem 2: The postulates ($\otimes$1) and ($\otimes$4)-($\otimes$6) are the special case of the postulates (N1), (N3)-(N5), respectively. ($\otimes$2) is implied by (M-S). For ($\otimes$3), if $F \cup K$ is consistent, then (N2) implies $K \otimes F = K + F$; if $F \cup K$ is inconsistent, $K \otimes F \subseteq K + F$ holds trivially. To prove ($\otimes$ IBR), we know that $(K \otimes F_1) \otimes (F_1 \cup F_2) = N_2(F_1 \cup F_2, K \otimes F_1)$. Since $F_1 \subseteq F_1 \cup F_2$, it follows from (M-S) that $F_1 \subseteq N_2(F_1 \cup F_2, K)$. Therefore $F_1 \subseteq (F_1 \cup F_2) \cap N_2(F_1 \cup F_2, K)$. By (N7) we yield $N_2(F_1 \cup F_2, N_2(F_1, K)) = N_2(F_1 \cup F_2, K) = K \otimes (F_1 \cup F_2)$. ($\otimes$7) and ($\otimes$8) are implied by ($\otimes IBR$).

Conversely, if $\otimes$ is an iterated revision function which satisfies ($\otimes$1)-($\otimes$8) and ($\otimes$IBR), then the defined negotiation function satisfies obviously (N1)-(N6).

To prove (N7), assume $F \subseteq Cn(Y) \cap N_1(X,Y)$. Then $Cn(F) \ominus X \subseteq Cn(F) \subseteq N_1(X,Y) \subseteq Cn(Y) \otimes X$. Hence $N_1(N_1(X,F),Y) = Cn(Y) \otimes N_1(X,F) = Cn(Y) \otimes (Cn(F) \otimes X) = Cn(Y) \otimes (X \cup (Cn(F) \ominus X)) = (Cn(Y) \otimes X) + (Cn(F) \ominus X) = Cn(Y) \otimes X = N_1(X,Y)$. Similarly we have $N_2(N_1(X,F),Y) = N_2(X,Y)$.

Again assume $F \subseteq Cn(X) \cap N_2(X,Y)$. Then $N_1(X, N_2(F,Y)) = N_1(X, Cn(Y) \otimes F) = (Cn(Y) \otimes F) \otimes X = Cn(Y) \otimes (F \cup X) = Cn(Y) \otimes X = N_1(X,Y)$. $N_2(X, N_2(F,Y))$ is exactly the same. Therefore (N7) is valid. □

Proof of Lemma 2: The item 1 and 2 are straightforward.

For 3, let $F = Cn(X) \cap Cn(Y)$. Obviously (F,F) is a deal over (X,Y). Assume that (Ψ_1, Ψ_2) dominates $(Cn(X) \cap Cn(Y), Cn(X) \cap Cn(Y)) = (F,F)$. Then $\Psi_1 \subseteq Cn(\Psi_1 \cup \Psi_2) = (Cn(X) \otimes_1 \Psi_2) \cap Cn(Y) \otimes_2 \Psi_1) = (Cn(X) \otimes_1 F) \cap (Cn(Y) \otimes_2 F) = F$. Thus $\Psi_1 \subseteq F$. Since (Ψ_1, Ψ_2) dominates (F,F), $\Psi_1 \not\subset F$. Similarly we have $\Psi_2 \not\subset F$, a contradiction. So $(F,F) \in (X,Y)$.

For 4, in order to prove the first equation, we only need to verify $Cn(X) \cap (Cn(Y) \otimes_2 \Psi_1) \subseteq \Psi_1$. Assume that there is a sentence A such that $A \in Cn(X) \cap (Cn(Y) \otimes_2 \Psi_1)$ but $A \notin \Psi_1$. Since $A \in Cn(Y) \otimes_2 \Psi_1$, it follows from ($\otimes$7) and ($\otimes$8) that $Cn(Y) \otimes_2 (\Psi_1 \cup \{A\}) = Cn(Y) \otimes_2 \Psi_1$. Therefore $(Cn(X) \otimes_1 \Psi_2) \cap (Cn(Y) \otimes_2 (\Psi_1 \cup \{A\})) = (Cn(X) \otimes_1 \Psi_2) \cap (Cn(Y) \otimes_2 \Psi_1) = Cn(\Psi_1 \cup \Psi_2) \subseteq Cn(\Psi_1 \cup \{A\} \cup \Psi_2)$. On the other hand, since (Ψ_1, Ψ_2) is rational, $Cn(X) \cap (Cn(Y) \otimes_2 \Psi_1) \subseteq Cn(X) \otimes_1 \Psi_2$. Thus $A \in Cn(X) \otimes_1 \Psi_2$. It follows that $Cn(\Psi_1 \cup \{A\} \cup \Psi_2) \subseteq (Cn(X) \otimes_1 (\Psi_1 \cup \{A\})) \cap (Cn(Y) \otimes_2 \Psi_2)$. Therefore $(\Psi_1 \cup \{A\}, \Psi_2)$ is also a deal over (X,Y), which obviously dominates (Ψ_1, Ψ_2). Thus (Ψ_1, Ψ_2) is not pareto optimal, a contradiction. The other one is similar. □

Proof of Theorem 3: The proof of (N1)-(N6) is straightforward. To verify (N7), assume that $F \subseteq Cn(Y) \cap N_1(X,Y)$. Then $F \subseteq Cn(Y) \cap (Cn(X) \otimes_1 \gamma_2(X,Y))$. It follows from Lemma 2 that $F \subseteq \gamma_2(X,Y)$. According to the construction of negotiation function, $N_1(N_1(X,F),Y) = N_1(X,F) \otimes_1 \gamma_2(N_1(X,F),Y) = (Cn(X) \otimes_1 \gamma_2(X,F)) \otimes_1 \gamma_2(Cn(X) \otimes_1 \gamma_2(X,F),Y)$. Let $F_2 = \gamma_2(X,F)$. Thus $F_2 \subseteq Cn(F) \subseteq \gamma_2(X,Y)$. Since γ is downwards compatible, we have $\gamma_2(Cn(X) \otimes_1 F_2, Y) = \gamma_2(X,Y)$. Therefore $N_1(N_1(X,F),Y) = (Cn(X) \otimes_1 F_2) \otimes_1 \gamma_2(Cn(X) \otimes_1 F_2, Y) = (Cn(X) \otimes_1 F_2) \otimes_1 \gamma_2(X,Y)$. By $(\otimes IBR)$ we know $(Cn(X) \otimes_1 F_2) \otimes_1 \gamma_2(X,Y) = Cn(X) \otimes_1 \gamma_2(X,Y) = N_1(X,Y)$. Put them together we yield $N_1(N_1(X,F),Y) = N_1(X,Y)$. For agent 2, similarly we have $N_2(N_1(X,F),Y) = Cn(Y) \otimes_2 \gamma_1 (N_1(X,F),Y) = Cn(Y) \otimes_2 \gamma_1(Cn(X) \otimes_1 \gamma_2(X,F),Y) = Cn(Y) \otimes_2 \gamma_1(Cn(X) \otimes_1 F_2, Y)$. By the downward compatibility of γ, we have $\gamma_1(Cn(X) \otimes_1 F_2, Y) = \gamma_1(X,Y) + F_2$. Since $F_2 \subseteq Cn(Y) \otimes_2 \gamma_1(X,Y)$, Therefore, $Cn(Y) \otimes_2 (\gamma_1(X,Y) + F_2) = Cn(Y) \otimes_2 \gamma_1(X,Y)$. These give us $N_2(N_1(X,F),Y) = Cn(Y) \otimes_2 (\gamma_1(X,Y) + F_2) = Cn(Y) \otimes_2 \gamma_1(X,Y) = N_2(X,Y)$. We have proved $N(N_1(X,F),Y) = N(X,Y)$. The other half of the postulate is symmetric. □

Proof of Theorem 4: The postulates (N1)-(N5) are straightforward except that the limit case when X or Y is inconsistent needs to be specified. Now we prove (N6). Assume that the negotiation ended at n round. $\Psi_2^1 \subseteq \Psi_2^2 \subseteq \cdots \subseteq \Psi_2^n$ and $\Psi_1^1 \subseteq \Psi_1^2 \subseteq \cdots \subseteq \Psi_1^n$ are all concessions made by agent 1 and agent 2, respectively. Note that some '$\subseteq$'s may be equal if the agent stood still at that round. Then by $(\otimes IBR)$ the final demand sets after the negotiation terminated are:

$$(((Cn(X) \otimes_1 \Psi_2^1) \otimes_1 \cdots) \otimes_1 \Psi_2^n = Cn(X) \otimes_1 \Psi_2^n$$
$$(((Cn(X) \otimes_2 \Psi_1^1) \otimes_2 \cdots) \otimes_2 \Psi_1^n = Cn(Y) \otimes_2 \Psi_1^n$$

Meanwhile, for i=1,···,n, $Cn(X) \cap (Cn(Y) \otimes_2 \Psi_1^i) \subseteq Cn(X) \otimes_1 \Psi_2^i$ and $Cn(Y) \cap (Cn(X) \otimes_1 \Psi_2^i) \subseteq Cn(y) \otimes_2 \Psi_1^i$. Therefore (N6) holds. □

References

[1] E. Alchourrón, P. Gärdenfors and Makinson, On the logic of theory change: partial meet contraction and revision functions, *J. Symbolic Logic,* 50(2)(1985), 510-530. 378

[2] R. Booth, A negotiation-style framework for non-priorised revision, *TARK'01*, 137-150. 386

[3] T. X. Bui and M. F. Shakun, Negotiation processes, evolutionary systems design, and NEGOTIATOR. In Melvin F. Shakun edt, *Negotiation Processes,* Kluwer Academic Publishers, 39-53, 1996. 377

[4] A. Darwiche and J. Pearl, On the logic of iterated belief revision. *Artificial Intelligence,* 89(1-2):1-29, 1997. 381

[5] P. Faratin, C. Sierra and N. R. Jennings, Negotiation decision functions for autonomous agents, *Robotics and Autonomous Systems,* 24(1998):159–182, 1998. 377

[6] N. Friedman and J. Y. Halpern, Belief revision: A critique, *KR'96,* 421-431, 1996.

[7] S. O. Hansson, E. Fermé, J. Cantwell, and M. Falappa, Credibilit-limited revision, *J. Symbolic Logic*, 66:1581-1596, 2001. 380, 386

[8] D. Lehmann, Belief revision, revised, *IJCAI-95,* 1534-1540, 1995. 381

[9] P. Liberatore and M. Schaerf, Arbitration (or How to merge knowledge bases), *IEEE Transactions on Knowledge and Data Engineering*, 1(10):76-90, 1998. 386

[10] N. E. Kfir-Dahav and M. Tennenholtz, Multi-agent belief revision, *TARK'96*, 175-194, 1996. 386

[11] S.Kraus, K.Sycara, A.Evenchik, Reaching agreements through argumentation: a logical model and implementation, *Artificial Intelligence*, 104(1998), 1-69, 1998. 377, 386

[12] S. Konieczny and R. Pino-Pérez, On the logic of merging, *KR'98*, 488-498, 1998.

[13] B. Malheiro, N. R. Jennings, and E. Oliveira, Belief revision in multi-agent systems, *ECAI-94*, 294-298, 1994. 386

[14] P. Maynard-Reid II and Y. Shoham, Belief fusion: aggregating pedigreed belief states, *J. of Logic, Language and Information* 10 (2):183-209, 2001. 381

[15] R. van der Meyden, Mutual belief revision (Preliminary Report), *KR'94*, 595-606, 1994. 386

[16] A. Nayak, Iterated belief change based on epistemic entrenchment, *Erkentnis* 41, 353-390,1994. 380

[17] S. Parsons, C. Sierra, N. Jennings: Agents that reason and negotiate by arguing, J. of Logic and Computation 8(3): 261-292, 1998. 377, 386

[18] D. G. Pruitt,*Negotiation Behavour*, Academic Press, 1981. 377

[19] P. Z. Revesz, On the semantics of arbitration, *Intern. J. of Algebra and Computation,* 7(2):133-160, 1997. 386

[20] J. S. Rosenschein and G. Zlotkin, *Rules of Encounter*, MIT Press, 1994. 377, 385

[21] K. Sycara, Persuasive arguementation in negotiation, *Theory and Decision*, 28:203-242, 1990. 377, 386

[22] D. Zhang, S. Chen, W. Zhu and Z. Chen, Representation theorems for multiple belief changes, *IJCAI-97*, 89-94, 1997. 380

[23] D. Zhang and N. Foo, Infinitary belief revision,*J. of Philosophical Logic*, 30(6):525-570, 2001.

A Probabilistic Line Breaking Algorithm

Remco R. Bouckaert

Xtal Mountain Information Technology and Computer Science Department
University of Waikato, New Zealand
rrb@xm.co.nz, remco@cs.waikato.ac.nz

Abstract. We show how a probabilistic interpretation of an ill defined problem, the problem of finding line breaks in a paragraph, can lead to an efficient new algorithm that performs well. The graphical model that results from the probabilistic interpretation has the advantage that it is easy to tune due to the probabilistic approach. Furthermore, the algorithm optimizes the probability a break up is acceptable over the whole paragraph, it does not show threshold effects and it allows for easy incorporation of subtle typographical rules. Thanks to the architecture of the Bayesian network, the algorithm is linear in the number of characters in a paragraph. Empirical evidence suggests that this algorithm performs closer to results published through desk top publishing than a number of existing systems.

1 Introduction

With the advance of information systems, the amount of automatically formatted text increases daily. However, judging from a software patent search, not a lot of attention is spent on increasing the quality of layout. New interest in this area is driven by the standardization of style specifications through CSS and XSL-FO [8]. Limitations of automatic typesetting systems are partly due to the fact that the underlying algorithms have been designed with resource restrictions in mind. For example, once TEX [3] has sufficient material to complete a page, the page is output to free up memory. If many floating objects like figures and tables are present, these can be placed in undesirable locations far from the place where they are referenced.

Considerations of limited resources were quite valid in the time that these algorithms were developed. However, they do not apply today and this opens a way to improve various typesetting algorithms. In typesetting a wide range of decisions need to be made in the areas of hyphenation, line breaking, page breaking, floating object placement, footnotes, headers and footers, page numbering, reference resolution, etc. In this paper, we consider line breaking only, but the principles presented should be extendible to page breaking, floating object placement and other areas as well.

A reasonable approach seems to be to ask professional publishers what makes a good line break-up. However, this knowledge is very difficult to elicit. Typically only examples of what is acceptable and what is not can be given, with

T.D. Gedeon and L.C.C. Fung (Eds.): AI 2003, LNAI 2903, pp. 390–401, 2003.

disagreement on borderline cases. So finding good line break-ups is an ill specified problem in which decisions need to be made with uncertainty of the function to optimize.

In this paper we take a probabilistic approach to line breaking in which we model the probability that a particular break-up is acceptable. The probability of acceptability can be maximized and the break-up with the highest probability chosen. A dynamic graphical model is used to capture acceptability.

The next section introduces concepts of line breaking and related work. In Section 3 a probabilistic algorithm is presented for breaking lines. Section 4 compares the performance of some popular typesetting systems with data published through desk top publishing together with an empirical comparison of the new algorithm. We finish with concluding remarks and suggestions for further extensions in Section 5.

2 The Line Breaking Problem

Breaking a paragraph into lines consists of selecting the break points in a paragraph. To determine how much space a line takes, each character (or rather glyph, see comment on ligatures below) in the paragraph is modeled as a box with a specific width, height and elevation from the baseline. These properties are determined by the font being used and the font size. The length of a line is the sum of the widths of the boxes in the line. The width of a box is not only determined by the width of the character, but is also influenced by the following character. If the width of a character would not be *kerned*, the result may look rather unpleasing. For example, compare these unkerned and kerned words:

BAYES
BAYES

The latter has less space between the A and Y, and the width of the A is modeled less wide then when an A is followed by, say, another A.

Some fonts treat certain combinations of characters as a single character. Examples of such so called *ligatures* is the combination of f and i and the combination of two f's and an i. The following line illustrates the difference between separate characters and their ligatures:

efficient fix
efficient fix

So, the first step in breaking a line is determining the width of its characters, taking kerning and ligatures in account. The next step is determining possible break points. Typical break points in Latin alphabet languages are whitespaces, hyphen characters and points where hyphenation is allowed.

Finding these hyphenation points by maintaining a wordlist is cumbersome because of the many variations of appearances of a word (e.g. singular and plural). Furthermore, maintaining such a word list in areas where new words are invented regularly (e.g. brand names) is a huge task. An elegant alternative is to use a hyphenation algorithm [3, 6] which essentially works as follows. Some words are stored in an exception list (for example, as-so-ciate). For words not in the exception list, a pattern list is used, which is generated from a word list containing hyphenated words. Patterns contain characters and numbers, for example 'y3thin'. A word is compared with the characters in the patterns and every character is assigned the highest number in the pattern. The word 'anything' matches 'ythin' in the pattern 'y3thin' and the y gets assigned the number 3. Any odd numbered character is a hyphenation point. So 'any-thing' is allowed. It is claimed [3] that more than 95% of the allowable hyphenation points in US English can be found by this algorithm (taking word frequencies in account).

Hyphenation points at the first λ and last ρ characters are discarded to prevent unacceptable hyphenation points. For exampe, consider the hyphenation points found when not suppressing the first and last points in the words 'a-ha', 'Bal-i', 's-tu-den-t' or 'A-pu-li-a'. By default, λ is 2 and ρ is 3 for English.

A broken line may not completely fit into the available space between the left and right margin. To make it fit, space may be distributed among the whitespaces in the line, or some whitespace may be taken out if the text exceeds the text width. We assume that there is a single target width for the complete paragraph and the paragraph is adjusted. An indent that applies to the first line of the paragraph can be modeled using a fixed width unbreakable space (possibly with a negative value).

Now that we have a flavor of the problem, we can specify some terms more formally. A *paragraph* p is a sequence of $n > 0$ characters c_i, $1 \leq i \leq n$. A *break point candidate* in p is an index of a character in p for which it is allowed to break a line. Typical break point candidates are space and hyphen characters and hyphenation points in words. We are interested in finding a sequence of character indexes $s_0, s_1, \ldots, s_k$, $\forall_{1 \leq j \leq k} 1 \leq s_j \leq n$ and $s_{j-1} < s_j$ where each s_j is a break point candidate. Each of the indices s_j indicates a *break point* in p. Index s_0 is added for convenience of further definitions and by convention $s_0 = 0$ and $s_k = n$. A *line* in p with break points $s_0, \ldots, s_k$ is the sub-sequence of characters between breakpoints, more specific for $1 \leq j \leq k$, $l_j = \{c_{s_{j-1}+1} \ldots c_{s_j}\}$ (note $s_0 = 0$ so $l_1 = \{c_1 \ldots c_{s_1}\}$). So, a break point s_j is the last character on line l_j.

The *character width* cw_i $1 \leq i \leq n$ of a character c_i in a paragraph p with break points $s_0, \ldots, s_k$ is the width of the character taking the font properties in account. Furthermore, the character width takes breakpoints in account with the following rules:

- A character not on a break point may have its width corrected due to kerning.
- A space character at the beginning or end of a line has width 0.
- A non-space and non-hyphen character not at the end of a word that is a break point has width of the character plus the width of a hyphen.
- If c_i is part of a ligature, and the ligature is not broken, cw_i is the width of the ligature divided among the characters that make up the ligature.

The *natural line width* nlw_i, $0 \leq i \leq n$ at character c_i is the sum of widths of characters on to the closest previous breakpoint s_j. More specific, nlw_0 is the indent, and if $s_j = 0$, $nlw_i = nlw_0 + \sum_{d=1}^{i} cw_d$, otherwise $nlw_i = \sum_{d=s_j+1}^{i} cw_d$. The *natural line width* of a line l_j, $1 \leq j \leq k$ is nlw_{s_j}. The *number of spaces* in a line l_j is $nsp_{s_j} = \sum_{i=s_{j-1}+1}^{s_j-1}\{1|c_i = \text{space}\}$, so spaces at the start and end of a line are not counted (it is assumed no space follows another space).

The *line breaking problem* for a paragraph p for a desired *text width* and *indent* is finding a set of break points of p that look 'nice'. We will consider fixed text width, tw, throughout this paper with the exception of the first line, which can have a (possibly negative) indent in.

A simple approach to the line breaking problem is to consider lines one by one, like Unix's Troff. Simply keep adding words as long as the natural line width does not exceed the text width. If a word makes the line overflow, try hyphenate the word (using a word list or Liang's hyphenation algorithm) and add as many parts (possibly zero) of the word to the line as long as the text width is not exceeded. Output the line and place the remainder on the following line. In this approach, line breaks are only optimized locally, without regard to the breaks in the remainder of the paragraph.

TEX's line breaking algorithm [4] on the other hand optimizes line breaks on the level of the paragraph. We give a short description of TEX's line breaking algorithm here (see [3] for an extensive description). TEX determines character widths taking kerning and ligatures in account and uses a three phased process.

In phase 1, no hyphenation points are considered for line breaking, only white spaces. For a paragraph broken in k lines, for each line $j = 1 \ldots k$ a badness b_i is calculated, using $b_j = 100(\frac{nlw_{s_j} - tw}{nsp_{s_j} \cdot f})^3$ where nlw_{s_j} the natural line width, tw the text width, nsp_{s_j} the number of spaces of line j and f is a factor that differs depending on whether the line needs to be stretched or shrunk. Note that $nlw_{s_j} - tw$ represents the amount that the line needs to be stretched or shrunk. Depending on the value of b_j, the line is classified as tight, decent, loose or very loose. If none of the line's badnesses exceed a 'pretolerance' limit, the paragraph is accepted.

Otherwise, in phase 2, the hyphenation candidates are calculated for the paragraph and a new set of breakpoints is calculated. If all of the line's badnesses are below a tolerance level (which differs from the pretolerance level) a demirit for the paragraph is calculated as a combination of line badnesses, roughly as

$$\sum_{j=1}^{n}(c + b_j)^2 + p \cdot |p| \tag{1}$$

where c is a constant line penalty and p a penalty term. A tight line next to a loose line increases the penalty p. If two consecutive lines are hyphenated, or if the last broken line is hyphenated, the penalty is increased. A paragraph with a demerit below a threshold is accepted.

Otherwise, in phase 3, the steps in phase 2 are repeated but now with an emergency stretch that allows lines to shrink or expand more. If this last phase does not succeed, TEX outputs overly long lines, due to the thresholds in the algorithm.

3 A Probabilistic Line Breaking Algorithm

We start with deriving the line breaking algorithm and illustrate how to design a simple but efficient probabilistic algorithm based on a dynamic Bayesian network. Then, this model is generalized and we show how the behavior of the network can be influenced by tuning network parameters. Because the network is easy to comprehend, the resulting change in behavior is intuitively satisfying.

3.1 A Simple Probabilistic Line Breaking Algorithm

We approach the line breaking problem by defining a probability distribution that represents the acceptability of a set of breakpoints.

Let p be a paragraph with n characters $c_1, \ldots, c_n$, text width tw, indent in and a set of break points $S = \{s_0, \ldots, s_k\}$. Let a_i, $1 \leq i \leq n$ be a boolean representing whether the break-up up to character i is acceptable, and let a denote a_n. Then we are interested in $P(a, S, p, in)$ where we leave tw and cw_i implicit. We make the following two assumptions

1. the acceptability of the first i characters in the break-up is independent of the break-up of the remainder, provided nlw_{i-1} is known for the remainder.
2. we have no initial preference for any breakup, that is, $P(S|p, in)$ is constant for any S.

The first assumption is the base for the efficiency of TEX's line breaking algorithm. The last assumption is for the sake of simplifying the derivation of the probabilistic algorithm, and will be relaxed in the following section. Now, $P(a, S, p, in) = P(a|S, p, in)P(S|p, in)P(p, in)$. Since p and in are given, and using the second assumption, $P(p, in)$ and $P(S|p, in)$ are constant, so $P(a, S, p, in) \propto P(a|S, p, in)$. The first assumption lets us decompose $P(a|S, p, in)$ as $P(a_i|S\backslash\{i+1, \ldots, n\}, p, in) \cdot P(a|S\backslash\{1, \ldots, i\}, p, nlw_i, in)$, where the first term is the acceptability of the break-up of the first i characters, and the second term the acceptability of the remainder given the natural line width nlw_i of the last line of the first i characters. Now note $P(a_i|S\backslash\{i+1, \ldots, n\}, p, in) = P(a_i|S\backslash\{i+1, \ldots, n\}, p\backslash\{c_{i+1}, \ldots, n\}, in)$. Therefore, we can write $P(a|S, p, in)$ as the product of the acceptability of the individual characters.

That is $P(a|S, p, in)$ equals $P(a_1|S\backslash\{2, \ldots, n\}, p, in) \cdot P(a|S\backslash\{2, \ldots, n\}, p\backslash \{c_2, \ldots, c_n\}, nlw_1, in)$. Observe, in the first term $S\backslash\{2, \ldots, n\}$ can be interpreted as the boolean $1 \in S$ and that in the latter term the indent does not give any extra information once nlw_1 is known. So, we can write $P(a_1|1 \in S, p, in) \cdot P(a|S\backslash\{2, \ldots, n\}, p\backslash\{c_2, \ldots, c_n\}, nlw_1)$. Repeating this process and realising $nlw_0 = in$ gives $P(a|S, p, in) = \prod_{i=1}^{n} P(a_i|i \in S, p, nlw_{i-1})$.

Further, the natural line width at character i, $0 \leq i \leq n$ is calculated as

$$nlw_i = \begin{cases} cw_i & |i-1 \in S \\ nlw_{i-1} + cw_i & |i-1 \notin S \\ in & |i=0 \end{cases}$$

Now we are left with determining $P(a_i|i \in S, p, nlw_{i-1})$. If c_i is not a break point candidate (i.e., it is not a space, hyphen, or hyphenation point) and $i \in S$, the break-up is not acceptable under any circumstances, and $P(a_i|i \in S, p, nlw_{i-1}) = 0$. If not breaking at c_i means that the natural line width remains within the text width, the line break-up is just as acceptable as the accaptability of the remainder of the paragraph, $P(a_i|i \in S, p, nlw_{i-1}) = 1$. However, not breaking may be unacceptable as well, if this means extending the natural length beyond the text width. Breaking before the natural line width has reached the text width also decreases acceptability. In summary, $P(a_i|i \in S, p, nlw_{i-1})$ is defined as

$$\begin{array}{ll} 0 & |i \text{ is not a break point candidate and } i \in S \\ 1 & |nlw_{i-1} + cw_i \leq tw \text{ and } i \notin S \\ P^a(cw_i|nlw_{i-1} - tw) & |nlw_{i-1} + cw_i > tw \\ P^a(cw_i|nlw_{i-1} - tw) & |nlw_{i-1} + cw_i \leq tw \text{ and } i \in S \end{array}$$

where P^a a distribution representing acceptability for breaking not earlier than at c_i instead of c_{i-1}. For the choice of this distribution, we let ourselves inspire by the empirical distribution of what seems to be acceptable as shown in Figure 2 by the graph for published line lengths. This shows a distribution which is roughly normal, with different variances for lines exceeding and undercutting the text width. So, we define

$$P^a(cw_i|x) = \begin{cases} N(x + cw_i, \sigma^+)/P^a(x) & |x + cw_i \geq 0 \\ N(x + cw_i, \sigma^-)/P^a(x) & |x + cw_i < 0 \end{cases}$$

where $N(x, \sigma)$ the standard normal distribution $N(x, \sigma) = \frac{1}{2\pi\sigma} e^{\frac{x^2}{\sigma^2}}$, $P^a(x) = N(x, \sigma^+)$ if $x \geq 0$ and $N(x, \sigma^-)$ otherwise and σ^+ and σ^- two variances, which are the two parameters to choose. Note that the contribution of line j to $P(a|S, p, in)$ is $\prod_{i=s_{j-1}+1}^{s_j} P(i \in S|p, nlw_{i-1})$ and that this reduces to $N(nlw_{s_j} - tw, \sigma^+)$ if line j's natural width exceeds the text width ($nlw_{s_j} > tw$) and $N(nlw_{s_j} - tw, \sigma^-)$ if $nlw_{s_j} \leq tw$.

What we are after is of course selecting a set of break points that maximizes $P(a|S, p, in)$. Note that what we have done so far is derive a dynamic Bayes network with a structure as pictured in Figure 1. The character widths cw_i are given (refer to the rules as described in Section 2 for the actual details). Node b_i indicates whether breaking is allowed at character c_i. The conditional distribution for nlw_i and a_i are described in this section above.[1] What remains is selecting a value for the boolean variable indicated by $i \in S$ in the figure. Straightforward inference e.g. as described by Pearl [7] or Lauritzen [5], for finding the most likely configuration for b_i suffices to find the breakpoints.

It is interesting to see how this algorithm relates to TeX's line breaking algorithm. It can be shown[2] for a paragraph p that a set of break points $S =$

[1] Instead of implementing one slice per character, one slice per break point candidate can be used. Of course, a correction for the characters widths between break point candidates need to be implemented then.

[2] See technical report version of this article for details.

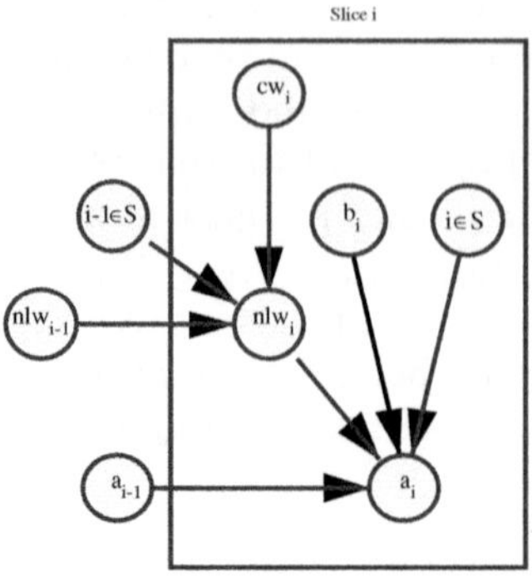

Fig. 1. Network representing simple probabilistic line breaking algorithm

$\{s_0, \ldots, s_k\}$ maximizes $P(a, S, p, in)$ iff S maximizes

$$\sum_{j \in S+} -(nlw_{s_j} - tw)^2 + \sum_{j \in S-} (-(nlw_{s_j} - tw)^2 C_1 + C_2) - k.C_3 \qquad (2)$$

where S^+ the breakpoints $j \in S$ such that $nlw_{s_j} > tw$, $S^- = S \backslash S$, $C_1 = \frac{\sigma^{+2}}{\sigma^{-2}}$, $C_2 = \sigma^{+2} \log \frac{\sigma^+}{\sigma^-}$, and $C_3 = \sigma^{+2} \log(2\pi\sigma^+)$. In words, maximizing the probability of acceptability $P(a, S, p, in)$ by choosing a set of break points S results in the same set of breakpoints as maximizing the sum over the lines longer than the text width tw (first sum term) and lines shorter than tw (second sum term). Comparing this with the formula that TeX maximizes, equation (1), we see that TeX uses the square of a sum of cubic functions, where we derived a quadratic one. Both have a term penalizing the number of lines (the term $-kC_3$ in (2), taking the reasonable assumption that $\log(2\pi\sigma^+) > 0$). Further, TeX is concerned with the amount that spaces need to be expanded or squeezed, instead of looking at the whole line width. In Section 4 we will see that this does not matter too much.

Benefits of this simple probabilistic approach are that it allows for tuning with intuitively attractive parameters. Further, by choosing a smooth probability distribution for P^a with infinite domain, no threshold effects leading to overly long lines occur (as happens in TeX's approach, resulting in overfull hboxes). Finally, note that when instantiating the network with values of S, cw etc., the functional dependencies between nlw_i and nlw_{i-1} have no impact on the inference of a. So, we are left with a network that has a poly-tree structure for which finding the most likely configuration of b_i is linear in the size of the network [7].

3.2 Extending the Probabilistic Line Breaking Algorithm

The second assumption in the previous section can be relaxed to take inter-line effects in account. This allows discouraging multiple hyphenated lines in a row and preventing tight lines following loose lines. This information can be incorporated by defining new metrics and updating $P(S|p, in)$, the preference of a break-up.

There are two issues that need to be considered when factoring in more metrics. First, to ensure it is possible to find the break points S that are most likely to be accepted efficiently, the network structure that would model the newly created metrics should not create large cycles. It would be sufficient to ensure that the metrics are calculable for each character c_i from the information in slice i and $i-1$ in the network. So when we have m metrics $f_1, \ldots, f_m$, we can write $P(S, f_1, \ldots, f_m|p) = \prod_{i=1}^{n} P(i \in S, f_{i,1}, \ldots, f_{i,m}|p)$, where $f_{i,j}$ is the metric f_j calculated at character i. Second, a choice need to be made modeling the interoperations of the metrics on the probability of acceptability. A pragmatic choice is to assume that the metrics have independent impact on out preference for a break-up S at a particular character c_i. Then we can write $P(i \in S, f_{i,1}, \ldots, f_{i,m}|p) = P(i \in S|p) \prod_{j=1}^{m} P(f_{i,j}|i \in S, p) \propto \prod_{j=1}^{m} P(f_{i,j}|i \in S, p)$.

Some metrics that could be calculated for each character are: tightness of previous line (one of 'tight', 'normal', 'loose'), number of previously hyphenated lines (one of 0, 1, 2, many), the amount spaces in a line need to be expanded to make the line fit the text width, and desirability of breakpoint (distinguishing desirability of spaces, hyphens, and hyphenation points). The latter requires the number of spaces at character i to be calculated as well.

A similar analysis as in the derivation of (2) reveals that each metric contributes as a seperate sum, i.e., $\arg\max_S P(a, S, p, in) = \arg\max_S \sum_{j=1}^{k} m_{1,s_j} + \sum_{j-1}^{k} m_{2,s_j} + \ldots + \sum_{j-1}^{k} m_{f,s_j}$. There is a similarity with (1) in that subtle effects impact additive. The difference is that TEX's penalties are difficult to interpret, especially because effects are multiplied (the term $p \cdot |p|$ in (1)).

This extension has all the benefits of the simple probabilistic approach: intuitively tunable parameters, complexity linear in the number of character, and no threshold effects. Furthermore, it captures subtle interline effects and can be extended easily to capture even more factors.

4 Empirical Results

To get an impression of the behavior of various line breaking algorithms, we looked at a set of existing publications and compared these with the output of some popular typesetting systems. Data was obtained from various scientific journals in the pharmaceutical area. Because the information was provided in PDF format, quite a bit of processing was required to get it into a useable format. Using the XPDF library[3], text, font and location information was extracted from the PDF files and strings glued together to lines, taking super and subscripts in account. Using simple pattern matching rules, any non-paragraph text, like headers and footers, titles, references etc., was removed and text lines glued to

[3] Available from http://www.foolabs.com/xpdf.

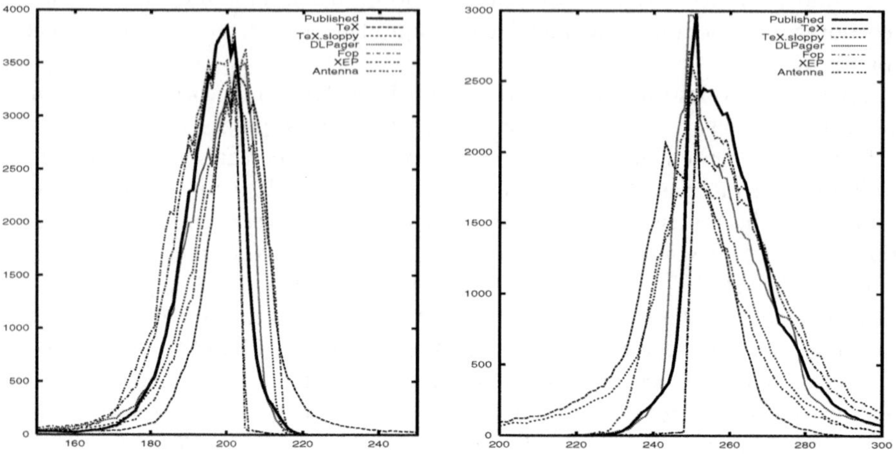

Fig. 2. Natural line widths (left) and space widths (right) for published paragraphs and TEX and DLPager output. Left horizontal axis shows the line length in points (204pt is optimal), right the space width in 100th of a point (250 is optimal). Vertical axis shows the number of lines with the particular property

paragraphs. Because of tables and figures floating through the text, this was not 100% successful and a manual post-processing step was required to ensure broken paragraphs were glued or split. The database contained 5.459 paragraphs with 74.120 lines in total. Natural line widths were calculated taking kerning in account.

We compared the published paragraphs with TEX to establish two benchmarks. The TEX input was obtained by collecting text from each of the paragraphs, mapping special characters (like $\leq$, $\pm$, © etc) and generating a TEX file with a pagebreak after each paragraph. The TEX input uses the same font and paragraph settings as in the published paragraphs. Then LATEX was run to generate PDF and the same process as for published PDF was applied. However, this time paragraphs could be reliably extracted because each of them was output on its own page. TEX was run both with default settings and with sloppy paragraphs. The same process was repeated using DL Pager, a commercial composition engine from Datalogics[4].

An informal experiment[5] was performed where publication managers were asked to rank randomly selected paragraphs comparing published text with Pager and our algorithm. Paragraphs were presented side by side on a page to make comparison easier and the paragraphs were randomly ordered on the page. The result of this experiment was that publication managers tended to prefer the published result, with our new algorithm close by and Pager last. However, all three types of paragraphs were ranked first and last some of the

[4] See http://www.datalogics.com for details

[5] See technical report version of this article for details of the experiment.

time. Overall, this result is encouraging because it shows that Bayesian networks can actually help in increasing the typesetting quality.

An instant quantitative impression was obtained by measuring the distribution of the natural line widths and space widths of published material and lines generated by the typesetting systems. Though this does not show subtle interline behavior (such as appearance of multiple hyphenated lines), it still gives a good initial assessment of a system since looseness of lines is mentioned by publication managers as the primary indication of the quality of a line.

Figure 2 shows the distribution of the natural line widths for the published paragraphs and the TEX output. Only middle lines are shown, not the first line (due to its indent) or last line. The ideal line is 204 points wide, if it is more, spaces have to be shrunk, or if it is less they have to be stretched. Figure 2 show clearly that handcrafted lines tend to be closer to the ideal length than TEX lines. In fact default TEX outputs a large number of lines that exceed the line length so much that there is not enough space in the line to shrink them, and the line exceeds the right boundary. Making paragraphs sloppy largely solves this problem, but there are even more looser lines than with the default setting. DL Pager output tends to have more lines closer to the ideal output compared to default TEX. However, in cases where lines are loose, they tend to be very loose, making the tail on the left in the left side graph rather thick.

Also in Figure 2, natural line widths of three XSL-FO processors are shown. FOP version 0.20.4[6] is an open source processor that uses a line breaking algorithm inspired by TEX. XEP version 3.2[7] and XSL Formatter version 2.4[8] are commercial XSL-FO processors using unknown line breaking algorithms. All three XSL-FO processors use hyphenation rules inspired by TEX's hyphenation files, indicating that all of them use Liang's hyphenation algorithm [6]. The XSL-FO standard [8] leaves the implementation of the line breaking rules to the XSL-FO processor, within bounds of common sense line breaking rules as specified in the Unicode standard [2]. Consequently, there is very little a user can do to influence the layout of paragraphs when using XSL-FO. Figure 2 shows that all three processors are quite far from the published output. FOP and XSL-Formatter both are very reluctant to expand a space, as indicated by the sharp drop in the number of lines longer than the optimal 204 point width.

The right side of Figure 2 shows the distribution of space widths for the systems under investigation. Lines can be made a bit longer or shorter to justify by expanding or squeezing spaces. For example, in a line with 3 spaces and a natural line width of 201 point 1 point may be added to each of the spaces to expand the line to 204 point. The space width for a line is calculated as (natural line width − text width)/number of spaces where the paragraph width is 204 point and lines without spaces are ignored. The distribution for published text is more peaked and has tails that are less fat than any of the systems. Sloppy TEX has a wider distribution than plain TEX and in fact any of the systems.

[6] Available from http://xml.apache.org.

[7] Available from http://www.renderx.com.

[8] Available from http://www.antennahouse.com.

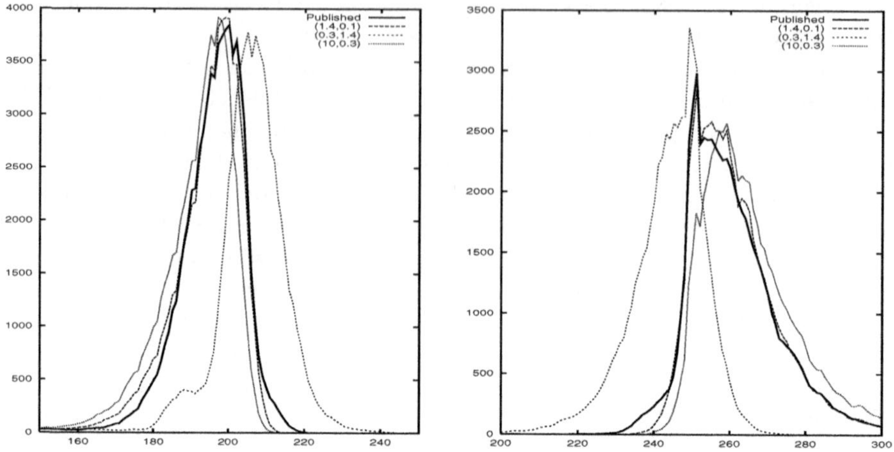

Fig. 3. Natural line widths (left) and space widths (right) for published paragraphs and simple algorithm with various values of σ^- and σ^+. Axis as in Figure 2

The distribution of DLPager comes closest to the published results. Figure 2 illustrates once more the reluctance of Fop and Antenna House's tendency to squeeze spaces. XEP shows least hesitation in squeezing spaces of all the systems. Figure 2 shows that there is considerable difference in behavior and indicate that there is room for improvement in automatic line breaking.

Figure 3 shows the distribution of natural line widths (left) and spaces (right) of our probabilistic line breaking algorithm for different settings of $\sigma-$ and σ^+. The published text is shown as reference. Interestingly, for $\sigma^- = 1.4$ and $\sigma^+ = 0.3$ we get a distribution that is especially close to the published text, closer than any of the other systems examined. By changing the values to $\sigma^- = 0.3$ and $\sigma^+ = 1.4$, we choose to accept tighter lines over loose lines and the line width distribution shifts to the right (so spaces get squeezed and its distribution goes to the left). Setting variances to $\sigma- = 10$ and $\sigma^+ = 0.3$ gives a result left from the two. Remarkably, the probability of acceptability depends on the natural line widths only, not on the space width, like in TEX's algorithm. Figure 3 shows that this does not result in overly tight lines tough.

We compared the simple probabilistic algorithm with the one extended with some of the terms of Section 3.2 by creating graphs similar to Figure 3. The graphs showed that the distributions of natural line widths and space widths are very close. Informal inspection of the output shows that the choice of break points indeed is influenced as desired by tuning parameters of the network.

5 Conclusions

We showed how a probabilistic interpretation of an ill defined problem, the problem of finding line breaks in a paragraph, can lead to an efficient new algorithm. The graphical model that results from the probabilistic interpretation has the advantage that it is easy to tune. Thanks to the architecture of the network structure being a polytree, the algorithm is linear in the number of characters in a paragraph. Line breaks are optimized taking the whole paragraph in account. We demonstrated how adding metrics to capture subtle typographical rules can be done without impacting the efficiency. The resulting algorithm does not show threshold effects (like TEX) and it allows for easy incorporation of subtle typesetting properties. Empirical evidence suggests that this algorithm performs closer to results published through desk top publishing (DTP) than a number of existing systems.

There are a few limitations to our work. We only considered English text with US-English hyphenation rules in our experiments. However, different languages have other hyphenation rules and some even have different breaking rules, like east Asian langauges [2]. Further, we only considered batch composition systems, but line breaking is routinely applied in word processors and DTP packages as well. These systems deal with incremental line breaking and the algorithm can be easily extended to deal with these limitations.

A bigger task is dealing with page breaks. It has been observed that page breaking can be interpreted as vertical version of the line breaking problem [1], so a probabilistic interpretation of this problem could be applied straigthforwardly to find a new algorithm. An extra complication, however, is placement of floating objects, like figures, tables and footnotes. Also, there is the interaction between line breaks and paragraph breaks, making this a more challenging problem.

References

[1] Jonathan Fine. Line breaking and page breaking. TUGBoat, vol 21 no 3, 210–221, 2000. 401

[2] Asmus Freytag. Line Breaking Properties Unicode Standard Annex #14 (part of the Unicode Standard). Technical Report, 2002. 399, 401

[3] D. E. Knuth. Computers & Typesetting Volume A, The TeXbook. Reading, Massachusetts: Addison-Wesley, 1984. 390, 392, 393

[4] D.E Knuth and M. F. Plass. Breaking Paragraphs into Lines. Software—Practice and Experience 11, p.1119–1184, 1981. 393

[5] S. L. Lauritzen and D. J. Spiegelhalter. Local computations with probabilities on graphical structures and their applications to expert systems (with discussion). Journal of the Royal Statistical Society B, 50:157–224, 1988. 395

[6] Franklin M. Liang. Word Hy-phen-a-tion by Com-put-er. Ph.D. Thesis, Department of Computer Science, Stanford University, August 1983. 392, 399

[7] J. Pearl. Probabilistic Reasoning in Intelligent Systems, Networks of Plausible Inference. Morgan Kaufmann, 1998. 395, 396

[8] Extensible Stylesheet Language (XSL). Version 1.0, W3C Recommendation, 15 October 2001. 390, 399

Semiring-Valued Satisfiability

Katarina Britz[1] and Johannes Heidema[2]

[1] Department of Computer Science and Information Systems
University of South Africa, Pretoria, South Africa
[2] Department of Mathematics, Applied Mathematics, and Astronomy
University of South Africa, Pretoria, South Africa
{britzk,heidej}@unisa.ac.za

Abstract. In this paper we show that a certain type of satisfiability problem for a set of propositional sentences with a many-valued semantics can be transformed into an equivalent commutative semiring-based constraint satisfaction problem (CSP). This is in analogy to the classical satisfiability problem (SAT) being an instance of a CSP. The characteristic matrix of the logic is a De Morgan lattice, seen as a commutative semiring. The levels of satisfiability is determined by the natural partial order on the semiring. The aim of a many-valued satisfiability problem may be to maximize the level of satisfaction, or to find a state in which a predetermined minimum level of satisfaction is attained. These aims can both be formulated as constraint problems.
Many-valued satisfiability problems occur naturally in an epistemic context of knowledge and beliefs about the state of a system S. The possible states of S are determined by the ontological constraints on S. The states are partially ordered by the beliefs of an agent observing the system, and trying to determine its current state. Unlike the ontological constraints on S, which are not defeasible, epistemic constraints are assigned degrees of plausibility, reflecting the reliability of the source or observation. The aim of the agent may be to find a possible state in which the degree of plausibility of the combination of its epistemic constraints is maximized, or to find a sufficiently plausible state. Such a state represents a best approximation of the current state of the system.

Keywords: Constraints; Logic; Uncertain reasoning

1 Introduction

Consider the representation of a physical system S as a knowledge system. The possible states of S are determined by the physical constraints of the system, as described by a set of sentences. These may be viewed as ontological constraints, since they describe the physical attributes of S, but they can also be viewed as background knowledge about S. The states of S are partially ordered by the beliefs of an agent observing the system. These beliefs constitute the epistemic constraints that the agent places on the system. They may be formulated syntactically, like the ontological constraints of S, but they may also be formulated

T.D. Gedeon and L.C.C. Fung (Eds.): AI 2003, LNAI 2903, pp. 402–413, 2003.

semantically. Unlike the ontological constraints of S, which are not defeasible, epistemic constraints are assigned degrees of plausibility, reflecting the reliability of the source or observation. The aim of the agent is to find a possible state in which the degree of plausibility of the combination of its epistemic constraints is maximised. Such a state represents a best approximation of the current state of the system. In this paper we show that this problem can be described and solved as a satisfiability problem in a many-valued logic, formulated as an instance of a semiring-based constraint satisfaction problem.

Related to the problem of finding a maximally plausible state, given a set of constraints, is the problem of partitioning the state space into reachable and unreachable states (or, possible and impossible worlds). Generating representative reachable states is a central aspect of model-checking, where the aim is to ensure that the given constraints are satisfied in all reachable states [1]. This, in turn, can be viewed an intermediary goal in the more ambitious enterprise of scientific discovery. Here, the scientific agent engages in a process of scientific enquiry in order to partition the state space, and then formulate a scientific theory (i.e. a set of constraints) which gives a complete (i.e. sound and adequate) specification of the partition. The focus of this paper is on the more modest aim of the truth-seeking agent outlined in the previous paragraph.

In the classical propositional satisfiability problem statement (SAT), n clauses in m variables are given. The aim is to find an assignment of truth values to the variables that satisfies all the clauses. This problem has been generalized to, for example, the MAX SAT problem, where the aim is to maximise the number of clauses that are satisfied. In this paper, we employ a notion of many-valued satisfiability, in which the aim is to maximise the level of satisfaction instead of the number of satisfied clauses. Working with multiple levels of satisfaction requires a logical framework with a many-valued semantics. The value of a proposition in a state or world indicates the extent to which the proposition is satisfied in that state.

SAT is an instance of a constraint satisfaction problem (CSP). In the classical CSP statement, n relations in m variables, ranging over some domain D, are given. The aim is to find an assignment of domain values to the variables such that each relation, or constraint, is satisfied. A number of mappings between SAT and CSPs have been investigated. In the standard non-binary encoding of SAT as a CSP the clauses are constraints, and a solution to the problem is an assignment of truth values to the variables such that all the constraints are satisfied. Not all CSP's have solutions. To accommodate such cases the CSP framework has to be extended to allow for the expression of a notion of partial satisfiability. The semiring constraint satisfaction problem framework has proved successful as a unifying framework for the characterization of a range of constraint satisfaction problems, including MAX CSPs, valued CSPs, fuzzy CSPs and weighted CSPs [2].

We define satisfiability in terms of a binary function which assigns to each proposition a truth value in each state, indicating the plausibility of the proposition in that state. We show that this corresponds to the notion of partial constraint satisfaction used in semiring constraint satisfaction problems. This extends the non-binary encoding of the classical satisfiability problem as a CSP. In Section 2, we give the relevant background to many-valued logic, and show that many-valued logic can be used to formulate the problem statement given above. In Section 3, we define the relevant concepts in semiring CSPs, reformulated in order to clarify the connection with many-valued logic. In Section 4, we represent the many-valued satisfiability problem as a commutative semiring CSP, and present an agent's satisfiability problems in a commutative semiring CSP framework. Section 5 summarizes our main result, and links it to related and future work.

2 Many-Valued Logic

Let L be a propositional language built in the standard way from a denumerable set of atoms L_0, and connectives $\wedge$, $\vee$ and $\neg$. Let A be a finite domain of truth values, and $\mathcal{A} = (A, \wedge, \vee, \neg)$ an abstract algebra with operations $\wedge$, $\vee$ and $\neg$ of the same arity as the corresponding connectives in L. A *valuation* is a function $v : L \to A$ which preserves the connectives of L, viewed as operations on the free algebra generated by L_0. That is, $v(\phi \wedge \psi) = v(\phi) \wedge v(\psi)$, $v(\phi \vee \psi) = v(\phi) \vee v(\psi)$, and $v(\neg\phi) = \neg v(\phi)$.

We assume in this paper that $\mathcal{A}$ is a De Morgan lattice, as in Definition 2 below:

Definition 1. *A distributive lattice is a structure* $(A, \wedge, \vee)$ *such that:*

1. *A is a non-empty, partially ordered set;*
2. *For each $a, b \in A$, their meet $a \wedge b$ and join $a \vee b$ exist, where the meet of two elements is their greatest lower bound, and the join of two elements is their least upper bound;*
3. *$\wedge$ and $\vee$ distribute over each other.*

Definition 2. *A De Morgan lattice is a structure* $\mathcal{A} = (A, \wedge, \vee, \neg, 0, 1)$ *such that:*

1. *$(A, \wedge, \vee)$ is a distributive lattice;*
2. *$\neg$ is an involution, i.e. $\neg\neg a = a$;*
3. *De Morgan's laws hold, i.e. $\neg(a \wedge b) = \neg a \ \vee \neg b$ and $\neg(a \vee b) = \neg a \ \wedge \neg b$;*
4. *0 is the zero element and 1 the unit element of $\mathcal{A}$, i.e. for every $a \in A$, $a = a \vee 0$ and $a = a \wedge 1$.*

A *filter* in $\mathcal{A}$ is a nonempty subset of A which is closed upwards under $\leq$, and closed under $\wedge$. Let $F = \{F_i : i \in I\}$ be a set of filters. Each filter partitions A into a set of designated and undesignated elements. Each tuple $\langle \mathcal{A}, F_i \rangle$ determines a semantic consequence relation on L: ϕ is a logical consequence of the set of premises Γ iff, whenever each element of Γ takes on a designated value in $\langle \mathcal{A}, F_i \rangle$, then ϕ also takes on a designated value in $\langle \mathcal{A}, F_i \rangle$. This definition generalizes the notion of logical consequence in classical propositional logic, where

ϕ is a logical consequence of Γ iff, whenever each element of Γ is true, then ϕ is also true. The set of determining matrices $\mathcal{M} = \{\langle \mathcal{A}, F_i \rangle : i \in I\}$ also gives rise to a semantic consequence relation on L, namely the intersection of the semantic consequence relations determined by each $\langle \mathcal{A}, F_i \rangle \in \mathcal{M}$ [3]:

Definition 3. *Let $\mathcal{M} = \{\langle \mathcal{A}, F_i \rangle : i \in I\}$ be a class of matrices. The semantic consequence relation $\models_{\mathcal{M}}$ is defined as follows: For any $\Gamma \subseteq L$ and $\phi \in L$,*

$$\Gamma \models_{\mathcal{M}} \phi \text{ iff for every } \langle \mathcal{A}, F_i \rangle \in \mathcal{M} \text{ and } v : L \to A \text{ , if } v(\Gamma) \subseteq F_i \text{ then } v(\phi) \in F_i.$$

A *state* is defined as a function $s : L_0 \to A$. A state is a *model* of a set of propositions Γ if its extension to a valuation $v : L \to A$ assigns a designated value to each element of Γ in each determining matrix of the logic. A logical *theorem* is a proposition which takes on a designated value in each state in each determining matrix. For notational convenience, we will also write $\phi(s)$ to refer to the truth value $s(\phi)$ of ϕ in state s. $\phi(s)$ may be viewed as the *level of satisfaction* of ϕ in s, indicating how close s is to being a model of ϕ. More generally, we define the level of satisfaction of a set of propositions Γ in a state s as the meet

$$\Gamma(s) = \bigwedge \{\phi(s) \mid \phi \in \Gamma\}.$$

Each determining matrix partitions the set of truth values into a set of designated values and a set of undesignated values. The choice of determining matrices determines the logical consequence relation of the resulting logic. These range from those consequence relations induced by a single filter in $\mathcal{A}$, to the consequence relation induced by the set of all filters in $\mathcal{A}$. The latter relation reflects the partial order on $\mathcal{A}$: ψ is a logical consequence of ϕ iff ψ is satisfied to at least the same level as ϕ in each state. That is, $(\forall s : L_0 \to A)\phi(s) \leq \psi(s)$.

For example, consider the three-valued Partial Logic, with characteristic algebra the Kleene lattice $\mathcal{K}_3 = \{\{t, u, f\}, \wedge, \vee, \neg\}$ shown in Figure 1 [4]. The third, non-classical truth value u is used to indicate that the truth or falsity of an atom is unknown. Negation is defined as follows: $\neg t = f$; $\neg f = t$; $\neg u = u$.

$\mathcal{K}_3$ is a De Morgan lattice (see Definition 2 above). The filters in $\mathcal{K}_3$ are $\{t\}$, $\{t, u\}$ and $\{t, u, f\}$. Choosing all three these filters as designated sets yields the

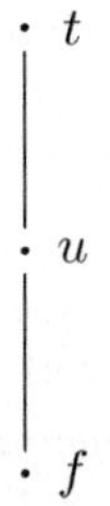

Fig. 1. The partial truth order on $\mathcal{K}_3$

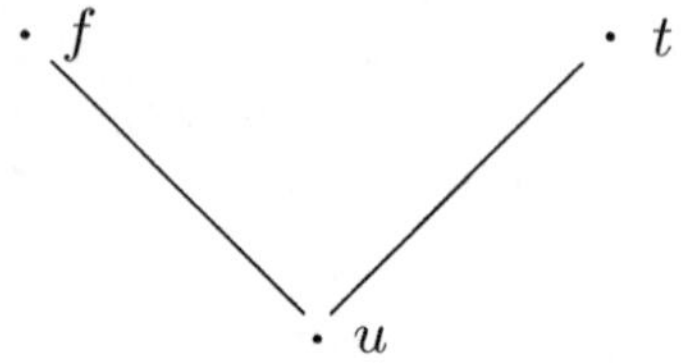

Fig. 2. The partial knowledge order on $\mathcal{K}_3$

logical consequence relation of Partial Logic, which can also be characterized as follows:

$$\phi \models_{\mathrm{PL}} \psi \text{ iff } (\forall s : L_0 \to A)\phi(s) \leq \psi(s).$$

The characteristic algebra of a many-valued logic is often endowed with a second partial order representing an increase in information, or a decrease in abstraction. This induces a pointwise order on states, yielding an information-ordered state space. The maximal states are the only realizable states. States lower down in the information order are partial descriptions, or abstractions, of realizable states. For example, the information order in Figure 2 is associated with the three-valued Kleene lattice:

The maximal elements in the state space induced by this information order are those states in which the truth of all atoms in the language are known, that is, no atom is assigned the third, non-classical truth value u.

3 Semiring-Based Constraint Satisfaction

Many constraint satisfaction schemes can be described using a *semiring*, defined below. The elements of the semiring indicate levels of compliance of states with constraints on them, and the operations of the semiring are used to combine constraints. In this section we give the necessary background on the semiring constraint satisfaction framework; for a more detailed account, see [2]. For technical reasons, and in order to simplify subsequent definitions and results, our definition of the domain of a constraint differs from that presented in [2]. We also restrict our attention to commutative semirings.

Definition 4. *A semiring is a tuple* $(A, \times, +, 0, 1)$ *such that*

1. *A is a set with* $0, 1 \in A$*;*
2. *$+$ is a closed, commutative, associative binary operation on* A*;*
3. *0 is the unit element for* $+$*, i.e.* $a + 0 = a = 0 + a$*;*
4. *$\times$ is a closed, associative binary operation on* A*;*
5. *1 is the unit element for* $\times$*, i.e.* $a \times 1 = a = 1 \times a$*, and* 0 *is its absorbing element, i.e.* $a \times 0 = 0 = 0 \times a$*;*
6. *$\times$ distributes over* $+$*, i.e.* $a \times (b + c) = (a \times b) + (a \times c)$*.*

Definition 5. *A commutative semiring is a semiring* $(A, \times, +, 0, 1)$ *such that*

1. $+$ *is idempotent, i.e.* $a + a = a$*;*
2. $\times$ *is commutative;*
3. 1 *is the absorbing element for* $+$*.*

Definition 6. *A constraint system is a tuple* $CS = \langle \mathcal{A}, D, V \rangle$*, where* $\mathcal{A} = (A, \times, +, 0, 1)$ *is a commutative semiring, and* V *is a set of variables ranging over a finite domain* D*.*

A *state* is an assignment of a domain element to each variable, i.e. states are elements of the function space $S = \{s : V \rightarrow D\} = D^V$. Let $W \subseteq V$. $s|_W$ denotes the restriction of s to W. For any state s, $s|_W$ is a function assigning a domain value to each element of W. W therefore partitions the state space into sets of states that are W-equivalent:

$$s \simeq_W t \text{ iff } s|_W = t|_W.$$

Our introduction of the relation $\simeq_W$ simplifies subsequent definitions dealing with constraints. We define a *constraint* c over W as a $\simeq_W$-preserving assignment of semiring values to states, with equivalent states assigned the same semiring value. (In [2], the domain of a constraint over W is defined as the function space $\{s : W \rightarrow D\} = D^W$, as opposed to the entire state space.)

Definition 7. *Let* $CS = \langle \mathcal{A}, D, V \rangle$ *be a constraint system, with state space* $S = \{s : V \rightarrow D\}$*. A constraint over* $W \subseteq V$ *is a* $\simeq_W$*-preserving function* $c : S \rightarrow A$*.*

Let $Var(c) = \bigcap\{W : c \text{ is a constraint over } W\}$. $Var(c)$ is called the set of constrained variables of c.

The additive $+$ operation of the commutative semiring induces a partial preference order on the semiring: $a \leq b$ if and only if $a + b = b$. This, in turn, induces a pointwise partial *constraint order*: $c_1 \sqsubseteq c_2$ if and only if for any state s, $c_1(s) \leq c_2(s)$. If $c_1 \sqsubseteq c_2$ and $c_2 \sqsubseteq c_1$, then $c_1 = c_2$.

Given a constraint c over W, $+$ is used to form a new constraint which only takes into account a subset of the variables in W:

Definition 8. *Given a constraint* c *over* W*, and* $U \subseteq W$*, the projection of* c *onto* U *is the constraint* $c\Downarrow_U$ *over* U*, where* $c\Downarrow_U (t) = \sum\{c(t') : t \simeq_U t'\}$*.*

The projection of a set of constraints onto a subset of variables is used in the formulation of local consistency search techniques. If $U \subset W$, then $\simeq_U$ is a larger equivalence relation than $\simeq_W$. For each $\simeq_U$-equivalence class of states, the projection of c onto U picks the best value amongst those values that c take in states that may not be $\simeq_W$-equivalent, but are $\simeq_U$-equivalent.

The multiplicative operation $\times$ is used to form the combination of two constraints:

Definition 9. *Given two constraints* c_1 *and* c_2 *over* U *and* W *respectively, their combination* $c_1 \otimes c_2$ *is the constraint* c *over* $U \cup W$*, where* $c(t) = c_1(t) \times c_2(t)$*.*

The complete solution to a set of constraints is their combination, which is an assignment of an element of the semiring to each element of the state space. However, generating a complete solution is typically prohibitive in terms of both time and space requirements. The problem statement may only require finding an element of the state space for which a maximum element in the preference ordering on the semiring is obtained, or finding an element of the state space for which a predetermined minimum preference value is exceeded.

We next turn to the notion of k-consistency, which is used in search techniques that look for inconsistencies in partial solutions by considering only constraints over k variables, in order to prune the search space at an early stage of the search, and so reduce the search time. A set of constraints C is called k-consistent if, for any set W of $k-1$ variables, and any kth variable x, the combination of all the constraints over W in C is equal to the projection onto W of the combination of all the constraints over $W \cup \{x\}$ in C.

Definition 10. *A set of constraints C is called k-consistent if, for any $W \subset V$ such that $|W| = k$, and $x \in V - W$,*

$$\otimes\{c : c \in C \text{ and } Var(c) \subseteq W\} \sqsubseteq (\otimes\{c : c \in C \text{ and } Var(c) \subseteq (W \cup \{x\})\})\Downarrow_W .$$

The converse of the inequality in this definition always holds [2]. The most common k-consistency techniques used are unary, or node consistency, and binary, or arc consistency, but the more general notion of k-consistency is also employed by certain local consistency search algorithms.

4 Many-Valued Satisfiability

The satisfiability problem in a many-valued logic can be defined as for classical logic, i.e. as the problem of finding a model of a given set of clauses. As with the classical satisfiability problem, this can be generalized to the maximal satisfiability problem (MAX-SAT), i.e. the problem of finding a state in which the maximum number of clauses are satisfied.

We propose an alternative generalization of the satisfiability problem, namely that of finding a state in which a maximal value in the partial truth order on the characteristic matrix of the logic, is obtained. The aim is to maximise the minimum level of satisfiability of the clauses, as opposed to finding a model, or maximising the number of satisfied clauses. This problem requires a logic with a many-valued semantics, to account for different levels of satisfiability. It is also the natural definition of satisfiability in a logic where the consequence relation reflects the partial truth order on the characteristic matrix, instead of being defined in terms of models as is the case in classical logic. We shall show that this formulation is an instance of a semiring constraint satisfaction problem, just as SAT is an instance of a CSP, and MAX-SAT is an instance of MAX-CSP.

In order to view the many-valued satisfiability problem as a constraint satisfaction problem, the domain of the constrained variables have to coincide with the domain of preference values in the semiring, i.e. in the constraint system

$CS = \langle \mathcal{A}, D, V \rangle$, we must have $A = D$. For example, in hardware verification, Belnap's four-valued logic has been used to model both the state space and the values that the specification of a circuit can take [5, 1]. The specification language used is a four-valued temporal logic. Statements in the logic are temporal constraints specifying the behaviour of a circuit over time. We also make the following assumptions about the characteristic algebra $\mathcal{A}$ of the logic:

- $\mathcal{A}$ is a De Morgan lattice;
- Each filter in $\mathcal{A}$ is a designated set.

The first assumption ensures that the algebra is a commutative semiring, with meet $\wedge$ as multiplicative operation, and join $\vee$ as additive operation corresponding to the conjunctive and disjunctive connectives $\wedge$ and $\vee$ in the language of the logic. It also ensures that propositions can be written in conjunctive normal form (CNF). Since $\mathcal{A}$ is a De Morgan lattice, its meet operation is idempotent. This ensures that the notion of local consistency is well-defined.

The second assumption ensures that the partial order on propositions induced by the consequence relation of the logic coincides with the constraint order. The consequence relation obtained is the intersection of the consequence relations obtained by taking each filter separately as designated set. Omitting this assumption allows for a coarser relation of logical strength on propositions than the constraint order in a CSCSP. This is appropriate when a solution with a given minimum level of satisfaction is sought, instead of an optimal solution.

The syntactic characterization of the combination of two constraints is their conjunction. The following theorem characterizes the projection of a constraint in CNF syntactically, instead of semantically as in Definition 8. Its proof is also an algorithm to construct the projection of a given constraint.

Theorem 1. *Let $W = U \cup \{P\}$. Let the sentence $c = c_1 \wedge \ldots \wedge c_n$ be a constraint over W, where each conjunct c_i is a disjunction of literals (i.e. a clause). The conjuncts of $c \Downarrow_U$ in CNF are as follows:*

- *Each conjunct c_i of c such that neither P nor $\neg P$ occurs in c_i, will also be a conjunct of $c \Downarrow_U$.*
- *If c_i and c_j are conjuncts of c ($i \neq j$) such that P occurs unnegated in c_i and $\neg P$ occurs in c_j, say $c_i = d_1 \vee P$ and $c_j = d_2 \vee \neg P$, then $d_1 \vee d_2 \vee k$ will be a conjunct of $c \Downarrow_U$, where k is the constant proposition with value the least upper bound of values that $P \wedge \neg P$ can assume in any state.*
- *No other conjuncts will occur in $c \Downarrow_U$.*

Proof. 1. Suppose that, for some i, where $1 \leq i \leq n$, neither P nor $\neg P$ occurs in the conjunct c_i. For any state t,

$$\begin{aligned} c_i(t) &= \sum \{c_i(t') : t \simeq_W t'\} \\ &= \sum \{c_i(t') : t \simeq_U t'\}. \end{aligned}$$

Projecting c onto U therefore has no effect on the value of c_i in any state t. The conjunct c_i is therefore also a conjunct of the projection of c onto U.

2. Next, suppose P occurs unnegated in c_i for some i, where $1 \le i \le n$, and $\neg P$ does not occur in any c_j, where $j \ne i$. For any state t, let t' be the state such that $t = t'$, except that $P(t') = 1$. Then $t \simeq_U t'$ and $c_i(t') = 1$. Therefore $\sum\{c_i(t') : t \simeq_U t'\} = 1$. The conjunct c_i must therefore be omitted in the formation of the projection of c onto U.
3. Similarly, suppose $\neg P$ occurs in c_i for some i, where $1 \le i \le n$, and P does not occur unnegated in any c_j, where $j \ne i$. For any state t, let t' be the state such that $t = t'$, except that $P(t') = 0$. Then $t \simeq_U t'$ and $c_i(t') = 1$. Therefore $\sum\{c_i(t') : t \simeq_U t'\} = 1$. The conjunct c_i must therefore be omitted in the formation of the projection of c onto U.
4. Suppose both P and $\neg P$ occur in c_i for some i, where $1 \le i \le n$. For any state t, let t' be the state such that $t = t'$, except that $P(t') = 1$. Then $t \simeq_U t'$ and $c_i(t') = 1$. Therefore $\sum\{c_i(t') : t \simeq_U t'\} = 1$. The conjunct c_i must therefore be omitted in the formation of the projection of c onto U.
5. Finally, suppose both P and $\neg P$ occur in c, say in conjuncts $c_i = d_1 \vee P$ and $c_j = d_2 \vee \neg P$, where $i \ne j$. For any state t,

$$\begin{aligned}(c_i \wedge c_j)\Downarrow_U (t) &= \sum\{(c_i \wedge c_j)(t') : t \simeq_U t'\} \\ &= \sum\{((d_1 \vee P) \wedge (d_2 \vee \neg P))(t') : t \simeq_U t'\} \\ &= \sum\{((d_1 \wedge d_2) \vee (d_1 \wedge \neg P) \vee (d_2 \wedge P) \vee (P \wedge \neg P))(t') : \\ &\quad t \simeq_U t'\} \\ &= d_1(t) \vee d_2(t) \vee \sum\{(P \wedge \neg P)(t') : t \simeq_U t'\}.\end{aligned}$$

The last equality above is derived as follows: Let t' be the state such that $t = t'$, except that $P(t') = 1$. Then $(c_i \wedge c_j)(t') = d_2(t') = d_2(t)$. Similarly, if t' is the state such that $t = t'$, except that $P(t') = 0$, then $(c_i \wedge c_j)(t') = d_1(t') = d_1(t)$. Therefore $(c_i \wedge c_j)\Downarrow_U (t) \ge d_1(t) + d_2(t)$. Further, $\sum\{(P \wedge \neg P)(t') : t \simeq_U t'\}$ is a constant, which is the least upper bound of values that $P \wedge \neg P$ can assume in any state. Call this constant k. The remaining inequalities needed follow from the observation that $(c_i \wedge c_j)\Downarrow_U (t) \ge k$, and that $c_i \wedge c_j$ is lower in the constraint order than $d_1 \vee d_2 \vee k$. (We assume that the language has an atom k for each truth value in the De Morgan lattice.)
Therefore, for each pair of conjuncts $c_i = d_1 \vee P$ and $c_j = d_2 \vee \neg P$ in c, $d_1 \vee d_2 \vee k$ is a conjunct in the projection of c onto U. □

We summarise the main claim of this paper as follows: Any satisfiability problem for a set of propositional sentences (of signature $\{\wedge, \vee, \neg\}$) with respect to a many-valued semantics using a De Morgan lattice as its matrix of truth values, can be transformed into an equivalent commutative semiring-based constraint satisfaction problem. The transformation is managed as follows:

1. Syntactically, we have the propositional language L, the free algebra of signature $\{\wedge, \vee, \neg\}$ generated by the denumerable ordered set L_0 of atoms. Semantically, we assign to atoms, and by homomorphic extension, to sentences, truth values in the finite De Morgan lattice $\mathcal{A} = (A, \wedge, \vee, \neg, 0, 1)$.

Since satisfaction is a semantic notion, sentences to be satisfied (to some degree) may be replaced by semantically equivalent sentences (i.e. which are assigned the same truth value in A by all assignments of truth values to atoms). Since $\mathcal{A}$ satisfies De Morgan's laws, every sentence has equivalent conjunctive and disjunctive normal forms. The satisfiability problem under consideration, formulated in terms of a set of sentences Γ involving only atoms in $W \subseteq L_0$, may be finding at least one truth assignment satisfying Γ to at least some specified degree.

2. The reduct $\mathcal{A}^- = (A, \times, +, 0, 1) = (A, \wedge, \vee, 0, 1)$ of the De Morgan lattice $\mathcal{A}$, obtained by simply ignoring its unary operation $\neg$, is a commutative semiring. Since this has no effect on the set of truth values A, the truth values of sentences involving $\neg$ are handled exactly as before.
3. We now construct the constraint system $CS = \langle \mathcal{A}^-, A, L_0 \rangle$, with state space $S = \{s : L_0 \to A\}$ the set of all truth value assignments to atoms, hence to sentences. Sentences are now seen as constraints with respect to CS. Suppose ϕ, with set of atoms U, is one of the sentences in Γ, where $U \subseteq W \subseteq L_0$. If s and t are two states such that $s|_U = t|_U$, then $s(\phi) = t(\phi)$. Hence, if we now see ϕ as a function $\phi : S \to A$, with $\phi(s) = s(\phi) \in A$, then ϕ is a constraint over U (and also W).
4. Finally, the conjoined single logical constraint $\bigwedge \Gamma$ coincides with the combination (or product) of these constraints in the commutative semiring framework. The disjoined logical constraint $\bigvee \Gamma$ does not feature prominently in the semiring framework, but the semantic projection of a constraint in a commutative semiring CSP coincides with the logical constraint obtained in Theorem 1.

Consider again the problem presented in the introduction to this paper: An agent's knowledge of a system consists of a specification of the physical properties of the system, and acquired information, for example from another agent or by observation of the system. Acquired information, or beliefs, are assigned a degree of plausibility, reflecting the reliability of the source or observation. The physical properties of the system are specified as a set of hard constraints, and are formulated syntactically. Beliefs are soft constraints, and can be formulated either syntactically or semantically. Indeed, not all semantic information acquired by the agent is expressible syntactically. For example, it is not possible to formulate the belief that a given state is definitely excluded, i.e. that it should have the lowest plausibility, syntactically. It is also impossible to formulate the belief that state s is more plausible than state t syntactically. This does not present a problem in a semiring constraint satisfaction framework, since searching the state space for a maximal solution is a semantic procedure.

The aim of the agent is to find a realizable state, i.e. a state which is maximal in the abstraction order, which satisfies all the hard constraints, and maximises the degree of plausibility of the combination of the soft constraints. Such a state represents a best approximation of the current state of the system. Although all the constraints may be satisfied in some of the abstract states, they are not eligible as solutions, since they only contain partial information about the state of the system.

The connectives $\wedge$, $\vee$ and $\neg$ are monotone with respect to the abstraction order on the state space. That is, given a constraint c, and states s and t such that $s \leq t$ in the abstraction order, then also $c(s) \leq c(t)$ (again in the abstraction order). This property can be used to prune the search space, and reduce search time, since a subtree can be pruned as soon as the set of plausibility values above $c(s)$ in the abstraction order, $\{x : c(s) \leq x\}$, does not contain an acceptable plausibility for a solution.

5 Conclusion and Related Work

We have shown that, for the class of many-valued logics with characteristic algebra a De Morgan lattice, the many-valued satisfiability problem can be transformed into an instance of a commutative semiring constraint satisfaction problem. Such a logic also provides a suitable formal framework within which to represent information that an agent is presented with in a knowledge system, including partial information, different levels of epistemic entrenchment, and semantic constraints. These can all be represented, either syntactically or semantically, as propositions in a suitably chosen many-valued logic. Propositions in the logic can then be viewed as constraints in a commutative semiring constraint satisfaction problem.

A number of mappings between SAT and CSPs have been investigated, for example in [6]. Our proposal extends the non-binary encoding of SAT as a CSP. Other mappings include the standard dual and hidden variable encodings of non-binary CSPs as binary CSPs. We briefly consider the dual variable encoding of SAT as a CSP here: A dual variable D_i is associated with each clause c_i. The domain of D_i consists of those states that satisfy the clause c_i. Binary constraints are posted between dual variables (i.e. clauses) that share propositional variables. These constraints ensure that value assignments are done consistently. In the dual encoding of many-valued SAT as a semiring CSP, the domain of each dual variable D_i is the state space. A unary constraint is posted on each D_i, indicating the degree to which c_i is satisfied in each state. The unary constraints are therefore functions from the state space into the De Morgan lattice of truth values. The binary constraints are two-valued hard constraints, as in the encoding of SAT as a CSP, and are posted between dual variables (i.e. clauses) that share propositional variables. The hidden variable encoding of many-valued SAT as a semiring CSP is similar to the dual encoding.

The approach of this paper has been representational, rather than algorithmic. Both a theoretical and an empirical investigation into the algorithms applicable to the problem domain presented here, and their complexity, are needed in order to properly evaluate its usefulness.

References

[1] S. Hazelhurst and C. H. Seger. Model checking lattices: Using and reasoning about information orders for abstraction. *Logic Journal of the IGPL*, 7(3):375–411, 1999. 403, 409

[2] S. Bistarelli, U. Montanari, F. Rossi, T. Schiex, G. Verfaillie, and H. Fargier. Semiring-based CSPs and valued CSPs: Frameworks, properties, and comparison. In *Over-constrained Systems*, volume 1106 of *Lecture Notes in Computer Science*, pages 111–150. Springer-Verlag, 1996. 403, 406, 407, 408

[3] G. Malinowski. *Many-Valued Logics.* Oxford University Press, 1993. 405

[4] S. Blamey. Partial logic. In D. Gabbay and F. Guenthner, editors, *Handbook of Philosophical Logic*, volume III, pages 1–70. D. Reidel, Dordrecht, 1986. 405

[5] N. D. Belnap. A useful four-valued logic. In G. Epstein and J. M. Dunn, editors, *Modern Uses of Multiple-Valued Logic*, pages 7–73. D. Reidel, Dordrecht, 1977. 409

[6] T. Walsh. SAT v CSP. In *Proceedings of CP-2000*, volume 1894 of *Lecture Notes in Computer Science*, pages 441–456. Springer-Verlag, 2000. 412

A Defeasible Logic of Policy-Based Intention*

Guido Governatori and Vineet Padmanabhan

School of Information Technology & Electrical Engineering
The University of Queensland, Brisbane, QLD, Australia
{guido,vineet}@itee.uq.edu.au

Abstract. Most of the theories on formalising intention interpret it as a unary modal operator in Kripkean semantics, which gives it a monotonic look. We argue that policy-based intentions [8] exhibit non-monotonic behaviour which could be captured through a non-monotonic system like defeasible logic. To this end we outline a defeasible logic of intention. The proposed technique alleviates most of the problems related to logical omniscience. The proof theory given shows how our approach helps in the maintenance of intention-consistency in agent systems like BDI.

1 Introduction

Formalising cognitive states like intention has received much attention in the AI community [7, 17, 18, 23]. All these theories are based on Normal Modal Logics (NMLs), where intention is formalised into a modal operator on the framework of kripkean possible world semantics. Due to this restriction, these theories suffer from the *logical-omniscience* problem [10, 22]. One of the solutions suggested to overcome this problem is to adopt a non-kripkean semantics as shown in [5]. In that work intention is interpreted in terms of its *content* and the intention consequence relation is explained based on the content of two intentions. There is also a *representationalist* theory of intention [11] that employs the minimal model semantics [4] to interpret the intention operator. Work has also been done relating intention to preferences [20] as well as commitments [6]. However none of these theories have explicitly addressed the need for a non-monotonic theory of intention and we argue that to capture the properties involved in *policy-based* intention we need such a non-monotonic setup.

Our claim is based on Bratman's [8] classification of intention as *deliberative, non-deliberative, policy-based* and we show that *policy-based* intention is non-monotonic (i.e. has a defeasible nature). Though, many of the theories mentioned above is based on Bratman's work, they fail to recognize the non-monotonic component involved in intention. In this paper we adopt a particular non-monotonic system, (*defeasible logic*), to study the properties involved in *policy-based* intention and show how one can relate it with an intentional system like BDI [17]. The reason for defeasible logic is due to its computational efficiency [13] and

* This research was partially supported by the University of Queensland under the grant UQRSF ITEE-03/2002001335.

T.D. Gedeon and L.C.C. Fung (Eds.): AI 2003, LNAI 2903, pp. 414–426, 2003.

easy implementation [15]. We are unaware of any existing work relating reasoning about intention with non-monotonic reasoning to the best of our knowledge. We believe that our approach helps in bridging the gap between non-monotonic reasoning and reasoning about intention.

The proposed method provides solutions to the problem of *logical-omniscience* which usually accompanies intention-formalisms based on normal modal logics. The use of non-monotonic logics in intention reasoning allows the agent to reason with partial knowledge without having a complete knowledge of the environment. This also helps the agent in avoiding a complete knowledge of the consequences. Moreover, we outline a proof-theory whereby one can reason about ways of maintaining intention consistency in agent systems like BDI. The new approach facilitates the designer of an agent system like BDI in describing rules for constructing intentions from goals and goals from knowledge. This is important as it is in alliance with the *commitment* axioms of Rao and Georgeff [17] and also provides an explanation on the practical nature of intentional systems like BDI. In this paper we don't want to recast the whole BDI theory but focus on the intention part supplemented by the factual knowledge and its underlying theory. Moreover similar considerations can be applied to the GOAL component.

In the next section we make the case for a non-monotonic theory of intention based on Bratman's classification of intention. In the third section we outline the problem of logical omniscience and in the fourth we give an overview of defeasible logic. The fifth section argues for a defeasible logic of intention. In the final section we make a comparison between our work and the work in policy-based reasoning

2 The Case for Non-monotonic Reasoning

An important classification of intention that is useful in computer science is that of *intending* versus *doing intentionally*, where the former involves the true intentions or preferences of the agent whereas the latter applies to the actions or states that the agent performs or brings about but not with any prior intention to do so. Based on this division Bratman classifies intentions as *deliberative, non-deliberative and policy-based.* When an agent i has an intention of the form $\mathrm{INT}_i^{t_1}\varphi, t_2$ (read as *agent i intends at t_1 to φ at t_2*) as a process of *present* deliberation, then it is called *deliberative intention.* On the other hand if the agent comes to have such an intention not on the basis of present deliberation, but at some earlier time t_0 and have retained it from t_0 to t_1 without reconsidering it then it is called *non-deliberative.* There can be a third case when intentions can be general and concern potentially recurring circumstances in an agent's life. Such general intentions constitute *policy-based intentions*, and is defined as follows: *when the agent i has a general-(policy/intention) to φ in circumstances of type ψ and i notes at t_1 that i am (will be) in a ψ-type circumstance at t_2, and thereby arrive at an intention to φ at t_2.* The difference here is that there is no *present* deliberation concerning the action to be performed as the agent already has a general intention to do a particular action (*doing intentionally*). Whether the agent is able to perform that action or not depends on the circumstances.

When dealing with such general policies/intentions (hereafter intention), we have to take into account two cases. General intentions could be either (1) *periodic* or (2) *circumstance-triggered.* They are *periodic* in the sense that their occasion for execution is guaranteed by the mere passage of a specific interval of time. For instance, the general intention of patching up and rebooting the Unix server, *hobbit* in our department on every friday at 7pm. In contrast to this, general intention could be *circumstance triggered* as in the case of being *Root if one is Super-user.* Its occasion is not guaranteed by the mere passage of time but require that certain specific circumstances obtain. In both cases one can find that the general intention has an underlying defeasible nature. The defeasible nature is explained as follows. Consider the above example for *circumstance-triggered* general intention:

$$SU(X) \Rightarrow Root(X) \tag{1}$$

which means, (super-users are typically root). Suppose, there exists an agent i (a software program) that monitors tasks related to giving root permissions as and according to whether a user is a normal-user (NU) or Super-User (SU) and i has a general intention like (1). This general intention has a defeasible nature in the sense that, if i knows that X is a SU then i may conclude that X is Root, unless there is other evidence suggesting that X may not be root (for instance, when X has only read and write permissions but not execute permission). But this does not mean that the agent i should know all such conditions but, only those he considers necessary to the intended outcome and that he/she isn't confident of their being satisfied. Hence our definition of general intention boils down to:

> *An agent intends all the necessary consequences of his performing his general intention and he isn't confident of their being satisfied.*

In order to intend the necessary consequence the agent has to make sure that all the evidence to the contrary has been defeated which basically is a defeasible logic conclusion. This is different from the usual NML interpretation where the agent intends all the consequences.

The formation of such general policies helps in extending the influence of deliberation as it is a partial solution to the problems posed by our limited resources for calculation and deliberation at the time of action. General policies also facilitate co-ordination. It may sometimes be easier to appreciate expectable consequences (both good and bad) of general ways of acting in recurrent circumstances than to appreciate the expectable consequences of a single case.

3 Logical Omniscience and Non-monotonicity

As we mentioned before, most of the theories based on NML's interpret intention as a unary modal operator in Kripkean semantics which makes it vulnerable to the problem of *logical-omniscience.* The problem in its general form as stated in [22] is as follows: (where **X** could represent a mental state like intention (INT)

1. $\models \mathbf{X}\varphi \wedge \mathbf{X}(\varphi \rightarrow \psi) \Rightarrow \mathbf{X}\psi$ (*side-effect problem*)
2. $\models \varphi \rightarrow \psi \Rightarrow \models \mathbf{X}\varphi \rightarrow \mathbf{X}\psi$ (*side-effect problem*)
3. $\models \varphi \Leftrightarrow \psi \Rightarrow \models \mathbf{X}\varphi \Leftrightarrow \mathbf{X}\psi$ (*side-effect problem*)
4. $\models \varphi \Rightarrow \models \mathbf{X}\varphi$ (*transference-problem*)
5. $\models (\mathbf{X}\varphi \wedge \mathbf{X}\psi) \rightarrow \mathbf{X}(\varphi \wedge \psi)$ (*unrestricted combining*)
6. $\models \mathbf{X}\varphi \rightarrow \mathbf{X}(\varphi \vee \psi)$ (*unrestricted weakening*)
7. $\models \neg(\mathbf{X}\varphi \wedge \mathbf{X}\neg\varphi)$

None of these properties except for (7) is valid when we take intention into consideration. For instance, consider a situation where an agent i goes to the bookstore with the intention of buying a *paper-back* and also with the intention of buying a *magazine* because he has a general intention to buy them.[1] Hence according to (5) it could be formally given as:

$$\mathrm{INT}_i(\mathit{paperback}) \wedge \mathrm{INT}_i(\mathit{magazine}) \rightarrow \mathrm{INT}_i(\mathit{paperback} \wedge \mathit{magazine})$$

But this general intention is *defeasible* in the sense that at the bookstore the agent might find that he doesn't have enough money to buy both of them and hence drops intention to buy each of them and now only intends to buy one of them. NMLs fail to account for such type of reasoning. In Sugimoto [20] an extra notion of *preference* is added and an ordering among the preferences is done to capture the desired effect. But we argue that, in general, such intentions are defeasible and hence a non-monotonic reasoning system would be more efficient for such occasions. The above example could be stated in a non-monotonic setup as

$$\begin{aligned}&(1)\ \ \mathit{paper\text{-}back}(X) \Rightarrow \mathit{buy}(X),\\&(2)\ \ \mathit{magazine}(X) \Rightarrow \mathit{buy}(X),\\&(3)\ \ \mathit{costly}(X) \rightsquigarrow \neg \mathit{buy}(X);\end{aligned}$$

where (1) and (2) are premises which reflects the agents general intention of buying a paper-back and magazine unless there is other evidence like (3) suggesting that he/she may not be able to buy. When intention is formalised in the background of NMLs it is often the case that the agent has to have a complete description of the environment before-hand or has to be omniscient in the sense of knowing all the consequences. Classically the logical omniscience problem amounts to say that an agent has to compute all consequences of its own theory. It is obvious that some of the consequences are not intended as shown above. Moreover in classical NML the set of consequences is infinite. Hence we need a system like DL (defeasible logic) which is easily implementable and where the set of consequences consists of the set of literals occurring in the agent theory i.e. in the knowledge base, which is finite.

4 Overview of Defeasible Logic

As shown in the previous section, reasoning about general intention has a defeasible nature (in the sense that it is fallible) and hence we need an efficient

[1] The example is a slightly modified one as given in [20].

and easily implementable system to capture the required defeasible instances. Defeasible logic, as developed by Nute [16] with a particular concern about computational efficiency and developed over the years by [3, 2, 1] is our choice. The reason being easy implementation [15], flexibility [1] (it has a constructively defined and easy to use proof theory) and it is efficient: It is possible to compute the complete set of consequences of a given theory in linear time [13]. We do not address any semantic issues in this paper but the *argumentation semantics* as given in [9] could be straightforwardly extended to the present case.

We begin by presenting the basic ingredients of DL. A defeasible theory contains five different kinds of knowledge: facts, strict rules, defeasible rules, defeaters, and a superiority relation. We consider only essentially propositional rules. Rules containing free variables are interpreted as the set of their variable-free instances.

Facts are indisputable statements, for example, "Vineet is a System Administrator". In the logic, this might be expressed as $SA(vineet)$.

Strict rules are rules in the classical sense: whenever the premises are indisputable (e.g., facts) then so is the conclusion. An example of a strict rule is "System-Administrators are Super-Users". Written formally: $SA(X) \rightarrow SU(X)$.

Defeasible rules are rules that can be defeated by contrary evidence. An example of such a rule is "Super-Users are typically root"; written formally: $SU(X) \Rightarrow Root(X)$. The idea is that if we know that someone is a super-user, then we may conclude that he/she is root, *unless there is other evidence suggesting that it may not be root.*

Defeaters are rules that cannot be used to draw any conclusions. Their only use is to prevent some conclusions. In other words, they are used to defeat some defeasible rules by producing evidence to the contrary. An example is "If a user is normal-user then he might not be a root". Formally: $NU(X) \leadsto \neg Root(X)$. The main point is that the information that a user is NU is not sufficient evidence to conclude that he/she is not root. It is only evidence that the user *may* not be able to become root. In other words, we don't wish to conclude $\neg root$ if NU, we simply want to prevent a conclusion *Root*.

The *superiority relation* among rules is used to define priorities among rules, that is, where one rule may override the conclusion of another rule. For example, given the defeasible rules $r : SU \Rightarrow Root$ and $r' : RW \Rightarrow \neg Root$ which contradict one another, no conclusive decision can be made about whether a Super-User with a read & write permission can be root. But if we introduce a superiority relation $>$ with $r' > r$, then we can indeed conclude that the Super-User cannot be root. The superiority relation is required to be acyclic. It turns out that we only need to define the superiority relation over rules with contradictory conclusions.

It is not possible in this short paper to give a complete formal description of the logic. However, we hope to give enough information about the logic to make the discussion intelligible. We refer the reader to [16, 3, 2] for more thorough treatments.

A *rule* r consists of its *antecedent* (or *body*) $A(r)$ ($A(r)$ may be omitted if it is the empty set) which is a finite set of literals, an arrow, and its *consequent* (or *head*) $C(r)$ which is a literal. Given a set R of rules, we denote the set of all strict rules in R by R_s, the set of strict and defeasible rules in R by R_{sd}, the set of defeasible rules in R by R_d, and the set of defeaters in R by R_{dft}. $R[q]$ denotes the set of rules in R with consequent q. If q is a literal, $\sim q$ denotes the complementary literal (if q is a positive literal p then $\sim q$ is $\neg p$; and if q is $\neg p$, then $\sim q$ is p).

A *defeasible theory* D is a triple $(F, R, >)$ where F is a finite set of facts, R a finite set of rules, and $>$ a superiority relation on R.

A *conclusion* of D is a tagged literal and can have one of the following four forms:

$+\Delta q$, meaning that q is definitely provable in D (using only facts and strict rules).
$-\Delta q$, meaning that we have proved that q is not definitely provable in D.
$+\partial q$, meaning that q is defeasibly provable in D.
$-\partial q$ meaning that we have proved that q is not defeasibly provable in D.

Provability is based on the concept of a *derivation* (or proof) in $D = (F, R, >)$. A derivation is a finite sequence $P = (P(1), \ldots P(n))$ of tagged literals satisfying four conditions (which correspond to inference rules for each of the four kinds of conclusion). $P(1..i)$ denotes the initial part of the sequence P of length i

$+\Delta$: If $P(i+1) = +\Delta q$ then
(1) $q \in F$ or
(2) $\exists r \in R_s[q]\ \forall a \in A(r) : +\Delta a \in P(1..i)$

$-\Delta$: If $P(i+1) = -\Delta q$ then
(1) $q \notin F$ and
(2) $\forall r \in R_s[q]\ \exists a \in A(r) : -\Delta a \in P(1..i)$

The definition of Δ describes just forward chaining of strict rules. For a literal q to be definitely provable we need to find a strict rule with head q, of which all antecedents have been definitely proved previously. And to establish that q cannot be proven definitely we must establish that for every strict rule with head q there is at least one antecedent which has been shown to be non-provable.

$+\partial$: If $P(i+1) = +\partial q$ then either
(1) $+\Delta q \in P(1..i)$ or
(2.1) $\exists r \in R_{sd}[q] \forall a \in A(r) : +\partial a \in P(1..i)$ and
(2.2) $-\Delta \sim q \in P(1..i)$ and
(2.3) $\forall s \in R[\sim q]$ either
(2.3.1) $\exists a \in A(s) : -\partial a \in P(1..i)$ or
(2.3.2) $\exists t \in R_{sd}[q]$ such that $t > s$ and $\forall a \in A(t) : +\partial a \in P(1..i)$.

$-\partial$: If $P(i+1) = -\partial q$ then
(1) $-\Delta q \in P(1..i)$ and
(2.1) $\forall r \in R_{sd}[q]\ \exists a \in A(r) : -\partial a \in P(1..i)$ or
(2.2) $+\Delta \sim q \in P(1..i)$ or
(2.3) $\exists s \in R[\sim q]$ such that
(2.3.1) $\forall a \in A(s) : +\partial a \in P(1..i)$ and
(2.3.2) $\forall t \in R_{sd}[q]$ either $t \not> s$ or $\exists a \in A(t) : -\partial a \in P(1..i)$.

Let us work through this condition. To show that q is provable defeasibly we have two choices: (1) We show that q is already definitely provable; or (2) we need to argue using the defeasible part of D as well. In particular, we require that there must be a strict or defeasible rule with head q which can be applied (2.1). But now we need to consider possible "attacks", that is, reasoning chains in support of $\sim q$. To be more specific: to prove q defeasibly we must show that $\sim q$ is not definitely provable (2.2). Also (2.3) we must consider the set of all rules which are not known to be inapplicable and which have head $\sim q$ (note that here we consider defeaters, too, whereas they could not be used to support the conclusion q; this is in line with the motivation of defeaters given earlier). Essentially each such rule s attacks the conclusion q. For q to be provable, each such rule s must be counterattacked by a rule t with head q with the following properties: (i) t must be applicable at this point, and (ii) t must be stronger than s. Thus each attack on the conclusion q must be counterattacked by a stronger rule. In other words, r and the rules t form a team (for q) that defeats the rules s.

The purpose of the $-\partial$ inference rules is to establish that it is not possible to prove $+\partial$. This rule is defined in such a way that all the possibilities for proving $+\partial q$ (for example) are explored and shown to fail before $-\partial q$ can be concluded. Thus conclusions tagged with $-\partial$ are the outcome of a constructive proof that the corresponding positive conclusion cannot be obtained.

Sometimes all we want to know is whether a literal is *supported*, that is if there is a chain of reasoning that would lead to a conclusion in absence of conflicts. This notion is captured by the following proof conditions:

$+\Sigma$: if $P(i+1) = +\Sigma p$ then	$-\Sigma$: if $P(i+1) = -\Sigma p$ then
(1) $+\Delta p \in P(1..i)$ or	(1) $-\Delta p \in P(1..i)$ and
(2) $\exists r_{sd}[p] : \forall a \in A(r) + \Sigma a \in P(1..i)$.	(2) $\forall r_{sd}[p] \exists a \in A(r) : -\Sigma a \in P(1.i)$

The notion of support corresponds to monotonic proofs using both the monotonic (strict rules) and non-monotonic (defeasible rules) parts of defeasible theories.

5 Defeasible Logic for Intentions

As we have seen in section 3 NMLs have been put forward to capture the intensional nature of mental attitudes such as, for example, intention. Usually modal logics are extensions of classical propositional logic with some intensional operators. Thus any classical (normal) modal logic should account for two components: (1) the underlying logical structure of the propositional base and (2) the logic behavior of the modal operators. Alas, as is well-known, classical propositional logic is not well suited to deal with real life scenarios. The main reason is that the descriptions of real-life cases are, very often, partial and somewhat unreliable. In such circumstances classical propositional logic might produce counterintuitive results in so far as it requires complete, consistent and reliable information. Hence any modal logic based on classical propositional logic is doomed to suffer from the same problems.

On the other hand the logic should specify how modalities can be introduced and manipulated. Some common rules for modalities are Necessitation and RM [4]. Consider the necessitation rule of normal modal logic which dictates the condition that an agent knows all the valid formulas and thereby all the tautologies. Such a formalisation might suit for the knowledge an agent has but definitely not for the intention part. Moreover an agent need not be intending all the consequences of a particular action it does. It might be the case that it is not confident of them being successful. Thus the two rules are not appropriate for a logic of intention.

A logic of policy-based intention should take care of the underlying principles governing such intentions. It should have a notion of the direct and indirect knowledge of the agent, where the former relates to facts as literals whereas the latter to that of the agent's theory of the world in the form of rules. Similarly the logic should also be able to account for general intentions as well as the policy-based (derived ones) intentions of the agent.

Accordingly a defeasible intention theory is a structure $(F, R^K, R^I, >)$ where, as usual F is a set of facts, R^K is a set of rules for knowledge (i.e., $\rightarrow_K$, $\Rightarrow_K$, $\leadsto_K$), R^I is a set of rules for intention (i.e., $\rightarrow_I$, $\Rightarrow_I$, $\leadsto_I$), and $>$, the superiority relation, is a binary relation over the set of rules (i.e., $> \subseteq (R^K \cup R^I)^2$).

Intuitively, given an agent, F consists of the information the agent has about the world and its immediate intentions; R^K corresponds to the agent's theory of the world, while R^I encodes its policy and $>$ its strategy (or its preferences). The policy part of a defeasible theory capture both intentions and goals. The main difference is the way the agent perceives them: goals are possible outcomes of a given context while intentions are the actual goals the agent tries to achieve in the actual situation. In other words goals are the choices an agent has and intentions are the chosen goals; in case of conflicting goals (policies) the agent has to evaluate the pros and cons and then decide according to its aims (preferences), which are encoded by the superiority relation.

In what follows we provide the appropriate inference rules for intentions, and we identify strong intentions – i.e., intentions for which there are no alternatives – using $\pm\Delta_I$; goals using $\pm\Sigma_I$, and intentions using $\pm\partial_I$.

In order to correctly capture the notion of intention we extend the signature of the logic with the modal operator INT; thus if l is literal then INTl and $\neg$INTl are modal literals. However we impose some restrictions on the form of the rules: modal literals can only occur in the antecedents of rules for intention.

Derivability for knowledge ($\pm\Delta_K$, $\pm\partial_K$) has the same conditions as those given for derivability in Section 4. It is true that the complete and accurate definition of the inference conditions is cumbersome but the intuition is natural and easy to understand. The conditions for deriving an intention are as follows:

$+\Delta_I$: if $P(i+1) = +\Delta_I p$ then
(1) $\text{INT}p \in F$ or
(2) $\exists r \in R^K_s[p] \forall a \in A(r) : +\Delta_I a \in P(1..i)$ or
(3) $\exists r \in R^I_s[p]$ such that
  (3.1) $\forall \text{INT}a \in A(r) : +\Delta_I a \in P(1..i)$ and
  (3.2) $\forall a \in A(r) : +\Delta_K a \in P(1..i)$.

$-\Delta_I$: if $P(i+1) = -\Delta_I p$ then
(1) $\text{INT}p \notin F$ and
(2) $\forall r \in R^K_s[p]$
  (2.1) $\exists a \in A(r) : -\Delta_K a \in P(1..i)$ or
  (2.2) $\exists a \in A(r) : -\Delta_I a \in P(1..i)$; and
(3) $\forall r \in R^I_s[p]$ either
  (3.1) $\exists \text{INT}a \in A(r) : -\Delta_I a \in P(1..i)$ or
  (3.2) $\exists a \in A(r) : -\Delta_K a \in P(1..i)$.

To prove a strong intention, we need either that the intention is unconditional (1), or that we have a strict rule for intention (an irrevocable policy) whose antecedent is indisputable (3). However we have another case (2): if an agent knows that B is an indisputable consequence of A, and it strongly intends A, then it must intend B. This is in contrast with the NML interpretation whereby the agent has to intend all the consequences of his/her intention.

To prove that a strong intention A does not hold ($-\Delta_I A$), first, A should not be a basic intention (1); then we have to discard all possible reasons in favour of it. If A is a definite consequence of B, that is $B \rightarrow_K A \in R^K$, we can disprove it if we can show that (2.1) B is not the case (i.e., $-\Delta_K B$) or (2.2) B is not strongly intended (i.e., $-\Delta_I B$). In case of strict policies for A (3), such as, for example the strict rule for intention $\mathrm{INT}B, C \rightarrow_I A$, we have to show that either B is not strongly intended (3.1), or the fact triggering the policy is not the case (3.2).

At the other extreme we have goals: literals supported by evidence and basic intentions.

$+\Sigma_I$: if $P(i+1) = +\Sigma_I p$ then
(1) $\mathrm{INT}p \in F$ or
(2) $\exists r \in R^K_s[p] \forall a \in A(r) : +\Sigma_I a \in P(1..i)$ or
(3) $\exists r \in R^I_s[p]$ such that
  (3.1) $\forall \mathrm{INT}a \in A(r) : +\Sigma_I a \in P(1..i)$ and
  (3.2) $\forall a \in A(r) : +\Sigma_K a \in P(1..i)$.

$-\Sigma_I$: if $P(i+1) = -\Sigma_I p$ then
(1) $\mathrm{INT}p \notin F$ and
(2) $\forall r \in R^K_s[p]$
  (2.1) $\exists a \in A(r) : -\Sigma_K a \in P(1..i)$ or
  (2.2) $\exists a \in A(r) : -\Sigma_I a \in P(1..i)$; and
(3) $\forall r \in R^I_s[p]$ either
  (3.1) $\exists \mathrm{INT}a \in A(r) : -\Sigma_I a \in P(1..i)$ or
  (3.2) $\exists a \in A(r) : -\Sigma_K a \in P(1..i)$.

The inference conditions for goals are very similar to those for strong intentions; essentially they are monotonic proofs using both the monotonic part (strict rules) and the supportive non-monotonic part (defeasible rules) of a defeasible theory.

On the other hand to capture intentions we have to use the superiority relations to resolve conflicts. Thus we can give the following definition for the inference rules for $\pm\partial_I$.

$+\partial_I$: if $P(i+1) = +\partial_I p$ then
1) $+\Delta_I p \in P(1..i)$ or
2.1) $-\Delta_K {\sim}p, -\Delta_I {\sim}p \in P(1..i)$ and
2.2) either
  .1) $\exists r \in R^K_{sd}[p] \forall a \in A(r) : +\partial_I a \in P(1..i)$, or
  .2) $\exists r \in R^I_{sd}[p]\ \forall \mathrm{INT}a, b \in A(s) :$
    $\partial_I a, +\partial_K b \in P(1..i)$; and
2.3) $\forall s \in R[{\sim}p]$ either
  .1) if $s \in R^K[{\sim}p]$ then
    $\exists a \in A(s) : -\partial_I a \in P(1..i)$ and
    $\exists b \in A(s) : -\partial_K b \in P(1..i)$; and
   if $s \in R^I[{\sim}p]$ then either
    $\exists \mathrm{INT}a \in A(s) : -\partial_I a \in P(1..i)$ or
    $\exists a \in A(s) : -\partial_K a \in P(1..i)$; or
  .2) $\exists t \in R[p]$ such that $t > s$ and
   if $t \in R^K[p]$ then $\forall a \in A(t) : +\partial_K a$ or
    $\forall a \in A(t) : +\partial_I a$; and
   if $t \in R^I[p]$ then $\forall a \in A(t) : +\partial_K a$ and
    $\forall \mathrm{INT}a \in A(t) : +\partial_I a$.

$-\partial_I$: if $P(i+1) = -\partial_I p$ then
1) $-\Delta_I p \in P(1..i)$ and
2.1) $+\Delta_K {\sim}p$ or $+\Delta_I {\sim}p \in P(1..i)$ or
2.2) both
  .1) $\forall r \in R^K_{sd}[p]\ \exists a \in A(r) : -\partial_K a \in P(1..i)$, and
    $\exists a \in A(r) : -\partial_I a \in P(1..i)$; and
  .2) $\forall r \in R^I_{sd}[p]\ \exists \mathrm{INT}a \in A(s) : -\partial_I a \in P(1..i)$ or
    $\exists a \in A(s) : -\partial_K a \in P(1..i)$; or
2.3)
  .1) $\exists s \in R^K[{\sim}p]\ \forall a \in A(s) : +\partial_K a$ or
    $\forall a \in A(s) : +\partial_I a$, or
   $\exists s \in R^K[{\sim}p]\ \forall a \in A(s) : +\partial_K a$ and
    $\forall \mathrm{INT}a \in A(s) : +\partial_I a$; and
  .2) $\forall t \in R[p]$ either $t \not> s$ or
   if $t \in R^K[p]$ then $\exists a \in A(t) : -\partial_K a$ and
    $\exists b \in A(t) : -\partial_I b$; and
   if $t \in R^I[p]$ then $\exists a \in A(t) : -\partial_K a$ or
    $\exists \mathrm{INT}a \in A(t) : -\partial_I a$.

The conditions for proving defeasible intentions are essentially the same as those given for defeasible derivations in Section 4. The only difference is that at each stage we have to check for two cases, namely: (1) the rule used is a rule for an intention; (2) the rule is a rule for knowledge. In the first case we have to verify that factual antecedent are defeasibly proved/disproved using knowledge ($\pm\partial_K$), and intentional antecedent are defeasibly proved/disproved using intention ($\pm\partial_I$). In the second case we have to remember that a conclusion of a factual rule can be transformed in an intention if all the literals in the antecedent are defeasibly intended. The intuition behind the definition of $-\partial_I$ is a combination of the motivation for $-\partial$ and the intuition of $-\Delta_I$.

We want to illustrate some of the aspects of derivability by means of examples. If it does not rain we intend to play cricket, and if we intend to play cricket we intend to stay outdoor. This example can be formalized as follows

$$\neg rain \Rightarrow_I cricket \qquad \mathrm{INT}cricket \Rightarrow_I outdoor$$

Once the fact $\neg rain$ is supplied we can derive $+\partial_I cricket$, and then the intention of staying outdoor ($+\partial_I outdoor$). However the same intention cannot be derived if the fact *cricket* is given.

If Vineet intend to travel to Italy then he intend to travel to Europe since Italy is in Europe. This argument can be formalized by the rule $Italy \rightarrow_K Europe$ plus the basic intention INT*Italy*. The conclusion $+\Delta_I Europe$ follows from clause (2) of $+\Delta_I$.

Most of the BDI systems are able to express positive and negative introspection of belief and intentions. Those notions are encoded, respectively, by the following axioms.

$$\text{INT}\phi \rightarrow \text{BEL}(\text{INT}\phi) \qquad \neg\text{INT}\phi \rightarrow \text{BEL}(\neg\text{INT}\phi)$$

One of the main effect of positive (resp. negative) introspection is the ability of using established (resp. rejected) intentions in epistemic contexts to derive (resp. prevent the derivation of) other intentions. But this is what is done in Clause 2 of $+\Delta_I$, Clause 2.2.1 of $+\partial_I$, for positive introspection, and Clause 2.2 of $-\Delta_I$ and Clause 2.2.1 of $-\partial_I$ for negative introspection.

The purpose of the $-\Delta$ and $-\partial$ inference rules is to establish that it is not possible to prove a corresponding tagged literal. These rules are defined in such a way that all the possibilities for proving $+\partial p$ (for example) are explored and shown to fail before $-\partial p$ can be concluded. Thus conclusions with these tags are the outcome of a constructive proof that the corresponding positive conclusion cannot be obtained.

As a result, there is a close relationship between the inference rules for $+\partial$ and $-\partial$, (and also between those for $+\Delta$ and $-\Delta$, and $+\Sigma$ and $-\Sigma$). The structure of the inference rules is the same, but the conditions are negated in some sense. This feature allows us to prove some properties showing the well behaviour of defeasible logic.

Theorem 1. *Let* $\# = \Delta_K, \partial_K, \Sigma_K, \Delta_I, \partial_I, \Sigma_I$, *and* D *be a defeasible theory. There is no literal* p *such that* $D \vdash +\#p$ *and* $D \vdash -\#p$.

The intuition behind the above theorem states that no literal is simultaneously provable and demonstrably unprovable, thus it establishes the coherence of the defeasible logic presented in this paper.

Theorem 2. *Let* D *be a defeasible theory, and* $M \in \{K, I\}$. $D \vdash +\partial_M p$ *and* $D \vdash +\partial_M {\sim} p$ *iff* $D \vdash +\Delta_M p$ *and* $D \vdash +\Delta_M {\sim} p$.

This theorem gives the consistency of defeasible logic. In particular it affirms that it is not possible to obtain conflicting intentions ($+\partial_I p$ and $+\partial_I {\sim} p$) unless the information given about the environment is itself inconsistent. Notice, however, that the theorem does not cover goals (Σ_I). Indeed, it is possible to have conflicting goals.

Let D be a defeasible theory. With Δ_K^+ we denote the set of literals strictly provable using the epistemic (knowledge) part of D, i.e., $\Delta_K^+ = \{p : D \vdash +\Delta_K p\}$. Similarly for the other proof tags.

Theorem 3. *For every defeasible theory D, and $M \in \{K, I\}$*

$$1.\ \Delta_M^+ \subseteq \partial_M^+ \subseteq \Sigma_M^+; \qquad 2.\ \Sigma_M^- \subseteq \partial_M^- \subseteq \Delta_M^-.$$

This theorem states that strict intentions are intentions ($\Delta_I^+ \subseteq \partial_I^+$), and intentions are goals ($\partial_I^+ \subseteq \Sigma_I^+$), which corresponds to the BDI principle INT$\phi \rightarrow$ GOALϕ. At the same time, we have that $\Delta_K^+ \subseteq \partial_K^+$. Thus if we assume that Δ_K corresponds to knowledge and ∂_K corresponds to belief we obtain KNOW$\phi \rightarrow$ BELϕ, the standard BDI axiom relating the two epistemic notions.

The proposed theory of intention satisfies many of the properties outlined by Bratman in [8]. The role of intention as a *conduct-controlling* pro-attitude rather than *conduct-influencing* is clearly illustrated in the elaborate proof-theory outlined for the types of intention. The proposed theory supports the fact that the rationality of an agent for his intention depends on the rationality of the relevant processes leading to that intention where the relevant processes includes using superiority relations to resolve conflicts as well as satisfying the rules of inclusion as shown in Theorem 3. The new approach provides a good formalisation as to the relation between *guiding intention* and *intentional action* termed as *historical principle of policy-based rationality* in [8]. The problem in general is to account for the rationality of an agent in performing a particular policy-based intention from a general policy. In our approach the defeasibility of general policies makes it possible to block/not block the application of the policy to the particular case without abandoning the policy.

6 Conclusion and Discussion

Based on Bratman's classification of intention, we have outlined a *policy-based* theory of intention which differs from the usual NML-based approaches in the sense of having a non-monotonic nature. To capture the properties involved in such intentions we adopted *defeasible logic* as the non-monotonic reasoning mechanism due to its efficiency and easy implementation as well as the defeasible nature of policy-based intentions. The new approach alleviates most of the problems related to logical-omniscience. We pointed out that some of the problems related to intention re-consideration could be easily understood through such an approach.

The approach outlined in this paper could be extended in at least two different directions.

The first is in alliance with the work done in [19, 12]. Here they outline a policy description language called *PDL* and use logic programs to reason about the policies. The main concern in that work is in tracing the *event* history that gives rise to an *action* history based on stable model semantics. In a similar manner our approach could be developed using the appropriate semantics (Kunen [14] or argumentation [9]) and developed from a logic programming point of view. The advantage in our approach is the use of the superiority relation (>) whereby we can mention a hierarchy between the rules and this is absent in other works.

The second direction in which our work could be extended is to define various rules required for constructing goals from beliefs, intentions from goals, intentions from beliefs etc. and giving a superiority relation among these rules. The recent work on BDI [21] seems to take this direction. On the other hand many new applications in emerging information technologies have advanced needs for managing relations such as authorization, trust and control among interacting agents (humans or artificial). This necessitates new models and mechanisms for structuring and flexible management of those relations. The issues of automated management of organisations in terms of policies and trust relations in highly dynamic and decentralised environments has become the focus in recent years.

Finally, as we have alluded to many semantics have been devised for defeasible logic and can be adapted straightforwardly to the extension proposed here. The method developed in [14] gives a set-theoretic fixed-point construction for $\Delta^+, \partial^+, \ldots$, which leads to a logic programming characterisation of defeasible logic. Programs corresponding to defeasible theories are sound and complete wrt Kunen semantics. The same technique is applicable in the present case with the obvious adjustments; however, it does not offer further insights on defeasible logic for BDI, because of the almost one-to-one correspondence between the inference conditions and the steps of the fixed-point construction. However semantics for defeasible BDI logic remains an interesting technical problem.

References

[1] G. Antoniou, D. Billington, G. Governatori, and M. Maher. A flexible framework for defeasible logics. In *AAAI'2000*, pages 401–405. AAAI/MIT Press, 2000. 418

[2] G. Antoniou, D. Billington, G. Governatori, and M. J. Maher. Representation results for defeasible logic. *ACM Transactions on Computational Logic*, 2(2):255–287, April 2001. 418

[3] D. Billington. Defeasible logic is stable. *Journal of Logic and Computation*, 3:370–400, 1993. 418

[4] B. F. Chellas. *Modal Logic, An Introduction.* Cambridge University Press, Cambridge, 1980. 414, 420

[5] X. Chen and G. Liu. A logic of intention. In *ICJAI'99*, 1999. 414

[6] P. R. Cohen and H. J. Levesque. Persistence, intention and commitment. In *In proceedings Timberline workshop on Reasoning about plans and actions*, pages 297–338, 1986. 414

[7] P. R. Cohen and H. J. Levesque. Intention is choice with commitment. *Artificial Intelligence*, 42(3), 1990. 414

[8] M. E. Bratman. *Intentions, Plans and Practical Reason.* Harvard University Press, Cambridge, MA, 1987. 414, 424

[9] G. Governatori and M. J. Maher. An argumentation-theoretic characterisation of defeasible logic. In *ECAI-2000*, pages 469–473, 2000. 418, 424

[10] J. Hintikka. *Knowledge and Belief.* Cornell University Press, 1962. 414

[11] M. E. Pollock and K. Konolige. A representationalist theory of intention. In *IJCAI-93*, pages 390–395, 1993. 414

[12] J. Lobo, R. Bhatia, and S. Naqvi. A policy description language. In *AAAI-99*. AAAI/MIT Press, 1999. 424

[13] M. J. Maher. Propositional defeasible logic has linear complexity. *Theory and Practice of Logic Programming*, 1(6):691–711, 2001. 414, 418
[14] M. J. Maher and G. Governatori. A semantic decomposition of defeasible logic. In *AAAI-99*, 1999. 424, 425
[15] M. J. Maher, A. Rock, G. Antoniou, D. Billignton, and T. Miller. Efficient defeasible reasoning systems. *International Journal of Artificial Intelligence Tools*, 10(4), 2001. 415, 418
[16] D. Nute. Defeasible logic. In *Handbook of Logic in Artificial Intelligence and Logic Programming*, volume 3, pages 353–395. Oxford University Press, 1987. 418
[17] A. S. Rao and M. P.Georgeff M. P. Modelling rational agents within a BDI-architecture. In *KR'91*, pages 473–484. Morgan Kaufmann, 1991. 414, 415
[18] M. P. Singh. Semantical considerations on intention dynamics for BDI agents. *Journal of Experimental and Theoretical Artificial Intelligence*, 1998. 414
[19] T. C. Son and J. Lobo. Reasoning about policies using logic programa. AAAI-spring symposium on answer set programming, March 26-28 2001. 424
[20] T. Sugimoto. A preference-based theory of intention. In *PRICAI-2000*, Springer-Verlag, 2000. 414, 417
[21] J. Thanagrajah, L. Padgham and J. Harland. Representation and reasoning for goals in BDI agents. In *Australasian Conference on Computer Science*, 2002. 425
[22] B. Van Linder. *Modal Logic for Rational Agents.* PhD thesis, Department of Computer Science, Utrecht University, 19th June 1996. 414, 416
[23] R. Zamparelli. Intentions are plans plus wishes (and more). In *AAAI Spring symposium-93*, 1993. 414

Dynamic Agent Ordering in Distributed Constraint Satisfaction Problems*

Lingzhong Zhou, John Thornton, and Abdul Sattar

School of Information Technology
Griffith University, Gold Coast Campus, Australia
{l.zhou,j.thornton,a.sattar}@griffith.edu.au

Abstract. The distributed constraint satisfaction problem (CSP) is a general formalisation used to represent problems in distributed multi-agent systems. To deal with realistic problems, multiple local variables may be required within each autonomous agent. A number of heuristics have been developed for solving such multiple local variable problems. However, these approaches do not always guarantee agent independence and the size of problem that can be solved is fairly limited.
In this paper, we are interested in increasing search efficiency for distributed CSPs. To this end we present a new algorithm using unsatisfied constraint densities to dynamically determine agent ordering during the search. The independence of agents is guaranteed and agents without neighbouring relationships can run concurrently and asynchronously. As a result of using a backtracking technique to solve the local problem, we have been able to reduce the number of nogoods stored during the search, leading to further efficiency gains. In an empirical study, we show our new approach outperforms an equivalent static ordering algorithm and a current state-of-the-art technique both in terms of execution time and memory usage.

Keywords: Constraints, Distributed Intelligence, Intelligent Agents, Search

1 Introduction

The constraint satisfaction paradigm is a well recognised and challenging field of research in artificial intelligence, with many practical and important applications. A constraint satisfaction problem (CSP) is a problem with a finite number of variables, each of which has a finite and discrete set of possible values, and a set of constraints over the variables. A solution of a CSP is an instantiation of all variables for which all the constraints are satisfied.

When the variables and constraints of a CSP are distributed among a set of autonomous and communicating agents, this can be formulated as a distributed constraint satisfaction problem (distributed CSP), where agents autonomously

* The authors gratefully acknowledge the financial support of the Australian Research Council, grant A00000118, in the conduct of this research.

T.D. Gedeon and L.C.C. Fung (Eds.): AI 2003, LNAI 2903, pp. 427–439, 2003.

and collaboratively work together to get a solution. A number of heuristics have been developed for solving distributed CSPs, such as synchronous backtracking, asynchronous backtracking (ABT) [4], asynchronous weak-commitment search (AWC) [4] and the distributed breakout algorithm (DB) [5]. However, these algorithms can only handle one variable per agent. In [1], Armstrong and Durfee use dynamic prioritisation to allow agents with multiple local variables in distributed CSPs. Here, each agent tries to find a local solution, consistent with the local solutions of higher priority agents. If no local solution exists, backtracking or modification of the prioritisation occurs. The approach uses a centralised controller, where one agent controls the starting and ending of the algorithm, and a nogood processor which records all nogood information. However, these centralised mechanisms are often not appropriate for realistic distributed CSPs. In [6], Yokoo and Hirayama extended AWC search to deal with multiple local variables in distributed CSPs. However, their approach requires a large space to store nogoods during the search.

In this paper, we propose a new *Dynamic Agent Ordering* (DAO) algorithm, which uses unsatisfied constraint density measures to locally compute a *degree of unsatisfaction* for each agent. These values are used to dynamically set the order in which agents are allowed to change their particular variable instantiations. In effect, the agents' orders are decided naturally by their unsatisfied constraint densities during the search. As each local computation is independent from other agents, the benefits of parallelism are retained, resulting in an approach that is suitable for agent oriented design and efficient in terms of memory cost.

In the rest of the paper, we formalise the definition of a distributed CSP. Then, we describe the new algorithm and investigate its performance in an empirical study. Finally, we discuss the possibility of using the new algorithm to solve other variants of distributed CSPs.

2 Distributed Constraint Satisfaction Problems

A distributed constraint satisfaction problem is defined as a CSP, in which variables and constraints are distributed among multiple autonomous and communicating agents. The agents may be distributed in different locations or in the same location but among different processes. Each agent contains a subset of the variables and tries to instantiate their values. Constraints may exist between the variables of one agent or between different agents. The final instantiations of the variables must satisfy all constraints. In this paper, we consider that all constraints are binary.

2.1 Formalisation

In a distributed constraint satisfaction problem:

1. There exists an agent set A :

$$A = \{A_1, A_2, ..., A_n\},\ n \in Z^+;$$

2. Each agent has a variable set X_i and domain set D_i,

$$X_i = \{X_{i1}, X_{i2}, ..., X_{ip_i}\};$$
$$D_i = \{D_{i1}, D_{i2}, ..., D_{ip_i}\},\ \forall i \in [1,n],\ p_i \in Z^+;$$

3. There are two kinds of constraints over the variables among agents:
   (a) Intra-agent constraints, which are between variables of same agent.
   (b) Inter-agent constraints, which are between variables of different agents.
   Agent A_i knows all constraints related to its variables. A variable may involve both intra-agent and inter-agent constraints.
4. A solution S, is an instantiation for all variables that satisfies all intra-agent and inter-agent constraints.

Since agents are distributed in different locations or in different processes, each agent only knows the partial problem associated with those constraints in which it has variables. A global solution then consists of a complete set of the overlapping partial solutions for each agent. Communication among agents is necessary and important in distributed CSPs, since each agent only knows its variables, variable domains and related intra-agent and inter-agent constraints. Hence, to evaluate a search algorithm, we not only need to measure the search speed but also to consider the communication cost.

Example:

Consider the following distributed CSP, shown in Figure 1:

1. Agents: A_1 and A_2;
2. Variable sets: $\{X_1, Y_1, Z_1\}$ in agent $A_1$and $\{X_2, Y_2, Z_2\}$ in agent A_2;
3. Domain sets: $\{D_{X_1} = \{1,3\}, D_{Y_1} = \{1,2,3\}, D_{Z_1} = \{2,4\}\}$ in agent $A_1$and $\{D_{X_2} = \{1,2,\}, D_{Y_2} = \{2,3\}, D_{Z_2} = \{2,3\}\}$ in agent A_2;
4. Intra-agent constraints: $\{X_1 \neq Y_1, Y_1 = Z_1\}$ and $\{X_2 = Y_2, Y_2 \neq Z_2\}$;
5. Inter-agent constraints: $\{X_1 \neq X_2, Y_1 \neq Z_2\}$;
6. Solution: $S = \{X_1 = 1, Y_1 = 2, Z_1 = 2, X_2 = 2, Y_2 = 2, Z_2 = 3\}$

In this example, each arc represents one constraint. If we reduce the number of the agents to one, a distributed CSP would become a local CSP. So we may consider a distributed CSP as a combination of several local CSPs. Compared with local CSPs, a distributed CSP has to deal with communication costs and delay, privacy, cooperation, extra computation, consistency, asynchronous changes, infinite processing loops, and all the basic problems of distributed computing.

3 The Dynamic Agent Ordering Algorithm

3.1 Motivation

In a CSP, the order in which values and variables are processed significantly affects the running time of an algorithm. Generally, we instantiate variables

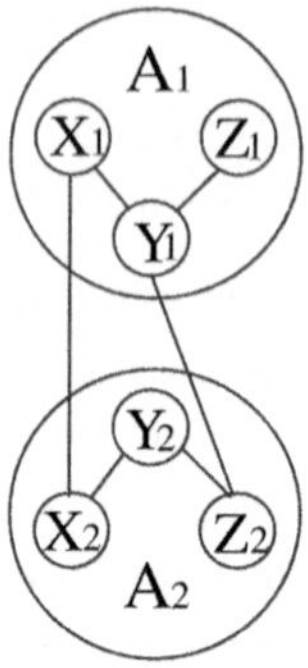

Fig. 1. An example of a distributed CSP

that maximally constrain the rest of the search space. For instance, selecting the variable with the least number of values in its domain tends to minimise the size of the search tree. When ordering values, we try to instantiate a value that maximises the number of options available for future instantiations.

The efficiency of algorithms for distributed CSPs is similarly affected by the order of value and variable selection. In the case where agents control multiple variables, the order in which agents are allowed to instantiate shared variables also becomes important. Agent communication and external computation (instantiating variables to be consistent with inter-agent constraints) is more costly than local computation (instantiating variables to be consistent with intra-agent constraints), and wrong or redundant computation can occur as a result of inappropriate agent ordering. It is therefore worth investigating agent orderings in order to develop more efficient algorithms.

The task of ordering agents is more complex than ordering variables, as more factors are involved, i.e. not only constraints and domains but also the structure of neighbouring agents. Deciding on agent ordering is analogous to granting a priority to each agent, where the priority order represents a hierarchy of agent authority. When the priority order is static, the order of agents is determined before starting the search process, and the efficiency of the algorithm is highly dependent on the selection of variable values. If the priority order is dynamic, this can be used to control decision making for each agent and the algorithm is more able to flexibly exploit to the current search conditions.

We propose a new algorithm which uses unsatisfied constraint density (related to both intra-agent and inter-agent constraints) to order agents in a distributed CSP. When a search becomes stuck (i.e. an inconsistency is found), the algorithm calculates the unsatisfied constraint densities and the degree of unsatisfaction for each agent, and the agent that is most unsatisfied is reassigned. As a backtracking search is used for each local computation, agents can still run asynchronously and concurrently. The algorithm also reduces the size of the nogood store (in comparison to AWC), and so allows larger problems to be solved.

3.2 Agent Ordering

To develop a dynamic agent ordering algorithm requires the specification of those features of the search space that should determine the ordering. In this study we develop a measure of the *degree of unsatisfaction* for each agent, such that the agent with the highest degree of unsatisfaction has the highest priority. In a standard CSP, the degree of unsatisfaction can simply be measured as the number of constraints unsatisfied divided by the total number of constraints. However, in a distributed CSP, we have the additional consideration of the relative importance of intra- versus inter-agent constraints. As inter-agent constraints affect variables in more than one agent, and these variables in turn can affect the intra-agent problems, we decided to develop separate measures for the intra- and inter-agent problems, such that the inter-agent constraints are given greater importance. To do this we looked at two problem features: (i) the degree of interconnectedness between constraints (or unsatisfied constraint density) and (ii) the degree of interconnectedness between inter-agent constraints and the intra-agent local problem.

To measure constraint density, we firstly divided the problem for a particular agent into an intra-agent constraint problem and an inter-agent constraint problem:

Intra-Agent Constraint Density. For the intra-agent problem, the maximum constraint density is simply defined as the ratio of the number of constraints over the number of variables, i.e. for agent i:

$$intraDensity_i = \frac{|intraC_i|}{|intraV_i|}$$

where $intraC_i$ is the set of intra-agent constraints for agent i and $intraV_i$ is the set of variables constrained by $intraC_i$. Assuming that each constraint has the same *tightness*[1], then we would expect a larger density to indicate a more constrained and hence more difficult problem.

Inter-Agent Constraint Density. The constraint density measure for the inter-agent problem contains two additional features which increase the relative importance of the inter-agent measure in comparison to the intra-agent measure. Firstly, for agent i, instead of dividing by the total number of variables constrained by i's inter-agent constraints $interC_i$, we divide only by the number of variables that are constrained by $interC_i$ *and* controlled by i, i.e. $|interV_i|$.

In addition, when counting agent i's jth inter-agent constraint, $c_{i,j}$, we also count the number of *intra*-agent constraints $m_{i,j}$ that share a variable with $c_{i,j}$. This means the more interconnected $c_{i,j}$ is with the intra-agent problem, the larger the value of $m_{i,j}$ and the greater the effect of $c_{i,j}$ on the overall inter-agent constraint density, given by:

[1] i.e. the ratio of the number unsatisfying assignments over the total number of possible assignments.

$$interDensity_i = \frac{|interC_i| + \sum_{j=1}^{|interC_i|} m_{i,j}}{|interV_i|}$$

The sum $staticDensity_i = intraDensity_i + interDensity_i$ now provides a general measure of the overall density of the problem for a particular agent. The greater this measure, the more constrained or difficult we would consider the problem to be.

Dynamic Constraint Density. The dynamic constraint density for a particular agent is based on the static density measure, except that only unsatisfied constraints are counted in the numerator. In this way the density of a current level of constraint violation during a search can be measured. Using the functions $intraUnsat(i, j)$, which returns one if the jth intra-agent constraint for agent i is unsatisfied, zero otherwise, and $interUnsat(i, j)$, which returns one if the jth inter-agent constraint for agent i is unsatisfied, zero otherwise, we define the following measures:

$$intraUnsat_i = \frac{\sum_{j=1}^{|intraC_i|} intraUnsat(i, j)}{|intraV_i|}$$

and

$$interUnsat_i = \frac{\sum_{j=1}^{|interC_i|} (interUnsat(i, j) \times (m_{i,j} + 1))}{|interV_i|}$$

These measures then range from a value of zero, if all constraints are satisfied, to $intraUnsat_i = intraDensity_i$ and $interUnsat_i = interDensity_i$ if all constraints are unsatisfied. Combining these measures, we define:

$$dynamicDensity_i = intraUnsat_i + interUnsat_i$$

and

$$degreeUnsat_i = \frac{dynamicDensity_i}{staticDensity_i}$$

$degreeUnsat_i$ now ranges from a value of zero, if all constraints for agent i are satisfied, to one, if all constraints are unsatisfied, while embodying the increased importance of inter-agent constraints in the overall evaluation. It is this measure we use to dynamically decide agent priority in our proposed algorithm.

Example. To further clarify the details of these measures, we use a distributed 3-colouring problem shown in Figure 2. The goal of the problem is to assign colours to each node so that nodes connected by the same arc have different colours[2]. In Figure 2 (a), Agent 1 has three inter-agent constraints ($interC_1 = \{C_{25}, C_{24}, C_{34}\}$) and two intra-agent constraints ($intraC_1 = \{C_{12}, C_{23}\}$). When Agent 1 attempts to satisfy the inter-agent constraint C_{25} by changing variable V_2, its instantiation affects two other intra-agent constraints C_{12} and C_{23}. Using definition of $interDensity_i$, this equates to $m_{1,1} = 2$. Similarly, Agent 1's second inter-agent constraint C_{24} is also connected to both of Agent 1's intra-agent constraints (C_{12} and C_{23}), making $m_{1,2} = 2$, and finally Agent 1's third inter-agent constraint C_{34} is connected to a single intra-agent constraint C_{23}, making $m_{1,3} = 1$, and giving:

$$\sum_{j=1}^{3} m_{1,j} = 2 + 2 + 1 = 5$$

As Agent 1 has three inter-agent constraints ($|interC_1| = 3$), two intra-agent constraints ($|intraC_1| = 2$), three intra-agent variables ($intraV_1 = \{V_1, V_2, V_3\}$, $|intraV_1| = 3$) and two inter-agent variables ($interV_1 = \{V_2, V_3\}$, $|interV_1| = 2$), Agent 1's intra- and inter-constraint densities are given by the following:

$$intraDensity_1 = \frac{|intraC_1|}{|intraV_1|} = \frac{2}{3}$$

and

$$interDensity_1 = \frac{|interC_1| + \sum_{j=1}^{3} m_{1,j}}{|interV_1|} = \frac{3+5}{2} = 4$$

Now considering Figure 2 (b), we can see that all Agent 1's intra-agent constraints are satisfied (i.e. each pair of colours on each arc is different) and all inter-agent constraints are satisfied except C_{24} where $V_2 = V_4 = \text{Y}$. This means the expression $\sum_{j=1}^{2} intraUnsat(1, j)$ evaluates to zero, (i.e. $intraUnsat(1, 1) = 0$ as C_{12} is satisfied and $intraUnsat(1, 2) = 0$ as C_{23} is satisfied). Similarly each term in $\sum_{j=1}^{3} interUnsat(1, j)$ will evaluate to zero, except $interUnsat(1, 2)$, corresponding to the unsatisfied inter-agent constraint C_{24}. In this case $m_{1,2} + 1 = 3$ as C_{24}'s variable V_2 is connected to two intra-agent constraints. From this it follows:

$$intraUnsat_1 = \frac{\sum_{j=1}^{2} intraUnsat(1, j)}{|intraV_1|} = \frac{0}{3} = 0$$

and

$$interUnsat_1 = \frac{\sum_{j=1}^{3} interUnsat(1, j) \times (m_{1,j} + 1)}{|interV_1|} = \frac{3}{2}$$

Putting these measures together we can now determine the degree of unsatisfaction for Agent 1:

[2] B = blue, R= red, Y = yellow.

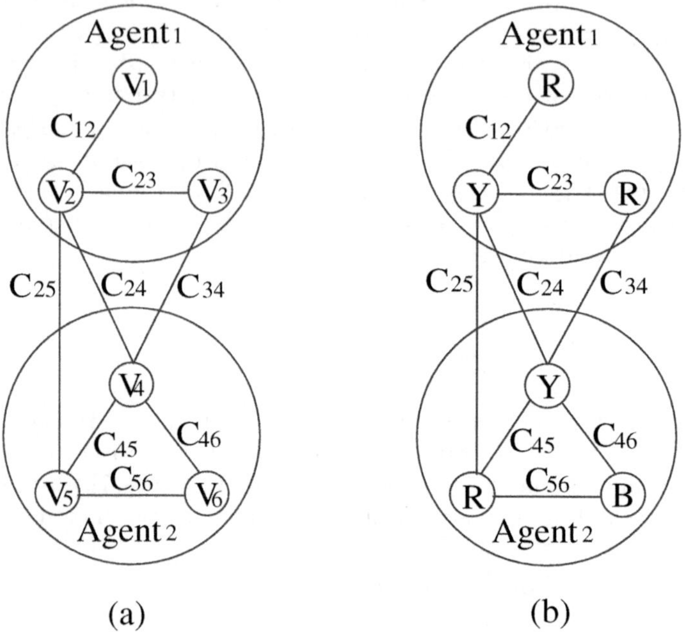

Fig. 2. A Distributed 3-colouring Problem

$$degreeUnsat_1 = \frac{intraUnsat_1 + interUnsat_1}{intraDensity_1 + interDensity_1} = \frac{0 + \frac{3}{2}}{\frac{2}{3} + 4} = \frac{9}{28}$$

Performing the same series of calculations for Agent 2, we obtain a degree of unsatisfaction of:

$$degreeUnsat_2 = \frac{intraUnsat_2 + interUnsat_2}{intraDensity_2 + interDensity_2} = \frac{0 + \frac{3}{2}}{\frac{3}{3} + \frac{9}{2}} = \frac{3}{11}$$

As $\frac{9}{28} > \frac{3}{11}$, it follows that $degreeUnsat_1 > degreeUnsat_2$, giving Agent 1 priority over Agent 2, and hence the authority to perform the next instantiation. In this case that would mean setting V_2 to B and hence finding a global solution to the problem. Note that the alternative of allowing Agent 2 to move would have resulted in several further instantiations before a global solution could be found. The reason for preferring Agent 1 in this situation can be expressed as follows: both Agent 1 and 2 have a single inter-agent constraint unsatisfied (C_{24}) for which their *dynamicDensity* measures are equal ($\frac{3}{2}$). However, Agent 1's overall static density measure is less than Agent 2 ($4\frac{2}{3}$ versus $5\frac{1}{2}$), because Agent 1's intra-agent problem is easier. Consequently the *dynamicDensity* value of $\frac{2}{3}$ has a greater effect on Agent 1, giving it priority. In effect, Agent 1 was preferred because its sub-problem was less dense (i.e. less constrained), meaning it would have the greater probability of finding a satisfying instantiation.

Overall, the *degreeUnsat* measure has the ability to show the complexity of the problem and the constraint strength in each agent. This further allows autonomous agents make decisions about who should change their variable values, without relying on a centralised evaluation mechanism.

3.3 Algorithm Implementation

The Dynamic Agent Ordering (DAO) algorithm was implemented as follows:

1. In the initial state, each agent concurrently instantiates their variables to construct a local solution, while checking consistency to guarantee that all intra-agent constraints are satisfied. Each agent then sends its local solution to its neighbouring agents (i.e. those with which it shares at least one inter-agent constraint);
2. Each agent then starts to construct a local solution which attempts to satisfy both intra- and inter-agent constraints. Assuming the overall problem is satisfiable, if an agent is unable to satisfy its partial solution, an inter-agent constraint must be involved. In this case, the two agents that share the constraint compare their *degreeUnsat* values, and the agent with greater value has priority and is allowed to reassign its variable. If the values are the same[3], we assign priority according to a predefined order; if an agent's *degreeUnsat* is less than its neighbours and its local problem is satisfied, then it simply waits for messages. If there is no suitable value for a local variable, the local agent tries to satisfy as many constraints as possible. This state will then be recorded as a nogood and sent to the related agents for the completeness of the algorithm;
3. After assigning its own variables, an agent sends messages to neighbouring agents. These messages contain the *degreeUnsat* value and the local instantiation for the agent.
4. The search will stop when each agent detects its and all its neighbouring agents' *degreeUnsat* values are equal to zero.

The DAO algorithm is shown in more detail in Algorithm *Dynamic Agent Ordering*. Here, an *optimal_local_partial_solution* is a partial solution for local variables that satisfy a maximal number of intra-agent and inter-agent constraints related to lower *degreeUnsat* agents, and against local *nogood_set*. If all constraints are satisfied, the *optimal_local_partial_solution* is equal to the local solution. *Culprit_variables* are the variables related to unsatisfied constraints that prevent a partial solution from being expanded further. These *culprit_variables* may be in different agents.

---

[3] Normally, this rarely happens in a distributed CSPs, since the density value not only depends on the number of unsatisfied constraints and variables in a local agent, but also on the structure of neighbours, the distribution of intra-agent and inter-agent constraints, and so on.

Algorithm *Dynamic Agent Ordering*

```
while received(Sender_id, variable_values, degreeUnsat) do
        calculate local_degreeUnsat;
        if local_degreeUnsat and all other agents' degreeUnsats = 0
          then the search is terminated;
          else add (Sender_id, variable_value, degreeUnsat) to agent_view;
               if local_degreeUnsat ¿ degreeUnsat
                 then assign_local_variables;
                      calculate local_degreeUnsat;
                      send(Sender_id, variable_values, local_degreeUnsat)
                      to neighbouring agents;
```

Algorithm *assign_local_variables*

```
Backtracking to construct optimal_local_partial_solution;
if optimal_local_partial_solution is not local_solution
  then assign remaining variables to satisfy as many constraints as possible;
       add the culprit_variables with their values, agent_ids to the nogood_set;
       if the nogood is new
         then record the new nogood;
              send nogood_set message to the related agents;
```

3.4 Experimental Evaluation

We evaluated the Dynamic Agent Ordering algorithm on a benchmark set of 3-colouring problems and against two other algorithms. The first, Asynchronous Weak-commitment (AWC) search, is recognised as the state-of-the-art for distributed CSPs, where each agent has control over multiple variables [6, 2]. We implemented the latest version of AWC which uses nogood learning and obtained comparable results to those reported in [2]. In addition, we implemented a version of DAO with the dynamic variable ordering switched off, called Static Agent Ordering (SAO). In this case, agent priority is determined statically before the search is commenced using the *staticDensity* measure defined in Section 3.2.

To simulate an autonomous agent environment we used an agent oriented design, implementing threads in FreeBSD that allow agents to run asynchronously and concurrently. All experiments were run on a Dell OptiPlex GX240 with a 1.6GHz P4 CPU and 256MB of PC133 DRAM. We used the same 3-colouring problem generator described in [3] and improved in [6] to evaluate the performance of our algorithms. We chose this domain as the 3-colouring problem has been used in many other studies, and this type of problem is often used in connection with scheduling and resource allocation problems. To build the problem set, we randomly generated 100 $\times$ (50-variable) and 40 $\times$ (100-variable) problems in the hard region of 3-colouring with a constraint to variable ratio of 2.7, assigning 50% of constraints as inter-agent and 50% as intra-agent constraints (within

Table 1. Results for distributed 3-colouring problems with 10 agents

Problem	Method	Checks	Nogoods	Local Instantiations	Time (seconds)
50 variables 100 runs	DAO	1,860.3	87.6	138.8	1.7254
	AWC	2,129.5	224.3	98.3	3.9624
	SAO	3,928.2	175.8	349.5	6.2387
100 variables 40 runs	DAO	17,357.4	734.8	1,277.0	33.4567
	AWC	23,617.6	1,927.2	825.4	56.9256
	SAO	44,998.2	1,256.0	2,774.8	245.4677

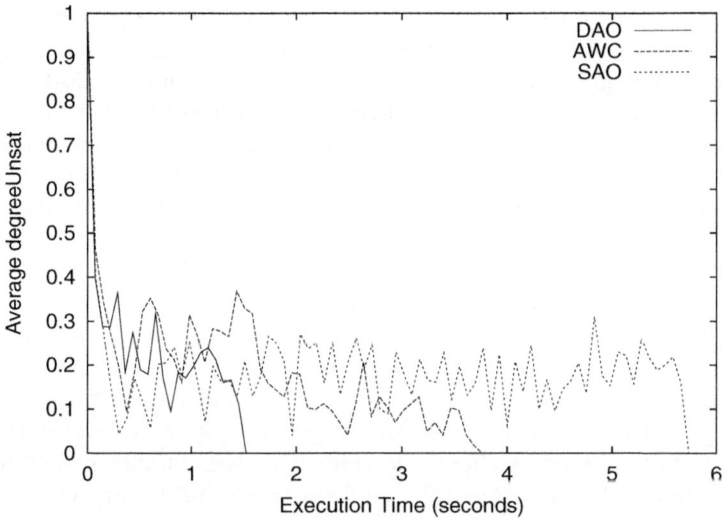

Fig. 3. The average *degreeUnsat* plotted against time

each problem). Each problem had ten agents, with each agent constrained to have at least one inter-agent constraint.

Table 1 shows the the number of average checks, the number of total nogoods produced, the number of total local instantiations broadcast and the total running time for all agents over the complete problem set. An individual problem run that most closely matched the aggregate results of the 50-variable problems in Table 1 was selected for each algorithm. The average degreeUnsat over time is shown in Figure 3 for this individual run. From these results it is clear that the new algorithm is considerably more efficient than AWC and SAO in terms of execution time. Although DAO produces more local instantiations, the size of its nogood store is significantly smaller. Also, unlike AWC, an optimal local solution can be sent by each agent during the search. As a result, a local agent has more options to instantiate its variables. The main disadvantage of DAO is that more communication costs are incurred, nevertheless, these costs are more than compensated for by the smaller number of nogoods recorded and the faster search times.

4 Conclusion and Future Work

We have demonstrated a new algorithm that uses constraint density to dynamically order agents and increase the search speed in distributed CSPs. We argue that our algorithm is more feasible and offers greater agent independence than the existing algorithms for distributed CSPs, especially for situations with multiple local variables in each agent.

Since agent independence is guaranteed, DAO can be used to solve dynamic distributed CSPs and distributed over-constrained CSPs. Dynamic distributed CSPs are common in realistic problems, where agents may be lost or added over time. By using our algorithm, real-time calculations are able to build new relations among agents, and constraints and/or variables in one agent will not affect other agents' local computations. In fact, it is not necessary to modify the algorithm to handle dynamic distributed CSPs. When a distributed CSP has no solutions, it is over-constrained. To deal with this kind of problem, we can setup a gate value (between 0 and 1) for the $degreeUnsat$ values. After all $degreeUnsat$ values reach the gate value, the problem is solved.

Finally, for problems where individual constraints have varying degrees of tightness, we can amend our constraint density measures to consider tightness directly. Currently we *count* the number of intra- and inter-agent constraints for each agent when calculating density. Alternatively, we can sum the tightness of these constraints, where tightness is defined as the number of possible unsatisfying assignments for a constraint divided by the total number of possible assignments.

References

[1] Aaron Armstrong and Edmund Durfee. Dynamic prioritization of complex agents in distributed constraint satisfaction problems. In *The Fifteenth International Joint Conference on Artificial Intelligence*, pages 620–625, 1997. 428

[2] Katsutoshi Hirayama and Makoto Yokoo. The effect of nogood learning in distributed constraint satisfaction. *The 20th International Conference on Distributed Computing Systems (ICDCS 2000)*, April 2000. 436

[3] S. Minton, M. D. Johnston, A. B. Philips, and P. Laird. Minimizing conflicts: a heuristic repair method for constraint satisfaction and scheduling problems. *Artificial Intelligence*, pages 161–205, 1992. 436

[4] Makoto Yokoo, Edmund H. Durfee, Toru Ishida, and Kazuhiro Kuwabara. The distributed constraint satisfaction problem: Formalization and algorithms. *IEEE Transaction on Knowledge and Data Engineering*, 10(5):673–685, 1998. 428

[5] Makoto Yokoo and Katsutoshi Hirayama. Distributed breakout algorithm for solving distributed constraint satisfaction problems. *Proceedings of the Second International Conference on Multiagent Systems (ICMAS-96)*, pages 401–408, 1996. 428

[6] Makoto Yokoo and Katsutoshi Hirayama. Distributed constraint satisfaction algorithm for complex local problems. In *the Third International Conference on Multiagent Systems (ICMAS-98)*, pages 372–379, 1998. 428, 436

On Why Discretization Works for Naive-Bayes Classifiers

Ying Yang and Geoffrey I. Webb

School of Computer Science and Software Engineering
Monash University, Melbourne, VIC 3800, Australia
{yyang,webb}@csse.monash.edu.au

Abstract. We investigate why discretization can be effective in naive-Bayes learning. We prove a theorem that identifies particular conditions under which discretization will result in naive-Bayes classifiers delivering the same probability estimates as would be obtained if the correct probability density functions were employed. We discuss the factors that might affect naive-Bayes classification error under discretization. We suggest that the use of different discretization techniques can affect the classification bias and variance of the generated classifiers. We argue that by properly managing discretization bias and variance, we can effectively reduce naive-Bayes classification error.

1 Introduction

Naive-Bayes classifiers are simple, effective, efficient, robust, and support incremental training. These merits have seen them employed in numerous classification tasks. Holding the *attribute independence assumption*, naive-Bayes classifiers need to estimate probabilities about individual attributes. An attribute can be either qualitative or quantitative. For a qualitative attribute, its probabilities can be estimated from corresponding frequencies. For a quantitative attribute, either probability density estimation or discretization can be employed to estimate its probabilities. Probability density estimation requires an assumption about the form of the probability distribution from which the quantitative attribute values are drawn. Discretization creates a qualitative attribute X_i^* from a quantitative attribute X_i. Each value of X_i^* corresponds to an interval of values of X_i. X_i^* is used instead of X_i for training a classifier.

In practice, discretization is more popular than probability density estimation. Pazzani [21] observed that naive-Bayes classifiers with probability density estimation were sometimes much less accurate than other learners. He concluded that unsafe assumptions about quantitative attributes' probability distributions was a major contributor to the poor performance. Dougherty et al. [7] conducted an empirical comparison of naive-Bayes learning with discretization and with probability density estimation. They observed that discretization could sometimes significantly improve the naive-Bayes classification accuracy. They hypothesized that this was because discretization did not make assumptions about the forms of quantitative attributes' probability distribution.

T.D. Gedeon and L.C.C. Fung (Eds.): AI 2003, LNAI 2903, pp. 440–452, 2003.

A number of previous authors have mentioned that discretization is critical to naive-Bayes learning's success, and have observed that different discretization techniques can result in different learning accuracy [10, 13, 20, 21, 23]. Hsu et al. [11, 12] proposed a theoretical analysis of discretization's effectiveness in naive-Bayes learning, based on an assumption that each X_i^* has a Dirichlet prior. Because 'perfect aggregation' holds for Dirichlet distributions, the probability estimation of X_i^* can be estimated independent of the shape of the curve of X_i's probability density function. However, this analysis suggested that 'well-known' discretization methods were unlikely to degrade the naive-Bayes classification accuracy, an *unconditional* excellence of discretization's effectiveness that we doubt. Besides, Hsu et al.'s analysis only addressed one-attribute classification problems, and suggested that the analysis could be extended to multi-attribute applications without indicating how this might be so. However, we believe that the analysis involving only one attribute differs from that involving multiple attributes, since the final choice of the class is decided by the product of each attribute's probability in the later situation. Further, their analysis does not lead to criteria to guide selection between alternative discretization methods.

Kononenko [17] presented a proof that so long as the estimate of the posterior probability of the class given the attribute is the same for the undiscretized attribute and its discretized counterpart, discretization will result in identical naive-Bayes classification accuracy to that obtained by use of the true probability density function. However, his proof requires that the attributes be *unconditionally* independent of each other. This assumption is much stronger than the naive-Bayes attribute independence assumption embodied in (3), below, and hence Kononenko's proof fails to account for the success of discretization for naive-Bayes.

Because of naive-Bayes classifiers' popularity and discretization's influential role in naive-Bayes learning, we believe it important to understand why discretization can be effective. We expect this understanding to help devise more effective discretization techniques for naive-Bayes classifiers, and thus to improve classification accuracy. To this end we seek to improve the current state of theoretical analysis of the factors that affect discretization effectiveness for naive-Bayes classifiers. In particular, we prove a theorem that explains why discretization can be effective. This theorem extends Kononenko's [17] theorem to the case of conditional independence. We suggest that discretization can affect the classification bias and variance of the generated naive-Bayes classifiers. We supply insights that link the interval size and interval number formed by a discretization method to its discretization bias and variance and introduce two new discretization methods informed by these insights.

The rest of this paper is organized as follows. Section 2 defines naive-Bayes classifiers. Section 3 proves a theorem that explains why discretization can be effective for naive-Bayes learning. Section 4 analyzes the factors that might affect discretization effectiveness, and proposes the bias-variance characteristics of discretization. It introduces two new discretization techniques that aim at managing discretization bias and variance. Section 5 presents the conclusion.

2 Naive-Bayes Classifiers

In naive-Bayes learning, each *instance* is described by a vector of *attribute* values and its *class* can take any value from some predefined set of values. A set of instances with their classes, the *training data*, is provided. A *test instance* is presented. The learner is asked to predict its class according to the evidence provided by the training data. We define C as a random variable denoting the class of an instance; $\mathbf{X}{<}X_1, X_2, \cdots, X_k{>}$ as a vector of random variables denoting the observed attribute values (an instance); c as a particular class label; $\mathbf{x}{<}x_1, x_2, \cdots, x_k{>}$ as a particular observed attribute value vector (a particular instance); and $\mathbf{X}{=}\mathbf{x}$ as shorthand for $X_1{=}x_1 \wedge X_2{=}x_2 \wedge \cdots \wedge X_k{=}x_k$.

Expected classification error under zero-one loss can be minimized by choosing $argmax_c(p(C{=}c \,|\, \mathbf{X}{=}\mathbf{x}))$ for each $\mathbf{x}$ [8]. We start with Bayes' theorem:

$$p(C{=}c \,|\, \mathbf{X}{=}\mathbf{x}) = \frac{p(C{=}c)p(\mathbf{X}{=}\mathbf{x} \,|\, C{=}c)}{p(\mathbf{X}{=}\mathbf{x})}. \tag{1}$$

Since the denominator in (1) is invariant across classes, it does not affect the final choice and can be dropped:

$$p(C{=}c \,|\, \mathbf{X}{=}\mathbf{x}) \propto p(C{=}c)p(\mathbf{X}{=}\mathbf{x} \,|\, C{=}c). \tag{2}$$

The probabilities $p(C{=}c)$ and $p(\mathbf{X}{=}\mathbf{x} \,|\, C{=}c)$ need to be estimated from the training data. Unfortunately, since $\mathbf{x}$ is usually a previously unseen instance that does not appear in the training data, it may not be possible to directly estimate $p(\mathbf{X}{=}\mathbf{x} \,|\, C{=}c)$. So a simplification is made: if attributes $X_1, X_2, \cdots, X_k$ are conditionally independent of each other given the class, then:

$$p(\mathbf{X}{=}\mathbf{x} \,|\, C{=}c) = p(\wedge_{i=1}^{k} X_i{=}x_i \,|\, C{=}c) = \prod_{i=1}^{k} p(X_i{=}x_i \,|\, C{=}c). \tag{3}$$

From (2) and (3), one can further estimate the most probable class by using:

$$p(C{=}c \,|\, \mathbf{X}{=}\mathbf{x}) \propto p(C{=}c) \prod_{i=1}^{k} p(X_i{=}x_i \,|\, C{=}c). \tag{4}$$

Classifiers using (4) are *naive-Bayes classifiers*. The assumption embodied in (3) is the *attribute independence assumption*. The probability $p(C{=}c \,|\, \mathbf{X}{=}\mathbf{x})$ denotes the conditional probability of a class c given an instance $\mathbf{x}$. The probability $p(C{=}c)$ denotes the prior probability of a particular class c. The probability $p(X_i{=}x_i \,|\, C{=}c)$ denotes the conditional probability that an attribute X_i takes a particular value x_i given the class c. For naive-Bayes learning, the class C is qualitative, and an attribute X_i can be either qualitative or quantitative. Since quantitative data have characteristics different from qualitative data [3, 22], the practice of estimating probabilities in (4) when involving quantitative data is different from that when involving qualitative data.

2.1 Calculating Frequency for Qualitative Data

The class, as well as a qualitative attribute, usually takes a small number of values [3, 22]. Thus there are usually many instances of each value in the training data. The probability $p(C{=}c)$ can be estimated from the frequency of instances with $C{=}c$. The probability $p(X_i{=}x_i \,|\, C{=}c)$, when X_i is qualitative, can be estimated from the frequency of instances with $C{=}c$ and the frequency of instances with $X_i{=}x_i \wedge C{=}c$. These estimates are strong consistent estimates according to the strong law of large numbers [5, 14]. In practice, a typical approach to estimating $p(C{=}c)$ is to use the Laplace-estimate [6]. A typical approach to estimating $p(X_i{=}x_i \,|\, C{=}c)$ is to use the M-estimate [6].

2.2 Probability Density Estimation for Quantitative Data

When it is quantitative, X_i usually has a large or even an infinite number of values [3, 22]. Thus the probability of a particular value x_i given the class c, $p(X_i{=}x_i \,|\, C{=}c)$ can be infinitely small. Accordingly, there usually are very few training instances for any one value. Hence it is unlikely that reliable estimation of $p(X_i{=}x_i \,|\, C{=}c)$ can be derived from the observed frequency. Consequently, in contrast to qualitative attributes, each quantitative attribute is modelled by some continuous probability distribution over the range of its values; and one can substitute a probability density function $f(X_i{=}x_i \,|\, C{=}c)$ for $p(X_i{=}x_i \,|\, C{=}c)$ for each quantitative X_i in (4) [14]. Since f is usually unknown for real-world data, *probability density estimation* is used to construct $\hat{f}$, an estimate of f from the training data. Typical approaches to estimating $f(X_i{=}x_i \,|\, C{=}c)$ are assuming f to have a Gaussian distribution [7, 18] or kernel density estimation [14].

2.3 Discretization

Discretization provides an alternative to probability density estimation for naive-Bayes learning with quantitative attributes. Under probability density estimation, if the assumed density is not a proper estimate of the true density, the naive-Bayes classification accuracy tends to degrade [7, 14, 21]. Since the true density is usually unknown for real-world data, unsafe assumptions unfortunately often occur. Discretization can circumvent this problem. Under discretization, a qualitative attribute X_i^* is formed for X_i. Each value x_i^* of X_i^* corresponds to an interval $(a_i, b_i]$ of X_i. Any original quantitative value $x_i \in (a_i, b_i]$ is replaced by x_i^*. All relevant probabilities are estimated with respect to x_i^*. Since probabilities of X_i^* can be properly estimated from corresponding frequencies as long as there are enough training instances, there is no need to assume a probability density function. However, because qualitative data have a lower level of measurement scale than quantitative data [22], discretization might suffer information loss.

Two important concepts involved in our study of discretization are *interval size* and *interval number*. Interval size is the frequency of training instances in an interval formed by discretization. Interval number is the total number of intervals formed by discretization.

3 Why Discretization Can Be Effective

We here prove Theorem 1 that suggests that discretization can be effective to the degree that $p(C{=}c\,|\,\mathbf{X}^*{=}\mathbf{x}^*)$ is an accurate estimate of $p(C{=}c\,|\,\mathbf{X}{=}\mathbf{x})$, where instance $\mathbf{x}^*$ is the discretized version of instance $\mathbf{x}$.

Theorem 1 *Assume the first l of k attributes are quantitative and the remaining attributes are qualitative*[1]. *Suppose instance $\mathbf{X}^*{=}\mathbf{x}^*$ is the discretized version of instance $\mathbf{X}{=}\mathbf{x}$, resulting from substituting qualitative attribute X_i^* for quantitative attribute X_i $(1{\leq}i{\leq}l)$. If $\forall_{i=1}^{l}(p(C{=}c\,|\,X_i{=}x_i) = p(C{=}c\,|\,X_i^*{=}x_i^*))$, and the naive-Bayes attribute independence assumption (3) holds, we have $p(C{=}c\,|\,\mathbf{X}{=}\mathbf{x}) \propto p(C{=}c\,|\,\mathbf{X}^*{=}\mathbf{x}^*)$.*

Proof: According to Bayes theorem, we have:

$$p(C{=}c\,|\,\mathbf{X}{=}\mathbf{x})$$
$$= p(C{=}c)\frac{p(\mathbf{X}{=}\mathbf{x}\,|\,C{=}c)}{p(\mathbf{X}{=}\mathbf{x})};$$

since the naive-Bayes attribute independence assumption (3) holds, we continue:

$$= \frac{p(C{=}c)}{p(\mathbf{X}{=}\mathbf{x})}\prod_{i=1}^{k} p(X_i{=}x_i\,|\,C{=}c);$$

using Bayes theorem:

$$= \frac{p(C{=}c)}{p(\mathbf{X}{=}\mathbf{x})}\prod_{i=1}^{k} \frac{p(X_i{=}x_i)p(C{=}c\,|\,X_i{=}x_i)}{p(C{=}c)}$$
$$= \frac{p(C{=}c)}{p(C{=}c)^k}\frac{\prod_{i=1}^{k} p(X_i{=}x_i)}{p(\mathbf{X}{=}\mathbf{x})}\prod_{i=1}^{k} p(C{=}c\,|\,X_i{=}x_i);$$

since the factor $\frac{\prod_{i=1}^{k} p(X_i=x_i)}{p(\mathbf{X}=\mathbf{x})}$ is invariant across classes:

$$\propto p(C{=}c)^{1-k}\prod_{i=1}^{k} p(C{=}c\,|\,X_i{=}x_i)$$
$$= p(C{=}c)^{1-k}\prod_{i=1}^{l} p(C{=}c\,|\,X_i{=}x_i)\prod_{j=l+1}^{k} p(C{=}c\,|\,X_j{=}x_j);$$

since $\forall_{i=1}^{l}(p(C{=}c\,|\,X_i{=}x_i){=}p(C{=}c\,|\,X_i^*{=}x_i^*))$:

$$= p(C{=}c)^{1-k}\prod_{i=1}^{l} p(C{=}c\,|\,X_i^*{=}x_i^*)\prod_{j=l+1}^{k} p(C{=}c\,|\,X_j{=}x_j);$$

[1] The order of attributes does not matter. We make this assumption only to simplify the expression of our proof. This does not at all affect the theoretical analysis.

using Bayes theorem again:

$$= p(C{=}c)^{1-k} \prod_{i=1}^{l} \frac{p(C{=}c)p(X_i^*{=}x_i^* \mid C{=}c)}{p(X_i^*{=}x_i^*)} \prod_{j=l+1}^{k} \frac{p(C{=}c)p(X_j{=}x_j \mid C{=}c)}{p(X_j{=}x_j)}$$

$$= p(C{=}c) \frac{\prod_{i=1}^{l} p(X_i^*{=}x_i^* \mid C{=}c) \prod_{j=l+1}^{k} p(X_j{=}x_j \mid C{=}c)}{\prod_{i=1}^{l} p(X_i^*{=}x_i^*) \prod_{j=l+1}^{k} p(X_j{=}x_j)};$$

since the denominator $\prod_{i=1}^{l} p(X_i^*{=}x_i^*) \prod_{j=l+1}^{k} p(X_j{=}x_j)$ is invariant across classes:

$$\propto p(C{=}c) \prod_{i=1}^{l} p(X_i^*{=}x_i^* \mid C{=}c) \prod_{j=l+1}^{k} p(X_j{=}x_j \mid C{=}c);$$

since the naive-Bayes attribute independence assumption (3) holds:

$$= p(C{=}c)p(\mathbf{X}^*{=}\mathbf{x}^* \mid C{=}c)$$
$$= p(C{=}c \mid \mathbf{X}^*{=}\mathbf{x}^*)p(\mathbf{X}^*{=}\mathbf{x}^*);$$

since $p(\mathbf{X}^*{=}\mathbf{x}^*)$ is invariant across classes:

$$\propto p(C{=}c \mid \mathbf{X}^*{=}\mathbf{x}^*). \quad \square$$

Theorem 1 ensures that as long as the attribute independence assumption holds, and discretization forms a qualitative X_i^* for each quantitative X_i such that $p(C{=}c \mid X_i^*{=}x_i^*) = p(C{=}c \mid X_i{=}x_i)$, discretization will result in naive-Bayes classifiers delivering probability estimates directly proportional to those that would be obtained if the correct probability density functions were employed. Thus, naive-Bayes classifiers with discretization can estimate $p(C{=}c \mid \mathbf{X}{=}\mathbf{x})$ without making any assumptions about the form of the probability density functions.

4 What Can Affect Discretization Effectiveness

In our analysis, the effectiveness of a discretization method is represented by the classification performance of naive-Bayes classifiers that are trained on data pre-processed by this discretization method. According to Theorem 1, we believe that the accuracy of estimating $p(C{=}c \mid X_i{=}x_i)$ by $p(C{=}c \mid X_i^*{=}x_i^*)$ is crucial in this issue. Two factors, *decision boundaries* and the *error tolerance of probability estimation*, have influence on the estimation accuracy. How discretization deals with these factors can affect the classification bias and variance of the generated classifiers, effects we name *discretization bias* and *variance*. According to (4), the prior probability of each class $p(C{=}c)$ also affects the final choice of the class. To simplify our analysis, we assume that each class has the same prior probability. That is, $p(C{=}c)$ is identical for each c. Thus we can cancel the effect of $p(C{=}c)$. However, our analysis extends straightforwardly to non-uniform cases.

4.1 Classification Bias and Variance

The performance of naive-Bayes classifiers discussed in our study is measured by their classification *error*. The error can be partitioned into a *bias* term, a *variance* term and an *irreducible* term [4, 9, 15, 16, 24]. Bias describes the component of error that results from systematic error of the learning algorithm. Variance describes the component of error that results from random variation in the training data and from random behavior in the learning algorithm, and thus measures how sensitive an algorithm is to changes in the training data. As the algorithm becomes more sensitive, the variance increases. Irreducible error describes the error of an optimal algorithm (the level of noise in the data). Consider a classification learning algorithm A applied to a set S of training instances to produce a classifier to classify an instance $\mathbf{x}$. Suppose we could draw a sequence of training sets $S_1, S_2, ..., S_l$, each of size m, and apply A to construct classifiers. The error of A at $\mathbf{x}$ can be defined as: $Error(A, m, \mathbf{x}) = Bias(A, m, \mathbf{x}) + Variance(A, m, \mathbf{x}) + Irreducible(A, m, \mathbf{x})$. There is often a 'bias and variance trade-off' [15]. All other things being equal, as one modifies some aspect of the learning algorithm, it will have opposite effects on bias and variance. A good learning scheme must have both low bias and low variance [19].

4.2 Decision Boundary

A *decision boundary* of a quantitative attribute X_i given an instance $\mathbf{x}$ is a point on the number line for X_i's value at which the most probable class of $\mathbf{x}$ changes.

Consider a simple learning task with one quantitative attribute X_1 and two classes c_1 and c_2. Suppose $X_1 \in [0, 2]$, and suppose that the probability distribution function for each class is $p(C{=}c_1 \mid X_1) = 1 - (X_1 - 1)^2$ and $p(C{=}c_2 \mid X_1) = (X_1 - 1)^2$ respectively, which are plotted in Fig. 1. The consequent decision boundaries are labelled as DB_1 and DB_2 respectively. The most-probable class for an instance $\mathbf{x}{=}{<}x_1{>}$ changes each time x_1's location crosses a decision boundary. Assume a discretization method to create intervals I_i

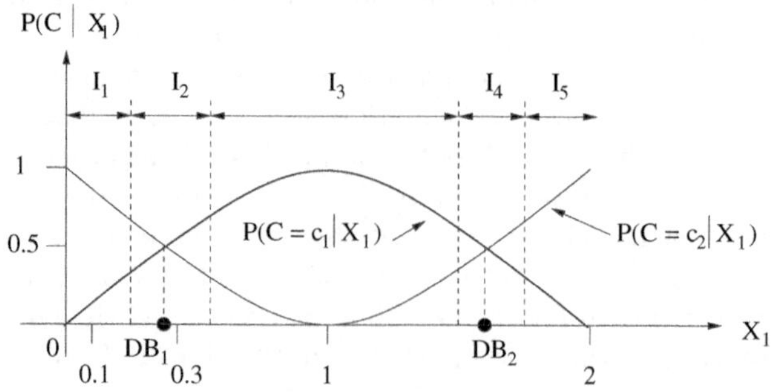

Fig. 1. Probability distribution in one-attribute problem

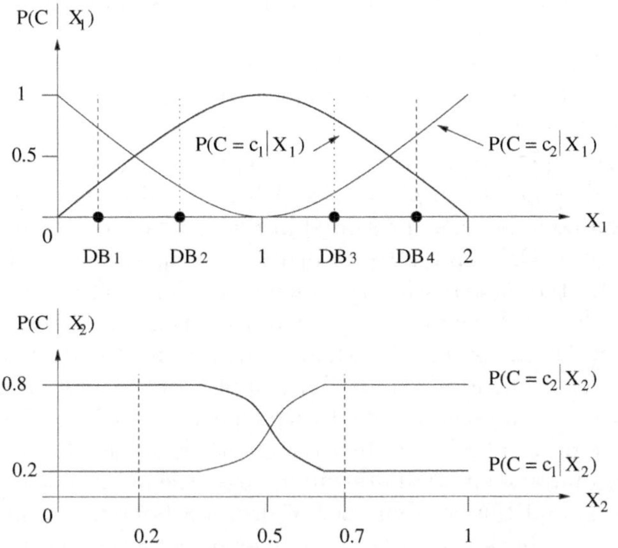

Fig. 2. Probability distribution in two-attribute problem

(i=1, ···, 5) as in Fig. 1. I_2 and I_4 contain decision boundaries while the remaining intervals do not. For any two values in I_2 (or I_4) but on different sides of a decision boundary, the optimal naive-Bayes learner under zero-one loss should select a different class for each value[2]. But under discretization, the values in a single interval can not be differentiated and all will have the same class probability estimate. Consequently, naive-Bayes classifiers with discretization will assign the same class to all of them, and thus values at one of the two sides of the decision boundary will be misclassified. The larger the interval size, the more likely that the value range of the interval is larger, thus the more likely that the interval contains a decision boundary. The larger the interval containing a decision boundary, the more instances to be misclassified, thus the higher the discretization bias.

In one-quantitative-attribute problems, the locations of decision boundaries of attribute X_1 depend on the distribution of $p(C \mid X_1)$ for each class. However, for a multi-attribute application, the decision boundaries of an attribute, say X_1, are not only decided by the distribution of $p(C \mid X_1)$, but also vary from test instance to test instance depending upon the precise values of other attributes. Consider another learning task with two quantitative attributes X_1 and X_2, and two classes c_1 and c_2. The probability distribution of each class given each attribute is depicted in Fig. 2, of which the probability distribution of each class given X_1 is identical with that in the above one-attribute context.

[2] Note that optimal classification may misclassify some instances. An optimal classifier *minimizes* error under zero-one loss. Hence even though optimal, it still can misclassify instances on both sides of a decision boundary.

We assume that the attribute independence assumption holds. We analyze the decision boundaries of X_1 for example. If X_2 does not exist, X_1 has decision boundaries as depicted in Fig. 1. However, because of the existence of X_2, those might not be decision boundaries any more. Consider a test instance **x** with $X_2{=}0.2$. Since $p(C{=}c_1 \,|\, X_2{=}0.2){=}0.8 > p(C{=}c_2 \,|\, X_2{=}0.2){=}0.2$, and $p(C{=}c \,|\, \mathbf{x}) \propto \prod_{i=1}^{2} p(C{=}c \,|\, X_i{=}x_i)$ for each class c according to Theorem 1, $p(C{=}c_1 \,|\, \mathbf{x})$ does not equal $p(C{=}c_2 \,|\, \mathbf{x})$ when X_1 falls on any of the single attribute decision boundaries as presented in Fig. 1. Instead X_1's decision boundaries change to be DB_1 and DB_4 as in Fig. 2. Suppose another test instance with $X_2{=}0.7$. By the same reasoning X_1's decision boundaries change to be DB_2 and DB_3 as in Fig. 2. When there are more than two attributes, each combination of values of the attributes other than X_1 will result in corresponding decision boundaries of X_1. Thus in multi-attribute applications the decision boundaries of one attribute can only be identified with respect to each specific combination of values of the other attributes. Increasing either the number of attributes or the number of values of an attribute will increase the number of combinations of attribute values, and thus the number of decision boundaries. In consequence, each attribute may have a very large number of potential decision boundaries. Nevertheless, for the same reason as we have discussed in one-attribute context, intervals containing decision boundaries have the potential negative impact on discretization bias.

Consequently, discretization bias can be reduced by identifying the decision boundaries and setting the interval boundaries close to them. However, identifying the correct decision boundaries depends on finding the true form of $p(C \,|\, X_1)$. Ironically, if we have already found $p(C \,|\, X_1)$, we can resolve the classification task directly; thus there is no need to consider discretization at all. Without knowing $p(C \,|\, X_1)$, an extreme solution is to set each value as an interval. Although this most likely guarantees that no interval contains a decision boundary, it usually results in very small interval size. The smaller the interval size, the fewer training instances per interval for probability estimation, thus the more likely the variance of the generated classifiers increases since even a small change of the training data might totally change the probability estimation. As a result, the estimation of $p(C \,|\, X_1)$ might be so unreliable that we can not identify the truly most probable class even if there is no decision boundary in the interval. Hence, smaller interval size can lead to higher discretization variance. A possible solution to this problem is to require that the size of an interval should be sufficient to ensure stability in the probability estimated therefrom. This raises the question, how reliable must the probability be? That is, when estimating $p(C{=}c \,|\, X_1{=}x_1)$ by $p(C{=}c \,|\, X_1^*{=}x_1^*)$, how much error can be tolerated without altering the classification. This motivates our following analysis.

4.3 Error Tolerance of Probability Estimation

To investigate this factor, we return to our example depicted in Fig. 1. We suggest that different values have different error tolerance of their probability estimation. For example, for a test instance $\mathbf{x}{<}X_1{=}0.1{>}$ and thus of class c_2, its true class probability distribution is $p(C{=}c_1 \,|\, \mathbf{x}){=}p(C{=}c_1 \,|\, X_1{=}0.1){=}0.19$ and $p(C{=}c_2 \,|\, \mathbf{x}){=}p(C{=}c_2 \,|\, X_1{=}0.1){=}0.81$. According to naive-Bayes learning, as long as $p(C{=}c_2 \,|\, X_1{=}0.1){>}0.50$, c_2 will be correctly assigned as the class and the classification is optimal under zero-one loss. This means, the error tolerance of estimating $p(C \,|\, X_1{=}0.1)$ can be as big as $0.81 - 0.50{=}0.31$. However, for another test instance $\mathbf{x}{<}X_1{=}0.3{>}$ and thus of class c_1, its probability distribution is $p(C{=}c_1 \,|\, \mathbf{x}){=}p(C{=}c_1 \,|\, X_1{=}0.3){=}0.51$ and $p(C{=}c_2 \,|\, \mathbf{x}){=}p(C{=}c_2 \,|\, X_1{=}0.3){=}0.49$. The error tolerance of estimating $p(C \,|\, X_1{=}0.3)$ is only $0.51 - 0.50{=}0.01$. In the learning context of multi-attribute applications, the analysis of the tolerance of probability estimation error is even more complicated. The error tolerance of a value of an attribute affects as well as is affected by those of the values of the other attributes since it is the multiplication of $p(C{=}c \,|\, X_i{=}x_i)$ of each x_i that decides the final probability of each class.

The lower the error tolerance a value has, the larger its interval size is preferred for the purpose of reliable probability estimation. Since all the factors that affect error tolerance vary from case to case, there can not be a universal, or even a domain-wide constant that represents the ideal interval size, which thus will vary from case to case. Further, the error tolerance can only be calculated if the true probability distribution of the training data is known. If it is not known, then the best we can hope for is heuristic approaches to managing error tolerance that work well in practice.

4.4 Summary

By this line of reasoning, optimal discretization can only be performed if the probability distribution of $p(C{=}c \,|\, X_i{=}x_i)$ for each pair of c and x_i, given each particular test instance, is known; and thus the decision boundaries are known. If the decision boundaries are not known, which is often the case for real-world data, we want to have as many intervals as possible so as to minimize the risk that an instance is classified using an interval containing a decision boundary. By this means we expect to reduce the discretization bias. On the other hand, however, we want to ensure that the intervals are sufficiently large to minimize the risk that the error of estimating $p(C{=}c \,|\, X_i^*{=}x_i^*)$ will exceed the current error tolerance. By this means we expect to reduce the discretization variance.

However, when the number of the training instances is fixed, there is a trade-off between interval size and interval number. That is, the larger the interval size, the smaller the interval number, and vice versa. Because larger interval size can result in lower discretization variance but higher discretization bias, while larger interval number can result in lower discretization bias but higher discretization variance, low learning error can be achieved by tuning interval size and interval number to find a good trade-off between discretization bias and variance. We argue that there is no universal solution to this problem, that is, the optimal trade-off between interval size and interval number will vary greatly from test instance to test instance.

Our analysis has been supported by the success of two new discretization techniques that we have recently developed: *proportional k-interval discretization*

(PKID) and *equal size discretization* (ESD) [25, 26]. To discretize a quantitative attribute, PKID equally weighs discretization bias and variance by setting interval size and interval number equal. It further sets both interval size and number proportional to the training data size. Hence as the training data increase, both discretization bias and variance tend to decrease. Bias can decrease because the interval number increases, thus the decision boundaries of the original quantitative values are less likely to be included in intervals. Variance can decrease because the interval size increases, thus the naive-Bayes probability estimation is more stable and reliable. ESD sets a *sufficient interval size*, m. It discretizes the values into intervals of size m. By introducing m, ESD aims to ensure that the interval size is sufficient so that there are enough training instances in each interval to reliably estimate the naive-Bayes probabilities. Thus ESD can prevent discretization variance from being very high. By not limiting the number of intervals formed, more intervals can be formed as the training data increases. This means that ESD can make use of extra data to reduce discretization bias. We have compared PKID and ESD against five existing key discretization methods on 29 natural datasets from UCI machine learning repository [2] and KDD archive [1]. With frequency significant at the 0.05 level, PKID and ESD each better reduce naive-Bayes classification error than previous methods. Also PKID and ESD each achieve lower mean and geometric mean of naive-Bayes classification error across all the datasets compared with previous methods.

5 Conclusion

In this paper, we prove a theorem that provides a new explanation on why discretization can be effective for naive-Bayes classifiers by showing that discretization will not alter the naive-Bayes estimate as long as discretization results in $p(C{=}c \mid X_i^*{=}x_i^*) = p(C{=}c \mid X_i{=}x_i)$. We explore the factors that can affect discretization effectiveness in terms of the classification bias and variance of the generated classifiers. We name the effects discretization bias and variance. We have argued that the analysis of the bias-variance characteristics of discretization provides insights into discretization effectiveness. In particular, we have obtained new understandings of how discretization bias and variance can be manipulated by adjusting interval size and interval number. In short, we want to maximize the number of intervals in order to minimize discretization bias, but at the same time ensure that each interval contains sufficient training instances in order to obtain low discretization variance.

Another illuminating issue arising from our study is that since the decision boundaries of a quantitative attribute value depend on the values of other quantitative attributes given a particular test instance, we can not develop optimal discretization by any apriori methodology, that is, by forming intervals prior to the classification time. However, even if we adopt a lazy methodology [27], that is, taking into account the values of other attributes during classification time, we still cannot guarantee optimal discretization unless we know the true probability distribution of the quantitative attributes. These insights reveal that, while discretization is desirable when the true underlying probability density function is not available, practical discretization techniques are necessarily heuristic in nature. The holy grail of an optimal universal discretization strategy for naive-Bayes learning is unobtainable.

References

[1] Bay, S. D. The UCI KDD archive [http://kdd.ics.uci.edu], 1999. Department of Information and Computer Science, University of California, Irvine. 450

[2] Blake, C. L., and Merz, C. J. UCI repository of machine learning databases [http://www.ics.uci.edu/~mlearn/mlrepository.html], 1998. Department of Information and Computer Science, University of California, Irvine. 450

[3] Bluman, A. G. *Elementary Statistics, A Step By Step Approach.* Wm.C.Brown Publishers, 1992. 442, 443

[4] Breiman, L. Bias, variance and arcing classifiers. *Tech. Rep., Statistics Department, University of California, Berkerley* (1996). 446

[5] Casella, G., and Berger, R. L. *Statistical Inference.* Pacific Grove, Calif., 1990. 443

[6] Cestnik, B. Estimating probabilities: A crucial task in machine learning. In *Proc. 9th European Conf. Artificial Intelligence* (1990), pp. 147–149. 443

[7] Dougherty, J., Kohavi, R., and Sahami, M. Supervised and unsupervised discretization of continuous features. In *Proc. 12th Int. Conf. Machine Learning* (1995), pp. 194–202. 440, 443

[8] Duda, R., and Hart, P. *Pattern Classification and Scene Analysis.* John Wiley & Sons, 1973. 442

[9] Friedman, J. H. On bias, variance, 0/1-loss, and the curse-of-dimensionality. *Data Mining and Knowledge Discovery 1*, 1 (1997), 55–77. 446

[10] Gama, J., Torgo, L., and Soares, C. Dynamic discretization of continuous attributes. In *Proc. 6th Ibero-American Conf. AI* (1998), pp. 160–169. 441

[11] Hsu, C.-N., Huang, H.-J., and Wong, T.-T. Why discretization works for naive Bayesian classifiers. In *Proc. 17th Int. Conf. Machine Learning* (2000), pp. 309–406. 441

[12] Hsu, C.-N., Huang, H.-J., and Wong, T.-T. Implications of the Dirichlet assumption for discretization of continuous variables in naive Bayesian classifiers. *Machine Learning* (in press). 441

[13] Hussain, F., Liu, H., Tan, C. L., and Dash, M. Discretization: An enabling technique, 1999. Tech. Rep., TRC6/99, School of Computing, National University of Singapore. 441

[14] John, G. H., and Langley, P. Estimating continuous distributions in Bayesian classifiers. In *Proc. 11th Conf. Uncertainty in Artificial Intelligence* (1995), pp. 338–345. 443

[15] Kohavi, R., and Wolpert, D. Bias plus variance decomposition for zero-one loss functions. In *Proc. 13th Int. Conf. Machine Learning* (1996), pp. 275–283. 446

[16] Kong, E. B., and Dietterich, T. G. Error-correcting output coding corrects bias and variance. In *Proc. 12th Int. Conf. Machine Learning* (1995), pp. 313–321. 446

[17] Kononenko, I. Naive Bayesian classifier and continuous attributes. *Informatica 16*, 1 (1992), 1–8. 441

[18] Mitchell, T. M. *Machine Learning.* McGraw-Hill Companies, 1997. 443

[19] Moore, D. S., and McCabe, G. P. *Introduction to the Practice of Statistics.* Michelle Julet, 2002. Fourth Edition. 446

[20] Mora, L., Fortes, I., Morales, R., and Triguero, F. Dynamic discretization of continuous values from time series. In *Proc. 11th European Conf. Machine Learning* (2000), pp. 280–291. 441

[21] Pazzani, M. J. An iterative improvement approach for the discretization of numeric attributes in Bayesian classifiers. In *Proc. 1st Int. Conf. Knowledge Discovery and Data Mining* (1995), pp. 228–233. 440, 441, 443

[22] Samuels, M. L., and Witmer, J. A. *Statistics For The Life Sciences, Second Edition.* Prentice-Hall, 1999. page10-11. 442, 443

[23] Torgo, L., and Gama, J. Search-based class discretization. In *Proc. 9th European Conf. Machine Learning* (1997), pp. 266–273. 441

[24] Webb, G. I. Multiboosting: A technique for combining boosting and wagging. *Machine Learning 40*, 2 (2000), 159–196. 446

[25] Yang, Y., and Webb, G. I. Proportional k-interval discretization for naive-Bayes classifiers. In *Proc. 12th European Conf. Machine Learning* (2001), pp. 564–575. 450

[26] Yang, Y., and Webb, G. I. Discretization for naive-Bayes learning: Managing discretization bias and variance. *Tech. Rep. 2003/131 (submitted for journal publication), School of Computer Science and Software Engineering, Monash University* (2003). 450

[27] Zheng, Z., and Webb, G. I. Lazy learning of Bayesian rules. *Machine Learning 41*, 1 (2000), 53–84. 450

Adjusting Dependence Relations for Semi-Lazy *TAN* Classifiers

Zhihai Wang[1], Geoffrey I. Webb[1], and Fei Zheng[2]

[1] School of Computer Science and Software Engineering
Monash University, Clayton Campus, Clayton, Victoria, 3800, Australia
{zhihai.wang,webb}@infotech.monash.edu.au
[2] Department of Computer Science
Nanchang University, Jiangxi Province, 330029, China
zhengfei_z@163.com

Abstract. The naive Bayesian classifier is a simple and effective classification method, which assumes a Bayesian network in which each attribute has the class label as its only one parent. But this assumption is not obviously hold in many real world domains. Tree-Augmented Naive Bayes (*TAN*) is a state-of-the-art extension of the naive Bayes, which can express partial dependence relations among attributes. In this paper, we analyze the implementations of two different *TAN* classifiers and their tree structures. Experiments show how different dependence relations impact on accuracy of *TAN* classifiers. We present a kind of semi-lazy *TAN* classifier, which builds a *TAN* identical to the original *TAN* at training time, but adjusts the dependence relations for a new test instance at classification time. Our extensive experimental results show that this kind of semi-lazy classifier delivers lower error than the original *TAN* and is more efficient than Superparent *TAN*.

1 Introduction

Classification learning seeks to build a classifier that can assign a suitable class label to an unlabelled instance described by a set of attributes. The naive Bayesian classifier is widely used in interactive applications due to its computational efficiency, competitive accuracy, direct theoretical base, and its ability to integrate prior information with data sample information [1, 2, 3, 4, 5, 6, 7]. It is based on Bayes' theorem and an assumption that all attributes are mutually independent within each class. Assume X is a finite set of instances, and $A = \{A_1, A_2, \cdots, A_n\}$ is a finite set of n attributes. An instance $x \in X$ is described by a vector $\langle a_1, a_2, \cdots, a_n \rangle$, where a_i is a value of attribute A_i. C is called the class attribute. Prediction accuracy will be maximized if the predicted class

$$Label(\langle a_1, a_2, \cdots, a_n \rangle) = argmax_c(P(c \mid \langle a_1, a_2, \cdots, a_n \rangle)). \qquad (1)$$

Unfortunately, unless $\langle a_1, a_2, \cdots, a_n \rangle$ occurs many times within X, it will not be possible to directly estimate $P(c \mid \langle a_1, a_2, \cdots, a_n \rangle)$ from the frequency with

T.D. Gedeon and L.C.C. Fung (Eds.): AI 2003, LNAI 2903, pp. 453–465, 2003.

which each class $c \in C$ co-occurs with $\langle a_1, a_2, \cdots, a_n \rangle$ within a given set of training instances. Bayes' theorem provides an equality that might be used to help estimate the posterior probability $P(c_i \mid x)$ in such a circumstance:

$$P(c_i \mid x) = \frac{P(c_i)P(\langle a_1, a_2, \cdots, a_n \rangle \mid c_i)}{P(\langle a_1, a_2, \cdots, a_n \rangle)} \tag{2}$$

$$= \alpha \cdot P(c_i) \cdot P(\langle a_1, a_2, \cdots, a_n \rangle \mid c_i) \tag{3}$$

where $P(c_i)$ is the prior probability of class $c_i \in C$, $P(\langle a_1, a_2, \cdots, a_n \rangle \mid c_i)$ is the conditional probability of $x \in T$ given the class c_i, and α is a normalization factor. According to the Bayes Theorem and the chain rule, equation 3 can be written as:

$$P(c_i \mid x) = \alpha \cdot P(c_i) \cdot \prod_{k=1}^{n} P(a_k \mid a_1, a_2, \cdots, a_{k-1}, c_i) \tag{4}$$

Therefore, an approach to Bayesian estimation is to seek to estimate each $P(a_k \mid a_1, a_2, \cdots, a_{k-1}, c_i)$.

If the n attributes are mutually independent within each class value, then the conditional probability can be calculated in the following way:

$$P(\langle a_1, a_2, \cdots, a_n \rangle \mid c_i) = \prod_{k=1}^{n} P(a_k \mid c_i). \tag{5}$$

Classification selecting the most probable class as estimation using formula 3 and formula 5 is the well-known naive Bayesian classifier.

The attribute independence assumption makes the application of Bayes' theorem to classification practical in many domains, but this assumption rarely holds in real world problems. Where some dependence relations do exist among attributes, the probability estimate of the naive Bayesian classifier may be incorrect. In such circumstances, comparing equation 4 with equation 5, we cannot use $P(a_k \mid c_i)$ instead of $P(a_k \mid a_1, a_2, \cdots, a_{k-1}, c_i)$, where $k = 1, 2, \cdots, n$. Notwithstanding Domingos and Pazzani analysis that demonstrates that some violations of the independence assumption are not harmful to classification accuracy [1], it is clear that many are, and there is an increasing body of work developing techniques to retain naive Bayesian classifiers' desirable simplicity and efficiency while delivering improved accuracy [2, 3, 4, 5, 8, 9, 10, 11, 12].

Of numerous proposals to improve the accuracy of naive Bayesian classifiers by weakening its attribute independence assumption, Tree Augmented Naive Bayes(TAN) [9, 3, 4] has demonstrated remarkable error performance [7]. Friedman, Geiger and Goldszmidt [9, 3] compared the naive Bayesian method and Bayesian network, and showed that using unrestricted Bayesian networks did not generally lead to improvements in accuracy and even reduced accuracies in some domains . They presented a compromise representation, called tree-augmented naive Bayes (TAN, called Friedman's TAN in our paper), in which the class node directly points to all attribute nodes and an attribute node can have only at most one additional parent to the class node. Based on this representation, they utilized a scoring measurement, called conditional mutual information, to efficiently find a maximum weighted spanning tree as a classifier. Keogh & Pazzani [4] took a different approach to constructing tree-augmented Bayesian networks(called Keogh and Pazzani's *TAN* in our paper). They used the same representation, but used leave-one-out cross validation to estimate the classification accuracy of the network when an arc was added. The two methods mainly differ in two aspects. One is the criterion of attribute selection used to select dependence relations among the attributes while building a tree-augmented Bayesian network. Another is the structure of the classifiers. The first one always tends to construct a tree including all attributes, the second one always tends to construct a tree with fewer dependence relations among attributes and better classification accuracy.

Friedman's *TAN* and Keogh & Pazzani's *TAN* are eager classifiers. They build tree-augmented Bayesian classifiers based on a given set of training instances at training time, and classify a new unlabelled instance directly using the classifiers at classification time.We analyze these two different approaches to *TAN* classifiers and their tree classifier structures. We show experimentally how different dependence relations impact on the accuracy of *TAN* classifiers. As a result of this study we present a new semi-lazy *TAN* classification algorithm. At training time, it builds a *TAN* identical to Friedman's *TAN*, but at classification time we adjust the dependence relations for each new test instance. Different Bayesian networks may apply to different unlabelled instances. Therefore, this approach can be thought of as a semi-lazy or partially-lazy classifier. Our extensive experimental results have shown that this kind of semi-lazy classifier has better accuracy than the previous *TAN* classifiers.

2 Restricted Bayesian Network Classifiers

Bayesian network classification is a probability classification method that can describe probability distributions over the training data. However, learning unrestricted Bayesian networks is very time consuming and quickly becomes intractable as the number of attributes increases [3, 13]. Previous research also shows that some scoring metrics used in learning unsupervised Bayesian networks do not necessarily optimize the performance of the learned networks in classification [9, 3]. Therefore, restricting the structure of Bayesian networks has become an active research area. The naive Bayesian classifier can be regarded as a highly restricted Bayesian network, which assumes that each attribute has the class label as its only one interdependent variable. TAN classifiers allow each attribute to depend on the class and at most one additional attribute. In this section, we will more formally describe *TAN* classifiers and show some issues in the implementations of the *TAN* classifiers.

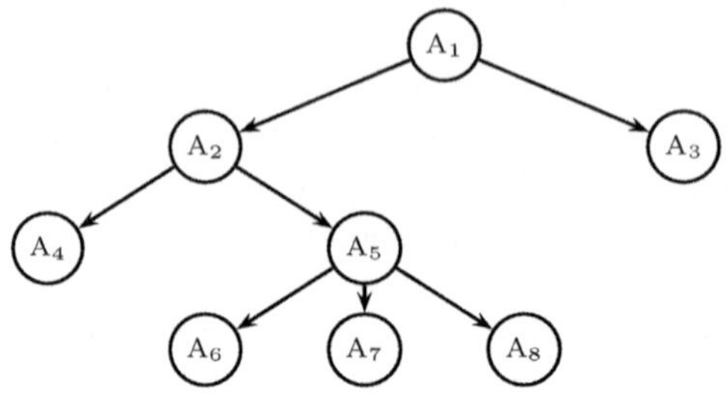

Fig. 1. A *TAN* classifier's tree structure

2.1 The Basic *TAN* Classifiers

A basic *TAN* classifier, i.e. a Friedman's *TAN* classifier, is a restricted Bayesian network classification model, which uses a tree-structure imposed on the naive Bayesian structure. In its Bayesian network, the class node is the root and has no parents, i.e. $\Pi(C) = \emptyset$ (here $\Pi(A_i)$ represents the set of parents of variable or attribute A_i). The class variable is a parent of each attribute variables, i. e. $C \in \Pi(A_i)$. And except for the class node, each attribute variable node has at most one other attribute variable node as its a parent, i.e. $|\Pi(A_i)| \leq 2$. Therefore $P(a_k \mid a_1, a_2, \cdots, a_{k-1}, c_i)$ in equation 4 can be simplified as follows.

$$P(a_k \mid a_1, a_2, \cdots, a_{k-1}, c_i) = P(a_k \mid \pi(a_k)). \tag{6}$$

Note that $\Pi(A_i)$ represents the set of parents of attribute A_i. $\pi(a_i)$ represents the set of parents of attribute value a_i. An example of Bayesian network structure for a *TAN* structure is shown in Figure 1, where class variable node C and all dependences from it to all attribute nodes A_i are omitted for simplicity.

The algorithm for building a basic *TAN* classifier consists of five main steps [3]:

1. Compute conditional mutual information $I_T(A_i, A_j \mid C)$ between each pair of attributes as follows($i \neq j$):

$$I_T(A_i, A_j \mid C) = \sum_{A_i, A_j, C} P(A_i, A_j \mid C) log \frac{P(A_i, A_j \mid C)}{P(A_i \mid C) \cdot P(A_j \mid C)} \tag{7}$$

2. Build a complete undirected graph in which the vertices are attributes $A_1, \cdots, A_n$. Annotate the weight of an edge connecting A_i to A_j by $I_T(A_i, A_j \mid C)$.

3. Build a maximum weighted spanning tree.

4. Transform the resulting undirected tree to a directed one by choosing a root variable and setting the direction of all edges to be outward from it.

5. Build a *TAN* model by adding a vertex labelled by C and adding an arc from C to each A_i.

Table 1. Descriptions of Data

	Domain	♯Instances	♯Classes	♯ Attributes
1	Adult	48842	2	14
2	Annealing Processes	898	6	38
3	Breast Cancer(Wisconsin)	699	2	9
4	Credit Screening(Australia)	690	2	15
5	German	1000	2	20
6	Glass Identification	214	7	10
7	Heart Disease(Cleveland)	303	2	13
8	Hepatitis Prognosis	155	2	19
9	House Votes 84	435	2	16
10	Hypothyroid Diagnosis	3163	2	25
11	Iris Classification	150	3	4
12	LED 24(noise level=10%)	1000	10	24
13	Letter Recognition	20000	26	16
14	Liver Disorders(bupa)	345	2	6
15	Lung Cancer	32	3	56
16	New-Thyroid	215	3	5
17	Pen Digits	10992	10	16
18	Pima Indians Diabetes	768	2	8
19	Pioneer	9150	57	36
20	Post-Operative Patient	90	3	8
21	Promoter Gene Sequences	106	2	57
22	Segment	2310	7	19
23	Solar Flare	1389	3	10
24	Sonar Classification	208	2	60
25	Soybean Large	683	19	35
26	Splice Junction Gene Sequences	3177	3	60
27	Syncon	600	6	60
28	Tic Tac-Toe End Game	958	2	9
29	Vehicle	846	4	18
30	Zoology	101	7	16

2.2 Some Issues in the Implementation

All the experiments in this paper are performed in the *Weka* system [14]. Now, we discuss some extended issues in our implementation of the basic *TAN* classifier.

One issue is related to the probability estimation assumption. In *TAN*, for each attribute we assess the conditional probability given the class variable and another attribute. This means that the number of instances used to estimate the conditional probability is reduced as it is estimated from the instances that share three specific values (the class value, the parent value and the child value). Thus it is not surprising to encounter unreliable estimates, especially in small datasets. Friedman, Geiger and Goldszmidt dealt with this problem by introducing a smoothing operation [3]. The estimation formula used by them is as follows.

$$\theta^s(x \mid \pi(x)) = \frac{N \cdot \widehat{P}_T P(\pi(x))}{N \cdot \widehat{P}_T P(\pi(x)) + N^0} \cdot \widehat{P}_T(x \mid \pi(x) + \frac{N^0}{N \cdot \widehat{P}_T P(\pi(x)) + N^0} \cdot \theta^0(x \mid \pi(x)) \quad (8)$$

where $\theta^0(x \mid \pi(x))$ is the prior estimate of $P(x \mid \pi(x))$, and N^0 is the confidence associated with that prior. In their experiments, $N^0 = 5$. A problem arises when an attribute value does not occur in the training data, a situation often occurs in cross validation tests. In this case the value of the estimate will be zero. To address this problem, in our implementation, we also apply a normal Laplace estimation to $\widehat{P}_T P(x)$. We use both smoothing functions to estimate any conditional probability, and only Laplace estimation to estimate a non-conditional probability.

Secondly, regarding the problem of missing values, in TAN classifiers, instances with missing values were deleted from the set of training instances by Friedman, et al. We keep all the instances, but ignore missing values from the counts for attribute values. Also, when we estimate a conditional probability $P(a_k \mid c_i)$, for a prior probability of class value c_i we exclude the occurrences of class value c_i with missing values on attribute A_k. Obviously, this makes the estimation of the condition more reliable while estimating $P(a_k \mid c_i)$.

Thirdly, although the choice of root variable does not change the log-likelihood of the TAN network, we have to set the direction of all edges for classification. When each edge (Ai, Aj) is added to the current tree structure, we always set the direction from Ai to Aj $(i < j)$ at once.

2.3 Keogh and Pazzani's TAN Classifiers

Keogh and Pazzani present another approach to learn TAN classifiers [4], called $SuperParent$,which searches heuristically for a TAN guided by cross-validation accuracy. They show that their algorithm consistently predicts more accurately than naive Bayes. It consists of two steps. The first step searches for a super parent that has the best cross-validation accuracy. A super parent is a node with arcs pointing to all others nodes without a parent except for the class label. The second step determines one favorite child for the super parent chosen at the first step, again based on the cross-validation accuracy. After this iteration of the two steps, one arc is added on the tree structure, and this process repeats until no improvement is achieved, or $n - 1$ arcs are added into the tree.

We also implemented Keogh and Pazzani's TAN classifiers in $Weka$ system. They follow Friedman and Goldszmidt's assumption about missing values [3], i.e., instances with missing values were deleted from the datasets. In their experiments, they replace zero probabilities with a small epsilon (0.0001). However, for consistency, our implementation utilises the smoothing techniques described above for our implementation of the basic TAN classifiers.

3 Adjusting Dependence Relations in Semi-Lazy Way

In this section, we discuss how to select dependence relations among attribute values given a test instance based on the basic TAN network. Experimentally we demonstrate that the tree structure is a useful description for the given set of training instances, and on the other hand, most of these conditional probabilities have extremely high variance and lead to poor predictions. We investigate a method of adjusting the dependence relation for a given conditional probability. At training time, we derive the basic TAN classifiers as Friedman's TAN. At classification time, we reinterpret the dependence relations for a given unlabelled instance. As the interpretation is done at classification time, we can regard this kind of classification model as a semi-lazy or partially-lazy classifier. Finally, we also experimentally demonstrate that building a TAN classifier in a totally lazy way is not effective. Before describing the new algorithms, we first describe the data sets used in our experiments and our experimental methodology.

3.1 Experimental Domains and Methodology

The thirty natural domains used in our experiments are shown in Table 1 [15]. All the experiments were performed in the Weka system [14], which provides a workbench that includes full and working implementations of many popular learning schemes that can be used for practical data mining or for research. The error rate of each classification model on each domain is determined by running 10-fold cross validation on a dual-processor 1.7Ghz Pentium 4 Linux computer with 2Gb RAM. All the data sets were used in previous research [9, 4, 5]. We also use the default discretization method "weka.filters.DiscretizeFilter" to discretize continuous attributes, which is based on Fayyad and Irani's method [16].

3.2 Applying Higher-Order Conditional Probabilities

A TAN structure is a kind of restricted Bayesian network, which combines some of Bayesian networks' ability to represent dependence relations with the simplicity of naive Bayes. A TAN structure means the way of using $P(a_k \mid c_i)$ *or* $P(a_k \mid \pi(a_k))$ instead of $P(a_k \mid a_1, a_2, \cdots, a_{k-1}, c_i)$ in equation 4. It is clear that in some situations, it would be useful to model correlations among attributes that cannot be captured by a TAN structure. This will be significant when there is a sufficient number of training instances to robustly estimate higher-order conditional probabilities [9]. However, estimating higher-order conditional probabilities will cost much more computation. Our alternative is to seek to find a better estimate instead of $P(a_k \mid \pi(a_k))$ as the estimate of $P(a_k \mid a_1, a_2, \cdots, a_{k-1}, c_i)$ based on a known TAN structure. Each node in a TAN structure always has the strongest mutual information with its parent. The performance of estimation for node A_k depends on the estimate for its parent node. This suggests using all the ancestors as the condition of node A_k. The result is an algorithm for classification based on the following equation:

$$P(a_k \mid a_1, a_2, \cdots, a_{k-1}, c_i) = P(a_k \mid Ancestors(a_k)). \qquad (9)$$

Table 2. Applying Higher-Order Conditional Probabilities

Domain	TAN^sT_1	TAN^sT_2	$HOCPT_1$	$HOCPT_2$
1 Adult	15.96	16.31	14.70	16.27
2 Annealing Processes	3.90	4.34	3.79	4.14
3 Breast Cancer(Wisconsin)	0.86	3.58	0.00	6.01
4 Credit Screening(Australia)	11.88	14.20	4.35	24.64
5 German	16.90	27.70	1.90	27.20
6 Glass Identification	0.93	9.34	0.93	7.94
7 Heart Disease(Cleveland)	10.56	18.48	4.29	24.09
8 Hepatitis Prognosis	3.87	18.71	1.29	21.94
9 House Votes 84	5.29	7.13	1.38	5.98
10 Hypothyroid Diagnosis	2.09	2.53	1.58	2.81
11 Iris Classification	2.67	10.00	2.00	12.67
12 LED 24(noise level=10%)	24.80	26.30	24.30	26.30
13 Letter Recognition	15.86	19.35	4.26	20.66
14 Liver Disorders(bupa)	20.29	39.71	6.09	40.58
15 Lung Cancer	9.37	46.88	0.00	43.75
16 New-Thyroid	2.33	6.98	1.40	10.23
17 Pen Digits	3.34	5.06	0.32	16.59
18 Pima Indians Diabetes	13.93	25.78	8.46	29.43
19 Pioneer	2.96	4.71	2.91	4.79
20 Post-Operative Patient	23.33	41.11	16.67	35.56
21 Promoter Gene Sequences	0.00	18.86	0.00	34.91
22 Segment	11.68	13.20	1.13	10.22
23 Solar Flare	0.94	1.01	0.86	0.94
24 Sonar Classification	2.40	29.81	0.00	44.71
25 Soybean Large	5.27	8.49	3.22	12.59
26 Splice Junc. Gene Sequences	2.80	4.60	0.03	41.33
27 Syncon	0.00	5.17	0.00	64.33
28 Tic-Tac-Toe End Game	22.23	26.10	21.82	26.20
29 Vehicle	23.52	32.39	1.30	37.35
30 Zoology	0.00	6.93	1.98	5.94

Note that this implies that the attribute subscripts are ordered so that $\forall a_j \in Ancestors(a_k)$, $j < k$, an ordering that need only be imposed implicitly. In Table 2, we list the experimental results of our implementation of Friedman's TAN and the classifier based on above formula 9. TAN^sT_1 represents the results of the basic TAN algorithm classifying all the training instances. TAN^sT_2 represents the results of the basic TAN algorithm using 10-fold cross validations. $HOCPT_1$ represents the results of applying higher-order conditional probabilities formula shown in equation 9 to classify all the training instances. $HOCPT_2$ represents corresponding results using 10-fold cross validations. The results are surprising. Most of the results of new estimation using 10-fold cross validations are worse than the TAN's, but most of the results of new estimation classifying the training instances are better than the TAN's. That tells us the new algorithm is overfitting.

3.3 Adjusting Dependence Relations Algorithm

Finding the dependence relations among the attributes is an important way to relax the attribute independent assumption made by naive Bayes. The main difference among Bayesian classification models of this kind is in way they calculate $P(a_k \mid a_1, a_2, \cdots, a_{k-1}, c_i)$. The above results experimentally show there is some possibility to find another attribute value instead of the value of the parent attribute. If $P(a_k \mid \pi(a_k))$ is a poor choice for $P(a_k \mid a_1, a_2, \cdots, a_{k-1}, c_i)$, we should first try to use $P(a_k \mid \pi(\pi(a_k)))$, because this attribute has the strongest dependence relations with its parent attribute. We use the following equation for adjusting the original dependence relation in the TAN structure.

$$P(a_k \mid a_1, a_2, \cdots, a_{k-1}, c_i) = P(a_k \mid MAX\{Ancestors(a_k)\}, c_i). \quad (10)$$

where $MAX\{Ancestors(a_k)\}$ represents the attribute value of its ancestors which has the maximum conditional mutual information with attribute value a_k. In this case, the conditional mutual information between two attribute values can be calculated as follows:

$$I_T(a_k, a_j \mid c) = \sum_c P(a_k, a_j \mid c) log \frac{P(a_k, a_j \mid c)}{P(a_k \mid c) \cdot P(a_j \mid c)} \quad (11)$$

because we are interested only in the specific values, no the full range of values for each attribute. At training time, we still build a TAN model in the same way, but at classification time, we will use formula 10 to classify an unlabelled instance. This is a semi-lazy classifier. Our algorithm also tests the tree structure. When a TAN structure is a single chain, we always use naive Bayes directly.

We compare the classification performance of four learning algorithms by running 10-fold cross validations. In the Table 3, we list the experimental results. We use the naive Bayes classifier implemented in the $Weka$ system, simply called $Naive$. We implemented in Weka Friedman's smoothed TAN, called TAN^s, Keogh and Pazzani's TAN, called SP, and our semi-lazy $S - Lazy$. The mean accuracy and running time over all data sets for each algorithm is also given in Table 3. Table 4 presents the WIN/LOSS/DRAW records for the semi-lazy TAN model together with the result of a binomial sign test which indicates the probabilities of obtaining the observed result or more extreme if WINS and LOSSES were equiprobable. This is a record of the number of data sets for which the nominated algorithm achieves lower, higher, and equal error to the comparison algorithm, measured to two decimal places. The semi-lazy TAN demonstrates significantly better classification performance than the original TAN models, and worse(albeit not significantly) than Keogh and Pazzani's TAN models, but is much more efficient than Keogh and Pazzani's TAN models.

Table 3. Experimental Results of Comparing Algorithms

	Naive	TAN^s		S-Lazy		SP	
Domain	Error	Error	Time	Error	Time	Error	Time
Adult	18.03	16.31	1.05	16.23	1.02	15.77	178.59
Annealing Processes	5.46	4.34	0.12	4.57	0.19	4.01	1.49
Breast Cancer(Wisconsin)	2.58	3.58	0.13	2.86	0.02	2.58	0.13
Credit Screening(Australia)	15.07	14.20	0.03	15.22	0.04	14.35	0.98
German	24.60	27.70	0.06	25.70	0.06	24.80	3.14
Glass Identification	11.68	9.35	0.03	7.01	0.06	6.07	0.10
Heart Disease(Cleveland)	16.50	18.48	0.01	17.49	0.02	15.19	0.23
Hepatitis Prognosis	16.13	18.71	0.02	11.61	0.02	16.13	0.11
House Votes 84	9.89	7.13	0.01	9.66	0.01	6.90	0.65
Hypothyroid Diagnosis	2.94	2.53	0.13	2.53	0.13	2.88	15.51
Iris Classification	6.00	10.00	0.01	6.00	0.01	6.00	0.00
LED 24(noise level=10%)	26.20	26.30	0.01	26.10	0.01	25.90	0.23
Letter Recognition	29.99	19.35	1.23	23.15	1.77	16.47	464.11
Liver Disorders(bupa)	36.81	39.71	0.01	39.13	0.01	40.29	0.05
Lung Cancer	46.88	46.88	0.11	46.88	0.18	50.00	1.18
New-Thyroid	8.37	6.98	0.00	6.05	0.02	7.44	0.02
Pen Digits	12.92	5.06	0.53	6.03	0.69	3.50	105.77
Pima Indians Diabetes	25.00	25.78	0.01	25.26	0.02	25.39	0.21
Pioneer	9.77	4.71	4.44	5.42	8.91	3.66	1256.99
Post-Operative Patient	28.89	41.11	0.00	33.33	0.00	30.00	0.03
Promoter Gene Sequences	8.49	18.87	0.15	14.15	0.19	8.49	7.35
Segment	11.08	13.20	0.20	9.26	0.31	6.28	17.06
Solar Flare	3.89	1.01	0.02	0.86	0.02	1.01	0.49
Sonar Classification	25.48	29.81	0.53	25.00	0.81	23.56	21.59
Soybean Large	7.17	8.49	0.18	7.76	0.27	7.03	6.24
Splice Junc. Gene Sequences	4.66	4.60	1.40	4.60	1.32	4.69	217.70
Syncon	3.00	5.17	1.61	3.00	2.44	3.00	64.71
Tic-Tac-Toe End Game	29.54	26.10	0.01	24.95	0.01	28.81	0.71
Vehicle	39.48	32.39	0.10	35.22	0.15	31.68	11.24
Zoology	5.94	6.93	0.00	4.95	0.01	5.94	0.05
The Mean	16.41	16.49	0.40	15.33	0.62	14.59	79.22

Table 4. Comparison of $Semi-LazyTAN$ to others

	WIN	$LOSS$	$DRAW$	P
$Naive$	18	9	3	0.122
TAN^s	20	7	3	0.019
SP	10	18	2	0.184

3.4 Building TAN Structures in Totally-Lazy Ways

Previous experimental results have shown that, in most cases, adjusting dependence relations can improve the performance of the basic TAN classifiers. Can we get lower error using a totally-lazy way? Can we build a better TAN structure for a given test instance? For many classification tasks classifier accuracy is more important than consideration of computational expense. In such a circumstance, building a classifier in a lazy way may be a better choice. To evaluate the promise of truly lazy TAN, we also implemented two ways for building a TAN structure using the measurement of conditional mutual information between attribute values. One is based on a given class value, another is based on all class values. Neither of them reduces error. Our previous research [7] also showed the implementation of Keogh and Pazzani's TAN in a lazy way did not improve classification performance. Probably, there should be some different measurement to show dependence relations among attribute values.

4 Conclusions

There are several contributions in this paper. The first one is that we have examined and implemented two different TAN classifiers and their tree classifier structures. Secondly, we experimentally show how different dependence relations impact on the accuracy of TAN classifiers. Thirdly, we mainly present a semi-lazy TAN classification model, which builds the same tree structure as the basic TAN model at training time, but adjusts the dependence relations for a new test instance at classification time. This approach can be thought of as a semi-lazy or partially-lazy method. Our extensive experimental results have shown that these semi-lazy classifiers have higher accuracy than the original TAN and are more efficient than Keogh and Pazzani's TAN.

It is remarkable that all our research is based on the assumption that the conditional mutual information can really reflect dependence relations among attributes. Because the measurements of conditional mutual information between attributes do not specify the direction of the dependence, this is also a reason that we can improve classification performances by adjusting dependence relations among attribute values. Although the semi-lazy TAN demonstrates better classification performance than the original TAN models, it is worse than Keogh and Pazzani's TAN models. These results may suggest a way to better restrict dependence relations based on the TAN structure.

References

[1] P. Domingos and M. Pazzani.: Beyond Independence: Conditions for the Optimality of the Simple Bayesian Classifier. In: Proceedings of the Thirteenth International Conference on Machine Learning, Morgan Kaufmann Publishers, Inc., San Francisco, CA(1996)105–112, 453, 454

[2] Kohavi, R.: Scaling up the Accuracy of Naive-Bayes Classifiers: A Decision-Tree Hybird, In: Simoudis, E.,Han, J. W., Fayyad, U. M.(eds.): Proceedings of the Second International Conference on Knowledge Discovery and Data Mining, The AAAI Press, Menlo Park, CA(1996)202-207 453, 454

[3] Friedman, N., Geiger, D., Goldszmidt, M.: Bayesian Network Classifiers. Machine Learning, Kluwer Academic Publishers, Boston, 29 (1997) 131–163 453, 454, 455, 456, 458

[4] Keogh, E. J., Pazzani, M. J.: Learning Augmented Bayesian Classifiers: A Comparison of Distribution-Based and Classification-Based Approaches. In: Proceedings of the Seventh International Workshop on Artificial Intelligence and Statistics. (1999) 225-230 453, 454, 455, 458, 459

[5] Zheng, Z., Webb, G. I.: Lazy Learning of Bayesian Rules. Machine Learning, Kluwer Academic Publishers, Boston, 41(2000) 53-84 453, 454, 459

[6] Webb, G. I., Boughton, J., Wang, Z: Averaged One-Dependence Estimators: Preliminary Results. In: S. J. Simoff, G. J. Williams and M. Hegland (Eds.). Proceedings of Australian Data Mining Workshop(ADM 2002), Canberra, Australia: University of Technology Sydney, (2002) 65-73 453

[7] Wang. Z., Webb, G. I.: Comparison of Lazy Bayesian Rule and Tree-Augmented Bayesian Learning. In: Kumar, V., Tsumoto, S., Zhong, N., Yu, P. S., Wu, X. (eds.): Proceedings of 2002 IEEE International Conference on Data Mining, IEEE Computer Society, Los Alamitos, CA(2002) 490-497 453, 454, 463

[8] Kononenko, I.: Semi-Naive Bayesian Classifier. In: Proceedings of European Conference on Artificial Intelligence, (1991) 206-219 454

[9] Friedman, N., Goldszmidt, M.: Building Classifiers Using Bayesian Networks. In: Proceedings of the Thirteenth National Conference on Artificial Intelligence, The AAAI Press, Menlo Park, CA, 1996 (1277-1284) 454, 455, 459

[10] Langley, P., Sage, S.: Induction of Selective Bayesian Classifiers. In: Proceedings of the Tenth Conference on Uncertainty in Artificial Intelligence, Morgan Kaufmann Publishers, Seattle, WA, 1994 (339–406) 454

[11] Sahami, M.: Learning Limited Dependence Bayesian Classifiers. In: Proceedings of the Second International Conference on Knowledge Discovery and Data Mining, AAAI Press, Portland, OR, 1996 (335–338) 454

[12] Webb, G. I., Pazzani, M. J.: Adjusted Probability Naive Bayesian Induction. In: Proceedings of the Eleventh Australian Joint Conference on Artificial Intelligence, Springer-verlag, Brisbane, 1998 (285–295) 454

[13] Chickering D. M.: Learning Bayesian Networks is NP-Hard. Technical Report MSR-TR-94-17, Microsoft Research Advanced Technology Division, Microsoft Corporation, 1994 455

[14] Witten, I. H., Frank, E.: Data Mining: Practical Machine Learning Tools and Techniques with Java Implementations. Seattle, WA: Morgan Kaufmann Publishers. (2000) 457, 459

[15] Merz, C., Murphy, P., Aha, D.: UCI Repository of Machine Learning Databases. Department of Information and Computer Science, University of California, Irvine. http://www.ics.uci.edu/mlearn/ MLRepository.html 459

[16] Fayyad, U., Irani, K.: Multi-Interval Discretization of Continuous-valued Attributes for Classification Learning. In: Proceedings of the 13th International Joint Conference on Artificial Intelligence, Seattle, WA: Morgan Kaufmann Publishers, (1993), 1022-1027 459

Reduction of Non Deterministic Automata for Hidden Markov Model Based Pattern Recognition Applications

Frederic Maire[1], Frank Wathne[1], and Alain Lifchitz[2]

[1] Smart Devices Laboratory, Faculty of Information Technology,
Queensland University of Technology,
2 George Street, GPO Box 2434, Brisbane Q4001 Australia
f.maire@qut.edu.au,
frankwathne@hotmail.com
[2] Laboratoire d'Informatique de Paris 6, Université P6 & CNRS (UMR 7606)
8, rue du Capitaine Scott, F-75015 Paris France
alain.lifchitz@lip6.fr

Abstract. Most on-line cursive handwriting recognition systems use a lexical constraint to help improve the recognition performance. Traditionally, the vocabulary lexicon is stored in a trie (automaton whose underlying graph is a tree). In a previous paper, we showed that non-deterministic automata were computationally more efficient than tries. In this paper, we propose a new method for constructing incrementally small non-deterministic automata from lexicons. We present experimental results demonstrating a significant reduction in the number of labels in the automata. This reduction yields a proportional speed-up in HMM based lexically constrained pattern recognition systems.

1 Introduction

Since the pioneering work of Vintsyuk [16] on automatic speech recognition systems, Hidden Markov Models (HMM) [12] and Dynamic Programming (DP) [3], [11], have provided a theoretical framework and practical algorithms for temporal pattern recognition with lexical constraints (even for large vocabularies). The techniques initially developed for speech recognition are also applicable to on-line handwriting recognition (especially if auto-segmentation from word to letter is used). Most on-line cursive handwriting recognition systems use a lexical constraint to help improve the recognition performance. Traditionally, the vocabulary lexicon is stored in a trie (automaton whose underlying graph is a tree). We have previously extended this idea with a solution based on a more compact data structure, the Directed Acyclic Word Graph (DAWG) [9]. In this paper, we propose a new construction algorithm that allows an incremental building of small non-deterministic automaton. Moreover, this new automaton is more compact than previously proposed automata. After recalling briefly the basics of lexically constrained pattern recognition problems in Section 2, we will describe taxonomy of automata in Section 3. In Sections 4 and 5, we review

T.D. Gedeon and L.C.C. Fung (Eds.): AI 2003, LNAI 2903, pp. 466-476, 2003.

standard reduction techniques for automata. In Section 6, we propose new heuristics to reduce node-automata. Experimental results demonstrating significant improvements are presented in Section 7. Our notation is standard and follows [12].

2 Lexically Constrained Pattern Recognition

A number of pattern recognition problems like hand gesture recognition, on-line hand writing recognition and speech recognition can be solved by performing an elastic matching between an input pattern and a set of prototype patterns. In all these applications, an a posteriori probability of a word given a sequence of frames (feature vectors) is computed using a HMM.

2.1 Word-HMM, Letter-HMM and Viterbi Algorithm

A word-HMM is made of the concatenation of the letter-HMM's corresponding to each letter of the word. We can abstract each word-HMM as an automaton whose underlying graph is a chain. Each transition of the automaton is labelled with a letter (or variant, namely allograph) of the word. That is each transition corresponds to a letter-HMM. At the letter scale, HMM states correspond to feature stationarity of frames (subunits of letter, namely graphemes). The objective of the lexically constrained pattern recognition problem is, given a sequence of frames and a lexicon, find the word with the largest a posteriori probability in this lexicon. The computation of this a posteriori probability of a word reduces to a matching of elastic patterns. In the framework of the so-called maximum approximation, an efficient DP algorithm, namely Viterbi Algorithm [17], [4], is used. A lexical constraint significantly helps to obtain better performance; practical experiments on a neuro-Markovian pattern recognition software called REMUS [6], [18], [19], shows that the recognition of words increases from 20% to 90%-98%, depending on the size of the vocabulary, when a lexical constraint is applied. Practical applications use lexicons with sizes ranging from 10 (digits recognition) to some 10^6 words (e.g. postcode dictionary, vocal dictation) [7]. Exhaustive application of Viterbi Algorithm to each word of the lexicon is only tractable for small and medium size lexicon, as the computational cost grows approximately linearly with the number of letters in lexicon.

2.2 Factorization of HMMs into Non-Deterministic Automata

If two words have a common prefix then the DP computations of the a posteriori probabilities can be factorized. Hence, a speed-up and reduction in storage can be obtained simply by using a trie (a well known tree-like data structure) [5]. Each edge/node in the trie corresponds to a letter. Thanks to the sharing of intermediate results, the running time has to improve dramatically compared to the trivial approach consisting in running Viterbi Algorithm independently on each word-HMM.

A trie eliminates the redundant computation/storage for common prefixes present in natural languages and is easy to implement. The trie structure is a good trade-off between simplicity and efficiency, and is widely used in practice. Unfortunately we were disappointed [9] by the poor compression ratio, from 1.5 to 4.2, dependent on languages (English/French) and vocabularies size (10^3 – 10^5 words), we got experimentally. Since practical applications, with large vocabulary, require very efficient processing, both in term of speed and storage, it is important to go further and extend the use of Viterbi Algorithm to more compact and complex lexicon structures, like DAWG. That is, use both prefix and suffix commonality [2], [14]. Lacouture et al. [8], and more recently Mohri et al. [9], have worked on similar problems with Finite State Automata (FSA) for Automatic Speech Recognition.

The automata that we build are not traditional deterministic automata. This choice is motivated by the following observations; traditional automata are graphs whose arcs have labels. Each arc is labelled with a letter. The nodes/states of the automata are not labelled. The nodes correspond to languages. It is natural to wonder whether putting the labels in the nodes instead of in the arcs would improve the compactness of the automata. The main computational cost of running Viterbi algorithm on a graph is a function of the number of labels in the graph. Hence the importance of finding a representation that minimizes the number of labels. Moreover, Viterbi algorithm does not require a deterministic automaton. We call *node-automaton* a directed graph whose nodes can be labelled with letters. The arcs of a node-automaton are transitions with no label. A transition of a node-automaton is just a routing device. Node-automata are better for HMM factorization because in a node-automaton the processing is done in the node and the routing is done with the arcs. Whereas with traditional automata (that we call *edge-automata*), these two tasks are not separated. Experimental results demonstrate a clear superiority of node-automata over edge-automata with respect to the computational cost of running Viterbi Algorithm on a whole lexicon (see Section 7).

3 Automata Taxonomy

We have experimented with two types of acyclic automata. They differ only in that the edge automata labels are stored in the edges, whereas in node-automata the labels are stored in the nodes. In order to describe the reduction algorithms we recall here the standard definitions we need throughout the rest of the paper.

A *finite state automaton* is a quintuple (Q,E,Σ,I,T) where Q is a set of states (nodes), Σ is an alphabet (a set of symbols), E is a set of *transitions* (*directed edges* or *arcs*), I is a set of *initial states* and T is a set of *terminal* (*accepting*) states. The automata discussed in this paper will have a single initial state called the *root* node. For some of the algorithms presented here, there will be a single terminal node referred to as the *sink* node. A quintuple $(Q,E,\Sigma,\{r\},\{s\})$ will denote an automaton with a single root r and a single terminal node s. A path of length k is a sequence of nodes $(n_0,n_1,\dots,n_k)$ where each successive pair of nodes in the sequence is connected by a transition. If (n_i,n_j) is a transition, then we say that n_i is a predecessor of n_j, and n_j a successor of n_i. We will denote the set of successors of n_i by

$\text{succ}(n_i)$. Similarly, $\text{pred}(n_i)$ will denote the set of predecessors of n_i. The node n_j is *reachable* from the node n_i, if there exists a path from n_i to n_j. In this case, we also say that n_j is a *descendant* of n_i.

Each ordered pair of nodes (n_i, n_j) implicitly defines a *language* (a set of words) denoted by $L_Q(n_i, n_j)$. Each sequence of labels encountered along a path from n_i to n_j makes up a word. $L_Q(n_i, n_j)$ is the language generated by all possible paths between n_i to n_j. More generally, $L_Q(A, B)$ will denote the language defined by $L_Q(A,B) = \bigcup_{(n_i,n_j)\in A\times B} L_Q(n_i,n_j)$. The language recognized by the automaton is $L_Q(I,T)$.

4 Automata Reduction

A key concept for automata minimization is the *contraction* (or *merging*) of equivalent nodes. We will first characterize useful equivalence relations, and then explain how they allow the merging of nodes.

4.1 Equivalence Relation and Node Contraction

Many equivalence relations can be defined on Q. Recall that an equivalence relation R on Q can be viewed as a partition of Q. Two nodes are equivalent with respect to R if and only if they belong to the same part of the partition of Q. Obviously, to be of any interest, the reduction operation must preserve the language of the automaton. That is, the reduced automaton should generate the same language as the original automaton. A sufficient condition for this to happen is that any two equivalent nodes with respect to R generate the same language. If an equivalence relation R satisfies this sufficient condition, we will say that R is *admissible*.

Formally, the contraction of two equivalent nodes n_i and n_j with respect to R will preserve the language of the automaton provided that

$$n_i R n_j \Rightarrow L_Q(n_i,T) = L_Q(n_j,T)$$

Whenever two nodes satisfy $L_Q(n_i,T) = L_Q(n_j,T)$, we will use the standard terminology, and say that the nodes n_i and n_j are *indistinguishable*. Our NFA minimization algorithm computes an equivalence relation between nodes that entails indistinguishability.

The contraction of a set of nodes A is a new node a' obtained by merging all nodes of A. The new automaton (after contraction of the set of nodes A) is obtained by removing all nodes belonging to A, then inserting a new node a', and connecting a' to all successors and predecessors of the nodes that were in A. Formally, $\mathrm{succ}(a') = \bigcup_{n_i \in A} \mathrm{succ}(n_i)$ and $\mathrm{pred}(a') = \bigcup_{n_i \in A} \mathrm{pred}(n_i)$. The merging of two nodes n_i and n_j is just a special case of set of nodes contraction where $A = \{n_i, n_j\}$.

We can extend the equivalence relation to sets of nodes. Two sets of nodes A and B are said to be equivalent if the two nodes a' and b' are equivalent.

4.2 Other Relevant Relations Defined on the Nodes

The type of equivalence relations we have defined so far is also referred to as *down-equivalence* because they only consider the descendants. We may also define *up-equivalence* and *up-indistinguishability* in a similar way by considering the reversed automaton. The reversed automaton is obtained by first swapping the set of initial states and the set of terminal states, then reversing all the transitions. That is, (n_i, n_j) is a transition in the reversed automaton if and only if (n_j, n_i) is a transition in the original automaton. The reversed automaton generates the mirror language of the original automaton.

When two nodes have exactly the same successors (which implies that they are indistinguishable), we say that these nodes are *similar*. *Subsumption* is a relationship more general than similarity. The node n_i subsumes the node n_j if and only if $\mathrm{succ}(n_i) \supseteq \mathrm{succ}(n_j)$. That is, every successor of n_j is also a successor of n_i. For node-automata we add the requirement that n_i and n_j have the same label. Two nodes are said to be *comparable* if one subsumes the other.

Automata minimization algorithms need sometimes to consider the *height* and the *depth* of a node. The height of a node is the length of the longest path from the node down to a leaf node. The depth of a node is the length of the longest path from the root to the node. A *level* is the set of all nodes of a given height. Levels are important because indistinguishable nodes must have the same height.

4.3 Contraction and Split Operators

In order to reduce the automata, we use mainly two node operators that work on pairs of nodes. The contraction operator merges two nodes and was defined in Section 4.1. The *split* operator is used for splitting a node into two nodes to prepare the automaton for further reduction. The new automaton obtained by splitting a node n is derived from the original automaton by replacing n with two nodes n' and n'' such that $\mathrm{pred}(n) = \mathrm{pred}(n') \cup \mathrm{pred}(n'')$ and $\mathrm{succ}(n') = \mathrm{succ}(n'') = \mathrm{succ}(n)$.

It is easy to see that

- Split operations do not change the language of the automaton,
- Contraction operations do not change the language of the automaton provided they are applied on equivalent nodes with respect to an admissible equivalence relation,
- After a split operation the new automaton will always be non deterministic.

5 Previous Works

We review in this section the main algorithms to minimize deterministic and non-deterministic automata.

5.1 Minimal Deterministic Automata

The classical algorithm for deterministic automata minimization [1] first computes the indistinguishable equivalence classes, and then contracts separately all these equivalence classes. In the special case of acyclic automata defined by a lexicon, a faster algorithm to build the minimal deterministic automaton was proposed by Revuz [13].

The algorithm of Revuz starts with the construction of the trie of the lexicon. Then the trie is traversed either level-by-level, starting with the leaves, or in a post order fashion. During the traversal the similarity classes are determined: the node currently considered is either equivalent to some other already visited node, or will become the representative of a new equivalence class. In the former case the current node is merged with the representative of that class. The bottom-up traversal ensures that all successor nodes of the current node have already been reduced. Therefore during the execution of the algorithm, two nodes on a same level will be indistinguishable if and only if they are similar (that is have the successors in the current automaton).

We have implemented and tested this algorithm for the classical edge-automata as well as for the node-automata.

5.2 Compact Non-deterministic Automata

The algorithm proposed by Sgarbas, Fakotakis and Kokkinakis [15] incrementally build a small non-deterministic automaton. We will describe the main ideas of this algorithm. The original algorithm was created for edge-automata, but we have adapted it to node-automata. The automaton returned by the algorithm is called a *compact NFA* (*CNFA*). The main feature of a CNFA is that it does not contain similar nodes.

The algorithm expects as input a NFA with a single root and a single sink node. Words are inserted incrementally in the automaton. A word insertion is made in two phases. In the first phase, a chain of new nodes corresponding to the new word is added to the automaton. The chain starts at the root node and finishes at the sink node. All the intermediate nodes are new. In the second phase, similar nodes are detected then merged in order to keep the automata compact. This is done by first traversing the automaton starting from the sink node to identify nodes that are down-similar with nodes of the newly inserted word. When no more such nodes can be found the procedure is repeated from the root, this time looking for up-similar nodes. When no more up-similar nodes are found, we continue looking from the bottom again. This procedure is repeated while we find either down-similar or up-similar nodes. The resulting automaton does not contain any similar nodes.

The CNFA algorithm provides good compression results, but further reduction can be obtained with the heuristics that we will describe in the next section.

6 Compressed Non-deterministic Automata

We propose a new heuristic, generalization of the one of [9], relying on the split and contraction operators, to further reduce NFAs. The equivalence relation R_k that we use for the contraction operator is original. For deterministic automata, we have seen that we can determine easily whether two nodes are indistinguishable by comparing their successors. For non-deterministic automata, the situation is more complicated. Let $\mathrm{succ}(n,x)$ denote the successors of node n with the label x. $\mathrm{succ}(n,x)$ can be a large set whereas it contains at most one node for deterministic automata. Next, we define an admissible equivalence relation with a given look-ahead depth. Then, we will outline our NFA compression algorithm.

6.1 The Equivalence Relation R_k

We define recursively the following equivalence relations R_k:

- R_0 is the similarity relation between nodes (their successors are required to be the same),
- Node n_i and node n_j are in relation with respect to R_k if and only if all the following conditions are satisfied
  - The nodes n_i and n_j have the same height,
  - The nodes n_i and n_j are both either terminal or non terminal.
  - The nodes n_i and n_j have the same set of labels for their outgoing transitions (but they are allowed to have a different number of transitions for a given label),
  - For each common transition label x, $\mathrm{succ}(n_i,x)$ is equivalent to $\mathrm{succ}(n_j,x)$ with respect to R_{k-1}.

The last condition concerns the equivalence between two sets of nodes. This was defined in Section 4.1.

6.2 The NFA Compression Heuristic

Our heuristic shuttles through the graph level-by-level starting at the deepest one. The skeleton of the heuristic is;

```
Loop until no change in the automaton
   Compute the heights of all nodes
   For each level
      Contract all equivalent nodes
      Separate all comparable nodes
   End for
End loop
```

In the deterministic case, the above algorithm will produce the unique minimal DFA. For non-deterministic automata, we know that there may be several minimal automata. Moreover the order in which we contract the nodes does matter. Blindly removing equivalent nodes in a NFA may lead to a sub-optimum automaton. The more reduced is level $(k+1)$, the more difficult it will be to reduce level k. To increase the likelihood of further contractions, we use the split operator to separate comparable nodes (see Section 4.2). Although splitting nodes increases the number of nodes on the current level, it makes more likely the creation of equivalent nodes in the predecessor level. Overall, there is a significant reduction in the number of nodes as demonstrated with the results of next section.

7 Experimental Results

We have run the different reduction algorithms on two families of artificial lexicons that have a known minimal NFA and DFA. Another purpose of these lexicons was to validate the implementation of the algorithms. We have also tested the algorithms on edge and node automata with different size English and French lexicons. Table 1 gives some details on the lexicons used.

Table 1. Features of the lexicons

Lexicons						
Name	**# words**	**# letters**	**Alphabet size**	**Mean length**	**Max length**	**Type**
DNA_4_8_1.txt	87380	669924	4	7.5	8	Artificial
HWR_4_8_1.txt	13120	98416	4	7.7	8	Artificial
Lex1000.txt	1000	6966	26	7	13	English
Lex10645.txt	10645	78197	26	7.3	21	English
Lex20233.txt	20233	149129	26	7.4	22	English
Lex65536f.txt	65536	631422	26	9.6	25	French
Lex130499.txt	130499	1256938	28	9.6	25	French

DNA_4_8_1 is the language of all words from 1 to 8 letters, on a alphabet of 4 letters, consecutive letters being required to be distinct. The HWR language differs from the DNA language by relaxing the constraint on consecutive letters.

Table 2. Comparison of edge-automata

	Edge Automata					
	Compact NFA (Edge Automata)		Compressed NFA (Edge Automata)		Minimal DFA (Edge Automata)	
Lexicon	Nodes	Transitions	Nodes	Transitions	Nodes	Transitions
DNA_4_8_1.txt	30	128	30	128	30	88
HWR_4_8_1.txt	9	60	9	60	9	32
Lex1000.txt	1433	2433	1427	2424	1462	2459
Lex10645.txt	7220	17864	7675	17119	9076	18294
Lex20233.txt	9463	29523	10460	26867	13685	28741
Lex65536f.txt	14102	68740	17157	60479	20949	44980
Lex130499.txt	15331	99808	17237	68783	20928	47732

Table 2 shows that for edge-automata, compressed NFA have fewer labels (transitions) than compact NFA. But for large lexicons minimal DFA contains fewer labels.

Table 3. Comparison of node-automata

	Node Automata					
	Compact NFA (Node Automata)		Compressed NFA (Node Automata)		Minimal DFA (Node Automata)	
Lexicon	Nodes	Transitions	Nodes	Transitions	Nodes	Transitions
DNA_4_8_1.txt	33	127	33	127	33	88
HWR_4_8_1.txt	33	162	33	162	33	116
Lex1000.txt	2100	3077	2099	3076	2105	3081
Lex10645.txt	10693	21309	10569	21174	11974	21248
Lex20233.txt	14142	34229	13781	33825	17282	32579
Lex65536f.txt	20559	72733	18251	65605	25075	50439
Lex130499.txt	21932	102276	17476	81304	24968	53255

Table 3 shows that for node-automata, compressed NFA have significantly fewer labels (nodes) than the other types of automata. With respect to the complexity of the compression algorithm, it is clear that the process of finding all indistinguishable states may sound computationally costly as the algorithm recursively checks nodes down the automaton. However, we discovered that if you are scanning an automaton with few indistinguishable states, the recursion rarely goes deeper than 2 or 3 levels. In fact, a practical solution consists in first building a compact NFA, then further reducing the NFA with our compression algorithm.

8 Conclusions

Figure 1 (below) illustrates clearly the benefit of using node-automata instead of edge-automata; the number of labels of the node-automata is a fraction of the number of labels of the edge-automata. Although our minimization method does not claim to return the optimal non-deterministic automaton, the heuristics proposed in this paper leads to a significant improvement for real-world vocabularies. Personal Digital Assistants (PDA) and other smart handheld devices have too modest resources (a relatively small storage capacity and slow CPU) to allow features like advanced user interfaces (natural interactivity). Nevertheless efficient use of these limited resources will permit sophisticated speech or hand writing recognition. Some recognition systems, especially for mobile computers, need the functionality of incremental updating of vocabulary (add/remove words). Our NFA construction allows such adaptive update, avoiding the recomputation from scratch of the minimized NFA of the slightly modified lexicon (so far, we have only implemented the addition of new words).

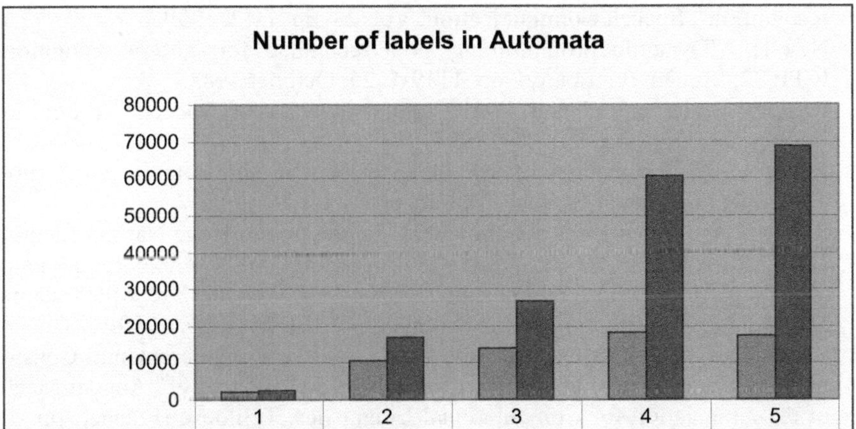

	1	2	3	4	5
Node Automata	2099	10569	13781	18251	17476
Edge Automata	2424	17119	26867	60479	68783

Fig. 1. Comparison of the numbers of labels for compressed real world lexicons

References

[1] Aho A. V., Hopcroft J. E. and Ullman J. D., "The Design and Analysis of Computer Algorithms", Addison-Wesley, Reading M.A., 1974.

[2] Appel A. W. and Jacobson G. J., "The world's fastest scrabble program", Communications of the ACM, Vol. 31, No. 5, pp. 572-578 & 585, May 1988.

[3] Bellman R., "Dynamic Programming", Princeton University Press, 1957.

[4] Forney Jr D. G., "The Viterbi Algorithm", Proceedings of the IEEE, Vol. 61, No 3, pp. 268-278, March 1973.

[5] Fredkin E., "Trie Memory", Communications of the ACM, Vol. 3, No 9, pp. 490-499, September 1960.
[6] Garcia-Salicetti S., "Une approche neuronale prédictive pour la reconnaissance en-ligne de l'écriture cursive", Thèse de Doctorat Paris 6, Spécialité: Informatique, 17 décembre 1996.
[7] Kosmala A., Willett D. and Rigoll G., "Advanced State Clustering for Very Large Vocabulary HMM-based On-Line Handwriting Recognition", ICDAR'99, Bangalore (India), pp. 442-445, 20-22 September 1999.
[8] Lacouture R. and De Mori R., "Lexical Tree Compression", Eurospeech'91, Genova (Italy), pp. 581-584, September 1991.
[9] Lifchitz, A. and Maire F., “A Fast Lexically Constrained Viterbi Algorithm For Online Handwriting Recognition”, In: L.R.B. Schomaker and L.G. Vuurpijl (Eds.), Proceedings of the Seventh International Workshop on Frontiers in Handwriting Recognition, Amsterdam, ISBN 90-76942-01-3, Nijmegen: International Unipen Foundation, pp 313-322, September 11-13 2000.
[10] Mohri M. and Riley M., "Network optimizations for large-vocabulary speech recognition", Speech Communication, Vol. 28, pp. 1-12, 1999.
[11] Ney H., "Dynamic programming as a technique for pattern recognition", ICPR'82, Munich (Germany), pp. 1119-1125, October 1982.
[12] Rabiner L. R. and Juang B.-H., "Fundamentals of Speech Recognition", Ed.Prentice Halls, pp. 321-389, 1993.
[13] Revuz D., "Minimization of acyclic deterministic automata in linear time"., Theoretical Computer Science, Vol 92, pp. 181-189, 1992.
[14] Ristov S. and Laporte E., "Ziv Lempel Compression of Huge Natural Language Data Tries Using Suffix Arrays", Proceedings of Combinatorial Pattern Matching, 10th Annual Symposium, Warwick University, UK, M.Crochemore and M.Paterson (editors), Berlin: Springer, pp. 196-211, July 1999.
[15] Sgarbas K. N., Fakotakis N. D. and Kokkinakis G. K., “Incremental Construction of Compact Acyclic NFAs”. Proceedings ACL-2001, 39th Annual Meeting of the Association for Computational Linguistics, Toulouse (France), pp. 474-481, July 6-11 2001.
[16] Vintsyuk T. K., "Recognition of words in spoken speech by dynamicprogramming methods", Kibernetika, Vol. 4, No 1, pp. 81-88, January 1968.
[17] Viterbi A. J., "Error bounds for convolutional codes and an asymptotically optimum decoding algorithm", IEEE Transactions on Information Theory, IT-13, pp. 260-269, April 1967.
[18] Wimmer Z., "Contribution à la lecture de documents papiers manuscrits: reconnaissance des champs numériques et cursifs par méthodes neuronales markoviennes", Thèse de Docteur-Ingénieur ENST de Paris, Spécialité: Informatique, 28 septembre 1998.
[19] Wimmer Z., Garcia-Salicetti S., Lifchitz A., Dorizzi B., Gallinari P., Artières T.,"REMUS", January 1999a http://www-poleia.lip6.fr/CONNEX/HWR/

Unsupervised Learning of Correlated Multivariate Gaussian Mixture Models Using MML

Yudi Agusta and David L. Dowe

Computer Science & Software Eng.
Monash University, Clayton, 3800 Australia
{yagusta,dld}@bruce.csse.monash.edu.au

Abstract. Mixture modelling or unsupervised classification is the problem of identifying and modelling components (or clusters, or classes) in a body of data. We consider here the application of the Minimum Message Length (MML) principle to a mixture modelling problem of multivariate Gaussian distributions. Earlier work in MML mixture modelling includes the multinomial, Gaussian, Poisson, von Mises circular, and Student t distributions and in these applications all variables in a component are assumed to be uncorrelated with each other. In this paper, we propose a more general type of MML mixture modelling which allows the variables within a component to be correlated. Two MML approximations are used. These are the Wallace and Freeman (1987) approximation and Dowe's MMLD approximation (2002). The former is used for calculating the relative abundances (mixing proportions) of each component and the latter is used for estimating the distribution parameters involved in the components of the mixture model. The proposed method is applied to the analysis of two real-world datasets - the well-known (Fisher) Iris and diabetes datasets. The modelling results are then compared with those obtained using two other modelling criteria, AIC and BIC (which is identical to Rissanen's 1978 MDL), in terms of their probability bit-costings, and show that the proposed MML method performs better than both these criteria. Furthermore, the MML method also infers more closely the three underlying Iris species than both AIC and BIC.

Keywords: Unsupervised Classification, Mixture Modelling, Machine Learning, Knowledge Discovery and Data Mining, Minimum Message Length, MML, Classification, Clustering, Intrinsic Classification, Numerical Taxonomy, Information Theory, Statistical Inference

1 Introduction

Mixture modelling [14, 17, 27] - generally known as unsupervised classification or clustering - models, as well as partitions, a dataset with an unknown number of components (or classes or clusters) into a finite number of components. The problem is also known as intrinsic classification, latent class analysis or numerical

T.D. Gedeon and L.C.C. Fung (Eds.): AI 2003, LNAI 2903, pp. 477–489, 2003.

taxonomy. Mixture modelling is widely acknowledged as a useful and powerful method to perform pattern recognition - as well as being useful in other areas, such as image and signal analysis.

In this paper, we discuss, in particular, an unsupervised classification that models a statistical distribution by a mixture (a weighted sum) of other distributions. The likelihood function - or objective function (see also Sec. 4, part 1d) - of the mixture modelling problem takes the form of:

$$f(x|M,\pi_1,\cdots,\pi_M,\widetilde{\theta}_1,\cdots,\widetilde{\theta}_M) = \sum_{m=1}^{M} \pi_m \times f_m(x|\widetilde{\theta}_m),$$

where there are M components, π_m is the relative abundance or mixing proportion of the mth component, and $f_m(x|\widetilde{\theta}_m)$ is the probability distribution of mth component (given the component distributional parameters).

There are two processes involved in performing mixture modelling. These are model selection for the model that best describes the dataset and point estimation for the parameters required. The former includes the selection of the most appropriate number of components. The problem we often face in choosing the best model is keeping the balance between model complexity and goodness of fit. In other words, the best model selected for a dataset must be sufficiently complex in order to cover all information in the dataset, but not so complex as to over-fit. Here, we apply the Minimum Message Length (MML) principle simultaneously for both parameter estimation and model selection.

The MML principle [27, 33, 30, 31] was first proposed by Wallace and Boulton [27] in 1968. It provides a fair comparison between models by stating each of them into a two-part message which, in turn, encodes each model (H) and the data in light of that model (D given H). Various related principles have also been stated independently by Solomonoff [24], Kolmogorov [15], Chaitin [4], and subsequently by Rissanen [20]. For a more comprehensive overview, see [30, 31].

Previous applications of MML to the problem of mixture modelling [27, 26, 28, 29, 32, 1, 2] includes the multinomial, Gaussian, Poisson, von Mises circular, and Student t distributions. In these applications, all variables in a component are assumed to be uncorrelated with one another.

For the correlated multivariate problem, various methods have also been proposed including AutoClass by Cheeseman *et. al.* [5], EMMIX (using AIC, BIC, and one other approach) by McLachlan *et. al.* [18], MCLUST (using BIC [22] - which is also the 1978 MDL [20]) by Fraley and Raftery [13] and MULTIMIX by Jorgensen *et. al.* [14]. (Relatedly, see also [7].) Figueiredo and Jain [8] also proposed a mixture modelling method for the same problem using an MML-like criterion. In their method, non-informative Jeffreys priors were utilised for the parameters estimated. A discussion of the appropriateness or otherwise of the Jeffreys prior can be found in [31] and the references therein.

Beginning with an elaboration of the MML principle and its approximations in Section 2, this paper proposes an MML mixture modelling method of correlated multivariate Gaussian distributions, where the variables within a component are assumed to be correlated with one another. This involves elaborations on point estimations for multinomial and multivariate Gaussian distributions (Section 3) and the coding scheme for MML mixture modelling of multivariate Gaussian distributions (Section 4). The proposed method is then applied to the analysis of two real-world datasets, the well-known (Fisher) Iris dataset and a diabetes dataset. The modelling results are compared with those obtained using two other commonly used criteria, BIC [22] (which is also the 1978 MDL [20]) and AIC (see Section 5), in terms of their probability bit-costings (see [25] and references therein). Comparisons in terms of the resulting number of components and the structure of the resulting components are also provided.

2 MML Principle and Its Approximations

The Minimum Message Length (MML) principle is an invariant Bayesian point estimation and model selection technique based on information theory. The basic idea of MML is to find a model that minimises the total length of a two-part message encoding the model, and the data in light of that model [27, 33, 30, 31].

Letting D be the data and H be a model with a prior probability distribution $P(H)$, using Bayes's theorem, the point estimation and model selection problems can be regarded simultaneously as a problem of maximising the posterior probability $P(H) \cdot P(D|H)$. From the information-theoretic point of view, where an event with probability p is encoded by a message of length $l = -\log_2 p$ bits, the problem is then equivalent to minimising

$$\text{MessLen} = -\log_2(P(H)) - \log_2(P(D|H)) \tag{1}$$

where the first term is the message length of the model and the second term is the message length of the data in light of the model.

In dealing with the mixture modelling problem of multivariate Gaussian distributions, it is required to perform parameter estimations of the multi-state and the multivariate Gaussian distributions. The parameter estimation of the multi-state distribution can be performed using the MML approximation proposed by Wallace and Freeman (1987) [33]. However, for the correlated multivariate Gaussian distribution, some mathematical challenges arise when using the 1987 MML approximation [33]. As an alternative more tractable approach, Dowe's recent MMLD approximation [16, 11, 10] is applied. These two approximations differ in the way they determine the optimal coding region of the possible models for a given dataset.

Given data x and parameters $\boldsymbol{\theta}$, let $h(\boldsymbol{\theta})$ be the prior probability distribution on $\boldsymbol{\theta}$, $f(x|\boldsymbol{\theta})$ the likelihood, $L = -\log f(x|\boldsymbol{\theta})$ the negative log-likelihood and

$$F(\boldsymbol{\theta}) = \det\Big\{E\Big(\frac{\partial^2 L}{\partial\boldsymbol{\theta}\partial\boldsymbol{\theta}'}\Big)\Big\}, \tag{2}$$

the Fisher information - i.e., the determinant of the matrix of expected second derivatives of the negative log-likelihood. Using (1), and expanding the negative log-likelihood, L, as far as the second term of the Taylor series about $\boldsymbol{\theta}$, the message length for the 1987 MML approximation is then given by [33, 32, 30]:

$$\text{MessLen} = -\log\left(\frac{h(\boldsymbol{\theta})}{\sqrt{\kappa_D^D F(\boldsymbol{\theta})}}\right) + L + \frac{D}{2} = -\log\left(\frac{h(\boldsymbol{\theta}) f(x|\boldsymbol{\theta})}{\sqrt{F(\boldsymbol{\theta})}}\right) + \frac{D}{2}(1 + \log \kappa_D) \tag{3}$$

where D is the dimension of the dataset and κ_D is a D-dimensional lattice constant [33] with $\kappa_1 = 1/12$ and $\kappa_D \leq 1/12$. The MML estimate of $\boldsymbol{\theta}$ can be obtained by minimising (3).

In Dowe's MMLD approximation, on the other hand, the optimal coding region, R, is determined by specifying the total two-part message length as follows [16, 11, 10]:

$$\text{MessLen} = -\log \int_R h(\theta')d\theta' - \frac{1}{\int_R h(\theta')d\theta'} \int_R h(\theta') \log f(x|\theta')d\theta' \tag{4}$$

Minimising (4) with respect to θ results in the following expression:

$$-\log f(x|\theta')\Big|_{\partial R} = -\frac{1}{\int_R h(\theta')d\theta'} \int_R h(\theta') \log f(x|\theta')d\theta' + 1 \tag{5}$$

where the first term on the right hand side of the equation above represents the second part of the message length. The expression above is known as the MMLD boundary rule - which means that the negative log-likelihood at the boundary, ∂R, of the optimal coding region is equal to the expected negative log-likelihood (with respect to the prior) throughout the region, R, plus one. However, using this approximation, it can be difficult to find the exact optimal coding region, R. Therefore, it is necessary to apply the approximation numerically.

The MMLD message length has been numerically approximated in [11], where the use of the importance sampling distribution was proposed. The importance sampling distribution is useful when we know the most likely area from where the possible models will be derived. With the posterior probability as the importance sampling distribution and applying Monte Carlo integration, the MMLD message length expression (4) can be numerically approximated as follows [11]:

$$\text{MessLen} = -\log\left(\frac{\sum_{\theta\in Q} f(x|\theta)^{-1}}{\sum_{\theta\in S} f(x|\theta)^{-1}}\right) - \left(\frac{\sum_{\theta\in Q} f(x|\theta)^{-1}\log f(x|\theta)}{\sum_{\theta\in Q} f(x|\theta)^{-1}}\right) \tag{6}$$

where S is the parameter space for the sampling distribution and Q is a subset of the optimal coding region, R.

Considering that the Gaussian is a continuous distribution, a finite coding for the message can be obtained by acknowledging that all recorded continuous data and measurements must only be stated to a finite precision, ϵ. In this way, a constant of $N\log(1/\epsilon)$ is added to the message length expression above, where N is the number of data [32, p74] [2, Sec. 2] [28, p38].

3 Parameter Estimation by Minimum Message Length

As mentioned earlier, for a mixture modelling problem involving multivariate Gaussian distributions, we need to perform parameter estimations of the multi-state distribution for the relative abundances of each component and the multivariate Gaussian distribution for the distribution parameters of each component. Unlike most previous MML mixture modelling works, it is assumed here that the data in each component are required to be normally distributed but that also the variables within each component can be correlated with one another.

3.1 Multi-state Distribution

For a multi-state distribution with M states (and sample size, N), the likelihood of the distribution is given by:

$$f(n_1, n_2, \cdots, n_M | p_1, p_2, \cdots, p_M) = p_1^{n_1} p_2^{n_2} \cdots p_M^{n_M},$$

where $p_1 + p_2 + \cdots + p_M = 1$, for all m: $p_m \geq 0$ and $n_1 + n_2 + \cdots + n_M = N$.

The distribution parameters are estimated using the 1987 MML approximation [33]. It follows from (2) that $F(p_1, p_2, \cdots, p_M) = N^{(M-1)}/p_1 p_2 \cdots p_M$. The derivation is also shown elsewhere for $M = 2$ [32, p75].

Assuming a uniform prior of $h(p) = (M-1)!$ over the $(M-1)$-dimensional region of hyper-volume $1/(M-1)!$, and minimising (3), the MML estimate $\hat{p}_m$ is obtained by [25, sec. 4.2]:

$$\hat{p}_m = (n_m + 1/2)/(N + M/2) \tag{7}$$

Substituting (7) into (3) provides the following two-part message length [27, p187 (4)] [27, p194 (28)] [32, p75 (6)] [1, p291 (5)]:

$$-\log(M-1)! + ((M-1)/2)(\log(N\kappa_{M-1}) + 1) - \sum_{m=1}^{M} (n_m + 1/2) \log \hat{p}_m \tag{8}$$

3.2 Multivariate Gaussian Distribution

The multivariate Gaussian distribution has a likelihood function:

$$f(x|\mu, \Sigma) = \frac{1}{(2\pi)^{\frac{n}{2}} |\Sigma|^{\frac{1}{2}}} e^{-\frac{1}{2}(x-\mu)^{\mathrm{T}} \Sigma^{-1} (x-\mu)}$$

where μ is the vector of means, Σ is the covariance matrix of the distribution (allowing correlations), and n is the number of variables in the dataset.

As explained in Section 2, the parameter estimation for this distribution is to be performed using the MMLD approximation. The parameter estimation in this approximation is conducted numerically using the following algorithm:

1. Sample a number of models from the importance sampling distribution.
2. Sort the models according to their likelihood values in decreasing order.

3. Apply the MMLD boundary rule (5) and select models that lie in the region.
4. Find the estimates using the Minimum Expected Kullback-Leibler (minEKL) distance method (weighted by the posterior [6], instead [11, 10] of the prior).

In the first step, the posterior probability is chosen as the importance sampling distribution. For this purpose, two prior probabilities on both parameters, μ and Σ, are required. Here, an improper uniform prior on μ over the $[-\infty, \infty]$ n-dimensional real space ($\Re^n$) and an improper conjugate prior, $|\Sigma|^{(-\frac{n+1}{2})}$, on Σ are considered [21, Sec. 5.2.3]. Both priors are the limiting form of the conjugate normal-inverted Wishart prior [21, Chapter 5]. We notice here that it is impossible to normalise both priors. However, as shown in (6), our numerical message length calculation only involves the priors via the posterior - which is proper. Therefore, it at least appears that we do not need to explicitly normalise the (possibly improper) priors. With these priors, the posterior probability becomes:

$$\mu|\Sigma, X \sim N(\bar{x}, N^{-1}\Sigma) \tag{9}$$

$$\Sigma|X \sim W^{-1}(N-1, (NS)^{-1}) \tag{10}$$

where N is the number of data and S is the data covariance matrix. Utilising the Gibbs sampling method, the possible models are sampled from (9) and (10).

Once the models are sampled, they are sorted according to their likelihood values in decreasing order, starting with the model at the Maximum Likelihood solution. We then apply the MMLD boundary rule (5) and select the models that lie inside the optimal coding region. Referring to equation (6), the following algorithm is utilised in simultaneously selecting models lying in the coding region and calculating the first and second parts of the resulting message length. This algorithm is a variation of that proposed in [11].

```
BEGIN ALGORITHM
 //Setting the first model (ML model) into the selected models
 Allocate first model into selected models;
 //Setting each expression involves in the message length
 Set FIRSTPARTNUMERATOR = 1.0;
 Set SECONDPARTNUMERATOR = minusLogLikelihood of first model;
 Set SECONDPARTDENOMINATOR = 1.0;
 //Calculating the second part of the message length
 Set SECONDPART = SECONDPARTNUMERATOR/SECONDPARTDENOMINATOR;
 Move to next model;
 While(not reaching the end of the sorted models) {
  //Applying the MMLD boundary rule for the rest of the models
  If(minusLogLikelihood of current model<=SECONDPART+1.0) {
   //Setting the model into the selected models, if it is inside
   Allocate current model into selected models;
   Set likelihood = exp(minusLogLikelihood of current model -
    minusLogLikelihood of first model);
   //Updating each expression involves in the message length
```

```
    FIRSTPARTNUMERATOR += likelihood;
    SECONDPARTNUMERATOR += minusLogLikelihood of current model *
     likelihood;
    SECONDPARTDENOMINATOR += likelihood;
    //Calculating the second part of the message length
    SECONDPART = SECONDPARTNUMERATOR/SECONDPARTDENOMINATOR;
   }
   Else -> Exit loop;
   Move to next model;
  }

  //Calculating the denominator of the first part until
  //the last model of the sorted models
  Set FIRSTPARTDENOMINATOR = FIRSTPARTNUMERATOR;
  While(not reaching the end of the sorted models) {
   Set likelihood = exp(minusLogLikelihood of current model-
    minusLogLikelihood of first model);
   FIRSTPARTDENOMINATOR += likelihood;
   Move to next model;
  }
  //Calculating the first part of the message length
  FIRSTPART = -log(FIRSTPARTNUMERATOR/FIRSTPARTDENOMINATOR);
END ALGORITHM
```

From the selected models, estimates are derived using the Minimum Expected Kullback-Leibler (minEKL) distance point estimation method, which is numerically calculated by taking the maximum likelihood of the (posterior-weighted) future samples, randomly sampled from the selected models [6]. This point estimation method is statistically invariant under 1-1 re-parameterisation.

4 MML Mixture Modelling

In order to apply MML to a mixture modelling problem, a two-part message conveying the mixture model needs to be constructed (in principle). Recall that from Section 1, the encoding of the mixture model hypothesis comprises several concatenated message fragments [27, 26, 28, 29, 32], stating in turn:

1a The number of components: Assuming that all numbers are considered as equally likely up to some constant, (say, 100), this part can be encoded using a uniform distribution over the range.

1b The relative abundances (or mixing proportions) of each component: Considering the relative abundances of an M-component mixture, this is the same as the condition for an M-state multinomial distribution. The parameter estimation and the message length calculation of the multi-state distribution have been elaborated upon in subsection 3.1.

1c For each component, the distribution parameters of the component attribute. In this case, each component is inferred as a multivariate correlated Gaussian distribution as in subsection 3.2.

1d For each thing, the component to which the thing is estimated to belong. (Part 1d is typically included and discussed in MML mixture modelling but is often omitted in other mixture modelling literature.)

For part (1d), instead of the total assignment as originally proposed in [27], partial assignment is used to approximate the improved cost. Further discussion of this can be found in [26, Sec. 3] [28, Sec. 3.2] [29, Sec. 3.2] [32, pp. 77-78][2, Sec. 5]. Once the first part of the message is stated, the second part of the message will encode the data in light of the model stated in the first part of the message.

5 Alternative Model Selection Criteria - AIC and BIC

In order to justify the proposed MML method, two criteria are considered for comparison. These are the *Akaike Information Criterion* (AIC) and Schwarz's *Bayesian Information Criterion* (BIC).

AIC, first developed by Akaike [3], is given by:

$$\text{AIC} = -2L + 2N_p$$

where L is the logarithm of the likelihood at the maximum likelihood solution for the investigated mixture model and N_p is the number of parameters to be estimated in the model. For the multivariate Gaussian mixture, N_p is set equal to $k-1+k[n+n(n+1)/2]$ as explained by Sclove [23] ($k-1$ mixing proportions, and $n+n(n+1)/2$ parameters for the n means and the matrix per component), where k is the number of components and n is the number of variables in the dataset. The model which results in the smallest AIC is the model selected.

The second criterion, BIC, first introduced by Schwarz [22] is given by:

$$\text{BIC} = -2L + N_p \log N$$

where L is the logarithm of the likelihood at the maximum likelihood solution for the investigated mixture model, N_p is the number of independent parameters to be estimated, and N is the number of data. For the multivariate Gaussian mixture, N_p is again equal to $k-1+k[n+n(n+1)/2]$ ($k-1$ mixing proportions, and $n+n(n+1)/2$ parameters per component), where k is the number of components and n is the number of variables in the dataset. (The number of components is not considered an independent parameter for the purposes of calculating the BIC as explained by Fraley and Raftery [12].) (The BIC model selection criterion is formally, not conceptually, the same as the 1978 Minimum Description Length (MDL) criterion proposed by Rissanen [20].) The model which results in the smallest BIC is selected as the best model.

6 Experiments

6.1 Iris Dataset

The Iris dataset was first analysed in 1936 by Fisher [9]. It comprises 150 iris plants belonging to three species, namely Iris Setosa (S), Iris Versicolour (Ve), and Iris Virginica (Vi). Four variables measuring sepal and petal length and width of the species are involved. Each group is represented by 50 plants. The measurement accuracies, ϵ, in this dataset (see Sec. 2) were set to 1.0 for all variables.

The analysis here is performed by dividing the original dataset into training and test datasets with proportions of 135:15. We first find the model for the training dataset and then fit the test dataset to the selected model. The latter is performed by measuring the probability bit-costing, $-\log(P(x))$, of each datum x in the test dataset (see [25] and the references therein). This process was repeated 20 times. The resulting averages ($\pm$ the standard deviations) of the probability bit-costings for the three criteria, MML, AIC and BIC, are 21.37 ($\pm$ 5.8), 23.14 ($\pm$ 10.1), and 21.83 ($\pm$ 6.2) nits (1 nit $= \log_2 e$ bits), respectively. These results suggest that MML performs better than both AIC and BIC.

In a further analysis using the proposed MML method, MML grouped the entire dataset into three components. Fig. 1 shows the original (Fisher) Iris dataset and the resulting three-component MML mixture of the dataset, plotted with the first two principal components as axes. Here, the relative abundances of the resulting MML components were 0.333:0.339:0.328, which were almost the same as those of the true model (0.333:0.333:0.333), and the MML fit appeared pleasing. The entire dataset was also analysed using AIC and BIC. The modelling using AIC resulted in a four-component mixture, whereas the modelling using BIC resulted in a two-component mixture in which the highly overlapping Versicolour and Virginica iris groups were modelled into one component. The second best model for BIC was a three-component mixture. However, the relative abundances in this model were 0.333:0.436:0.231, which were substantially different from those of the true model.

6.2 Diabetes Dataset

This diabetes dataset was first reported in 1979 by Reaven and Miller [19] and comprises 145 samples with three variables measuring glucose area, insulin area and the steady state plasma glucose response (SSPG). The modelling reported in [19] was performed based on the groupings established using conventional clinical criteria. In this conventional classification, diabetes was grouped into three categories: Normal, Chemical and Overt. A subsequent analysis which also resulted in a three-component mixture has been reported by Fraley and Raftery [12]. In the latter analysis [12], BIC was used to select the number of components. In the present application, we aimed to compare these earlier results [19, 12] with the analysis obtained using the proposed MML method.

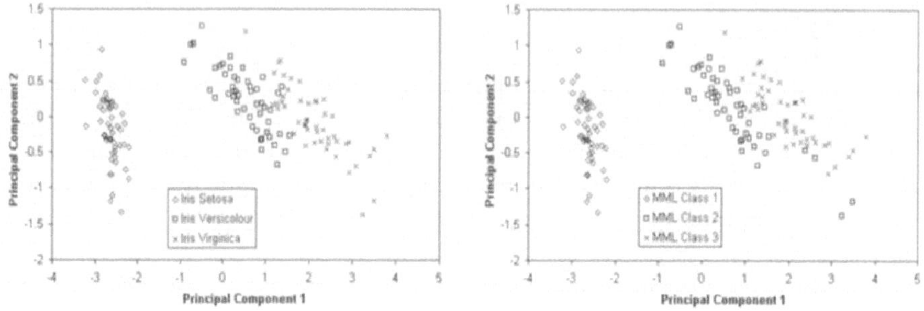

(a) Original Iris Data (b) MML Model

Fig. 1. The original (Fisher) Iris dataset and the resulting three-component MML mixture, plotted with the first two principal components as axes

The measurement accuracies, ϵ, in this modelling were set equal to 1.0 for all three variables.

We applied the same analysis as in Sec. 6.1, where the proportions of the training and test datasets are set equal to 130:15. The experiment was repeated 20 times. The averages ($\pm$ the standard deviations) of the probability bit-costings on this diabetes dataset for MML, AIC and BIC were 235.94 ($\pm$ 8.9), 237.49 ($\pm$ 12.4), and 236.54 ($\pm$ 10.4) nits (1 nit $= \log_2 e$ bits), respectively. Again, MML performed better than AIC and BIC.

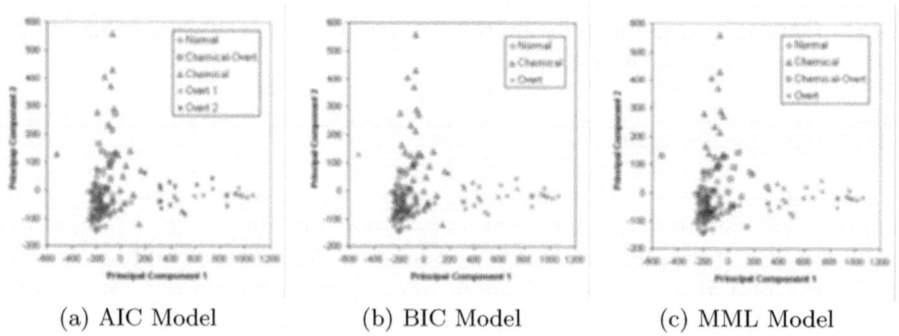

(a) AIC Model (b) BIC Model (c) MML Model

Fig. 2. Modelling using AIC, BIC and MML (plotted on 2 principal component axes)

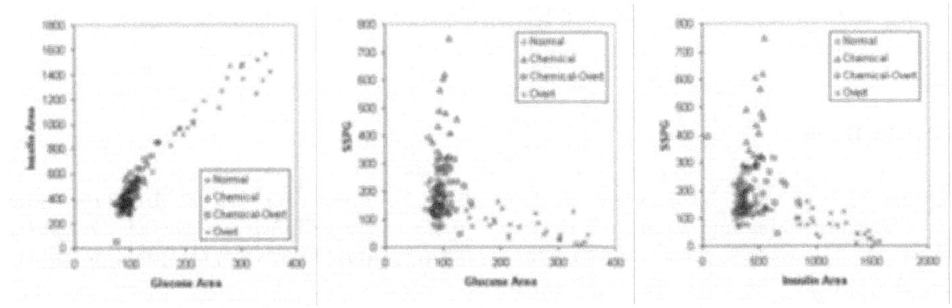

Fig. 3. Modelling using the proposed MML method (the results are pair-plotted)

We further analysed the original diabetes dataset using AIC, BIC and MML, with the results (plotted with the first two principal components as axes) being shown in Fig. 2. The modelling using AIC (see Fig. 2(a)) resulted in five components with two components dividing the Overt group, and one component overlapping with the Chemical and Overt groups. The modelling using BIC resulted in three components, which are the same as those reported in [12] and shown in Fig. 2(b).

In the modelling using the proposed MML method, a four-component mixture resulted: this is plotted against the first two principal components in Fig. 2(c) and its ${}^3C_2 = 3$ cross-sectional pair plots are shown in Fig. 3. The additional component to the original classification appears to highly overlap with the Chemical and Overt groups, and consists of members that originally belonged to both groups. Although the results are different from the original classification, this does not imply that the proposed method has modelled the dataset incorrectly. As mentioned earlier, the original groupings used to justify the analysis in [19] (and possibly also in [12]) were performed based on conventional clinical criteria. Thus, no true model exists which can be used to justify which classification is correct. Conversely, the results obtained here and the performance of the proposed MML method compared to both AIC and BIC in terms of the probability bit-costings might suggest an alternative diabetes classification by the addition of a Chemical-Overt group.

7 Conclusion

In conclusion, we draw the attention of the reader to the following results:

1. The proposed method broadens the scope of problems handled by MML mixture modelling, by now modelling correlated multivariate data. This provides flexibility since most real-world datasets contain variables that are correlated within each component in their mixture models.

2. The proposed MML method performs better than two other modelling criteria, AIC and BIC (or 1978 MDL), as shown in the analysis of the probability bit-costings for both the (Fisher) Iris and diabetes datasets.

References

[1] Y. Agusta and D. L. Dowe. Clustering of Gaussian and t Distributions using Minimum Message Length. In *Proc. Int'l. Conf. Knowledge Based Computer Systems - KBCS-2002*, pp. 289-299, Mumbai, India, 2002. Vikas Publishing House Pvt. Ltd. 478, 481

[2] Y. Agusta and D. L. Dowe. MML Clustering of Continuous-Valued Data Using Gaussian and t Distributions. In *Lecture Notes in Artificial Intelligence*, vol. 2557, pp. 143-154, 2002. Springer-Verlag, Berlin. 478, 480, 484

[3] H. Akaike. A new look at the statistical model identification. *IEEE Transactions on Automatic Control*, AC-19(6):716–723, 1974. 484

[4] G. J. Chaitin. On the length of programs for computing finite sequences. *J. the Association for Computing Machinery*, 13:547–569, 1966. 478

[5] P. Cheeseman and J. Stutz. Bayesian Classification (AutoClass): Theory and Results. In *Advances in Knowledge Discovery and Data Mining*, pp. 153-180, 1996. AAAI Press/MIT Press. 478

[6] D. L. Dowe, R. A. Baxter, J. J. Oliver, and C. S. Wallace. Point Estimation using the Kullback-Leibler Loss Function and MML. In *Lecture Notes in Artificial Intelligence*, vol. 1394, pp. 87-95, 1998. Springer-Verlag, Berlin. 482, 483

[7] R. T. Edwards and D. L. Dowe. Single factor analysis in MML mixture modelling. In *Lecture Notes in Artificial Intelligence*, vol. 1394, pp. 96-109, 1998. Springer-Verlag, Berlin. 478

[8] M. A. T. Figueiredo and A. K. Jain. Unsupervised Learning of Finite Mixture Models. *IEEE Trans. on Pattern Analysis and Machine Intelligence*, 24(3):381–396, 2002. 478

[9] R. A. Fisher. The use of multiple measurements in taxonomic problems. *Annals of Eugenics*, 7:179–188, 1936. 485

[10] L. J. Fitzgibbon, D. L. Dowe, and L. Allison. Change-Point Estimation Using New Minimum Message Length Approximations. In *Lecture Notes in Artificial Intelligence*, vol. 2417, pp. 244-254, 2002. Springer-Verlag, Berlin. 479, 480, 482

[11] L. J. Fitzgibbon, D. L. Dowe, and L. Allison. Univariate Polynomial Inference by Monte Carlo Message Length Approximation. In *Proc. 19th International Conf. of Machine Learning (ICML-2002)*, pp. 147-154, Sydney, 2002. Morgan Kaufmann. 479, 480, 482

[12] C. Fraley and A. E. Raftery. How Many Clusters? Which Clustering Method? Answers Via Model-Based Cluster Analysis. *Computer J.*, 41(8):578–588, 1998. 484, 485, 487

[13] C. Fraley and A. E. Raftery. MCLUST: Software for Model-Based Cluster and Discriminant Analysis. Technical Report 342, Statistics Dept., Washington Uni., Seattle, USA, 1998. 478

[14] L. A. Hunt and M. A. Jorgensen. Mixture model clustering using the Multimix program. *Australian and New Zealand Journal of Statistics*, 41(2):153–171, 1999. 477, 478

[15] A. N. Kolmogorov. Three approaches to the quantitative definition of information. *Problems of Information Transmission*, 1:4–7, 1965. 478

[16] E. Lam. Improved approximations in MML. Honours Thesis, School of Computer Science and Software Engineering, Monash Uni., Clayton 3800 Australia, 2000. 479, 480
[17] G. J. McLachlan and D. Peel. *Finite Mixture Models*. John Wiley, NY, 2000. 477
[18] G. J. McLachlan, D. Peel, K. E. Basford, and P. Adams. The EMMIX software for the fitting of mixtures of Normal and t-components. *J. Stat. Software*, 4, 1999. 478
[19] G. M. Reaven and R. G. Miller. An Attempt to Define the Nature of Chemical Diabetes Using a Multidimensional Analysis. *Diabetologia*, 16:17–24, 1979. 485, 487
[20] J. Rissanen. Modeling by shortest data description. *Automatica*, 14:465–471, 1978. 478, 479, 484
[21] J. L. Schafer. *Analysis of Incomplete Multivariate Data*. Chapman & Hall, London, 1997. 482
[22] G. Schwarz. Estimating the dimension of a model. *Ann. Stat.*, 6:461–464, 1978. 478, 479, 484
[23] S. L. Sclove. Application of model-selection criteria to some problems in multivariate analysis. *Psychometrika*, 52(3):333–343, 1987. 484
[24] R. J. Solomonoff. A formal theory of inductive inference. *Information and Control*, 7:1–22, 224–254, 1964. 478
[25] P. J. Tan and D. L. Dowe. MML Inference of Decision Graphs with Multi-way Joins. In *Lecture Notes in Artificial Intelligence*, vol. 2557, pp. 131-142, 2002. Springer-Verlag, Berlin. 479, 481, 485
[26] C. S. Wallace. An improved program for classification. In *Proc. 9th Aust. Computer Science Conference (ACSC-9)*, vol. 8, pp. 357-366, Monash Uni., Australia, 1986. 478, 483, 484
[27] C. S. Wallace and D. M. Boulton. An information measure for classification. *Computer J.*, 11(2):185–194, 1968. 477, 478, 479, 481, 483, 484
[28] C. S. Wallace and D. L. Dowe. Intrinsic classification by MML - the Snob program. In *Proc. 7th Aust. Joint Conf. on AI*, pp. 37-44, 1994. World Scientific, Singapore. 478, 480, 483, 484
[29] C. S. Wallace and D. L. Dowe. MML Mixture Modelling of Multi-State, Poisson, von Mises Circular and Gaussian Distributions. In *Proc. 6th International Workshop on Artificial Intelligence and Statistics*, pp. 529-536, Florida, 1997. 478, 483, 484
[30] C. S. Wallace and D. L. Dowe. Minimum Message Length and Kolmogorov Complexity. *Comp. J.*, 42(4):270–283, 1999. Special issue on Kolmogorov Complexity. 478, 479, 480
[31] C. S. Wallace and D. L. Dowe. Refinements of MDL and MML Coding. *Computer J.*, 42(4):330–337, 1999. Special issue on Kolmogorov Complexity. 478, 479
[32] C. S. Wallace and D. L. Dowe. MML clustering of multi-state, Poisson, von Mises circular and Gaussian distributions. *Statistics and Computing*, 10:73–83, Jan. 2000. 478, 480, 481, 483, 484
[33] C. S. Wallace and P. R. Freeman. Estimation and Inference by Compact Coding. *J. Royal Statistical Society (B)*, 49(3):240–265, 1987. 478, 479, 480, 481

Cooperative Learning in Self-Organizing E-Learner Communities Based on a Multi-Agents Mechanism

Fan Yang[1], Peng Han[1], Ruimin Shen[1], Bernd J. Kraemer[2], and Xinwei Fan[1]

[1] Department of Computer Science and Engineering
Shanghai JiaoTong University
Shanghai, 200030, China
{fyang,phan,rmshen,xwfan}@mail.sjtu.edu.cn
[2] Faculty of Electrical and Information Engineering
FernUniversitaet Hagen, Germany
bernd.kraemer@fernuni-hagen.de

Abstract. Web based learning provides an unprecedented flexibility and convenience to both learners and instructors. However, it also creates a great number of lonely learners. Hence, there is an urgent need for finding an efficient way to help the learners share their learning experiences and exchange their learning materials during the learning process. This paper reports on an attempt to construct a flexible and effective self-organization system that groups similar learners according to their preferences and learning behaviors. We use a multi-agent mechanism to manage and organize learners and learner groups. Furthermore, we present effective award and exchange algorithms, so eventually learners with similar preferences or interests can be clustered into the same community. Experiments based on real learner data have shown that this mechanism can organize learners properly, and has sustainably improved speed and efficiency of searches for peer students owning relevant knowledge resources.

Keywords: Intelligent agents

1 Introduction

Recently, web based learning technology enables many more students to have access to e-learning environments, which provides students and teachers with an unprecedented flexibility and convenience. On the other hand, this development from classroom teaching to blended or e-learning creates too many lonely learners. While a lot of research has been pursued to provide collaborative learning environments for geographically dispersed learner groups [1], web-based lectures allow instructors and learners to share information and ideas with the entire class, supplemented by multimedia resources, electronic mailing lists and digital video links. But this teacher-centered learning mode bears inherent limitations such as learner passiveness or lack

T.D. Gedeon and L.C.C. Fung (Eds.): AI 2003, LNAI 2903, pp. 490-500, 2003.

of interaction. As a result, there is an urgent need for finding an efficient way to help learners share their learning experiences and insights and exchange learning materials during the learning process. An intuitive way to accomplish this objective is to group learners with similar preferences into the same community and help them learn collaboratively.

A web based learning environment usually involves a great number of active members, including information providers, information users, consultants, tutors or administrative staff. Because the learners involved are generally widely spread and often don't know each other personally, a great challenge is how to find suitable and timely information resources in a distributed environment. To cope with this challenge, middle agents are often used in distributed information systems to facilitate services among users [2], such as the InfoSleuth [3] or the Open Agent Middleware (OAM) of IDIoMS [4]. However, in these distributed systems, the relationship between user agents and middle agents is always pre-defined by a human designer [5]. They have, however, difficulties in handling the dynamism inherent in such open environments. To achieve a good performance and a high scalability, the organizational structure of a socio-technical information system should be both self-organizing and adaptive [6].

In this paper, we present a self-organization model that relies on earlier work by Wang Fang [5]. Around this model we have implemented our own self-organization method to cluster learners automatically and quickly. Section 2 briefly introduces our work, including the underlying conceptual framework and the architecture of our prototype system. The detailed design patterns and the group formation algorithm are presented in Section 3. Section 4 describes some experiments we conducted to demonstrate the formation of user communities and evaluate the efficiency of this mechanism. Section 5 concludes this paper and provides an outlook on future research work.

2 Conceptual Framework

This paper describes the construction of an automatic and effective organization of learners and group agents in distributed e-learning systems. We refer to the concept "E-Learner Community" as a group of learners who share common preferences and mutually satisfy each other's requirements in terms of relevant knowledge resources.

We generated a Learner Agent (LA) acting on behalf of a real learner. LA is in charge of restorations and updates of learning resources. It is well-known that the behaviors of learners are very complex. An LA also handles the requests of a learner and seeks corresponding learning sources from the system.

During the learning process, learners will browse online courses, submit questions or assignments and perform exercises. All of these actions represent the learning interest and intent of the learners. Generally, we view all of them as different resource requests. For instance, browsing courses can be looked at as many http request flows of learning content. The submission of questions represents a request for specific subjects, and so on.

To make our conceptual model more precise, we provide a few formal definitions on which we will also rely in the definition of our group formation algorithm presented in Section 3.2.

Let **G** and **L** be disjoint sets of group and learner names with typical elements g and l, respectively, and let **P** and **R** be sets of preferences and resources. For $P \subseteq \mathbf{P}$ and $R \subseteq \mathbf{R}$ the mapping s: $R \rightarrow P$ models the fact that preference $p \in P$ is *supported by* resource $r \in R$ if $s(\mathrm{r}) = p$. By $R_p = \{r \in R \mid s(\mathrm{r}) = p\}$ we denote the set of *all resources in R supporting preference p.* Now we can define a learner agent acting on behalf of a learner l by combining the learner name with the learner's preferences and the resources supporting these preferences.

Definition 1. A learner agent is a triple $A_l = (l, P_l, R_l)$ with $P_l \subseteq \mathbf{P}$ and

$$R_l = \bigcup_{p \in P_l} R_p \tag{1}$$

As the community of learners typically becomes pretty large, it would be a performance bottleneck if the LAs would send requests directly to other LAs. To avoid traffic overload and increase the efficiency of searches, we propose another kind of agent, called Group Agent (GA), to serve as the broker for requests from a smaller community of LAs. A GA is responsible for locating providers of resources and managing the association of learners to communities and it can interact with both the local LAs in its management domains and the other GAs.

The GAs are distributed and only manage local communities of learners, they can enhance the whole system's robustness as there is still some probability of component failure dynamically adjusting learners of the provider sends neighboring GAs on real learners. Furthermore, some GAs may take charge of two or more categories of documents and some of them may lose all of their members during the community formation.

A *multi-agent structure* can be modeled by associating LAs and GAs through the mapping m: $L \rightarrow G$, where $m(l)=g$ denotes the fact that learner l is a *member of the group* managed by g. All *LAs managed by* g are then defined by the set:

$$A_g = \{l \in L \mid m(l) = g\} \tag{2}$$

and the set of resources maintained by the community of LAs managed by g is defined by:

$$R_g = \bigcup_{l \in A_g} R_l \tag{3}$$

Therefore, we propose a two-layer multi-agent structure. Figure 1 illustrates a schematic view of it.

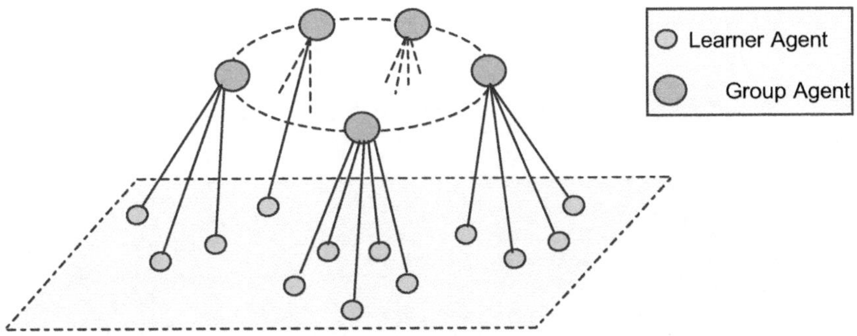

Fig. 1. A schematic view of the two-layer multi-agent structure

The LAs submit resource requests to their GA, and each GA will take care of finding suitable learning resources providers and return the requested resources to the real learner. The interaction between LAs and GAs is transparent to the human learner. Our learner behavior model is based on clustering of learners according to dynamic resource requests and does not require laborious human interference.

3 Core Mechanism and Group Formation Algorithm

We have divided the process of group formation into two main parts:

- Initialization and automatic registration
- Dynamic adjustment and self-organization.

We shall discuss the details of each part in the following subsections.

3.1 Initialization and Automatic Registration

In our open E-Learning platform, preferences and behaviors of learners can be collected from web servers, automatic question–and-answering system, assignment system, examination system, newsgroups or other resources. In order to analyze the behavior of learners, we have to generate reports that cover a large period of time and gather all relevant learning information. Therefore, we have defined the Learner Markup Language, which offers the flexibility to combine all learning information to produce the profiles of learners [7]. As discussed above, we only focus on the preferences and resources attributes. The XML document in Fig.2 shows a typical user profile.

The **users** element is the root element of a User Profile valid document. A **users** element can contain zero or more **user** elements. All **user** elements have global attributes ***ID*** and ***interest***. The ***ID*** attribute is a unique number to identify the **user** element and the ***interest*** are strings denoting the preference of the user. The **resources** element is the container of a **resource** element, which describes the title of a resource owned by a learner object.

```xml
<?xml version="1.0" encoding="gb2312"?>
<users>
    <user ID="149" interest="Undergraduate Computer Science">
        <resources>
            <resource>Software Develope Methods</resource>
            <resource>Program Design</resource>
            <resource>C Language Program Shortcut</resource>
            <resource>Computer  and Information Processing</resource>
            <resource>Data Structure Tutorials</resource>
            <resource>Network Tutorials</resource>
            <resource>C Program Design Instances</resource>
        </resources>
    </user>
    <user ID="148" interest="Computer">
        <resources>
            <resource>Foundation of Computer Application</resource>
            <resource>Program Design</resource>
        </resources>
    </user>
    <user ID="147" interest="Business Administration">
        <resources>
            <resource>Foundation of Probability Theory and Administrative Analysis</resource>
            <resource>Management Behabvior Cases</resource>
        </resources>
    </user>
    <user ID="146" interest="E-Commerce">
        <resources>
            <resource>Customer Relationship Managment</resource>
            <resource>E-Commerce Application Develope Techniques</resource>
            <resource>E-Commerce Application Tutorials</resource>
            <resource>E-enterprise Management</resource>
            <resource>E-Commerce Security</resource>
        </resources>
    </user>
    ......
</users>
```


Fig. 2. Example of a learner profile

In the initialization process, the system employs three procedures:

1. First, an LA is generated for every real learner *l* automatically. Its task is to maintain the profile of *l*. It provides a succinct and valid way to save relevant leaner information.
2. Second, one or more GAs are generated together with an initial multi-agent structure *M* specifying which LAs are managed by which GA. In this step, every GA generates two tables to maintain the information of local learners and other GAs respectively. These tables implement the mappings *s* and *m* introduced in Section 2.
3. We also provide an operation that allows a Learner Agent *l* to de-register from its GA *g* and register with another GA *g'* when required. The effect of this operation is that the pair *(l, g)* is removed from *M* and the pair *(l, g')* is added to *M*. In order to meet the requirement of frequently changeable relationships between LAs and

GAs, the system maintains a table of all group members and their relationships. Every GA is associated with a mutable table to maintain the information shared by all learners in a local group (we will discuss this in section 3.2). The initialization and interaction of a Learner Agent with a Group Agent is transparent to the real learner.

3.2 Self-Organizing Learner Communities

In our former research, we have constructed a collaborative learning platform based on a multi-agent model [8]. Some investigations have been pursued on the infrastructure and interaction language between multi-agents. In this paper, we focus on automatically grouping learners sharing similar preferences and dynamically adjust learners according to their changeable behaviors.
The main algorithm implementing this search strategy is shown in Figure 3. Besides the preferences and resources of the actual learner community, the algorithm takes variables **Gnum, MaxRequestTime,** and **topSearch** as inputs. They are needed to constrain the number of searches across large communities. Each LA and GA maintains a variable **award**, which is used to maintain information about matching preferences. The local variable "Provider" refers to an LA, while variable "Requester" refers to both LAs and GAs.

INPUT:

1. 1.A **learner.xml** file containing a learner profile for each learner in a community (including the learners' preferences and information about their resources)
2. 2.The number **Gnum** of GAs to be generated
3. 3.The maximum **MaxRequestTime** of request times per learner
4. 4.The maximum **topSearch** of searching times

OUTPUT: A self-organized learner community

PROCEDURE

```
Self-organizing (learner.xml, Gnum, MaxRequestTime, topSearch)
{ Find-Provider();                    //find the resource provider
  if (isSucceed=true) then
  { LocateResource ();                //send the resource to searcher
    Award(requester);
    Award(provider);
  }
  if provider.GA<>requester.GA then
  Exchange(provider, requester);
}
```

Fig. 3. Self-organizing Algorithm

The search schema helps GAs to find suitable learning resources. Here, the requests are titles of learning resources, answers are names of identified providers if there were any matching resources titles. The main search process **Find-Provider()** can be divided into several steps as follows:

- **Judge the type of requester**
  When receiving an information query message from an agent, the GA will judge the type of the agent. Here, a GA can only communicate with local LAs and other GAs. It can not communicate with other LAs. So a GA can only receive two kinds of requests. One is from another GA, the other is from local LAs.
- **If the type of requester is GA**
  If the requester is another Group Agent, it only searches the local group and delivers the provider information if it exists. Because every GA maintains a table that records the information registered by all of its own LAs, the GA can easily find out whether the required information is owned by one of its LAs. Considering the communication problem, only the GA of a provider sends the confirm message and the GA of a requester only chooses the first returned provider.
  If the type of requester is LA, then the GA will search in a local group first. However, if the information is not available locally, the GA seeks help by forwarding the request to other GAs. These other GAs then check their own databases for a match and deliver any positive feedback to the requesting GA, which passes the feedback directly on to the original requester.
- **If the search does not succeed**
  If no positive answer can be found by interacting with neighbored GAs, the middle agent of the requester stops searching and declares that the search has failed. The **topSearch** here is an upper limit for the number of attempts to prevent endless searches in a large and distributed e-learning environment. Since every LA is registered with a GA randomly in the initialization process, one group may have learners with different preferences. Hence, we adopted Wang's award and exchange schemas [5] aiming at recognizing learner behavior and reorganizing learners accordingly.
- **Award schema**
  When a GA relays search results to a LA, it examines whether the search is successful. If the search is successful, that is, if there is an information provider matching a request, the award of the requester and provider LA are increased by 1. This is because both the requester and provider have made a good contribution to the system.
- **Exchange schema**
  After that, the GA of the requester determines whether the requester and provider are both in its group. If they are not in the same group, the GA of the requester contacts the GA of the provider for a membership exchange such that both the requester and provider - who have similar interests - are registered with the same group. The rule is to move the LA with lower award towards the GA managing the LA with higher award. This hypothesis is based on the belief that a learner with high award usually means that it has either requested or provided useful information to other users. The highly awarded LA is always acting on behalf of the main interest of the group. It is called the

authority learner. Hence an attraction to an authority LA can drive other LAs with similar interests to join the same group quickly.

By means of the award scheme, it is possible for GAs to differentiate between learners who are contributing to their community and those who are passive. This information plays an important role in generating the authority learner and the formation of learner communities. By using the exchange schema, GAs can quickly cluster the LAs with the same preferences, which is essential for a quick decrease of the search time and an increase of the success rate.

4 Experiments

The experiments presented here are based on the real learner from the Network Education College of Shanghai Jiao Tong University. We focus on the evaluation of the usefulness and efficiency of this self-organizing mechanism. For simplicity in the experimental system, we made two assumptions:

- **Hypothesis1**: the resources owned by learners are documents and can be classified into certain categories according to their context.
- **Hypothesis2**: every learner only has one preference or interest and owns zero or more documents of relevant category.

In the experimental setting, we chose 1500 users, and the number of owned documents are 1000 (one document can be owned by one or more learners), the preference category is 10.

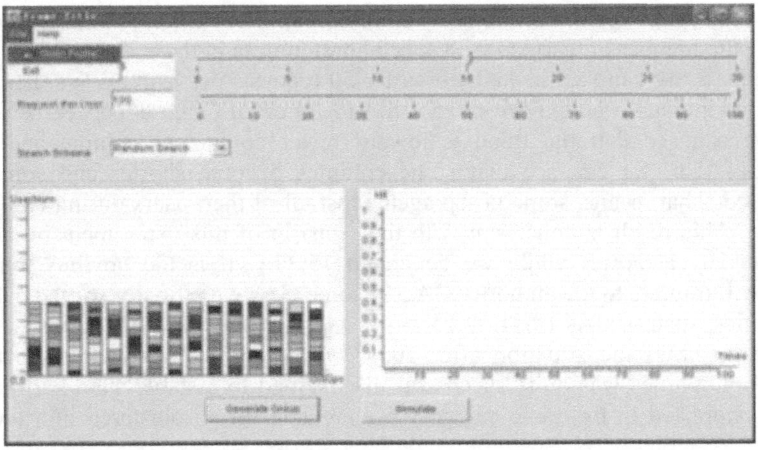

Fig. 4. Introduction of the test platform and system initialization

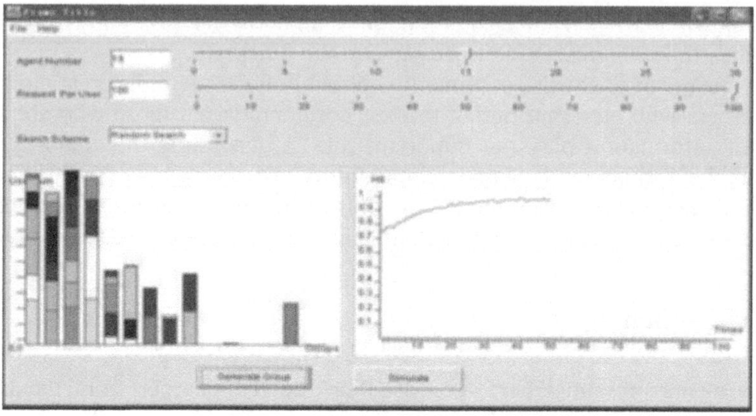

Fig. 5. System situation after 50 requests per learner

Figure 4 illustrates the main test platform of this system. In the top panel, we can define the number of Group Agents (**Gnum**) and the request times per learner (**maxRequestTime**). The left graph at the bottom shows the learner distribution in every group, whilst the right one shows the statistic analysis of the request success rate. When the experiments started, the system generated 1500 Learner Agents on behalf of the real learners and 15 Group Agents. LAs randomly registered with one GA together with the summarized registration information. The GAs kept all information of learners in this group. Figure 4 shows the initial situation in which the colors of every column are mixed and the distribution is almost average.

In the Learner Distribution Graph, every column represents a group, and the colors of the rectangles represent different preferences. The height of each rectangle illustrates the number of learners with special preferences in the corresponding group.

In Figure 5, we can see the situation after 50 requests per learner. We can see the success rate increased quickly from 75% to 91% after 20 requests per learner. In this figure we can see that the trend is toward fewer colors per column and longer rectangles. Also, we can see that some columns become shorter and some even disappeared. That means, some group agents lost all of their users during community formation. This result is consistent with the scenario of this experiment because we only have 10 categories while we generated 15 GA. It is the obvious trend that learners will migrate to the authority GA and some GAs were finally shifted out of the communities, such as GAs 10,11,12,14.

Figure 6 shows the situation after 100 requests per learner. The formation of learner communities was quite successful. It is settled to a stable state with learners who are interested in the same category preferences are all clustered into the same group.

From the RequestSuccess-Rate Graph, we can see that when users are not well organized, the success rate is low. This is because Group Agents often can not find available resources locally and need to send requests to other Group Agents. Since we limited the number of searched GAs, the success rate is lower during the first ten requests per learner. Once learner communities have started forming, however, the system exhibits an obviously improved success rate and greater efficiency. Because learners who have matching requests and results are gradually grouped together, GAs can more easily find correct answers in their own groups. As a result there is an increase in the success rate of search results. Figures 4 – 6 show that the system's ability to find correct answers to requests was obviously improving, as more and more requests were initiated in the system. And the success rate approached 1 after the learner communities were set up. Meanwhile, the average search time for a request was greatly decreased.

The experiments have been executed using varying numbers of learners and different kinds of learners. The formation of learner communities was always quite successful and can be scaled with the increased number of learners quite effectively.

5 Conclusion

This paper has described a multi-agent environment to self-organize learner communities according to users' preferences and capabilities. Furthermore, we presented effective award and exchange algorithms whose effect is that eventually learners with similar preferences or interests can be clustered into the same community. We evaluated this system in the real e-learning environment operational at Jiao Tong University. The experimental results illustrate that this method can cluster the learners quickly and facilitate appropriate organization between learner agents and group agents. In particular, this method achieves higher search success rates and a sharp decrease of search time.

Our further work will be devoted to extend the adjustment mechanism to handle multiple preferences since learners usually have multiple interests and information resources. Furthermore, we will consider more complex behaviors than mere resource requests. We also plan to run an evaluation with students using the system to verify whether our measurements correlate with the students' satisfaction.

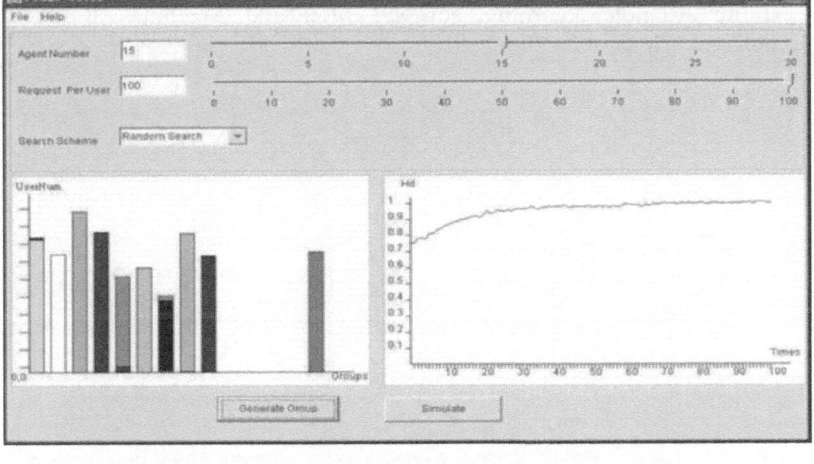

Fig. 6. System situation after 100 requests per learner

References

[1] Suthers, D. Collaborative Representations: Supporting Face-to-Face and Online Knowledge Building Discourse. In: Proceedings of the 34th Hawaii International Conference on the System Sciences (HICSS-34), January 3-6, 2001, Maui, Hawaii.

[2] K. Decker, K. Sycara and M. Williamson, Middle-agents for the internet. IN: Proceedings of the Fifteenth International Joint Conference on Artificial Intelligence, Japan, 1997, pp.578-583.

[3] R.J. Bayardo Jr., W. Bohrer, R. Brice, A. Cichocki, J. Fowler, A. Helal, V. Kashyap, T. Ksiezyk, G. Martin, M. Nodine, M. Raschid, M. Rusinkiewicz, R. Shea, C. Unnikrishnan, A. Unruh, and D. Woelk. Infosleuth: agent-based semantic integration of information in open and dynamic environments. In: Proceedings of the ACM International Conference on Management of Data, 1997, pp.195-206.

[4] T. Mohri and Y. Takada, Virtual Integration of Distributed Database by Multiple Agents. IN: Proceedings of the First International Conference on Discovery Science, pp. 413-414, 1998.

[5] F.Wang, Self-organising communities formed by middle agents. In: Proceedings of the First International Conference on Autonomous Agents and Multi-Agent Systems, Bologna, Italy, July 2002, pp.1333-1339.

[6] P.J. Turner and N.R. Jennings, Improving the scalability of multi-agent systems, IN: Proceedings of the First International Workshop on Infrastructure for Scalable Multi-Agent Systems, 2000, pp.246-262.

[7] Ruimin Shen, Peng Han, Fan Yang, Qiang Yang, Zhexue Huang. An Open Framework for Smart and Personalized Distance Learning. In: Proceedings of International Conference on Web-based Learning ICWL 2002, pp19-30, HongKong, China, Aug 2002.

[8] Dazheng Wang, Ruimin Shen, Liping Shen, Collaborative Learning based on Multi-agent Model. IN: Proceedings of International Conference on Web-based Learning: Remote Learning between Men and Machines ICWL 2002, HongKong, China, Aug 2002.

The Effects of Material, Tempo and Search Depth on Win-Loss Ratios in Chess

Peter Smet, Greg Calbert, Jason Scholz, Don Gossink,
Hing-Wah Kwok, and Michael Webb

Command and Control Division
Defence Science and Technology Organisation
Edinburgh, South Australia 5011
Australia
peter.smet@dsto.defence.gov.au

Abstract. What exactly is it that gives one player a decisive advantage over the other in an adversarial contest? We have played tens of thousands of simulated chess games to compare how different kinds of advantage contribute to overcoming an opponent. This has allowed us to quantify the effects of search depth, material strength, and tempo on game outcomes. The results indicate that tempo and material are both advantageous, but that the degree of advantage is highly dependent on skill level. Skill level, as reflected by search depth, leverages small inequalities in material or tempo. In addition, searching one move deeper than an otherwise equally matched opponent equates to an advantage of between 85%-98%.

1 Introduction

Chess is a game of perfect information often used to model conflicts and decision-making strategies. The playing of chess involves looking ahead at hypothetical move progressions, and assessing the values of positions that arise from these. Strong players are able to look many moves ahead, and take into account a great number of positional nuances. We have set out to quantify the extent to which various factors contribute to chess-playing strength, and to examine the relationship between these factors.

The factors of initial interest to us were material, search depth, and tempo. Material asymmetries were created by removing pieces from one of the players, while search depth was explored by allowing a player to search one ply deeper than their opponent. Tempo advantages took the form of extra moves, executed at predefined intervals. Our strategy was to play agents with various strengths and weaknesses off against each other. This approach has several advantages. The trials can be repeated many times under controlled conditions. Further, by making the agents non-deterministic, statistically rigorous evaluation and quantification is possible.

Previous studies have examined the effects of tempo and material changes on human chess performance [8]. These studies indicate that material and tempo

T.D. Gedeon and L.C.C. Fung (Eds.): AI 2003, LNAI 2903, pp. 501–510, 2003.

gains are highly advantageous, but tempered by the level of uncertainty. Uncertainty was modelled by hiding an opponent's last one or two moves. In addition, material advantages were degraded to a far greater extent by uncertainty than were tempo advantages. While innovative, these human studies are limited in the number of games that can be evaluated, variable performance, and side-effects of learning from repeated trials. We were therefore inspired to examine these issues in greater detail, using six different tempo conditions, four levels of material advantage, and seven levels of search depth.

This paper outlines the manner in which search depth, material, and tempo advantages were implemented in section 2. Section 3 examines the outcomes of games played with these various advantages, and section 4 discusses the implications of the results. The work is summarized in section 5.

2 Methods

Agents were played against each other, using a straightforward minimax algorithm with a lookahead range from 0 to a maximum of 7 ply. The 0 ply lookahead uses no search at all, and simply plays random legal moves. The evaluation function examines material only, using a standard set of weightings [9]. A queen is valued at nine units, a rook at five, bishops and knights at three. Pawns are worth one unit. A board position is scored by subtracting the summed value of the adversary's pieces from the summed value of the player's pieces. The evaluation function also detects checkmate.

A draw results if the game exceeds 500 moves, or if one of the players is in a stalemate position. Draws by three-fold repetition were ignored. Such draws typically go on to exceed the move limit by n-fold repetition, and can so be classified using our simpler heuristic. The move limit was chosen because preliminary trials indicated that less than 1% of games were won or lost in more than 500 moves. Here, 'move' is defined as a single turn by one player.

All agents were non-deterministic in their move selection. This enabled us to examine conditions over thousands of games, and so produce valid statistical inferences. Random moves were picked by generating and scoring all legal moves. Moves were sorted best-first, and one of the top-scoring moves was chosen randomly. A linear congruential algorithm [7] was used to generate random numbers, seeded from the system clock at the start of each set of trials.

Games were scored according to the rules of the International Chess Federation. A player scores one point for a win, half a point for a draw, and nothing for a loss [4]. To neutralize the first move advantage that white traditionally has (approximately 4%, [10]), black and white alternate in opening games.

The statistical analysis was conducted as follows. Variance was calculated according to the binomial theorem. var $= (p * q)/n$. Here, p is the probability of white winning, q is the probability of white losing $(1 - p)$, and n represents the number of games played. p is calculated using a modification of the maximum likelihood estimate, where $p = \text{wins} + (0.5 * \text{draws})/n$.

Table 1. The effect of search depth on win-loss ratios. n denotes the number of games played for each trial condition

Trial	Score	Standard Deviation	Average Game Length	n
1 versus 0	95.48%	0.66%	103	1000
2 versus 1	97.50%	0.49%	72	1000
3 versus 2	98.00%	0.45%	90	1000
4 versus 3	94.70%	0.45%	130	2425
5 versus 4	94.42%	0.73%	172	976
6 versus 5	91.12%	0.97%	205	861
7 versus 6	85.93%	1.85%	313	373

3 Results

3.1 The Effect of Search Depth on Game Outcomes

For these experiments agents with *depth* ply lookahead were matched against agents using $depth - 1$ ply lookahead. The results are summarized in the table below. Scores are expressed as the percentage of games won by the player with the greater lookahead depth. It is clear that, for fixed depth searches, an extra ply of lookahead represents a consistent and overwhelming advantage (85%-98%).

It is interesting that the number of moves required to win games increases with search depth. Average game length is perhaps a more sensitive measure of dominance than the simple win or loss outcomes. If we assume that stronger agents can overcome their opponents in fewer moves, the advantage of an extra search ply appears diminished at higher search levels. Agents of search depth seven took nearly twice as long as agents of search depth five to overcome their depth - 1 opponents. This effect is borne out by the win-loss ratios, which seem to decrease slightly towards the higher ply levels.

The results suggest the effect of an extra ply may be 'diluted' at greater search depths. For example, at one-ply, an extra ply represents a 100% improvement in lookahead relative to your opponent. At five-ply, an extra ply represents only a 20% improvement in lookahead.

3.2 The Effect of Tempo on Game Outcomes

Tempo was simulated by granting white an extra move at predefined intervals. Six conditions were examined, in which an extra move was given each 1, 5, 10, 25, 50 or 100 moves. In the second condition, for example, the agent is able to move again immediately after making its 5th, 10th, 15th... and so forth, moves. Players and their opponents were always matched on search depth, which ranged from 0 to 6 ply across the various trials.

The lookahead algorithm was adapted to anticipate extra moves, for both players. Figure 1 shows how this was implemented. It was also necessary to

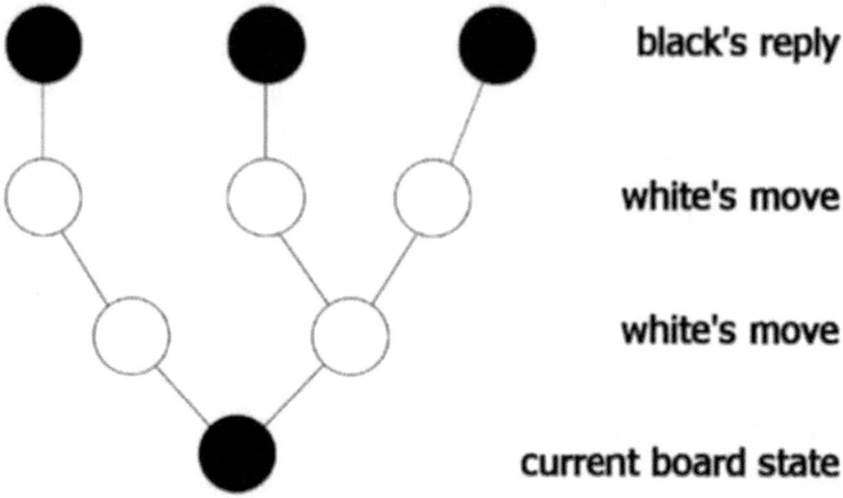

Fig. 1. Hypothetical search tree evaluating the best move for white when an extra move is due. At two-ply, white examines it's two consecutive moves, and black's counter move. Conversely, black examines it's move and white's two consecutive replies. The search tree thus defines and scores all possible paths involving extra moves

modify the rules of chess to accomodate the extra moves. If either king was captured as a result of a two move sequence, this was deemed checkmate for the player making the capture. Figure 2 shows the results of various tempo advantages to white. Players with deeper search depths exploit extra moves to a greater extent. The relationship between search depth and the advantage gained from tempo is not a straightforward one, and appears to be non-linear in nature.

3.3 The Effect of Material on Game Outcomes

Material advantage was tested by removing either black's pawn, knight, bishop, rook, or queen. The removed piece always came from the queen side of the board. The pawn was removed from square b7. The effects of losing a piece on win-loss ratios are shown for various lookahead levels in figure 3. Again, players and their opponents were matched on search depth. The results bear out the traditional ranking of the pieces, and are in line with the relative values accorded to the pieces by human judgement. White's queen advantage has a greater effect than the rook advantage, while the rook advantage is more important than that of

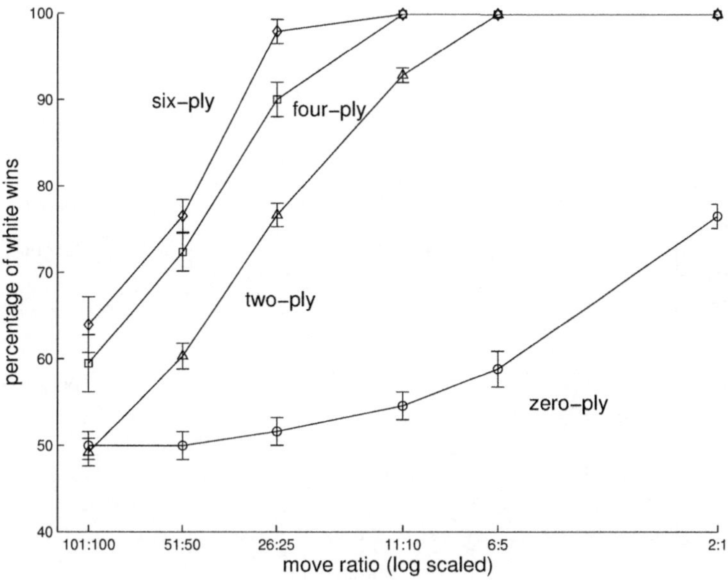

Fig. 2. The effect of extra moves on win-loss ratios for white at various search depths. Each data point represents the mean of between 100-1000 games. Error bars represent the standard deviation

the knight, bishop and pawn. This was true across all skill levels. The chart further shows that a material advantage is never absolute, and depends on the skill level of the match. As with tempo, players of greater skill exploit material to a greater extent. Once again, the relationship between search depth and the amount of advantage gleaned from a material imbalance is discrete.

4 Discussion

4.1 Search Depth

The advantage of looking an extra move ahead of your opponent ranged from 85% to 98%. Others have reported a 70%-80% advantage for an extra level of search depth in chess [5, 6, 11]. The discrepancy with the current result is likely to be attributable to the evaluation function. A strong evaluation function can compensate for a lack of search depth, by accurately predicting the future value of board positions at the search limit. Our evaluation function is far simpler than most others. With such an unsophisticated evaluation function, search depth assumes a greater importance in determining the outcome of a game. These considerations suggest agents with sophisticated evaluation functions are better able to compensate for the handicap of a lower search depth than those with simple evaluation functions.

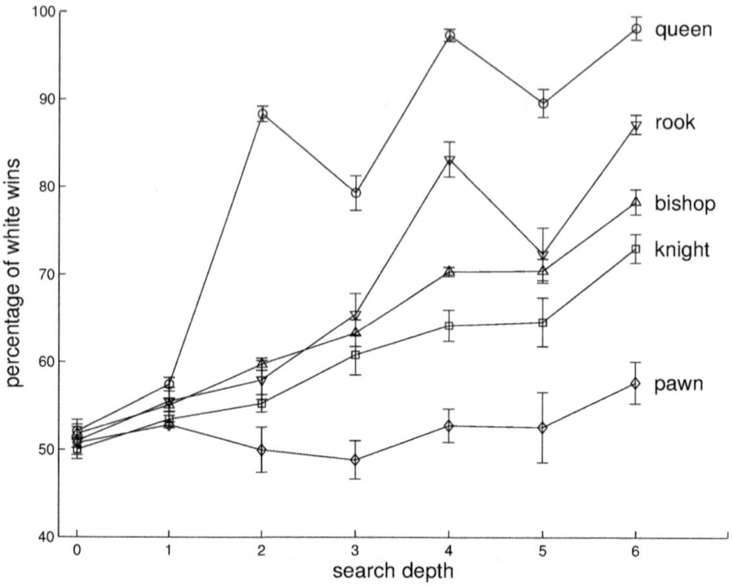

Fig. 3. The effect of extra material on win-loss ratios for white at various search depths. Each data point represents the mean of between 100-1000 games. Error bars represent the standard deviation

Our agents furthermore use a fixed-depth search strategy. This introduces the well-known 'horizon effect' [2], where the consequences of bad moves cannot be seen when these lie beyond the search depth limit. In the lookahead trials, the $depth - 1$ ply agents are likely to suffer from the horizon effect far more than their *depth* ply counterparts. Strong chess programs use a variable-depth search strategy to offset the shortcomings of a fixed-ply search strategy.

There is a linear relationship between the proportion of wins and Elo score differences separating human chess players [3, 10]. Each percentage gain in excess of a 50% win-ratio represents an approximate 8.6 Elo point advantage over an opponent. Using this simplified formula to calculate the relative Elo differences between our agents [10], we estimate that each ply adds approximately 299-412 Elo rating points to an agent. Others have estimated that agents gain between 100-250 Elo points when their search depth is increased by a single ply [5, 11]. These differences simply reflect the greater effect of search depth on win-ratios which we have found.

4.2 Tempo

Extra moves deliver a powerful advantage in chess. Our observations of played games indicated that double moves are often used to capture pieces, or indeed the king. The defending player either cannot defend all exposed pieces, or is vulnerable to a 'hit and run'. Chess assumes pieces can be protected against capture, normally by recapture of the attacker. With double moves, a defended piece can be captured for free, since the second move provides an escape for the aggressor.

Interestingly, extra moves were a strong advantage even to players with no goal. Agents playing random legal moves won in the order of 76% of games when allowed to move twice as often as their opponents. At higher search depths players with extra moves won a large percentage of games. Even with an advantage as small as one extra move per hundred, players still eked out a 60%-64% advantage. Considering most of these games are only 120 or so moves in length, this represents only one extra move in the entire game.

Tempo was more advantageous than material, in the sense that it was exploited better by players with low skill levels.

4.3 Material

Figure 3 shows that the absolute advantage of a material imbalance is highly dependent on skill level. This bears out the intuitive notion that pieces have more power in the hands of skilful players, as expressed in the maxim:

"A weapon is only as strong as the hand that wields it."

Below a particular skill level, material has only the smallest of effects. This is particularly obvious in the random player, where even a queen advantage yields only a 52.14% proportion of wins. Removing both black rooks and the queen extends this proportion to only 54.5% (data not shown). This contrasts with extra moves, which produce robust advantages. An extra move each single turn, or each five turns, yield 76% and 59% advantages, respectively (Figure 2). At the lower skill levels (0-1 ply), the *maximum* obtainable advantage is therefore much higher for tempo than it is for material.

According to the commonly held values for chess pieces, a knight or bishop should give three times as much advantage as a pawn, and a queen three times as much value as a knight. Our results partially support these notions. At six-ply, for example, a pawn gives a 7.7% advantage over the expected 50% outcome for evenly matched players. If we normalize this advantage to one point, then our results indicate a knight is worth 3.0 points, a bishop 3.7, a rook 4.9, and a queen 6.3 points. This compares to the actual values of 3, 3, 5, and 9. The estimate of the value of the queen is almost certainly an underestimate. The trials with black's queen removed produced a 98% winning advantage to white, and are limited by a 'ceiling effect'.

While the program therefore produced an approximation of the values commonly accorded to the pieces, it must be kept in mind that these values were derived by skilled human players, and are perhaps most applicable in that context. Our simple agents may underestimate the true power of the various chess pieces. Trials using temporal difference learning have shown that the 'correct' values of the chess pieces can vary widely. The values for a knight range from

2.2-2.9, the bishop from 2.9-3.4, those of the rook from 4.0-5.0, and the queen from 7.5-9.75 [1].

It would be interesting to recalculate our piece values with an 8-12 ply lookahead and a more sophisticated evaluation function.

It is apparent from figure 3 that even-ply searches perform better than odd-ply searches. This is particularly obvious when one agent has either a rook or queen advantage, and is an indirect consequence of the horizon effect. Even ply searches always examine an opponent's move at their search limit, odd ply searches examine their own moves last. Even ply searches therefore tend towards conservative play, and do not leave pieces where these can be captured. Odd ply searches play aggressively, and will often capture an opponent's piece if negative consequences are, or can be pushed, over their search horizon.

We examined odd-ply games where white lost, despite a rook or queen advantage, in detail. Typically, white's queen or rook was placed in an attacking position. White's lookahead then played a hypothetical 'optimal' move sequence where the queen captures a significant black piece as it's last move. This move sequence was invariably bad, due to the impending recapture of the queen. Sometimes the move sequence involved the loss of minor pieces for white, which was offset by the anticipated final move capture. It was such lines of play that led to the loss of material, and ultimately, the game. This artifact of the fixed-depth search causes the oscillations between odd and even ply lookahead trials.

4.4 Exchange Rate of Different Advantages

Having quantified various advantages, we can now say something about the 'equivalence point', at which one type of advantage can match, or neutralize, another. For example, at six-ply, a queen advantage, an extra level of lookahead, and one extra move every twenty five all produce identical advantages (95%-100%). Put another way, 9 units of material $\approx$ 1 search depth level $\approx$ 26:25 move ratio. It may be more informative to examine what produces a half-maximal advantage. Since a 50% win ratio indicates no side has an advantage, and a maximal advantage produces 100% wins, the half-maximal advantage point is defined as a win rate of 75%. This avoids the problems of diminishing returns and ceiling effects associated with near-maximal advantages. At six-ply, a knight or bishop advantage, or an extra move every fifty, create a 73%-78% advantage. That is, 3 units of material $\approx$ 51:50 move ratio.

Figure 4 attempts to generalize this concept for material and tempo. It plots the move ratio that produces an equivalent advantage to any given material advantage.

5 Conclusions and Future Work

We have shown that an imbalance in material, tempo, or search depth leads to improved win-ratios in chess. The extent of a material or tempo advantage is

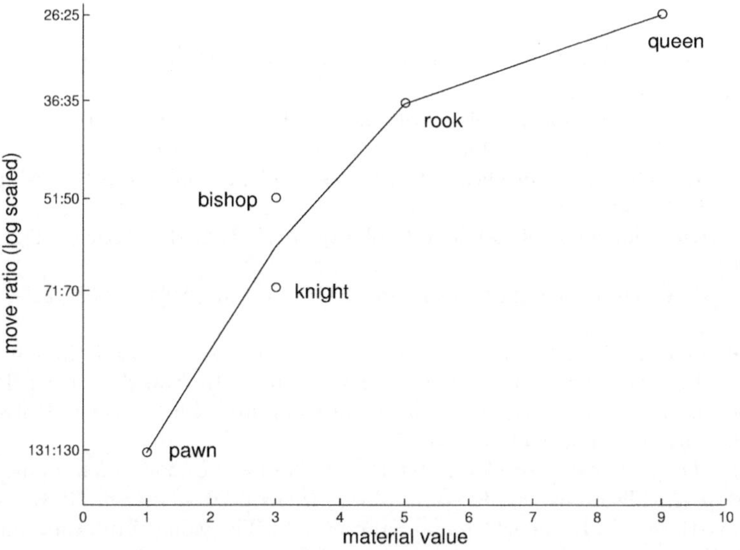

Fig. 4. Equivalence relation between material and tempo. The x-axis shows material gain, using the standard chess values of the pieces. The y-axis shows the move-ratio that produces an advantage identical to a given material gain. Games were all played at a search depth of six-ply

greatly enhanced by search depth, in a non-linear manner. In addition, the *maximum* advantage gained from tempo is greater than that gained from material, but only at the lower skill levels. Finally, search depth is a strong determinant of game outcomes, in that agents that search a single ply deeper than their opponents win 85%-98% of their games.

It will be important in future work to determine whether quantitatively similar advantages are truly equivalent. For example, is an agent with a tempo advantage that yields a 75% win rate evenly matched against an agent with a material advantage that also produces a 75% win rate? We intend to verify the equivalence of material, tempo, and search depth advantages by playing agents with quantitatively identical but qualitatively distinct advantages off against each other.

Acknowledgements

The authors would like to thank Don Perugini for an elegant solution to the problem of implementing a search tree with extra moves, and Mofeed Shahin for insightful discussions. Sam Hutchison and Matthew Thyer were extremely helpful in getting the chess trials running on a linux cluster.

References

[1] Beal, D. F. and Smith, M. C.: Learning Piece Values Using Temporal Differences. ICCA Journal 20(3) (1997) 147-151 508
[2] Berliner, H. J.: Chess as Problem Solving: The Development of a Tactics Analyzer. Ph.D. Thesis, Carnegie-Mellon University, Pittsburgh, (1974) 506
[3] Elo, A. E.: The rating of chess players past and present. Arco Publishing, New York (1978) 506
[4] http://www.fide.com/official/handbook.asp?level=EE101 Article 11: scoring (2000) 502
[5] Heinz, E. A.: DarkThought Goes Deep. ICCA Journal, 21(4) (1998) 228-244 505, 506
[6] Junghanns, A., Schaeffer, J., Brockington, M., Bjornsson, Y. and Marsland, T. A.: Diminishing returns for additional search in chess. In: van den Herik, H. J. and Uiterwijk, J. W. H. M. (eds.): Advances in Computer Chess, Vol 8. University of Maastricht, Maastricht (1997) 53-67 505
[7] Knuth, D. E.: The Art of Computer Programming, Volume 2 Seminumerical Algorithms, Addison-Wesley, Reading Mass., ISBN 0-201-03822-6. (1981) 502
[8] Kuylenstierna, J., Rydmark, J. and Fahraeus, T.: The value of Information in War: Some Experimental Findings. International Command and Control Research and Technology Symposium, Canberra (2000) 501
[9] Levy, D.:Computer Gamesmanship: Elements of Intelligent Game Design. Simon & Schuster, New York, ISBN 0-671-49532-1. (1984) 502
[10] Sonas, J.: The sonas rating formula - better than Elo? www.chesscenter.com/twic/event/sonas/sonasrat.html. (2003) 502, 506
[11] Thompson, K. Computer chess strength. In: Clarke, M. R. B. (ed.): Advances in Computer Chess, Pergamon (1982) 55-56 505, 506

Using Multiple Classification Ripple Down Rules for Intelligent Tutoring System's Knowledge Acquisition

Yang Sok Kim[1], Sung Sik Park[1], Byeong Ho Kang[1], and Joa Sang Lim[2]

[1]School of Computing
University of Tasmania, Australia
{yangsokk,sspark,bhkang}@.utas.edu.au
[2] School of Media Technology,
University of Sangmyung, Korea
jslim@smu.ac.kr

Abstract: This research focuses on the knowledge acquisition (KA) for an intelligent tutoring system (ITS). ITSs have been developed to provide considerable flexibility in presentation of learning materials and greater abilities to respond to individual students' needs. Our system aims to support experts who want to accumulate the classification knowledge. Rule based reasoning has been widely used in ITSs. Knowledge acquisition bottleneck is a major problem in ITSs as it is known in AI area. This problem hinders the diffusion of ITSs. MCRDR is a well known knowledge acquisition methodology and mainly used in classification domain. MCRDR is used to acquire knowledge for the classification of learning materials (objects). The new ITS is used to develop a part of online education system for the people who learn English as a second language. Our experiment results show that the classification of learning materials can be more flexible and can be organized in multiple contexts.

1 Introduction

Various Artificial intelligence researches have been applied to ITS development. [1], [2], [11], [12], [14], [15], [25] ITSs that use AI technology are very effective in many domains. For example, measures taken at Carnegie-Mellon University have demonstrated that the intelligent tutor usage allowed an improvement of quality of learning by 43% and a reduction in the length of learning 30%. [19] Other more spectacular experiences have been undertaken by the SHEROCK system that is a tutoring system for the detection of plane breakdowns. The research result shows that employees who learn an apprenticeship about 20 hours by using this system get a comparable expertise to that of employees having 4 years of experience. [18]

However, ITSs still are not widely used because of the high development costs. According to Woolf, ITSs require at least 200 hours of development per hour of instruction. [26] This, in our view, is caused by traditional knowledge acquisition (KA) methods. Traditionally the construction of ITS's knowledge base (KB) is

T.D. Gedeon and L.C.C. Fung (Eds.): AI 2003, LNAI 2903, pp. 511-519, 2003.

founded on the rule based approaches. [3] However, this approach is very difficult because it is hard to uncover the hundreds of rules that the expert uses to solve problems. This problem became known as the KA bottleneck. [13] Moreover, the rule-based systems are very brittle since all their knowledge was recorded only in terms of rules. If a problem did not match any of the rules, the system could not solve it. It also makes maintenance work very difficult. To solve this problem, new KA approaches have been suggested by researchers (e.g Case Based Approach, Knowledge Mapping Approach). [4], [15], [20], [22]

2 Multiple Classification Ripple Down Rules

MCRDR was developed to deal with the problems found in maintenance of the medical expert system GARVANES1. [5] One of the key observations in this was that experts never gave a comprehensive explanation for their decision making. Rather, they justified that the conclusion was correct and the justification was created for and shaped by the context in which it was given. [7], [8] This view is in turn consistent with Popper's falsification approach to the development of knowledge. Knowledge is always a hypothesis which can never be proven correct. At most it can be proven incorrect and be replaced by another perhaps "less false" hypothesis. The hypothesis is a model and as such is intrinsically different from reality. Obviously the model is also constructed within a particular paradigm and particular context and so will have to changed or replaced to be cope with other points of view. [20] From this point, the KA tools must support incremental KA and store the context information. The most common form of KA is correcting errors of interpretation for particular cases. In such maintenance the justification for a new conclusion will be presented at least in terms of the features of the case that suggest a different interpretation of the data from the one given. That is, the context in which this justification is given is that a particular incorrect conclusion has been made for a specific case. In MCRDR, a rule consists of the conjunction of features in the case identified by the expert and the rule is added to the KBS so that it will only be used when the same wrong conclusion is reached through the same pathway of rules (i.e. in the same context in which the error was made). MCRDR provides a further validation step in that the case is stored with the rule, and if a further correction rule is added, the expert must select some features that distinguish the new case from the stored case(s) for which the given interpretation was correct. In MCRDR, this maintenance strategy is used to build the entire knowledge base. [9]

3 Intelligent Tutoring System with MCRDR

3.1 System Overview

A classification system for English test questions, which is a component of ITS for the students who learn English, are implemented with Java program language and MCRDR methodology. Test questions are called ***cases*** in our system. We use two types of multiple choice questions. One is sentence structure question, and the other is a written expression question. These types are typical questions in a TOEFL test. Each question contains following data; question type, difficulty level, question explanation, related words and idioms, and ***keyword***. We divided keywords into three types; ***case keyword, rule condition keyword, and abstract keyword.*** Firstly, ***case keywords*** are the metadata provided by experts. Experts can add, delete and modify the keyword while they construct a knowledge base. Case keywords are used to create rule condition and if the case is a cornerstone case, they will be used to validate new rule. Secondly, ***rule condition keywords*** represent the condition of a rule. They can be either case keyword or not. Each rule has one or many keywords. Our system uses these keywords when it inferences a conclusion. Thirdly, ***abstract keyword*** represents a set of rule keywords. Experts give this keyword directly or select from rule condition keywords. They can directly see this keyword from the system user interface.

3.2 Knowledge Base Structure

The expert's knowledge is stored in a tree structure. Each node is a rule of the form if *A* then *B*, where *A* (condition) typically consists of a conjunction of attribute-value tests and *B* (conclusion) could be anything from a nominal value to a string in a natural language. Fig. 1. shows the rule structure of our system's knowledge base. Condition is a set of keywords and conclusion is an abstract word. Each rule has one or more cornerstone cases that are used when new rule is validated. Both the condition keyword(s) and cornerstone case(s) represent the relationship information of learning contents.

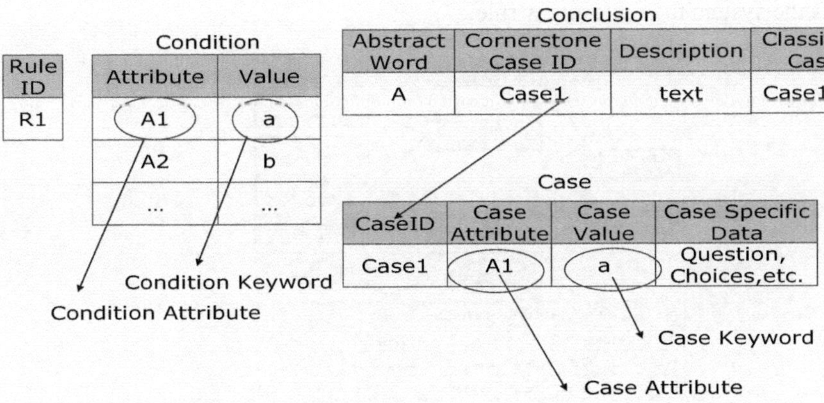

Fig. 1. Each case (question) has following data; question type, difficulty level, question explanation, related words and idioms, and keyword. Keyword has three types; case keyword, rule condition keyword, and abstract keyword

3.3 Knowledge Acquisition Process

3.3.1 Importing Cases

The process begins with importing new cases (questions). The system shows the unclassified cases in the case tree pane. Each case has case keywords, question, choices, and answer. The expert can choose one case from the case tree and edit (add, delete, and modify) the case keywords. These case keywords are used the in inference and knowledge acquisition processes.

3.3.2 Inference

The inference process begins when the expert demands a conclusion from a given case. The boxes in bold (Fig.2) represent rules that are satisfied for the test case of a set of keywords with {***a, b, c, d, e, f, g***}. The system evaluates all the rules in the first level of the tree for the given case (rule1, 2, 3 and 5 in Fig. 2.). Then, it evaluates the rules at the next level that are refinements of the rule satisfied at the top level and so on. The process stops when there are no more children to evaluate or when none of these rules can be satisfied by the case in hand. If rules 2, 3 and 5 are evaluated as satisfied by the case, the next rules to be evaluated are rules 6, 7, 8, 9 and 10 (refinement of rule 2, 3, and 5). The process will stop when there are no more child nodes to be evaluated or none of these refinement rules are unsatisfied by the case. In this instance, there exist 4 rule paths and 3 classifications (class 2, 5, and 6). Therefore, it will end up with more than one path for a particular case. The only job that the expert has to do is determine whether the recommendations are correct or incorrect. If the expert satisfies a recommendation, he/she accepts the result and then the system classifies this case into the recommended rule nodes. If not, the expert requests the system to create a new rule.

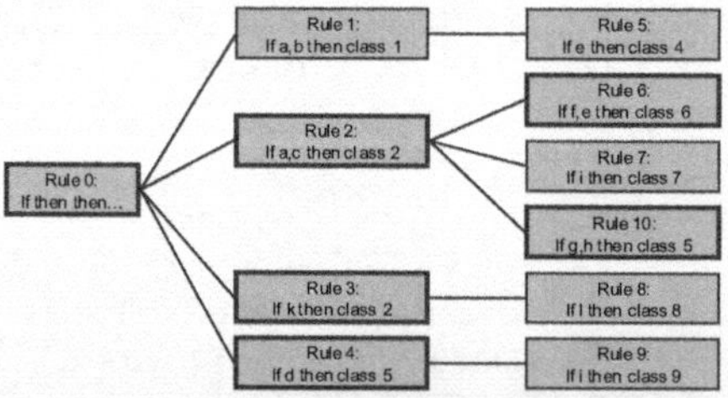

Path 1	[(Rule 0,...), (Rule2, class2), (***Rule6, class6***)]	info 1
Path 2	[(Rule 0,...), (Rule2, class2), (***Rule10, class5***)]	info 2[4]
Path 3	[(Rule 0,...), (***Rule3, class2***)]	info 3
Path 4	[(Rule 0,...), (***Rule5, class5***)]	info 4[2]

Fig. 2. The highlighted boxes represent rules that are satisfied for the case {a, c, d, e, f, h, k} and pathways through the knowledge base. The rules producing conclusions are highlighted. Info n [...] indicates other rule numbers with the same classification

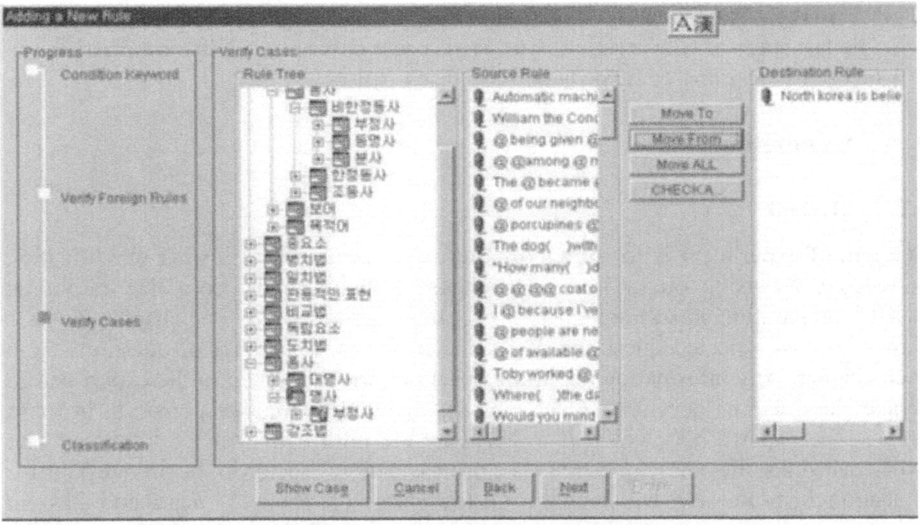

Fig. 3. User create new rule (infinitive). Two foreign rules are validated by user. All cases are shown in the "Source Rule" pane. Expert can validate each case very easily

3.3.3 Knowledge Acquisition

KA and inference are inextricably linked in MCRDR methodology, so some of KA steps depend on the inference structure and vice versa. When a case has been classified incorrectly or is missing a classification, KA process is required.

- **Select Rule Creation Method**
  There are three rule creating methods; refinement rule creation, stop rule creation, and independent rule creation. If the expert selects one method and requests the system to create new rule, then the new rule creation process begins.
- **Choose Condition Keyword**
  In the first stage, the expert chooses one or more keywords from a case keyword list. The selected keywords are used as condition keyword. If the expert needs another keyword, he/she also can create new case keyword.
- **Verify Foreign Rules**
  The system shows conflicting rules by highlighting the nodes. In the original MCRDR, only the parent rule and its child rule (the sibling rules of new rule) are selected as verification rules. However in our approach, the system validates the rules that have the same name because we consider both the folder structure and the condition keywords as expert's knowledge. This is a variation from the original MCRDR algorithm. The expert can choose to access all conflicting rules or only some of the conflicting rules to validate the cases.
- **Verify Cases**
  The system shows all the cases that need verification. The expert simply decides whether a certain case moves from conflicting rules to the new rule. This is a very simple process because the expert focuses on only each case, not the whole knowledge base.

- **Write Rule Description**
  The last stage of rule creation is to write a rule description. This information can be used as learning contents later.

4 Experiments

4.1 Method

The aim of experiment is to show how the expert accumulates his/her classification knowledge by using our system. We choose 900 questions from the traditional TOEFL preparation book. Each question is categorized into only one chapter. Table 1 shows how we selected questions. We selected the same number of questions from each chapter. Any information about the questions was not given to the expert before he/she classifies the questions. Questions are inputted by another person in order to prevent the expert's bias which is affected by the information in the textbook. The expert who classifies these questions is a professional instructor in the private TOEFL education institute. He has taught TOEFL class for about 5 years and has no information about our system. We trained him for about 3 hours about how to use the system. Because we only focus on the classification knowledge, the expert does not write the rule description when he classifies questions.

Table 1. Question Sets for Experiment

Chapter Name	Sentence Structure Question	Written Expression Question	Total
Noun	50	50	100
Pronoun	50	50	100
Adjective	50	50	100
Adverb	50	50	100
Verb	50	50	100
Verbals	50	50	100
Relative Pronoun	50	50	100
Agreement	50	50	100
Parallel structure	50	50	100
Total	450	450	900

Table 2. Classification Results(1)

Depth		Depth1	Depth2	Depth3	Depth4	Depth5	Total
Total	NOR	2	21	35	20	18	96
	NOC	0	200	1,248	489	397	2,334
	ACR	0.00	9.52	35.66	24.45	22.06	24.31
Sentence Structure Questions	NOR	1	10	18	7	6	42
	NOC	0	66	597	174	127	964
	ACR	0.00	6.60	33.17	24.86	21.17	22.95
Written Expression Questions	NOR	1	11	17	13	12	54
	NOC	0	134	651	315	270	1,370
	ACR	0.00	12.18	38.29	24.23	22.50	25.37

NOR: Number of Rules, NOC: Number of Cases, ACR: Average Case per Rule

4.2 Results

4.2.1 Classification Results

The expert classified the questions into 96 rules. The sentence structure question has 42 rules and the written expression question has 54. Total number of cases that classified each rule is 2,334 and the average case per rule (ACR) is 24.31 cases. Maximum depth of rule is 5, and ACR increases from depth 0 to depth 3 and decreases from depth 4 to depth 5 because the rule in the lower depth is more general and the rule in the deep depth rule is a more specialized rule (Table2). Each case classified into an average 2.59 rules (Table 3). 61.9% of all cases are classified into 2 or 3 rules. (Table 4) This means the learning contents have been classified into multiple categories. However, multiple classification property depends on the characteristics of learning objects. Our experiment supports this. The written expression questions are classified into more rules (3.04) than that of the structure expression questions (2.14). This fact also agrees with actual education experience - experts usually explain that one question can be solved by other rules.

4.2.2 Classification Times

The expert consumed 26 hours 45minutes to classify all questions. The average consumed time per question is 1.78 minute. This shows that classification is very easy when an expert uses our system. Classification time changes according to the property of the classification task (from 15 seconds to 378 seconds). If questions are only classified by inference results, the classification time is very small. But in the case of rule creation, more time is needed to classify questions.

Table 3. Number of Classification (2)

Number	Number of Classification (A)	Case Number (B)	Average (A/B)
Sentence Structure Question	964	450	2.14
Written Expression Question	1,370	450	3.04
Total	2,334	900	2.59

Table 4. Number of Classification (3)

Number of Classification	Total	1	2	3	4	5 ~
Number of Cases	900	125	331	226	134	84
Ratio	100%	13.9%	36.8%	25.1%	14.9%	9.3%

Table 5. Classification Time

	Time (second)
Total Classification Time (TCT)	96,300
Average Classification Time (TCT/Question Number)	107
Minimum Classification Time	15
Maximum Classification Time	378

5 Conclusion

The main problem of traditional ITSs is KA bottleneck. We used MCRDR methodology to solve this problem. Our system focuses on the classification knowledge of learning contents, which are specialized in TOEFL questions. KA processes begin when the expert gets no sufficient inference result by using current KB. If the expert specifies the condition keyword of new rule, system shows the rules and the cases that need verification. The expert only validates the rules and cases that are suggested by system, not the whole system. This makes the KA process very simple. Our experiment results show that the classification of learning materials can be more flexible and can be organized in multiple contexts.

References

[1] Ainsworth, S., Grimshaw, S., Underwood, J.: Teachers Implementing Pedagogy through REDEEM in Computer and Education, Vol 33 (1999) 171-187

[2] Beck, J., Stern, M., Haugsjaa, E.: Applications of AI in Education

[3] Buchanan, B., Shortliffe, E.(eds.) : Rule based Expert System: The MYCIN Experiments of the Stanford Heuristic Programming Project (1984)

[4] Burke, R.: Conceptual Indexing and Active Retrieval of Video for Interactive Learning Environments in Knowledge-Based Systems, Vol 9 (1996) 491-499

[5] Compton, P.J. and Jansen, R.: A Philosophical Basis for Knowledge Acquisition in 3rd European Knowledge Acquisition for Knowledge-Based Systems Workshop, Paris (1989) 75 – 89

[6] Compton, P.: Insight and Knowledge in AAAI Spring Symposium: Cognitive aspects of knowledge acquisition, 1992 (Stanford University, 1992) (1992) 57-63

[7] Compton, P., Edwards, G., Kang, B., Lazarus, L., Malor, R., Menzies, T., Preston. P., Srinivasan, A., Sammut, C.: Ripple Down Rules: Possibilities and Limitations. 6th Bannf AAAI Knowledge Acquisition for Knowledge Based Systems Workshop, Banff (1991)

[8] Compton, P., Kang, B., Preston, P., Mulholland, M.: Knowledge Acquisition without Analysis. Knowledge Acquisition for Knowledge Based Systems. Lecture Notes in AI (723). N. Aussenac, G. Boy, B. Gaineset al. Berlin, Springer Verlag (1993) 278-299

[9] Compton, P. Richard, D.: Extending Ripple Down Rules in The 4th Australian Knowledge Acquisition Workshop, Sydney, University of NSW (1999) 87-101.

[10] Kang, B., Compton, P., Preston, P.: Multiple Classification Ripple Down Rules : Evaluation and Possibilities. Proceedings of the 9th AAAI-Sponsored Banff Knowledge Acquisition for Knowledge-Based Systems Workshop, Banff, Canada, University of Calgary (1995)

[11] Evens, M.W., Brandle, S., Chang, R., Freedman, R., Glass, M., Lee, Y.H, Shim, L.S., Woo, C.W., Zhang, Y., Zhou, Y., Michael, J.A., Rovick, A.A.: CIRCSIM-Tutor: An Intelligent Tutoring System using Natural Language Dialogue in 12th Midwest AI and Cognitive Science Conference, Oxford OH (2001) 16-23

[12] Frasson, C., Aimeur, E.: Designing a Multi-Strategic Intelligent Tutoring System for Training in Industry in Computers in Industry, Vol 37 (1998) 153-167

[13] Hayes-Roth, F., Waterman, D.A., Lenat, D.B. (eds.) : Building Expert Systems. Reading, MA: Addison-Wesley (1983)

[14] Heffernan, N.T., Koedinger, K.R.: An Intelligent Tutoring System Incorporating a Model of an Experienced Human Tutor

[15] Kim, Y.C., Evens, M.W., Michael, J. A., Trace, D.: Physiology Tutorials Using Casual Concept Mapping in 13th Midwest AI and Cognitive Science Conference, Chicago IL (2002) 61-64

[16] Kolodner, J.L., Jona, M.Y.: Case-Based Reasoning: An Overview in Technical Report #15 (1991)

[17] Larkin, J., Chabay, R., Sheftic, C. (eds) .: Computer-assisted Instruction and Intelligent Tutoring Systems: Establishing Communication and Collaboration. Erlbaum, Hillsdale, NJ (1990)

[18] Nussbaum, M., Rosas, R., Peirano, I., Cardenas, F.: Development of Intelligent Tutoring Systems Using Knowledge Structures in Computer and Education, Vol36 (2001) 15-32

[19] Polson, M.C., Richardson J.J.(eds).: Foundations of Intelligent Tutoring Systems. Erlbaum, Hillsdale, NJ (1988) pp. 21–53

[20] Popper, K.: Conjectures and Refutations. London, Routledge and Kegan Paul (1963)

[21] Schank, R.C.: Case-Based Teaching: Four Experiences in Educational Software Design in Interactive Learning Environments, Vol 1:4 (1990) 231-253

[22] Schulze, K.G., Shelby, R.N., Treacy, D.J., Wintersgill, M.C.: Andes: A Coached Learning Environment for Classical Newtonian Physics in 11th International Conference on College Teaching and Learning

[23] Slade, S.: Cased-Based Reasoning: A Research Paradigm in AI Magazine, Vol 12:1 (1991) 42

[24] Sokolnicki, T.: Towards Knowledge-Based Tutors: a Survey and Appraisal of Intelligent Tutoring Systems in The Knowledge Engineering Review, Vol 6:2 (1991) 59-95

[25] Zhou, Y., Freedman, R., Glass, M.: Delivering Hints in a Dialog-Based Intelligent Tutoring System in Proceedings of 16th National Conference on Artificial Intelligence(AAAI-99), Orlando (1999)

[26] Woolf, B.: Theoretical Frontiers in Building a Machine Tutor in Kearsley, G (ed.), Artificial Intelligence and Instruction-Applications and methods Addison-Wesley (1987)

Model-Based Reinforcement Learning for Alternating Markov Games

Drew Mellor

School of Electrical Engineering and Computer Science
The University of Newcastle
Callaghan, 2308, Australia
Telephone: (+612) 4921 6034
Facsimile: (+612) 4921 6929
dmellor@cs.newcastle.edu.au

Abstract. Online training is a promising technique for training reinforcement learning agents to play strategy board games over the internet against human opponents. But the limited training experience that can be generated by playing against real humans online means that learning must be data-efficient. Data-efficiency has been achieved in other domains by augmenting reinforcement learning with a model: model-based reinforcement learning. In this paper the Minimax-MBTD algorithm is presented, which extends model-based reinforcement learning to deterministic alternating Markov games, a generalisation of two-player zero-sum strategy board games like chess and Go. By using a minimax measure of optimality the strategy learnt generalises to arbitrary opponents, unlike approaches that explicitly model specific opponents. Minimax-MBTD is applied to Tic-Tac-Toe and found to converge faster than direct reinforcement learning, but focussing planning on successors to the current state resulted in slower convergence than unfocussed random planning.

Keywords: Game playing, machine learning, planning, reinforcement learning, search.

1 Introduction

The field of reinforcement learning contains a broad group of methods for finding an optimal policy based on trial-and-error interaction with the problem domain [21]. Two classes of methods are the direct (model-free) and indirect (model-based) methods. Model-free reinforcement learning samples the problem domain directly but does not attempt to store the experience in a model. Model-based reinforcement learning, introduced by Sutton with the Dyna framework [17, 19, 20], saves the experience in a model, which it uses as another source of input into the underlying learning mechanism. The extra processing of modelled state transitions - called *planning* - accelerates convergence per interaction with the problem domain, and is particularly beneficial for systems where the cost of interacting with the environment is high relative to the computational

T.D. Gedeon and L.C.C. Fung (Eds.): AI 2003, LNAI 2903, pp. 520–531, 2003.

expense. In this paper, model-based reinforcement learning is extended to deterministic alternating Markov games, a generalisation of two-player zero-sum strategy board games like chess and Go.

Efficiency in reinforcement learning can be measured by the number of observations made from the environment (data-efficiency), or by the number of applications of the learning rule (computational-efficiency). In direct reinforcement learning, these measures reflect each other, but for model-based approaches they can be very different. In systems that can be modelled well, data-efficiency can be high as the number of observations required by the system is small. The computational efficiency though, will be equivalent to that of a direct reinforcement learning approach, since the computational requirements of the problem have not decreased - in fact model-based systems may apply the learning rule less optimally than direct approaches and can be less computationally efficient.

Strategy board games are a prime candidate for model-based methods because a model does not have to be learnt - the state transitions can be inferred from the game rules, and opponent's moves can selected by application of the minimax heuristic. Previously, data-efficiency has not been an issue when learning strategy board games, as the typical training method, self-play, generates training examples very cheaply. Self-play is an attractive method of training since it requires almost no domain specific knowledge, and because it has achieved spectacular success in the domain of backgammon [22, 4]. However, further research has shown that the nature of backgammon itself facilitates the use of self-play [14], and that the method performs poorly for deterministic games like chess [2].

A promising alternative to self-play, is to register the program on an Internet game server and train against real humans. This online training method has produced a strong chess player, KnightCap [2], (and also solves the problem of domain knowledge representation - by embodiment in human form). Interestingly, incremental training was achieved naturally, since as the program's rating improved it attracted higher caliber opponents.

Training against real opponents is more costly than training against simulated opponents however, and data-efficiency is a necessity. Whereas the backgammon programs could afford to learn from hundreds of thousands of training matches, and sometimes even millions, KnightCap had to make equivalent progress over only a few hundred matches. KnightCap was able to reduce convergence time by employing a knowledge intensive approach, particularly by initialising it's linear evaluation function approximator with good weight values. Finding a set of good initial weights is less likely to be the case for non-linear approximators, such as multi-layer perceptrons. Model-based techniques are a more general way of achieving data-efficiency.

The remainder of this paper is organised as follows. Section 2 reviews model-free methods for learning strategy board games, and ends with comments about their suitability as methods for doing planning backups in model-based approaches. Section 3 discusses some issues that arise when extending model-based methods to strategy board games, and presents a new algorithm Minimax-

MBTD. Section 4 describes some experiments where Minimax-MBTD was applied to Tic-Tac-Toe. The paper ends with some conclusions and ideas for further research.

2 Model-Free Reinforcement Learning for Strategy Board Games

The planning component of model-based reinforcement learning selects target states from the model, which are then backed up. A *backup* is an application of the underlying learning rule to a target state. Model-free methods are, in a sense, the trivial case for model-based methods when the number of planning backups is zero. This section presents two model-free algorithms that have been used to learn strategy board games, then discusses their suitability as methods for doing planning backups in model-based approaches. The first, TD(0), is based on the framework of Markov decision processes and has become very popular. The second and lesser known, Minimax-TD, is grounded in the framework of alternating Markov games.

Markov Decision Processes: A Markov Decision Process (MDP) [7] consists of a decision making agent operating within an environment. More precisely, it is a finite discrete set of states, $\mathcal{S}$, and controls, $\mathcal{A}$, a state transition function, $\mathcal{P}$, and a reward function, $\mathcal{R}$. At time t, the agent finds itself in state $s_t \in \mathcal{S}$, it chooses control $a_t \in \mathcal{A}$ and moves to state s_{t+1} with probability $\mathcal{P}^{a_t}_{s_t s_{t+1}}$, and is given some "reward" $\mathcal{R}^{a_t}_{s_t s_{t+1}}$. It is the agent's task to maximise the amount of reward it receives, that is, to find an optimal policy mapping states to controls that maximises the expected sum of discounted reward, $E(\sum_{t=0}^{\infty} \gamma^t \mathcal{R}^{a_t}_{s_t s_{t+1}})$. The discount factor, $\gamma \in [0, 1)$, ensures that the sum is bounded and also makes the agent prefer to be rewarded sooner rather than later.

The optimal value function, V^*, gives the total expected discounted reward for each state when following the optimal policy. It satisfies the following Bellman equation

$$V^*(s) = \max_a \sum_{s'} \mathcal{P}^a_{s,s'} [\mathcal{R}^a_{s,s'} + \gamma V^*(s')] \qquad (1)$$

which expresses a recursive relation between the optimal value of a state and it's successors. Once V^* is known, the optimal policy is found by choosing the control, a, that satisfies (1) given the state, s.

Temporal Difference Learning: A widely used algorithm for finding the optimal value function is TD(0), the method of temporal differences [18]. The agent maintains a table, $\hat{V}$, that stores an estimate of V^*, the optimal value function. It samples the state space according to the estimate of the optimal policy derived from $\hat{V}$, and at each time step, t, the algorithm shifts the estimate of the value function for the current state, $\hat{V}(s_t)$, to be more consistent with the sum of the immediate reward, r_t, and the discounted estimate of the value of the next state, $\hat{V}(s_{t+1})$, as expressed in the following *temporal difference* rule

$$\hat{V}(s_t) \leftarrow \hat{V}(s_t) + \alpha(r_t + \gamma\hat{V}(s_{t+1}) - \hat{V}(s_t)) \tag{2}$$

where $\alpha \in [0,1]$ is a learning rate. It can be shown that the TD(0) algorithm converges to an optimal policy given that all states are visited infinitely often [18]. In order to ensure that all states are adequately explored, at every time step a control is chosen uniformly at random with some small probability, ϵ, the exploration threshold.

The method can be applied to sequential games by treating the opponent as part of the environment, or more precisely, as part of the state transition function $\mathcal{P}$. For each training match played, the sequence of positions for the learning player is observed, $s_1, s_2, \ldots, s_N$, where s_i is the position of the game immediately after the learning agent has made their ith move, and N is the total number moves that they played during the match. Starting with s_1, the temporal difference rule (2) is applied to each state in turn, with the exception of s_N, when $\hat{V}(s_{t+1})$ is replaced with zero. The reward function is usually determined by the win-loss relation, for example, the following reward function is commonly used for many games

$$r_t = \begin{cases} 1 \; s_t \text{ wins} \\ -1 \; s_t \text{ looses} \\ 0 \text{ otherwise.} \end{cases}$$

For some games, such as those that involve capturing territory, like Go and Othello, a more expressive reward can be given at the conclusion of the match based on the size of the territory captured or the piece difference between the players.

Alternating Markov Games: The domain of alternating Markov games [10], sometimes called sequential games, generalises MDPs to capture the essence of strategy board games like chess, Go and backgammon.

The domain has a discrete, finite set of states, $\mathcal{S}$, and two opponents, agent 1 and agent 2, which in each turn of the game alternately choose controls from their respective control sets, $\mathcal{A}_1, \mathcal{A}_2$. The sequence of events in one complete turn is shown by $\ldots s \overset{u\in\mathcal{A}_1}{\rightarrow} x \overset{v\in\mathcal{A}_2}{\rightarrow} s' \ldots$, a portion of a trajectory through a hypothetical match. For any turn, the probability that it ends in state s' given the initial state s, and the control choices u and v, is $\mathcal{P}^{u,v}_{s,s'} = \sum_x \mathcal{P}^u_{s,x}\mathcal{P}^v_{x,s'}$ where $\mathcal{P}^u_{s,x}$ and $\mathcal{P}^v_{x,s'}$ are the transitional probabilities for the state transitions occurring during the turn. Let the corresponding reward for each turn be $\mathcal{R}^{u,v}_{s,s'}$, which one agent attempts to maximise and the other agent attempts to minimise, resulting in diametrically opposed goals for the two agents.

Like Markov Decision Processes, the agent's aim is maximise (minimise) the expected sum of discounted reward, $E(\sum_{t=0}^{\infty} \gamma^t \mathcal{R}^{u_t,v_t}_{s_t s_{t+1}})$. But this time there is a complication, since the accrued reward depends not only on the agent's choice of controls, but the opponent's strategy too. Which opponent strategy should be used to determine the accumulated reward? Typically, the resolution is to use the opponent that makes the agent's policy look the worst. This choice leads to a conservative measure of optimality, where strategies that consistently draw are preferred to strategies that give big wins but give big losses too. Using this "minimax" definition of optimality, the Bellman equation for alternating Markov games is given by

$$V^*(s) = \max_u \min_v \sum_{s'} \mathcal{P}^{u,v}_{s,s'}[\mathcal{R}^{u,v}_{s,s'} + \gamma V^*(s')]$$

assuming agent 1 is the maximiser (if agent 1 is the minimiser then negate the reward function $\mathcal{R}$). For deterministic games there is only one successor for every state, so (3) can be simplified by removing the summation, as follows

$$V^*(s) = \max_u \min_v [\mathcal{R}^{u,v}_{s,s'} + \gamma V^*(s')]. \tag{3}$$

A review of solution methods for alternating Markov games can be found in [10, 3].

Minimax Temporal Difference Learning: I now present Minimax-TD, an algorithm for solving deterministic alternating Markov games. Like the TD(0) algorithm, the agent maintains a table, $\hat{V}$, that stores an estimate of the optimal value function, and is used to generate it's policy. For simplicity, in this work the policy will be generated by greedily selecting the successor position with the best estimated value[1], but there is nothing to prevent the use of game searches over $\hat{V}$. The value function estimate is improved over a series of training matches as follows. Let s_t be the board position directly after the agent has made their tth move. At each turn in the game, t, the agent applies the consistency relation (3) to $\hat{V}(s_t)$, using the following backup rule

$$\hat{V}(s_t) \leftarrow \hat{V}(s_t) + \alpha(r_t + \gamma \max_u \min_v \hat{V}(T(s_t, u, v)) - \hat{V}(s_t)), \tag{4}$$

where $T(s, u, v)$ is the transition function that gives the position succeeding s when the opponent selects control v then the agent chooses control u, with the proviso that u can be a *null move* if the opponent's move v has ended the match. The assumption that the game is deterministic is required to ensure that T returns a unique move. Minimax-TD has been applied to Othello [23], and generalised for Markov Games with simultaneous turns [9, 6].

Comparison of TD(0) and Minimax-TD: This section presented two model-free algorithms that have been used to learn strategy board games. The first, TD(0), forms strategies that are predictive and exploit weaknesses in the opponent's play, but the flip side is that they are biased by the training partner and may not generalise well to arbitrary opponents.

[1] Technically, I use ϵ-greedy policies, which ensure that positions estimated to be sub-optimal, perhaps incorrectly, are also occasionally visited.

The second, Minimax-TD, is based on a formalisation of strategy board games called alternating Markov games. It uses a minimax measure of optimality, resulting in conservative strategies that assume the opponent will make optimal moves irrespective of their actual behaviour. It's advantage is that it plays well against arbitrary opponents, therefore this rule will be used for the planning backups of Minimax-MBTD.

3 Model-Based Reinforcement Learning for Strategy Board Games

To extend model-based reinforcement learning to strategy board games, some further issues must be considered. Model-based reinforcement learning relies on a model of the environment to predict the state transitions. In strategy board games, the state transitions depend on the behaviour of the opponent as well as the rules of the game. The game rules are completely known to the agent, but the behaviour of the opponent usually isn't. So the first issue is: how should the opponent to be modelled?

The planning component of model-based methods involves a selection process, which chooses target states to be backed up. Careful choice of target states can increase computational efficiency. The second issue is: which game positions should be selected as backup targets?

The remainder of this section examines these two issues, and proposes answers that will form the basis for a new algorithm Minimax-MBTD.

Modelling the Opponent: An opponent model can be acquired by *model-learning*, that is, by observing the frequency of the opponent's moves for each state. For example, given a set of board positions, the opponent's decision in each position, and a feature decomposition of the game, a set of constraints can be constructed over the co-efficient vector for the features, which can be solved using linear programming techniques [5]. We do not pursue this approach because it "overfits" to the training opponent - when playing against new opponent, the model has to be re-learnt.

A more general approach would be to model the opponent as an optimal player. Since the optimal strategy is not known it must be estimated, for example by using the minimax heuristic [16]. If our evaluation function reflects the true game theoretic values, then the heuristic actually gives opponent's optimal response. Of course, at the beginning of training, the evaluation function estimate will be far from reflecting the true game theoretic values, leading to suboptimal predictions for the opponent's moves. As training continues and the evaluation function improves, the predictions should improve. The current research focuses on this approach.

Focussing Planning Backups: The simplest way to select target states for the planning component would be to choose them at random. The Dyna framework is an example of this approach [17, 19, 20]. Unsurprisingly, random selection does not make best use of the model, and better performance has been reported for methods that employ more focussed approaches to selection, such as prioritized sweeping [11], Queue-Dyna [13] and trajectory sampling [8].

The trajectory sampling method looks forward along the trajectory given by the current estimate of the optimal policy, and selects the next k successor states. These successor states then form the target states for planning, and are backed up in reverse order, i.e. furthest away first. The heuristic behind this approach is that attention is focused on states that are likely to be encountered in the near future. The current research generalises the trajectory sampling method by focusing on *all* the k successor states, covering future possibilities more broadly. This way of selecting target states for the planning backups is inspired by suggestions in Sutton and Barto's book ([21], see Section 9.7).

The Minimax-MBTD Algorithm: I now give the Minimax-MBTD algorithm (see Figure 1). For each position encountered during training, the algorithm generates the tree of k successors to use as targets for planning (not counting transpositions and the opponent's positions), then backs them up using Minimax-TD. Within a layer, sibling successors are ordered by their evaluation function estimate so that positions with higher estimated importance are selected before positions with a lower estimate (see Figure 2). The order of backups is from the leaves towards the root, which propagates the leaf values towards the root. Note that when $k = 0$ the algorithm reduces to a model-free version of the algorithm, which is Minimax-TD.

- Let $s_1, s_2, \ldots, s_N$ be the N positions occurring during the match after each of the agent's moves. For $t = 1, 2, \ldots, N$ let r_t be the reward corresponding to position s_t.
- Let $H(s_t)$ be the game tree rooted at s_t after all transpositions are removed, and siblings are ordered s.t. if x and y are siblings and they are positions after the agent has moved and $\hat{V}(x) > \hat{V}(y)$, then x is to the left of y; else if x and y are siblings and they positions after the opponent has moved and $\hat{V}(x) < \hat{V}(y)$, then x is to the left of y.
- Let s_t^l be the lth node in $H(s_t)$, where nodes are numbered in the same order as visited by a breath-first search, the opponent's positions are not counted, and $s_t^0 = s_t$.
- For $t = 1$ to N do
  - Generate $H(s_t)$ to depth(s_t^k)
  - For $l = k$ downto 0 do
    - $\hat{V}(s_t^l) \leftarrow \hat{V}(s_t^l) + \alpha(r_t + \gamma \max_u \min_v \hat{V}(T(s_t^l, u, v)) - \hat{V}(s_t^l))$

Fig. 1. The Minimax-MBTD algorithm

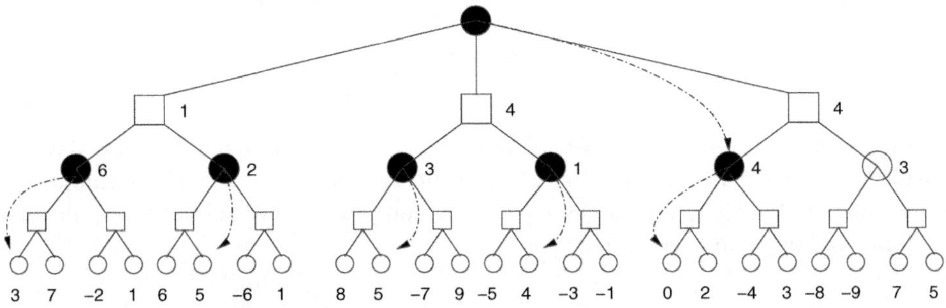

Fig. 2. A partial game tree generated by Minimax-MBTD when $k = 5$. The target positions for planning are shown as filled black circles and the backups are indicated by dashed arrows. The values next to the nodes are returned by the value function estimate, that is, they have not been propagated up the tree from the leaves like a minimax search would do

4 Experiments

In this section I apply Minimax-MBTD to Tic-Tac-Toe, also called noughts and crosses. The aim is firstly to see if the model-based approach will converge faster than direct reinforcement learning in a strategy board game domain, and secondly to see if there is an advantage to focussing the planning steps on successors to the current state. In all experiments, the learning rate, α, was set at 0.2 and annealed during training; the discount factor, γ, was set at 1 (no discounting); the exploration threshold, ϵ, was set at 5%; and the table entries were initialised to random values drawn uniformly from the interval [0.1,-0.1].

Generalisation techniques (such as neural networks) were not used because Minimax-TD is an off-policy rule. *On-policy* rules sample the problem domain according to the current estimation policy, whereas *off-policy* rules sample using a different distribution. TD(0) is an on-policy rule, and is frequently combined with neural networks when learning board games, but off-policy rules are not guaranteed to converge when used with function approximation [15], in fact simple problems exist for which they never converge [1]. Tic-Tac-Toe has 6045 unique legal positions which is easily small enough to store in memory, and to explore without using a generalisation method, so using tables avoided complications arising from the combined use of function approximation and off-policy backups.

Training and Evaluation: Training was by self-play, with the two opponents alternating as the first player every match. Every 1,000 matches a set of 100 evaluation matches were played against two fixed strategy challengers, during which learning was switched off, and no exploratory moves were made. After each evaluation set is played, a match equity score is computed as follows

$$\text{equity}(e_i) = \frac{\text{wins}(p, e_i) - \text{wins}(c, e_i)}{\text{total}(e_i)}$$

where e_i is the ith evaluation set, total(e_i) is the total number of games played during e_i - always 100 in the current experiments, wins(x, e_i) is the number of wins counted during e_i for player $x \in \{p, c\}$, and p and c are the learning player and the challenger respectively.

Challengers: The first challenger is a *semi-random player*, who firstly tries to complete a winning line, then secondly to block the opponent from winning on the next turn; or if neither of these options are possible, it chooses a move from all legal candidates uniformly at random. The semi-random challenger will cover the state space broadly due to the randomness in its decision making.

The other challenger is an *heuristic player*, who selects from all the immediate legal candidate moves, that which maximises the following *max-lines* heuristic [12]. All of the eight groupings of three cells lying in a straight line are examined, and the number of friendly singlets, S_f, and doublets, D_f; and enemy singlets, S_e, and doublets, D_e, are counted. A singlet contains one marker only, a doublet contains two by the same player only. In addition, if a Tic-Tac-Toe is found a flag is set. An evaluation for the board position, s, is computed as

$$V(s) = \begin{cases} 1 & \text{Tic-Tac-Toe made} \\ -1 & D_e > 0 \\ \frac{2D_f + Sf - S_e}{6} & \text{otherwise} \end{cases}$$

If more than one move has the greatest evaluation, then one of those moves is made at random. The heuristic challenger plays a strong game, although it can be beaten as the second player by constructing a fork.

Results: The first experiment compares the model-based approach ($k > 0$) with the model-free ($k = 0$). The results of the experiment are shown in Figure 3. Against both challengers all model-based approaches converge by about 5,000 training matches, whereas for the model-free case it is closer to 10,000 matches. Interestingly, better strategies were also found as the number of planning steps increases.

To test whether focused backups are better than unfocused, the previous experiment was repeated, only this time the target positions for the planning steps were selected uniformly at random from all legal positions. The results are shown in Figure 4. In all cases convergence occurs by 1000 training matches, much faster than the focused case. In addition, the strategies learnt are also more optimal than those found using focussed planning.

The strong performance of the unfocussed approach is due to the size of the problem domain relative to number of planning backups. The state space of Tic-Tac-Toe is smaller than the total number of planning backups, therefore

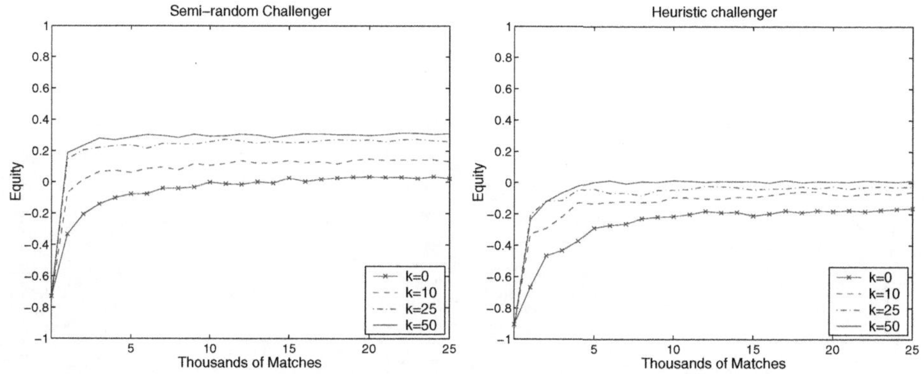

Fig. 3. Learning curves for model-based Minimax-MBTD ($k > 0$) versus model-free Minimax-TD ($k = 0$). Each curve is the average of 100 training episodes

a random distribution of states will cover it more completely than the distribution focused on successors to the current state. However, focussed approaches are likely to scale better, because the state space cannot be covered completely for larger games, and a heuristic is needed to restrict attention to the more "interesting" states.

5 Conclusions

This paper outlines a way of extending model-based reinforcement learning to strategy board games, and presents Minimax-MBTD, an algorithm based on these ideas. A key finding is that it converges faster than direct methods per observation, which is consistent with other studies of model-based methods. More interestingly, as the number of planning steps increased, the final strategy improved too. Unfocussed backups were found to perform better than focused, nevertheless focussed approaches may scale better to larger games.

Future work could concentrate on scaling up to larger games. The comparatively small state space of Tic-Tac-Toe meant that the entire value function could be stored in a lookup table, however, most games have too many states for a table based approach to work. The issue is deeper than the limit that physical memory places on the size of the evaluation function; more importantly, it is the need to recognise when new situations resemble previously encountered ones, so that best use can be made of limited training experience. Therefore, when scaling up to games with larger state spaces it is typical to use a generalisation technique, such as a neural network.

Minimax-TD is off-policy and therefore could be susceptible to the divergence frequently observed for the combination of off-policy methods and function approximation [1]. An intuitive explanation for the divergence is that approximator error is compounded by the min and max operators of Minimax-TD (or just max

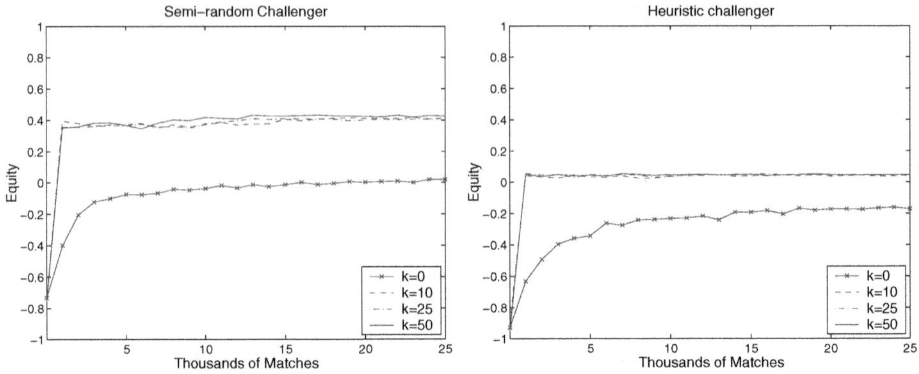

Fig. 4. Learning curves when the k target positions for planning are selected at random. Each curve is the average of 100 training episodes

in the case of Q-Learning), propagated back to the approximator in the training signal, where it causes further approximation error, and so on, ultimately leading to divergence. More optimistically, Minimax-TD has been combined with non-linear gradient descent RBF networks to train an Othello playing program [23], though my preliminary efforts at combining Minimax-TD with backpropagation MLP networks for Tic-Tac-Toe have been discouraging.

References

[1] Leemon C. Baird. Residual algorithms: Reinforcement learning with function approximation. In *Proceedings of the 12th International Conference on Machine Learning*, pages 30–37. Morgan Kaufmann, 1995. 527, 529

[2] Jonathan Baxter, Andrew Tridgell, and Lex Weaver. TDLeaf(λ): Combining temporal difference learning with game-tree search. *Australian Journal of Intelligent Information Processing Systems ISSN 1321-2133*, 5(1):39–43, 1998. 521

[3] D. P. Bertsekas and J. N. Tsitsiklis. *Neuro-Dynamic Programming*. Athena Scientific, 1996. 524

[4] Justin A. Boyan. Modular neural networks for learning context-dependent game strategies. Master's thesis, University of Cambridge, 1992. 521

[5] David Carmel and Shaul Markovitch. Model-based learning of interaction strategies in multiagent systems. *Journal of Experimental and Theoretical Artificial Intelligence*, 10(3):309–332, 1998. 525

[6] Fredrik A. Dahl and Ole Martin Halck. Minimax TD-learning with neural nets in a markov game. In Ramon López de Mántaras and Enric Plaza, editors, *Proceedings of the 11th European Conference on Machine Learning*, pages 117–128. Springer-Verlag, 2000. 524

[7] Ronald A. Howard. *Dynamic Programming and Markov Processes*. The MIT Press, Cambridge, MA, 1960. 522

[8] L. Kuvayev and R. Sutton. Model-based reinforcement learning with an approximate, learned model. In *Proceedings of the Ninth Yale Workshop on Adaptive and Learning Systems*, pages 101–105, 1996. 526

[9] Michael L. Littman. Markov games as a framework for multi-agent reinforcement learning. In *Proceedings of the 11th International Conference on Machine Learning (ML-94)*, pages 157–163, New Brunswick, NJ, 1994. Morgan Kaufmann. 524

[10] Michael L. Littman. *Algorithms for Sequential Decision Making.* PhD thesis, Brown University, 1996. 523, 524

[11] Andrew W. Moore and Christopher G. Atkeson. Prioritized sweeping: Reinforcement learning with less data and less time. *Machine Learning*, 13:103–130, 1993. 526

[12] Daniel Kenneth Olson. Learning to play games from experience: An application of artificial neural networks and temporal difference learning. Master's thesis, Pacific Lutheran University, Washington, 1993. 528

[13] J. Peng and R. J. Williams. Efficient learning and planning within the dyna framework. In *Proceedings of the 2nd International Conference on Simulation of Adaptive Behavior, Hawaii*, 1993. 526

[14] Jordan B. Pollack and Alan D. Blair. Co-evolution in the successful learning of backgammon strategy. *Machine Learning*, 32(1):225–240, 1998. 521

[15] Doina Precup, Richard S. Sutton, and Sanjoy Dasgupta. Off-policy temporal-difference learning with function approximation. In *Proc. 18th International Conf. on Machine Learning*, pages 417–424. Morgan Kaufmann, San Francisco, CA, 2001. 527

[16] C. Shannon. Programming a computer for playing chess. *Philosophical Magazine*, 41(4):256–275, 1950. 525

[17] R. S. Sutton. Integrated architectures for learning, planning, and reacting based on approximating dynamic programming. In *Proceedings of the Seventh International Conference on Machine Learning*, pages 216–224. Morgan Kaufmann, 1990. 520, 526

[18] Richard S. Sutton. Learning to predict by the method of temporal differences. *Science*, 3(1):9–44, 1988. 522, 523

[19] Richard S. Sutton. DYNA, an Integrated Architecture for Learning, Planning and Reacting. In *Working Notes of the AAAI Spring Symposium on Integrated Intelligent Architectures*, pages 151 155, 1991. 520, 526

[20] Richard S. Sutton. Planning by incremental dynamic programming. In *Proceedings of the Eighth International Workshop on Machine Learning*, pages 353–357. Morgan Kaufmann, 1991. 520, 526

[21] Richard S. Sutton and Andrew G. Barto. *Reinforcement Learning: An Introduction.* The MIT Press, 1998. 520, 526

[22] G. Tesauro. TD-Gammon, a self-teaching backgammon program, achieves masterlevel play. *Neural Computation*, 6(2):215–219, 1994. 521

[23] Taku Yoshioka, Shin Ishii, and Minoru Ito. Strategy acquisition for the game "Othello" based on reinforcement learning. *IEICE Transactions on Information and Systems E82-D*, 12:1618–1626, 1999. 524, 530

HLabelSOM: Automatic Labelling of Self Organising Maps toward Hierarchical Visualisation for Information Retrieval

Hiong Sen Tan

University of South Australia
Mawson Lakes SA 5095, Australia
tan@cs.unisa.edu.au

Abstract. The self organising map technique is an unsupervised neural network that is able to cluster high-dimensional data and to display the clustered data on a two-dimensional map. However, without the help of a visualisation technique or labels assigned by an expert, users without some prior knowledge will find it difficult to understand the clusters on the map. In this paper, we present the HLabelSOM method to automatically label the self organising map by utilising the features the nodes on the map hold after the training process. Users will be able to see the clusters on the labelled map since the neighbouring nodes have similar features and the labels themselves reveal the cluster boundaries based on the common features held by neighbouring nodes. Further, the HLabelSOM method produces several maps that can be utilised to create hierarchical visualisation for information retrieval. We demonstrate the applicability of the HLabelSOM method in mining medical documents from the Internet and visualising the information in hierarchical maps.

1 Introduction

The self organising map (SOM), proposed by Kohonen [1], is a clustering technique with the ability to map data from high-dimensional input space into low-dimensional output space, usually two-dimensional map. This SOM technique has been used in the information retrieval arena, not only because of its ability to preserve, as faithfully as possible, the relation of data in the input space, but also because its two-dimensional output map can be utilized to visualise the information for retrieval purposes. Lin, et al. [2] are probably the pioneers, based on our literature review, of using SOM in information retrieval setting. Later work [3-5] shows several researchers and their research groups following Lin's path, and indicates the acceptance of using SOM in information retrieval.

The ability of the SOM to cluster data without supervision is undoubted, but the ability of a user to understand the clusters on the map is depend on the user's prior knowledge of the input data. Users without some prior knowledge of the input data

T.D. Gedeon and L.C.C. Fung (Eds.): AI 2003, LNAI 2903, pp. 532-543, 2003.

will find difficult to figure out the similarities and dissimilarities of the clusters. Experts usually assign labels to the map to help the users to understand the clusters.

Several visualisation techniques have been proposed to detect the cluster boundaries of the SOM. U-matrix [6] detects the cluster boundaries based on the distances between neighbouring nodes and produces a three-dimensional landscape where the valleys indicate similarities in the input data and the altitudes show dissimilarities. The Cluster Connection technique [7] uses a similar method to U-matrix where the distances among weight vectors and a set of thresholds are used to determine whether the nodes belong to the same clusters and should be connected. The Adaptive Coordinate technique [8] creates an additional map to capture the movement on the weight vectors during the training process. The clusters on this additional map identify the clusters on the output map.

The examples of visualisation techniques mentioned above certainly enhance the usability of the SOM. However, our interest is to automatically label the SOM by utilising the characteristic the nodes hold after the training process. The labels themselves then form the clusters which can be seen by the users by noticing the similarities of the features the neighbouring nodes hold and the common features shared by the neighbouring nodes. Lin, et al. [2] actually labelled the map automatically, and Rauber and Merkl proposed the LabelSOM method [3, 9] to automatically label the map. In this paper, we propose a novel method for automatic labelling nodes of the SOM, which can also be used for hierarchical visualisation in information retrieval.

The remainder of the paper is organised as follows: In section 2, we give a brief introduction of the SOM, its architecture and its algorithm. We apply the algorithm to cluster a well-known animal data set and show the difficulty in understanding the clusters if the users do not have prior knowledge of the input data. Next, in section 3, we briefly discuss the previous automatic labelling methods, Lin's method (to refer the automatic labelling method used by Lin et al.) and LabelSOM, and present the results of applying these techniques on animal data set before we introduce our method to automatically label the map and produce hierarchical maps. We apply the HLabelSOM method to a medical document collection, as an example of text mining application, in section 4. In section 5, we present the user interface for information retrieval created by utilising the hierarchical maps that HLabelSOM produces. Finally, the conclusion is presented in section 6.

2 The Self Organising Map

The SOM is an artificial neural network that is able to perform a data clustering. One of the most important aspects of the SOM is that it is primarily a visualisation method for the clustering. Unlike statistical methods or other ways of grouping data, the SOM provides a topological ordering where the relationships between data items are made apparent. In fact, relationships between data of high-dimensionality are reduced to relationships on a two-dimensional surface.

The SOM (Fig. 1) has a two-dimensional matrix of processing elements (the output layer) and a scalar array of processing elements (the input layer). Each node in the output layer is connected to every node in the input layer by weights. A training process is necessary to change the initially random weights into values that respond to input data with minimum distance[1)] (see below). The two-dimensional output map effectively orders the input data once the training is complete.

The self organising map algorithm is as follows:

1. Initialise the weights from the nodes in the output layer to each node in input layer with small, usually random, values.
2. Present random choice from the training data to input nodes.
3. Compute the distances from the input to all output nodes.
4. Select a winning node that is the output node that has the minimum distance[1)] to the input.
5. Update the weights of the winning node and its neighbours.
6. Repeat step 2 to step 5 until the map converges or for a number of training cycles.

The distances of the input to the output nodes are computed based on the Euclidean distance function:

$$D_j = \sqrt{\sum_{i=0}^{N-1} \left(X_i(t) - W_{ij}(t)\right)^2} \tag{1}$$

N is the number of nodes in the input layer, $X_i(t)$ is the input to node i at time t and $W_{ij}(t)$ is the weight from input node i to output node j at time t.

The weights of the winning node and its neighbours are updated by the following function:

$$W_{ij}(t+1) = W_{ij}(t) + \eta(t)(X_i(t) - W_{ij}(t)) \tag{2}$$

$\eta(t)$ is the learning rate at time t. The neighbourhood of the winning node shrinks in size over time. We use a simple topological neighbourhood function, as shown in Fig. 2, to reduce the size of the neighbourhood over time. The details of the SOM algorithm can be found in [10], and the practical version in [11].

To show the capability of the SOM to cluster data we use a well-known animal data set, which is also used in [9, 12]. The data set consists of 16 animals with 13 features to describe the animals, is given in Table 1, and used as the input of 3 by 3 SOM. The 16 inputs - which are actually 14 distinct inputs since horse and zebra, and hawk and owl have the same features - have to be fitted onto a map with 9 nodes.

After the training of the SOM is complete, all the inputs are mapped to the nodes on the map. Each input is mapped to a node that has minimum distance from it. The users with some prior knowledge of the input data can easily figure out the clusters on the result map, given in Fig. 3. For examples, animals that have two legs (owl, hawk, eagle, dove, hen, duck, and goose) are at the top of the map, and animals that have four legs (cat, fox, dog, wolf, tiger, lion, zebra, horse and cow) are at the bottom of the map. In the four-leg-animal cluster, there are three sub-clusters: small size animal (cat), medium size animals (fox, dog and wolf) and big size animals (tiger, lion, horse, zebra, and cow). Further, in the big-size-animal cluster the animals that like to hunt (tiger and lion) are separated from the animals that do not (horse, zebra and cow).

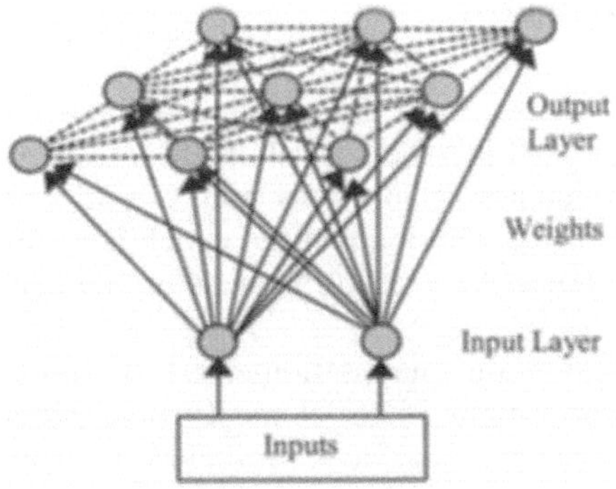

Fig. 1. Architecture of the Self Organising Map. The network consists of two layers: input layer and output layer. Each node in output layer is connected to every node in input layer by weights

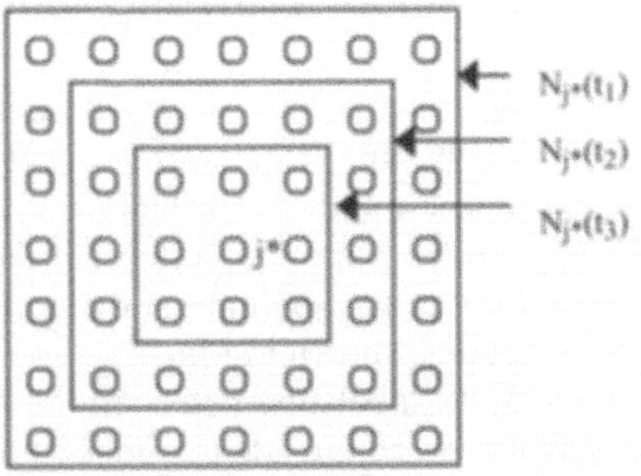

Fig. 2. The topological neighbourhood, where j* is the winning node and $N_{j*}(t)$ is a group of nodes considered as the neighbours of j* including j* at a discrete time t (shown within a square). The size of neighbourhood decreases in time

However, how will the users without prior knowledge of the animals mentioned above figure out the clusters on the map? What is the similarity between tiger and lion that makes them in the same cluster? What is the similarity between tiger-and-lion cluster and horse-zebra-and-cow cluster so that these two clusters are neighbours?

Table 1. The animal data set. The row has value 1 if the animal has the particular feature and value 0 otherwise

Attribute		Dove	Hen	Duck	Goose	Owl	Hawk	Eagle	Fox	Dog	Wolf	Cat	Tiger	Lion	Horse	Zebra	Cow
Is	small	1	1	1	1	1	1	0	0	0	0	1	0	0	0	0	0
	medium	0	0	0	0	0	0	1	1	1	1	0	0	0	0	0	0
	big	0	0	0	0	0	0	0	0	0	0	0	1	1	1	1	1
Has	2 legs	1	1	1	1	1	1	1	0	0	0	0	0	0	0	0	0
	4 legs	0	0	0	0	0	0	0	1	1	1	1	1	1	1	1	1
	hair	0	0	0	0	0	0	0	1	1	1	1	1	1	1	1	1
	hooves	0	0	0	0	0	0	0	0	0	0	0	0	0	1	1	1
	mane	0	0	0	0	0	0	0	0	0	1	0	0	1	1	1	0
	feathers	1	1	1	1	1	1	1	0	0	0	0	0	0	0	0	0
Likes to	hunt	0	0	0	0	1	1	1	1	0	1	1	1	1	0	0	0
	run	0	0	0	0	0	0	0	0	1	1	0	1	1	1	1	0
	fly	1	0	0	1	1	1	1	0	0	0	0	0	0	0	0	0
	Swim	0	0	1	1	0	0	0	0	0	0	0	0	0	0	0	0

3 Automatic Labelling and HLabelSOM

Before we introduce our proposed method, we will examine the previous methods for automatic labelling, Lin's method and LabelSOM. Lin's method labels the nodes by comparing each node to all feature vectors, which are vectors consisting of only one feature, and labelling the node with the name of the winning feature, which is a feature vector that has the closest distance to the node. As a result the areas on the map are continuous, shown in Fig. 4.

LabelSOM labels the nodes on the map based on the quantisation errors of all features that are accumulated distance between the weight vector elements of all input mapped to the nodes. The quantisation errors that are close to 0 or below a given threshold (λ_1) indicate the features best characterise the input mapped to the nodes. Further, we need another threshold, especially for the applications where we find a high number of input vectors that have values of 0, e.g. text documents. These zero values mean the features are not present in the particular input. The features selected should have weight vectors above the second threshold (λ_2). Fig. 5 shows the result map when LabelSOM is used with λ_1=0.05 and λ_2 = 0.4.

Lin's method and LabelSOM can help the users to understand the clusters, but each of them has its own drawback. Lin's method labels the nodes with fewer features on average than LabelSOM. The more features on the node, the easier the understanding of the inputs mapped to the node. It makes it easier for the users to understand the input data represented by the nodes and will be useful especially in information retrieval to find the specific information needed. The drawback of the LabelSOM method is that we have to define the values of the thresholds.

When applying the LabelSOM, we found out that for binary input vectors, at least in this particular experiment setting, actually the node is labelled by all features of the input document if only one input is mapped to the node and by the common features of the input documents if more than one input mapped to the node. In this case, we do not need to calculate the accumulated quantisation error and define the thresholds.

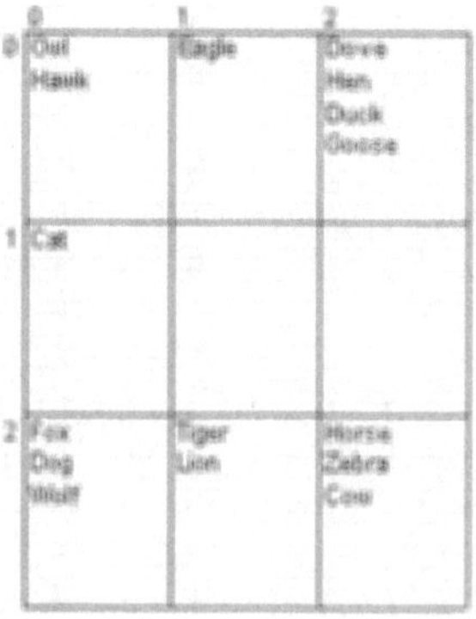

Fig. 3. The result map of using standard representation of the SOM. After training is complete all inputs are mapped to the nodes on the map

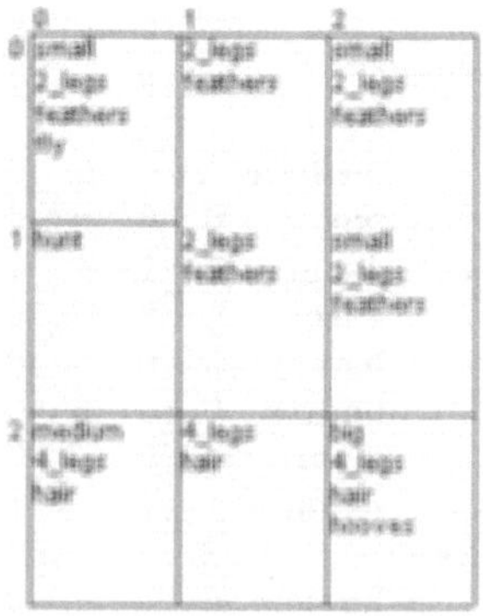

Fig. 4. The map is labelled by using Lin's method. The areas on the map show continuity

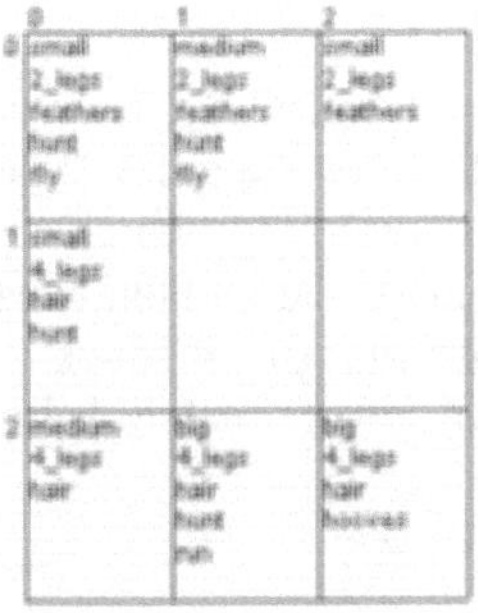

Fig. 5. The map is labelled by using the LabelSOM method. Two nodes are not labelled since there are no inputs mapped to the node

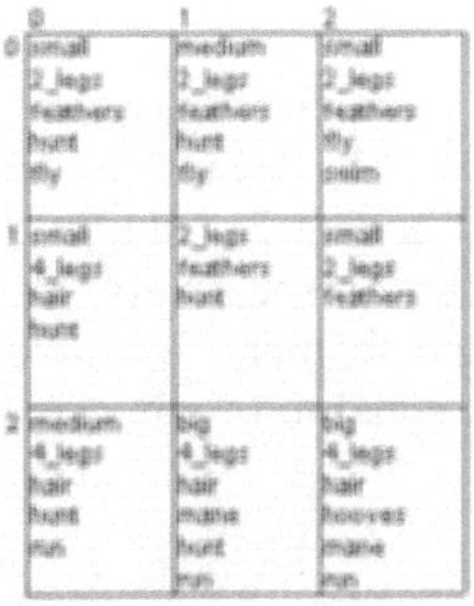

Fig. 6. The detail map. The map is labelled by using HLabelSOM method. The nodes are labelled by the features that have weight values >= 0.5

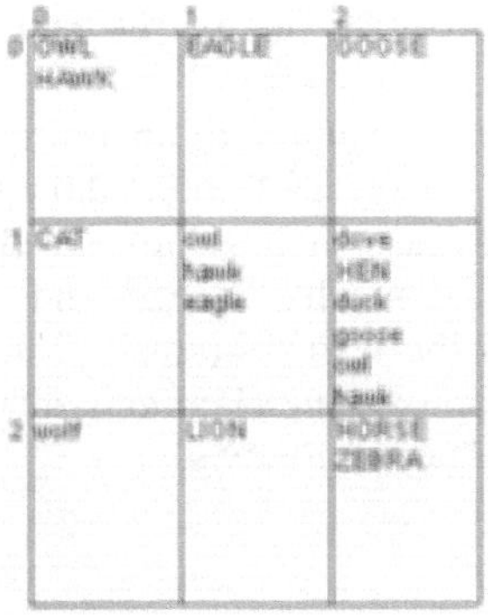

Fig. 7. The inputs are mapped onto the map in Fig. 6. Animal names with capital letters represents exact matches, otherwise approximate matches

Now, we introduce our method, which is very simple, for labelling the map. The nodes are labelled with the features if the node weight value for the particular feature is greater or equal to 0.5. The selection of value 0.5 as a threshold is based on the fact that the document vectors are binary vectors that are either 0 or 1. If after training the weight value is above or equal to 0.5 that means closer to 1 than to 0, the node should have the feature, otherwise it should not. The result map of labelling the SOM using HLabelSOM is shown in Fig. 6.

By using the HLabelSOM, we obtain a result map with richer features on the nodes in average than that using LabelSOM and Lin's method. We also do not need to define the threshold as in LabelSOM since we use a fixed value 0.5. But, using HLabelSOM some nodes are labelled with features that are absent in the inputs mapped to these particular nodes. For example, in Fig. 6, node(2,2) is labelled with big, 4_legs, hair, hooves, mane and run, but from Fig. 3 we know that besides horse and zebra, which have all these features, cow is also mapped to this node. However, cow does not have a mane and does not like to run. Here is the significant different between HLabelSOM and the previous methods, especially LabelSOM. In LabelSOM, the inputs are mapped to the nodes before the nodes are labelled. In HLabelSOM, the nodes are labelled first, then the inputs are mapped to the nodes if the nodes and the inputs have exactly the same features (exact match) or at least all the features on the nodes are present on the inputs (approximate match).

An exact match is defined as having a Hamming distance between the input and the node equal to 0. An example of an exact match is between 'hen' and node(1,2), where both have vectors that contain [1,0,0,1,0,0,0,0,1,0,0,0,0], which represents features 'small', '2_legs', and 'feathers'. The Hamming distance function is as follows.

$$D_H(A,B) = \sum_{i=1}^{N} |A_i - B_i| \tag{3}$$

N is the number of features in the vector, A is the document input vector and B is the node vector.

An example of an approximate match is between 'duck' and node(1,2). The vector of 'duck' is [1,0,0,1,0,0,0,0,1,0,0,0,1], which represents features 'small', '2_legs', 'feathers' and 'swim', and that of node(1,2) is [1,0,0,1,0,0,0,0,1,0,0,0,0], which represents 'small', '2_legs', and 'feathers', respectively. It is an approximate match because all the features (small, 2_legs and feathers) on node(1,2) are present in 'duck'. The Approximate Hamming distance equal to 0 indicates the approximate match. It is calculated as follows.

$$D_{AH}(A,B) = \sum_{i=1}^{N} \begin{cases} 1, \text{ if } B_i > A_i \\ 0, \text{otherwise} \end{cases} \tag{4}$$

The result of mapping the inputs onto the labelled map in Fig. 6 is shown in Fig. 7. The animals named in capital letters show the exact matches, the others are approximate matches. The map labelled by using HLabelSOM has more exact matches, 9 exact matches (from labelled map in Fig. 6 and animal map in Fig. 7), than the map labelled by using LabelSOM, which has 3 exact matches (from labelled map in Fig. 5 and animal map in Fig. 3) and than the map labelled by using Lin's method, which has no exact match at all (from labelled map in Fig. 4 and animal map in Fig. 3). The more exact matches, the better the map, because the nodes on the map exactly represent the features on the inputs mapped to the node. The exact matches make it easier for the users to find specific data.

We also find that several animals, i.e., fox, dog, and cow, are missing from the map in Fig. 7. This problem will be solved with the hierarchical characteristic of the HLabelSOM method where more general maps are produced so that all inputs can be mapped to the nodes either with exact matches or at least approximate matches.We refer to our novel method as the HLabelSOM, where H is stand for Hierarchical since our purpose is to have a hierarchical visualisation. We believe creating user interface that follows the information visualisation mantra: "overview first, zoom and filter, then details on demand" [13] can lead to an effective information retrieval system. A variation of the SOM, the HSOM [14, 15], will be able to achieve this purpose as well. But, using HSOM will require several trainings for different SOMs. We can also produce hierarchical maps by combining four adjacent nodes on the map into one node and naming the new node with the common features the four nodes share if the common features exist, otherwise with all the features the four nodes have [16]. The absence of the common features in the four adjacent nodes leads to a drawback of this method because renaming a new node with all features that the four nodes have does not really represent the documents the node contains.

In the HLabelSOM method, the map is re-labelled to produce other maps. In the new maps the nodes are labelled with the features if the node weight value for the particular feature is greater or equal to 0.5 and only the n greatest weight values of features are selected. We can choose the values of n from 1 to a certain value that will lead to the production of the detail map (Fig. 6) as a result. Fig. 8 shows the labelled map for $n = 3$ and Fig. 9 shows the animals mapped to the labelled map in Fig. 8. Fig. 10 shows labelled map for $n = 1$ and Fig. 11 shows the animals mapped to the labelled map in Fig. 10. We can now use the maps in Fig. 6, Fig. 8, and Fig. 10 as hierarchical maps: the detail map, more general (middle) map and most general (overview) map, respectively. The overview map produced, in this experiment setting, is the same as the map labelled by Lin's method, (Fig. 4).

4 Applying HLabelSOM to Medical Document Collection

We also apply this HLabelSOM method to a medical document collection. The medical documents are fetched from the internet. The keywords of diseases included in Australia's six national health priority areas: “cancer”, “diabetes”, “arthritis”, and “asthma”, were entered to the ‘Google' search engine and a number of URLs is returned in a linear list. From the results ten documents for each disease were selected, 40 documents in total, and the keywords of the documents, not only from the title of the document but also from the whole text in document, were determined by using automatic indexing software. Twelve keywords that had the most frequent occurrences in the 40 documents were selected to be the features of the documents.

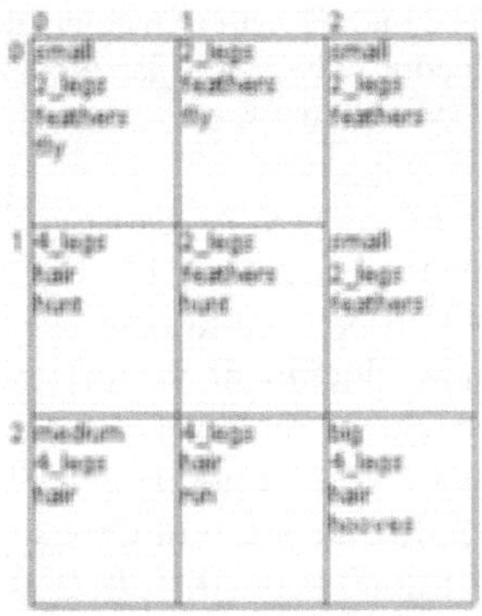

Fig. 8. The middle map. The map is labelled by using HLabelSOM with n = 3

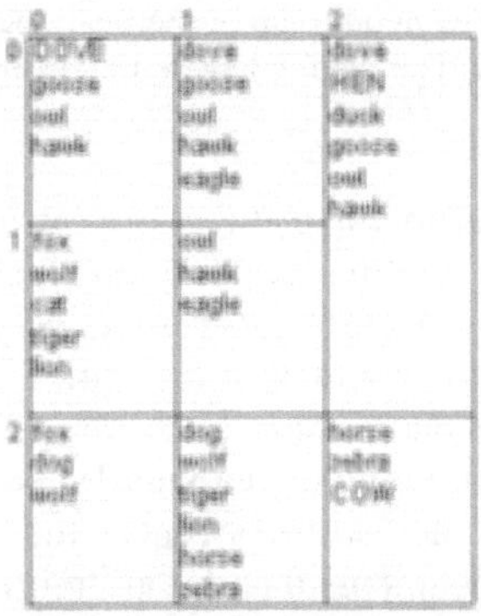

Fig. 9. The animals are mapped to the map in Fig. 8 for exact and approximate matches

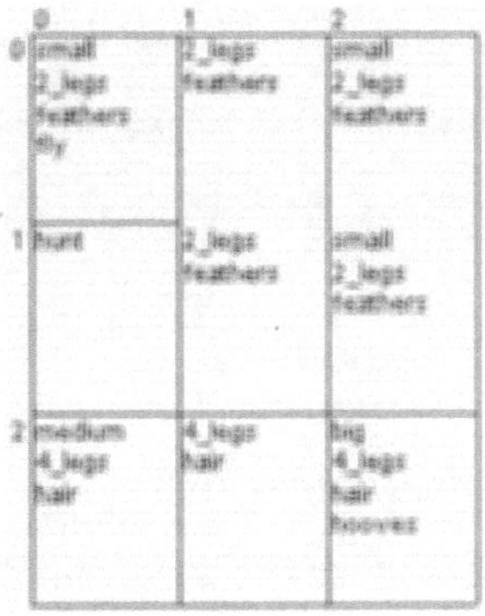

Fig. 10. The overview map. The map is labelled by using HLabelSOM with n = 1

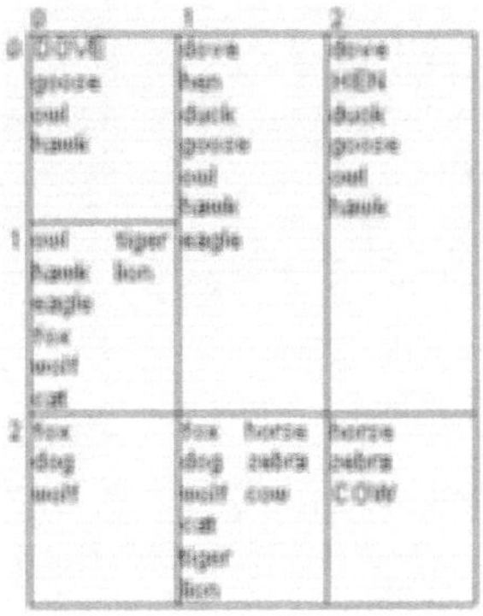

Fig. 11. The animals are mapped to the map in Fig.10 for exact and approximate matches

Fig. 12. The detail map of medical document collection. The map is labelled by using HlabelSOM

Fig. 13. The middle map of medical document collection. The map is labelled by using HLabelSOM with n = 2

The document features are ‘arthritis', ‘asthma', ‘cancer', ‘care', ‘cause', ‘diabetes', ‘disease', ‘health', ‘joint', ‘pain', ‘patient', and ‘treatment' (10, 9, 9, 8, 11, 10, 12, 11, 8, 8, 9 and 11 occurrences respectively). These features will be used to label the map.

Each document is transformed to a twelve element document vector (the vector represents the features [arthritis, asthma, cancer, care, cause, diabetes, disease, health, joint, pain, patient, treatment]). We use a binary vector that contains 1 in the given column if the feature occurs as the keyword in the document, and 0 otherwise. For example, documents 0 and 1 are represented as follows: doc0 = [0, 0, 1, 0, 0, 0, 1, 0, 0, 0, 0, 1] and doc1 = [0, 0, 0, 0, 1, 1, 1, 1, 0, 0, 0, 1].

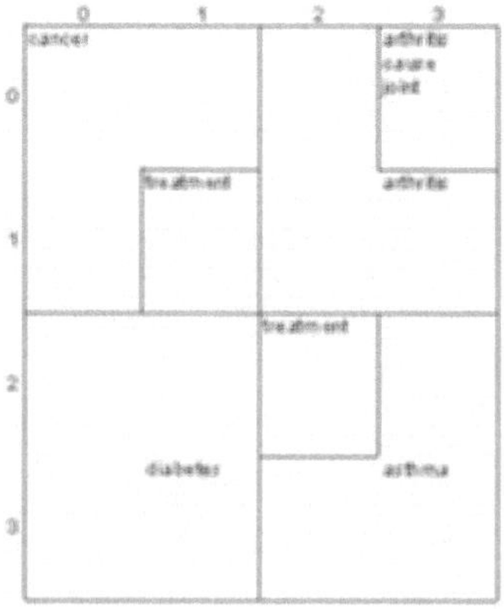

Fig. 14. The overview map of medical document collection. The map is labelled by using HlabelSOM with n = 1

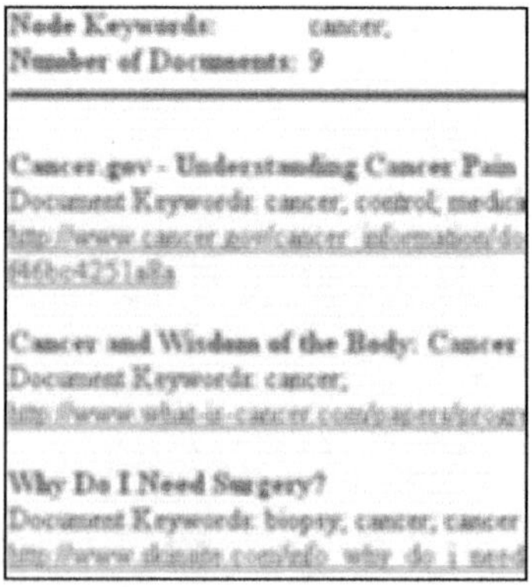

Fig. 16. The result of clicking the areas having the keyword ‘cancer' in Fig. 15. The documents' titles, keywords and URLs are given

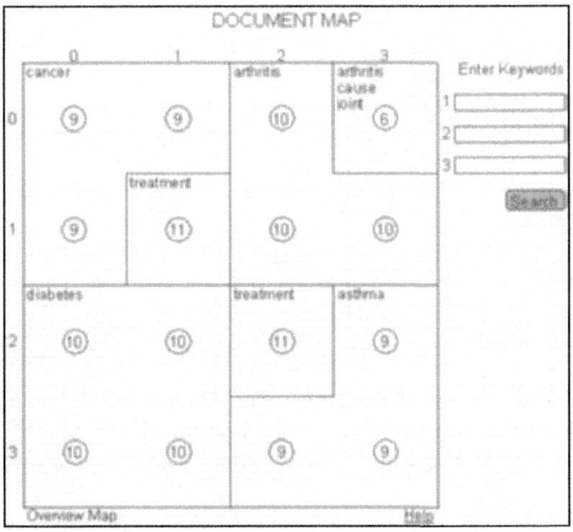

Fig. 15. The overview map of document collection with the possibility for the users to enter the keyword as in traditional searching. The numbers in the circle represent the numbers of documents mapped to the nodes

Then, a 4 by 4 SOM is trained by using the 40 feature vectors of the documents. We chose to use 4 by 4 SOM on the output map as a model where a larger number of cases, i.e. 40 documents with 31 unique cases, has to be fitted into a smaller space, i.e. 16 nodes on the output node. This model mirrors future real situations where larger amounts of data, e.g. 1000 documents with 100 features, have to be fitted into a smaller space, e.g. a 10 by 10 SOM.

After the training completes, the HLabelSOM method is applied to produce the result maps: the detail, middle and overview maps, which are shown in Fig. 12, Fig. 13 and Fig. 14, respectively. The overview map is again the same map as if we label the map using Lin's method.

5 User Interface for Information Retrieval

The user interface for information retrieval can be created by utilising the hierarchical maps in Fig. 12, Fig. 13, and Fig. 14 that are detail map, middle map, and overview map, respectively. First, the overview is given, with the possibility to enter the keywords for users who prefer traditional search to use of the maps, shown in Fig. 15.
The numbers on the circles represent the number of documents mapped onto the nodes. The user can click the areas to retrieve the documents with these keywords, or zoom in by clicking the nodes on the map to see more detailed maps with more keywords on the nodes.

Fig. 16 shows the result when the user clicks an area with keyword 'cancer'. The titles, keywords and URLs of the documents mapped to the node are given so that the user can follow the links to retrieve the documents he/she is interested in.

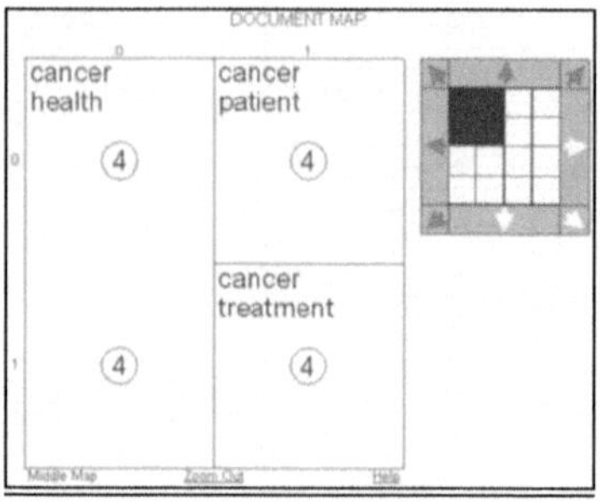

Fig. 17. The partial middle map of document collection. The result of zooming in node(0,0) on overview map in Fig 15

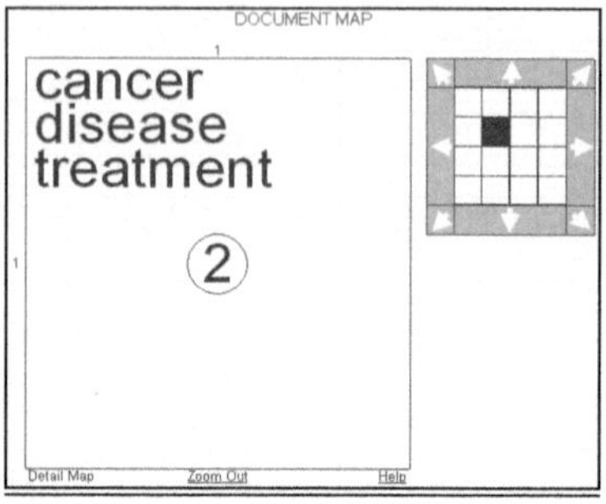

Fig. 18. The partial detail map of document collection. The result of zooming in node(1,1) on middle map in Fig 17

Fig. 17 shows the result when we zoom in by clicking node(0,0) that has keyword 'cancer' in Fig. 15. The keywords are extend to 'cancer and health', 'cancer and patient', and 'cancer and treatment'. If we still want to extend the keywords of 'cancer and treatment', we can click node (1,1) on map in Fig. 17 that will take us to the detail map in Fig. 18. The navigation tool is also provided in the detail map and middle map so that the user will not get lost while exploring the map. The facilities to zoom out (in addition to zooming in) and to get help increase the flexibility and the user friendliness of the interface.

The interactive version of this user interface can be found at http://cis213818.levels.unisa.edu.au:8080/docmap/ver1/overviewMap.jsp.

6 Conclusion

We have presented the HLabelSOM method to automatically label a trained SOM. Like other previous automatic labelling methods, Lin's method and LabelSOM, the HLabelSOM can help the users to understand the clusters on the map since the labels show that neighbouring nodes have similar features. The cluster boundaries are also revealed through the common features shared by neighbouring nodes. HLabelSOM has some advantages over the previous methods. First, HLabelSOM labels the detail map with richer features than Lin's method does and has more exact matches than Lin's method and LabelSOM. The richer features on the nodes and the more exact matches make it easier for the users to find specific information. Second, we do not need to define thresholds as in LabelSOM, since HLabelSOM uses a fixed value of 0.5 as threshold. HLabelSOM and LabelSOM also are significantly different in the time of mapping the inputs to the nodes. In HLabelSOM, the inputs are mapped to the nodes on the maps after the nodes are labelled, but in LabelSOM the inputs have to be mapped to the nodes before the nodes can be labelled. Furthermore, HLabelSOM generates hierarchical maps that can be utilised for information retrieval user interface with overview, zoom and filter, then detail options on demand. The next tasks are to evaluate the user interface and to apply the HLabelSOM to larger amounts of data than that in the experiments presented in this paper.

References

[1] Kohonen, T.: Self-Organized Formation of Topologically Correct Feature Maps. Biological Cybernetics, Vol 43. (1982) 59-69

[2] Lin, X., et al.: A Self-Organizing Semantic Map for Information Retrieval. The 14th Annual Int'l ACM/SIGIR Conf. on Research and Development in Info. Retrieval (1991)

[3] Rauber, A., Merkl, D.: Automatic Labeling of Self-organizing Maps for Information Retrieval. Journal of System Research Info. System, Vol. 10:10 (2001) 23-45.

[4] Kohonen, T., et al.: Self Organization of a Massive Document Collection. IEEE Transactions on Neural Networks, Vol. 11:3 (2000) 574-585

[5] Houston, A. L., et al.: Medical Data Mining on the Internet: Research on a Cancer Information System. Artificial Intelligence Review, Vol. 13 (1999) 437-466

[6] Ultsch, A.: Self-Organizing Neural Networks for Visualisation and Classification. Info. and Classification: Concepts, Methods and App. Springer, Berlin (1993)

[7] Merkl, D., Rauber, A.: Cluster Connections: A Visualzation Technique to Reveal Cluster Boundaries in Self-Organizing Maps. The 9th Italian Workshop on Neural Nets. Vietri sul Mare, Italy (1997)

[8] Merkl, D, Rauber, A.: On the Similarity of Eagles, Hawks, and Cows: Visualization of Semantic Similarity in Self-Organizing Maps. Int'l Workshop Fuzzy-Neuro-Systems. Soest, Germany (1997)

[9] Rauber, A.: LabelSOM: On the Labeling of Self-Organizing Maps. Int'l Joint Conference on Neural Networks. Washington (1999)

[10] Kohonen, T.: Self-Organizing Map. 2nd Ed. Springer-Verlag, Berlin (1995)

[11] Lippmann, R. P.: An Introduction to Computing with Neural Nets. IEEE Acoustics, Speech, and Signal Processing Society (1987) 4-22

[12] Ritter, H., Kohonen, T., Self-Organizing Semantic Maps. Biological Cybernetics, Vol. 61 (1989) 241-254

[13] Shneiderman, B.: The Eyes Have It: A Task by Data Type Taxonomy for Information Visualizations. IEEE Symposium on Visual Languages. Boulder, CO, USA (1996)

[14] Miikkulainen, R.: Script Recognition with Hierarchical Feature Maps. Connection Science, Vol. 2:1 (1990) 83-102

[15] Merkl, D.: Exploration of Text Collections with Hierarchical Feature Maps. Proc. of the 20th Annual Int'l ACM SIGIR Conf. on Research and Development in Info. Retrieval (1997)

[16] Tan, H. S., George, S. E.: Multi-Media Based Web Mining for an Information Resource. The 3rd Int'l Conf. on Data Mining Methods and Databases for Engineering, Finance and Other Fields, Bologna, Italy (2002)

Using Images to Compare Two Constructive Network Techniques

Ross Hayward

Smart Devices Lab
Centre for Information Technology Innovation
Queensland University of Technology
Box 2434 GPO, Brisbane 4001 Qld, Australia

Abstract. A common theme in answering the question of how distinct concepts are represented in neural networks is to associate meaning to the activity of an individual, or set of neurons. While such models may provide an explanation of how reasoning may occur, they rarely suggest how structures come to exist, often relying on some form of the innateness conjecture. This paper addresses the issue of concept formation by using the principle of structural risk minimization during network construction. This paper describes two related methods for incrementally constructing networks from smaller network architectures. The methods are applied to an image prediction problem that provides a sufficiently difficult learning task, while allowing both quantitive and qualitative analysis of performance. Experiments suggest that incrementally introducing structure during learning in a principled manner can assist in overcoming the bias/variance problem as, for the given problem, the methods show no sign of over-fitting after prolonged training.

1 Introduction

Given a set of m labeled n-bit examples there is no efficient way to determine whether a hypothesis space H_n contains a hypothesis consistent with the examples [1]. The implication for multilayer perceptrons is that choosing a sufficient number of hidden units for a particular problem is largely guesswork. In contrast, increasing the expressiveness of a network by increasing the number of hidden units, and thereby admitting functions with less approximation error has the effect of increasing the estimation error. This characteristic is referred to as the *Bias-Variance dilemma* [4], which supports the use of Occam's Razor as a principle in machine learning - given a finite number of training examples, the smallest network will yield a function with the smallest estimation error. Interestingly, experimental results using *Boosting* techniques have shown a decrease in the error for unseen examples as the complexity of the underlying classifier increases [8, 6].

This paper describes two similar constructive neural network techniques called TowerNet and CascadeNet, which incrementally expand the space of hypotheses during learning by identifying a supportive hypothesis in each expansion, which is then used to bias subsequent learning. The algorithms are applied

T.D. Gedeon and L.C.C. Fung (Eds.): AI 2003, LNAI 2903, pp. 544–553, 2003.

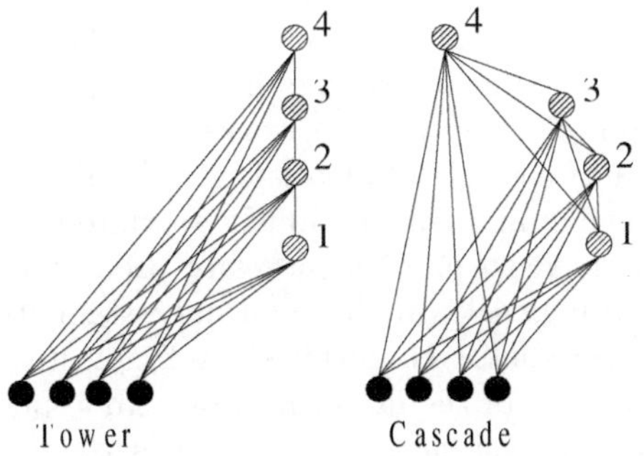

Fig. 1. The structure of Tower and Cascade Networks

to the problem of constructing images by predicting a red, green and blue value for a given pixel location. Problems of this type are useful, as for a given image size and number of possible colors the number of possible images will be finite, and at any stage during learning an image can be generated to provide a qualitative analysis of the networks search through the space of hypotheses.

The paper is organized as follows. This Section introduces background to the algorithms, while Section 1.1 describes the algorithms in more detail. Section 3 results of several experiments and Section 4 discusses the results. The paper concludes with Section 5.

1.1 Dynamically Constructing Networks

The Tower algorithm [3, 5] and Cascade Correlation algorithm [2] are examples of algorithms that dynamically construct neural networks. These networks have the unique feature that whenever a new unit is recruited, it's weights are frozen, and if we assume that the set of training examples is fixed, components within the training examples can be manipulated or added, enabling portions of the network structure to be discarded prior to further training. That is, instance descriptions can be modified to express network structure.

A significant drawback with constructive algorithms lies in the requirement that: as units are added sequentially to construct a larger network, there must be some criteria that identifies when training a particular unit should cease and the training of a new unit begin. Gallant, in showing that the Tower algorithm will converge to find a function $f : \{0,1\}^n \rightarrow \{0,1\}$ over a set of m examples, identifies *optimal units* as those that correctly classify the most examples. The search for an optimal unit is terminated after a unit fails to classify more examples in some given time, this unit then becomes the new output for the network. The search for the next optimal unit then begins by creating a unit (with Heaviside activation function) having inputs from the input space and the last (possibly optimal) unit inserted into the network (see Figure 1).

The Cascade Correlation algorithm uses a similar technique for regression problems, albeit a little more complex. A network is said to stagnate if the error is not reduced by a specified amount in a given number of epochs. Training the network consists of only altering weights in the output layer and when this training stagnates, a pool of candidate networks are trained. Each network in the candidate pool contains a single hidden layer and unit, with the unit having inputs from the input space and all previously recruited units. The networks in this pool are then trained to predict the error seen when constructing the larger network. Once the networks in the pool have stagnated, the unit in the hidden layer of the best predictor is recruited by fixing it's input weights and then training the output layer of the larger network.An appealing characteristic of the Tower algorithm over Cascade Correlation is that recruitment can be avoided altogether by initially fixing structure and training units in parallel.

Structural Risk Minimization: It is interesting to consider these algorithms, in particular the Tower algorithm, with respect to Structural Risk Minimization [9]. The principle of Structural Risk Minimization (SRM) is based upon the identification of a bound on the difference between empirical and actual risks.

Vapnik shows that, with probability $1-\eta$, the following bound holds on how well the empirical risk ($R_{emp}(\alpha)$) approximates the actual risk ($R(\alpha)$) for a set of parameters α implementing a classifier. The bound is:

$$R(\alpha) \leq R_{emp}(\alpha) + VC_{conf}(VC(H), m, \eta). \tag{1}$$

Note that the second term on the right hand side, the *VC confidence* depends upon the number of training instances m, the confidence level $1 - \eta$ and the VC-dimension of hypothesis space H. As the empirical and actual risk only depend upon α (the parameters identified during learning) the VC confidence is dependent on the set of functions used for learning[1].

For a set of m training examples the training error $R_{emp}(\alpha)$ will decrease as the VC-dimension increases, but the VC-Confidence always increases with increasing VC-dimension. Thus, the VC-dimension can be used as a controlling variable by constructing a graded set of hypothesis spaces:

$$H_1 \subset H_2 \subset \cdots \subset H_n \subset \cdots \tag{2}$$

where, for finite VC dimension, we have

$$VC(H_1) \leq VC(H_2) \leq \cdots \leq VC(H_n) \leq \cdots \tag{3}$$

The SRM principle implies that the function identified by a learner with the smallest empirical error, selected from a set of functions with the smallest VC-dimension, will have the smallest difference between the actual and empirical error.

[1] The VC-dimension of a hypothesis space H is indicative of the classification power of hypotheses in H

Table 1. The GTower Algorithm

– Assume a matrix $\mathbf{x}_0$ is a set of m samples, $(x_1, \cdots, x_m)$, as column vectors. Each $x_i \in X$ is assumed to have been drawn randomly and independently from a distribution D on X. Let $\mathbf{y}$ be a matrix of labels for each vector in $\mathbf{x}_0$ generated by a target function $g : X \rightarrow Y$. Create a set of examples $E_0 = (\mathbf{x}_0, \mathbf{y})$

1. Given examples $E_t, t \in \mathbb{N}$ train l networks $NN_{1t}(\mathbf{x}_t), \cdots, NN_{lt}(\mathbf{x}_t)$ to approximate the target function g. Stop training when the training error has not decreased by more than a set amount α in K epochs.
2. If the error for the ith network is given by $err_i = |NN_{it}(\mathbf{x}_t) - \mathbf{y})|$ then choose j so that $j = \underset{1 \geq i \geq l}{argmin}(err_i)$ and store NN_{jt}
3. If $err_j \geq bestError$ Go to 1 else set $bestError = err_j$.
4. If err_j is sufficiently small or some other criteria then halt.
5. Create a new set of m examples $(x'_1, \cdots, x'_m)$ using:

$$\mathbf{x}_{t+1} = \begin{bmatrix} \mathbf{x}_0 \\ NN_i(\mathbf{x}_t) \end{bmatrix}$$

6. Go to 1, with $E_{t+1} = (\mathbf{x}_{t+1}, \mathbf{y})$.

With respect to the Tower algorithm, the first unit represents a hypotheses space H_1 with the VC-dimension of a single threshold unit of $n+1$, identifying an optimal unit in this space, and adding this into the Tower network, reposes the problem so that finding the next optimal unit is a search through H_2 where $(H_2 \supset H_1)$. Hence, the Tower algorithm searches through a graded set of Hypothesis spaces until a solution is found.

2 GTower and GCascade

The GTower and GCascade algorithms are similar to the Tower algorithm, but aimed at regression problems. The GTower algorithm creates a similar structure to a Tower network, however the individual units in the Tower algorithm are replaced by the hidden and output layers of a multi-layer perceptron (a subnetwork). The GCascade algorithm generates networks that differ slightly in structure by adding subnetworks with inputs from all previously added subnetworks, analogous to those generated by the Cascade correlation algorithm.

Where the Tower algorithm searches for an optimal unit, the GTower and GCascade algorithms search for an optimal subnetwork - a network with the lowest possible mean square error over the training data. Where the Tower algorithm ceases to look for an optimal unit if classification does not improve in a specified number of epochs, the algorithms use the cascade correlation stagnation criteria.

Figure 1 depicts the structure of GTower and GCascade networks after four subnetworks have been added. In this case the black nodes indicate network inputs and the hatched nodes subnetworks. The subnetworks consist of a number of hidden layers and an output layer specified by the problem domain. The subnetworks have been numbered in the order in which they have been inserted into the network. Note that the input space for successively added subnetworks for the GCascade networks increases in dimension by the number of output units, so that the nth subnetwork will have $i + o(n-1)$ inputs, where i and o are the dimensions of the input and output space respectively. In contrast, the input space for GTower is i for the first subnetwork and then $i + o$ for all successive subnetworks.

The GTower algorithm, shown in Table 1, proceeds by training a fixed number of candidate MLP's in parallel on a given task. The subnetworks stagnate when the training error has not improved by an amount α over K epochs. When all the candidate networks have stagnated, the one with the smallest training error is selected and recruited into the larger network structure if it's error over the training set is less than that obtained from the most recently recruited subnetwork.

After an initial subnetwork has been recruited, another set of candidate MLP's are generated, each with an input layer composed from the original problem space and the outputs of the most recently added subnetwork. Recruitment of the initial subnetwork is equivalent to extending the dimensionality of the training vectors by adding components generated by the most recently added subnetwork. The recruitment of subsequent subnetworks simply requires the recalculation of these components. The algorithm in Table 1 is presented in this manner (in contrast with Figure 1 that shows the structure of the network). This training/extending process is repeated until some criteria is met.

The GCascade algorithm differs from the GTower by extending the set of input vectors for every subnetwork added. When candidate subnetworks are trained they have inputs from the input space and all, if any, previously recruited subnetworks. Hence GCascade can be described by amending step 5 in Table 1 to:

5 Create a new set of m examples $(x'_1, \cdots, x'_m)$ using

$$\mathbf{x}_{t+1} = \begin{bmatrix} \mathbf{x}_t \\ NN_i(\mathbf{x}_t) \end{bmatrix}$$

The following experiments investigate the difficulty associated with identifying subnetworks with increasingly smaller training error.

3 Experiments

Consider the situation where, given a pair of integers between 0 and 255 inclusive, the goal is to predict 3 numbers from the same set. Any regression estimation technique capable of solving such a problem is capable of predicting any 256 ×

Fig. 2. The left image corresponds to the training data (uniformly sampled 25% of image) and the right image test data

256 pixel image with 16777216 different colors just from pixel coordinates. The number of possible images provides a concrete example of the hypothesis space required to identify a mapping $f : (x, y) \to (r, g, b)$.

What is interesting about such a problem is that a learning algorithm over the space of hypotheses can be evaluated in two ways. Firstly, by the generalization error, and secondly, by a visual inspection of the predicted image. A coloured image (see Figure 2) was cropped to 256 pixels high by 256 pixels wide with the x and y pixel locations used for training, scaled to range between 0.0 and 1.0. The colour of each pixel is represented by 256 red, green and blue values in the range 0 to 1. The training set was a uniform sample consisting of 25% (16384) of the examples.

3.1 Methodology and Parameters

Samples from the image were used to train two GTower and two GCascade networks. The two experiments for each network type were differentiated by the number of hidden units in each of the subnetworks. Subnetworks consisted of either 3 or 50 hidden units in a single layer and were all constructed using the same set of examples. The stagnation parameters α and K were constant over all experiments at 10^{-6} and 100 epochs respectively.

For each experiment three candidate subnetworks were trained concurrently using the RPROP learning algorithm [7] and after all had stagnated, the subnetwork with the smallest training error was selected for recruitment into the larger network. Training errors in the following experiments are the mean square error of the networks prediction over training set, and the generalization error refers to the mean square error of the networks prediction over all the pixels in an image.

On the occasion that none of the stagnated candidate subnetworks have a smaller training error than the most recently recruited, their weights are randomized and training repeated. However, although discarded, the results of the

Table 2. Result summary for GTower and GCascade networks recruiting subnetworks consisting of 3 and 50 hidden units

		GCascade		GTower	
Units in hidden layer		3	50	3	50
A-Generalization error-initial	($\times 10^{-2}$)	7.68	3.87	7.85	3.71
B-Generalization error-final	($\times 10^{-2}$)	4.49	2.06	5.08	2.24
C-Training error - final	($\times 10^{-2}$)	4.34	1.82	5.01	2.10
D-Candidate pools trained		41	10	44	18
E-Subnets recruited		34	10	28	6
F-Final number of weights		5763	9780	831	2568
G-Training Epochs	($\times 10^{5}$)	1.41	8.66	10.0	6.84
H-Weight Epochs	($\times 10^{8}$)	8.00	7.61	0.80	4.64

experiments include these details to provide information concerning the difficulty in locating subnetworks with smaller training error.

3.2 Network Weights

The number of weights in GTower subnetworks remains constant after the first network. For example, a task with i inputs and o outputs and using subnetworks with h units in a single hidden layer will have $h(i+1)+o(h+1)$ weights in the first subnetwork and $h(i+o+1)+o(h+1)$ weights in subsequent networks. In contrast, the nth subnetwork in a GCascade network will have $h(i+1+no)+o(h+1)$ weights. Due to the increase in the number of weights with additional subnetworks, the results are reported in terms of weight epochs to provide a better indication of the training times required. A weight epoch for a network is simply the number of weights in the network multiplied by the number of epochs the network was trained.

3.3 Results

Table 2 indicates the results obtained from constructing the networks. Row A shows the generalization error of the first subnetworks trained on just the pixel locations and gives some indication of how standard MLP performs on the problem. The generalization error of the last subnetwork recruited is recorded in row B. In all cases the generalization error reduced with each subnetwork recruited. Row C indicates the training error of the last subnetwork, row D indicates the total number of candidate pools trained, and row E the number of recruited subnetworks. Row F indicates the number of weights in the resulting network, row G the total number of epochs to train the subnetwork with the smallest training error from all candidate pools and row H the weight epochs for these same subnetworks.

Figure 3 shows results plotted for each of the networks where the steps in each graph refer to the error over all image pixels and the line graph the training

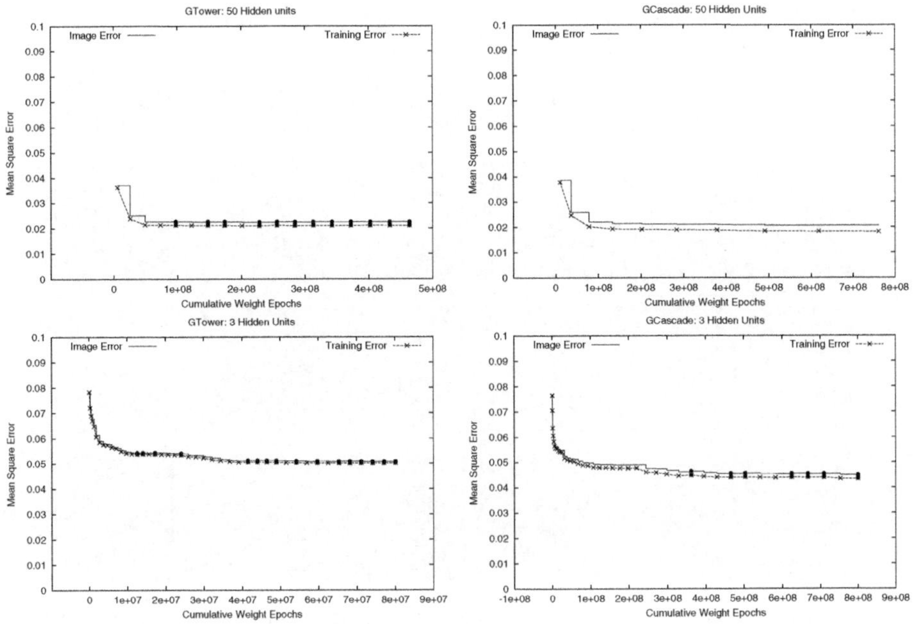

Fig. 3. Training results for the networks. The top row shows results from GTower and GCascade networks consisting of subnetworks with 50 hidden units and the bottom row with 3 hidden units

error. Solid points in the graphs indicate results for the best subnetworks from a candidate pool when none were recruited. For example in the case of GTower with subnetworks containing 50 units in the hidden layer, the last 9 subnetworks are not recruited.

4 Discussion

A significant observation is that none of the subnetworks recruited ever led to an increase in the generalization error no matter how small the improvement in the training error. However, it is clear that this can not, in general, always be true. A simple example would be an image where the left half was white and the right half black, sampling the image would induce a much more complicated decision boundary and while an initial subnetwork (hypothesis space) may be restricted to providing the perfect generalization, subsequent expansions of the hypothesis space would yield approximations with increasingly worse generalization errors.

In all the networks constructed, the first few subnetworks recruited reduced the training and generalization errors the most. Usually after which additional subnetworks only reduced the training error by a small degree. However, the results for experiments where subnetworks with 3 hidden units are recruited indicates that occasionally sequences of subnetworks are introduced that decrease

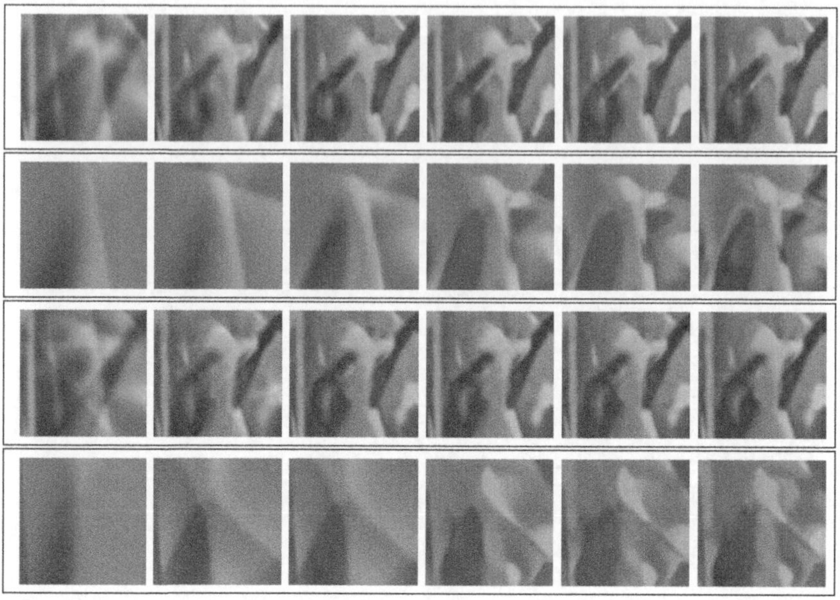

Fig. 4. Subnetwork predictions of image for four different networks. From top row: GCascade 50 hidden, GCascade 3 hidden, GTower 50 hidden, GTower 3 hidden

the error at an increased rate. There exists a question of whether a number of subnetworks are required prior to this increased reduction, or whether it is simply a matter of not having previously identified optimal networks at every recruitment.

In each of the above cases the GCascade networks converged towards better solutions with a much higher recruitment rate than their counterparts. It seems reasonable to speculate that as the GCascade networks expand the space of possible hypotheses at a much faster rate than the GTower networks, then there is an increased liklihood of recruitment. However the difficulty that GTower networks seem to have in finding subnetworks with less error may be indicative of a decreased liklihood in identifying optimal subnetworks. Each subnetwork in a GCascade network takes into account the prediction of all previously added subnetworks and is more representative of a committee machine than GTower networks that only take into account the previous best prediction.

5 Conclusion

The experiments were conducted to explore differences in the two algorithms and do not address the flexibility that constructing networks offers during learning. During learning tasks, whether regression or classification, we generally feel free to provide as many descriptive attributes as necessary to solve the task. The algorithms above take the same approach by introducing functions of the input space as descriptive attributes that might be considered useful in determining a network with good generalization. A significant point is that these functions need not be represented by subnetworks, any function will suffice. For example, we might introduce a rule that x and y values in certain ranges should yield particular RGB values. In this way approximate problem domain knowledge can be introduced easily during learning and integrated with the overall approximation.

This perspective views individual attributes as concepts of the problem domain, and and makes no distiction as to whether the concepts are represented by subnetworks or prior knowledge; both are just methods of producing attribute/value pairs. An explanation how a network reaches a conclusion can be stated in terms of the descriptive attribute/value pairs, some of which are represented by subnetworks. Consider the functions generated by each subnetwork in Figure 4, the first subnetwork recruited provides some initial structure for the image, which is subsequently added to by later subnetworks. Simply subtracting the output of the first subnetwork from the second can give yield information about what the second subnetwork actually does.

While the results in Table 2 show that single networks with 50 hidden units provide a better generalization error than those constructed of subnetworks each with 3 hidden units, the latter provides a gradual progression (seen in Figure 4) that helps explain how the final result is reached.

References

[1] M. Anthony and N. Biggs. *Computational Learning Theory. An Introduction.* Number 30 in Cambridge Tracts inTheoretical Computer Science. Cambridge University Press, 1992. 544

[2] S. Fahlman and C. Lebiere. The cascade-correlation learning architecture. In R. Lippman, J. Moody, and D. Touretzky, editors, *Advances in Neural Information Processing Systems*, volume 3, pages 190–196, San Mateo, 1990. Morgan-Kaufmann. 545

[3] S Gallant. Perceptron based learning algorithms. *IEEE Transactions on Neural Networks*, 1(2):179–191, June 1990. 545

[4] S. Geman, E. Bienenstock, and R Doursat. Neural networks and the bias/variance dilemma. *Neural Computation*, 4:1–58, 1992. 544

[5] JP Nadal. Study of a growth algorithm for a feedforward network. *International Journal of Neural Systems*, 1(1):55–59, 1989. 545

[6] J. Quinlan. Bagging boosting and c4.5. In *Proceedings of the Thirtenth National Conference on Artificial Intelligence*, pages 725–730, 1996. 544

[7] Martin Riedmiller and Heinrich Braun. A direct adaptive method for faster backpropagation learning: The rprop algorithm. In *Proceedings of the IEEE International Conference on Neural Networks*, pages 586–591, San Francisco, 1993. 549

[8] R. Schapire, Y. Fruend, P. Bartlett, and W. S. Lee. Boosting the margin: A new explanation for the effectiveness of voting methods. *The Annuals of Statistics*, 26(5):1651–1686, 1998. 544

[9] V. Vapnik. *The Nature of Statistical Learning Theory.* Springer-Verlag, 1995. ISBN 0 387 94559 8. 546

Pareto Neuro-Ensembles

Hussein A. Abbass

Artificial Life and Adaptive Robotics (A.L.A.R.) Lab
School of Information Technology and Electrical Engineering
Australian Defence Force Academy, University of New South Wales
Canberra, ACT 2600, Australia
h.abbass@adfa.edu.au

Abstract. The formation of a neural network ensemble has attracted much attention in the machine learning literature. A set of trained neural networks are combined using a post-gate to form a single super-network. One main challenge in the literature is to decide on which network to include in, or exclude from the ensemble. Another challenge is how to define an optimum size for the ensemble. Some researchers also claim that for an ensemble to be effective, the networks need to be different. However, there is not a consistent definition of what "different" means. Some take it to mean weakly correlated networks, networks with different bias-variance trade-off, and/or networks which are specialized on different parts of the input space.

In this paper, we present a theoretically sound approach for the formation of neural network ensembles. The approach is based on the dominance concept that determines which network to include/exclude, identifies a suitable size for the ensemble, and provides a mechanism for quantifying differences between networks. The approach was tested on a number of standard dataset and showed competitive results.

1 Introduction

The formation of a mixture of predictors (ensemble) can improve the prediction accuracy [10]. There have been many studies on ensembles of neural networks, see for example [10, 11, 12, 13] and [17] for a good review. In some studies, the networks are trained separately then combined to form an ensemble. In some other studies a group of networks is trained simultaneously while an explicit term is added to the error function to make them different. For example, in *negative correlation learning* (NCL), proposed by Liu and Yao [12], a mathematical penalty term describing the negative correlation between the networks is added to the conventional mean square error of each network and traditional *backpropagation* (BP) [16] is used for network training. In [14], an analysis of NCL revealed that the penalty function in NCL acts to maximize the average difference between each network and the mean of the population; while the intended aim of anti-correlation mechanisms is to maximize the average difference between pairs of population members, which is not necessarily the same thing. In [13], an *evolutionary algorithm* (EA) is used in conjunction with NCL to

T.D. Gedeon and L.C.C. Fung (Eds.): AI 2003, LNAI 2903, pp. 554–566, 2003.

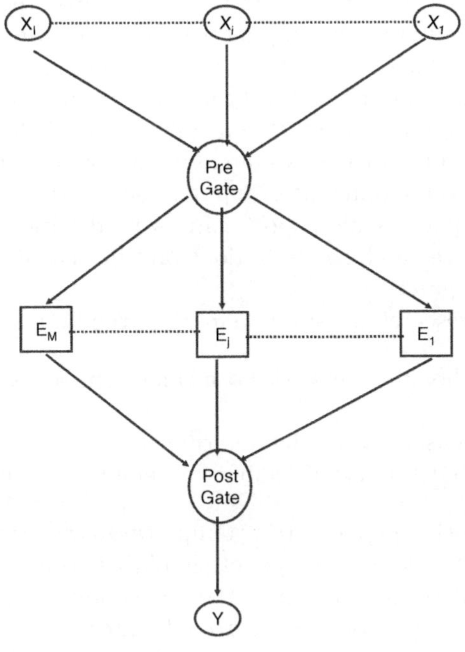

Fig. 1. A generic description to ensemble research

evolve the ensemble. The k-means algorithm was used to cluster the individuals in the population. The fittest individual in each cluster is used as a member in the ensemble. The authors found no statistical difference between the use of the population as a whole in the ensemble and the use of a subset. It is not clear however how to choose the value of k.

Figure 1 presents a generic representation of research in ensembles of learning machines. Two types of gates can be identified: pre-gate and post-gate. A pre-gate is used as a mechanism to decide on which input instance affects which member of the ensemble. Research in this area is mainly concerned with getting members of the ensemble to specialize on different parts of the input space. For example, we can see a pre-gate as a result of clustering the input space into k clusters, with the pre-gate deciding on which member of the k clusters being assigned to which members of the M learning machines in the ensemble. The study of this pre-gate falls under the umbrella of competitive-learning, where the main objective of the learning machine is to compete against other learning machines on the inputs.

A post-gate is used to decide on how to combine the different outputs made by the members in the ensemble. The majority of research in ensemble-learning is concerned with the post-gate, while attempting to implicitly construct a pre-gate. The motivation to this learning approach comes from the area of expert systems, where a committee of experts is used to make decisions. If one and only one committee member responses to each situation, there will be no conflict within the committee and the problem becomes similar to the use of pre-gates with one-to-one or many-to-one mapping between the clusters in the input space and the ensemble members. However, if a group of committee members responses differently to the same input instance, the problem would then be how to combine the different opinions of the committee members. In this last case, the research would fall into the study of the post-gate. Obviously, we can have a pre-gate as well as a post-gate, where different group of experts get specialized on different parts of the input space, while a post gate is used within each group. Up to our knowledge, little research has been done on this last situation. This paper is concerned with post-gates.

Three key open research questions remain in the literature:

1. On what basis should a network be included in, or excluded from the ensemble?
2. How should the ensemble size be determined?
3. How to ensure that the networks in an ensemble are different?

The objective of this paper is to attempt to answer these three questions. This is a challenging task and this paper should be seen as only an attempt to provide a theoretically-sound answer to these questions. The rest of the paper is organized as follows: In Section 2, background materials are covered followed by an explanation of the methods in Section 3. Results are discussed in Section 4 and conclusions are drawn in Section 5.

2 Multiobjective Optimization and Neural Networks

Consider a *multiobjective optimization problem* (MOP) as presented below:-

$$\textbf{Optimize } F(\boldsymbol{x} \in \Upsilon) \tag{1}$$

$$\textbf{Subject to: } \Upsilon = \{\boldsymbol{x} \in R^n | G(\boldsymbol{x}) \leq 0\} \tag{2}$$

Where $\boldsymbol{x}$ is a vector of decision variables $(x_1, \ldots, x_n)$ and $F(\boldsymbol{x} \in \Upsilon)$ is a vector of objective functions $(f_1(\boldsymbol{x} \in \Upsilon), \ldots, f_K(\boldsymbol{x} \in \Upsilon))$. Here $f_1(\boldsymbol{x} \in \Upsilon), \ldots, f_K(\boldsymbol{x} \in \Upsilon)$, are functions on R^n and Υ is a nonempty set in R^n. The vector $G(\boldsymbol{x})$ represents a set of constraints.

The aim is to find the vector $\boldsymbol{x}^* \in \Upsilon$ which optimizes $F(\boldsymbol{x} \in \Upsilon)$. Without any loss of generality, we assume that all objectives are to be minimized. We note that any maximization problem can be transformed to a minimization one by multiplying the former by -1.

The core aspect in the discussion of a *multiobjective optimization problem* (MOP) is the possible conflict that arises in the case of optimizing the objectives simultaneously. At a certain point, one objective can just be improved at the expense of at least another objective. The principle of dominance (Figure 2) in MOP allows a partial order relation that works as follows: a solution does not have an advantage to be included in the set of optimal solutions unless there

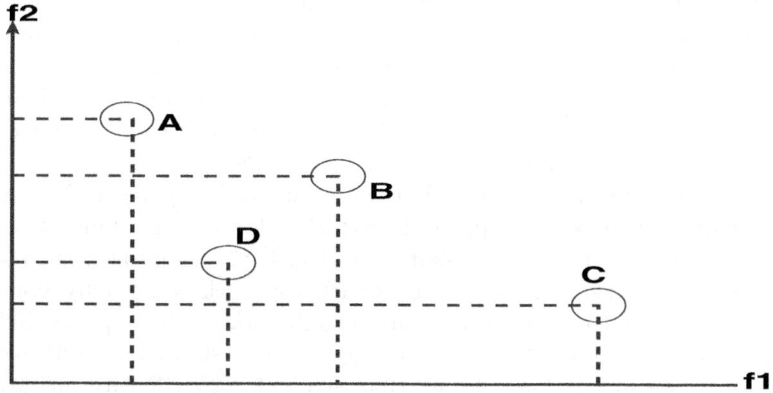

Fig. 2. The concept of dominance in multiobjective optimization. Assuming that both f_1 and f_2 are to be maximized, D is dominated by B since B is better than D when measured on all objectives. However, A, B and C are non-dominated since none of them is better than the other two when measured on all objectives

is no solution that is better than the former when measured on all objectives. A non–dominated solution is called Pareto. To formally define the concept of non–dominated solutions in MOPs, we need to define two operators, $\ncong$ and $\precsim$ and then assume two vectors, $\boldsymbol{x}$ and $\boldsymbol{y}$. $\boldsymbol{x} \ncong \boldsymbol{y}$ iff $\exists\, x_i \in \boldsymbol{x}$ and $y_i \in \boldsymbol{y}$ such that $x_i \neq y_i$. And, $\boldsymbol{x} \precsim \boldsymbol{y}$ iff $\forall\, x_i \in \boldsymbol{x}$ and $y_i \in \boldsymbol{y}, x_i \leq y_i$, and $\boldsymbol{x} \ncong \boldsymbol{y}$. $\ncong$ and $\precsim$ can be seen as the "not equal to" and "less than" operators over two vectors, respectively. We can now define the concepts of local and global optimality in MOPs.

Definition 1. Neighborhood or Open Ball *The open ball (ie. a neighborhood centered on $\boldsymbol{x}^*$ and defined by the Euclidean distance) $B_\delta(\boldsymbol{x}^*) = \{\boldsymbol{x} \in R^n \mid ||\boldsymbol{x} - \boldsymbol{x}^*|| < \delta\}$.*

Definition 2. Local Efficient (Non–inferior/ Pareto–Optimal) Solution *A vector $\boldsymbol{x}^* \in \Upsilon$ is said to be a local efficient solution of MOP iff $\nexists\, \boldsymbol{x} \in (B_\delta(\boldsymbol{x}^*) \cap \Upsilon)$ such that $F(\boldsymbol{x}) \precsim F(\boldsymbol{x}^*)$ for some positive δ.*

Definition 3. Global Efficient (Non–inferior/ Pareto–Optimal) Solution *A vector $\boldsymbol{x}^* \in \Upsilon$ is said to be a global efficient solution of MOP iff $\nexists\, \boldsymbol{x} \in \Upsilon$ such that $F(\boldsymbol{x}) \precsim F(\boldsymbol{x}^*)$.*

Definition 4. Local Non–dominated Solution *A vector $\boldsymbol{y}* \in F(\boldsymbol{x})$ is said to be local non–dominated solution of MOP iff its projection onto the decision space, $\boldsymbol{x}^*$, is a local efficient solution of MOP.*

Definition 5. Global Non–dominated Solution *A vector $\boldsymbol{y}* \in F(\boldsymbol{x})$ is said to be global non–dominated solution of MOP iff its projection onto the decision space, $\boldsymbol{x}^*$, is a global efficient solution of MOP.*

A MOP can be solved in different ways. Evolutionary algorithms (EAs) [7, 8], being population based, they are able to generate a set of near-Pareto solutions in a single run. In addition, they do not require assumptions of convexity, differentiability, and/or continuity as traditional optimization problems do. EAs with local search are usually used to improve the performance of EAs to get closer to the actual optimal or, the Pareto set in case of MOPs.

In EANN, an evolutionary algorithm is used for training the ANN. A major advantage to the evolutionary approach over BP alone is the ability to escape a local optimum. The major disadvantage of the EANN approach is that it is computationally expensive, as the evolutionary approach is normally slow.

Recently, the problem of simultaneous optimization of the network architecture and the corresponding training error has been casted as a multiobjective optimization problem [4]. It was found that by combining BP with an EMO algorithm, a considerable reduction in the computational cost can be achieved [4]. The multiobjective optimization outputs a set of solutions, where no solution in the set is better than all others when compared on all objectives. The Pareto set provided a new insight into the learning problem, where an obvious question that emerged from the study is how to utilize the information embedded in the set as a whole. One answer to this question is to form an ensemble, which is the focus of this paper.

2.1 Evolutionary Artificial Neural Networks

The following notations will be used for a single hidden layer ANN:

- I and H are the number of input and hidden units respectively.
- $\boldsymbol{X}^p \in \mathbf{X} = (x_1^p, x_2^p, \ldots, x_I^p), p = 1, \ldots P$, is the pth pattern in the input feature space $\mathbf{X}$ of dimension I, and P is the total number of patterns.
- Without any loss of generality, $\mathbf{Y}_o^p \in \mathbf{Y}_o$ is the corresponding scalar of pattern $\mathbf{X}^p$ in the hypothesis space $\mathbf{Y}_o$.
- w_{ih} and w_{ho}, are the weights connecting input unit i, $i = 1 \ldots I$, to hidden unit h, $h = 1 \ldots H$, and hidden unit h to the output unit o respectively. The number of outputs is assumed to be 1 in this paper.
- $\Theta_h(\boldsymbol{X}^p) = \sigma(a_h)$; $a_h = \sum_{i=1}^{I} w_{ih} x_i^p$, $h = 1 \ldots H$, is the hth hidden unit's output corresponding to the input pattern $\mathbf{X}^p$, where a_h is the activation of hidden unit h, and $\sigma(.)$ is the activation function taken in this paper to be the logistic function $\sigma(z) = \frac{1}{1+e^{-Dz}}$, with D the function's sharpness or steepness and is taken to be 1.
- $\hat{Y}_o^p = \sigma(a_o)$; $a_o = \sum_{h=1}^{H} w_{ho} \Theta_h(\boldsymbol{X}^p)$ is the network's output and a_o is the activation of output unit o corresponding to the input pattern $\boldsymbol{X}^p$.

In this paper, we use the quadratic error function by squaring the difference between the predicted and actual output. We will also make use of the traditional BP algorithm as a local search method. For a complete description of BP, see for example [9].

3 Formation of Neural Networks Ensembles

3.1 The Multiobjective Learning Problem

In casting the learning problem as a multiobjective problem, a number of possibilities appear to be useful. A successfully-tested possibility that was presented in [4] is to formulate the problem as follows:

Prob1

$$\textbf{Minimize } f_1 = \Psi \tag{3}$$

$$\textbf{Minimize } f_2 = \sum_o \sum_p (\hat{Y}_o^p - Y_o^p)^2 \tag{4}$$

$$\textbf{Subject to } \Theta_h(\mathbf{X}^p) = \sigma(a_h); a_h = \sum_{i=1}^{I} w_{ih} x_i^p, \quad h = 1 \dots H \tag{5}$$

$$\hat{Y}_o^p = \sigma(a_o); a_o = \sum_{h=1}^{H} w_{ho} \Theta_h(\boldsymbol{X}^p) \tag{6}$$

where Ψ is a measure for the architecture's complexity. This measure can be a counting of the number of synapses in the network, the number of hidden units, or a combination. We may note that the VC-dimension, which is a measure for the network capacity, for a feedforward neural network with sigmoid activation function is $O(\Gamma^2)$; Γ is the number of free parameters in the network [9]. By minimizing the number of connections or nodes, we control the VC-dimension.

Another alternative that we will use in this paper is to divide the training set into two subsets using stratified sampling methods. Let $p1$ and $p2$ be the number of patterns/instances in the two subsets $Sub1$ and $Sub2$ respectively. The learning problem can be formulated as

Prob2

$$\textbf{Minimize } f_3 = \sum_o \sum_{p1} (\hat{Y}_o^{p1} - Y_o^{p1})^2 \tag{7}$$

$$\textbf{Minimize } f_4 = \sum_o \sum_{p2} (\hat{Y}_o^{p2} - Y_o^{p2})^2 \tag{8}$$

$$\textbf{Subject to } \Theta_h(\mathbf{X}^p) = \sigma(a_h); a_h = \sum_{i=1}^{I} w_{ih} x_i^p, \quad h = 1 \dots H, \quad p \in p1 \cup p2 \tag{9}$$

$$\hat{Y}_o^p = \sigma(a_o); a_o = \sum_{h=1}^{H} w_{ho} \Theta_h(\boldsymbol{X}^p), \quad p \in p1 \cup p2 \tag{10}$$

There are advantages of the formulation presented by **Prob2** over the one presented by **Prob1** [4]. To demonstrate these advantages, let us assume that we are able to generate the actual Pareto frontier for both Prob1 and Prob2. The Pareto frontier for **Prob1** will contain by definition very inferior networks in terms of training error. That is because a very small network will have an advantage in terms of f_1; therefore the frontier will contain networks which have VC-dimension less than that required to capture the underlying function in the training set. It becomes a problem in defining which network should be included in the ensemble since some of these networks are very inferior and will introduce a lot of noise in the ensemble. It is also difficult to draw a line between good and bad networks on the frontier. Apparently, one can leave it to another algorithm to decide on which network to include in the ensemble but this turns to be not simpler than training a single network that generalizes well. The formulation presented by **Prob2**, however, does not suffer from this problem. The worst training error that a network will have on $Sub1$, for example, is paired with the best training error the network will have on $Sub2$ by virtue of the Pareto definition. This entails, as we will see in the next subsection, that the range of networks on the actual Pareto frontier is bounded from right and left with the bias-variance trade-off of $Sub1$ and $Sub2$ respectively.

3.2 Bias-Variance Tradeoff

To understand the significance of the Pareto concept, let us assume that the actual Pareto frontier $Pareto_{actual}$ in a problem is composed of the following networks $\Omega_1, \ldots, \Omega_i, \ldots, \Omega_m$, where Ω_i is the ith network in a Pareto set of m networks. Let us assume that these networks are in an increasing order from left to right in terms of their performance on the first objective; that is, Ω_{i+1} has higher prediction accuracy on $Sub1$ than Ω_i, therefore the former has lower prediction accuracy on $Sub2$ than the latter by virtue of the Pareto definition.

Proposition 1. *If $Pareto_{actual}$ is the actual Pareto frontier for Prob2, then all networks on the frontier are a subset of the bias/variance trade-off set for Prob2.*

Let us define $error(\Omega_i, Sub1)$, $bias(\Omega_i, Sub1)$ and $variance(\Omega_i, Sub1)$ to be the training error, bias and variance of Ω_i corresponding to $Sub1$ respectively. The same can be done for $Sub2$. If $Pareto_{actual}$ is the actual Pareto frontier for this problem, then $error(\Omega_m, Sub1) < error(\Omega_{m-1}, Sub1)$ and $error(\Omega_m, Sub2) > error(\Omega_{m-1}, Sub2)$. Let us assume that the network's architecture is chosen large enough to have a zero bias for both $Sub1$ and $Sub2$. Then $variance(\Omega_m, Sub1) < variance(\Omega_{m-1}, Sub1)$ and $variance(\Omega_{m-1}, Sub2) < variance(\Omega_m, Sub2)$. Note that the variance represents the inadequacy of the information contained in $Sub1$ and $Sub2$ about the actual function from which $Sub1$ and $Sub2$ were sampled. Therefore, the Pareto frontier for $Prob2$ is also a bias/variance trade-off.

3.3 Assumptions

In $Prob2$, there is no guarantee that the ensemble size would be more than 1. Take the case where a network is large enough to over-fit on both $Sub1$ and $Sub2$, where the training error on each of them is zero. In this case, the Pareto set would have a single solution. It is therefore our assumption that we choose a network size small enough not to over-fit the union of *Sub*1 and *Sub*2, while it is large enough to learn each of them independently.

Our second assumption relates to the algorithm. Getting to the actual Pareto frontier is always a challenge for any algorithm. We do not assume that we have to get to the actual Pareto frontier for this method to work. What we assume instead is that there is no guarantee that the Pareto set returned by our method is the optimal ensemble as being suggested by this paper.

3.4 Formation of the Ensemble

The ensemble is formed from all networks on the Pareto frontier found by the evolutionary method. We have used three methods for forming the ensemble's gate. In voting, the predicted class is the one where the majority of networks agreed. In winner-take-all, the network with the largest activation in the ensemble is used for prediction. In simple averaging, the activations for all networks in the ensemble are averaged and the class is determined using this average.

3.5 The MPANN Algorithm

Recently, we developed the *Pareto differential evolution* (PDE) method [5]. The algorithm is an adaptation of the original *differential evolution* (DE) introduced by Storn and Price [18] for optimization problems over continuous domains. There is a large number of *evolutionary multiobjective optimization* (EMO) algorithms in the literature [7, 8], but we selected PDE in our experiments because it improved on twelve evolutionary algorithms for solving MOPs.

PDE generally improves over traditional differential evolution algorithms because of the use of Gaussian distribution, which spreads the children around the main parent in both directions of the other two supporting parents. Since the main parent is a non-dominated solution in the current population, it is found [3] that the Gaussian distribution helps to generate more solutions on the Pareto-front; which is a desirable characteristic in EMO. PDE was also tested successfully for evolving neural networks, where it is called *Memetic Pareto Artificial Neural Network* (MPANN) algorithm [1, 2, 4].

1. Create a random initial population of potential solutions. The elements of the weight matrix Ω are assigned uniformally distributed random values $U(0,1)$.
2. Apply BP to all individuals in the population.
3. Repeat
   (a) Evaluate the individuals in the population and label the non-dominated ones.
   (b) If the number of non-dominated individuals is less than 3 repeat the following until the number of non-dominated individuals is greater than or equal to 3 so that we have the minimum number of parents needed for crossover:
      i. Find a non-dominated solution among those who are not marked.

   ii. Mark the solution as non-dominated.
   (c) Delete all dominated solutions from the population.
   (d) Repeat
      i. Select at random an individual as the main parent α_1, and two individuals, α_2, α_3 as supporting parents.
      ii. **Crossover:** For all weights do

$$\omega_{ih}^{child} \leftarrow \omega_{ih}^{\alpha_1} + N(0,1)(\omega_{ih}^{\alpha_2} - \omega_{ih}^{\alpha_3}) \tag{11}$$

$$\omega_{ho}^{child} \leftarrow \omega_{ho}^{\alpha_1} + N(0,1)(\omega_{ho}^{\alpha_2} - \omega_{ho}^{\alpha_3}) \tag{12}$$

      iii. Apply BP to the child then add the child to the population.
   (e) Until the population size is M
4. Until termination conditions are satisfied.

4 Experiments

Similar to [13], we have tested MPANN on two benchmark problems; the Australian credit card assessment problem and the diabetes problem, available by anonymous ftp from ice.uci.edu [6].

The Australian credit card assessment dataset contains 690 patterns with 14 attributes; 6 numeric and 8 discrete (with 2 to 14 possible values). The predicted class is binary - 1 for awarding the credit and 0 for not. To be consistent with the literature [15], the dataset is divided into 10 folds where class distribution is maintained in each fold. Cross–validation is used where we run the algorithm with 9 out of the 10 folds for each data set, then we test with the remaining one. Similar to [13], the number of generations is 200, the population size 25, the learning rate for BP 0.003, the number of hidden units is set to 5, and the number of epochs BP was applied to an individual is set to 5.

The diabetes dataset has 768 patterns; 500 belonging to the first class and 268 to the second. It contains 8 attributes. The classification problem is difficult as the class value is a binarized form of another attribute that is highly indicative of a certain type of diabetes without having a one-to-one correspondence with the medical condition of being diabetic [15]. To be consistent with the literature [15], the dataset was divided into 12 folds where class distribution is maintained in each fold. Cross–validation is used where we run the algorithm with 11 out of the 12 folds for each dataset and then we test with the remaining one. Similar to [13], the number of generations is 200, the population size 25, the learning rate for BP 0.003, the number of hidden units is set to 5, and the number of epochs BP was applied to an individual is set to 5.

Figure 3 shows the accuracy achieved over each of the two training subsets averaged over all networks on the Pareto frontier found in each evolutionary generation for one of the ten runs. For the Australian credit card dataset, it is clear that convergence occurred early from generation 20 while for the diabetes dataset, convergence occurred from generation 90 and onwards. It is also interesting to see that the accuracy on both subsets for both datasets does not meet,

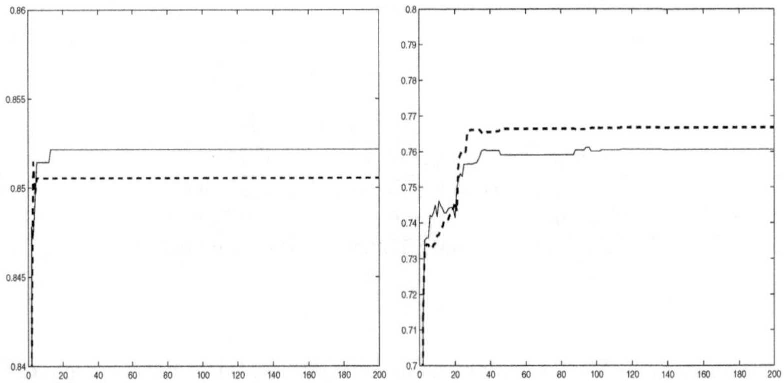

Fig. 3. The average accuracy rate for networks on the Pareto frontier found in each evolutionary generation. Left figure represents the Australian credit card dataset while the right figure represents the diabetes dataset. The x-axis is the generation number and the y-axis is the accuracy rate. The dotted line represents the first training subset while the dashed line represents the second subset

Table 1. Accuracy rates of MPANN2 for the Australian Credit Card and the diabetes datasets

	Simple averaging		Majority voting		Winner-rakes-all	
Accuracy rate	Training	Testing	Training	Testing	Training	Testing
Australian credit card dataset						
Mean	0.849	0.865	0.854	0.862	0.847	0.858
SD	0.019	0.043	0.018	0.049	0.018	0.044
Min	0.803	0.797	0.812	0.783	0.805	0.797
Max	0.877	0.928	0.879	0.928	0.876	0.913
Diabetes dataset						
Mean	0.769	0.777	0.771	0.779	0.767	0.777
SD	0.023	0.032	0.023	0.033	0.022	0.037
Min	0.737	0.723	0.737	0.723	0.737	0.723
Max	0.808	0.84	0.81	0.84	0.807	0.84

which may imply that one subset is usually difficult to learn than another. This implies that the Pareto set is not empty. On the average, there was 4 networks on the Pareto front for the Australian credit card dataset and 6 networks on the Pareto front for the diabetes dataset. The minimum number of networks on the Pareto front was 3 for both datasets and the maximum was 5 for the Australian credit cards and 11 for the diabetes. The final ensemble was formed from 4 and 6 networks for the Australian credit card and the diabetes datasets respectively.

The results of this experiment are presented in Table 1. It is interesting to see that the training error is slightly worse than the test error. In a normal situation,

Table 2. Comparing MPANN against other work for the Australian credit card and diabetes data sets. The three rates for MPANN and Liu et. al. represent the performance using simple average, vote, winner-takes-all strategies, respectively

	Accuracy Rate on Test Set	
Algorithm	Australian	Diabetes
This paper	0.865,0.862,0.858	0.777,0.779,0.777
Backprob[15]	0.846	0.749
Liu et.al. [13]	0.855,0.857,0.865	0.766,0.764,0.779

this might be interpreted as under-training. The situation here is very different indeed. Let us first explain how this training error is calculated. Recall that we have divided the training set into two subsets using stratified samples. By virtue of the Pareto definition, the Pareto frontier would have networks which are over-fitting on the first subset but not on the second and other networks which are over-fitting on the second subset but not on the first. In a strategy like "simple averaging", this will be translated into taking the average activation of all networks on the frontier when deciding on an input instance. However, the sum of the bias of a machine trained on the first subset and another machine trained on the second subset would be less than a machine trained on the union of the two subsets. Therefore, in a conventional situation, where the whole training set is used as a single training set, a biased machine would usually perform better on the training and worse on the test. In our case, this is unlikely to occur since the bias and the variance are averaged over the entire Pareto-set.

We can also observe in Table 1 that the standard deviation on the training set is always smaller than on the test. This should be expected given that the networks are trained on two subsets which constitute the entire training set.

4.1 Comparisons with Other Work

We compare our results against Backprob[15] and Liu et.al. [13]. In Table 2, we find that the Pareto-based ensemble is better than BP and equivalent to Liu et.al. with smaller ensemble size. Although it is desirable in a new method to perform better than other methods on some dataset, the main contribution from our perspective here is that the decision on which network to include in, or exclude from, the ensemble is determined automatically by the definition of the Pareto frontier. Also, the ensemble's size is automatically determined.

5 Conclusion

In this paper, we cast the problem of training artificial neural networks as a multiobjective optimization problem and use the resultant Pareto frontier to form an ensemble. The method provides a theoretically-sound approach for formation of neuro-ensembles while answering three main questions in the neural network ensemble regarding the criteria for including a network in, or excluding it from, the ensemble, the Pareto set defines the size of the ensemble, and the Pareto concept ensures that the network in the ensemble are different.

References

[1] H. A. Abbass. A memetic pareto evolutionary approach to artificial neural networks. In M. Stumptner, D. Corbett, and M. Brooks, editors, *Proceedings of the 14th Australian Joint Conference on Artificial Intelligence (AI'01)*, pages 1–12, Berlin, 2001. Springer-Verlag. 561

[2] H. A. Abbass. An evolutionary artificial neural network approach for breast cancer diagnosis. *Artificial Intelligence in Medicine*, 25(3):265–281, 2002. 561

[3] H. A. Abbass. The self-adaptive pareto differential evolution algorithm. In *Proceedings of the IEEE Congress on Evolutionary Computation (CEC2002)*, volume 1, pages 831–836, Piscataway, NJ, 2002. IEEE Press. 561

[4] H. A. Abbass. Speeding up back-propagation using multiobjective evolutionary algorithms. *Neural Computation*, to appear, 2003. 558, 559, 561

[5] H. A. Abbass, R. A. Sarker, and C. S. Newton. PDE: A pareto-frontier differential evolution approach for multi-objective optimization problems. In *Proceedings of the IEEE Congress on Evolutionary Computation (CEC2001)*, volume 2, pages 971–978, Piscataway, NJ, 2001. IEEE Press. 561

[6] C. L. Blake and C. J. Merz. UCI repository of machine learning databases, `http://www.ics.uci.edu/~mlearn/mlrepository.html`. *University of California, Irvine, Dept. of Information and Computer Sciences*, 1998. 562

[7] C. A. Coello, D. A. Van Veldhuizen, and G. B. Lamont. *Evolutionary Algorithms for Solving Multi-Objective Problems.* Kluwer Academic, New York, 2002. 558, 561

[8] K. Deb. *Multi-objective optimization using evolutionary algorithms.* John Wiley & Sons,, New York, 2001. 558, 561

[9] S. Haykin. *Neural networks - a comprehensive foundation.* Printice Hall, USA, 2 edition, 1999. 558, 559

[10] Anders Krogh and Jesper Vedelsby. Neural network ensembles, cross validation, and active learning. In G. Tesauro, D.S. Touretzky, and T.K. Leen, editors, *Advances in Neural Information Processing Systems*, volume 7, pages 231–238. MIT Press, 1995. 554

[11] Y. Liu and X. Yao. Ensemble learning via negative correlation. *Neural Networks*, 12(10):1399–1404, 1999. 554

[12] Y. Liu and X. Yao. Simultaneous training of negatively correlated neural networks in an ensemble. *IEEE Transactions on Systems, Man, and Cybernetics, Part B: Cybernetics*, 29(6):716–725, 1999. 554

[13] Y. Liu, X. Yao, and T. Higuchi. Evolutionary ensembles with negative correlation learning. *IEEE Transactions on Evolutionary Computation*, 4(4):380–387, 2000. 554, 562, 564

[14] R. McKay and H. A. Abbass. Anti-correlation: A diversity promoting mechanisms in ensemble learning. *The Australian Journal of Intelligent Information Processing Systems (AJIIPS)*, 7(3/4):139–149, 2001. 554

[15] D. Michie, D. J. Spiegelhalter, and C. C. Taylor. *Machine learning, neural and statistical classification.* Ellis Horwood, 1994. 562, 564

[16] D. E. Rumelhart, G. E. Hinton, and R. J. Williams. Learning internal representations by error propagation. In J. L. McClelland D. E. Rumelhart and the PDP Research Group Eds, editors, *Parallel Distributed Processing: Explorations in the Microstructure of Cognition., Foundations, 1, 318,*. MIT Press Cambridge, 1986. 554

[17] A. J. C. Sharkey. On combining artificial neural nets. *Connection Science*, 8:299–313, 1996. 554

[18] R. Storn and K. Price. Differential evolution: a simple and efficient adaptive scheme for global optimization over continuous spaces. Technical Report TR-95-012, International Computer Science Institute, Berkeley, 1995. 561

Predicting the Distribution of Discrete Spatial Events Using Artificial Neural Networks

Andrew Skabar

School of Information Technology, Deakin University
221 Burwood Highway, Burwood, Victoria, 3125, Australia
andrews@deakin.edu.au

Abstract. Although the development of geographic information system (GIS) technology and digital data manipulation techniques has enabled practitioners in the geographical and geophysical sciences to make more efficient use of resource information, many of the methods used in forming spatial prediction models are still inherently based on traditional techniques of map stacking in which layers of data are combined under the guidance of a theoretical domain model. This paper describes a data-driven approach by which Artificial Neural Networks (ANNs) can be trained to represent a function characterising the probability that an instance of a discrete event, such as the presence of a mineral deposit or the sighting of an endangered animal species, will occur over some grid element of the spatial area under consideration. A case study describes the application of the technique to the task of mineral prospectivity mapping in the Castlemaine region of Victoria using a range of geological, geophysical and geochemical input variables. Comparison of the maps produced using neural networks with maps produced using a density estimation-based technique demonstrates that the maps can reliably be interpreted as representing probabilities. However, while the neural network model and the density estimation-based model yield similar results under an appropriate choice of values for the respective parameters, the neural network approach has several advantages, especially in high dimensional input spaces.

1 Introduction

In this paper, the term 'discrete spatial event' is used to refer to a localized event that occurs within some spatial area. Examples of discrete spatial events include occurrences such as the sighting of an endangered species, the location of a rare mineral deposit, the outbreak of a bush fire, etc. In general, we expect that the likelihood of observing such a discrete event at some location is dependent to some degree on factors such as habitat, atmospheric conditions, basement geology, and so on. These factors are referred to as *indicator variables* and the selection of these variables

T.D. Gedeon and L.C.C. Fung (Eds.): AI 2003, LNAI 2903, pp. 567-577, 2003.

clearly depend on the nature of the problem domain. Thus, the prediction task can be thought of as the problem of discovering a mapping by which the values of various relevant input variables are combined in some way to produce an estimate of the likelihood of occurrence of a discrete event of the given type. When this likelihood is mapped over each grid element of the spatial region, the result will be referred to as a *likelihood map*.

The task of producing a likelihood map for discrete spatial events differs in several important respects from classification-based spatial mapping tasks. Consider a typical spatial classification task—land-cover classification using remote-sensed data. The problem is to classify pixels into one of several mutually exclusive land-cover classes (vegetation, desert, rock, etc.) on the basis of the pixel's radiance value in various spectral bands [1][2][3]. This can be done by selecting a set of examples representative of each of the land-cover classes, and searching for decision boundaries in input space that minimize the classification error on these training examples. The search for decision boundaries is based on the assumption that pixels belonging to the same class will exhibit similarities in their attribute values, and the problem is to discover these patterns of similarity. Once these decision boundaries have been discovered, other pixels can then be classified depending on which side of the boundaries they fall.

While the classes in the land-cover classification domain are clear and predefined, this is not the case when producing likelihood maps. The domains for which we wish to produce likelihood maps are inherently probabilistic, and we cannot, for example, speak of a pixel *belonging* to the class `mineralization_present` in the same sense in which we can say that a pixel *belongs* to the class `vegetation_cover`. In the case of producing likelihood maps, class-membership can only be interpreted as representing a probability, or some other measure of likelihood (belief, confidence, etc). [1]

A second difference between classification-based mapping tasks and likelihood mapping tasks concerns the nature of the data to be used for training. Classification algorithms normally require that the training set consists of examples representative of each of the classes in the domain, and that the attribute vectors for these examples are accompanied with a *label* indicating to which class the example belongs. However, in the case of likelihood mapping, class membership is generally unknown; aside from a those pixels which are known (perhaps from historical records) to correspond to an occurrence of the discrete spatial event, all other pixels are in this sense *un*labeled.

It is also significant to note that tasks such as land-cover classification can also be performed using *unsupervised* learning algorithms; i.e., algorithms which do not require the class membership of training examples to be supplied [4]. These algorithms form *clusters* of examples based on some measure of the proximity of the examples in attribute space, and these clusters can then be identified with land-cover classes in a post-clustering operation, thus forming a classifier. Direct clustering approaches are not expected to be of much use in the prediction of (rare) spatial events such as the presence of mineral deposits, as a pixel is likely to be more similar in its attribute values to pixels in its immediate neighbourhood than it is to other mineralized pixels

---

[1] Unless stated otherwise, *likelihood* will be interpreted as *probability*.

[5]. This means that the clusters will not necessarily be good indicators of mineralization potential.

Thus, from an inductive learning perspective, the problem of producing likelihood maps for discrete spatial events can be characterized as one that is inherently probabilistic, and in which the training set is made up of a small number of labelled positives accompanied by a large corpus of unlabeled data. Because the search mechanism in most conventional classification and function approximation algorithms is based either directly or indirectly on some measure of the performance of the current hypothesis in classifying or approximating the training data, these conventional approaches are not generally suitable for this task.

This paper describes a data-driven approach by which Artificial Neural Networks (ANNs) can be trained to represent a mapping function characterising the probability that an instance of a discrete spatial event will occur over some grid element of the spatial area under consideration. The paper is set out as follows. Section 2 provides a formal specification of the task. Section 3 describes the neural network structure, and specific training requirements. Section 4 presents a case study in which the technique is applied to the task of mineral prospectivity mapping in the Castlemaine region of Victoria. Section 5 concludes the paper.

2 Specification of Task

In the context of GIS, likelihood mapping can be seen as a process whereby a set of input maps, each representing a distinct indicator variable, are combined using some function to produce a single map which ranks areas according to their potential to host an occurrence of the discrete spatial event. There are two general approaches by which such a mapping function can be derived: (i) a *knowledge-based* approach; and (ii) a *data-driven* approach. In the knowledge-based approach, the input maps are combined on the basis of expert knowledge. This knowledge would normally be expressed in terms of symbolic rules (e.g., a rule-based system). In contrast, the *data driven* approach attempts to discover, or *learn*, the function by measuring in some way the association between mapped predictor variables and a response map that indicates the locations of known occurrences of the discrete spatial event. The major assumption made in the data-driven approach is that the known occurrences constitute an adequate and unbiased sample of the true occurrences in the region, and that by discovering signatures for these known occurrences, other regions in which there is a high likelihood of occurrence will also be highlighted [6].

The likelihood mapping technique described in this paper is based on the following assumptions:

1. occurrences of the discrete spatial event are distributed according to some unknown distribution;
2. the discrete event occurrences which have been observed are random selections from the above distribution; and
3. indicator variables provide clues to finding other occurrences of the event.

In general these indicator variables can themselves be thought of as maps; that is, it is assumed that each pixel is assigned a value for each indicator variable.

The general task of producing a likelihood map for discrete spatial events can be expressed as follows:

Given:

1. Background information provided by m layers of data, each of which represents the value of a distinct indicator variable x_i at each pixel p;
2. A subset of pixels, each of which is known from historical data to contain one or more occurrences of the spatial event;

Find:

A function f that assigns to each pixel p in the study area a continuous value on the interval [0,1] that represents the probability that pixel p contains one or more of the spatial events, given the evidence supplied by the background information.

Thus, assuming that the evidence for a pixel p is described by a vector $\boldsymbol{x} = (x_1, \ldots, x_m)$, the objective is thus to learn a function f: $X \rightarrow [0,1]$, where $f(\boldsymbol{x})$ represents the posterior probability that p contains an occurrence of the discrete event. This task is depicted schematically in Figure 1.

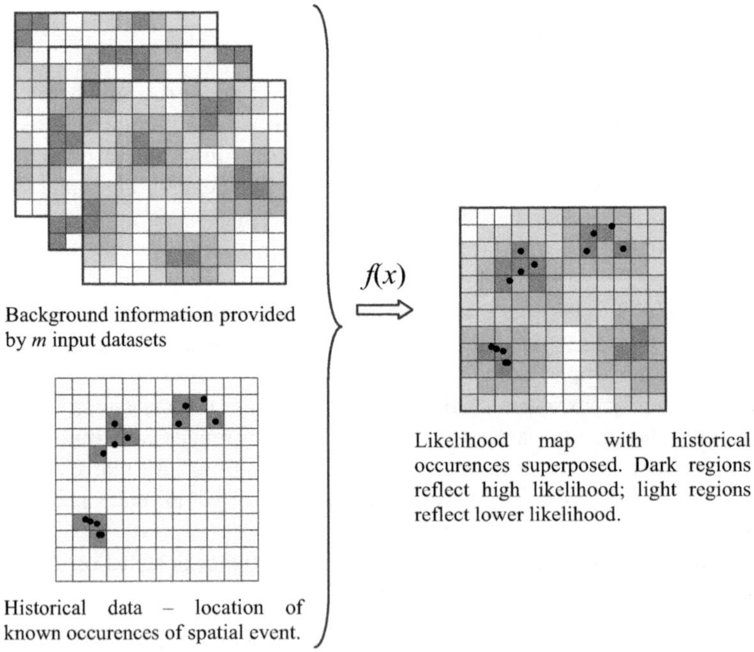

Fig. 1. Constructing a likelihood map. The output map represents the likelihood that a grid element will host an occurrence of the spatial event, and should reflect the location of known occurrences

It is assumed that each input layer takes the form of a *raster* image, i.e., a rectangular array of grid elements, where the intensity of each element represents the value of some variable over the physical region corresponding to the pixel. Because the raw values for input data may be in a variety of formats, due in large part to the method with which the data were acquired (e.g., point-sampled geochemical data, remote-sensed magnetics data, etc), conversion of raw input data into this form will generally require pre-processing operations including interpolation, scaling, vector-to-raster conversion, etc. The next section describes how a feedforward neural network can be trained to represent the mapping function $f(x)$.

3 Likelihood Mapping Using Artificial Neural Networks

The domain characteristics described in the previous section have two very important implications for the use of neural networks to model the mapping function—both in regard to the selection of error function used for neural network training, and also for the selection of training set examples. This section demonstrates that the success of the task depends critically on these two factors. The first part describes the network structure and substantiates the use of cross-entropy error for back-propagation training. The second part addresses the issue of training set selection.

3.1 Selection of Error Reduction Function

Recall from Section 2 that the target function $f(\mathbf{x})$ represents the posterior probability that a cell contains at least one occurrence of the spatial event. The function $f(\mathbf{x})$ can be discovered as follows. Assume the existence of a *binary* function $g(\mathbf{x})$ that represents the presence or absence of a *known* occurrence of the event in a pixel p with attribute vector $\mathbf{x} = (x_1, x_2, \ldots, v_m)$, where x_k is the value of the kth variable for pixel p and m is the number of input variables. If pixel p contains a known occurrence, then $g(\mathbf{x}) = 1$, otherwise $g(\mathbf{x}) = 0$. Now let $h(\mathbf{x})$ be a *probabilistic* function whose output is the *probability* that $g(x) = 1$. The objective is to learn the function h: $X \rightarrow [0,1]$, such that $h(\mathbf{x}) = P(g(\mathbf{x}) = 1)$.

The function $h(x)$ is to be represented by a feed-forward neural network. Because the network is required to produce only a single value for each input example, only one output unit is required. Because the output at this unit is to represent a probability, the output values should be bounded between 0 and 1, and the simplest way of arranging this is to use a sigmoidal activation function on the output node.

The function we seek to discover is the maximum likelihood function (or *maximum likelihood hypothesis*), i.e., the hypothesis h_{ML} that results in the highest probability of observing the given data D [7][8]. Suppose that an example $\mathbf{x}_i$ is drawn randomly from the training set. By the definition of $h(\mathbf{x})$, the probability that $\mathbf{x}_i$ has a target output of 1 is $h(\mathbf{x}_i)$, and the probability that is has a target output of 0 is $1 - h(\mathbf{x}_i)$. The probability of observing the correct target value, given hypothesis h, can therefore be expressed as

$$P(d_i \mid h, \mathbf{x}_i) = h(\mathbf{x}_i)^{d_i} (1 - h(\mathbf{x}_i))^{1-d_i} \tag{1}$$

where $\mathbf{x}_i = (x_{i1}, x_{i2}, \ldots, x_{im})$ is the attribute vector for pixel p_i, $h(\mathbf{x}_i)$ is the value of the hypothesis h applied to vector $\mathbf{x}_i$ and $d_i = 1$ if pixel p_i contains a known deposit, and $d_i = 0$ otherwise. Assuming that the examples in the training set are independent, the probability of observing the correct value for *all* the examples is given by

$$P(D \mid h) = \prod_{i=1}^{m} h(\mathbf{x}_i)^{d_i} (1 - h(\mathbf{x}_i))^{1-d_i}) \tag{2}$$

where m is the number of examples. h_{ML} is the function for which $P(D|h)$ is a maximum:

$$h_{ML} = \max_{h \in H} \prod_{i=1}^{m} h(\mathbf{x}_i)^{d_i} (1 - h(\mathbf{x}_i))^{1-d_i}) \tag{3}$$

Taking the natural logarithm, and using the fact that $\ln(h_{ML})$ is a monotonic function of h_{ML}, this can be shown to be equivalent to

$$h_{ML} = \max_{h \in H} \sum_{i=1}^{m} \{d_i \ln h(\mathbf{x}_i) + (1 - d_i) \ln(1 - h(\mathbf{x}_i))\} \tag{4}$$

Alternatively, the maximum likelihood function can be expressed as a minimization:

$$h_{ML} = \min_{h \in H} \left[-\sum_{i=1}^{m} \{d_i \ln h(\mathbf{x}_i) + (1 - d_i) \ln(1 - h(\mathbf{x}_i))\} \right] \tag{5}$$

The error function to be minimized is commonly referred to as *cross-entropy* [9]. Alternatively, cross-entropy can be expressed as

$$E = -\sum_{i=1}^{m} \ln\left(1 - \left| d_i - h(\mathbf{x}_i) \right|\right) \tag{6}$$

which makes the interpretation of distance between d_i and $h(\mathbf{x}_i)$ more intuitive [10].

In regard to back-propagation training of the network, it can be shown that the output node delta term for cross-entropy error reduction is $\delta_{cross\_entropy} = (d_i - h(\mathbf{x}_i))$, which differs from the delta term for quadratic error reduction which contains an additional factor $h(\mathbf{x}_i)\,(1 - h(\mathbf{x}_i))$.

3.2 Training Set Selection

Up to this point, the network configuration and training scheme described are no different to that used in many binary classification tasks to which neural networks are commonly applied. However, one of the requirements of the map we wish to produce is that it represents the *posterior probability* of a pixel containing an occurrence of spatial event of the given type. In other words, the sum of probabilities over all pixels in the map should sum to the actual number of known occurrences. This has implications regarding the choice of examples used for training. To illustrate, consider a network trained using only a *subset* of the pixels in the study region. The sum of network outputs over this subset will be equal to the number of pixels in the subset which contain known occurrences of the event. Clearly, the output of the trained network will depend on the particular subset of pixels used for training. This is not necessarily a problem for the interpretation of the output values as posterior probabilities when applied to the entire study region, as the outputs can be simply rescaled to sum to the number of known occurrences. The problem, rather, is that the mapping function itself will be highly sensitive to the choice of training examples, especially if the number of training examples is small. This is a major problem, since we cannot be sure that a network trained using a subset of pixels will be representative of the true posterior probability that a pixel contains an occurrence of the event.

To avoid this problem the network should be trained using *all* examples. Thus, rather than using the traditional approach of dividing the available data into disjoint training and test sets (and possibly also a validation set), the approach taken here is to use all pixels in the study region for training. Note however, that this does not preclude the possibility of testing the model. Rather than holding out a set of examples for testing, we hold out only the *label* of these examples. That is, we select a set of known positives (i.e., pixels known to contain an occurrence of the event) and assign these examples a target label of 0 rather than 1. We then test to see whether the trained network assigns a high probability to these examples.

To have outputs sum to the number of historically known occurrences may at first sight appear to be at odds with the assumption that the network output represents the posterior probability of an occurrence. This is because we don't know the true *a priori* probability of an occurrence. However, this is consistent with the assumption for data-driven mapping that the known occurrences constitute an adequate and unbiased sample of the true occurrences in the region, and that by discovering signatures for these known occurrences, other regions of high potential will also be highlighted.

It was shown previously that the appropriate error reduction function for this task is cross-entropy. The use of cross-entropy error is, in fact, critical. While quadratic error reduction can often approximate the posterior probability of class membership on binary classification tasks [8], its use in this domain will lead to results that cannot be interpreted as probabilities (in fact the network output may not even approximate probabilities). The use of quadratic error reduction is justified (under maximum likelihood considerations) on problems in which the training data can be modelled by normally distributed noise added to the target function value, an assumption that usually holds when approximating continuous-valued functions (i.e., regression problems). However, in the spatial predictivity domain, target values are binary and noise follows a binomial, and not Gaussian, distribution. The problem is further compounded by the fact that there is usually a gross imbalance between the number of examples known to contain an historical occurrence and the remaining examples (e.g., mineralization is a rare event).

We now summarize the important points from this section. Firstly, because the objective is to estimate probabilities of (binary) class membership, the appropriate error-reduction function is *cross-entropy* error, and not *quadratic* error. Secondly, because the sum of output values over the map should equal the number of historically known occurrences in the study region, it is necessary to discard the traditional approach of using separate training and test sets. Rather, all examples in the study region should be used for training.

4 Case Study: Mineral Potential Mapping

Mineral potential mapping is the process of producing a map that depicts the favourability of mineralization occurring over a specified region. The map should reflect the location of known mineral occurrences and also predict the distribution of areas of high mineral potential where little or no mining activity currently exists. Mineral potential mapping is typically used in the early stages of mineral exploration to identify regions favourable for follow-up exploration activity [6]. This section describes the application of the technique described in Sections 2 and 3 to the production of a mineral potential map for a study area near the Castlemaine region of Victoria, using a range of geological, geophysical and geochemical indicator variables.

The study region extends from a Northwest corner with coordinates 251,250mE, 5,895,250mN, to a Southeast corner with coordinates 258,250mE, 5,885,000mN. Input data consisted of 16 magnetic, radiometric, geochemical and geological variables made available by the Department of Natural Resources and Environment, Victoria. Based on a grid-cell resolution of 50m by 50m, the study region described above can be represented by a rectangular grid consisting of 141 cells in the horizontal direction and 206 cells vertically. The number of documented known reef gold deposits in the study area is 148. In addition, 938 other gold deposits locations have been recorded, but the historical information on these does not indicate the type of occurrence (i.e., reef or alluvial) or their significance. Additional information on Victorian geology can be found in [11] and [12]. The Castlemaine Goldfield is described in [13]. Full details of results presented here, including pre-processing input data etc., can be found in [14].

In order to test whether the output map indeed represents posterior probabilities, it useful to compare the technique with an alternative technique for estimating probabilities. For these purposes, the output map can be compared with maps produced using a non-parametric density estimation-based approach. This density estimation-based method consists in estimating the class-conditional and class unconditional probability density functions ($P(\mathbf{x}|D)$ and $p(\mathbf{x})$ respectively), and combining these using Bayes' theorem. The results presented here are based on estimating density functions using the Parzen window method with Gaussian kernels [7].

It is useful to compare the two predictive models on the basis of the favourability *ranking* that they assign to pixels. A convenient means of performing such a comparison is to plot a graph of cumulative deposit frequency versus cumulative area. Such a graph can be constructed by ranking pixels according to their assigned posterior probability value, and plotting the cumulative frequency of deposits (either predicted or observed) against cumulative area as the posterior probability is increased from its minimum to its maximum value. Fig. 2(a) shows the curve produced using a neural network with a single hidden layer containing 6 units. Fig. 2(b) was produced using the Parzen window approach with a σ value of 0.2. (σ is the smoothing parameter that determines the width of the Gaussian kernel surrounding each point). As a guide to interpreting the graphs, consider the dark solid curve of Fig. 2(a), which represents prediction results for the 148 training deposits using the neural network model. It can be seen that a cumulative area value of 80% corresponds to a cumulative deposits value of approximately 25% (this is shown on the graph). This is equivalent to stating that 75% of the deposits occur in the 20% of the map depicted as being most favourable to host mineralization. However, of more interest is the predictive performance on the 938 holdout test deposits (represented by the grey curves), and it can be seen that both models yield similar performance in predicting these deposits. The broken curves simply represent the cumulative sum of posterior probabilities converted to a percentage, and it is useful to compare these with the grey curves. Note that some of the 938 holdout examples are alluvial, i.e., transported by water from original mineralization location. Predictive performance on test deposits would be improved significantly if the alluvial deposits were removed, however sufficient information on the type of these deposits is not available.

The fact that two models provide (almost) identical prediction curves does not imply that the two maps are identical, as the curves only provide a statistical measure of predictive performance. In order to compare the actual maps, the product moment correlation coefficient, *r*, can be calculated. Table 1 shows the correlation between maps produced using various numbers of hidden units (for neural network models) and sigma values (for the density estimation-based method). A relatively high correlation of 0.59 occurs between a Parzen window model with σ=0.2 and neural network models with 6 and 9 hidden layer units. This suggests that the two models do indeed yield similar results, and we can be reasonably confident that these do represent mineralization probability.

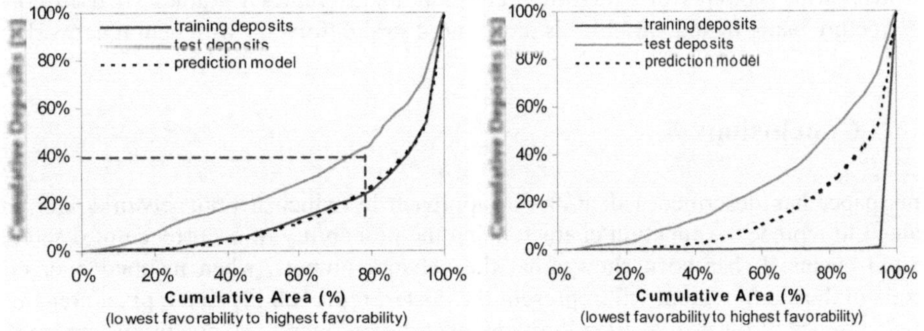

Fig. 2. Cumulative deposits versus cumulative area and cumulative deposits versus posterior probability graphs. (a) Neural network approach (6 hidden units); (b) Non-parametric density estimation approach (σ=0.2)

Table 1. Correlation between networks trained using CE error reduction and Parzen window models

Hidden	σ (smoothing parameter for Parzen

	Window model)					
units	0.05	0.10	0.15	0.20	0.25	0.50
3 hlu	0.13	0.20	0.31	0.39	0.36	0.13
6 hlu	0.33	0.44	0.56	0.59	0.46	0.14
9 hlu	0.34	0.46	0.58	0.59	0.45	0.12

An obvious question arising out of these experimental results is why we should bother with neural network models at all if non-parametric density estimation-based models yield similar results. While the neural network model and the density estimation-based model will provide similar results under an appropriate choice of values for the respective parameters, the neural network approach has several advantages:

(i) *Fewer free parameters to be determined subjectively*. The only free parameter requiring subjective determination in the neural network model is the number of hidden layer units. In the general case, the density estimation-based approach requires estimating a smoothing parameter for each variable;
(ii) *Feature weighting*. While the density estimation-based approach implicitly assumes that all variables are equally relevant, neural networks make no such assumption; however, it cannot in general be determined how a network will react to irrelevant variables;
(iii) *Better able to deal with high-dimensional input spaces*. Neural networks are more reliable than density estimation-based methods when applied to datasets containing large numbers of input variables;
(iv) *Storage*. The neural network stores the mapping function; the density estimation-based method requires that all examples be stored in memory;
(v) *Better able to deal with mixed data*. While density estimation-based methods can deal with mixtures of categorical and continuous-values variables (if used correctly), using mixed variables is much more straightforward with neural networks.

5 Conclusions

This paper has described a data-driven approach by which neural networks can be trained to represent a function characterising the probability of occurrence of discrete spatial events. It has been shown that the network outputs, when mapped over all pixels in the study region, will represent the posterior probability that a pixel contains an occurrence of the spatial event provided that the following two conditions are satisfied: (i) all examples in the study region are used for training, and (ii) *cross-entropy*, not *quadratic*, error reduction is used. A case study has described the application of the technique to the task of mineral prospectivity mapping in the Castlemaine region of Victoria using a range of geological, geophysical and geochemical input variables. Comparison of the maps produced using neural networks with maps produced using a density estimation-based technique demonstrates that the maps can reliably be interpreted as representing probabilities. However, while the neural network model and the density estimation-based model yield similar results under an appropriate choice of values for the respective parameters, the neural network approach has several advantages, especially in high dimensional input spaces.

References

[1] D.L. Civco, "Artificial neural networks for land-cover classification", *International Journal of Geographical Information Systems*, vol. 7, no. 2, 173-86, 1990.

[2] Miller, D.M., Kaminsky, E.J. & Rana, S. 1995, 'Neural network classification of remote-sensing data', *Computers and Geosciences*, vol. 21, no. 2, pp. 377-86.

[3] Mohanty, K.K. & Majumdar, T.J. 1996, 'An artificial neural network (ANN) based software package for classification of remotely sensed data', *Computers & Geosciences*, vol. 22, No. 1, pp. 81-87.

[4] B. Everitt, *Cluster Analysis*, London, Heinemann, 1980.

[5] F.P. Agterberg, *Geomathematics: Mathematical Background and Geo-Science Applications*, Elsevier Scientific Publishing Company, Amsterdam, 1974.

[6] G.F. Bonham-Carter, Geographic Information Systems for Geoscientists: Modelling with GIS, Elsevier Science Ltd, U.K., 1994

[7] R.O. Duda & P.E. Hart, *Pattern Recognition and Scene Analysis*, John Wiley & Sons, New York, 1973.

[8] C. Bishop, *Neural Networks for Pattern Recognition*, Oxford University Press, Oxford, 1995.

[9] E.B. Baum & F. Wilczek, "Supervised learning of probability distributions by neural networks", in *Neural Information Processing Systems*, ed. D.Z. Anderson, American Inst. of Physics, New York, pp. 52-61, 1988.

[10] M. Schumacher, R. Rossner & W. Vach, "Neural networks and logistic regression: part 1", *Computational Statistics & Data Analysis*, vol. 21, pp. 661-82, 1996.

[11] G.W. Cochrane, G.W. Quick & D. Spencer-Jones (eds), *Introducing Victorian Geology*, 2nd edn, Geological Society of Australia Incorporated (Victorian Division), Melbourne, Australia, 1995.

[12] I. Clark & B. Cook (eds), *Victorian Geology Excursion Guide*, Australian Academy of Science, Canberra, Australia, 1988.

[13] C.E. Willman, Castlemaine Goldfield: Castlemaine-Chewton, Fryers Creek 1 : 10 000 Maps Geological Report, Geological Survey Report 106, Energy and Minerals Victoria, 1995.

[14] A. Skabar, *Inductive Learning Techniques for Mineral Potential Mapping*, PhD Thesis, School of Electrical and Electronic Systems Engineering, Queensland University of Technology, Australia, 2000.

Learning Action Selection Network of Intelligent Agent

Eun-Kyung Yun and Sung-Bae Cho

Department of Computer Science, Yonsei University
134 Shinchon-dong, Sudaemoon-ku, Seoul 120-749, Korea
ekfree@candy.yonsei.ac.kr, sbcho@cs.yonsei.ac.kr
http://sclab.yonsei.ac.kr

Abstract. Behavior-based artificial intelligent system is to derive the complicated behaviors by selecting appropriate one from a set of basic behaviors. Many robot systems have used behavior-based systems since the 1980's. In this paper, we propose new method to create the sequences of behaviors appropriate to the changing environments by adding the function of learning with Learning Classifier System to P. Maes' action selection network. Links of the network need to be reorganize as the problem changes, because each link is designed initially according to the given problem and is fixed. Learning Classifier System is suitable for learning of rule-based system in changing environments. The simulation results with Khepera robot simulator show the usefulness of learning in the action selection network by generating appropriate behaviors.

1 Introduction

In behavior-based artificial intelligent system, the architecture is not to compose of sequential modules, such as perception, modeling, planning, and task execution, but to derive more complicated behaviors by selecting appropriate one from a set of basic behaviors [1] [2]. Tinbergen has proposed Hierarchical Decision Structure for action selection [3]. There are some related publications such as Brooks' Subsumption Architecture [4], Tyrrell's hierarchical structure [5], and Mataric's group studying learning via human behavior imitation [6].

Learning in robots is known to be a difficult problem, due to the uncertainty in sensing and action, partial observability of the robot's environment, and non-stationarity of the world. Learning in behavior-based system focuses on the efficiency improved in short time instead of long optimization [7]. Effective action selection is a matter of primary concern in behavior-based system. It can be easily supported by using the well-known reinforcement learning whose policy selects the action that maximizes rewards. Some examples of reinforcement learning in the context of behavior-based system demonstrated hexapod walking [1] and box-pushing [8]. Besides in multi-robot systems, M. Dorigo and M. Colombetti [9] applied the idea of Shaping to the original reinforcement learning.

In this paper, we propose the new dynamic action selection network adding Learning Classifier System (LCS) in order to adapt to the changing environment

T.D. Gedeon and L.C.C. Fung (Eds.): AI 2003, LNAI 2903, pp. 578–589, 2003.

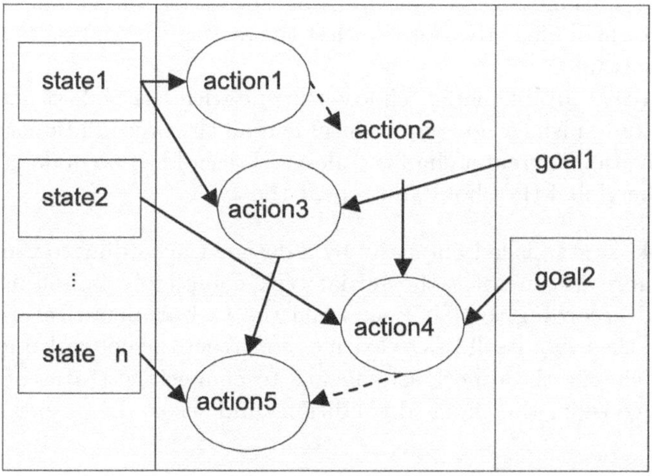

Fig. 1. The structure of the action selection network

like the real world. Maes' network has a shortcoming of selecting inappropriate actions when the task changes because it is difficult to modify the structure. In order to solve this problem, we improve the network appropriate to the changing environment by applying the function of learning.

2 Action Selection Network

Maes' action selection network is a distributed, non-hierarchical network [10]. It consists of action nodes, motivation or goal nodes, and environmental state nodes (see Fig. 1). These nodes have the different types of links that encode various relationships and stimulate each other. The compositions of a node are preconditions, add list, delete list, activation level, and the code that is run if the node is executed. Fig. 2 shows the elements of a node in action selection network.

If the proposition X about environment is true and X is in the precondition list of node A, there is an active link from the state node X to action node A. If goal Y has an activation greater than zero and Y is in the add list of node A, there is an active link from goal node Y to action node A. Internal links include predecessor links, successor links, and conflicter links. Fig. 3 shows the conditions of each internal link.

The following is the procedure to select an action node to be executed at each step.

1. Calculate the excitation coming in from the environment and the motivations.

2. Spread excitation along the predecessor, successor, and conflicter links, and normalize the node activations so that the average activation becomes equal to the constant π.
3. Check any executable nodes, choose the one with the highest activation, execute it, and finish. A node is executable if all the preconditions are true and if its activation is greater than the global threshold. If no node is executable, reduce the global threshold and repeat the cycle.

Every link is established manually by a designer according to the initial task given. Therefore it can solve the initial task completely but it needed to be reconstructed according to that task when the task changes. In order to solve this problem, the robot itself has to learn to construct the network and reorganize all links. To change the robot tasks means to change the status of robot. The robot has to consider the changed conditions and select the proper actions.

3 Learning Action Selection Network with LCS

In machine learning, neural networks and genetic algorithm are widely used. However, in this paper, learning classifier system is used for learning of the action selection network. It is the learning algorithm for rule-based system and is appropriate to this case. The state nodes and goal nodes, which are elements of action selection network, can be considered as a kind of conditions. Therefore these are easy to convert to the rule-based system. This is the reason that we select the learning classifier system for learning.

3.1 LCS (Learning Classifier System)

LCS, a kind of machine learning technique, was introduced by Holland and Reitman in 1978 [12]. LCS has two different learning procedures. One is to learn via mixing given rules (Credit Assignment System) and the other is to learn via creating useful rules as possible (Rule Discovery System). It is very appropriate

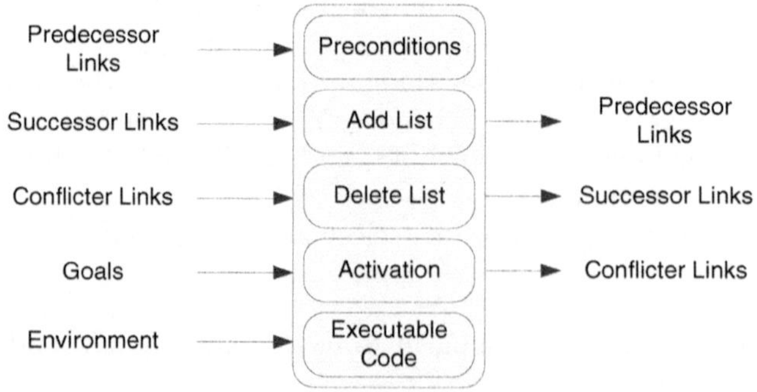

Fig. 2. The elements of a node in action selection network

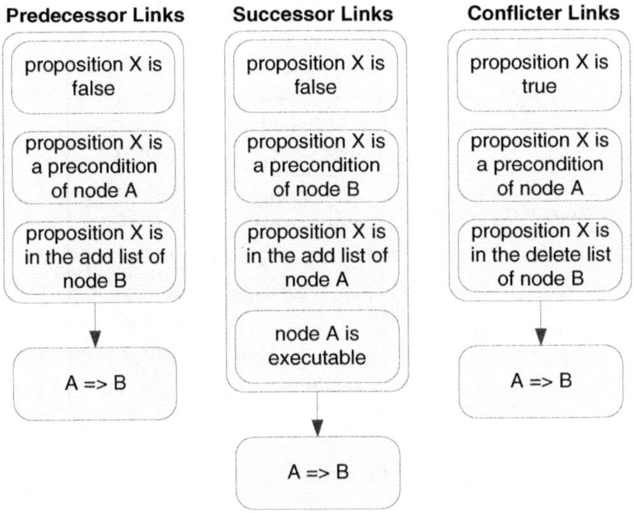

Fig. 3. The internal links

to be adapted to a changing environment. Classifier system consists of several rules, so-called classifiers. One classifier has one or more condition parts and one action part. The condition of a classifier consists of ternary elements {0, 1, #} and the action part consists of {0, 1}. The character '#' means "don't care" and can take either '0' or '1'.

In the competition of classifiers, the value of strength gives a measure of the rule's past performance. That is, the higher a classifier's strength the better it has performed. In addition, each classifier has the value of specificity, which is the number of non-# symbols in the condition part. LCS consists of three modules as shown in Fig. 4.

- Classifier System: The system compares input messages with the condition part of all classifiers in the classifier list and performs matching. This acts by bit-to-bit. The matching classifiers enter competition for the right to post their output messages, and only the winner of the competition actually posts messages. The measure of each classifier is the value of bid as follows:

$$bid = c \times specificity \times strength$$

  where c: a constant less than 1, specificity: specificity of the classifier condition, condition's length minus the number of '#' symbols, and strength: the measure of confidence in a classifier. When feedback comes from its environment, the strength of the winner is recomputated.
- Credit Assignment System: When the rewards from the environment transmit, they cause the strength of the classifiers to change in order to reflect their relevance to the system performance. This system is to classify rules

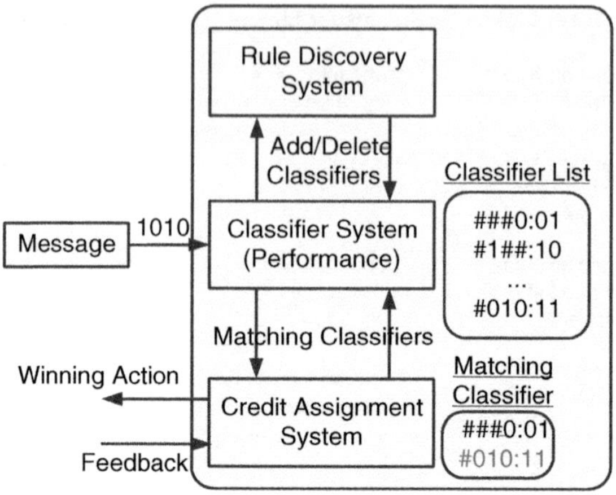

Fig. 4. The structure of LCS

in accordance with their usefulness. The classical algorithm used for this purpose is the Bucket Brigade algorithm [12].

- Rule Discovery System: The system uses the genetic algorithms to discover new rules. Genetic algorithms are stochastic algorithms that have been used both as optimization and as rule discovery mechanism. They work modifying a population of solutions (in LCS, a solution is a classifier). Solutions are properly coded and a fitness function is defined to relate solutions to performance. The value from this function is a measure of the solution quality. The fitness of a classifier is set by its usefulness calculated with a credit apportionment system instead of a fitness function.

In general, while classifier system and credit apportionment system are interested in the classifiers with better performances, rule discovery system is interested in the classifiers with worse performances. If the only classifiers with high scores survive, the system cannot discover new classifiers.

3.2 Learning with LCS

Our goal is to improve the robot's ability to solve problems using LCS, a kind of learning methods, in the action selection network. Firstly the rules of LCS, the basic elements, have to be defined for learning links in the network. The rules include the state nodes, the extent of the problems, and links between the nodes. They do not include the action nodes. When the rule is fixed, or the network structure is decided, it calculates the activation level and selects the action node with the highest activation value. Fig. 5 shows the structure of the rule defined. The extent of the problems and state nodes are determined through the robot's current states from environment.

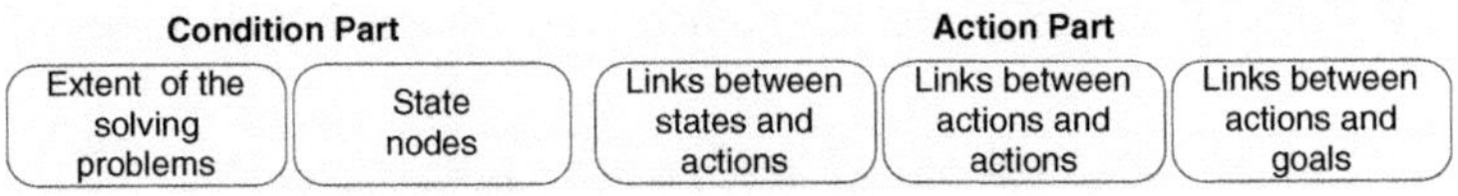

Fig. 5. The structure of the rule

While original action selection network includes three kinds of links between the nodes, we redefine the excitatory links for LCS. Excitatory links indicate predecessor links and successor links because they have similar meaning. In addition, the classifier lists are initialized in accordance with the purpose of tasks, not randomly. That is, the initial structure of network is converted into rules.

4 Experimental Results

4.1 Experimental Environments

The task of the robot is to reach the specific goal position in the shortest path. The network has only two actions of going straight and avoiding obstacles initially. Therefore the robot is going to complete the task using only two actions regardless of the goal position. However as shown in Fig. 6, other actions are better than the action of going straight if the robot considers where it exists. The initial network is converted into the rules, which form the classifier list, and the network is improved using LCS in order to select the proper action even though the position of robot changes.

Initial network is constructed as shown in Fig. 7 and Table 1 shows preconditions, add list elements, and delete list elements for each action node.

The learning of action selection network is tested on the Khepara robot simulator [13]. It has 8 distance sensors and light sensors. The distance sensors return the value ranging between 0 and 1023. 0 means that no object is perceived while 1023 means that an object is very close to the sensor. Intermediate values indicate approximate distance between the sensor and the object. Therefore we define the state node,'obstacle around' is true if the distance value is more than 900, and nothing around is true otherwise.

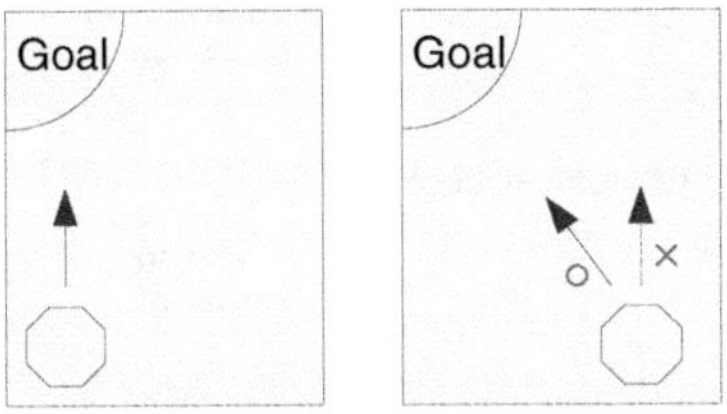

Fig. 6. The action selection for the position of robot

Table 1. Preconditions, add list elements, and delete list elements for each action node

Action node	Preconditions	Add list	Delete list
Avoiding obstacles	Obstacles around	Nothing around	Obstacles around
Going straight	Nothing around	Obstacles around	Nothing around
Turn left an angle of 45 degrees and go straight	Nothing around	Obstacles around	Nothing around
Turn right an angle of 45 degrees and go straight	Nothing around	Obstacles around	Nothing around

Table 2 shows the composition of a classifier in LCS. Firstly the extent of solving problems is set by checking the location of the goal robot reaches. That is, 4 bits indicate up,down,right,left. For example, 1000 means the goal is upper and 1010 means the goal is left. In LCS, genetic operators act every 100 iterations.

4.2 Results

Before learning, the robot always begins to move by going straight as shown in Fig. 8. Therefore, it is impossible to reach the goal in the shortest path whenever the initial position of the robot changes. The goal of robot is the left upper corner and robot's initial position is set randomly in the experiments. The goal of robot is the left upper corner and robot's initial position is set randomly in the experiments. Many of the actions of going straight were selected before learning (see Fig. 8) but through the learning, other actions rather than the

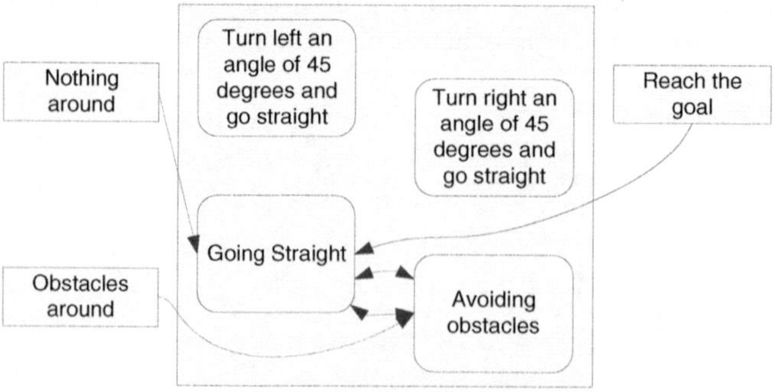

Fig. 7. The structure of initial network: Dotted lines mean inhibitory links and solid lines mean excitatory links in the internal links

Table 2. The composition of a classifier

	Bit	Descriptions	
Condition part (6 bits)	C1	Position of goal = upper	
	C2	Position of goal = under	
	C3	Position of goal = left	
	C4	Position of goal = right	
	C5	Nothing around	
	C6	Obstacles around	
Action part (17 bits)	A1	Links between "Nothing around" node and action nodes, 11: Going straight, 10: Turn left, 01: Turn right	External links (from sensors of the environment, or state nodes)
	A2		
	A3	Links between "Obstacles around" node and "Avoiding obstacles" action node	
	A4	Link from "Going straight" to "Avoiding obstacles"	Internal links (00: no link, 01: excitatory links, 10: inhibitory links)
	A5		
	A6	Link from "Avoiding obstacles" to "Going straight"	
	A7		
	A8	Link from "Turn left" to "Avoiding obstacles"	
	A9		
	A10	Link from "Avoiding obstacles" to "Turn left"	
	A11		
	A12	Link from "Turn right" to "Avoiding obstacles"	
	A13		
	A14	Link from "Avoiding obstacles" to "Turn right"	
	A15		
	A16	Links between goal nodes and action nodes 11: Going straight, 10: Turn left, 01: Turn right	External links (from goals)
	A17		

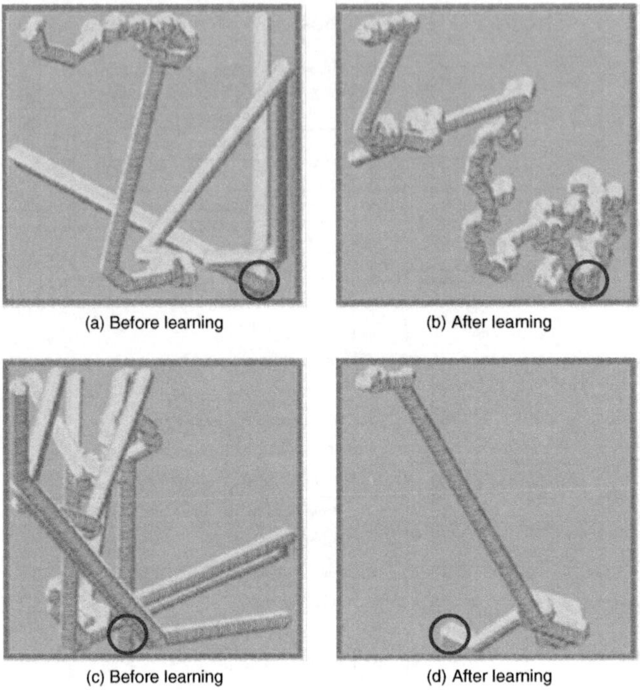

Fig. 8. Action results according to different initial positions of the robots

action of going straight were selected. Fig. 8 shows the results of learning when the initial position of robot is same. While in Fig. 8(a) the robot reached the goal in 5533 steps, it reached the goal in 1457 steps after learning (see Fig. 8(b)). Also as shown in Fig. 8(c) and Fig. 8(d), while the robot reached goal in 5435 steps before learning, it reached in 448 steps after learning.

Fig. 9 shows an example of the networks with higher fitness after robot's learning. As comparing with the initial network (see Fig. 7), learning procedures make the links between nodes adapt to the changes of task. Firstly the robot moved only by going straight, but after learning it moved by selecting the actions of reaching the goal faster as shown as Fig. 8 and Fig. 9.

The learning mode performance measures how well the LCS is learning to perform the correct behavior in an environment. In order to evaluate its performance, we use the simple measurement of learning performance, the ratio of the number of correct responses during epoch to the epoch length, as follows [14]:

$$\frac{Number\ of\ correct\ responses\ during\ epoch}{Epoch\ length}$$

Fig. 10 shows the average number of correct responses at every epoch. It proves that the robot can select appropriate behaviors via learning. Table 3 shows the average steps of robot's reaching the goal. It means the robot reached the goal much faster after learning.

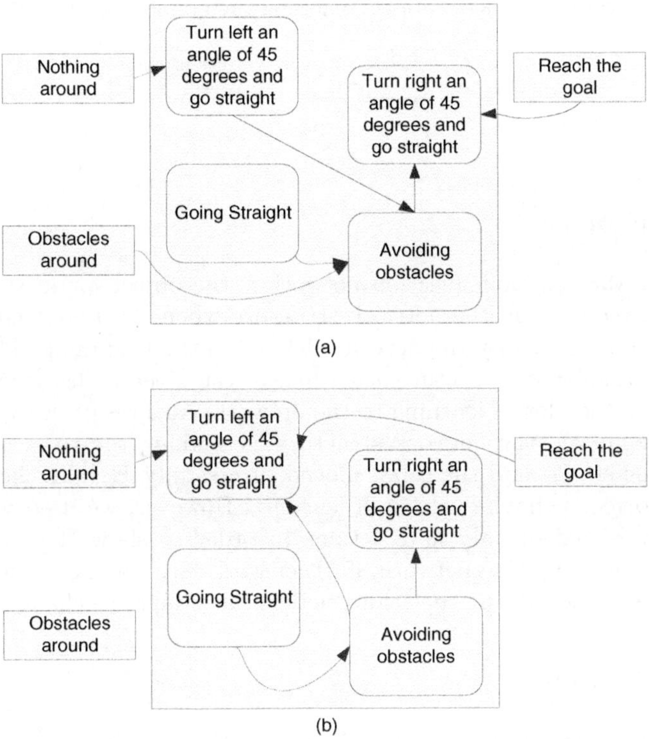

Fig. 9. Changes in action selection network after learning: Dotted lines mean inhibitory links and solid lines mean excitatory links in the internal links

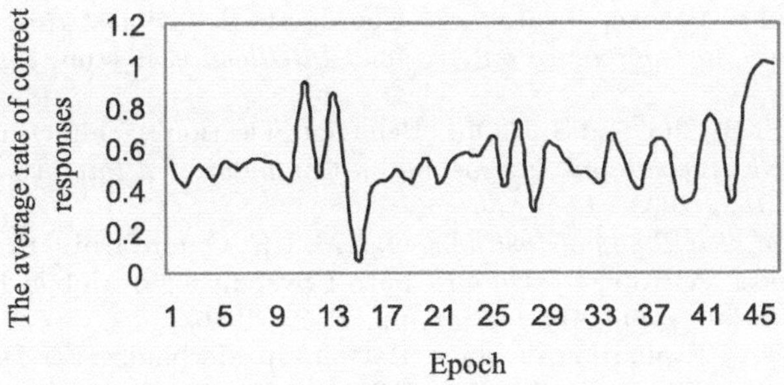

Fig. 10. The average rate of correct responses with respect to epochs

Table 3. The average steps of robot's reaching the goal before and after learning

	Before learning	After learning
Average steps	7936.7	4421.8

5 Conclusions

In this paper, Maes' action selection network is the object for learning. It has a weakness of redesigning the network structure when the task changes. That is, it is impossible to modify the network defined in the beginning. The network needs tobe reorganize its structure according to the given tasks. In order to do this, we applied the idea of learning to the network. As a result, we verified that the robot reorganized the network structure by making the rules appropriate to the changing states, and the robot moved successfully. Besides the robot can select more correct behaviors through learning. However, we used very simple model that consisted of only four actions in order to show the usefulness of learning the action selection network. Further works are going on studying more complicated network with more various nodes that adapts to changes.

Acknowledgements

This work was supported by Korea Research Foundation Grant (2002-005-H20002).

References

[1] P. Maes, and R. Brooks, "Learning to Coordinate Behaviors," *Proceedings AAAI-90: the American Conference on Artificial Intelligence*, Boston, MA, pp.796-802, 1990. 578

[2] K.-J. Kim, S.-B. Cho, and S.-R.Oh, "Behavior Selection Architecture for Tangible Agent," *IEEE Internationl Symposium on Computational Intelligence in Robotics and Automation*, 2003. 578

[3] N. Tinbergen, *The Study of Instincts*, Oxford, UK: Oxford Univ. Press, 1996. 578

[4] R. A, Brooks, "A Robust Layered Control System for a Mobile Robot," *IEEE Journal of Robotics and Automation*, pp. 14-23, 1986. 578

[5] T. Tyrrell, "An Evaluation of Maes' Bottom-up Mechanism for Bahavior Selection," *Adaptive Behavior*, vol. 2, pp. 307-348, 1994. 578

[6] M. J. Mataric, "Behavior-Based Control: Examples from Navigation, Learning, and Group Behavior," *Journal of Experimental and Theoretical Artificial Intelligence, Special Issue on Software Architectures for Physical Agents*, pp. 323-336, 1997. 578

[7] R. Arkin, *Behavior-Based Robotics*, MIT Press, Boston, MA, 1998. 578

[8] S. Mahadevan, and J. Connell, "Scaling Reinforcement Learning to Robotics by Exploiting the Subsumption Architecture," *Eighth International Workshop on Machine Learning*, Morgan Kaufmann, pp. 328-337, 1991. 578

[9] M. Dorigo, and M. Colombetti, *Robot Shaping: An Experiment in Behavior Engineering*, MIT Press, MA, 1997. 578
[10] P. Maes, "How To Do the Right Thing," *Connection Science Journal*, vol. 1, no. 3, pp. 291-323, 1989. 579
[11] D.E. Goldberg, *Genetic Algorithms in Search, Optimization, and Machine Learning*, Addison-Wesley, MA, 1989.
[12] L. Booker, D.E. Goldberg, and J.H. Holland, "Classifier Systems and Genetic Algorithms," *Artificial Intelligence*, vol. 40, pp. 235-282. 1989. 580, 582
[13] O. Michel, "Khepera Simulator Package version 2.0: Freeware Mobile Robot Simulator Written by Oliver Michel," `http://diwww.epfl.ch`, 1996. 583
[14] R.A. Richards, "Zeroth-Order Shape Optimization Utilizing a Learning Classifier System," *Ph. D. Dissertation*, Mechanical Engineering Department, Stanford University, 1995. 586

A Dynamic Self-Organizing E-Learner Communities with Improved Multi-agent Matchmaking Algorithm

Ruimin Shen, Fan Yang, and Peng Han

Department of Computer Science and Engineering
Shanghai Jiao Tong University, Shanghai, 200030, China
{rmshen,fyang,phan}@mail.sjtu.edu.cn

Abstract. Learning experiences and resources sharing is becoming one of the most important research fields of the advanced E-Learning techniques. This paper proposed a dynamic self-organizing community mechanism with an improved matchmaking algorithm, which makes the systems having the ability of self-organizing community that groups similar agents according to their preferences and capabilities. How to represent, evaluation, exchange as well as how to matchmaking during the self-organizing process are discussed in the algorithms. A prototype is also built to verify the algorithms. Experiments based on rea learner data have shown that this mechanism can organize learners properly, and has sustainably improved speed and efficiency of searches for peer students owning relevant knowledge resources.

Keywords: Intelligent agents

1 Introduction

Learning experiences and resources sharing is becoming one of the most important research fields of the advanced E-Learning techniques. An intuitive way to accomplish this objective is to group learners with similar preferences and reciprocal capabilities into the same community and help them learn collaboratively. Due to the distributed and dynamically changing characteristics of learners, middle agents are used to search suitable service providers in response to learner requests. Furthermore, the performance of middle agents relies heavily on the matchmaking algorithms used.

The earliest matchmaker we are aware of is the ABSI facilitator [1]. Sycara et al. proposed an agent capability description language, called LARKS [2]. In [3], it introduced a HTML-like Service Description Language (SDL) and used the find_nn algorithm to find all service names within a specified distance *d*. For a more comprehensive survey of matchmaking and brokering, see [4].

However, in these distributed systems, the relationship between user agents and middle agents is always pre-defined by a human designer [5]. They have, however, difficulties in handling the dynamism inherent in such open environments. To achieve

T.D. Gedeon and L.C.C. Fung (Eds.): AI 2003, LNAI 2903, pp. 590-600, 2003.

a good performance and a high scalability, the organizational structure of a socio-technical information system should be both self-organizing and adaptive [6].

In E-Learning environment, the objective of community-organizing is to organize the learners which can help each other into a group. Here, we consider the term 'help each other' as 'similar preferences' or 'reciprocal capabilities'. Since the characteristics and behaviors are very complex of the real e-learners, for instance, learners will browse online courses, submit questions or assignment and perform exercises. Hence, we did some investigations on the learning process, both of the learning log files and the learning behaviors. Then we adopted a method to describe the capability of a learner and find the matchmaking learner agents referring to the SDL and find_nn algorithm mentioned in [3]. Otherwise, we modified the method to suit for the knowledge concept hierarchy and grouped agents in the self-organizing E-Learner community environment.

In this paper, we present a self-organization model with an improved matchmaking algorithm. We also conducted some experiments to demonstrate the formation of user communities and to evaluate the efficiency and scalability of this mechanism.

2 Conceptual Framework of the Self-Organizing Model

2.1 Conceptual Framework

This paper describes the construction of an automatic and effective organization of learners and group agents in distributed e-learning systems. We refer to the concept "E-Learner Community" as a group of learners who share common preferences and mutually satisfy each other's requirements in terms of relevant knowledge resources.

We generated a Learner Agent (LA) acting on behalf of a real learner. A LA is in charge of not only the advertisement of capabilities, but also the restorations and updates of learning resources. Besides, a LA also handles the requests of a learner and seeks corresponding learning sources from the system.

To avoid traffic overload and increase the efficiency of search, we propose another kind of agent, called Group Agent (GA), to serve as the broker for requests from a smaller community of LAs. A GA is responsible for locating providers of services and managing the association of learners to communities. Furthermore, it can interact with both the local LAs in its management domains and the other GAs. To make our conceptual model more precise, we provide a few formal definitions.

Definition 1. A learner agent l is a **3-tuple** **(C_l,CET, PT)** with $C_l \subseteq$**C.** Here, **C** is set of Capabilities the learner has, CET is a Capability Evaluation Table and PT is a Preference Table. (See as Section 2.2 and 2.3)

Definition 2. A GA g maintains three data structures: the Type set **T**, **Σ_C-hierarchies** and a **Local-Agent Table** (See as Section 2.2 and 2.3)

Let **G** and **L** be disjoint sets of group and learner names with typical elements g and l. Now we can define the relationship between the g and the local agents.

Definition 3. A ***agent-community structure*** can be modeled by associating LAs and GAs through the mapping $m: L \rightarrow G$, where $m(l)=g$ denotes the fact that learner l is a *member of the group* managed by **g**. All *LAs* ***managed*** *by* **g** are then defined by the set:

$$A_g = \{l \in L \mid m(l) = g\} \quad (1)$$

2.2 Definition of Capability

During the learning process, learners will browse online courses, submit questions or assignments and perform exercises. All of these actions represent the learning interest, intent and capability of the learners. Generally, we view all of them as different requests of learning contents (each content) belonging to different knowledge points. For instance, the preferences can be looked at as many http request flows of learning content. The submission of questions represents a request for specific knowledge points of one subject. The marks of each homework or exercise also show the mastery-degree or capability of the relevant knowledge points, and so on.

Here we propose a method to define the learning capability of a learner. We consider the learning capability name can be described as (type: knowledge point), where the Type is the kinds of requests and the knowledge point is the knowledge discussed in the learning content. We will give the formal definition as the following:

Definition 4. Σ_K-Node is a finite set $K = \{k_1, k_2, ..., k_n\}$ of all possible knowledge points of capability learners may learn.

Definition 5. Σ_K-Hierarchy is a directed acyclic graph *KH*=(V, E) such that

- Each vertex of V is a noun $k_i \in K$
- For each edge e={k_i, k_j}, there is a $\prec$ relationship between two vertexes, we write $k_j \prec k_i$. That is, the concept level of k_i is more general than k_j and we call k_j is the child of k_i.

Definiton 6. $T = \{t_1, t_2, ..., t_m\}$: a finite set of all possible types the requests should be.

Definition 7. If $t_i \in T$ and $k_i \in K$, call (t_i, k_i) a Capability Term. Let $c_i=(t_i, k_i)$, then the Capability Set C={ c_i | $c_i \in T \times K$}.

Definition 8. The set of capabilities maintained by the community of LAs managed by g is defined by:

$$C_g = \bigcup_{l \in A_g} C_l \quad (2)$$

Definition 9. The matching function $f: C \times C \to \{0, 1\}$ with:

$$f(c_i, c_j) = \begin{cases} 1, & if\,(c_i, c_j)\; is\; a\; matching\; pair \\ 0, & otherwise \end{cases}$$

Here c_i is a pair (t_i, k_i), if $t_i = t_j$ and $k_i = k_j$, then we consider (c_i, c_j) is a matching pair.

2.3 Definition of Evaluation Record

As discussed above, a LA can advertise its capabilities and preferences when it registered to the GA. But the question is that there are always more than one service provider agents claim that they have the same or very similar capabilities to accomplish a task in an application. But it is not totally consistent with the actual performance of provider agents. So we propose two tables to track the performances and preferences respectively.

Table 1 is an example of the Capability Evaluation Table (CET). The evaluation records of CET consist of 3-tuples with a form as [t, k, evaluation]. The t and k parameters in the 3-tuple are the type and knowledge point of the evaluated capability separately. And the evaluation parameter is satisfactory degree returned by the agent received the service. For example, if it is the (algorithm, Apriori) capability of the agent was selected and the feedback evaluation is '8', then a 3-tuple, [algorithm, Apriori, 8], will be added to the CET table of the agent has the capability, which is in the database of the GA. Such a representation is easy to process and calculate the evaluation of each kind of capability.

In the same way, we also define the Preference Table (PT). The preference records of PT consist of 3-tuples with a form as [t, k, preference]. Different with the CET, the third item of the record is the preference, which on behalf of the preference award of the learner. Bigger the number is, higher the preference is.

3 Community-Organizing Algorithm with Matchmaking

In this paper, we focus on automatically grouping reciprocal learners together and dynamically adjust learners according to their changeable behaviors. Here we consider 'reciprocal' as the learners with similar preferences and most helpful capabilities. The main algorithms implementing this search strategy are discussed as following.

Table 1. Capability Evaluation Table

Type	Knowledge	Evaluation
t_1	k_2	8
t_2	k_2	7
t_3	k_4	6
t_2	k_3	4.5
t_1	k_2	7
...	...	...

3.1 Community-Organizing Algorithm

The algorithm takes variables **Requester, t, k** and **n** as inputs. They are needed to constrain the number of searches across large communities. Each LA and GA maintains **CET** and **PT**, which is used to maintain information about capability and preferences evaluations during the matchmaking process. Matching Agent Table, which is called MATable in the rest of this paper, is the list of matching agents after the performance of the matching subroutines. The local variable "Provider" refers to an LA, while variable "Requester" refers to both LAs and GAs.

INPUT:
1.The parameter **Requester** of Agent seeking for help
2.The type **t** of requesting capability
3.The knowledge point **k** of requesting capability
4.The number **n** of maximum searching times per matching
OUTPUT: Learner Communities

PROCEDURE
Community-organize (requester,t,k,n)

```
/* find the most promise providers and return to the requester */
/* insert the evaluation and preference record to provider.CET and requester.PT separately */
/* dynamic exchange the learners according to both of the CET and PT */
{
      MATable=∅;
      IF (t,c)∉ T×C return NoAnswer;
      Else
      {
            If (Requester.type=LA) THEN
            {
                  MATable= matchingLocally (t,c,n);
                  num=count(MATable);
                  if num<n then
                    MATable = PrioriSelect(matchingGlobally(t,c,n-num,MATable))
            }
            else MATalbe=matchingLocally(t,c,n,MATable);
      }
    if MATable<>NIL then
    {
        Rank_promise(MATable);
        Community_organizing();
    }
    else return NoAnswer;
}
```

The search schema helps GAs to find suitable learning resources. Here, the requests are capability-term of learns may have, answers were matching agent table (MATable). The MATable includes not only the information of the matching agent, but also the sum-up evaluation of the matching capability in the CET. The main search process can be divided into several steps as follows:

1. Judge the type of requester

When receiving a request from an agent, the GA will judge the type of the agent. Here, a GA can only communicate with local LAs and other GAs. It can not communicate with other LAs. So a GA can only receive two kinds of requests. One is from another GA, the other is from local LAs.

- **If the type of requester is LA**
  If the type of requester is LA, then the GA will call the subroutine **matchingLocally()** to search in a local group firstly. However, if the information is not available locally, the GA seeks help by forwarding the request to other GAs. These other GAs then check their own databases for a match and deliver any positive feedback to the requesting GA, which passes the feedback directly on to the original requester. Considering the communication problem, only the GA of provider send the confirm message and the GA of requester only chooses the first returned MATables. We call this the Priori Selection. (As shown in Section 3.2)
- **If the type of requester is GA**
  If the requester is another Group Agent, it only searches the local group and returns the matching agent table (MATable) if it is not NIL. Because every GA maintains the a table that records the information registered by all of its own LAs, the GA can easily find whether the required capability is owned by one of its LAs.

2. Judge if the search is success

- **If the search is success**
  If the search is success, it will call **Rank_promise(MATable)** to rank the matching agent according to the average evaluations, and suggest the most promise Agent.
- **If the search is not success**
  If no positive answer can be found by interacting with neighbored GAs, the GA of the requester stops searching and declares that the search has failed. We embedded a **topSearch** parameter as an upper limit for the number of attempts to prevent endless searches in a large and distributed e-learning environment.

In Section 3.2 and 3.3, we will discuss the core subroutines of the main algorithm.

3.2 Search Matchmaking Agents Algorithm

Given a capability term **(t, k)** and an integer **m**, returns the set of all agents, **MATable**, that have the capabilities. It is composed of three main subroutines.

- **CreateTccT(t, k)**. To extend the capability term (t,k) according to the Σ_K-Hierarchy. The returned capability-term set is the children terms of (t,k) if (t,k) is not the leaf node.
- **sql_agent(t,k)**. To match the agents that matching the capability pair (t,k) abide by the matchmaking function (see Definition 9). It executes the SQL query from the Agent Table maintained by the GA as:
  Select Agent **From** [GA-Agent] **Where** type=t, kp=k

- **sql_CET (LA,t,k)**. To calculate and return the average-evaluation of agent LA with the capability (t,k).
  Select evaluation **As** 'average-evaluation'
  From [LA-CEV] **Where** type=k, kp=t

PROCEDURE:
matchingLocally(t, k, m)

```
/*find the m agents owned the capabilities same or closest to (t, c), and output them with */
/*the average evaluations*/
    TccT=createTccT(t, k);
    isfinish=false;
    SqlA=sql_agent(t,k);
    SqlEV=sql_CET(SqlA, t, k);
    while ¬isfinish do
    {
        insert(SqlA, SqlEV, MATable);
        n=num (MATable);
        if n≥m then isfinish = true
        else
        {
            (t',k')=NEXT(TccT);
            if (t',k') = NIL then isfinish=true
            else
            {
            (t,k)=(t',k');
            MATable=search_agent_local(t,k,m-n, MATable);
            }
        }
    }//end of while
    return(MATable);
}
```

3.3 Evaluation and Exchange Schemas

Since every LA is registered with a GA randomly in the initialization process, one group may have learners with different preferences and capabilities. Hence, we proposed an evaluation record and dynamic exchange mechanism aiming at recognizing learner behavior and reorganizing learners accordingly.

1. Award Schema

When a GA relays search results to a LA, it examines whether the search is successful. If the search is successful, that is, there is a service provider matching a request, then firstly it will insert a new evaluation record from the requester (t,k,evaluation') into the CET of the provider, secondly it will add 1 to the preference item of the requester's PT, where the preference record is (t,k,preference).

2. Exchange Schema

After that, the GA of the requester determines whether the requester and provider are both in its group. If they are not in the same group, the GA of the requester contacts the GA of the provider for a membership exchange such that both the requester and provider - who can help each other - are registered with the same group.

The exchange rules are:

1) Calculations

- Calculate the number the records of the provider (NE), where the value of evaluation > threshold α (which is an empirical number that on behalf of the minimum value of a valid capability).
- Calculate the award of preferences of the requester (AP).

2) Conditions

- If AP is the highest one in the PT and higher than a threshold β (which is an empirical number), that is, it is the primary preference of the requester;
- If NE >γ, which is decided by the total number of the capability records of the provider, that is, the provider did has the relative capability which can satisfy the requirement of the requester.

3) Exchange

- If above conditions are all satisfied, then move the requester LA towards the GA managing the provider LA. This hypothesis is based on the belief that a learner with high evaluated capability usually means that it can provide useful information to other learners. On the other hand, the highest preference award almost shows the primary preference of the learner. The LA with highly NE is always acting on behalf of the main interest of the group. It is called the *authority learner*. Hence an attraction to an authority LA can drive other LAs with similar preferences to join the same group quickly.

4 Experimental Results and Evaluations

The experiments presented here are based on the real learners from the Network Education College of Shanghai Jiao Tong University (www.nec.sjtu.edu.cn). We focus on the evaluations of the usefulness and efficiency of this self-organizing mechanism.

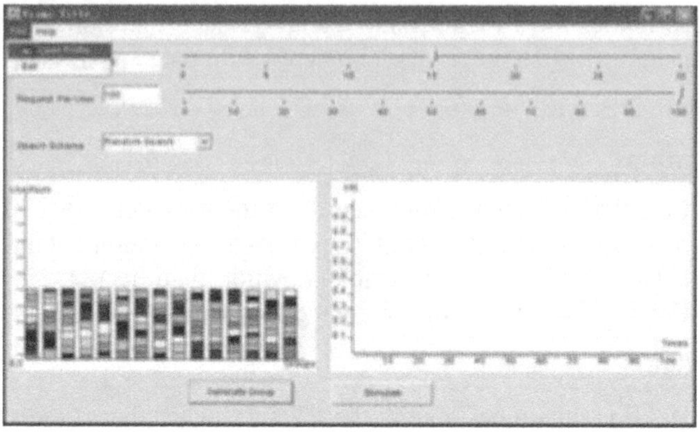

Fig. 1. Introduction of the test platform and system initialization

In the experimental setting, we chose 1500 users, and the number of category of C is 10. Figure 1 illustrates the main test platform of this system. In the top panel, we can define the number of Group Agents (**Gnum**) and the request times per learner (**maxRequestTime**). The left graph at the bottom shows the learner distribution in every group, whilst the right one shows the statistic analysis of the request success rate. When the experiments started, the system generated 1500 Learner Agents on behalf of the real learners and 15 Group Agents. LAs randomly registered with one GA together with the summarized registration information. The GAs kept all information of learners in this group. Figure 1 shows the initial situation, in which the colors of every column are mixed and the distribution is almost average.

In the Learner Distribution Graph, every column represents a group, and the colors of the rectangles represent different preferences. The height of each rectangle illustrates the number of learners with special preferences in the corresponding group. As shown in Figure 2, we can see that after 20 requests per learner, there was an obvious extension of color rectangles and the success rate increased quickly from 75% to 91%. After 50 requests per learner, the trend is toward fewer colors per column and longer rectangles. Also, we can see that some columns become shorter and some even disappeared. That means, some group agents lost all of their users during community formation. This result is consistent with the scenario of this experiment because we only have 10 categories while we generated 15 GA. After 100 requests per learner, the formation of learner communities was quite successful. It is settled to a stable state with learners who are interested in the same category preferences are all clustered into the same group.

From the Request Success-Rate Graph, we can see the success rate is lower during the first ten requests per learner. Once learner communities have started forming, however, the system exhibits an obviously improved success rate and greater efficiency. Because learners who have matching requests and results are gradually grouped together, GAs can more easily find correct answers in their own groups. And the success rate approached 1 after the learner communities were set up. Meanwhile, the average search time for a request was greatly decreased.

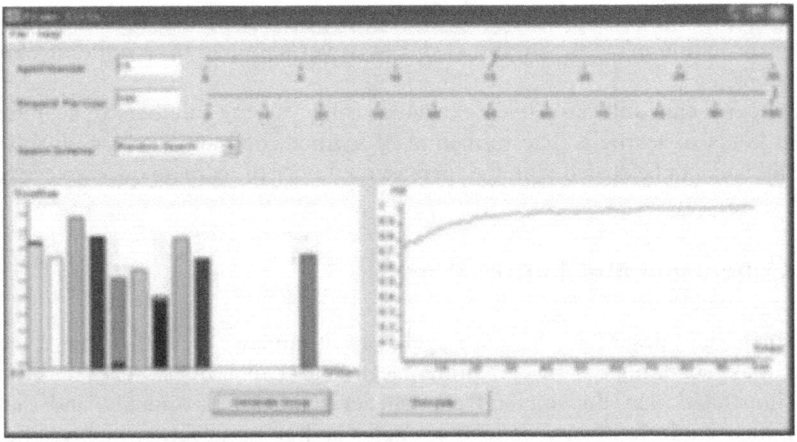

Fig. 2. System situation after 100 requests per learner

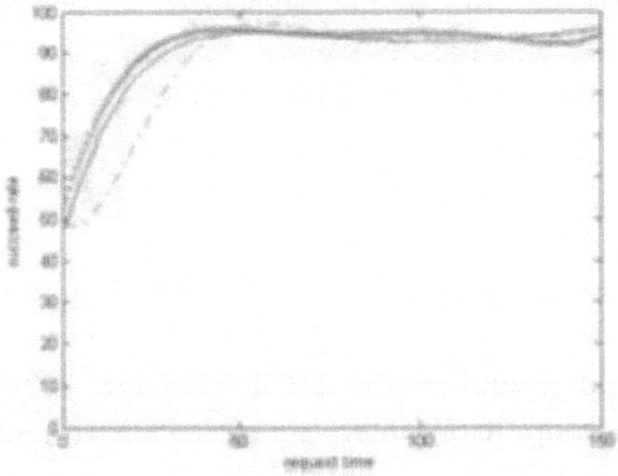

Fig. 3. Success rates of query results

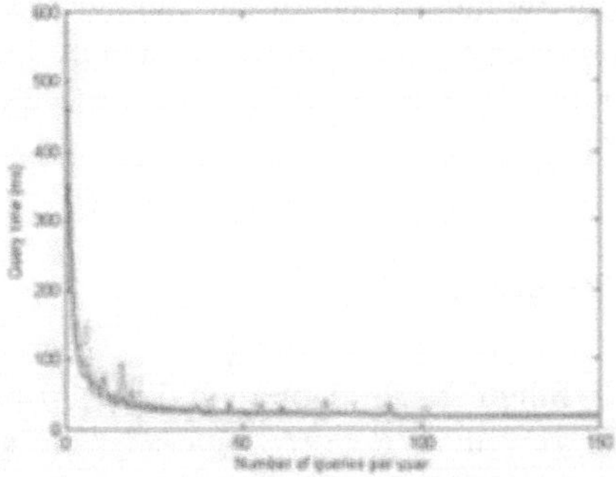

Fig. 4. Average Query Time

Scalability is an important issue in large information systems. The systems should work properly even when more and more users and learning knowledge join the systems. In this paper, we mainly investigate the scalability of self-organizing communities associated with increased categories and learning contents. Self-organizing communities have been tested with varied numbers of categories in order to examine their scalability.

Figure 3 shows that the system's ability to find correct answers to queries was obviously improving, as more and more queries were initiated in the system. Figure 4 portrays the continuously decreasing search time spent on each query. The green and '.' curve in Figure 3 and Figure 4, shows the success rate and query time with 1000 categories. The blue and '-' curve shows the situation with 5000 categories. The red and '.-' curve shows the situation with 10000 categories.

According to Figure 3 and Figure 4, we can see that the user communities were found after every learner launched about 46 queries. With the formation of communities, the actual success rate approached 1. Meanwhile, the average search time for a query was greatly decreased from a huge amount of 575 milliseconds. Due to the communication and processing delay, the query search time oscillated a little after the formation of communities. It did however stabilize to a value of about 20 milliseconds.

The experiments still have been executed using varying numbers of learners and different kinds of learners. The formation of learner communities was always quite successful and can be scaled with the increased number of learners quite effectively.

5 Conclusion and Future Work

This paper has described a multi-agent environment to self-organize learner communities according to users' preferences and capabilities. The experimental results illustrated that this method can cluster the learners quickly and facilitate appropriate organization between learner agents and group agents, which verified that this method can achieve higher search success rates and a sharp decrease of search time. In particular, the self-organizing communities scaled with the increased number of categories.

Our further work will be devoted to extend the adjustment mechanism to handle multiple relationships between the GA and LAs. Meanwhile, we will research how to help the learners learning cooperatively based on the Self-Organizing E-learner Communities.

References

[1] Singh N., A Common Lisp API and Facilitator for ABSI: Version 2.0.3, *Technical Report Logic-93-4*, Logic Group, Computer Science Department, Stanford University, 1993.

[2] K. Sycara, J. Lu, M. Klusch, and S. Wido., Matchmaking Among Heterogeneous Agents on the Internet, *Proceedings of AAAI Spring Symposium on Intelligent Agents in Cyberspace*, Stanford, USA, 1999, *http://www.cs.cmu.edu/~softagents/publications.html.*

[3] V. Subrahmanian, P. Bonatti, J. Dix, et al., *Heterogeneous Agent Systems*, The MIT Press, 2000.

[4] M. Klusch and K. Sycara, Brokering and Matchmaking for Coordination of Agent Societies: A Survey, in: A. Omicini, F. Zambonelli, M. Klusch, and R. Tolksdorf (Eds.), *Coordination of Internet Agents: Models, Technologies and Applications*, Springer, 2001, Chapter 8.

[5] F.Wang, Self-organising communities formed by middle agents. In: Proceedings of the First International Conference on Autonomous Agents and Multi-Agent Systems, Bologna, Italy, July 2002, pp.1333-1339.

[6] P.J. Turner and N.R. Jennings, Improving the scalability of multi-agent systems, IN: Proceedings of the First International Workshop on Infrastructure for Scalable Multi-Agent Systems, 2000, pp.246-262.

Learning to Survive: Increased Learning Rates by Communication in a Multi-agent System

Paul Darbyshire[1] and Dianhui Wang[2]

[1] School of Information Systems, Victoria University
PO Box 14428 Melbourne City MC, Melbourne Victoria, 8001, Australia
Tel (03) 9688-4393, Fax (03) 9688-5024
Paul.Darbyshire@vu.edu.au

[2] Department of Computer Science and Computer Engineering
Latrobe University, Bundoora, Melbourne, Victoria 3086, Australia
Tel (03) 9479 3034, Fax (03) 9479 3060
dhwang@cs.latrobe.edu.au

Abstract. In a hostile multi-agent environment, a team of learning agents utilizing a reinforcement-learning paradigm may not learn at a sufficient rate for the agents to adapt and survive. If we can increase the learning rate of the individual agents, they may learn earlier to select more optimal actions and hence increase the chance of survival. Implementing a simple communication structure between the agents and allowing them to communicate their learning events with team members at each time step considerably increases the learning rate. This increased learning rate can significantly improve the team's chances of survival in the agent's environment, but alone is not a guarantee of better performance. The type of information communicated also plays a crucial role in determining increased survival chances. This paper reports on some experimental results from simulating two teams of combative agents, one utilizing reinforcement learning as a control paradigm using communication to increase the measured learning rate.

1 Introduction

A Multi-Agent System (MAS) in which there are a number of agents interacting, with each affecting the actions of the others essentially constitutes a complex system. Performing and completing tasks in such an environment can be extremely difficult. Indeed, some tasks (depending on the MAS) may be beyond the capabilities of individual agents and require a group of such agents to successfully complete. Adding an extra dimension of a hostile environment compounds the problem further as individual agents may die, weakening the group as a whole and reducing its chances of success, and perhaps its chance of survival.

T.D. Gedeon and L.C.C. Fung (Eds.): AI 2003, LNAI 2903, pp. 601-611, 2003.

In such cases, the implementation of individual learning paradigms by the agents in the group can help the agents to adapt to the hostile nature of the environment and complete the task. However, depending on the environment, this may still not be sufficient for the group to survive. Agents implementing individual learning paradigms are essentially individual learners. One way for the group to increase its chances of success is for the agents within the group to communicate with each other to accelerate the learning process, thus increasing the learning rate of the group as a whole. By increasing the learning rate, agents can begin to choose optimal actions at an earlier stage, which in turn can enhance the group's chances of success.

This paper reports on results from a simulation where a group of agents implementing a reinforcement-learning paradigm communicated their learning events to each other in order to significantly increase the learning rate of the group. The simulation environment was that of a combat distillation used to study land combat at a conceptual level. Initially, the agents were implemented as individual learners to provide a benchmark scenario against which to measure the effects of the inter-group communication. In the following sections a brief background is given, followed by a discussion on the simulation environment and the structure of the communication implemented between the agents. The results from a number of simulation trials are then present, along with a short analysis of the results.

2 Background

Reinforcement-learning addresses the problem of how an autonomous agent that senses and acts in its environment learns by interaction with that environment [1]. Reinforcement Learning is generally based on the premise that the underlying environment is a Markov Decision Process (MDP), and in fact, most of the current theory is restricted to MDPs [2]. However, a MAS, while generally Markov-like in nature is essentially non-Markov decision process, as the interaction between the agents themselves and concurrent learning affect the state transition probabilities [3]. Even though the underlying state signal is non-Markov, it is still appropriate to approximate it with a Markov state signal, and thus a reinforcement learning approach [2]. Mundhe and Sen [4] report promising results on convergence in a MAS even though the conditions for the underlying convergence proofs for reinforcement-learning algorithms are violated in such systems.

A reinforcement learning technique often used to incorporate intelligent behaviour into agents is Q-Learning [5-8]. This research, utilizes Q-Learning as it comes from a class of techniques known as Temporal Difference algorithms. This is appropriate as the simulation is viewed as a non-episodic learning problem, hence agents learn from direct experience as the simulation progresses. The equation representing the Q-Learning algorithm implemented is given below in Eqn(1) [2].

$$Q(s_t,a_t) \leftarrow Q(s_t,a_t) + \alpha\left[r_{t+1} + \gamma \max_a Q(s_{t+1},a) - Q(s_t,a_t)\right] \tag{1}$$

In Eqn(1), $Q(s_t, a_t)$ represents the current estimate of the action-value pair, action a in state s at time t. This is updated from the current value of $Q(s_t, a_t)$, the reward r_{t+1} received from taking action a while in state s at time t, (received at time $t+1$), and the maximum action-value (over all possible actions) of the resultant state s_{t+1} at time $t+1$. The values α and γ represents the values of the step-size, and discount factors respectively.

In a multi-agent system using Q-Learning, the agents each learn individually from interaction with the environment, and each other. However, there are a number of reasons why we might want our agents to cooperate with each other during the learning process, and these include scalability, speed fault tolerance and encapsulation [9]. One of the criticisms of incorporating intelligent agents into military simulations by [10], was that often the systems were not scalable. Using a multi-agent system where the intelligent agents cooperate in learning could allay these concerns.

Of particular interest is the speed of the learning process. If the agents in a multi-agent system can cooperate in the learning process, we can effectively parallelize the learning and hopefully achieve an increase in the speed of learning. This research looks at communication between agents within a team to effectively speed up the learning process. As the research will demonstrate, a group of individual Q-Learning agents placed in the hostile environment of a distillation cannot learn fast enough to survive a combat scenario with a team of non-learning agents utilizing a human control strategy.

Dayan and Hinton discuss speeding up the reinforcement learning process using a multiple resolution approach [11]. In this approach, a managerial hierarchy is set up using reinforcement learning where high-level agent managers learn how to sub-delegate tasks to others. However, in this combat simulation, we don't want to specifically set up such a hierarchy. This is a type of emergent behavior we would ideally like to see emerge as a natural consequence of the simulation. Wolpert and Tumer discuss the emergence of Collective Intelligence multi-agent systems (COINS) [12]. In these multi-agent systems, every agent runs its own reinforcement learning algorithm, and the questions revolve around the reward function best suited to achieve high reward gains. While these are closely related to the approach in this paper, COINS are typically viewed as a control problem for real-world distributed applications.

Very closely related work is presented in [8], however there are some significant differences. Tan considered the limited hunter/prey problem in a small 10×10 grid world with a maximum of 2 hunters and 2 prey. This research utilizes a distillation with a 500×500 world with two teams of 15 agents. Tan also considered the learning problem as an episodic learning problem, and at the end of each episode, an agent communicated its entire episode-sequence to the other agent. This research considers the learning task as non-episodic, and communications occur at each time step throughout the simulation. The agents within this research are also placed within a hostile environment where they can be killed if the learning rate is insufficient.

3 Agent Environment

The agent environment represents a military distillation, which was designed to study land combat at a simple conceptual level. The environment consists of a continuous 500 by 500 unit square landscape with each position coordinate approximated by a pixel location. There are two teams of agents, 15 agents each, who combat for survival within the limited landscape. The two teams begin at opposite ends of the environment and slowly move towards each other until initial contact is made, when they battle each other for survival.

The teams are designated *blue* and *red* teams, with the behaviour of the agents in *blue* team governed by a Q-learning reinforcement learning algorithm. These agents are given no initial training data and all behaviour is learned as the combat progresses. While recognizing the existence of other agents, these agents would still be classed as 0-level agents [13]. The behaviour of the agents in *red* team is governed by a rigid control paradigm, controlled by a series of weighted values for various behavioural actions. This is a typical agent control paradigm prevalent in distillations. The *red* team then represents a highly trained unit whose actions are based on instinctual behaviours rather than any cognitive reasoning.

The simulation engine utilizes a discrete time-step progression; with the agents thinking processes separated from the determined actions so as to not perpetuate and unwanted side effects of the simulation itself. At each simulated time step, the agents are each given unlimited time to perceive their environment, team members and hostile forces to determine a course of action. Before the actions are applied, the sequence is randomized to remove bias due to pre-determined action sequences.

Each agent begins with certain level of health, which is reduced proportionately with each 'hit' it receives as the combat progresses. When the health level reaches zero, the agent is removed from the simulation and plays no further part. Thus as the combat progresses, the effectiveness of each team is reduced as it incurs losses. The simulation proceeds until a team is completely decimated or until a pre-determined simulation time limit is reached. This limit is preset to 1000 time units.

4 Increasing Learning Rates via Communication

In this instance, we define a learning event (l_e) as a single update to a Q-learning agent's action-value matrix. The learning rate ($L_r(a)$) of agent a is then the number of learning events per time unit (l_e/t). This will vary over time as the simulation progresses, depending on the spatial position of the agent in the environment and its proximity to other agents. However, if an agent can receive a maximum of m, learning events per time-step, then A group (G) of n agents described thus far has a maximum learning rate of:

$$L_r(G) = \frac{m l_e}{t} n \quad (2)$$

In the discrete time-step simulation described above, each Q-learning agent experiences one learning event per time-step. An agent receives a single reward from the environment at time *t+1* after all actions have been processed at time *t*. This reduces the maximum learning rate per agent to $m = 1$ event per time unit. Thus, for the group of *n* Q-learning agents in the simulation, the learning rate of the group is bounded by the simple linear function:

$$L_r(G) = n \quad (3)$$

Increasing the learning rate for individual agents is difficult. Each agent receives a total reward from the environment, and that reward is made up of a number of smaller positive and negative amounts depending on damage inflicted and taken from various sources. While these could be separated, still the action value matrix is updated based on the single *state* and *action* taken in the previous time-step. However, as each other agent in the group is also receiving rewards for past *states* and *actions* taken, we can increase the learning rate of the group by allowing each individual agent to share their learning experiences with the rest of the group.

If each of the *n* agents experiences a learning event at time *t*, this can be passed on to each of the other *n-1* agents in the group at time *t+1*. The maximum learning rate for the group at time *t+1* would then occur when each of the *n* agents experiences another learning event, as well as receiving and processing reward data from the other *n-1* agent's learning events from time *t*. In such a case, the group-learning rate is bounded by the function:

$$L_r(G) = n + n(n-1) = n^2 \quad (4)$$

This upper bound to the learning rate assumes that the communication of the Q-learning agents is global and not restricted by range. If range restrictions apply, then the above maximum rate is conditional upon the proximity of agents to each other during the simulation, and would be reduced accordingly.

In order for an agent a_i to process a learning event from agent a_j, it needs not only the reward that was given to a_j from the simulation, but it must be placed in context by the corresponding state-action pair that generated it, along with the resultant state. Thus the structure of the messages required to allow a group of Q-learning agents to share learning experiences is of the form $\langle s,a,r,s' \rangle$. Where $s \in S$ (set of states), $a \in A_S$ (set of actions for state S), *r* is the reward received and $s' \in S$ is the resulting state.

These messages carry enough information about learning events from other group members for the agent to be able to update its own action-value matrix accordingly. These secondary learning events will be slightly older than an agent's own, but will be only one time step behind. Thus, the age of the extra input signal should not be a significant concern. The consequential change to the Q-learning algorithm specified in Eqn (1),can be seen in the updated form of Eqn (5).

$$\forall < s,a,r,s' > \in \{S, A_s, r, S\} : Q(s,a) \leftarrow Q(s,a) + \alpha\left[r + \gamma \max_a Q(s',a) - Q(s,a)\right] \quad (5)$$

By applying each of the $\langle s, a, r, s' \rangle$ learning events to its own action-value matrix through the update function in Eqn (5), each individual agent is incorporating the learning events of its team members. Thus at each time-step, an agent can potentially increase its learning rate based on the team learning. This technique is applicable given that the team of agents is homogeneous, each having the same set of states and actions available.

5 Increasing Group Success via Learning Rate?

The following reports on the results obtained from four comparative simulation-runs. These are summarized in Table 1 below. The data for each of the four comparative simulation-runs was obtained by averaging 1000 simulations for each run under the same conditions. Each simulation was allowed to proceed for 1000 simulated time-steps before it was terminated. In many cases, the simulations finished well before the time limit and averages were only computed where data was available.

As previously mentioned, the *blue* team consisted of the Q-learning agents, and these agents were allowed to communicate over the different simulation-runs. The *red* team parameters were kept constant throughout. In all runs, the communication was global and no range restrictions on communication were enforced. The difference between the experimental runs was in the learning events that were communicated. The average learning rate of the group ($L_r(G)$) for each of these simulation-runs is plotted against the learning rate for the benchmark Q-learning team (no communication) and the results can be seen in Fig 1.

Table 1. Brief description of the four comparative simulation trials to measure the effects of communication on group learning and success

Simulation-run	Description
benchmark	agents do not communicate and are individual learners
$comm_1$	agents communicate positive reward learning events only
$comm_2$	agents communicate negative reward learning events only
$comm_3$	agents communicate all learning events

The results shown in Fig 1 were wholly not unexpected, and demonstrate a significant difference in the learning rate between the Q-learning agent teams that communicate and the benchmark Q-learning team that does not. Interestingly enough, the learning rates for all four experimental runs peak at approximately the same time throughout the simulation. This was unexpected as it was assumed that an increased learning rate might bring forward the learning rate peak somehow. However, this may be more due to the spatial positioning of the agents as the two groups meet and hostilities begin.

The effect of the increased learning rate can be viewed a number of ways, in particular by plotting the health of the group against time as the simulation progresses. A number of the damage vs time plots for these simulations were presented in [14] but the effects can be summarized Fig 2.

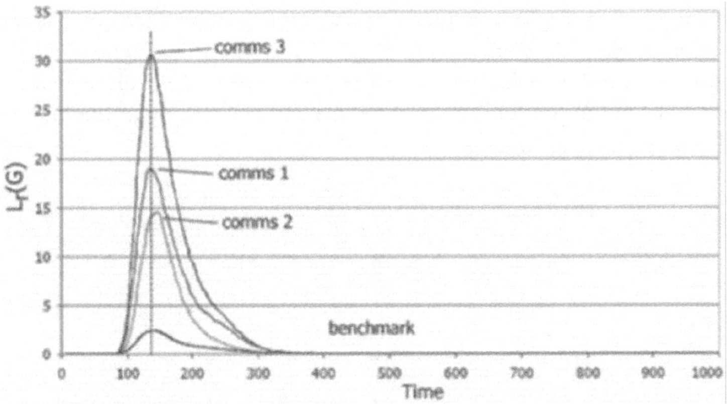

Fig. 1. Comparative group learning rates for the *blue* team, Q-learning agents. The horizontal axis represents simulated time-steps, while the vertical axis is the number of learning events

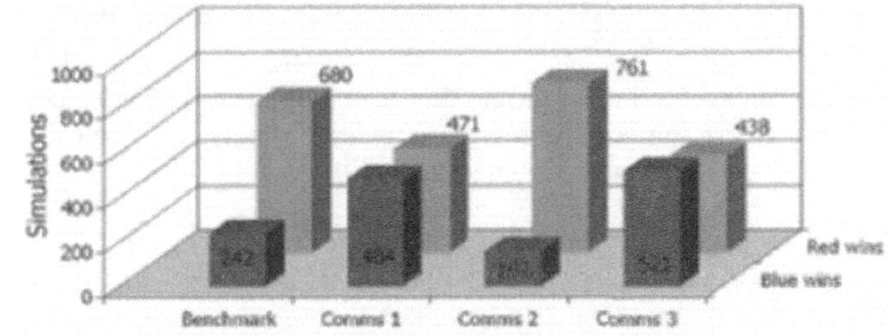

Fig. 2. Comparative bar chart showing number of victories for the blue and red teams in each of the simulation-runs. The vertical axis represents the number of decisive victories from the 1000 simulations. A decisive victory is when the opposing team is completely eradicated

In the *benchmark* simulation run, the Q-learning agents could only manage to survive as a group in approximately 24% of the simulation runs. By allowing the agents in the group to communicate their learning events in the $comm_1$ simulation run, the learning rate for the group was increased (Fig 1) and the group significantly improved its chances of survival (as in Fig 2) to approximately 48%, doubling its chance.

Fig 2 shows the simulation-run for $comm_2$ indicating a significantly worse performance for group survival than the *benchmark.* The Q-learning team does approximately 50% worse in the number of victories and this is despite the significantly increased learning rate for comm_2 shown in Fig 1. An explanation could lie in the non-deterministic nature of this type of simulation. In the same state, the same action chosen may produce different rewards at different time steps. Thus by processing only the negative rewards, the actions producing them could be actions which might normally produce positive rewards at other times. By processing only the negative learning events, the negative reinforcement of these actions would tend to make them be selected less in future thereby restricting chances for positive reward. However, in the light of the significant gains made from the communication of positive learning events, the option needed investigating.

Fig 2 also shows a further increase in the *blue* groups survival chances for the $comm_3$ simulation-run. As indicated in Fig 1, $comm_3$ has the highest learning rate of all the simulation-runs, and includes both the positive and negative learning events. However, in light of the previous discussion it wasn't obvious that the highest learning rate would provide the best outcome.

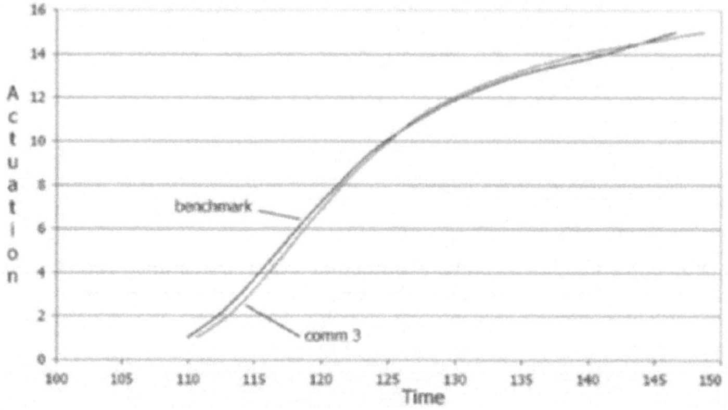

Fig. 3. Actuation rate of agents in the Q-learning team for the *benchmark* compared to the $comm_3$ simulation-run. Horizontal axis represents simulation time-steps, vertical axis represents the accumulated number of agents in the team active

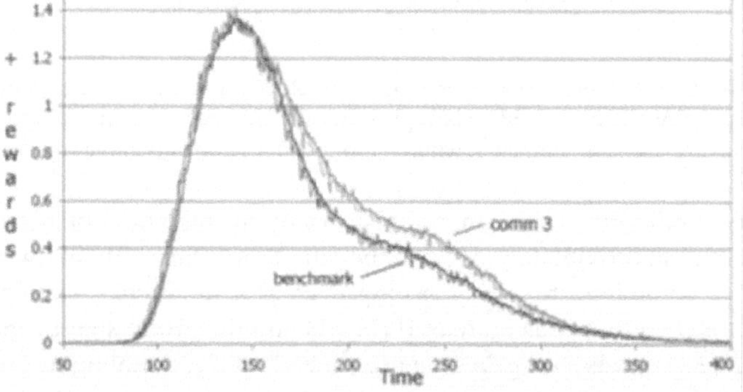

Fig. 4. Graph of the rate of positive rewards received from the simulation for the Q-learning *benchmark* team compared to the $comm_3$ team. Horizontal axis is in simulated time-steps, vertical axis is the positive reward rate for individual agents per simulated time-step

Given the significant difference in the simulation outcomes between the *benchmark* and $comm_3$ simulation-runs it would be of interest to try and identify other metrics significantly mutated by inter-group communication, other than the group learning rate. One metric to investigate is the actuation rate of the agents in the Q-learning team. That is, the accumulative measurement over time of when different agents begin to positively participate throughout the simulation. If the actuation rate increases, then more damage to the opposing team could be inflicted early giving a positive advantage as the simulation progresses. The actuation rate of *comm₃* is charted against that of the *benchmark* and shown in Fig 3.

As indicated in Fig 3, there is little if any difference in the actuation rates between the two Q-learning teams. Thus, the agent communication and increased learning rate do not seem to have affected this metric. As discussed earlier though, the spatial positioning of the agents within the environment may influence this to some degree, but it seems to remain un-affected. Two other items to investigate are the rate of the positive and negative and rewards received from the simulation as time progresses. This could give an indication of the activity of the Q-learning group during this time period.

Fig 4 and Fig 5 graph the reward rate (positive and negative respectively) during simulation time-steps 50-400. Time-steps outside this range were not graphed, as they didn't show any significant activity. There are observable differences in the reward rates. In Fig 4, after the apex in the positive reward rate, the *comm₃* Q-learning team sustains a higher rate of rewards for approximately 160 simulation time steps. In Fig 5, it can be observed that the *comm₃* Q-learning team maintains a lower rate of negative rewards for the first part of the conflict. However, there is a crossover at about time-step 180 where the benchmark Q-learning team maintains a slightly lower negative reward rate for some time.

The observable differences in the above figures are not very large, however the accumulative effect of both the increased positive reward rate and the lower negative reward rate must be enough to account for the differences in the final outcomes. The jump in the success rate of the Q-learning team from 24% to 52% (Fig 2) is significant but the observable changes in some of the metrics investigates above seem small in comparison.

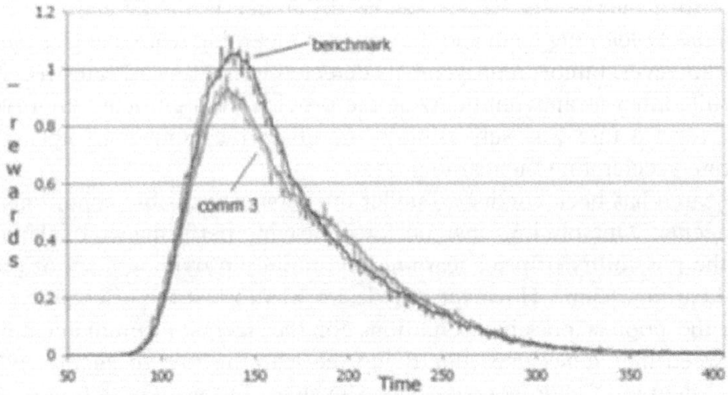

Fig. 5. Graph of the rate of negative rewards received from the simulation for the Q-learning *benchmark* team compared to the *comm₃* team. Horizontal axis is in simulated time-steps, vertical axis is the negative reward rate for individual agents per simulated time-step

6 Conclusions

Without any form of cooperation, the agents using a Q-learning control paradigm are each learning as individuals, not as a group. The group itself has a very poor chance of survival in the simulation environment, and is thus not able to achieve its objective. The group itself is not enhanced significantly by the presence of an agent, and adding another agent will slightly add to the groups capability, but do nothing directly to the learning capability of the other group members. Consequently, the group is basically the sum of its parts.

When agents in the same group communicate their learning events to other group members this allows the team to parallelize the learning task. Thus enabling all members of the team to accelerate the learning process and develop a synergy with other team members. While this results in a significant increase in learning rates achieved, it was seen that improved learning rate alone did not automatically lead to improved performance. Communication of the negative learning events improved the learning rate, but resulted in significantly worse performance for the group. The type of communication also plays a vital role. Significant improvement of the group's success resulted from the communication of all learning events, that is, with no filtering of the types of events. This indicates that over time, those actions which tend to lead to positive rewards get positively rewarded more often despite the non-determinism of the simulation. The negative learning events obviously do play an important part, but only when coupled with the positive events as well.

While the $comm_3$ simulation did significantly improve the chances of team success, it was difficult to pinpoint significant changes in some of the metrics we could use to measure performance levels. There was no difference in the actuation rates between the *benchmark* Q-learning team and the $comm_3$ Q-learning team using communication. There was however, minor improvements detected in the reward rates received by the agents. While this was apparently enough to provide the significant improvement as a whole, the reward rate was surprisingly low given the number of agents in the Q-learning team accelerating the learning rates.

This research has been conducted under the harshest possible conditions for the Q-Learning agents. One obvious method for improving performance of the agents is to allow for the possibility of prior learning, by initially providing a set of training data for the Q-Learning team. However, the desire was to test the effects of communication given the poorest possible conditions for the agents. Communicating learning events between the agents does improving the learning rate of agents and promotes some initial degree of team cooperation. However, further research is required as to the type of information communicated, and whether localized communication would produce better results, given that spatially close agents are more likely to have similar states.

References

[1] T. M. Mitchell, *Machine Learning*: McGraw-Hill International Editions, 1997.

[2] R. S. Sutton and A. G. Barto, *Reinforcement Learning: An Introduction*: MIT Press, Cambridge, MA, 1998.

[3] S. Arai, K. Sycara, and T. Payne, "Multi-agent Reinforcement Learning in a Dynamic Environment," 1999.

[4] M. Mundhe and S. Sen, "Evaluating concurrent reinforcement learners," University of Tulsa, Department of Mathematical & Computer Sciences 2000.

[5] S. A. Gugel, D. R. Pratt, and R. A. Smith, "The Implementation of Multiple Behavioural Approaches in the CFG Test-bed," presented at 10th Conference on Computer Generated Forces & Behavioral Representation, Norfolk, VA, 2001.

[6] S. A. Gugel, D. R. Pratt, and R. A. Smith, "Using Q-Learning to Control Individual Combatants," presented at 10th Conference on Computer Generated Forces & Behavioral Representation, Norfolk, VA, 2001.

[7] B. D. Lapin, R. L. Kolbe, J. Langworthy, V. Tran, and R. Kang, "Semi-Automated Forces that Learn from Experience," presented at 10th Conference on Computer Generated Forces & Behavioral Representation, Norfolk, VA, 2001.

[8] M. Tan, "Multi-Agent Reinforcement Learning: Independent vs. Cooperative Agents," presented at 10th International Conference on Machine Learning (ICML), Amherst, MA, USA, 1993.

[9] S. Sian, "Extending Learningto Multiple Agents: Issues and a Model for Multi-Agent Machine Learning," presented at Machine Learning - EWSL-91, Porto, Portugal, 1991.

[10] M. D. Petty, "Do We Really Want Computer Generated Forces That Learn?," presented at 10th Conference on Computer Generated Forces & Behavioral Representation, Norfolk, VA, 2001.

[11] P. Dayan and G. E. Hinton, "Feudal Reinforcement Learning," presented at Advances in Neural Information Processing Systems 5, Denver USA, 1993.

[12] D. H. Wolpert and K. Tumer, "An Introduction to Collective Intelligence," NASA NASA tech rep NASA-ARC-IC-99-63, 2000.

[13] J. M. Vidal and E. H. Durfee, "Agents learning about agents: A framework and analysis," in *Working Notes of the AAAI-97 workshop on Multiagent Learning*, 1997, pp. 71-76.

[14] P. Darbyshire and B. McKay, "Using Communication to Increase Learning in a Hostile Multi-Agent Environment," presented at Complex Systems 2002, Tokyo, Japan, 2002.

An Infrastructure for Agent Collaboration in Open Environments

Kenichi Yoshimura[1], Lin Padgham[1], and Wei Liu[2]

[1] School of Computer Science and Information Technology
RMIT University, Australia
{kenichi,linpa}@cs.rmit.edu.au
[2] School of Computer Science and Software Engineering
The University of Western Australia, Australia
wei@csse.uwa.edu.au

Abstract. In this paper, we present a service composition tool which enables specification of composite services for users with limited programming skills. We view composite services as a loose form of *teamwork* of service provider agents coordinated by a personal user agent. Such view enables us to use an existing hierarchical model of teamwork as the basis for service composition. An extension to the current model of teamwork is also proposed to achieve *flexible* and *robust* execution of composite services in an open environment.

Keywords: Intelligent Agents

1 Introduction

The World Wide Web is undergoing a significant evolution since Berners-Lee et al. unleashed the semantic web possibilities in 2001 [2]. The ultimate aim of the semantic web is to revolutionise today's web (a human friendly and dependent information repository) into a machine (agent) friendly backbone connecting automated service providers across the Internet.

The existence of content markup and service description languages (e.g. DAML-S and WSDL [12]) alone is not enough to realize the semantic web vision. Many have identified the importance of *service compositions* [5, 13] for developing more complex services using existing services available on the network.

Agents and agent technologies are well recognized as the key players and the active components in the semantic web world [2]. They are to interpret, manipulate and execute web services in an open environment.

Viewing web services as agent services, as far as we are concerned, a service composition problem naturally turns into an agent coordination and collaboration issue. In this paper, we present a novel infrastructure that uses teamwork framework to realize *flexible* and *robust* service composition.

It is well known that the success of the current Web attributes to its ease of use and the fault tolerance nature of web browsers. If the Semantic Web is to enjoy a similar level of acceptance by the large majority of normal end users,

T.D. Gedeon and L.C.C. Fung (Eds.): AI 2003, LNAI 2903, pp. 612–623, 2003.

intuitive service composition tools should be developed to enable users browsing and combining services in a familiar way as browsing and using the Web content today. We have implemented a prototype system in order to explore such vision.

The Web is an *open environment* where services are deployed by different organizations using various technologies. One of the key characteristics of an open environment is that services are not available reliably all the time. The system evolves over time and new services may be added at an ongoing basis. A successful system operating in such environment must be both *flexible* and *robust* to cope with the uncertain nature of the environment.

Previous study of Beliefs, Desires, and Intentions (BDI) architecture has shown it is one of most successfully implemented platforms (e.g. JACK [4] and JAM [9]) for developing systems in dynamic environment [7, 14]. In addition, work on *teamwork* addresses issues of how to improve robustness of systems consist of numerous agents (discussed further in Section 4). For this reason, we use a BDI architecture, JACK agent development framework, which also provides an extension for teamwork for the deployed prototype system. We present a generalizable extension of such a teamwork model that is particularly suitable for representing composite services.

Our prototype system is built for, and deployed in, the Agentcities[1] network. Agentcities is an international initiative, which aims at facilitating the exploration of a worldwide open network of platforms hosting a variety of interoperable agent services. Agentcities grew out of the work of the Foundation for Intelligent Physical Agents (FIPA)[2], which is a standards body focussing on standards and infrastructure for rational interaction and cooperation among heterogenous agents.

In this paper, we present:

- A novel view of service composition as a loose form of teamwork, and an exploration of a team programming environment as a base for service composition;
- The use of a goal directed BDI framework to assist in achieving robustness;
- Development of mechanisms for repairing a plan in the event of a team member (i.e. a service provider agent) becoming incapacitated or failing to provide the needed service.

In the following sections we present a general approach for realizing our vision of a service composition tool and explore key characteristics of the design of our prototype system. It is known as **ASKIT**, **A**gent **S**ervice **K**omposition **I**nterface **T**ool [3]. In section 2, an overall architecture is described. Section 3 introduces the key concepts of a BDI team oriented programming environment, JACK Teams. Section 4 explores the team view of service composition, followed by Section 5 describing our extension to the current team plan model. Section 6 compares our work with others and the paper concludes in section 7.

[1] `http://www.agentcities.org`

[2] `http://www.fipa.org`

[3] The prototype of ASKIT for demonstration purposes is available at `http://agentcities.cs.rmit.edu.au:7198/agentcities/login.jsp`.

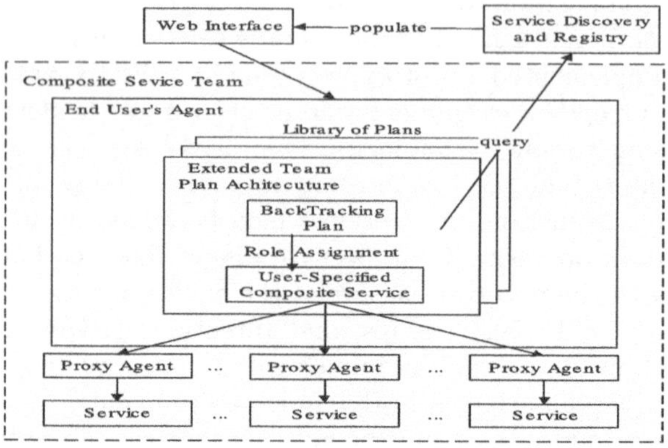

Fig. 1. A service composition Infrastructure

2 An Overall Architecture

Figure 1 depicts the key components and data flow of an infrastructure that supports automatic FIPA compliant agent service composition. It consists of three core components:

- *Web interface* is responsible for collecting data from end users.
- *Service discovery and registry module* keeps service provider details, service types and service APIs. It is also responsible for populating the Web interface and display available service APIs in an intuitive manner to the end user.
- *Team plan generation module* is responsible for generating abstract composite service specification from the Web interface data, generate abstract team plans, backtracking upon plan failures, and generating *proxy agents* for interacting with service providers.

In the context of research presented in this paper, a service is modelled as a service type and a list of interfaces associated with the service. Each interface has a list of expected input parameters and a list of expected output values. For example, a bank service type has interfaces to open an account, it takes your name, address and passport number as input values, and it returns your account number as an output value of the interface[4]. Multiple instances of the same type of service can be available on the network simultaneously. For example, it is possible that several CD seller services are available on the network, offering competitive prices. This is one of the key characteristics of the open environments where services may appear or disappear by the interest of service providers, and services are not reliably available all the time.

The interface also allows a user to specify the control sequence of a composite service by combining the existing services sequentially, or in parallel, or combination of both.

[4] Due to the limit of pages, interested user are referred to the ASKIT Web site for further details.

One of the key characteristics of our architecture is composite services is specified using *service types* rather than *service providers*. When composite services are executed, the infrastructure identifies appropriate service providers depending on the configuration of the environment. This enables us to implement a flexible and adaptive service composition tool which in particular suits for the dynamic and evolving Web environment.

The current model of services makes simplifying assumptions about service description and Ontology issues. Currently, the system only interacts with a known set of services using directory facilitators for *agent service discovery*. However, this should not influence the demonstration of our service composition ideas. We expect to extend the Service Discovery and Registry module to accommodate the current industry standard such as UDDI[5].

The infrastructure automatically generates the source code for a team *proxy agents* that are responsible for interacting with each service provider agent so as to fulfill the specified composite service. This is based on our view of composite services as *teamwork* among service provider agents. Section 4 has further discussion on this, and Section 5 describes how the current model of teamwork can be extended to improve the overall flexibility and robustness of the system.

3 JACK Teams

JACK Intelligent Agents^{TM} is an extension of the Java programming language designed to provide additional programming constructs for developing applications [6]. It is based on the Beliefs, Desires, and Intentions model [3] and previous practical implementations of such systems, for example, JAM [9].

The BDI agent model is an event-driven execution model providing both reactive and proactive behaviour. In this model, an agent has certain beliefs about the environment, has goals (desires) to achieve, and has plans (intentions) describing how to achieve goals. One of the key strengths of the BDI architectures is their ability to recover from failures by selecting alternative plans to achieve the same goal.

JACK Teams is an extension of JACK Intelligent Agent^{TM} for *Team Oriented Programming*. It is a nuance of *Agent Oriented Programming* aimed at developing teams of agents. Programming of teamwork amongst agents is an area that has been recognised as causing a range of challenges and difficulties.

JACK Teams does not operate with the traditional constructs of mutual belief, joint goals, and joint intentions [6]. Rather it has a hierarchical team model where the team entity at each level is reified and contains its own beliefs, goals and intentions, giving a model of team. JACK Teams takes the approach of modelling the team as a specific entity, separate from its members. Having a team entity explicitly modelled allows for reasoning to be done at the level of the team, rather than at the level of team members. Thus reasoning about

[5] Universal Description, Discovery ad Integration of Web Services at `http://www.uddi.org`

[6] An evaluation package is available for download from *www.agent-software.com.au*

which team plan to use, based on the situation, is encapsulated within the team entity.

Roles are conceptual entities which are defined by the events that are required to be handled by that role, events that can be generated by that role - with the expectation that the team entity will handle these and beliefs which are propagated to and from the role.

A team plan is used to describe how to go about achieving a particular goal of the team executing the plan. A team plan declaration includes a set of team members required to achieve its goal (called *role positions*). The strategy within a team plan typically involves co-ordination of its team members. It is described in terms of *role positions* associated with the team plan using programming constructs such as *@team_achieve(...)* (sets out a goal for a team member) and *@parallel(...)* (executes multiple tasks concurrently) statements.

A team entity declaration includes what roles the team uses. Once a team entity is instantiated, it is possible to assign members to *role containers* which can be thought of as groupings of team members, organised according to the kind of function, i.e. role, they can have within the particular team. At this stage there is no commitment for any particular team member to play a part within a given team plan. Once the team entity chooses a plan to execute in the service of one of its goals, the role-positions required by the plan must be filled by particular agents from the role containers. It is only when a member accepts a role position in a particular plan that it commits to acting on behalf of the team.

JACK Teams is especially well suited to modelling larger organisations where there is a hierarchical structure which can readily be modelled as teams within teams. In these contexts the concept of the team as an entity which can itself be a team member is very powerful.

4 Composite Services and Teamwork

In this section, we present our approach of viewing composite services as a form of *teamwork*. A comparison study is also presented to differentiate our approach from the existing teamwork literature.

4.1 A Team Plan View of Service Composition

In the work presented in this paper, we view a composite service as a particular form of hierarchical *teamwork* involving an agent pursuing goals on behalf of its user and service providers. In this model, only the user agent views its service providers as its team members, and the service providers do not have any awareness of their roles played within the team.

The standard view of agent interactions in a system such as Agentcities is that interaction between agents happens according to protocols which are agreed in advance and which form the basis of interaction patterns between participating agents. One of the difficulties of the protocol based view of agent interactions is that the protocols must be decided in advance, prior to the development of all agents who will use these protocols. In this context, we refer to an interaction protocol as a description of exact sequences of permitted messages between the interacting agents (service providers). It is a useful implementation level description of interactions, and developers of the system should follow the specification. It is important to note that such a specification includes all possible scenarios of the conversation including successful and failure cases of the interaction.

Our view is that while some basic protocols are necessary at the implementation level, it is useful to be able to specify and program customized and extended conversation using the existing services without detailed specification of individual conversation. That is, such a specification of composite services should only capture the new scenarios of composite services without showing every possible conversation with every individual services. In our approach, the translation process of such a specification of composite services deals with the detailed specification of individual conversations.

This is one of key characteristics of our system which enables the end users with limited programming experience to compose new services using the services available on the network. From the end users' perspective of our system, the specification of a composite service is only a matter of specifying synchronization and co-ordination of existing services and input values for each service.

Such an abstract representation of composite services captures the aspects of a composite service that are concerned with the co-ordination and synchronization of messages between the service providers and the initiating agent. Team plans as described in the previous section are concerned with synchronization and co-ordination of activities done by different agents. Consequently we see team plans as an ideal tool to use for both coordinating the actions and interactions of agents and for composing the services offered by other agents.

In JACK Teams, the separation of *teams* from individual agents enables an abstract specification of coordination strategies as team plans. An individual agent then figures out how to achieve its assigned task from the team and reports the outcome of the assigned task once it finishes the execution. In fact, such an approach of teamwork model enables an implementation of service composition model which we proposed above.

In this model, team plans are used to capture the order of execution of composite services. It only captures the sequence of interaction which the end user of our system specified. An individual interaction with a particular service provider is represented as a task for a team member to achieve, and it is the team member's responsibility to ensure coherent interaction with the service provider.

4.2 Roles of Proxy Agents

A service composition involves more than a simple exchange of messages between service providers and a user agent. For example, an interaction involves discovery of service providers, management of different interaction protocols, and message delivery over the network just to name a few issues involved. Furthermore, service providers are not aware of their roles within a team. For example, a service provider could disappear any time during the interaction, and it does not agree to meet any responsibilities as a member of a team.

Proxy agents are responsible for the management of individual interaction with service providers. Proxy agents hide most of the environment specific requirements (FIPA standards in this case) and ensure reliable interaction with unreliable services. Key roles of a proxy agent are summarized below.

- Failure Management: *Proxy agents* deal with failure of various kinds, eg. a restaurant booking service is unable to make a booking due to its restaurant is fully booked.
- Time out management: An *proxy agent* monitors the progress of an interaction. It declares failure if the service provider associated with the proxy agent does not reply within a reasonable time.
- Interaction protocol management: Currently each proxy agent knows *FIPA-QUERY* and *FIPA-REQUEST* interaction protocols. Depending on the interaction protocol used by a service provider, the proxy agent prepares appropriate messages.
- Content language management: Conversion of content language to/from an end user friendly format.
- Message delivery management: Delivers messages over the network and reports any failures.

The proxy agents implement the basic requirements to play a part in a team, and it enables our view of composite service as a form of teamwork transparently to service providers.

4.3 A Comparison with the Teamwork Literature

Over the past decade, theoretical analyses of teamwork [6, 8, 11] have lead to a variety of practical approaches to facilitating the implementation of coordinated agent activity, e.g. [1, 16]. It is aimed at decreasing the complexity of developing a team of collaborative agents and improve the overall robustness of the system. Aspects of this have been well illustrated by empirical work, e.g. [16].

In the teamwork literature, collaborative activities are typically achieved by team members having certain beliefs about the team while executing team plans. For example, [6] argue that teamwork involves more than team members obeying certain rules, but also team members having certain commitments as a member of the team such as being responsible to make its knowledge about the progress of the current goal mutually known when it knows something different from what was known previously.

In comparison with the previous work on teamwork, our view of composite services as a form of teamwork has a much looser notion of *teams*. Our approach realizes that a service composition is analogous to a specification of team plan from the perspective of a user who is composing the services. Furthermore, the implementation of the service composition infrastructure based on the teamwork framework would complement each other since both domains inherently share the similarities.

Despite the fact that our approach does not comply with the previous work on teamwork in several aspects, we can assume the outcomes of co-operative tasks, i.e. composite services, will not suffer from constant failures due to lack of understanding as a team amongst team members. It is based on the assumption that service providers are typically co-operative once they agree to provide services although the services can be expensive [7].

In the service composition domain, we believe a reliable discovery of services and a smart recovery strategy from failed interaction are important to achieve the robust execution of composite services. Section 5 describes our approach to improve the current model of teamwork to address these issues.

5 Extending Team Plans for Robust Execution

In any form of teamwork, an outcome of joint effort largely depends on the participating team members' performance as well as the environmental influences. For example, one of the major motivations of Tambe's work [16] on his general model of teamwork is to improve the overall robustness of executing joint plans by having generalised model of teamwork as a part of the underlying infrastructure. A large portion of the underlying teamwork model is concerned with the behaviours of failed team members. Furthermore, additional work has been done to improve the robustness by exploring the possibilities of monitoring the progress of team members [10].

The uncertain nature of the open environments introduce a new problem to the current teamwork models which has not been addressed in the current practical work on BDI platforms and teamwork models. Typically, the failure recovery approach available in such systems is the infrastructure support to select an alternative plan to achieve the same goal. This process continues until either the goal is achieved or all applicable plans failed to achieve the goal. Such an approach to failure recovery potentially imposes computational redundancy especially in the service composition type of teamwork. For example, unless a smart team member selection algorithm is implemented in the system, an alternatively selected plan could select a completely different set of team members and the effort of previously selected team members is completely wasted. In other words, in face of a certain sub-team failure, we do not want to reconstruct the whole team again, we want to be able to replace only the failed sub-team or team member.

We propose a new approach of failure recovery which extends the current model of teamwork as an additional means to achieve the robust execution of team plans. In this model, a failed plan is analyzed to identify the cause of failure, finds a replacing team member for the failed step, performs *backtracking* and the failed plan resumes with a new set of team members.

This approach is based on our observation of composite services in open environments where the failure of a particular service does not often indicate the

[7] We are currently not concerned with the security issues of open environments.

failure of the entire composite service. This approach is suitable given the fact that there are multiple instances of the same type of service available, and services are typically independent of each other since they are deployed by different organizations. Also bear in mind, we are trying to develop an easy to use service composition tool for end users who have limited programming experiences. Therefore, it is more desirable to ensure the robust execution of specified composite services, rather than relying upon the users to specify alternative plans to achieve the same goal.

In our approach, a generated team plan is executed until either the plan succeeds, or all combination of team members (service providers) are attempted and have failed to fulfil the requirements. This is an additional level of *backtracking* we built on the existing BDI model of failure recovery. The backtracking process involves five steps:

1. Find a failed step in the team plan
2. Find an alternative service provider of the same type and replace the failed service provider
3. Mark all the interfaces associated with the failed service provider as *not executed*
4. Identify any other steps which depend on the output of failed steps and mark as *not executed* (perform this recursively)
5. Re-execute the plan

To implement the *backtracking* strategy, we implemented an additional data structure which keeps track of the current execution state of a team plan. The data structure holds information required to perform the *backtracking*, which includes information such as an *execution-state flag* (true if it was successfully executed), *dependency with other steps* in the plan, and *assignment of a team member to a role*. Dependencies amongst the steps of a team plan are analyzed at the compilation time of the user specified composite services.

A hypothetical example illustrating the need for smart backtracking is a situation where a user wishes to purchase a hi-fi unit for his office, and a CD, but also wants to check that the hi-fi unit chosen is one for which there is a suitable power supply amongst those available at the workplace. Assume there are agents selling hi-fi units, an information agent providing information about available power supply units and agents that sell CDs. It is necessary to select a hi-fi unit in order to check availability of the power supply, as this is different for different units. However the user does not want to commit to purchase until it is confirmed that an appropriate power supply is available. Figure 2 illustrates the interaction.

Assume that the first three steps execute successfully, but the fourth step fails as the agent has now sold the hi-fi that was available. This necessitates going back to redo the first step. The third step will also need to be re-done, as it is dependent on the output of the first step. The second step on the other hand is independent and does not need to be repeated. Our system takes care of these dependencies, backtracking to redo only necessary steps.

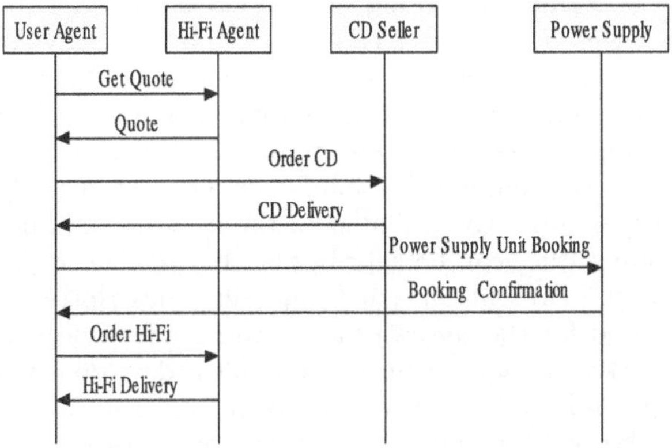

Fig. 2. Hi-Fi Ordering Example

The additional level of *backtracking* we developed could easily be extended to implement a generalized model for other domains. Our approach potentially provide a useful basis for development of multi-agent systems, in particular where a selection of team members is important.

6 Related Work

In this section, we discuss three pieces of related work that are strongly relevant to our work presented in this paper.

eFlow [5] is a service composition platform designed for specification, management, and execution of composite services using the existing Web-based services, primarily designed for the future *e-business* applications. The project aims at developing an enabling technology for more customisable and dynamic composite services. A composite service is modeled as a graph which defines the order of execution. The nodes in a graph includes service selection rules, which enables a dynamic selection of service providers. The system is flexible to incorporate new services and *broker agents* as the configuration of environment changes over the time.

eFlow and the ASKIT share great deal of similarities. In particular, both approaches enable specification of sequential and parallel execution, dynamic allocation of service providers, and provides support to access every instances of the same type of service just to name a few common characteristics. *eFlow* focuses on the development of infrastructure support for organization deploying services and manage their services easily. In comparison, our project aims at developing a service composition tool that enables end users with limited programming experiences compose *flexible* and *robust* services using services available on the network.

SWORD [13] is another implemented service composition tool. In SWORD, a service is represented as a simple rule using its expected input values and output values. Those rules of existing services are stored in a rule-based expert system and they are used to determined if a new composite service can be realized using the existing services given the expected input and output values of the new service. The primary contribution of their work is their approach of automatically generating new composite services. Our approach differs significantly in that our interface enables a user compose customized services by directly manipulating a set of service types currently available on the network. Ultimately, these user specified composite service can be kept in plan libraries for re-use.

Sycara et. al. [15] present an agent capability description language called *LARKS* that is used for the *matchmaking* process of services available on the Internet. Their work is primarily concerned with the development of a service description language and the implementation of the efficient and adaptive matchmaking process. Work on service description languages such as LARKS and WSDL is important for the service registration and discovery aspect of our system.

7 Conclusions

In this paper, we present a service composition tool-ASKIT, which enables users with limited programming experiences to compose customized services using the existing services available on the Agentcities network. We extended the current teamwork model in that a failed team plan is repaired by only replacing the failed team member, i.e. service provider, but not discard the whole team. *Backtracking* is performed so that the failed plan resumes its execution without any redundancy. We demonstrate such a failure recovery method using an fictional example composite service. Aspect of key features of our system are demonstrated through the web interface.

Being a prototype system, ASKIT has demonstrated the feasibility of implementing composite services using team plans. Moreover, the prototype system is developed on Agentcities network and has live interaction with other agent services. Currently, we are working on simple hierarchical team structures, where the team coordination is heavily imposed upon the personal user agent (the abstract team). Future work will look into other team architectures, taking into account joint intention and commitment using belief propagation among team members. We will also look at the Web services in the commercial world to compare the agent and teamwork approach with the standard industry approach.

Acknowledgements

The project is funded by Australian Research Council (ARC) Linkage Grant LP0218928 (2002-2004) with Agent Oriented Software Pty. Ltd. A shorter version of the paper is presented at Autonomous Agents and Multi-Agent Systems 2003's workshop - the Agentcities: Challenges in open agent environments.

References

[1] AOS. JACK Teams User Guide. Technical report, Agent Oriented Software Pty. Ltd., 2002. 618

[2] T. Berners-Lee, J. Hendler, and O. Lassila. The semantic web. *Scientific American*, 2001. 612

[3] M. E. Bratman. *Intentions, Plans, and Practical Reason.* Harvard University Press, Cambridge, MA, 1987. 615

[4] P. Busetta, R. Rönnquist, A. Hodgson, and A. Lucas. JACK Intelligent Agents - components for intelligent agents in Java. Technical report, Agent Oriented Software Pty. Ltd., 2001. 613

[5] F. Casati, S. Ilnicki, L. Jin, V. Krishnamoorthy, and M. Shan. Adaptive and dynamic service composition in eFlow. Technical report, Software Technology Lab, Hewlett-Packard Laboratories, 2000. 612, 621

[6] P. Cohen and H. Levesque. Teamwork. Technical Note 504, AI Center, SRI International, Menlo Park, CA, March 1991. 615, 618

[7] M. Georgeff, B. Pell, M. Pollack, M. Tambe, and M. Wooldridge. The Belief-Desire-Intention model of agency. In *Proceedings of Agents, Theories, Architectures and Languages (ATAL)*, 1999. 613

[8] B. Grosz and S. Kraus. Collaborative plans for complex group action. *Artificial Intelligence*, 86:269–357, 1996. 618

[9] M. J. Huber. JAM: a BDI-theoretic mobile agent architecture. In *Proceedings of the Third International Conference on Autonomous Agents (Agents'99)*, pages 236–243, Seattle, USA, May 1999. 613, 615

[10] G. Kaminka and M. Tambe. Robust agent teams via socially attentive monitoring. *Journal of Artificial Intelligence Research (JAIR)*, 2000. 619

[11] D. Kinny, M. Ljungberg, A. Rao, E. Sonenberg, G. Tidhar, and E. Werner. Planned team activity. In C. Castelfranchi and E. Werner, editors, *Artificial Social Systems*, volume 830 of *Lecture Notes in Computer Science*, pages 227–256. Springer Verlag, 1994. 618

[12] S. McIlraith, T. Son, and H. Zeng. Semantic web services. *IEEE Intelligent Systems (Special Issue on the Semantic Web*, 2001. 612

[13] S. Ponnekanti and A. Fox. SWORD: A developer toolkit for web service composition. In *Proceedings of The Eleventh World Wide Web Conference*, 2002. 612, 621

[14] A. Rao and M. Georgeff. An abstract architecture for rational agents. In *Proceedings of the Third International Conference on Principles of Knowledge Representation and Reasoning*, 1991. 613

[15] Katia P. Sycara, Matthias Klusch, Seth Widoff, and Jianguo Lu. Dynamic service matchmaking among agents in open information environments. *SIGMOD Record*, 28(1):47–53, 1999. 622

[16] M. Tambe and W. Zhang. Towards flexible teamwork in persistent teams: extended report. *Journal of Autonomous Agents and Multi-agent Systems, special issue on "Best of (ICMAS) 98"*, 2000. 618, 619

Fingerprint Images Segmentation Using Two Stages Coarse to Fine Discrimination Technique

T. S. Ong, T. B. J. Andrew, N. C. L. David, and Y. W. Sek

Faculty of Information Science and Technology (FIST), Multimedia University
Jalan Ayer Keroh Lama, Bukit Beruang, 75450, Melaka, Malaysia
{tsong,bjteoh,david.ngo,ywsek}@mmu.edu.my

Abstract. Segmentation of fingerprint image is necessary to reduce the size of the input data, eliminating undesired background, which is the noisy and smudged area in favor of the central part of the fingerprint. In this paper, an algorithm for the segmentation which uses two stages coarse to fine approach is presented. The coarse segmentation will be performed at first using the orientation certainty values that derived from the blockwise directional field of the fingerprint image. The coarse segmented image will be carry on to the second stage which consist Fourier based enhancement and adaptive thresholding. Orientation certainty values of the resultant binarized image are calculated once again to perform the fine segmentation. Finally, binary image processing is applied as postprocessing to further reduce the segmentation error. Visual inspection shows that the proposed method produce accurate segmentations result. The algorithm is also evaluated by counting the number of false and missed detected center points and compare with the fingerprint image which have no segmentation and with the proposed method without postprocessing. Experiments show that the proposed segmentation method perform well than others.

1 Introduction

An acquitted fingerprint image usually consists of two components, which are called the foreground and background. The foreground is the component that originated from the contact of a fingertip with the fingerprint scanner. The noisy and smudged area at the border of the image is called the background [1]. Therefore, segmentation is the process of discriminating the foreground and background area in the fingerprint image. In segmentation, areas corresponding to the background are determine and discarded from the rest of the processing. Segmentation is served for two purposes, which are reducing the size of the input data and reliable extraction of features like center point.

Several techniques have been found in the literature. In [2], the fingerprint image is partitioned into 16 x 16 pixels subblocks. Then, each block is classified according to

T.D. Gedeon and L.C.C. Fung (Eds.): AI 2003, LNAI 2903, pp. 624-633, 2003.

their distribution of the gradients. [3] extended this method by setting a threshold value and compare to a gray value variance. Regions in the image for which the variance is less than the threshold will be excluded. In [4], the output of a set of Gabor filters is used as input to a clustering algorithm that constructs spatially compact clusters. [5] segmented fingerprint images based on the coherence, while morphology is used to obtain smooth regions while [1] uses coherence, mean, variance, supervised linear classifier and morphology for the segmentation.

A good segmentation technique should be insensitive to the contrast to the original image, be independent of whether the input image is enhanced or raw and should give consistent results for a variety of images used in application. This motivated us to proposed a two stages coarse to fine discrimination technique which able to segment the fingerprint image efficient and reliably. The coarse segmentation is performed using the orientation certainty values. However, non homogeneous illumination and the uniform regions may yield the errors. Hence, second stage refinement that consists of Fourier based enhancement and adaptive thresholding are applied. Orientation certainty values of the resultant binarized image are calculated once again to perform the fine segmentation. Finally, binary image processing is applied as postprocessing.

2 Coarse Segmentation

As shown in Fig. 1, the information carrying features in a fingerprint are the line structures, called ridges and furrows. In this figure, the ridges are black and the furrows are white.

The elementary orientations in the image are given by the gradient vector $[G_x\ G_y]^T$, which is defined as

$$\begin{bmatrix} G_x \\ G_y \end{bmatrix} = sign(G_x)\nabla I(x,y) \quad (1)$$

where $I(x,y)$ represents the gray-scale image. In order to describe the orientation of the ridge-furrow structures, which is a much coarser scale, a blockwise averaged squared gradient field that shown in Fig. 2 is derived. Here, gradients of each pixel is calculated and squared and then combines the entire squared gradient within the block W x W to get a blockwise averaged squared gradient field that given by

Fig. 1. Ridges and furrows in the fingerprint

Fig. 2. The orientation field of the fingerprint image

$$\begin{bmatrix} G_{x,s} \\ G_{y,s} \end{bmatrix} = \begin{bmatrix} G_{xx} - G_{yy} \\ 2G_{xy} \end{bmatrix} \tag{2}$$

In this expression, $G_{xx} = \Sigma_W G_x^2$, $G_{yy} = \Sigma_W G_y^2$ and $G_{xy} = \Sigma_W G_x G_y$. are estimate for the variance and crossvariance of G_x and G_y, averaged over the window W.

Now, the orientation certainty (OC) is defined as:

$$\text{OC} = \sqrt{\frac{(G_{xx} - G_{yy}) + G_{xy}}{W^2 (G_{xx} + G_{yy})}} \tag{3}$$

The expression above will form values that normalized into the interval from 0 and 1 [6]. The magnitude of OC representing the flow directions of ridges in a block are used as certainties for the extracted information on flow direction for the block.

At this stage, the image regions that has small OC (<0.01) will be discarded as shown in Fig. 3. However, uneven contact pressure fingerprint with the sensor causes non homogeneous illumination and the uniform regions also may lead to the high OC, but these regions are supposed to be segmented in the background segmentation process. In order to compensate these kinds of pitfall, the second stage of fine segmentation is needed to extract the tightly bounded fingerprint region from the image

Fig. 3. Coarse segmentation of the fingerprint image

3 Fine Segmentation

In order to eliminate the effect of non homogeneous illumination and high OC values in the uniform region, a series of processing are used to fulfill these requirements. These steps included Fourier based image enhancement, adaptive thresholding and OC values are calculated again on the binarized image. Finally, binary image processing is applied as post processing to further reduce the segmentation error. At the second stage, the input image is the coarse segmented and reduced size fingerprint image. This will help the refinement to be done efficiently and more reliably.

3.1 Fourier Based Image Enhancement

The main objective of the image enhancement is to increase the contrast in the, many time pretty rough, grayscale fingerprint image from the sensor. That is, making the non-continuous ridges-valleys of the fingerprint continuous and extracts the highly interesting foreground.

Directional information that conveyed by the line structure fingerprint image, $I(x,y)$ is contained in the magnitude of the frequency domain. In the blockwise operation, by choosing the window size, W (16 x 16 in default) so that the image contains two or three roughly parallel ridges the dominant frequencies are presumed to correspond to the ridges in the window. The target is to thicken ridges and also to separate parallel ridges more finely, and thus increase the overall contrast in the fingerprint [8]. In order to do this, 2D Fourier transform of each window, W is taken according to

$$F_W(u,v) = \sum_{x=0}^{W-1}\sum_{y=0}^{W-1} I(x,y)e^{-j2\pi\left(\frac{ux}{W}+\frac{vy}{W}\right)} \tag{4}$$

for $u = 0, 1, 2, \ldots, W\text{-}1$ and $v = 0, 1, 2, \ldots, W\text{-}1$. Subsequently $F_W(u,v)$ is multiplied with its powered magnitude and finally the inverse 2D Fourier transform, F_W^{-1} of the product is calculated as follow:

$$I_{W,enh}(x,y) = F_W^{-1}(F_W(u,v)|F_W(u,v)|^k) \tag{5}$$

where $I_{Wenh}(x,y)$ is the resulting enhanced window and k is the enhancement power factor, an experimentally determined constant. Higher k can improve the appearance of the ridges and filling up small holes in ridges whereas having too high k can result in false joining of ridges. Thus a termination minutia might become a bifurcation which is unallowable. This process will repeat W x W times for a fingerprint image. Fig. 4 shows the enhanced image with $k = 0.24$.

3.2 Adaptive Thresholding

As mentioned before, the OC approach fails for uniform regions and non homogeneous illumination on the fingerprint image, therefore adaptive thresholding is designed in order to alleviate these problems. Firstly grayscale means are computed for each window, W of the enhanced image. The mean values of a window (k,l) is computed as follows:

$$\mathrm{M}_W(k,l) = \frac{1}{W^2}\sum_{i-1}^{W}\sum_{j=1}^{W} f(kW+i, lW+j) \tag{6}$$

Fig. 4. Enhanced Image using Fourier based approach with $k = 0.24$

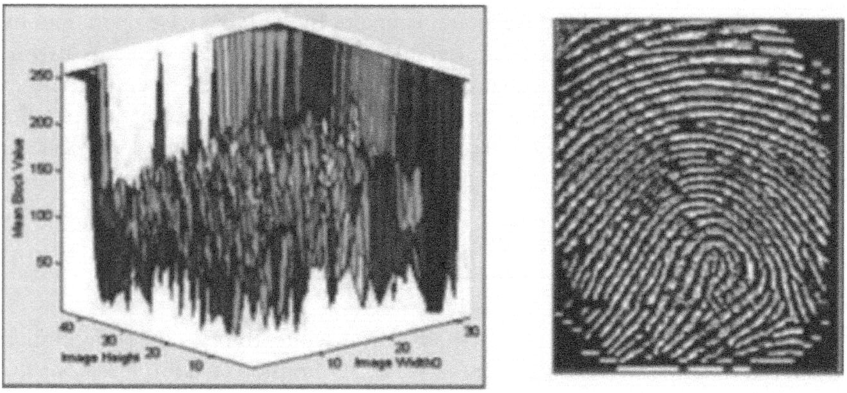

Fig. 5. (a) Windowed Mean Value Map and (b) Thresholded fingerprint image

Mean values of those windows in uniform regions approach 255 (white background is assumed). So, by setting a proper threshold, τ, the uniform regions in the image for which the M_W is more than τ can be segmented.

A mean values map, MP can be constructed according to (4) as below:

$$\mathrm{MP} = \bigcup_i M_{Wi} \tag{7}$$

From this map, another threshold value, τ' can be estimated from the border MP in order to eliminate the non homogeneous illumination regions. The MP and the thresholded fingerprint image are shown in Fig. 5(a) and Fig. 5(b) respectively.

Once again, OC values are calculated over this binarized image and image regions that has small OC (<0.01) will be assigned value 1 and 0 otherwise. Therefore, a tightly bound binarized image that acts as a mask is drawn. By multiply the binary mask with the grayscale input image, a fine segmented image will be obtained. Fig.6(a) and Fig. 6(b) show the binary mask and the final segmented gray scale image.

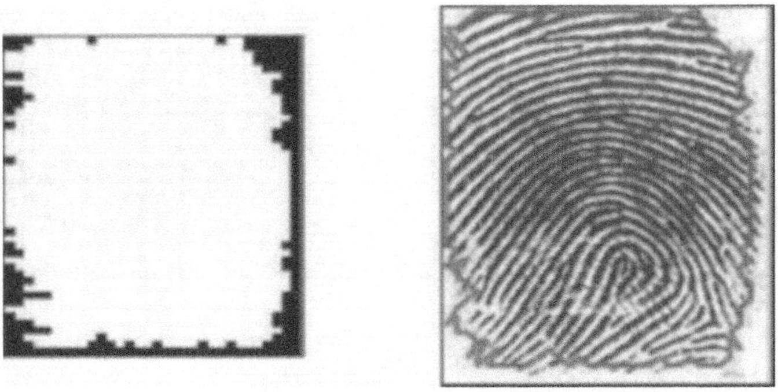

Fig. 6. (a) Binary mask of the fingerprint and (b) Final segmented gray scale image

3.3 Postprocessing

Even though the previous mentioned procedure is able to produce a quite satisfying result for most of the cases. However, in some cases small illumination difference in within non-homogeneous regions will cause poor segmentation due to the estimation difficulty of the threshold value, τ'. This may lead to a not tightly bounded region for the segmented image as shown in Fig. 7.

This problem can be solved by apply the binary image processing techniques into the binary mask as depicted in Fig. 8. At first, a labeling algorithm is used to assign the same label to all pixel connected blobs, then each connected blob that contained the same label will be isolated and eliminate if the blob sizes is smaller than 100.

Then, the holes in the binarized image are filled and every row in the image will be scanned through and if their size is smaller than 10, all the pixel in that particular row will be turn off. Same procedure is also applied for every column in the image. This leads to the result as shown in Fig. 8(b). Finally, only the biggest blob will be reserved and acts as a mask in order to get the tightly bounded segmented image as shown Fig.9.

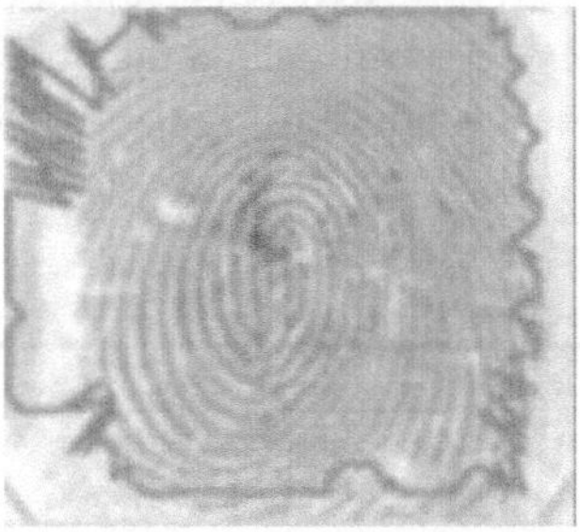

Fig. 7. Poor segmented image

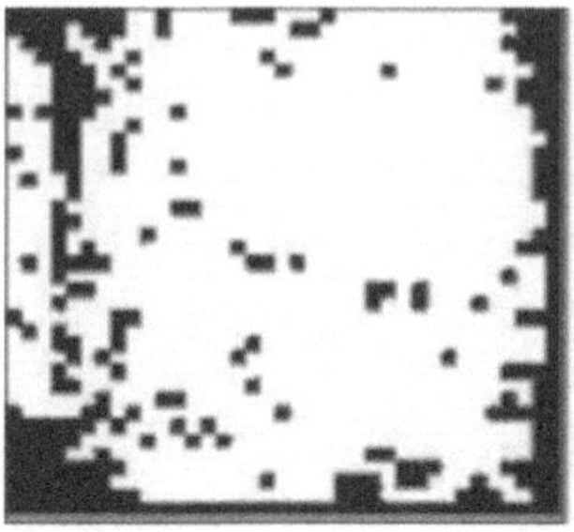

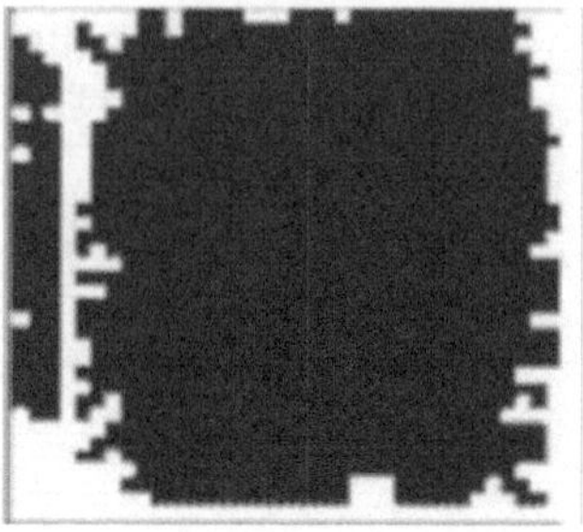

Fig. 8. (a) Binary mask of the sample in Fig. 7. and (b) Result obtained after apply the binary image processing

Fig. 9. Final segmented image

4 Experimental Results

Experiments are performed using fingerprint samples obtained from FVC2000 (Set B) Database 2 and Database 3 [9]. There are total in 160 samples. Visual inspection shows that the algorithm provides satisfactory results. Some sample results from Database 2 and Database 3 are shown in Figure 10 and in Fig. 11 respectively.

Apart from the visual inspection, the proposed algorithm also evaluated by counting the number of false and missed detected center points (CP) and compare with the fingerprint image which have no segmentation and the proposed method without postprocessing. The results for the center point, using the methods presented in [10], are shown in Table 1.

Segmentation method	False CPs	Missed CP
PM without postprocessing	11.88%	1.25%
PM with postprocessing	6.25%	2.5%

PM: Proposed Method

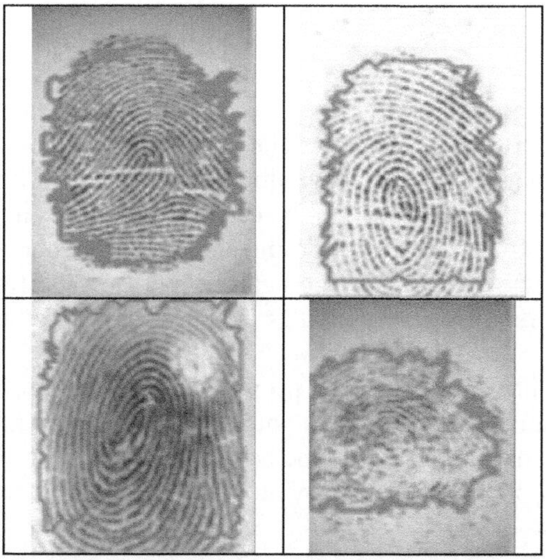

Fig. 10. Result examples from Database 2 FVC 2000 (Set B)

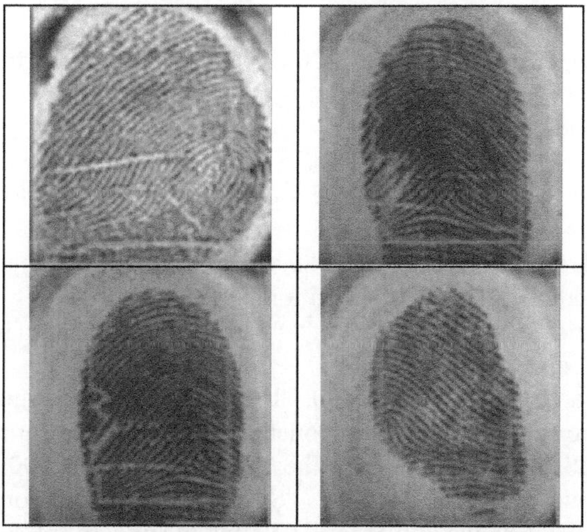

Fig. 11. Result examples from Database 3 FVC 2000 (Set B)

From the Table 1, can be observed that the proposed method with postprocessing rejects more false center points, which is just 6.25% compare to the 11.88% and 46.87% for one without processing and when no segmentation is applied respectively. This is because the proposed method without postprocessing may cover wider area and thus introduce more false center points. However, the percentage of miss detected center point for the proposed algorithm with postprocessing is slightly higher than the without postprocessing one. This is due to the over-segmentation of few bad quality fingerprint images and hence causes the region that contain center point of the image was excluded.

5 Conclusion

A fingerprint segmentation algorithm has been presented in this paper. This method uses two stages coarse to fine approach. The coarse segmentation is performed using the certainty values that derived from the blockwise directional field of the fingerprint image. The coarse segmented image will be carry on to the second stage which consist Fourier based enhancement and adaptive thresholding. Certainty values of the resultant binarized image are calculated once again to perform the fine segmentation. Finally, binary image processing is applied as post processing to further reduce the segmentation error.

Visual inspection has shown that the proposed method provides accurate segmentation results. Alternative experiments by counting the false and missed center point of the fingerprint show that the proposed method performs well in rejecting false center point compare to the proposed method without postprocessing and when no segmentation is applied.

References

[1] Asker. M., Bazen and Sabih H. G. 2001. "Segmentation of Fingerprint Images". ProRISC 2001 workshop on Circuit, System and Signal Processing, Veldhoven. The Netherlands. November 2001.

[2] Mehtre, B.M., Murthy, N. N. Kapoor, S. and Chatterjee, B. 1987. "Segmentation of .fingerprint images using the directional image". Pattern Recognition, 20(4): 429–435.

[3] Mehtre, B.M. and Chatterjee, B. 1989. "Segmentation of .fingerprint images - a composite method". Pattern Recognition, 22(4): 381–385.

[4] Jain, A. K. and Ratha, N. K. 1997. "Object detection using Gabor filters", Pattern Recognition. 30(2): 295–309.

[5] Bazen,A.M. and Gerez, S. H. 2000. "Directional field computation for fingerprints based on the principal component analysis of local gradients". Proceedings of ProRISC2000, Veldhoven, The Netherlands, Nov.2000.

[6] Bazen,A.M. and Gerez, S. H. 2002. "Systematic Methods for the Computation of the Directional Fields and Singular Points of Fingerprint". IEEE Transaction on Pattern Analysis and Machine Intelligence. 24(7): 905-918.

[7] Halici, U., Ongun, G. 1996. "Fingerprint Classification through Self Organization Maps Modified to Treat Uncertainties". Proceeding of the IEEE. 84(10): 1497-1512.

[8] Willis, A. J., Myers, L. 2001. "A Cost-Effective Fingerprint Recognition System for use with Low-Quality Prints and Damaged Fingerprints," Pattern Recognition 34, 255 - 270, 2001.

[9] Maio, D., Maltoni, D., Cappelli, C., Wayman, J. L. and Jain, A.K. 2000. "FVC2000:Fingerprint verification competition." Biolab internal report, University of Bologna, Italy. http://bias.csr.unibo.it/fvc2000/.

[10] Rao, A. R. 1990. "A Taxonomy for Texture Description and Identification", SpringerVerlag, New York, 1990.

Automatic Fingerprint Center Point Determination by Using Modified Directional Field and Morphology

T. B. J. Andrew, T. S. Ong, N. C. L. David, and Y. W. Sek

Faculty of Information Science and Technology (FIST), Multimedia University
Jalan Ayer Keroh Lama, Bukit Beruang, 75450, Melaka, Malaysia
{bjteoh,tsong,david.ngo,ywsek}@mmu.edu.my

Abstract. Accurate and reliable center point determination of the fingerprint is an essential pre-processing step for the image based fingerprint recognition approach. This step is particularly important since a reference point is required to cancel the variation in position of the reference and testing fingerprint image. This paper illustrated an automatic center point determination algorithm that couples the modified averaged square directional field (MASDF) with morphological operation, in order to pinpoint the center point of the fingerprint efficiently and accurately. The experiment result shows only 3.13% rejection rate for the false center point by using the proposed method.

1 Introduction

Today, there are various approaches of automatic fingerprint matching that have been proposed which include minutiae-based approaches and image-based approaches. Minutiae-based approaches as the most popular ones being included in almost all contemporary fingerprint recognition systems. Image based approaches on the other hand usually applied directly onto the gray scale with lesser pre-processing compare to minutiae-based approaches, and hence they may achieve higher computation efficiency than minutiae methods [1]. For both image-based and minutiae–based systems, the center point (or core point, which differentiate from singular points that consist core and delta point) of the fingerprint is important to find for reference in obtaining features [1, 3, 5]. Particularly for the image-based approach, its main disadvantage is the limitation ability to track with variation in position, scale and orientation angle. Usually the variation in position between the two fingerprints is cancelled by choosing a center point in each fingerprint.

Until now, a number of methods to detect the center points in fingerprint images have been proposed [2-7]. The most popular one is the Poincare Index (PI) that developed by [2], was adopted by [3] and [5] and was enhanced by [6] and [7], which is derived from continuous curves. However, two problems with this method are fingerprints of the Arch type do not have any singularity in terms of PI and also more time are consumed to smooth the directional image iteratively in order to calculate the PI.

T.D. Gedeon and L.C.C. Fung (Eds.): AI 2003, LNAI 2903, pp. 633-640, 2003.

In this paper, an algorithm that couples the modified averaged square directional field (MASDF) with morphological operation is proposed to resolve the mentioned problems.

2 Modified Averaging Squared Directional Field (MASDF)

The information carrying features in a fingerprint are the line structures, called ridges and furrows.

The elementary orientations in the image are given by the gradient vector $[G_x\ G_y]^T$, which is defined as

$$\begin{bmatrix} G_x \\ G_y \end{bmatrix} = sign(G_x)\nabla I(x,y) \tag{1}$$

where $I(x,y)$ represents the gray-scale image. In order to describe the orientation of the ridge-furrow structures, which is a much coarser scale, gradient field is derived from the gradients by performing some averaging operation on the gradients, involving pixels in some neighborhood. However, gradient cannot directly be averaged in some local neighborhood since opposite gradient will then cancel each other, although they indicate the same furrow-ridge orientation. This is caused by the fact that local ridge-furrow structures remain unchanged when rotated over 180 degrees. A solution to this problem is proposed by doubling the angles of the gradients vectors before averaging [8]. After doubling the angles, opposite gradient vectors will point in the same direction and therefore, will reinforce each other, while perpendicular to the direction of the average gradient vector. The resultant averaged squared gradient field that defined in some neighborhood, using a uniform window, W is therefore given by

$$\begin{bmatrix} G_{x,s} \\ G_{y,s} \end{bmatrix} = \begin{bmatrix} G_{xx} - G_{yy} \\ 2G_{xy} \end{bmatrix} \tag{2}$$

In this expression, $G_{xx} = \Sigma_W G_x^2$, $G_{yy} = \Sigma_W G_y^2$ and $G_{xy} = \Sigma_W G_x G_y$. are estimates for the variance and crossvariance of G_x and G_y, averaged over the window W.

Now, the average gradient direction, φ, with $-\frac{1}{2}\pi < \varphi \leq \frac{1}{2}\pi$, is given by:

$$\varphi = \frac{1}{2} tan^{-1}\left(\frac{2G_{xy}}{G_{xx}-G_{yy}}\right) \tag{3}$$

and the average rigde-furrow direction ϕ or also known as averaged squared directional field (ASDF), with $-\frac{1}{2}\pi < \phi \leq \frac{1}{2}\pi$, is perpendicular to φ:

$$\phi = \varphi + \frac{1}{2}\pi \tag{4}$$

Fig. 1(a) and Fig. 1(b) show the gradient field and the ASDF respectively.

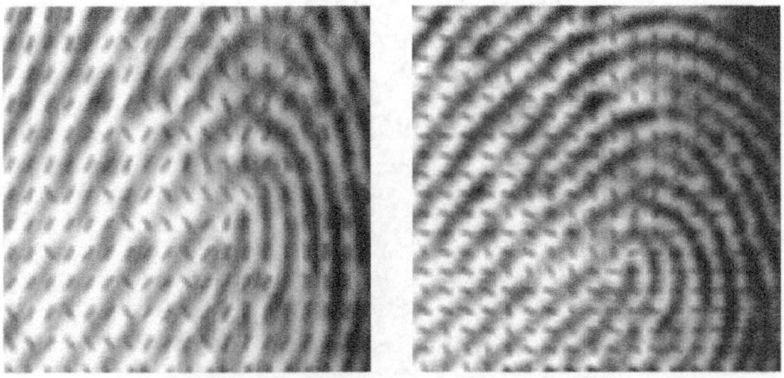

(a) (b)

Fig. 1. Detail area in a fingerprint: (a) gradient field and (b) ASDF

Fig. 2. Modified Averaged Squared Direction Field

However, estimated ASDF may not always be correct due to the corrupted ridge and furrow structures that caused by noise. Since local ridge orientation varies slowly in a local neighborhood where no singular point appears, Wierner filter had been applied to modify the incorrect local ridge orientation. In order to perform the filtering, the orientation image needs to convert into a continuous vector field, which can be defined as follow:

$$\Phi_x = \cos(2\phi) \quad \text{and} \quad \Phi_y = \sin(2\phi) \tag{5}$$

With the resulting vector field, the filtering then is performed as follows:

$$\Phi'_x = \Sigma\Sigma_W H\Phi_x \quad \text{and} \quad \Phi'_y = \Sigma\Sigma_W H\Phi_y \tag{6}$$

where H is a 2D Wiener low-pass filter. After all, MASDF, ϕ' is calculated using

$$\phi' = \frac{1}{2} tan^{-1}(\frac{\Phi'_y}{\Phi'_x}) \tag{7}$$

Thus, a fairly smooth directional field can be obtained. Fig. 2 shows an example of the MASDF.

Fig. 3. Illustration of the pattern. White area indicates the slop values that ranging from 0 to ½ π

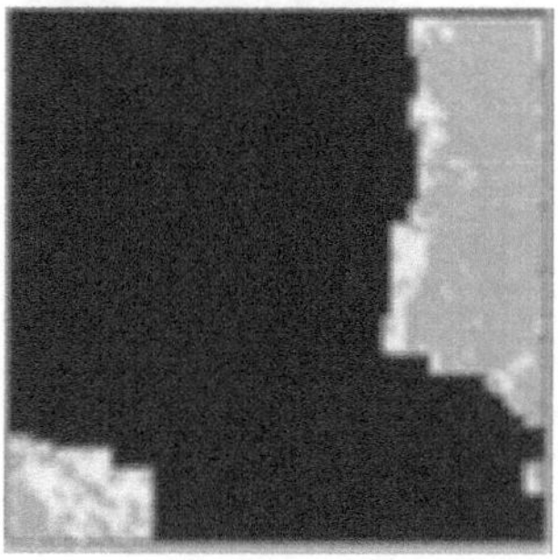

Fig. 4. Binary pattern after performing close and open morphological operation

3 Algorithm

In this algorithm, center point of a fingerprint is defined as the point of maximum curvature in the fingerprint image [5]. Firstly, an input fingerprint will be divided into non-overlapping blocks of size 5 x 5. In each block, the x and y magnitudes of the gradient, G_x and G_y at each pixel in each block were determined. With each block, slope that perpendicular to the local gradient orientation of each block will be computed using equation (4), (5), (6) and (7) to produce a MASDF that had described in section 2.0.

In the MASDF, the blocks with slopes value ranging from 0 to $\frac{1}{2}\pi$ will be sought and marked as '1' otherwise '0' was assigned. The illustration of this step is shown in Fig. 3.

However, the binary pattern shows a lot spurious points or cluster area that may lead to the false detection center point. In order to rectify this problem, mathematical morphological open and close operations are performed. These processes are commonly used to smooth, fill in, and/or remove objects in a binary image. Morphologic closing is equivalent to a dilation followed by erosion with a structuring element. Similarly, morphologic opening is equivalent to erosion followed by dilation. The choice of a structuring element depends on the size and shape of the objects to be modified. In this case, a disk-shaped structuring element with a radius of 3 pixels is created. This step 'repairs' the estimate by removing small areas, thus creating more compact cluster and thus this also causes less time to perform the searching process. Fig. 4 shows the binary pattern after performing close and followed by the open operation.

By looking only at blocks which consists '1', a path is traced down until a slope that is not ranging from 0 to $\frac{1}{2}\pi$ be encountered and marked. Fig. 5 shows the map of the point that indicates the possible location for a center point.

Finally, the block that has the highest number of marks will compute the slope in negative y direction and location of center point can be pinpointed. The marked map result is shown in Fig. 6.

4 Experiments and Results

Experiments are conducted by using fingerprint samples obtained from FVC2000 (Set B) Database 1 and Database 2 [9]. There are 160 samples in total and the acquiring method for these databases was using low-cost optical sensor and capacitive sensor respectively. In Fig. 7, the pinpointed center points of the five Henry classes [10] are shown. This indicates that this approach has no problem to detect center point of all classes of the fingerprint if compare to the PI approach.

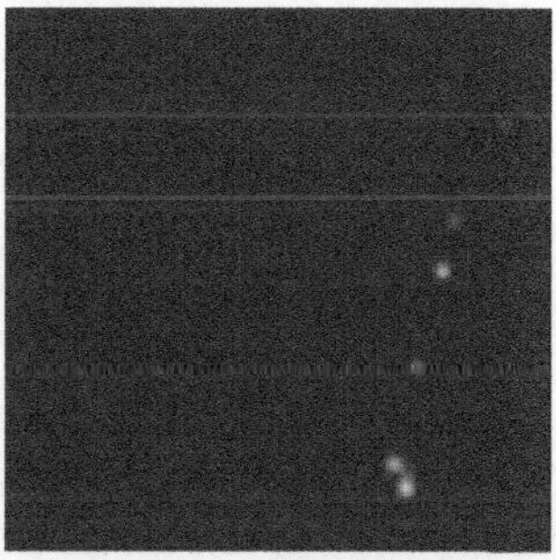

Fig. 5. Map of possible fingerprint center point

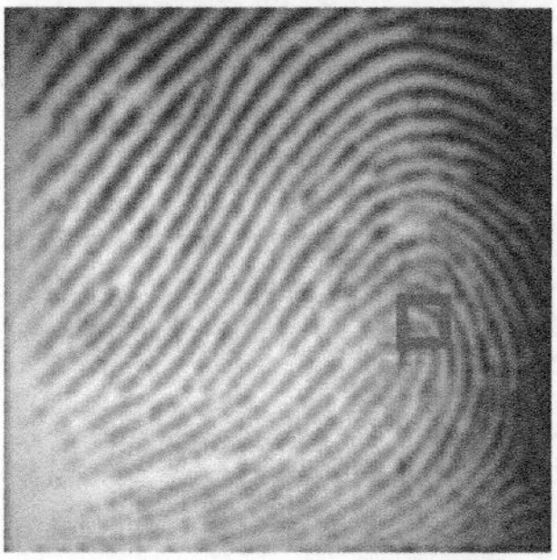

Fig. 6. Center point is located

The experiments are performed by counting the number of false detected center points in four cases: (a) just applying ASDF which originally proposed by [4], (b) ASDF with morphology, (c) MASDF alone without morphology and (d) MASDF with morphological operation. The results are shown in Table 1.

From Table 1, method (b) is performed better than method (a) with the rejection rate 5.00% compare to the 6.25%. This indicates that morphology operation is able to remove the spurious points or clusters effectively that may lead to the false center points.

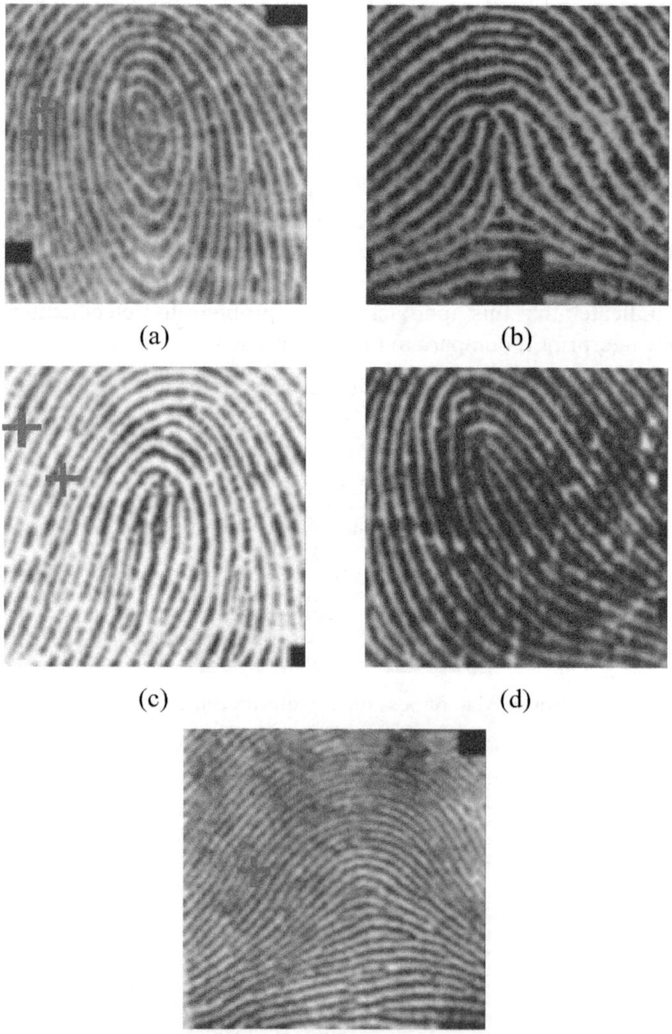

(a) (b) (c) (d) (e)

Fig. 7. Center point extraction for example from each of the five Henry classes with. (a) Whorl. (b) Tented arch. (c) Left Loop. (d) Right Loop and (e) Arch

Table 1. Result of center point extraction

Method	False CPs
a. ASDF [4]	6.25%
b. ASDF with morphology	5.00%
c. MASDF	5.00%
d. MASDF with morphology	3.13%

Table 2. Performances comparison between the proposed method and the PI approach

Method	False CPs	Speed Factor
MASDF with morphology	3.13%	0.7
Poincare Index	5.63%	1

On the other hand, method (c) still gave the better result over method (a), this shows the modified version of ADSF is able to correct the corrupted local ridge orientation that was caused by noise and thus reveal the area that contain the maximum curvature. However, when the results from method (b) and method (c) were observed, both have contributed 5% of the rejection rate, which indicates that modified ASDF and morphology are perform equally well.

The best result has been obtained by combine the MASDF and the morphology as shown in method (d). This proposed method rejects more false center points, which is 3.13% of rejection rate, compare to the 5.00%, 5.00% and 6.25% when applied method (c), method (b) and method (a) respectively.

Comparison method (d) with the method based on Poincare Index (PI) which adopted in [3] also been performed in terms of rejection rate and execution rate. The computation speed of two approaches is evaluated using speed factor relative to the computation speed of the PI. The PI, being the slowest is assigned a speed factor of one. For faster method, it will have a speed factor of less than one. Rejection rate and speed factor for each of the method are given in Table 2.

Apparently, the proposed method is outperformed Poincare Index approach in terms of the rejection rate and the speed execution as well.

5 Conclusion

A fingerprint center point determination algorithm that couples MASDF and mathematical morphological operation has been presented in this paper.

Experiment results show the proposed method is able to detect center point of all classes of the fingerprint and performs well in rejecting false center point. This exhibit the use of this algorithm into the image based fingerprint recognition system as a pre-processing step to detect the center point accurately and efficiently is promising.

References

[1] Tico, M.; Immonen, E.; Ramo, P.; Kuosmanen, P.; Saarinen, J.: 'Fingerprint recognition using wavelet features'. The 2001 IEEE International Symposium on Circuits and Systems, 2001, 2: 21 -24.

[2] Kawagoe, M. and Tojo, A.: 'Fingerprint Pattern Classification', Pattern Recognition, 1984, 17(3): 295-303.

[3] Karu, K. and Jain, A. K.: 'Fingerprint Classification', Pattern Recognition, 1996, 29(3): 389-403.

[4] Rao, A. R.: 'A Taxonomy for Texture Description and Identification', SpringerVerlag, New York, 1990.

[5] Jain, A. K., Prabhakar, S. and Hong, L.: 'A Multichannel Approach to Fingerprint Classification'. IEEE Transaction on Pattern Analysis and Machine Intelligence. 1999, 21(4):348-359.

[6] Cho, B.H., Kim, J.S., Bae, J.H., Bae, I.G. and Yoo, K.Y.: 'Core-based Fingerprint Image Classification'. ICPR'00, Spain, 2000, vol 2.

[7] Ching T.H., Zhuang Y.L., Tan C.L. and Kung C.M.: 'An effective method to extract fingerprints Y ingular point.': Proc. The Fourth International Conference/ Exhibition on High Performance Computing in the Asia-Pacific Region, 2000, pp. 696 -699 vol.2.

[8] Kass, M. and Witkin, A.: 'Analyzing Oriented Patterns', Computer Vision, Graphic and Image Processing, 1990. 37(3): 362-385.

[9] Maio, D., Maltoni, D., Cappelli, C., Wayman, J. L. and Jain, A.K.: 'FVC2000:Fingerprint verification competition', Biolab internal report, University of Bologna, Italy, 2000. http://bias.csr.unibo.it/fvc2000/.

[10] Henry E. R.: 'Classification and Uses of Fingerprints'. George Routledge and Sons, London. 1990.

Convolutional Neural Networks for Image Processing: An Application in Robot Vision

Matthew Browne and Saeed Shiry Ghidary

GMD-Japan Research Laboratory, Collaboration Center
2-1 Hibikino, Wakamatsu-ku, Kitakyushu-city
matthew.browne@gmd.gr.jp, saeed.shiry@gmd.gr.jp

Abstract. Convolutional neural networks (CNNs) represent an interesting method for adaptive image processing, and form a link between general feed-forward neural networks and adaptive filters. Two dimensional CNNs are formed by one or more layers of two dimensional filters, with possible non-linear activation functions and/or down-sampling. CNNs possess key properties of translation invariance and spatially local connections (receptive fields). We present a description of the convolutional network architecture, and an application to practical image processing on a mobile robot. A CNN is used to detect and characterize cracks on an autonomous sewer inspection robot. The filter sizes used in all cases were 4x4, with non-linear activations between each layer. The number of feature maps used in the three hidden layers was, from input to output, 4, 4, 4. The network was trained using a dataset of 48x48 sub-regions drawn from 30 still image 320x240 pixel frames sampled from a pre-recorded sewer pipe inspection video. 15 frames were used for training and 15 for validation of network performance. Although development of a CNN system for civil use is on-going, the results support the notion that data-based adaptive image processing methods such as CNNs are useful for image processing, or other applications where the input arrays are large, and spatially / temporally distributed. Further refinements of the CNN architecture, such as the implementation of separable filters, or extensions to three dimensional (ie. video) processing, are suggested.

1 Introduction

The term *convolutional network* (CNN) is used to describe an architecture for applying neural networks to two-dimensional arrays (usually images), based on spatially localized neural input. This architecture has also been described as the technique of shared weights or local receptive fields [1, 2, 3] and is the main feature of Fukushima's *neocognitron* [4, 5]. Le Cun and Bengio [6] note three architectural ideas common to CNNs: local receptive fields, shared weights (weight averaging), and often, spatial down-sampling. Processing units with identical weight vectors and local receptive fields are arranged in an spatial array, creating an architecture with parallels to models of biological vision systems [6].

T.D. Gedeon and L.C.C. Fung (Eds.): AI 2003, LNAI 2903, pp. 641–652, 2003.

A CNN image mapping is characterized by the strong constraint of requiring that each neural connection implements the same local transformation at all spatial translations. This dramatically improves the ratio between the number of degrees of freedom in the system and number of cases, increasing the chances of generalization [7]. This advantage is significant in the field of image processing, since without the use of appropriate constraints, the high dimensionality of the input data generally leads to ill-posed problems. To some extent, CNNs reflect models of biological vision systems [8]. CNNs take raw data, without the need for an initial separate pre-processing or feature extraction stage: in a CNN the feature extraction and classification stages occur naturally within a single framework.

In the CNN architecture, the 'sharing' of weights over processing units reduces the number of free variables, increasing the generalization performance of the network. Weights are replicated over the spatial array, leading to intrinsic insensitivity to translations of the input - an attractive feature for image classification applications. CNNs have been shown to be ideally suited for implementation in hardware, enabling very fast real-time implementation [9]. Although CNN have not been widely applied in image processing, they have been applied to handwritten character recognition [2, 9, 10, 11] and face recognition [7, 8, 12]. CNNs may be conceptualized as a system of connected feature detectors with non-linear activations. The first layer of a CNN generally implements non-linear template-matching at a relatively fine spatial resolution, extracting basic features of the data. Subsequent layers learn to recognize particular spatial combinations of previous features, generating 'patterns of patterns' in a hierarchical manner. If down-sampling is implemented, then subsequent layers perform pattern recognition at progressively larger spatial scales, with lower resolution. A CNN with several down-sampling layers enables processing of large spatial arrays, with relatively few free weights.

The present paper presents an application of CNN to an applied problem associated with the development of the visual system of a KURT2 robot [13]. This robot is used in present experiments on autonomous sewer inspection and is equipped with an infra-red video camera, as well as other sensors. The task of the system is the online detection of cracks and other faults in sewer pipes. The cracks are defined by a relatively distinctive space-frequency structure. However, they are often embedded in a variety of complex textures and other spurious features. Lighting and reflection effects present an additional source of difficulty. The implementation and performance of the CNN architecture is discussed in terms of this application. The CNN is also applied to certain artificial image processing problems.

Figure 1 shows the architecture of a CNN with two layers of convolution weights and one output processing layer. Neural weights in the convolution layers are arranged in an 2-D filter matrices, and convolved with the preceding array. In figure 1, a single layer 1 neural filter is shown operating at two different spatial translations on the input array. The output (shaded pixel) forms a spatially distributed feature map, which is processed by the second convolutional layer,

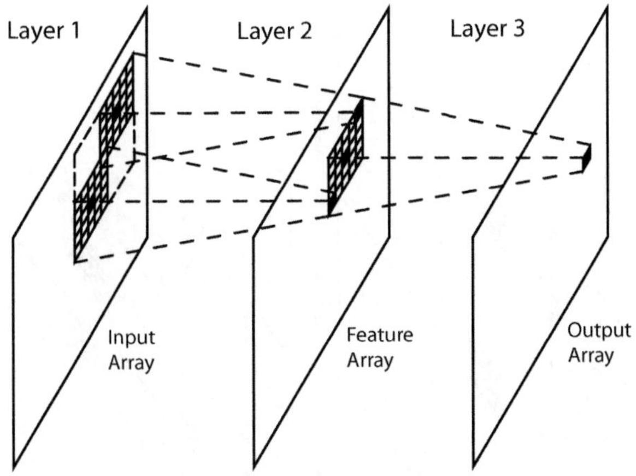

Fig. 1. Architecture of a CNN with a single convolutional neuron in two layers. A 5x5 filter at two different translations is shown mapping to shaded pixels in the first feature array. Shaded pixels in turn are part of the local feature information which is mapped to the output (or second feature) array by a second 5x5 filter

and passed to the output array. down-sampling of the feature array may be implemented between the convolution layers. For fixed filter sizes, this has the effect of increasing the spatial range of subsequent layers, while reducing the level of spatial resolution. As with most neural networks, the exact architecture of a CNN should depend on the problem at hand. This involves determination of: the mixture of convolutional and down-sampling layers (if any), filter-weight sizes, number of neurons, and interconnectivity between layers.

Figure 2 shows an example CNN with two hidden layers and multiple neurons in each layer, with down-sampling being implemented by the second layer.

Figure 3 displays the application of a simple CNN to a problem of detecting road markers from a camera image mounted above an intersection. It may be seen that the feature arrays in layer 1 capture simple features, while the feature arrays in the second hidden layer capture have more complex responses. The final output of the system does a good job of detecting road markers despite low degrees of freedom, a high degree of noise, and the presence of many distractors such as cars or road signs.

2 Convolutional Neural Networks

CNNs perform mappings between spatially / temporally distributed arrays in arbitrary dimensions. They appear to be suitable for application to time series, images, or video. CNNs are characterized by:

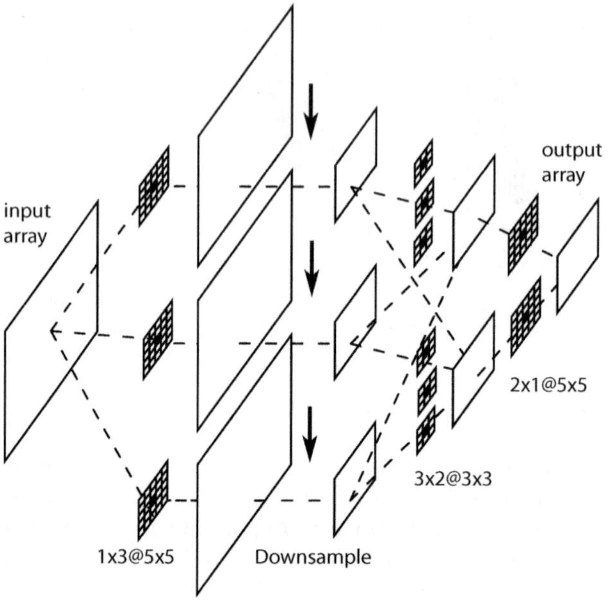

Fig. 2. Architecture of a fully interconnected CNN with multiple neurons in the convolution layer, and a factor of two translation in the second layer

- translation invariance (neural weights are fixed with respect to spatial translation)
- local connectivity (neural connections only exist between spatially local regions)
- an optional progressive decrease in spatial resolution (as the number of features is gradually increased).

These constraints make a CNN operate like a system of interconnected filters, and profitable comparisons may be made between other filtering systems, since the neural weights of a CNN operate like the taps of a system of finite impulse response (FIR) or wavelet filters. Thus a trained CNN may be thought of trainable filter system, custom made for a certain function mapping application. Finally, CNNs allow the processing of large spatially distributed arrays without a correspondingly large number of free parameters, increasing the chances of minima avoidance and generalization.

Initially, we will describe the case of one dimensional input and a single hidden layer. Extensions may then be made to multiple dimensions and multiple layers, with possible operation of down-sampling. We wish to obtain the formula for changing the weight and bias parameters of a CNN given a training set of input-output pairs $\xi^{\mu}_{k,r}, \zeta^{\mu}_{i,p}$. The indexes $\{i, j, k\}$ refer respectively to neuron arrays in the output, hidden, and input layers. In a CNN neurons are 'replicated' with respect to spatial translation, although they share the same weight and bias

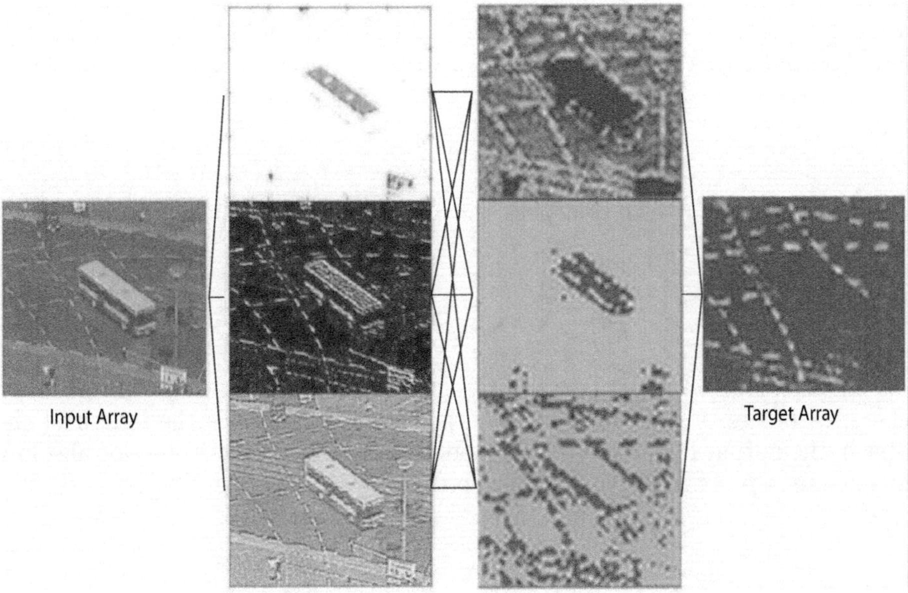

Fig. 3. Input, feature, and output arrays of a convolution network applied to detecting road markers

Table 1. Indexes and array terms, organized with respect to layer and data type

	output	hidden	input
array label	O	V	ξ
array index	i	j	k
spatial index	p	q	r
weight index	s	t	

vectors. Indices $\{p, q, r\}$ are to used as a spatial index for each layer. A property of CNNs is that translated neurons $h^{\mu}_{j,q}$ receive only *local* connections from the previous layer: a CNN is not a fully interconnected network. So, when calculating the net input to a particular translated neuron, it is convenient to index the spatially distributed weights separately, using indices s, t, u. The weights of a CNN are invariant to spatial translation, so it is natural to think of the set of weights $w^{t}_{j,k}, t = \{-T, .., 0, ..., T\}$ as a filter that connects input array $\xi^{\mu}_{k,r}$ to feature array $V^{\mu}_{j,q}$. $2T + 1$ is the size of the region surrounding each translation point for which network weights exist, ie. the filter size.

A summary of the indices used in the present paper is shown in table 1.

Given input pattern μ hidden unit j, q receives net input

$$h^{\mu}_{j,q} = \sum_{k} \sum_{t} w^{t}_{j,k} \xi^{\mu}_{k,q+t} + b_j \tag{1}$$

where the index to ξ_k^μ is clamped to spatially local positions, centered at translation q, by setting $r = q + t$. The term b_j refers to the usual constant bias. The neural output forms the hidden feature arrays, produced by the transfer function

$$V_{j,q}^\mu = g\left(h_{j,q}^\mu\right). \tag{2}$$

The neuron at translation p in the ith array in the output layer receives net input

$$h_{i,p}^\mu = \sum_j \sum_s w_{i,j}^s V_{j,p+s}^\mu + b_i \tag{3}$$

where, as before, $s = \{-S, .., 0, ..., S\}$, and $2S + 1$ describes the length of the filter in the output layer, and relative indexing has been substituted for absolute indexing; $q = p + s$. Final output of the network is

$$O_{i,p}^\mu = g\left(h_{i,p}^\mu\right) = g\left(\sum_j \sum_s w_{i,j}^s V_{j,p+s}^\mu + b_i\right). \tag{4}$$

Identical to the effect of applying two convolution filters sequentially, CNNs with hidden layers result in progressively larger portions of the input contributing to the function, $O_p = f(\xi_{p-S-T}, ..., \xi_p, ..., \xi_{p+S+T})$.

3 Delta Rule for CNNs

Although CNNs make use of the same weight update rule as normal neural networks, some care should be taken in implementation, particularly with differential indexing of spatial translation and feature maps.

As with normal ANNs, the cost function of a CNN is

$$E = \frac{1}{2} \sum_{\mu,i,p} \left[\zeta_{i,p}^\mu - O_{i,p}^\mu\right]^2. \tag{5}$$

We find the derivative of the error with respect to the sth weight of the filter connecting the jth feature array to the ith output array

$$\frac{\partial E}{\partial w_{i,j}^s} = -\sum_{\mu,p} \left[\zeta_{i,p}^\mu - g\left(h_{i,p}^\mu\right)\right] g'\left(h_{i,p}^\mu\right) V_{j,p+s}^\mu \tag{6}$$

which, used in combination with the familiar gradient descent weight update rule yields

$$\Delta w_{i,j}^s = \eta \sum_{\mu,p} \delta_{i,p}^\mu V_{j,p+s}^\mu \tag{7}$$

where

$$\delta^{\mu}_{i,p} = \left[\zeta^{\mu}_{i,p} - g\left(h^{\mu}_{i,p}\right)\right] g'\left(h^{\mu}_{i,p}\right). \tag{8}$$

In order to find the weight change of the input to hidden connections $w^{s}_{j,k}$ the delta rule is applied with a change of indices

$$\Delta w^{s}_{j,k} = \eta \sum_{\mu,q} \delta^{\mu}_{j,q} \xi^{\mu}_{k,q+t} \tag{9}$$

where

$$\delta^{\mu}_{j,q} = \sum_{s} g'\left(h^{\mu}_{j,q}\right) \sum_{i} \delta^{\mu}_{i,q-s} w^{s}_{i,j}. \tag{10}$$

Bias terms b_i and b_j are treated as normal weights with constant input, and may be likewise updated using (10) and (11).

4 Down-Sampling

Often when applying CNNs we wish to progressively reduce spatial resolution at each layer in the network. For example, a CNN may be used for classification where an image is mapped to a single classification output. Given fixed filter sizes, reducing spatial resolution has the effect of increasing the effective spatial range of subsequent filters. In a CNN with down-sampling in each layer, the outcome is a gradual increase in the number of features used to describe the data, combined with a gradual decrease in spatial resolution. Because the change in coordinate system is accomplished in a nonlinear, incremental, hierarchical manner, the transformation can be made insensitive to input translation, while incorporating information regarding the relative spatial location of features. This provides an interesting contrast to methods such as principle components analysis, which make the transition from normal coordinate space to feature space in a single linear transformation.

We can rewrite the previous formulas for calculating the output of the network, given that both layers incorporate spatial down-sampling. This has been previously accomplished using a separate 'averaging' layer with fixed neural weights [2]. However, it is described below by increasing the shift indexes by a factor of two, thus combining adaptive and down-sampling functions. Since the averaging layer in the method of Le Cun [2] may be specified by a 'double shift' layer with a filter size of 2, it may be shown that the present formalism is essentially similar, albeit more general, and allowing for adaption of previously fixed averaging weights.

$$h^{\mu}_{j,q} = \sum_{k} \sum_{t} w^{t}_{j,k} \xi^{\mu}_{k,2q+t} + b_j \tag{11}$$

$$h^{\mu}_{i,p} = \sum_{j} \sum_{s} w^{s}_{i,j} V^{\mu}_{j,2p+s} + b_i \tag{12}$$

The output of the network is then

$$O_{i,p}^{\mu} = g(\sum_j \sum_s w_{i,j}^s g(h_{j,2p+s}^{\mu}) + b_i) \tag{13}$$

$$= g(\sum_j \sum_s w_{i,j}^s g(\sum_k \sum_t w_{j,k}^t \xi_{k,4p+2s+t}^{\mu} + b_j) + b_i). \tag{14}$$

For a general CNN with N layers, being some combination of non-down-sampling and down-sampling layers, and filter sizes being given by F_n, the local region of input contributing to the output is given by the recursive formulas

$$R_{n+1} = R_n + F_n - 1 \text{ if non} - \text{down} - \text{sampling} \tag{15}$$

$$R_{n+1} = 2(R_n + F_n) - 3 \text{ if down} - \text{sampling} \tag{16}$$

given $R_1 = F_1$. Given fixed filter sizes F_n, it is clear that the input 'window' of CNN may grow rapidly as the number of down-sampling layers increase. Most wavelet transforms utilize a similar nested series of down-sampling operations, achieving significant computational savings. In fact, given a tree-like connectivity between feature arrays, the sub-sampled CNN may be profitably conceptualized as a wavelet transform with adaptable filters.

For down-sampling layers, the weight update rules are identical to [7] and [9], with the shift indices p and q increased by a multiple of two.

5 Method

CNNs are investigated in the current work for use as an image processing system on an autonomous mobile robot (see [14, 15] for details). The task of the system is autonomous detection and characterization of cracks and damage in sewer pipe walls. The robot scans the pipe wall using a monochrome CCD camera, which is digitally converted at a resolution of 320x240 pixels per frame. The task of the CNN is to perform filtering of the raw pixel data, identifying the spatial location of cracks, enabling subsequent characterization of the length, width, etc of the damage. Figure 4 displays a sample input frame, along with the ideal output of the CNN. Although the cracks are easily identifiable by eye, the image processing task is quite complex, as variability in lighting and orientation, width and type of crack, along with the presence of other crack-like structures (such as joins between pipe sections), combine to make a challenging computer vision task.

A representative data-set of 30 frames from an on-line recording session were manually classified for training and validation of the network. A training data set was generated using 15 of the images, sampling 737 pixel locations. Although all pixel locations had associated targets, not every pixel was used for training because of: a) computational expense, and b) the low proportion of 'crack' to 'clean' training samples tended to bias the network towards classifying all samples as 'clean'. Alternatively, it would be possible to use a biased learning procedure, where the error generated by the rarer class would be weighted in inverse proportion to the ratio of occurrence. A validation set of 676 pixel locations drawn from a separate set of 15 video frames was also generated.

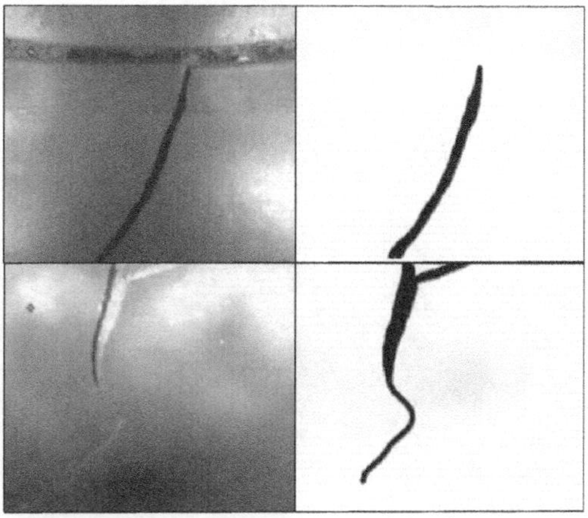

Fig. 4. Example input and target images for large cracks on a concrete pipe. Note the horizontal feature in the upper left image is a pipe joint, not a crack. Differentiating between pipes and joints, accounting for shadows and lighting effects are significant challenges for a detection / filtering system

The CNN architecture used involved a total of five layers: a single input and output map, and three hidden layers. The filter sizes used in all cases were 4x4, and the activation function used was a tan-sigmoid for the hidden layers and log-sigmoid for the output layer. The number of feature maps used in the three hidden layers was, from input to output, 4, 4, 4. The number of neural weights to be optimized was 680 while the input to the network was a square region with side lengths of 48 pixels, yielding a total of 2304 pixel inputs to the network. These figures are indicative of the degrees-of-freedom saving achieved by the weight sharing method. Training was conducted using standard weight updated rules, as described about, for 10,000 epochs, using gradient descent with an adaptive learning rate and momentum, requiring approximately three hours of compute time. The network was implemented in C++.

6 Results

All pixels in the validation set were correctly classified except one. Final classification on the training set was 98.1% and classification of the validation set was 95.1%. We note that some of the misclassified pixels were non-cracks, spatially adjacent to crack pixels. Thus, there is some inherent uncertainty in the classification task. Figure 5 displays three example validation frames, that were representative of the data set, including crack present, no crack and pipe joint, and crack and joint together. The network appears to have successfully ignored the presence of joints, and attenuated lighting effects while enhancing the cracks. In the context of the application to sewer pipe defect detection and character-

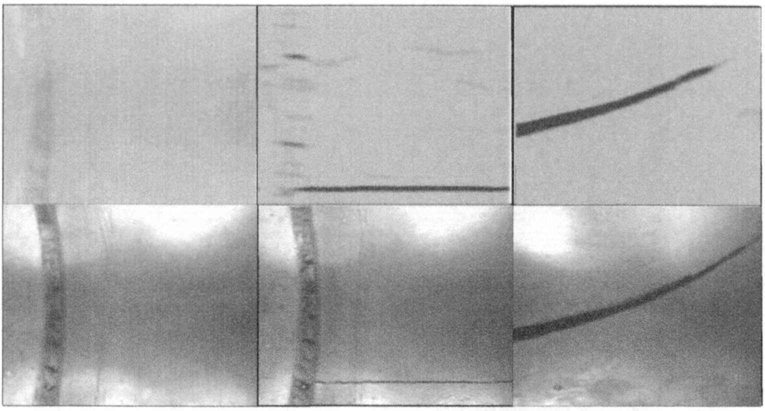

Fig. 5. Example input and CNN output frames. Note that because of the 48 pixel window required by the network, the output image represents a subregion of the input

ization, the output may be profitably used by a subsequent crack detection algorithm. However, the present system is in its early stages, and we believe that further refinement may provide better results.

7 Discussion

We have described and applied a general CNN architecture to a 'real-world' problem of some interest to the civil robotics community. CNNs may be expected to achieve significantly better results than standard feed-forward networks for many tasks because they impose appropriate constraints on the way the function mapping is learnt. The key characteristics of local connectivity, and translation invariant weight sharing, are appropriate when the input data is spatially or temporally distributed. In addition, implementing down-sampling allows the network to progressively trade-off resolution for a greater input range.

In testing the present CNN, some issues became apparent. Firstly, our next task is to implement training using Levenburg's method, which should improve the quality of results considerably. Training using bitmap type images results in an over abundance of training sample points. It appears that, in order to maximize the available computational resources, that not all pixel-centre points of the training set should be used. Rather, a representative sub-sample would be more appropriate. In the present application, this might mean over-sampling joints as opposed to flat-wall regions, or thin cracks as opposed to thick cracks. This may be done manually by hand-coding multiple targets, and sampling each one equally. Alternatively, some statistical criteria might be developed for selection of a representative data subset. Also, we have not yet done work on deciding the best CNN architecture for a given application - the present architecture was simply chosen as appearing reasonable for the current application. Determination of the architecture of the CNN, like many applications of neural networks, was somewhat arbitrary, based on user judgment rather than well defined rules.

As yet, the utility of CNNs does not appear to have been fully realized, and applications to a wide variety of data-types and function mapping problems (ie. physiological recordings, financial time-series analysis, automatic satellite image processing) remain to be explored. In particular, through the implementation of 3-D filters, CNNs may represent a computationally feasible method of adaptive video processing. Refinements in the CNN architecture remain to be explored. For example, sequential CNN layers comprising 1xN and Nx1 filters may be used to learn separable NxN filter functions. There are clear links between CNNs and finite impulse response filters, adaptive filters, and wavelet transforms, and theoretical work bridging these disciplines would be of significant interest. As a final point, the development of methods for adapting the CNN architecture via discriminant or entropy-based cost functions, similar to those developed for wavelet packet analysis, would be of significant interest.

References

[1] D. E. Rumelhart, G. E. Hinton, and R. J. Williams, "Learning internal representation by error propagation," in *Parallel Distributed Processing: Explorations in the Microstructure of Cognition*, D. E. Rumelhart and J. L. McClelland, Eds., vol. 1, pp. 318–362. Cambridge, MA: MIT Press, 1986. 641

[2] Y. B. Le Cun, J. S. Boser, D. Denker, R. E. Henderson, W. Howard, W. Hubbard, and L. D. Jackel, "Backpropagation applied to handwritten zip code recognition," *Neural Computation*, vol. 4, no. 1, pp. 541–551, 1988. 641, 642, 647

[3] K. J. Lang and G. E. Hinton, "Dimensionality reduction and prior knowledge in e-set recognition," in *Advances in Neural Information Processing Systems*, D. S. Touretzky, Ed., pp. 178–185. Morgan Kauffman, San Marteo, CA, 1990. 641

[4] K. Fukushima, S. Miyake, and T. Ito, "Neocognitron: a neural model for a mechanism of visual pattern recognition," *IEEE Transactions on Systems, Man, and Cybernetics*, vol. 13, pp. 826–834, 1983. 641

[5] Kunihiko Fukushima, "Neocognitron: A hierachical neural network capable of visual pattern recognition," *Neural Networks*, vol. 1, no. 2, pp. 119–130, 1988. 641

[6] Y. Le Cun and Y. Bengio, "Convolutional networks for images, speech, and time series," in *The Handbook of Brain Theory and Neural Networks*, M. A. Arbib, Ed., pp. 255–258. MIT Press, Cambridge, MA, 1995. 641

[7] Steve Lawrence, C. Lee Giles, A. C. Tsoi, and A. D. Back, "Face recognition: A convolutional neural network approach," *IEEE Transactions on Neural Networks*, vol. 8, no. 1, pp. 98–113, 1997. 642

[8] B. Fasel, "Robust face analysis using convolutional neural networks," in *Proceedings of the International Conference on Pattern Recognition (ICPR 2002), Quebec, Canada*, 2002. 642

[9] E. Sackinger, B. Boser, J. Bromley, and Y. LeCun, "Application of the anna neural network chip to high-speed character recognition," *IEEE Transactions on Neural Networks*, vol. 3, pp. 498–505, 1992. 642

[10] Y. Le Cun, "Generalization and network design strategies," Tech. Rep. CRG-TR-89-4, Department of Computer Science, University of Toronto, 1989. 642

[11] Y. Bengio, Y. Le Cun, and D. Henderson, "Globally trained handwritten word recognizer using spatial representation, convolutional neural networks, and Hidden Markov Models," in *Advances in Neural Information Processing Systems*, Jack D. Cowan, Gerald Tesauro, and Joshua Alspector, Eds. 1994, vol. 6, pp. 937–944, Morgan Kaufmann Publishers, Inc. 642

[12] Beat Fasel, "Facial expression analysis using shape and motion information extracted by convolutional neural networks," in *Proceedings of the International IEEE Workshop on Neural Networks for Signal Processing (NNSP 2002), Martigny, Switzerland*, 2002. 642

[13] F. Kirchner and J. Hertzberg, "A prototype study of an autonomous robot platform for sewerage system maintenance," *Autonomous Robots*, vol. 4, no. 4, pp. 319–331, 1997. 642

[14] M.Browne, M. Dorn, R. Ouellette, and S. Shiry, "Wavelet entropy-based feature extraction for crack detection in sewer pipes.," in *6th International Conference on Mechatronics Technology, Kitakyushu, Japan*, 2002. 648

[15] M. Browne, S. Shiry, M. Dorn, and R. Ouellette, "Visual feature extraction via pca-based parameterization of wavelet density functions," in *International Symposium on Robots and Automation, Toluca, Mexico*, 2002. 648

Towards Automated Creation of Image Interpretation Systems

Ilya Levner, Vadim Bulitko, Lihong Li, Greg Lee, and Russell Greiner

University of Alberta, Department of Computing Science
Edmonton, Alberta, T6G 2E8, CANADA
{ilya,bulitko,lihong,greglee,greiner}@cs.ualberta.ca

Abstract. Automated image interpretation is an important task in numerous applications ranging from security systems to natural resource inventorization based on remote-sensing. Recently, a second generation of adaptive machine-learned image interpretation systems have shown expert-level performance in several challenging domains. While demonstrating an unprecedented improvement over hand-engineered and first generation machine-learned systems in terms of cross-domain portability, design-cycle time, and robustness, such systems are still severely limited. This paper inspects the anatomy of the state-of-the-art Multi resolution Adaptive Object Recognition framework (MR ADORE) and presents extensions that aim at removing the last vestiges of human intervention still present in the original design of ADORE. More specifically, feature selection is still a task performed by human domain experts and represents a major stumbling block in the creation process of fully autonomous image interpretation systems. This paper focuses on minimizing such need for human engineering. After discussing experimental results, showing the performance of the framework extensions in the domain of forestry, the paper concludes by outlining autonomous feature extraction methods that may completely remove the need for human expertise in the feature selection process.

Keywords: Computer Vision, Machine learning, Reinforcement learning.

1 Introduction & Related Research

Image interpretation is an important and highly challenging problem with numerous practical applications. Unfortunately, hand engineering an image interpretation system requires a long and expensive design cycle as well as subject matter and computer vision expertise. Furthermore, hand-engineered systems are difficult to maintain, port to other domains, and tend to perform adequately only within a narrow range of operating conditions atypical of real world scenarios. In response to the aforementioned problems, various *automated* ways of constructing image interpretation systems have been explored in the last three decades [10].

T.D. Gedeon and L.C.C. Fung (Eds.): AI 2003, LNAI 2903, pp. 653–665, 2003.

Based on the notion of "goal-directed vision" [9], a promising approach for autonomous system creation lies with treating computer vision as a control problem over a space of image processing operators. Initial systems, such as the Schema System[9], had control policies consisting of ad-hoc, hand-engineered rules. While presenting a systemic way of designing image interpretation systems, the approach still required a large degree of human intervention. In the 1990's the second generation of control policy-based-image interpretation systems came into existence. More than a systematic design methodology, such systems used theoretically well-founded machine learning frameworks for automatic acquisition of control strategies over a space of image processing operators. The two well-known pioneering examples are a Bayes net system [15] and a Markov decision process (MDP) based system [8].

Our research efforts have focused on automating the latter system, called ADaptive Object REcognition system (ADORE), which learned dynamic image interpretation strategies for finding buildings in aerial images [8]. As with many vision systems, it identified objects (in this case buildings) in a multi-step process. Raw images were the initial input data, while image regions containing identified buildings constituted the final output data; in between the data could be represented as intensity images, probability images, edges, lines, or curves. ADORE modelled image interpretation as a Markov decision process, where the intermediate representations were continuous state spaces, and the vision procedures were actions. The goal was to learn a dynamic control policy that selects the next action (i.e., image processing operator) at each step so as to maximize the quality of the final image interpretation.

As a pioneering system, ADORE proved that a machine learned control policy was much more adaptive that its hand-engineered counterparts by outperforming any hand-crafted sequence of operators within its library. In addition, the system was effortlessly ported to recognize stationary (staplers, white-out, etc.) in office scenes and again was shown to outperform operator sequences designed by human domain experts [7]. However, the framework of ADORE was still limited in a number of ways and left several directions for future work and improvement. This paper discusses perhaps the biggest stumbling block to realization of autonomous image interpretation systems, namely, the need for hand-crafted features. The project that investigates approaches to fully autonomous object recognition systems is named MR ADORE for Multi-Resolution ADaptive Object REcognition [3].

The rest of the paper is organized as follows. First, we review the requirements and design of MR ADORE, in order to demonstrate the critical assumptions made and the resulting difficulties. We then present framework extensions that minimize the need for hand-crafted features and present experimental results of applying the upgraded system to the domain of forestry. The paper concludes with future research directions on how to completely replace domain experts with automated feature selection methods.

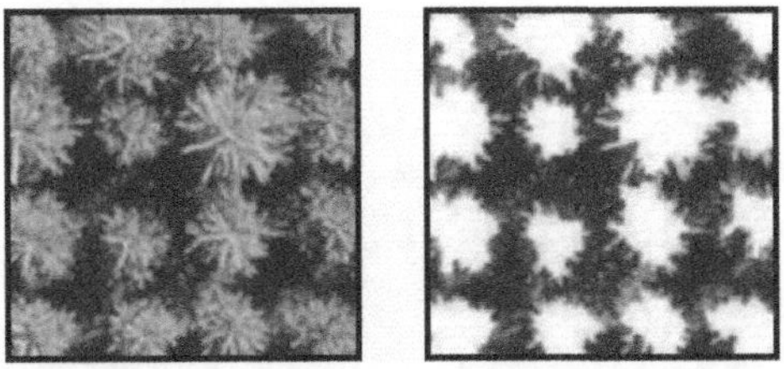

Fig. 1. Artificial tree plantations result in simple forest images. Shown on the left is an original photograph. The right image is the desired labeling provided by an expert as part of the training set

2 MR ADORE Design Objectives

MR ADORE was designed with the following objectives as its target: (i) rapid system development for a wide class of image interpretation domains; (ii) low demands on subject matter, computer vision, and AI expertise on the part of the developers; (iii) accelerated domain portability, system upgrades, and maintenance; (iv) adaptive image interpretation wherein the system adjusts its operation dynamically to a given image; and (v) user-controlled trade-offs between recognition accuracy and utilized resources (e.g., run-time).

These objectives favor the use of readily available off-the-shelf image processing operator libraries (IPLs). However, the domain independence of such libraries requires an intelligent policy to control the application of library operators. Operation of such control policy is a complex and adaptive process. It is *complex* in that there is rarely a one-step mapping from input images to final interpretation; instead, a series of operator applications are required to bridge the gap between raw pixels and semantic objects. Examples of the operators include region segmentation, texture filters, and the construction of 3D depth maps. Figure 2 presents a partial IPL operator dependency graph for the forestry domain. Image interpretation is an *adaptive* process in the sense that there is no fixed sequence of actions that will work well for most images. For instance, the steps required to locate and identify isolated trees are different from the steps required to find connected stands of trees. The success of adaptive image interpretation systems therefore depends on the solution to the control problem: for a given image, what sequence of operator applications will most effectively and reliably interpret the image?

3 MR ADORE Operation

MR ADORE starts with a Markov decision process (MDP) [16] as the basic mathematical model by casting the IPL operators as MDP **actions** and the results of their applications (i.e., data tokens) as MDP **states**. However, in the context of image interpretation, the formulation frequently leads to the following

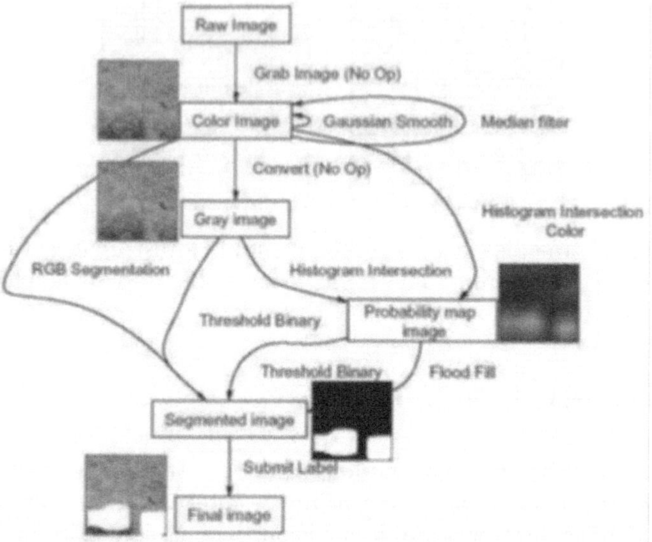

Fig. 2. Partial operator graph for the domain of forest image interpretation. The nodes and the corresponding example images depict that data processing layers, which in turn describe the *type* of MDP states present with MR ADORE. The edges represent vision routines, typically ported from the Intel OpenCV and IPL libraries, that transform one state to another (i.e., the MDP actions)

challenges absent from typical search settings and standard MDP formulations: (i) Standard machine learning algorithms cannot learn from raw pixel level data since the individual states are on the order of several mega-bytes each. Selecting optimal features as state descriptions for sequential decision-making is a known challenge in itself; (ii) The number of allowed starting states (i.e., the initial high-resolution images) alone is effectively unlimited for practical purposes. Additionally, certain intermediate states (e.g., probability maps) have a continuous nature; (iii) In addition to the relatively high branching factor possible, due to large image processing operator libraries, some of the more complex operators may require hours of computation time each; (iv) Unlike standard search, typically used to find the shortest sequence leading to a goal state, the goal of MR ADORE is to find/produce the best image interpretation. In that respect the system solves an optimization problem rather than one of heuristic search. Therefore, since goal states are not easily recognizable as the target image interpretation is usually not known *a priori*, standard search techniques (eg IDA* [14]) are inapplicable.

In response to these challenges MR ADORE employs the following off-line and on-line machine learning techniques. First, the domain expertise is encoded in the form of training data. Each training datum consists of 2 images, the input image, and its user-annotated counterpart allowing the output of the system to be compared to the desired image labeling (typically called ground-truth). Figure 1 demonstrates a training pair for the forestry image interpretation domain. Second, during the off-line stage the state space is explored via limited depth expansions of all training image pairs. Within a single expansion all sequences of IPL operators up to a certain user-controlled length are applied to a training image. Since training images are user-annotated with the desired output, terminal rewards can be computed based on the difference between the produced labeling and the desired labeling. System **rewards** are thus defined by creating a scoring metric that evaluates the quality of the final image interpretation with respect to the desired (used-provided) interpretation[1]. Then, dynamic programming methods [2] are used to compute the value function for the explored parts of the state space. We represent the value function as $Q : S \times A \rightarrow R$ where S is the set of states and A is the set of actions (operators). The true $Q(s,a)$ computes the maximum cumulative reward the policy can expect to collect by taking action a in state s and acting optimally thereafter. (See Figure 3 for an illustration of the process)

Features (f), used as **observations** by the on-line system component, represent relevant attributes extracted from the unmanageably large states (i.e., data tokens). Features make supervised machine learning methods practically feasible, which in-turn are needed to extrapolate the sampled Q-values (computed by dynamic programming on the explored fraction of the state space) onto the entire space (see Figure 4).

Finally, when presented with a novel input image, MR ADORE exploits the machine-learned heuristic value function $Q(f(s),a)$ over the abstracted state space, f(S), in order to intelligently select operators from the IPL. The process terminates when the policy executes the action `Submit(`$\langle$*labeling*$\rangle$`)`, which becomes the final output of the system. (see Figure 5).

3.1 Learning Control Policies

The purpose of the off-line learning phase within MR ADORE is to automatically construct the on-line control policy. The policies studied to date can be conceptually represented via the following unifying framework.

First, data tokens (i.e., states) can be split into disjoint "processing levels". The initial input image is a data token of the top (first) data level and the final submitted image interpretation is a data token from the bottom (last) data level. Figure 2 illustrates six processing levels as follows: *raw image, color image, gray image, probability map image, segmented image,* and *final image.*

[1] For all experiments presented, the intersection over union scoring metric, $\frac{A \cap B}{A \cup B}$ is used. Typically used in vision research, this pixel-based scoring metric computes the overlap between the set of hypothesis pixels produced by the system A and the set of pixels within the ground-truth image B. If set A and B are identical then their intersection is equal to their union and the score/reward is 1. As the two sets become more and more disjoint the reward decreases, indicating that the produced hypothesis corresponds poorly to the ground-truth.

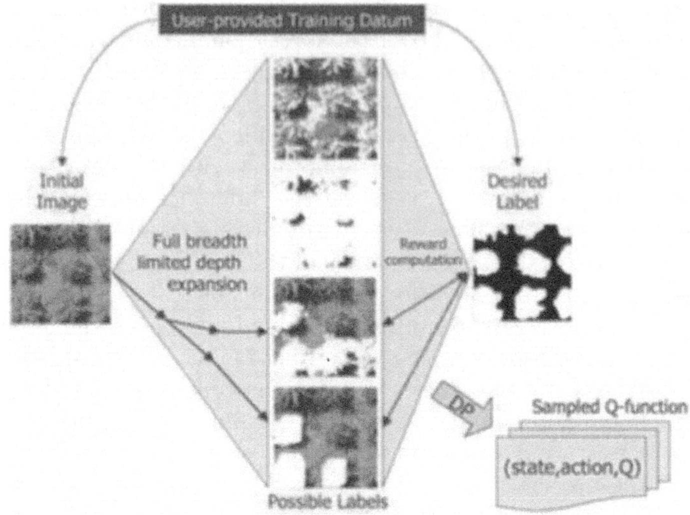

Fig. 3. Off-line operation

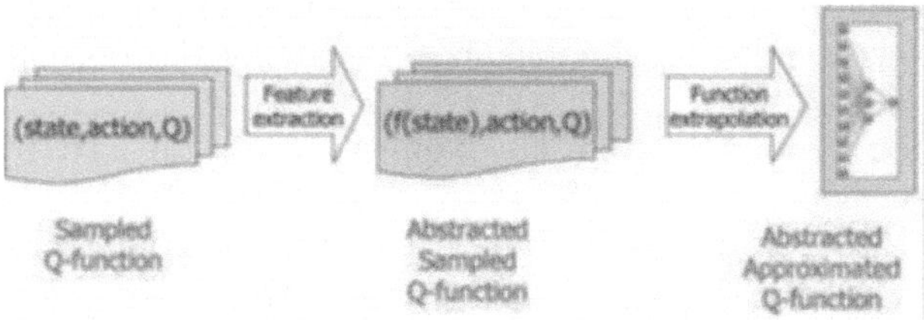

Fig. 4. Feature Extraction and Q-function approximation

Consequently, the set S of data tokens can be partitioned into subsets $S_1, \ldots, S_n$ where S_i contains data tokens belonging to processing level i. Likewise, the set A of operators can be represented as a union of (disjoint) operator subsets A_i, where action $a_i \in A_i$ is applicable at level i. Therefore, any control policy $\pi : S \rightarrow A$ can be conceptually divided into n control policies: $\pi_1, \ldots, \pi_n$ where π_i operates over S_i only, that is $\pi_i : S_i \rightarrow 2^{A_i}$. There are several different classes of π_i policies possible, including:

fixed: $\pi_i(s) = \{a_i\}$ for all states s. Such policies blindly apply the same action regardless of the data token s at hand.

full expansion: $\pi_i(s) = A_i$: all applicable operators within the set A_i are executed.

value-based: $\pi_i(s) = \{\arg\max_{a \in A_i} Q_i(s,a)\}$ where a machine-learned approximator is used for the Q_i-function. Such policies greedily select actions to be executed with the highest Q-values. This policy is specified by the machine-learned approximator and the feature selection function.

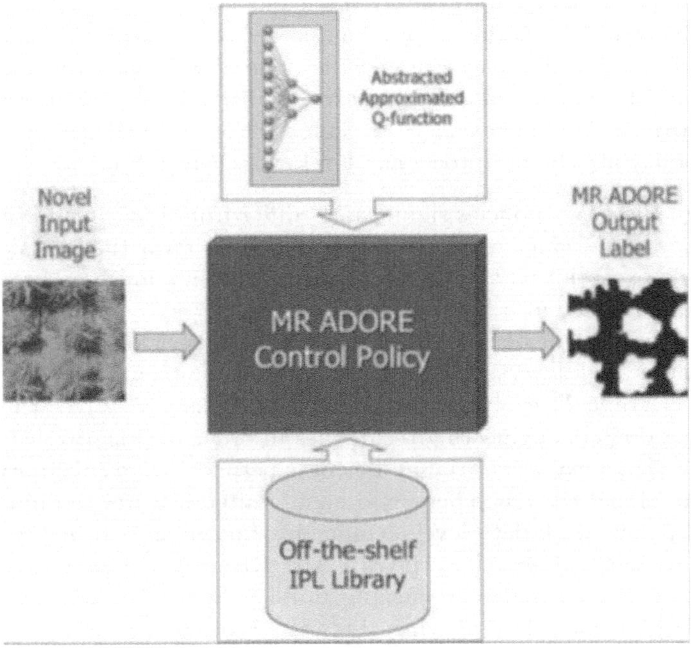

Fig. 5. On-line operation

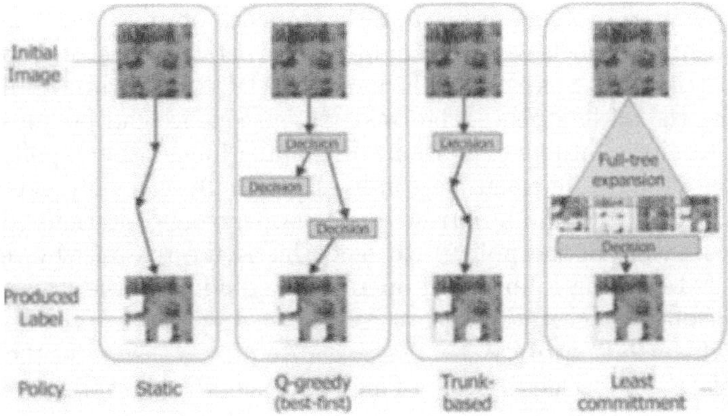

Fig. 6. Four types of on-line polices

In addition to the three aforementioned types of processing level policies, two special policy types are possible within MR ADORE that do not conform to the standard policy definition presented above.

path-selector: $\pi_i(s) = a^i \in A_i, a^{i+1} \in A_{i+1}, \dots, a^j \in A_j$: This policy selects a (sub)sequence of actions, the first of which is executed and the rest are passed on to the next processing level(s). This policy is really a derivative of the value-based policy in the sense that it too is a Q-greedy policy defined by a machine-learned function approximator and the feature selection function. The most notable difference is unlike the value-based policy, the path-selector selects a (sub)sequence of actions to execute rather than a (sub)set of actions.

follow-through: This policy $\pi_j(s)$ executes action a_j specified by a path-selector policy at a higher processing level, π_i, where $1 \leq i < j$.

Although these two policies significantly differ from the three initial polices, all five policy types can be mixed and matched to form the global policy π. (One obvious exception is that a follow-through policy must be proceeded by a path-selection policy at a higher level.)
Thus, any policy π can be described as an n-ary vector of π_i, where n is the number of processing layers. Selecting the optimal combination of π_i's type and parameters is a difficult problem. Pre-ADORE systems have solved this problem by employing domain engineering techniques in order to design π, which virtually always calls for extensive trial-and-error experiments and substantial human expertise. In [8] and [7] researchers used hand-crafted features to implement the value-based policy at all data levels (commonly known as best-first/greedy policy) within ADORE. Research in the MR ADORE project casts policy selection as a challenging open machine learning problem with the elements of meta-learning. To that end, the following three types of control policies have been explored to date.

A **static** policy, π_{static}, uses a single sequence of operators and therefore consists of only fixed π_i's. Research within the field of computer vision has produced numerous systems using a static sequence of operators (e.g. [6]). As previously mentioned such systems require extensive trial-and-error experiments, domain expertise, etc. Since ADORE and MR ADORE perform a full-breadth search during the off-line phase, the **best** static sequence can be automatically determined from the off-line expansions of training images. Such policies, while being computationally inexpensive (on-line), generally can only achieve near-optimal performance within a narrow range of operating conditions (due to the non-adaptive nature of the policy). In fact, this is the reason why researchers initially turned to using a library of operators and adopted the "goal-directed" paradigm for object recognition.

A **"trunk-based" policy** π_{trunk} makes its only decision at the top processing level (S_1) by selecting an n-long sequence of operators based on the input image only. The policy maintains a library of *optimal*[2] sequences (called "trunks") collected during the off-line stage. Therefore, π_1 is a path-selector over the space of trunks while $\pi_2, \ldots, \pi_n$ are all of the follow-through type.

While best-first policies are theoretically capable of much more flexibility than static or (semi-static) trunk-based policies, they depend crucially on (i) data token features for *all* levels and (ii) adequate amounts of training data to train the Q-functions for *all* levels. Experience has shown that it is substantially harder to select features at the earlier levels where the data tokens exhibit less

[2] An optimal sequence is the one yielding the greatest reward over all other sequences with respect to a given input image.

structure [10]. To make the problem worse, a single user-labeled training image delivers exponentially larger numbers of ⟨ state, action, reward ⟩ tuples at later processing levels. Unfortunately, the first processing level gets the mere $|A_1|$ tuples per training image since there is only one data token (the input image itself). As a net result, best-first control policies have been known to backtrack frequently [8] as well as produce highly suboptimal interpretations [5], due to poor decision making at the top processing layers. In fact the trunk-based policy suffers from the same phenomenon, a lack of high-quality features and sparseness of training examples.

Rather than making control decisions at every level based on the frequently incomplete information provided by imperfect features, the **least-commitment policies** postpone their decisions until more structured and refined data tokens are derived. That is: $\pi_j = A_j$ for $1 \leq j \leq n-1$ and π_n is value-based. In other words, we apply all operator sequences up to a predefined depth and only then engage the machine-learned control policy to select the appropriate action. Doing so allows the control system to make decisions based on high-quality informative features, resulting in an overall increase in the quality increases. As a side benefit, the machine learning process is greatly simplified since feature selection and value function approximation are performed for considerably fewer processing levels while benefiting from the largest amount of training data. (Figure 6 illustrates the outlined policies.) In order to determine the merit of each policy we tested MR ADORE with static, trunk-based and least-commitment policies along with two base-line policies: random and random trunking. A random policy is simply a least-commitment policy that randomly assigns rewards to data tokens rather than using a machine-learned function approximator. Similarly, a random trunk-based policy is a trunk-based policy that selects an operator sequence (i.e., a trunk) at random rather than using a machine-learned function approximator to assign Q-values to each of the possible sequences. As previously mentioned, trunk-based policies suffer from lack of informative features and training examples. The performance evaluation was conducted on 35 images of forest plantations (cf. Figure1). Using a leave-one-out cross-validation strategy (i.e, 34 for training and 1 for testing), each of the five control policies was ran on the set of forestry images. Table 1 shows the experimental results. As expected, since the least-commitment policy uses a function approximator trained on data tokens at bottom processing level that contains an exponentially many more examples than preceding levels, this policy outperforms all other policies[3] The trunk-based policy follows next[4]. As mentioned previously, the lack of information at the top processing layer clearly makes the trunk-based policy

[3] The results presented here used an artificial neural network [12] as the function approximator and an HSV histogram for input features. It should be noted that other function approximation methods were used including k-nearest neighbors[11] and sparse networks of Winnows (SNoW) [1]. In addition a number of feature sets were tried as well.

[4] Here we used KNN with a number of features and distance metrics. The best results show in the table were obtained by using texture/shape features in the form of local binary patterns (lbp) together with a Euclidean distance measure.

Table 1. Performance of five different policies on the forestry data. 35 images of forest plantation scenes were used in the experiment. Results for each policy were averaged over the 35 leave-one-out cross-validation runs. **% optimal** refers to the fact that each policy is evaluated relative to the best off-line interpretation achievable given a particular IPL

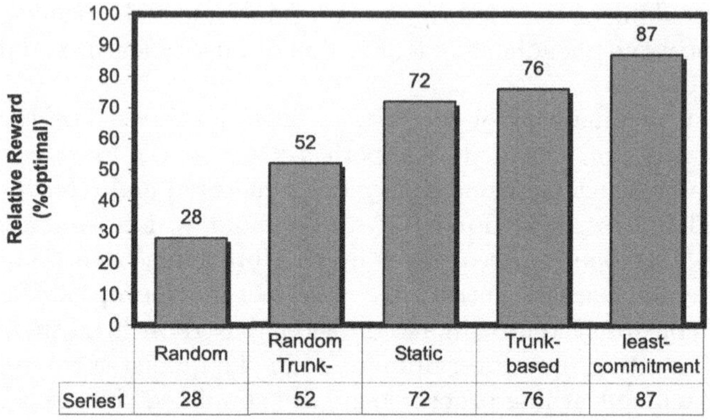

perform much worse that the least-commitment policy. The results are consistent with previous work. In [10] the researchers (informally) report that as less and less informative features are used to train the best-first policy, it converges to a static policy (i.e., the best-first policy applies the same sequence of actions regardless of input). Thus the performance of the static policy is almost as good as the best trunk-based policy. Not surprisingly the random policies perform the worst. It should be noted that the probability of performing optimally for this experiment (ie randomly choosing an optimal sequence for each image) can be easily calculated. There are 118 possible sequences for each image and 35 images, but only one optimal sequence per image. Thus $P(\pi^{optimal}) = (\frac{1}{118})^{35} \approx 3 * 10^{-72}$.

4 Future Research

One of the problems with both the trunk-based and least-commitment policies is the need for high-quality features. While both policies (trunk-based and least-commitment) reduce the need for hand-crafted features by requiring only a single set (respectively at either the top or bottom processing level), the need for a domain expert is only reduced and **not** eliminated. Principal component analysis (PCA) can be used to reduce the vast dimensionality of the raw data tokens by projecting them into an eigen space [13]. This results in an automatically generated set of features (typically one set per processing level). Lazy instance-based learning (k-nearest-neighbors) [11] can then be used to approximate the Q_i function at level i. Additionally, the features can take the state history into account insomuch as Q_i is computed on abstracted previous states:

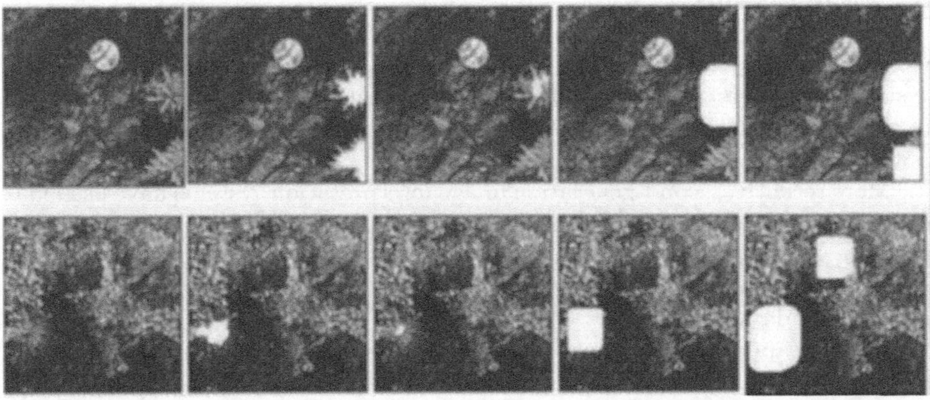

Fig. 7. Each row from left to right: the original image, desired user-provided labeling, optimal off-line labeling, best static policy of length 4 labeling, best-first policy labeling (using KNN and automatically extracted features via PCA). The adaptive policy has been observed to outperform best static policy (top row) and sometimes the human experts as well (bottom row)

$[f_i(s_i), f_{i-1}(s_{i-1}), \ldots, f_1(s_1)]$ where $s_1, \ldots, s_i$ are the states on the path from the input image s_1 to the current image at hand, s_i. Figure 7 demonstrates the potential of using KNN with PCA-based features. An alternative to using the PCA procedure to compress the images is to simply down-sample the images to a size that can be managed by the modern machine learning methods. Unfortunately, both of these methods (PCA and down-sampling) assume that the intermediate data levels are indeed representable as images. If an intermediate processing level consists, for example, of contour points, both PCA and down-sampling cannot be readily applied. Thus the most probable solution to the feature extraction problem will perhaps come in the form of feature extraction libraries (similar to the operator libraries). Just as MR ADORE learned efficient operator selection, next generation object recognition systems will need to *select* which features are relevant [4] for a given processing layer through off-line trail-and-error processes based, perhaps, on the very same MDP framework used today to efficiently select operator sequences.

Automated selection of optimal π_is for all processing levels is an open-ended problem. The particular difficulty lies with the tightly coupled nature of individual policies and the exponentially large space of all combinations of π_is. An initial investigation into bottom-up construction of π via filling the later levels with value-based π_is for $i = n, n-1, \ldots$ one-by-one, while leaving the earlier processing levels populated with the full expansion $\pi_j = A_j$ policies, appears promising.

5 Conclusions

Conventional ways of developing image interpretation systems often require that system developers posses significant subject matter and computer vision expertise. The resulting knowledge-engineered systems are expensive to upgrade, maintain, and port to other domains.

More recently, second-generation image interpretation systems have used machine learning methods in order to (i) reduce the need for human input in developing an image interpretation system or port it to a novel domain and (ii) increase the robustness of the resulting system with respect to noise and variations in the data.

In this paper we presented a state-of-the-art adaptive image interpretation system called MR ADORE and demonstrated several exciting machine learning and decision making problems that need to be addressed. We then reported on the progress achieved on reducing the need for feature selection and went on to propose ideas on how to completely eliminate the need for hand-crafted features.

Acknowledgements

Bruce Draper participated in the initial MR ADORE design stage. Lisheng Sun, Yang Wang, Omid Madani, Guanwen Zhang, Dorothy Lau, Li Cheng, Joan Fang, Terry Caelli, David H. McNabb, Rongzhou Man, and Ken Greenway have contributed in various ways. We are grateful for the funding from the University of Alberta, NSERC, and the Alberta Ingenuity Center for Machine Learning.

References

[1] S. Agarwal and D. Roth. Learning a sparse representation for object detection. In *7th European Conference on Computer Vision*, volume 4, pages 113–130, Copenhagen, Denmark, 2002. 661

[2] A. G. Barto, S. J. Bradtke, and S. P. Singh. Learning to act using real-time dynamic programming. *Artificial Intelligence*, 72(1):81–138, 1995. 657

[3] V. Bulitko, B. Draper, D. Lau, I. Levner, and G. Zhang. MR ADORE : Multiresolution adaptive object recognition (design document). Technical report, University of Alberta, 2002. 654

[4] Vadim Bulitko, Greg Lee, and Ilya Levner. Evolutionary algorithms for operator selection in vision. In *Proceedings of the Fifth International Workshop on Frontiers in Evolutionary Algorithms*, 2003. 663

[5] Vadim Bulitko and Ilya Levner. Improving learnability of adaptive image interpretation systems. Technical report, University of Alberta, 2003. 661

[6] Michael C. Burl, Lars Asker, Padhraic Smyth, Usama M. Fayyad, Pietro Perona, Larry Crumpler, and Jayne Aubele. Learning to recognize volcanoes on venus. *Machine Learning*, 30(2-3):165–194, 1998. 660

[7] B. Draper, U. Ahlrichs, and D. Paulus. Adapting object recognition across domains: A demonstration. In *Proceedings of International Conference on Vision Systems*, pages 256–267, Vancouver, B. C., 2001. 654, 660

[8] B. Draper, J. Bins, and K. Baek. ADORE: adaptive object recognition. *Videre*, 1(4):86–99, 2000. 654, 660, 661

[9] B. Draper, A. Hanson, and E. Riseman. Knowledge-directed vision: Control, learning and integration. *Proceedings of the IEEE*, 84(11):1625–1637, 1996. 654

[10] Bruce A. Draper. From knowledge bases to Markov models to PCA. In *Proceedings of Workshop on Computer Vision System Control Architectures*, Graz, Austria, 2003. 653, 661, 662

[11] Richard O. Duda and Peter E. Hart. *Pattern Classification and Scene Analysis.* John Wiley & Sons, New York, 1973. 661, 662

[12] S. Haykin. *Neural Networks: A Comprehensive Foundation.* Macmillian College Pub. Co., 1994. 661

[13] Michael Kirby. *Geometric Data Analysis: An Emprical Approach to Dimensionality Reduction and the Study of Patterns.* John Wiley & Sons, New York, 2001. 662

[14] Richard E. Korf. Real-time heuristic search. *Artificial Intelligence*, 42(2-3):189–211, 1990. 656

[15] R. Rimey and C. Brown. Control of selective perception using bayes nets and decision theory. *International Journal of Computer Vision*, 12:173–207, 1994. 654

[16] R.S. Sutton and A.G. Barto. *Reinforcement Learning: An Introduction.* MIT Press, 2000. 655

Dealing with Decision Costs in CBR in Medical Applications

Monica H. Ou, Geoff A.W. West, and Mihai Lazarescu

Department of Computing, Curtin University of Technology
GPO Box U1987, Perth 6845, Western Australia, Australia
{ou,geoff,lazaresc}@cs.curtin.edu.au

Abstract. This paper presents an approach for developing a Web-based decision support system that aids general practitioners in diagnosing dermatological problems using case-based reasoning. Our research aims to address the lack of decision support in current store-and-forward telemedicine systems and the need to deal with different costs of decisions. We propose a cost calculation method for minimizing the total misclassification cost. Two different techniques of measuring cost are evaluated. A series of tests is carried out on different sets of cost matrices for comparing the two methods. Our results show taking max values gives a lower misclassification cost and still manages to achieve a slightly higher accuracy compared to the summing method of instance-weighting.

1 Introduction

A patient with a dermatological problem usually sees a general practitioner (GP) first. If the GP is unsure, the patient is referred to a consultant dermatologist. This is expensive and inconvenient for patients from rural areas because they must travel large distances to a main city. These problems can be avoided by the development of teledermatology systems. Teledermatology is defined as the practice of dermatological services at a distance [15]. There are two main methods available for teledermatology namely live video conferencing and store-and-forward [9].

Although the development of teledermatology has increased the provision of medical services to rural communities and the efficiency of medical practices, current systems are lacking in decision support abilities. In medical applications, case-based reasoning (CBR) and machine learning have been used for decision support with considerable success for tasks such as patient diagnosis. Case-based reasoning is an artificial intelligence technique that attempts to solve new problems by using the solutions applied to similar cases in the past. This paper focuses on the retrieval stage of the CBR cycle, which mainly deals with the retrieval of similar cases and reusing the existing knowledge in that case to solve new problems.

Our aim is to develop a Web-based CBR system that can be used to provide decision support to the GPs for diagnosing patients with dermatological problems. In this paper, we address the issue of specifying costs of misclassification

T.D. Gedeon and L.C.C. Fung (Eds.): AI 2003, LNAI 2903, pp. 666–676, 2003.

to improve the performance of CBR in terms of benefit and risk of death due to misdiagnosis. This is important in minimizing the high cost of misdiagnosis. We also explore CBR and supervised machine learning techniques to enable learning and the classification of new cases.

1.1 System Design and Functionalities

The CBR system involves asking questions for appropriate symptoms, and returning the most likely diagnosis for that particular case. The CBR system is implemented so that the processing of each stage is visible to the users. This has the advantage of letting the user know the effect/outcome of each decision they make. In developing the system, we require a convenient way of storing all the past cases (i.e. symptoms and diagnosis) and allow the new cases to be added easily to the case-base. We selected a modern relational database to store our data instead of a flat or customized file structure, since a relational database allows cases to be added, updated, and deleted independently.

1.2 Case Retrieval and Misclassification Costs

We use the C4.5 decision tree as the main classification tool. The classifier is built on past cases which are stored in the case-base. C4.5 is chosen because it can generate questions on the fly, and hence dynamically change the order of the questions presented to the user when the tree is updated. Another major advantage of using decision trees is that it makes retrieval more efficient in comparison to having cases organized as a list. Rather than having to attempt matches to all cases in the case-base, it considers only a subset of cases.

The generated questions are represented by the tree nodes. To retrieve a case from a tree, a breadth-first search is used. The input from the GP (i.e. symptoms) is matched against the contents of each node from the root of the tree. The best-matching node is chosen. If it is a leaf node, the diagnosis is returned. Otherwise, if it is an internal node, the process is repeated among its descendants. This continues until a diagnosis is returned.

Techniques using inductive decision trees for case retrieval have been widely applied in CBR systems due to the simplicity of the algorithms and good classification performance. However, such classifiers have not take into consideration the associated risks/costs involve for each decision when performing classification.

Therefore, the important aspect of this paper is the need to fully consider the costs of misdiagnosis. Different decisions have different outcomes and hence costs will effect how GPs and consultants use the system. Importantly for CBR, costs have to be integrated into such methods as decision tree classifiers. It enables the classifier to avoid predicting high cost errors by moving decision boundaries. This is very important in medical diagnosis because of the consequences of making a wrong decision. The cost of misclassification involves: performing unnecessary diagnosis/surgery, delaying a diagnosis, a mixture of social and economic factors, and death.

This paper is organized as follows. In section 2, a brief review of the related literature is given. Section 3 discusses the cost-sensitive learning techniques and the data used. An analysis of the results can be found in Section 4. Section 5 presents the conclusions.

2 Related Work

CBR has been successfully applied to medical diagnosis. There are some well-known medical CBR systems such as PROTOS [7] and CASEY. PROTOS was developed in the domain of clinical audiology. It learns to classify hearing disorders by using the patients' symptoms, histories, and test results. CASEY is a system that diagnoses heart failure. It combines case-based reasoning and causal models [13]. Additionally, there are large European CBR projects on induction and reasoning from cases (e.g. INRECA and INCRECA-II) [3].

There are a number of learning algorithms in the literature that are used for case retrieval: Nearest Neighbor, Decision Trees and Bayesian networks. Most commercial CBR tools support case retrieval using Nearest Neighbor or Inductive Decision Trees due to the simplicity of the algorithms and good classification performance [13]. Other methods of retrieval used in the CBR system are syntactic and semantic similarities [1].

Most research into machine learning has concentrated on error classification rather than cost-sensitive classification, that is assume all misclassifications have equivalent cost. A better performance might be obtained by taking cost information into account during learning. The cost-sensitive learning method involves using a cost matrix C. Let i and j be the class labels of the instances. The contents of $C(i,j)$ specify the cost incurred when a case is classified to be in class i when in fact it belongs to class j. It deals with the situations in which different types of incorrect prediction have different costs. For example, in medical diagnosis, the cost of diagnosing someone as healthy when one has a life threatening disease is usually considered to be much higher than a false alarm (diagnosing someone as having a disease when one is in fact healthy). Other well known examples include credit risk evaluation [10], fraud detection [2], and marketing analysis [4]. Although the cost of misclassification errors has been recognized as being an issue for the last four decades, it has only been recently investigated [14, 5]. Of relevance to our work is the recent research of incorporating the cost of misclassification into decision trees [11, 12].

Ting [11] has introduced an instance-weighting method to induce cost-sensitive trees directly from training data. Let N be the total number of instances, N_j be the number of class j instances, $N(t)$ and $N_j(t)$ be the number of instances and class j instances respectively in node t of a decision tree. The standard divide-and-conquer procedure is used when node t contains instances that belong to a mixture of classes. The commonly used criterion is entropy — a threshold is set in a feature such that the samples at the node are derived into the classes that have a better classification (refer to Quinlan [8]).

Let $C(j)$ be the cost of misclassifying a class j instance, then the weight of a class j instance can be computed as

$$w(j) = C(j)\frac{N_j(t)}{\sum_i N_i(t)}, \text{ where} \tag{1}$$

$$C(j) = \sum_i^I cost(i,j) \tag{2}$$

Finally, the expected misclassification cost for predicting class i with respect to the example x is given by:

$$EC_i(x) \propto \sum_j W_j(t(x))cost(i,j) \tag{3}$$

where $t(x)$ is the leaf of the tree in which instance x falls into, and $W_j(t)$ is the total weight of class j training instances in node t. $EC^i(x)$ is computed for every class when attempting to classify a new example x, then x is assigned to class i with the smallest value for $EC^i(x)$. This method works in the classic two-class case, but it creates problems for multiple-classes because it transforms the cost matrix of size $J \times J$ to J (i.e. classifies an example of class j, irrespective of class predicted). For the two-class case, the Equation 2 summation only occurs over the one misclassified entry (i,j). For more than two classes it is the sum over all the misclassified states. The issue is how to choose the right misclassification state when the misclassification entries contain one high and one low cost. It is obviously reasonable if the misclassification costs are similar.

3 Cost-Sensitive Learning Techniques

The experiment involves using the instance-weighting method to induce cost-sensitive trees proposed by Ting [11], which minimize the number of high cost and total misclassification cost errors. The instance-weighting method changes the class distribution, and will produce a decision structure that is skewed toward the avoidance of high cost errors. Suppose in a two-class situation, the cost of classifying "no" when the true class is "yes" is penalized ten times more heavily than predicting "yes". In this case, the classifier that uses instance-weighting will often predict "yes", given the heavy penalty for erroneously predicting "no". This mimics the situation often found in medical diagnosis (see example mentioned in Section 2). Our research aims to improve the performance of the instance-weighting method by reducing the total misclassification cost errors.

3.1 Approaches to Risk Calculations

Our approach to risk calculations employs the cost matrix conversion of the form *cost(i,j)* to cost vector *cost(j)* in order to use Equation 1, where *C(j)* is the cost of misclassifying an example of class *j*, irrespective of the class predicted. In

Table 1. An example of confusion matrix entries

Predicted class			
Melanoma	Mole		
a	b	Melanoma	Actual class
c	d	Mole	

instance-weighting, the total costs of misclassification is calculated by summing each row — called the *Summing Method.*

We propose an alternative method for minimizing the total misclassification cost. The method involves taking the *maximum* value of the misclassification costs across each row — called the *Max Method.* This enables the classifier to be biased towards predicting high cost errors by moving decision boundaries. Since some diseases cost more than others if a misdiagnosis occurs, this method is good at calculating the number of high cost misclassification errors among different classes. To determine which method gives the minimum total cost, we set up a series of tests on both methods.

The average misclassification cost is calculated by taking the sum of the incorrectly classified examples and divided it by the total number of examples. Additionally, we define the definition and calculation of accuracy, precision and recall. For instance, lets start with a simple illustration using a confusion matrix for two classes which have the entries:

- a is the number of *correctly* classified examples as Melanoma.
- b is the number of *incorrectly* classified examples as Mole.
- c is the number of *incorrectly* classified examples as Melanoma.
- d is the number of *correctly* classified examples as Mole.

Accuracy: The accuracy is the percentage of the total number of predictions that were correct.

$$AC = \frac{a+d}{a+b+c+d}100\% \tag{4}$$

Precision: The percentage of the predicted mole or melanoma examples that were correct, as calculated using the equations:

$$P = \frac{d}{b+d}100\%, or P = \frac{a}{a+c}100\% \tag{5}$$

Recall: The percentage of the mole or melanoma examples that were correctly classified, as calculated using the equations:

$$R = \frac{d}{c+d}100\%, or R = \frac{a}{a+b}100\% \tag{6}$$

3.2 The Data

The data we used in our experiments consisted of patient records for the diagnosis of erythemato-squamous diseases. The data set "Dermatology database" is made available on the Web [1] by Prof. H. Altay Guvenir of Bilkent University. The data set contains 34 attributes and 366 instances. The data set was collected by clinical evaluation of 12 features of the patients. Skin samples were taken for the evaluation of 22 histopathological features, in which the values are determined by an analysis of the samples under a microscope. Every feature (clinical and histopathological) was given a degree in the range of 0 to 3. The value of 0 indicates that the feature was not present, 1 and 2 indicate the relative intermediate values, and 3 indicates the largest amount possible.

3.3 Cost Matrices

The sets of cost matrices were generated by a consultant dermatologist heuristically assigning costs based on his perception of the total relative costs of misdiagnosis. Table 2 shows the relative misclassification costs for a patient to obtain diagnosis in descending order from the most expensive to the least expensive. The classes are assigned different cost values to reflect the seriousness of misclassifying the class. For example, it costs a lot more to misclassify CronicDermatitis than PityriasisRosea. CronicDermatitis is a skin disease with inflammation in the skin, whereas PityriasisRosea is a common skin disorder which is usually mild and will go away in several months without treatment [6].

The cost matrix shown in Table 3 is derived from the relative cost indicated in Table 2. In our experiments, cost matrix rows correspond to actual classes, while columns correspond to predicted classes. The off-diagonal entries contain the costs of misclassifications. The normal convention is to assign zero for the cost of correct classifications because the right decision has been made. Sometimes, these entries can be viewed as benefits instead of costs with the values of non-zero. However, costs/benefits of correct classifications is outside the scope of this paper.

4 Results and Analysis

Different sets of cost matrices are used to illustrate the effect the cost entries have on the classification decision and total misclassification cost. The total cost of misclassification is defined as the sum of the off-diagonal elements in the confusion matrix . A series of tests is carried out for comparing the two cost calculation methods described in Section 3.1. At each stage of the test, the confusion matrix, the performance measure (i.e. accuracy), the total misclassification cost and error are examined. Total misclassification error is the sum of the off-diagonal entries.

[1] www.cormactech.com/neunet/download.html

Table 2. Relative ranking of misclassification costs

Rank	Costs (thousands)	Class
1	100	CronicDermatitis
2	50	PityriasisRubra
3	20	LichenPlanus
4	10	Psoriasis
5	5	Seboretic
6	2	PityriasisRosea

Table 3. Cost Matrix 1

classified as →	a	b	c	d	e	f
Psoriasis = a	0	5	20	50	100	2
Seboretic = b	10	0	20	50	100	2
LichenPlanus = c	10	5	0	50	100	2
PityriasisRubra = d	10	5	20	0	100	2
CronicDermatitis = e	10	5	20	50	0	2
PityriasisRosea = f	10	5	20	50	100	0

Table 4. Confusion matrix using a default cost matrix and Summing Method

classified as →	a	b	c	d	e	f	Recall (%)
Psoriasis = a	108	2	2	0	0	0	96.4
Seboretic = b	2	56	0	1	0	2	91.8
LichenPlanus = c	1	1	69	0	1	0	95.8
PityriasisRubra = d	2	1	0	17	0	0	85.0
CronicDermatitis = e	0	0	0	0	52	0	100
PityriasisRosea = f	0	3	0	0	0	46	93.9
Precision (%)	95.6	88.9	97.2	94.4	98.1	95.8	95.1

Note, this research involves using the Waikato Environment for Knowledge Analysis (Weka) written by Witten and Frank [14]. We use J48 which is the Weka's implementation of C4.5 decision tree [8]. Additionally, we extend the CostSensitiveClassifier in the Weka system to incorporate our proposed Max Method which initially uses the cost weighting method proposed by Ting [11].

4.1 Effects of Applying Cost Matrix 1

Table 4 presents the result of applying the default cost matrix to the data set using J48 in the Weka tool set with the Summing Method proposed by Ting [11]. The default cost matrix has all misclassification cost entries set to 1. The results indicate the average misclassification cost to be around 4.9%, with an accuracy of 95.1%.

Table 5 presents the result of applying cost matrix 1 to the data set. The results show that the classifier tends to favor the class with more weight compared to the results of applying the default cost matrix (i.e. no cost). It avoids making errors with the high weighted classes, so the penalty is as low as possible. For example, no instances are misclassified as CronicDermatitis due to the high cost compared to the results of applying a cost matrix with equal misclassification cost. The predicted precision has increased from 98.1% to 100% after applying cost matrix 1. On the other hand, more instances are predicted as Seboretic (i.e. a skin disease that often affects those with a tendency to dandruff) due to the

Table 5. Confusion matrix using Cost Matrix 1 and Summing Method

classified as →	a	b	c	d	e	f	Recall (%)
Psoriasis = a	105	4	2	1	0	0	93.8
Seboretic = b	2	55	1	2	0	1	90.2
LichenPlanus = c	1	0	71	0	0	0	98.6
PityriasisRubra = d	2	2	0	16	0	0	80.0
CronicDermatitis = e	0	0	0	0	52	0	100
PityriasisRosea = f	0	3	0	0	0	46	93.9
Precision (%)	95.5	85.9	95.9	84.2	100	97.9	94.3

Table 6. Confusion matrix using Cost Matrix 1 and Max Method

classified as →	a	b	c	d	e	f	Recall (%)
Psoriasis = a	107	2	2	1	0	0	95.5
Seboretic = b	2	55	1	1	0	2	90.2
LichenPlanus = c	1	1	70	0	0	0	97.2
PityriasisRubra = d	2	1	0	17	0	0	85.0
CronicDermatitis = e	0	0	0	0	52	0	100
PityriasisRosea = f	0	3	0	0	0	46	93.9
Precision (%)	95.5	88.7	95.9	89.5	100	95.8	94.8

low cost of misclassification with a slight drop in precision and recall to 85.9% and 91.8% respectively. The reason for the decrease is because the cost of misdiagnosing Seboretic is low compared to other classes. The results also indicate that the average misclassification cost has increases from 4.9% (given no cost) to around 5.8%.

The second test is performed on the same cost matrix using the Max Method. The results are then compared with the results of the Summing Method. Table 6 shows the confusion matrix generated when applying the Max Method. The classifier produces an average misclassification cost of 5.2% compared to a higher average cost of 5.8% for the Summing Method. The Max Method gives a lower total misclassification cost and still manages to achieve a slightly higher accuracy of 94.8% compared to the previous method. The results indicate that the Max Method outperforms the Summing Method.

4.2 Effects of Applying Cost Matrix 2

Another series of tests is carried out using a completely different cost matrix (shown in Table 8) on the same data set to show the above analysis holds for other cases. This time the ranking order of the misclassification costs shown in Table 2 is reversed as shown in Table 7. As can be seen, CronicDermatitis has the least misclassification cost while PityriasisRosea has the most cost. Note that the relative cost in this table is just for testing purposes, and does not apply to real costs of a diagnosis.

Table 7. Reversed relative ranking of misclassification costs

Rank	Costs (thousands)	Class
6	2	CronicDermatitis
5	5	PityriasisRubra
4	10	LichenPlanus
3	20	Psoriasis
2	50	Seboretic
1	100	PityriasisRosea

Table 8. Cost Matrix 2

classified as →	a	b	c	d	e	f
Psoriasis = a	0	50	10	5	2	100
Seboretic = b	20	0	10	5	2	100
LichenPlanus = c	20	50	0	5	2	100
PityriasisRubra = d	20	50	10	0	2	100
CronicDermatitis = e	20	50	10	5	0	100
PityriasisRosea = f	20	50	10	5	2	0

Table 9. Confusion matrix using Cost Matrix 2 and Summing Method

classified as →	a	b	c	d	e	f	Recall (%)
Psoriasis = a	110	0	2	0	0	0	98.2
Seboretic = b	4	53	0	1	0	3	86.9
LichenPlanus = c	2	1	68	0	1	0	94.4
PityriasisRubra = d	0	0	0	19	1	0	95.0
CronicDermatitis = e	0	0	0	0	52	0	100
PityriasisRosea = f	1	12	0	0	0	36	73.5
Precision (%)	94.0	80.3	97.1	95.0	96.3	92.3	92.3

Table 9 shows the results generated using cost matrix 2. It is important to see that there are less instances being classified as PityriasisRosea in this confusion matrix than the one shown in Table 5 and 6. This is due to the fact that a higher misclassification cost (100) is assigned to PityriasisRosea compared to the one shown in Table 3. As can be seen from Table 9, the classifier tends to classify more instances as PityriasisRubra and CronicDermatitis, because they have the lowest predicted cost compared to other classes. The average misclassification cost for this classifier is 7.7%, with a high accuracy of 92.4%. However, the recall rate for PityriasisRosea is low compared to other classes. This is affected by the high misclassification cost of the class.

The results of the Max method are shown in Table 10. This method gives the minimum misclassification cost with an average of 7.1%. Additionally, the rate of accuracy has increased slightly to around 92.9%. This method again has the advantage of avoiding misclassification with high cost diagnoses and still manages to achieve high precision and recall rate. For instance, the predicted PityriasisRosea has a higher recall at 93.9% using the Max Method as compared to 73.5% of the Summing Method.

5 Conclusions

In this paper we present a CBR approach that provides decision support to the GP for diagnosing dermatological problems using machine learning technique (decision trees) and cost-sensitive approach (instance weighting) for classifying

Table 10. Confusion matrix using Cost Matrix 2 and Max Method

classified as →	a	b	c	d	e	f	Recall (%)
Psoriasis = a	110	0	2	0	0	0	93.8
Seboretic = b	1	56	0	1	0	3	90.2
LichenPlanus = c	1	3	68	0	0	0	98.6
PityriasisRubra = d	1	1	0	18	0	0	80.0
CronicDermatitis = e	0	0	0	0	52	0	100
PityriasisRosea = f	0	13	0	0	0	36	93.9
Precision (%)	97.3	73.7	97.1	94.7	100	92.3	92.9

cases. The CBR system is integrated into a Web-based application that enables dynamic interaction between the CBR engine and the users across the Internet.

The results show that instance weighting method usually avoids classifying the instances that have high misclassification cost. However, in seeking to minimize the errors with high cost, it will usually increase the number of low cost errors as a result. Sometimes the rate of accuracy can be very good and still give a high average misclassification cost. In addition, our results show that our proposed Max Method performs better than Summing Method in providing a lower misclassification cost, and still manages to achieve a slightly higher accuracy.

Future work will investigate more objective methods to decide on costs of misdiagnosis, such as the age (i.e. quality of life because of life expectancy) and health condition of the patient.

References

[1] A. Aamodt and E. Plaza. Case-Based Reasoning: Foundational Issues, Methodological Variations, and System Approaches. *AI Communications*, 7(1):39–59, March 1994. 668

[2] D. W. Abbott, P. Matkovsky, and Elder IV. J. F. An Evaluation of High-End Data Mining Tools for Fraud Detection. In *Proceedings of IEEE International Conference on systems, man and Cybernetics*, pages 2836–2841, 1998. 668

[3] R. Bergmann. Highlights of the European INRECA Projects. In David W. Aha and Ian Watson, editors, *Proceedings of the 4th International Conference on Case-Based Reasoning, ICCBR-2001*, pages 1–15, Vancouver, Canada, July/August 2001. Springer. 668

[4] V. Ciesielski and G. Palstra. Using a Hybrid Neural/Expert System for Database Mining in Market Survey Data. In *Proceedings of the 1st International Conference on Knowledge Discovery and Data Mining*, pages 38–43, 1996. 668

[5] P. Domingos. MetaCost: A General Method for Making Classifiers Cost-Sensitive. In *Proceedings of the 5th International Conference on Knowledge Discovery and Data Mining*, pages 155–164, San Diego, CA, 1999. 668

[6] J. A. A. Hunter, J. A. Savin, and M. V. Dahl. *Clinical Dermatology.* Blackwell Scientific Pulications, Australia, 1989. 671

[7] B. W. Porter and E. R. Bareiss. PROTOS: An Experiment in Knowledge Acquisition for Heuristic Classification Tasks. In *Proceedings of the First International*

Meeting on Advances in Learning (IMAL), pages 159–174, France, 1986. Les Arcs. 668

[8] J. Ross. Quinlan. *C4.5 Programs for Machine Learning.* Morgan Kaufmann Publishers, USA, 1993. 668, 672

[9] C. P. Tait and C. D. Clay. Pilot Study of Store and Forward Teledermatology Services in Perth, Western Australia. *Australian Journal of Dermatology*, 1999. Research Report. Royal Perth Hospital, Australia. 666

[10] A. C. Tessmer. What to Learn From Near Misses: An Inductive Learning Approach to Credit Risk Assessment. *Decision Sciences*, 1:105–120, 1997. 668

[11] K. M. Ting. Inducing Cost-Sensitive Trees via Instance Weighting. In *Proceedings of the 2nd European Symposium on Principles of Data Mining and Knowledge Discovery, LNAI-1510*, pages 139–147. Springer-Verlag, 1998. 668, 669, 672

[12] K. M. Ting and Z. Zheng. Boosting Cost-Sensitive Trees. In *Proceedings of the 1st International Conference on Discovery Science, LNAI-1532*, pages 244–255, Berlin, 1998. Springer-Verlag. 668

[13] I. Watson. *Applying Case-Based Reasoning: Techniques for Enterprise Systems.* Morgan Kaufmann Publishers, USA, 1997. 668

[14] I. Witten and E. Frank. *Data Mining: Practical Machine Learning Tools and Techniques with Java Implementations.* Morgan Kaufmann Publishers, USA, 2000. 668, 672

[15] R. Wootton and A. Oakley, editors. *Teledermatology.* Royal Society of Medicine Press Ltd, London, UK, 2002. 666

A Case Study in Feature Invention for Breast Cancer Diagnosis Using X-Ray Scatter Images

Shane M. Butler[1], Geoffrey I. Webb[1], and Rob A. Lewis[2]

[1] School of Computer Science and Software Engineering
Bldg 26, Monash University Vic. 3800, Australia
{sbutler,webb}@infotech.monash.edu.au
[2] School of Physics and Materials Engineering
Bldg 19, Monash University Vic. 3800, Australia
rob.lewis@spme.monash.edu.au

Abstract. X-ray mammography is the current clinical method for screening for breast cancer, and like any technique, has its limitations. Several groups have reported differences in the X-ray scattering patterns of normal and tumour tissue from the breast. This gives rise to the hope that X-ray scatter analysis techniques may lead to a more accurate and cost effective method of diagnosing beast cancer which lends itself to automation. This is a particularly challenging exercise due to the inherent complexity of the information content in X-ray scatter patterns from complex hetrogenous tissue samples. We use a simple naïve Bayes classier as our classification system. High-level features are extracted from the low-level pixel data. This paper reports some preliminary results in the ongoing development of this classification method that can distinguish between the diffraction patterns of normal and cancerous tissue, with particular emphasis on the invention of features for classification.

Keywords: Knowledge discovery and data mining; Applications.

1 Introduction

The current clinical method used for screening for breast cancer is X-ray mammography. In this method X-rays are directed through the breast and a radiograph is recorded. A radiologist then examines the photograph for evidence of cancer and the appropriate action taken.

Mammography has been successful at reducing breast cancer mortality [7], but like all diagnostic techniques has its limitations. Mammography is non-specific since X-ray attenuation has no direct connection with the presence of disease. As a result it features a high false-positive rate, with less than 20% of women recalled following a suspicious mammogram actually having breast cancer [1]. It also has a significant false-negative rate, with an overall sensitivity (the proportion of cancers successfully detected) of 90% [6]. The sensitivity is

T.D. Gedeon and L.C.C. Fung (Eds.): AI 2003, LNAI 2903, pp. 677–685, 2003.

further reduced in younger women, in those women with a dense background pattern and in women on hormone replacement therapy [8, 9].

Many women recalled require percutaneous or surgical breast biopsy which is subject to pathological analysis to confirm the nature of their breast lesion. Unfortunately, biopsy analysis is also subject to errors, both from sampling errors, where the sample does not contain part of the lesion, and because the interpretation of a substantial fraction of biopsies is difficult. Whilst the screening programme is undoubtedly successful at detecting cancer, it has important limitations, missing some abnormalities and providing insufficient information for the classification of others. Any technique capable of reducing the need for breast biopsy and/or aiding the analysis of biopsy specimens, especially in the presence of sampling error would be highly advantageous.

Small Angle X-Ray Scattering (SAXS) is a X-ray diffraction based technique where a narrow collimated beam of X-rays are focused on to a sample and the scattered X-rays recorded by a detector. The pattern of the scattered X-rays carries information on the molecular structure of the material. The technique is particularly successful when combined with a synchrotron [12] which produces monochromatic X-ray beams of sufficient intensity to allow scatter patterns to be recorded in a few seconds. The Australian Synchrotron will be operational in 2007.

Small angle X-ray scattering is particularly useful where the material possesses a well ordered molecular structure such as is the case with several biological materials. In particular, collagen which is the most common protein in the human body, is a major constituent of breast tissue and yields an immediately recognisable scatter pattern.

Several groups have applied X-ray scattering to breast tissue [4, 5, 3] and all have detected differences in the scattering patterns produced by normal and tumour tissue. The hope is that this will lead to a more accurate and cost effective method of diagnosing beast cancer.

The challenge remains, however, for a method of automatic classification of the SAXS images. This paper reports on some preliminary results in developing a classification method that can distinguish between diffraction patterns of normal and cancerous tissue, with particular emphasis on the invention of features for classification.

2 Data Description

The data used in this paper has previously been published by Lewis *et al* [5]. Samples were obtained from patients via cosmetic breast reduction and core cuts of invasive breast carcinomas. Diffraction data was then collected at the Synchrotron Radiation Source at Daresbury Laboratory in the UK. All samples were examined and classified by a pathologist. See Lewis *et al* [5] for details of samples and the experimental protocol.

The subset of the data used in this paper consisted of 512 pixels by 512 pixels diffraction images obtained from normal and invasive malignant tumours. Each

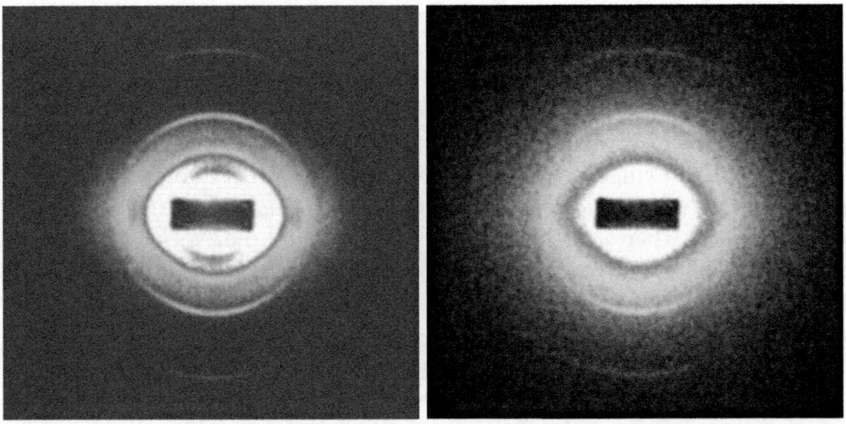

(a) Normal (b) Tumour

Fig. 1. Diffraction images were obtained using a synchrotron

pixel has a real number associated with it that represents the number of photons hitting a point on the detector. The classes used were Normal (which consisted of 20 instances) and Tumour (which consisted of 22 instances).

Some example diffraction images are shown in Figure 1. The dark area in the centre is created by the beam stop employed to prevent the direct beam from impinging upon the detector. The scatter pattern radiates from the beam centre which is not necessarily the centre of the beam stop. The well structured collagens found in healthy breast tissue produces clear arcs such as those visible in Figure 1(a). It is believed that many breast tumours express enzymes that degrade collagen [2]. This is consistent with the diffraction patterns from tumour tissue such as that shown in Figure 1(b) where the arcs are much less prominent.

3 Feature Extraction

Whilst each pixel could have been treated as an attribute, it is unlikely that a classifier built from such a low-level description would be able to classify the images. Furthermore, there is inherit uncertainty in the scatter of the diffracted X-rays and hence the values at individual pixels cannot be considered as precise. High-level features therefore need to be extracted from the diffraction images. We report here on preliminary work identifying such features.

In order to develop useful features which can subsequently be related to the changes from which the differences arise it is important to develop an understanding of the physics underlying the problem. The angles to which X-rays are scattered depends on the molecular properties of the material. As a direct result, not only does the pattern change but the amount of X-ray energy hitting the detector varies according to this structure. The first feature seeks to use this possible difference between classes for classification. The value of this feature was sum of all of the pixel intensities across an entire scatter image. We designated it SumIntWhole.

Note that this feature does not make use of the diffraction patterns. It is hypothesised that if the structure of the tissue has been degraded due to cancer, the intensity of the rings or arcs at specific angles will be lower. Locating the rings is relatively straight forward in the case where the structure is intact, but is more problematic when the rings are less well defined. For this reason, and also to search for other significant features, we chose not to directly locate and assess the rings. Rather, we seek to segment the images in such a way as to localise the regions in which the ring may occur. Whether or not one observes a complete ring or an arc such as seen in Figure 1(a) is dependent upon the degree of orientation of the tissue. In the experiments care was taken to align the tissue in the same way from sample to sample but minor variations in orientation will always exist.

Given the orientation is known to be approximately vertical down the centre of the image, a vertical slice down this line will locate the points at which the highest intensities are expected. There is a further problem however, in that the centre of the beam does not necessarily correspond to the centre of the image or centre of the beam stop. For technical reasons it is difficult to determine where the centre of the beam falls but an attempt has been made to do this, of unknown accuracy. Due to the experimental controls the expectation is that the same location should be the centre of the beam for all samples.

The first method of segmentation used a slice down the diffraction image from $(260, 0)$ to $(260, 512)$ as shown in Figure 2(a). $(260, 259)$ is the estimated beam centre. Since individual intensity values recorded as a pixel by the detector can be spurious, a simple averaging function was employed to ensure that finding the maximum value was less heavily influenced by noise. The calculated value was the average of the pixel and the pixels immediately surrounding it, such that the result was an average of 9 pixel intensities (with the exception of pixels located on the edge of the image that were averages of fewer pixel intensities). The slice was split into either 5 or 10 sections and the maximum (smoothed) value of a pixel in a section_s was recorded in a feature designated MaxSliceSection_s. The y-axis location of the maximum for section_s was also recorded as another feature, $\text{YlocMaxSliceSection}_s$. These features should pick up the differences in intensity between the peaks.

The rings can vary in intensity, distance from the beam centre and how complete they are. To capture this information, a radial segmentation technique was employed. Images were segmented into either 5 or 10 circular regions radiating from the beam centre. This is shown in Figure 2(b). From each region_r several features were gathered. $\text{SumCircularRegion}_r$ and $\text{MaxCircularRegion}_r$ are the sum of all intensity values in the region and maximum of the intensity values for these regions, respectively. Intensity values were again averaged for determining the maxima. Following the detection of the maximum intensity

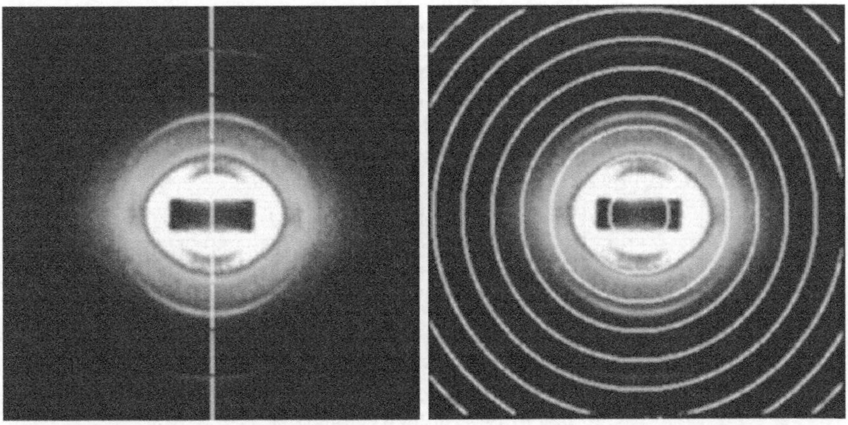

(a) Centre slice (b) Circular regions

Fig. 2. Segmentation used among some of the features

for a region$_r$, the distance from that point to the beam centre is designated RlocMaxCircularRegion$_r$.

As the correct beam centre is uncertain, RlocMaxCircularRegion$_r$ may be compromised. The features YlocMaxCircularRegion$_r$ and XlocMaxCircularRegion$_r$ were therefore included as alternate maximum location measures (using x and y co-ordinate values) as they may be less sensitive to this error. Note: Throughout the remainder of this paper feature variable names represent all of the variables relating to that feature, unless otherwise noted.

By sampling both the slice sections and circular regions at different frequencies of 5 and 10 we hoped to create features that included the significant information.

4 Classification

A naïve Bayes classifier was selected as the initial base classifier since it is known to be a simple, efficient and effective classifier. These features are all numeric, so we were faced with the issue of whether to use probability density estimation or discretization, and if the latter, which discretization to employ. Due to the small amount of data we deemed probability density estimation inappropriate as accurate estimation of a probability density function may require large volumes of data. Influenced by Yang and Webb [11] we instead used Equal Frequency Discretization (EFD) with the number of buckets set to 5.

The Waikato Environment for Knowledge Analysis (WEKA) [10] software version 3.3.6 was selected for this analysis as it encompasses both the chosen

Table 1. Leave-one-out naïve Bayes classification accuracy

Feature #	Name	Accuracy
1	SumIntWhole	45.24%
2-11	MaxSliceSections (10)	76.18%
12-21	YlocMaxSliceSections (10)	78.57%
22-31	SumCircularRegions (10)	90.48%
32-41	MaxCircularRegions (10)	85.71%
42-51	YlocMaxCircularRegions (10)	69.05%
52-61	XlocMaxCircularRegions (10)	80.95%
62-71	RlocMaxCircularRegions (10)	73.81%
72-76	MaxSliceSections (5)	61.90%
77-81	YlocMaxSliceSections (5)	76.19%
82-86	SumCircularRegions (5)	80.95%
87-91	MaxCircularRegions (5)	95.24%
92-96	YlocMaxCircularRegions (5)	42.86%
97-101	XlocMaxCircularRegions (5)	57.14%
102-106	RlocMaxCircularRegions (5)	90.48%

classifier and discretization method required. This meant that it was necessary for the input data to be written in the WEKA ARFF file format.

The settings used in the experiments were as follows. The weka.classifiers.bayes.NaiveBayes classifier was used in conjuntion with weka.filters.unsupervised.attribute.Discretize. The number of bins was set to 5 and the useEqualFrequency option enabled. The leave-one-out cross-validation test method was used and all other settings were the WEKA program defaults.

5 Results

When the naïve Bayes classifier was run on the various features it produced some very interesting results, shown in Table 1. The first column gives the feature a feature number. Most features are the result of splitting some area into either 5 or 10 separate sections. These will have a range in the feature number column instead of just one number, with each value in the range representing a feature corresponding to one section. The second column is the feature name. The accuracy using a naïve Bayes classifier with EFD discretization on only the single feature or group of features is reported in the next column.

The individual features performed very differently. SumIntWhole was the worst performing feature. Similarly, the x and y features at samples of both 5 and 10 sections also delivered very low accuracy. Some of the features that stand out are the 10 SumCircularRegions, the 5 MaxCircularRegions and the 5 RlocMaxCircularRegions, although interestingly for no set of features was equivalent performance obtained using both 5 and 10 partitions.

Our expectation was that effective classification would require a combination of features. We first tried the combination of all 10 slice section features and the 10 circular region features. This delivered extremely high accuracy, 97.62%. This

Table 2. Combinations of the 10 slice sections and 10 circular regions

Feature #	Excluded Feature	Accuracy
12-71	MaxSliceSections (10)	97.62%
2-11, 22-71	YlocMaxSliceSections (10)	92.86%
2-21, 32-71	SumCircularRegion (10)	95.24%
2-31, 42-71	MaxCircularRegion (10)	95.24%
2-41, 52-71	YlocMaxCircularRegion (10)	92.86%
2-51, 62-71	XlocMaxCircularRegion (10)	95.24%
2-61	RlocMaxCircularRegion (10)	95.24%
2-71	(None excluded)	97.62%

Table 3. Combinations of the 5 slice sections and 5 circular regions

Feature #	Excluded Feature	Accuracy
77-106	MaxSliceSections (5)	90.48%
72-76, 82-106	YlocMaxSliceSections (5)	85.71%
72-81, 87-106	SumCircularRegions (5)	92.85%
72-86, 92-106	MaxCircularRegions (5)	88.10%
72-91, 96-106	YlocMaxCircularRegions (5)	90.48%
72-96, 102-106	XlocMaxCircularRegions (5)	90.48%
72-101	RlocMaxCircularRegions (5)	90.48%
72-106	(None excluded)	90.48%

means that only one of the examples was misclassified by leave-one-out cross-validation. To evaluate whether all of the 10 slice section and the 10 circular region features were important for this outcome, we next tried all combinations of these features with a single group omitted. Omitting the MaxSliceSections group did not affect the accuracy obtained, but omitting any other single group did, each one resulting in one or two more misclassifications. These results are presented in Table 2.

We also tried the combination of all 5 slice section features and the 5 circular region features in a similar fashion. This delivered high accuracy, 90.48% which equates to four misclassified examples. To evaluate whether all of the 5 slice section and the 5 circular region features were important for this outcome, we next tried all combinations of these features with a single group omitted. Omitting most of the groups did not affect the accuracy obtained, but when the groups YlocMaxSliceSections and MaxCircularRegions were individually omitted they resulted in several more misclassifications. Omitting the SumCircularRegions group actually resulted in slightly improved accuracy, with one extra correctly classified case. These results are presented in Table 3.

We also tried the combinations of all features at a sample size of both 5 and 10 sections. Again, we found that omitting some groups of features yielded high accuracy results, with all results being over 90%. These results are shown in Table 4.

Table 4. Combinations of both the 5 & 10 slice sections and the 5 & 10 circular regions

Feature #	Excluded Feature	Accuracy
12-106	MaxSliceSection (10)	92.85%
2-11, 22-106	YlocMaxSliceSection (10)	90.48%
2-21, 32-106	SumCircularRegion (10)	95.24%
2-31, 42-106	MaxCircularRegion (10)	95.24%
2-41, 52-106	YlocMaxCircularRegion (10)	92.85%
2-51, 62-106	XlocMaxCircularRegion (10)	90.48%
2-61, 72-106	RlocMaxCircularRegion (10)	92.85%
2-71, 77-106	MaxSliceSection (5)	92.85%
2-76, 82-106	YlocMaxSliceSection (5)	90.48%
2-81, 87-106	SumCircularRegion (5)	92.85%
2-86, 92-106	MaxCircularRegion (5)	95.24%
2-91, 97-106	YlocMaxCircularRegion (5)	95.24%
2-96, 102-106	XlocMaxCircularRegion (5)	95.24%
2-101	RlocMaxCircularRegion (5)	95.24%
2-106	(None excluded)	95.24%

6 Conclusions and Future Research

Herein we present preliminary results from an exploratory study that seeks to identify features for diagnosing breast cancer from synchrotron X-ray scatter data. This is a particularly challenging exercise due to the inherent complexity of the information content in X-ray scatter patterns from complex hetrogenous tissue samples. Nonetheless, we are encouraged by the high accuracy achieved by our initial simple features.

Some caution is required in interpreting these results. With many different experiments performed, it should be expected that some observed accuracies are unrealistically high due to chance variation. That we obtained above 90% accuracy on all combinations of features relating to 10 slice sections and 10 circular regions does however provide evidence that we are capturing significant regularities in the data.

Our next challenge is to refine the features. A first step will be to more accurately locate the centre of the beam for each image and then more precisely locate and measure each of the rings.

We are also seeking to increase the size of the data set by collecting further samples. This will allow better evaluation of the features. The ultimate test will be to evaluate a specific classifier on a new data set of previously unsighted cases. However, the research is still a long distance short of this objective.

Our preliminary results provide some hope to the prospect of developing X-ray based techniques for detecting and diagnosing cancer that are more accurate and less intrusive than those currently existing. If such techniques could be developed they may lead to further decreases in mortality coupled with a decrease in the intrusiveness and uncertainty to which large numbers of women are currently subjected.

Acknowledgements

We would like to acknowledge Marcus Kitchen for his invaluable consultation in interpreting the SAXS data.

We also wish to thank the authors of the WEKA machine learning environment [10], which was used in these experiments, for making their software freely available.

References

[1] N. Bjurstam, L. Bjorneld, S. W. Duffy, T. C. Smith, E. Cahlin, O. Eriksson, L. O. Hafstrom, H. Lingaas, J. Mattsson, S. Persson, C. M. Rudenstam, and J. Save-Soderbergh. The gothenburg breast screening trial. *Cancer*, 80(11):2091–2099, 1997. 677

[2] L. M. Coussens and Z. Werb. *Chemistry and Biology*, 3:895–904, 1996. 679

[3] M. Fernandez, J. Keyrilainen, R. Serimaa, M. Torkkeli, M-L. Karjalainen-Lindsberg, M. Tenhunen, W. Thomlinson, V. Urban, and P. Suortti. Small-angle x-ray scattering studies of human breast tissue samples. *Physics in Medicine and Biology*, 47:577–592, 2002. 678

[4] G. Kidane, R. D. Speller, G. J. Royle, and A. M. Hanby. X-ray scatter signatures for normal and neoplastic breast tissues. *Physics in Medicine and Biology*, 44:1791–1802, 1999. 678

[5] R. A. Lewis, K. D. Rogers, C. J. Hall, E. Towns-Andrews, S. Slawson, A. Evans, S. E. Pinder, I. O. Ellis, C. R. M. Boggis, A. P. Hufton, and D. R. Dance. Breast cancer diagnosis using scattered x-rays. *Journal of Synchrotron Radiation*, 7:348–352, 2000. 678

[6] A. I. Mushlin, R. W. Kouides, and D. E. Shapiro. Estimating the accuracy of screening mammography: a meta-analysis. *Am J Prev Med*, 14(2):143–53, 1998. 677

[7] L. Nystrom, L. Rutqvist, and S. Wall et al. Breast cancer screening with mammography: overview of swedish randomised trials. *Lancet*, 341:973–978, 1993. 677

[8] R. D. Rosenberg, W. C. Hunt, M. R. Williamson, F. D. Gilliland, P. W. Wiest, C. A. Kelsey, C. R. Key, and M. N. Linver. The effect of age, density, ethnicity, and estrogen replacement therapy on screening mammography sensitivity and cancer state: A review of 183,134 screening mammograms in Albuquerque, New Mexico. *Radiology*, 209:511–518, 1998. 678

[9] D. M. Sibbering, H. C. Burrell, A. J. Evans, L. J. Yeoman, R. M. Wilson, and J. F. Robertson. Mammographic sensitivity in women under 50 years presenting symptomatically with breast cancer. *Breast*, 4(2):127–129, 1995. 678

[10] I. H. Witten and E. Frank. *Data Mining: Practical Machine Learning with Java Implementations*. Morgan Kaufmann, Sydney, 2000. 681, 685

[11] Y. Yang and G. I. Webb. Discretization for naive-bayes learning: managing discretization bias and variance. Technical Report 2003/131, School of Computer Science and Software Engineering, Monash University, 2003. 681

Effectiveness of A Direct Speech Transform Method Using Inductive Learning from Laryngectomee Speech to Normal Speech

Koji Murakami[1], Kenji Araki[1],
Makoto Hiroshige[1], and Koji Tochinai[2]

[1] Hokkaido University, Graduate school of Engineering
North 13, West 8, Kita-ku, Sapporo, 060-8628, Japan
{mura,araki,hiro}@media.eng.hokudai.ac.jp
http://sig.media.eng.hokudai.ac.jp/~mura
[2] Hokkai Gakuen University, Graduate school of Business Administration
Asahimachi 4-1-40, Toyohira-ku, Sapporo, 062-8605, Japan
tochinaik@econ.hokkai-s-u.ac.jp

Abstract. This paper proposes and evaluates a new direct speech transform method with waveforms from laryngectomee speech to normal speech. Almost all conventional speech recognition systems and other speech processing systems are not able to treat laryngectomee speech with satisfactory results. One of the major causes is difficulty preparing corpora. It is very hard to record a large amount of clear and intelligible utterance data because the acoustical quality depends strongly on the individual status of such people.

We focus on acoustic characteristics of speech waveform by laryngectomee people and transform them directly into normal speech. The proposed method is able to deal with esophageal and alaryngeal speech in the same algorithm. The method is realized by learning transform rules that have acoustic correspondences between laryngectomee and normal speech. Results of several fundamental experiments indicate a promising performance for real transform.

Keywords: Esophageal speech, Alaryngeal speech, Speech transform, Transform rule, Acoustic characteristics of speech

1 Introduction

Speech is a perfect medium and most common for human-to-human information exchange because it is able to be used without hands or other tools, being a fundamental contribution for ergonomic multi-modality[1]. Much research has also been developed to realize such advantages for human-machine interaction. Many applications have been produced and they are contributing to human life.

On the other hand, many people who are unable to use their larynxes are not able to benefit from such advances in technology although such assistance is hoped for. Both esophageal and alaryngeal speech, which laryngectomee people

T.D. Gedeon and L.C.C. Fung (Eds.): AI 2003, LNAI 2903, pp. 686–698, 2003.

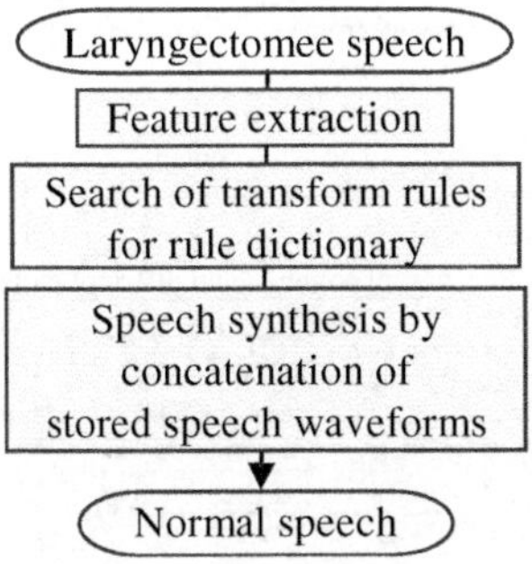

Fig. 1. Processing of the proposed method

practice to enable conversation, are understandable and enables adequate communication. However, conventional speech processing systems are not able to accept them as inputs because almost all acoustic models in the current systems are trained by intelligible utterances spoken by normal people. It is easy to find a lot of corpora high in both quality and quantity in many languages. However, there are not many resources of laryngectomee or other disordered speech because it is very difficult to sample a number of intelligible and clear utterances. One of the major causes is dependence on individual status of speech.

We only focus on laryngectomee speech waveforms themselves to transform them into normal speech. Many studies have attempted to transform laryngectomee speech to normal speech, for example: re-synthesizing fundamental frequency or formant of normal speech[2], or by utilizing a codebook[3]. We propose a radically different speech transform approach which handles only acoustic characteristics. The concepts of the method have been applied to realize a speech translation method and provided promising effectiveness[4, 5]. Fig.1 shows the processing stages of our method. The proposed method is realized by dealing with only correspondences of acoustic characteristics between both speech waveforms. Our basic conception is based on belief that even laryngectomee utterances have certain contents although these are inarticulate and quite different from normal speech waveforms. At first, acoustic common and different parts are extracted by comparing two utterances within the same speech side. These parts should have correspondences of meaning between two different types of speech. Then we generate transform rules and register them in a transform dictionary. The rules also have the location information of acquired parts for speech synthesis on time-domain. The transform rules are obtained by comparing not only speech samples but also acquired transform rules themselves using Inductive Learning Method[6], still keeping acoustic information within the rules. Deciding the correspondence of meaning between two speech sides is the unique condition necessary to realize our method.

In a transform phase, when an unknown utterance of laryngectomee speech is applied to be transformed, the system compares this sentence with the acoustic information of all rules within the speech side. Then several matched rules

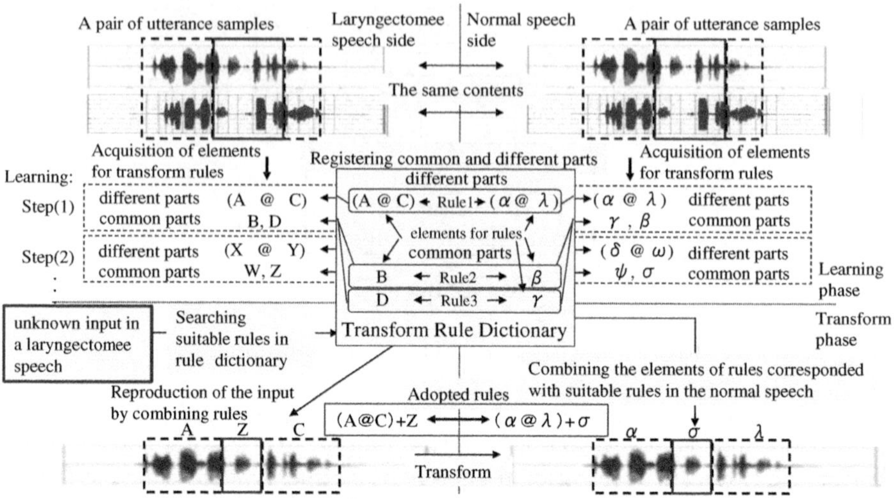

Fig. 2. Processing structure

are utilized and referred to their corresponding parts of the normal speech side. Finally, we obtain roughly synthesized normal speech utterance by simply concatenating several suitable parts of rules in the normal speech side according to the information of location. Fig.2 explains an overview of the processing structure of our method.

Although the method deals with speech waveforms, the boundaries of word, syllable, or phoneme are not important for our method because the proposed method does not prepare any acoustically correct interpretation, and transform rules are designed by acoustic characteristics of speech from utterances. Therefore, these rules will be able to adapt the speaker's characteristics and habits by recursive learning. We evaluate effectiveness of the transform rules through fundamental experiments and offer discussion on behaviors of the system.

2 Laryngectomee Speech

Laryngectomee people try to acquire esophageal or alaryngeal speech as second speech to enable them to once again communicate effectively in society. The characteristics of these types of speech are explained in this section. Fig. 3, 4, and 5 show normal, alaryngeal and esophageal speech, respectively. Each Figure contains (A)waveform itself, (B)RMS power and (C)fundamental frequency.

2.1 Alaryngeal Speech

Alaryngeal speech has an unnatural quality and is significantly less intelligible than normal speech. The utterances spoken using an artificial larynx, are not able to contain any intonation, accent and emotion despite the speakers intention.

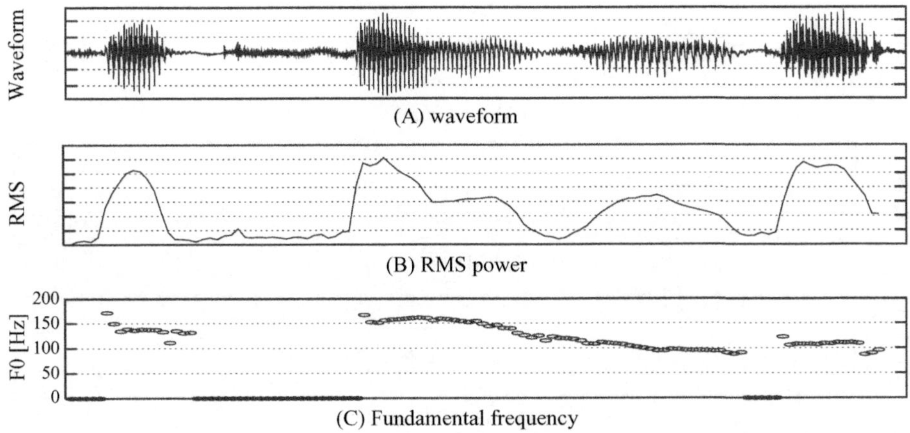

Fig. 3. Normal speech

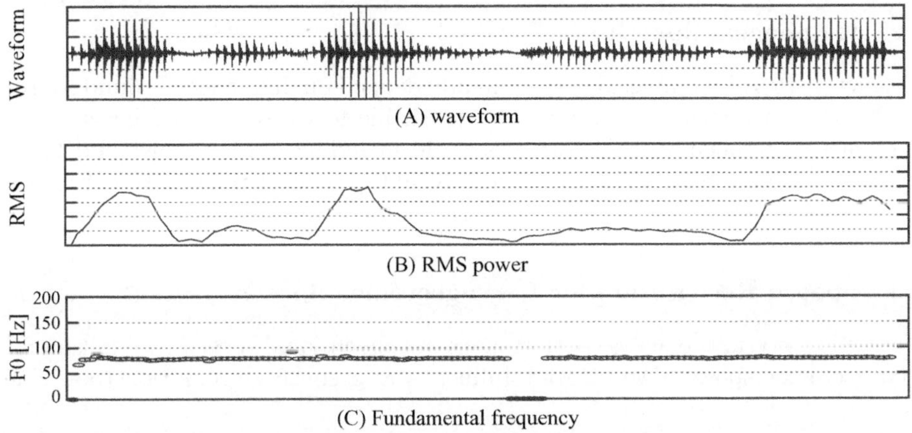

Fig. 4. Alaryngeal speech

Table 1. Result of speech recognition

Type of speech	Number of utterance	Accuracy of correct words
Normal Speech	80	65.82%
Alaryngeal Speech	119	29.61%
Esophageal Speech	107	24.32%

The cause is that this device is only able to vibrate fixed impulse source. Some research has improved the performance and quality of speech[9]. Fig. 4 shows a sample of alaryngeal speech.

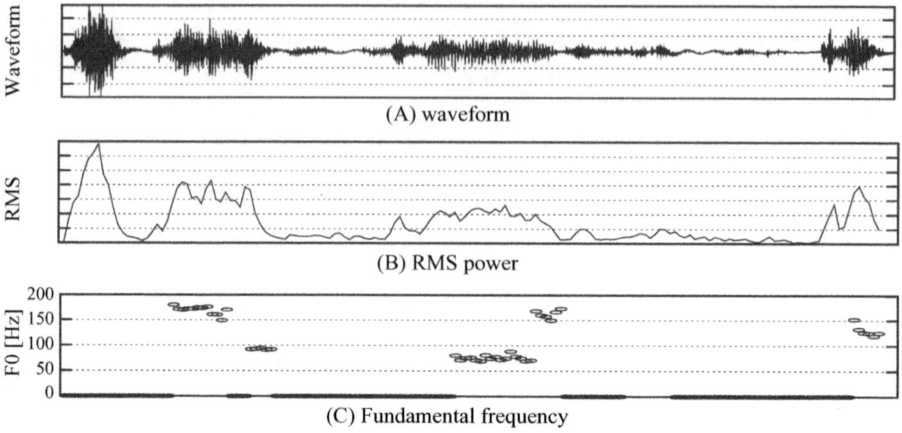

Fig. 5. Esophageal speech

2.2 Esophageal Speech

Characteristics of esophageal speech mainly depend on difference of sound source mechanism as shown Fig. 5. Several remarkable features are as follows: lower fundamental frequency than normal speech, including a lot of noise and lower volume[7]. Moreover, differences on prosody and spectral characteristics of speech are also reported[8].

2.3 Speech Recognition for Laryngectomee Speech

We need to reveal the actual performance of conventional speech recognition for laryngectomee speech. We utilized Julius[10] as a speech recognition tool. The

Table 2. Experimental conditions of speech processing

Size of frame	30msec
Frame cycle	15msec
Speech window	Hamming Window
AR Order	14
Cepstrum Order	20

Table 3. Information of speakers

Type of speech	Age/Gender	Speaker's feature
Normal Speech	24/Male	student
Alaryngeal Speech	70/Male	operation in 1990
Esophageal Speech	65/Male	operation in 1994

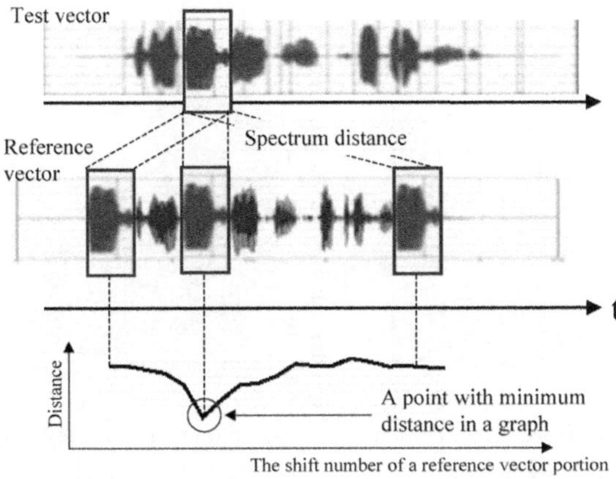

Fig. 6. Comparison of vector sequences

acoustic and language models in the system were constructed by the learning of normal speech utterances. Table 1 explains the result of recognition performance. It is obvious that the system is not able to treat laryngectomee speech without rebuilding the acoustic model for esophageal and alaryngeal speech utterances.

3 Speech Processing

3.1 Speech Data and Spectral Characteristics

Various acoustic parameters specific to disordered speech have been developed and applied to many studies[11]. Our study has succeeded to show acoustic differences by a clustering method using these values between normal and disordered female voices[12]. However, we have focused on results of comparison experiments using only spectral analysis[8].

We recorded utterance data with 16bit and 48kHz sampling rate, and downsampled to 16kHz. These data were spoken by three people whose usual speech is normal, esophageal and alaryngeal, respectively. Table 2 shows parameters adopted for speech processing, and Table 3 shows these speaker's characteristics. In this report, LPC Cepstrum coefficients were chosen as spectral parameter, because we could obtain better results than the other representations of speech characteristics[4].

3.2 Searching for the Start Point of Parts between Utterances

When speech samples were being compared, we had to consider how to normalize the elasticity on time-domain. We meditated upon suitable methods that would be able to give a result similar to dynamic programming[13] to execute time-domain normalization. We adopted a method to investigate the difference

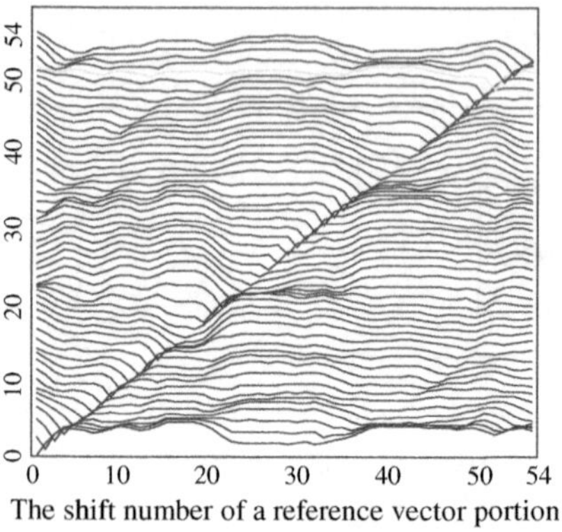

(A) "*KOKODE HITOWO MIRU* (I watch a man.)."

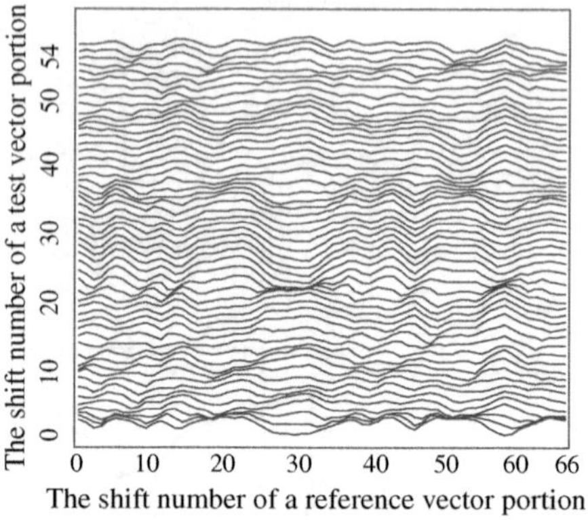

(B) "*KOKODE HITOWO MIRU* (I watch a man.)."-"*WATASHIWA KARATEWO SURU* (I play Karate.)."

Fig. 7. Difference between utterances

between two characteristic vectors of speech samples for determining common and different acoustic parts. We also adopted the Least-Squares Distance Method for the calculation of the similarity between these vectors. Two sequences of characteristic vectors named "test vector" and "reference vector" are prepared. The "test vector" is picked out from the test speech by a window that has definite length. At the time, the "reference vector" is also prepared from the reference speech. A distance value is calculated by comparing the present "test vector" and a portion of the "reference vector". Then, we repeat the calculation between the current "test vector" and all portions of the "reference vector" that are picked out and shifted in each moment with constant interval on time-domain. When a portion of the "reference vector" reaches the end of the whole reference vector, a sequence of distance values is obtained as a result. The procedure of comparing two vectors is shown in Fig. 6. Next, the new "test vector" is picked out by the constant interval, then the calculation mentioned above is repeated until the end of the "test vector". Finally, we should get several distance curves as the result between two speech samples.

Fig. 7 shows two examples of the difference between two utterances. Italic characteristics express Japanese. These applied speech samples are spoken by the same esophageal speaker. The contents of the compared utterances are the same in Fig. 7(A), and are quite different in Fig. 7(B) The horizontal axis shows the shift number of reference vector on time-domain and the vertical axis shows the shift number of test vector, i.e., the portion of test speech. In the figures, a curve in the lowest location has been drawn by comparing the head of the test speech and whole reference speech. If a distance value in a distance curve is obviously lower than other distance values, it means that the two vectors have much acoustic similarity.

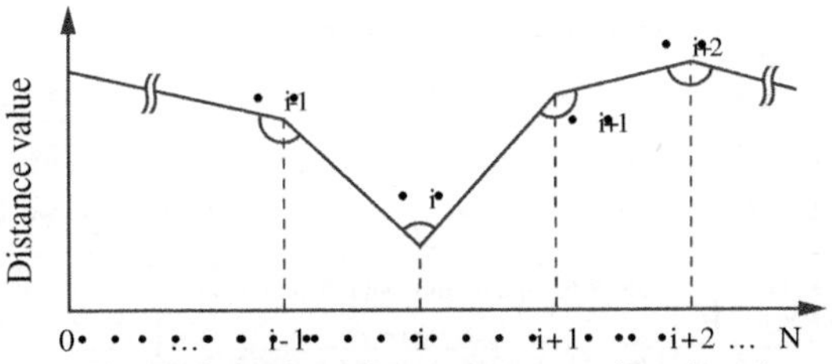

Fig. 8. Angle calculation

As shown in Fig. 7(B), the obvious local minimum distance point is not discovered even if there is the lowest point in each distance curve. On the other hand, as shown in Fig. 7(A), when the test and reference speech have the same content, the minimum distance values are found sequentially in distance curves. According to these results, if there is a position of the obviously smallest distance point in a distance curve, that point should be regarded as a frame in the "common part" by evaluating the point by a decision method in our previous research[4]. Moreover, if these points sequentially appear among several distance curves, they will be considered a common part. At the time, there is a possibility that the part corresponds to several semantic segments, longer than a phoneme and a syllable.

3.3 Evaluation of the Obvious Minimal Distance Value

To determine that the obviously lowest distance value in the distance curve is a common part, we need to employ a method based on a criterion. We chose the angles formed where two lines cross each other. These lines are linked by two distance points. Figure 8 shows angle calculation within the distance curve. The angles θ_i are determined by a focused point(i) and two neighboring distance points($i-1, i+1$) excluding the start and end points of the distance curve. If these two points have the minimal distance in a distance curve, we evaluate them with heuristic techniques. A common part is detected if the lowest degree of an angle θ_i is formed by a minimal distance point(i) and its neighboring points($i-1, i+1$), because the portion of reference speech has much similarity with the "test vector" of the distance curve at a point.

If several common parts are decided continuously, we deal with them as one common part because we want to partition the utterance into few parts, and the first point in this part will be the start point finally. In our method, the acoustic similarities evaluated by several calculations based on Least-Square Distance Method are the only factor for judgment in classifying common or different parts in the speech samples.

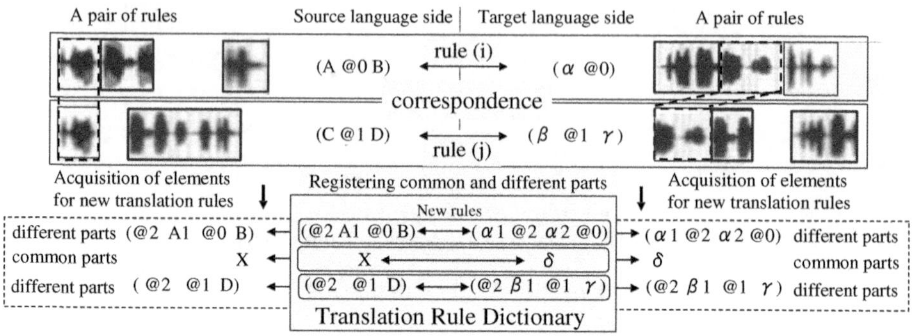

Fig. 9. Rule acquisition using Inductive Learning Method

4 Inductive Learning Method

Inductive Learning Method[6] acquires rules by extracting common and different parts through the comparison between two examples. This method is designed from an assumption that a human being is able to find out common and different parts between two examples although those contents are unknown. The method is also able to obtain rules by repetition of the acquired rules registered in the rule dictionary. At the time, a sentence form rule is also acquired with a different part rule by each comparison of utterances, and registered in the dictionary. This type of rule consists of different parts and a common part which is replaced with the variable "@". Therefore, the rule should represent the abstracted sentence without losing its meaning and be able to restore its context structure in the learning stage and the transform stage, respectively. This approach has been applied to many areas of natural language processing. The effectiveness of the method was also confirmed in machine translation. Therefore, we applied this method to acoustic characteristics of speech instead of character strings for realizing a direct speech transform approach.

Figure 9 shows an overview of recursive rule acquisition by this learning method. Two rules acquired as rule(i) and rule(j) are prepared and compared to extract common and different acoustic parts similar to comparisons between speech utterances. Then, these obtained parts are designed as new rules. If the compared rules consist of several common or different parts, the calculation is repeated within each part. It is assumed that these new rules are much more reliable for transform.

If several rules are not useful for transform, they will be eliminated by generalizing the rule dictionary optimally to keep a designed size of memory. The ability of optimal generalization in Inductive Learning Method is an advantage, as less examples have to be prepared beforehand. Much sample data is needed to acquire many suitable rules with conventional approaches.

5 Generation and Application of Transform Rule

5.1 Acquisition of Transform Rules

Acquired common and different parts are applied to determine the rule elements needed to generate transform rules. At the time, there are three cases of sentence structure as the "rule types". If two compared utterances were almost matching or did not match at all, several common or different parts are acquired, respectively. And the other case is that these utterances have both parts at the time. Combining sets of common parts of both normal and laryngectomee speech become elements of the transform rules for rule generation. The set of common parts extracted from the laryngectomee speech, which have a correspondence of meaning with a set of common parts in normal speech, are kept. The sets of different parts become elements of the transform rules as well. Finally, these transform rules are generated by completing all elements as below. It is very important that the rules are acquired if the types of sentences in both speech sides are the same. When the types are different, it is impossible to obtain the transform rules and register them in the rule dictionary because we are not able to decide the correspondence between two speech sides uniquely. Information that a transform rule has are as follows:

- rule types as mentioned above
- index number of an utterance in both speech sides
- sets of start and end points of each common and different part

5.2 Transform and Speech Synthesis

When an unknown laryngectomee speech is applied to be transformed, acoustic information of acquired parts in the transform rules are compared in turn with the unknown speech, and several matched rules become the candidates to transform. The inputted utterance should be reproduced by a combination of several candidates of rules. Then, the corresponding parts of the normal speech in candidate rules are referred to obtain transformed speech. Although the final synthesized normal speech may be produced roughly, speech can directly be concatenated by several suitable parts of rules in the normal speech side using the location information on time domain in the rules.

6 Evaluation Experiments

All data in experiments were achieved through several speech processes as explained in 3.1. We applied 80 utterances of each speaker. The contents of input speech were 54 three-digit numbers and 26 simple sentence included 8 of *"WATASHIWA * WO SURU."*(I play * .) and 5 of *"SOKODE * WO MIRU."*(I watch a *.). Italic fonts express Japanese. The system was prepared with the same parameters throughout the experiments for both between esophageal or alaryngeal and normal speech to evaluate the generality of the system. The

Table 4. Conditions for experiments

Frame length of test vector	120msec
Frame rate of both vectors	60msec
Margin of time delay	±180ms(±120ms)
The rate of agreement for adopting rules	95%

Table 5. Comparison of correspondences of acquired rules

Speech data	Number of Data	Number of acquired rules	±180ms	±120ms
Alaryngeal-Normal	80	2,284	1,665(73.9%)	1,315(57,6%)
Esophageal-Normal	80	1,378	1,055(76.6%)	646(61.4%)

conditions shown in Table 4 were also adopted in these experiments. The rule dictionary had no rule or initial information at the beginning of learning.

We evaluated that the system could obtain a number of useful transform rules created by only the calculation of acoustic similarity between both esophageal and normal speech, and alaryngeal and normal speech. Any other criterion was adopted to limit to acquire transform rules throughout the experiments. When several common parts were found in the calculation result in comparing utterances, the one with the longest match was acquired as the transform rule. Moreover, location of parts on time-domain was also evaluated because this characteristic expressed the accuracy of correspondence of parts to those in another speech side. We allowed a margin for parts appearing in time domain for difference in elasticity of individual utterances. Two margins, ±180ms and ±120ms were applied because they corresponded with 1 and 1.5 mora in the Japanese speech rate, respectively. When corresponding parts between two speech sides in a rule appeared in appropriate location on time-domain with suitable length, the rule included these parts was regarded as a correct rule because the correspondences were able to be decided uniquely.

Table 5 shows a number of transform rules acquired by only acoustic similarity. The system could also obtain many rules that have appropriate correspondences without any limitation. Percentages in parentheses show the ratio of total acquired rules to appropriate rules. These rules imply that it is possible to acquire correspondences between both speech sides by only calculating of acoustic similarity.

7 Discussion

Many appropriate rules are obtained in both experiments through the same parameters. The results shows common and different parts appear approximately close location on time-domain independent of speech type. They also indicate that calculation of acoustic similarity is able to be a criterion to partition laryngectomee utterances although these are not clear and intelligible and are not able to be deal with in conventional speech recognition. So, these rules show promising possibilities for transform experiments. The number of appropriate rules from esophageal speech is lower than from alaryngeal speech. Noises accrued from injecting volumes of air into the esophagus are one of the major causes. The table also show corresponding parts were adequately acquired in close location on time-domain between normal and laryngectomee speech. However, the injecting volumes causes the other problem that esophageal utterances have a tendency to be longer than other types of speech. Therefore, longer margin should be needed to deal with esophageal speech.

In both experiments, the system is able to obtain more than 50% of suitable rules with our unique criterion, location of time-domain. This limitation is indispensable to manage the number of rules. We need other criteria to keep a small number of rules in the dictionary, for example, stricter limitation on time-domain, or checking length of extracted common or different parts, and so on.

We need to increase the number of speech utterances to obtain more suitable transform rules, and it is also necessary to consider the contents of utterances for more effective rule acquisition and application.

8 Conclusion and Future Works

In this paper, we have described the proposed method and have evaluated rule acquisition without being parameter tuning specific for esophageal and alaryngeal speech. We have confirmed that appropriate acoustic information is able to be extracted by calculation of acoustic similarity and that rules have been generated.

We will consider adopting DP matching method to decrease calculation cost because the method described in 3.2 needs a large amount of calculation.

We will have to implement transform experiments with a large amount of data, and confirm the synthesized speech in normal speech by listening.

References

[1] J. Müller and H. Stahl. 1999. *Speech understanding and speech transform by maximum a-posteriori semantic decoding. Proceedings of Artificial Intelligence in Engineering*, pages 373–384. 686

[2] Wen Ding and N. Higuchi. *A voice conversion method based on complex RBF network. Proceedings of the 1997 autumn meeting of ASJ(Japanese)*, pp.335–336. 1997. 687

[3] Oyton Turk and Levent M.Arslan. *Subband based Voice Conversation. Proceedings of ICSLP2002, pp.289–292.* 2002. 687

[4] K. Murakami, M. Hiroshige, K. Araki and K. Tochinai. 2002. *Evaluation of direct speech transform method using Inductive Learning for conversations in the travel domain. Proceedings of ACL-02 Workshop on Speech-to-Speech Translation*, pages .45–52. 687, 691, 693

[5] K. Murakami, K. Araki, M. Hiroshige, and K. Tochinai. 2003. *Evaluation of the rule acquisition on a direct speech translation method with waveforms using Inductive Learning for nouns and noun phrases. Proceedings of Pacific Association for Computational Linguistics(PACLING)'03*, pp.121–130. 687

[6] K. Araki and K. Tochinai. 2001. *Effectiveness of natural language processing method using inductive learning. Proceedings of Artificial Intelligence and Soft Computing(ASC)'01*, pages 295–300. 687, 694

[7] K. Matsui and E. Noguchi. *Enhancement of esophageal speech. proceedings the 1996 autumn meeting of the ASJ(Japanese)*, pp.423–424. 1996. 690

[8] J. Lu, Y. Doi, S. Nakamura and K. Shikano. *Acoustical Characteristics of Vowels of Esophageal Speech. Technical report of IEICE, SP96-126*, pp.233–240. 1997. 690, 691

[9] Carol Y. Espy-Wilson, Venkatesh R. Chari and Caroline B. Huang. *Enhancement of Alaryngeal Speech by Adaptive Filter. Proceedings of ICSLP '96*, pp764–767. 1996. 689

[10] A. Lee, T. Kawahara and K. Shikano. *Julius – a Open Source Real-Time Large Vocabulary Recognition Engine. Proceedings of EUROSPEECH '01*, pp.1691–1693. 2001. 690

[11] Y. Katoh. *Acoustic characteristics of speech in voice disorders. The 2000 spring meeting of the ASJ(Japanese)*, pp.309–310. 2002. 691

[12] Daniel Callan, Ray D. Kent, Nelson Roy and Stephen M. Tasko. *Self-organizing Map for the Classification of Normal and Disordered Female Voices. Journal of Speech, Language, and Hearing Research*, Vol.43, pp.355–366. 1999. 691

[13] H. F. Silverman and D. P. Morgan. 1990. *The application of dynamic programming to connected speech recognition. IEEE, ASSP Magazine*, pp.6–25. 691

Robustness for Evaluating Rule's Generalization Capability in Data Mining

Dianhui Wang, Tharam S. Dillon, and Xiaohang Ma

Department of Computer Science and Computer Engineering
La Trobe University, Melbourne, VIC 3083, Australia
Phone: +61-3-94793034, Fax: +61-3-94793060
http://myprofile.cos.com/dhwang

Abstract. The evaluation of production rules generated by different data mining algorithms currently depends upon the data set used, thus their generalization capability cannot be estimated. Our method consists of three steps. Firstly, we take a set of rules, copy these rules into a population of rules, and then perturb the parameters of individuals in this population. Secondly, the maximum robustness bounds for the rules is then found using genetic algorithms, where the performance of each individual is measured with respect to the training data. Finally, the relationship between maximum robustness bounds and generalization capability is constructed using statistical analysis for a large number of rules. The significance of this relationship is that it allows the algorithms that mine rules to be compared in terms of robustness bounds, independent of the test data. This technique is applied in a case study to a protein sequence classification problem.

1 Introduction

Construction of if-then rule classifiers can be viewed as a class of data mining tasks, which is concerned with the exploitation of information inherent in databases. This is often achieved by clustering data points that are close to one another according to some metric or criteria. In the past few years, a variety of techniques have been proposed to address the issue of classification rules miming. Amongst the techniques, decision tree and soft computing based rule extraction methods are representative [1,2,4]. There are various concerns with data mining techniques, for example, rule representation, feature selection, condensed rule extraction and optimization, knowledge insertion, rules adaptation, and rule quality evaluation [6,8]. While specific data mining algorithms extract different sets of rules from the same set of data, there currently exists no method for evaluating which rule set is better. Then, one will be interested in knowing which rule/rule-set is more suitable for performing the classification task. To answer this question, some criteria for rule/rule-set quality evaluation are needed [3].

T.D. Gedeon and L.C.C. Fung (Eds.): AI 2003, LNAI 2903, pp. 699-709, 2003.

Robustness can be defined as the ability of a system to maintain its performance when subjected to noise in external inputs, changes to internal structure and/or shifts in parameters. This important concept has been studied extensively in control engineering [7]. However, it has not received attention in data mining and intelligent systems. For rule-based classification systems, the robustness evaluation plays a significant role for judging the quality of the systems. Some benefits for building advanced classification systems can be obtained from this quantitative analysis of robustness, for example, the rule set may be further optimized by swapping rules extracted from different trials. As a core part of artificial intelligence, generalization capability of intelligent systems should be addressed in data mining techniques. Issues related to rule evaluation have been discussed and argued in the literature [3,5,6,8]. Most of them are concerned with rule size, classification accuracy, coverage rate, and predictive quality. The predictive quality is quantified by generalization accuracy, which is a popular measure for evaluating the goodness of learning systems. Usually, generalization accuracy is computed using a set of test data that is, theoretically, independent of the training data. Unfortunately, the measure obtained by this approach is not reliable due to the limitations of examples in test data set. It should be realized that the commonly used estimation techniques like cross-validation or bootstrap cannot provide direct information on generalization capability at rule or rule set level. As a matter of fact, these evaluation approaches are only applicable at algorithm level. Therefore, it is inappropriate to evaluate the generalization power (GP) of rules extracted by specified data mining or machine learning approaches based on the estimation techniques. It is believed that links exist between GP and robustness for some basic classification performance like classification rate and coverage rate. Thus, it will be very helpful and useful to measure the goodness of the rules generalization accuracy through robustness bounds, which is independent of the test data.

This paper aims to explore the relationships between the GP and the maximal robustness bounds associated with the misclassification rate for single rules and rule sets. This research mainly contributes the following aspects: (i) formulate the problem in a generic framework; (ii) develop a GA based searching algorithm for computing the robustness bound; and (iii) explore the relationships between robustness and generalization by a set of rules extracted by decision tree techniques for classifying protein sequences. The remainder of this paper is organized as follows: Section 2 gives the problem formulation with two remarks. Section 3 presents a GA based algorithm for computing the robustness bound. Section 4 carries out an empirical study on the relationships between GP and robustness using a protein sequence data set. We conclude this work in the last section.

2 Problem Formulation

Consider a set of production rules described by

$$\text{If}(A_1 \le Y_1 \le B_1) \wedge ..(A_m \le Y_m \le B_m) \text{ Then } X = [X_1, X_2, \cdots, X_n] \in C_r, m \le n, \quad (1)$$

where $\wedge$ is "and" logic operator, Y_j represents a subset of the feature X_j, A_j and B_j are the bounds of the rule, C_r represents the *r-th* class. The conditional terms in (1) can be simplified as the bounds take negative or positive infinity.

In this paper, our discussions will focus on the robustness analysis with respect to parameter perturbations. Without loss of generality, we assume that all of the bound parameters in (1) are finite, and denote the nominal parameter vector by $P_0 = [A_1, B_1, \cdots, A_m, B_m]^T$. The performance index vector for the nominal rule is denoted by $Q(P_0)$. We use $\|*\|$ to represent a vector norm.

Definition 1 (ε *-Robustness*):
For a given positive real number $\varepsilon > 0$, the classification rule (1) is said to be ε - *Robust* with respect to rule parameter perturbations, if there exists a positive real number $\delta > 0$ such that $\| Q(P) - Q(P_0) \| \leq \lambda_P \varepsilon$ holds for all admissible parameters P in the set $A_P = \{P : \| P - P_0 \| < \delta\}$, where $\lambda_P > 0$ is a real number related to the rule.

Definition 2 (ε *-Robustness Bound)*:
Let $\bar{\delta} = \sup\{\delta : \| P - P_0 \| < \delta, s.t. \| Q(P) - Q(P_0) \| \leq \lambda_P \varepsilon\}$. Then, $\bar{\delta}$ is called the ε - *Robustness Bound.*

Remark 1: The performance index vector can be comprised of several independent classification performance metrics, for example, coverage rate and misclassification rate. Also, the components of the performance index vector may be some form of combination of the individual classification performance metric.

Remark 2: The norms used in the parameter space and performance index space can be different from each other. Also, weighted norms will have more practical meanings. We here simply apply the standard 2-norm.

3 Determination of Robustness Bounds

The robustness property studied in this paper reflects the performance sensitivity of rules with respect to the internal structural parameters. Generally the classifier performance index is a discontinuous function of parameters, which characterize the classification rules. Therefore, robustness bounds computations are equivalent to solving an optimization problem in parameter space (see Definition 2). It is difficult to apply numerical optimization techniques to this problem. Genetic algorithms (GAs) can be used for this purpose. It is very time consuming to get the robustness bound because of the large search space. To reduce the computational burden, this paper proposes a strategy and it is summarized as follows.

Robustness Bound Searching Algorithm (RBSA)

Step 1. Set step indicator k=0, initial bound $\delta(k) = \delta_0 > 0$, initial bound step $\Delta(k) = \Delta_0 > 0$ and parameters for GA algorithm.

Step 2. k=k+1.

Step 3. Generate a set of populations (chromosomes) using unit vectors, that is,

$$P_i(k) = P_0 + \frac{\delta(k)V_i}{\| V_i \|}, i = 1,2,\ldots,q \,, \tag{2}$$

where V_i is a non-zero vector with the same dimension as P_0.

Step 4. Apply GA operations to maximize the objective function $E(P) = \| Q(P) - Q(P_0) \|$ subject to $\| P - P_0 \| = \delta(k)$ until the process meets some termination criteria.

Step 5. Calculate $E_{\max}(k) = \max\{\| Q(P_i(k)) - Q(P_0) \|, i = 1,2,\ldots,q\}$.

Step 6. If $E_{\max}(k) \le \varepsilon$, set $\Delta(k) = \Delta(k-1), \delta(k) = \delta(k-1) + \Delta(k)$;
Else $\delta(k) = \delta(k-1) - \Delta(k), \Delta(k) = 0.5 * \Delta(k-1)$.

Step 7. If $| \delta(k) - \delta(k-1) | < \sigma$ (a very small positive real number), Then Stop;
Else go to Step 2.

Note that in Step 6 the next step remains the same as the previous one in the forward search process, but it takes half of the previous step in the further search process. Mathematically, we can prove that this enables us to find the bound. To obtain an accurate bound, a sufficiently large population and running step for the GA algorithm are necessary. It has been realized that some advanced GA programming techniques with spatial constraints will be useful in speeding up the optimization process. The following example illustrates the effectiveness of our proposed RBSA algorithm. Also, it addresses an issue arising from the use of the GA optimization technique.

Example: Consider a discontinuous function $f(x,y)$ defined on $D = [0,3] \times [0,3]$ and three nominal points $P_1 = (1,2.8), P_2 = (1.5,1.5)$ *and* $P_3 = (2.1,0.8)$, respectively. Let $f(x,y) = -1$ if the points are out of *D*, and

$$f(x,y) = \begin{cases} 0.0, & if\,(x,y) \in R_{21} \cup R_{31} \cup R_{32} \\ 0.5, & if\,(x,y) \in R_{13} \cup R_{23} \cup R_{33} \cup R_{12} \cup R_{11} \\ 1.0, & if\,(x,y) \in R_{22} \end{cases}$$

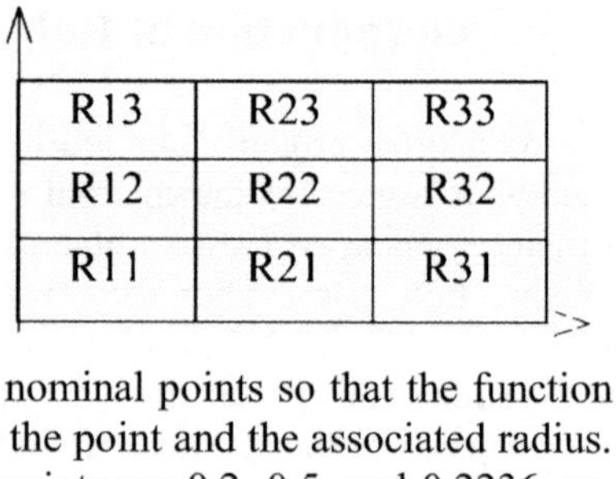

The task is to find the maximum radius for the nominal points so that the function keeps the same value within the circle specified by the point and the associated radius. It is easy to show that the ideal bounds for these points are 0.2, 0.5, and 0.2236, respectively.

These bounds can be obtained using our proposed algorithm with 500 populations. Figure 1 below shows a result obtained by the RBSA for P3. It demonstrates that a sufficiently large of population in GA programming for computing the robustness bounds is essential. For example, if we use 40 populations in our GA optimization calculation, the obtained bound for P3 is not reliable.

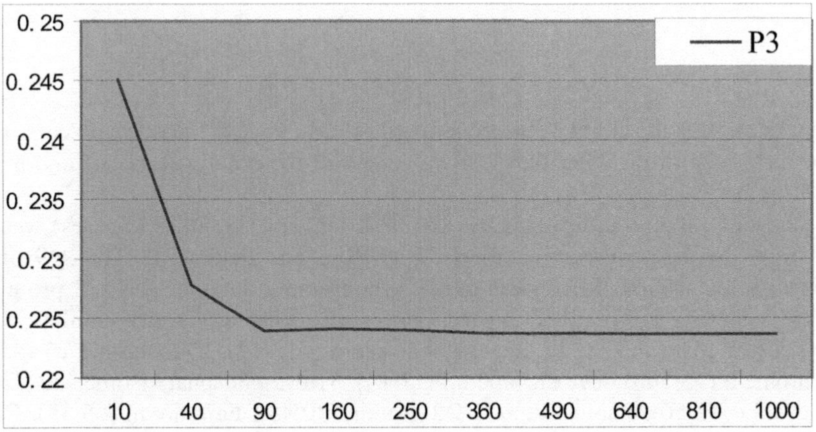

Fig. 1. Maximum robustness bounds vs. the number of populations

4 Robustness vs. Generalization: An Empirical Study

The protein sequences are transformed from DNA sequences using the predefined genome code. Protein sequences are more reliable than DNA sequence because of the redundancy of the genetic code. Two protein sequences are believed to be functionally and structurally related if they show similar sequence identity or homology. These conserved patterns are of interest for the protein classification task.

A protein sequence is made from combinations of variable length of 20 amino acids Σ = {A, C, D, E, F, G, H, I, K, L, M, N, P, Q, R, S, T, V, W, Y}. The n-grams or k-tuples [9] features will be extracted as an input vector of the classifier. The n-gram features are a pair of values (v_i, c_i), where v_i is the feature i and c_i is the counts of this feature in a protein sequence for i = 1… 20^n. In general, a feature is the number of occurrences of an animal in a protein sequence. These features are all the possible combinations of n letters from the set Σ. For example, the 2-gram (400 in total) features are (AA, AC,…,AY, CA, CC,…,CY,…,YA, …,YY). Consider a protein sequence VAAGTVAGT, the extracted 2-gram features are {(VA, 2), (AA, 1), (AG, 2), (GT, 2), (TV, 1)}. The 6-letter exchange group is another commonly used piece of information. The 6-letter group actually contains 6 combinations of the letters from the set Σ. These combinations are A={H,R,K}, B={D,E,N,Q}, C={C}, D={S,T,P,A,G}, E={M,I,L,V} and F={F,Y,W}. For example, the protein sequence VAAGTVAGT mentioned above will be transformed using 6-letter exchange group as EDDDDEDDD and their 2-gram features are {(DE, 1), (ED, 2), (DD, 5)}. We will use e_n and a_n to represent n-gram features from a 6-letter group and 20 letters set. Each sets of n-grams features, i.e., e_n and a_n, from a protein sequence will be scaled separately to avoid skew in the counts value using the equation below:

$$\bar{x} = \frac{x}{L-n+1}, \tag{3}$$

where x represents the count of the generic gram feature, $\bar{x}$ is the normalized x, which will be the inputs of the classifier; L is the length of the protein sequence and n is the size of n-gram features.

In this study, the protein sequences covering ten super-families (classes) were obtained from the PIR databases comprised by PIR1 and PIR2 [10]. The 949 protein sequences selected from PIR1 were used as the training data and the 533 protein sequences selected from PIR2 as the test data. The ten super-families to be trained/classified in this study are: Cytochrome c (113/17), Cytochrome c6 (45/14), Cytochrome b (73/100), Cytochrome b5 (11/14), Triose-phosphate isomerase (14/44), Plastocyanin (42/56), Photosystem II D2 protein (30/45), Ferredoxin (65/33), Globin (548/204), and Cytochrome b6-f complex 4.2K(8/6). The 56 features were extracted and comprised by e_2 and a_1.

Table 1. Maximum robustness bounds with respect to misclassification rate

RuleSet 1	Class	Bounds
Rule 4:	1	0.000314062
Rule 25:	1	0.00101563
Rule 11:	2	0.0005
Rule 35:	3	0.00134531
Rule 30:	3	0.000801563
Rule 31:	4	1.56E-06
Rule 8:	4	0.00170313
Rule 10:	5	1.56E-06
Rule 26:	6	0.00119063
Rule 22:	6	0.0001
Rule 20:	6	0.00308125
Rule 36:	7	0.000598438
Rule 13:	8	0.00120156
Rule 23:	8	0.00276406
Rule 17:	9	0.0193328
Rule 1:	9	0.00473437
Rule 29:	10	0.002

RuleSet 2	Class	Bounds
Rule 27:	1	0.00161094
Rule 60:	1	0.00148906
Rule 12:	2	0.00095156
Rule 13:	2	0.0009
Rule 56:	2	1.56E-06
Rule 28:	3	0.0028
Rule 4:	4	0.00060156
Rule 30:	5	0.00061875
Rule 80:	6	0.0010125
Rule 40:	6	0.00291094
Rule 11:	6	0.0002
Rule 6:	7	0.00020063
Rule 73:	8	0.00150156
Rule 33:	8	0.00289844
Rule 57:	9	0.0087625
Rule 42:	9	0.01
Rule 105:	9	0.0030125
Rule 45:	9	0.00030938
Rule 37:	10	0.0125109

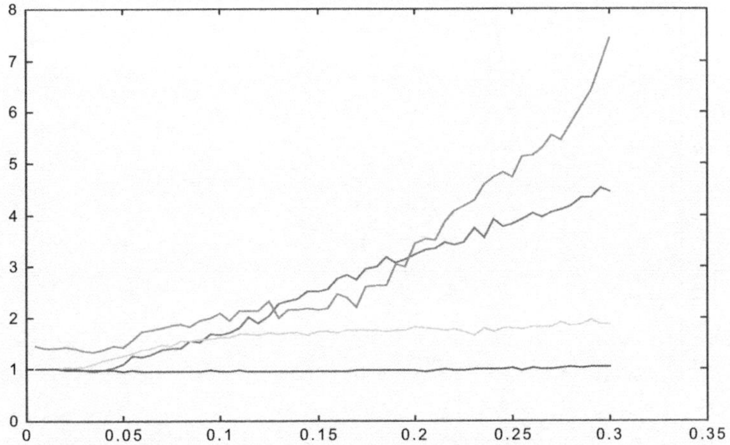

Fig. 2. Relative misclassification ratio vs. noise level for R1

Using C4.5 program and the given training data set, we produce two sets of production rules (see Appendix A), denoted by RuleSet1 (R1) and RuleSet2 (R2) respectively with a single trial and 10 trials correspondingly. In this work, we only consider the correlation between generalization accuracy and the maximum ε-robustness bound, where the performance index vector Q in Definition 1 takes a scalar value of misclassification rate. The relevant parameters in the RBSA algorithm are taken to be as follows: population number=5000, crossover rate=0.9, Mutation rate=0.01, generation step=10, termination criteria $\sigma = 0.000001$, and $\varepsilon = 0.05$, $\lambda_P = MS_P * CR_P$, and MS_P and CR_P are the local misclassification rate of Rule p and the number of total examples covered by Rule p, respectively. Table 1 gives the maximum ε-robustness bounds for each rule in R1 and R2. To measure the generalization power of each rule, we first perturbed for a 100 times the original whole data set comprising of the training and test data sets with different levels of random noise, and then computed the misclassification rate using the contaminated (unseen) data. The variation of the misclassification rate will keep changing at a slow pace with the change of the noise level if the rule has stronger generalization capability. Thus, we use the relative ratio of the misclassification rates calculated using the unseen data and the nominal misclassification rate to characterize the generalization power. Our empirical studies were carried out using the rules that classify one class alone from R1 and R2.

Figure 2 and 3 depict the ratio changes of the misclassification rate alone with the varying noise level. Refer to the maximum robustness bounds given in Table 1, the correlation between generalization power and robustness bounds can be examined. In Figure 2, the corresponding mappings between colors and rules are: blue line-R29, yellow line-R36, red line-R11, and green line-R10. In Figure 3, the corresponding mappings between colors and rules are: purple line-R37, deep-blue line-R28, green line-R30, red line-R4, and light-blue-R6. The following observations can be made:

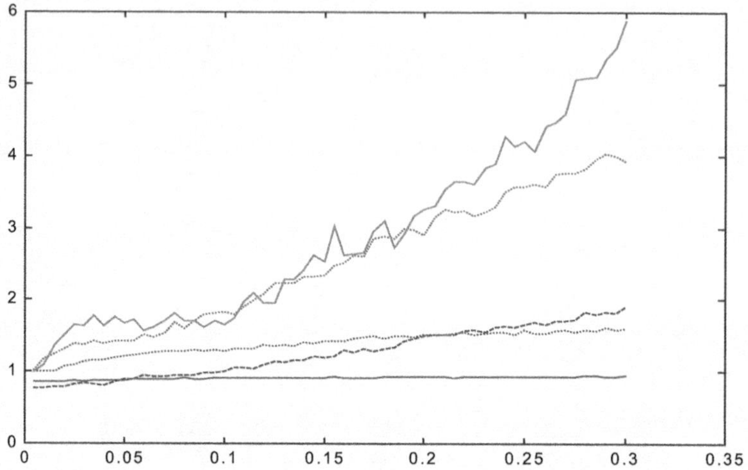

Fig. 3. Relative misclassification ratio vs. noise level for R2

- The proposed method for generalization estimation is dependent on the noise level;
- There exists a nonlinear relationship between the GP and the maximum robustness bound although the overall trend conforms to a general law, i.e., rules with larger robustness bounds will have stronger generalization power in terms of specific classification performance.

It is interesting to see that the relative misclassification ratio becomes less than one for some rules with larger robustness bounds as the noise level varies in a small interval. This implies that the characterization of generalization accuracy using finite examples is not reliable and true.

5 Conclusion

In practice, it is hard to characterize the value of classification rules that have very weak generalization power. Therefore, it is meaningful and significant to establish direct or indirect approaches for measuring this quantitatively. Previously, researchers have explored ways to estimate this important measure for learning systems, however more of them had some limitations due to the restricted amount of data used in their methods. The existing estimation techniques for evaluating rules generalization are only applicable at a higher level, that is, they compare rule sets extracted using different algorithms. To the best of the authors' knowledge, we have not seen papers that address this issue in a manner that is independent of the data set used.

This paper attempts a pioneering study on measuring the rules' generalization power through robustness analysis in data mining. Our primary results from this empirical study show that (i) in general, production rules for classification extracted by a specified data mining algorithm or machine learning technique will have stronger generalization capability, if they have larger robustness bounds; (ii) The relationship between the maximum robustness bound and the generalization accuracy is nonlinear, which may depend upon many other relevant factors, for example, the degree of noise in generalization estimation, rules nature and the distribution of data to be classified.

References

[1] Quinlan J.R, C4.5: Programs for Machine Learning, San Mateo, CA: Morgan Kaufmann, 1994.

[2] Dunham, M. H., Data Mining: Introductory and Advanced Topics, Prentice Hall, 2003.

[3] Kononenko and I. Bratko, Information based evaluation criterion for classifier's performance, Machine Learning, 6(1991) 67-80.

[4] S. R. Safavian and D. Landgrebe, A survey of decision tree classifier methodology, IEEE Transactions on Systems, Man, and Cybernetics, 21(1991) 660-674.

[5] M. E. Moret, Decision tree and diagrams, Computing Survey, 14(1982) 593-623.

[6] R. W. Selby and A. A. Porter, Learning from examples: generation and evaluation of decision trees for software resource analysis, IEEE Transactions on Software Engineering, 14(1988) 1743-1757.

[7] K. M. Zhou and J. C. Doyle, Essentials of Robust Control, Prentice Hall, 1997.

[8] S. Mitra, K. M. Konwar and S. K. Pal, "A fuzzy decision tree, linguistic rules and fuzzy knowledge-based network: generation and evaluation", IEEE Systems, Man and Cybernetics, Part C: Application and Reviews, 32(2002) 328-339.

[9] C. H. Wu, G. Whitson, J. McLarty, A. Ermongkonchai, T. C. Changc, PROCANS: Protein classification artificial neural system, Protein Science, (1992) 667-677.

[10] Protein Information Resources (PIR), http:// pir.Georgetown.edu.

Appendix A: Two Sets of Production Rules for Classifying Protein Sequence

RuleSet1 (R1)	RuleSet2 (R2)

Rule 35:
Att19 > 0.0274
Att50 > 0.0106
Att52 <= 0.0105
->class 3[97.9%]

Rule 30:
Att5 > 0.0391
Att8 > 0.0801
Att19 > 0.009
Att19 <= 0.027
->class 3[85.7%]

Rule 31:
Att19 > 0.0274
Att36 <= 0.0028
Att37 <= 0.0044
Att50 <= 0.0106
->class 4[84.1%]

Rule 8:
Att3 > 0.0625
Att5 <= 0.0391
Att41 <= 0.0106
Att55 > 0.0194
->class 4[63.0%]

Rule 36:
Att19 > 0.0274
Att52 > 0.0105
->class 7[95.3%]

Rule 13:
Att2 > 0.0259
Att5 <= 0.0391
Att6 <= 0.125
Att41 > 0.0106
->class 8[97.6%]

Rule 23:
Att6 <= 0.099
Att19 <= 0
Att28 > 0.0602
->class 8[96.0%]

Rule 4:
Att5 <= 0.0391
Att20 > 0.0245
Att33 > 0.0078
Att36 <= 0.01
Att41 <= 0.0106
Att53 <= 0.0029
Att55 <= 0.0194
->class 1[96.9%]

Rule 25:
Att5 > 0.0391
Att6 > 0.099
Att8 <= 0.0801
Att18 <= 0.0833
Att33 > 0.0085
->class 1[94.4%]

Rule 11:
Att2 <= 0.0259
Att5 <= 0.0391
Att36 <= 0.0147
Att41 > 0.0106
->class 2[92.9%]

Rule 10:
Att5 <= 0.0391
Att56 > 0
->class 5[73.1%]

Rule 29:
Att5 > 0.0391

Rule 27:
Att7 <= 0.0388
Att9 > 0.0421
Att18 <= 0.073
Att31 <= 0.0515
Att33 > 0.0078
Att41 <= 0.0106
->class 1[98.7%]

Rule 60:
Att2 > 0.0085
Att2 <= 0.0203
Att9 > 0.0962
Att10 <= 0.0826
->class 1[98.6%]

Rule 28:
Att9 > 0.0028
Att9 <= 0.0421
Att49 > 0.0795
->class 3[98.1%]

Rule 4:
Att5 <= 0.0407
Att33 <= 0.0077
Att41 <= 0.0106
Att54 <= 0
->class 4[88.2%]

Rule 30:
Att7 <= 0.0388
Att18 > 0.073
Att21 > 0.0109
Att33 > 0.0078
Att40 > 0.0714
Att41 <= 0.0106
->class 5[90.6%]

Rule 80:
Att5 > 0.0391
Att9 > 0.0351
Att10 <= 0.0833
Att18 > 0.0841
Att55 <= 0.0114
->class 6[94.6%]

Rule 40:
Att7 <= 0.0388
Att9 > 0.0421
Att33 > 0.0078
Att53 > 0.0029
->class 6[77.7%]

Rule 11:
Att2 > 0.0213
Att2 <= 0.0259
Att26 > 0.0027
Att41 > 0.0106
->class 6[63.0%]

Rule 6:
Att5 > 0.0407
Att9 <= 0.0351
Att19 > 0.0227
Att30 > 0.0486
->class 7[95.5%]

Rule 73:
Att2 > 0.0259
Att5 <= 0.0417
Att6 <= 0.125
Att41 > 0.0106
->class 8[97.8%]

Rule 33:
Att41 > 0.0276
->class 8[97.3%]

Att8 > 0.0801
Att19 <= 0.009
->class 10[61.2%]

Rule 26:
Att5 > 0.0391
Att6 > 0.099
Att8 <= 0.0801
Att18 > 0.0833
Att40 <= 0.0722
->class 6[87.9%]

Rule 22:
Att19 <= 0.0274
Att53 > 0.0072
->class 6[77.7%]

Rule 20:
Att5 > 0.0391
Att6 <= 0.099
Att37 <= 0.0029
Att42 > 0.1887
->class 6[61.2%]

Rule 17:
Att5 > 0.0391
Att6 <= 0.099
Att7 > 0.0134
Att8 <= 0.0801
Att11 <= 0.0438
Att19 <= 0.0274
Att28 <= 0.0602
Att42 <= 0.1887
Att53 <= 0.0072
->class 9[99.5%]

Rule 1:
Att19 <= 0.0274
Att33 <= 0.0078
Att36 <= 0.0028
Att55 <= 0.0194
->class 9[99.0%]

Rule 12:
Att1 > 0.1111
Att5 <= 0.0407
Att32 <= 0
Att43 > 0.0706
Att52 <= 0.0046
->class 2[89.0%]

Rule 13:
Att2 <= 0.0259
Att5 <= 0.0407
Att41 > 0.0106
->class 2[87.0%]

Rule 56:
Att1 > 0.1376
Att9 > 0.0421
Att10 > 0.0826
Att36 > 0.0077
->class 2[85.7%]

Rule 37:
Att9 <= 0.0028
Att49 > 0.0795
->class 10[73.1%]

Rule 57:
Att1 > 0.0687
Att9 > 0.0421
Att10 > 0.0826
Att13 <= 0.0617
Att36 <= 0.0077
->class 9[99.7%]

Rule 42:
Att5 > 0.0323
Att7 > 0.0388
Att9 > 0.0421
->class 9[99.7%]

Rule 105:
Att2 <= 0.0072
Att5 > 0.0391
Att9 > 0.0351
Att14 > 0.0135
->class 9[98.9%]

Rule 45:
Att2 <= 0.0268
Att5 > 0.0407
Att9 <= 0.0421
Att19 <= 0.0238
Att34 <= 0.0071
Att49 <= 0.0795
->class 9[89.9%]

Choosing Learning Algorithms Using Sign Tests with High Replicability

Remco R. Bouckaert

Xtal Mountain Information Technology and Computer Science Department
University of Waikato, New Zealand
rrb@xm.co.nz
remco@cs.waikato.ac.nz

Abstract. An important task in machine learning is determining which learning algorithm works best for a given data set. When the amount of data is small the same data needs to be used repeatedly in order to get a reasonable estimate of the accuracy of the learning algorithms. This results in violations of assumptions on which standard tests are based and makes it hard to design a good test.
In this article, we investigate sign tests to address the problem of choosing the best of two learning algorithms when only a small data set is available. Sign tests are conceptually simple and no assumption about underlying distributions is required. We show that simplistic sample generation can lead to flawed test outcomes. Furthermore, we identify a test that performs well based on Type I error (showing a difference between algorithms when there is none), power (showing a difference when it indeed exists) and replicability.
Replicability is a novel measure of a quality of a test that gives an indication how likely it is that the test outcome will be the same when the same test on the same data with the same sampling scheme and same pair of algorithms is executed, but with a different randomization of the data. A new definition of replicability is provided and its benefits highlighted. Empirical evidence is provided to show the test is robust under a varied range of circumstances.

1 Introduction

Choosing learning algorithms for classification tasks when a single data set is given is an area marked with unexpected pitfalls [3, 10]. As a result, many methods can indicate one algorithm outperforms another while the experimental data actually does not support such a claim. If the algorithms are closely related, standard techniques such as minimum description length, information criteria or Bayesian techniques [4] can be used for comparing the algorithms. However, if the algorithms are based on a range of varied techniques such as naive Bayes [5], C4.5 [8], or nearest neighbor [11], those techniques cannot be applied straightforwardly. A commonly used technique is to repeatedly split the data set in different training/test sets, train the algorithms on the traing set and use the classification accuracies of the test sets as input for a hypothesis test.

T.D. Gedeon and L.C.C. Fung (Eds.): AI 2003, LNAI 2903, pp. 710–722, 2003.

Inevitably, the various samples of accuracy thus obtained are somewhat dependent because they are partially based on the same data. Consequently, the thus far widely used t-test [6] based on resampling was found to have a high probability of resulting in a wrong recommendation [3]. Also the popular repeated cross validation test [11] suffers from this problem [2].

This article tries to draw out a small part of the map by studying sign tests for choosing between two learning algorithms when a small data set is given (question 8 in Dietterich's taxonomy of problems [3]). Sign tests are attractive because they are conceptually simple: when multiple samples agree on the outcome more than a threshold number of times, we decide to choose the preferred algorithm, otherwise we accept the null hypothesis that the algorithms perform equally well. Furthermore, sign tests require no assumptions about the underlying distribution, unlike t-tests. The only assumption underlying sign tests is that samples are independent. In the remainder of this paper we will see how easily this assumption can be violated.

Replicability of experiments is routinely addressed in areas like the social sciences and physics. However, in experimental computer science and machine learning in particular not a lot of attention is given to this issue. In a machine learning setting, it became recently clear that many popular tests give different outcomes when repeating the test on the same data, but with different randomizations [2]. Clearly, this is an undesirable property of a test, because a researcher or data analyst can be misled unintentionallys in thinking one method outperforms another, while the data may not support such claim except for the specific randomization that was used in the experiment. In this article, we consider replicability when judging the quality of a test.

This article is organized as follows. We start with an introduction to sign tests and explain how to apply such tests to learning algorithm selection. In Section 3, we discuss how to determine the quality of these tests, with special emphasis on replicability. Section 4 describes experimental evaluation under a range of circumstances. We conclude with some final remarks, recommendations and directions for further research.

2 Sign Tests for Selecting Learning Algorithms

The null hypothesis of tests for determining which of two learning algorithms performs best is that both algorithms performs the same. A sample to perform the test is typically obtained by performing an experiment where the two algorithms are repeatedly trained on one part of the data and tested on another part of the data. Based on this sample (details follow later in this section), the probability of the sample assuming the hypothesis is true can be calculated, the so called p-value. If the p-value is below the significance level, which is a predetermined threshold, we reject the null hypothesis and decide that one of the algorithms outperforms the other, otherwise we conclude that there is no support for such a claim.

The benefit of a sign test is that it does not make any assumption about the underlying distribution and is very well suited to data from matched pairs. Matched pairs naturally occur when comparing two algorithms A and B on n data sets giving a pair of accuracies $P_{A,i}$ and $P_{B,i}$ $(1 \leq i \leq n)$ on the test set. Using the sign of the difference $x_i = P_{A,i} - P_{B,i}$, we can count the number of plusses and minuses. When accuracies $P_{A,i}$ and $P_{B,i}$ are the same, which occurs quite often when two algorithms are performing very similarly, we count this as half a plus and half a minus. The number of plusses and minusses can be used to test the null hypothesis H_0 that both algorithms perform the same. If this is true, the probability of generating a plus or a minus is 0.5, in other words $H_0 : p = 0.5$.

Assuming the samples are independent and the null hypothesis is true, the number of plusses and minuses follows a binomial distribution [9]. The probability of observing k plusses in n comparisons is $P(k) = \binom{n}{k} p^k (1-p)^{n-k}$, which with $p = 0.5$ is $P(k) = \binom{n}{k} \frac{1}{2}^n$. We accept the null hypothesis as long as the probability that the number of plusses is less than k or larger than $n-k$ is more than α.

A potential drawback of sign tests is that they do not exploit the magnitute of the difference, just the sign of the difference, thus ignoring potentially valuable information. The Wilcoxon sign-ranked test may aleviate this problem, but is out of scope of this article.

Note that the one and only assumption made for performing a sign test is that the samples are drawn independently. Unfortunately, we do not have the luxury of getting independent samples because we are forced to reuse the data in generating samples due to the small size of the data set. There are various ways to obtain these samples. We describe the popular methods of resampling, k-fold cross validation, and various ways to benefit from repeated cross validation. These make obvious candidates to provide input for sign tests because they have been used extensively in t-tests before.

Resampling. Resampling is repeatedly splitting the data randomly r times in a training and a test set. The algorithms A and B learn on the training set and accuracies $P_{A,i}$ and $P_{B,i}$, $1 \leq i \leq r$ are obtained by classifying instances on the accompanying test set. The sign of the difference $x_i = P_{A,i} - P_{B,i}$ is used for the sign test with $n = r$ comparisons.

k-fold Cross Validation. Cross validation splits the data D in k approximate equal parts $D_1, \ldots, D_k$, and learn on the data $D \backslash D_i$, $1 \leq i \leq k$ with one part left out. The part D_i left out is used as test set. This gives k accuracy differences $x_i = P_{A,i} - P_{B,i}$ from which the sign is used for a sign test with $n = k$ comparisons.

Use All Data. Repeating k-fold cross validation r times and using the results is a way to increase replicability [2]. Such an experiment gives $r \times k$ differences x_{ij},

$1 \leq i \leq r$, $1 \leq j \leq k$. The signs can be used for a sign test with $n = r \times k$ comparisons.

Average over Folds. Averaging over folds in a repeated cross validation experiment, the recommended method for Weka [11], may be expected to give a better estimate of the difference of accuracies and thus a way to improve the Type I error. To perform a sign test, take the sign of the average difference $x_{i.} = \sum_{j=1}^{k} x_{ij}$ (where x_{ij} as for the use all data test) and apply a sign test with $n = r$ comparisons.

Average over Runs. Averaging over folds can be interpreted as an improved way of doing resampling. The natural extension is performing an improved way of k-fold cross validation, and instead of averaging over folds, average over runs. To perform a sign test, take the sign of the average difference $x_{.j} = \sum_{i=1}^{r} x_{ij}$ and apply the test with $n = k$ comparisons.

Average over Sorted Runs. Averaging over runs combines results from different runs rather arbitrary. One gets better estimates of a k-fold cross validation experiment by first sorting the results for the individual k-fold cross validation experiments and then taking the average. This way, the estimate for the minimum value is calculated from the minimum values in all folds, the one but lowest from the one but lowest results in all folds, etc. Let $x_{\theta(ij)}$ be the jth highest value of accuracy difference $x_{i'j'}$ of run i. Then, take the sign of the average difference $x_{.j} = \sum_{i=1}^{r} x_{\theta(ij)}$ and apply a sign test with $n = k$ comparisons.

Figure 1 illustrates the difference between the data used for the tests. The figure shows an example of 3x3 fold cross validation outcomes in the box at the left half (though in practice a 10x10 fold cross validation is more appropriate). All the data in the box in Figure 1 is used for the "use all data" test. For resampling, essentially only the first column is required when performing a 2/3-1/3 split of training and test data. Cross validation uses only the first run, that is, the first row of a 3x3 fold cross validation outcome. Averaging over folds and runs is essentially summing over columns and rows respectively. For getting sorted means, first the results have to be sorted over folds, giving the table at the right of Figure 1. Then the means are obtained by summing over rows.

3 Quality of a Test

There are various ways to judge the quality of a test. Firstly, the *Type I error* is the probability that a test rejects the null hypothesis while the hypothesis is true. If the Type I error is lower than the significance level, the test is called conservative. The *power* is the probability a test rejects the null hypothesis when the null hypothesis is indeed false. The more power, the better the test.[1]

[1] The Type II error (which explains Type I error's name) is the probability a test incorrectly does not reject the null hypothesis and it is related to the power through Type II error = 1 - power.

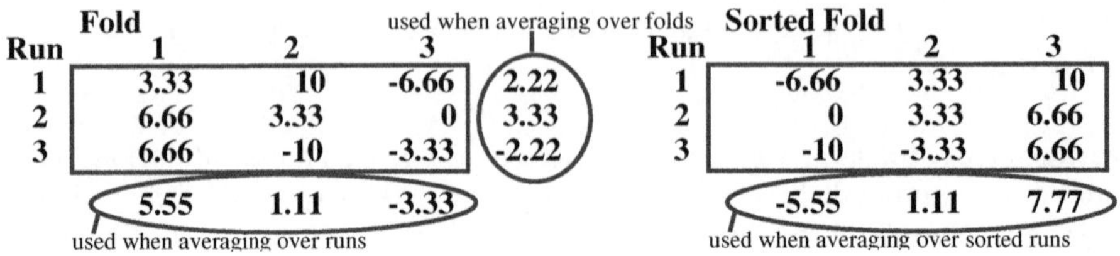

Fig. 1. Example illustrating the data used for the various tests. It shows the data involved in doing the various tests if the data of a 3x3 fold cross validation experiment were available. The 'use all data' test uses all points in the top rectangle

Typically, Type I error and power are related: the more power, the higher the Type I error. Ideally, the Type I error should coincide with the significance level and the power as high as possible.

Often, a test is only judged on its Type I error and power, but when a test is based on repeatedly using the same small data set the issue of replicability arises [2]. Replicability is the ability of a test to produce the same outcome on the same data with the same pair of algorithms but with a different randomization of the data. Obviously, this is an important property since it guarantees that an experiment performed by a researcher or data analyst can be repeated by others and that the outcome is unambiguous.

In [2], a rather ad hoc measure of replicability is proposed that demonstrates that replicability is indeed an issue. The measure proposed is obtained by counting how often a test gives the same outcome ten times when performing the test with ten different randomizations of a data set. By generating 1000 data sets from a single data source and counting the proportion of data sets for which the test produces the same outcome in 10 different runs, an indication of replicability for a particular data source is obtained. This procedure can be repeated for different data sources. Minimal replicability tend to occur for those data sources the hypothesis test has difficulty in deciding which outcome to prefer. This is exactly the area where machine learning researchers typically work and where data analysts make their most difficult decisions. So, the minimum replicability over various data sources is determined to be the overall replicability of a test in [2].

The problem with the proposed definition is that it depends on the number of times a test is performed on a data set. If the number of tests is increased from 10 to say 100, the probability that at least one test outcome differs from the others increases and consequently the replicability decreases. So, replicability cannot be mutually compared when different numbers of tests are performed on a data set. We propose a new definition of replicability that does not depend on the number of times a test is performed.

Definition: *Replicability* of a hypothesis test is the probability two runs of the test on the same data set, with the same pair of algorithms and the same method of sampling the data produce the same outcome minimized over data sets.

To distinguish between this definition and the one in [2], we will refer to the latter as *10x consistency.* The outcome of a hypothesis test is two valued (reject or not reject). Therefore, the probability two tests produce the same outcome is at worst 50%. So, replicability is valued between 50% and 100%, unlike 10x consistency which lies between 0% and 100%. In fact, our definition can be interpreted as an upper bound of 10x consistency.

A naive approach to estimating the replicability would be to generate pairs of randomization of a data set and run the test to obtain samples with which to estimate replicability. However, assuming randomizations of a data set are uniformly distributed we can perform a test n times and take every pair of outcomes as a sample. There are $\binom{n}{2}$ such pairs. Suppose k $(0 \leq k \leq n)$ tests reject the null hypothesis and the remaining $n-k$ do not. Then $\binom{k}{2}$ pairs of rejecting pairs and $\binom{n-k}{2}$ pairs of non rejecting pairs can be formed. This gives an estimate of replicability as $R(k,n) = (\binom{k}{2} + \binom{n-k}{2})/\binom{n}{2} = \frac{k(k-1)+(n-k)(n-k-1)}{n(n-1)}$. By drawing m data sets from a data source and performing the test n times on each data source, we can estimate replicability for the data source as follows. Let i_k $(0 \leq k \leq n)$ be the number of times the test agrees k times (so $\sum_{k=0}^{n} i_k = m$). Then, replicability is $R = \sum_{k=0}^{n} \frac{i_k}{m} R(k,n)$. By comparing replicability for various data sources a better indication of replicability of a test can be obtained.

4 Simulation Study

We performed some experiments to study the behavior of the sign tests by measuring Type I error, power and replicability under different circumstances. A random data source was created over 10 binary variables with a class probability of 50% and all attributes independent. From this data source, 1000 data sets of 300 instances were drawn and one test set of 10.000 instances. Training naive Bayes [5] and C4.5 [8] as implemented in Weka version 3.3[2] [11] on the 1000 data sets and averaging the accuracy measured on the test set showed that both algorithms have a 50% accuracy, as expected. Another three data sources were created by randomly generating Bayes networks over 10 binary variables and generating 1000 data sets and one test set similarly. Table 1 shows the average accuracies of those data sets. For sets 2, 3 and 4, C4.5 outperforms naive Bayes with an increasing margin. The results for Set 1 are used to measure Type I error because it is a data source for which both algorithm perform exactly the same on average, guaranteeing that the null hypothesis is true. For the other three sets we know that the null hypothesis is false and should be rejected. The proportion of datasets for which this happens is a measure of the power of a test.

The input for the sign tests was obtained by training naive Bayes and C4.5 on each of the 4000 data sets using 10 fold cross validation repeated 1000 times.

[2] Available from http://www.cs.waikato.ac.nz/ml.

Table 1. Average accuracies (in percentage) on test data by training on the 1000 data sets and measuring on the test set

Algorithm	Set 1	Set 2	Set 3	Set 4
Naive Bayes:	50.0	87.84	71.92	81.96
C4.5:	50.0	90.61	77.74	93.23
Difference	0.0	2.77	5.83	11.27

For each of the tests, a subset of these 1000x10 accuracies was used. For example, for a 10x10 use all data test, only the first 10 runs of 10 fold cv was used. Default options for naive Bayes and C4.5 in Weka were used. Table 2 shows the results for ten trials of each test on a data set for the 1000 data sets at 5% significance level. In each test, naive Bayes or C4.5 can be preferred or alternatively the null hypothesis that both perform equal can be accepted. So, a total of 10.000 wins can be reached maximally for an algorithm.

The last two columns of Table 2 shows replicability for the tests. The first is 10x consistency as defined in [2], which is calculated by determining the proportion of data sets for which the outcome of a test is the same with 10 different randomizations. The minimum of the four data sources is shown. The last column is replicability as defined in Section 3. Note that the same ranking of tests is obtained by both forms of replicability.

Some observations on Table 2:

- The Type I error for resampling is just over 15% at the 5% significance level. This elevated Type I error is considerably lower than the over 50% reported on resampled t-test [3], but still higher than the expected 5%. The power is reasonable, but the replicability is lowest of all test considered, 65.2% (or even 16% more dramatically highlighted through 10x consistency), which is realized on Set 3. This means that the probability of getting the same outcome twice is only increased from 50% to 65%. Alternatively, this means that for 84% of the data sets in Set 3 this test can provide different outcomes

Table 2. Type I error (for Set 1), power (for Set 2, 3 and 4) and replicability of various sign tests at 5% significance level (all numbers in percentages)

test	Set 1 Type I	Set 2 Power	Set 3 Power	Set 4 Power	10x consistency	Replicability
Resampling	15.23	25.36	45.45	93.77	16.0	65.2
k-fold cv	11.30	29.79	43.31	95.64	28.8	71.9
Use all data	48.35	68.49	86.64	100.00	43.1	78.3
Average over folds	61.80	78.86	90.22	100.00	34.7	75.3
Average over runs	53.74	69.07	86.28	100.00	32.6	74.2
Average over sorted runs	5.00	21.24	48.61	99.05	66.7	87.6

depending on the particular randomization of the data set when running the test twice.
- The k-fold cv test performs slightly better than the resampling test. The Type I error is elevated at just over 11% due to the dependence between samples because they are based on the same data. This increase is comparable to the elevated Type I error for t-test where typical values are 10% [3, 7]. The power is quite reasonable, however the replicability (see last column of Table 2) is rather low at almost 72%, which is considerably worse than for tests based on repeated cross validation.
- The use all data test has an unacceptable high Type I error due to the independence assumption underlying the sign test being violated. The dependence between the various outcomes of a k-fold cv experiment is already high, but increases when performing a repeated cross validation experiment.
- Also the Type I error for the average over folds and average over runs is unacceptable high at over 50%. The reason is that the signs are based on means over a number of cases and due to the dependence between the various cases, these means will have the same sign very often.
- The Type I error for the average over sorted runs is exactly on the mark at 5.00% and the power is quite acceptable as well. Sorting the folds before averaging effectively breaks the dependence between the various cases. Also notable is that the replicability is highest of all the tests considered. The ranking of tests is in fact the same according to the two definitions of replicability, however our definition makes it easier to compare replicability with other experiments.

Table 2 shows that the Type I error for some tests is so bad at the 5% significance level that we looked into trying to repair the test by changing the threshold value artificially by rejecting the null hypothesis only if all signs are the same. However, even then the Type I error for average over folds and average over runs was unacceptable high (over 15%) while the use all data test lost its power.

The overall conclusion is that judged on Type I error, power and replicability, only the average over sorted runs test shows any sign of promise and the others will not be considered further.[3]

4.1 Varying the Number of Runs

Another experiment was set up to measure the impact of the number of runs and Table 3 shows the result for 5, 10, 15, 20, 25, 30, 40, 60 and 100 runs. Note, only the outcomes of the sorted runs test is shown. Type I error for set 1 and power for sets 2, 3 and 4 are listed and only one test per data set counts, so the numbers are based on a sample of 1000. The last two rows list 10x consistency and replicability.

[3] Actually, the numbers were calculated for the tests for following tables, but it only served to confirm the flaws just mentioned.

Table 3. Type I error (for Set 1), power (for Set 2, 3 and 4) and replicability for average over sorted runs test with 10 folds at 5% significance level for various numbers of runs (in percentages)

test	5	10	15	20	25	30	40	60	100
Set 1	5.4	5.1	4.7	4.5	4.9	4.4	4.6	4.9	4.6
Set 2	20.4	21.7	22.2	22.1	22.0	22.7	22.6	23.1	23.3
Set 3	48.0	49.7	49.8	50.0	49.9	49.7	49.4	49.6	50.6
Set 4	98.4	99.0	99.4	99.2	99.4	99.6	99.4	99.3	99.4
10x consistency	54.1	66.7	72.5	74.9	79.3	80.1	82.2	85.3	89.4
Replicability	83.2	87.6	90.1	91.2	92.3	93.1	93.7	94.6	96.3

Table 3 suggests that there is a slight improvement of Type I error and power with increasing numbers of runs. More interestingly, the replicability increases to a point where it exceeds the highest known replicability of the acceptable t-tests [2] according to 10x consistency. The 10x consistency of almost 90% for 100 runs comes at the cost of extra computational effort. It is also comforting to realize that with 100 runs, the probability of getting the same outcome using this test is over 96%.

4.2 Varying Class Probability

Data sources similar to the one used for Set 1 were created with class probabilities ranging from 10% to 50% and 1000 data sets generated with each of them (Set 1 was used for the 50% data source). The Type I error measured based on a sample of 1000 was 0.0, 0.0, 1.3, 3.8 and 5.1% for class probabilities of 10, 20, 30, 40 and 50% respectively. Note that the Type I error decreases with decreasing class probability because the classifiers tend to become more similar and predict the majority class only. The Type I error is very acceptable for the desired significance level.

4.3 Varying Significance Level

In the experiments, the significance level of the test was varied ranging from 1 to 11 percent for the sorted runs test only and results are shown in Table 4. For Set 1, the Type I error is listed and for Set 2, 3 and 4 the power is listed. Only the first of the ten tests on each data set counts in this experiment, so the Type I error/power is based on a sample of 1000 (which explains the slight difference in Type I error at 5% in this table and Table 2).

Table 4 shows that the Type I error is acceptably close to the desired level for 1%, 5% and 10% levels, but jumps to almost 25% where 11% is desired (increasing the number of runs shows the same effect shown in a separate experiment). Note that there is a threshold effect: for ten folds, an algorithm is recommended if all 10 signs are the same at $\alpha = 0.1\%$, if 9 or more signs are the same at

Table 4. Type I error (for Set 1) and power (for Set 2, 3 and 4) for average over sorted runs test with 10 runs and 10 folds at various significance levels (in percentages)

test	1%	5%	10%	11%
Set 1	0.5	5.1	5.6	24.3
Set 2	7.2	21.7	26.7	48.2
Set 3	19.9	49.7	52.9	75.6
Set 4	92.6	99.0	99.3	100.0

$\alpha = 1.1\%$, and 8 or more at $\alpha = 5.5\%$. So when putting the significance level at say 2.5% the number of signs at which an algorithm is preferred over the other is the same as at a 5% significance level for 10 runs. This threshold effect is inevitable when using the small sample of 10 folds.

4.4 Varying Number of Folds

Table 5 lists the results when varying the number of folds based on a sample of 1000. This table highlights the aforementioned threshold effect and shows that one has to be careful in selecting an appropriate number of folds.

4.5 Varying Algorithms and Variable Cardinality

To see the behavior of the sign test with other algorithms, we trained nearest neighbor, tree augmented naive Bayes and support vector as implemented in Weka [11] with default parameters. To get an impression of how well non-binary data is handled, two new data sources were used: one with ten ternary variables and one with ten four-valued variables with all variables independent and uniformly distributed. Using the outcomes of 10 times repeated 10 fold cross validation, pair-wise sign tests were performed at 5% significance level. Table 6 shows the Type I error for pair-wise comparisons of algorithms (based on a sample of 1000). Weka's support vector implementation works on binary variables only, so the outcome for ternary and four valued variables is omitted.

Table 5. Type I error (for Set 1) and power (for Set 2, 3 and 4) for average over sorted runs test with 10 runs at 5% significance level for various numbers of folds (in percentages)

test	5	10	20	30	40	50
Set 1	0.4	5.1	9.4	11.2	9.8	13.8
Set 2	9.4	21.7	22.6	20.1	14.5	14.7
Set 3	25.3	49.7	48.7	46.5	43.8	44.5
Set 4	97.0	99.0	98.7	97.6	96.6	96.6

Table 6. Type I error (in percentage) for average over sorted runs test with 10 runs and 10 folds on a binary (Set 1), ternary and four valued data for various algorithms (nb = naive Bayes, nn=nearest neighbor, tan = tree augmented naive Bayes, sv = support vector)

	Binary				Ternary			Four valued		
	C4.5	nb	nn	tan	C4.5	nb	nn	C4.5	nb	nn
nb	5.1				3.9			4.1		
nn	2.7	4.9			2.3	5.3		2.4	4.7	
tan	5.4	2.1	3.7		3.8	0.2	5.6	4.0	0.0	4.8
sv	1.6	0.6	0.8	0.7						

Remarkably, all Type I errors do not exceed the 5% significance level substantially, suggesting that this is a reasonable conservative test for comparing pairs of algorithms. Also, tests with closely linked underlying mechanisms, like naive Bayes and tree augmented naive Bayes, show the dependency of algorithms in the very low Type I error.

4.6 Replicability on Real Datasets

In order to see how the test behaves on real world data sets, we performed an experiment running naive Bayes (NB), C4.5 and nearest neighbor (NN) using 27 UCI [1] datasets[4]. Though this does not allow us to measure Type I error or power, since it is not known whether the null hypothesis is true or not, it makes it possible to get an indication of replicability of a hypothesis test. Ten times running a 10x10 sorted runs test as before, 20 instances returned consistent outcomes in all ten runs for NB against C4.5 and against NN. Comparing C4.5 against NN even gives consistent outcomes for 24 datasets. Calculating replicability by pooling the outcomes of the runs gives 89% for NB vs C4.5, 92% for NB vs NN and 95% for C4.5 vs NN. So, the sorted runs test has a rather acceptable level of replicability on real live dataset as well.

5 Conclusions

We demonstrated how to successfully apply a sign test to selecting between two learning algorithms when a small data set is available. The benefits of sign tests are that they are conceptually simple and based on very few assumptions. We investigated various methods for generating samples from a single small data set. Most of these methods are well known for generating samples used in t-tests, but surprisingly all these methods turned out to have serious flaws. Only one test

[4] Namely anneal, arrhythmia, audiology, autos, balance-scale, breast-cancer, credit-rating, ecoli, German credit, glass, heart-statlog, hepatitis, horse-colic, Hungarian heart disease, ionosphere, iris, labor, lymphography, pima-diabetes, primary-tumor, sonar, soybean, vehicle, vote, vowel, Wisconsin breast cancer and zoo.

among the investigated tests exhibited an acceptable Type I error, reasonable power and a good level of replicability .

This test is based on a repeated 10 fold cross-validation experiment where the folds are sorted and the accuracy of each fold is averaged. Then the signs of the 10 averaged accuracies can be used for a standard sign test. Increasing the number of runs improves replicability, which can reach over 96% at 100 runs. This compares favorably with results for t-tests [2]. Though replicability is already reasonably good at 10 runs, the number of runs should only be limited by computational considerations in order to guarantee replicability. Changing the number of folds is not recommened due to threshold effects inherent to the sign test. Empirical evidence suggests this test performs robustly under a varied range of circumstances.

Another contribution of this paper is a refined definition of replicability. Its benefits are highlighted and it is demonstrated that various tests are undesirable due to their low replicability.

There are two obvious areas in which to extend the tests presented here. One is in selecting the best out of multiple algorithms instead of selecting the best of just two algorithms. Another one is in selecting the best of two algorithms when data sets from multiple domains are available since often machine learning algorithms are compared on benchmark data sets like from the UCI repository. Combined, those two problems form the final goal of this exploration. All these questions pose various multiple comparison problems; the probability that a pair-wise comparison indicates a preference for an algorithm, while in reality there is none, increases when comparing multiple algorithms and multiple data sets. Given the issues highlighted in this paper for the simple case of just two algorithms, it is worth studying those problems in greater detail.

Acknowledgements

I thank Eibe Frank for his stimulating discussions on this topic.

References

[1] C. L. Blake and C. J. Merz. UCI Repository of machine learning databases. Irvine, CA: University of California, 1998. 720
[2] R. R. Bouckaert. Choosing between two learning algorithms based on calibrated tests. International Conference on Machine Learning, 51–58, 2003. 711, 712, 714, 715, 716, 718, 721
[3] T. G. Dietterich. Approximate Statistical Tests for Comparing Supervised Classification Learning Algorithms. Neural Computation, 10(7) 1895–1924, 1998. 710, 711, 716, 717
[4] T. Hastie, R. Tibshirani and J. Friedman. The elements of statistical learning. Springer Series in Statistics. Springer-Verlag, New York, 2001. 710
[5] G. H. John and Pat Langley. Estimating Continuous Distributions in Bayesian Classifiers. Uncertainty in Artificial Intelligence, 338–345, 1995. 710, 715
[6] T. Mitchell. Machine Learning. McGraw Hil, 1997. 711

[7] C. Nadeau and Y. Bengio. Inference for the generalization error. Advances in Neural Information Processing Systems 12, MIT Press, 2000. 717

[8] R. Quinlan. C4.5: Programs for Machine Learning. Morgan Kaufmann Publishers, San Mateo, CA, 1993. 710, 715

[9] C. J. Wild and G. A. F. Weber. Introduction to probability and statistics. Department of Statistics, University of Auckland, New Zealand, 1995. 712

[10] S. Salzberg. On Comparing Classifiers: Pitfalls to Avoid and a Recommended Approach. Data Mining and Knowledge Discovery 1:3, 317-327, 1997. 710

[11] I. H. Witten and E. Frank. Data mining: Practical machine learning tools and techniques with Java implementations. Morgan Kaufmann, San Francisco, 2000. 710, 711, 713, 715, 719

Evaluating a Nearest-Neighbor Method to Substitute Continuous Missing Values

Eduardo R. Hruschka[1], Estevam R. Hruschka Jr.[2], and Nelson F. F. Ebecken[3]

[1] Universidade Católica de Santos (Unisantos)
Rua Carvalho de Mendonça, nº 144, CEP 11.070-906, Santos, SP, Brasil
erh@unisantos.br

[2] COPPE / Universidade Federal do Rio de Janeiro
Bloco B, Sala 100, Caixa Postal 68506, CEP 21945-970, Rio de Janeiro, RJ, Brasil
estevamr@terra.com.br

[3] COPPE / Universidade Federal do Rio de Janeiro
Bloco B, Sala 100, Caixa Postal 68506, CEP 21945-970, Rio de Janeiro, RJ, Brasil
nelson@ntt.ufrj.br

Abstract. This work proposes and evaluates a Nearest-Neighbor Method to substitute missing values in datasets formed by continuous attributes. In the substitution process, each instance containing missing values is compared with complete instances, and the closest instance is used to assign the attribute missing value. We evaluate this method in simulations performed in four datasets that are usually employed as benchmarks for data mining methods - Iris Plants, Wisconsin Breast Cancer, Pima Indians Diabetes and Wine Recognition. First, we consider the substitution process as a prediction task. In this sense, we employ two metrics (Euclidean and Manhattan) to simulate substitutions both in original and normalized datasets. The obtained results were compared to those provided by a usually employed method to perform this task, i.e. substitution by the mean value. Based on these simulations, we propose a substitution procedure for the well-known K-Means Clustering Algorithm. Then, we perform clustering simulations, comparing the results obtained in the original datasets with the substituted ones. These results indicate that the proposed method is a suitable estimator for substituting missing values, i.e. it preserves the relationships between variables in the clustering process. Therefore, the proposed Nearest-Neighbor Method is an appropriate data preparation tool for the K-Means Clustering Algorithm.

1 Introduction

Knowledge discovery in databases (KDD) is the non-trivial process of identifying valid, novel, potentially useful, and ultimately understandable patterns in data [1]. In this context, data mining is a step in this process that centers on the automated discov-

T.D. Gedeon and L.C.C. Fung (Eds.): AI 2003, LNAI 2903, pp. 723-734, 2003.

ery of new facts and relationships in data and it consists of three basic steps: data preparation, information discovery and analysis of the mining algorithm output [2]. The data preparation step has a major importance in the whole process and it is used as a tool to adjust the databases to the information discovery step. Thus, when it is performed in a suitable way higher quality data are produced, and the KDD outcomes can be improved. In spite of its importance, the data preparation process became an effervescent research area only in the last few years.

The substitution of missing values is an important subtask in the data preparation step. The absence of values in a dataset is a common fact in real world applications and, further than, it may generate bias in the data, affecting the quality of the KDD process. One of the most used methods to deal with the missing values problem is the mean or mode imputation [3], but this method can bring bias in the data and is not adequate in all situations. This work shows a missing value substitution method using an algorithm based on the instance-based learning method [4]. More specifically, we propose and evaluate a missing values substitution method, based on the nearest-neighbor approach, for the well-known K-Means Clustering Algorithm.

The next section presents works related to the missing values problem, whereas Section 3 presents our proposed substitution method. This method is evaluated in simulations performed in four datasets that are benchmarks for data mining methods - Iris Plants, Wisconsin Breast Cancer, Pima Indians Diabetes and Wine Recognition. The prediction simulations, described in Section 4, allow us to compare the performance of two different distance metrics – Euclidean and Manhattan. Besides, we also evaluate the influence of normalized datasets in the substitution process and all these results helped us to parameterize the Nearest-Neighbor Method, which is then evaluated as a data preparation tool for the K-Means Algorithm - Section 5. Finally, Section 6 describes the conclusions and points out some future works.

2 Related Work

The missing values problem is an important issue in data mining. Thereby there are many approaches to deal with it [5]: i) Ignore objects containing missing values; ii) Fill the gaps manually; iii) Substitute the missing values by a constant; iv) Use the mean of the objects in the same class as a substitution value; and v) Get the most probable value to fill the missing values. The first approach usually wastes too much data, whereas the second one is unfeasible in data mining tasks. The third approach assumes that all missing values represent the same value, probably leading to considerable distortions. The substitution by the mean value is common and sometimes can even lead to reasonable results. However, we believe that the best approach involves trying to fill missing values with the suitable, most probable ones.

The literature that deals with the problem of missing values describes several works. For example, when working with decision trees, some practical results can be found in [6], which just ignore the objects with missing values. Another approach involves replacing the missing values by the most frequent value [7] - *majority method.* In the *probability method* [8], a decision tree is constructed to determine the missing values of each attribute, using the information contained in other attributes and ignoring the class. The *dynamic path generation* [6] and the *Lazy decision tree approach* [9] do not generate the whole tree, but only the most promising path.

Some works about missing value classification tasks, using committee learning approach, are *Boosting* [10], *Bagging* [11], *Sasc* (Stochastic attribute selection committee) [12] and *SascMB* (Stochastic attribute selection committee with Multiple Boosting) [13], which applies decision trees to the learning task.

Considering Bayesian methods and supposing a missing at random data [16], a way to find the posterior distribution of the variable joint probabilities and the marginal probability of the variable having missing values is treating the missing values as unknown parameters, applying a Monte Carlo Markov Chain method [17]. When the missing data mechanism is not ignorable [15], an *imputation-based* analysis can be used [16]. However, these methods have some disadvantages [14]. First, they need information about the missing values mechanism. Second, the sampling variability and the non-response variability are mixed. Third, they have a high computational cost. Some approaches that try to solve these problems can be found in [18, 19].

In multivariate analysis, some works apply the Multiple Imputation (MI) [20] method to handle missing data. MI methods provide good estimations of the sample standard errors, and several analyses can be applied. However, the data must be missing at random in order to generate a general-purpose imputation.

The EM (Expectation-Maximization) algorithm [21] can be applied when the model belongs to an exponential family, but it has a slow convergence rate. For example, the MS-EM (Model Selection – Expectation Maximization) [22], which plays relatively few iterations to find the best network with incomplete data, implements a version of the EM and uses a metric to choose the best Bayesian model. Other methods applying the EM algorithm can be seen in [15, 23].

Instance-based (IB) learning methods [4] are part of another class of algorithms that can be applied to the missing values substitution process. There are some classical IB learning algorithm classes as the *nearest neighbor* (k-NN) [3], the *locally weighted linear regression* [24], and the *case based reasoning* (CBR) [25]. One of the most important characteristics of these methods is that they do not generate a model to describe the data. In other words, they do not have the training step as the other learning methods do. Thus, instead of consulting a generated model to estimate the best value to substitute the missing one (for each substitution), these algorithms search the whole dataset to find the best instance to be used. This characteristic produces a high computational cost when working with many attributes. On the other hand, as the learning process is specific to each query, it may be more accurate.

Few works deal with the problem of substituting missing values in clustering problems. This paper proposes and evaluates a Nearest Neighbor Method to substitute missing values - similar to that employed in [3,31,32] for algorithms C4.5 and CN2 – to be employed as a data preparation tool for the K-Means Algorithm.

3 Nearest-Neighbor Method

The proposed substitution method considers that missing values can be substituted by the corresponding attribute value of the most similar complete instance (object) in the dataset. In other words, we employ a K-nearest-neighbor method [4], using K=1 and two different distance functions (e.g. Euclidean and Manhattan). More specifically, let us consider two objects *i* and *j*, both described by a set of N continuous attributes $\{x_1,x_2,...,x_N\}$. The distance between objects *i* and *j* will be here called d(i,j). Besides, let us suppose that the *k*-th attribute value ($1 \leq k \leq N$) of the object *m* is missing. Thus, the Nearest-Neighbor Method (NNM) will compute the distances d(m,i), for all $i \neq m$, according to the Euclidean or the Manhattan distance, i.e. respectively:

$$d(m,i)_E = \sqrt{(x_1^m - x_1^i)^2 + ... + (x_{k-1}^m - x_{k-1}^i)^2 + (x_{k+1}^m - x_{k+1}^i)^2 + ... + (x_N^m - x_N^i)^2}\ . \quad (1)$$

$$d(m,i)_M = \left|x_1^m - x_1^i\right| + ... + \left|x_{k-1}^m - x_{k-1}^i\right| + \left|x_{k+1}^m - x_{k+1}^i\right| + ... + \left|x_N^m - x_N^i\right|. \quad (2)$$

One observes that we are not taking into account the attribute x_k, because it is missing. After computing all distances, we choose the smallest one, which refers to the most similar object in respect to *m*. This object is here called *s,* which is the nearest neighbor. In this way, one observes that d(m,s) = min d(m,i) for all $i \neq m$, and x_k^m is substituted by x_k^s. The proposed method can be easily adapted to datasets formed by discrete attributes. To do so, one can just substitute the Euclidean/Manhattan distance function by the Simple Matching Approach [26].

4 Simulation Results in a Prediction Task

We performed simulations in four datasets that are benchmarks for data mining methods: Iris Plants, Wisconsin Breast Cancer, Pima Indians Diabetes, and Wine Recognition. These datasets were chosen because they are formed only by numeric attributes and because they are available at the UCI Machine Learning Repository [27]. All these datasets describe classification problems. However, we are mainly interested in evaluating the proposed method in the context of clustering problems, i.e. as a preprocessing tool for clustering algorithms. This fact leads us to investigate the performance of the proposed method in an *unsupervised way*, i.e. applying the method in the dataset formed by examples of all classes. In this way we can simulate the substitution process in the dataset to be clustered. Thus, in all the experiments we applied the substitution process in the datasets formed just by the attribute values (without the *class value*). Besides, we also compare these simulation results with those obtained by means of a simple and usual substitution method, which consists in substituting the missing values by the mean of the attribute values.

Basically, our simulations consider that there is just one missing value at a time. Let us consider that one has a dataset formed by L objects $i=\left(x_1^i, x_2^i, ..., x_N^i\right)$. First, we simulate that x_1^1 is missing and it is consequently substituted. Second, x_2^1 is missing and it is consequently substituted. This process is repeated until x_N^1 is substituted. After that, we simulate that x_1^2 is missing and it is consequently substituted. In summary, this procedure is repeated for all x_k^i (i=1,...,L; k=1,...,N). In this way the simulations can be easily reproduced, i.e. they are not influenced by the choice of random samples. Besides, if there is a set of objects whose distances d(m,i) are equal the substituted value comes from the first object of this subset, in the sense that the algorithm starts from the first object in the dataset and goes until the last one. After the substitution process, one has two datasets (the original one and the substituted one) and it is possible to verify how similar the substituted values are compared to the original ones. In this sense, we calculate the absolute difference between the substituted value and the original one, obtaining an average error for each attribute – considering all possible substitutions. These errors are shown by means of graphics.

Obviously the method is sensitive to the distance function choice. Therefore, in this paper we compare the results obtained by two distance functions that are commonly used: the Euclidean (1) and the Manhattan (2). Another important issue is about normalization, which is merely an option that may or may not be useful in a given application [26]. In order to evaluate its influence in the substitution process, we compare the results obtained in the original values with those obtained with the normalized ones. In this sense, we convert all the attribute values into the range [0,1], using a linear interpolation.

4.1 Iris Plants

This database consists of 3 classes (Setosa, Versicolour and Virginica), each one formed by 50 examples of plants. There are 4 attributes (sepal and petal length and width). The class Setosa is linearly separable from the others, whereas the classes Versicolour and Virginica are not linearly separable from each other. Figures 1 and 2 show that, in all experiments, the Nearest-Neighbor Method (NNM) provided lower prediction errors than the substitution by the mean. Considering the normalized data, the Euclidean distance provided better results than the Manhattan distance in attributes A3 and A4, whereas in the original data the Euclidean distance provided better results in attributes A1 and A3. If one compares the results obtained by means of normalized and original data, the normalized data provided better results in attributes A1 and A2 (Euclidean), as well as in A3 and A4 (Manhattan).

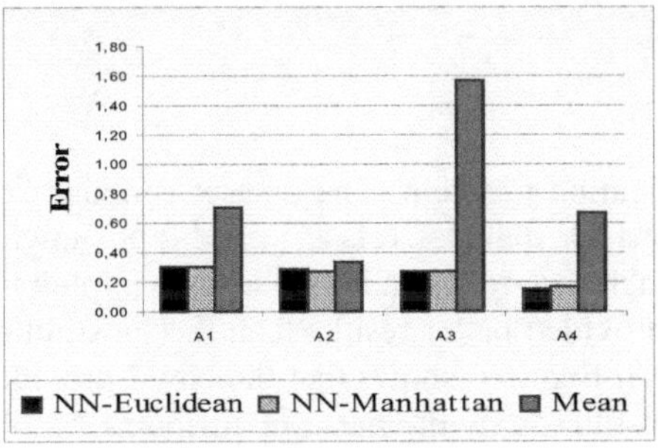

Fig. 1. Iris Plants Normalized

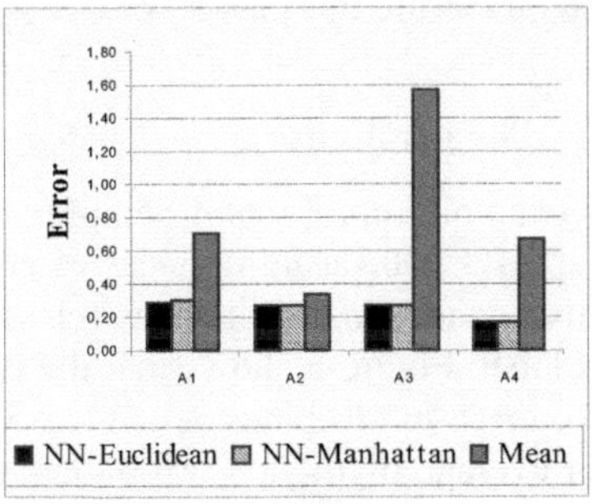

Fig. 2. Iris Plants Original

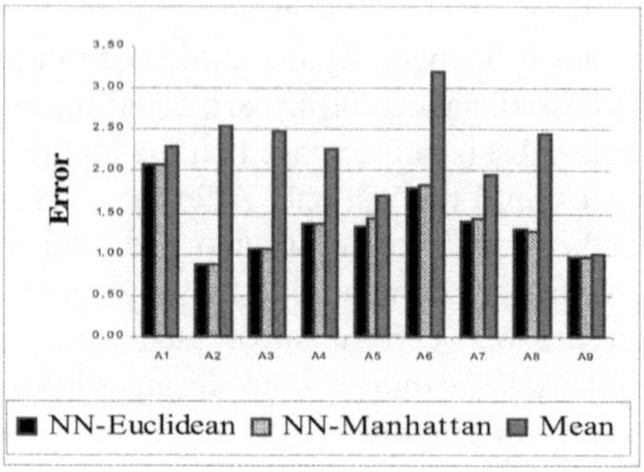

Fig. 3. Wisconsin Normalized.

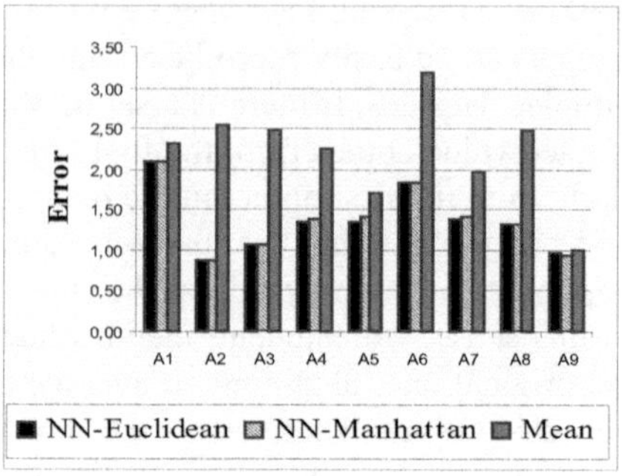

Fig. 4. Wisconsin Original.

4.2 Wisconsin Breast Cancer

In this database each object has 9 attributes and an associated class label (benign or malignant). The two classes are known to be linearly inseparable. The total number of objects is 699 (458 benign and 241 malignant), of which 16 have a single missing feature. We removed those 16 objects and used the remaining ones to simulate the substitution of missing values. Figures 3 and 4 show that, in all experiments, the NNM provided better results (lower prediction errors) than the substitution by the mean. Considering the normalized data, the Euclidean distance provided better results than the Manhattan distance in attributes A5, A6 and A7, whereas in the original data the Euclidean distance provided better results in attributes A1, A4, A5, A6 and A7. If one compares the results obtained by means of normalized and original data, the normalized data provided better results in attributes A3, A5, A6 and A8 (Euclidean), as well as in A1, A3, A4, A6, A7 and A8 (Manhattan). It is also necessary to say that, in theory, the results obtained in the original and in the normalized dataset should be equal, because all attribute values in this dataset belong to the set {1, 2,..., 9}. However, small differences were observed due to *rounding errors* that occur in the normalization process. In fact, the NNM can also be employed in datasets formed by discrete ordinal attributes [26], and the simulations performed in the Wisconsin Breast Cancer dataset illustrate this property.

4.3 Pima Indians

This example represents a complex classification problem. The dataset contains 768 examples – 500 meaning negative conditions for diabetes (class 1) and 268 showing positive conditions of diabetes (class 2). Each example contains 8 attributes plus the class label. Figure 5 shows that the NNM provided better results than the substitution by the mean in attributes A1, A4, A5 and A8. Figure 6 shows that the NNM provided lower average prediction errors than the substitution by the mean in attributes A4, A5 and A8 (just for Manhattan distance). The substitution errors for A7 do not appear in any figure because they are very low (less than 0.33). Considering the normalized data, the Euclidean distance provided better results than the Manhattan distance in attributes A1, A4 and A7, whereas in the original data the Euclidean distance provided better results in attributes A3-7. If one compares the results obtained by normalized and original data, the normalized data provided better results in A1, A2, A4, A7 and A8 (Euclidean), as well as in A1, A3-8 (Manhattan).

4.4 Wine Recognition

In this database each object has 13 attributes and an associated class label (1, 2 or 3). The total number of objects is 178 (59 – class 1, 71 – class 2, 48 – class 3). Figure 7 shows that the NNM provided better results than the substitution by the mean in attributes A1-2, A4 (just for Manhattan distance), A5-13. Figure 8 shows that the NNM provided lower average errors than the substitution by the mean in attributes A1 (just for Manhattan distance), A6-7, A10-11 (just for Manhattan distance), A12-13. Considering the normalized data, the Euclidean distance provided better results than the Manhattan distance in attributes A1, A5-7, A10-13, whereas in the original data the Euclidean distance provided better results just in attribute A2. If one compares the results obtained by means of normalized and original data, the normalized data provided better results in all attributes when the Euclidean distance is applied, as well as in 12 attributes (less A13) when the Manhattan distance is applied.

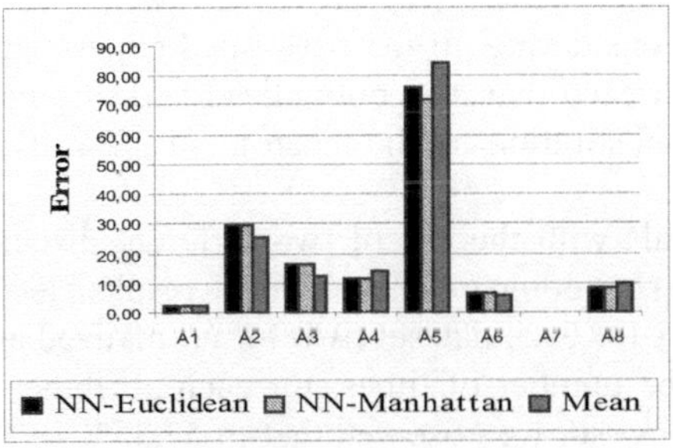

Fig. 5. Pima Indians Normalized

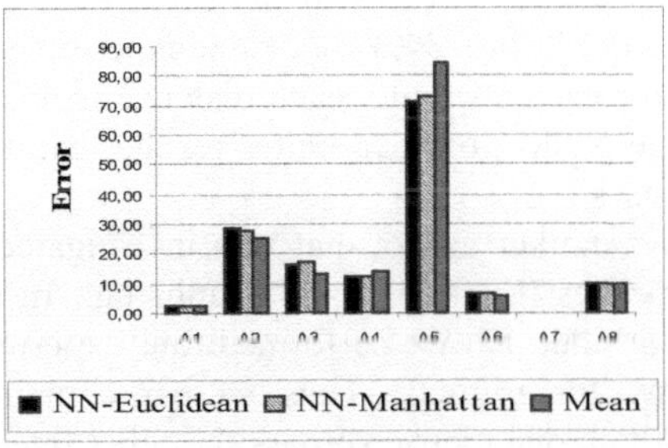

Fig. 6. Pima Indians Original

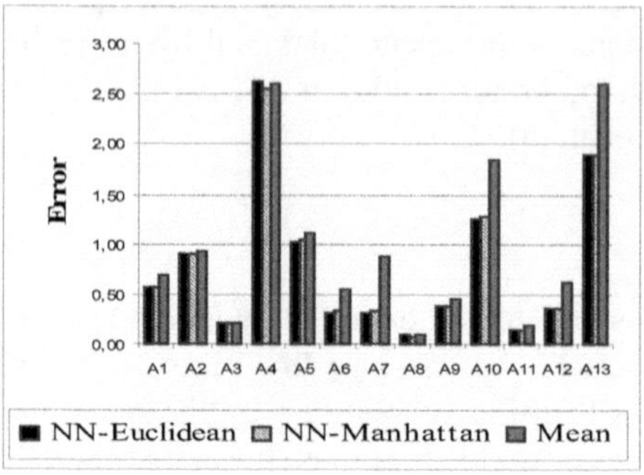

Fig. 7. Wine Recognition Normalized

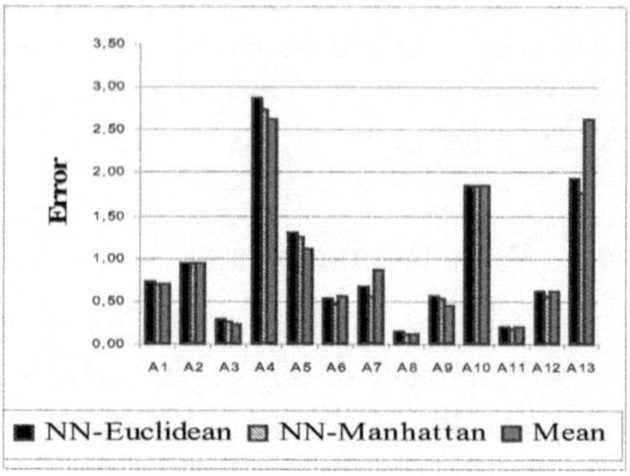

Fig. 8. Wine Recognition Original

4.5 Discussion

In general, the prediction simulation results showed that the NNM provided better results than the use of the mean. In fact, the NNM provided better results in all experiments involving the datasets Iris Plants and Wisconsin Breast Cancer. In the Wine Recognition dataset the NNM provided lower average errors in 65.38% of the experiments, whereas in the Pima Indians Diabetes the NNM provided lower average errors in just 34.38% of the experiments. Considering all the performed experiments (for each attribute, each distance metric and each dataset - normalized and original), the NNM provided better results than the substitution by the mean in 71.32% of the cases.

Another aspect that was investigated deals with the use of two different distance metrics: Euclidean and Manhattan. In order to compare the obtained results, let us consider that we performed four experiments for each dataset (two for normalized and two for original). Thus, we can compute the number of times, for each attribute, in which one metric surpass the other one. Doing so, we observed that in 57.35% of the experiments the Manhattan Distance provided lower errors than the Euclidean one. Besides, the Manhattan distance is computationally more efficient and should be preferred. Finally, we also compared the results obtained in normalized datasets with those obtained in the original datasets and we observed that normalized datasets provided better results in 75% of the experiments.

In summary, we verified that in most simulations: (i) the NNM provided better results than the substitution by the mean; (ii) the Manhattan distance is a better metric that the Euclidean one; (iii) it is better to employ normalized datasets. Although the prediction evaluation is relevant and valid, it is not the only important issue to be analyzed. In this sense, one of the most important aspects is that the substitution method must generate values that least distort the original characteristics of the original sample [28]. This being the case, we are also interested in evaluating if the substitution process preserves the between-variable relationships, which, in our study, are defined by the clustering process. This aspect is approached in the next section.

5 Simulation Results in a Clustering Process

The substituted values should preserve the between-variable relationships. In a clustering process, it means that the *natural* clusters should be preserved, i.e. the imputed values should not change the group that the object *really* belongs to. In order to evaluate this aspect, one has to suppose that the *natural* clusters are *a priori* known. If real-world datasets are considered, it is a hard task to find the *natural, correct* clusters. As previously mentioned, we have employed classification datasets in our simulations, considering that the classes form the *natural* clusters. Thus, one can verify to what extent the K-Means is capable of finding the correct clusters, which are defined by each class. In this sense, we propose to compare the A*verage Classification Rates (ACRs)* obtained by the K-Means Algorithm in the original dataset with those obtained in the substituted datasets, which are formed by the procedure described in Section 4. When the NNM is concerned, we have employed the Manhattan distance and normalized datasets, which, in our experiments, provided better results when one evaluates the substitution method as a prediction task.

Our clustering simulations were performed by means of the WEKA System [29], using the K-Means Algorithm with the default parameters and considering that the number of clusters is equal to the number of classes, defined by each employed dataset. Table 1 shows the simulation results both in the original and in the substituted datasets. In most cases, the ACRs obtained in the original datasets were very similar to the ones obtained in the substituted datasets, what indicates that the NNM provides unbiased estimates of the missing values. Bad results were just observed in the Pima Indians Dataset, which represents a very difficult classification problem [30]. However, our main goal was not to evaluate the performance of the K-Means Algorithm, which is well known. Instead, our objective was to evaluate how suitable is the NNM – as a data preparation tool - to the K-Means Algorithm. In this sense, one important aspect to be observed is the ACR difference between the original and the substituted dataset - 14.46%, which can be considered an estimate of the *inserted bias* in this dataset.

6 Conclusions and Future Work

This paper described and evaluated a Nearest-neighbor Method (NNM) to substitute missing values in datasets formed by continuous attributes. The proposed method compares each instance containing missing values with complete instances, using a distance metric, and the closest complete instance is used to assign the missing attribute value.

We evaluated the proposed method by means of simulations performed in four datasets that are benchmarks for data mining methods. First, we considered the substitution process as a prediction task. In that manner, we employed two metrics (Euclidean and Manhattan) to simulate substitutions both in original and normalized datasets. These substitutions were compared with a usually employed method, i.e. using the mean value. Besides, we evaluated the efficacy of the proposed method both in original and normalized datasets. In summary, these simulations showed that in most cases: (i) the NNM provided better results than the substitution by the mean; (ii) the Manhattan distance is a better metric that the Euclidean one; (iii) it is really better to employ normalized datasets. Thus, our simulations suggest that a NNM based on the Manhattan distance and on normalized datasets provide better results. In this sense, the Euclidean metric is more influenced by outliers than the Manhattan one, and some attributes can have undue weights if the dataset is not normalized.

Table 1. Average Classification Rates (ACR): K-Means Algorithm

Dataset	ACR (%) Original	ACR (%) Substituted	(Original – Substituted)
Iris Plants	88.00	89.33	-1.33 %
Wisconsin Breast Cancer	96.19	95.90	+0.29 %
Pima Indians Diabetes	66.80	52.34	+14.46 %
Wine Recognition	94.38	95.51	-1.13 %

Although the prediction results are relevant, they are not the only important issue to be analyzed. In fact, one of the most important aspects is that the substitution method must generate values that least distort the original characteristics of the original sample [28], i.e. the substituted values should preserve the between-variable relationships. In our work, we evaluated this aspect in the context of the K-Means Algorithm, performing clustering simulations and comparing the results obtained in the original datasets with the substituted ones - using the Manhattan distance and normalized datasets. These results indicated that the proposed method is a suitable estimator for missing values, i.e. it preserves the relationships between variables in the clustering process. Therefore, the NNM is a suitable data preparation tool for the K-Means Algorithm.

Considering our future work, there are many aspects that can be further investigated. One important issue is to evaluate the best number of neighbors, i.e. the best K value in the K-nearest-neighbor method. In this sense, we are also going to evaluate the substitution process in the context of other learning algorithms. Besides, we are going to evaluate the efficacy of the proposed method in datasets with more missing values, comparing the NNM results with those obtained with other substitution methods. Another important aspect is to investigate the results of the substitution process when the method is applied in the examples of each class separately, because this methodology can be useful in classification tasks. Besides, we are going to evaluate the NNM in discrete datasets, applying the Simple Matching Approach [26] as the distance metric.

References

[1] Fayyad, U. M., Shapiro, G. P., Smyth, P. "From Data Mining to Knowledge Discovery : An Overview". In: Advances in Knowledge Discovery and Data Mining, Fayyad, U.M., Piatetsky-Shapiro, G., Smyth, P., Uthurusamy, R., Editors, MIT Press, pp. 1-37, 1996.

[2] Bigus, J. P., Data Mining with Neural Networks , First edition, USA, McGraw-Hill, 1996.

[3] Batista, G. E. A. P. & Monard, M. C., An Analysis of Four Missing Data Treatment Methods for Supervised Learning, Proceedings of the First International Workshop on Data Cleaning and Preprocessing, IEEE International Conference on Data Mining, Maebashi, Japan, 2002.

[4] Mitchell, T. M. Machine Learning. McGraw-Hill, 1997.

[5] Han, J. & Kamber, M. - Data Mining: Concepts and Techniques. Morgan Kaufmann, 2001.

[6] Quinlan, J., Induction of Decision Trees. Machine Learning, 1,81-106, 1986.

[7] Kononenko, I., Bratko, I. & Roskar, E., Experiments in Automatic Learning of Medical Diagnostic Rules. Technical Report, Jozef Stefan Institute, Ljubjana, Yogoslavia, 1984.

[8] Quinlan, J. R. Unknown Attribute Values in Induction. Proceedings of 6th International Workshop on Machine Learning, 164-168, Ithaca, NY, 1989.

[9] Friedman, H. F., Kohavi, R. & Yun, Y., Lazy Decision Trees. In Proceedings of the 13th National Conference on Artificial Intelligence, pp. 717-724, AAAI Press/MIT Press, 1996.

[10] Schapire, R. E., Freund, Y., Barlett, P. & Lee, W. S., Boosting the Margin: A New Explanation for the Effectiveness of Voting Methods. Proceedings of the 14th International Conference on Machine Learning, Morgan Kaufmann, pp. 352-330, 1997.

[11] Breiman, L. Bagging Predictors. Machine Learning, 24, 123-140, 1996.

[12] Zheng, Z. & Webb, G. I., Stochastic Attribute Selection Committees. In Proceedings of the 10th Australian Joint Conference of Artificial Intelligence. Berlin: Springer-Verlag, 1998.

[13] Zheng, Z. & Webb, G. I., Stochastic Attribute Selection Committees with Multiple Boosting: Learning more Accurate and more Stable Classifier Committees. In Proceedings of the 3rd Pacific-Asia Conference on Knowledge Discovery and Data Mining, Berlin: Springer-Verlag, 1999.

[14] Gilks W. R., Richardson, S. & Spiegehalter, D. J., Markov Chain Monte Carlo in Practice. Chapman and Hall, London, 1996.

[15] Gelman, A., Carlin, J. B., Stern, H. S. & Rubin, D. B., Bayesian Data Analysis. Chapman and Hall, London, 1995.

[16] Sebastiani, P. & Ramoni, M., Bayesian Inference with Missing Data Using Bound and Collapse. Technical Report KMI-TR-58, KMI, Open University, 1997.

[17] Little, R. & Rubin, D. B., Statistical Analysis with Missing Data. Wiley, New York, 1987.

[18] Hruschka Júnior, E. R., Ebecken, N. F. F. Missing Values prediction with K2. Intelligent Data Analysis.(IDA).IOS Press, Netherlands: , v.6, n.6, 2002.

[19] Hruschka Júnior, E. R., Hruschka, E. R., Ebecken, N. F. F. A Data Preparation Bayesian Approach for a Clustering Genetic Algorithm, In: Frontiers in Artificial Intelligence and Applications, A. Abraham et al. (Eds), Soft Computing Systems: Design, Management and Applications, pp. 453-461, IOS Press, 2002.

[20] Rubin, D. B., Multiple Imputation for non Responses in Surveys. New York, John Wiley & Sons, 1987.

[21] Dempster, A. P., Laird, N. M. & Rubin, D. B., Maximum Likelihood from Incomplete Data via the EM algorithm. Journal of the Royal Statistical Society B, 39, 1-39, 1977.

[22] Friedman, N., Learning Belief Networks in the presence of Missing Values and Hidden Variables. Proceedings of the 14th International Conference on Machine Learning, 1997.

[23] Lauritzen, S. L., The EM Algorithm for Graphical Association Models with Missing Data. Computational Statistics and Data Analysis, 19, 191-201, 1995.

[24] Atkeson, C. G., Moore, A. W., Schaal, S. A., Locally Weighted Learning for Control. AI Review, 1997.

[25] Aamodt, A. & Plazas, E., Case-Based Reasoning: Methodological Variations, and System Approaches. AI Communications, 7(1), 39-52, 1994.

[26] Kaufman, L., Rousseeuw, P. J., Finding Groups in Data – An Introduction to Cluster Analysis, Wiley Series in Probability and Mathematical Statistics, 1990.

[27] Merz, C.J., Murphy, P.M., UCI Repository of Machine Learning Databases, http://www.ics.uci.edu, Irvine, CA, University of California, Department of Information and Computer Science.

[28] Pyle, D., Data Preparation for Data Mining. Academic Press, 1999.

[29] Witten, I. H., Frank, E., Data Mining – Practical Machine Learning Tools and Techniques with Java Implementations, Morgan Kaufmann Publishers, USA, 2000.

[30] Nakhaeizadeh, G., Kressel, U., Keh, S. et al., Dataset Descriptions and Results, In: Machine Learning Neural and Statistical Classification, D. Michie, D.J. Spiegelhalter and C.C. Taylor editors, pp. 131-174, Ellis Horwood Series in Artificial Intelligence, Bookcraft, Midsomer Norton, 1994.

[31] Batista, G.E.A.P.A. and Monard, M.C., A Study of K-Nearest Neighbor as an Imputation Method. In Second International Conference on Hybrid Intelligent Systems, Santiago, Chile, Soft Computing Systems: Design, Management and Applications, pp. 251-260, IOS Press, 2002.

[32] Batista, G.E.A.P.A. and Monard, M.C., A Study of K-Nearest Neighbor as a Model-Based Method to Treat Missing Data, Proceedings of Argentine Symposium on Artificial Intelligence, v. 30, pp. 1-9, Buenos Aires, 2001.

Single-Class Classification Augmented with Unlabeled Data: A Symbolic Approach

Andrew Skabar

School of Information Technology, Deakin University
221 Burwood Highway, Burwood, Victoria, 3125, Australia
andrews@deakin.edu.au

Abstract. Supervised machine learning techniques generally require that the training set on which learning is based contain sufficient examples representative of the target concept, as well as known counter-examples of the concept; however, in many application domains it is not possible to supply a set of labeled counter-examples. This could be due to either the nature of the problem domain, or the expense of obtaining labels for the training data. This paper presents a technique that can be used for learning symbolic concept descriptions from a dataset consisting of labeled positive examples together with a corpus of unlabeled positive and negative examples. The technique uses evolutionary search to explore the space of concept descriptions, guided by an evaluation function that seeks to achieve a balance between the generalization and specialization capacities of individuals. One of the features of the technique is that it requires empirical determination of a bias weighting factor that controls the relative balance between the generalization and specialization of hypotheses. The advantage of incorporating a bias weighting factor is that it can be used to guide search towards the discovery of hypotheses in which the emphasis may be on achieving either a low false negative rate, or a low false positive rate. This is a useful property, because in many application domains misclassification costs are not equal. Although determining an appropriate bias weighting may require some degree of subjective judgement, heuristics are presented to assist in this task. The technique is able to cope with noise in the training set, and is applicable to a broad range of classification and pattern recognition problems.

1 Introduction

In this paper the term 'single-class classification' is used to denote the problem of discovering a classifier from a set of training examples in which only examples of the target class (i.e., *positive* examples) are present. From an inductive learning perspective, single-class learning is difficult because in the absence of counter-examples it is

T.D. Gedeon and L.C.C. Fung (Eds.): AI 2003, LNAI 2903, pp. 735-746, 2003.

not clear how unrestricted over-generalization can be prevented. Put simply, the role of positive examples is to induce generalizations, and the role of negative examples is to rule out over-generalizations [1]. Because every inductive learner must achieve a balance between specialization and generalization, and because the responsibility for specialization rests so heavily on negative examples, the very idea of inductive learning without counter-examples appears strange.

This paper addresses the problem of learning a classifier from a dataset consisting of labeled examples belonging to the target class (i.e., positive examples) together with a corpus of unlabeled (positive and negative) examples. More formally:

Given:

LP: a set of *positive examples* of the target concept

UL: a set of *unclassified examples* from the domain (i.e. *un*labeled positive and negative examples)

Find:

H: an (intentional) description of the concept (i.e. hypothesis) which

(i) covers all (or at least a high proportion of) the examples in *LP* as well as similar examples in *UL*

(ii) excludes examples in *UL* that are not similar to examples in *LP*.

Put slightly differently, we would like to discover a hypothesis sufficiently general to describe the known (i.e., labeled) positive examples, but yet sufficiently specific to exclude unlabeled examples dissimilar to the labeled positives. The technique presented in this paper is based on the use of genetic search to explore the space of concept descriptions described in the VL1 language. It can be described as a partially supervised learning algorithm based on a generate-and-test search procedure.

The paper is structured as follows. Section 2 describes the VL1 concept description language and outlines how genetic search can be used to explore a VL1 concept description space. Section 3 formally defines the notions of *specificity* and *generality* as applied to concept descriptions, and proposes metrics for these. The hypothesis evaluation function is also described in Section 3. Results of applying the technique are presented in Section 4 as a case study in which various features of the technique are explored and discussed. Section 5 concludes the paper.

2 Genetic Search through VL1 Concept Description Space

This section describes the knowledge representation scheme and search mechanism on which the proposed technique is based.

2.1 The VL1 Concept Description Language

The VL1 concept description language [2] is a multiple valued extension to propositional logic that allows concepts to be expressed in disjunctive normal form (DNF). According to this formalism, a concept description C is a disjunction of one or more *complexes* c_i,

$$c_1 \lor c_2 \lor c_3 \lor \ldots \lor c_k \Rightarrow C$$

where each complex c_i is a conjunction of *selectors*, and a selector is a triplet of the form (*attribute relation set-of-values*). As an example, consider an attribute space consisting of the four attributes: *Shape* ∈ {*sphere, cube, cone*}, *Colour* ∈ {*red, green, blue, yellow*}, *Weight* ∈ [0, 100], and *Size* ∈ [0, 100]. Examples of complexes in this space could include:

$$\langle Shape \in sphere\rangle \land \langle \text{Colour} \in \{\text{red, green}\}\rangle \land \langle \text{Size} > 0.5\rangle$$
$$\langle \text{Shape} \in \{\text{sphere, cube}\}\rangle \land \langle \text{Weight} < 50.0\rangle$$

An example belongs to the concept if its attribute values satisfy one or more of the complexes appearing in the concept description. In this case we say that the instance is *covered* by the concept description, and that it belongs to the *positive* class. If an example's attribute values do not satisfy any of the complexes, then the example would be classified as belonging to the *negative* class. Note that it is not necessary that all attributes appear in a complex. If an attribute does not appear in a complex, then that attribute is irrelevant to determination of membership according to that complex.

2.2 Binary Representation of VL1 Expressions

In order to apply genetic search a mapping must be established between concept descriptions (expressed in the concept description language) and individuals in the search population. The mapping between hypotheses expressed in the VL1 concept description language and variable length binary strings in the genetic search population is depicted in Figure 1. The '⟨' and '⟩' indicate respectively the start and end of a complex, and the '|' is used only to indicate the separation between selectors within a complex. Note that individuals can vary in length, but that the total length must be a multiple of the length of a complex, which is fixed, and determined by the number and type of selectors in the problem domain. The selector representation depends on the type of attribute being encoded. Two cases are considered here: (i) nominal attributes, and (ii) continuous attributes.

Selectors for nominal attributes are represented by n-bit binary strings, where n is the number of values in the domain of the respective attribute. There is a one-to-one correspondence between each bit in the binary vector and the values from the attribute's domain. For example, using the domain from the example above, the selector *Shape* = *sphere* is represented by the 3-bit string 100. In order for an example to be covered by this selector, the value for attribute *Shape* must be *sphere*. Similarly, the selector *Colour* = {*red, yellow*} would be represented as 1001. Note that because a selector containing all 1's will always cover that attribute, it is effectively a *don't care value*.

Continuous attributes can be represented by mapping an unsigned integer encoding linearly from $[0, 2^l]$ to a specified interval $[E_{min}, E_{max}]$, where l is the number of bits used in the representation. Thus E_{min} is mapped to the binary representation of 0, E_{max} maps to the binary representation of the integer $2^l - 1$, and other values map linearly in between.

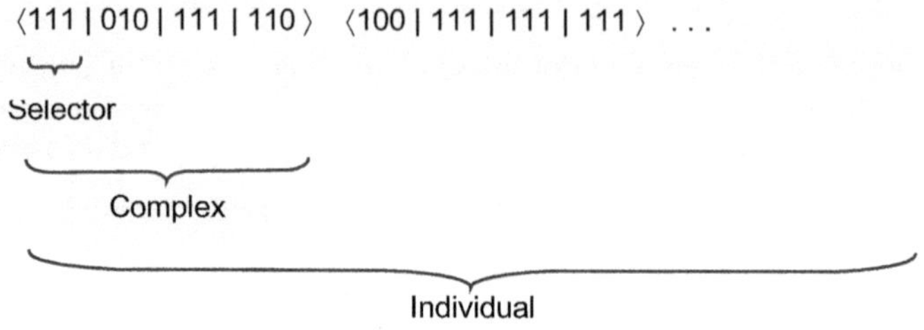

Fig. 1. Mapping scheme between VL1 expressions and variable length binary strings

$$E_{min} \rightarrow 0\,0\,0\,0\,0\,0\,0\,0 \quad (0)$$
$$E_{max} \rightarrow 1\,1\,1\,1\,1\,1\,1\,1 \quad (2^l - 1)$$

In the case of continuous attributes, we often require to represent a range of values, bounded by some upper or lower bound. This can be achieved by incorporating an extra bit to represent the type of bound (i.e. upper or lower). Assuming that 1 and 0 represent lower and upper bounds respectively, and that the values of attribute A_l range between 0.0 and 100.0, a range of values on A_l such as $A_l > 25.0$ can be represented as:

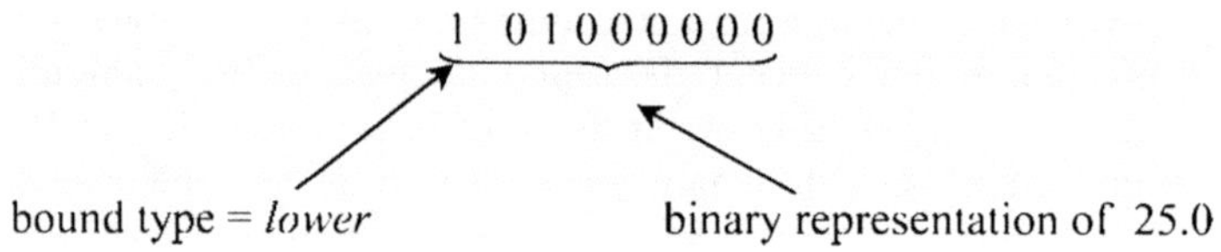

2.3 Genetic Search

The variable-length encoding scheme requires that the standard low-level genetic crossover operator be modified slightly. The modified crossover operator (for two-point crossover) works as follows. Two crossover points are first randomly selected for the first individual in the crossover pair. Crossover points for the second individual are then randomly selected, but subject to the constraint that the distances from the leftmost and rightmost bit of a complex are the same as for the first individual. This ensures that the representational structure of individuals is preserved through crossover.[1]

Search is performed as per standard genetic search. An initial population of individuals is generated. These individuals are then evaluated according to a fitness metric based on the evaluation function that is described in the next section, and the fitter

---

[1] This is the approach De Jong & Spears take in GABIL [5]. A somewhat different approach is taken by Janikow [6], whose GIL system uses 14 specialized, task-dependent genetic operators (rule specialization, rule generalization, etc.) that take advantage of the particular representation being used. The rationale for this is that exploiting knowledge about the problem being solved allows a more efficient search with faster convergence to the desired solution. Janikow acknowledges that the operators are really just special cases of the traditional mutation and crossover operators.

individuals are chosen to undergo reproduction, crossover, and mutation operations in order to produce a population of children for the next generation. This procedure is continued until either convergence is achieved, or a sufficiently fit individual has been discovered. Further information on genetic algorithms can be found in [3] and [4]. For details on the variable length string approach, see [5] and [6].

3 Hypothesis Evaluation

Genetic search through the space of VL1 concept descriptions must be guided by an evaluation function which evaluates competing hypotheses. This evaluation function must rank hypotheses according to some objective measure of the performance of the hypothesis in classifying the examples. This section defines the notions of hypothesis *specificity* and *generality*, provides metrics for these, and proposes an objective function suitable for the problem of learning in the absence of labeled counter-examples.

3.1 Defining Specificity and Generality

As described above, we would like to discover a hypothesis sufficiently general to describe the known (i.e. labeled) positive examples, but yet sufficiently specific to exclude unlabeled examples dissimilar to the labeled positives. Before proceeding to provide measures for these quantities, it is useful clarify the meanings of these terms.

There are at least two distinct meanings that can be associated with the use of the words *specific* and *general* as applied to concept descriptions (hypotheses):

Definition 1:
A hypothesis is specific/general if the hypothesis covers a small/large subset of the instance space.

Definition 2:
A hypothesis is specific/general if it covers a small/large subset of the total set of supplied examples from the problem domain.

The problem with defining *specificity* and *generality* in terms of the instance space is that it assumes that all combinations of attribute values are equally likely. However, in many application domains this assumption is not justified. Many problem domains simply do not include examples representing every combination of attribute values. For example, in an animal classification domain one would not expect to observe an instance in which `has_tusks = TRUE`, `has_fins = TRUE` and `method-of-locomotion = FLYS`. Moreover, even if every possible combination of attribute values did appear, it would be highly unlikely that each of these occurs with equal frequency. A set of examples—irrespective of whether the individual examples are labeled or unlabeled—provides a *context* for the learning problem, and it is from within this context that the *specificity* and *generality* of a hypothesis must be evaluated. Thus, through the remainder of this paper the assumed meanings of *specificity* and *generality* are those of Definition 2 above.

3.2 Generality and Specificity Metrics

A hypothesis should be sufficiently general to cover all (or most) of the labeled positives. This suggests that the generality of a hypothesis Φ be measured in terms of the proportion of labeled positive examples that it covers:

$$generality(\Phi) = \frac{1}{P}\sum_{i=1}^{P}\delta(x_i, \Phi) \tag{1}$$

where

P is the number of labeled positive examples, and

$$\delta(x_i, \Phi) = \begin{cases} 1 & \text{if the } i^{\text{th}} \text{ example is covered by } \Phi \\ 0 & \text{if the } i^{\text{th}} \text{ examples is not covered by } \Phi \end{cases}$$

It is clear that a measure of generality on its own is inadequate as a measure of classifier performance. If classification performance is measured solely on the capacity of the concept description to cover labeled positives, a hypothesis that predicts every instance as positive will be assigned a measure of performance at least as high as that of any other hypothesis.

A hypothesis should be sufficiently specific such that it excludes those unlabeled examples that are not similar to the labeled positives. In order to develop the argument, it is convenient to temporarily leave aside the issue of *similarity*, and define a measure of specificity analogous to the measure of generality as defined above. This specificity measure is based on the proportion of unlabeled examples excluded by the hypothesis:

$$specificity(\Phi) = \frac{1}{U}\sum_{i=1}^{U}\delta(x_i, \Phi) \tag{2}$$

where:

U is the number of examples in the corpus of unlabeled data, and

$$\delta(x_i, \Phi) = \begin{cases} 1 & \text{if the } i^{\text{th}} \text{ example is } \textit{not} \text{ covered by } \Phi \\ 0 & \text{if the } i^{\text{th}} \text{ examples is covered by } \Phi \end{cases}$$

According to this measure of specificity, a hypothesis Φ_1 is more specific than a hypothesis Φ_2 if it covers fewer unlabeled examples. A hypothesis that covers none of the unlabeled examples will be assigned a specificity measure of 1, and a hypothesis which covers all of the unlabeled examples will be assigned a specificity value of 0.

The major problem with defining the specificity of a hypothesis in this way is that it does not take into account the distance of an example from the decision boundary represented by a concept description, and in this sense is an overly crude measure of specificity. This problem can be solved by defining specificity as follows:

$$specificity(\Phi) = \frac{1}{nU} \sum_{i=1}^{U} \sum_{j=1}^{n} \delta(x_i^j, \Phi) \tag{3}$$

where:

U is the *total* number of unlabeled examples;
n is the number of attributes, and

$$\delta(x_i^j, \Phi) = \begin{cases} 1 & \text{if the } j^{th} \text{ attribute of } x_i \text{ is } \textit{not} \text{ covered by } \Phi \\ 0 & \text{if the } j^{th} \text{ attribute of } x_i \text{ is covered by } \Phi \end{cases}$$

3.3 Combining Specificity and Generality into an Objective Function

The metrics defined above must be combined into an hypothesis evaluation function that can be used to guide search through the hypothesis space. A hypothesis should receive a high evaluation only if it is both (i) sufficiently general to cover a high proportion of the labeled positives, and (ii) sufficiently specific to exclude unlabeled examples that are not similar to the positives. The simplest way of ensuring that only a hypothesis that meets both these criteria receives a high evaluation is to define the fitness of a hypothesis as the product of these two measures:

$$Fitness\ (\Phi) = generality\ (\Phi) \times specificity\ (\Phi)$$

However, defining fitness as the simple product of hypothesis generality and specificity does not provide scope for balancing specialization and generalization. The balance between specificity and generality will depend on a variety of factors including the problem domain, the specific training examples, as well as additional factors such as whether there are irrelevant variables. The following analytic fitness function form incorporates a bias factor k that can be used to adjust the balance between the generalizational and specializational capacity of a hypothesis:

$$Fitness(\Phi) = generality\ (\Phi) * [\ 1.0 + k \times specificity\ (\Phi)]$$

which is equivalent to

$$Fitness(\Phi) = generality(\Phi) + [k \times generality(\Phi) \times specificity\ (\Phi)] \tag{4}$$

As can be seen, the fitness is now defined as the sum of two terms: *generality*(Φ), and a second term that involves the product of *generality* (Φ) and *specificity* (Φ). Observe that the bias towards specificity now actually increases as the generality of the hypothesis increases. This is an appealing property, because it means that in the early stages of search, the emphasis will be on discovering general hypotheses, with the role of the specificity bias increasing as more general hypotheses are found.

4 Case Study: Australian Credit Dataset

This section reports on the application of the technique described in Section 2 and 3 to the Credit approval dataset. The Credit approval dataset, due to Quinlan [7], is a widely used dataset that contains a good mix of attributes. The dataset contains 690 examples (307 positives and 383 negatives) over 14 attributes (6 continuous, 8 nominal). The nominal variables range from 2 values up to 14 values.

The technique was applied as follows. For each trial, 30% of the positive examples were randomly selected to appear with class labels. These examples constitute the set of *labeled positives*. All remaining examples were treated as *unlabeled* (i.e. their class labels are hidden from the algorithm during the learning phase). Genetic search was implemented using the VL1 encoding scheme as described in Section 2, with fitness of individuals evaluated using the function given in Equation 4. At the completion of training, the classifier represented by the fittest member of the population was used to assign class labels to the unlabeled examples, and the following statistics were collected: accuracy on labeled positives; accuracy on unlabeled positives; accuracy on unlabeled negatives, and overall accuracy on unlabeled data.

4.1 Dependency of Classification Performance on Value of *k*

Because the parameter *k* in the evaluation function determines the relative weighting given to the specificity measure in determining overall fitness, it plays an important role in explicitly determining the overall inductive bias of the technique. If the value of *k* is too small, then the resulting concept description will be too general. If the value of *k* is too high, then the thrust towards specialization will be too strong. To illustrate, consider Figure 2, which shows the effect that the bias weighting factor, *k*, has on classification performance on the Credit dataset.

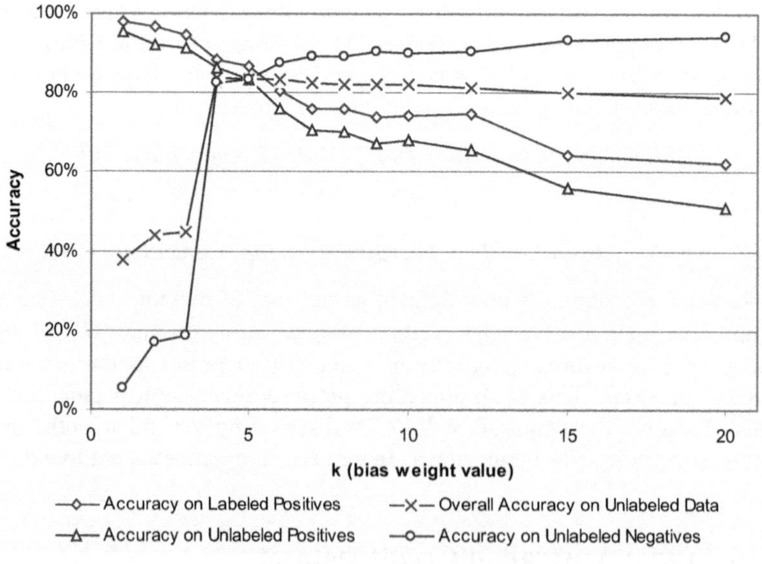

Fig. 2. Variation of classifier performance with bias weighting factor *k* for Credit dataset. There is a *critical* value for *k* (approx. 4.5) for which the specificity bias is sufficient to prevent over-generalization

The following observations can be made:

Low k Values. For low k values the accuracy on both labeled and unlabeled positives is high, and accuracy on unlabeled negatives is very low. This is due to the bias weighting being too small to produce sufficient specialization, thus resulting in over-generalization.

Critical k Value. There appears to be a critical value of k (approximately 4.5) at which there is a sharp increase in the classification accuracy on unlabeled negative examples. No such sudden change is observed for classification accuracy on positive examples. At this point, the specialization bias is sufficient to prevent most of the negative examples from being covered by the hypothesis. The critical k value will henceforth be referred to as k_c.

Values of k Greater than Critical Value. As the value of k increases beyond k_c there is a drop in classification accuracy on labeled and unlabeled positives, and there is a corresponding increase in the classification accuracy on unlabeled negatives. Note however, that these changes are gradual, and that there are no sudden leaps. Note also that the overall classification accuracy on unlabeled data remains relatively constant as the value of k is increased beyond the critical value.

4.2 Sensitivity and Predictive Value

These observations suggest that for a given problem domain, there may be a range of k values which result in acceptable classification performance. As demonstrated in Figure 2, as the value of k is increased beyond the critical value, k_c, there is a drop in the proportion of positive examples that are covered by the hypothesis. This means that the number of positive examples misclassified by the hypothesis increases. In other words, there is a fall in the *true positive* rate. However, as the value of k is increased there is a corresponding increase in the proportion of negative examples that are correctly predicted by the hypothesis. This means that the number of negative examples misclassified as positive decreases. In other words, there is a fall in the *false positive rate*.

These observations are consistent with an intuitive understanding of the role of the weighting factor k. Increasing the value of k leads to more specific (i.e. less general) hypotheses. This means that a higher proportion of positives will fall outside of the boundary defining positive class membership, thus increasing the number of false negatives (i.e., reducing the true positive rate). It also means that the chance of a negative being covered is decreased, thus reducing the number of false positives.

4.3 Heuristics for Determining k_c

The critical value for k will depend on a variety of domain specific factors. Since k_c cannot be determined analytically, it must be determined empirically for each dataset. Heuristics that can be used to indicate when the threshold value of k has been reached include the following:

1. Sudden drop in the number of unlabeled examples covered.

Recall that on the Credit dataset there was a sudden drop in the number of unlabeled negatives covered by the hypothesis as the critical value was reached, but no sudden change was observed in the number of unlabeled positives that were covered. This means that the critical value may be able to be determined by consideration of the number of unlabeled examples covered. This is shown for the CREDIT dataset in Figure 3. Note that as k approaches 4.5, there is indeed a sharp fall in the proportion of unlabeled examples that are predicted as positive.

2. 'a priori' knowledge of proportion of examples expected to be positive.

In some problem domains, there may be some *a priori* expectation of the approximate proportion of positives in the dataset. For example, in a credit screening domain, there may be an expectation that the proportion of customers characterized as high risk credit customers is on average, say, 20%. If the total number of examples classified as *high risk* significantly exceeds this value, then this indicates that the specialization bias is insufficient, and that k_c has not been reached. If such knowledge is available, then the k value can be determined by increasing k until the proportion of examples covered by the hypothesis is reduced to a reasonable value.

3. 100% accuracy on labeled positives.

Most real world domains contain some degree of noise. Therefore, it is quite possible that there may be misclassified negatives amongst the set of labeled positives. This suggests that the experimenter should be wary of any classifier that achieves 100% accuracy on labeled positives as this signifies that over-generalization may be occurring. In this case, the value of k should be increased to beyond the point at which all labeled positives are classified as positive.

These heuristics are general *rules of thumb* and require considerable subjective human judgement. In practice, determining the value of k_c requires careful consideration of the dataset, and may involve considerable trial and error. At best, it is an educated guess.

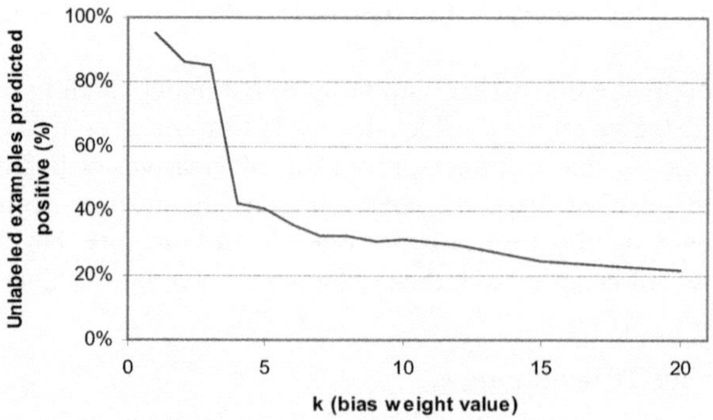

Fig. 3. Proportion of unlabeled examples covered by hypothesis as k is increased on Credit dataset. As the value of k is increased, a sharp drop in the proportion of unlabeled examples classified as positive is observed corresponding to a k-value of approximately 4.5

Table 1. Average predictive accuracy (%) and standard deviation on training and test sets for C4.5 (10-fold cross-validation)

	Train (%)	Test (%)
C4.5 decision trees (*no pruning*)	94.73 (0.17)	82.97 (0.70)
C4.5 decision trees (*pruning*)	90.23 (0.09)	85.57 (0.62)
C4.5 Rules	90.83 (0.19)	85.45 (0.57)

Table 2. Average predictive accuracy (%) and standard deviation on training and test sets for genetic search of VL1 concept descriptions. Results are averaged over 10 trials

	Train (%)	Test (%)
1-DNF	89.35 (1.83)	83.49 (0.89)
3-DNF	88.70 (2.95)	84.80 (0.82)

4.4 Benchmark Testing

In order to evaluate the overall performance of the technique, it is useful to have some benchmark according to which this performance can be compared. Quinlan's C4.5 [8] is a suitable technique because it is a well-known and well-understood algorithm that has been shown to give good performance on a large variety of classification tasks involving noisy data.

Table 1 summarises results of applying C4.5 to the Credit dataset. These results are based on 10-fold cross-validation; that is, on each trial the training set contained 90% of the examples (positive and negative—all labeled), and the remaining 10% of examples constituted the test set. Table 2 provides results for the technique proposed in this paper. Note that performance in this case also depends on the maximum number of disjuncts allowed, and that the first row of the table corresponds to concept descriptions consisting of only a single disjunct, and the second row to concept descriptions up to a maximum of 3 disjuncts. Limiting the number of disjuncts allowed is a form of representational bias.

It can be seen from these results that the performance of the proposed technique compares favourably with that of C4.5. This is despite the fact that C4.5 uses 90% of examples (both positively labeled and negatively labeled) for training, in comparison to only 30% of positives and no negatives appearing labeled in the case of the proposed technique. The latter does of course use all of the examples for training, but in a very different way to conventional supervised approach.

5 Conclusions

A technique has been presented for learning symbolic concept descriptions from a dataset containing labeled examples of a target concept (i.e., positive examples) together with a corpus of unlabeled positive and negative examples. The technique is based on a generate-and-test procedure in which the classification performance of hypotheses is evaluated using an evaluation function that combines measures of the specificity and generality of hypotheses.

One of the features of the technique is that it requires empirical determination of a bias weighting factor that controls the relative balance between the generalization and specialization of hypotheses. The advantage of incorporating a bias weighting factor is that it can be used to guide search towards the discovery of hypotheses in which the emphasis may be on achieving either a low false negative rate, or a low false positive rate. This is a useful property, because in many application domains misclassification costs are not equal. For example, in medical domains *sensitivity* is often the critical measure of the ability of a test to accurately screen patients for some condition, whereas in other domains *predictive value* (i.e. the probability that a predicted positive is a true positive) may be the more important measure of classifier performance. Although determining an appropriate bias weighting may require some degree of subjective judgement, heuristics are available to assist in this task.

Although this paper has only presented results for the Credit dataset, applications of the technique to the well-known Cleveland Heart Disease and Wisconsin Breast Cancer datasets yield similar results and identical conclusions [9].

References

[1] Mitchell, T.M. 1982, 'Generalization as search', *Artificial Intelligence*, vol. 18, no. 2, pp. 203-226.

[2] Michalski, R.S., Mozetic, I, Hong, J. and Lavrac, N., *The AQ15 Inductive Learning System: An Overview and Experiments*, University of Illinois, Urbana-Champaign, 1986.

[3] Holland, J.: Adaptation in Natural and Artificial Systems, University of Michigan Press, Ann Arbor, (1983).

[4] Goldberg, D.: Genetic algorithms in search, optimization, and machine learning, Addison-Wesley, Reading, MA (1989).

[5] De Jong, K.A., Spears, W.M., and Gordon, D.F.: "Using genetic algorithms for concept learning", *Machine Learning*, 13 (1993) 161-188.

[6] Janikow, C.Z.: "A Knowledge Intensive Genetic Algorithm for Supervised Learning". *Machine Learning, **13***, (1993) 198-228.

[7] Quinlan, J.R.: "Simplifying decision trees", *International Journal of Man-Machine Studies*, 27, (1987), 221-234.

[8] Quinlan, J.R. *C4.5 Programs for Machine Learning*, Morgan Kaufmann Publishers, San Francisco (1993).

[9] A. Skabar, *Inductive Learning Techniques for Mineral Potential Mapping*, PhD Thesis, School of Electrical and Electronic Systems Engineering, QUT, Australia (2000).

C3: A New Learning Scheme to Improve Classification of Rare Category Emails

Jie Yang[1], Joshua Zhexue Huang[2], Ning Zhang[1], and Zhuoqun Xu[1]

[1] Department of Computer Science and Technology
Peking University, Beijing, 100871, P.R.China
{yangjie,nzhang}@ebusiness.pku.edu.cn, zqxu@pku.edu.cn
[2] E-Business Technology Institute
University of Hong Kong, Pokfulam Road, Hong Kong
jhuang@eti.hku.hk

Abstract. This paper[1] proposes $C3$, a new learning scheme to improve classification performance of rare category emails in the early stage of incremental learning. $C3$ consists of three components: the *chief-learner*, the *co-learners* and the *combiner*. The chief-learner is an ordinary learning model with an incremental learning capability. The chief-learner performs well on categories trained with sufficient samples but badly on rare categories trained with insufficient samples. The co-learners that are focused on the rare categories are used to compensate for the weakness of the chief-learner in classifying new samples of the rare categories. The combiner combines the outputs of both the chief-learner and the co-learner to make a finial classification. The chief-learner is updated incrementally with all the new samples overtime and the co-learners are updated with new samples from rare categories only. After the chief-learner has gained sufficient knowledge about the rare categories, the co-learners become unnecessary and are removed. The experiments on customer emails from an e-commerce company have shown that the $C3$ model outperformed the Naive Bayes model on classifying the emails of rare categories in the early stage of incremental learning.

1 Introduction

Many applications of text classification systems follow the following scenario. Given a set of samples collected before time t_0, first classify the samples into m categories or classes (usually manually) according to some defined business criteria. Then divide the samples into a training set and a test set, use a classification algorithm to build a model from the training set and test it with the test samples. After that, install the model to classify new samples coming after time t_0 and verify the results of classification through other means. Finally update the model with new samples. A typical example is automatic classification and routing of customer emails that is adopted in many companies. In these systems,

[1] Supported by the National Natural Science Foundation of China (No 60003005) and IBM.

T.D. Gedeon and L.C.C. Fung (Eds.): AI 2003, LNAI 2903, pp. 747–758, 2003.

a model is first built from selected customer emails and then applied to classify new customer emails in real time and route them to corresponding people to handle. The people who process these emails identify the misclassified emails, correct the classification results and save the results in the email database. These identified emails are later used to update the model.

These real world applications require the learning model to possess the following capabilities:

1. *Incremental learning.* Since the misclassified emails are continuously fed into the email database, the model must be able to incrementally learn the new knowledge embedded in these emails to avoid future mistakes.
2. *Adaptive to new categories.* Some new emails may not fall into any of existing categories in the model. The knowledge of the new categories can be added incrementally to the model without rebuilding it.
3. *Learning quickly from few examples.* In many application domains, the distribution of samples of different categories is uneven. Some categories have much less samples than others. It is difficult for the learning model to learn the characteristics of these categories due to insufficient samples.

A few learning algorithms [2], such as Naive Bayes, neural networks and super vector machines, possess the incremental learning capability. Adding new categories to a model is straightforward to the Naive Bayes algorithm and achievable in neural networks and super vector machines. The challenge to a learning algorithm is the speed of incremental learning. If new samples come slowly, most learning algorithms take long time to learn sufficient knowledge of the rare categories, because their incremental learning is done either through modification of the model architecture such as a neural network architecture [6] or readjustment of model parameters during the learning process such as the weights of a neural network or probabilities of a Naive Bayes model [4][5][7]. When classifying new samples of rare categories, the model usually performs badly until sufficient knowledge is learnt about the categories.

In this paper, we propose a new learning scheme, called $C3$, that uses three learning components to improve the learning performance on rare categories. The first component *chief-learner* is an ordinary learning model with the incremental learning capability. In the initial model building stage the chief-learner can learn sufficient knowledge on the categories with enough training samples and perform well on classification of new samples of these categories. However, it performs badly on some rare categories because it is insufficiently trained. As time goes and more new samples are fed to the model incrementally, the knowledge about the rare categories augment and its classification accuracy on the new samples of the rare categories increases.

To compensate for the weakness of the chief-learner in classifying samples of rare categories in the early stage, we use a group of *co-learners* to quickly gain the knowledge of these categories. The co-learners contain only the knowledge of rare categories and therefore are more sensitive to the new samples of rare categories. The outputs of the chief-learner and the co-learners are combined in the combiner in such a way that if the received new sample is of a rare category, the output of one co-learner will boost its chance to be classified correctly. The new samples of rare categories are fed to both the chief-learner and the co-learners to incrementally update the models. The more knowledge about the rare categories is learnt by the chief-learner, the less contribution to the classification is made by the co-learners. After the chief-learner has accumulated enough knowledge about the rare categories and is able to classify their samples independently, the co-learners become unnecessary and dead. The age of a co-learner depends on the learning speed of the chief-learner on that category.

We have implemented the $C3$ scheme using the Naive Bayes model as the chief-learner and instance-based learning technique as the co-learners. The combiner is designed as $(1 - f(t))P_{C_f} + f(t)P_{C_o}$ where P_{C_f} and P_{C_o} are the outputs of the chief-learner and co-learners respectively and $f(t)_{t \to \infty} \to 0$. Experiments on classification of customer emails from an e-commerce company have shown that the new learning scheme raises classification accuracy of the rare email categories in the early stage of the model.

The reminder of this paper is structured as follows: In Section 2, we introduce the framework of $C3$ and explain its incremental learning process and combination method. The implementation is addressed in Section 3. Section 4 presents the experimental results. Finally, we give a brief conclusion and point out our future work in Section 5.

2 C3 Scheme

2.1 C3 Framework

The $C3$ scheme is a variant of the ensemble learning algorithms [8] that combine the outputs of multiple base learners to make the final classification decision. Unlike most ensemble algorithms who see no differences among their base classifiers, $C3$ employs two kinds of base classifiers aiming at improving the classification performance at the early stage of incremental learning. The learning system of $C3$ consists of three components: a chief-learner, a set of co-learners and a combiner. Figure 1 shows the relationships of these three components. The chief-learner $C_f(t)$ is an ordinary classifier that has an incremental learning capability. Let $D(t_0)$ be a training data set collected at time t_0 and classified into m categories. $C_f(t_0)$ is a classifier built from $D(t_0)$ with a learning algorithm. Due to the uneven distribution of categories in the training data set, $C_f(t_0)$ may not perform on some rare categories whose samples are not sufficient in $D(t_0)$.

The co-learners $C_o^i(t_0)$ are built to compensate for the weakness of $C_f(t_0)$ in classifying the new samples of the rare categories. Each $C_o^i(t_0)$ is built only from the samples of the rare category c_i. It turns to produce a strong output when it classifies a new sample of category c_i and a weak output if the sample is in other categories.

The combiner $C_m(t)$ combines the outputs from $C_f(t)$ and $C_o^i(t)$ to determine the final classification on a new sample. If a new sample is in a rare

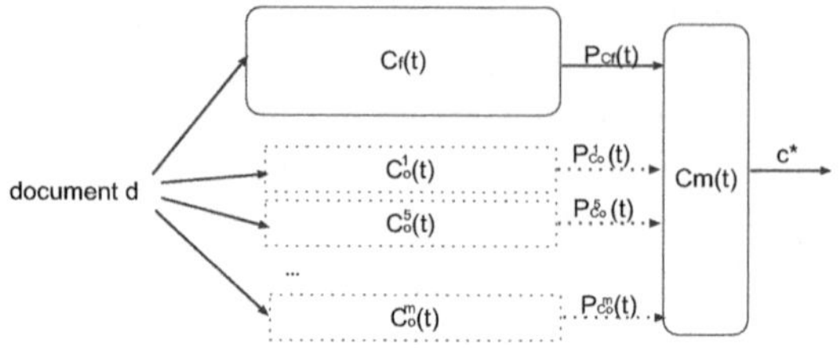

Fig. 1. The framework of $C3$ scheme. The combiner $C_m\,(t)$ combines the outputs from the chief-learner $C_f\,(t)$ and the co-learners $C_o^i\,(t)$

category c_i, $C_f(t)$ tends to produce a weak output while the co-learner $C_o^i(t)$ will generate a strong output. Let $P_{C_f}^i(t)$ and $P_{C_o^i}(t)$ be the outputs of $C_f(t)$ and $C_o^i(t)$ given input sample d from category c_i respectively. The output of the combiner $C_m(t)$ is calculated as

$$c^* = argmax_{i \in C}((1 - f_i(t))P_{C_f}^i(t) + f_i(t)P_{C_o^i}(t)) \tag{1}$$

where C is the set of all categories, $f_i(t)_{t \to \infty} \to 0$ is a decay function and $f_i(t) = 0$ if c_i is not a rare category.

From (1) we can see that if a new sample d belongs to a category c_j that is not a rare category, $P_{C_f}^j$ for category c_j will be large while P_{C_f} for other categories will be small. All P_{C_o} will be small. Therefore, the final classification will be determined only by $P_{C_f}^j$. If the new sample is in a rare category c_i, the chief-learner output $P_{C_f}^i$ may not be greater than those of other categories and the sample is likely misclassified by the chief-learner. However, the co-learner for category c_i will generate a large $P_{C_o^i}$ that will compensate for the weakness of the chief-learner result through the combiner (1) and a correct classification can be obtained.

As more examples of a rare category c_i are learnt by the chief-learner $C_f(t)$ overtime, $P_{C_f}^i(t)$ for category c_i increases. Because the decay function $f_i(t)$ decreases as more samples for category c_i are received, the effect of $P_{C_o^i}(t)$ also decreases to avoid bias on category c_i. At time t_n when sufficient knowledge about category c_i has been learnt by the chief-learner, $f_i(t_n) \to 0$. The co-learner for category c_i becomes unnecessary and dead. This process indicates that the co-learners are dynamic. When a sample of a new category is presented to the system, a co-learner for that category is created. When enough knowledge of a rare category is learnt by the chief-learner, the corresponding co-learner is removed. The life cycles of co-learners are determined by the decay functions $f_i(t)$.

2.2 Incremental Learning Process

The incremental learning process of a $C3$ learning system is illustrated in Figure 2. At the initial stage time t_0, a set of training samples from a learning domain

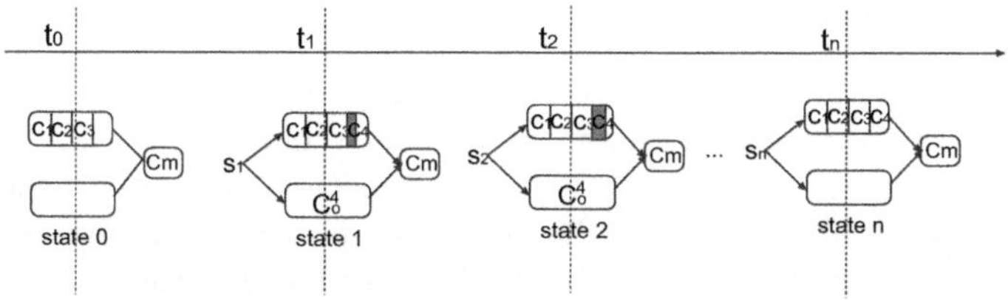

Fig. 2. The incremental learning process of $C3$. The new category c_4 is added at time t_1. At time t_n, the chief-learner is well trained on this category and the co-learner C_o^4 is removed

is collected. Assume the samples are classified into 3 categories $\{c_1, c_2, c_3\}$ and all categories have enough samples. The initial model of the chief-learner is built from the initial samples and no co-learner is created. At time t_1, few samples of a new category c_4 are identified. The chief-learner model built at time t_0 is unable to classify them. The knowledge of category c_4 is learnt incrementally by the chief-learner. A co-learner for this category is created from the few samples. The updated learning system now contains partial knowledge about category c_4, therefore can partially process new coming samples of that category. At time t_2, more c_4 samples have been present to the system, some being classified correctly and some classified wrongly. These samples are fed to the learning system and the chief-learner has learnt more complete knowledge about c_4, therefore, being able to classify new coming samples with a greater confidence. At time t_n, enough c_4 samples have been learnt by the chief-learner that has gained sufficient knowledge about c_4 through the incremental learning process and is able to classify new coming c_4 samples correctly. The co-learner for category c_4 has become unnecessary, therefore being removed from the learning system.

The above process describes the life cycle of one co-learner. In a real dynamic learning process, a few rare categories can be present in the same or different time periods. At any time t_i, old co-learners can be removed and new co-learners can be created. Also the life cycles of different co-learners are different, depending on the coming speed of the new samples of the rare categories. However, the $C3$ learning system is adaptive to this dynamic process. One important part of the dynamic learning process is the identification of samples of rare categories and feeding of the correct samples to the learning system. In real applications, this part is often performed by human being.

3 An Implementation

In this section, we present an implementation of the $C3$ scheme that uses the Naive Bayes model as the chief-learner and the instance-based techniques as the co-learners. We use the learning curve generated from the training samples to determine the combiner.

3.1 The Naive Bayes Chief-Learner

We choose the Naive Bayes model [1] as the chief-learner in our $C3$ implementation for its simplicity, capability in incremental learning and performance in text document classification which is our application domain. In the Bayesian learning framework, it is assumed that the text data is randomly generated by a parametric model (parameterized by θ). The estimates of the parameters $\hat{\theta}$ can be calculated from the training data. The estimated parameters then can be used to predict the class of new documents by calculating the posterior probability of the new documents belonging to the existing classes in the model.

Let D be the domain of documents that has $|C|$ categories and $p(d_i|\theta)$ a model that randomly generates a document $d_i \in D$ according to

$$p\left(d_i|\theta\right) = \sum_{j=1}^{|C|} p\left(c_j|\theta\right) p\left(d_i|c_j;\theta\right) \tag{2}$$

where $p(c_j|\theta)$ is the probability of category c_j and $p(d_i|c_j;\theta)$ is the prior probability of document d_i belonging to category c_j. In this paper we assume that $p(d_i|c_j;\theta)$ is a multinomial model [1]. Let W be the word vocabulary, document d_i is represented as a vector $\left\langle w_{d_{i,1}}, w_{d_{i,2}}, \ldots, w_{d_{i,|d_i|}} \right\rangle$, where $w_{d_{i,j}} \in W$ is the jth word in document d_i. With the Naive Bayes assumption on the independence of the word variables, the probability of document d_i in class c_j is given by

$$p\left(d_i|c_j;\theta\right) = p\left(|d_i|\right) \prod_{k=1}^{|d_i|} p\left(w_{d_{i,k}}|c_j;\theta\right) \tag{3}$$

where $\theta = (\theta_j, \theta_{j,t})$, $\theta_j \equiv p\left(c_j|\theta\right)$, $\theta_{j,t} \equiv p\left(w_t|c_j;\theta\right)$ and $\sum_{t=1}^{|W|} \theta_{j,t} = 1$. Let $N\left(w_t, d_i\right)$ be the count of word w_t in document d_i, and define $P(c_j|d_i) \in \{0,1\}$, as given by the documents class label. Given training documents set D, $\theta_{j,t}$ can be estimated by

$$\hat{\theta}_{j,t} = \frac{1 + \sum_{i=1}^{|D|} N\left(w_t, d_i\right) P\left(c_j|d_i\right)}{|W| + \sum_{s=1}^{|W|} \sum_{i=1}^{|D|} N\left(w_s, d_i\right) P\left(c_j|d_i\right)} \tag{4}$$

and θ_j is estimated by

$$\hat{\theta}_j = \sum_{i=1}^{|D|} P\left(c_j|d_i\right) / \left|D\right| \tag{5}$$

Given $\hat{\theta}_j$ and $\hat{\theta}_{j,t}$, the posterior probability of document d_i belonging to class c_j can be calculated by

$$p\left(c_j|d_i;\hat{\theta}\right) = \frac{p\left(c_j|\hat{\theta}\right) \prod_{k=1}^{|d_i|} p\left(w_{d_{i,k}}|c_j;\hat{\theta}\right)}{\sum_{r=1}^{|C|} p\left(c_r|\hat{\theta}\right) \prod_{k=1}^{|d_i|} p\left(w_{d_{i,k}}|c_r;\hat{\theta}\right)} \tag{6}$$

Incremental learning is performed by re-calculating new $\hat{\theta}$ based on previous $N(w_t, d_i)$, $P(c_j|d_i)$ and the new training sample. To reduce the vocabulary size and increase the stability of the model, feature selection is conducted by selecting words that have the highest mutual information with the class variable.

3.2 Instance-Based Co-learners

A co-learner for a rare category c_j is built from $|D_j|(\geq 1)$ samples where D_j is the set of samples from category c_j. Let $d_i = \langle w_{i,1}, w_{i,2}, \ldots, w_{i,|W|} \rangle$ be a document in D_j where $w_{i,j}$ is the frequency of word $w_j \in W$ in document d_i and $1 \leq i \leq |D_j|$. The co-learner for category c_j is modeled by center of the training samples as

$$\langle v_{j,1}, v_{j,2}, \ldots, v_{j,|W|} \rangle \tag{7}$$

where

$$v_{j,k} = \sum_{i=1}^{|D_j|} w_{i,k} / |D_j| \tag{8}$$

Given a new document $d = \langle w_1, w_2, \ldots, w_{|W|} \rangle$, we first calculate the Euclidean distance between the document and the center as

$$d_{v_j} = \sqrt{\sum_{k=1}^{|W|} (w_k - v_{j,k})^2} \tag{9}$$

We then map d_{v_j} to its probability of the normal distribution as

$$g(d_{v_j}) = \frac{e^{-(d_{v_j} - \mu)^2} / 2\sigma^2}{\sigma\sqrt{2\pi}} \tag{10}$$

where $\mu = 5.5$ and $\sigma = 1$ in our settings.

3.3 The Combiner

We use a decay function $f(t)$ to balance the influence of the chief-learner and the co-learners in the combiner when making the final classification. In learning knowledge of a new category, the co-learner learns much faster than the chief-learner.

Figure 3 shows the learning curves of the chief-learner and a co-learner. The learning speed of the co-learner $C_o(t)$ is faster than the chief-learner $C_f(t)$ in the early stage of the model. The co-learner is more accurate than the chief-learner in classification of a rare category samples. Therefore, more weight should be put on the co-learner. As time goes and more knowledge about the rare category is learnt by the chief-learner, its classification accuracy increases. Then the weight on the co-learner decreases in order to avoid bias on this category and result in overfitting. In this work we define the decay function as

$$f_j(t) = \frac{1}{\sqrt[4]{age_j(t)}}, \qquad t \geq 1 \tag{11}$$

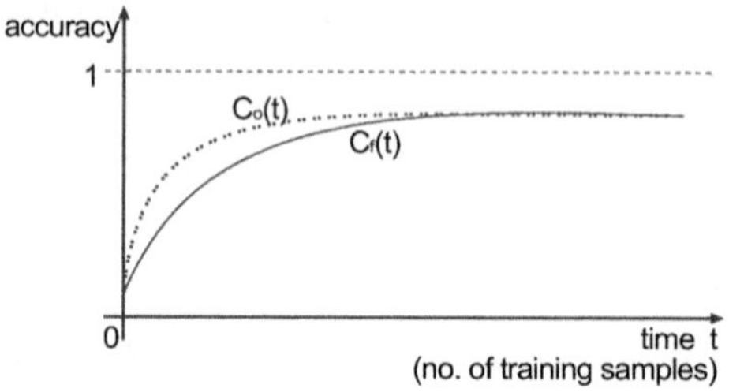

Fig. 3. Learning curves of the chief-learner $C_f(t)$ and the co-learner $C_o(t)$

where $age_j(t)$ is defined as the total number of samples on rare category c_j fed to $C3$ at time t. Thus the combiner of (1) is implemented as

$$c^* = argmax_{i \in C}((1 - \frac{1}{\sqrt[4]{age_i(t)}})P^i_{C_f}(t) + \frac{1}{\sqrt[4]{age_i(t)}}P_{C^i_o}(t)) \qquad (12)$$

4 Experimental Results

The implementation of the $C3$ scheme was tested with customer email data collected from a small e-commerce company in Hong Kong that sells products to global customers through emails. The company receives daily about 100 customer emails that order products, pay with credit cards, query about products, prices and services, and return products. The emails are written in different languages. Currently, an email categorization system is employed to classify incoming emails and route them to the corresponding people to process. All classification results are saved in an email database. If an email has been classified wrongly, the person who processes it will reclassify it and save the right result in the email database for model update.

4.1 E-mail Data Sets

971 customer emails received in 10 consecutive days in September 2002. These mails were manually classified to 17 categories according to some business criteria. In this experiment we threw away the junk emails and selected only 8 categories with more than 20 emails. We divided these emails into three groups. Group 1 contains categories of "Discount Question", "Item Question", "Order Address Amend". Group 2 contains categories of "Credit card Refused", "Over Charge", "Product Inquiry" and "Order Amend". Group 3 contains categories of "Order Status Question", "Over Charge" and "Product Inquiry". Two categories appeared in two groups without any particular reason.

We divided emails in each group into a training set and a test set. For the training set we chose one category as the rare category and assumed that emails in this category arrived in separate days. The emails in other categories collected before the date when the model was built. The settings of training emails are

Table 1. Training sets. Categories marked with * were the rare categories that arrived in separate days

Group1

Category	day0	day1	day2	day3	day4	day5	day6	day7	day8
Discount Question	14								
Item Question	13								
Order Address Amend*		1	6	11	16	21	26	31	36

Group2

Category	day0	day1	day2	day3	day4	day5	day6	day7	day8
Credit Card Refused	17								
Over Charge	31								
Product Inquiry	60								
Order Amend*		1	6	11	16	21	26	31	36

Group3

Category	day0	day1	day2	day3	day4	day5	day6	day7	day8	day 9
Order Status Question	149									
Over Charge	43									
Product Inquiry*		1	11	21	31	41	51	61	71	81

Table 2. Test sets

Group1

category	sample no.	category	sample no.
Discount Question	13	Item Question	11
Order Address Amend	18		

Group2

category	sample no.	category	sample no.
Credit Card Refused	15	Over Charge	30
Product Inquiry	15	Order Amend	14

Group3

category	sample no.	category	sample no.
Order Status Question	63	Over Charge	18
Product Inquiry	23		

shown in Table 1. We used one test data set for each group due to lack of sample emails. The same test set was used to test all updated models obtained from different dates. The number of samples for each category is given in Table 2.

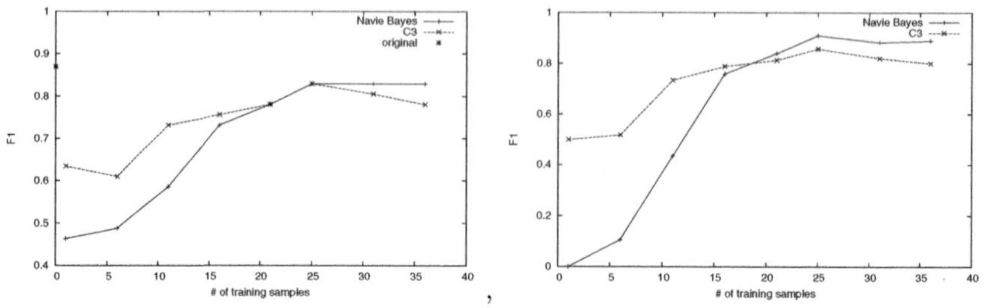

Fig. 4. Experimental results on Group1. Left side shows F_1 obtained from all the testing samples in Group1 and the right side that of the rare(Order Address Amend) category

4.2 Model Building and Incremental Learning

We first used training emails on Day 0 to build a Naive Bayes model and tested it with the test emails without the rare category. Then, we updated the model with emails of the rare category on Day 1 and tested the updated model with all test emails in a group. Since a new category was added on Day 1, we created a co-learner for the rare category to build a $C3$ model. We then tested the $C3$ model with the test emails. We continued this process for all the consecutive days and evaluated the classification performance of each updated model with the test emails.

The classification performance of each updated model was measured by

$$F_\beta = \frac{(\beta^2 + 1) \times p \times r}{\beta^2 p + r} \tag{13}$$

where p and r are the precision and recall [3] calculated from the test data set. In our approach, we regard precision and recall equally important, so we set $\beta = 1$. Since precision and recall are not isolated from each other, this measure avoids bias on either.

4.3 Result Analysis

Figure 4 shows the experiment results of Group 1 data. The horizontal axis represents the number of the rare category emails accumulatively fed to the model in consecutive dates. Each point on the curves represents a particular date when the model was updated incrementally and tested. The solid curve is the result of the Naive Bayes model and the dotted curve is the result of the $C3$ model. Each curve is the average result of three independent experiments on the Group 1 data.

From the left side of Figure 4, one can see the chief-learner Naive Bayes model performed very well on Day 0 when no email of the rare category was

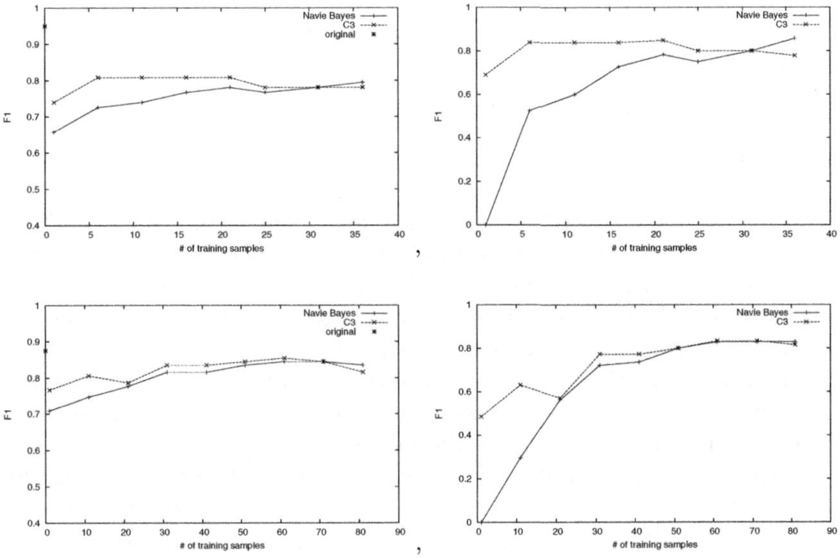

Fig. 5. Experimental results on Group2(top) and Group3(bottom)

present. On Day 1 when one rare category email was learnt, the performance of the Naive Bayes model dropped dramatically, the classification accuracy of the $C3$ model could still achieve more than 60% even if only one email was present in the co-learner. The effectiveness of the co-learner in the early stage of learning a rare category was obvious. On Day 2, 5 emails of the rare category were learnt by the models, the performance of the Naive Bayes model improved a little but the performance of the $C3$ model decreases slightly due to the disturbance of the data. On Day 3, 5 more emails of the rare category were learnt again. Both the Naive Bayes model and the $C3$ model showed a significant increase in classification performance because more knowledge has been accumulated in both models. However, the $C3$ model performed better than the Naive Bayes.

On Day 6, 26 emails of the rare category had been learnt. The two models achieved the best performance because the sufficient knowledge on that category had been accumulated. After Day 6, the performance of the Naive Bayes model remained stable. However, the performance of the $C3$ model dropped slightly. This is due to the overfitting of the $C3$ model caused by the co-learner. When this point arrived, the function of the co-learner should stop because the Naive Bayes could perform the function alone. The right side of Figure 4 shows the only result of the rare category emails. The advantage of the $C3$ model on classifying the rare category emails in the early learning stage is quite clear.

Figure 5 shows the results of Group 2 and Group 3 data sets. Similar trends can be observed, which demonstrated that the co-learner was effective in enhancing the classification accuracy of the rare category in the early stage of the model. However, the parameters of the co-learner and the combiner have to be well adjusted. In this experiment, the parameters of the normal distribution distance scaling of the co-learner was set as $\mu = 5.5$ and $\sigma = 1$. The decay function on co-learner was $f_i(t) = \frac{1}{\sqrt[4]{age_i(t)+0.2}}$.

5 Conclusions

In this paper we have described $C3$, a new learning scheme to use co-learners to boost the classification performance of a classification model on the rare categories in the early stage of incremental learning. We have discussed the implementation of the $C3$ scheme with the Naive Bayes model as the chief-learner and instance-based learning techniques as the co-learners. Our preliminary experiments on classification of limited customer emails have demonstrated that the co-learners can improve the classification performance of the rare category emails in the early learning stage of the model.

Although the initial results are encouraging, more experiments on diverse data from different domains to further test the effectiveness of the new scheme are necessary. We will continue experimenting with more customer emails in more categories and further study the parameters in the co-learners and the combiner. Furthermore, we will study other implementations with different learning models such as super vector machines and k-nearest neighbors.

References

[1] A. McCallum and K. Nigam, A Comparison of Event Models for Naive Bayes Text Classification, *AAAI-98 Workshop on Learning for Text Categorization* (1998). 752

[2] C. Giraud-Carrier, A Note on the Utility of Incremental Learning, *AI Communications* 13(4) (2000), 215-223. 748

[3] F. Sebastiani, Machine Learning in Automated Text Categorization, *ACM Computing Surveys* 34(1) (2002), 1-47 756

[4] J. D. M. Rennie, ifile: An Application of Machine Learning to E-Mail Filtering, *Proceedings of the KDD-2000 Workshop on Text Mining* (2000). 748

[5] R. B. Segal and J. O. Kephart, Incremental Learning in SwiftFile, *Proceedings of The 17th International Conference on Machine Learning* (2000), 863-870. 748

[6] S. E. Fahlman and C. Lebiere, The Cascade-Correlation Learning Architecture, *Advances in Neural Information Processing Systems* 2 (1990), 524-532. 748

[7] S. K. Chalup, Incremental Learning in Biological and Machine Learning Systems, *International Journal of Neural Systems* 12(6) (2002), 447-465. 748

[8] T. G. Dietterich, Machine-Learning Research: Four Current Directions, *AI Magazine* 18(4) 1997, 97-136 749

A New Approach for Scientific Citation Classification Using Cue Phrases

Son Bao Pham and Achim Hoffmann

School of Computer Science and Engineering
University of New South Wales, Australia
{sonp,achim}@cse.unsw.edu.au

Abstract. This paper introduces a new method for the rapid development of complex rule bases involving cue phrases for the purpose of classifying text segments. The method is based on Ripple-Down Rules, a knowledge acquisition method that proved very successful in practice for building medical expert systems and does not require a knowledge engineer. We implemented our system KAFTAN and demonstrate the applicability of our method to the task of classifying scientific citations. Building cue phrase rules in KAFTAN is easy and efficient. We demonstrate the effectiveness of our approach by presenting experimental results where our resulting classifier clearly outperforms previously built classifiers in the recent literature.

1 Introduction

With an ever increasing number of scientific papers and other documents available, it is impossible for researchers to read everything that might be relevant. It is useful to have a system that assists researchers in dealing with this problem possibly by suggesting only a small number of specific papers that they should read. Building a citation map is a good way of summarizing comments on a paper by other authors. A citation map is a directed graph with papers as nodes. A directed edge between two nodes indicates that one paper cites the other. The type of edges depend on their corresponding citation types. A citation map is useful for experienced researchers as well as for those that are new to the field. The newcomers could use the citation map, for instance, to find the fundamental work in the field. These usually include papers on which others *base* their work. Papers that are *supported* by a large number of other papers are also a good starting point.

Building robust natural language processing systems remains a challenging engineering task. A reason for the persistent difficulty with building more sophisticated NLP systems is the extraordinary variety in which certain content may be presented in natural language. Capturing the vast variety of possible natural language patterns which would allow the same system response, such as classifying entire texts or text segments, appears to require substantial task-specific knowledge.

T.D. Gedeon and L.C.C. Fung (Eds.): AI 2003, LNAI 2903, pp. 759–771, 2003.

In this paper, we introduce a new approach for capturing such task-specific linguistic patterns from human experts (most humans who are fluent in a language would qualify as experts).

Our approach builds on the basic idea of Ripple-Down Rules [3], which strongly support the knowledge acquisition in the context of its actual use. For the tasks in natural language processing that means that an expert monitors the system's performance on individual pieces of text. As soon as the system makes a mistake due to some rule R, the expert tells the system how the current context, i.e. the given text or text segment, differs from previous contexts in which rule R performed satisfactorily. This kind of knowledge acquisition approach proved to be substantially more effective than machine learning techniques applied to the same data [5]. Reason being that the human is in the loop who provides important guidance for building a knowledge base which learning techniques can only achieve by the use of much more data. Since the data needs to be manually classified, the overall involvement of the human can be significantly reduced using a knowledge acquisition approach instead of a learning approach.

We developed KAFTAN, our Knowledge Acquisition Framework for TAsks in Natural language, which is capable of rapidly acquiring cue phrases for classifying individual citations in a text. Recently CiteSeer, a citation visualization tool which performs Autonomous Citation Indexing, has been built [6]. References together with surrounding text are automatically extracted from all papers available to CiteSeer. However, CiteSeer does not provide users with the types of citations between papers. The user has to read the citation context in order to work out the type of citation. KAFTAN would therefore benefit CiteSeer substantially.

2 Related Work

In this section we present the basic idea of Ripple-Down Rules in 2.1 upon which our approach was based. In 2.2 we discuss related work on our application domain of citation classification.

2.1 Knowledge Acquisition with Ripple Down Rules

Ripple Down Rules (RDR) is an unorthodox approach to knowledge acquisition. RDR does not follow the traditional approach to knowledge based systems (KBS) where a knowledge engineer together with a domain expert perform a thorough domain analysis in order to come up with a knowledge base. Instead a KBS is built with RDR incrementally, while the system is already in use. No knowledge engineer is required as it is the domain expert who repairs the KBS as soon as an unsatisfactory system response is encountered. The expert is merely required to provide an explanation for why in the given case, the classification should be different from the system's classification.

This approach resulted in the expert system PEIRS used for interpreting chemical pathology results [3]. PEIRS appears to have been the most comprehensive medical expert system yet in routine use, but all the rules were added by a pathology expert without programming or knowledge engineering support or skill whilst the system was in routine use. The basic idea has been extended into a number of directions - none of it addressing the specific problems present in NLP applications.

Ripple-Down Rules and some further developments are now successfully exploited commercially by a number of companies.

Single Classification Ripple Down Rules A single classification ripple down rule (SCRDR) tree is a finite binary tree with two distinct types of edges. These edges are typically called *except* and *if not* edges. See Figure 1. They are used for evaluating cases and will be discussed later. Associated with each node in a tree is a *rule*. A rule has the form: *if* α *then* β where α is called the *condition* and β the *conclusion.*

Cases in SCRDR are evaluated by passing a case to the root of the tree. At any node in the tree, if the example entails the condition of the node, the node is said to *fire*. If a node fires, the example is passed to the next node following the *except* branch. Otherwise, the case is passed down the *if not* branch. This determines a path through a SCRDR tree for an example. The conclusion given by this process is the conclusion from the last node which fired on this path. To ensure that a conclusion is always given, the root node typically contains a trivial condition which is always satisfied. This node is called the *default* node.

A new node is added to an SCRDR tree when the evaluation process returns the wrong conclusion. The new node is attached to the last node evaluated in the tree. If the node has no exception link, the new node is attached using an exception link, otherwise an *if not* link is used. The case causing the new node being added (call this example e) is associated with the new node and is called the *cornerstone case* for that node. To determine the rule for the new node, the expert formulates a rule which is satisfied by e but not by the cornerstone case for the last node which fired in the evaluation path.

While the process of incrementally developing knowledge bases will eventually lead to a reasonably accurate knowledge base, provided the domain does not drift and the experts are making the correct judgements, the time it takes to develop a good knowledge base depends heavily on the appropriateness of the used language in which conditions can be expressed by the expert. For example, if the target function to be represented is a linear threshold function in the numerical input space and the conditions an expert can use allow only axis-parallel cuts, it will take very long until most of the relevant objects will be classified correctly.

Despite a number of different application areas in which RDR systems have been employed, serious applications in natural language processing have not yet been tackled with Ripple-Down Rules. For natural language tasks it is not obvious what a suitable language for conditions will be. In this paper, we tried rather simple conditions as discussed in 3.2 and they seem to be sufficient for our task.

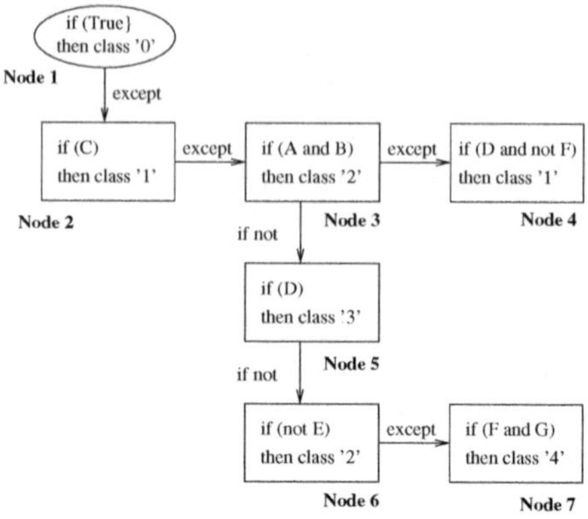

Fig. 1. An example SCRDR tree. Node 1 is the default node. A case for which only 'A' is true is classified as '0' by node 1. If only A, B, and C are true, it is classified as '2' by node 3, while it is classified as '1' by node 4 if on top of A,B, and C also D was true. If only A, C, and D are true it is classified as '3' by node 5. If only A and C are true, the case is classified as '2' by node 6, etc.

2.2 Work on Citation Classification and Linguistic Patterns

Linguistic patterns or cue phrases have a long history as features for determining global importance of the sentences containing them. In Edmundson's approach [2], sentences containing positive cue phrases that express confidence or importance (e.g. *absolutely*, *important*) were candidates for inclusion in the summary. On the other hand, sentences containing stigma words (e.g. *unclear*, *hardly*) were unlikely to be included in the summary. More recent works include Burstein et al. [1] where cue phrases were used to derive rhetorical relations. Teufel [8] also used long cue phrases (e.g. *similar to those of*) for her Argumentative Zoning work.

Nanba and Okumura [7] introduce a supporting tool for writing survey papers. They use cue phrases to classify citations in text into one of three categories i.e type B (Basis), type C (Comparison or Contrast) and type O (other). Inspecting Nanba & Okumura's cue phrases after they were selected manually by using n-gram models suggests that they could easily be expanded to generalize better if our notion of word group were used in those phrases. For example, a sentence having the cue phrase

however, ... they

would be classified into type C. But Nanba & Okumura's list of cue phrases does not contain the phrase

however, ... he/she

which is a similar pattern. This is probably because the latter kind of phrase did not appear often enough in the used corpus (before formulating cue phrases Nanba & Okumura analyzed their corpus by searching for frequently occurring patterns). However, including that phrase would clearly classify more sentences outside the corpus correctly. This problem is addressed in our approach by using word groups which will be described in section 3.4.

Teufel [8] tries to classify sentences into Argumentative Zones which she argues could be used for building citation maps as well as for text summarization. The seven zones consist of *Background*, *Other*, *Own*, *Aim*, *Textual*, *Contrast*, and *Basis*. Among those, only *Contrast* and *Basis* are relevant to our task.

Different to Nanba & Okumura, Teufel [8] proposes to use meta-discourse features which include formulaic expressions, agent patterns and semantic verb classes. Formulaic expressions are bundled into 20 major semantic groups (e.g. *CONTRAST*, *GAP*, *PROBLEM*, etc.). To determine which group a sentence belongs to, 396 formulaic patterns were manually constructed. An example of a *GAP* formulaic pattern is

as far as @SELF_NOM know

where @SELF_NOM is an entry in a manually constructed concept lexicon consisting of lists of words. Similarly, there are 168 agent patterns and semantic verb classes composed of 365 verbs constructed manually. The patterns manually constructed by Teufel are similar to our patterns used in the condition part of our rules, though Teufel's rules are less expressive. The lexicon is also very much like our word groups.

Both Nanba & Okumura's as well as Teufel's work show that the skillful use of cue phrases can lead to useful tools for purposes such as building citation maps. However, the manual development of suitable cue phrases is obviously a rather time-consuming task. Furthermore, certain cue phrases are bound to contradict each other, i.e. a single sentence will be matched by multiple cue phrases which may well belong to different classes. Resolving such conflicts manually is a time-consuming activity.

Our approach provides automatic support for eliminating such conflicts from the beginning by structuring the knowledge base in a suitable way and by presenting potential conflict-causing sentences to the expert as soon as they are encountered. This makes it very easy for the expert to provide a rule that resolves such conflicts. The integration of the rule into the knowledge base without causing any inconsistencies is also done automatically.

3 KAFTAN Design

Our Knowledge Acquisition Framework for TAsks on Natural language (KAFTAN) allows users to easily and quickly develop a knowledge base for natural language sentence classification. It is in general hard for the expert to formulate a good rule from one example. With KAFTAN, the expert has merely to justify why the current example is of a particular category. The expert will then be guided through a simple process to expand the current rule so that it could cover a larger number of cases. Our KAFTAN system offers the user the following features.

- KAFTAN supports an effective user machine interaction by first highlighting those parts of a sentence that led to a misclassification. Based on that it is

usually easy for an expert to come up with a suitable choice of one or more words in the current sentence that justifies its classification into a different class.
- Following that initial choice of words, KAFTAN offers generalizations of each selected word to a word group composed of synonyms that are appropriate in the context.
- To support the quick selection of synonyms, KAFTAN colour-codes its lists of synonyms according to the various meanings of the original word, which KAFTAN takes from WordNet [4].

3.1 Knowledge Bases Built with KAFTAN

KAFTAN allows users to define a number of classes into which a sentence can be classified. For each class C an SCRDR tree will be acquired which determines whether a sentence is classified as class C or not. Every SCRDR tree is composed of a number of nodes each containing a KAFTAN rule. Rules in KAFTAN are composed of a condition part and a conclusion part. The condition is to be evaluated on an individual sentence. The conclusion part is either a class label or a text segment selector. KAFTAN allows users to build two types of rules. One is for plain sentence classification, i.e. it uses a class label as conclusion. The other type is an information extraction rule, which uses a selector conclusion to select parts of a sentence to be filled into a corresponding field of a template. Both types of rules use the same kind of conditions.

3.2 Conditions in KAFTAN Rules

Regular expression was considered to be used as the condition for KAFTAN rules initially because of its expressiveness. In order to take this advantage however, the user has to predict sentences similar to the one at hand to formulate the desired conditions that cover more cases. This puts a burden on users which is against the motivation of our system. KAFTAN is built to help users minimize their effort in formulating new rules. The user is only required to differentiate the case at hand from the cornerstone cases without worrying about unseen cases. It is therefore desirable to have a condition syntax that is easy and intuitive to use while expressive enough. A simplified version of regular expression turns out to suffice for KAFTAN's objectives.

A condition in KAFTAN is a simple pattern consisting of a sequence of an arbitrary number of word groups (see section 3.4 for details). Between two word groups there may be a gap whose maximum length can be specified (or left unlimited) within the condition. A gap represents effectively a wild-card which is matched by any sequence of words of up to the maximum number of words. Inter-punctuation within a sentence is ignored at this stage. Each word group represents a set of alternative words that may occur in a matching sentence at the respective position. An example condition may be the following:

(approach|method) (gap 5) (is|was) (good)

This condition would match sentences containing "approach" or "method" followed by "is" or "was" with at most 5 words apart (gap 5), followed immediately by "good". For example, it would match the sentence

Our approach of using knowledge acquisition is good because it makes the development of intelligent systems much easier.

To increase the expressiveness of the condition language, we also allow negation to be used. That is one can require that a particular word, word group or multiple such words or word groups do *not* match the sentence. (This is an important feature when formulating exception rules to a rule.)

3.3 Automatic Check for Consistency

As discussed in section 2.1 our approach has the objective of ensuring the incremental improvement of a system's capabilities. This implies that all sentences which KAFTAN had already classified to the satisfaction of the user need to be classified in the same way after any modification to KAFTAN's knowledge base. The only type of modification of KAFTAN's knowledge base that is allowed is the addition of further (exception) rules to its rule base.

To ensure this and to support the user to add suitable rules, KAFTAN checks automatically for any potential inconsistencies with previous classifications of sentences for each candidate rule a user enters. I.e. KAFTAN stores all sentences it sees along with their classification which was endorsed by the user. For any new rule R a user wants to enter, KAFTAN checks whether R would misclassify any previously classified sentences. If a rule would misclassify such a sentence, this is indicated to the user and KAFTAN asks to modify that rule or to start from scratch to form a new rule. I.e. after a rule has been accepted by KAFTAN, it does not need to be revisited again and will remain in the knowledge base.

3.4 Word Groups

A specific sequence of words used in a condition in a KAFTAN rule would apply to the sentence that triggered the user to enter the rule (the cornerstone case) and may also apply to a number of other sentences. However, in many cases there will be a number of different ways of expressing essentially the same content. Hence, to avoid entering essentially the same rules over and over again with some slight variation in the particular words being used, KAFTAN allows users to use a set of words from which only one has to match a sentence. The set of words can be chosen by the user as needed in order to accommodate synonyms or other related words which are commonly used in place of the particular word in the cornerstone case.

Furthermore, KAFTAN supports the user rather strongly in finding a suitable set of words in the following way:

– WordNet [4] is used to provide an initial set of possible synonyms and hypernyms.

- The synonyms and hypernyms in WordNet are ordered according to different meanings of the query word. These different groups of synonyms and hypernyms are colour-coded when presented to the user.
- The user can form their own sets of related words which can be tailored to suit specific purposes not catered for in WordNet.

For each of the groups *Noun*, *Adjective*, *Verb*, and *Adverb* synonyms in different "contexts" are highlighted in different colours. If there is no synonym in a particular group, there is no display of that group at all.

Within a group of the same "context", synonyms and hypernyms are also highlighted differently. In the event that the suggested lists and sublists of synonyms are not suitable for forming patterns, the user can also form their own word groups. These word groups can also be effectively reused by browsing quickly through existing word groups which are presented together with the synset-based word groups taken from WordNet.

Giving the user full control, as we do, does not appear to put an unreasonable burden on the user. By doing that, we also eliminate all potential problems which might come with an automatic preselection based on, e.g. a Part-of-Speech tagger. With the provided interface the user can easily select those words from one of the lists he or she deems appropriate. This is usually accomplished within 10s of seconds.

3.5 Templates

To build systems for more sophisticated tasks than simple sentence classification, KAFTAN allows users to develop concurrently for each category an information extraction knowledge base that fills the fields of a template with the respective information provided in the sentence. For each category the template can have different fields. What part of a sentence is filled into which slot is determined by a separate RDR-style knowledge base. For each slot in each template a separate RDR-tree is built. The knowledge for filling the templates is also incrementally acquired. Each time a sentence is correctly classified the user can check the content of the fields in the corresponding template. If a slot is incorrectly filled, the user will enter an exception rule to the rule responsible for the mistake. This gives us flexibility to refine any previously entered rule for both the classification task and the information extraction task. The conditions KAFTAN allows for the template extraction RDR node are the same as the condition in the classification RDR node.

The concept of templates for each class allows many different applications. In the case of citation classification which we discuss in this paper, it can be used to address the following problem: Sometimes it is impossible to tell what category a sentence belongs to just by looking at that sentence. This may for example be due to the fact that the authors introduce the cited work in one (or more) sentence(s) and describe the relationship later (in a sentence that does not contain any citations). In this situation, we can "inspect" sentences around the citations and use templates to check if they talk about the citation.

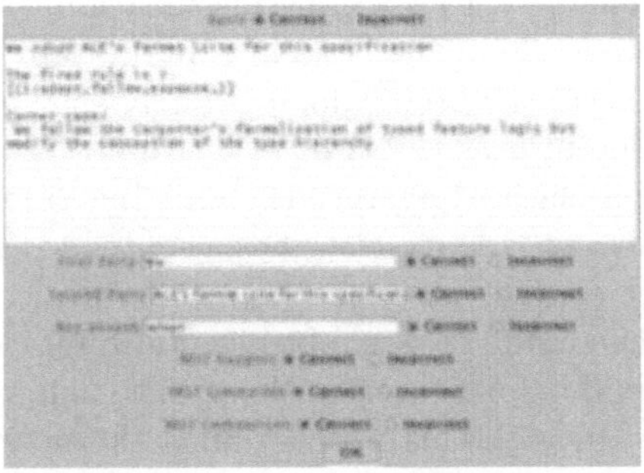

Fig. 2. A screen shot of a KAFTAN window allowing users to provide the correct classification of a sentence. If a sentence is classified into a particular type, the fired rule and its cornerstone case are shown. The extracted phrases of all the slots in the corresponding template are also displayed. For each classification result that the user deems incorrect, s/he can tick the corresponding *Incorrect* box to enter a new exception rule

For example:

Taylor describes the Naive Bayes Classifier in \cite. In this paper, we adopt the Naive Bayes Classifier to do our classification task.

We cannot tell which category the first sentence belongs to. However the second sentence can be classified as *Basis* by the following rule:

(We|I|he|she) (adopt)

and the extracted fields for the template are:

- First party: *In this paper, we*
  Second party: *the Naive Bayes Classifier to do our classification task*
- Phrase: *adopt*

Using simple word matching, we can see that *Naive Bayes Classifier* mentioned in the first sentence appears in the *Second party* field which would suggest that it is a basis of the work in the *First party* field.

3.6 Building Citation Classifiers

The user is presented with the text (paragraph). He can either select a sentence with citation manually by highlighting the sentence with the mouse or ask the system to pick one automatically.

The system automatically classifies the selected sentence and show the result to the user. The result includes the classifications as well as the extracted fields for every category into which the sentence was classified (Figure 2). The user then verifies the displayed result. If an item was incorrect, the fired rule is presented so the user can see why the misclassification or incorrect information extraction has occurred.

To help the user decide how to fix a misclassification, the phrases that match the condition of the firing rule in the current sentence are highlighted. Together with the cornerstone case of the fired rule, it is fairly easy to come up with a new exception rule that differentiates the two.

The new rule is then checked against the existing rule base for consistency. There is a conflict if the new rule misclassifies a sentence that has been correctly classified before. In the conflict case, the new rule is rejected and the user has to construct a different condition possibly by modifying the previous one. The conflicting sentence would be presented to the user to help revising the condition.

The user has to go through this process until the new rule is accepted into the rule base. The sentence then is added into the list of previously seen cases for future consistency checks.

4 Experiments with KAFTAN

In this paper, we used four citation types, namely *Basis*, *Support*, *Limitation* and *Comparison*. In [9] 15 different citation types were distinguished. Our citation types appear to be the most relevant among those 15 types for the purpose of a citation map that is designed to support obtaining an overview of a field of research.

- *Basis*: One work is based on another work.
- *Support*: One work is supported by another work.
- *Limitation*: One work has been criticized to have some limits or weaknesses.
- *Comparison*: Two approaches are compared.

These citation types arguably reflect the most important relationships between papers that users are interested to know. They had been chosen before we acquired the corpus used in our experiment.

For each of these four categories, we have developed one SCRDR tree. We classify sentences and deem the citations occurring in those sentences to belong to the category of the sentence. Since a sentence could belong to more than one category we developed one RDR tree for each category so that one sentence may be classified into multiple categories.

Experiments In our experiments, we divided the data into two groups. The first group has 482 sentences while the second has 150 sentences. Initially, we used sentences in the first group to incrementally build our knowledge base(KB) using KAFTAN. As the result, our KB has 70 rules for the *Basis* class, 24 rules for the *Support* class, 23 rules for the *Limitation* class and 54 rules for the *Comparison* class. Each rule took about 2-3 minutes to be created by the user. The created KB is then expanded using 150 sentences from the second data group. At the end of this phase, the rule base ended up having 79 rules for the *Basis* class, 28 rules for the *Support* class, 30 rules for *Limitation* class and 59 rules for the *Comparison* class.

Table 1. Nanba & Okumura's results

		reference type identified by rules			accuracy for each type(%)
		C	B	O	
correct reference type	C	12	0	4	75.0
	B	2	25	5	78.1
	O	1	5	46	88.5

Every additional rule added was a result of one misclassification. Therefore the number of rules added reveals the number of error the rule base makes on second data group. To see how well KAFTAN performs, we compare our results against Nanba and Okumura's results which used the same corpus.

Nanba & Okumura used 282 reference areas for making rules and 100 for evaluation. A reference area in their approach usually contains multiple sentences. Their results are shown in Table 1.

Because Nanba & Okumura allow single classification and they have only two categories (excluding *Other*) we compare their accuracy against ours in classifying citations into a single class. Our *Limitation* and *Comparison* roughly corresponds to their TYPE C class whereas the BASIS category corresponds to the TYPE B class. Our *Support* category has no correspondence in their approach.

To make Nanba & Okumura's numbers comparable to our results we perform the following calculation which actually increases their accuracy numbers: When classifying citations into TYPE C, their system misclassified 4 TYPE C cases as seen in Table 1. Additionally, 2 TYPE B cases and 1 TYPE O case were classified incorrectly as TYPE C. This resulted in 7 misclassified cases out of 100 cases which is equivalent to an accuracy of 93%. Similarly, 7 cases of TYPE B were misclassified and 5 cases not of TYPE B were misclassified into TYPE B. Therefore the accuracy for TYPE B is (100-12) = 88%.

The accuracy of KAFTAN in classifying sentences into one category is calculated as follows. For the BASIS category, 9 rules were added which is equivalent to making 9 errors out of 150 cases. This results in an accuracy of (150-9)/150 = 94%. KAFTAN performed favourably in comparison to Nanba & Okumura's system (see Table 2).

Inspecting our knowledge base reveals that there are 47 exception rules not counting exception rules of the default rule. Every exception rule is the result of a misclassification of an existing rule. In approaches using a list of linguistic patterns, the rule that caused the misclassification needs to be modified. This might cause conflict as the modified rule usually has already correctly classified more than one sentence. This proves the usefulness of the rules structure in KAFTAN over the "flat" list of linguistic patterns.

Table 2. Our results using KAFTAN compared against Nanba and Okumura's results on the same set of documents

	# of rules with 482 sentences	# of errors in 150 citations	Accuracy
BASIS	70	9	94.0%
SUPPORT	24	4	97.3%
LIMITATION	23	7	95.3%
COMPARISON	54	5	96.7%
Nanba and Okumura's results			
TYPE B			88%
TYPE C			93%

5 Conclusion

This paper presented our system KAFTAN (Knowledge Acquisition Framework for Tasks in Natural language). KAFTAN allows users to rapidly develop knowledge bases for classification of sentences along with extracting information relevant for the particular sentence class.

Our results are very encouraging as the achieved classification accuracy is higher than a previous approach from the literature and the effort required with KAFTAN seems less than that in the comparison case in [7]. It should also be noted that while our measured accuracy is based on 'training data', the data was independent from the data used to build the bulk of the knowledge base. The accuracy measurements are taken from the last batch of 'training data' which did not result in a significant change of the knowledge base, i.e. if the knowledge base had not been changed on that last batch of data the same accuracy would have been measured. Hence our measurement is as good as measuring the accuracy on independent test data.

Further research will investigate the feasibility to automatically propose a list of possible conditions for discriminating the given sentence from cornerstone sentences. This would make the task of entering rules for the user even easier and we hope that the rules could become more general, so that the developed knowledge base can reach high accuracy even sooner.

Sentence classification was an essential subtask for our citation classification. But sentence classification is a recurring theme in other advanced tasks as well, including text summarization and information extraction.

We believe that the approach we took with KAFTAN is widely applicable for building powerful natural language processing tools. It represents a serious alternative to supervised machine learning approaches for NLP. While supervised machine learning approaches require a sizeable set of training instances which need to be classified, usually manually, we strongly believe that we can achieve the same or better results with significantly less data using the advice of a human who provides crucial information, e.g. in form of cue phrases, etc. to the system. This results in utilizing the time of the human in a more direct way to building the system. In fact, it looks like a detour to ask the human to classify training instances manually so that a program can find rules, instead of asking the human for the rules directly. While we haven't conducted rigorous studies in the NLP field as yet to confirm this point, studies in the medical domain involving fixed-length attribute vectors to be classified have been conducted and strongly support our point [5]. In NLP we expect this effect to be even more pronounced as the set of potential attributes to be used in conditions is almost infinitely larger than in the studies mentioned and hence, will hamper the automatic learning substantially, as is known from theoretical studies, such as the VC dimension, as well as practical studies.

Acknowledgements

We thank Hidetsugu Nanba for providing us with the corpus of scientific papers from the E-print archive which we used in our experiments and comparisons with Nanba & Okumura's work.

References

[1] J. Burstein, D. Marcu, S. Andreyev, and M. Chodorow. Towards automatic classification of discourse elements in essays. In *Proceedings of ACL-2001*, Toulouse, France, 2001. 762

[2] H. Edmundson. New methods in automatic extracting. *The Association for Computing Machinery*, 16(2):264–285, 1969. 762

[3] G. Edwards, P. Compton, R. Malor, A. Srinivasan, and L. Lazarus. PEIRS: a pathologist maintained expert system for the interpretation of chemical pathology reports. *Pathology*, 25:27–34, 1993. 760

[4] C. Fellbaum, editor. *WordNet - An electronic lexical database.* MIT PRESS, CAMBRIDGE, MA, 1998. 764, 765

[5] B. Kang, P. Compton, and P. Preston. Multiple classification ripple down rules: Evaluation and possibilities. In *Proceedings of the 9th AAAI-sponsored Banff Knowledge Acquisition for Knowledge Based Systems Workshop*, pages 17.1–17.20, 1995. 760, 771

[6] S. Lawrence, C. L. Giles, and K. Bollacker. Digital libraries and autonomous citation indexing. *IEEE Computer*, 32(6):67–71, 1999. 760

[7] H. Nanba and M. Okumura. Towards multi-paper summarization using reference information. In *Proceedings of the International Joint Conference on Artificial Intelligence*, Stockholm, Sweden, 1999. 762, 770

[8] S. Tuefel. *Argumentative Zoning: Information Extraction from Scientific Text.* PhD thesis, Edinburgh, 1999. 762, 763

[9] N. Weinstock. Citation indexes. In A. Kent, editor, *Encyclopedia of Library and Information Science*, volume 5, pages 16–41. Marcel Dekker, New York, 1971. 768

Automatic Dialogue Segmentation Using Discourse Chunking

T. Daniel Midgley and Cara MacNish

School of Computer Science and Software Engineering
Discipline of Linguistics, University of Western Australia
{fontor,cara}@csse.uwa.edu.au

Abstract. This study explains a method for arranging dialogues into discourse chunks. Discourse chunking is a simple way to segment dialogues according to how dialogue participants raise topics and negotiate them. It has been used successfully to improve performance in dialogue act tagging, a classification task where utterances are classified according to the intentions of the speaker.
Earlier work showed that discourse chunking improved performance on the dialogue act tagging task when the chunk information was correct and hand-coded. The goal for the current study is two-fold: first, to investigate how accurately dialogues can be marked with discourse chunks automatically, and second, to determine the effect of the discourse chunk information on dialogue act tagging. We present evidence which shows that discourse chunking improves the performance of the dialogue act tagger, even when the chunk information is imperfect.
The dialogue act tagger for this study uses case-based reasoning, a machine learning technique which classifies utterances by comparing their similarity to examples from a knowledge base.

Keywords: dialogue act tagging, case-based reasoning, evolutionary algorithms, language understanding, language generation, machine learning

1 Dialogue Act Tagging

Dialogue has an observable structure. When humans engage in purposeful dialogue, they structure their utterances in ways that help them accomplish their conversational goals. A knowledge of this structure can be helpful in natural language processing tasks, such as dialogue act tagging.

A dialogue act (hereafter DA) is a coarse-grained representation of a speaker's intentions in dialogue—what the speaker is trying to accomplish by saying something. Dialogue act tagging is a text classification task whereby utterances in a dialogue are categorised according to the intentions of the speaker. In this task, human annotators tag utterances in a training corpus with the most appropriate DA from a tagset. With this information, machine learning techniques are used to determine the most appropriate DA tag for the set of utterances in a test corpus.

T.D. Gedeon and L.C.C. Fung (Eds.): AI 2003, LNAI 2903, pp. 772–782, 2003.

DA tagging is a difficult task. Even human annotators agree with each other only about 84 percent of the time, as noted in [15]. This is partly because intentions cannot be examined directly; they must be inferred from the context of the dialogue. To do this, a listener uses linguistic and real-world knowledge, and makes subtle assessments about the plans, goals, beliefs, and knowledge of the speaker. Computational linguists are making headway in these areas, but the problem remains a difficult one.

Another difficulty in DA tagging is that speakers frequently perform more than one dialogue act with the same utterance. For instance, the utterance "yeah" could indicate that the speaker is agreeing to a proposition, giving positive feedback, or simply following the conversation—or many of these at once [3]. In these situations, it may be difficult to select the one 'best' speech act.

Despite these difficulties, DA tagging is an important area of natural language processing, with application in speech recognition and language generation. On a theoretical level, DA tagging represents an initial step toward intention recognition, which is critical for true natural language understanding.

This DA tagging project uses Verbmobil-2, a corpus of appointment scheduling dialogues, along with its accompanying tagset, shown in Table 1.

2 Discourse Chunking

There has traditionally been a lack of true dialogue-level information in automated dialogue act tagging. Past work [13, 14, 15, 16] has focussed on lexical information (the words or n-grams in an utterance), and to a lesser extent syntactic and phonological information (as with prosody). Discourse-level information is typically limited to surrounding DA tags [13, 14]. While such information can be useful, knowledge of prior DA tags is not always an accurate guide to the current one, especially when this information is imperfect. Information about the structure of dialogue would be the 'right tool for the job' of dialogue act tagging.

Theories about the structure of dialogue (for example, centering [6], and more recently Dialogue Macrogame Theory [9]) have not been applied to the DA tagging task. Incorporating these theories into the DA tagging task would constitute a separate tagging task of its own, with the concomitant time-consuming corpus annotation.

Discourse chunking is a method of dialogue segmentation that gives information about patterns of topic raising and negotiation in dialogue, and where an utterance fits within these patterns. It is also able to use existing DA tag information, without the need for separate annotation. In previous work [10], we showed a rule-based method for marking discourse chunks, and showed that applying the rule-derived discourse chunk information to the test corpus improved the performance of a dialogue act tagger. However, these rules relied on knowledge of DA tags—information that would be unavailable for a new corpus. The goal of the current study, then, is twofold: first, to investigate whether discourse chunks can be recognised in a test corpus without using DA tag information

Table 1. The tagset from Verbmobil-2, with examples

Tag	Example
ACCEPT	sounds good to me
BACKCHANNEL	mhm
BYE	see you
CLARIFY	I said the third
CLOSE	okay <uhm> so I guess that is it
COMMIT	I will get that arranged then
CONFIRM	well I will see you <uhm> at the airport on the third
DEFER	and I will get back to you on that
DELIBERATE	so let us see
DEVIATE_SCENARIO	oh I have tickets for the opera on Friday
EXCLUDE	January is basically shot for me
EXPLAINED_REJECT	I am on vacation then
FEEDBACK	gosh
FEEDBACK_NEGATIVE	not really
FEEDBACK_POSITIVE	okay
GIVE_REASON	because that is when the express flights are
GREET	hello Miriam
INFORM	<uhm> I I have a list of hotels here
INIT	so we need to schedule a trip to Hanover
INTRODUCE	Natalie this is Scott
NOT_CLASSIFIABLE	and <uh>
OFFER	<uhm> would you like me to call
POLITENESS_FORMULA	good of you to stop by
REFER_TO_SETTING	want to step into your office since we are standing right outside of it
REJECT	no that is bad for me unfortunately
REQUEST	you think so?
REQUEST_CLARIFY	I thought we had said twelve noon
REQUEST_COMMENT	is that alright with you
REQUEST_COMMIT	can you take care of <uhm> arranging those reservations
REQUEST_SUGGEST	do you have any preference
SUGGEST	we could travel on a Monday
THANK	okay thanks John

in a new corpus; and second, whether this automatically-recognised discourse chunk information, though imperfect, can be used to improve performance on the DA tagging task.

To test this, we divided the Verbmobil-2 corpus into two parts: a training corpus consisting of 50 dialogues (7197 utterances) from the Verbmobil-2 corpus, and a test corpus of 9 dialogues (1202 utterances). In the first stage of the experiment, we used a set of hand-built rules to tag the training corpus with discourse chunk information. These rules, described in detail below, rely on the DA tag information supplied by human annotators. Machine learning is then

Table 2. A sample of dialogue from Verbmobil, before dialogue chunking

LMT	so there is a noon flight	SUGGEST
LMT	and then there is a three o'clock flight	SUGGEST
LMT	so you want to	NOT_CLASSIFIABLE
KNT	okay	ACCEPT
KNT	I think we better shoot for the noon	CONFIRM
LMT	the noon yeah	REQUEST_CLARIFY
LMT	okay	FEEDBACK_POSITIVE
KNT	yeah	FEEDBACK_POSITIVE
LMT	right	FEEDBACK_POSITIVE

used to learn to mark the test corpus for discourse chunks without using the DA tags, and the results are compared to the results obtained by applying the hand-built rules. In the second stage of the experiment, the performance of the DA tagger is compared with and without the output of the discourse chunk learning system as a feature.

2.1 Observations about Dialogue

Discourse chunking grew from a set of observations about actual dialogues from Verbmobil-2. Participants in these dialogues raise topics and negotiate them in orderly and predictable ways. Different kinds of speech acts can be observed at different phases of the negotiation process.

To illustrate, consider the one-word utterance "okay." On its own, this is a difficult utterance to tag because of its low semantic content and its ability to appear in a wide variety of contexts. It can be used to accept a proposition (usually marked by the ACCEPT tag), to respond favourably (marked FEEDBACK_POSITIVE), to ask for feedback (REQUEST_COMMENT), or to signal that the speaker is listening (BACKCHANNEL). Less commonly, it can be used to respond unfavorably (FEEDBACK_NEGATIVE), or even to say goodbye (BYE).

Dialogue participants appear to use "okay" differently at different phases of the negotiation process. If the "okay" appears close to the beginning of the negotiation cycle, it is likely to be tagged ACCEPT. If it appears closer to the end of the cycle, it is likely to be tagged FEEDBACK_POSITIVE or one of the other tags.

This pattern is not limited to one-word utterances. In general, an ACCEPT speech act tends to appear close to the SUGGESTion that occasions it, while utterances marked FEEDBACK_POSITIVE tend to appear later on as the topic winds down, as seen in Table 2.

Note how the dialogue participants, having raised and negotiated the topic, finish up with some general FEEDBACK_POSITIVE statements. Then a new topic is raised, and the negotiation cycle continues.

Discourse chunking is designed to use information about where an utterance appears within the negotiation cycle. Such information tells more about what is happening in the dialogue than traditional features such as patterns of words or previous DA tags.

2.2 Chunking Rules

Dialogue chunks can be found by employing a simple set of rules. These rules come from observation; optimality is not guaranteed. However, the rules are guided by a principled assumption: the raising of a topic or a proposition entails the beginning of a chunk.

The dialogue acts are explained and defined in [1]. By these definitions, only four DA's contain or may contain a topic or proposition. These are `INIT`, `EXCLUDE`, `REQUEST_SUGGEST`, and `SUGGEST`.

The chunking rules are as follows:

1. The first utterance in a dialogue is always the start of a chunk.
2. A tag marked `INIT` or `SUGGEST` or `REQUEST_SUGGEST` or `EXCLUDE` in a dialogue is the start of a chunk.
3. If the previous utterance is also the start of a chunk, and if it is spoken by the same person, then this utterance is considered to be a continuation of the chunk, and is not marked.
4. The first `BYE` is the start of a chunk.

These rules were used to mark discourse chunk boundaries in the training corpus, and this information served as a standard when finding the chunk boundaries automatically in the test corpus.

3 The Case-Based Reasoning (CBR) Tagger

A thorough discussion of our CBR tagger goes beyond the scope of this paper, but a few comments are in order.

Case-based reasoning [7] is a form of machine learning that differs from commonly used rule-induction systems such as Transformation-Based Learning [2] or decision trees [12]. In rule-induction systems, the tagger attempts to derive rules from the training data and apply these rules when tagging the test corpus. By contrast, a case-based reasoner does not attempt to induce any such rules; it simply compares instances from the test corpus (in this case, utterances) against a database of correctly-tagged instances. Each new instance is marked with the same tag as its "nearest neighbour" (that is, the closest match) from the database. A k-nearest neighbour approach selects the closest k matches from the database to be committee members, and the committee members "vote" on the correct classification. In our implementation, each committee member gets a vote equal to its similarity to the test utterance. Values of k ranged from 1 to 13, and odd numbers were used to avoid "tie" situations.

Not all features are equally important, and so an Evolutionary Programming algorithm (adapted from [4]) was used on a subset of the data to weight the features. Weightings were initially chosen randomly for each member of a population of 20, and the 5 best performers were allowed to "survive" and "mutate" their weightings by a Gaussian random number. This was repeated for 20 generations, and the weightings from the highest performer were used for the CBR tagging runs.

Removing low-content function words (or *stopwords*) from some kinds of analysis helped to improve our results. We chose ten stopwords (the, of, and, a, an, in, to, it, is, was), the ten most common words from the British National Corpus [8]. These stopwords were removed when considering word similarity, but not n-gram similarity, since these words are useful for distinguishing sequences of words that would otherwise be very similar.

3.1 Finding the Discourse Chunks

The first part of this project was to discover how accurately discourse chunk boundaries could be found in the test corpus, using the information from the training corpus. We decided to rework the chunking task into a tagging task by simply changing the tagset. The well-known part-of-speech tagging task has been successfully transformed into a phrase-level chunking task using this method [11], and this work has served as a model for the current study.

For this part of the experiment, the DA tagset has been exchanged for a discourse chunk tagset. The tagset for discourse chunk recognition consists of three tags: **B** for an utterance at the beginning of a chunk, **I** for an utterance anywhere else, and * for the blank lines between dialogues. These lines serve to separate different dialogues from each other in the training corpus.

The discourse chunker makes a classification for an utterance u by examining every utterance in the database and calculating its similarity to u, based on the feature set.

The feature set used for this part of the experiment largely follows the set used in [14], and is as follows:

- Speaker change
- Word number
- Word similarity
- n-gram similarity
- Previous DA tag

and the following feature not included in that study,

- 2-previous DA tag

Inclusion of this feature enables more complete analysis of previous DA tags. However, the DA tagger was not given the correct information about the previous DA tags—that information was assumed to be unavailable. Instead, the DA tagger had to rely on its 'best guess' as to what the previous tags were, using the tagger's own results as it went through the corpus utterance by utterance. This method allows us to test performance even with imperfect DA tag information.

Utterances were also compared with regard to the *cue phrases* contained in each. A cue phrase (sometimes called a *discourse marker*) is a word or a phrase that serves as a special indicator of intention and discourse structure [5]. For example, the phrase "how about" is strongly indicative of a `SUGGEST` dialogue act. Cue phrases for this project were derived experimentally from the corpus.

The discourse chunking phase uses four committees of 1, 5, 9, and 13 members. We examined the accuracy for chunk starts for each of these groups, using the well-known measures of precision (P), recall(R), and f-measure(F), where

Table 3. Discourse chunk start accuracy

	Precision	Recall	F-measure
$k = 1$	0.435	0.857	0.577
$k = 5$	0.460	0.852	0.597
$k = 9$	0.475	0.847	0.609
$k = 13$	0.482	0.852	0.616

$$P = \frac{number\ of\ correct\ answers}{total\ number\ of\ answers}$$

and

$$R = \frac{number\ of\ correct\ answers}{number\ in\ the\ corpus}$$

F-measure is the harmonic mean of precision and recall, or

$$F = \frac{2(PR)}{P + R}$$

Table 3 shows accuracy for the chunk-finding phase of the experiment.

Precision and recall usually have a see-saw effect on each other; when the system makes more conservative guesses, it tends to become more accurate in those guesses, but finds fewer of the targets. In this implementation, however, recall holds steady while precision increases. With 13 committee members, the chunker was able to find 85 percent of the discourse chunk starts, while being accurate about half the time.

Having found the most likely candidates for chunk starts, the question now becomes whether this information, though imperfect, can help the DA tagger improve its performance.

3.2 DA Tagging with Dialogue Chunking

For the next phase of the experiment, the discourse chunking information was used as a feature in the CBR tagger to help in the DA tagging task. The discourse chunk tags (B, I, *) were converted into numbers. The first utterance in a discourse chunk (B) was marked with a 1. All other utterances were numbered depending on their position within the chunk with values from 1 to 100 (the last utterance in the chunk). This normalizes the chunk distances to facilitate comparison between utterances. A sample is shown in Table 4.

Discourse chunk similarity was added as a feature. Similarities for this tag were computed by dividing the smaller discourse chunk number by the larger. Comparing two "chunk starter" utterances would give the highest similarity of 1, and comparing a chunk starter (1) to a chunk-ender (100) would give a lower similarity (.01).

Table 4. A sample of dialogue from Verbmobil-2, after dialogue chunking

LMT	so there is a noon flight	1	SUGGEST
LMT	and then there is a three o'clock flight	13.375	SUGGEST
LMT	so you want to	25.75	NOT_CLASSIFIABLE
KNT	okay	38.125	ACCEPT
KNT	I think we better shoot for the noon	50.5	CONFIRM
LMT	the noon yeah	62.875	REQUEST_CLARIFY
LMT	okay	75.25	FEEDBACK_POSITIVE
KNT	yeah	87.625	FEEDBACK_POSITIVE
LMT	right	100	FEEDBACK_POSITIVE

Table 5. DA Tagger accuracy, in percentages

Committees used for DA tagging	Committees used for finding discourse chunks					
	k=0 (no chunks)	$k = 1$	$k = 5$	$k = 9$	$k = 13$	Correct information
$k = 1$	47.17	47.25	47.75	48.41	48.25	52.49
$k = 5$	47.83	49.41	49.91	50.41	50.24	55.32
$k = 9$	46.58	49.83	50.66	50.99	51.16	55.32
$k = 13$	47.50	50.83	51.66	51.91	52.24	56.65

4 Results and Discussion

A baseline was obtained by running the CBR tagger with no discourse chunk information, obtaining a maximum score of 47.83 percent. As a check, the DA tagger also performed a run using the rule-derived discourse chunk tags. It found the correct DA tag 56.65 percent of the time. This suggests an upper and lower boundary for the tagger using this feature set and corpus.

The results are shown in Table 5. Performance improves as the number of committee members increases, reaching a maximum of 52.24 percent accuracy, a five percent increase over tagging without discourse information. This result is statistically significant for $p \leq 0.05$. For comparison, the accuracy that could be obtained by simply tagging every utterance INFORM (the most common tag) is 21.71 percent.

As mentioned before, human performance has been estimated at about 84 percent. Other DA tagging work has achieved anywhere from 47 to 75 percent (see [15] for a more thorough review), but comparing these projects can be difficult because they use different corpora of different sizes, and different tagsets. For example, the 75.12 percent score reported by [14] is something of a record, but that project used a more coarse-grained tagset (18 tags) than the current study (32 tags).

It could be argued that since discourse chunks are based on DA tags (at least for the training corpus), the gains in performance could be explained by the extra information that discourse chunks provide—as though the chunk tags

Table 6. DA Tagger accuracy, by DA tag

	Without discourse chunks			With discourse chunks		
Tag	Precision	Recall	**F-measure**	Precision	Recall	**F-measure**
THANK	1.000	1.000	**1.000**	1.000	1.000	**1.000**
GREET	0.923	0.923	**0.923**	0.850	0.920	**0.880**
POLITENESS_FORMULA	0.846	0.786	**0.815**	0.850	0.610	**0.710**
BYE	0.778	0.667	**0.718**	0.850	0.600	**0.700**
FEEDBACK_POSITIVE	0.822	0.542	**0.653**	0.960	0.550	**0.700**
BACKCHANNEL	0.958	0.442	**0.605**	0.780	0.610	**0.680**
FEEDBACK_NEGATIVE	0.500	0.667	**0.571**	0.820	0.480	**0.600**
INFORM	0.488	0.485	**0.487**	0.450	0.530	**0.490**
EXPLAINED_REJECT	0.333	0.667	**0.444**	0.310	0.630	**0.420**
SUGGEST	0.405	0.472	**0.436**	0.370	0.450	**0.410**
INIT	0.313	0.500	**0.385**	0.310	0.420	**0.360**
REQUEST_COMMENT	0.263	0.500	**0.345**	0.290	0.440	**0.350**
REQUEST	0.323	0.323	**0.323**	0.270	0.430	**0.330**
ACCEPT	0.293	0.348	**0.318**	0.210	0.570	**0.310**
DELIBERATE	0.269	0.350	**0.304**	0.170	1.000	**0.290**
COMMIT	0.222	0.400	**0.286**	0.190	0.500	**0.280**
GIVE_REASON	0.211	0.364	**0.267**	0.160	0.380	**0.220**
REQUEST_CLARIFY	0.208	0.217	**0.213**	0.150	0.240	**0.190**
REQUEST_SUGGEST	0.182	0.222	**0.200**	0.110	0.330	**0.170**
NOT_CLASSIFIABLE	0.152	0.175	**0.163**	0.120	0.230	**0.160**
CLARIFY	0.120	0.176	**0.143**	0.080	0.180	**0.110**
EXCLUDE	0.080	0.500	**0.138**	0.070	0.140	**0.100**
REJECT	0.071	0.250	**0.111**	0.000	0.000	**0.000**

were providing an unfair hint as to the identity of the DA tag. This is unlikely, as the discourse chunking for the test corpus was done with no recourse to DA tag information. Further, if the rule-based chunk information were responsible for the performance gain, we would expect that discourse chunking would have the greatest positive effect on the four tags that serve as indicators to chunk starts (SUGGEST, INIT, REQUEST_SUGGEST, and EXCLUDE). However, when the best DA tagging results were broken down by tag, performance for the each of the four "chunk start" tags actually *declined* slightly with discourse chunks. As seen in Table 6, the greatest performance gains were made with the tags FEEDBACK_POSITIVE, FEEDBACK_NEGATIVE, and BACKCHANNEL. This suggests that discourse chunking is in fact helpful in recognising a greater range of speech acts than just the "chunk starters."

5 Conclusion and Future Work

This study shows that discourse chunks can be recognised automatically from simple features, and that these chunks do improve performance in DA tagging, even when found with only moderate accuracy. Better automatic chunk recognition will be a primary focus in future work.

In this experiment, we used a set of rules for breaking up dialogues into chunks. We chose these rules because the boundaries they define seem to correspond to topic boundaries. Although our results seem fairly intuitive and do produce some improvement in the DA tagging task, no claims can be made with regard to the psychological reality of these boundaries for human speakers. Future work will focus on how well these rules correspond to discourse boundary judgements made by human subjects.

Acknowledgements

This work was supported by an Australian Postgraduate Award. Thanks to Shelly Harrison for supervision and advice. Many thanks to Verbmobil for generously allowing use of the corpus which formed the basis of this project.

References

[1] J. Alexandersson, B. Buschbeck-Wolf, T. Fujinami, E. Maier, N. Reithinger, B. Schmitz, and M. Siegel. 1997. *Dialogue Acts in Verbmobil-2.* Verbmobil Report 204. 776

[2] E. Brill 1995. Transformation-based error-driven learning and natural language processing: a case study in part-of-speech tagging. *Computational Linguistics* 21(4):543-565. 776

[3] M.G. Core, and J.F. Allen. 1997. Coding dialogs with the DAMSL annotation scheme. In *Working Notes of the AAAI Fall Symposium on Communicative Action in Humans and Machines.* Cambridge, MA. 773

[4] D. Fogel. 1994. An introduction to evolutionary computation. *Australian Journal of Intelligent Information Processing Systems*, 2:34–42. 776

[5] B. J. Grosz, C. L. Sidner. 1986. Attention, intention, and the structure of discourse. *Computational Linguistics*, 12(3):175–204. 777

[6] B.J. Grosz, A.K. Joshi, and S. Weinstein. 1995. Centering: A framework for modelling the local coherence of discourse. *Computational Linguistics*, 21(2):203–225. 773

[7] J. Kolodner. 1993. *Case-Based Reasoning.* Academic Press/Morgan Kaufmann. 776

[8] G. Leech, P. Rayson, and A. Wilson. 2001. *Word Frequencies in Written and Spoken English: based on the British National Corpus.* Longman. 777

[9] W. Mann. 2002. Dialogue Macrogame Theory. In *Proceedings of the 3rd SIGdial Workshop on Discourse and Dialogue*, pages 129–141, Philadelphia PA. 773

[10] D. Midgley. 2003. Discourse chunking: a tool for dialogue act tagging. In *ACL-03 Companion Volume to the Proceedings of the Conference*, pages 58–63, Sapporo, Japan. 773

[11] M. Osborne. 2000. Shallow Parsing as Part-of-Speech Tagging. In *Proceedings of CoNLL-2000 and LLL-2000*, pages 145–147, Lisbon, Portugal. 777

[12] J.R. Quinlan. 1993. *C4.5: Programs for Machine Learning.* Morgan Kaufmann, San Mateo, California. 776

[13] N. Reithinger and M. Klesen. 1997. Dialogue act classification using language models. In G. Kokkinakis, N. Fakotakis, and E. Dermatas, editors, *Proceedings of the 5th European Conference on Speech Communication and Technology*, volume 4, pages 2235–2238, Rhodes, Greece. 773

[14] K. Samuel. 2000. *Discourse learning: An investigation of Dialogue Act tagging using transformation-based learning.* Ph.D. thesis, University of Delaware. 773, 777, 779

[15] A. Stolcke, K. Ries, N. Coccaro, E. Shriberg, R. Bates, D. Jurafsky, P. Taylor, R. Martin, C. Van Ess-Dykema, M. Meteer. 2000. Dialogue act modeling for automatic tagging and recognition of conversational speech. *Computational Linguistics*, 26(3):339–373. 773, 779

[16] H. Wright. 1998. Automatic utterance type detection using suprasegmental features. In *ICSLP (International Conference on Spoken Language Processing) '98.* Sydney, Australia. 773

On Using Prototype Reduction Schemes and Classifier Fusion Strategies to Optimize Kernel-Based Nonlinear Subspace Methods*

Sang-Woon Kim[1] and B. John Oommen[2]

[1] Dept. of Computer Science and Engineering
Myongji University, Yongin, 449-728 Korea
kimsw@mju.ac.kr
[2] School of Computer Science
Carleton University, Ottawa, ON, K1S 5B6, Canada
oommen@scs.carleton.ca

Abstract. In Kernel based Nonlinear Subspace (KNS) methods, the length of the projections onto the principal component directions in the feature space, is computed using a kernel matrix, K, whose dimension is equivalent to the number of sample data points. Clearly this is problematic, especially, for large data sets. To solve the problem, in [9] we earlier proposed a method of reducing the size of the kernel by invoking a Prototype Reduction Scheme (PRS) to reduce the data into a smaller representative subset, rather than define it in terms of the entire data set. In this paper we propose a new KNS classification method for further enhancing the efficiency and accuracy of the results presented in [9]. By sub-dividing the data into smaller subsets, we propose to employ a PRS as a *pre-processing* module, to yield more refined representative prototypes. Thereafter, a Classifier Fusion Strategies (CFS) is invoked as a *post-processing* module, so as to combine the individual KNS classification results to derive a consensus decision. Our experimental results demonstrate that the proposed mechanism *significantly* reduces the prototype extraction time as well as the computation time without sacrificing the classification accuracy. The results especially demonstrate that the computational advantage for *large* data sets is significant when a parallel programming philosophy is applied.

Keywords: Kernel based Nonlinear Subspace (KNS) Method, Prototype Reduction Schemes (PRS), Classifier Fusion Strategies (CFS)

1 Introduction

This paper presents a novel scheme by which we can enhance the efficiency of a pattern classification system. The novelty in the scheme lies in the fact that

* The work of the first author was done while visiting at Carleton University, Ottawa, Canada. The first author was partially supported by KOSEF, the Korea Science and Engineering Foundation, and the second author was partially supported by NSERC, Natural Sciences and Engineering Research Council of Canada.

T.D. Gedeon and L.C.C. Fung (Eds.): AI 2003, LNAI 2903, pp. 783–795, 2003.

the data is subdivided into subsets, each of which trains a nonlinear kernel-based subspace method. The results of the individual trained classifiers are then merged using classical fusion methods to yield an enhanced classifier that is superior to each individual one. Thus, the results presented here essentially describe a technique wherein the various advantages of three fields, (namely, prototype reduction, nonlinear kernel methods, and fusion-based methods), are used to augment each other. In that sense, these results are both novel, and of a pioneering sort.

The subspace method of pattern recognition is a technique in which the pattern classes are not primarily defined as bounded regions or zones in a feature space, but rather given in terms of linear subspaces defined by the basis vectors, one for each class [13]. The length of a vector projected onto a class subspace, or the distance of the vector from the class subspace, is a measure of similarity or degree of membership for that particular class, and serves as the discriminant function of these methods. More specifically, the following steps are necessary for the methods: First, we compute the covariance matrix or scatter matrix. Then, we compute its eigenvectors and normalize them by the principal components analysis or the Karhuen-Loève (KL) expansion technique. Finally, we compute the projections (or distances) of a test pattern vector onto (or from) the class subspaces spanned by the subset of principal eigenvectors. All of the subspace classifiers have employed the Principal Components Analysis (PCA) to compute the basis vectors by which the class subspaces are spanned. Since the PCA is a linear algorithm, it is, clearly, beyond its capabilities to extract nonlinear structures from the data, which essentially limits the use of such classifiers.

To overcome the above limitation, a kernel Principal Components Analysis (kPCA) was proposed in [15]. The kPCA provides an elegant way of dealing with nonlinear problems in an input space R^N by mapping them to linear ones in a feature space, F. That is, a dot product in space R^N corresponds to mapping the data into a possibly high-dimensional product space F by a nonlinear map $\Phi : R^N \rightarrow F$, and taking the dot product in the latter space.

Recently, using the kPCA, kernel-based subspace methods, such as the Subspace method in Hilbert Space [18], the Kernel based Nonlinear Subspace (KNS) method [12], and the Kernel Mutual Subspace method [14], have been proposed. All of them solve a similar linear eigenvalue problem and utilize the *kernel trick* to obtain the kernel PCA components. The only difference is that the size of the problem is decided by the *number* of data points, and not by the dimension, since the dimension of the kernel (covariance) matrix is equivalent to the number of data points, n, which is not desirable when a large number of observations are available. Nonlinear subspace methods are reportedly better than their linear counterparts.

To solve the computational problem, a number of methods, such as the techniques proposed by Achlioptas and his co-authors [1], [2] (detailed presently), the power method with deflation [15], the method of estimating K with a subset of the data [15], the Sparse Greedy matrix Approximation (SGA) [16], the Nystom method, and the sparse kernel PCA method based on the probabilistic feature-space PCA concept [17], have been proposed. Details of these methods are omitted here in the interest of brevity.

Pioneering to the area of reducing the complexity of KNS are the works of Achlioptas [2] and his co-authors, whose techniques can be classified into two main categories. The first category includes the strategy of artificially introducing sparseness into the kernel matrix, which, in turn, is achieved by *physically* setting some randomly chosen values to *zero*. The other alternative, as suggested in [1], proposes the elimination of the underlying data points themselves. *This is the strategy we advocate* - this is indeed, the theoretical basis for our present solution for optimizing KNS. However, rather than eliminate the "unnecessary" underlying data points in a randomized manner, we propose that this be done in a systematic manner, namely in a way by which the distribution of the points is maintained, albeit, approximately. Indeed, we propose to solve the problem by reducing the size of the design set without sacrificing the performance, where the latter is achieved by using a Prototype Reduction Scheme (PRS). The PRS is a way of reducing the number of training vectors while simultaneously insisting that the classifiers built on the reduced design set perform as well, or nearly as well, as the classifiers built on the original design set.

The PRS has been explored for various purposes, and has resulted in the development of many algorithms [3], [4]. One of the first of its kind, was a method that led to a smaller prototype set, the Condensed Nearest Neighbor (CNN) rule [5]. Since the development of the CNN, other methods have been proposed successively, such as the Prototypes for Nearest Neighbor (PNN) classifiers and its modified version, the Vector Quantization (VQ) technique, and a new hybrid method (HYB) [7] etc[1].

On the other hand, Classifier Fusion Strategies (CFS) have been investigated in the literature, because of their potential to improve the performance of pattern recognition systems. The basic idea is to solve each recognition problem by designing a *set* of classifiers, and then combining the designs in some way to achieve reduced classification error rates. Therefore a choice of an appropriate fusion method can further improve on the performance of the combination. Various CFSs have been proposed in the literature - excellent studies are found in [10], [11].

In this paper we propose to utilize a PRS and a CFS in succession to yield a new KNS classification method. We highlight the salient contributions of the paper below.

1.1 Contributions of the Paper

The primary contribution of this paper is the augmented strategy used to enhance kernel-based nonlinear subspace (KNS) methods. This is achieved by employing two pre-processing modules, and a post-processing module - *neither of which are related to KNS methods at all.* The first of the pre-processing modules

[1] Bezdek *et al* [3], who have composed an excellent survey of the field, report that there are "zillions!" of methods for finding prototypes (see page 1459 of [3]).

sub-divides the data into subsets. The second invokes a suitable PRS, by which the prototypes are selected or created for each of the latter subsets. Finally, the post-processing is achieved by invoking a relevant fusion-based scheme, which represents a strategy of combining the results of the individual classifiers in a suitable way so as to enhance their collective classification. Thus, the results presented here essentially describe a technique wherein the various advantages of three fields, (namely, prototype reduction, nonlinear kernel methods, and fusion-based methods), are used to augment each other. In that sense, these results are both novel, and of a pioneering sort.

The second contribution of this paper is the method by which we have reduced the computational burden of *each* KNS method, namely, by utilizing a PRS. The speed is thus enhanced, (as addressed earlier in [9]) because we work with a smaller matrix, leading to less expensive matrix-related operations.

The third contribution of the paper is a parallelization of the proposed mechanism. If the data subsets were processed in parallel (as demonstrated later), the processing CPU-times could be reduced significantly when compared to the times required for sequential computations. All of these concepts are novel to the field of designing kernel-based nonlinear subspace methods.

2 Kernel Based Nonlinear Subspace (KNS) Method

In this section, we provide a *brief* overview to the kernel based nonlinear subspace method. Since the eigenvectors are computed linearly by the Principle Component Analysis (PCA), these classifiers are very effective for linear problems, but not good for non-linear feature distributions. This has naturally motivated various developments of nonlinear PCA's, including the kernel PCA (or kPCA). The details of the kPCA are omitted (as they are essentially, irrelevant) but can be found in the well-known literature, and also in [6] and [9].

The Kernel-based Nonlinear Subspace (KNS) method is a subspace classifier, where the kPCA is employed as the specific nonlinear feature extractor. Given a set of n N-dimensional data samples $X = (\boldsymbol{x}_1, \cdots, \boldsymbol{x}_n)^T \in \omega_i$ and a kernel function $k(\cdot, \cdot)$, the column matrix of the orthogonal eigenvectors, $\boldsymbol{U} = (\boldsymbol{U}_1, \cdots, \boldsymbol{U}_{p_i'})$, and the diagonal matrix of the eigenvalues, $\boldsymbol{\Lambda}$, can be computed from the kernel matrix $K_{ij} = k(\boldsymbol{x}_i, \boldsymbol{x}_j), 1 \le i, j \le n$. Then, it can be shown [6], that the length of the projection of a new test pattern $\boldsymbol{z}$ onto the principal directions of $\boldsymbol{U}$ in F is

$$\|P_i'\|^2 = \sum_{j=1}^{p_i'} \sum_{i=1}^{n} (\alpha_{ji} k(\boldsymbol{x}_i, \boldsymbol{z}))^2 = \|\frac{1}{\sqrt{\boldsymbol{\Lambda}}} \boldsymbol{U}^T k(X, \boldsymbol{z})\|^2. \quad (1)$$

Thus, the analogous decision rule as that of linear subspace classifiers, classifies each $\boldsymbol{z}$ into the class on whose "class subspace" it has the longest projection as follows:

$$\forall j \neq i, \ \|P_i'\|^2 > \|P_j'\|^2 \Rightarrow \boldsymbol{z} \in \omega_i. \quad (2)$$

In KNS, the subspace dimensions, $p'_i, i = 1, \cdots, C$, have a strong influence on the performance of subspace classifiers. In order to get a high classification accuracy, we have to select a large dimension. However, designing with dimensions which are too large leads to low performances due to the overlapping of the resultant subspaces. Also, since the speed of computation for the discriminant function of subspace classifiers is inversely proportional to the dimension, the latter should be kept small.

Various dimension selection methods have been reported in the literature [13]. The simplest approach is to select a global dimension to be used for all the classes. A more popular method is that of selecting different dimensions for the various classes based on the cumulative proportion, α, which is defined as $\alpha(p'_i) = \sum_{i=1}^{p'_i} \lambda_i / \sum_{i=1}^{n} \lambda_i$.

In this method, the p'_i of each class for the data sets is determined by considering the cumulative proportion, $\alpha(p'_i)$. For each class, the kernel matrix K is computed using the available training samples for that class. The eigenvectors and eigenvalues are computed, and the cumulative sum of the eigenvalues, normalized by the trace of K, is compared to a fixed number.

The overall procedure for kernel based nonlinear subspace classifiers is summarized as follows:

1. For every class, j, compute the kernel matrix, K, using the given training data set $\{\boldsymbol{x}_i\}_{i=1}^n$ and the kernel function $k(\cdot, \cdot)$;
2. For every class, j, compute the normalized eigenvectors of K in F, and select the subspace dimension, p'_j;
3. For all the test patterns, compute the projections of the pattern onto the eigenvectors, and classify it by the decision rule of (2).

In the above, the computation of principal component projections for a pattern requires evaluation of the kernel function with respect to *all* the training patterns. This is unfortunate, because, in practice, the computation is excessive for large data sets. We overcome this limitation by reducing *each subset* of the training set using a PRS.

3 Schema for the Overall Proposed Solution

Let us, for the present, assume that we have sub-divided the training samples into mutually exclusive subsets of size 'n' each. Our reasoning below is for each of these subsets.

The fundamental problem we encounter is that of computing the solution to the kernel subspace classifier from the set of training samples. This, in turn, involves four essential phases, namely that of computing the kernel matrix, computing *its* eigenvalues and eigenvectors, extracting the principal components of the kernel matrix from among these eigenvectors, and finally, projecting the samples to be processed onto the reduced basis vectors. We observe, first of all, that all of these phases depend on the size of the data set. In particular, the most time consuming phase involves computing the eigenvalues and eigenvectors of the kernel matrix.

There is an underlying irony in the method, in itself. For a good and effective training phase, we, typically, desire a large data set. But if the training set is large, the dimensionality of the kernel matrix is correspondingly large, making the computations extremely expensive. On the other hand, a smaller training set makes the learning less accurate, although computations are correspondingly less.

There are a few ways by which the computational burden of the kernel subspace method can be reduced. Most of the reported schemes [1], [2], [16] and [17], resort to using the specific properties of the underlying kernel matrix, for example, its sparseness. Our technique is different.

The method we propose is by reducing the size of the training set. However, we do this, by not significantly reducing the accuracy of the resultant classifier. This is achieved by resorting to a PRS. The theoretical basis for this will be explained in a subsequent section.

The question now is essentially one of determining which of the training points we should retain. Rather than deciding to discard or retain the training points, we permit the user the choice of either *selecting* some of the training samples using methods such as the CNN, or *creating* a smaller set of samples using the methods such as those advocated in the PNN, VQ, and HYB. This reduced set effectively represents the new "training" set. Additionally, we also permit the user to migrate the resultant set by an LVQ3-type method to further enhance the quality of the reduced samples.

The PRS serves as a preprocessor to the n N-dimensional training samples to yield a subset of n' potentially new points, where $n' << n$. The "kernel" is now computed using this reduced set of points to yield the so-called *reduced-kernel* matrix. The eigenvalues and eigenvectors of *this* matrix are now computed, and the principal components of the kernel matrix are extracted from among *these* eigenvectors of smaller dimension. Notice now that the samples to be tested are projected onto the reduced basis directions represented by *these* vectors. More specifically, the data set D_j per each class, j, is randomly divided into small subsets, $D_{ij}, i = 1, \cdots, M$. Subsequently, each subset D_{ij} is reduced into prototype vectors, P_{ij}, by a PRS algorithm. The number of prototype vectors, $n' = |P_{ij}|$, is significantly smaller than that of the original subset, $n = |D_{ij}|$, since is $n' \ll n$. After that, the KNS classifier, C_{ij}, is designed using the $n' \times n'$ kernel matrix, K, instead of the $n \times n$ matrix.

To investigate the computational advantage gained by resorting to such a PRS preprocessing phase, we observe, first of all, that the time used in determining the reduced prototypes is *fractional* compared to the time required for the expensive matrix-related operations. Once the reduced prototypes are obtained, the eigenvalue/eigenvector computations are significantly smaller, since these computations are now done for a much smaller set, and thus for an $n' \times n'$ matrix.

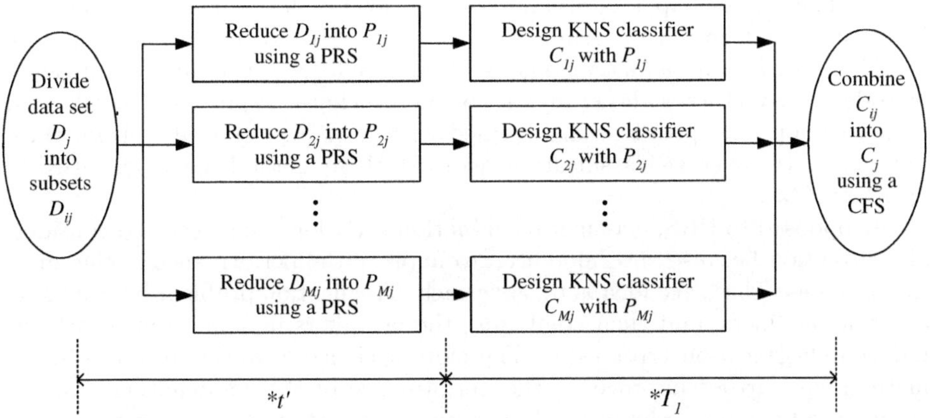

Fig. 1. The process of the proposed mechanism. Here, the D_j is a data set of class j, and the $D_{ij}, i = 1, \cdots, M$ is a subset divided randomly. The P_{ij} is a prototype set reduced by a PRS. The C_{ij} is an KNS classifier designed with the reduced kernel matrix, and the C_j is a KNS classifier combined by a CFS

Once the outputs of the individual KNS schemes are obtained, they have to be coalesced in a systematic manner. This is achieved by resorting to a fusion-based philosophy, which compensates for the decreased efficiency caused by the data set division. The latter essentially combines the decisions of the individual KNS classification results to derive a consensus decision from the M classifiers. Figure 1 shows the overall process of the proposed mechanism.

Another important issue that we highlight (as is clear from the figure), is the inherent parallelism in the process. If the modules of the algorithm are processed in parallel, as shown in the figure, the times taken for the PRS reduction and the KNS classifier design are $*t'$ and $*T_1$, respectively. In this case, the sequential computation times, t' and T_1, for both steps almost can be reduced by the factor of $\frac{1}{M}$. The net result of these two reductions is reflected in the time savings we report in the next section in which we discuss the experimental results obtained for artificial and real-life data sets.

4 Prototype Reduction (PRS) and Fusion Schemes (CFS)

Various data reduction methods have been proposed in the literature - two excellent surveys are found in [3], [4]. The most pertinent ones are reviewed in [6] in two groups: the conventional methods and a newly proposed hybrid method. Among the conventional methods, the CNN and the SVM are chosen as representative schemes of *selective* methods. The former is one of first methods proposed, and the latter is more recent. As opposed to these, the PNN and VQ (or SOM) are considered to fall within the family of prototype-*creative* algorithms.

The unabridged paper [6] also describes a newly-reported hybrid scheme, the Kim_Oommen Hybridized technique, [7], which is based on the philosophy of invoking *creating* and *adjusting* phases. First, a reduced set of initial prototypes or code-book vectors is chosen by any of the previously reported methods, and then their optimal positions are learned with an LVQ3-type algorithm, thus, minimizing the average classification error. All these methods have been used to optimize KNS.

As opposed to PRS, classifier combination ("Fusion") has received considerable attention because of its potential to improve the performance of classification systems. The basic idea is to solve each classification problem by designing a *set* of classifiers, and then combining the classifiers in some way to achieve reduced classification error rates. Therefore a choice of an appropriate fusion method can further improve on the performance of the combination. Various CFSs have been proposed in the literature - excellent studies are found in [10], [11]. We summarize the CFS's decision rules of [10] here briefly.

Consider a pattern recognition problem where pattern z is to be assigned to one of the C possible classes, $\omega_1, \cdots, \omega_C$. Assume that there are M classifiers each representing the given pattern by a distinct measurement vector. Denote the measurement vector used by the ith classifier by $\boldsymbol{x}_i, i = 1, \cdots, M$. In this case, the Bayesian decision rule computes the *a posteriori* probability $p(\omega_k | \boldsymbol{x}_1, \cdots, \boldsymbol{x}_M)$ using the Bayes theorem as follow:

$$p(\omega_k | \boldsymbol{x}_1, \cdots, \boldsymbol{x}_M) = \frac{p(\boldsymbol{x}_1, \cdots, \boldsymbol{x}_M | \omega_k) P(\omega_k)}{\sum_{j=1}^{C} p(\boldsymbol{x}_1, \cdots, \boldsymbol{x}_M | \omega_j) P(\omega_j)}. \tag{3}$$

Let us assume that the representations used are conditionally statistically independent. Then the joint probability distribution of the measurements extracted by the classifiers can be rewritten as follow:

$$p(\boldsymbol{x}_1, \cdots, \boldsymbol{x}_M | \omega_k) P(\omega_k) = \prod_{i=1}^{M} p(\boldsymbol{x}_i | \omega_k), \tag{4}$$

where $p(\boldsymbol{x}_i | \omega_k)$ is the measurement process model of the ith representation.

Based on (3) and (4), we obtain commonly used decision rules such as *Product*, *Sum*, *Max*, *Min*, *Median*, and *Majority vote* rules. Their details can be found in [10] and [11]. Although all of them can be used in a CFS, we mention, explicitly, the one that we have used in our experiments, namely the *Majority vote* rule which operates under the assumption of equal priors, as follows :

$$\sum_{i=1}^{M} \Delta_{ji} = \max_{1 \le k \le C} \left\{ \sum_{i=1}^{M} \Delta_{ki} \right\} \Rightarrow \boldsymbol{z} \in \omega_j, \tag{5}$$

$$\Delta_{ki} = \begin{cases} 1, \text{ if } p(\omega_k | \boldsymbol{x}_i) = \max_{1 \le j \le C} \{p(\omega_j | \boldsymbol{x}_i)\}, \\ 0, \text{ otherwise.} \end{cases} \tag{6}$$

Here, for each class ω_k, the sum of Δ_{ji} simply counts the votes received for this result from the individual classifiers. Thus the class which receives the largest number of votes is then selected as the majority decision.

5 Experimental Verification

The proposed method has been rigorously tested and compared with many conventional KNSs. This was done by performing experiments on both "artificial" and "real-life" data sets. In our experiments, the two artificial data sets "Non_normal 2" and "Non_normal 3", were generated with different sizes of testing and training sets of cardinality 500 and 5,000 respectively. The data sets "Arrhythmia" (in short, "Arrhy") and "Adult4", which are real benchmark data sets, are cited from the UCI Machine Learning Repository[2]. Their details can be found in the latter site, and also in [6], [8] and [9], and omitted here in the interest of compactness.

The data set named "Non_normal", which has also been employed as a benchmark experimental data set [6], was generated from a mixture of four 8-dimensional Gaussian distributions whose means are $\mu_{11} = [0, 0, \cdots, 0]$, $\mu_{12} = [6.58, 0, \cdots, 0]$, $\mu_{21} = [3.29, 0, \cdots, 0]$, $\mu_{22} = [9.87, 0, \cdots, 0]$, and covariance matrices are the *8*-dimensional *Identity* matrix.

In all the data sets examined, all of the vectors were normalized within the range $[-1, 1]$ using their standard deviations. For every class j, the data set for the class was randomly split into two subsets of equal size for training and testing respectively, and the roles of these sets were later interchanged. The parameters used for the PRS and KNS are found in the unabridged version of this paper [6].

5.1 Experimental Results

We report below the run-time characteristics of the proposed algorithm for the artificial and real-life data sets as shown in Table 1. In this table, the KNS represents the method of calculating the kernel matrix with the entire data set. The SGA-KNS[3], the CNN-KNS, and the HYB-KNS are obtained by calculating the kernel matrix using the prototypes obtained with each respective method. On the other hand, the rCNN-KNS and rHYB-KNS are obtained by calculating the kernel matrix using the prototypes obtained with each respective recursive method [8]. The above two methods were experimented with a hope of reducing the prototype extraction time. Then, the KNSc is obtained by using a combined KNS classifier, where each KNS classifier is designed with the one of the divided subsets. That is, in this experiment, the whole data set is divided into three subsets, $M = 3$ (refer to Figure 1), and the *Majority vote* rule (as specified by Equation (5)) is employed as an ensemble strategy. The CNN-KNSc and HYB-KNSc are obtained by using the combined KNS classifiers.

[2] http://www.ics.uci.edu/mlearn/MLRepository.html

[3] The authors are very grateful to Dr. Alex Smola, of the Australian National University, for his assistance in this study.

Table 1. The experimental results of the proposed method for the artificial data set, "Non_normal 3", and real-life data sets, "Arrhy" and "Adult4". The values marked with $*$ of the column T_1 and t', that is, $*T_1$ and $*t'$, are the processing CPU-times when a parallel programming technique is applied

Data Sets	Methods	*Acc*	T_1	(p_1, p_2)	n'	t'
Non_3	KNS	94.81	12,464.00	(1,1), (1,1)	5,000	0
	SGA-KNS	92.53	7.90	(1,1), (1,1)	8	1,305.35
	CNN-KNS	94.81	360.29	(1,1), (1,1)	491	58.57
	HYB-KNS	94.76	488.87	(1,1), (1,1)	641	58.38
	rCNN-KNS	94.79	299.47	(1,1), (1,1)	408	15.06
	rHYB-KNS	94.77	479.32	(1,1), (1,1)	627	46.61
	KNSc	94.84	4,797.80	{(1, 1) (1, 1) (1, 1)},	{1666, 1667, 1667}	0
			*1,612.10	{(1, 1) (1, 1) (1, 1)}		*0
	CNN-KNSc	94.77	440.79	{(1, 1) (1, 1) (1, 1)},	{216, 184, 199}	11.38
			*146.93	{(1, 1) (1, 1) (1, 1)}		*5.69
	HYB-KNSc	94.74	469.54	{(1, 1) (1, 1) (1, 1)},	{232, 204, 206}	10.15
			*156.52	{(1, 1) (1, 1) (1, 1)}		*3.38
Arrhy	KNS	98.24	18.29	(5,10), (13,19)	226	0
	SGA-KNS	98.46	18.83	(25,28), (28,32)	206	213.05
	CNN-KNS	95.80	2.72	(15,16), (14,14)	30	0.84
	HYB-KNS	98.90	7.12	(23,32), (21,25)	67	3.50
	rCNN-KNS	92.26	1.77	(6,8), (9,8)	16	1.75
	rHYB-KNS	96.46	3.93	(10,22), (11,27)	44	3.67
	KNSc	97.57	16.79	{(40, 35) (41, 35) (41, 34)},	{75, 76, 75}	0
			*6.79	{(41, 34) (41, 35) (41, 34)}		*0
	CNN-KNSc	93.15	4.08	{(6, 6) (9, 10) (5, 6)},	{13, 18, 11}	1.88
			*1.36	{(10, 4) (10, 7) (6, 4)}		*0.63
	HYB-KNSc	98.68	8.40	{(6, 8) (9, 18) (5, 9)},	{33, 37, 34}	1.34
			*2.80	{(15, 18) (7, 11) (9, 17)}		*0.45
Adult4	KNS	85.78	15,558.00	(13,12), (13,12)	4,168	0
	SGA-KNS	81.80	244.85	(27,18), (17,19)	301	108,135.50
	CNN-KNS	90.10	536.38	(8,8), (8,8)	755	237.69
	HYB-KNS	91.96	287.70	(11,11) (11,11)	440	5,453.64
	rCNN-KNS	88.24	367.74	(9,9), (8,8)	520	38.61
	rHYB-KNS	91.75	91.31	(10,9), (9,9)	143	80.01
	KNSc	90.60	5,482.30	{(20, 15) (19, 13) (21, 14)},	{1389, 1389, 1389}	0
			*1,836.10	{(14, 13) (19, 13) (21, 14)}		*0
	CNN-KNSc	89.76	543.65	{(15, 14) (13, 12) (16, 12)},	{246, 266, 267}	25.85
			*181.22	{(18, 14) (15, 14) (13, 11)}		*8.62
	HYB-KNSc	93.91	314.27	{(9, 8) (13, 11) (10, 8)},	{155, 161, 167}	1,095.99
			*104.76	{(9, 8) (9, 8) (9, 8)}		*365.33

The proposed methods, KNSc, CNN-KNSc and HYB-KNSc can be compared with the traditional versions, such as the KNS (which utilizes the entire set), SGA-KNS, CNN-KNS, and HYB-KNS as well as the rCNN-KNS and rHYB-KNS, using four criteria, namely, the classification accuracy (%), *Acc*, the processing CPU-time (in seconds), T_1, the number of prototypes extracted, n', and the prototype extraction time, t'. We report below a summary of the results obtained for the case when one subset was used for training and the second for testing. The results when the roles of the sets are interchanged are almost identical. In the table, each result is the averaged value of the training and the test sets, respectively.

In KNS, the subspace dimensions have a strong influence on the performance of subspace classifiers. Therefore, the subspace dimensions, the number of prototypes employed for constructing the kernel matrix, and their extraction times in the PRSs and the SGA, should be investigated. To assist in this task, we display the results of the *Acc* and T_1, the subspace *dimensions* of the two classes, $(p_{11}, p_{12}), (p_{21}, p_{22})$, or their divided three subsets, the final number of prototypes, $n' = \frac{1}{2}(n'_1 + n'_2)$, and their extraction times, $t' = \frac{1}{2}(t'_1 + t'_2)$.

From Table 1, we can see that the processing CPU-time and the prototype extraction time of the rCNN-KNS and rHYB-KNS methods can be further reduced by employing a recursive PRS such as the recursive CNN and the recursive HYB. It should be mentioned that the processing time T_1, can be decreased *dramatically* when a parallel programming technique is applied. In the table, the values of the columns T_1 and t' are indicated with a '$*$'.

Consider the KNS, HYB-KNS, KNSc and HYB-KNSc methods for the "Non_3" data sets. First of all, the classification accuracy (%) of these methods, *Acc*, is almost the same, namely, 94.81, 94.76, 94.84, and 94.74. However, the processing CPU-time (seconds), T_1, is significantly different, which are 12,464.00, 488.87, 4,797.80 (*1,612.10), and 469.54 (*156.52) respectively. The number of prototypes, that is, the dimension of the kernel matrix, n' are 5,000, 641, $\{1666, 1667, 1667\}$ and $\{232, 204, 206\}$, and their extraction times are 0, 58.38, 0 (*0), and 10.15 (*3.38) respectively. Additional remarks about the experimental results can be found in [6].

When a parallel programming technique is applied to the HYB-KNSc scheme, the processing times of T_1 and t', 469.54 and 10.15 seconds, are reduced by those of $*T_1$ and $*t'$, 156.52 and 3.38 seconds, respectively. Identical comments can also be made about the CNN-KNS and CNN-KNSc schemes, and about the experimental results obtained for the real-life data set captioned *Adult*4. One final concluding remark is that the HYB-KNSc is uniformly superior to the others in terms of the classification accuracy, *Acc*, while requiring a little additional prototype extraction time, t'. We thus propose this method as the most ideal candidate for enhancing a KNS by combining a PRS and a CFS.

6 Conclusions

Kernel based nonlinear subspace methods suffer from a major disadvantage for large data sets, namely that of the excessive computational burden encountered by processing all the data points. In this paper, we suggest a computationally superior mechanism to solve the problem, which incorporates the techniques akin to many different pattern recognition philosophies. Rather than define the kernel matrix and compute the principal components using the entire data set, we propose that the data be reduced into smaller prototype subsets (which lead to "reduced-kernel" matrices) using a Prototype Reduction Schemes (PRS) followed by a Classifier Fusion Strategy (CFS). The CFS classifier can compensate for the decreased accuracy caused by the data set subdivision.

The proposed method has been tested on artificial and real-life benchmark data sets, and compared with the conventional methods. The experimental results for large data sets demonstrate that the proposed schemes can improve the computation speed of the KNS classification by an order of magnitude, while yielding almost the same classification accuracy. Also, if invoked in parallel, the times taken for the PRS reduction and the KNS classifier design are significantly reduced, rendering the overall classifier an order-of-magnitude superior to the previous traditional counterparts.

References

[1] D. Achlioptas and F. McSherry, "Fast computation of low-rank approximations", in *Proceedings of the Thirty-Third Annual ACM Symposium on the Theory of Computing*, Hersonissos, Greece, ACM Press, pp. 611 - 618, 2001. 784, 785, 788

[2] D. Achlioptas, F. McSherry and B. Schölkopf, "Sampling techniques for kernel methods", in *Advances in Neural Information Processing Systems 14*, MIT Press, Cambridge, MA, pp. 335-342, 2002. 784, 785, 788

[3] J. C. Bezdek and L. I. Kuncheva, "Nearest prototype classifier designs: An experimental study", *International Journal of Intelligent Systems*, vol. 16, no. 12, pp. 1445 - 11473, 2001. 785, 789

[4] B. V. Dasarathy, *Nearest Neighbor (NN) Norms: NN Pattern Classification Techniques*, IEEE Computer Society Press, Los Alamitos, 1991. 785, 789

[5] P. E. Hart, "The condensed nearest neighbor rule", *IEEE Trans. Inform. Theory*, vol. IT-14, pp. 515 - 516, May 1968. 785

[6] S.-W. Kim and B. J. Oommen, "On using prototype reduction schemes and classifier fusion strategies to optimize kernel-based nonlinear subspace methods". *Unabridged version of this paper.* 786, 789, 790, 791, 793

[7] S. -W. Kim and B. J. Oommen, "Enhancing prototype reduction schemes with LVQ3-type algorithms", *Pattern Recognition*, vol. 36, no. 5, pp. 1083 - 1093, 2003. 785, 790

[8] S.-W. Kim and B. J. Oommen, "Recursive prototype reduction schemes applicable for large data sets", *Proceedings of Joint IAPR International Workshops*, Windsor, Canada, Springer, pp. 528-537, Aug. 2002. 791

[9] S.-W. Kim and B. J. Oommen, "On using prototype reduction schemes to optimize kernel-based nonlinear subspace methods". *Submitted for publication.* The preliminary version can be found in the *Proceedings of AI'02, the 2002 Australian Joint Conference on Artificial Intelligence*, Canberra, Australia, pp. 155-166, December 2002. 783, 786, 791

[10] J. Kittler, M. Hatef, R. P. W. Duin and J. Matas, "On combining classifiers", *IEEE Trans. Pattern Anal. and Machine Intell.*, vol. PAMI-20, no. 3, pp. 226 - 239, Mar. 1998. 785, 790

[11] L. I. Kuncheva, "A theoretical study on six classifier fusion strategies", *IEEE Trans. Pattern Anal. and Machine Intell.*, vol. PAMI-24, no. 2, pp. 281 - 286, Feb. 2002. 785, 790

[12] E. Maeda and H. Murase, "Multi-category classification by kernel based nonlinear subspace method", in the *Proceedings of the IEEE International Conference on Acoustics, Speech, and Signal Processing (ICASSP 99)*, IEEE Press, 1999. 784

[13] E. Oja, *Subspace Methods of Pattern Recognition*, Research Studies Press, 1983. 784, 787

[14] H. Sakano, N. Mukawa and T. Nakamura, "Kernel mutual subspace method and its application for object recognition", *IEICE Trans. Information & Systems*, vol. J84-D-II, no. 8, pp. 1549 - 1556, Aug. 2001. (*In Japanese*). 784

[15] B. Schölkopf, A. J. Smola, and K. -R. Müller, "Nonlinear component analysis as a kernel eigenvalue problem", *Neural Comput.*, vol. 10, pp. 1299 - 1319, 1998. 784
[16] A. J. Smola and B. Schölkopf, "Sparse greedy matrix approximation for machine learning", *Proceedings of ICML'00*, Bochum, Germany, Morgan Kaufmann, pp. 911 - 918, 2000. 784, 788
[17] M. Tipping, "Sparse kernel principal component analysis", in *Advances in Neural Information Processing Systems 13*, MIT Press, Cambridge,MA, pp. 633 - 639, 2001. 785, 788
[18] K. Tsuda, "Subspace method in the Hilbert space", *IEICE Trans. Information & Systems*, vol. J82-D-II, no. 4, pp. 592 - 599, April 1999. (*In Japanese*). 784

Noise Tolerance of EP-Based Classifiers

Qun Sun[1], Xiuzhen Zhang[2], and Kotagiri Ramamohanarao[1]

[1] Department of Computer Science and Software Engineering
The University of Melbourne, Vic 3010, Australia
{qun,rao}@cs.mu.oz.au
[2] School of Computer Science and Information Technology
RMIT University, Melbourne, Vic 3001, Australia
zhang@cs.rmit.edu.au

Abstract. Emerging Pattern (EP)-based classifiers are a type of new classifiers based on itemsets whose occurrence in one dataset varies significantly from that of another. These classifiers are very promising and have shown to perform comparably with some popular classifiers. In this paper, we conduct two experiments to study the noise tolerance of EP-based classifiers. A primary concern is to ascertain if overfitting occurs in them. Our results highlight the fact that the aggregating approach in constructing EP-based classifiers prevents them from overfitting. We further conclude that perfect training accuracy does not necessarily lead to overfitting of a classifier as long as there exists a suitable mechanism, such as an aggregating approach, to counterbalance any propensity to overfit.

Keywords: Knowledge Discovery and Data Mining, Classification, EP-based classifier, Noise, Overfitting

1 Introduction

Real datasets are usually affected by noise. Reliable classifiers should be tolerant to a reasonable level of noise. Sensitivity to noise within classifiers can usually be attributed to overfitting. Overfitting refers to the phenomenon that a classifier performing perfectly on training, is nevertheless poor in predictive accuracy. The explanation is that noise has been fit into the classification model during training.

Emerging Patterns (EPs) [3] are itemsets whose occurrence in one dataset varies significantly from that of another. The *growth rate* of an itemset X from the *background* dataset D_1 to the *target* dataset D_2 is $\frac{supp_{D_2}(X)}{supp_{D_1}(X)}$. X is a (ρ) EP from D_1 to D_2 (or EP of D_2), if its growth rate $\geq \rho$, where ρ is a given threshold greater than 1. X is a *Jumping Emerging Pattern* (JEP) from D_1 to D_2 (or JEP of D_2), if $supp_{D_1}(X) = 0$ and $supp_{D_2}(X) \neq 0$, i.e., the growth rate of X is infinity.

Example 1. Table 1 [11] is a small training dataset on a Saturday morning activity. There are 14 instances, labelled by one of the two classes, P or N, in-

T.D. Gedeon and L.C.C. Fung (Eds.): AI 2003, LNAI 2903, pp. 796–806, 2003.

Table 1. Saturday Morning Activity (A Small Training Dataset)

N	P
sunny,hot,high,false	overcast,hot,high,false
sunny,hot,high,true	rain,mild,high,false
rain,cool,normal,true	rain,cool,normal,false
sunny,mild,high,false	overcast,cool,normal,true
rain,mild,high,true	sunny,cool,normal,false
	rain,mild,normal,false
	sunny,mild,normal,true
	overcast,mild,high,true
	overcast,hot,normal,false

dicating participation or non-participation. All the instances with label P constitute dataset D_P, others belong to dataset D_N. Three of the many EPs are: {*Temperature=cool, Humidity=normal*} is an EP of D_P with a growth rate of $\frac{5}{3}$, as its support in D_N is $\frac{1}{5}$ and in D_P is $\frac{1}{3}$; {*Outlook=overcast*} is a JEP of D_P, as its support in D_N is 0 and in D_P is $\frac{4}{9}$; {*Outlook=sunny, Windy=false*} is an EP of D_N with a growth rate of $\frac{18}{5}$, as its support in D_P is $\frac{1}{9}$ and in D_N is $\frac{2}{5}$.

Research on EPs mainly includes the efficient mining of (J)EPs [3, 1], and most importantly, EP-based classifiers, which classify by aggregating the differentiating power of mined EPs. The CAEP (Classification by Aggregating Emerging Patterns) classifier [4, 13] and the JEP classifier [9] are the two most significant ones. Others such as iCAEP [14] and eJEP-classifier [5] are variants developed to improve the efficiency and/or classification accuracy. Experimental results on datasets from the UCI Repository [2] show that EP-based classifiers are powerful. Their accuracies are comparable with the most popular classifiers available.

In this paper, we conduct two experiments to study the noise tolerance of EP-based classifiers. A primary concern is to ascertain if overfitting occurs in them.

Building upon CAEP, in our first experiment, we propose the CAEEP (Classification by Aggregating Essential EPs) classifier. Here two additional constraints are augmented to those native to CAEP. These additional constraints provide mechanisms of regulating the consistency and generality of the mined EPs. While consistency control has been used as a means of avoiding overfitting in other classifiers, our results indicate that CAEP and CAEEP share a similar tolerance to noise and thus demonstrate their ability to withstand overfitting.

The JEP classifier, a classifier utilising those EPs with a growth rate of infinity, is also reviewed. For the JEP classifier, the fact that any training instance can only contain JEPs of its class label, meaning it will always be classified correctly, ensures 100% training accuracy. (Note this is true when every training instance contributes at least one JEP). The interest here surrounds its susceptibility to overfitting given its guarantee of a 100% training accuracy. In our second experiment, after comparing the noise tolerance levels of various classifiers using learning curves, its high ranking suggests that the JEP classifier is also not prone to overfitting.

Our results highlight the fact that the aggregating approach in constructing EP-based classifiers prevents them from overfitting and makes them noise tolerant.

2 Background on EP-Based Classifiers

The essential idea of EP-based classifiers is to classify by aggregating the differentiating power of the mined individual EPs (for the JEP classifier, it uses JEPs).

EP-based classifiers proceeds as follows:

- Step 1: The training dataset D with m classes is partitioned into subsets of $D_1, ..., D_m$. Each subset contains instances sharing the same class label.
- Step 2: Mine EPs or JEPs from $\bigcup_{j\neq 1}^{m} D_j$ to D_1, from $\bigcup_{j\neq 2}^{m} D_j$ to D_2, ..., and from $\bigcup_{j\neq m}^{m} D_j$ to D_m.
  The difference among EP-based classifiers is mainly in this step.
  In the initial version of CAEP [4], the border-based algorithm was applied to mine EPs. All the EPs had to be extracted from their border representations and had their supports and growth rates later calculated, often at great expense. ConsEPMiner [13] was proposed later to mine EPs even at low support on large high-dimensional datasets. It prunes the EP search space with constraints and, directly mines EPs with their supports and growth rates calculation inherent in the mining procedure. In this paper, any reference to CAEP implies the one with ConsEPMiner as the EP miner.
  For the JEP classifier, at this step, only the *most expressive JEPs* [9] are mined. They are the JEPs whose subsets can not be JEPs. Also, a border-based algorithm was first employed [9], followed by a tree-based mining algorithm [1]. This tree-based mining algorithm was reported to be typically around 5 times faster than the border-based one.
- Step 3: Score all the test instances in an aggregating approach.
  Both the scoring of the CAEP and JEP classifiers are characterised by the aggregating approach, although they have different scoring functions. For a test instance, all the (J)EPs of a particular class contained in it contribute to the aggregate score of this class. Calculate the aggregate score for each class. The class with the highest score wins.
  Given a test instance T, the aggregate score of class i (of m classes) for CAEP is given by: $\sum_{(X\in E(D_i)\ and\ X\subseteq T)} \frac{growthRate(X)}{growthRate(X)+1} \times supp_{D_i}(X)$, where $E(D_i)$ is the set of EPs from $\bigcup_{j\neq i}^{m} D_j$ to D_i [4]. (The score is then normalised before comparison). For the JEP classifier: $\sum_{(X\in MEJEP(D_i)\ and\ X\subseteq T)} supp_{D_i}(X)$, where $MEJEP(D_i)$ is the set of the most expressive JEPs from $\bigcup_{j\neq i}^{m} D_j$ to D_i [8].

3 Examining the Noise Tolerance of EP-Based Classifiers

We carry out two experiments to study the noise tolerance of EP-based classifiers and determine whether they suffer from overfitting. In the first experiment, we study how consistency control, a means of overfitting avoidance, works on CAEP. In the second one, we compare the noise tolerance of EP-based classifiers with 3 others by studying learning curves.

We experiment with 3 kinds of noise, attribute, label and mixed noise. Attribute noise only affects the attribute values, label noise distorts the class label descriptions, and mixed noise refers to the kind applied to both attribute values and class labels. Should we apply a $n\%$ noise, there is a $n\%$ chance that the value is replaced by a random value in its domain. Specifically, we first decide if the original attribute value or class label will be exposed to noise. If the answer is yes, for continuous attributes, a new value is chosen randomly between the minimum and the maximum. For categorical attributes or class labels, a random value is selected from these available. When applying additional noise to a classifier, if the amount is too low, its influence may not be clear. On the contrary, if there is too much noise applied, the classifier may be totally disabled. In our study, in order to conduct a reasonable study, a 40% attribute noise, a 40% label noise and a 40% mixed noise systems are employed. Also note that noise is only applied to the training dataset, not the test dataset.

Short descriptions of the parameters set for all the 6 classifiers used in our experiments are as follows. In NB and EP-based classifiers, scoring ties are broken by classifying with the majority class.

- **C5.0:** C5.0 is the successor of C4.5. We experiment with Release 1.12, RuleQuest Research Pty Ltd. We use the default settings with only one option $-f$ indicating names of the training and test datasets.
- **k-Nearest Neighbour classifier:** We choose k to be 3 in our implementation. The Euclidean distance function is used to measure similarity. Continuous attributes are first normalised to the range 0 to 1.
- **Naive Bayesian classifier:** In our implementation of NB, when an attribute is absent in a transaction, we assign 0.0001 as its count.
- **JEP classifier:** We adopt the JEP classifier with the tree-based JEP mining algorithm [1].[1] It is usually much faster than the one with the border-based JEP mining algorithm.
- **CAEP classifier:** We employ the CAEP classifier as in [4, 13].[3] The settings on the thresholds are as follows: minimum target support = 0.01, minimum growth rate = 5, minimum relative growth-rate improvement = 1.01, and maximum number of EPs = 100,000.
- **CAEEP classifier:** The settings on the thresholds are: minimum target support for non-JEPs = 0.01, minimum growth rate = 5, minimum relative growth-rate improvement = 1.01, maximum background support for non-JEPs = 0.04, minimum target support for JEPs = 0.05, and maximum number of EPs = 100,000.

[1] Thank the authors for providing the source code.

3.1 Experiment One

We propose CAEEP, a classifier derived from CAEP, by modifying its EP mining component from ConsEPMiner to ConsEEPMiner. ConsEEPMiner differs from ConsEPMiner by employing two extra constraints, i.e., consistency and generality controls. Consistency Control means that for those non-JEPs that satisfy all the constraints from ConsEPMiner, we choose only those having small background supports. The rationale is that such EPs are regarded here consistent enough to be taken as a JEP, i.e., admitting a bounded number of inconsistencies within JEPs. We think the concept of JEP is overly specific regarding the background support and, no longer force it to be strictly zero. Generality Control refers to the procedure that we demand JEPs to have enough target supports. In the definition of JEP, while the background support is zero, the target support is required to be any non-zero value. That is to say, the target support could be very small. We believe this is overly general regarding the target support. A JEP with a low target support has less discriminating power over class labels and also can be incurred by noise. CAEEP retains the most important feature of CAEP, the aggregate scoring approach. It has the same scoring function as CAEP.

We compare CAEEP with CAEP. The experiments are carried out on some datasets from the UCI Repository, using 10-fold Cross Validation.

Experimental Results. Table 2 is a summary of the resultant classification accuracies of CAEEP and CAEP on 23 datasets, listing the number of times one classifier outperforms the other. In this table, CAEEP obviously wins over CAEP on clean datasets, but they are almost tied in noisy situations. We also do paired t-tests (2-tailed) to compare CAEEP and CAEP. The results are shown in Table 3. The small p-value of 0.0215 when the datasets are clean points out a significant difference between CAEEP and CAEP, while all the other 3 p-values show similar performances. As indicated by the 2 tables, using consistency and generality controls, with 10-fold Cross Validation, CAEEP is able to achieve a considerably better accuracy than CAEP on clean datasets. However, when the training datasets have noise added, their discrepancy in accuracy is reduced, showing that CAEEP and CAEP perform similarly in noisy environments. Since consistency control is one of the methods to avoid overfitting[12], this result shows that CAEEP and CAEP do not overfit. We attribute this feature of not overfitting to the aggregating approach of the classifiers.

3.2 Experiment Two

In this experiment, we systematically compare the noise tolerance of 6 classifiers, namely JEP, CAEP, CAEEP, C5.0, kNN and NB, by analysing their learning curves. The main question is that if CAEP and CAEEP do not overfit, what about the JEP classifier? The JEP classifier attains 100% training accuracy. Our first conjecture therefore would be that overfitting is prevalent or, JEP is noise sensitive.

Table 2. CAEEP and CAEP Classification Accuracy Comparison with 10-fold Cross Validation

Classifier	Clean	Attribute Noise	Label Noise	Mixed Noise
CAEEP	17	12	10	10
CAEP	5	11	11	9

Table 3. p-values of Paired t-tests (2-tailed)

Clean	Attribute Noise	Label Noise	Mixed Noise
0.0215	0.1387	0.9225	0.2539

In order to improve the comparison of performances of classifiers by interpreting their learning curves, we propose the concept of *learning ability* of a classifier. For a given dataset, we think a classifier has a good learning ability, if it achieves a good overall classification accuracy with a smaller size of training dataset. We also suggest using the area below the learning curve to measure the learning ability of a classifier. Larger area indicates a better learning ability. As a result, a classifier whose learning curves are more often above another illustrates better learning ability. This criterion also provides us an explicit means of comparing the noise tolerance of classifiers from their learning curves. Classifiers having better learning abilities with noisy datasets are more noise tolerant.

Experiments are carried out on 22 datasets from the UCI Repository. We define an accuracy set as the one containing accuracies with different training dataset size percentages. For any of the 6 classifiers, 22 datasets, we generate 4 accuracy sets, resulting from clean dataset, and those with attribute, label and mixed noise. Each accuracy set forms a learning curve. Algorithm 1 outlines the process to calculate one accuracy set (with averaged values). In our implementation, we set the times of repetition (variable *times* in the algorithm) to 5, i.e., there are actually 5 accuracy sets generated before averaging. Each of them has the accuracies with training dataset size percentages of 20%, 40%, 60%, 80% and 100%. To obtain the accuracy of any one of the above percentage value, a 10-fold Generalised Cross Validation[6] is employed (line 5). 10-fold Generalised Cross Validation is usually applied to assess performance where different percentages of the training dataset are used. Instead of taking the remaining 9 subsets as training dataset, we take a certain percentage from them.

Normally, in order to plot one learning curve, 250 ($= 5\ (repetitions) \times 10\ (general\ cross\ validataions) \times 5\ (percentages\ of\ training\ datasets)$) classifications are conducted. This makes our total number of classifications in the experiments almost 132,000 ($= 250 \times 6\ (classifiers) \times 22\ (datasets) \times 4\ (versions\ of\ datasets)$). (There are several cases when a classifier on a dataset

Algorithm 1 Function GetAccuracies

```
GetAccuracies(dataset D, repeatTimes times)
     ;; Return the accuracy set A containing the average
     ;; accuracies with training dataset size percentages
     ;; of 20%, 40%, 60%, 80% and 100%.
     ;; A_p refers to the accuracy with training dataset size
     ;; percentage of p%.
     ;; A_{p,n} refers to the accuracy with training dataset
     ;; size percentage of p% in the n^th repetition.
1)   n ← 1;
2)   while n ≤ times do {
3)        p ← 20;
4)        while p ≤ 100 do {
5)             A_{p,n} ← 10-foldGeneralisedCrossValidation(D, p);
6)             p ← p + 20;
7)        };
8)        n ← n + 1 ;
9)   }
     ;; Take the average values.
10)  p ← 20;
11)  while p ≤ 100 do {
12)       A_p = (Σ_{n=1}^{times} A_{p,n}) / times
13)       put A_p into set A;
14)       p ← p + 20;
15)  };
16)  return A;
```

takes a very long time to finish, therefore we perform the 10-fold Generalised Cross Validation only once.)

Experimental Results. The classification results are used to plot 88 (= 22×4) figures (Figure 1 to 4 are examples of them on dataset ionosphere), each of them shows learning curves corresponding to different classifiers, on one particular dataset, clean or with noise. We then analyse the learning curves to make final judgement on the noise tolerance of the classifiers.

Generally speaking, the resultant learning curves reinforce the trend of a typical learning curve. The classification accuracies increase as the percentage of the original training dataset grows. This increase gradually declines. The learning curves rise steeply early on, they then tend to reach plateaus, although many of them in our figures do not. As for noise, it almost always lowers the learning ability of a classifier. Also, we make the following observations:

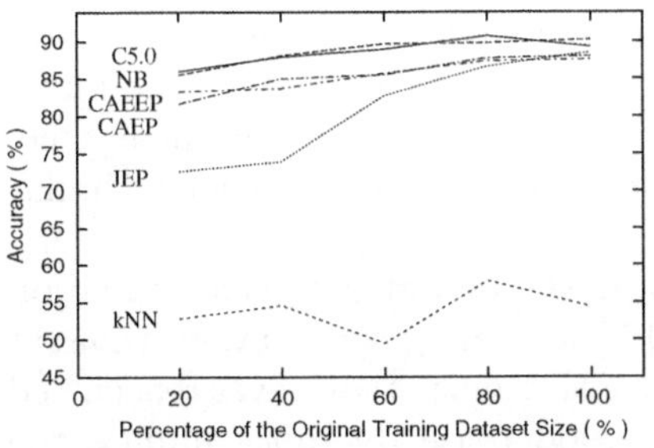

Fig. 1. ionosphere-clean

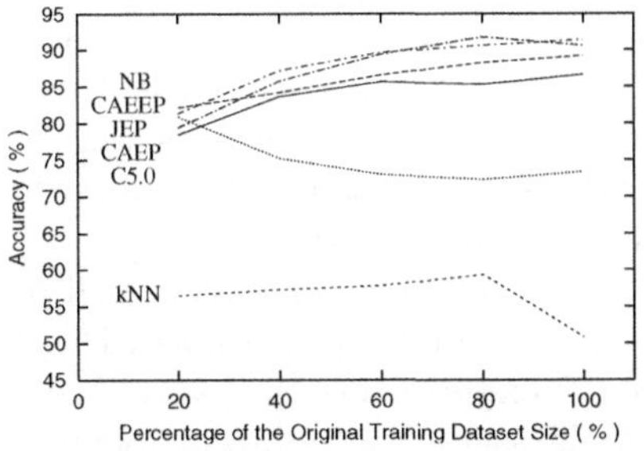

Fig. 2. ionosphere-noisyAttr

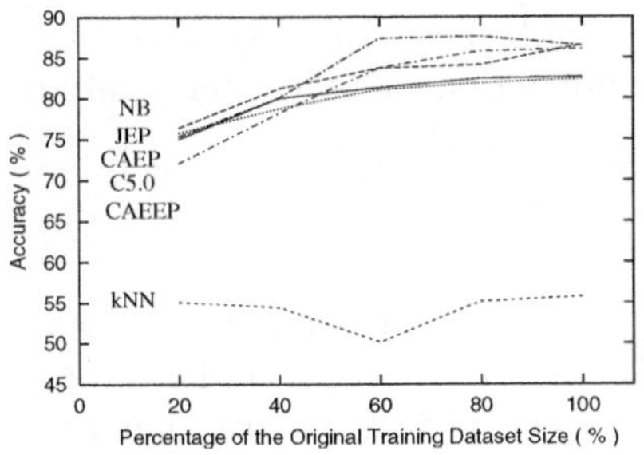

Fig. 3. ionosphere-noisyLabel

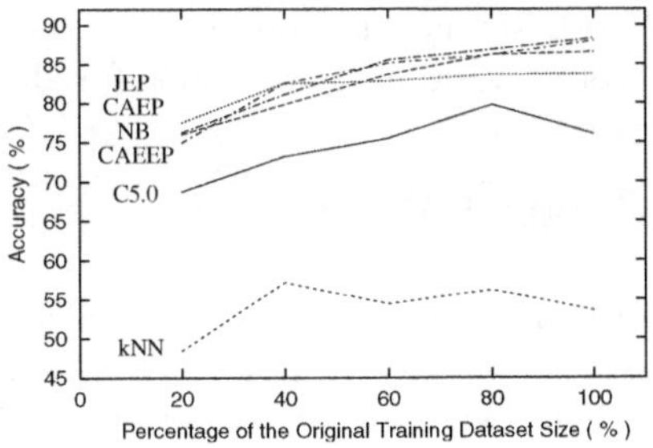

Fig. 4. ionosphere-noisyMixed

1. Under the noise levels in our experiments, classifiers are more tolerant to attribute noise than label noise, and classifiers are more robust to label noise than mixed noise.
2. The greater the number of class labels, the more pronounced the impact of label noise. This is due to the way we manufacture label noise. Under the same level of label noise, the more class labels in a dataset, the more likely class labels are changed.

We record the classifier with the best learning ability in each figure. We call such a classifier the top classifier. If the top several learning curves in one figure are so close that their average accuracy difference is smaller than 1%, then all the corresponding classifiers are regarded as top classifiers. Table 4 lists the number as top classifiers for each classifier, under different noise conditions. The performances of the 6 classifiers are finally ranked in Table 5. It shows that NB has the best performance with noise, kNN ranks last and, with JEP, C5.0, CAE(E)P ordered in between.

NB has been regarded as inherently noise tolerant due to its collection of class and conditional probabilities [7]. kNN is the worst under any type of noise despite its good performance on clean datasets. We conclude that kNN's feature of relying on a few closest instances, which may become contaminated, makes it sensitive to noise.

CAEP and CAEEP are last with regard to clean datasets and rank in the middle when considering noise. Table 6 is the result if we only compare the two from their learning curves. It lists the number of times one classifier outperforms the other. This result accords well with those obtained in Experiment One. CAEEP is superior to CAEP before the application of noise, while being similar following its introduction.

The JEP classifier performs well on clean datasets and surprisingly well on datasets with all 3 types of noise. Due to its 100% training accuracy, we thought before the experiments that some JEPs representing noise that never reoccurred in test datasets would cause overfitting. Its good noise tolerance implies the opposite. We believe, like CAEP and CAEEP, it is the aggregating approach that prevents it from overfitting. All the JEPs of one class label contribute collectively to the final score of this class label. Those JEPs representing noise are usually small both in quantity and target supports. Their influence within the aggregated final score is minimised and therefore has little impact on classification accuracy.

4 Related Work

Perfect training accuracy is regarded as a source of overfitting. Overfitting makes a classifier sensitive to noise. The essence of all the overfitting avoidance algorithms is to avoid achieving perfect training accuracy. Overfitting avoidance has

Table 4. Number of Times as Top Classifier

	JEP	CAEP	CAEEP	C5.0	NB	*k*NN
clean	7	4	4	10	7	7
Attribute Noise	10	4	8	4	7	1
Label Noise	7	3	3	8	7	1
Mixed Noise	7	7	3	3	15	0
Total	31	18	18	25	36	9

Table 5. Ranks of Classifiers

	Clean	Attribute Noise	Label Noise	Mixed Noise	Noise Overall
1	C5.0	JEP	C5.0	NB	NB
2	JEP,NB,*k*NN	CAEEP	JEP,NB	JEP,CAEP	JEP
3	-	NB	-	-	C5.0
4	-	CAEP,C5.0	CAEP,CAEEP	CAEEP,C5.0	CAEP,CAEEP
5	CAEP, CAEEP	-	-	-	-
6	-	*k*NN	*k*NN	*k*NN	*k*NN

been achieved by stopping the induction earlier according to some criteria. Another strategy is to tolerate the overfitting in the induction phase, but later dealing with it by reducing the complexity of the induction results. Overfitting avoidance can also be accomplished by relaxing the consistency requirement until prediction time. The Disjunctive Version Space (DiVS) approach [12] is an example. In the DiVS approach, induction is achieved by building DiVS in which all consistent hypotheses covering at least one training instance are represented. The prediction is performed while allowing us to control the degrees of consistency and generality. We apply this idea of consistency and generality controls in our first experiment. The difference is that we use them in the induction phase.

Contrary to the previous belief, our experiments show that the JEP classifier, despite its 100% training accuracy, is not overfitting. All the EP-based classifiers demonstrate no sign of overfitting, we attribute this to the aggregating approach employed in them.

Note the aggregating approach in EP-based classifiers is different from that in bagging and boosting. In EP-based classifiers, each EP is too weak to be taken as a classifier. An individual EP is usually sharp in telling the class of only a small fraction, e.g. 3%, of instances [4]. In bagging and boosting, multiple classifiers are generated and then aggregated into a single classifier by voting. Also, there is still controversy on the effect of bagging and boosting on overfitting.

Both EP-based classifiers and association-based classifiers, such as CBA [10], use composite tests on itemsets as classification rules. However, they are different in which itemsets are used and how they are used for classification.

5 Conclusion

Based on the two experiments, we conclude that EP-based classifiers on the whole do not overfit. We believe the aggregating approach implicit in EP-based classifiers is the primary reason. A perfect training accuracy is generally regarded by researchers as the source of overfitting for a classifier, and thus should be avoided. In this paper, we challenge this view by concluding that perfect training accuracy, even as high as 100%, does not necessarily lead to overfitting of a classifier as long as there exists a suitable mechanism, such as an aggregating approach, to counterbalance any propensity to overfit.

Table 6. CAEEP and CAEP Classification Accuracy Comparison with 5 Repetitions of Generalised 10-fold Cross Validation

Classifier	Clean	Attribute Noise	Label Noise	Mixed Noise
CAEEP	13	10	10	9
CAEP	4	6	11	9

References

[1] James Bailey, Thomas Manoukian, and Kotagiri Ramamohanarao. Fast algorithms for mining emerging patterns. In *Proc. of the 6th European Conference on Principles and Practice of Knowledge Discovery in Databases*, pages 39–50, 2002. 797, 798, 799

[2] C. L. Blake and C. J. Merz. UCI repository of machine learning databases. *Department of Information and Computer Sciences, University of California, Irvine*, 1998. [http://www.ics.uci.edu/~mlearn/MLRepository.html]. 797

[3] Guozhu Dong and Jinyan Li. Efficient mining of emerging patterns: Discovering trends and differences. In *Proc. of the 5th ACM SIGKDD International Conference on Knowledge Discovery and Data Mining*, pages 43–52, 1999. 796, 797

[4] Guozhu Dong, Xiuzhen Zhang, Limsoon Wong, and Jinyan Li. CAEP: Classification by aggregating emerging patterns. In *Proc. of the 2nd International Conference on Discovery Science*, pages 30–42. Springer-Verlag, 1999. 797, 798, 799, 805

[5] Hongjian Fan and Kotagiri Ramamohanarao. An efficient single-scan algorithm for mining essential jumping emerging patterns for classification. In *Proc. of the 6th Pacific-Asia Conference on Knowledge Discovery and Data Mining*, pages 456–462, 2002. 797

[6] Carl Myers Kadie. *SEER: Maximum Likelihood Regression For Learning-speed Curves*. PhD thesis, University of Illinois at Urbana-Champaign, 1995. 801

[7] Pat Langley and Stephanie Sage. Induction of selective Bayesian classifiers. In *Proc. of the 10th Conference on Uncertainty in Artificial Intelligence*, pages 399–406. Morgan Kaufmann, 1994. 803

[8] Jinyan Li. *Mining Emerging Patterns to Construct Accurate and Efficient Classifiers*. PhD thesis, The University of Melbourne, 2001. 798

[9] Jinyan Li, Guozhu Dong, and Kotagiri Ramamohanarao. Making use of the most expressive jumping emerging patterns for classification. In *Proc. of the 4th Pacific-Asia Conference on Knowledge Discovery and Data Mining*, pages 220–232, 2000. 797, 798

[10] Bing Liu, Wynne Hsu, and Yiming Ma. Integrating classification and association rule mining. In *Proc. of the 4th International Conference on Knowledge Discovery and Data Mining*, pages 80–86, 1998. 805

[11] John Ross Quinlan. Induction of decision trees. *Machine Learning*, 1:81–106, 1986. 796

[12] Michele Sebag. Delaying the choice of bias: A disjunctive version space approach. In *Proc. of the 13th International Conference on Machine Learning*, pages 444–452. Morgan Kaufmann, 1996. 800, 805

[13] Xiuzhen Zhang, Guozhu Dong, and Kotagiri Ramamohanarao. Exploring constraints to efficiently mine emerging patterns from large high-dimensional datasets. In *Proc. of 6th ACM SIGKDD International Conference on Knowledge Discovery and Data Mining*, pages 310–314, 2000. 797, 798, 799

[14] Xiuzhen Zhang, Guozhu Dong, and Kotagiri Ramamohanarao. Information based classification by aggregating emerging patterns. In *Proc. of the 2nd International Conference on Intelligent Data Engineering and Automated Learning*, pages 48–53, 2000. 797

Guided Operators for a Hyper-Heuristic Genetic Algorithm

Limin Han and Graham Kendall

Automated Scheduling, Optimisation and Planning Research Group
School of Computer Science and IT University of Nottingham NG8 1BB, UK
{lxh,gxk}@cs.nott.ac.uk

Abstract. We have recently introduced a hyper-heuristic genetic algorithm (hyper-GA) with an adaptive length chromosome which aims to evolve an ordering of low-level heuristics so as to find good quality solutions to given problems. The guided mutation and crossover hyper-GA, the focus of this paper, extends that work. The aim of a guided hyper-GA is to make the dynamic removal and insertion of heuristics more efficient, and evolve sequences of heuristics in order to produce promising solutions more effectively. We apply the algorithm to a geographically distributed training staff and course scheduling problem to compare the computational result with the application of other hyper-GAs. In order to show the robustness of hyper-GAs, we apply our methods to a student project presentation scheduling problem in a UK university and compare results with the application of another hyper-heuristic method.

1 Introduction

Since their introduction by Bremermann [2] and Fraser [12] and the seminal work of Holland [17] genetic algorithms have been used to solve a whole range of problems including the travelling salesman problem [25], bin packing problems [11] and scheduling problems [24].

Personnel scheduling problems have also successfully been solved using GAs. Aickelin and Dowsland [1] used a GA for a nurse rostering problem in a large UK hospital. The result of their approach was a fast, robust implementation and the approach proved flexible and able to solve the large rostering problem in the hospital with a range of objectives and constraints. Easton and Mansour [10] also used a GA for deterministic and stochastic labour scheduling problems. Their approach is a distributed genetic algorithm, which runs in parallel, on a network of workstations. Their procedure uses a combination of feasibility and penalty methods to help exploit favourable adaptations in infeasible offspring so as to maintain the search near the feasible region. They applied this approach to three different sets of published test suites of labour scheduling problem. They compared the results to those problems

T.D. Gedeon and L.C.C. Fung (Eds.): AI 2003, LNAI 2903, pp. 807-820, 2003.

solved by other meta-heuristics and conventional heuristics and found, on average, the genetic algorithm outperformed other methods.

In addition to GAs with a direct chromosome representation, indirect genetic algorithms have been studied widely. For example, Terashima-Marin, Ross and Valenzuela-Rendon [23] designed an indirect GA to solve an examination timetabling problem. They encode strategies for the problem as parameters into a 10-position array, thus the chromosome represents how to construct a timetable rather than representing the timetable itself. The use of an indirect chromosome representation avoids the limitation of a direct chromosome, which is known as the coordination failure between different parts of a solution when solving examination timetabling problem. Corne and Ogden [4] compared their indirect and direct GA for a Methodist preaching timetabling problem and found the former was more efficient.

Although genetic algorithms have been shown to be effective in solving a range of problems, they are often time intensive and require domain knowledge. Normally the chromosome of a genetic algorithm is either the solution of the target problem or a structure of the solution. This means problem specific knowledge is essential in the design of chromosome. The heavy dependence of domain knowledge makes it difficult to reuse on different problems. In order address this problem, and have a reusable, robust and fast-to-implement approach, applicable to a wide range of problems and instances, we have designed a genetic algorithm using an indirect chromosome representation based on evolving a sequence of heuristics which are applied to the problem. In this case the problem is a personnel scheduling problem that can be regarded as the allocation of staff to timeslots and possibly locations [26]. The motivation behind this approach is a hyper-heuristic, which we introduce in the next section.

2 Hyper-Heuristics

Hyper-heuristics [3] are an approach that operate at a higher level of abstraction than a meta-heuristic. It is described in [6] thus: "*The hyper-heuristics manage the choice of which lower-level heuristic method should be applied at any given time, depending upon the characteristics of the heuristics and the region of the solution space currently under exploration.*" Their hyper-heuristic consists of a set of low level heuristics and a high level heuristic selector. They defined a general framework for the hyper-heuristic which selects which low-level heuristics to apply at each decision point. A state is maintained within the hyper-heuristic which records the amount of CPU time taken by each low-level heuristic and the change in the evaluation function. A hyper-heuristic only knows whether the objective function is to be maximised or minimised and has no information as to what the objective function represents. There is no domain knowledge in the hyper-heuristic, and each low level heuristic communicates with the hyper-heuristic using a common, problem independent, interface architecture [6]. In addition to the general framework, they also have a *choice function* which decides which heuristic to call next. This is calculated based on information derived from recently called low-level heuristics. The three levels of data they maintain are the recent improvement of each individual heuristic, the recent improvement of each pair of heuristics, and the CPU time used by each heuristic. Their hyper-heuristic approach was applied to a sales summit scheduling problem

[7], a project presentation scheduling problem [8], and a nurse scheduling problem [9], and they found that it solved these problems effectively. The choice function was further improved in [8].

A hyper-heuristic method was also developed by Hart, Ross and Nelson [16]. They developed an evolving heuristically driven schedule builder for a real-life chicken catching and transportation problem. The problem was divided into two sub-problems and each was solved using a separate genetic algorithm. The two genetic algorithms evolved a strategy for producing schedules, rather than a schedule itself. All the information collected from the company is put into a set of rules, which were combined into a schedule builder by exploiting the searching capabilities of the genetic algorithm. A sequence of heuristics was evolved to dictate which heuristic to use to place a task into the schedule. The approach successfully solved a highly constrained real-life scheduling problem of a Scottish company that must produce daily schedules for the catching and transportation of a huge number of live chickens. They also compared the performance of the approach to other methods such as hill-climbing and simulated annealing and found their approach to be superior.

Randall and Abramson [22] designed a general purpose meta-heuristic based solver for combinatorial optimisation problems, where they used linked list modelling to represent the problem. The problem was specified in a textual format and solved directly using meta-heuristic search engines. The solver worked efficiently and returned good quality results when being applied to several traditional combinatorial optimisation problems. Nareyek [21] proposed an approach that was able to learn how to select promising heuristics during the search process. The learning was based on weight adaptation. The configuration of heuristics was constantly updated during the search according to the performance of each heuristic under different phases of the search. The results showed that the adaptive approach could improve upon static strategies when applied to the same problems. Gratch and Chien [13] developed an adaptive problem solving system to select heuristic methods from a space of heuristics after a period of adaptation and applied it successfully to a network scheduling problem.

In our previous work we have designed an indirect genetic algorithm hyper-heuristic approach, hyper-GA, which may be regarded as a hyper-heuristic that uses a GA to select low-level heuristics [5] and further improved this approach to an adaptive length chromosome hyper-GA (ALChyper-GA), which is more parameter-adaptive [14] than our previous work. We investigated the behaviour of the algorithms for a trainer scheduling problem and believe that given an appropriate set of low-level heuristics and an evaluation function the hyper-GA approach may be applied to a wide range of scheduling and optimisation problems.

In this paper, we investigate a guided mutation and crossover hyper-GA. The motivation for this work is to find good heuristic combinations that we can use in a crossover operator in order to guide the search. We also want to identify badly performing sequences of heuristics and use mutation to remove them from a chromosome. In previous work [14] we explicitly controlled the length of the chromosome. We hope this new approach will maintain the chromosome length to reasonable limits by a method of self-adaptation. This is similar to the idea of "*select programming components based on the evolution of the program*" of genetic programming [19].

3 Geographically Distributed Course Scheduling Problem

The problem is to create a timetable of geographically distributed courses over a period of several weeks using geographically distributed trainers. We wish to maximise the total priority of courses which are delivered in the period, while minimising the amount of travel for each trainer. To schedule the events, we have 25 staff, 10 training centres (or locations) and 60 timeslots. Each event is to be delivered by one member of staff from the limited number who are competent to deliver that event. Each staff member can only work up to 60% of his/her working time (i.e. 36 timeslots). Each event is to be scheduled at one location from a limited list of possible locations. Each location, however, can only be used by a limited number of events in each timeslot because of room availability at each site. The start time of each event must occur within a given time window. The duration of each event varies from 1 to 5 time slots. Each course has a numerical priority value. Each member of staff has a home location and a penalty is associated with a staff member who must travel to an event. The objective function is to maximise total priority for scheduled courses minus total travel penalty for trainers. A mathematical model for the problem is shown in figure 1, where we have

E: *the set of events;* ***S****: the set of staff members;* ***T****: the set of timeslots;*
L: *the set of locations;* ***dur****$_i$: the duration of event e_i;*
d_{sl}: *the distance penalty for staff member s delivering a course at location l;*
w_i: *the priority of event e_i;* ***c****$_l$: the number of room at location l*

Objective

$$MaxW = \sum_{i\in E}(w_i * \sum_{s\in S}\sum_{t\in T}\sum_{l\in L} y_{istl}) - \sum_{s\in S}\sum_{l\in L} d_{sl} \sum_{i\in E}\sum_{t\in T} y_{istl}$$

Subject

$$\sum_{s\in S}\sum_{t\in T}\sum_{l\in L} y_{istl} \le 1 \quad (i \in E) \qquad (1)$$

$$\sum_{i\in E}\sum_{l\in L}\sum_{t\in T} x_{istl} \le 1 \quad (s \in S) \qquad (2)$$

$$\sum_{i\in E}\sum_{s\in S}\sum_{t\in T} x_{istl} \le c_l \quad (l \in L) \qquad (3)$$

$$x_{istl} <= \sum_{j=1}^{t} y_{istl} \quad (i\in E)(s\in S)(t\in T)(l\in L) \qquad (4)$$

$$\sum_{s\in S}\sum_{t\in T}\sum_{l\in L} x_{istl} = dur_i * \sum_{s\in S}\sum_{t\in T}\sum_{l\in L} y_{istl} \quad (i\in E) \qquad (5)$$

$$x_{istl} <= \sum_{\substack{j\in T \\ 0<=t-j<dur_i}} y_{istl} \quad (i\in E)(s\in S)(t\in T)(l\in L) \qquad (6)$$

Fig. 1. Mathematical model for the geographically distributed trainer scheduling problem

Variable $y_{istl} = 1$ when event e_i is delivered by staff s at location l commencing at timeslot t, or 0 otherwise. Variable $x_{istl} = 1$ when event e_i is delivered by staff s at location l, or 0 otherwise. Constraint (1) ensures that one event can happen at most once. Constraint (2) ensures that each staff member is only required to deliver at most one event in each timeslot. Constraint (3) ensures that each location has sufficient room capacity for the event scheduled. Constraints (4), (5), and (6) link the x_{istl} and y_{istl} variables, which ensures that if one event is delivered, its duration must be consecutive.

4 Low-Level Heuristics

A hyper-heuristic consists of a set of low-level heuristics and a high level selector. We have developed fourteen problem-specific low-level heuristics, which accepted a current solution, and modify it in an attempt to return an improved solution. At each generation the hyper-GA could call the set of low-level heuristics and apply them in any sequence. These low-level heuristics may be considered in four groups: (1) *add,* (2) *add-swap,* (3) *add-remove,* and (4) *remove.* The add heuristics comprise five methods which can be sub-divided into two groups. *Add-first, add-random* and *add-best* try to add unscheduled events by descending priority, and *add-first-improvement* and *add-best-improvement* consider the unscheduled list in a random order. There are four add-swap heuristics: *add-swap-first* and *add-swap-randomly* try to schedule one event but if there is a conflicting event when considering a particular timeslot, staff member and location, we will consider all swaps between that conflicting event and other scheduled events to see if the conflict can be resolved; *add-swap-first-improvement* and *add-swap-best-improvement* do the same but consider the unscheduled list in a random order. The mechanism of the third group (*add-remove-first, add-remove-random,* and *add-remove-worst*) is: select one event from the unscheduled event list by descending priority, if the event is in conflict with event(s) in the timetable (none of the event's possible staff members are able to work in the possible timeslots), and the event's fitness is higher than the fitness(es) of the conflicting event(s), delete the conflicting event(s) and add the unscheduled event. The last group has two heuristics: *remove-first* and *remove-random.* These two heuristics try to remove events from the schedule. They are likely to return a worse solution but will hopefully lead to an improvement later on, after other heuristics have been applied.

We list all our 14 low-level problem specific heuristics as follows:

0. Add-first
1. Add-random
2. Add-best
3. Add-swap-first
4. Add-swap-randomly
5. Add-remove-first
6. Add-remove-random
7. Add-remove-worst
8. Add-first-improvement
9. Add-best-improvement
10. Add-swap-first-improvement
11. Add-swap-best- improvement
12. Remove-first
13. Remove-random

The integer in front of each heuristic is the integer used in the chromosome.

2	3	1	5	0	7	9	8	11	1	10	9	12	13

Fig. 2. Example of hyper-GA

5 Hyper-GA and Guided Operators

5.1 Hyper-GA

Hyper-GA [5] and the adaptive length chromosome hyper-GA (ALChyper-GA) [14] are hyper-heuristics that use a GA to select low-level heuristics to produce high quality solutions for the given problem. The GA is an indirect GA with the representation being a sequence of integers each of which represents a single low-level heuristic. Each individual in a GA population gives a sequence of heuristic choices which tell us which low-level heuristics to apply and in which order to apply them. Figure 2 is an example of hyper-GA. Integer 2 represents low-level heuristic *add-best,* 3 refers *add-swap-first* and so on.

5.2 Adaptive Length Chromosome Hyper-GA (ALCHyper-GA)

The adaptive length chromosome hyper-GA assumes that a fixed length chromosome is not always the optimal length and aims to evolve good combinations of low-level heuristics without explicit consideration of the chromosome length. Each heuristic may work efficiently at one moment but work poorly during other periods. A heuristic such as *add-random* helps to add a few courses to the schedule during the initial generations, but it becomes less helpful when the schedule is "full". Thus, the behaviour of each heuristic can be different in different chromosomes at different times of the evolutionary process. Because the behaviour of a given low-level heuristic, or a combination of low-level heuristics, within a chromosome, could be very promising, while another low-level heuristic or combination could perform poorly, we use removal of poor-performing heuristics from a chromosome or the injection of efficient heuristics from one chromosome to another in order to try and improve the quality of solutions. As a result, the length of the chromosomes in each generation changes as genes are inserted or removed. A new crossover operator and two new mutation operators were designed for this algorithm. The *best-best crossover*, will select the best group of genes (the call of low-level heuristics by these genes that gives the most improvement of the objective function) in either selected chromosome, and exchanges them. One new mutation operator, *removing-worst mutation*, will remove the worst group of genes (the call of low-level heuristics by these genes which gives the largest decrease in the objective function, or which is the longest group giving no improvement to the objective function) in the selected chromosome. Another mutation, *inserting-good mutation,* inserts the best group of genes from a randomly selected chromosome to a random point of the desired chromosome.

The reason why we combine good groups of genes and remove bad genes is that we hope the work of particular heuristic or combination of heuristics can be helpful with in other chromosome. See [14] for detailed description of these operators.

5.3 Guided Operators

The adaptive length chromosome, hyper-GA, produced promising results with our test data sets. However, we hypotheses the algorithm should work more efficiently if removal and injection of genes can be better guided. The ALChyper-GA has the ability to identify good groups and poor groups of genes in each chromosome, but it still needs to identify whether the chromosome needs new genes to enhance the search or redundant genes need to be removed. For example, the ALChyper-GA should have the ability to know that new genes need to be added to very short chromosomes and some genes need to be removed from a chromosome when it becomes too long (and thus increases the computation time). In order to add this ability to ALChyper-GA, we wanted to design a strategy to guide the algorithm to adapt the length of chromosome more effectively and efficiently. We wanted the strategy to be able to identify when it is appropriate to remove genes and when it should inject genes. There are two points in the strategy:

a. When the chromosome is longer than the average length of chromosomes over previous generations, remove the worst-block of genes.
b. When the chromosome is shorter than the average length of chromosomes over previous generations, inject the best-block of genes from another chromosome.

These two strategies are designed to help the dynamic injection/removal of genes, and should also maintain a reasonable length chromosome. The average length should not become too short, because this will not help the solution. On the other hand, long chromosomes will be computationally expensive to evaluate.
The pseudo-code for the guided operator hyper-GA is shown below:

1. Generate an initial solution (S) randomly.
2. Generate 30 initial chromosomes (length of 14), put them into a pool
3. For each chromosome k ($0 \leq k < 30$),
   a. Apply low-level heuristics in the order given in the chromosome to S
   b. Record the solution S_k (k is the position of the chromosome in the pool)
   c. Record the change each single gene makes to the objective function
4. Compare each S_k to S: if $S_k > S$, then $S = S_k$.
5. Select parents. For each pair of parents: decide which crossover operator to use (either *best-best crossover* or *one-point crossover*, choosing them with equal probability), if *best-best crossover* is used, select best groups of genes in each parent, and exchange them.
6. Select chromosomes for mutation and for each selected chromosome: decide which mutation operator to use (*inserting-good mutation* or *removing-worst mutation*). If *a* (see previous page) can be used, using *removing-worst mutation* to remove the worst group of genes from the chromosome; if *b* can be used, use *inserting-good mutation* to inject the best group of genes from a randomly selected chromosome to the desired chromosome.
7. Add all new chromosomes and 10 best chromosomes in current pool to a new pool. If the stopping criteria is met, stop the evolution, else, go to 3.

We have 4 versions of hyper-GA, two with adaptive parameters and two with non-adaptive parameters. In the adaptive versions, the mutation rate and crossover rate adapt according to the change in fitness in each generation. When there is no improvement in average fitness over 3 generations, the mutation rate will be increased using the following formula:

$$\textit{New Mutation Rate} = (\textit{Old Mutation Rate} + 1)/2 \tag{7}$$

and the crossover rate will be decreased using :

$$\textit{New Crossover Rate} = \textit{Old Crossover Rate}/2. \tag{8}$$

If the average fitness has improved over 3 generations, the mutation rate will be decreased using:

$$\textit{New Mutation Rate} = \textit{Old MutationRrate}/2 \tag{9}$$

and the crossover rate will be increased using:

$$\textit{New Crossover Rate} = (\textit{Old Crossover Rate} + 1)/2 \tag{10}$$

The aim of the modification to the crossover/mutation rates is to observe the effect these have on the performance of hyper-GA.

There are two fitness functions in our algorithm. One uses total priority minus total travelling penalty. The formula is:

$$\sum \Pr iority \; - \sum \textit{Travelling Penalty} \tag{11}$$

The other uses total priority minus total travelling penalty divided by the CPU time, so that improvement per unit time is the fitness. The formula for this objective function is:

$$(\sum \Pr iority - \sum \textit{TravellingPenalty}) / (\textit{CPU Time in Chromosome}) \tag{12}$$

The consideration of CPU time is so as to easily compare the efficiency of each individual sequence of low-level heuristic. The comparison of these four versions can test the robustness of hyper-GA under a range of conditions.

The four versions of hyper-GA, according to the objective function and the context of parameters for mutation and crossover rate, are as follows:

- PPPN uses (11) as the objective function.
- PPPA uses same objective function as PPPN, and the crossover and mutation rate are adapted using (7)-(10).
- FTPN, uses (12) as the objective function.
- FTPA, whose objective function is the same as FTPN, and the crossover and mutation rate are adapted using (7)-(10).

PPPN, PPPA, FTPN, and FTPA are simply mnemonic names.

Table 1. Results of guided ALChyper-GA applied to the basic data set [5] (percentage of shorter than average length over the generation in row, percentage of longer than average length over the generation in column. Objective (maximise)/CPU time (secs))

	10%	20%	30%	40%	50%
10%	1973/1042	1971/1021	1967/1035	1965/1024	1964/1039
20%	1972/1053	1967/1060	1969/1040	1966/1053	1968/1042
30%	1969/1031	1968/1043	1965/1031	1964/1035	1969/1052
40%	1971/1046	1969/1045	1963/1052	1965/1059	1969/1047
50%	1970/1038	1967/1049	1966/1061	1966/1049	1968/1051
100%	1970/1041	1965/1057	1967/1060	1968/1074	1967/1072
200%	1962/1074	1963/1080	1962/1079	1964/1083	1962/1076

6 Results

All algorithms were implemented in C++ and the experiments were conducted on an AMD 800MHZ with 128MB RAM running under Windows 2000. We used five data sets, which describe realistic problem instances, each having differing degrees of difficulty [5]. Each data set contains more than 500 events. The events in each data set are generated randomly, based on the characteristics of a real staff trainer scheduling problem at a large financial institution.

In order to ascertain the correct length at which to remove and inject genes, we tried to remove genes when the chromosome was 10%, 20%, 30%, 40%, 50%, 100%, or 200% longer than the average length of chromosome in one generation, and inject genes when the chromosome was 10%, 20%, 30%, 40%, 50% shorter than the average length in our experiment. We also combined these rates so as to find the effect of them on the evolution of hyper-GA. Table 1 shows the results.

All combinations find relatively good results (the best result in our previous work [14] is *1961/1357*), and the CPU time is less than our previous work. However, trying to find the best combination is computationally extensive, since the sum of CPU time in table 1 is large. In order to avoid tuning parameters manually, we tried to evolve the guiding rates within the algorithm itself.

To evolve rates, we add two more genes to each chromosome, one for the rate of removing genes, the other one for the rate of injection. These two genes are real numbers in the range 0 to 1, they don't exchange information with other genes during the evolution. The evolution of these parameters is carried out by adding a random number. The result of the self-evolving rate experiment is **1972/1184**. The objective function is not better than the best result (*1973/1042*) in the fixed rate group, and the CPU time is greater. However, the chromosomes are kept to a reasonable length and the guiding parameter does not need to be manually tuned. This is important as different problems, or even problem instances could require different parameters to produce good quality solutions.

In some experiments in table 1, because the length of chromosomes drops sharply during the evolution (the average length becomes 1 or 2 after about 100 generations), we designed a new mutation, (*add-remove-mutation)*. This operator adds a *best block of genes* from a list of *best blocks* to the selected chromosome and replaces the *worst-block of genes* in that chromosome. This heuristic is added to the original 14 and is used in the experiments discussed below.We compare each of our hyper-heuristic approaches over the five problem instances (each result is the average of 5 runs).

The results of the self-evolving guiding rate hyper-GA tests on the 5 data sets are shown in table 2 along with the results of our previous work: application of hyper-GA and ALChyper-GA to the same data. We also compare with the results of the application of genetic algorithm and memetic algorithm (defined by Moscato [20]). The upper bound is calculated by solving a relaxed knapsack problem [18] where we ignore travel penalties. We can see from the table that the results of our new algorithm are all better than previous results. In addition the computation al time is less. We find that the result of guided operator hyper-GA for the basic data set is *1972/1184*, which is not better than the best value in table 1, however, the self-evolving guiding rate hyper-GA avoids manually tuning the guiding rate.

Results from our hyper-TGA [15] are also included in table 2. Although the hyper-TGA consumes less processing time, the guided-operator hyper-GA produces superior results. We suspect this is due to the increased computation required to identify the best/worst genes in each chromosome by the guided-operator hyper-GA.

Table 2. Comparison of Guided Operator Hyper-GA with Other Algorithms (Objective (maximise)/Time)

Heuristics	*Basic data*	*Very few staff*	*Few staff 1*	*Few staff 2*	*Non -restricted*
Upper bound (priority)	2261	2179	2124	2244	2179
GA (30, 100) [5]	1796/1628	1633/1629	1589/1641	1706/1721	1644/1699
MA (30, 100) [5]	1832/2064	1678/2054	1617/2129	1769/2254	1698/2133
Hyper-GA PPPN [5]	1959/1456	1780/1387	1749/1404	1858/1496	1742/1422
Hyper-GA PPPA [5]	1939/1448	1754/1461	1712/1306	1854/1475	1814/1571
Hyper-GA FTPN [5]	1943/1411	1770/1437	1673/1436	1803/1422	1774/1434
Hyper-GA FTPA [5]	1951/1420	1731/1424	1738/1436	1769/1427	1770/1419
ALCHyper-GA PPPN [14]	1961/1357	1788/1250	1816/1163	1831/1591	1822/1437
ALCHyper-GA PPPA [14]	1933/1638	1757/1644	1795/1325	1862/1506	1804/1638
ALCHyper-GA FTPN[14]	1949/1450	1780/1365	1781/1277	1821/1638	1813/1488
ALCHyper-GA FTPA1[14]	1954/1526	1764/1496	1766/1364	1799/1583	1799/1419
Guided Hyper-GA PPPN	**1972/1184**	**1792/1139**	**1819/1087**	**1869/1257**	**1826/1194**
Guided Hyper-GA PPPA	1960/1208	1780/1158	1796/1135	1849/1306	1814/1270
Guided Hyper-GA FTPN	1964/1223	1786/1164	1802/1158	1852/2286	1811/1248
Guided Hyper-GA FTPA	1969/1215	1785/1162	1807/1143	1865/1324	1816/1186
Hyper-TGA PPPN [15]	1966/972	1789/911	1820/834	1866/1004	1824/941
Hyper-TGA PPPA [15]	1959/958	1782/931	1804/864	1852/996	1809/970
Hyper-TGA FTPN [15]	1960/963	1784/933	1799/856	1852/1012	1814/982
Hyper-TGA FTPA [15]	1965/985	1782/942	1811/892	1857/997	1804/930

7 Student Project Presentation Scheduling Problem

From table 2 it is apparent that the guided-operator hyper-GA outperforms other algorithms across all problem instances for the trainer scheduling problem. In order to further demonstrate the effectiveness of the guided-operator hyper-GA, and to test the robustness and generality of our hyper-GAs, we applied the hyper-heuristic to a student project presentation scheduling problem, which was solved by Cowling et al [8].

7.1 Problem Description

Every final year BSc student in the School of Computer Science and IT at the University of Nottingham has to give a 15-minute presentation to describe his/her project. A 4-week time slot is allocated for all presentations. Each student chooses a project topic and works under the supervision of an assigned staff member. Project presentations are then organised and each student must present his work in front of a panel of three staff members (a first marker, a second marker and an observer), who will mark the presentation. Ideally, the project's supervisor should be involved in the presentation (as the first marker or the observer) but this is rarely the case in practice. Once every student has been assigned a project supervisor, the problem is to schedule all individual presentations, i.e. determine a panel of three staff members for each presentation, and allocate both a room and a timeslot to the presentation. The presentations are organised in sessions, each of which contains up to six presentations. Typically the same markers and observers will see all presentations in a particular session. So the problem can be seen as the search of (student, 1st marker, 2nd marker, observer, room, timeslot) tuples, and has the to following constraints:

(1) Each presentation must be scheduled exactly once;
(2) No more than 6 presentations in each session;
(3) Only one session in one room at one time;
(4) No marker can be scheduled to be in 2 different rooms within the same session.

Moreover, presentations can only be scheduled in a given session when both the academic members of staff and the room assigned to those presentations are available during that session. There are four objectives to be achieved:

(A) Fair distribution of the total number of presentations per staff member;
(B) Fair distribution of the total number of sessions per staff member;
(C) Fair distribution of the number of inconvenient sessions (before 10 am and after 4 pm) per staff member;
(D) Optimise the match between staff research interest and project themes, and try to ensure that supervisors attend presentations for projects they supervise.

There are 151 students, 26 staff members, 60 sessions and 2 rooms involved in this problem. For further details of the problem and its formulation please refer to [8].

Table 3. Application of hyper-GAs to the student project presentation scheduling problem

	A	B	C	D	Obj (E)	Time (CPU sec)
Choice Function	344.40	14.60	17.70	-1637.00	-1444.99	600
Hyper-GA	360.50	19.10	10.90	-1622.00	-1419.38	984
ALCHyper-GA	347.20	15.10	9.40	-1631.80	-1440.28	975
Guided Operator Hyper-GA	**325.00**	**14.80**	**5.60**	**-1632.30**	**-1453.32**	**924**

7.2 Application of Hyper-GAs to the Problem

Cowling et al [8] designed a hyper-heuristic choice function to select 8 low-level heuristics to solve the problem. They ran their algorithm for 600 CPU seconds. In our application, we use the same objective function and same low-level heuristics. The choice function, however, is replaced by the hyper-GAs. The initial length of each chromosome is 8. The same parameter rates and population size as for the trainer scheduling problem were used. We ran our program over 200 generations. All experimental results were averaged over 10 runs. Table 3 presents the result.

The A, B, C, D in table 3 represent the evaluation results of A, B, C and D in 7.1. The objective function is try to minimise $E(x) = 0.5A + B + 0.3C - D$.

From table 3 we find that the guided operator hyper-GA gives the best result, though it takes longer time than Cowling et al' hyper-heuristic choice function. The result also demonstrates that the guided operator hyper-GA is superior to our other versions of hyper-GA.

8 Conclusion

The guided adaptive length chromosome hyper-GA is a further improvement of our previous work. It is a promising approach for solving personnel scheduling problems and other optimisation problems. The removal of poorly performing heuristics and the injection of promising heuristics as a guiding strategy appears to help the search. The strategy presented in this paper also helps to reduce the CPU time. We have tried many combinations of guiding rates. However, trying to find suitable parameters is computationally expensive. Therefore, we designed a new strategy that is able to evolve the removal/injection rates. The use of this strategy can evolve good quality solutions on all our testing data sets, and it improves upon our previous work. The comparison of hyper-GAs to the hyper-heuristic choice function further shows that the guided operator hyper-GA can achieve better results than other versions of hyper-GA, and hyper-GAs are robust across a range of problem instances.

References

[1] Aickelin, U, Dowsland, K, Exploiting Problem structure In A Genetic Algorithm Approach To A Nurse Rostering Problem, 2000, *Journal Of Scheduling*, vol 3, pp.139-153.

[2] Bremermann, H.J. 1958. The Evolution of Intelligence. The Nervous System as a Model of its Environment. Technical Report No. 1, Contract No. 477(17), Dept. of Mathematics, Univ. of Washington, Seattle.

[3] Burke, E., Kendall, G., Newall, J., Hart, E., Ross, P., and Schulenburg, S., *Handbook of metaheuristics*, chapter 16, Hyper-heuristics: an emerging direction in modern search technology, pp. 457–474. Kluwer Academic Publishers, 2003.

[4] Corne, D, Ogden, J, Evolutionary Optimisation of Methodist Preaching Timetables, Lecture Notes in Computer Science, PATAT1995: 142-155.

[5] Cowling, P.I., Kendall, G., and Han, L.. An investigation of a hyperheuristic genetic algorithm applied to a trainer scheduling problem. Proceedings of the Congress on Evolutionary Computation 2002, CEC 2002. Morgan Kaufman, pp. 1185-1190, 2002.

[6] Cowling, P.I., Kendall, G., Soubeiga, E., Hyperheuristic Approach to Scheduling a Sales Summit, Selected papers of *Proceedings of the Third International Conference of Practice And Theory of Automated Timetabling,* Springer LNCS vol 2079, pp. 176-190.

[7] Cowling, P.I., Kendall, G., Soubeiga, E., A Parameter-free Hyperheuristic for Scheduling a Sales Summit, 2001,*Proceedings of the Third Metaheuristic International Conference (MIC 2001)*, pp. 127-131

[8] Cowling, P.I., Kendall, G., Soubeiga, E., Hyperheuristics: A Tool for Rapid Prototyping in Scheduling and Optimisation, 2002, European Conference on Evolutionary Computation (EvoCop 2002), Springer LNCS.

[9] Cowling, P.I., Kendall, G., and Soubeiga, E., Hyperheuristics: A robust optimisation method applied to nurse scheduling, 2002, Seventh International Conference on Parallel Problem Solving from Nature, PPSN2002, Springer LNCS, pp. 851-860.

[10] Easton, F, Mansour, N, A Distributed Genetic Algorithm For Deterministic And Stochastic Labor Scheduling Problems, 1999,*European Journal of Operational Research*, pp. 505-523.

[11] Falkenauer, E., A Hybrid Grouping Genetic Algorithm for Bin Packing, 1996, Journal of Heuristics, vol 2, No. 1, pp. 5-30.

[12] Fraser, A.S., Simulation of genetic systems by automatic digital computers. II, 1957, Effects of linkage on rates under selection. Australian J. of Biol Sci, vol 10, pp 492-499

[13] Gratch, J., Chien, S., Adaptive Problem-Solving for Large-Scale Scheduling Problems: A Case Study, 1996, *Journal of Artificial Intelligence Research*, vol. 4, pp. 365-396.

[14] Han, L., Kendall, G., and Cowling, P., An adaptive length chromosome hyperheuristic genetic algorithm for a trainer scheduling problem, SEAL2002: 267-271.

[15] Han, L., Kendall, G., Investigation of a Tabu Assisted Hyper-Heuristic Genetic Algorithm, 2003, accepted by CEC'03, Perth, Australia.
[16] Hart, E, Ross, P, Nelson, J, Solving a Real-World Problem Using an Evolving Heuristically Driven Schedule Builder, 1998, *Evolutionary Computation* vol. 6, No.1, pp. 61-80.
[17] Holland, J. H. (1975). Adaptation in Natural and Artificial Systems, Ann Arbor, MI: University of Michigan Press.
[18] Martello, S., Toth, P., *Knapsack Problems Algorithms and Computer Implementations*, 1990, John Wiley & Son Ltd, Chichester, England.
[19] Mitchell, M.. *An introduction to genetic algorithms,* 1996, MIT Press, Cambridge.
[20] Moscato, P., 1989, On Evolution, Search, Optimisation, Genetic Algorithms and Martial Arts: Towards Memetic Algorithms, report 826, Caltech Concurrent Computation Program, California Institute of Technology, Pasadena, California, USA.
[21] Nareyek, A., Choosing Search Heuristics by Non-Stationary Reinforcement Learning, 2001, in Resende, M.G.C., and de Sousa, J.P. (eds.), *Metaheuristics: Computer Decision-Making*, Kluwer Academic Publishers, pp.523-544.
[22] Randall, M, Abramson, D, A General Meta-Heuristic Based Solver for Combinatorial Optimisation Problems, 2001,*Computational Optimisation and Applications*, vol. 20, pp.185-210.
[23] Syswerda, G., Schedule Optimisation Using Genetic Algorithm, 1991, Handbook of Genetic Algorithms, Edited by Davis, L., International Thomson Computer Press.
[24] Terashima-Marin, H., Ross, P., Valenzuela-Rendon, M., Evolution of Constraint Satisfaction Strategies in Examination Timetabling, 1999, *Proceedings of the Genetic and Evolutionary Computation Conference (GECCO99)*. pp. 635-642.
[25] Whitley, D., Starkweather, T., and Shaner, D., The Travelling Salesman and Sequence Scheduling: Quality Solutions Using Genetic Edge Recombination, 1991,, Handbook of Genetic Algorithm, Edited by Davis, L., International Thomson Computer Press.
[26] Wren, A. Scheduling, Timetabling and Rostering - a Special Relationship? 1995, in: *ICPTAT'95- Proceedings of the International Conference on the Practice and Theory of Automate Timetabling,* pp. 475-495 Napier University

Translating Novelty of Business Model into Terms of Modal Logics

Hiroshi Kawakami[1], Ryosuke Akinaga[2], Hidetsugu Suto[3], and Osamu Katai[1]

[1] Graduate School of Informatics, Kyoto University
Sakyo-ku, Kyoto, 606-8501, Japan
{kawakami,katai}@i.kyoto-u.ac.jp
http://www.symlab.sys.i.kyoto-u.ac.jp/
[2] Matsushita Electric, Kadoma, Osaka, 1006, Japan
[3] Department of Industrial Design, Akita Municipal Junior College of Arts and Craft
12-3 Okawamachi, Araya, Akita, 010-1632, Japan

Abstract. By employing some modalities as bases for modeling business processes, the present paper shows that a conventional way of analyzing *Petri nets* also turns out to be a method for specifying the novelty of business processes. This novelty is translated into terms of *modal logics*. Most conventional methods for modeling business processes represent "structures" of their processes, which do not show directly their "novelty." The proposed method, on the other hand, enables us to specify their novelty. From the viewpoint of alethic, deontic and temporal modalities, business processes are modeled by interactions among three layers, i.e., top, main and base layers. The possible actions and states represented by using Petri nets in the main layer are controlled by designers' intentions represented in the top layer, and by what we call *social causalities* represented in the base layer. It is well known that a conventional method for checking the reachabilities of Petri nets results in a *coverability tree* of the net. We show that the coverability tree of the net in the main layer can be interpreted as a transition tree of *possible worlds* in terms of the modal logic, and that the type of tree specifies some kind of novelty of a business process.

Keywords: Modal Logic, Petri Nets, Business Process Model, Novelty, Modality

1 Introduction

The ban on obtaining patents for business processes has been removed [1]. Superficially, some of these processes seem to be trivial (not novel). In other words, some conventional business processes may now be monopolized. Most conventional tools for modeling business processes [2, 3] represent "structures" of their processes, which do not directly show their "novelty". In order to specify the novelty of a business process, we have to analyze its patent claim, which details the novelty in various ways. The present paper proposes a method for modeling

T.D. Gedeon and L.C.C. Fung (Eds.): AI 2003, LNAI 2903, pp. 821–832, 2003.

business processes based on several modalities, which enables us to specify their novelty in a uniform manner, i.e., in terms of the modal logic.

In section 2 of this paper, we propose a method for modeling business processes that represents the process of business models and relevant social constraints by the use of interactions among three layers, i.e., top, main and base layers. By focusing on the modalities of processes and states of actors in business models, the contents of those three layers are strictly defined. In the top layer, deontic and temporal logic formulae represent designers' intentions. In the main layer, Petri nets represent possible actions and states of actors in a business process. In the base layer, what we call *social causalities* are represented. The three layers are closely related with each other, and the interactions among the layers show the relationship between a business model and its surroundings.

In section 3, the proposed model is applied to a method for specifying the novelty of a business model. The method is based on a conventional way of checking the reachability of Petri nets.

2 A Model for Representing Business Processes

2.1 Modalities in Business Processes

We discuss several modalities in business processes and their surroundings, e.g., technologies, cultures, social criteria and so on.

A Viewpoint of Alethic Modality (□: necessity v.s. ◇: possibility) Considering *possibility* and *necessity* on business processes, actors (seller, buyer, broker, distributor and so on) are to act to satisfy their own desires [8], and there are several *possibilities* for their actions. However, from the viewpoint of the designer of a business model, it is *necessary* to restrict actors' choices in order to organize the whole business process. Actors cannot widely deviate from a designed sequence of processes. Generally, necessity is either teleological or causal. This kind of restriction can be interpreted as the result of *teleological necessities.*

Actors are also restricted by causalities that we call social causalities. Their actions cause certain results governed by social constraints, e.g., legal restraints, contracts, ethical restraints and so on. For example, after a buyer notifies his/her decision to purchase to a seller, the only party who reserves the right to cancel it is the buyer. This type of constraint can be interpreted as *causal necessities.*

A Viewpoint of Deontic Modality (□: obligation v.s. ◇: permission) There are several types of business processes in which actors take actions governed by the designers' intentions. That is, several restrictions fixed by designers from the viewpoint of teleological necessities turn out to be *obligations* for actors. On the other hand, possible alternatives of actions for actors are interpreted as designers' *permission.* For example, if a buyer wants to buy something, it is obliged to follow the purchase process designed by the designer, but as long as

Table 1. Relations between Modalities and Contents to be Represented

	Alethic modality	Teleological -Causal	contents	Deontic modality	Temporal modality
Top layer	Necessity	Teleological	intention	Obligation	○
Main layer	Possibility	Free	process	Permission	○
Base layer	Necessity	Causal	social causality	n.a.	×

he/she follows the obligations, he/she is permitted to buy anything anytime. In this paper, we use two symbols $\mathcal{O}(=\Box)$ and $\mathcal{P}(=\Diamond)$ to denote deontic modes.

Table 1 shows the relation between alethic and deontic modalities, and what we use to represent those modes.

A Viewpoint of Temporal Modality ($\Box$: $\mathcal{F}$ v.s. $\Diamond$: $\mathcal{G}$) When designing business processes, it is essential to consider temporal modalities. In the case where business processes are called "business flows," fragmentary processes are linked to form a chain. This is the most rigid case from the viewpoint of temporal modalities. Even if the temporal constraints are set up as loosely as possible, the process will be at most quasi-ordered. Actions (operations), their pre-state and post-state are also temporally ordered in nature. Therefore, in business processes, temporal modalities have to be taken into account when designers' intentions, actions and state transitions are represented, as shown by open circles in Table 1.

Sometimes, causalities are also discussed with the notion of temporal order. It has been said that "it is obvious that each causal relation involves a time sequence in nature" [9]. However, several problems arise when setting the temporal order as the reason for distinction between "cause" and "result." For example, even though an event A causes event B, A might continue longer than B. In this case, we cannot say that A precedes B. Of course, if A did not exist in this world, neither would B, but this fact is irrelevant to the temporal order [10]. Therefore, as indicated by a symbol "×" in Table 1, social causalities do not always involve temporal modalities.

Employing an axiom system "K_T," the following modal operators ($\mathcal{T}$, $\mathcal{G}$, $\mathcal{F}$, $\mathcal{U}$) are introduced:

$\mathcal{T}A$: A will be true at the next moment,
$\mathcal{G}A$: A will be true forever, i.e., at any future time point,
$\mathcal{F}A$: A will be true at some time in the future,
$A\mathcal{U}B$: B is true at the current moment or A will be true at all times until the first moment when B will be the case,

where A, B denote atomic formulae.

Regarding state transitions in the future, there are two aspects, i.e., b (branching) and l (linear), thus the modes $\mathcal{G}$, $\mathcal{F}$ are more precisely defined as [6]:

$\mathcal{G}_b A$: A will be necessarily persistent in the future,
$\mathcal{G}_l A$: A will be possibly persistent in the future,
$\mathcal{F}_b A$: A will be possibly the case in the future,
$\mathcal{F}_l A$: A will be necessarily the case in the future,

as shown in Figure 1. In the figure, each circle denotes a state, each arc denotes a state transition, and the letter A means that A is true in that state.

2.2 Three-Layered Model for Representing Business Processes

Based on the discussions of alethic, deontic and temporal modalities in business processes, we propose three layers (i.e., main, top and base layers) to represent a business process and its surroundings.

Main Layer In the *main layer*, actors' possible actions and their pre-state and post-state are represented by Petri nets. Since many actors are involved in a business model, their actions and state transitions are concurrent and distributed.

A set of local states represents the conditions of an event. Events occur either by actions or by causalities when their conditions are satisfied. Each condition has a binary value, i.e., true or false. Employing Petri nets, each local state and event are represented by a `place` and a `transition`, respectively. Therefore, the Petri nets we use are 1-bounded, i.e., the upper limit of the number of tokens in each `place` is 1. When a token is in a `place`, the business process is in the state denoted by the `place`. Each arc from a `place` to a `transition` means that the `place` denotes one of the necessary conditions of an event denoted by the `transition`. Firing of a `transition` means the occurrence of an event, while each arc from a `transition` to a specific `place` means that firing the `transition` alters the local state into what the `place` denotes. Markings of Petri nets show the state of a business process as a whole.

When it is possible for multiple `transitions` to fire, Petri nets *permits* any of the `transitions` to fire. In other words, it is *possible* to take any of the actions denoted by the `transitions`.

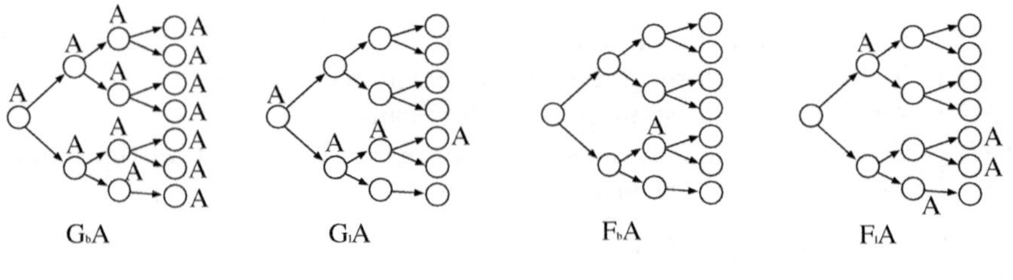

Fig. 1. World Transition Trees

Table 2. The Claim of US Patent 5,794,207

0. inputting into the computer a conditional purchase offer which includes an offer price
1. inputting into the computer a payment identifier specifying a credit card account, the payment identifier being associated with the conditional purchase offer
2. outputting the conditional purchase offer to the plurality of sellers after receiving the payment identifier
3. inputting into the computer an acceptance from a seller, the acceptance being responsive to the conditional purchase offer
4. providing a payment to the seller by using the payment identifier

Top Layer In the top layer, designers' requests are represented by combinations of temporal and deontic logic formulae. For example,

$$\mathcal{OF}A \cdots \text{ it is obligatory that } A \text{ will be true in the future}$$
$$\mathcal{PG}A \cdots \text{ it is permitted that } A \text{ is always true in the future}$$
$$\mathcal{O}(A\mathcal{U}B) \cdots \; A \text{ must always be true before } B \text{ can be true}$$

We have confirmed that each combination of temporal and deontic operators can be translated into extended Petri nets that we call "task unit graphs" [11]. By translating logic formulae into Petri nets, the top layer and the main layer can be connected in a natural manner.

"Business model patent claim" is a kind of design specification of business processes, which represents designers' intentions. Table 2 shows an example of patent claim (A portion of US Patent 5,794,207 [12]). Designers' (inventors') intentions are implied by means of a sequence of processes, which can be represented by modal logic formulae and then translated into extended Petri nets in the top layer.

Base Layer In the base layer, what we call social causalities are represented. When a certain action is taken in a business process, the corresponding `transition` fires in the main layer, which affects the surrounding of the business process. As a result, governed by social causalities, some actions become enabled or disabled. Social causalities are governed by legal, physical, political, cultural, or technical restrictions and so on.

Social causalities are represented by networks, as shown in Fig. 2. In the figure, an event "buyer received response" enables "the buyer cancels it" or "his/her deputy cancels it" and disables "other people cancels it," governed by contracts.

2.3 Interactions between layers

Figure 3 shows an overview of the main, top and base layers and the interactions between layers. These interactions represent relations between business processes, designers' intentions and social causalities.

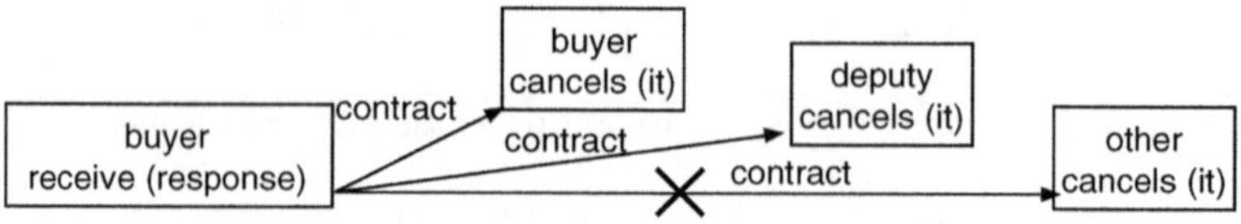

Fig. 2. An Example of Social Causality

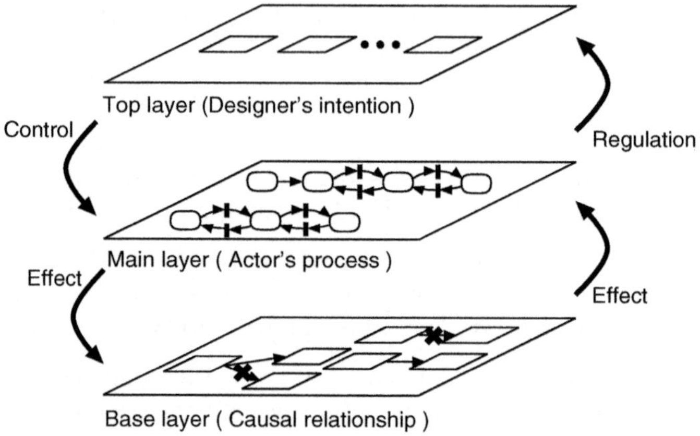

Fig. 3. Overview of Three-Layered Model

Main and Top Layers Actors involved in a business model are permitted to take actions, but the designers' intentions (patent claims) limit the alternatives of these actions. In the three-layered model, the main layer notifies the current state to the top layer, and the top layer forbids or forces certain `transitions` to fire in the main layer according to the current state.

Main and Base Layers Some local states denoted by `places` in the main layer are unified with some "cause events" in the base layer. When the `place` receives a token in the main layer, social causalities derive "result events" in the base layer. Each "result event" enables and/or disables certain `transitions` to fire in the main layer.

3 A Method for Specifying Novelty of Patent Claim

In this section, an example of business model patent (US Patent 5,794,207) is represented by the three-layered model, and analyzed for its novelty by comparing it with a conventional "reverse-auction business model."

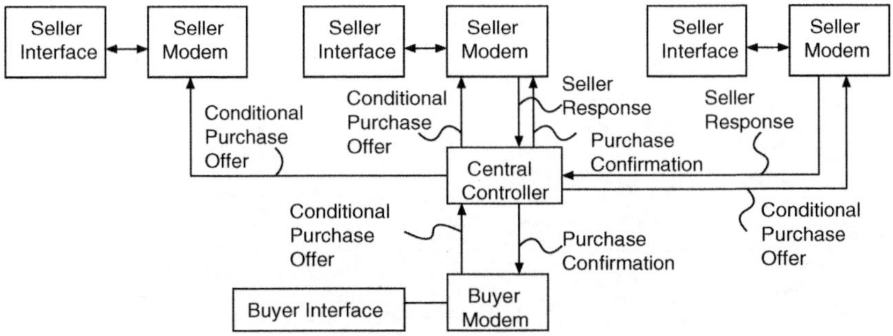

Fig. 4. The Structure of Reverse-Auction Business Process [12]

3.1 Representing Patent Claim by Three-Layered Model

Conventional Reverse-Auction Generally, an auction means that buyers propose the prices of an article presented by a seller, but the reverse-auction requires presentation of an article that a buyer wants in the first place, then sellers propose the prices.

Figure 4 shows the static structure of an internet auction [12], which is common to normal and reverse-auction except for exchanging the "seller" and the "buyer". From this structure, possible actions and local states can be derived, which are encoded into Petri nets and represented in the main layer.

An Example of Patent Reverse-Auction The main part of the patent claim, shown in table 2, is a sequence of processes. The temporal order of these processes is represented by modal logic formulae, then translated into extended Petri nets.

Figure 5 shows the "patent reverse-auction process" represented by the three-layered model. The upper part of the figure shows a modal logic formula represented by extended Petri nets[1]. The middle part of the figure shows that this business model involves a buyer, a broker, sellers and credit cards, while the lower part of the figure shows some social causal relations.

Note that the top layer (designer's intention) forces the transition "`notice_proof`" (the buyer notifies a payment identifier to the broker) to fire just after the transition "`offer`" (the buyer offers purchase) fires. When comparing with the normal reverse-auction processes represented by the three-layered model, the difference between normal and patent processes comes to light.

3.2 Analyzing Reachability to Goal State

Analyzing the Petri nets represented in the main layer enables us to examine the novelty of the business process. In this section, we focus on the process from the

[1] There are more temporal constraints that are derived from the patent claim, but they are omitted in the figure.

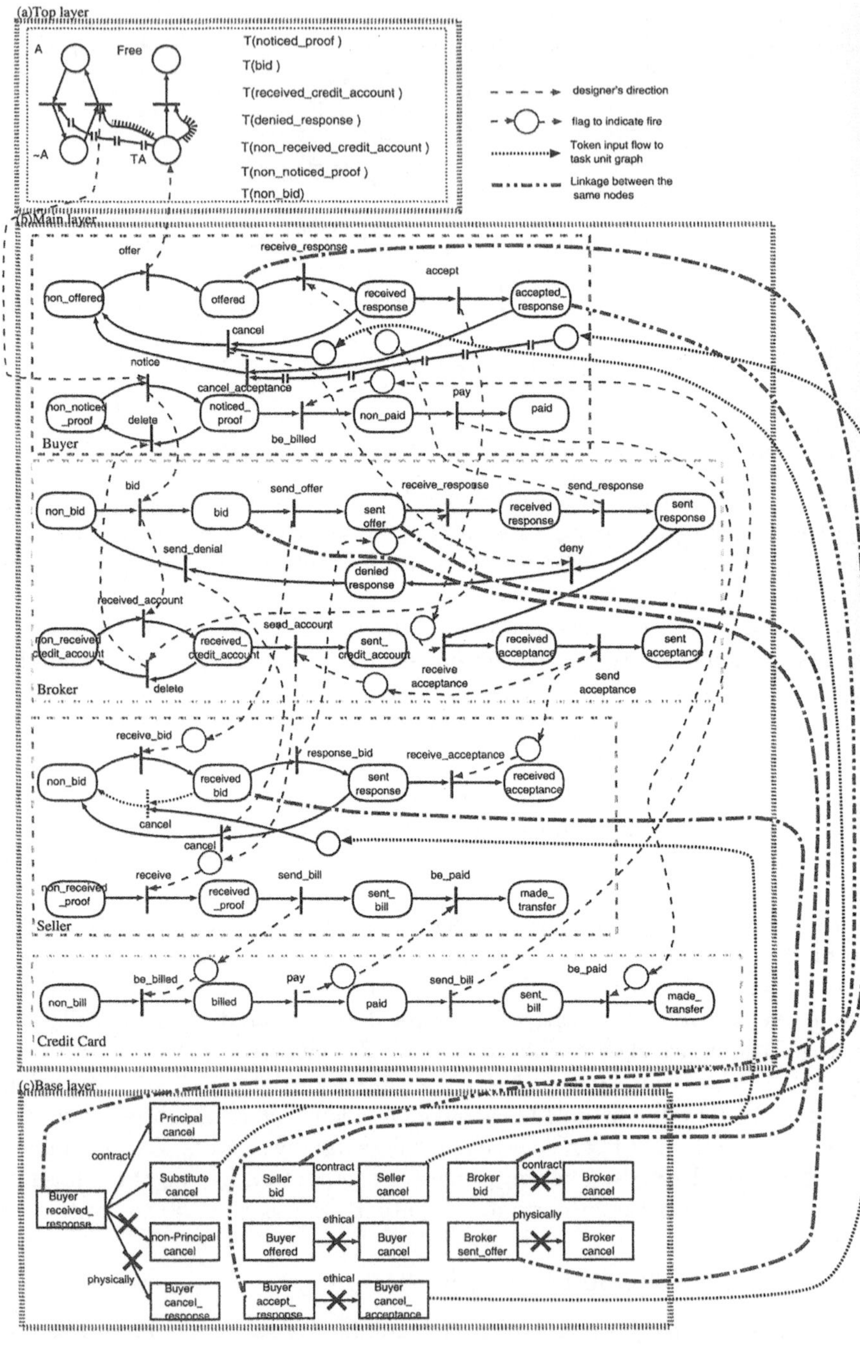

Fig. 5. An Example of Business Model Represented by the Three-Layered Model

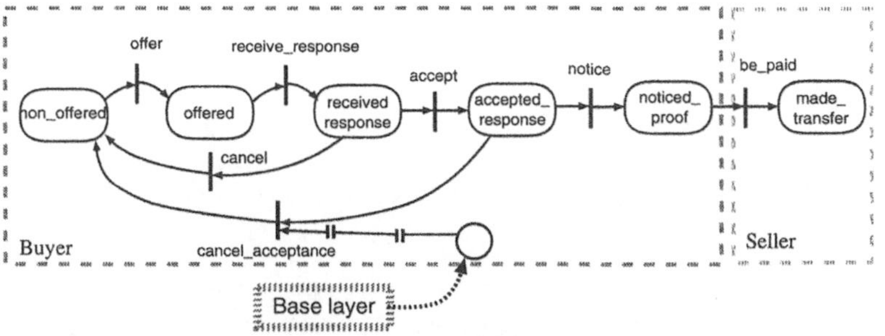

Fig. 6. A Portion of Petri Nets for Normal Process

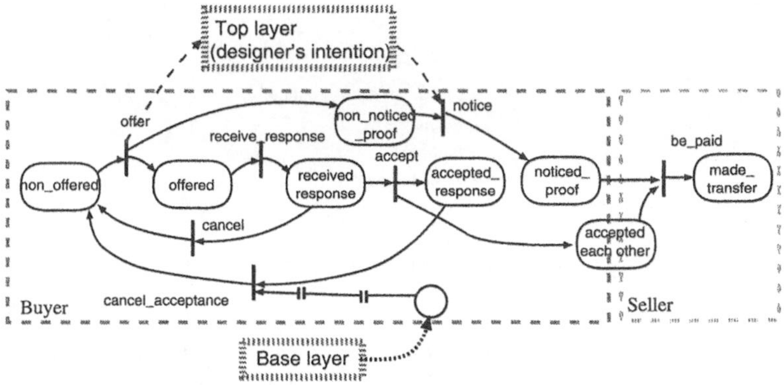

Fig. 7. A Portion of Petri Nets for Patent Process

action "`purchase_offer`" to the goal state "`paid`." Figure 6 shows Petri nets extracted from the main layer of the normal process model, and Fig. 7 shows the counterpart of the patent process model.

Generally, once the buyer notifies the broker of acceptance of purchase, ethical restraints inhibit the buyer from canceling. In Figs. 6 and 7, the arc from the place "`accepted_response`" to the transition "`cancel_acceptance`" exists because the patent claim does not explicitly forbid it. The special arc from the base layer to a blank place represents the ethical restraints.

Assuming a world where ethical restraints have no power, let us compare the reachabilities between Figs. 6 and 7. For the initial marking, both Petri nets have a token in the place "`non_offered`", and no other places have tokens. The goal state is "`made_transfer`," which denotes the resultant state of payments.

It is well known that a conventional way for checking the reachabilities of a Petri nets results in a *coverability tree* of the nets. Figure 8 (a) shows coverability trees of the normal process, and (b) shows that of the patent process, with the sequences of actors' actions shown from left to right. The branches represent alternative actions. Each broken line from right to left denotes state regressions.

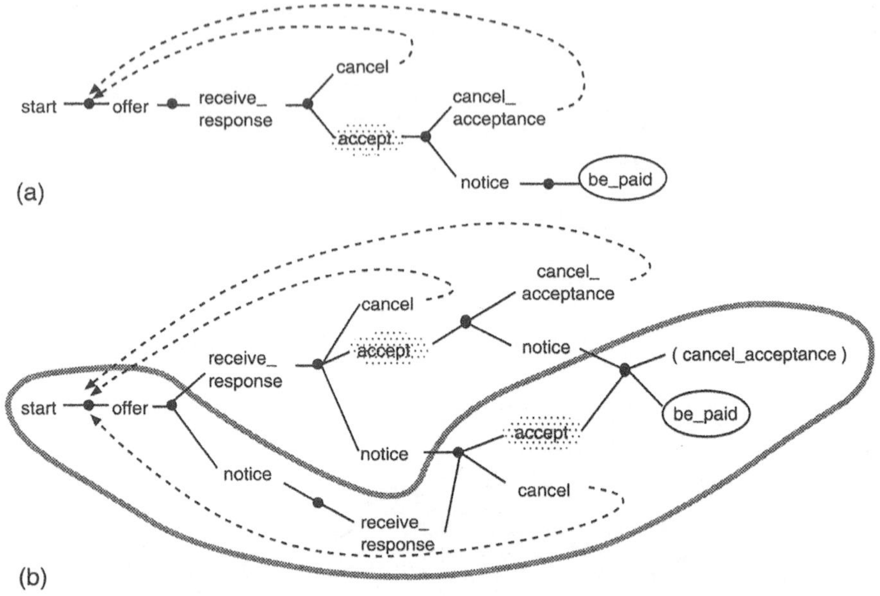

Fig. 8. Coverability Trees of Petri Nets

Comparing Fig. 8 (a) with (b) clarifies the *novelty* of the patent process. After accepting a purchase (denoted by the label "`accept`" in the figure), the buyer has a chance to cancel it (denoted by the label "`cancel`") in Fig. 8 (a), and the state regresses to an almost initial state.

On the other hand, Fig. 7 shows that the designer's intention represented in the top layer enforces the transition "`notice`" to fire just after firing the transition "`offer`." Consequently, the coverability tree shown in Fig. 8 (b) is confined to the area circled by a thick line. Within that area, after accepting a purchase, there are only two possible alternative actions: the buyer pays or the seller cancels it. In other words, the structure of the business process implicitly prevents "canceling after an acceptance by buyer" without support from ethical constraints.

3.3 Temporal Mode of State

A coverability tree shows all the possible state (marking) transitions, and all the possible world (in terms of modal logics) transitions can be represented by a tree as shown in Fig. 1. Each marking of Petri nets corresponds to a possible world of modal logics. Therefore, we can interpret coverability trees as world-transition trees.

The novelty implied by the structure of a business process is formulated into the structure of a coverability tree, which is then translated into temporal mode $\mathcal{F}_l$ shown in Fig. 1.

"Canceling after acceptance" is ethically inhibited, although sometimes buyers do so for fun or offer to purchase just for fun. In this case, sellers go to a lot of trouble for nothing, and brokers lose their attraction for sellers and ethical buyers. On the other hand, the patent process guarantees that sellers can get their money. The novelty of the patent process can be explained that way in natural language, but in terms of modal logics, the novelty can be explained in more formal manners, e.g., "$\mathcal{F}_l A$ is true" where A denotes a goal state.

4 Discussions

4.1 Supporting Design Process of Business Model

The three-layered model represents the relation between business processes and their surroundings, and a conventional method for checking the reachability of Petri nets clarifies temporal modes of a certain state. Therefore, by modeling a business process with the use of three-layered model, we can check "what will happen if we change the flow of an existing business process" by altering the content of the top layer. Furthermore, we can check "what will happen if social common sense shifts" by altering the content of the base layer. There may also be more ways to support designers (inventors) of business processes.

4.2 Implementation and Limitation of the Proposed Model

The proposed method consists of an "extended Petri nets simulator (EPNS)," a "causal chain generator (CCG)" and their mediator.

The EPNS was implemented by modifying a common algorithm for generating coverability trees of Petri nets. The variation of "marking" of a normal 1-bounded Petri nets is at most 2^n, where n denotes the number of `places` in the net. Forbiddance and enforcement by the main and the base layers can be either implemented as limitations of possible transitions from specific markings to others. Consequently, for the Petri nets in the main layer, the number of its marking, that is the number of nodes on its coverability tree, is finite (less than 2^n). The CCG was implemented [14] based on CMS (Clause Management System).

When the "degree" of states (e.g., amount of stockpile, circulating rate of goods, etc.) plays a significant role in a business process, some difficulties arise in modeling it when the proposed method is used. To represent such a degree or quantity of states by schemes for discrete events, e.g., Petri nets, the states have to be divided into some "value fixed sub-states," which in turn explode the size of Petri nets. Removing the ban on the capacity of each `place`, on the other hand, seems to be one way to reduce the size of the net, but the coverability tree does not show how many tokens are in each place; i.e., each node of the tree shows only whether each `place` has tokens or not. Therefore, the "degree" of states, i.e., the number of tokens, cannot be taken into account.

5 Conclusions

In this paper, focusing on modalities of actions and states involved in business processes, we proposed a three-layered model for representing business processes and their surroundings.

Designers (inventors) implement their ideas into a sequence of processes, which is represented by the combination of temporal and deontic logic formulae. The static structure of a business process is broken down to possible actions and their pre-states and post-states, which are represented by Petri nets, and the arbitrality of transition firings corresponds to the possibility of actions permitted by designers. The temporal order of state transition is represented by a firing sequence, and the temporal modes of each local state can be analyzed with the reachability of the net.

In this paper, we only showed that $\mathcal{F}_l A$ exists in the case of representing novelties, but there may be other temporal characteristics of novelty. For example, when $\mathcal{G}_b A$ is the case, anytime actors are guaranteed to take an action, in which the only necessary condition is A. We are now investigating some other novelties that can be represented by temporal modes.

References

[1] Makino, K., Sidney, H.W. and Kawamura, K.: Business Model Patent (in Japanese), Nihon Keizai Shinbun, Inc., Tokyo (2000) 821
[2] Scheer, A.W.: ARIS -Business Process Modeling, Springer-Verlag Berlin Heidelberg (1998) 821
[3] Yoshihara, K.: An Introduction to Business Model, Part 1, Chap. 4 (in Japanese), Kogyo Chosakai Publishing co.,ltd, Tokyo, Japan (2000) 821
[4] Kripke, S.A.: Semantical Analysis of Modal Logic I -Normal Modal Propositional Calculi-, Zeitschrift fur Mathematische Logik und Grundlagen der Mathematik, 9 (1963) 67-96
[5] Hughes, G.E. and Cressewll, M.J.: An Introduction to Modal Logic, Methuen and Co. Ltd., London (1968)
[6] Katai, O.: Modal Logic and System Control (in Japanese), Trans. Soc. Instrum. & Control Eng., vol.22, no.12 (1983) 1005-1011 823
[7] Reisig, W., Rozenberg, G.: Lectures on Petri Nets I, Springer LNCS 1491 (1998)
[8] MIT Process Handbook Project, `http://ccs.mit.edu/ph` 822
[9] Hiromatsu, W., et al., eds.: Encyclopedia of Philosophy (in Japanese), Iwanami Shoten, Publishers (1998) 823
[10] Bunge, M.: Causality, Harvard University Press, Cambridge (1959) 823
[11] Katai, O, et al.: Decentralized Control of Discrete Event Systems based on Extended Higher-Order Petri Nets, Proc. of the 1st Asian Control Conference, 2 (1994) 897-900 825
[12] United States Patent, 5,794,207, `http://164.195.100.11/` (1998) 825, 827
[13] Suto, H., Kawakami, H. and Katai, O.: A model for representing artifacts based on the modality of operations and states (in Japanese), Trans. Soc. Instrum. & Control Eng., 37, 11 (2001) 1078-1086
[14] H. Kawakami, et al.: A causal explanation of physical systems for supporting synthesis process (in Japanese), Trans. Soc. Instrum. & Control Eng. 37, 8, (2001) 786-794 831

An eNegotiation Framework

John Debenham

University of Technology, Sydney,
Faculty of Information Technology,
PO Box 123, NSW 2007, Australia
debenham@it.uts.edu.au

Abstract. Negotiation between two trading agents is a two-stage process. First, the agents exchange offers whilst acquiring and exchanging information. Second, they attempt to reach a mutually satisfactory agreement in the light of that information. This process aims to reach informed decisions in eMarket bargaining by integrating the exchange of offers and the acquisition and exchange of information drawn from the environment. Negotiation proceeds by a loose alternating offers protocol that is intended to converge when the agents believe that they are fully informed. The eNegotiation framework is a prototype for developing agents that exploit an information-rich environment in one-to-one negotiation. This work is part of a program that is investigating deeper issues of market evolution and business network development in an immersive, virtual worlds eMarket environment.

1. Introduction

The economic (game theoretic) and multiagent interest in negotiation became fused in "Agent Mediated Electronic Commerce" which has attracted increasing attention since the mid-1990s. Much of this work has studied economically rational strategies and mechanisms in an eMarket context; for example, the use of auction mechanisms for multi-unit and multi-attribute trade [1]. From an artificial intelligence perspective humans do not always strive to be utility optimisers. Sometimes they do so, for example, traders taking short-term positions in exchanges act to make short-term profits. And sometimes they do not, for example when buying a house or an automobile an intelligent human may have intelligent, justifiable but completely irrational motives. The study of negotiation between artificially intelligent agents should not be bound by economic rationality; see also [2].

Here, two agents reach a mutually beneficial agreement on a set of issues because they have chosen to become well informed. The aim is to reach *well-informed* decisions rather than *economically rational* decisions. "Good negotiators, therefore, undertake integrated processes of knowledge acquisition combining sources of knowledge obtained at and away from the negotiation table. They learn in order to plan and plan in order to learn" [3]. The work described here attempts to encapsulate this intimate link between the negotiation itself and the knowledge generated both by and because of it. During a negotiation, an agent may actively acquire information that it may, or may not, choose to place on the negotiation table. Information is a strategic weapon in competitive interaction [4].

The eNegotiation framework is a prototype for developing negotiation agents that exploit an information-rich environment in one-to-one negotiation. The

T.D. Gedeon and L.C.C. Fung (Eds.): AI 2003, LNAI 2903, pp. 833-846, 2003.

negotiation agents are responsible for conducting the negotiation on behalf of another. They are purely self-interested agents. Eventually, the negotiation architecture will consist of two further generic agent types: mediation agents and observer agents. *Mediation agents* are impartial observers of negotiations and develop expertise in how to lead multi-attribute negotiations to a successful conclusion [5]. *Observer agents* analyse unsuccessful negotiations. A failed negotiation is a missed business opportunity and so may be a clue to the discovery of an innovative, evolutionary way of doing business.

The eNegotiation framework is being built by members of the eMarkets research group in the Faculty of IT at UTS in Sydney: http://research.it.uts.edu.au/emarkets/ where the project is named the "Curious Negotiator" [6]. Minghui Li, a PhD student in the group is building the eNegotiation framework. To date a version has been roughed up in Java. Woralak Kongdenfha, another PhD student is developing negotiation strategies for the framework. Bots that access the World Wide Web have been designed and built by Dr Debbie Zhang, a Senior Researcher in the group. A fully integrated demonstrable version of the framework will be operational in July when it will be trialed on an exemplar application described below and made publicly available. Plans are to then re-engineer the framework using the Australian Jack agent building-environment and to apply it to a variety of richer negotiation situations.

The following *exemplar application* will be used in initial trials of the framework. An agent has owned a digital camera for three years and wants to upgrade to a better model. This agent has set some money aside, and has a pretty good idea of the features that she wants in her next digital camera. She is in no hurry to buy. She has let it be known that she will consider buying a second-hand camera as long as it is in as-new condition. On receiving an opening offer from a seller to buy a camera she then negotiates with that seller—hopefully leading to a mutually satisfactory deal. This problem is simple in that the actual negotiation is single-attribute—namely the price of the camera. It involves an assessment of the extent to which the camera offered is suited to her requirements—this involves multi-attribute reasoning. It also involves an assessment of the market value of the camera. Both of these two issues are resolved by reference to information on the World-Wide-Web that is accessed by information gathering bots. The negotiation uses a "loose" alternating offers protocol in which each agent may make an offer at any time but is under no contractual obligation until a deal is made [7]. During a negotiation the agent is expected to be able to feed information to its opponent such as "that model does not have the features that I am looking for" and "your price is higher than I could buy it for on the Web". This exemplar application has been chosen to enable us to make the whole negotiation apparatus work in a satisfactory way on a moderately simple problem. Future plans are to use the augmented negotiation framework in the group's two long-term projects that are in eMarket evolution and in the development of business networks in an eMarket environment.

Game theoretic analyses of *bargaining* are founded on the notion of agents as utility optimisers in the presence of complete and incomplete information about their opponents [8]. To deal with incomplete information, agents are typically assumed to know the probability distribution of that which is not known completely [9]. Likewise, game theoretic analyses of auction mechanisms in *markets* are typically base on assumptions concerning the probability distributions of the utilities of the agents in both the private and public value cases [10]. Further, game theoretic analyses of *exchanges* focus on price formation, strategies and mechanisms for agents with complete and incomplete information. Game theory tells us what to do, and what outcome to expect, in many well-known negotiation situations, but these strategies and expectations are derived from assumptions. For example, in auction theory, the most restrictive of these are that the players are utility opimisers with SIPV (symmetric, independent, private valuations) [10].

Impressive though the achievements in game theory are, the significance of that work in describing how humans negotiate—and so too how artificially intelligent systems should negotiate—is questionable. When confronted with the SIPV formula for an optimal bid in a first-price sealed-bid auction a colleague once remarked: "Look, I may simply want to buy it". The sense in this remark is that utility optimisation is not always the motivation for a negotiation.

The work here is based on the notion that when an intelligent agent buys a hat, a car, a house or a company she does so because she *feels comfortable* with the general terms of the deal. This "feeling of comfort" is achieved as a result of information acquisition and validation. This information includes both that which the agent is given and information that which the agent chooses to acquire. Here negotiation is as much of an information exchange process as it is an offer exchange process. The exchange of information is equally as important as the exchange of offers—one feeds off the other. In so far as game theory aims to reach a *rational* solution, the aim here is to reach an *informed* solution.

2. Managing an Information-Rich Environment

The form of negotiation considered is between two agents in an information-rich environment. That is, the agents have access to general information that may be of use to them. They also exchange information as part of the negotiation process. The two agents are called "me" and my opponent ω. The environment here is the Internet, in particular the World Wide Web, from which information is extracted on demand using special purpose 'bots'. So my agent, "me", may receive information either from ω or from one of these bots. In a 'real life' negotiation, the sort of information that is tabled during a negotiation includes statements such as "this is the last bottle available", "you won't get a better price than this", and so on. To avoid the issue of natural language understanding, and other more general semantic issues, the interface between each of the negotiating agents and these information sources is represented using the language of first-order, typed predicate logic, and a set of pre-agreed, pre-specified predicates. All information passed to "me" is expressed in this way.

As well as being unambiguous, the use of first-order typed predicate logic has the advantage of admitting metrics that describe, for example, how "close" two statements are. These metrics are useful in enabling the agent to manage the information extracted from the environment in a strategic way. The terms predicate, term, function, variable and constant are used in their well-accepted sense. In typed logic a term, constant or variable appearing as part of a proposition belongs to a certain domain determined by the argument in which it occurs in the predicate of that proposition. If functions can be avoided then these metrics are simpler. Functions are not used in this discussion, although their inclusion does not introduce any technical problems beyond making the Herbrand universe unbounded. The notation: <variable>/<term> denotes the substitution in which the named variable, the *subject*, is replaced by the term wherever that variable occurs in an expression.

Using the usual notation and terminology for Horn clauses, given a set of facts—ie: unit clauses with just one positive predicate—the following defines a partial ordering of those facts:

$P(\underline{x}) \geq_S P(\underline{y})$ if there exists set of substitutions J whose subjects are variables in $\underline{x}$ such that: $\underline{y} = \underline{x} / J$

where $\underline{x}$ and $\underline{y}$ are complete sets of arguments of P. If this holds then $P(\underline{x})$ *subsumes* $P(\underline{y})$. This ordering between two literals captures the notion of one being *more general* than the other.

Given two positive propositions, p_1 and p_2, both of which have the same predicate symbol, then the *unification* of those two propositions is denoted by $p_1 \cap p_2$, and the *anti-unification*—using the terminology of Reynolds—by $p_1 \cup p_2$. [The anti-unification of two positive literals is the unique literal that subsumes both of them and of all such literals is minimal with respect to the partial order $\geq_S$.]

Suppose that G is a finite set of ground literals that is closed with respect to unification and anti-unification. G is a non-modular lattice where unification and anti-unification are the meet and join respectively. Given a literal c, the function λ_G : {the set of all literals} → {the positive integers} is defined as $\lambda_G(c)$ = the number of literals in G with which c is resolvable. $\lambda_G(c)$ is a measure of the "generality" of c with respect to unification over G. It is a measure of how "general" c is over G. Given an isotone valuation v on the vertices of G, the function:

$\delta_1(c_1, c_2) = v[c_1] - v[c_2]$

is a measure of how much "more general" c_1 is compared with c_2. Further

$\delta_2(c_1, c_2) = v[c_1 \cup c_2] - v[c_1 \cap c_2]$

is a measure of the "distance" between c_1 and c_2. There is a problem with δ_2 when the meaning of the partial ordering is such that: if $c_1 \leq c_2$ then knowing c_2 means that knowing c_1 brings no new information—this is the case with the partial ordering $\geq_S$. In this case the function:

$\delta_3(c_1, c_2) = v[c_1 \cup c_2] - v[c_2]$

is a measure of how much more "novel" c_1 is compared with c_2. [Another possibility is $v[c_1] - v[c_1 \cap c_2]$ but that is not so useful when the intersection is likely to be empty—ie: at the "bottom" of the lattice—if $c_1 \leq c_2$ then $\delta_3(c_1, c_2) = 0$.

Suppose that J is any subset of G, given a literal c, then:

$\delta_i(c, J) = \min_{s \in J} \delta_i(c, s)$ for i = 1, 2, 3.

are generalisations of the measures between two vertices to measures between a vertex and a set of vertices. Suppose that S is the set of literals that agent "me" knows, let H be set of all possible literals obtained by instantiating the predicates in $\{c\} \cup S$ over the Herbrand universe for $\{c\} \cup S$, then the above two measures are:

$\delta_1(c, S) = \min_{s \in S} [\lambda_H[c] - \lambda_H[s]]$

$\delta_2(c, S) = \min_{s \in S} [\lambda_H[c \cup s] - \lambda_H[c \cap s]]$

$\delta_3(c, S) = \min_{s \in S} [\lambda_H[c \cup s] - \lambda_H[s]]$

If there are functions present then the Herbrand universe will be non-finite. In this case H may be defined as the set of such instantiations up to a certain chosen function nesting depth. As long as that chosen nesting depth is greater than that occurring in the literals in $\{c\} \cup S$ the resulting measures are generally useful.

This leads to a general approach to managing information in negotiation:

- develop a predicate logic representation
- introduce a partial ordering on that representation that captures the meaning of "generalisation" in a way that is relevant to the application
- define lattice meet and join operations
- define an isotone valuation on the lattice that is consistent with the ordering—this valuation need not necessarily treat each argument (domain) in the same way (as in the example above)—it may be useful to introduce weights that give some arguments more significance than others.

and then apply the ideas above. The introduction of two additional measures, of "cost" of acquiring the information and of "belief" in its validity, complete the machinery required to manage the acquisition and maintenance of an information base in a strategic way.

To illustrate this, in the exemplar system one job that the information bots do is to find the cheapest camera that has a certain set of features. This is achieved by scraping the sites of various information providers $\{P_i\}$. My agent knows what sites are available to it. To my agent, the bot that scrapes site P_j is represented as:

Cheapest(P_j:$\{P_i\}$, f_1:F_1,..., f_n:F_n; c:C, p:\$) [1]

where f_i is the i'th feature (such as, whether the camera has an on-board, electronic flash), F_i is the domain (ie: all possible values) for each of these features, c is the cheapest camera with those features (according to source P_j), and p is its price (according to source P_j). The ";" separates the "inputs" from the "outputs". It is assumed that each domain F_i contains the features dc_i meaning "I don't care what f_i is". So to my agent the bots appear as a set of as yet uninstantiated literals. If my agent chooses to activate the j'th bot then it does so by setting the P_j and all of the f_i to particular values and the result of the scraping exercise is then, for example:

Cheapest(P_2, "no flash",..., dc_n; Kodak123, \$54)

meaning that "according to site P_2 the Kodak123 camera is the cheapest camera available with the set of features specified and it costs \$54". In so far as that information is valid, it can only be assumed to be valid at the time that bot is activated. Here it is assumed that information is valid for the period of the negotiation. There is also the matter of how reliable the information is—with what degree of belief should the agent believe it to be true? This is achieved by attaching a certainty factor to each retrieved tuple that is based solely on an assessment of the general reliability of the information provider P_j.

The "don't care" value, dc, may be a value for any or all of the $\{f_i\}$. [Sites vary in their accuracy and completeness—some do not contain information for every such combination.] At some time after activation a bot, information will be received by my agent also in form [1] but with the "output" values (ie: values for c_x and p_x) specified. This completes the first step—the predicate logic representation. Introduce the partial order defined by the values of the first (n+1) arguments only. If the first (n+1) arguments of expressions e_1 and e_2 are the same except that e_1 contains dc values where e_2 does not then $e_1 \geq_D e_2$. $\geq_D$ is the partial ordering. If e_1 and e_2 can be "unified" by setting dc input values in one expression to the corresponding value in the other then define $e_1 \cap e_2$ to be the expression with that "unified" set of input values, [output values remaining undefined as they play no part in this ordering], otherwise $e_1 \cap e_2$ is the empty expression. In defining this ordering the constant dc is treated rather like a logic variable in conventional unification, but *it is a logical constant*. The expression $e_1 \cup e_2$ is defined similarly—ie: to reflect the sense of "generality" in the "Cheapest" predicate. Let S be the set of ground literals that the agent believes to be true, then define H as above by treating dc as a "variable", and λ_H as above.

After all of this the measures of "how general" statements are [11], and how "close" statements are, may be used to strategically gather information. In the exemplar system this amounts to managing the "information lattice" in a fairly common sense manner. It enables the agent to explore "nearby" alternatives, and to obtain a "more general feeling" for price. More importantly, it has been shown that this approach may be used for first-order resolution logic—it is proposed that this approach may be applied to any application that can be expressed in first-order logic, including those of greater complexity, to strategically manage the acquisition of information. This approach relies on the hand crafting of the lattice structure and the lattice ordering for each predicate involved.

3. Accepting an Offer

A mechanism decides whether to accept an offer or to cease the negotiation—neither of these may occur in which case either one of these two agents has to "think again". This mechanism is the first step towards automatic negotiation in which agents aim to reach comfortable agreements through managing both the exchange of offers and requests for information in an integrated way. The goal here is to design something that works. The solution described does work, but is not claimed to be the best, or even a particularly good, way representing "comfort". The following section embeds this mechanism into a negotiation framework.

Suppose that a negotiation between two agents, my opponent ω and me, stands alone—that it is not part of a related sequence of negotiations. Each agent aims to reach an agreement on a deal. A *deal* Δ is a commitment for me to do something, τ (my "terms"), subject to the other agent agreeing to do something, υ, $\Delta = (\tau, \upsilon)$. Suppose my negotiation with ω is conducted in the light of information, ι, (ie: ι is the information available to me, but not necessarily to ω) then:

$\text{Acc}(\omega, \tau, \upsilon, \iota)$

is a proposition in which the predicate "Acc" means "taking account of information ι, if I agree to do τ, subject to ω agreeing to do υ, then I will feel comfortable with the deal". This raises the two questions of defining what it means to say that I feel comfortable with a deal, and determining the point in time at which that determination should be made. It makes acceptable sense to the author to state "looking back on it, I made the right decision at the time"—those two questions are not addressed further here. The information, ι, consists of any information tabled by either party as part of the negotiation process, and any information extracted by me from the environment as described in the previous section. The commitments, τ and υ, may be quite complex and may include staggered payments or related "sweeteners"—in general they will contain multiple attributes [12]. A deal is normally associated with a time, or time frame, when the respective commitments of the two agents are to be performed. In addition, the "Acc" predicate could contain an assessment of the guarantees, assurances, escrow services and "after sales service", γ—but this has been omitted for simplicity. In addition to the Acc predicate, a "Cea" predicate aims to detect when a negotiation should cease—possibly because it appears to be a "waste of time". The Cea predicate is not described here.

A probability is attached to the "Acc" proposition so that it may be used as the basis for making a decision on whether or not to accept the terms τ. This probability aims to represent the likelihood that the proposition will turn out to be true in due course—see the comment above concerning the point in time at which "I may feel comfortable". A fuzzy set is introduced to define "Accept" to some level of confidence χ. Its membership function μ_A: $[0, 1] \rightarrow [0, 1]$ is monotonic increasing. μ_A is applied to $P(\text{Acc}(\omega, \tau, \upsilon, \iota))$. This function is a matter for each agent to specify, and in general may be a function of the terms, τ, and maybe υ, and so there is no claim that this work is leading to context-independent super-negotiator. See Fig. 1 for an example of a piece-wise linear fuzzy membership function μ_A. The μ_A function just scales the values—it accepts offers when $P(\text{Acc}) \geq \mu_A^{-1}(\chi)$. It is included so that the framework will support experiments with cooperative agents seeking to reach a compromise deal with equivalent certainty—the μ_A function enables agents to have different "types" analogous to Kasbah's "anxious", "cool-headed", and "frugal" [13].

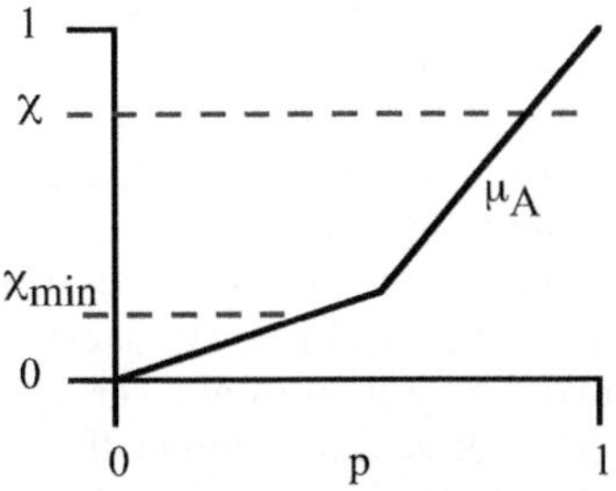

Figure 1. Fuzzy Accept

The value χ_{min} is the greatest value of confidence χ for which both:

$P(\text{Acc}(\omega, \tau, \upsilon, \iota)) \in$ Accept, and
$P(\text{Cea}(\omega, \tau, \upsilon, \iota)) \in$ Cease

are true. χ_{min} is the confidence value at which the agent will be equally disposed to accept the offer *and* to cease negotiation—values of χ at or below χ_{min} are of no interest here. The level of confidence χ is vaguely related to the notion of "risk averseness" in game theory. If, to confidence χ,

$P(\text{Acc}(\omega, \tau, \upsilon, \iota)) \notin$ Accept, and
$P(\text{Cea}(\omega, \tau, \upsilon, \iota)) \notin$ Cease

then the agent continues to negotiate as described in the following section.

The probability attached to the "Acc" proposition—and similarly for the "Cea" proposition—is derived from probabilities attached to four other propositions:

$\text{Suited}(\iota, \upsilon)$, $\text{Good}(\iota, \omega)$, $\text{Fair}(\iota, \tau, \upsilon)$, and $\text{Me}(\iota, \tau, \upsilon)$

meaning respectively: "taking account of information ι, terms υ are perfectly *suited* to my needs", "taking account of information ι, ω will be a *good* agent for me to be doing business with", "taking account of information ι, τ are generally considered to be *fair* terms in exchange for ω agreeing to do υ given that ω is impeccable", and "independent of what everybody else believes, on strictly subjective grounds, taking account of information ι, τ are fair terms in exchange for ω agreeing to do υ given that ω is impeccable". The last two of these four explicitly ignore the suitability of υ and factors out the appropriateness of the opponent ω. If ω's reputation is doubtful then the mechanism will compensate for these doubts by looking for better terms than would be comfortable for me if I had dealt with an impeccable opponent. The difference in meaning between the third and fourth proposition is that the third captures the concept of "a fair market deal" and the fourth a strictly subjective "what υ is worth to me". The "Me" proposition is closely related to the concept of private valuations in game theory.

To deal with "Suited" depends on what the terms υ are. In the exemplar system the terms will be a second-hand camera if I am buying, and money if I am selling.

There are a large number of sites on the World Wide Web that may be used to estimate the extent to which a digital camera will suit me. For example:
`http://www.dpreview.com/reviews/compare.asp`
Those sites are used in the exemplar system to attach a level of confidence in the "Suited" proposition. This is described in detail in the following section. In the digital camera market area there are a variety of such advisory sites $\{S_1,..,S_n\}$. These sites differ by the questions that they ask. If the agent's preferences are known sufficiently well to answer some of the questions on one of these sites then it is a matter of seeing whether the camera, υ, is amongst the recommended choices. So each of the sites $\{S_i\}$ may be used to generate evidence for "Suited". These sites are designed to help a buyer choose a camera—they are used here for a rather different task: to check that a given camera is suitable for me. A Bayesian revises the *a priori* probability attached to "Suited" in the light of evidence obtained from the $\{S_i\}$. For any seller it is assumed here that ω is a commitment to deliver money, and that the probability of money being "suitable" is 1.0. That is, P(Suited("money")) = 1.0. Another approach to dealing with the suitability of a deal is used in some B2B e-procurement reverse auctions [14]. There the buyer specifies various packages and attaches a factor to them as a real number ≥ 1.0. The sellers bid on the various packages. The bids are then multiplied by the factor associated with the relevant package to determine the winner—ie: the lowest bid.

Attaching a probability to "Good" involves an assessment of the reliability of the opposing agent. For some retailers (sellers), information—of possibly questionable reliability—may be extracted from sites that rate them. For individuals, this may be done either through assessing their reputation established during prior trades in a multi-agent environment [15], or through the use of some intermediate escrow service that in turn is deemed to be "reliable" in which case the assessment is of the individual *and* the escrow service. In the exemplar system this factor has been ignored as the camera is assumed to be offered for sale by an individual. That is, P(Good(ω)) = 1.0.

Attaching a probability to "Fair" is achieved by reference to an appropriate market. When a deal is to purchase retail goods for money this may be achieved by using bots to scrape a set of sites. In the exemplar system the second-hand camera, υ, is assumed to be in "as new" condition, and so its fair market value, α_υ, is obtained by discounting the best price, β_υ, that is found for that item in new condition, say by 30%. The sites scraped are those merchant sites that are linked from the sites used for "Suited" above. These are shown as $\{\bar{S}_1,..,\bar{S}_n\}$. There is no causal link between the $\{S_i\}$ and the $\{\bar{S}_i\}$. If the terms τ specify that money is to be exchanged for the camera then the set $\{ \tau : \text{Fair}(\tau , \omega) \}$ is defined by a fuzzy membership function such as that shown in Figure 2.

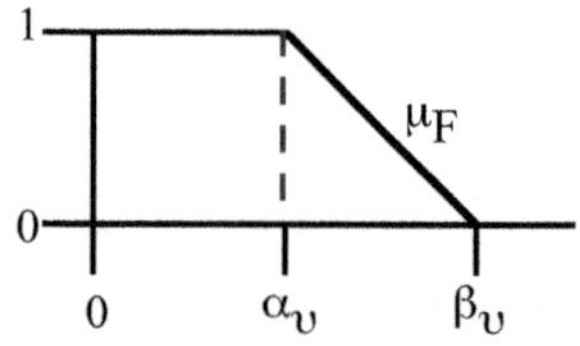

Figure 2. Fuzzy Fair

Attaching a probability to "Me" is a purely subjective matter. For a single-attribute, real-valued τ it is dealt with in a similar way to "Fair" above using a fuzzy membership function that is zero at my upper limit, $\bar{\sigma}_\upsilon$, and 1.0 at the greatest price that "I would be delighted to pay", $\underline{\sigma}_\upsilon$. Figure 3 shows a possible fuzzy membership function for "Me" for such a τ. That function aims to encapsulate the notion of "τ

are personally fair terms to me in exchange for ω doing υ". My upper limit, $\bar{\sigma}_\upsilon$, is my "valuation" in game theoretic terms—it is a link between this informed-decision approach and economically rational approaches.

The whole "accept an offer" apparatus is shown in Figure 4. There is no causal relationship between the four propositions, with the possible exception of the third and fourth. So to link the probabilities associated with the five propositions, the probabilities are treated as epistemic probabilities and form the nodes of a simple Bayesian net. The weights on the three arcs of the resulting Bayesian net are meaningful in practical terms. They are a subjective, high-level representation of what "comfortable" means to an agent. More important, the resulting net divides the problem of assessing "Accept" into four more convenient sub-problems.

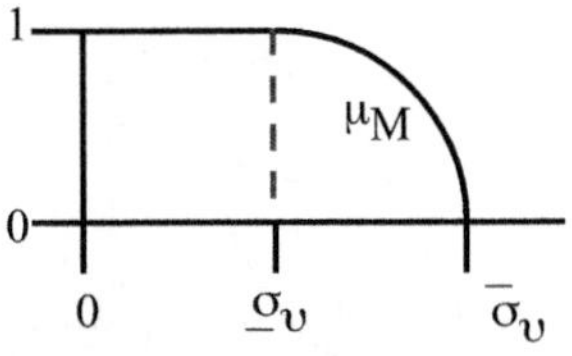

Figure 3. Fuzzy Me

Figure 4 shows the structure of the hybrid net for the general case—ie: including the machinery for "Good". The top half of the hybrid net combines data obtained from the Internet through the "fuzzy filters" represented by •'s on that figure. There "Good" is derived as a result of scraping a set of sites $\{M_1,..,M_m\}$ as would most likely be the case in a 'real' electronic business application. The product of these "fuzzy filters" are treated as probabilities and attached to the bottom half of the hybrid net that is a conventional Bayesian belief network. The probability attached to "Acc" is then passed through the fuzzy filter μ_A also denoted by a • on that figure. The conditionals on the Bayesian network are subjective—they are easy to specify because 12 of them are zero—for the cases in which I believe that either Fair or Me is "false". With fairly optimistic priors of 0.9 on each of the four evidence nodes, and the conditionals set to:

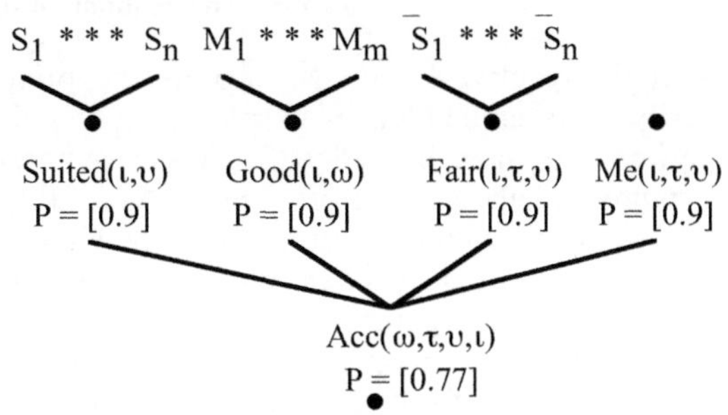

Figure 4. Hybrid net to accept an offer

P(Acc | Suited, Good, Fair, Me) = 1.0, P(Acc | ~Suited, Good, Fair, Me) = 0.7,
P(Acc | Suited, ~Good, Fair, Me) = 0.8, P(Acc | ~Suited, ~Good, Fair, Me) = 0.6
the probability P(Acc) = 0.77, and so the membership function, μ_A, for the fuzzy set "Accept" and the confidence value χ should be defined so that $\mu_A(0.77) << \chi$. It then remains to access the information from the available sources to, hopefully, increase P(Acc) so that the deal is acceptable.

4. The Negotiation Process

The ideas in the previous two sections are summarised in Figure 5. My agent is a hybrid agent. Its beliefs are derived first from incoming messages in the "Information received" box, expressed in pre-determined predicates, from ω (and from "John")—these beliefs activate the deliberative machinery. They are time-stamped on arrival as they include offers from ω. Second, beliefs are also derived from information that arrives because a plan has activated a macro tool—these beliefs

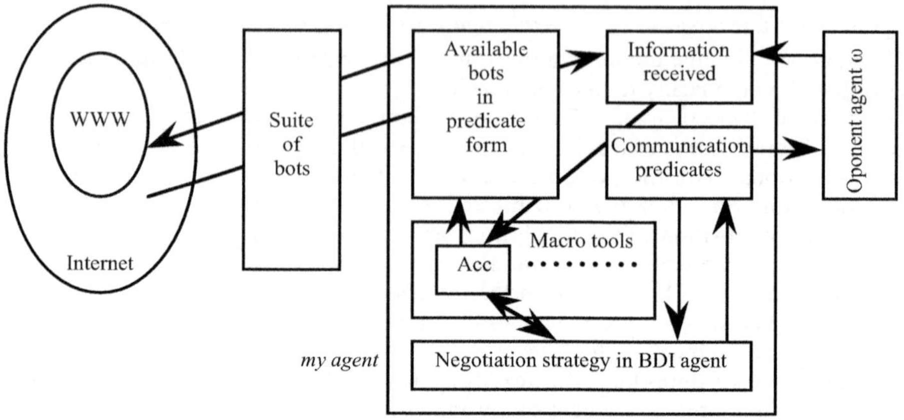

Figure 5. The negotiation framework

trigger the reactive logic. My agents' negotiation strategy—which has yet to be described—is embedded in its plans. It is guided by the lattice structure that is superimposed on each predicate. Those structures are built by hand and are intended to capture something of the meaning of "generality" for each predicate.

The background of information in which a negotiation takes place will in general be continually changing. This dynamic nature of information greatly complicates its management—dealing with belief revision is one consequence of this. For example, most people who contemplate upgrading their personal computing equipment are confronted with the problem "shall I buy now or wait to see what becomes available next month". In the markets for stocks and bonds the background information is continually changing, and only experienced traders are able to detect deep patterns in the information shift on which to build their trading strategies. To simplify the discussion here we assume that the background information remains constant during the period of each negotiation. Specifically we assume that any information derived from information sources during a negotiation remains valid for the period of the negotiation.

A negotiation here is a sequence of offers tabled together with attendant information. The offers in this sequence may originate from the two agents in turn—the alternating offers model—or the agents may table offers at random. As time progresses during a negotiation my agent will have access to an increasing amount of information—shown in the lower half of Figure 6. The source of that information will be the information bots, information tabled by my opponent ω, and the deals themselves. Of this information, only the terms offered by my agent, τ, are certain. The validity of the information extracted by the bots will be uncertain, as will the information tabled by ω, and even ω's terms, υ, may be uncertain in that if I accept an offer, (τ, υ), ω may not meet her commitments—she may "renege" on the deal.

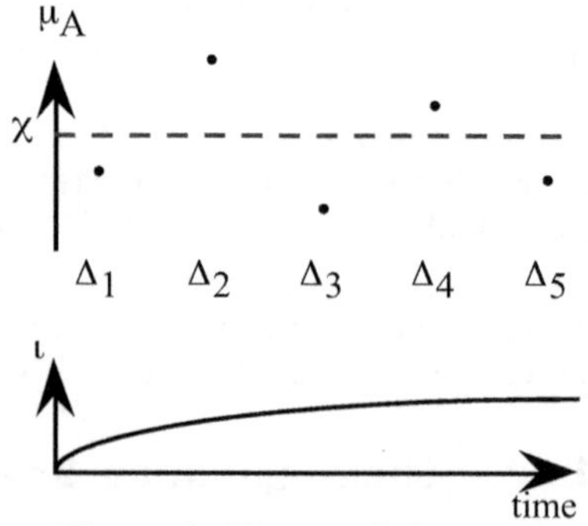

Figure 6. The negotiation process

The μ_A value of acceptable offers will be above my confidence level χ—five offers are shown in the top half of Figure 6.

The negotiation process is in two stages—first, the exchange of offers and information, and, second, the end-game. As the information is assumed to be static, there will come a time during the exchange of offers and information when the agents believe that have all the information that they require. They are then "fully informed", and should be in a position to take a final position. The development of the offer sequence, the construction of counter-offers [16], and the tabling of information are all part of the negotiation game. This can be a game of bluff and counter-bluff in which an agent may not even intend to close the deal if one should be reached. Strategies for playing the "negotiation game" in an information rich environment are the subject of on-going research. The aim here is to present the machinery that those strategies will have at their disposal.

4.1 Stage 1: Offer, counter-offer, re-offer and information exchange

The current version of the exemplar application has two pre-defined bot predicates in addition to the "Cheapest" predicate (described above):

Features(c:C, P_j:{P_i}; f_1:F_1,..., f_n:F_n, r:*s, p:$)

Cameras(P_j:{P_i}, f_1:F_1,..., f_n:F_n, r:*s, p:$; {c_k:C})

where the meaning of Features and Cameras is self-evident—in Features "r:*s" is a rating in *s (typically one to five). The lattice structure for Cameras is defined similarly to that for Cheapest. The structure for Features is null—it could be given a trivial structure by including "dc" in C but this serves no useful purpose.

There are four pre-defined interaction predicate templates:

Message("John", "me", IWant(f_1:F_1,..., f_u:F_u; c:C))

Message(a_1:A, a_2:A, Offer(τ:{ $, C }, υ:{ $, C }))

Message(a_1:A, a_2:A, Accept τ:{ $, C }, υ:{ $, C }))

Message(a_1:A, a_2:A, Unacceptable((τ:{ $, C }, υ:{ $, C }) , <reason>)

where the meaning of IWant is self-evident, "A" is the domain of agent names (eg: me or ω), and <reason> is derived here by identifying the evidence node with lowest probability in the Accept Bayesian net. A "Message" is from agent a_1 to agent a_2 and here contains either an "Offer", an "Accept", or a reason for an offer being "Unacceptable". At present my agent does not understand received "Unacceptable" messages but it sends them if wishes to do so. It should not do so as a matter of course as that would then give ω a sure way to discover my threshold. My agent is presently unable to receive tabled information from ω.

If an agent's confidence in accepting an offer is close to her threshold then she may decide to seek further information—either from the bots or from ω—in the hope that she will be able to accept it. An issue here is what information to request, particularly when costs are involved.

In 'real' negotiation it is common for an agent to table information in an attempt to encourage the opposition to see her point of view: "I really wanted a blue one". Advising an agent that its terms are not quite what I wanted is not difficult, but we can do more subtle things than that. In the exemplar system, when assessing an offer my agent may ask its information bots "to find if there is a camera whose features nearly match those of the one on offer but which sells new for less than the offer price". If such a camera exists then it is used as fodder for the Unacceptable predicate.

In the exemplar application my agent's offer strategy is to make a, possibly empty, sequence of counter offers. The price in the initial counter offer, if any, is the greatest such that both P(Fair) = P(Me) = 1.0 and the confidence in Accept > χ. Then the subsequent prices are chosen so that the confidence in Accept tends asymptotically to χ so as to more or less reach χ is 5 steps. This sequence will be empty if P(Suited) and P(Good) are such that confidence in Accept does not exceed χ for any positive price. If ω were to discover this simple strategy then it should take advantage of it. This is not seen as a weakness in these experiments that aim primarily at trialing the information management machinery.

More important, it is proposed that more complex applications may be supported in the same way as long as the information in the application is expressed in first-order predicates with lattice orderings for each predicate that capture "generalisation". For example, the same machinery could in principle handle a negotiation to purchase a house where features could include the number of bedrooms, the size of the house, the "aspect" (sunny or otherwise), the extent to which there are views from the property, the prime construction method, number of car spaces and so on.

4.2 Stage 2: Fully informed agents and the end-game

If my agent believes that it has all the information it requires to make an informed decision—with the exception of the position of the opposing agent—and no matter what the opposing agents position is it will make no difference to my agents position then my agent is *fully informed*. An agent may believe that it is fully informed prior to any negotiation taking place, or it may come to believe that it is fully informed during the exchange of offers and information.

If both agents in a negotiation believe that they are fully informed then they should be prepared to enter a final, single round of the negotiation process that determines a final, definitive outcome of either "no deal" or "a deal will be conducted under these terms..." [17].

If:

$$\{ (\tau, \upsilon) \mid P(Acc_{me}(\omega, \tau, \upsilon, \iota_{me})) \in Accept_{me} \} \cap$$
$$\{ (\tau, \upsilon) \mid P(Acc_{\omega}(me, \upsilon, \tau, \iota_{\omega})) \in Accept_{\omega} \} = PosDeals$$

is non-empty then a deal is possible. There will typically be a large or unbounded number of such possible deals. The structure of the admissible deals in this intersection may be quite complex if the negotiation involves multiple attributes. The comfort level of any acceptable deal will be no less than my level of confidence χ_{me}, and clearly no greater than 1.0. It will be less than 1.0 as the deal must also be acceptable to my opponent ω. The best confidence level that I may hope for, β_{me}, for will be determined by ω's confidence threshold χ_{ω}, and *vice versa*. This is illustrated in Figure 7. The [χ, β] intervals are vaguely related to the idea of players having different "types" in the game-theoretic analysis of bargaining [9].

There are no ready answers to the following questions. Could some analogue of the Nash Bargaining Solution of 1950 [9] be used here? In so far as the Nash solution yields a fair distribution of the surplus utility, perhaps the basis for a final definitive outcome could be based on a fair distribution of "comfort"? Perhaps some mechanism that split each agent's range of feasible confidence in fair proportions would satisfy an analogue of the Myerson-Satterthwaite [1983] result [19] for the linear equilibrium in "split the difference" bargaining? The linear equilibrium in the raw form of "split the difference" bargaining yields the highest expected payoff of any mechanism *ex ante*, but it is not truth-revealing—there are associated truth-revealing mechanisms—this mechanism is not optimal *ex post*. These questions require some serious analysis, and have yet to be addressed. Two possible ways in which a final, single round of negotiation could be managed are now described.

First, two fully informed agents table their final, proposed deals simultaneously. If either the terms that I propose are unsatisfactory to ω, or *vice versa*, then there is no deal. Otherwise a trusted intermediary sets about constructing a compromise deal based on the agents' two proposals. If the difference between the deals concerns one continuous attribute—such as an amount of money—then the intermediary may simply "split the difference"—although this will not lead to truth-revealing behaviour by the agents. If there are multi-attributes, or attributes with discrete values then this intermediation will be more complex, although the second method following may be employed.

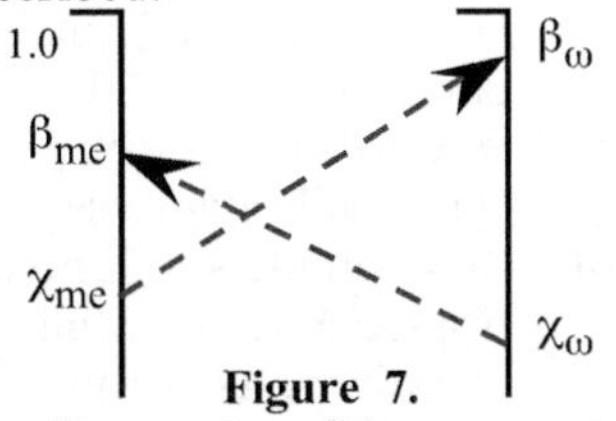

Figure 7.
Expected confidence range

Second, two fully informed agents make their whole reasoning apparatus available to a trusted intermediary who determines whether the set PosDeals is empty. If it is not then the intermediary attempts to find a deal that divides the two agent's $[\beta, \chi]$ intervals into equal proportions π:1. If this proves impossible, as may be the case if one of the deal attributes is non-continuous, then a deal is selected at random from the pair of deals that most closely match with one deal's μ_A value above and one below the π:1 division.

5. Conclusion

The eNegotiation framework is the first step in the design of agents that can exploit information-rich environments in one-to-one negotiation. Such agents should be adept at handling both the exchange of information and the exchange of offers. The negotiation protocol used is a loose exchange of offers and information in an initial stage, followed by an end-game mechanism when the agents believe that they have become fully informed. This protocol assumes that sufficient information is available and that the currency of the information is sustained throughout the negotiation process. Semantic issues [18] are avoided by introducing pre-defined, pre-agreed predicates. The information is managed by representing it in these predicates and then by overlaying a lattice structure with isotone valuations on each predicate. It is proposed that this method may be applied to a wide range of negotiation situations, and if so then general negotiation strategies may be developed to manage both the offers and the information. To date this work has raised more problems than it has solved. Initial indications are that it appears to work and so it is expected to lead to the design of sound negotiation agents that strive to make informed decisions rather than economically rational decisions.

References

[1] Martin Bichler. The Future of E-Markets: Multi Dimensional Market Mechanisms. Cambridge University Press, 2001.

[2] Felix Brandt and Gerhard Weiß. Antisocial Agents and Vickrey Auctions. in Intelligent Agents VIII proceedings ATAL 2001. Springer Verlag, 2002.

[3] Watkins, M. Breakthrough Business Negotiation—A Toolbox for Managers. Jossey-Bass, 2002.

[4] S.S. Fatima, M. Wooldridge and N. R. Jennings. The Influence of Information on Negotiation Equilibrium. In: Agent Mediated Electronic Commerce IV, Springer-Verlag, 2002.

[5] Howard Raiffa. Negotiation Analysis: The Science and Art of Collaborative Decision Making. Harvard, 2002.

[6] Rob Saunders. Curious Design Agents and Artificial Creativity: A Synthetic Approach to the Study of Creative Behaviour. Ph.D. Dissertation. University of Sydney, Department of Architectural and Design Science, 2001.

[7] Sarit Kraus. Strategic Negotiation in Multiagent Environments. MIT Press, 2001.

[8] Muthoo, A. Bargaining Theory with Applications. Cambridge UP, 1999.

[9] Martin J. Osborne and Ariel Rubinstein. Bargaining and Markets. Academic Press, 1990.

[10] Elmer Wolfstetter. Topics in Microeconomics. Cambridge UP, 1999.

[11] Hilderman, R.J. and Hamilton, H.J. "Evaluation of Interestingness Measures for Ranking Discovered Knowledge." In Cheung, D., Williams, G.J., and Li, Q. (eds.), Proceedings of the 5th Pacific-Asia Conference on Knowledge Discovery and Data Mining (PAKDD'01), Hong Kong, April, 2001, pp. 247-259.

[12] Gerding, E.H., van Bragt, D.D.B. and La Poutre, J.A. Multi-issue negotiation processes by evolutionary simulation: validation and social extensions. In proceedings Workshop on Complex Behavior in Economics. Aix-en-Provence, France, May 4-6, 2000.

[13] Anthony Chavez, Daniel Dreilinger, Robert Guttman and Pattie Maes. A Real-Life Experiment in Creating an Agent Marketplace. in proceedings of the Second International Conference on the Practical Application of Intelligent Agents and Multi-Agent Technology (PAAM97). London, UK, April 1997.

[14] Birgit Burmeister, Tobias Ihde, Thomas Kittisteiner, Benny Moldavanu and Jörg Nikutta.. A Practical Approach to Multi-Attribute Auctions. in proceedings Third International Workshop on Negotiations in Electronic Markets, Aix-en-Provence, France, 2002.

[15] S. Ramchurn, N. Jennings, C. Sierra and L. Godo. "A Computational Trust Model for Multi-Agent Interactions based on Confidence and Reputation". in proceedings workshop on "Proc. 5th Int. Workshop on Deception, Fraud and Trust in Agent Societies" AAMAS-2003, Melbourne, Australia, July 2003.

[16] Peyman Faratin, Carles Sierra and Nick Jennings. Using Similarity Criteria to Make Issue Trade-Offs in Automated Negotiation. Journal of Artificial Intelligence. 142 (2) 205-237, 2003.

[17] Jeffrey S. Rosenschein and Gilad Zlotkin. Rules of Encounter. MIT Press, 1998.

[18] Steven Wilmott, Monique Calisti and Emma Rollon. Challenges in Large-Scale Open Agent Mediated Economies. In: Agent Mediated Electronic Commerce IV, Springer-Verlag, 2002.

[19] R. Myerson and M. Satterthwaite. Efficient Mechanisms for Bilateral Trading. Journal of Economic Theory, 29, 1–21, April 1983.

Teaching Computational Intelligent Techniques with Real-Life Problems in Stock Trading

Jiabin Li[1] and Chun Che Fung[2]

[1] Department of Electrical and Computer Engineering
Curtin University of Technology
GPO Box U1987, Perth WA 6845, Australia
[2] School of Information Technology, Murdoch University, South Street
Murdoch, WA 6150, Australia

Abstract. Artificial Intelligence (AI) and Computational Intelligence (CI) techniques have formed part of the core or elective studies in many computing and engineering courses. On the other hand, there is a need for non-commerce graduates to appreciate and be aware of the business and financial environments. The analysis of financial market is a time-consuming and complicated process. It is proposed that such problem can be used as a vehicle to teach and encourage students on the use of intelligent techniques to solve real-life problems. The participants are required to investigate the problem and to develop a system to assist an investor on acquisition, analysis, and selection of financial market investments. The possible intelligent techniques to be considered are fuzzy expert systems, artificial neural networks and evolutionary computation. In this paper, a fuzzy expert system modeling vague trading rules is used as an example to demonstrate the feasibility of such approach.

1 Introduction

The IEEE Computer Society and the Association for Computing Machinery (ACM) have jointly produced a final report on Computing Curricula in December 2001. In the report, the topic area of Intelligent Systems (IS) has been clearly established as one of the core topics in the body of knowledge in the computing discipline. From the survey conducted during the preparation of the report, most of the schools have either included such topic as a core or as an elective unit. The teaching of artificial intelligence as a set of tools to solve problems that are difficult or impractical to solve with other traditional methods is now commonly recognized. At the end of the teaching module, the student should be able to determine when an intelligent approach is appropriate for a given problem, and to be able to select and implement the most suitable intelligent methodology.

On the other hand, the same report also suggested a "discipline-based model" to provide flexibility in the course design in order to broaden the student's horizon and to

T.D. Gedeon and L.C.C. Fung (Eds.): AI 2003, LNAI 2903, pp. 847-856, 2003.

enhance a student's overall education. In addition, opportunities for computing graduates may exist in other fields such as psychology, sociology, economics, biology, business, or any of the science or engineering disciplines. Hence, from an education prospective, it is logical to include specific application areas and practical problems in the teaching of intelligent systems.

An example of real life problems is the prediction of the performance of specific stocks within the financial market. Firstly, the stock market environment is dynamic, rapid changing and highly non-linear. Such phenomena result in a highly complicated process for stock price prediction. Secondly, a number of known and unknown economic factors influence the movement of the stock prices. Such uncertainties generally cannot be coped with traditional methods easily. Furthermore, uncertainties sometimes arise from the participation of human investors with specific set of trading rules and strategies, which affect the behaviours of the stock market. Finally, due to dynamic nature of the stock market, real-time analysis is vital to an investor who has high stakes in short-term investments.

Although many market indicators have been used by financial experts to assist analysis of the stock market, such trading rules, are in general ambiguous, inconsistent, confused, and even conflicted to each other. Intelligent systems present as viable alternative solutions to the problem. For example, it is possible to implement the trading rules in a fuzzy expert system to aid the decision making process. As each investor follows his/her own set of goals, rules, or strategies, it is necessary for an intelligent system to have the capability in providing an explanation on the choice of trading rules and interpretation of the analysis results.

This paper proposes a competition based on real-time problems to teach AI or CI techniques in the Intelligent Systems unit. The stock market has been considered as the target problem due to:

- practical characteristics of a real-life problem,
- availability of data,
- possibility to apply various intelligent techniques,
- flexibility of implementation,
- simple input and output structures, and
- challenging.

Stock prices of most of the stock markets are now readily and freely available. A stock market predictor can be based on relatively simple input and output structures. The inputs are typically historical stock prices. The outputs are recommendations of *sell*, *buy* or *hold*. Another factor for consideration will be transition volumes. The implementation of the predictor is therefore flexible and private which makes the problem suitable for undergraduate courses. The predictor could be built based on either one or a combination of CI techniques implemented in a hybrid fashion.

In this paper, two artificial intelligent techniques, fuzzy expert systems and artificial neural networks, (ANN) are evaluated and comparison of two techniques is reported. They are used to illustrate typical solutions which could be implemented by the students.

2 Financial Market Analysis

Technical analysis is a class of techniques for numerical evaluation of market prices with an assumption that market price is a reflection of human investors' attitude towards the changing market factors. Basically, there are three types of analyses techniques in technical analysis [1]:

- Trend Analyses – used to indicate whether a consistent trend has appeared in the data or whether an existing trend has finished.
- Momentum Analyses – measure the rate of change of prices as opposed to the actual price levels.
- Volatility Analyses – measure the rate of random change in market prices.

As mentioned previously, trading rules are to be defined according to these indicators to help an inventor to make decision regarding the trading. Theses trading rules could be based on heuristic rules extracted from financial experts. As these rules are generally vague and contradict, an example is to use fuzzy logic to translate the trading rules into IF-THEN statements in an Expert System. The design and implementation of these rules are illustrated in the latter section.

3 Financial Market Predictor

The main purpose of the proposed financial market predictor is to assist investors on analysis of trend of stock price. Fig. 1 illustrates the overall system. There are two inputs to the system: (1) quantitative data, or market indices, which are used to perform technical analysis; and (2) trading rules provided by user. After processing the inputs, the system will generate one of three signals: *buy*, *sell*, or *hold*, to user. Based on these output signals, user decides action to be taken to maximise his/her return. For the sake of simplicity, the following assumptions were made:

- Future trend of the stock market is based on its historical performance; and
- Trading rules are means to determining investors' behaviours.

As dynamic nature of the stock market, the prediction system is crucial to provide real-time decision to user, or in other word, computation time is a constraint to the system. Furthermore, the system is capable to provide reliable information to user despite mistakes and errors occurs in the trading rules and qualitative data made by experts. Consequently, any technique to be implemented in the system must meet all the constraints.

4 Fuzzy Logic in Market Prediction

As mentioned previously, experts in financial market create a great amount of trading rules, based on different market indicators, to assist market analysis. As a result, expert systems, which are capable to store and represent human experts' knowledge, can store these trading rules, in turn give decision to user according to these trading rules. Consider the following trading rules created by [3].

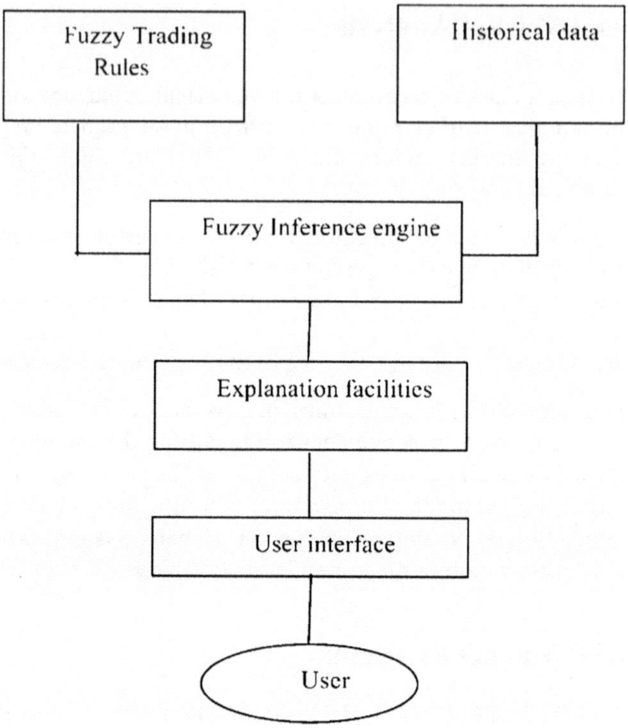

Fig. 1. Overall structure of the fuzzy expert system

- If percentage of investment capital is small, decision is *buy*.
- If percentage of investment capital is large, decision is *sell*.
- If the daily closing price crosses above the 5 days weighted moving average, decision is *buy*.
- If the daily closing price crosses below the 5 days weighted moving average, decision is *sell*.

These trading rules with fuzzy terms, such as small and large, cannot be represented in any rule-based system as these systems require crisp inputs. To cope with fuzzy values, fuzzy logic is used in an expert system. With the help of linguistic variables, a fuzzy expert system is able to describe a term or concept with vague or fuzzy values. Basically, these values are represented in fuzzy sets. As fuzzy expert system has the ability to model human expert's fuzziness trading rules, it is leading to deliver more human intelligent and accuracy of analysis to user. Advantages of using fuzzy logic listed in [4] are summarised as follows:

- Fuzzy logic is multi-valued as knowledge is represented as degrees of membership;
- Membership functions can be varied based on requirements of an application;
- The shape of fuzzy sets can be modified by using hedges to reflect human thinking, such as very, quite or extremely;
- Like crisp logic, fuzzy set can interact via operations.

However, as fuzzy logic needs complex computation to map crisp inputs to fuzzy variables, in turn to generate desire outputs, fuzzy expert systems usually run slower than other two types of expert systems do. Furthermore, the task of defining membership function for an application is time-consuming and subjective to the designer. It is thereby difficult to show the correctness of the design.

Despite all the weaknesses, fuzzy expert systems provide the most convenient and precise ways to implement those trading rules with ambiguous and contradictory. As the power of computers is increasing, complicate computation is no longer an issue.

4.1 Design of Fuzzy Trading Rules

It is obvious that these trading rules inherent fuzziness. Instead of defining precise values, the rules use vague words, above and below, as conditions to compare two quantities, such as daily closing price and 5 days moving average. To implement these rules, we first define three membership functions for each linguistic variables, *DMA(5)*, *DMA(5)-DMA(20)* and *decision*, which respectively represent daily moving average for 5 days, difference of daily moving average between 5 days and 20 days and output signals. Fig. 2 illustrates the membership function for the linguistic variables DMA(5).

Based on these linguistic variables, fuzzy trading rules thereby can be constructed from the original trading rules:

- If (DMA(5) is upcross) then (decision is short)
- If (DMA(5) is downcross) then (decision is long)
- If (DMA(5) – DMA(20) is upcross) then (decision is short)
- If (DMA(5) – DMA(20) is downcross) then (decision is long)

Similarly, the membership functions for the other two linguistic variables are also created, as shown in Fig. 3 and Fig. 4. It should emphasise that ranges of each linguistic variable are defined arbitrarily. Basically, there is no unique configuration for the system. User can adjust ranges of every linguistic variable to meet particular requirements. One of the possible solutions is to allow automatically modify these values by using artificial neural network (ANN), without user inference.

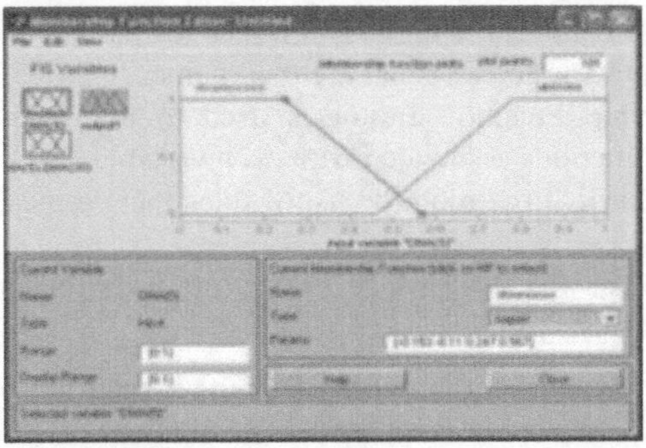

Fig. 2. Membership function defines two ranges of upcross and downcross for linguistic variable DMA(5)

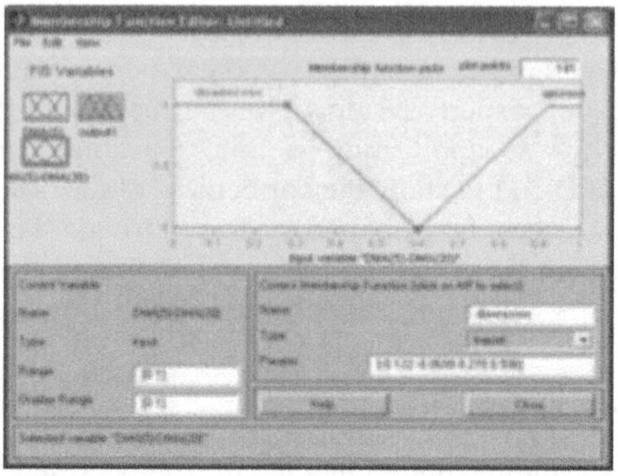

Fig. 3. The membership function is created for linguistic variable DMA(5)-DMA(20)

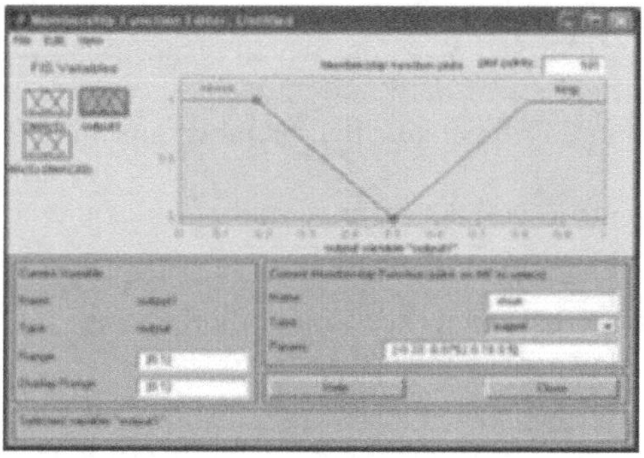

Fig. 4. The membership function of the output signal, decision

4.2 Implementation of the Fuzzy Logic Predictor

Matlab® was used to implement the financial market prediction system. There are five essential components in Matlab to implement fuzzy logic:

- The FIS Editor handles the high level issues for the system, such as how many input and output variables? What are their names?
- The Membership Function Editor is used to define the shapes of all the membership functions associated with each variable.
- The Rule Editor is for editing the list of rules that defines the behaviour of the system.
- The Rule Viewer and the Surface Viewer are used for looking at, as opposed to editing, the FIS.

It is important to note that the number of inputs should not be too large, because large number of inputs requires a great amount of memory of the target machine. The implementation of the prediction system is straightforward. It involves five major stages:

1. Define inputs and outputs using the FIS editor as shown in Fig. 5.
2. Define membership functions for each linguistic variable as shown in Fig. 2-4.
3. Implement fuzzy rules using the Rule editor as shown in Fig. 6.
4. Test the system using the Rule viewer and the Surface viewer as illustrated in Fig. 7.
5. Go back to step 2 if results are not satisfied.

It is not unusual that it requires a number of iterations to obtain a satisfactory result. Consequently, it is wise to use genetic algorithm (GA) to automate and optimise the implementing process.

5 Simulation Results

To demonstrate the performance of the prediction system, the daily high, low, and opening and closing prices, which are based on Telstra® shares in Australian ASX index in 1-year period (from 4/2/2002 to 31/12/2002), are used as simulation inputs as shown in Fig. 8 [5].

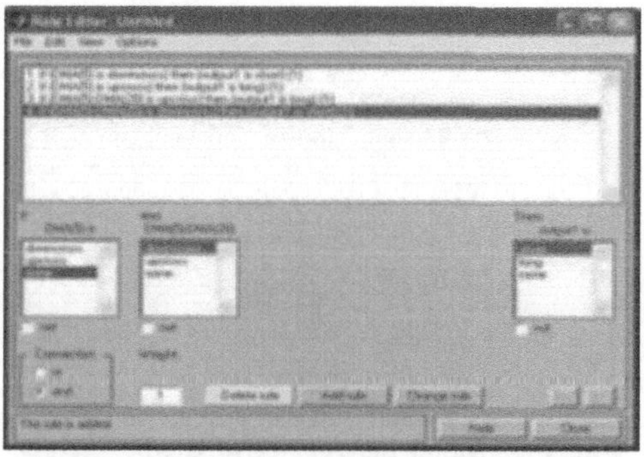

Fig. 5. Implementation of fuzzy trading rules

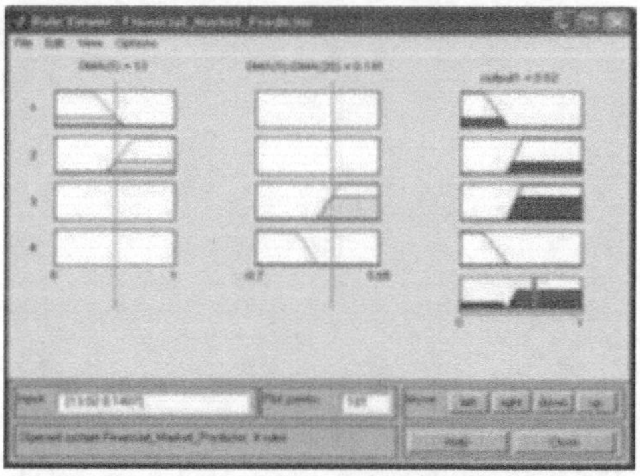

Fig. 6. Use the Rule viewer to test the fuzzy rules

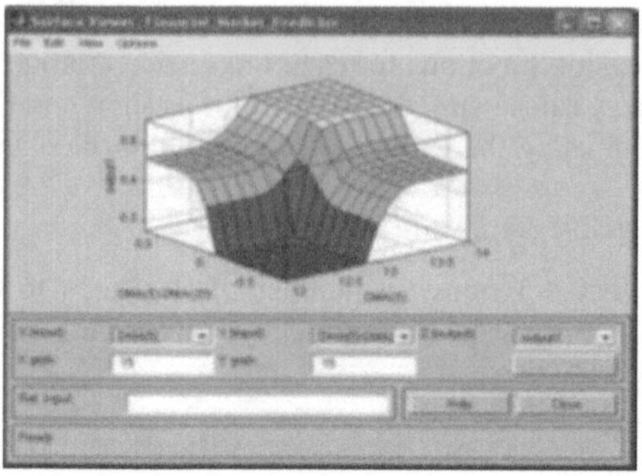

Fig. 7. 3-D plot of the output as functions of two inputs

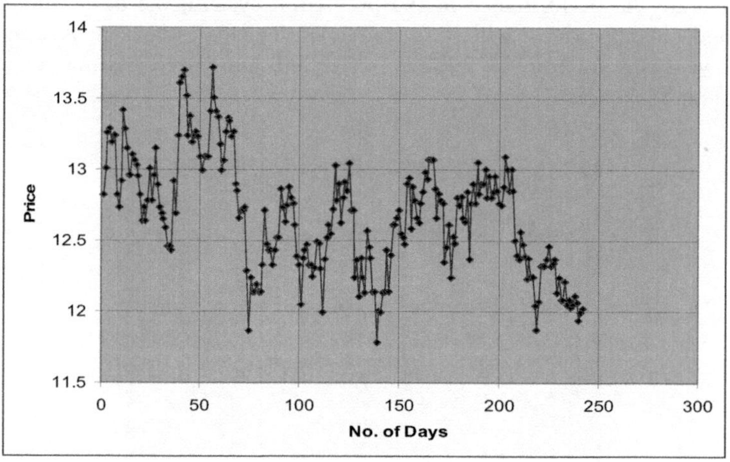

Fig. 8. Closing Price of Telstra Shares in 2002

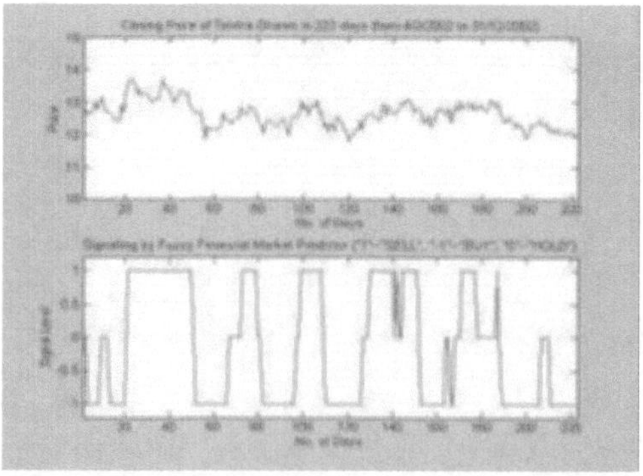

Fig. 9. Signalling in the stock prediction system

Table 1. The prediction system makes profit 3.05% at the end of the testing period

Initial Investment Capital	2000	Profit	3.05%
Total share value	1282		
Total	3282		
Investment Capital	316.9		
Total Share value	3065.1		
Total	3382		

With help of Matlab®, signals generated by the prediction system together with closing price are plotted as Fig. 9. Note that SELL signal is indicated by 1; BUY signal by -1; HOLD signal by 0.

It is obvious that buy signals generated by the prediction system occur when closing price is relative low, while sell signals are generated during high closing price. It thereby meets the buy-low-and-sell-high requirement as mentioned previously.

To further demonstrate the strength of the system, another simulation based on simulation results shown in Fig. 9 is carried out. Assume the initial investment capital 2000 AUS dollars and zero Telstra share. The amount of buy (sell) is 10 shares every time the output signal is -1 (1), otherwise, we'll do nothing – hold. After testing period, we compare the initial capitals, including cash and share, with the final capitals. To calculate the profit, we use the following formula:

$$\text{profit}=\frac{\text{finalcapital-initialcapital}}{\text{initialcapital}}\times 100\%$$

Table 1 indicates the system generates 3.05% return at the end of the testing period.

The usefulness of the financial market prediction system is verified through two simulations. The system is capable to (1) perform technical analysis by using fuzzy rules, (2) generate buy/sell signals at the right time, and (3) help investors make profit in the stock market. However, the weaknesses of the fuzzy logic predictor are (1) lacking of adaptability, (2) poor automated knowledge discovery and data mining, and (3) difficult maintaining of the fuzzified knowledge base.

6 Conclusion

This paper proposes a competition for teaching CI techniques which is based on the stock market prediction problem. Due to its typical characteristics, stock market is used as real life problems to introduce CI techniques to students. Furthermore, stock prices of most of the stock markets are readily available and free to download. Flexibility of implementations of a stock market predictor is another feature of the competition.

A simple structure of a stock market predictor is described in this paper. The advantages of the system include the simple structures and easy implementation. The limitations of the system, however, are also outlined in the paper. Nevertheless, in the real competition, the amount of historical data is expected to be larger. Qualitative data will also be included in the simulated stock market to provide more choices for challengers to build their predictors.

References

[1] 1. S. S. Lam, "A Genetic Fuzzy Expert System for Stock Market Timing," presented at Evolutionary Computation, 2001. Proceedings of the 2001 Congress on, Seoul , South Korea, April 9, 2003. 2001.

[2] V. K. Sagar and L. C. Kiat, "A Neural Stock Price Predictor Using Qualitative and Quantitative Data," presented at Neural Information Processing, 1999. Proceedings. ICONIP '99. 6th International Conference on, Perth, WA , Australia, April 9, 2003. 1999.

[3] C. J. Neely and P. A. Weller, "Technical Trading Rules in the European Monetary System," *Journal of International Money and Finance*, vol. 18, pp. 429-458, 1999.

[4] M. Negnevitsky, Artificial Intelligence: A Guide to Intelligent Systems: Addison Wesley, 2002.

[5] *Yahoo!Finance*, 2003, [Available] http://au.finance.yahoo.com/, 1 Jun, 2003.

Finding Optimal Architectures and Weights for ANN: A Combined Hierarchical Approach

Ranadhir Ghosh

School of Information Technology and Mathematical Sciences
University of Ballarat, PO Box 663, Victoria 3353, Australia
r.ghosh@ballarat.edu.au

Abstract. In this paper, we present a novel approach of implementing a combined hierarchical methodology to find appropriate neural network architecture and weights using an evolutionary least square based algorithm (GALS). This paper focuses on the aspects such as the heuristics of updating weights using an evolutionary least square based algorithm, finding the number of hidden neurons for a two layer feed forward neural network, the stopping criteria for the algorithm and finally comparisons of the results with error back propagation (EBP) algorithm.

1 Introduction

Earlier work by Verma and Ghosh [1], suggested a learning methodology, which uses a hybrid technique by using evolutionary learning [2-7] for the hidden layer weights and least square based solution method for the output layer weights. The proposed algorithm solved the problems of time complexity of the evolutionary algorithm by combining a fast searching methodology using least square technique. Also, by using a matrix based solution method, it could avoid further the inclusion of other iterative local search method such as error back propagation (EBP), which had its own limitations.

2 Research Methodology

Henceforth, the two separate modules for architecture and weight will be referred to as findArchitecture and GALS module respectively. A simple rule base can describe the working of the stopping criteria for the algorithm.

Rule 1: If the current output is satisfactory, then stop the algorithm, else check rule 2.

Rule 2: If the stopping criteria for weight search is met and the search is completely exhausted (in terms of number of iterations) then stop else check rule 3.

T.D. Gedeon and L.C.C. Fung (Eds.): AI 2003, LNAI 2903, pp. 857-865, 2003.

Rule 3: If the stopping criteria for weight search is met then go to rule 4, else continue.
Rule 4: If the stopping criteria for GALS is met then go to rule 1, else continue.

2.1 Issues for Combining GA and Leaset Square Method Area

How we can combine the evolutionary algorithm with the least square method is again a very important issue as there are many possibilities joining the two independent modules. Earlier experiments showed that the time / memory complexity highly depends on this factor of combination. To improve the memory complexity of our original GALS [1], we call the least square method after the convergence routine of the evolutionary algorithm is over. The methodology is described in Fig. 1.

We start the training process using number of hidden neurons obtained from the findArchitecture module. Then we initialize all the hidden layer weights using a uniform distribution of a closed interval range of [-1, +1]. We also encode the genotype, which represents the weights for the hidden layer with those values for all the population strings.

Then we apply the evolutionary algorithm. We create an intermediate population from the current population using a selection operator. We use roulette wheel selection. The method creates a region of a wheel based on the fitness of a population string. All the population strings occupy the space in the wheel based on their rank in fitness. A uniform random number is then generated within a closed interval of [0,1]. The value of the random number is used as a pointer and the particular population string that occupies the number is selected for the offspring generation Once the intermediate population strings are generated, we randomly choose two parents from the pool of the intermediate population, and apply the genetic operators (mutation only) to generate the offspring with some predefined probability distribution. We continue this process till such time the number of offspring population becomes same as the parent population. Once the new population has been created, we find the fitness for all the population based on the weights of the hidden weights (obtained from the population string) and the output layer weights (obtained from the least square method). We also normalize the fitness value to force the population string to maintain a pre-selected range of fitness.

After applying the evolutionary the output layer weights are computed using Least square method. After replacing the upper half gene from the best-selected candidate from the evolutionary algorithm using least square method, the fitness of the new transformed gene is calculated. If the fitness of the new transformed gene meets the error criteria then stop, else go to findArchitecture module to start the process again.

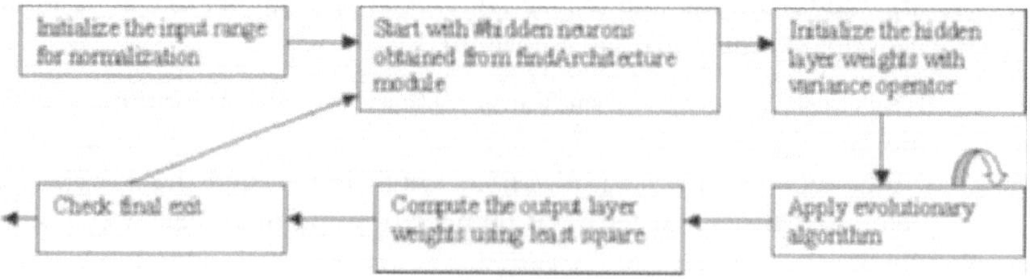

Fig. 1. Block diagram of GALS

2.2 Stopping Criteria for GALS

Stopping criteria for GALS is also based on few simple rules. All the rules are based on current train and test output and the maximum number of generation for the evolution algorithm. Following, we describe the stopping criteria for the convergence of the evolutionary algorithm.

If (best_RMS_error[1] < goal_RMS_error) then
    Stop
Else if (number_of_generation = total_number_of_generation[2]) then
    Stop
Else if (train_classification_error is increased in #m[3] consecutive generation) then
    Stop

2.3 Finding Optimal Number of Hidden Neurons (findArchitecture)

We used two different types of experiments – Linear incrementing for GALS (LIGALS) and binary tree search type for GALS (BTGALS) to find the number of hidden neurons-

1. Starting with a small number and then incrementing by 1 (LIGALS)
2. Using a binary tree search type (BTGALS)

2.3.1 Experiment A (LIGALS)

In experiment A, we start with a small number of hidden neurons and then increment by one. The stopping criteria for this experiment was as follows:

If (train_classification_error = 0) then
    Stop.
Else If (the test classification error is high in #n[4] consecutive generation) then
    Stop.

2.3.2 Experiment B (BTGALS)

In experiment B, we use a binary tree search type to find the optimal number. A pseudo-code of the algorithm is given below:

Step 1: Find the % test classification error & train_classification_error (error_min) for #min number of hidden neurons

    error_min = (train_classification_error (%)+ test_classification_error (%)) / 2

Step 2: find the % test classification error & train classification error (error_max) for #max number of hidden neurons

    error_max = (train_classification_error (%) + test_classification_error (%)) / 2

---

[1] The best_RMS_error is the best of the RMS error from the population pool

[2] Total number of generations is considered as 30

[3] m is considered as 3

[4] We use n = 3 for our experiments, which was determined by trial and error method

Step 3: Find the % test classification error & train classification error (error_mid) for #mid (mid = (min + max) / 2) number of hidden neurons

error_mid = (train_classification_error (%) + test_classification_error (%)) / 2

Step 4: if (error_mid < error_min) and (error_mid > error_max) then
            min = mid
            mid = (min + max / 2)
        else
            max = mid
            mid = (min + max / 2)
        end if

Step 5: Go to Step1, if (mid > min) and (mid < max)
        Else go to Step 6

Step 6: #number of hidden neurons = mid.

3 Analysis and Discussion of the Results

We tested our proposed approach on XOR, 10 bit Odd Parity, and some other real-world benchmark data sets such as handwriting character dataset from CEDAR, Breast cancer and Heart Disease from UCI Machine Learning repository. The results are shown in Table 1 – 3.

In terms of training classification accuracy for LIGALS and BTGALS outperforms the traditional EBP. The improvements are shown in Fig 2. For example in case of heart disease data set, the classification accuracy for LIGALS is 16.31% higher over EBP and for BTGALS the classification accuracy is 11.67% higher. For handwriting and breast cancer data set the average classification accuracies for LIGALS and BTGALS over EBP are 6.66% and 4% respectively.

Table 1. Experimental results for EBP

Data set	#time (minutes)	Classification Error	
		Train	Test
XOR	0.66	0	X
10 bit odd parity	127	15.40	8.20
Handwriting characters (CEDAR)	238	3.00	18.15
Breast cancer (Wisconsin)	121	3.50	11.40
Heart disease(Cleveland)	97	3.90	15.01
Heart disease(Hungary)	93	1.89	13.21
Heart disease(Switzerland)	71	2.00	13.31

Table 2. Experimental results for LIGALS

Data set	#time (minutes)	Classification Error	
		Train	Test
XOR	0.47	0	X
10 bit odd parity	115	10	5.80
Handwriting characters (CEDAR)	219	2.80	15.16
Breast cancer (Wisconsin)	102	3.10	9.16
Heart disease(Cleveland)	85	3.50	10.52
Heart disease(Hungary)	86	1.87	10.68
Heart disease(Switzerland)	56	1.90	10.28

Table 3. Experimental results for BTGALS

Data set	#time (minutes)	Classification Error	
		Train	Test
XOR	0.47	0	X
10 bit odd parity	115	10	5.80
Handwriting characters (CEDAR)	219	2.80	15.16
Breast cancer (Wisconsin)	102	3.10	9.16
Heart disease(Cleveland)	85	3.50	10.52
Heart disease(Hungary)	86	1.87	10.68
Heart disease(Switzerland)	56	1.90	10.28

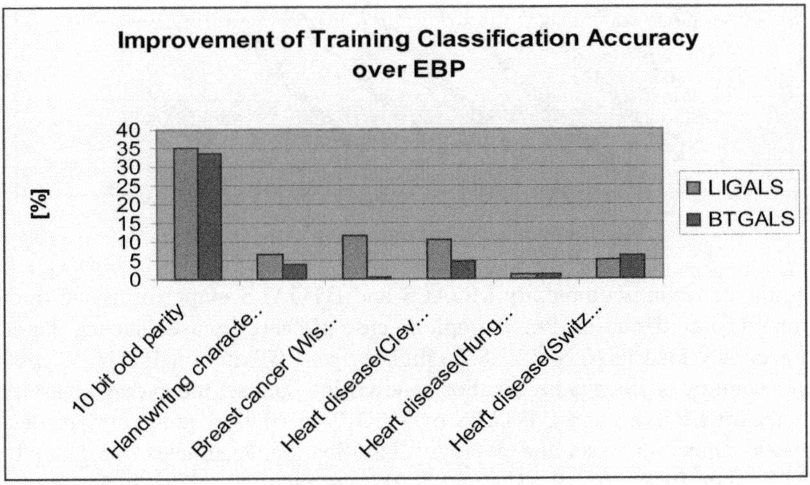

Fig. 2. Improvement of training classification accuracy over EBP

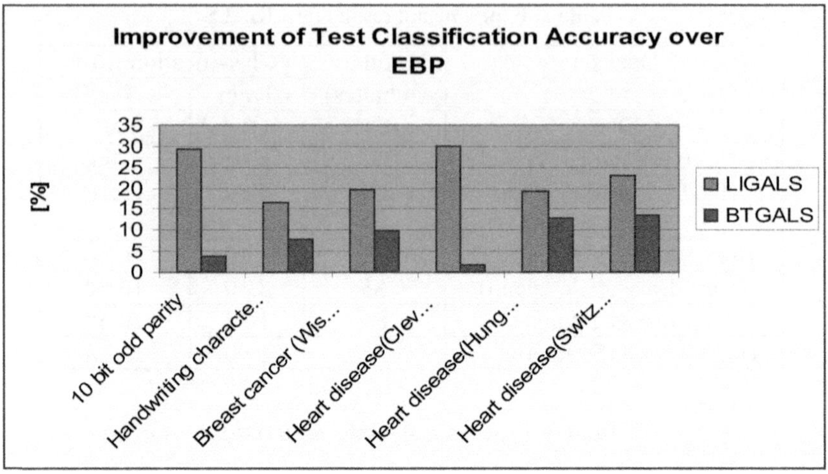

Fig. 3. Improvement of testing classification accuracy over EBP

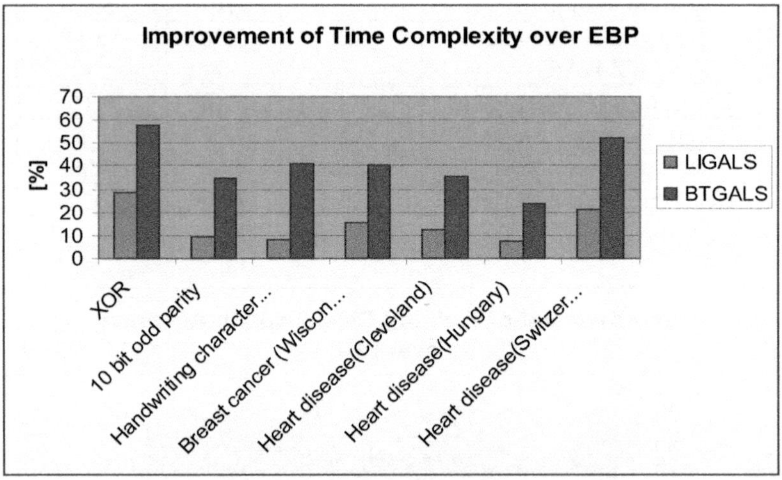

Fig. 4. Improvement of time complexity over EBP

In terms of testing complexity LIGALS and BTGALS outperforms the traditional EBP in almost all the cases. For example in case of heart disease data set, the classification accuracy for LIGALS is 71.83% higher than EBP and for BTGALS the classification accuracy is 28.04% higher. For handwriting data set the average classification accuracies for LIGALS and BTGALS over EBP are 16.47% and 7.71% respectively. For breast cancer data set the average classification accuracies for LIGALS and BTGALS over EBP are 19.64% and 9.91% respectively. The improvements are shown in Fig. 3.

The ratio of classification error for testing and training data set results show that the comparatively low ratio for LTGALS and BTGALS for most of the time shows that the optimal training obtained by the proposed hierarchical approach provides better generalization result than that of their manual EBP counterpart. When we linearly increase the number of hidden neurons, the ratio increases more or less in a linear fashion for LIGALS and BTGALS.

In terms of time complexity also, LTGALS and BTGALS are slightly better than the traditional EBP. The results are shown in Fig. 4. For all the data set the average improvement in time complexity for LTGALS over EBP are 28.78%, 9.44%, 7.98%, 15.7%, 12.37%, 7.52%, 21.12% respectively. For all the data set the average improvement in time complexity for BTGALS over EBP are 57.57%, 34.64%, 40.75%, 40.49%, 35.05%, 23.65%, 52.11% respectively. Some further training with some other initial weight set shows that the time taken by EBP for similar classification accuracy rate is much more than that of the obtained initial result.

The following figure (Fig. 5) shows the comparison of memory usage for the EBP, and the new algorithm[5]. Only the CEDAR dataset for the analysis with different data size is considered.

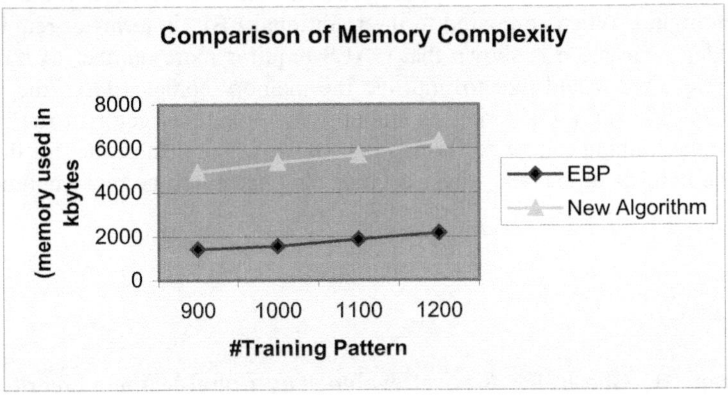

Fig. 5. Comparison of memory complexity

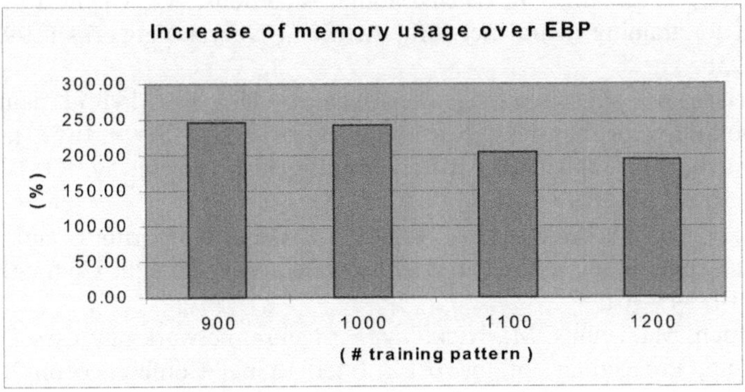

Fig. 6. Comparison of memory complexity

5 Proposed algorithm refers only to the findWeight module, as the findArchitecture module is not required to add in those cases.

The following figure (Fig. 6) shows the increment of memory usage over EBP. Only the CEDAR dataset with variable data length is considered for the analysis. It shows that, in case of EBP the increment is from 193% to 246%.

4 Conclusion and Further Research

In this paper, we proposed a novel approach for finding appropriate architecture and weights of an MLP. We have discussed with experimental results, how to achieve that with two different stopping criteria approaches. We also improved our earlier GALS algorithm, to reduce the memory complexity and also the choice of parameter setting for the algorithm. From the experimental results, we show that the new approach outperforms some other traditional approaches in terms of time, classification accuracy, etc., and it provides us with a guaranteed solution. It should be noted here that had we considered a fully such automated system for EBP, the training time would have even been much higher. When compared to the traditional EBP, in terms of required number of hidden neurons, it is shown that GALS requires more number of hidden neurons. In future, we would like to improve the memory complexity of the algorithm further by introducing a clustering technique for the feature vector of the input data set. So that the training can be performed for not the whole data set but for the data set equal to the number of classes, where each of the vector will be representing a single class.

References

[1] Verma, B., Ghosh, R.: A novel evolutionary neural learning algorithm, IEEE World Congress on Computational Intelligence, Honolulu, Hawaii, USA (2002), 1884-89.

[2] Petridis, V., Kazarlis, S., Papaikonomu, A., Filelis, A.: A hybrid genetic algorithm for training neural network, Artificial Neural Networks (1992), Vol. 2, 953-956.

[3] Likartsis, A., Vlachavas, I., Tsoukalas, L. H.: New hybrid neural genetic methodology for improving learning, Proceedings of the 9th IEEE International Conference on Tools with Artificial Intelligence, Piscataway, NJ, USA (1997), IEEE Press, 32-36.

[4] Whitley, D., Starkweather T., Bogart, C.: Genetic algorithms and neural networks - optimizing connections and connectivity, Parallel Computing (1990), Vol. 14, 347-361.

[5] Koeppen, M., Teunis, M., Nickolay, B.: Neural network that uses evolutionary learning, Proceedings of the IEEE International Conference on Neural Networks (1994), Part 5 (of 7), Piscstaway, NJ, USA, IEEE press, 635-639.

[6] Topchy, A.P., Lebedko, O.A.: Neural network training by means of cooperative evolutionary search, Nuclear Instruments & Methods in Physics Research, Section A: Accelerators, Spectometers, Detectors and Associated Equipment (1997), Vol. 389, no. 1-2, 240-241.

[7] Gutierrez, G., Isasi, P., Molina, J. M., Sanchis, A., Galvan, I. M.: Evolutionary cellular configurations for designing feedforward neural network architectures, connectionist models of neurons, learning processes and artificial intelligence, Jose Mira et al (Eds), Springer Verlag – Germany, LNCS 2084 (2001), 514-521.

Race Car Chassis Tuning Using Artificial Neural Networks

David Butler[1] and Vishy Karri[2]

[1] School of Engineering, University of Tasmania
PO Box 252-65, Hobart 7001, Tasmania, Australia
david.butler@utas.edu.au
[2] School of Engineering, University of Tasmania
PO Box 252-65, Hobart 7001, Tasmania, Australia
vishy.karri@utas.edu.au

Abstract. Proficient chassis tuning is critical to the overall performance of any race car. Determination of the optimum arrangement for specific track conditions can require a large amount of practical testing and, as such, any tools that reduce this expenditure will be of great value to the racing industry. Traditional computer modeling based on simplified vehicle dynamics has had a growing use in this field, but due to the extremely complex nature of the vehicle / driver / environment entity it has a number of practical limitations. Intelligent models, such as Artificial Neural Networks, on the other hand are not limited in this way and show a number of potential benefits. This study presents a simplified application of ANN to predict the optimum chassis arrangement for a steady state cornering condition for a Formula SAE race car to see if these benefits can be realised. The race car was equipped with a large sensor array, including engine speed, throttle angle, wheel speed, suspension position, steering angle, longitudinal and lateral acceleration and yaw rate, and chassis tuning was accomplished by varying caster, toe and front and rear tyre pressures. Data was collected for a total of six different chassis tuning combinations for the steady state cornering condition and a feed-forward back-propagation ANN model capable of predicting the lateral (centrifugal) acceleration of the vehicle for any given chassis tuning was produced. A numerical investigation was then completed with the ANN model to find the maximum lateral acceleration, and therefore speed, of the vehicle for each different possible chassis tuning combination. Each of the resulting 480 combinations were then ranked and compared against the optimal combination found from extensive practical vehicle testing. Despite a few problems encountered throughout the investigation that deteriorated ANN model accuracy, a high degree of correlation was found.

T.D. Gedeon and L.C.C. Fung (Eds.): AI 2003, LNAI 2903, pp. 866-877, 2003.

1 Introduction

It is obvious that the chassis properties have a large impact on the performance of most vehicles, racing cars especially. The suspension system defines how effectively the tyres can be utilised to obtain traction with the pavement, and also goes a long way in determining driver comfort and control. To be competitive, race car suspension must be designed and tuned to meet the demands of each track, each vehicle, and each driver [1..5].

The driver / vehicle / environment system is extremely complex, and an optimum suspension arrangement can be very difficult to develop, as Milliken [1] demonstrates. This is compounded by the fact that suspension design involves a number of compromises, and suspension requirements continuously change. Between races, between track conditions, between vehicle arrangements and between drivers the optimum suspension arrangement must be constantly redeveloped and refined. Short of installing a completely new suspension system for each change, these modifications are best done through chassis tuning.

Tuning allows the fine adjustment of many of the chassis parameters in an attempt to provide the most effective set up for the specific race conditions. Typically, these include camber angle, caster angle, toe, tyre choice and pressure, spring rate, damper resistance and stabiliser bar stiffness, or more depending on the application. Each of these parameters are interrelated, and their effects on vehicle dynamics are intricate and difficult to predict. A small change to one parameter may improve performance in one aspect, but impede it in many others. It is the goal of chassis tuning to provide an arrangement that gives optimal performance in critical situations (such as hard cornering), while still providing adequate performance in all other conditions, with the intent of providing the fastest consistent race times possible, as explained by Adams [2].

Because of the complexities of vehicle dynamics, and the fact that in most categories every vehicle is different to a large degree, there are no specific rules for chassis tuning. Instead, each vehicle must undergo extensive testing on the racetrack until an optimum arrangement is found through iteration and experience.

Obviously, this amount of track time can be expensive, in terms of both time and money. The amount of testing required to find the optimal solution can, however, be reduced with increased experience and the adoption of appropriate tools. Computer modeling has had a growing use in this field, and provides a powerful tool for testing tuning arrangements off-track. In this manner, different arrangements can be evaluated quickly, and without track testing. This means that when track testing is conducted, a high level of chassis tuning can be reached comparably quickly, with minimal expense.

The accuracy of the computer model then, clearly, has a large impact on how quickly the best arrangement can be found. The more parameters that the model takes into account, the better it will generally be. But the dynamics of a vehicle are very complex. Not only must the suspension geometries be considered, but also chassis movement and deflection, as well as tyre grip, deformation and slip - to name only a few. Since the development of mathematical models required for traditional computer modeling grows exponentially more difficult as additional parameters are considered (tyre dynamics alone are extremely complex [6,7]), it can be seen that there are a number of limitations in its use [8].

Artificial Neural Networks (ANN), on the other hand, offer an alternative to traditional computer modeling, and bring a number of potential benefits [9].

It is the goal of this study to conduct a simplified investigation into the use of ANNs in chassis tuning to discover if these benefits can be realised. As such, the investigation has been restricted to optimising caster, toe and front and rear tyre pressures for an approximately steady state cornering condition.

2 ANN Architecture

Artificial Neural Networks attempt to mimic the operation of the brain at a neural level [10]. As such, they exhibit some similar features, including the ability to learn and model process behaviour where *a priori* knowledge of the associated scientific principles is not available, or extremely difficult to obtain. This means that an ANN model can be programmed using just the input / output data (called training data) of a system, without the need to develop the complex mathematical representations that would otherwise be necessary to characterise the inner workings of the system. A by-product of this property is also that ANNs can include data from a large number of inputs. Therefore, because the ANN statistical model is more complex than conventional models, it can handle a wider variety of operating conditions [11].

The ability of an ANN to model process behaviour depends to a large extent on the network architecture. There are many proven architectures, in a range of applications, and new ones are continuously being developed. Among the most common is the multi-layed feedforward backpropagation (BP) ANN, which has been in use by the manufacturing industry for some time for systems control. This architecture is well known and well documented [15, 16]. It is very simple to construct, is robust and often provides reasonable accuracy - but can take a comparatively long time to train through error backpropagation, as discussed by Zeidenberg et al [12..15].

BP ANNs are structured as shown on Figure 1, with an input layer, one or more hidden layers and an output layer, with *i, j & k* processing units (called neurons, or nodes) in each respectively. The role of the input neurons is to simply pass forward the model input data to the hidden layer, and so the number of input neurons is equal to the number of model inputs. The hidden and output layers then perform all of the network processing, with each neuron performing an input (summation) function of the weighted inputs, an activation function and an output function, shown in Figure 2.

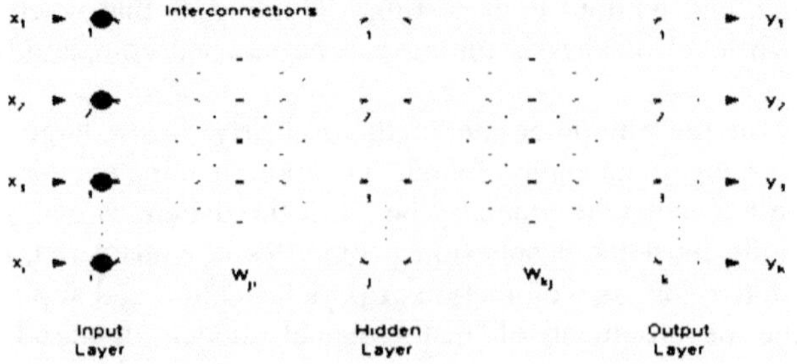

Fig. 1. Feedforward Architecture of a Basic BP ANN

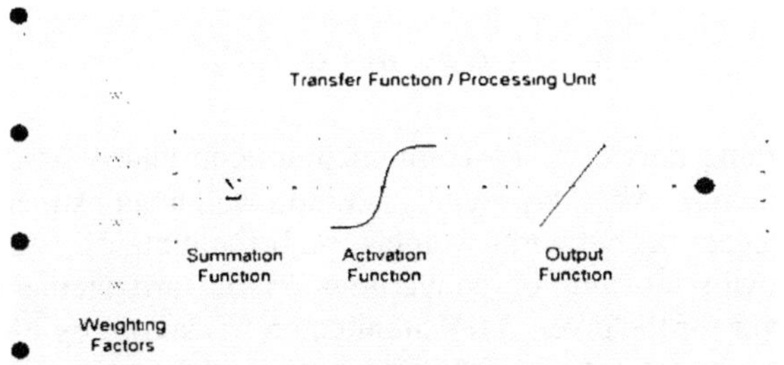

Fig. 2. Structure of the Artificial Neuron

The input function can thus be written as:

$$net_j = \sum_i x_i w_{ji} \tag{1}$$

Where; net_j=summation function result for neuron j, x_i=output from neuron i, w_{ji}=weight factor applied to x_i at neuron j

The goal of the activation function is then to perform a non-linear operation to the summation result. The most common is the sigmoidal function, which has the form:

$$f(net_j) = \frac{1}{1+\exp(-net_j)} \,. \tag{2}$$

The purpose of the output function is to condition the activation function result before it is passed to other neurons. Generally, this is not required, so the output function normally does not perform any operation.

Looking at the architecture, it can then be seen that the ANN characteristics can be altered by changing the neural weights it contains. Iteratively modifying these neural weights until a state of minimum error is achieved gives the ANN the ability to 'learn' a process, called Network Training. As the name 'backpropagation' suggests, this is done by comparing model predicted outputs (at the output layer) with the training data, and propagating this error value back through the network to the input layer, updating weight magnitudes on the way.

Therefore, the error values for the output layer neurons are given by:

$$\delta_k = (t_k - a_k) \cdot f'(net_k) \,. \tag{3}$$

Where; δ_k=error value for neuron k, t_k=target training value for neuron k, a_k=output value of neuron k

And the error values for the hidden layer neurons are determined using:

$$\delta_j = \left[\sum_k \delta_k W_{kj} \right] \cdot f'(net_j) \,. \tag{4}$$

Where; W_{kj}=weight factor to neuron k from neuron j

These error values can then be used to calculate the required weighting factor adjustments for the next training iteration, as shown:

$$\Delta W_{ji}{}^{h} = \eta \cdot \delta_j \cdot a_i + \alpha \cdot \Delta W_{ji}{}^{h-1} . \quad (5)$$

Where; η=learning rate, $0<\eta<1$, α=momentum constant, $0<\alpha<1$, $\Delta W_{ji}{}^{h}$=weight adjustment at iteration *h*, $\Delta W_{ji}{}^{h-1}$=previous iteration weight adjustment

Through this process the networks weights can be continuously adjusted during the training process, with the aim of converging on the arrangement that will give minimum RMS error for the given ANN architecture, as shown by Frost [16].

Typical leaning values used in training are $0.05<\eta<0.9$ and $0.05<\alpha<1$, with higher values improving convergence speed, and lower values enabling network 'fine tuning'. Values should be chosen to produce reasonable convergence speed while maintaining the ability to converge to a specific solution, avoiding false minima.

3 Parameter Selection and Data Acquisition

It was decided at an early stage to keep this initial investigation into the practically of ANNs as an aid in chassis tuning as simple a possible. As such, the testing conditions were limited to a flat uniform asphalt surface, and the test tack was constructed in a 'figure of 8' shape to the specifications of the Formula SAE skid pad [17]. Since the goal on this track is to achieve best performance in steady state cornering (in both directions), and transient responses are unimportant, it significantly simplifies the problem. Also, by assuming the vehicle (built to Formula SAE specifications) will corner in one direction in the same manner as the other, the ANN training data can be shortened to cover only the steady state cornering in one direction. The goal of the model, therefore, is simplified to the prediction of lateral acceleration (and velocity using the relationship $a_s = v^2 / r$) under steady conditions using a selection of measured parameters describing the vehicle/driver dynamics and chassis tuning.

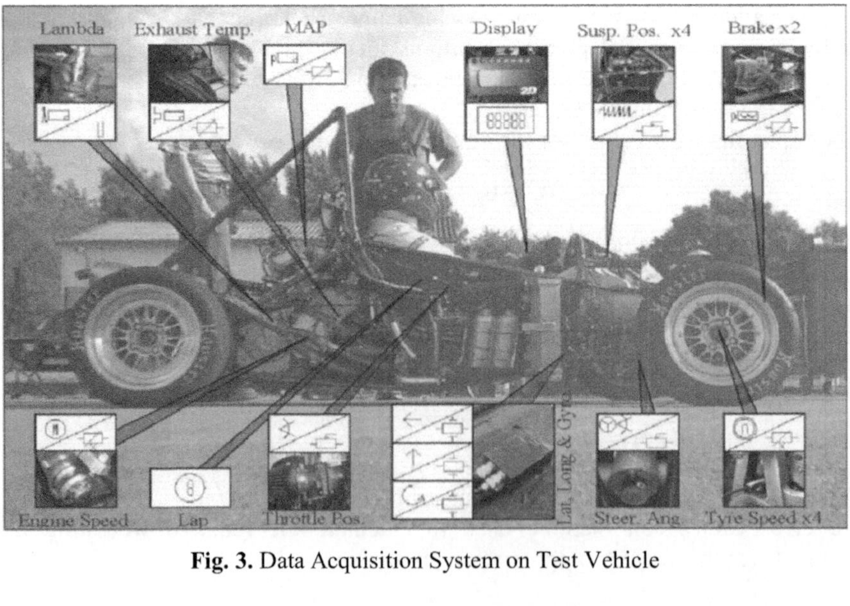

Fig. 3. Data Acquisition System on Test Vehicle

Table 1. Training chassis arrangements

Caster (o)	Toe (')	Front Tyre Press. (bar)	Rear Tyre Press. (bar)
5	-1.2	1	1
5	-1.2	0.8	0.6
5	-1.2	0.6	0.6
5	-1.2	0.4	0.6
5	2.5	0.8	0.6
0	2.5	0.8	0.6

The Formula SAE class race car built at the University of Applied Sciences, Stralsund, Germany for the 2002 Formula Student competition was used as the test vehicle. This racing car is equipped with a comprehensive data acquisition system, and can measure a variety of engine and chassis parameters as shown in Figure 3.

All of the parameters were measured during the data acquisition phase, with a mind to discard parameters that were found to have little or no effect on steady state cornering ability. Data was acquired for six different chassis arrangements, varying caster, front toe and front and rear tyre pressures, as shown in Table 1. It should also be noted that the data for the rear left wheel speed was later found to be erroneous, and so was omitted from the analysis.

4 Model Development

A number of different ANN arrangements were experimented with, and different architectures used, in the hope of finding a suitable model. The ANN programs were first written in LabVIEW [18], and a parameter importance analysis was conducted to see which of the vehicle parameters had significant effects. As anticipated from a previous study [9], the parameters associated with the engine and brakes were found to have negligible effects. As a result, the ANN inputs used to produce a prediction of lateral acceleration were chosen as:

- Caster,
- Front Toe Angle,
- Tyre Pressure (front and rear),
- Wheel Speed & Accel. (x3),
- Suspension Travel & Speed (x4),
- Steering Wheel Angle & Angular Velocity,
- Yaw Rate & Yaw Acceleration,
- Longitudinal Acceleration & Rate of Change.

Network training then went ahead, testing many different network architectures by varying the number of hidden layers and hidden layer neurons. In total over 20 000 lines of data (called patterns) were used to train each network, with about 2000 patterns randomly set aside for network testing and RMS error evaluation.

The 24 input, 16 first hidden layer, 0 second hidden layer and 1 output neuron architecture (Figure 4) proved the best arrangement for this case (using the method suggested by Frost et al [15]), exhibiting a full scale RMS error of only 4.0%.

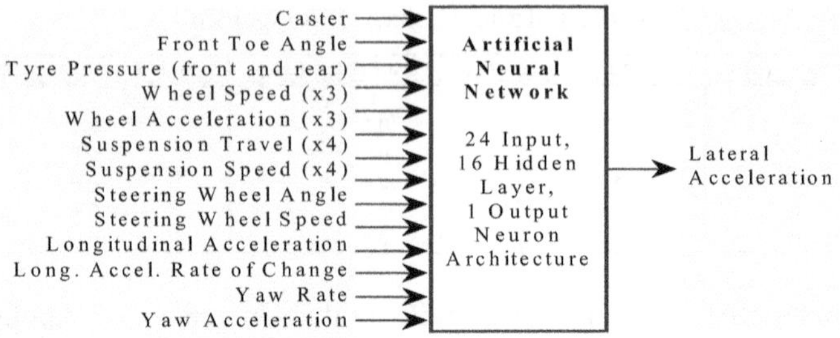

Fig. 4. ANN Model Used for Analysis

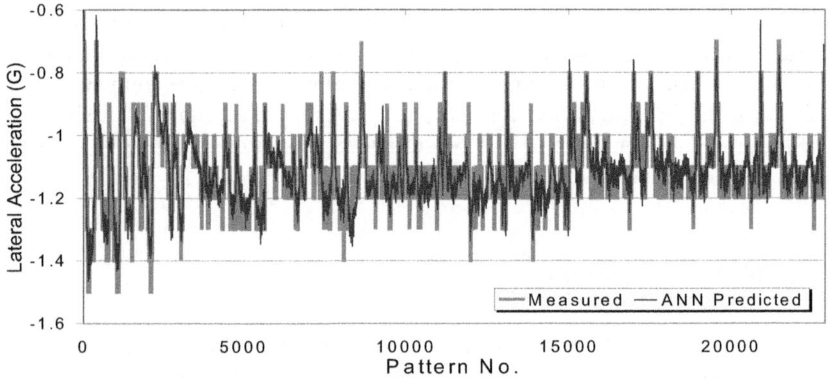

Fig. 5. Training ANN Prediction Results

Figure 5 shows the comparison between the actual measured values (grey) and the ANN predicted values. It can be seen that the measured values have been rounded to one decimal place when logged, which makes error comparison difficult. However, the ANN prediction results seem to follow the trend well and make very few, if any, large deviations from the original data. This is also reflected in the error distribution in Figure 6. Instances of large random errors are non-existent, and the curve follows a general bell shape. This also tends to support the theory that a large part of the error is the result of data rounding, and in reality is much lower.

Figure 7 and 8 follow the same format as above, except in this case just the testing data is displayed. The testing data was excluded from training and as a result remains an excellent tool to establish how well the ANN has 'learnt' the process. It can be seen that the results are almost identical to the training data. This not only shows that the ANN has learnt the vehicle dynamics very well, but also supports the theory that a significant contributor to the model error comes from data rounding.

Further investigation into this model then revealed the percent importance it placed on the input variables. Parameters that the model finds to have a large impact on the output are given high importance values, while unimportant ones are given low values. The results are shown in Figure 9.

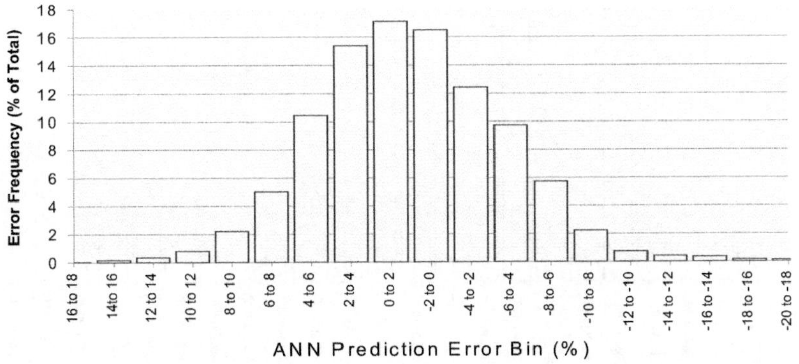

Fig. 6. Training Error Distribution

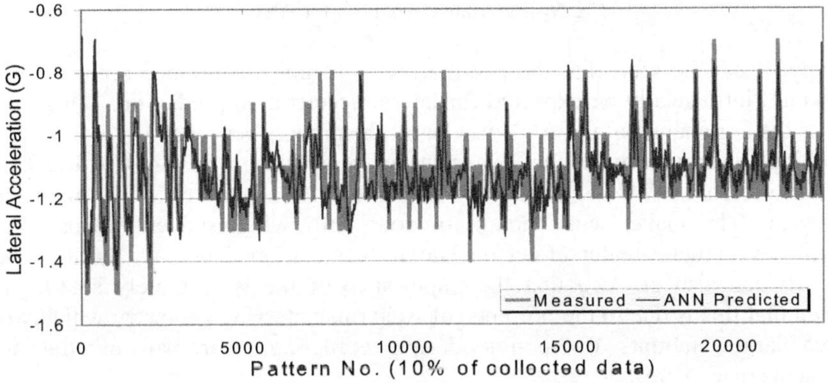

Fig. 7. Testing ANN Prediction Results

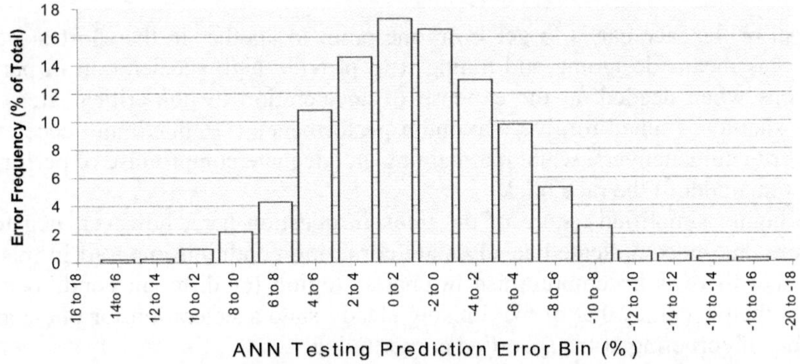

Fig. 8. Testing Error Distribution

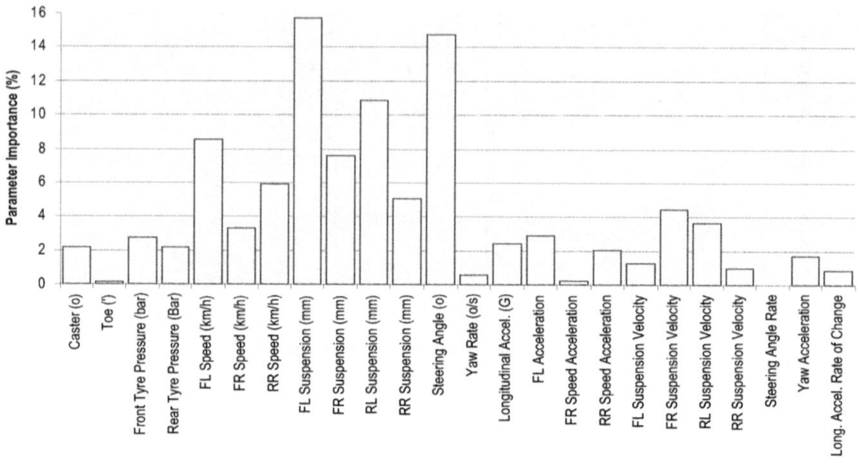

Fig. 9. Importance Analysis for ANN

Firstly, it can be seen that the parameters with high importance seem to follow what would intrinsically be expected for lateral acceleration prediction. This further supports the argument that the ANN has learnt the process very well.

Looking closer we can see that caster has an importance of just 2.16%, toe 0.13%, front tyre pressure 2.74% and rear tyre pressure 2.18% - just over 7% of the total importance. This makes sense though, obviously if a wheel suddenly stops rotating this will have much greater effect on lateral acceleration than, say, the amount of caster. It can also be seen that the importance of toe is extremely small. It is expected that this is due to the presence of Ackerman steering geometry which would produce large amounts of toe-out during cornering, marginalising the small adjustments made here.

5 Chassis Performance Prediction

The goal of the race car is to get from one point to another in the shortest possible time. This means designing and tuning it to provide high acceleration in particular conditions when needed, at the expense of acceleration in non-critical areas. The chassis should be tuned to give maximum performance (i.e. maximum acceleration) during critical maneuvers, while maintaining an adequate compromise of performance for the remainder of the race track.

The highly simplified nature of the track in question here, however, negates this problem. The uncomplicated steady state cornering condition imposed in this study means that there is no compromise in chassis tuning for different conditions. The problem then becomes that of maximising steady state acceleration for one condition only, that of cornering at the specified corner radius.

All things being equal, it is the chassis tuning arrangement that defines this maximum acceleration. Each arrangement of every tuning parameter produces a very large amount of possible combinations, each with their own effect on vehicle performance. In theory then, it should be possible to rank each of these combinations in order of performance for specific maneuvers, such as the steady state cornering condition discussed here.

With this in mind, the next stage of the investigation was to use the trained ANN model discussed above to predict the maximum obtainable lateral acceleration for any given combination of caster, toe and front and rear tyre pressure. It was decided that the simplest way to do this was to enter the desired tuning arrangement, set all of the 'rate of change' parameters defining steady state to zero and let a program (written in LabVIEW again) continuously enter random numbers into the remaining network inputs. This numerical investigation would then, therefore, explore all of the possible running conditions of the car, in the hope of finding the condition that produced the maximum lateral acceleration for the given chassis tuning arrangement. The program would then record the conditions that produced the highest lateral acceleration, and rank it against other chassis tuning arrangements.

Of note, however, is that while the method above should give steady state results, it has the potential of predicting a condition that would be very hard for a driver to control. An additional condition was also included to make sure that the vehicle and wheel velocities reflected a realistic driving condition and correlated with the predicted lateral acceleration. Variations due to slip were included to a degree however, with the introduction of a constant slip error term (1% slip) within the test.

Using this process it was possible to input any chassis arrangement (within the minimum and maximum bounds of the training data) into the ANN and obtain the maximum achievable velocity of the vehicle in steady state cornering at the designated corner radius. Finding the optimum chassis arrangement was then just a process of repeating this procedure for an array of arrangements and identifying the fastest. The different chassis arrangements used in this process were comprised of every combination (480 total) of the following:

- Caster – 0, 2, 3, 4 & 5°
- Toe = -1.2, 0, 1.2 & 2.5'
- Front and rear tyre pressure = 0.6, 0.7, 0.8, 0.9 & 1.0bar.

The fastest final results are given in Table 2, and show an optimum arrangement of 0° caster, -1.2' toe, 1.0bar front tyre pressure and 0.9bar rear tyre pressure.

Table 2. Top ten ANN chassis arrangements

Caster (o)	Toe (')	Front Tyre Press. (bar)	Rear Tyre Press. (bar)	FL Wheel Speed (km/h)
0	-1.2	1.0	0.9	43.414
0	-1.2	1.0	1.0	43.413
0	-1.2	0.8	0.9	43.402
0	-1.2	0.6	1.0	43.400
0	-1.2	0.8	1.0	43.390
0	-1.2	1.0	0.8	43.390
0	-1.2	1.0	0.7	43.385
0	-1.2	0.4	1.0	43.382
0	0	1.0	1.0	43.379
0	-1.2	0.6	0.9	43.376

6 Testing and Appraisal of ANN Model

The data acquisition required for this research was acquired over an eight hour period, with most of this time going into changing the chassis arrangements and re-tuning to suit. This was done in conjunction with the normal chassis tuning regiment that the Stralsund University of Applied Sciences was conducting to prepare for an upcoming competition. All of the data used to train and test the ANN model was recorded on the first day of testing. This was followed by continued chassis tuning for the same steady state cornering condition for a further two days to provide a comparison between practical experimentation and the ANN model.

The optimum arrangement found through practical experimentation was found to be 0° caster, -1.2' toe, 1.0bar front tyre pressure and 1.0bar rear tyre pressure. It can be seen that this arrangement was ranked second in the ANN model, by only 0.001km/hr, which suggests exceptional correlation between the two methods.

The degree of correlation is also even more impressive when considering the quality of the data used to train the ANN. The chassis arrangements used in the training data were taken when the University of Applied Sciences was predominantly investigating the effects of tyre pressure variation, and as a result did not cover a broad range of conditions. Comparing Table 1 and Table 2 shows that there is very little relationship between the training data and the ANN predicted optimum arrangements. In fact, the best training arrangement is ranked at 216th of the 480 ANN predictions. This meant that the ANN model had to extrapolate much of its information from obscure and largely unhelpful data to provide this highly accurate solution.

7 Conclusions

The conditions used in this research are obviously highly simplified when considering the complexities involved in race car chassis tuning. Nonetheless, in this case three days of rigorous track testing was effectively and accurately modeled using an ANN model based on only one day of testing. This was despite model errors induced by coarse measured data rounding and the large deviation between the chassis arrangements used in training and the final, optimal solution. This, strongly suggests a real benefit may exist in the use of ANN in chassis tuning.

Acknowledgements

This work is the result of financial support and collaboration between the University of Applied Sciences, Stralsund, Germany and the University of Tasmania, Australia. This arrangement included a research visit to Germany by David Butler for the duration of the study.

References

[1] W. MILLIKEN and D. MILLIKEN. "Race Car Vehicle Dynamics", Society of Automotive Engineers, 1995.
[2] H. ADAMS. "Chassis Engineering", HP Books, 1993.
[3] C. CAMPBELL. "Design of Racing Sports Cars", 1st Edition, Chapman and Hall, 1973.
[4] M. COSTIN and D. PHIPPS. "Racing and Sports Car Chassis Design", 2nd Edition, B. T. Batsford, 1966.
[5] L. TERRY and A. BAKER. "Racing Car Design and Development", 1st Edition, Robert Bentley, 1973.
[6] V. V. VANTSEVICH, M. S. VYSOTSKI and S. V. KHARITONT. "Control of Wheel Dynamics", International Congress & Exposition, Society of Automotive Engineers, 1998.
[7] A. van ZANTEN, W. RUF and A. LUTZ. "Measurement and Simulation of Transient Tire Forces", Robert Bosch GmbH, Society of Automotive Engineers, 1989.
[8] R. BANNATYNE. "Future Developments in Electronically Controlled Braking Systems", Motorola Inc, Transportation Systems Group, 1998.
[9] D. BUTLER. "Traction Control Using Artificial Neural Networks", MEngSci, University of Tasmania, 2002.
[10] M. ARBIB and J. ROBINSON. "Natural and Artificial Parallel Computation", Massachusetts Institute of Technology, 1990.
[11] "Artificial Neural Networks Technology", www.dacs.dtic.mil/techs/neural.
[12] M. ZEIDENBERG., "Neural Networks in Artificial Intelligence", University of Wisconsin, Ellis Horwood, 1990.
[13] M. CAUDILL and C. BUTLER. "Understanding Neural Networks", Volume 1, The MIT Press, 1994.
[14] V. KARRI. "ANN for performance estimation in wood turning", ICMA '97, Hong Kong, Apr 29, pp 618.
[15] V. KARRI and F. FROST. "Selecting Optimum Network Conditions In BP Neural Networks with respect to Computation Time and Output Accuracy", Comalco Aluminium Ltd, University of Tasmania, CR/01/1999.
[16] F. FROST. "Neural Network Applications to Aluminium Manufacturing", PhD, University of Tasmania, 2001
[17] "2001 Formula SAE Rules", Society of Automotive Engineers, 2000.
[18] "LabVIEW 6i", National Instruments Corporation, 2000.

Applications of Soft Computing for Musical Instrument Classification

Daniel Piccoli, Mark Abernethy, Shri Rai, and Shamim Khan

Murdoch University, Western Australia
{dpiccoli,mark.abernethy,smr,s.khan}@murdoch.edu.au

Abstract. In this paper, a method for pitch independent musical instrument recognition using artificial neural networks is presented. Spectral features including FFT coefficients, harmonic envelopes and cepstral coefficients are used to represent the musical instrument sounds for classification. The effectiveness of these features are compared by testing the performance of ANNs trained with each feature. Multi-layer perceptrons are also compared with Time-delay neural networks. The testing and training sets both consist of fifteen note samples per musical instrument within the chromatic scale from C3 to C6. Both sets consist of nine instruments from the string, brass and woodwind families. Best results were achieved with cepstrum coefficients with a classification accuracy of 88 percent using a time-delay neural network, which is on par with recent results using several different features.

Keywords: neural networks, musical instrument recognition

1 Introduction

With the advent of digital multimedia, there is an increasing need to be able to catalogue audio data in much the same way that books are catalogued. Most digital audio formats in use today such as MP3 and WAV contain limited meta-data about the actual recordings that they contain [1]. However, the MPEG-7 specification requires that meta-data, such as the types of musical instruments in a recording, should be stored in the file with the recording to enable effective cataloguing of files [2]. The classification and identification of important features of musical instruments in digital audio will be a step towards such a cataloguing system. The process of classification based on a set of features is often referred to as 'Content-Based Classification [1].'

Meta-data generated by such a system may be used by a search engine to allow users to find specific styles of music. For example, a music teacher may be interested in searching for audio files with certain instruments playing or music from a certain genre. Currently, musical search systems only have the ability to classify their music based on filenames.

The model discussed in this paper concentrates on identifying musical instruments in sound recordings whilst assessing the usefulness of different features

T.D. Gedeon and L.C.C. Fung (Eds.): AI 2003, LNAI 2903, pp. 878–889, 2003.

that can represent musical instruments for the purpose of classification. Meta-data gathered by an effective instrument classification system can be used to build databases based on MPEG-7 meta-data. Another possible use of such an audio classification system include the ability to automatically generate meta-data about music files and to fight the distribution of copyright audio material on the Internet.

The main aim of this research was to compare the usefulness of spectrum and cepstrum based features using an artificial neural network. Also, the performance of the time delay neural network and the multi-layer perceptron were compared. The results may lead to an improved cataloguing system that allows people to search through music databases more effectively.

2 Previous Work

Two major characteristics of an audio classification system are the features that are used to distinguish between the sounds and method in which the classifier is 'trained' to recognise the sounds. A trained human ear can easily identify musical instruments that are playing in a sound mix, even if each of these instruments share a similar note range. This is because each instrument has a set of auditory features that distinguishes it from other instruments. For example, many musical instruments have harmonics or overtones that can be heard at multiples of the fundamental frequency. These harmonics colour the sound making each instrument sound different. The features that enable one to distinguish musical instruments can be described as the timbre of the sound. At present, computerised audio classifiers lack the accuracy of human classifiers. As a result, research is being conducted in improving the features that are fed to audio classifiers as well as the audio classification engines themselves (eg. ANNs).

Audio Classification Techniques

Herrera [2] provides a survey of different techniques that have been used to classify musical instruments in monophonic (where one instrument plays only one note at any given time) sounds. Techniques discussed include K-nearest neighbours, Bayesian classifiers, binary trees, support vector machines and neural networks. Comparability between these techniques is difficult due to differing experimentation approaches and sound samples used. Also, many of these experiments have been based on a limited set of musical instruments. However, the attractiveness of the ANN comes from its ability to generalise after being trained with a finite set of sound samples. Herrera also alludes to the difficulty involved with using the same algorithms for identifying musical instruments in sound mixes.

Experiments performed using ANNs to classify musical instrument sounds have found that good results can be achieved for monophonic instrument samples [3, 4]. Cemgil [4] provides a comparison between a multi-layer perceptron, a time delay neural network and a hybrid Self-Organising Map/Radial Basis

Function (RBF) for classifying instrument sounds. All network types are presented sound in the form of a set of harmonic envelopes. The number of instrument harmonics required for good neural network generalisation performance is also discussed. The results indicated that generalisation improves when more harmonics are presented to the neural networks. The number of time windows had less bearing on the performance of each of the models. It was found that the spectral content in the attack portion of the sound contained most of the information needed by the network to make a classification.

Cemgil's experiment found that the performance of each of the models in order of success were: the Time Delay Neural Network with classification success of up to 100%; the Multi-layer perceptron with a classification success of up to 97%, and lastly the hybrid Self Organising Map (SOM)/Radial Basis Function with a classification success of up to 94%. However, the results shown by the SOM are promising because they show how the instruments are organised according to timbre. Unfortunately, these results are optimistic for the classification of real life sounds. The sound samples included a limited range of notes (one octave) and limited instrument articulation (eg. a cello string can be plucked or bowed).

Experiments detailed in this paper test the adequacy of the standard back-propagation Multi-Layer Perceptron (MLP) the Time Delay Neural Network (TDNN) for detecting the presence of a musical instrument in a monophonic source regardless of the note played or its volume. Each ANN is fed a series of spectral parameters based on the Fourier Transform.

Selection of Auditory Features

A digital audio signal must be converted into a suitable form before classification becomes possible. Auditory features can be classified as spectral or temporal. Spectral features are parameters that can be extracted from the frequency spectrum of a sound whereas temporal features relate to the timing of events over the duration of a note.

When using multiple features for classifying musical instrument, it is possible for one bad feature to destroy the classification results. It is therefore important to determine whether a certain feature allows musical instruments to be distinguished. Kostek's [3] work goes some way to identifying instrument distinguishing sound characteristics.

Brown [5] demonstrated the validity of cepstral coefficients for distinguishing between an oboe and a saxophone by using a K-means algorithm. Eronen [6] later verified their robustness in classifying a wider range of instruments.

Having surveyed the literature, it must be noted that different features may be suitable for classifying a small set of instruments. However, a universal set of parameters with the ability to distinguish between any musical instrument is non-existent. Most of the timbral features discussed are based on perceptual models of the ultimate sound classifier, namely the human being. However, the difficulty lies in the objective measurement of these properties using a computer [5]. The human auditory system is a highly non-linear device in which some harmonic components produced by musical instruments may be masked by spectral energy at nearby frequencies.

Extensive research has been completed on defining timbre in terms of human perception. The definition of timbre in terms of spectral features was apparent as early as 1954, when Helmholtz claimed that the relative amplitudes of the harmonic partials that compose a periodic tone is the primary determinant of a tone's sound quality [7, 8]. Temporal features were recognised as important determinant of musical instrument sounds as early as 1910 when Stumpf recognised the importance of the onset of a musical note for distinguishing musical sounds [8]. For this reason, experiments detailed in this paper involve presenting the ANN with representations of sounds beginning at the onset of a note.

The field of computational auditory scene analysis [9] attempts to mimic the ability of humans to conceptualise music. Abdallah [10] discusses how to extract features that make up our perception of musical sounds. His idea is to create a 'single unifying description of all levels of musical cognition from the raw audio signal to abstract musical concepts.'

This research could lead to improvements in how a neural network is fed sound features in the form of reduced redundancy.
In the meantime, techniques such as Mel scaling (passing audio through a set of band pass filters based on experimental results on human hearing) are used to approximate the human auditory system. Future work in musical classification based on perceptual models may be fruitful.

3 Methodology

Sound samples from a variety or woodwind, brass and string instruments were collected from the University of Iowa music samples web page [11]. Piano samples were also obtained from this site. A synthesised guitar was included in the sample set (from general MIDI) in order to have a representation of another stringed instrument. The note samples, which ranged from C3 to C6 on the chromatic scale, were separated into two sets. One of these sets was used for training and the other for testing. The testing set was used to assess the ability of the ANN to generalise based on a finite number of examples that were presented during training (the training set).

The note range from C3 to C6 was chosen because the instruments selected produce sounds that overlap in these frequencies. Notes above C6 were not included in either the training or test sets because the Nyquist theorem prevents the extraction of higher harmonics for higher notes. To enable the extraction of higher harmonics, a larger audio sampling rate must be used. However, due to processing time constraints, a sampling rate of 22 kHz was used.

It is important to use real life instrument samples for classification to ensure that the ANN can generalise regardless of how the instrument is played. The ANN training set contained a combination of soft notes, moderately loud notes and loud notes. The harmonic compositions of each note change dynamically (not proportionally) depending on how the note is articulated (see figure 1).

This makes the classification process more difficult for live instruments. For synthesized instruments the harmonic amplitudes are generally scaled in proportion to the loudness of the fundamental frequency. The training and testing sets contain examples of notes played vibrato (with modulation) and pizzicato (plucked) notes from the cello.

After the onset of each note is detected (by computer algorithm), 50 percent overlapping windows of 2048 Hann windowed samples were converted to the frequency domain using the fast Fourier transform (FFT). From the FFT coefficients, harmonic envelopes and cepstral coefficients were then calculated. The extracted features were either averaged over a number of frames or taken directly and the resultant feature vector was placed into a pattern file ready for presentation to the ANN. For the cepstral coefficients, only the first 128 coefficients were presented to the neural network. The lower coefficients correspond to the rough or coarse spectral shape. The rough spectral shape contains information of the formants or resonances of the instrument body [6], thus these were used as features. The calculation of cepstral coefficients generally involves taking the spectrum of a log spectrum [12].

These features were fed into both a multi-layer perceptron (MLP) and a time-delay neural network. A limitation of the multi-layer perceptron comes from the fact that the entire pattern has to be presented to the input layer at once. The nature of sound is that variations occur over time. For example, musical sounds have attack, sustain and decay transients that are difficult to represent. The time delay neural network is an extension of the MLP back propagation network developed by Alex Waibel [13] that has the advantage of being able to learn patterns that vary over time [14]. TDNNs have proven to be useful for musical instrument identification [4]. The time-delay neural network, like the perceptron, uses back-error propagation as its learning algorithm. However, the TDNN has a variety of time-delay components built into its architecture as described by Waibel [15, 13].

The convergence of each Neural Network towards the correct classification was assessed by monitoring the 'mean squared error' (MSE) of the training data and comparing it to the MSE of the test data after each epoch. Training was stopped once the MSE of the test no longer improved. The validity of the features (harmonic envelopes vs cepstral coefficients) was assessed by comparing the recognition accuracy of ANNs using each of these features.

4 Results

4.1 Instrument Classification with Neural Networks

MLPs Trained with Harmonic Amplitudes

An MLP was initially trained with examples of individual frames (no averaging over a number of frames) with 50% overlapping windows shifted across the entire sound file for each instrument. Therefore, the number of neurons in the input layer was equal to the number of harmonics presented to the neural network.

That is, the ANN was presented with the harmonic amplitudes of one frame at a time (not a harmonic envelope). This tests the ability of the ANN to recognise a musical instrument given just one frame.

Results were modest with a best classification score of 67%. However, these results suggest that the neural network struggled to identify musical instruments correctly given just one frame at a time. This may be due to the fact that harmonic content is not consistent across the entire length of a musical note, or across different notes or articulations produced by the same instrument. This makes it difficult for the ANN to generalise. Also, based on research regarding phenomena that influence timbre perception [16], frequency information in the attack transient of a musical note is important for classification. In the instances where the sliding window appears over a non-attack part of a note, instruments may be more difficult to distinguish.

There seems to be an indicative trend that by increasing the number of harmonics presented to the neural network, the ANN will be better able to distinguish between instrument sounds. Unfortunately, there were not enough test runs available to statistically verify this claim due to the Nyquist limitation (insufficient number of harmonics). However, this trend is consistent with findings indicated by Cemgil [4] and Kostek [3].

MLPs Trained with Harmonic Envelopes

The MLP was then presented with harmonic envelopes from several adjacent sliding windows starting from the onset of each musical note. In other words, the ANN was presented with harmonic amplitudes for several 50% overlapping frames beginning from the onset of each note. Each window was individually normalised (note that this is additional normalisation for each frame that was performed in addition to the normalisation of the entire note during pre-processing). It was hoped that a trend could be shown between the number of frames presented to the MLP and the ability of the MLP to generalise. However, the addition of new frames to the input vector failed to improve the classification results of the ANN. As shown in table 8 in appendix A, the best result attained by presenting multiple frames of harmonic amplitudes (harmonic envelope) was 65%. These results are well below those attained by Cemgil [4] using harmonic envelopes and are also inferior to the results obtained for this research for ANNs presented with harmonic amplitudes from only one frame.

These poor results indicate that the ANN failed to generalise to the general (test set) population given harmonic envelopes. There are two possible causes for this. Firstly, if the number of hidden layer nodes were excessive, the ANN is likely to learn the nuances of the training set rather than the major features that distinguish the instrument patterns. The second possible cause of the problem occurs when the training sample is not representative of the general population. For example, the trombone samples given in the training set may not be representative of all trombone sounds. Since the number of hidden layer nodes was carefully chosen, the former is unlikely to be the problem. Analysis of error during training revealed a disappointing relationship between test and training data. Ideally, if the ANN was learning to distinguish musical instruments based on their harmonic content, then both training and testing errors would have decreased at a similar rate.

TDNN Trained with Harmonic Envelopes

When the Time-Delay Neural Network (TDNN) was presented with the harmonic envelope as an input vector, the classification success for the best performing TDNN increased to only 68%. This result is well below the 100% achieved by Cemgil (Cemgil et al., 1997). However, this may be due to the fact that every individual frame was normalised. To improve this experiment, the harmonic envelope should have been normalised as a whole. By normalising each individual frame, important information regarding change in volume associated with the onset of a note may be lost.

4.2 Supervised Neural Networks Using Cepstral Coefficients as Features

MLP Trained with Cepstral Coefficients

Results for MLPs, which were disappointing with a classification success of just 53 percent. It became apparent that a multi-layer perceptron is unable to distinguish musical instruments very well when trained with just one set of Cepstral coefficients per note. Thus, in order to reduce the effect of noise when measuring the cepstrum of a note, the average of several frames over each individual note were taken. The best result attained from averaging the Cepstrum over a number of frames is depicted in table 1.

Improvements to the classification ability of the MLP as a consequence were excellent. After interpretation of the results, it became evident that the ANN is better able to generalise when the Cepstrum is averaged over an increasing number of frames. This implies that the average of a number of cepstral coefficients at the onset of each note is more representative of a particular instrument than cepstral coefficients in isolation. More research is needed to examine whether this phenomenon continues beyond 7 frames. However, classification results of up to 85% for the MLP using cepstral coefficients averaged over 12 frames (at the onset of each note) indicate that 128 cepstral coefficients are a valid and robust input vector option for neural networks.

TDNN Trained with Cepstral Coefficients

The TDNN presented with cepstral coefficients (the first 128 coefficients) proved to be the most successful architecture for classifying musical instruments. Each TDNN was configured to accept 10 fifty percent overlapping delayed frames containing 128 cepstral coefficients. The TDNN was given 8 such examples per note. As depicted in table 2, the TDNN successfully classified up to 88% of instruments correctly.

Table 1. Matrix of 9 instruments classified with an MLP using the first 128 cepstrum coefficients averaged over 12 frame(s) at the onset of each note

Actual Class/Target	Bass.	Flute	Clar.	Trom.	Horn	Oboe	Piano	Guit.	Cello
Bassoon	14	0	0	0	0	0	0	0	0
Flute	0	14	1	3	0	0	0	0	0
Clarinet	0	1	14	0	0	2	0	0	0
B. Tr	0	0	0	8	0	1	0	0	0
Horn	1	0	0	0	15	0	0	0	0
Oboe	0	0	0	4	0	12	0	0	0
Piano	0	0	0	0	0	0	8	0	0
Guitar	0	0	0	0	0	0	2	15	0
Cello	0	0	0	0	0	0	5	0	15
Total	15	15	15	15	15	15	15	15	15
Percentage Correct	**93**	**93**	**93**	**53**	**100**	**80**	**53**	**100**	**100**
Total Correct	**115**								
Total % Correct	**85**								

Table 2. Matrix of 9 instruments classified with a TDNN Using the first 128 cepstrum coefficients delayed over 10 frames at the onset of each note (best performance)

Actual Class/Target	Bass.	Flute	Clari.	Trom.	Horn	Oboe	Piano	Guit.	Cello
Bassoon	105	0	0	0	0	0	0	12	0
Flute	0	102	7	0	0	0	0	0	0
Clarinet	0	1	87	0	0	10	0	0	0
B. Tr	0	0	0	103	4	0	2	0	0
Horn	0	0	0	2	98	0	0	0	0
Oboe	0	0	0	0	0	95	0	25	0
Piano	0	0	0	0	0	0	85	15	0
Guitar	0	0	0	0	0	0	14	54	2
Cello	0	2	4	0	3	0	4	0	103
Total	**105**	**105**	**105**	**105**	**105**	**105**	**105**	**105**	**105**
Percentage Correct	**100**	**97**	**83**	**98**	**93**	**90**	**81**	**51**	**98**
Total Correct	**88**								
Total % Correct	**832**								

4.3 Discussion of Results

Comparison of Feature Vectors

From the above analysis, it appears that averaged cepstral coefficients (with classification success up to 88%) are a much more useful input vector to a classification algorithm (such as neural networks) than harmonic envelopes (with classification success of up to 68%). This may be indicative of the fact that harmonic information alone is insufficient for distinguishing between musical instruments.

For classification with harmonic amplitudes to be possible, the relationship between the harmonics must be consistent for a given instrument class (eg. bassoon) regardless of the note played or its articulation. Figure 1 below depicts the harmonic structure of two different bassoon notes used in the experiments:

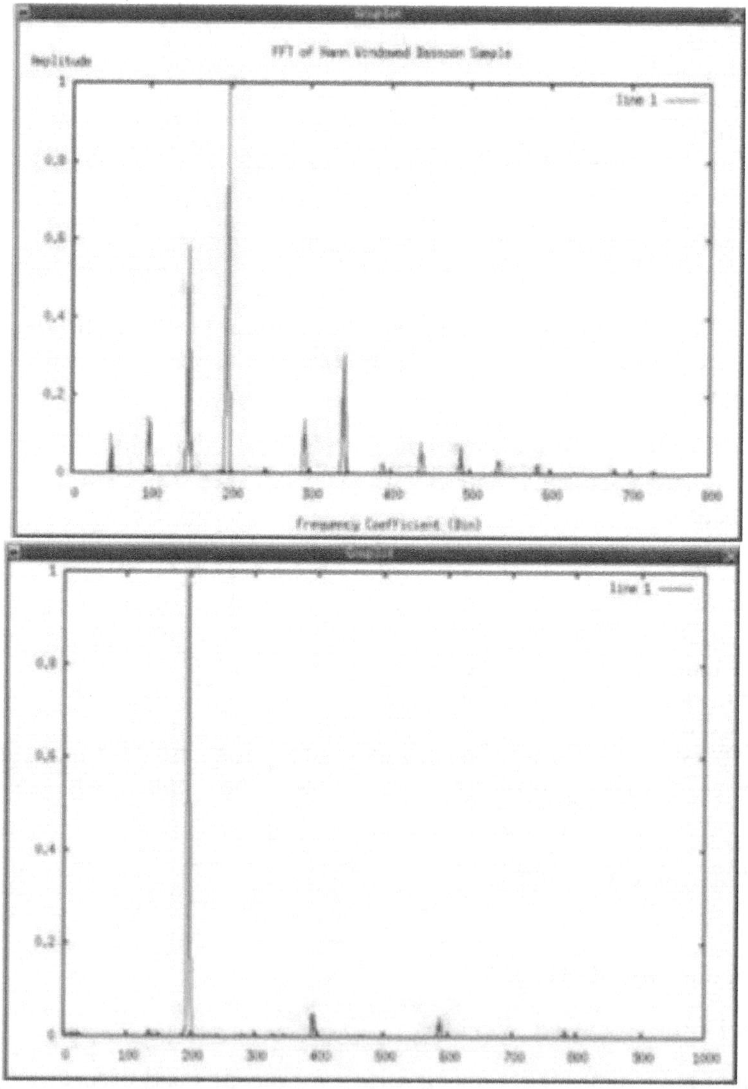

Fig. 1. Comparison of two bassoon notes with different articulation

The graphs in figure 1 depict harmonic peaks of a 2048-sample frame gathered from the onset of two bassoon notes. Each frame has been normalised to 1 so that the relationship between harmonic amplitudes for each note can easily be seen. From these diagrams, it is evident that the harmonic pattern is not consistent for each bassoon note. In the top diagram, the third harmonic has the highest amplitude whereas the second diagram shows the fundamental as having the highest amplitude. These inconsistencies may limit the ability of an ANN to distinguish between instruments using harmonic amplitudes exclusively. Future studies may be interested in testing whether the process of averaging harmonic envelopes improves the classification process as it did for cepstral coefficients.

The relatively poor results attained using harmonic envelopes are not comparable to results from Cemgil's [4] experiments. Differences in the training and test-set data may be the cause for these discrepancies. For example, it appears that Cemgil used sound samples from the standard AWE32 (sound card) set. This makes classification using harmonic envelopes alone feasible due to consistency in the harmonic envelopes of synthesised sounds. As stated by Kaminskyj [17], it is important for the sound classification research community to make their results as comparable as possible in future.

Comparison of Classifiers

In terms of the feature classifiers themselves, the TDNN consistently returned better results than the MLP. The TDNN had classification results of up to 68% using harmonic envelopes and 88% using cepstral coefficients while the MLP had classification results of up to 67% using harmonic envelopes and 85% using cepstral coefficients as features. This is indicative of the importance of temporal features for classifying musical sounds. Further research is needed to confirm this statistically.

One advantage of using the MLP as opposed to the TDNN came from the fact that it completed training much more quickly. However, this was influenced by the experiment technique used. For the MLP, cepstral coefficients were averaged over several frames. Therefore, the number of input layer nodes was equal to the window size (or 128 for the ANNs designed to accept the first 128 cepstrum coefficients). However, for the TDNN, 10 individual frames (delays) of 128 cepstral coefficients were applied to the input layer, totalling 1280 input layer nodes. This slowed the training process for the TDNN.

Identification of Timbral Families

One of the aims of these experiments was to identify timbral families and relationships between the musical instruments. This can reveal whether or not the computer 'perceives' musical instruments in a similar way to the human auditory system. Table 3 provides totals of the most common misclassifications with the TDNN architecture using cepstral coefficients as features:

This data can be used to reveal relationships inherent between the musical instruments. For example, the table reveals that the flute and the clarinet,

Table 3. Common Instrument Misclassifications

Instrument	Most Commonly Misclassified As	No. Of Instances
Bassoon	Bass Trombone	2
Flute	Clarinet	12
Clarinet	Flute	44
Trombone	Horn	90
Horn	B. Trombone	10
Oboe	Clarinet	29
Piano	Guitar	77
Guitar	Bassoon	59
Cello	B. Trombone	34

both woodwind instruments are commonly misclassified as one another. Also, the two brass instruments, namely the bass trombone and the horn are commonly misclassified. The only two instruments that significantly did not exhibit the expected misclassification pattern were the guitar (string), which was most commonly misclassified as a bassoon (woodwind), and the cello (string), which was most commonly misclassified as a trombone (brass). The fact that most instruments were misclassified as instruments within their own orchestral (timbral) family indicates that cepstral coefficients have some relationship with the perception of timbre by humans.

5 Conclusions and Future Research

The initial aim of this research was to compare the usefulness of cepstral coefficients and harmonic envelopes as inputs to a neural network for the purpose of classifying musical instruments. Results have indicated that cepstral coefficients may be more useful than harmonic envelopes for the purpose of distinguishing between musical instruments. This research has also demonstrated the usefulness of the MLP and TDNN as classification tools.

Future work may involve combining Cepstral coefficients with spectrum related parameters such as spectral brightness and odd/even harmonic components in order to produce a better classification model. Temporal features must also be carefully analysed for their usefulness in classifying musical sounds. Further experiments involving the human perception of sound may also be fruitful for devising a better way to present a classifier with sound data.

The results attained for this research were limited to monophonic sounds. For classification models discussed in this paper to be useful, prior separation of sounds is required. The ultimate goal of audio classification is to identify musical instruments that exist in polyphonic sounds. This may be achieved in the future by using a technique known as source (or stream) separation. Future research in this area will improve the viability of classifying musical instruments in polyphonic sounds.

References

[1] E. Wold, T. Blum, D. Keislar, and J. Wheaton, "Classification, search and Rereval of audio," *IEEE Multimedia Magazine, vol. 3,* no. *3, pp. 27-36, 1996.* 878

[2] P. Herrera, X. Amatraiain, B. E., and X. Serra, "Towards instrument segmentation for music content description: A critical review of instrument classification techniques," in *Proceedings of the International Symposium On Music Information Retreival, 2000.* 878, 879

[3] B. Kostek and R. Krolikowski, "Application of artificial neural networks to the recognition of musical sounds," *Archives of Acoustics, vol. 22,* no. 1, pp. *27-50, 1997.* 879, 880, 883

[4] A. Cemgil and F. Gurgen, "Classification of musical instrument sounds using neural networks," *Proceedings of SUI97, 1997.* 879, 882, 883, 887

[5] J. Brown, "Computer identification of musical instruments using pattern recognition with cepstral coefficients as features," *J. Acoust. Soc. Am., vol. 105, pp. 19331941, 1999.* 880

[6] A. Eronen and A. Klapuri, "Musical instrument recognition using cepstral coefficients and temporal features," *Proceedings of the IEEE International Conference an Acoustics, Speech and Signal Processing, 2000.* 880, 882

[7] H. Helmholtz, "On the sensations of tone as a physiological basis for the theory of music," *Dover, A. J. Ellis Trans, 1954.* 881

[8] K. Martin and Y. Kim, "Musical instrument identification: A pattern-recognition approach," in *136th Meeting of the Acoustical Society of America,* October 1998. 881

[9] A. S. Bregman, "Auditory scene analysis: The perceptual organisation of sound," *MIT Press, 1990.* 881

[10] S. Abdallah and M. Plumbley, "Unsupervised learning in neural networks for auditory perception and music cognition," *Cambridge Music Processing Colloquium,* Sept 1999. 881

[11] L. Fritts, "Musical instrument samples web page," tech. rep., University of Iowa, URL: http://theremin.music.niowa.edu/ web/sound/, 2002. 881

[12] R. Randall, *Frequency Analysis.* Naerum: Bruel and Kjaer, 1987. 882

[13] A. Waibel, T. Hanazawa, G. Hinton, K. Schikano, and K. Lang, "Phoneme recognition using time-delay neural networks," *IEEE Transactions in Acoustics, Speech and Signal Processing 37, 1989.* 882

[14] J. Dayhoff, *Neural Network Architectures - An Introduction.* New York: Van Nostrand Reinhold, 1990. 882

[15] A. Waibel, "Consonant recognition by modular construction of large pnonetic timedelay neural networks," *Neural Information Processing Systems, 215-223.* 882

[16] J. Grey, "Multidimensional perceptual scaling of musical timbres," *J. Acoust. Soc. Am. 61(5), pp. 1270-1271, 1977.* 883

[17] I. Kaminskyj, "Multi-feature musical instrument sound classifier w/user determined generalisation performance," in *Proceedings of ACMA02, 2002.* 887

Nonlinear Time Series Prediction Based on Lyapunov Theory-Based Fuzzy Neural Network and Multiobjective Genetic Algorithm

Kah Phooi Seng [1] and Kai Ming Tse [2]

[1] School of Engineering & Science, Monash University (Malaysia), Bandar Sunway, 46150 PJ, Malaysia
jasmine.seng@engsci.monash.edu.my
[2] School of Microelectronics, Griffith University Kessels Rd, Nathan QLD 411 1, Australia
k.tse@griffith.edu.au

Abstract. This paper presents the nonlinear time series prediction using Lyapunov theory-based fuzzy neural network and multi-objective genetic algorithm (MOGA). The architecture employs fuzzy neural network (FNN) structure and the tuning of the parameters of FNN using the combination of the MOGA and the modified Lyapunov theory-based adaptive filtering algorithm (LAF). The proposed scheme has been used for a wide range of applications in the domain of time series prediction. An application example on sunspot prediction is given to show the merits of the proposed scheme. Simulation results not only demonstrate the advantage of the neuro-fuzzy approach but it also highlights the advantages of the fusion of MOGA and the modified LAF.

1 Introduction

Time series prediction is a very important practical application with a diverse range of applications including economic and business planning, inventory and production control, weather forecasting, signal processing and control [1]. As a result, there has been considerable interest in the application of intelligent technologies such as neural networks (NNs) and fuzzy logic [2]-[3]. More recently, these two computationally intelligent techniques have been viewed as complementary leading to developments in fusing the two technologies [4], with a number of several successful neurafuzzy systems reported in the literature [5]. Such work has demonstrated the superior prediction capabilities of a fuzzy neural network as compared to the conventional neural network approach [5)-[6].

In this paper, we employ the fuzzy neural network (FNN) for nonlinear time series prediction. A new combination of the modified Lyapunov theory-base filtering algorithm (LAF) and multi-objective genetic algorithm (MOGA) [10] is used to adjust the network parameters. The proposed scheme provides not only the advantages of fuzzy

T.D. Gedeon and L.C.C. Fung (Eds.): AI 2003, LNAI 2903, pp. 890-898, 2003.

logic and NN but it also offer additional advantages offered by the modified LAF and MOGA. The weight parameters of FNN in the consequence part are adaptively adjusted by the modified LAF. The MOGA is used to tune the parameters of the membership functions (MBFs) in the premise part. Most real world problems require the simultaneous optimisation of multiple criteria/objectives. In this case, MOGA can provide the solution to these problems. In the proposed scheme, 2 types of error: instantaneous and a prior errors defined in later section are the multiple criteria to be solved by MOGA. The proposed scheme has been used for a wide range of applications in the domain of time series prediction. The theoretical prediction mechanism of the proposed scheme is further confirmed by the simulation example for real world data such as sunspot forecasting.

The paper is organized as follow: section 2 briefly describes the main features of the proposed FNN and the two criteria. Section 3 presents the fuzzy neurel network learning: structure learning and parameter learning. The parameter learning 1 – the modified LAF algorithm is presented in section 4. Section 5 describes the MOGA and the transformation of the multiobjective function into a new function so that single objective optimization methods can be used. The prediction results are presented in section 6. The finally section 7 concludes the paper with a discussion of the significance of the results.

2 Fuzzy Neural Network

Neural-fuzzy systems have been applied to many fields successfully since a decade ago. In this section, the fuzzy logic inference system can be implemented as a five-layer NN (Fig. 1). This type of architecture is the most common among neural fuzzy inference systems. Given the training input data x_n, $n = 1, 2 \ldots. N$, and the desired output d_m $m=1,2 \ldots. M$, the inference rules of simplified fuzzy reasoning can be established by experts or generated based on numerical data as proposed by Wan and Mendel [11]. The rule base contains the following form:

$$\begin{aligned} &R^i : IF\ x_1\ is\ A_1^i\ and \ldots\ x_N\ is\ A_N^i, \\ &\quad THEN\ y_1\ is\ w_1^i\ and \ldots\ y_M\ is\ w_M^i \end{aligned} \tag{2.1}$$

where i is a rule number, the A^i_N 's are MBF's of the antecedent part and w^i_M's are real numbers of the consequent part.

The operation of the this system can be described layer by layer as follows:

Layer 1: Fuzzification

This layer consists of linguistic variables. The crisp inputs x_n, $n=1,2 \ldots. N$ are fuzzified by using MBFs of the linguistic variables A^i_N. Usually, triangular, trapezoid, Gaussian or bell-shaped membership functions are used.

Layer 2. Rule Nodes

The second layer contains one node per each fuzzy if-then rule. Each rule node performs connective operation between rule antecedents (if-part). Usually, the minimurn or the dot product is used as intersection AND. The union OR is usually done using maximum operation. In our example case the firing strengths μ_i, of the fuzzy rules are computed according to

$$\mu_i = A_1^i(x_1) \cdot A_2^i(x_2) \cdot ....A_N^i(x_N) \tag{2.2}$$

Layers 3-5:Normalization, Consequence & Summation

In the third layer, the firing strengths of the fuzzy rules are normalized. Layer 4 is related to consequent fuzzy labels which are singletons in our case. The values of the singletons are multiplied by normalized firing strength. The final layer computes the overall output as the summation of the incoming signals. Therefore the output y_m of the fuzzy reasoning can be represented by the following equation:

$$y_m = \frac{\sum_i^Q \mu_i w_m^i}{\sum_i^Q \mu_i} \tag{2.3}$$

$$Y = [y_1, y_2, ..., y_M] \tag{2.4}$$

After the fuzzy logic rules and network structure have been established, the learning algorithm can then applied to adjust the parameters of the MBFs in the premise part and the weights in the consequence parts. In this paper, we proposed to use the modified LAF algorithm to adaptively adjust the weights in the consequence parts and MOGA to tune the parameters of MBFs.

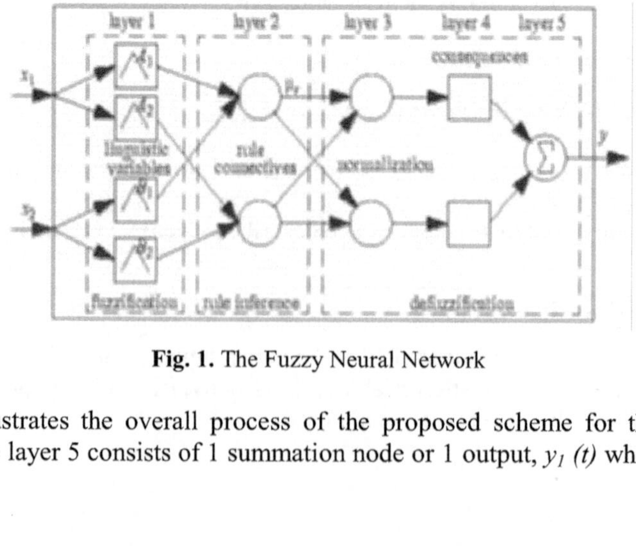

Fig. 1. The Fuzzy Neural Network

Fig. 1 illustrates the overall process of the proposed scheme for the prediction problem. The layer 5 consists of 1 summation node or 1 output, $y_1\ (t)$ which is defined as

$$y_1(t) = \frac{\sum_i^Q \mu_i w_m^i(t)}{\sum_i^Q \mu_i} \tag{2.5}$$

$y_2(t)$ is not another output node of FNN as shown in Fig. 2 and it is only computed using (2.6)

$$y_2(t) = \frac{\sum_i^Q \mu_i w_m^i(t-1)}{\sum_i^Q \mu_i} \tag{2.6}$$

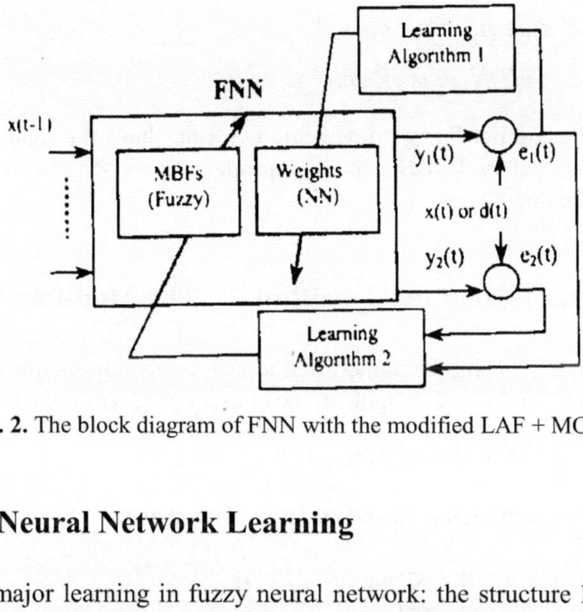

Fig. 2. The block diagram of FNN with the modified LAF + MOGA

3 Fuzzy Neural Network Learning

There are two major learning in fuzzy neural network: the structure learning and the parameter learning. The structure learning is to generate initial fuzzy rules when there is no initial fuzzy rule established by the experts. The initial fuzzy rules can generated based on numerical data. Wang and Mendal have proposed a simple and straightforward algorithm [11]. The main steps of the algorithm are described as follows:

First, we have to determine the input and output variables and construct the numerical data set. The kth numerical data set can be formed as

$$\{x'_1(t), x'_2(t), \ldots, x'_n(t)\} \rightarrow \{y'_1(t), y'_2(t), \ldots, y'_n(t)\}$$

where the left hand side and the right hand side specify the input and output numerical data. Second, calculate the membership degree for each variable by a row vector denoted by $\overline{x}_i(t)$ for inputs and $\overline{y}_j(t)$ for outputs as

$$\overline{x}_i(t) = [A_i^1(t), A_i^2(t),...., A_i^{N_i}(t)] \tag{3.1}$$

$$\overline{y}_j(t) = [B_j^1(t), B_j^2(t),...., B_j^{M_j}(t)] \tag{3.2}$$

$A_i^1(t), A_i^2(t),....,\ A_i^{N_i}(t)$ and $B_j^1(t), B_j^2(t),....,\ B_j^{M_j}(t)$ are the membership degrees for $x'_i(t)$ and $y'_i(t)$. Third, calculate the importance degree to each data pair as

$$\overline{h}(t) = \prod_{i=1}^{n} \max\{\overline{x}_i(t)\} \tag{3.3}$$

Fourth, construct the fuzzy rules for all data pairs. For instance, a fuzzy rule has this form:

$$\begin{aligned} R^{(t)} : IF\ x_1\ is\ A_1^2\ and\ .x_2\ is\ A_2^2 \\ THEN\ y_1\ is\ B_1^1\ and\ y_2\ is\ B_2^2 \end{aligned} \tag{3.4}$$

Fifth, delete the conflict fuzzy rules, while two rules have the same fuzzy set in IF part but a different set in THEN part. The proper rule is selected according to the highest importance degree.

4 Parameter Learning Algorithm 1 – The Modified LAF

The LAF algorithm in [7] has been modified to adaptively adjusts the weights of FNN in the consequence part. Those weights in the consequence part are updated as follow:

$$w_m^i(t) = w_m^i(t-1) + g_m^i(t)\alpha_m(t) \tag{4.1}$$

where g^i_m (t) is the adaptation gain and α_m (t) is defined as

$$\alpha_m(t) = d_m(t) - \frac{\sum_i \mu_i w_m^i(t-1)}{\sum_i \mu_i} \tag{4.2}$$

The adaptation is given by

$$g_m^i(t) = \frac{\mu_i(t)}{\|U(t)\|^2}\left(1 - k\frac{|e_m(t-1)|}{|\alpha_m(t)|}\right) \tag{4.3}$$

where $0 < k \le 1$. $U(t) = [\mu_1, \mu_1, ..., \mu_Q]$.

It is noticeable that the values of $U(t)$ and α_m in (4.3) may be zero and rise singularities problem. Therefore the adaptation gain may be modified as (4.4)

$$g_m^i(t) = \frac{\mu_i(t)}{\| U(t) \|^2 + \lambda_1}\left(1 - k\frac{| e_m(t-1) |}{| \alpha_m(t) | + \lambda_2}\right) \tag{4.4}$$

where $0 < k \leq 1$, and λ_1, λ_2 are small positive numbers.

5 Parameter Learning Algorithm 2 - MOGA

The MOGA is used to tune the parameters of the membership functions (MBFs) in the premise part. Without the need of linearly combining multiple attributes into a composite scalar objective function, evolutionary algorithms incorporate the concept of Pareto's optimality or modified selection schemes to evolve a family of solutions along the tradeoff surface. The manner of Parento optimality is one of the useful approaches to solve mutliobjective optimization problems. In this paper we employ the weighted sum-based optimization method.

5.1 Weighted Sum Based Optimization

In a weighted sum-based optimization, multiobjective $F=(f_1, \ldots, f_2)$ is transformed into $F_w = \sum_{i=1}^{k} w_i f_i$ so that single objective opirnization methods can be used. Preferences are used for specifying weights, With reference to Fig. 2, two fitness F_1, and F_2 can be evaluated from $e_1(t)$ and $e_2(t)$ $\forall t$. Thus, these two objective functions is transformed in an overall fitness function $F = w_1 F_1 + w_2 F_2$, where $w_1 + w_2 = 1$. For example, we may choose $w_1 = 0.3$ and $w_2 = 0.7$.

5.2 Computational Algorithms

The procedure of the MOGA algorithm is described as follows:

1. Initialization: Training data is clustered to generate 9 centroids based on which the Gaussian MBFs (mean and variance) are evaluated. 80 potential candidates P(t) are created by varying ± 20% of the MBF s.
2. Evaluate the overall fitness F. Select candidates proportional to their fitness relative to the others in *P(t)* using the Stochastic Universal Sampling technique.
3. Applying genetic operators, whole arithmetic: crossover, mutation, and adaptation with the best candidate by adding a perturbation to the relative best-fit candidate, to reproduce new candidates.
4. Combine all new candidates with the P(t) to form the new population for the next generation.
5. Repeat step 2, 3, 4 until termination condition is satisfied.

6 Simulation Example

In order to demonstrate the performance of the proposed method, we have applied the method to prediction of the sunspot data. Sunspot data is used as a benchmark for many years by researchers. Data file of the Sunspot times series is download from [9]. It consists the sunspot data from the year 1700 to 1999 (300 Samples).

Fig. 3 shows the plot of the sunspot time series. Fig. 4 shows the mean squared error of $e_2(t)$ giving MSE=0.0159 at the 30th generation. Fig. 3 show no distinct difference between the $y_2(t)$ and $x(t)$. Fig. 5 and 6 reveal some weight parameters of the FNN. A computer program developed in MATLAB software is used to implement the proposed scheme.

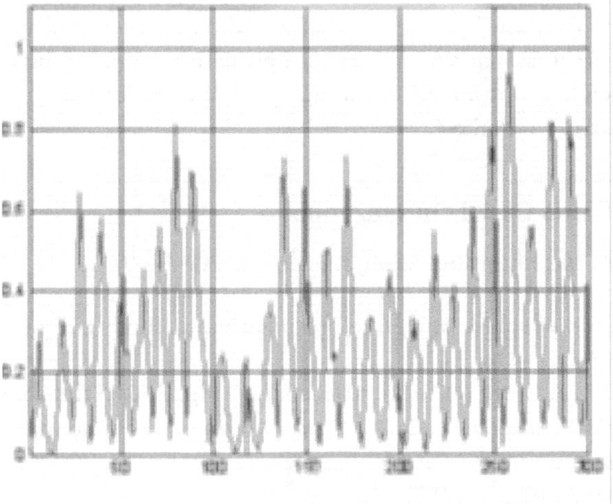

Fig 3. Normalized sunspot time series & the predictor output

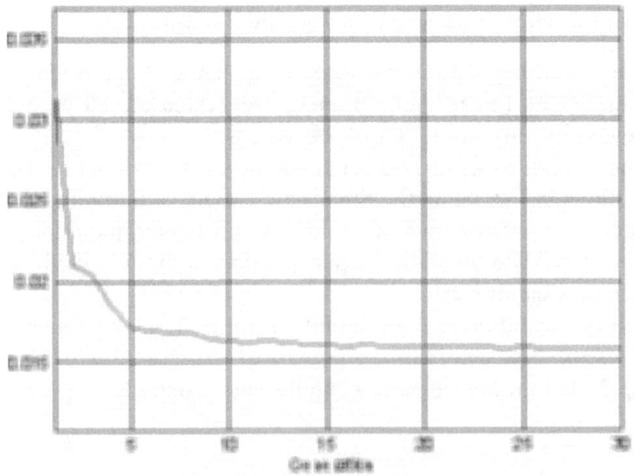

Fig. 4. MSE of $e_2(t)$

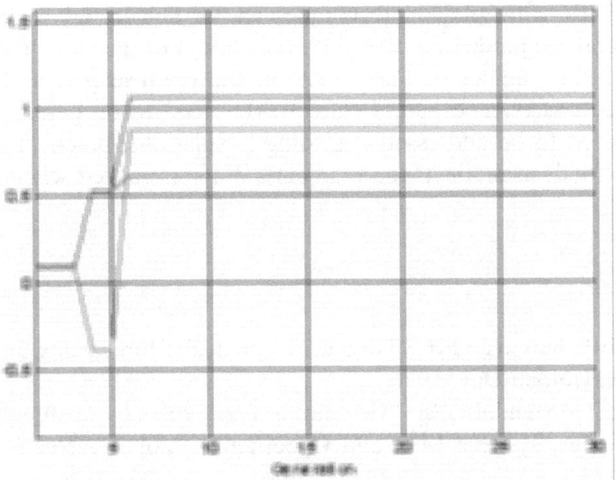

Fig. 5. Weight parameters of FNN

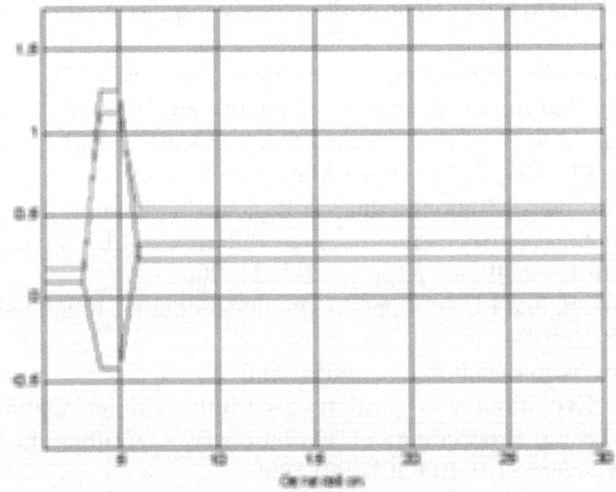

Fig. 6. Weight param8eters of FNN

7 Conclusion

This paper has presented a new approach in designing a FNN with MOGA and the modified LAF techniques.The previous section clearly demonstrate the performance of the proposed FNN for the prediction of nonlinear time series. The modified LAF has provided the fast error convergence to the training of FNN. On the other hand, MOGA has added advantage of global optimization to the FNN training based on two criteria/objectives. The results have emphasized the benefits of the fusion of fuzzy and NN technologies as well as the advantages of the fusion of the modified LAF and MOGA. This increases in transparency of the neurofuzzy approach overcomes the drawback of FNN with gradient techniques and/or GA in the conventional NNs or FNNs. In general the prediction capability (accuracy) of this system is proportional to its granularity (the number of fuzzy sets) in the premise part and the numbers of weights in the consequence part. Future works need to be conducted in this area. Many issues need to be addressed regarding simulations, practical implementations, and the further analysis on the theoretical parts of the proposed scheme.

Reference

[1] Box, GEP; Jenkins, GM: "Time series analysis: forecasting and control", Oakland CA: Holden-Day, 1976.

[2] Wang, LX,- Mendel. YM: "Generating fuzzy rules by leaming from examples", IEEE Trans. Systems Man and Cybernetics, Vol 2"'. No. 6, pp. 1414-1427, 1992.

[3] Moody, J; Darken, C: "Fast learning in networks of locally tuned processing units", Neural Computation, pp. 281-294, Vol. 1, 1989~

[4] Brown, M- Harris, C: "Neurofuzzy adaptive modelling and controi", Prentice-Hall, 1994.

[5] Jang, Roger JS: "Predicting chaotic time series with fuzzy IF-THEN rules" Proc. IEEE 2nd Int. Conf. on Fuzzy Systems, pp. 1079-1084 Vol. 2, 1993.

[6] Jang, Roger J S; Sun, C-T: "Neuro-fuzzy modelling and Control" Proc. of the IEEE, pp. 378-406, Vol 83, No 3 March 1995.

[7] Seng Kah Phooi, Zhihong Man, H.R. Wu, "Lyapunov Theory-based Radial Basis Function Networks for Adaptive Filtering", IEEE Transaction on Circuit and System 1, vol 49, no. 8, pp.1215-1221, 2002.

[8] Slotine, J-J. E. and Li, W. Applied. nonlinear control, Prentice-Hall, Englewood Cliffs, NJ, 19 9 1.

[9] http://www.astro.oma.be/SIDC/index.html

[10] Deb, K., "Evolutionary Algorithms for Mutli-Criterion Optimization in Engineering Design. Proceedings of Evolutionary Algorithms in Engineering and Computer Science (EUROGEN-99),1999.

[11] L. X. Wang and J. Mendel, "Generating fuzzy rules by learning form examples", IEEE transactions on systems, Man, Cybernetics, vol. 22, pp. 332-342, 1994.

A Unified Stochastic Architecture for Spoken Dialogue Systems

Owen Lamont and Graham Mann

Murdoch University, School of Information Technology,
Murdoch 6150, Western Australia
{olamont,G.Mann}@murdoch.edu.au

Abstract. Spoken Dialogue Systems (SDS) have evolved over the last three decades from simple single word command, speech recognition applications and technologies to the large vocabulary continuous systems used today. SDSs have long been reliant on hierarchies of stochastic architectures, in particular Hidden Markov Models (HMM), for their components and sub-components. In this paper, we examine the applications of HMMs in speech recognition including phoneme recognition, word recognition and stochastic grammars. Other applications of HMMs within SDSs are also covered including word tagging and semantic classification at the parsing level, dialogue management and strategy optimisation and stochastic reply generation. We then propose that the Hierarchical Hidden Markov Model (HHMM) of Fine, Singer and Tishby serve as replacement for many of these specialised HMMs, creating a more unified and consistent architecture. We explore the feasibility of this model within a specific information retrieval SDS and demonstrate ways HMM merging can be used with contextual and entropic clustering to dynamically generate HHMM data structures. Issues of training time and applicability of the HHMMs to natural language processing are also examined.

Keywords: AI architectures, Language understanding/generation, Machine learning

1 Introduction

Spoken Dialogue Systems (SDS) are an application of speech technology integrated with a database and other software modules and are used primarily for telephone customer service systems. Common uses of SDSs include directing customer calls, answering simple queries and 'form filling' applications such as booking tickets [1]. These SDSs involve specifying a few parameters with a finite number of discrete values that require little in-depth understanding on the part of the SDS.

Figure 1 shows the generic SDS architecture that serves as a basis for this development, which will be described in more detail in section 3. The *User* speaks to

T.D. Gedeon and L.C.C. Fung (Eds.): AI 2003, LNAI 2903, pp. 899-909, 2003.

the system via an audio stream that is processed into a sequence of words after passing through several sub-processes within the speech *Recogniser* (these processes are listed in more detail section 3). The *Parser* component discards or tags the words generated by the *Recogniser* with various syntactic, prosodic or semantic labels. In the case of a form filling application, such as an airline reservation system, semantic labels would consist of the field names of the form to be filled, eg. city names are labelled with a destination or departure tag. The *Dialogue Manager* acts upon the output from the *Parser*, filling forms or conducting queries for the user and then signalling the *Response Generator* to give an appropriate confirmation or counter query depending on how successful the *Dialogue Manager* was at interpreting the initial user input. Finally, the *Response Generator* sends a grammatical sequence of words to the *synthesizer* that speaks the response to the *User*. This process repeats until the end of the user's interaction with the system.

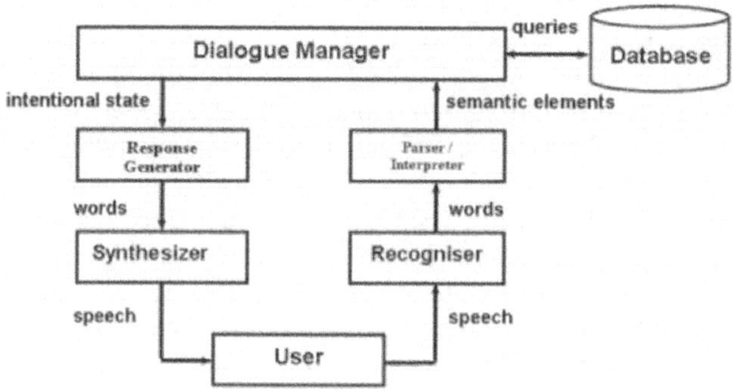

Fig. 1. Generic Spoken Dialogue System Arch itecture

2 Applications of Hidden Markov Models in Spoken Dialogue Systems

To give a brief overview, Hidden Markov Models (HMM) are finite state automata that have two levels of representation, a set of hidden states (vertices) and a set of observable symbols. Weighted links represent events in probability distributions and various algorithms exist to adjust and search the weightings to make predictions about observation sequence probabilities or modify HMM weightings to better model training sequences [2]. A simple left-right HMM is shown in Figure 2.

HMMs are frequently used to find a higher level of structure in data, to 'step up' one level of abstraction. Many such 'step ups' are involved in SDSs. The translation of sounds and their derived feature vectors into phonemes, the translation of phonemes into words, words into syntactic and prosodic tokens, syntactic and prosodic tokens into semantic categories and semantic tokens into actions have all been modelled in this way.

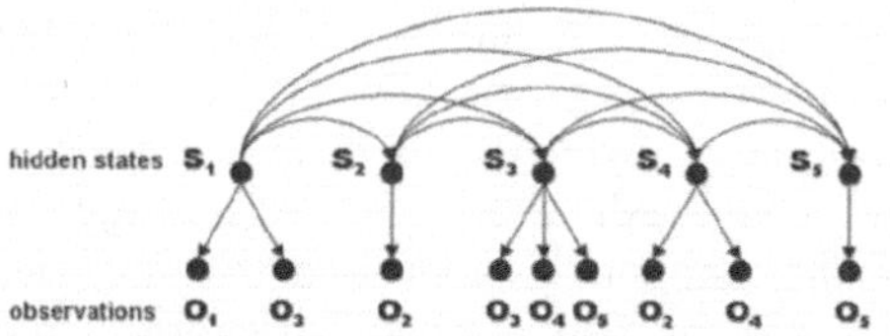

Fig. 2. Hidden Markov Model (weight values not shown for simplicity)

A simplifying assumption of SDSs is that the vocabulary to be used is defined in advance. A HMM is constructed for each word in the vocabulary, using the phonemes as observable symbols. These models are then used to determine the probability of each word given a particular sound. HMMs are also trained contextually with a corpus of training texts which can then be used to give the probability of a word occurring within a certain context in what's referred to as a language model or stochastic grammar. Superior models have been found but remain impractical for real time applications [3]. However, improving computer hardware is helping to alleviate this problem.

HMMs can also be used to classify words into prosodic, syntactic and semantic categories. Stochastic clustering has been used in models to make superior probability estimates as clustered words can better model relationships and sentences that have never been observed in the training corpus [3]. Speech tagging using Markovian models trained with large hand annotated corpuses has shown to be very successful when given new word sequences to annotate [4], [5]. Hidden Understanding Models (HUM) which are closely related to HMMs, represent semantic categories and accept processed word sequences as input and can be trained from annotated data to generate the most probable semantic classifications for each word in the input [6].

The use of stochastic models and more specifically Markov Decision Processes (MDP) has been applied to optimising dialogue strategies [3], [7], [8]. It applies expectation maximisation algorithms with a reward function to select the actions which lead to the most probable reward. Rewards are given for minimising the length of and successfully completing the dialogue.

The fact that alphabets, lexicons, syntax and semantics are continually evolving makes static or manually edited processing rules impractical for wider applications beyond the rigidly controlled template based SDSs used at present. Some elements of language are obviously more stable than others; fundamental elements such as phonemes and alphabets remain fairly constant while more abstract elements such as expressive phrases are ephemeral. The fact that dictionary editors are still in business attests to how dynamic language is, as new idioms and jargon words are incorporated officially into language each year. When analysing spoken dialogue, as opposed to written language, rules of grammar become often very loose if present at all [9], [10].

3 A Unified Model

An intuitive question to pose then is, given the abundant applications of specialised HMMs in SDSs, could a single HMM fulfill all or most of the functions presently performed by specialised HMMs? A unified model might have several advantages over individual HMMs. It would organise hierarchical structures in a systematic way rather than the arbitrary and awkward levels of abstractions used by the specialized HMMs at present. This would reduce the types of training data to a single level of abstraction and reduce the amount of supervised training needed to create a useful model. It would avoid human defined linguistic structures and categories, which limit the automation of current SDSs. The combined HMM could also create a more computationally feasible and efficient system.

3.1 Hierarchical Hidden Markov Model

But how can different HMMs working at different levels of abstraction be merged into a unified whole? Given the loose hierarchical organisation of HMMs in current SDSs, the Hierarchical Hidden Markov Model (HHMM) as conceived by Fine et al [11] is a possible candidate for a unified model. Fine and his colleagues propose a nested architecture where each state in a HMM can itself be HMM and so on for as many levels as required. Rather than just a single symbol, higher-level states can produce sub-sequences of observable symbols and therefore can cluster data at higher forms of abstraction. The HHMM has a number of advantages over conventional HMMs. It can model dependencies and interactions between more dimensions than would computationally tractable with a single layer HMM. It can cluster data into more levels of abstraction and model more complex relationships.

Many problems involving time series analysis have applications of HHMMs and their derivatives. These areas include: language modelling [11], [12], optical character recognition [11], task analysis [13], robot navigation [14], video analysis [15], gene identification [16], [12] and information extraction [12]. Fine and colleagues demonstrated that HHMMs can find intuitive hierarchical structures in time series data and this remains an active area of research [17] so there is no doubt HHMMs can model natural language effectively. However, to create a model of the scale we're proposing would require an architecture far larger than would be computational tractable to optimise with the current training algorithms and far too complex to construct by hand. We discuss methods for tackling these issues later in the paper.

3.2 Applying Hierarchical Hidden Markov Models to a Spoken Dialogue System

Figure 3 shows how a HHMM could conceivably replace the existing *dialogue manager*, *parser* and *response generator* of Figure 1, thereby eliminating much of the redundant functionality of each. The HHMM would be trained with query – answer pairs in the form of sequences of tokens where each token represents either a word or an action to be executed by the database management system. The stream of words from the speech recogniser would be converted into these unique tokens to be analysed by the HHMM. The most probable path of observations following the input observations could be calculated using the Viterbi algorithm to find the most probable hidden state where the input ends and the feed forward algorithm to generate the most probable remainder of the observation sequence. This architecture presupposes natural language processing can be treated as a pattern completion problem. The generated sequence of action and word tokens (the action/word distinction is meaningless to the HHMM itself) would be processed by hand coded script that extracts these action tokens converting them into database queries using the adjacent word tokens as parameters. The query results would be displayed on screen with the SDS response consisting of the remaining word tokens.

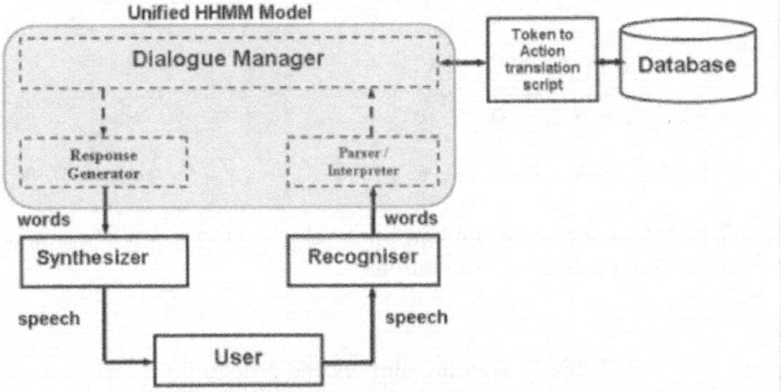

Fig. 3. Position of a Unified HHMM within a Spoken Dialogue System

This is an ambitious application for the HHMM, if it proved incapable of completely replacing those three components; it could still be used as a superior stochastic optimiser for the state transitions between the rule-based components of the *Dialogue Manager* itself. Even in this limited role, it would be a far more powerful and sophisticated solution than the crude stochastically optimised Dialogue Management systems implemented at present. If the HHMM does prove capable of replacing the *Dialogue Manager*, *Parser* and *Response Generator*, then new lower levels of abstraction can easily be added to the architecture to replace other components such as the speech recogniser and word spellings for written input.

3.3 Hierarchical Hidden Markov Model Architecture

Figure 4 shows a simple HHMM topology. Only some states produce observable symbols (although it is possible to attach observation symbol links to any state if need be, not just the bottom level as shown here). HHMMs consist of two types of states, the *internal states* that connect to other states without producing an observation and the *production states* that generate *observations*.

3.4 Hierarchical Hidden Markov Model Construction and Generalisation for Natural Language Modelling

A HHMM can act as a perfect auto associative memory recording every training sequence as a new state path, with each state producing one symbol with the probability of 1, as shown in Figure 5. We demonstrate the dynamic construction of a HHMM by showing a topology for predicting the following three phrases:

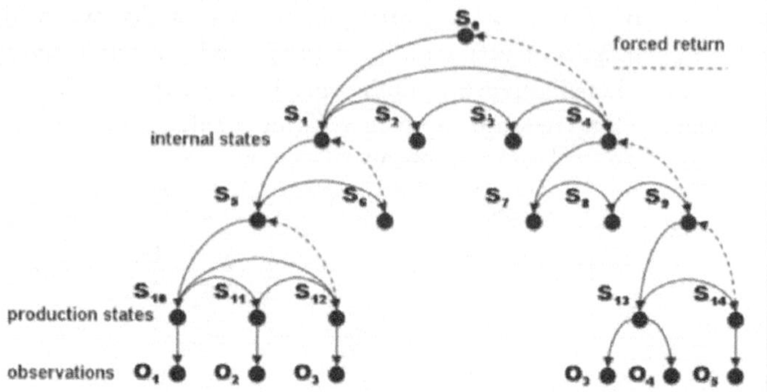

Fig. 4. Simple HHMM architecture showing internal and production states and how internal states can produce sub-sequences of observations

Table 1. Training phrases and probabilities

ID	Phrase	Probability
1	The cat sat on the mat	1/3
2	The cat sat on the hat	1/3
3	The cat ate the rat	1/3
	Total:	1

The construction process follows the details of HMM merging specified by Stolcke [18]. Figure 5 shows the initial model after all phrases have been modelled with an individual state path with a probability of 1 (excluding the first link), each starting and ending at the origin state, S_0. The symbol links have not been shown for simplicity sake (they all have a value of 1.0). In this model, phrases 1, 2 and 3 has a probability of 1/3 each, totalling 1. All other phrases, excluding sub-phrases of the three above, would have a probability of 0 under this model.

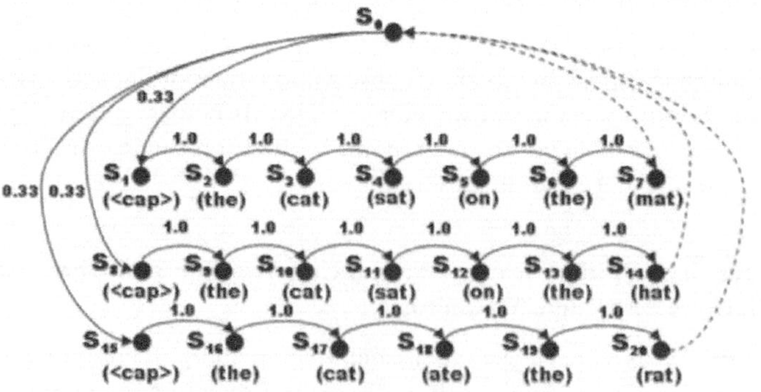

Fig. 5. HHMM of three phrases

Figure 6a shows the model with redundant states merged. It has a much simpler topology while retaining the same probabilistic properties of the previous topology, i.e. the values shown in table 1 are still true. Notice that *state 7*, now has two symbols (unlike the other states that have one each), each with a probability of 1/2.

In Figure 6b, the redundant 'the' producing *states 6 and 9* from Figure 6a have been merged. This has created three new possible state sequences and increased the generalisation of the model. The probabilities of phrases 1, 2 and 3, have changed from 1/3 each as shown in Table 1 and three new phrases are now possible with this model as shown in Table 2.

Table 2. Phrase Probabilities after Generalisation

	ID	**Phrase**	**Probability**
Training Phrases	1	The cat sat on the mat	2/9
	2	The cat sat on the hat	2/9
	3	The cat ate the rat	1/9
Generalised Phrases	4	The cat sat on the rat	2/9
	5	The cat ate the mat	1/9
	6	The cat ate the hat	1/9
	Total:		1

The training phrase probabilities in the Table 1 sum to exactly 1, showing that there is no generalisation. The amount by which this sum decreases below 1 gives a good indication of how much the model has generalised, as seen in Table 2.

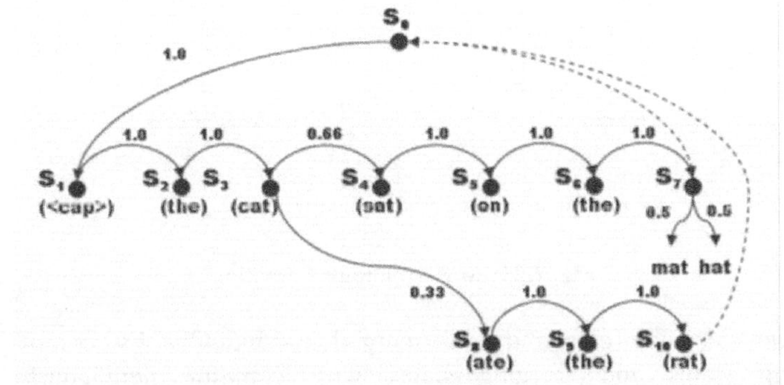

Fig. 6a. HHMM of three phrases after state merging

By merging states systematically and monitoring the change in the sum of the training set probabilities, a selected level of generalisation can be achieved. The topology examples in Figure's 5, 6a and 6b are still essentially single layer HMMs, *state 0* exists just as a start and end point. Merging states into a multi-level or pre-merged topology requires methods that are more complex.

An automated method of constructing and modifying topologies is essential to the creation of such a unified system. While this is a recognised HHMM problem [11], [17], few have tackled this. However, various data clustering methods from other induction systems would be applicable. The Baum-Welch algorithm used to train the weightings of HHMM can be used to modify the topology by pruning out connections with zero or near zero weight values [18]. State connections with significantly higher weightings, compared to their neighbouring connections, are indicative of higher structure, thus it may be appropriate to insert higher order states to represent both these lower states (shown in Figure 7).

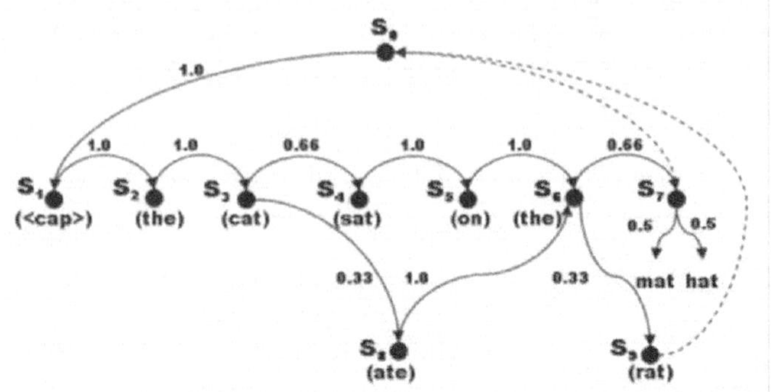

Fig. 6b. HHMM of three phrases after state merging with some generalisation

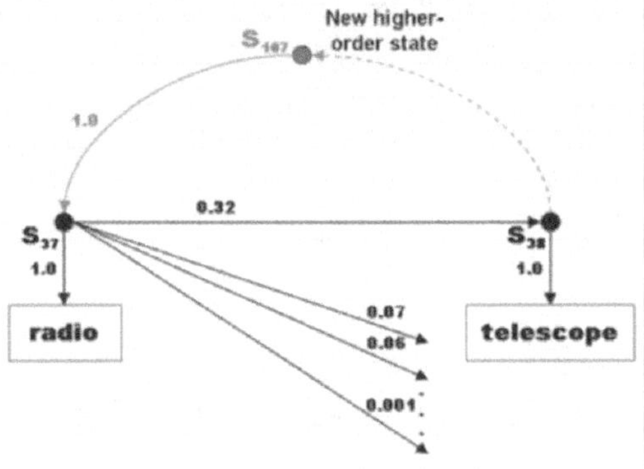

Fig. 7. HHMM Topology Alteration

States can also be clustered and merged contextually by comparing their neighbouring states and merging states with common neighbours. Initially, generalisation and topology modification would be kept to a minimum while data is being collected. State pruning and merging would be used once a significant body of data is collected, to prevent spurious relationships in inadequate training data being modelled.

4 Feasibility

Two main questions need to be considered for assessing the feasibility of this architecture. Firstly, are HHMMs able to perform as well as the specialised HMMs and conceptual systems currently used in SDSs and how would this performance be evaluated? Secondly, is the HHMM computationally efficient? We address these questions in sections 4.1 and 4.2 respectively.

4.1 Hierarchical Hidden Markov Model Applicability

Previous experiments with HHMMs have shown their ability to dynamically detect logical and non-redundant structures in natural language text as well as detect long-range interactions such as punctuation marks [11]. HHMM dynamic topology modification would be essential to create a HHMM of useful complexity, the methods described in 3.4 give some insights in to how this could be achieved and there are a multitude of other unsupervised clustering and abstraction techniques in existence that could be employed.

Methods by which the Unified HHMM model can be evaluated would have to be carefully considered. Holistic evaluations of the systems via user satisfaction especially if system use was voluntary would be paramount. However, user satisfaction is a highly abstract and imprecise measure. Simpler constructs correlated to user satisfaction such as feedback times could be measured as part of comprehensive testing plan [19]. Other issues such as the volume of data required to train the model and the overall development time of a system using the Unified HHMM. Lastly information theoretic measures such as perplexity can be used to test the quality of the HHMM as a predictor.

4.2 Hierarchical Hidden Markov Model Efficiency

HMMs have complex training algorithms and HHMMs even more so. The inference algorithm for a HHMM is of the order $O(T^3Q^D)$ where T is the length of the sequence, Q is the maximum number of states at each level of the hierarchy and D is the depth of the hierarchy [20]. Some research has been done in implementing HHMMs as Dynamic Bayesian Networks (DBN) with a inference algorithm of $O(TQ^{2D})$ worst case time complexity [20]. The amount of Baum-Welch training can be reduced by careful merging of individual HMM state paths into the HHMM, when merged the weightings will be optimal without any retraining (provided no generalisation is tolerated). Systematic merging of states followed by testing the probability of the training sequence would be of the order $O(TQ^4D^3)$. While this time complexity appears no better than Baum-Welch retraining, it reduces QD after each successful merge and only requires one iteration, unlike Baum-Welch training.

While these time complexity formulas appear daunting with exponential run times or worse, it should be remembered that the run times of an average topology will be up to several orders of magnitude better since the formulas assume the worst possible topology interconnections. Many additional shortcuts exist that can dramatically improve runtime with minimal performance degradation.

The generalised weight re-estimation methods proposed by Fine and colleagues involve recursive brute force state searches when calculating sub-sequence probability which leads to products of Q^D time complexity (since the collapsed state-space of an HHMM is up to Q^D in size) [15]. This approach is both contradictory and crippling in terms of computational efficiency given the hierarchical paradigm being used. By asserting that all sequences must begin at the top state of the HHMM hierarchy (a logical and intuitive preposition given this architecture) and by relaxing the constraint that a higher state transitions cannot occur until after it's sub-states have expressed their symbols, massive improvements in training time complexity can be achieved whilst still covering the state space efficiently and eliminating the need for the recursive generalised forward, backward, Viterbi and Baum Welch algorithms. With caching of intermediate variables a HHMM could be implemented with $O(T^2Q^3D^2)$ worst case time complexity using only slightly modified HMM algorithms. It remains to be seen if this assumption limits performance for practical purposes; which we are currently testing experimentally.

5 Conclusion

HHMMs have potential for SDSs because they have all the advantages of the conventional and widely applied HMMs along with the ability to model complex and multi-scaled relationships in time series data. Given that hierarchies of HMMs are already used in almost every aspect of SDSs, the amalgamation of the specialised HMMs seems highly appropriate. The processor intensive nature of HHMMs and the need for a manually edited topology have slowed their uses in practical applications. However, with carefully chosen clustering, merging and training algorithms, their use in a natural language interfaces should now be feasible. We are currently testing a HHMM model and will be creating an information retrieval SDS, called the Speech Librarian for intelligent document retrieval based on the Unified HHMM SDS architecture.

References

[1] Young, S. J.: Probabilistic methods in spoken dialogue systems, Philosophical Transactions of the Royal Society (Series A), Vol. 358, no 1769, (2000) 1389–1402 (1)

[2] Rabiner L. R.: A Tutorial on Hidden Markov Models and Selected Applications in Speech Recognition, Proceedings of IEEE, Vol 77, no. 2 (1989)

[3] Goodman J. T.: A bit of Progress in Language Modelling, Computer Speech and Language, no. 15, (2001) 403-434

[4] Brill E.: Unsupervised Learning of Disambiguation Rules for Part of Speech Tagging, Proceedings of 3rd Workshop on Very Large Corpora, MIT (1995)

[5] Dermatas, E., Kokkinakas, G., Automatic Stochastic Tagging of Natural Language Texts, Computational Linguistics Vol. 21, no. 2, (1995) 137-163

[6] Miller, S. Bobrow, R., Ingria, R., Schwartz R.: Hidden Understanding Models Of Natural Language, 32nd Annual Meeting of the Association for Computational Linguistics (1994)

[7] Levin, E., Pieraccini, R.: A stochastic model of computer-human interaction for learning dialogue strategies, Proceedings of 5th European Conference on Speech Communication and Technology, (1997) 1883-1886

[8] Goddeau D., Pineau J.: Fast Reinforcement Learning of Dialog Strategies, IEEE Int. Conf. on Acoustics, Speech, and Signal Processing. (2000)

[9] Mann, G.: Control of a Navigating Agent by Natural Language [PhD Thesis], University of New South Wales (1996)

[10] Glass, J.: Challenges for Spoken Dialogue Systems, Proceedings of the 1999 IEEE ASRU Workshop (1999)
[11] Fine, S., Singer, Y., Tishby N.: The Hierarchical Hidden Markov Model: Analysis and Applications, Machine Learning, Vol. 32, no. 1, (1998) 41-62
[12] Skounakis, M., Craven, M., Ray, S.: Hierarchical Hidden Markov Models for Information Extraction, 18th International Joint Conference on Artificial Intelligence (2003)
[13] Lühr, S., Bui H. H., Venkatesh, S., West, G.A.W.: Recognition of Human Activity through Hierarchical Stochastic Learning, First IEEE International Conference on Pervasive Computing and Communications (2003) 416
[14] Theocharous, G., Mahadevan S.: Approximate Planning with Hierarchical Partially Observable Markov Decision Process for Robot Navigation, IEEE International Conference on Robotics and Automation (2002)
[15] Xie, L., Chang, S. F., Divakaran, A., Sun, H.: Learning Hierarchical Hidden Markov Models for Video Structure Discovery, Tech. Rep. 2002-006, ADVENT Group, Columbia Univ., http://www.ee.columbia.edu/dvmm/ (2002)
[16] Hu, M., Ingram, C., Sirski, M., Pal, C., Swamy, S., Patten, C.: A Hierarchical HMM Implementation for Veterbrate Gene Slice Prediction, Technical report, Department of Computer Science, University of Waterloo (2000)
[17] Barto, A. G., Mahadevan, S.: Recent Advances in Hierarchical Reinforcement Learning, Special Issue on Reinforcement Learning, Discrete Event Systems, Vol 13, (2003) 41-77
[18] Stolcke, A.: Bayesian Learning of Probabilistic Language Models [PhD Thesis], University of Berkeley (1994)
[19] Litman, D. J. and Pan, S.: Empirically Evaluating an Adaptable Spoken Dialogue System, In Proceedings of the 7th International Conference on User Modelling, pp 55-64
[20] Murphy, K., Paskin M.: Linear Time Inference in Hierarchical HMMs, Proceedings of Neural Information Processing Systems (2001)

Evaluating Corpora for Named Entity Recognition Using Character-Level Features

Casey Whitelaw and Jon Patrick

Sydney Language Technology Research Group
Capital Markets Co-operative Research Centre
University of Sydney
{casey,jonpat}@it.usyd.edu.au

Abstract. We present a new collection of training corpora for evaluation of language-independent named entity recognition systems. For the five languages included in this initial release, Basque, Dutch, English, Korean, and Spanish, we provide an analysis of the relative difficulty of the NER task for both the language in general, and as a supervised task using these corpora. We construct three strongly language-independent systems, each using only orthographic features, and compare their performance on both seen and unseen data. We achieve improved results through combining these classifiers, showing that ensemble approaches are suitable when dealing with language-independent problems.

Keywords: computational linguistics, machine learning

1 Introduction

Named entity recognition is an important subtask of Information Extraction. It involves the recognition of named entity (NE) phrases, and usually the classification of these NEs into particular categories. Language-independent named entity recognition involves the construction of a single system that learns, through its application to a training corpus for a target language, how to identify NEs in that language. Supervised learners have been used on Chinese, English, and Spanish corpora as part of the Multilingual Entity Task (MET), and on Spanish and Dutch as the shared task of CoNLL 2002 [9]. Most recently, English and German corpora were the subject of comparison at the shared task of CoNLL-03 [10]. Unsupervised language-independent systems have been tested on small corpora in Romanian, English, Greek, Turkish, and Hindi [4]. While language-specific systems have reached high levels of performance in English, language-independent systems have much room for improvement, especially in broad domains.

For this paper, we have compiled corpora for English, Basque, Spanish, Dutch, and Korean. The Basque and Korean corpora have not been used previously, and this collection forms the largest set of cross-language NE-tagged corpora currently available. The corpora differ greatly in size and domain. One of the challenges of evaluating language-independent systems is to understand how

T.D. Gedeon and L.C.C. Fung (Eds.): AI 2003, LNAI 2903, pp. 910–921, 2003.

the specific corpora used effect the observed performance. We present a range of statistics to aid in this analysis.

We describe three different language-independent classification strategies that each use probabilistic representations. These include word-level classification through probabilistic learned lists, and character-level classification through letter n-grams and orthographic tries. These approaches are tested on the recognition of NEs, and we present details of both their individual performance and together as an ensemble system. Using only simple methods and less than thirty features, we obtain f-scores ranging from 62.2% in Korean to 86.9% in Spanish.

2 Corpora

The collected corpora were standardised to use the UNICODE character encoding and the IOB2 method of annotation [11]. Beyond the removal of extraneous information such as part-of-speech tags, no cleaning was performed.

Basque: The Basque data was provided by the IXA group at the University of the Basque Country, Donostia. It consists of articles about economics, taken from the Basque-language newspaper Egunkaria from January and February in 1999.

Dutch: The Dutch data was originally used in CoNLL 2002, where a full description is provided. The corpus consists of four editions of the Belgian newspaper, "De Morgen" from the year 2000. The annotation was undertaken by the University of Antwerp in Belgium.

English: The English data was the dataset used in MUC-7 NER task. The corpus consists of newswire articles from the New York Times News Service from 1996. Articles are solely on the specific topics of aircrashes and missile launches. The distribution is controlled by the Linguistic Data Consortium.

Korean: The Korean data was provided by the KORTERM group at the Korea Advanced Institute of Science and Technology (KAIST). It consists of a portion of their POS-tagged corpus that has had an NE annotation added manually, following the IREX tagging standard. The data consists of both newspaper articles and general texts, including academic and scientific texts.

Spanish: The Spanish data was also originally used in CoNLL 2002. The corpus is taken from a collection of newswire articles from the year 2000, from the Spanish EFE News Agency, and was annotated by University of Catalonia and the University of Barcelona.

Table 1. Dataset statistics

	characters		words		NEs		
	train	test	train	test	train	test	repeated
Basque	408945	105271	68046	17855	5353	1450	336
Dutch	1120542	329255	240692	68994	15960	3941	899
English	849930	293068	190791	63941	10786	4097	761
Korean	109375	40168	72029	25914	1936	872	167
Spanish	1417766	228888	317638	51533	23147	3558	957

Table 2. Character-level statistics

	total	O only	NE only
Basque	71	8	2
Dutch	100	7	3
English	82	11	0
Korean	1088	589	90
Spanish	91	6	5

3 Data Analysis

While the datasets for each language share their annotation for named entities, they may differ greatly in their suitability to evaluating named entity recognition (NER) and classification (NEC) tasks. A corpus must be rich enough to make apparent the differences between classification techniques, both in terms of raw performance, and the particular strengths and weaknesses of each system. While it may be possible to produce difficulty indices for a single language[3], we have chosen a range of statistics that will allow us to make hypotheses concerning the expected performance of different types of NER systems.

Dataset Statistics. Table 1 shows the sizes of training and test sets in terms of characters, words, and NEs. The training set for Spanish has over four times more words than the Basque corpus, and proportionately more NEs. The Korean dataset is smallest in terms of both number of characters and number of NEs. We would expect lower performance from the smaller corpora. In practice, this could be ameliorated by using semi-supervised techniques such as co-training in conjunction with larger unannotated corpora.

The final column in Table 1 shows the frequency of NEs in the test set that occur more than once. Many NER systems use global information such as other occurences of the same token as part of their classification strategy [2]. All datasets have repeated terms, so the benefit of such an approach could be tested using these corpora. The proportion of repeated terms places an upper limit on the observable performance gain for each corpora.

Character-Level Statistics. Table 2 shows the number of unique characters that are present in the corpus. All alphabetic languages have less than one hundred unique characters, including punctuation. This is starkly different to Korean, which has over one thousand characters represented in the corpus, a small sample of the language's combined phonetic *hangul* and logographic *hanja* alphabets. Table 4 shows that the test set contains 167 more, previously unseen, characters. Korean and other non-alphabetic languages pose a challenge for naive character-level feature representations. If a binary feature were assigned to the presence of each possible character in a word, each token would have a very large and sparse representation, which would deal poorly with unseen characters. This suggests that a direct representation of orthography may not be a suitable basis for performing machine learning.

All corpora contain characters that occur uniquely within one category. An expected example of this is punctuation characters, which would usually not occur within NEs, such as '?' or '!'. Even in alphabetic languages, there are characters that occur exclusively within NEs, both specific punctuation such as '&', and characters that are from foreign languages, such as 'X' (Dutch) and 'Q' (Basque). Korean contains many characters that are uniquely used in NEs. We will see that these differences can be exploited through techniques such as character n-grams.

Word- and Phrase-Level Statistics. Table 3 shows the average word length for non-entities and NEs. NEs have longer average length across all languages. Korean is set apart by its short average word length and by having almost no NEs with more than one token. Basque makes heavy use of suffixes which gives it longer average word length across all categories. The final column shows the percentage of tokens which appear both inside and outside of NEs; this ambiguity will be problematic for word-level techniques that rely on gazetteers or learned lists. From this we would expect English to perform worse than Dutch on seen words unless contextual clues are also used.

Unseen Data. A good test corpus should contain both NEs that are present in the training set and some that are not. Classifiers that work well within a small domain, where the set of NEs is limited, may be less suited to a domain such as newspaper articles, where new NEs arise constantly. Table 4 shows the number of previously unseen words, NEs, and characters in the test set. All corpora have a significant percentage of unseen NEs, so should give reliable performance estimates.

Recent work [5] has corroborated our belief that performance on unseen words is the major contributing factor to the differences in current NER systems. It is important that the unseen term performance is examined explicitly, rather than relying on holistic measures of precision and recall. In addition to overall performance, we report seen and unseen performance for all systems.

Table 3. Phrase-level statistics

	word length		NE length		O / NE
	O	NE	words	chars	ambiguity
Basque	5.8	7.6	1.4	10.6	8.2%
Dutch	4.5	6.4	1.4	9.2	3.9%
English	4.3	6.0	1.6	9.5	10.7%
Korean	1.5	3.0	1.1	3.2	11.1%
Spanish	4.2	6.0	1.7	10.5	7.0%

Table 4. Unseen data in test sets

	O	NE	chars
Basque	13.4%	43.0%	0
Dutch	6.6%	13.0%	2
English	8.3%	48.8%	2
Korean	9.8%	60.4%	167
Spanish	4.0%	29.3%	0

4 Language Independent Classification Techniques

When developing an NER system for a single language, one can use explicit linguistic knowledge in choosing suitable attributes, and in the choice of a feature representation. In this way, even simple features, such as capitalisation, can be very powerful.

When an NER system is language-independent, these features must be eschewed in favour of a representation that is equally applicable to all languages, regardless of its character set, and can *learn* to recognise important features, rather than having them provided explicitly. Most language-independent NER systems include features based explicitly around phenomena such as capitalisation, and use many thousands of very sparse features. We demonstrate three representation schemes that are all strongly language-independent, in that they do not rely on any specific phenomena, and further that the representation is identical across all languages. In addition, each uses only a small number of probabilistic attributes. Each approach aims to target a single linguistic phenomena through careful feature representation, rather than being made up of a wide variety of unrelated but potentially useful features.

We use decision trees for all classification tasks. While the use of other classifiers, such as decision graphs or SVMs, would be expected to improve performance, the relative performance of the various approaches would not be expected to vary significantly. The results should be taken as illustrative of each approach's strengths and weaknesses.

4.1 Baseline Word-Level Classification

To evaluate the performance of more advanced classifiers, it is necessary to establish a baseline using a simple classification system. Here, the baseline classifier tags a phrase as an NE if it appears as such in the training set. This will obviously perform very poorly on unseen tokens in the dataset, and even on seen tokens used in new contexts. Its performance is limited by the phrase-level ambiguity of the dataset (see Table 3) and the amount of unseen data in the test set. We consider this to be a gazetteer-like approach and expect it to give higher precision than recall.

4.2 Word-Level Classification Using Probabilistic Learned Lists

Many NER systems have used gazetteers, compiled lists of known entities, to aid the identification of named entities. While gazetteers are useful, they are not strictly language independent, as they rely on languages sharing (at least) their writing system. The cost of acquisition, construction, and maintenance of gazetteers remains high, and their usefulness in broad domains such as newswires remains doubtful[7].

An alternative to using pre-built gazetteers is to use *learned lists*, based on the training set. For each token w and each category c, one can use the training data to estimate

$$P(C(W_{i+offset}) = c \mid W_i = w)$$

for offsets of $-1, 0, 1$. The zero offset captures the probability of each word occurring in each category, much like a traditional gazetteer. An offset of -1 (or 1) captures the probability of a word occurring before (after) each category. This contextual information is not available in a usual gazetteer.

To use these probabilistic learned lists to make feature sets for a machine learner, each token can be represented by a set of real-value attributes $P(C(w_i) = c \mid W_{i+offset})$, for each category c and each offset. Unseen tokens are given a flat distribution. Using these three attributes per category results in nine attributes per token. Attributes for the preceding and following tokens were included using data windowing, to give a total of 27 real-valued attributes per token.

When classifying a previously seen token, performance would be expected to be good, limited for the most part by the ambiguity of the corpus (see Table 3). In classifying unseen tokens, classification depends on contextual evidence, the reliability of which will vary greatly from language to language. Since NEs are a specific type of noun phrase, this contextual evidence may also lead to over generalisation and the false identification of nouns as NEs.

4.3 Orthographic Classification Using Letter N-grams

Word-level classification suffers from the sparseness problem, especially in supervised systems. While there are a large and dynamic number of words, the character set for a language is always much more constrained (see Table 2). By moving to letter-level features, data sparsity can be largely overcome, but the problem of finding effective language independent representations remains.

Here we consider letter sequences of fixed length (*letter n-grams*) as possible indicators of class membership. For each n-gram g that occurs in the dataset, we estimate $P(c \mid g)$, the probability of it occurring in a word of each category c.

For a word composed of m n-grams $g_1, g_2, \ldots, g_m$, using full distribution information for all n-grams is an unwieldy representation, resulting in $m * c$ attributes. Instead, we produce aggregate statistics across all n-grams in the word:

$MAX(c)$	highest $P(c \mid g_i)$
$MIN(c)$	lowest $P(c \mid g_i)$
$AVE(c)$	$\sum_{i=1}^{m} \frac{P(c \mid g_i)}{m}$

Not only is this representation more compact, the number of attributes is independent of language, word length and n-gram size. Additionally, these aggregate statistics should give salient indicators of class membership. While $AVE(c)$ looks at the whole word, $MAX(c)$ and $MIN(c)$ highlight individual n-grams that may give strong (counter-)indications. These may be morphological constructs such as affixes or infixes with a strong affinity to a class (NE or non-NE), or orthographic artifacts such as capitalisation or foreign letters.

As with probabilistic learned lists, the three attributes per category were windowed to produce 27 real-valued attributes per token.

N-grams of different sizes should be sensitive to different phenomena. Unigrams are enough to capture capitalisation or punctuation, but cannot detect longer morphological constructs such as affixes. Moreover, since the importance of these phenomena will vary for each language, there is no reason to expect n-grams of any one length to perform consistently well. For this evaluation, n-grams of lengths 1, 2, and 3 were tested on all corpora. It is interesting to note that the function of n-grams changes as their length approaches average word length, as with 2- or 3-grams in Korean, when n-grams may be representing entire words.

4.4 Contextual Classification Using Orthographic Tries

Tries are an efficient data structure for capturing statistical differences between strings in different categories. In an orthographic trie, a path from the root through n nodes represents a string $a_1a_2 \ldots a_n$. The n-th node in the path stores the occurrences (frequency) of the string $a_1a_2 \ldots a_n$ in each category. These frequencies can be used to calculate probability estimates $P(c \mid a_1a_2 \ldots a_n)$ for each category c. Tries have previously been used in both supervised [8] and unsupervised [4] named entity recognition.

Given a string $a_1a_2 \ldots a_n$ and a category c an orthographic trie yields a set of relative probabilities $P(c \mid a_1)$, $P(c \mid a_1a_2)$, ..., $P(c \mid a_1a_2 \ldots a_n)$. The probability that a string indicates a particular class is estimated along the whole trie path, which helps to smooth scores for rare strings, and functions as a progressive backoff model from unseen words. The contribution of each level in the trie is governed by a linear weighting function, which may be fixed or be obtained experimentally using parameter optimisation [12].

A trie may be rooted at the beginning or end of a word, capturing commonality of prefixes or suffixes respectively. We build a forward and backward trie for the current word, and also for the words on each side, capturing contextual information in much the same way as probabilistic learned lists. The scores from each trie are obtained independently, then used as attributes for a machine learner. Using six tries, each producing a score for each category, gives 18 real-valued attributes.

Tries are highly language independent. They make no assumptions about character set, or the relative importance of different parts of a word or its context. Tries should deal well with unseen words, using their progressive back-off model and probability smoothing to give good estimates on the available data. Combining word-internal and contextual tries is expected to give a robust indicator of class membership. The probabilistic classifications generated by tries have also been used within a Hidden Markov Model framework, resulting in improvements in phrase modelling [13].

4.5 Ensemble Classification

It is not reasonable to expect any one of the above approaches to consistently outperform other methods for all languages. It is more likely that each method will have its own characteristic strengths and weaknesses. Word-level classification might be expected to have higher precision but lower recall, especially for unseen tokens. Letter n-grams and tries may have high recall but lower precision, since their intent is to generalise orthographic phenomena. These strengths can be harnessed, and weaknesses overcome, by combining the classifiers into a single system. To test the performance of an ensemble, we used a simple voting scheme, in which the classification with the highest vote was assigned.

5 Results

When examining the performance of these systems, it does not suffice to look at performance on the corpus as a whole. We provide separate results for those tokens that occur in the training set ("seen tokens"), and those that do not ("unseen tokens"), and include the ratio of unseen performance to seen performance. All results are presented on page 920. $F_{\beta=1}$ (f-score) is the harmonic mean of precision and recall.

Baseline Results (Table 5). The results for all European languages follow the expected pattern of higher precision and lower recall. In the baseline system, precision is harmed by words that occur as both NE and non-entity. There is a strong correlation to the phrase-level ambiguity reported in Table 3. Such ambiguity can be produced by the natural ambiguity of the language, such as the word "American" in the English dataset, which occurs as a valid NE through ellipsis in the phrase "on American" (Airlines). It can also be the product of tagging error, which is illustrated by the Spanish results. The Spanish training corpus contains a single example of the preposition "de" occurring as an NE phrase by itself. Due to the extremely high frequency of the word in the test set, the removal of this one erroneous entry from the corpus produced a forty percent gain in precision, bringing Spanish in line with other European languages. This pattern of precision and recall is broken in the Korean dataset. Both figures are low, but recall is nearly double precision.

Table 5. Baseline results

	precision	recall	$F_{\beta=1}$
Basque	86.41%	58.34%	**69.66**
Dutch	82.37%	48.95%	**61.40**
English	75.90%	48.11%	**58.89**
Korean	22.36%	36.47%	**27.72**
Spanish (orig)	35.51%	63.38%	**45.51**
Spanish (corr)	75.44%	63.38%	**68.89**

Table 6. Probabilistic learned list results

	all tokens			seen tokens			unseen tokens			
	precision	recall	$F_{\beta=1}$	precision	recall	$F_{\beta=1}$	precision	recall	$F_{\beta=1}$	seen%
Basque	83.32%	59.59%	**69.48**	86.35%	83.87%	**85.09**	14.89%	1.74%	**3.12**	3.7%
Dutch	67.07%	71.05%	**69.00**	86.59%	90.33%	**88.42**	33.35%	37.08%	**35.12**	39.7%
English	74.76%	72.93%	**73.83**	82.15%	88.05%	**85.00**	52.87%	43.26%	**47.58**	56.0%
Korean	76.59%	34.52%	**47.59**	84.18%	84.41%	**84.30**	18.75%	1.20%	**2.26**	2.7%
Spanish	77.31%	83.11%	**80.10**	82.42%	88.87%	**85.53**	47.00%	51.80%	**49.28**	57.6%

Table 7. Letter n-gram results

		all tokens			seen tokens			unseen tokens			
	n	precision	recall	$F_{\beta=1}$	precision	recall	$F_{\beta=1}$	precision	recall	$F_{\beta=1}$	seen%
Basque	1	70.43%	74.90%	**72.59**	74.89%	79.10%	**76.94**	63.38%	73.63%	**68.12**	88.5%
Dutch	1	72.24%	77.34%	**74.71**	71.38%	78.93%	**74.97**	74.20%	77.59%	**75.86**	101.2%
English	1	63.31%	71.30%	**67.06**	63.95%	75.24%	**69.14**	59.47%	63.14%	**61.25**	88.6%
Korean	3	69.82%	49.89%	**58.19**	80.79%	76.88%	**78.79**	56.25%	30.60%	**39.64**	50.3%
Spanish	1	81.84%	88.56%	**85.07**	81.77%	88.11%	**84.82**	81.66%	90.03%	**86.05**	101.5%

Probabilistic Learned Lists (Table 6). For all languages except Basque, learned lists performed better than the baseline, increasing f-value by up to 20 points. Performance on seen data was very high across the board. When dealing with an unseen token, the classification was made purely on the basis of the tokens on each side. This gives lower f-scores for English, Spanish, and Dutch, while the performance for Korean and Basque approaches zero. This can be explained by looking at the relative significance of word order for each language. Basque has free word ordering, so context is an unreliable source of information. The low unseen performance for all languages indicates that this approach, while generally an improvement over the baseline, is highly dependent on the training set, and suffers sparsity problems in broad domains.

Letter N-grams (Table 7). The performance profile of this character-level classifier is totally different to that of the word-level classifiers. The performance difference between seen and unseen tokens has been removed, with unseen performance actually being better for Dutch and Spanish. Under this representation,

Table 8. Orthographic tries results

	all tokens			seen tokens			unseen tokens			
	precision	recall	$F_{\beta=1}$	precision	recall	$F_{\beta=1}$	precision	recall	$F_{\beta=1}$	seen%
Basque	79.50%	83.17%	**81.29**	85.07%	88.65%	**86.82**	67.27%	74.13%	**70.53**	81.2%
Dutch	84.44%	87.82%	**86.09**	88.31%	92.92%	**90.56**	78.05%	81.17%	**79.58**	87.9%
English	76.81%	82.55%	**79.58**	84.03%	90.38%	**87.09**	60.67%	67.31%	**63.82**	73.3%
Korean	71.26%	55.16%	**62.18**	71.88%	64.45%	**67.96**	70.12%	37.58%	**48.94**	72.0%
Spanish	83.25%	89.52%	**86.27**	83.54%	89.67%	**86.50**	79.20%	85.63%	**82.29**	95.1%

there is conceptually no difference between seen and unseen data. The data sparseness problem is much smaller when dealing with the order of one hundred letters rather than thousands of words.

Unigrams are weak at identifying terms in all-capitals. From the training corpora for the European languages, capitals have a high probability of occurring in NEs, which in the case of an all-caps term will result in high $AVE()$, $MAX()$, and $MIN()$ scores. In the English corpus, 40% of all-caps terms were identified as NEs, an accuracy of 83%.

Korean was the only language that achieved better performance using a longer n-gram, in this case trigrams. Since the average length of words and NEs is much lower, a trigram may be capturing most or all of a token, in which case performance would be expected to be more like a word-level classifier. Indeed, the performance on seen tokens is significantly higher than unseen tokens, as would be expected if this were the case. Overall, performance was slightly worse than learned lists for seen tokens, but greatly improved for unseen tokens.

Orthographic Tries (Table 8). Prefix and suffix tries for the current token, plus one token of context on either side, were combined using a machine learner. Despite using less attributes than the n-gram representation, performance is boosted significantly for all languages. As with letter n-grams, recall is higher than precision, which suggests that while the orthographic representations discussed here are quite effective they have a tendency to overgeneralise.

Performance on seen tokens is significantly higher than for unseen tokens. This is to be expected, as tries are an explicit representation of the phenomena present in the training corpus. The performance on unseen tokens is still higher than that of n-grams, showing the strength of the progressive back-off model obtained through using tries.

All-caps terms were problematic for tries also, with an accuracy of 86% in the English corpus, down from over 97%. In subsequent work [13], we have shown that case restoration techniques can remove any performance loss from capitalisation artifacts. Other errors in English included ambiguously tagged words such as "defense", which occurs equally as a word and as an NE in the training corpus. For some words, such as "earth", the training and test corpora disagreed completely on the correct categorisation.

Table 9. Voting results

	precision	recall	$F_{\beta=1}$
Basque	83.32%	83.67%	**83.49**
Dutch	88.23%	85.58%	**86.88**
English	83.67%	82.57%	**83.12**
Korean	35.09%	90.00%	**50.50**
Spanish	89.49%	84.42%	**86.88**

Voting (Table 9). The combination of classification techniques improved performance for all languages except Korean, where the poor performance of each classifier contributed too much noise for voting to be successful. Using only word-level and letter-level features, we have obtained up to 86.9% on the NER task. From CoNLL 2002, the best results were 91.64% for Spanish and 90.70% for Dutch [1].

6 Conclusion

In addition to presenting new multilingual corpora for named entity recognition, we have presented an initial analysis of the suitability and difficulty of the data for each language. By identifying specific aspects such as the repetition and ambiguity of NEs, we can make predictions of the performance of different approaches, and the suitability of these corpora for their evaluation.

Orthographic classifiers, which use no global or external evidence, are the most fundamental types of classifier used in NER systems. We have presented three simple yet powerful language-independent techniques for representing both word-internal and contextual data using a small number of real-valued attributes. Our results, using under thirty attributes and simple machine learning are presented here in contrast to the trend of natural language learning systems to use ever-increasing numbers of features [6].

Evaluating performance on both seen and unseen data provides greater understanding of the characteristics of a system. Probabilistic learned lists function in much the same way as gazetteers, but can boost the performance on unseen data by incorporating contextual evidence. Aggregate statistics gathered from letter n-grams perform robustly across seen and unseen data, and were shown to be capable of capturing useful orthographic structure even across writing systems. Tries, especially a combination of multiple tries, exploit prefix and suffix information to give the most reliable single orthographic classification tested.

Given the wide range of possible linguistic cues that may or may not be present in a given language, it is not reasonable to expect a single set of attributes to always perform well. Instead, robust language independence will be obtained through using a range of general techniques, and combining them in an appropriate manner for each target language.

This paper has not dealt with the classification of named entities, only their recognition. The authors look forward to future recognition and classification work using this diverse collection of languages.

References

[1] Xavier Carreras, Lluís Màrques, and Lluís Padró. Named entity extraction using adaboost. In *Proceedings of CoNLL-2002*, pages 167–170. Taipei, Taiwan, 2002. 920

[2] Hai Leong Chieu and Hwee Tou Ng. Named entity recognition: A maximum entropy approach using global information. In *Proceedings of the 19th International Conference on Computational Linguistics (COLING 2002)*, pages 190–196. Taipei, Taiwan, 2002. 912

[3] Satoshi Sekine Chikashi Nobata and Jun'ichi Tsuji. Difficulty indices for the named entity task in japanese. In *Proceedings of the Annual Meeting of the Association for Computational Linguistics (ACL-2000)*. Hong Kong, China, 2000. 912

[4] S. Cucerzan and D. Yarowsky. Language independent named entity recognition combining morphological and contextual evidence, 1999. 910, 916

[5] Dan Klein, Joseph Smarr, Huy Nguyen, and Christopher D. Manning. Named entity recognition with character-level models. In *Proceedings of CoNLL-2003*, pages 180–183. Edmonton, Canada, 2003. 913

[6] James Mayfield, Paul McNamee, and Christine Piatko. Named entity recognition using hundreds of thousands of features. In *Proceedings of CoNLL-2003*, pages 184–187. Edmonton, Canada, 2003. 920

[7] A. Mikheev, M. Moens, and C. Grover. Named entity recognition without gazetteers, 1999. 915

[8] Jon Patrick, Casey Whitelaw, and Robert Munro. Slinerc: The sydney language-independent named entity recogniser and classifier. In *Proceedings of CoNLL-2002*, pages 199–202. Taipei, Taiwan, 2002. 916

[9] Erik F. Tjong Kim Sang. Introduction to the conll-2002 shared task: Language-independent named entity recognition. In *Proceedings of CoNLL-2002*, pages 155–158. Taipei, Taiwan, 2002. 910

[10] Erik F. Tjong Kim Sang and Fien De Meulder. Introduction to the conll-2003 shared task: Language-independent named entity recognition. In *Proceedings of CoNLL-2003*, pages 142–147. Edmonton, Canada, 2003. 910

[11] Erik F. Tjong Kim Sang and Jorn Veenstra. Representing text chunks. In *Proceedings of EACL'99*, pages 173–179. Bergen, Norway, 1999. 911

[12] Casey Whitelaw and Jon Patrick. Orthographic tries in language independent named entity recognition. In *Proceedings of ANLP02*, pages 1–8. Centre for Language Technology, Macquarie University, 2002. 916

[13] Casey Whitelaw and Jon Patrick. Named entity recognition using a character-based probabilistic approach. In *Proceedings of CoNLL-2003*, pages 196–199. Edmonton, Canada, 2003. 917, 919

Active Learning: Applying *RinSCut* Thresholding Strategy to Uncertainty Sampling

Kang Hyuk Lee[1], Judy Kay[1], and Byeong Ho Kang[2]

[1] School of Information Technologies, University of Sydney, NSW 2006, Australia
{kangl,judy}@it.usyd.edu.au

[2] School of Computing, University of Tasmania, Hobart, Tasmania 7001, Australia
bhkang@utas.edu.au

Abstract. In many supervised learning approaches to text classification, it is necessary to have a large volume of manually labeled documents to achieve a high level of performance. This manual labeling process is time-consuming, expensive, and will have some level of inconsistency. Two common approaches to reduce the amount of expensive labeled examples are: (1) selecting informative uncertain examples for human-labeling, rather than relying on random sampling, and (2) using many inexpensive unlabeled data with a small number of manually labeled examples. Previous research has been focused on a single approach and has shown the considerable reduction on the amount of labeled examples to achieve a given level of performance. In this paper, we investigate a new framework to combine both approaches for similarity-based text classification. By applying our new thresholding strategy, *RinSCut*, to the conventional uncertainty sampling, we propose a new framework which automatically selects informative uncertain data that should be presented to human expert for labeling and positive-certain data that could be directly used for learning without human-labeling. Extensive experiments have been conducted on Reuters-21578 dataset to compare our proposed scheme with random sampling and conventional uncertainty sampling schemes, based on micro and macro-averaged F_1. The results showed that if both macro and micro-averaged measures are concerned, the optimal choice might be our new approach.

1 Introduction

Text classification (or text categorization) is, simply, the task of automatically assigning arbitrary documents to predefined classes (or categories) based on their content. Because most natural languages are ambiguous and usually return a large number of input features (or attributes), most supervised learning approaches to text classification require a large number of labeled examples to give an accurate classification scheme. In many practical situations, such a large number of labeled data for learning is not really available, since manually labeling them is such a big burden on human

T.D. Gedeon and L.C.C. Fung (Eds.): AI 2003, LNAI 2903, pp. 922-933, 2003.

experts [1]. It is, therefore, important for the success of a text classification system to develop new learning methods that give reasonable performance with fewer labeled examples.

There have been approaches to reduce the amount of these expensive labeled examples to achieve a given level of performance. They include:

- Selecting informative uncertain data from a pool of unlabeled ones for human-labeling.
- Directly using many inexpensive unlabeled examples with a small number of labeled data.

One way of selecting uncertain examples is based on the uncertainty values of unlabeled ones and so-called uncertainty sampling [3]. In this way, uncertainty value is measured by comparing the estimated similarity score and threshold. Its motivation is based on the reasoning that the example that is most uncertain is the most informative one for training. Recent work showed that this approach can significantly reduce the amount of labeled data to achieve a given level of classification performance [2, 3]. Also, the latter approach has recently been applied to text classification [5, 6] and the results showed that learning algorithms could be improved by using unlabeled data, especially when the amount of labeled examples is small.

While many researchers have been successfully applying one of the approaches to text classification, few have combined both approaches. Our goal in this paper is to develop a framework to combine these two approaches for similarity-based text classification. For this goal, we applied *RinSCut* thresholding strategy [7, 8] to the previous conventional uncertainty sampling scheme.

This paper is organized as follows. Section 2 reviews the related work in similarity-based text classification and uncertainty sampling. Section 3 explains our new uncertainty sampling framework with *RinSCut*. Then, we give experimental setup and performance results in Section 4. Finally, conclusions and future work are given in Section 5.

2 Uncertainty Sampling in Similarity-Based Text Classification

2.1 Similarity-Based Text Classification

There has been a wide range of learning algorithms applied to text classification task. One group of algorithms that has been shown good performance is similarity-based. These algorithms assign relevance similarity scores to every document-class pair.

Before applying learning algorithm, raw documents should be transformed into machine-readable representation. The common representation adopted by most similarity-based algorithms is the "bag-of-words" representation. . In this representation, each document d is transformed to a vector $v = (x_1, x_2, \ldots, x_n)$. Here, n is the number of unique features and each x_i is the weighting value of the ith feature. This is calculated as a combination of two common weighting schemes: the term frequency, TF_i, is the number of times the ith feature occurs in document d and the inverse document frequency, IDF_i, is $log(|N|/DF_i)$, where DF_i is the number of documents that contain the ith feature and $|N|$ is the total number of documents in the training set. Then, x_i is

$TF_i{\times}IDF_i$. Because the document lengths may vary widely, a length normalization factor is applied to the term weighting function. The weighting equation that is used in this work is given in [9].

The main task of similarity-based learning algorithm is to give more weight to the informative features than it assigns to non-informative ones. Similarity-based algorithms can be grouped into two classes: profile-based and instance-based. Profile-based algorithms (see, for example, *Rocchio* [10] and Widrow-Hoff [11]) prepare a generalized profile set for each class in which each feature has a weight vector computed from a set of training examples. The *k*-NN [12] is one of the most widely used instance-based algorithms. It uses training examples directly in the similarity computation, and its assumption is that an example itself has more representative power than a generalized feature vector.

The mapping from a new (or test) example to relevant classes is achieved by thresholding the similarity scores. Yang [13] explained and evaluated on existing common thresholding techniques, and noted that combining the strengths of the existing strategies is a challenge in text classification.

2.2 Uncertainty Sampling

In most supervised learning settings, the more labeled training examples we provide, the better performance we have. Previous experiments in text classification literature showed that the system usually needs several thousands of human-labeled examples to get reasonable performance. However, given the human resource limitations, such a large number of labeled examples are often not freely available and obtaining them is expensive. This problem suggests an active learning [4] that controls the process of sampling examples.

- Create initial classifiers
- Until "there are no more unlabeled examples" or "human experts are unwilling to label more exa mples"
  - (1) Apply the current classifiers to each unlabeled example
  - (2) Find the *k* examples with the highest uncertainty values
  - (3) Have human experts label these *k* examples
  - (4) Train the new classifiers base on all labeled exa mples to this point

Fig. 1. Uncertainty sampling

Lewis and Gale in [3] proposed the uncertainty sampling method for active learning. Rather than relying on random sampling, uncertainty sampling selects unlabeled examples for human-labeling, based on the level of uncertainty about their correct class. A text categorization system with the uncertainty-sampling method examines unlabeled examples and computes the uncertainty values for the predicted class membership of all examples. Then, those examples with largest uncertainties are selected as the most informative ones for training and presented to human experts for labeling. The uncertainty of an example is typically estimated by comparing its numeric similarity score with the threshold of the category. The most uncertain (informative) example is the one whose score is closest to the threshold. Figure 1 shows the pseudo-code for the uncertainty sampling.

3 Uncertainty Sampling with *RinSCut*

For our goal in this paper, the important issue in the previous uncertainty sampling scheme is how to design a mechanism showing the boundaries of the threshold regions to which the uncertain and positive-certain examples should belong. With such a mechanism, we can use the positive-certain examples for training, without asking human experts for their correct classes. We achieved this mechanism by applying *RinSCut* thresholding strategy.

3.1 *RinSCut* Thresholding Strategy

In [7, 8], we proposed the rank-in-score threshold (*RinSCut*) by jointly using the score-based threshold (*SCut*) and rank-based threshold (*RCut*). The *SCut* finds a different similarity score for each class and so optimizes local performance while *RCut* gives one single rank to all classes by optimizing the global performance. *RinSCut* is designed to use strengths of two strategies and, as a result, to overcome weakness in both *SCut* and *RCut*. *RinSCut* finds two threshold scores, s_{top} and s_{bottom}, for each class. When s is the optimal score from *SCut*, let R be the set of negative examples with similarity scores above s and T be the set of positive ones with similarity scores below s. Then, two threshold scores are defined as follows:

$$s_{top} = s + \frac{\sum_{i \in R}(v_i - s)}{|R|} \qquad (1)$$

$$s_{bottom} = s - \frac{\sum_{i \in T}(s - v_i)}{|T|}$$

where v_i is the similarity value of document, $|R|$ and $|T|$ are the number of examples in R and T respectively. The range of similarity values between these two threshold scores is considered as the ambiguous zone: the classification decision for a test example to this class can not be made only with the similarity score. For the test documents that belong to this zone, the rank threshold is used to make the final decision. For an assignment decision on a test document d to a class c with the similarity score, $Sim(d, c)$, assigns a "YES" decision if $Sim(d, c) \geq s_{top}$ and a "NO" decision if $Sim(d, c) < s_{bottom}$. If $Sim(d, c)$ is between s_{top} and s_{bottom}, the assignment decision depends on the rank threshold t. The t threshold can be defined by two ways, one optimizing the global performance (*GRinSCut*) and the other optimizing the local performance of each category (*LRinSCut*). Our studies in [7, 8], similarity-based algorithms with *RinSCut* showed performance improvements over *SCut* strategy.

3.2 New Framework for Uncertainty Sampling

For our new scheme of uncertainty sampling, we use the *RinSCut* thresholding strategy. The key difference in our uncertainty sampling, from the previous conventional approach, is that our scheme distinguishes the uncertain, positive-certain, and negative-certain examples. In this paper, we do not use negative-certain examples. As explained in Section 3.1, *RinSCut* defines the ambiguous zone for a given class c using the $s_{top}(c)$ and $s_{bottom}(c)$. In our approach, this ambiguous zone is used for differentiating the uncertain, positive-certain, and negative-certain examples in the set of unlabeled examples.

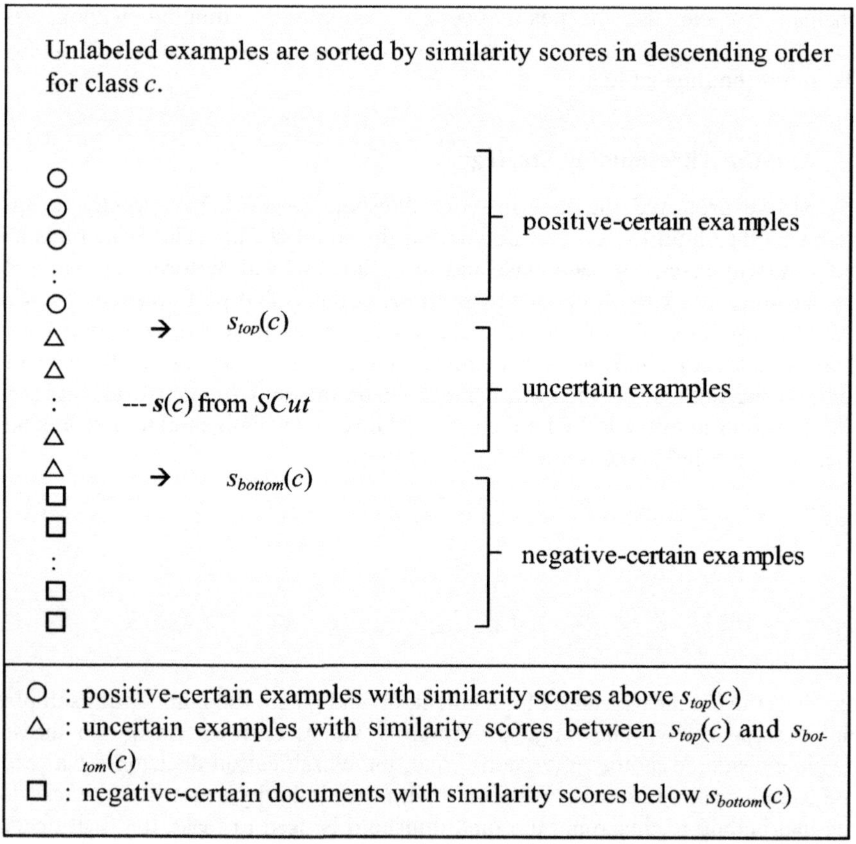

Fig. 2. Definition of uncertain and certain examples using $s_{top}(c)$ and $s_{bottom}(c)$ for a given class c

For a given class c, Figure 2 shows and defines the three ranges of similarity scores of unlabeled examples as follows:

$$s_{bottom}(c) \leq Sim(uncertain\ examples) < s_{top}(c)$$
$$Sim(positive\text{-}certain\ examples) \geq s_{top}(c) \quad (2)$$
$$Sim(negative\text{-}certain\ examples) < s_{bottom}(c)$$

In this figure, the uncertain examples for the category c are shown as Δ having similarity scores that belong to the ambiguous zone, the positive-certain examples are shown as O, having similarity scores above $s_{top}(c)$, and the negative-certain examples are represented as □ having similarity scores below $s_{bottom}(c)$. In our uncertainty sampling method, the classification system selects, for each iteration, a number of uncertain examples with the largest uncertainty values (i.e., with similarity scores closest to the threshold $\boldsymbol{s}$), but they also must be within the ambiguous zone. Then, the system presents them to human experts for their correct labels. Also, our selective sampling method automatically selects the positive-certain examples with similarity scores above s_{top}. Then, it uses them directly for training without asking human experts for their correct class information.

4 Experiments with KAN

4.1 Keyword Association Network (KAN)

Keyword association network (*KAN*) is a lazy linear learning algorithm that exploits feature co-occurrence information extracted from a set of training examples and discriminating power value of each feature in similarity computation. Unlike the profile-based linear algorithms, *KAN* does not fix the feature weights through a pre-learning process. The absolute weight for each feature in a certain class is determined by seeing a new example. This is why we call a lazy linear algorithm. In our previous work [7, 8], *KAN* has been shown to be effective in text classification. For its parameter settings, we used same values as used in [7].

4.2 Test Dataset

The dataset used in our experiments is the Reuters-21578 test collection [14]. We split the articles into training and test sets according to the Modified Apte ("ModApte") splitting method that has been the most widely used splitting method in text classification on this corpus. Instead of analyzing all 135 categories in "Topics" group, we choose the categories with at least 10 articles in both training and test sets. This gives a corpus of 6,984 training articles and 3,265 test articles across selected 53 categories. In this corpus, many articles in both training and test sets belong to multiple categories.

4.3 Experimental Setup

For the text preprocessing to convert raw articles to the representations, we applied a stop-list and Poter's stemming algorithm [16]. We used information gain for feature selection and took the same number of features, 50 in these experiments, for all 53 classes. The classifiers (i.e., learning algorithm + thresholding strategy) implemented and evaluated are *KAN+GRinSCut* and *KAN+LRinSCut*. In the experiments, we want to compare the following sampling methods:

- Random sampling (***RS***)
  The examples in the training set are randomly selected from the set of unlabeled examples and then, manually labeled by human experts.
- Uncertainty sampling of uncertain examples (***US-U***)
  For the training set, the most uncertain examples are selected based on their uncertainty scores and then, manually labeled by human experts.
- Uncertainty sampling of uncertain and certain examples (***US-U&C***)
  As well as the most uncertain examples, a set of positive-certain examples that is automatically labeled by the system is also added to the training set.

Uncertainty sampling methods (***US-U*** and ***US-U&C***) are based on the homogeneous approach that uses the same type of classifier as the one used for the classification of test examples. For the number of positive-certain examples used for training in the ***US-U&C***, we choose 250 and 500 examples, . Also, to obtain generalized and reliable results for the evaluation of the ***RS*** method, we conducted the experiments 3 times.

The classification performance was measured using the standard F_1 that is designed to balance recall and precision by giving them equal weight [17].

$$F_1 = 2 \times \text{Recall} \times \text{Precision} / (\text{Recall} + \text{Precision}) \tag{3}$$

4.4 Experimental Results

For the results in this section, we use the following experimental methodology. To build an initial-classifier, we randomly select the same number, n, of positive-training examples for each class (n = 2, 106 examples in total). Also, for each iteration, the same number of examples (i.e., 106 examples) is selected based on the adopted sampling method and added to the training set. For the ***US-U&C*** sampling methods, a set of an additional k automatically labeled positive-certain examples (250 and 500) is added to the training set. These k positive-certain examples are almost evenly distributed across categories. Then, the classifier rebuilt from the training set is evaluated against the test set.

The learning traces of *KAN+GRinSCut* and *KAN+LRinSCut* with four sampling methods are presented in Figures 3 and 4. In these charts, the curves of uncertainty sampling methods stop when they achieve the target performance of random sampling. In these figures, we can see that all the uncertainty sampling methods for the micro-averaged F_1 do not show any desirable effects over the random sampling. These results are mainly due to the fact that the documents in the Reuters-21578 corpus are very unevenly distributed across the categories. More than 50% of examples belong to two categories, "earn" and "acq", in both training and test sets. In this situation, the micro-averaged measure mainly depends on the performances on these two categories. In the ***RS***, the randomly selected documents for the training set of each trial might be mainly from "earn" and "acq" categories and the classifiers that are built from this unevenly distributed training set might be working well with the test examples of two frequent categories. By contrast, the selected documents in the uncertainty sampling methods are evenly distributed across categories.

In Figures 3, note that the ***US-U&C*** variants perform better than the ***US-U*** sampling for the micro-averaged performance of *KAN+GRinSCut.* This superior result of ***US-U&C***, against ***US-U***, is probably due to the large number of positive-certain examples added to the training set, since these examples can solve the problem of the sparse training examples for the frequent categories like "earn" and "acq". Unlike for the micro-averaged performance, the advantage of the ***US-U*** sampling is apparent for the macro-averaged measure in Figure 3. As shown in Table 1, ***US-U*** achieves 0.497 macro-averaged F_1 at 848 examples, while the random sampling needs 1,378 examples to achieve this level of performance. So, the ***US-U*** shows 38.4% saving on the number of labeled examples required, over the random sampling. The advantages of the ***US-U&C*** variants for the macro-averaged measure are not clear with the small numbers of labeled examples used. But, after 954 examples, both ***US-U&C***[500] and ***US-U&C***[250] work well and achieve 0.507 and 0.500 macro-averaged F_1 at 954 and 1,166 examples, respectively. These results show 30.7% saving for ***US-U&C***[500] and 15.3% saving for ***US-U&C***[250] in Table 1. Also, we can see that using 500 positive-certain examples achieves better performance than using 250 positive-certain examples.

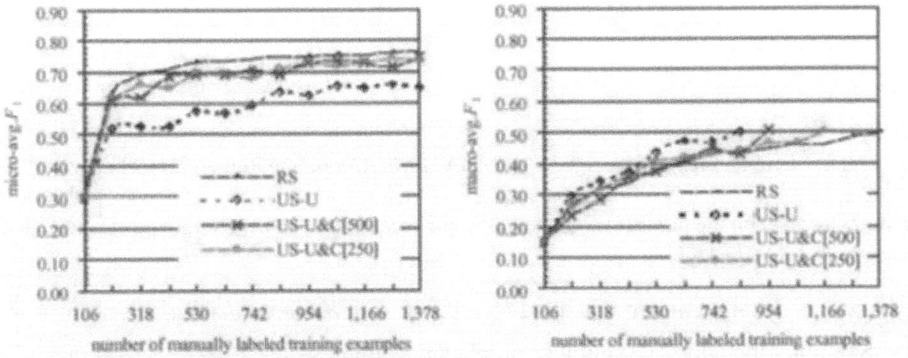

Fig. 3. Micro and macro-averaged F_1 of *KAN+GRinSCut* on the Reuters-21578 dataset

Table 1. Savings on the number of labeled examples required to achieve target macro-averaged measures in *KAN+GRinSCut*

target macro averaged F_1 = 0.497 with *KAN+GRinSCut*			
sampling method	number of labeled examples	F_1	savings
RS	1,378	0.497	0%
US-U	848	0.502	38.4%
US-U&C[500]	954	0.507	30.7%
US-U&C[250]	1,166	0.500	15.3%

Figure 4 depicts the micro and macro-averaged F_1 performance of another classifier, *KAN+LRinSCut* in which the locally optimized *RinSCut* used. The ***US-U*** shows somewhat unstable learning curves in the micro-averaged performance and, like *KAN+GRinSCut* in Figure 3, it failed to achieve 0.765 micro-averaged F_1 with less than ***RS*** at 1,378 examples. The ***US-U&C*** variants work better than the other sampling methods with small numbers of labeled examples. But, after 530 examples for ***US-U&C***[500] and 374 examples for ***US-U&C***[250], their performances are worse than ***RS***. For the macro-averaged performance of *KAN+LRinSCut*, all the uncertainty sampling methods consistently work better than the ***RS***. To achieve 0.500 macro-averaged F_1, while the ***RS*** needs 1,378 examples, the ***US-U*** needs 848 training examples representing 38.4% saving in the required number of examples. The ***US-U&C***[500] and ***US-U&C***[250] need 954 and 1,166 examples, showing 30.7% and 15.3% savings, respectively in Table 2.

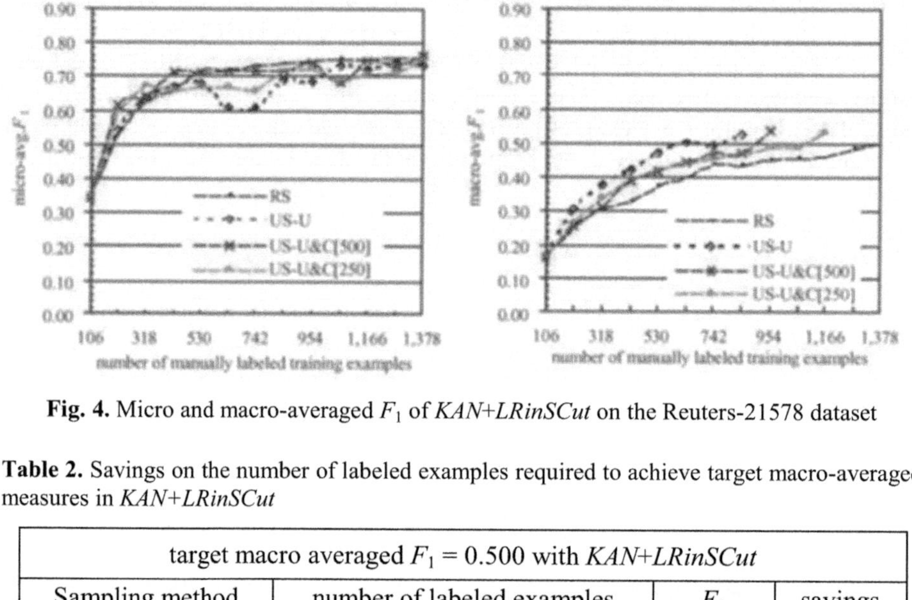

Fig. 4. Micro and macro-averaged F_1 of *KAN+LRinSCut* on the Reuters-21578 dataset

Table 2. Savings on the number of labeled examples required to achieve target macro-averaged measures in *KAN+LRinSCut*

target macro averaged F_1 = 0.500 with *KAN+LRinSCut*			
Sampling method	number of labeled examples	F_1	savings
RS	1,378	0.500	0%
US-U	848	0.528	38.4%
US-U&C[500]	954	0.538	30.7%
US-U&C[250]	1,166	0.533	15.3%

From these results on the Reuters-21578, we can see that when test examples are unevenly distributed across categories, the optimal choice for the sampling methods becomes difficult. And, the choice will depend on the user's needs. For example, if micro-averaged performance is the primary concern on the Reuters-21578, the choice of the optimal sampling method will be the ***RS***, otherwise, it will be the ***US-U*** sampling. If both micro and macro-averaged measures are concerned, the optimal choice seems to be the ***SS-U&C*** sampling method. Also, the results show that using more positive-certain examples (i.e., 500 for the Reuters-21578) works slightly better than the smaller number of positive-certain ones (250 examples). However, using a larger number of training examples would make overall text classification process slower.

5 Conclusions and Future Work

In similarity-based text classification, we have explored conventional uncertainty sampling method for the important goal of reducing the number of labeled examples to achieve a given level of performance. By applying *RinSCut* thresholding strategy, we developed a new framework for uncertainty sampling to exploit positive-certain unlabeled examples as well as uncertain ones.

The comparative experiments on sampling methods with *KAN+GRinSCut* and *KAN+LRinSCut* were conducted on the Reuters-21578 dataset. For the micro-averaged performance, all the uncertainty sampling methods failed to show a clear advantage. This result was mainly caused by an uneven distribution of test examples in this dataset. However, ***US-U&C*** variants showed much better performance than ***US-U*** and their micro-averaged performances were very close to ***RS***. By contrast, for the macro-averaged, there was a clear advantage over the ***RS*** sampling. The conclusions drawn from these results are (1) if micro-averaged performance is the primary concern, the ***RS*** sampling should be used, (2) otherwise, the ***US-U*** sampling could be the optimal choice, and (3) if both averaged measures are concerned, the choice for the sampling methods for training examples might be ***US-U&C*** since it did not show the worst performance on both micro and macro-averaged performances.

We allocated nearly the same number of uncertain examples (and positive-certain examples for the ***US-U&C***) to each category (i.e., kept an even distribution of recommended examples across categories). When the test documents in a given data set show an uneven distribution as in the Reuters-21578 corpus, varying the number of training examples recommended for each category in our uncertainty sampling methods may increase the micro-averaged performance. This suggests that it would be fruitful to explore effects of having an uneven distribution of training examples and exploring what proportion of these training examples is actually correct for a given category. More experiments are needed to evaluate our methods on other corpora like the Reuters Corpus Volume 1 [15]. Also, we note that applying our framework to the classification of other types of documents like web documents is plausible. More work is envisaged to investigate the performance of our uncertainty sampling method with other similarity-based learning algorithms.

References

[1] C. Apte, F. Damerau, and S. M. Weiss. Automated Learning of Decision Rules for Text Categorization. ACM Transactions of Information Systems, 12(3), pages 233-251, 1994.

[2] D. D. Lewis and J. Catlett, "Heterogeneous Uncertainty Sampling for Supervised Learning", In Proceedings of the Eleventh International Conference on Machine Learning, San Francisco, CA., Morgan Kaufman, pages 148-156, 1994.

[3] D. D. Lewis and W. Gale, "A Sequential Algorithm for Training Text Classifiers", In Proceedings of the 17th Annual International ACM SIGIR Conference on Research and Development in Information Retrieval, pages 3-12, 1994.

[4] D. Cohn, L. Atlas, and R. Lander, "Improving Generalization with Active Learning", Machine Learning, 15(2), pages 201-221, 1994.

[5] Blum and T. Mitchell, "Combining Labeled and Unlabeled Data with Co-Training", In Proceedings of the 11th Annual Conference on Computational Learning Theory, pages 92-100, 1998.

[6] K. Nigam, A. K. McCallum, S. Thrun, and T. M. Mitchell, "Text Classification from Labeled and Unlabeled Documents using EM", Machine Learning, 39, pages 103-134, 2000.

[7] K. H. Lee, J. Kay, and B. H. Kang, "Lazy Linear Classifier and Rank-in-Score Threshold in Similarity-Based Text Categorization" ICML Workshop on Text Learning (TextML'2002), Sydney, Australia, pages 36-43, 2002.

[8] K. H. Lee, J. Kay, B. H. Kang, and U. Rosebrock, "A Comparative Study on Statistical Machine Learning Algorithms and Thresholding Strategies for Automatic Text Categorization", The 7th Pacific Rim International Conference on Artificial Intelligence (PRICAI-02), Tokyo, Japan, pages 444-453, 2002.

[9] Buckley, G. Salton, J. Allan, and A. Singhal, "Automatic Query Expansion Using SMART: TREC 3", The Third Text Retrieval Conference (TREC-3). National Institute of Standards and Technology Special Publication 500-207. Gaithersburg, MD, 1995.

[10] J. Rocchio, "Relevance Feedback in Information Retrieval", In G. Salton (Ed.), The SMART Retrieval System: Experiments in Automatic Document Processing. Prentice-Hall, 1971.

[11] Widrow and S. D. Stearns, "Adaptive Signal Processing", Prentice-Hall Inc., Eaglewood Cliffs, NJ, 1985.

[12] Y. Yang, "Expert Network: Effective and Efficient Learning from Human Decisions in Text Categorization and Retrieval", In Proceedings of the 17th International ACM SIGIR Conference on Research and Development in Information Retrieval, pages 13-22, 1994.

[13] Y. Yang, "A Study on Thresholding Strategies for Text Categorization", In Proceedings of the 24th Annual International ACM SIGIR Conference on Research and Development in Information Retrieval (SIGIR'01), pages 137-145, 2001.

[14] Reuters-21578 collection, originally collected and labeled by Carnegie Group Inc and Reuters Ltd, may be freely available for research purpose only from, http://www.daviddlewis.com/resources/testcollections/reuters21578/

[15] The new Reuters collection, called Reuters Corpus Volume 1, has recently been made available by Reuters Ltd, may be freely available for research purpose only from, http://about.reuters.com/researchandstandards/corpus/

[16] M. F. Porter, "An Algorithm for Suffix Stripping", Program, 14(3), pages 130-137, 1980.

[17] Y. Yang, "An Evaluation of Statistical Approaches to Text Categorization", Journal of Information Retrieval, 1(1/2), pages 67-88, 1999.

The Effect of Evolved Attributes on Classification Algorithms

Mohammed A. Muharram and George D. Smith

School of Computing Sciences
UEA Norwich, Norwich, England
m.muharram@uea.ac.uk , gds@cmp.uea.ac.uk

Abstract. We carry out a systematic study of the effect on the performance of a range of classification algorithms with the inclusion of attributes constructed using genetic programming. The genetic program uses information gain as the basis of its fitness. The classification algorithms used are C5, CART, CHAID and a MLP. The results show that, for the majority of the data sets used, all algorithms benefit by the inclusion of the evolved attributes. However, for one data set, whilst the performance of C5 improves, the performance of the other techniques deteriorates. Whilst this is not statistically significant, it does indicate that care must be taken when a pre-processing technique (attribute construction using GP) and the classification technique (in this case, C5) use the same fundamental technology, in this case *Information Gain*.

Keywords: Evolutionary algorithms, Knowledge discovery and Data Mining.

1 Introduction

In this paper, we present the results of experiments in which a genetic program is used to evolve new attributes which are non-linear functions of the original attributes. The purpose of the attribute construction is to test the improvement in performance of a range of classifiers on a number of public domain data sets. This work should be viewed as complimentary to the work of [1], in which the authors also evolved attributes using a GP and compared the classification performance of C4.5 using the original attributes with that of C4.5 using the augmented attribute set which included a new evolved attribute.

The reason we have chosen to extend their work is that the GP used in [1] uses *Information Gain Ratio* as the fitness function, whilst the splitting criterion in the decision tree algorithm C4.5 is also based on information gain ratio, see [2]. The question arises therefore, will another classifier, whose induction technique is not based on information gain, also benefit from the inclusion of a new attribute evolved using a GP with information gain as a fitness function, or is the process biased towards C4.5?

The remainder of the paper is structured as follows: Section 2 presents a brief review of attribute construction and the use of GP to evolve features, or attributes. Section 3 describes the experimental methodology, the details of the

T.D. Gedeon and L.C.C. Fung (Eds.): AI 2003, LNAI 2903, pp. 933–941, 2003.

data sets used and the parameters and settings used in the GP. The results are presented in Section 4 and a summary in Section 5.

2 Attribute Construction

In data mining, when constructing classification models from data sets, the data is normally presented as a fixed number of features, or *attributes*, one of which is the discrete valued, dependent variable, or *class*. The purpose of classification is to find a description of the class variable based on some or all of the other predicting variables. The representation of this description varies depending on the particular induction technique used, and includes decision trees, rules, artificial neural networks, Bayesian classifiers, and many others, see [3].

In this paper, we are primarily addressing the performance of a decision tree classifier. A decision tree is typically constructed using a greedy, iterative process, wherein, during the induction stage, each internal decision node is associated with a test on one of the predicting attributes, the particular test being chosen to optimise a measure relating to the splitting criterion. Successors of the internal nodes are formed and the process is repeated at each successor node. In C4.5, for instance, the splitting criterion is the *Information Gain Ratio*, see [2], whilst in CART (Classification and Regression Trees), the splitting criterion is the Gini index [4].

The success of any classification algorithm depends on its ability to represent any inherent pattern in the data set, and hence depends on the set of predictive attributes available, or its attribute vector. Techniques such as tree induction typically assess the predictive ability of attributes on a one-by-one basis. In other words, at each internal node of the tree, each attribute is analysed in turn to measure its predictive power in terms of the class output. Any *combination* of predicting attributes which presents a much stronger prediction may therefore be missed, if the operators available to the induction process are insufficient to identify that combination.

One approach to overcome this problem is to allow the induction process the flexibility to identify and 'construct' these powerfully predictive combinations. For instance, in the work of [5], where the authors use a feed forward neural network to extract knowledge from corporate accounting reports, it is interesting to note that the first hidden layer of nodes were inclined to construct ratios from some of the raw accounting data. It is widely recognised that accounting ratios, rather than the basic accounting data, are more useful in terms of what can be deduced about a company's financial status. Turning to decision trees, OC1 is an oblique tree induction technique designed for the use with continuous real-valued attributes. During the induction stage, OC1 considers linear combinations of attributes, and partitions the data set into both oblique and axis-parallel hyperplanes, [6].

Another approach is to construct new attributes which are combinations of the original attributes, the objective of the construction technique being to identify highly predictive combinations of the original attribute set and hence, by including such combinations as new features (attributes), to improve the predictive power of the attribute vector, see [7]. In this paper, we restrict our attention to the construction of new attributes, and in particular, to the use of *genetic programming* to construct/evolve new attributes, see [8].

There are essentially two approaches to constructing attributes in relation to data mining; one method is as a separate pre-processing stage, in which the new attribute are constructed before any induction process, i.e. before the classification algorithm is applied to build the model. The second approach is an integration of construction and induction, in which new attributes are constructed within the induction process. The latter is therefore a hybrid induction algorithm.

In [9], the author integrates the tree induction process with the attribute construction process. At each internal node a new attribute is generated by one of four types of construction algorithms. If this attribute is better (in terms of the splitting criterion) than the original attributes and all previous constructed attributes, then it is selected as the test for that node. Zheng applied this hybrid to a range of data sets from the UCI repository, and found a general improvement in the performance in terms of accuracy achieved.

In [7] and [10], the authors have developed a GP to evolve new Boolean attributes from the original Boolean attribute vector. This was applied as a pre-processing stage prior to testing the inclusion of the new attributes on parity problems, using C4.5 and backpropagation [10] and quick-prop [7].

In a recent study [1], the authors use genetic programming as a pre-processing, attribute construction technique. For each data set used, and for each trial in a 10xCV test, a single new attribute was evolved using a GP with *Information Gain Ratio* as the fitness function of the GP. Classification using C4.5 was applied to the data set, both with and without the new attribute and the results showed an improvement in the performance of C4.5 with the use of the newly evolved attribute. However, in this study, it is notably that the objective function of the pre-processing technique, i.e. the attribute construction GP, and the induction technique C4.5, both use *Information Gain Ratio.* One is left asking, therefore, will a classifier whose induction process is not based on information gain benefit, or indeed benefit as much as C4.5, with the inclusion of an attribute which has been evolved by a GP with information gain as its fitness function?

This paper addresses this question.

3 Experimental Details

3.1 Data Sets

The experiments are performed on 5 data sets, 4 of which were the data sets used in [1] plus the BUPA Liver Disorder data set. All data sets are from the UCI data repository, see www.ics.uci.edu/~mlearn/MLRepository.html. Table 1 shows the number of cases, classes and attributes for each data set. Note that our

Table 1. Data Sets used in experimental work

Data Set	Cases	Classes	Attributes
Abalone	4177	28	8
Balance-scale	625	3	4
BUPA Liver Disorder	345	2	6
Waveform	300	3	21
Wine	178	3	13

GP considers all attributes to be real-valued variables. Thus the single Boolean attribute in the Abalone data set is considered as real-valued for the purposes of this study.

3.2 Methodology

The main aim of this work is to ascertain if classification using C5 (C4.5) is advantaged in any way by the inclusion of a constructed attribute which has been evolved by a GP which also uses information gain (as the fitness function).

Thus, in addition to using C5 to perform the classification, we use three other classification methods, two of which are also decision tree algorithms whilst the third is a feed forward neural net. The classification algorithms chosen are:

1. C5, a descendant of C4.5, a decision tree algorithm whose splitting criterion is based on information gain, see [2];
2. CART, a decision tree algorithm whose splitting criterion is based on the Gini index; see [4];
3. CHAID, a decision tree algorithm whose splitting criterion is based on achieving a threshold level of significance in a chi-squared test, see [11];
4. MLP ANN.

For each data set we use 10-fold cross validation to compute the error rate. Thus the data set is firstly partitioned into 10 subsets. For each trial, 9 of the 10 subsets are used as the training set whilst the remaining set is the test set.

The methodology is as follows: For each trial,

1. apply each classification algorithm to the training set (original attributes only) and evaluate resulting models on test set,
2. evolve a single, new attribute from the training set using GP,
3. apply each classification algorithms on augmented training set (original attribute set plus evolved attribute). Evaluate resulting model on the test set.

For both the classification using original attributes and that using the augmented set, we then determine the average error rate over the 10 trials for each algorithm. These are referred to respectively as *Original* and *Augmented* in the results tables in the following section.

Note that, like [1], the attribute construction is a pre-processing technique, whilst [9], for example, integrated the tree induction and attribute construction processes. Thus, in this work, for each data set and for each classification technique, the GP was run 10 times and the classification technique 20 times.

Note also that, for each trial, the attribute evolved by the GP may be different.

3.3 The GP

The GP is designed to construct real-valued attributes from the original (assumed) real-valued attributes of the data set. Thus the terminal set consists of all the original attributes plus the constant "1", whilst the function set consists of the arithmetic operators +,-,*,%.

The initial population is created using a ramped half-and-half method, and the size is fixed at 600. The GP was run for 100 iterations.

The selection method used is tournament, with a tournament size of 7. Mutation and crossover are fairly standard, with mutation replacing nodes with like nodes, and crossover swapping subtrees. Mutation rate is 50%, whilst crossover rate is 70%.

The fitness function adopted here is *Information Gain*, see [2] and [1].

As [1] showed, limiting the size of the tree, and hence the complexity of the constructed attribute, made little difference to the results. We also limit the size of the constructed trees, choosing an upper limit of 40 nodes. Indeed, it is worth mentioning here that some experiments were carried out without this restriction. For the larger data sets, trees of the order of 200 nodes were being generated. When we included the restriction, we noticed an improvement in the fitness of the constructed attribute. However, further experimentation is needed to validate this result.

4 Results

4.1 Error Rates

In Tables 2 to 5, we present the error rates (for the test sets) averaged over the 10xCV trials. In each table, the second column shows the error rate achieved by the induction technique using the original attribute set. The figure after the $\pm$ is the standard deviation of the accuracies over the 10 trials. In the final column, we show the error rates achieved by the induction technique on the augmented attribute set, i.e. the original attributes plus the single evolved attribute. [Note that, in each of the 10 runs necessary for 10xCV, a new attribute was evolved from scratch.]

For the purposes of comparison, we also show the results from [1], in which C4.5 is used as the base classifier. The result we report for the augmented attribute set for C4.5 is the best result from the 9 experiments carried out in [1] for each data set (3 different tournament sizes and 3 different maximum tree sizes).

Table 2. Error rates for the Abalone data set

Abalone	Original	Augmented
C5	78.18 ± 1.91	77.70 ± 2.45
CHAID	76.45 ± 1.61	74.77 ± 2.62
CART	74.53 ± 2.31	73.84 ± 2.98
ANN	75.83 ± 1.58	75.30 ± 2.46
C4.5 [1]	79.20 ± 0.37	79.16 ± 0.36

Table 3. Error rates for the Balance-scale data set

Balance-scale	Original	Augmented
C5	22.42 ± 6.20	0.00 ± 0.00
CHAID	28.39 ± 5.17	5.65 ± 2.55
CART	29.19 ± 5.76	5.49 ± 2.18
ANN	10.32 ± 5.49	7.96 ± 4.02
C4.5 [1]	22.42 ± 1.34	7.78 ± 0.66

Table 4. Error rates for the BUPA data set

BUPA	Original	Augmented
C5	36.47 ± 6.68	32.35 ± 7.20
CHAID	40.00 ± 9.73	30.00 ± 6.17
CART	42.35 ± 8.34	35.59 ± 6.12
ANN	44.41 ± 8.60	37.06 ± 13.38
C4.5	-	-

For the Abalone data set, Table 2 shows that, although all induction techniques show an improvement using the augmented attribute set, the results are barely significant. In fact, it has to be said that all classification algorithms perform rather badly with this data set, even with the evolved attribute. We include these results, however, in order to compare our results with those of [1].

On the other hand, when we look at the results for the Balance data set, in Table 3, we see a significant improvement in accuracy when we include the evolved attribute, particularly for C5, in which the error rate for the test set was 0 in all 10 trials. Indeed, the accuracies of all the decision tree techniques are improved by around 23%. The ANN, although having a better performance than the other classifiers with the original attributes, also improves performance with the augmented attribute set, but not so dramatically, and indeed less significantly if we take the standard deviations into account. It is also worth pointing out that, for the Balance-scale data set, not only are the error rates reduced significantly, but the standard deviations are also reduced showing a more consistent performance, particularly for the decision tree models.

Table 5. Error rates for the Waveform data set

Waveform	Original	Augmented
C5	22.86 ± 1.85	19.50 ± 1.67
CHAID	28.36 ± 1.81	24.92 ± 2.30
CART	29.00 ± 2.14	24.10 ± 1.37
ANN	18.76 ± 1.00	16.20 ± 1.87
C4.5 [1]	25.06 ± 0.66	22.22 ± 0.49

Table 6. Error rates for the Wine data set

Wine	Original Attributes	Original + New Attribute
C5	7.06 ± 8.23	5.29 ± 7.04
CHAID	17.06 ± 8.06	17.65 ± 8.77
CART	17.65 ± 8.77	19.41 ± 8.79
ANN	4.71 ± 6.08	5.66 ± 5.52
C4.5 [1]	6.48 ± 2.05	3.54 ± 1.30

Table 4 shows the average error rates for the BUPA data set. Once again, all classifiers have improved their performance with the augmented set. However, the relatively large standard deviations represent a more inconsistent performance all round, even with the augmented attribute set. The story is much the same with the Waveform data set, see Table 5, with all classifiers showing an improvement with inclusion of an evolved attribute, although only barely significantly.

Finally, turning attention to the Wine data set, the results in Table 6 show that, while the performance of C5 has marginally improved with the augmented attribute, the performance of each of the other classifiers has deteriorated.

Although not statistically significant, bearing in mind the standard deviations, the results in Table 6 are significant in that the performance of C5, a classifier based on information gain, has improved with the inclusion of an attribute evolved using information gain, whilst other classifiers have not.

The results shown in Tables 2 to 6 generally support those of [1], in that attribute construction using genetic programming can generate improvements in the performance of a classifier, sometimes significantly so, as in the case of the Balance-scale data set.

If we measure the absolute improvement in performance for each classifier, averaged out over all the data sets, then, using the evolved attribute, C5 manages an overall 6.43% improvement in accuracy, CHAID achieves a 7.46% improvement, CART 6.86% and ANN 2.37%. It cannot be concluded from these results, therefore, that C5 has any advantage over the other classifiers with the inclusion of a feature evolved using a GP with information gain as a fitness measure.

Table 7. Difference in error rates between the training and the test sets

	C5		CHAID		CART		ANN	
Data set	Orig.	Aug.	Orig.	Aug.	Orig.	Aug.	Orig.	Aug.
Abalone	45.41	48.15	6.54	5.22	5.41	5.78	0.98	0.97
Balance	12.21	0.00	10.73	0.94	11.78	0.73	1.60	0.54
BUPA	24.03	16.95	11.29	7.94	21.10	17.17	2.26	3.36
Wave	12.70	9.44	12.70	10.76	15.40	12.72	1.19	0.87
Wine	5.76	4.80	7.18	8.64	7.77	11.84	1.66	3.55
average	20.02	15.87	9.69	6.70	12.29	9.64	1.54	1.86

However, the results in Table 6 would suggest caution when including a new attribute that has been evolved using a GP with information gain as a fitness function.

4.2 Training and Test Error Comparison

We turn our attention now to the issue of the relative performance of the classifiers on the training and the test sets.

Each entry in Table 7 represents the difference in error rate between the training and test set, averaged over the 10xCV sets. The results for C5 on the Abalone data set are somewhat exaggerated due to the vast difference in accuracy between training and test sets. However, it is notable that all three tree induction methods suffer significantly from overtraining, but the inclusion of the evolved attribute reduces this (Aug. columns). The ANN suffers less from overtraining, with and without the evolved attribute.

5 Conclusions

In this paper, we have used a GP, with information gain as the fitness function, to evolve a new attribute which is a non-linear function of the original, real-valued attributes. This was done as a pre-processing stage in a data mining exercise to build classification models for a number of public domain data sets.

We compared the performance of 4 classifiers (C5, CHAID, CART and ANN) with and without the new attribute, to ascertain if C5 was benefiting more than the other classifiers on the basis that its splitting criterion is also based on information gain, see [1]. We have found no evidence that C5 has an advantage over the other classifiers, and that, in general, all classifiers benefit from the inclusion of an evolved attribute. Only in one data set did we notice an improvement for C5 whilst the performance of the other classifiers deteriorated when the evolved attribute was included.

Whilst this study was primarily aimed at decision tree algorithms, and their potential to benefit from the inclusion of an attribute evolved using a GP with information gain as a fitness function, a fuller study is underway to investigate the effect on classifier performance when attributes are evolved using a more generic fitness function.

Acknowledgements

The first author would like to acknowledge financial support from his parents and from the Yemen Education Ministry.

References

[1] Otero, F.E.B., Silva, M.M.S., Freitas, A.A., Nievola, J.C.: Genetic programming for attribute construction in data mining. In: Lecture Notes in Computer Science: Genetic Programming: Proc. 6th European Conference (EuroGP-2003). Volume 2610., Springer (2003) 384-393 933, 935, 937, 938, 939, 940
[2] Quinlan, JA.: C4.5: Programs for Machine Learning. Morgan Kaufmann, San Mateo (1993) 933, 934, 936, 937
[3] Witten, I.H., Frank, E.: Data Mining: Practical machine learning tools and techniques with Java. Morgan Kaufmann, CA (1999) 934
[4] Breiman, L., Friedman, J.H., Olshen, RA., Stone, C.J.: Classification and Regression Trees. Wadsworth, Inc. Belmont, California (1984) 934, 936
[5] Treigueiros, D., Berry, R-H.: The application of neural network based methods to the extraction of knowledge from accounting reports. In: Proceedings of 24th Annual Hawaii Int. Conf. on System Sciences IV. (1991) 137-146 934
[6] Murthy, S., Salzberg, S.: A system for induction of oblique decision trees. Journal of Artificial Intelligence Research 2 (1994) 1-32 934
[7] Kuscu, I.: A genetic constructive induction model. In Angeline, P.J., Michalewicz, Z., Schoenauer, M., Yao, X., Zalzala, A., eds.: Proc of Congress an Evolutionary Computation. Volume 1., IEEE Press (1999) 212-217 935
[8] Koza, J.: Genetic Programming: On the Programming of Computers by Means of Natural Selection. MIT Press (1992) 935
[9] Zheng, Z.: Effects of different types of new attribute an constructive induction. In: Proc of 8th Int. Conf. on Tools with Artifical Intelligence (ICTAI'96), IEEE (1996) 254-257 935, 937
[10] Bensusan, H., Kuscu, I.: Constructive induction using genetic programming. In Fogarty, T., Venturini, G., eds.: Proceedings of tut. Conf. Machine Learning, Evolutionary Computing and Machine Learning Workshop. (1996) 935
[11] Kass, G. V.: An exploratory technique for investigating large quantities of categorical data. Applied Statistics 29 (1980) 119-127 936

Semi-Automatic Construction of Metadata from a Series of Web Documents

Sachio Hirokawa[1], Eisuke Itoh[1], and Tetsuhiro Miyahara[2]

[1] Computing and Communications Center, Kyushu University
Hakozaki 6-10-1, Higashi-ku, Fukuoka, 812-8581, Japan
{hirokawa,itou}@cc.kyushu-u.ac.jp
[2] Faculty of Information Sciences, Hiroshima City University
Otsuka-Higashi 3-4-1, Asaminami-ku, Hiroshima, 731-3194, Japan
miyahara@its.hiroshima-cu.ac.jp

Abstract. Metadata plays an important role in discovering, collecting, extracting and aggregating Web data. This paper proposes a method of constructing metadata for a specific topic. The method uses Web pages that are located in a site and are linked from a listing page. Web pages of recipes, real estates, used cars, hotels and syllabi are typical examples of such pages. We call them a series of Web documents. A series of Web pages have the same appearance when a user views them with a browser, because it is often the case that they are written with the same tag pattern. The method uses the tag-pattern as the common structure of the Web pages.

Individual contents of the pages appear as plain texts embedded between two consecutive tags. If we remove the tags, it becomes a sequence of plain texts. The plain texts in the same relative position can be interpreted as attribute values if we presume that the pages represent records of the same kind.

Most of these plain texts in the same position vary page to page. But, it may happen that the same texts show up at the same relative position in almost all pages. These constant texts can be considered as attribute names. "Location", "Rating" and "Travel from Airport" are examples of such constant texts for pages of hotel information. If the frequency of a text is higher than a threshold, we accept it as a component of metadata.

If we mark a constant text with "N" and a variable text with "V", the sequence of plain texts forms a series of N's and V's. A page in a series contain two kinds of NV sequence pattern. The first pattern is $(NV)^n$, which we call vertical, where an attribute value follows the attribute name immediately. The second pattern is $N^n V^n$, which we call horizontal, where names occur in the first row and the same number of values follow in the next row. Thus we can understand the meaning of values and can construct records from a series of Web pages.

Keywords: Knowledge acquisition, Knowledge engineering, Knowledge discovery and Data Mining, Machine learning, Ontology.

T.D. Gedeon and L.C.C. Fung (Eds.): AI 2003, LNAI 2903, pp. 942–953, 2003.

1 Introduction

Due to the rapid spread of the Web, huge amounts of information in various formats are available on the Web. In order to extract useful information from huge Web pages, there are many research tasks such as extraction of knowledge from semistructured data or HTML files [2, 4, 9], topical crawling which automatically collects Web pages of user's topics [1, 3], and integration of semantically homogeneous information which is heterogeneous in representation.

We are constructing an information integration system utilizing Web pages on specific topics. In order to realize such an information integration system, we need to realize the following subtasks: collection of Web pages on a specific topic, classification of the collected Web pages, information extraction from Web pages, and construction of a database of extracted information. However, Web pages on a specific topic are available in multiple Web sites of different formats. To resolve differences between multiple databases, the database schema is necessary for each source and metadata is necessary as an integration target.

There have been many efforts to construct metadata for this purpose. But it is not a trivial task. In this paper we propose a new method for semi-automatic construction of metadata from a *series of Web pages* [14]. A series of Web pages are pages that are located in a site and are linked from a listing page in the site. Web pages of recipes, real estates, used cars, hotels and syllabi are typical examples of such pages. These series of pages can be searched with a phrase "list of" and a keyword of the topic.

In most cases, a series of Web pages are in the same site and are linked from a page of contents in the site. We call the source page of the links as a *page of type A*. The Web pages that are linked from the page of type A are called *pages of type B*

A key problem in Web Mining is the separation of the structure and the contents from an HTML file. A series of Web pages have the same appearance when a user views them with a browser, because it is often the case that they are written with the same tag-pattern. The proposed method uses the tag-pattern as the structure of the Web pages. Individual contents of the pages appear as plain texts embedded between two consecutive tags. If we remove the tags, it becomes a sequence of plain texts. The plain texts in the same relative position can be interpreted as attribute values if we presume that the pages represent records of the same kind. Most of these plain texts in the same position vary

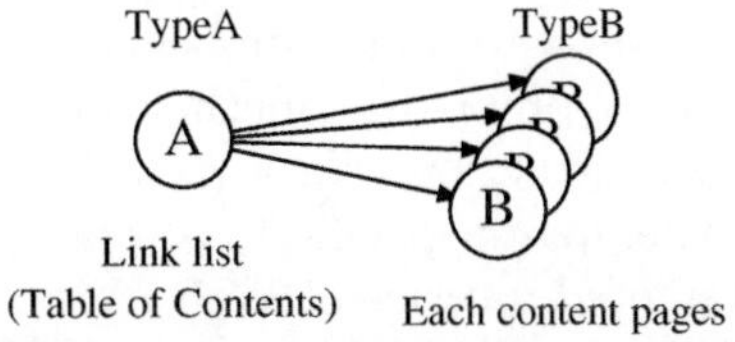

Fig. 1. Link Structure of a Series of Web Pages

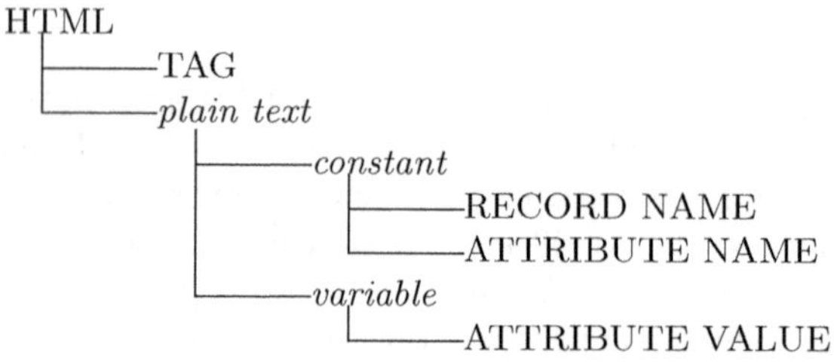

Fig. 2. Structure and Contents of HTML Files

page to page. But, it may happen that the same texts show up at the same relative position in almost all pages. These constant texts can be considered as attribute names. "Location", "Rating" and "Travel from Airport" are examples of such constant texts for pages of hotel information. If the frequency of a text is higher than a threshold, we accept it as a component of metadata. Our solution is to use (TAG, RECORD NAME, ATTRIBUTE NAME) as the structure, and ATTRIBUTE VALUE as contents (see Fig. 2).

If we mark a constant text with "N" and a variable text with "V", the sequence of plain texts forms a series of N's and V's. A page in a series contains two kinds of NV sequence pattern. The first pattern is $(NV)^n$, which we call vertical, where an attribute value follows the attribute name immediately. The second pattern is N^nV^n, which we call horizontal, where names occur in the first row and the same number of values follow in the next row. Thus we can understand the meaning of values and can construct records from a series of Web pages.

Making content-related metadata plays an important role in information integration on Web pages. As the extensive studies on the Semantic Web, RDF and RDF schema show, extraction of meaningful metadata describing the contents of Web pages is the key to realize an information integration system. However, almost all Web pages in the current WWW are HTML files and the acquisition of appropriate metadata is still a major problem to realize such a system. Hence we propose a new method for semi-automatic construction of metadata from a series of Web pages as a realistic method for implementing information integration on the Web pages.

In order to extract a template specific to a series of Web pages, our method uses a measure of structural similarity among Web pages. Measuring the structural similarity among semistructured data has been an active research topic [4, 5, 6, 8, 10] and it is fundamental to many applications such as integrating Web data sources. The authors have proposed a method for extracting a common tree structured pattern from semistructured data [11]. This extraction method can be applied to extracting a template specific to a series of Web pages.

Umehara et al. [14] targeted a series of Web pages. But their aim is not in generating metadata, but in transforming a series of HTML files into a series of corresponding XML files. Their method requires a user to prepare transformation examples. Stuckenschmidt et al. [17] proposed a knowledge-based approach for metadata validation and generation but their approach is within a framework of ontology. Handschuh et al. [16] proposed a general framework for creation of semantic metadata in the Semantic Web but the framework does not deal with raw data of Web pages. Arakas et al. [2] proposed an extraction method of metadata from Web pages in a Web site but do not consider information integration using metadata.

These works focus on some of the subtasks such as extracting a template or making metadata. But our aim is to construct a total system of information integration utilizing Web pages on specific topics. We believe that our method will be improved by using the methods of these related works.

This paper is organized as follows. In Section 2, we explain the idea with the Web documents of hotel information. In section 3, we propose a method for extracting records from a series of Web pages by transforming the pages into tag sequences. Then the metadata is constructed as a list of strings that appear in the same position of all the records. In section 4 and 5, we give a method for combining the names and the values of attributes. In Section 6 we conclude this paper.

2 Metadata for "Hotel"

In this section, we explain the idea of constructing metadata for "hotels". An online hotel reservation site "Bookingsavings" [1] has lists of hotels for many destinations. For example, a list of hotels in Perth is displayed in Fig. 3.

The list has 20 links to the pages of detailed information of the hotels, where "Rating", "Rates", "Room facilities", "Hotel facilities", "Location", "Travel from Perth Airport", "Travel from railway Station", "Children/extra bed" and "Places of interest nearby" are displayed in the same pattern as we see in Fig. 4.

The page for "Novotel Vines Resort" contains "Wineries", "Historic Sites" and "Local Attractions" but does not contain other fields. Nevertheless, 19 pages are written with the same pattern. We describe the pattern as the following tag-sequence, where "*" represents the positions of texts which vary hotel to hotel.

```
html head title * meta meta meta meta link /head body div table tr td
img /td /tr tr td div * * * /td /tr /table table tr td br table tr td
img /td /tr /table div * div * br a img /a br br span * * img img img
/span br span * * * br br div * br div * br div * br div br li b * *
br li b * * br li b * * br li b * * br li b * * br li b * * br li b *
* br br /td /tr tr td a * * * * * * * * /td /tr /table table tr td img
/td /tr /table a * /div /body /html
```

The pattern contains 36 such fields or texts. These 36 texts form a record of each hotel information. The fields are classified as common parts and individual

[1] http://www.bookingsavings.com/asia_pacific/australia/perth/index.shtml

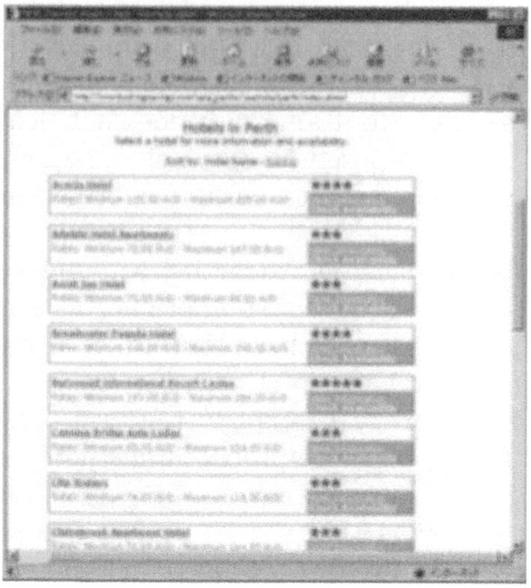

Fig. 3. A List of Hotels in Perth

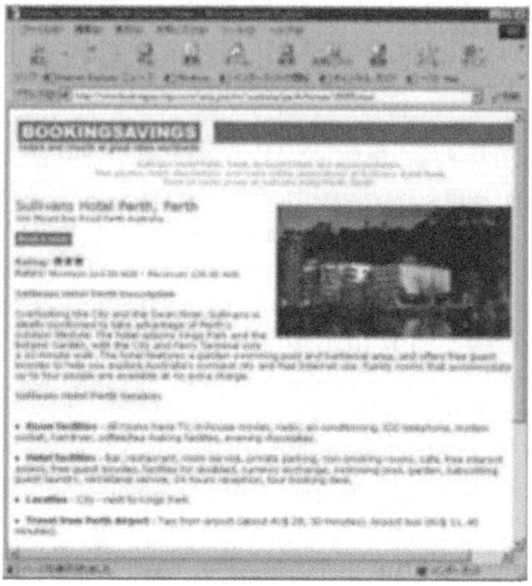

Fig. 4. A Page of Detailed Information of a Hotel

parts according to the frequency. The common parts appear as the names of fields of the record. A field with frequency 19 represents a name of an attribute. On the other hand, the texts in a field with frequency 1 vary according to each HTML page. There may be some exception, as in the 25th field of "Ascot Inn Hotel", where "Extra bed" is used instead of "Childbed/extra bed" and as in the 23rd field of "City Waters Hotel", where "Travel from Perth" is used instead of "Travel from Perth Airport".

3 Metadata Construction Algorithm

In this section, we describe an algorithm for constructing metadata given a keyword. Fig. 5 shows the algorithm.

In the first step, we send the keyword to a search engine for collecting pages related to a topic. Here we send the keyword augmented with the phrase "list of". Thus we obtain Web pages which are supposed to have many links to related pages. These are the pages of "type A" as we explained in the previous section. A page of type A contains many links to the desired pages, i.e., pages of "type B". But it may contain other kind of links, e.g., links to the top pages and links to famous pages. It is often the case that the pages of "type B" are located in the same directory in the same site. Such directory is calculated with the base URL. For example, the HTML files of the hotels in Section 2 are in `http://www.bookingsavings.com/asia_pacific/australia/perth/hotels/`. The pages which are linked from the page of type A and are in the directory are the pages of type B.

Table 1. Contents and Frequency

field	frequency	contents
1	1	Sullivans Hotel Perth - Perth Discount Hotels
2	1	Sullivans Hotel Perth, Perth discount hotels ...
3	1	View photos, ... at Sullivans Hotel Perth
4	1	Save on room prices at Sullivans Hotel Perth, Perth.
5	1	Sullivans Hotel Perth, Perth
6	1	166 Mount Bay Road Perth Australia
7	19	Rating:
8	19	
9	19	Rates:
10	19	
11	1	Minimum 110.00 AUD - Maximum 130.00 AUD
12	1	Sullivans Hotel Perth Description
13	1	Overlooking the City and the Swan River, ...
14	1	Sullivans Hotel Perth Services
15	16	Room facilities
16	1	- All rooms have TV, in-house movies, ...
17	19	Hotel facilities
18	1	- Bar, restaurant, room service, ..
19	19	Location
20	1	- City - nest to Kings Park.
21	7	Travel from Perth Airport
22	1	- Taxi from airport (about AU$ 28, 30 minutes)....
23	8	Travel from railway station
24	1	- Taxi from railway station (about AU$ 5, 5 minutes)....
25	6	Children/extra bed
26	1	- Maximum 2 children are allowed to stay ...
27	19	Places of interest nearby
28	1	Kings Park, Swan River.
29	19	Worldwide Hotels
30	19	–
31	19	Asia Pacific Hotels
32	19	–
33	19	Australia Hotels
34	19	–
35	19	Perth Hotels
36	19	Copyright, terms and conditions.

The second step is to obtain the common tag-pattern of pages of type B. It is obtained as the tag-sequence for a Web page which belongs to the maximal cluster with respect to the tag-sequence mapping. To use the tag-sequence is introduced in [12]. The tag-sequence mapping is a function from an HTML file to the tag-sequence of the file. It eliminates attributes of tags and deletes textual contents which appear outside of HTML tags. The *maximal cluster* is defined as follows. Let f be a mapping from a set X to a set Y. A *cluster* in X with respect to f is an inverse image of $y \in Y$ with respect to f, i.e., $f^{-1}(y) = \{x \in X \mid f(x) = y\}$. A cluster is *maximal* if the number of elements in the cluster is maximal.

The 3rd step is to extract the contents from pages of type B using the tag-sequence. We do not need pattern matching at this stage. Because, we already obtain the corresponding contents when we calculate the tag sequence of the HTML file. At this stage, the i-th page $a[i]$ of type B is represented as a list $a[i, 1], a[i, 2], ..., a[i, n]$ of strings. Since the pages of type B are supposed to have the same tag sequence, the length n of this strings is the same to all pages of type B. Thus a page of type B is transformed into a record with n-fields.

The final step is to classify the fields and distinguish the names of the attributes and the values of the attributes. If all j-th fields have the same string, we consider the field represents the name of the attribute whose contents follow in the sequel. Conversely, the attribute values vary one by one. So, we distinguish the names and the values of attribute by the frequency of the strings among the same fields.

4 Alignment of Names and Values

We can consider an HTML page to be a merged sequense $(T_1, P_1, T_2, P_2, \cdots, T_n, P_n, T_{n+1})$, where T_i is a tag and P_j shows a text. A series of Web pages $H_1, H_2, \cdots, H_m$ have the same tag-sequence, so that the sequence of plain texts $(P_1^i, P_2^i, \cdots, P_n{}^i)$ can be extracted uniformly from each page $H_i (i = 1, \cdots, m)$.

The plain text P_j^i in the j-th position may be an attribute name or an attribute value. In the previous section, we showed an algorithm to distinguish names and values according to the frequency of the text. In this section and next section, we explain how to name the values or how to bind names and values.

Consider a series of pages of "used cars" as an example. Characteristic keywords of these pages are "maker", "model", "year", "mileage", "color" and so on. These keywords are the attribute names and are displayed with attribute values aligned vertically or horizontally when we see the pages with a browser (Fig. 6). In both cases of alignment, a name and the corresponding value are displayed close to each other. Such display improves the visual effect and helps user's comprehension.

Imagine that we have a series of pages as in Fig. 7 and that we know that F_1 is a name. Where does a corresponding value appear? If F_3 is another instance of name, the value for F_1 should be F_2. If F_2 is a name, then the value for F_1 should be F_3. Thus, if names are aligned vertically the corresponding value appears horizontally next to the name. If names are aligned horizontally, the corresponding value appears vertically next to the name.

5 Binding Name and Value by NV Sequence

Fig. 8 shows a series of syllabus pages at "Anan National College of Technology"[2].

The page of type A in Fig. 8 (a) has 44 links. Three of them are the links to the top pages of the site, the college and the syllabi. The other 41 links are the links to course pages or pages of type B, e.g., "Applied mathematics"(Fig. 8 (b)), "Circuit theory", "Electromagnetics" and so on. These pages have the same template of 31 fields shown in Table 2. The second column "Freq." shows the frequency of the field text. For example, the 4th field text "course" appears in

[2] `http://www.anan-nct.ac.jp/gakka/syllabus/h13/curri_e.html`

```
procedure Pattern-and-Bs {
   Input: a listing page a;
   Output: tag-pattern;

    X = the list of pages linked from a;
    B = the maximal cluster in X w.r.t. base();
   D = d_1, ..., d_m = the maximal cluster in B w.r.t. tag();
   p = p_1.*.p_2.*....*.p_n.*.p_{n+1} the tag-sequence tag(d_1);
   return(p, B);
}

procedure ExtractFields {
   Input: a tag-pattern p = p_1.*.p_2.*....*.p_n.*.p_{n+1};
            an HTML file h;
   Output: a record a = (a[1], a[2], ..., a[n]);

   if tag(h)=p {
      a[i] = the i-th variable parts "*" in h;
   }
   return (a[1], a[2], ..., a[n]);
}

procedure NameValueSeparation {
   Input: a_1, ..., a_m : a list of records of the same fields,
      i.e., a_i = (a[i, 1], ..., a[i, n]);
      α : a threshold;
   Output: a list nv of "N" and "V" of length n;

   function g(i) = a[i,j] the j-th field of i-th record;
   for( j = 1; j <= n; j++ ) {
      weight[j] = the number of elements of maximal cluster w.r.t. g()
      in {1, 2, ···, m}
      if ( weight[j] > α) {
         nv[j] = "N";
      } else {
         nv[j] = "V";
      }
   }
   return nv;
}

main {
   Input: a keyword w;
   Output: a list of keywords;

   SearchResult = search("list of " + keyword);
   (p, B) = Pattern-and-Bs(SearchResult);
    C = map { λ(h) ExtractFields(p, h) } B;
   NV = NameValueSeparation(C);
   return the names of NV;
}
```

Fig. 5. Metadata Construction Algorithm.

41 pages, i.e., in all pages. A field text which appears in all pages is an attribute name or a common text to the site, such as a site name.

Field texts of Web documents in a series can be classified in two types.

N : Constant text that appears in almost all pages at the same position.
V : Variable text that varies page to page.

Maker	*Mitsubishi*
Model	*Lancer Evolution*
Year	*2003*
Mileage	*100*
Color	*Yellow*

(a) Vertical Alignment

Maker	Model	Year	Mileage	Color
Mitsubishi	*Lancer Evolution*	*2003*	*100*	*Yellow*

(b) Horizontal Alignment

Fig. 6. Name Value Alignment

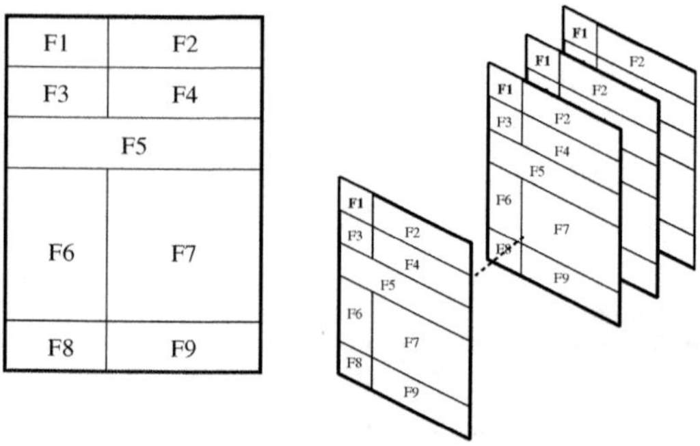

Fig. 7. Relative Position of Name and Value

(a) page of type A

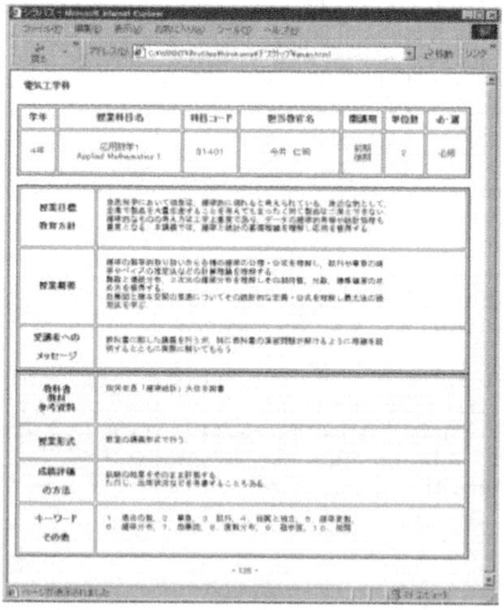

(b) page of type B

Fig. 8. A Syllabus Page of Anan National College of Technology

"N" stands for name and "V" stands for value. Then we can describe the alignment of a page with some *NV sequences* N^nV^n. The i-th value is obtained at the position of i-th V.

Table 2. Appearance frequency of fields.

Field No.	Freq.	Word(s) (translated)	Word(s) (in Japanese)
1	41	Syllabus	• • • • • • • • •
2	41	Dept. Elec. eng.	• • • • • • • • • • •
3	41	Grade	• • • •
4	41	Course	• • • • • • • • • • •
5	41	Code	• • • • • • • • • • •
6	41	Lecturer Name	• • • • • • • • • • •
7	41	Period	• • • • • •
8	41	Credit	• • • • • •
9	41	elective or required	• • • • • •
10	10	5th degree	• • • •
	15	4th degree	• • • •
14	28	First term, Second term	• • • • • • • • •
15	14	1	• •
	24	2	• •
16	17	elective (optional)	• • • •
	24	required (compulsory)	• • • •
17	41	Course Goal	• • • • • • • • • • • • • • • • • •
19	41	Abstract	• • • • • • • • •
21	41	Message to students	• •
23	41	Text books, Teaching Materials, Reference books	• •
25	41	Class type	• • • • • • • • •
27	41	Evaluation	• • • • • • • • • • • • • • • •
29	41	Keywords, MISC.	• • • • • • • • • • • • • • • • • •

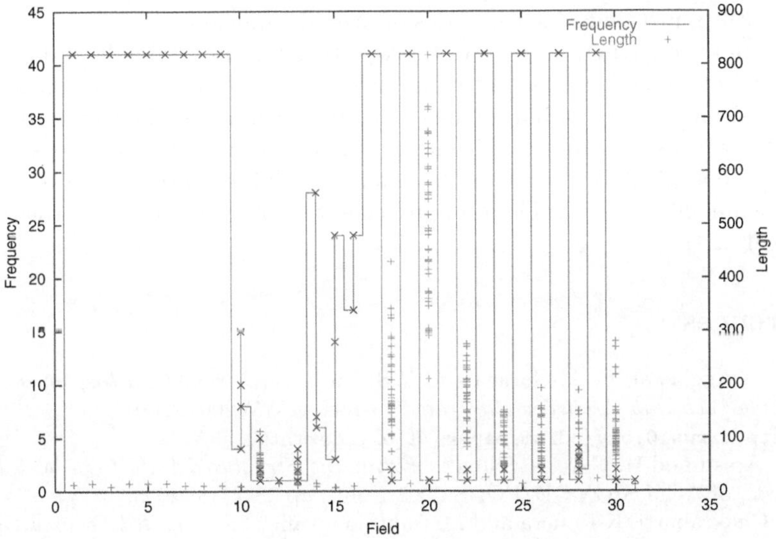

Fig. 9. Frequency and Length of Field Texts

Table 2 displays the names of high frequency. Note that another characteristic feature of value fields is that the length is relatively large.

Fig. 9 shows the frequency and the length of field texts. We can see that the text length is short and the frequency is high for the fields 1-9. On the other hand, the texts in the fields 10-16 are relatively long and have low frequency. If we see the fields 1-17 closely, we notice a discrepancy of the number of names (9) and the number of values (7). The first and second name fields are the cause of the mismatch. There is no value corresponding to these "N". This kind of name can be considered as a record name instead of an attribute name. Thus we have the NV sequence $N^2N^7V^7(NV)^7$ and we can confirm that the record has 14 attributes.

6 Conclusion

We proposed a method for constructing metadata from a series of Web documents. It is often the case that a site provides many Web pages for a specific contents. Moreover these pages look very similar, because they are written with the same template HTML file. We used the common tag-sequence as the template of these pages. The individual contents of the pages are surrounded by these tags. The i-th string enclosed by the i-th and $i+1$-st tags can be considered as the i-th field of the record. If all of the i-th field are identical and can be considered as a constant, it does not represent the value but the name of an attribute in the record. The method we proposed uses the frequency of the string in the same field to distinguish the name and the value. The fields of names form a metadata for the series of Web documents.

Metadata construction is one of the key steps for integrating Web documents. The core idea of the method is to use the frequency of texts that appear in the same position of the similar semi-structured documents. The tag-sequence is used in the present paper, but there are many other similarity measurements that can be applied for detecting the similarity and for extracting a template of Web documents [5, 6, 7]. These approaches will improve the robustness of the proposed method.

References

[1] C. C. Aggarwal, F. Al-Garawi and P. S. Yu : "*Intelligent Crawling on the World Wide Web with Arbitrary Predicates*", Proc. WWW2001, 2001. `http://www10.org/cdrom/papers/110/index.html` 943

[2] A. Arasu and H. Garcia-Molina : "*Extracting Structured Data from Web Pages,*" Proc. of ACM SIGMOD/PODS 2003 Conf., pp.337-348, 2003. 943, 945

[3] S. Chakrabarti, K Punera and M. Subramanyam : "*Accelerated Focused Crawling through Online Relevance Feedback*", Proc. WWW2002, 2002. `http://www2002.org/CDROM/refereed/336/index.html` 943

[4] C. H. Chang, S. C. Lui, Y. C. Wu : "*Applying Pattern Mining to Web Information Mining,*" Proc. PAKDD 2001, Spring LNAI 2035, pp.4-16, 2001. 943, 944

[5] I. F. Cruz, S. Borisov, M. A. Marks and T. R. Webb : "*Measuring Structural Similarity Among Web Documents: Preliminary Results*", Proc. EP 1998, Springer LNCS 1375, pp.513–524, 1998. 944, 952

[6] S. Flesca, G. Manco, E. Masciari, L. Pontieri and A. Pugliese : "*Detecting Structural Similarities between XML Documents*", Proc. WEBDB2002, 2002. `http://feast.ucsd.edu/webdb2002/papers/19.pdf` 944, 952

[7] J. Han, J. Pei and Y. Yin : "*Mining Frequent Patterns without Candidate Generation*", Proc. ACM SIGMOD Intl. Conf. Management of Data, pp.1–12, 2000. 952

[8] J.W. Lee, K. Lee, W. Kim : "*Preparations for semantics-based XML mining*," Proc. IEEE Int. Conf. od Data Mining (ICDM) 2001, pp.345-352, 2001. 944
[9] K. Lerman, C. Knoblock and S. Minton : "*Automatic Data Extraction from Lists and Tables in Web Sources*", `http://www.cs.waikato.ac.nz/~ml/publications/1999/99SJC-GH-Innovative-apps.pdf` 943
[10] H. Leung, F. Chung, S.C. Chan : "*A New Sequential Mining Approach to XML Document Similarity Computation*," Proc. PAKDD 2003, Spring LNAI 2637, pp.356-362, 2003. 944
[11] T. Miyahara, Y. Suzuki, T. Shoudai, T. Uchida, S. Hirokawa, K. Takahashi and H. Ueda : "*Discovery of Frequent Tag Tree Patterns in Semistructured Web Documents*", Proc. PAKDD 2003, Springer LNAI 2637, pp.430–436, 2003. 944
[12] T. Taguchi, Y. Koga and S. Hirokawa : "*Integration of Search Sites of the World Wide Web*", Proc. CUM, Vol2, pp.25–32, 2000. 947
[13] S. Yamada, Y. Matsunaga, E. Itoh and S. Hirokawa : "*A study of design for intelligent web syllabus crawling agent*" Trans. of IEICE D-I, Vol.J86, No.8, pp.566-574, 2003. (in Japanese)
[14] 943, 944
M. Umehara, K. Iwanuma, H. Nagai :f "A Case-Based Semi-automatic Transformation from HTML Documents to XML Ones –Using the Similarity between HTML Documents Constituting a Series–," Journal of JSAI, Vol.16, No.5, pp.408-416, 2001. (in Japanese)
[15] C.C. Marshall : "*Making metadata: a study of metadata creation for a mixed physical-digital collection DL '98*," Proc. of the 3rd ACM Int'l Conf. on Digital libraries, pp.162-171, 1998.
[16] S. Handschuh and S. Staab : "*Authoring and annotation of web pages in CREAM*," Proc. WWW2002, 2002.
`http://www2002.org/CDROM/refereed/506/index.html` 945
[17] H. Stuckenschmidt and F. van Harmelen : "*Ontology-based metadata generation from semistructured information*," Proc. of K-CAP'01, pp.440–444, 2001 945

Constructive Plausible Logic Is Relatively Consistent

David Billington and Andrew Rock

School of Computing and Information Technology
Nathan Campus, Griffith University
Brisbane, Queensland 4111, Australia
Telephone: +61 (0)7 3875 {5017,5016}
Facsimile: +61 (0)7 3875 5051
{d.billington,a.rock}@griffith.edu.au
http://www.griffith.edu.au

Abstract. An implemented, efficient, propositional non-monotonic logic, called Constructive Plausible Logic, is defined and explained. Several important properties enjoyed by this logic are stated. The most important property, relative consistency, means that whenever the set of axioms is consistent so is the set of all formulas proved using defeasible information. Hence the non-monotonic deduction mechanism is trustworthy. This is the first Plausible Logic which has been proved to be relatively consistent. Constructive disjunction is characterised by the property that a disjunction can be proved if and only if at least one of its disjuncts can be proved. Constructive Plausible Logic uses constructive disjunction. Moreover the ambiguity propagating proof algorithm is simpler than the one in Billington and Rock [4].

Keywords: Nonmonotonicity; Common-sense reasoning; Logic; Knowledge representation.

1 Introduction

In the late 1980s Nute [6] introduced a non-monotonic reasoning formalism called Defeasible Logic. It was a propositional logic which dealt with uncertain and incomplete information, as well as factual information. However the uncertainty and incompleteness was not represented by numbers. Moreover the reasoning was based on principles rather than on the manipulation of numbers. So probabilities and certainty factors have no place in Defeasible Logic. Defeasible Logic has many desirable properties, but perhaps the most important of these is a deterministic polynomial deduction procedure which enables a straightforward implementation of this logic. However, Defeasible Logic can neither represent nor prove disjunctions. In the late 1990s Billington [2] introduced Plausible Logic which was based on Defeasible Logic but which could represent and prove clauses. Both Defeasible Logic and Plausible Logic have a syntax which distinguishes between formulas proved from just the facts or axioms, and those proved using defeasible information.

T.D. Gedeon and L.C.C. Fung (Eds.): AI 2003, LNAI 2903, pp. 954–965, 2003.

Relative consistency means that whenever the set of axioms is consistent so is the set of all formulas proved using defeasible information. Relative consistency is important because it shows that the non-monotonic deduction mechanism does not create inconsistencies. The incorporation of disjunction makes proving relative consistency very difficult. Indeed relative consistency has not been proved for the Plausible Logic introduced in Billington and Rock [4].

Constructive disjunction is characterised by the property that $\bigvee L$ can be proved if and only if at least one element of L can be proved. With classical disjunction it is possible to prove $\bigvee L$ and not be able to prove any element of L. Experience with translating business rules and regulations into Plausible Logic indicates that often only constructive disjunction is needed. For example, the eligibility criteria for becoming a member of the IEEE is a disjunctive list, which means that if at least one criterion in the list is satisfied then the candidate is eligible. In general constructive disjunction is exactly what is needed for any disjunctive list of eligibility criteria. The fact that constructive disjunction is computationally simpler than classical disjunction means that constructive disjunction may be just the right compromise between computational efficiency and expressive power. Certainly Constructive Plausible Logic is much more expressive than Defeasible Logic.

The purpose of this paper is to define, and give some intuitions about, a Plausible Logic which is relatively consistent. All attempts at proving relative consistency for many different Plausible Logics with a non-constructive disjunction have failed. As this paper shows, when the simpler constructive disjunction is used then relative consistency can be proved.

A separate issue is that of ambiguity. An atom a is *ambiguous* if and only if neither a nor its negation, $\neg a$, can be proved. Suppose there is evidence for b. If a is ambiguous and a is evidence for $\neg b$ then what should be concluded about b? A logic is ambiguity blocking if it can conclude b; and it is ambiguity propagating if b is ambiguous, because the ambiguity of a has been propagated to b.

The Plausible Logic presented in this paper has a simplified ambiguity propagating proof algorithm compared to the Plausible Logic in Billington and Rock [4]. The ambiguity propagating proof algorithm of Billington and Rock [4] required an auxiliary proof algorithm. No such auxiliary algorithm is needed in this paper.

Section 2 of this paper defines and explains Constructive Plausible Logic. Section 3 presents the main results. Unfortunately space limitations allow the inclusion of only the proof of the main theorem on relative consistency. However all the proofs are in Billington [3]. Section 4 describes our implementation of Constructive Plausible Logic, and contains an example of a reasoning problem and its automated proof. Section 5 is the conclusion.

2 Constructive Plausible Logic

We begin by establishing our basic notation and terminology. We often abbreviate "if and only if" by "iff". X is a subset of Y is denoted by $X \subseteq Y$; the notation $X \subset Y$ means $X \subseteq Y$ and $X \neq Y$, and denotes that X is a *proper subset* of Y. The empty set is denoted by $\{\}$, and the set of all integers by $\mathbb{Z}$. If m and n are integers then we define $[m..n] = \{i \in \mathbb{Z} : m \leq i \leq n\}$. The cardinality of a set S is denoted by $|S|$. The length of a sequence P is denoted by $|P|$. Let $P = (P(1), P(2), \ldots, P(|P|))$ be a finite sequence. If $i \in [1..|P|]$ then $P[1..i] = (P(1), \ldots, P(i))$, and if $i = 0$ then $P[1..i] = ()$, the empty sequence. The notation $x \in P$ means that there exists j in $[1..|P|]$ such that $x = P(j)$. And $x \notin P$ means not($x \in P$). The concatenation of a sequence P onto the beginning of the sequence Q is denoted by $P\&Q$. Let S be any set. It is sometimes convenient to abbreviate "for all x in S" by "$\forall x \in S$". Also "there exists an x in S such that" is sometimes abbreviated to "$\exists x \in S$ such that", and sometimes to just "$\exists x \in S$". $\mathcal{P}(L)$ is the powerset of L. When convenient we abbreviate "for all I in $\mathcal{P}(S)$" by "$\forall I \subseteq S$". Also "there exists an I in $\mathcal{P}(S)$ such that" is sometimes abbreviated to "$\exists I \subseteq S$ such that", and sometimes to just "$\exists I \subseteq S$".

Our *alphabet* is the union of the following four pairwise disjoint sets of symbols: a non-empty finite set, *Atm*, of (propositional) atoms; the set $\{\neg, \bigwedge, \bigvee, \rightarrow, \Rightarrow, \rightharpoondown\}$ of connectives; the set $\{+, -, \delta, \gamma, \pi\}$ of proof symbols; and the set of punctuation marks consisting of the comma and both braces. By a *literal* we mean any atom, a, or its negation, $\neg a$. A clause, $\bigvee L$, is the disjunction of a finite set, L, of literals. $\bigvee\{\}$ is the *empty clause* or *falsum* and is thought of as always being false. If l is a literal then we regard $\bigvee\{l\}$ as another notation for l and so each literal is a clause. A clause $\bigvee L$ is a tautology iff both an atom and its negation are in L. A *contingent* clause is a clause which is not empty and not a tautology. A *dual-clause*, $\bigwedge L$, is the conjunction of a finite set, L, of literals. $\bigwedge\{\}$ is the *empty dual-clause* or *verum* and is thought of as always being true. If l is a literal then we regard $\bigwedge\{l\}$ as another notation for l and so each literal is a dual-clause. Thus $\bigwedge\{l\} = l = \bigvee\{l\}$. Neither the verum nor the falsum are literals. The verum is not a clause, and the falsum is not a dual-clause. A *cnf-formula*, $\bigwedge C$, is the conjunction of a finite set, C, of clauses. A *dnf-formula*, $\bigvee D$, is the disjunction of a finite set, D, of dual-clauses. If c is a clause then we regard $\bigwedge\{c\}$ as another notation for c. If d is a dual-clause then we regard $\bigvee\{d\}$ as another notation for d. Thus both clauses and dual-clauses are both cnf-formulas and dnf-formulas. By a *formula* we mean any cnf-formula or any dnf-formula. The set of all literals is denoted by *Lit*; the set of all clauses is denoted by *Cls*; the set of all dual-clauses is denoted by *DCls*; the set of all cnf-formulas is denoted by *CnfFrm*; the set of all dnf-formulas is denoted by *DnfFrm*; and the set of all formulas is denoted by *Frm*. *Frm* is finite.

We define the *complement*, $\sim f$, of a formula f and the *complement*, $\sim F$, of a set of formulas F as follows. If f is an atom then $\sim f$ is $\neg f$; and $\sim\neg f$ is f. If L is a set of literals then $\sim L = \{\sim l : l \in L\}$. If $\bigvee L$ is a clause then $\sim\bigvee L = \bigwedge\sim L$. If $\bigwedge L$ is a dual-clause then $\sim\bigwedge L = \bigvee\sim L$. So the complement of a clause is a dual-clause, and the complement of a dual-clause is a clause. In particular the falsum and the verum are complements of each other. If E is a set of clauses or a set of dual-clauses then $\sim E = \{\sim e : e \in E\}$. If $\bigwedge C$ is a cnf-formula then $\sim\bigwedge C = \bigvee\sim C$. If $\bigvee D$ is a dnf-formula then $\sim\bigvee D = \bigwedge\sim D$. So the complement of a cnf-formula is a dnf-formula, and the complement of a dnf-formula is a cnf-formula. If F is a set of formulas then $\sim F = \{\sim f : f \in F\}$. Both *Lit* and *Frm* are closed under complementation.

The information with which constructive plausible logic reasons is either certain or defeasible. All the information is represented by various kinds of rules and a priority relation on those rules. Define r to be a *rule* iff $r = (A(r), arrow(r), c(r))$ where $A(r)$ is a finite set of literals called the *antecedent* of r, $arrow(r) \in \{\rightarrow, \Rightarrow, \rightharpoonup\}$, $c(r)$ is a literal called the *consequent* of r, $c(r) \notin A(r)$, and $\sim c(r) \notin A(r)$. A rule r which contains the *strict arrow*, $\rightarrow$, is called a *strict rule* and is usually written $A(r) \rightarrow c(r)$. A rule r which contains the *plausible arrow*, $\Rightarrow$, is called a *plausible rule* and is usually written $A(r) \Rightarrow c(r)$. A rule r which contains the *defeater arrow*, $\rightharpoonup$, is called a *defeater rule* and is usually written $A(r) \rightharpoonup c(r)$. The antecedent of a rule can be the empty set. The set of all rules is denoted by *Rul*. *Rul* is finite.

Strict rules, for example $A \rightarrow l$, behave like the material conditional. If all the literals in A are proved then l can be deduced. Plausible rules, for example $A \Rightarrow l$, represent some of the aspects of a situation which are plausible. If all the literals in A are proved then l can be deduced provided that all the evidence against l has been defeated. So we take $A \Rightarrow l$ to mean that, in the absence of evidence against l, A is sufficient evidence for concluding l. A defeater rule, for example $A \rightharpoonup \sim l$, is evidence against l, but it is not evidence for $\sim l$. $A \rightharpoonup \sim l$ can be defeated by defeating $\bigwedge A$. Defeater rules can be used to prevent conclusions which would be too risky. For instance, given the rules $a \Rightarrow b$ and $b \Rightarrow c$ it may be too risky to conclude that things with property a usually have property c. In which case we could add the defeater rule $a \rightharpoonup \sim c$. In this case adding $a \Rightarrow \sim c$ would be wrong because having property a is not a reason for having property $\sim c$, indeed it is a weak reason for having property c.

Let R be any set of rules. The set of antecedents of R is denoted by $A(R)$; that is $A(R) = \{A(r) : r \in R\}$. The set of consequents of R is denoted by $c(R)$; that is $c(R) = \{c(r) : r \in R\}$. We denote the set of strict rules in R by R_s, the set of plausible rules in R by R_p, and the set of defeater rules in R by R_d. Also we define $R_{pd} = R_p \cup R_d$ and $R_{sp} = R_s \cup R_p$.

Let l be any literal. If C is any set of clauses define $C[l] = \{\bigvee L \in C : l \in L\}$ to be the set of all clauses in C which contain l. If R is any set of rules and L is any set of literals then define $R[l] = \{r \in R : l = c(r)\}$ to be the set of all rules in R which end with l; and $R[L] = \{r \in R : c(r) \in L\}$ to be the set of all rules in R which have a consequent in L.

Any binary relation, $>$, on any set S is *cyclic* iff there exists a sequence, $(r_1, r_2, \ldots, r_n)$ where $n \geq 1$, of elements of S such that $r_1 > r_2 > \ldots > r_n > r_1$. A relation is *acyclic* iff it is not cyclic. If R is a set of rules then $>$ is a *priority relation* on R iff $>$ is an acyclic binary relation on R such that $>$ is a subset of $R_p \times R_{pd}$. We read $r_1 > r_2$ as r_1 *beats* r_2, or r_2 is *beaten* by r_1. Notice that strict rules never beat, and are never beaten by, any rule. Also defeater rules never beat any rule. Let $R[l; s] = \{t \in R[l] : t > s\}$ be the set of all rules in R with consequent l that beat s.

A *plausible description* of a situation is a 4-tuple $PD = (Ax, R_p, R_d, >)$ such that PD1, PD2, PD3, and PD4 all hold.
(PD1) Ax is a set of contingent clauses.
(PD2) R_p is a set of plausible rules.
(PD3) R_d is a set of defeater rules.
(PD4) $>$ is a priority relation on R_{pd}.
The clauses in Ax, called *axioms*, characterise the aspects of the situation that are certain. The set of strict rules, R_s, is defined by $R_s = \{{\sim}(L - \{l\}) \rightarrow l : l \in L$ and $\bigvee L \in Ax\}$. Define $R = R_s \cup R_p \cup R_d$ to be the set of rules *generated* from PD. The ordered pair $(R, >)$ is called a *plausible theory.* If $T = (R, >)$ is a plausible theory then $Ax(T) = \{\bigvee(\{c(r)\} \cup {\sim}A(r)) : r \in R_s\}$ is the set of axioms from which R_s was generated.

Let S be a set of clauses. A clause C_n is *resolution-derivable from S* iff there is a finite sequence of clauses $C_1, \ldots, C_n$ such that for each i in $[1..n]$, either $C_i \in S$ or C_i is the resolvent of two preceding clauses. The sequence $C_1, \ldots, C_n$ is called a *resolution-derivation* of C_n *from S*. The set of all clauses which are resolution-derivable from S is denoted by $Res(S)$. Define $Rsn(S) = Res(S) - \{\bigvee\{\}\}$ to be the set of all non-empty clauses in $Res(S)$.

Let S be a set of sets. Define the set of *minimal elements* of S, $Min(S)$, to be the set of minimal elements of the partially ordered set $(S, \subseteq)$. That is, $Min(S) = \{Y \in S : \text{if } X \subset Y \text{ then } X \notin S\}$.

Let R be the set of rules generated from some plausible description $(Ax, R_p, R_d, >)$. Define $Inc(R) = \{{\sim}L : \bigvee L \in Rsn(Ax) \cup \{\bigvee\{k, {\sim}k\} : k \in c(R)\}\}$ to be the set of non-empty sets of literals which are *inconsistent with R*. Define $Inc(R, l) = Min(\{I - \{l\} : I \in Inc(R) \text{ and } l \in I\})$. Each member of $Inc(R, l)$ is a minimal set of literals which is *inconsistent with l*. Since Ax is finite and R is finite, $Res(Ax)$, $Rsn(Ax)$, $Inc(R)$, and $Inc(R, l)$ are finite.

Cnf-formulas can be proved at three different levels of certainty or confidence. The definite level, indicated by δ, is like classical monotonic proof in that more information cannot defeat a previous proof. If a formula is proved definitely then one should behave as if it is true. Proof at the general [respectively, plausible] level, indicated by γ [respectively, π], is non-monotonic and propagates [respectively, blocks] ambiguity. If a formula is proved at the general or plausible level then one should behave as if it is true, even though it may turn out to be false.

It is sometimes necessary to prove that a cnf-formula is impossible to prove. To signify the six possible cases of proof we use the following *tags*: $+\delta$, $-\delta$, $+\gamma$, $-\gamma$, $+\pi$, $-\pi$. If $\alpha \in \{\delta, \gamma, \pi\}$ then $+\alpha f$ indicates f is proved at the α level, and $-\alpha f$ indicates that we have proved that $+\alpha f$ is impossible prove. A *tagged* formula is a formula preceded by a tag; so all tagged formulas have the form $\pm\alpha f$ where $\pm \in \{+, -\}$, $\alpha \in \{\delta, \gamma, \pi\}$, and f is a formula.

In the following *inference conditions*, $P = (P(1), \ldots, P(|P|))$ is a finite sequence, $(R, >)$ is a plausible theory, C is a non-singleton set of clauses, L is a non-singleton set of literals, l is a literal, and $\alpha \in \{\delta, \gamma, \pi\}$.

$+\bigwedge$) If $P(i+1) = +\alpha\bigwedge C$ then $\forall c \in C$, $+\alpha c \in P[1..i]$.
$-\bigwedge$) If $P(i+1) = -\alpha\bigwedge C$ then $\exists c \in C$, $-\alpha c \in P[1..i]$.
$+\bigvee$) If $P(i+1) = +\alpha\bigvee L$ then $\exists l \in L$, $+\alpha l \in P[1..i]$.
$-\bigvee$) If $P(i+1) = -\alpha\bigvee L$ then $\forall l \in L$, $-\alpha l \in P[1..i]$.
$+\mathcal{L}$) If $P(i+1) = +\alpha l$ then either
  .1) $\exists r \in R_s[l], +\alpha\bigwedge A(r) \in P[1..i]$; or
  .2) both
    .1) $\alpha \in \{\gamma, \pi\}$ and $\exists r \in R_p[l], +\alpha\bigwedge A(r) \in P[1..i]$, and
    .2) $\forall J \in Inc(R, l)\ \exists j \in J\ \forall s \in R[j]$ either
      .1) $\exists t \in R_p[l; s], +\alpha\bigwedge A(t) \in P[1..i]$; or
      .2) $A(s) \neq \{\}$ and $+\alpha{\sim}\bigwedge A(s) \in P[1..i]$; or
      .3) $\alpha = \pi$ and $-\pi\bigwedge A(s) \in P[1..i]$.
$-\mathcal{L}$) If $P(i+1) = -\alpha l$ then
  .1) $\forall r \in R_s[l], -\alpha\bigwedge A(r) \in P[1..i]$, and
  .2) either
    .1) $\alpha = \delta$ or $\forall r \in R_p[l], -\alpha\bigwedge A(r) \in P[1..i]$; or
    .2) $\exists J \in Inc(R, l)\ \forall j \in J\ \exists s \in R[j]$ such that
      .1) $\forall t \in R_p[l; s], -\alpha\bigwedge A(t) \in P[1..i]$, and
      .2) $A(s) = \{\}$ or $-\alpha{\sim}\bigwedge A(s) \in P[1..i]$, and
      .3) $\alpha = \gamma$ or $+\pi\bigwedge A(s) \in P[1..i]$.

Let T be a plausible theory. A formal proof or *T-derivation* P is a finite sequence, $P = (P(1), \ldots, P(|P|))$, of tagged cnf-formulas such that for all i in $[0..|P|-1]$ all the inference conditions hold. So () is a T-derivation. If the plausible theory T is known or is irrelevant then we often abbreviate T-derivation to just derivation. Each element $\pm\alpha f$ of a derivation is called a *line* of the derivation.

First notice that the inference conditions are paired, one positive and one negative. The negative one is just the strong negation of the positive one. That is, the negative condition is obtained by negating the positive condition and then replacing $+X \notin P[1..i]$ by $-X \in P[1..i]$, and $-X \notin P[1..i]$ by $+X \in P[1..i]$. So only the positive inference conditions need explaining. $+\bigwedge$ says that to prove a conjunction at level α, all the conjuncts must have been proved at level α previously. $+\bigvee$ says that to prove a disjunction at level α, at least one of the disjuncts must have been proved at level α previously. This is just the definition of constructive disjunction.

$+\mathcal{L}$ shows how to prove a literal l at level α. $+\mathcal{L}$.1 says that if the conjunction of the antecedent of a strict rule has been previously proved at level α, then its consequent can be added to the derivation. $+\mathcal{L}$.1 is just modus ponens for strict rules. $+\mathcal{L}$.2.1 says that $+\mathcal{L}$.1 is the only way to prove a literal at level δ. Hence proof at level δ only uses strict rules, which were generated from the axioms.

$+\mathcal{L}$.2 gives another way of proving a literal at the defeasible levels. $+\mathcal{L}$.2.1 says there must be some evidence for the literal, and $+\mathcal{L}$.2.2 says that all the evidence against the literal must be defeated. $+\mathcal{L}$.2.1 is similar to $+\mathcal{L}$.1 but with plausible rules replacing strict rules. $+\mathcal{L}$.2.2 says that every set of literals which is inconsistent with l must contain an element j such that every rule s supporting j is defeated. Among other things this ensures that j cannot be proved at a defeasible level. $+\mathcal{L}.2.2.1$, $+\mathcal{L}.2.2.2$, and $+\mathcal{L}.2.2.3$ give the three ways in which a rule can be defeated. $+\mathcal{L}.2.2.1$ says that s is beaten by an applicable plausible rule, t, which supports l. $+\mathcal{L}.2.2.2$ says that the antecedent of s is not empty and that the complement of the conjunction of the antecedent of s has previously been proved. $+\mathcal{L}.2.2.3$ gives the π level an alternative way to defeat s, which is not available at the γ level. This is the only difference between these two defeasible levels. $+\mathcal{L}.2.2.3$ says that at the π level the conjunction of the antecedent of s has previously been proved to be not provable.

The notation $T \vdash \pm\alpha f$ means that $\pm\alpha f$ is in a T-derivation. If $\alpha \in \{\delta, \gamma, \pi\}$ then we define $T(+\alpha) = \{f : T \vdash +\alpha f\}$ and $T(-\alpha) = \{f : T \vdash -\alpha f\}$. A *constructive plausible logic* consists of a plausible theory and the inference conditions.

3 Results

This section contains some of the results which have been proved for Constructive Plausible Logic. The proofs of these and other results are in Billington [3].

Since a negative tag means that the corresponding positively tagged formula has been proved to be unprovable, it would be reassuring to know that the same formula cannot have both a positive and a negative tag. This property, called coherence, is proved in the first result.

Intuitively an ambiguity propagating proof procedure should be more reliable than an ambiguity blocking proof procedure. This is because in an ambiguity propagating proof procedure more evidence is required to defeat an attack than in an ambiguity blocking proof procedure. So it would be good to know that every formula that can be proved at the δ level (using only infallible information) can be proved at the γ level (which is ambiguity propagating), and every formula that can be proved at the γ level can be proved at the π level (which is ambiguity blocking). This hierarchy is established in the first result, as is a corresponding hierarchy for the negative tags.

Theorem 1 (Coherence and Hierarchy).
Let T be a plausible theory and suppose $\alpha \in \{\delta, \gamma, \pi\}$.
(1) (Coherence) $T(+\alpha) \cap T(-\alpha) = \{\}$.
(2) (Hierarchy) $T(+\delta) \subseteq T(+\gamma) \subseteq T(+\pi)$, and $T(-\pi) \subseteq T(-\gamma) \subseteq T(-\delta)$.
End

Sets in $Inc(R)$ are inconsistent with the rules in R. So it would be nice to show that it is not possible to prove every element of a set in $Inc(R)$. Unfortunately it is possible to prove every element of a set in $Inc(R)$, but only if the axioms were inconsistent. This is what lemma 3 says. To help prove lemma 3, we need lemma 2 which shows that if every element of a set in $Inc(R)$ is proved then the fault is confined to just the strict rules (which were derived from the axioms).

Lemma 2 (Inconsistency Needs Strict Rules).
Suppose $T = (R, >)$ is a plausible theory, $I \in Inc(R)$, and $\alpha \in \{\delta, \gamma, \pi\}$. Let P be a T-derivation such that for all $l \in I$, $+\alpha l \in P$. Then either

1. for all $l \in I$, $+\mathcal{L}.2$ fails, and so $+\mathcal{L}.1$ holds; or
2. $\alpha \in \{\gamma, \pi\}$ and there exists a literal k such that $+\alpha k$ and $+\alpha{\sim}k$ are in a proper prefix of P.

End

Let T be a plausible theory and F be a set of cnf-formulas. Define $F^{-\wedge} = \{c : \bigwedge C \in F \text{ and } c \in C\}$. Then F is *consistent* [respectively, T-*consistent*] iff $\bigvee\{\} \notin Res(F^{-\wedge})$ [respectively, $\bigvee\{\} \notin Res(F^{-\wedge} \cup Ax(T))$]. So T-consistent means consistent with the axioms of T. F is *inconsistent* iff F is not consistent.

Since $Res(F^{-\wedge}) \subseteq Res(F^{-\wedge} \cup Ax(T))$, if F is T-consistent then F is consistent. The converse is not always true. Consider the following counter-example. Let $Ax(T) = \{\bigvee\{a, b\}\}$, and $F = F^{-\wedge} = \{\neg a, \neg b\}$. Then F is consistent but F is not T-consistent.

Lemma 3 (Relative Consistency for Members of $Inc(R)$).
Suppose $T = (R, >)$ is a plausible theory and $\alpha \in \{\delta, \gamma, \pi\}$. If $I \in Inc(R)$ and there is a T-derivation P such that for all $l \in I$, $+\alpha l \in P$ then $Ax(T)$ is inconsistent.
End

Consider the following property. If two clauses can be proved then their resolvent (if it exists) can be proved. At least some variation of this "resolution property" seems to be necessary for relative consistency to be proved. Hence a resolution property is not only a useful result by itself, but also highly important. The exact form of the resolution property which Constructive Plausible Logic satisfies is given in lemma 4.

Lemma 4 (Resolution Property).
Suppose T is a plausible theory and $\alpha \in \{\delta, \gamma, \pi\}$. Let L and M be two sets of literals and let l be a literal. If $\bigvee(L \cup \{l\}) \in T(+\alpha)$ and $\bigvee(M \cup \{{\sim}l\}) \in T(+\alpha)$ then either $\bigvee(L \cup M) \in T(+\alpha)$ or $\{l, {\sim}l\} \subseteq T(+\alpha)$.
End

We would like to show that the set of all proved formulas was consistent. Unfortunately this is not true, because if the axioms are inconsistent then the set of all proved formulas may also be inconsistent. Our final result shows that if the axioms are consistent then the set of all proved formulas is not only consistent but also consistent with the axioms.

Theorem 5 (Relative Consistency).
If T is a plausible theory and $\alpha \in \{\delta, \gamma, \pi\}$ then $T(+\alpha)$ is T-consistent iff $Ax(T)$ is consistent.

Proof.
To prove this theorem we shall need the definition of a subclause and a technical lemma, Lemma TL, concerning only classical propositional resolution.

If $\bigvee L$ and $\bigvee M$ are two clauses then $\bigvee L$ is a *subclause* of $\bigvee M$ iff $L \subseteq M$.

Lemma TL.
Let S and S^* be two sets of clauses. Let L and X be two sets of literals.

1. If $S \subseteq S^*$ then $Res(S) \subseteq Res(S^*)$, and $Rsn(S) \subseteq Rsn(S^*)$.
2. If every clause in S^* has a subclause in S then every clause in $Res(S^*)$ has a subclause in $Res(S)$.
3. If $\bigvee X \in Res(S \cup L)$ and $\bigvee X \notin Res(L)$, then there is a finite subset K of L such that $\bigvee(X \cup {\sim}K) \in Res(S)$, and $X \cap {\sim}K = \{\}$.

EndTL

Let $Ax = Ax(T)$. Then $Ax^{-\wedge} = Ax$, and $T(+\alpha)^{-\wedge} = T(+\alpha) - \{\bigwedge C : \bigwedge C \in T(+\alpha)$ and $|C| \neq 1\}$. By lemma TL(1), $Res(Ax) \subseteq Res(T(+\alpha)^{-\wedge} \cup Ax)$, so if $T(+\alpha)$ is T-consistent then Ax is consistent.

Conversely, suppose $T(+\alpha)$ is not T-consistent. Then $\bigvee\{\} \in Res(T(+\alpha)^{-\wedge} \cup Ax)$. Let $T = (R, >)$. If there is a literal l such that $\{l, {\sim}l\} \subseteq T(+\alpha)^{-\wedge}$, then $\{l, {\sim}l\} \in Inc(R)$, and so by lemma 3, Ax is inconsistent. So suppose there is no literal l such that $\{l, {\sim}l\} \subseteq T(+\alpha)^{-\wedge}$. By lemma 4, $Rsn(T(+\alpha)^{-\wedge}) = T(+\alpha)^{-\wedge}$. If $\bigvee\{\} \in Res(T(+\alpha)^{-\wedge})$ then there is a literal k such that $\{k, {\sim}k\} \subseteq Rsn(T(+\alpha)^{-\wedge})$, and hence $\{k, {\sim}k\} \subseteq T(+\alpha)^{-\wedge}$. So suppose $\bigvee\{\} \notin Res(T(+\alpha)^{-\wedge})$. Let L be the set of all literals in $T(+\alpha)$. Then $L \subseteq T(+\alpha)^{-\wedge}$, and so by lemma TL(1), $\bigvee\{\} \notin Res(L)$. By $+\bigvee$, every clause in $T(+\alpha)^{-\wedge} \cup Ax$ has a subclause in $L \cup Ax$, and so by lemma TL(2), every clause in $Res(T(+\alpha)^{-\wedge} \cup Ax)$ has a subclause in $Res(L \cup Ax)$. Hence $\bigvee\{\} \in Res(L \cup Ax)$. By lemma TL(3), there is a finite subset K of L such that $\bigvee({\sim}K) \in Res(Ax)$. If ${\sim}K = \{\}$ then Ax is inconsistent. So suppose ${\sim}K$ is not empty. Then $\bigvee({\sim}K) \in Rsn(Ax)$ and so $K \in Inc(R)$. By lemma 3, Ax is inconsistent.
EndProof5

4 Implementation

Constructive Plausible Logic has been implemented as a tool, CPL, that attempts the proofs of tagged formulas. The tool is written in literate Haskell, and is fully listed and documented in Rock [9]. It is implemented in a manner emphasising correctness and simplicity, similar to that of previous versions of Plausible Logic [4, 8], and Defeasible Logic [5, 7], but does not at present contain as many speed optimisations to cope with hundreds of thousands of rules, and is simpler to use. It could be extended with more usage options and optimisations to match the previous systems. It presently consists of about 1300 logical lines of code, compared to 5000 for the previous implementation of Plausible Logic [8].

To demonstrate the tool, the following reasoning puzzle will be used. We know for certain that Hans is a native speaker of Pennsylvania-Dutch, *nspd*, that native speakers of Pennsylvania-Dutch are native speakers of German, $nspd \rightarrow nsg$ or $\bigvee\{\neg nspd,\ nsg\}$, and that persons born in Pennsylvania are born in the United States of America, $bp \rightarrow busa$ or $\bigvee\{\neg bp,\ busa\}$. We also know that *usually* native

```
/* file:    PennDutch.d
   purpose: Generalised competitors. */

% plausible description:
    nspd.
    \/{~nspd, nsg}.
    \/{~bp, busa}.
R1: nsg =>~busa.
R2: nspd => bp.
    R2 > R1.

% the requested proofs:
output{+d busa}.  output{+g busa}.  output{+p busa}.
output{-d busa}.  output{-g busa}.  output{-p busa}.
output{+d~busa}.  output{+g~busa}.  output{+p~busa}.
output{-d~busa}.  output{-g~busa}.  output{-p~busa}.
output{+d bp}.    output{+g bp}.    output{+p bp}.
output{-d bp}.    output{-g bp}.    output{-p bp}.
output{+d~bp}.    output{+g~bp}.    output{+p~bp}.
output{-d~bp}.    output{-g~bp}.    output{-p~bp}.
```

Fig. 1. CPL input file containing the Pennsylvania-Dutch plausible description and requests to output specific proofs

speakers of German are not born in the United States, $nsg \Rightarrow \neg busa$, but that *usually* native speakers of Pennsylvania-Dutch are born in Pennsylvania, $nspd \Rightarrow bp$. The latter rule, being more specific, should take priority over the former. These axioms, plausible rules and the priority form a plausible description.

$$\begin{array}{ll} & nspd \\ & \bigvee\{\neg nspd, nsg\} \\ & \bigvee\{\neg bp, busa\} \\ r_1\colon & nsg \Rightarrow \neg busa \\ r_2\colon & nspd \Rightarrow bp \\ & r_2 > r_1 \end{array}$$

We would like to know whether we can conclude that Hans was born in the USA. Figure 1 shows the plausible description coded for input to the CPL tool. $\bigvee$ is approximated by `\/`. `%`, `/*` and `*/` delimit comments as in Prolog. The input includes `output` directives to request proofs of tagged formulas. In these `d`, `g` and `p` stand for δ, γ and π respectively.

The CPL tool prints traces for all of the requested proofs and a summary table of the results. The trace of the proof of $+\pi busa$ is listed in Figure 2. The trace starts with the tagged formulas to be proved, the goal. Each line of each inference rule used is indicated by printing the label for that line, *e.g.* `+L.1` for line $+\mathcal{L}.1$, and `-/\` for line $-\bigwedge$. As variables scoped by $\forall$ and $\exists$ are instantiated their values are printed. The attempted proofs of subgoals are printed indented.

```
To Prove: +p busa
. +L.1
. r = {bp} -> busa
. To Prove: +p bp
. . +L.1
. . +L.2.1
. . r = R2: {nspd} => bp
. . To Prove: +p nspd
. . . +L.1
. . . r = {} -> nspd
. . . To Prove: +p /\{}
. . . . +/\
. . . Proved: +p /\{}
. . Proved: +p nspd
. . +L.2.2
. . J = [~busa]
. . j =~busa
. . s = R1: {nsg} =>~busa
. . +L.2.2.1
. . t = R2: {nspd} => bp
. . Proved previously: +p nspd
. . J = [~bp]
. . j =~bp
. . s = {~busa} ->~bp
. . +L.2.2.1
. . +L.2.2.2
. . Loop detected: +p busa
. . +L.2.2.3
. . To Prove: -p~busa
. . . -L.1
. . . -L.2.1
. . . r = R1: {nsg} =>~busa
. . . To Prove: -p nsg
. . . . -L.1
. . . . r = {nspd} -> nsg
. . . . To Prove: -p nspd
. . . . . -L.1
. . . . . r = {} -> nspd
. . . . . To Prove: -p /\{}
. . . . . . -/\
. . . . . Not proved: -p /\{}
. . . . Not proved: -p nspd
. . . Not proved: -p nsg
. . . -L.2.2
. . . J = [bp]
. . . j = bp
. . . s = R2: {nspd} => bp
. . . -L.2.2.1
. . . -L.2.2.2
. . . To Prove: -p~nspd
. . . . -L.1
. . . . r = {~nsg} ->~nspd
. . . . To Prove: -p~nsg
. . . . . -L.1
. . . . . -L.2.1
. . . . Proved: -p~nsg
. . . . -L.2.1
. . . Proved: -p~nspd
. . . -L.2.2.3
. . . Proved previously: +p nspd
. . Proved: -p~busa
. Proved: +p bp
Proved: +p busa
Goal count = 10
```

Fig. 2. CPL trace of the proof $+\pi busa$

Duplicated proofs of subgoals are avoided by maintaining a history of prior results of attempted proofs. This same history is used to detect loops, where a goal generates itself as a subgoal. Where a loop is detected the proof must be achieved by an alternate path if possible. The trace ends with a count of the goals and subgoals, a measure of the effort required for that proof.

5 Conclusion

A constructive Plausible Logic has been defined and explained and implemented. This is the first Plausible Logic which has been proved to be relatively consistent, an important property ensuring that the non-monotonic deduction mechanism is trustworthy. It also has the desirable properties of coherence, hierarchy, and resolution. Moreover its ambiguity propagating proof algorithm is simpler than the one in Billington and Rock [4]. Its implementation has been at least as straightforward and efficient as its predecessors.

In the past many Plausible Logics have been defined with a non-constructive disjunction, and then the proof of relative consistency has been attempted. In every case the attempt has failed. But now we can start with this constructive Plausible Logic and try to generalise the disjunction while maintaining relative consistency. Antoniou and Billington [1] showed that an ambiguity propagating version of Defeasible Logic could be embedded into Default Logic with Priorities. An attempt to embed an ambiguity propagating version of the Plausible Logic of Billington and Rock [4] into Default Logic with Priorities failed because the relative consistency of that Plausible Logic could not be proved. Now that constructive Plausible Logic is relatively consistent it is worthwhile to see if it can be embedded into Default Logic with Priorities.

References

[1] G. Antoniou and D. Billington. Relating defeasible and default logic. In *Proceedings of the 14th Australian Joint Conference on Artificial Intelligence*, volume 2256 of *Lecture Notes in Artificial Intelligence*, pages 13–24. Springer, 2001. 965

[2] D. Billington. Defeasible deduction with arbitrary propositions. In *Poster Proceedings of the 11th Australian Joint Conference on Artificial Intelligence*, pages 3–14, Griffith University, 1998. 954

[3] David Billington. Constructive Plausible Logic version 1.2 repository. Available from the author, 2003. 955, 960

[4] David Billington and Andrew Rock. Propositional plausible logic: Introduction and implementation. *Studia Logica*, 67:243–269, 2001. 954, 955, 962, 965

[5] Michael J. Maher, Andrew Rock, Grigoris Antoniou, David Billington, and Tristan Miller. Efficient defeasible reasoning systems. *International Journal on Artificial Intelligence Tools*, 10(4):483–501, 2001. 962

[6] D. Nute. Defeasible reasoning. In *Proceedings of the 20th Hawaii International Conference on System Science*, pages 470–477, University of Hawaii, 1987. 954

[7] Andrew Rock. *Deimos*: A query answering Defeasible Logic system. Available from the author, 2000. 962

[8] Andrew Rock. *Phobos*: A query answering Plausible Logic system. Available from the author, 2000. 962

[9] Andrew Rock. Implementation of Constructive Plausible Logic (version 1.2). Available from the author, 2003. 962

Heuristic Search Algorithms Based on Symbolic Data Structures

Albert Nymeyer and Kairong Qian

School of Computer Science, The University of New South Wales
UNSW Sydney 2052 Australia
Tel: 02 9385 7708, Fax: 02 9385 5995
{anymeyer,kairongq}@cse.unsw.edu.au

Abstract. Search algorithms in formal verification invariably suffer from combinational explosions in the number of states as complexity increases. One approach to alleviate this problem is to use symbolic data structures called BDDs (Binary Decision Diagrams). BDDs represent sets of states as Boolean expressions, which makes them ideal data structures for problems in AI and verification with large state spaces. In this paper we demonstrate a BDD-based A*algorithm, and compare the performance of A* with this algorithm using heuristics that have varying levels of informedness. We find that contrary to expectation, the BDD-based algorithm is not faster in certain circumstances, namely if the heuristic is strong. In this case, A* can be faster and will use a lot less memory. Only if the heuristic is weak will the BDD-based algorithm be faster, but even in this case, it will require substantially more memory than A*. In short, a symbolic approach only pays dividends if the heuristic is weak, and this will have an associated memory penalty. We base our performance comparisons on the well-known sliding tile puzzle benchmark.

Keywords: Heuristic search, shortest path, binary decision diagrams.

1 Introduction

Heuristic search methods have been employed to solve a variety of problems in areas such as AI planning, problem-solving tasks and recently bug-finding in verification domain. The heuristic search algorithm A* in particular has been the subject of much research. However, it has long been known that the usefulness of A* is handicapped by the fact that it processes the states of the system individually. As most real-world problems have enormous state spaces, A* is often not able to cope.

Research into making A* more efficient for problems with large state spaces has resulted in a number of variations of the algorithm. One body of research [1, 2, 3, 4] improves efficiency by modifying A* to construct a linear search tree. In other work [5, 6, 7], the search tree is pruned to reduce its size. Since the late 1980s, however, the most favoured technique for handling the large sets of states

T.D. Gedeon and L.C.C. Fung (Eds.): AI 2003, LNAI 2903, pp. 966–979, 2003.

has involved the use of Binary Decision Diagrams (BDDs). Introduced by Byrant in 1986 [8], a BDD is a data structure for the purpose of graphical management of Boolean functions. Due to their abilities to efficiently represent and process sets of states, BDDs were then developed for the purposes of formal verification [9, 10, 11] and have been successfully implemented in symbolic model checking to avoid the so-called "state explosion" problem [12]. BDDs are efficient because they represent sets of states as Boolean expressions, so computing successor states involves simple Boolean operations on these expressions. We omit the details of BDDs in this paper and interested readers may refer to [8, 13, 14, 15, 16].

The first study of BDD-based A* algorithm was called BDDA* [17, 18]. More recently in [19, 20], another two variants, ADDA* and setA* respectively, have been proposed. Although they are all variants of the conventional A* algorithm representing state space using decision diagrams, they use different *Open* set implementation and heuristic representation. Surprisingly, BDDA* and ADDA* do not seem to include mechanisms to record path information, therefore they can only be used to find the optimal search length of the goal given the heuristic is admissible. While these algorithms were tested in [17, 18, 19, 20], no performance comparison was made between the explicit-state A* and BDD-based A* algorithms when directed by different heuristics. We believe that the nature of the heuristic is crucial to the performance of the (BDD-based) algorithm, but in the above work, this aspect is generally ignored. Other very recent work, for example in [21, 22, 23], has investigated a variety of heuristics in a different setting. In particular, Groce et al. [21, 22] is concerned with Java software verification and uses a testing-based structural heuristic as well as others. In an attempt to applying heuristic search in process verification, Santone [23] develops inductive heuristic functions to evaluate each state in processes specified by CCS (Calculus for Communicating Systems). In this work, we use a similar approach to BDDA*, ADDA* and setA*, but additionally experiment with the role of the heuristic in the performance of the search algorithm. As such, our work falls in between the BDDA*-ADDA*-setA* work described above, and the work of Groce and Santone [21, 22, 23].

The rest of the paper is organised as follows: in Section 2 we show our version of BDD-based A* algorithm, called SA* (for *symbolic* A*), and demonstrate a trivial marker example using our algorithm. In section 3, we apply SA* to the sliding tile puzzle and the results of a number of experiments comparing the conventional A* and BDD-based A* algorithms are presented as well. Finally, in Section 4 we present our conclusions.

2 Symbolic A* Algorithm

Modifying the A* algorithm to use BDDs for its data structures requires the states of the system to be encoded as a Boolean function, called the *characteristic* function. We also need to compute the successor states (and hence a new characteristic function) using a Boolean function, called the *transition* function. Both functions can be represented as BDDs. As the BDD-based algorithm does

```
Procedure SA* (TP, HP, CF_start, CF_goal)
    Open.push() ← (0, h_start, CF_start)
    while (Open ≠ φ)
        (g, h, CF) ← Open.pop()
        if (CF ∧ CF_goal ≠ False)
            return SUCCESS
        Closed.push() ← (g, CF)
        for i = 0 to | TP | −1
            CF_next ← (∃x(CF ∧ TP(i)))[x′/x]
            if (CF_next ≠ False)
                for j = 0 to |HP| −1
                    CF_tmp ← (CF_next ∧ HP(j))
                    if (CF_tmp ≠ False)
                        g ← g + Cost(TP(i))
                        h ← Heu(HP(j))
                        Open.push() ← (g, h, CF_tmp)
        Open.union()
    if (Open = φ) then NoGoalExists
```

Fig. 1. The symbolic A* algorithm

not use the explicit states but a symbolic representation of these states in the form of Boolean functions, it can be referred to as the symbolic A* algorithm, which we will refer to as SA*. The SA* algorithm is shown in Figure 1.
The SA* algorithm has 4 input parameters:

TP The *transition partition TP* is partitioned according to the cost of a transition. The technique of partitioning the transition function has been widely used in symbolic model checking area and allows large transition relations, which cannot be manipulated as a single BDD, to be handled incrementally [20]. The ith partition of the relation is referred to by *TP(i)*. *Cost(TP(i))* is the cost of moving from state to state in the relation.

HP The *heuristic partition HP* partitions the states into sets that have equal heuristic cost. Note that such partition may be non-trivial to compute for some cases, but for those problems whose state spaces have high level description, this can often be easily accomplished. *Heu(HP(i))* is the heuristic value for the ith set of states.

$\mathbf{CF_{start}}$ and $\mathbf{CF_{goal}}$ The *characteristic function* is denoted by *CF*. This function encodes the set of states. The characteristic functions of the start and goal states comprise the third and fourth parameters of the algorithm.

The initialisation of SA* is the same as A* except that in SA* the start state has been encoded as the BDD CF_{start}. The elements in the lists *Open* and *Closed* in SA* are not single states, but sets of states nodes containing state BDDs. The elements in *Open* are still ordered by the cost f however, where f is the cost of each of the states represented in the BDD.

The algorithm iterates and test whether the *Open* list is empty or not: if it is not empty then the first element in *Open* is taken. As this element includes a BDD that may encode many states, we will need to check the conjunction of this BDD and the goal BDD in order to determine whether the goal has been

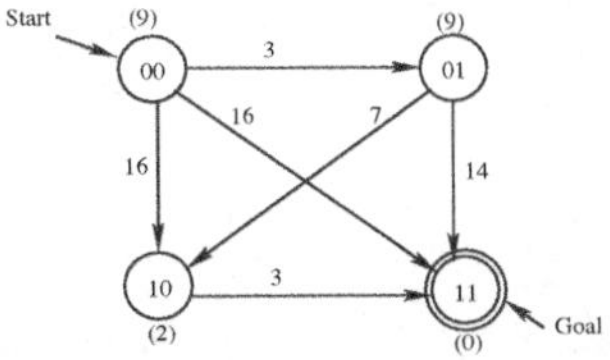

Fig. 2. The state space of the marker game on a 2×2 board

reached. If the goal has been found, the path can be extracted from the *Closed* list. Otherwise, we compute:

Successors The iteration in line 7 computes the successors of all the states represented by the BDD *CF*. The computation $(\exists x(CF \wedge TP(i)))[x'/x]$ (in line 8) is called the *relational production*. The characteristic function *CF* encodes the set of current states. The transition relation *TP(i)* is a BDD in which half the variables (denoted by x) encode the current state, and the other half of the variables (denoted by x') encode the successor states. To compute (the BDD of) the successors of *CF*, we compute the conjunction of *CF* and *TP(i)*, and apply existential quantification to the result. As well, we need to change the names of the variables x' to x and thereby generate a BDD for a new characteristic function.

Heuristic Costs The iteration in line 10 computes a new BDD that comprises the newly generated characteristic function and the heuristic cost. In line 15, this BDD and the actual and heuristic costs are added to *Open*.

The operation *Open.union()* merges all elements in *Open* that have the same cost $g + h$ and eliminate those non-optimal paths. Note that, in SA*, we only change the way that states are represented and processed. As a result, like A*, SA* guarantees it will find a least-cost path if the heuristic function is admissible.

Marker Game

We illustrate the computations performed by the SA* algorithm by determining a least-cost path in a simple game that moves a marker on a 2×2 board. The size of the state space in this game is 4, corresponding to the 4 board positions where the marker can sit. In the start state, the marker is in position $(0, 0)$, in the goal state it is in position $(1, 1)$. A graph depicting the state space is shown in Figure 2. Each state (node in the graph) is labeled with the position of the marker. The marker can move between states along the edges in the graph. The labels on the edges are the costs of moves, and the labels in parentheses on the states are the heuristic costs of reaching the goal from that state.

To encode the state space we need 2 Boolean variables, x_0, x_1. The characteristic function of the start state is $CF_{start} = \overline{x_0} \wedge \overline{x_1}$ and of the goal state is $CF_{goal} = x_0 \wedge x_1$. We need 4 variables, x_0, x_1, x'_0, x'_1, to encode the transition relation. The transition partitions are the following:

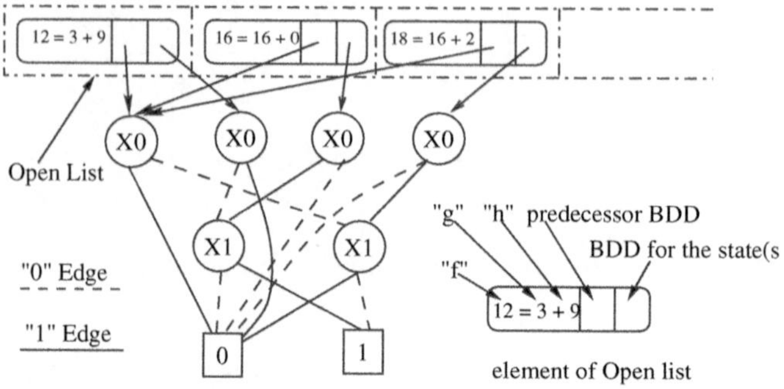

Fig. 3. The *Open* BDD after one iteration of state exploration

$\mathrm{TP}(0){=}(\overline{x_0} \wedge \overline{x_1} \wedge \overline{x_0'} \wedge x_1') \vee (x_0 \wedge \overline{x_1} \wedge x_0' \wedge x_1')$
$\mathrm{TP}(1){=}\overline{x_0} \wedge x_1 \wedge x_0' \wedge \overline{x_1'}$
$\mathrm{TP}(2){=}\overline{x_0} \wedge x_1 \wedge x_0' \wedge x_1'$
$\mathrm{TP}(3){=}(\overline{x_0} \wedge \overline{x_1} \wedge x_0' \wedge x_1') \vee (\overline{x_0} \wedge \overline{x_1} \wedge x_0' \wedge \overline{x_1'})$

The costs of these 4 partitions are 3, 7, 14 and 16 resp. The heuristic partitions are:

$\mathrm{HP}(0){=}x_0 \wedge x_1$
$\mathrm{HP}(1){=}x_0 \wedge \overline{x_1}$
$\mathrm{HP}(2){=}(\overline{x_0} \wedge \overline{x_1}) \vee (\overline{x_0} \wedge x_1)$

with heuristic values 0, 2 and 9 resp. In the first step of the algorithm, *Open* is set to the characteristic function of the start state CF_{start}, together with its cost $g = 0$ and heuristic value $h = 9$. In the outer **for** loop iteration, we compute the successors of the start node using $CF_{next} \leftarrow (\exists x(CF \wedge TP(i)))[x'/x]$ for i: 0 to 3. For $i = 0$ we have:

$$CF_0 \leftarrow \exists(x_0, x_1)(\overline{x_0} \wedge \overline{x_1}) \wedge ((\overline{x_0} \wedge \overline{x_1} \wedge \overline{x_0'} \wedge x_1') \vee (x_0 \wedge \overline{x_1} \wedge x_0' \wedge x_1'))[x'/x]$$

which simplifies to $CF_0 = \overline{x_0} \wedge x_1$. In the inner **for** loop, we calculate the heuristic cost of this successor. If $j = 2$ then $\exists x(\overline{x_0} \wedge x_1 \wedge HP(2))$ is true, so the heuristic cost h of CF_0 is given by $Heu(HP(2)) = 9$. The actual cost g is given by $Cost(TP(0)) = 3$. In line 15 we add (g, h, CF_0) to *Open*. After computing all the successors of the start node, the *Open* set looks like:

Successor	$f = g + h$	Characteristic Function
0	$12 = 3 + 9$	$CF_0 = \overline{x_0} \wedge x_1$
1	$16 = 16 + 0$	$CF_1 = x_0 \wedge x_1$
2	$18 = 16 + 2$	$CF_2 = x_0 \wedge \overline{x_1}$

and the *Open* BDD is shown in Figure 3. We pick the element in *Open* with minimum f and continue iterating using the **while** iteration until we find the goal node.

3 Sliding-Tile Experiment

To better appreciate the difference in performance between the A* and SA* algorithms we needed a problem that has a large but solvable state space. The largest solvable (by A*) $n^2 - 1$ sliding tile puzzle is the 8-puzzle. This puzzle consists of a 3×3 board and 8 tiles, numbered 1 to 8. Initially the tiles are placed in some configuration on the board. The aim is to re-arrange the tiles until some goal configuration is reached, making as few moves as possible.

To illustrate how different heuristics influence the performance of A* and SA* algorithm, we apply the search algorithms using 3 different heuristics:

Odd Tiles out of Position The first heuristic we apply is as follows:

$$h_1(n) = \sum_{i=1}^{8} \begin{cases} 1 \text{ if } p^i \neq g^i \text{ and } i \text{ is odd} \\ 0 \text{ otherwise} \end{cases}$$

where p^i is the position of tile i in the present configuration and g^i is the position of tile i in the goal configuration.

Tiles out of Position This heuristic is similar to the first one, except we now consider all tiles. The heuristic can also be expressed as:

$$h_2(n) = \sum_{i=1}^{8} \begin{cases} 1 \text{ if } p^i \neq g^i \\ 0 \text{ otherwise} \end{cases}$$

where p^i and g^i have the same meanings as in $h_1(n)$.

Manhattan Distance This heuristic can be expressed as:

$$h_3(n) = \sum_{i=1}^{8} \mid p_x^i - g_x^i \mid + \mid p_y^i \quad g_y^i \mid$$

where p_x^i is the x-coordinate of tile i in the present configuration (similarly p_y^i), and g_x^i is the x-coordinate of tile i in the goal configuration (similarly g_y^i). This heuristic computes the number of moves that each tile needs to make to reach its final position in the goal configuration, assuming there are no other tiles on the board. It never overestimates the actual number of moves that will be required.

Note that all 3 heuristics are admissible and therefore an optimal solution is guaranteed. Moreover, we can say:

$$h_1(n) < h_2(n) < h_3(n) < h(n)$$

where $h(n)$ is the minimum real cost from state n to the goal state. The Manhattan distance heuristic is of course the strongest heuristic, and the "odd tiles out of position heuristic" is the weakest one.

The first task is to encode all 9! configurations of the puzzle as a Boolean vector. Ideally we would like to use only 19 variables ($2^{18} < 9! < 2^{19}$) to encode the state space of the puzzle, but such an encoding requires a non-trivial computation. As each square can hold one of 8 tiles, in our implementation, we use 4 Boolean variables to represent one square and the binary assignment to them encode the tile number in it. For example, if we use x_0, x_1, x_2, x_3 to represent left-upper corner square, $\overline{x_0}, x_1, \overline{x_2}, x_3$ means tile number 5 is in that square. Although this kind of encoding will need more variables, it makes the transition function much easier to compute.

The next task is to partition the transition and heuristic functions.[1] (We let the cost of a transition (i.e. moving a tile) be constant.) Partitioning the heuristic function involves grouping all states that have the same cost. Note that the transition and heuristic functions are computed *a priori*. In the sliding tile puzzle, we keep all the transition relation and heuristic partitions in memory for convenience. In bigger applications, we would not do this as we only need this information for successor calculation and heuristic cost estimation.

Experimental Results

We used the BDD package CUDD[2] in this work. The package contains all standard BDD operations and reordering functions, as well as a set of statistical functions that help with experimentation. CUDD was written for symbolic model checking, and has been integrated with such tools as VIS[3] and NuSMV[4]. The work was carried out on an Intel PIII 933Hz CPU with 512MB RAM.

We tested 500 randomly generated solvable start configurations of the sliding tile puzzle. We used the aforementioned heuristics in the test, and we studied their relative informedness (i.e. the relative strengths of each heuristic), the effect of clustering in SA*, the relative speeds of A* and SA*, and their memory consumption.

The Relative Informedness of Heuristics A heuristic h_a is said to be more informed than heuristic h_b if, to find the optimal solution path, the algorithm directed by h_a explores a smaller proportion of the state space than directed by h_b. For each test, we collected the total number of states explored by the algorithm, and we did this for each of the 3 heuristics[5]. From the graph in Figure 4 we can clearly see that when the Manhattan distance heuristic (h_3) is used, the algorithm explores less than 10% of the state space in order to find the solution path for all 500 different instances of the puzzle, whereas in the case of the weakest heuristic (h_1), the algorithm needs to explore up to 95% of the state space. Note that the tests are ordered by the results of h_1 and this naturally results in the perfect linear line for h_1 in the graph.

[1] Partitioning is actually not necessary in the puzzle but we do it for the sake of generality.

[2] `http://vlsi.colorado.edu/~fabio/`

[3] `http://vlsi.colorado.edu/~vis/`

[4] `http://nusmv.irst.itc.it/`

[5] Note that both A* and SA* explore the same state space proportion if we apply the same heuristic.

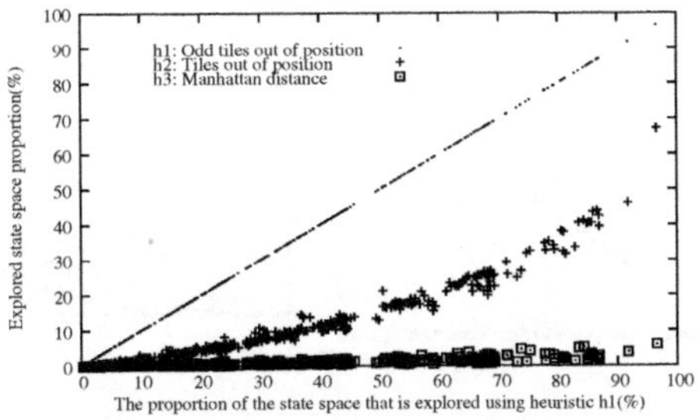

Fig. 4. The relative informedness

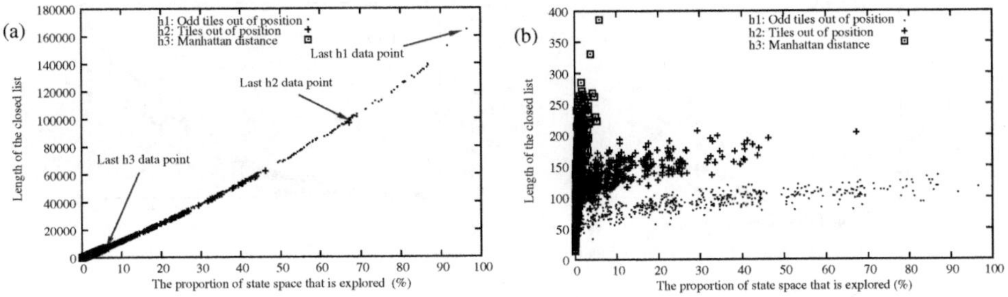

Fig. 5. (a) The length of the *Closed* list in A* for all heuristics (b) The length of *Closed* list in SA* for all heuristics

Clustering Effects In contrast to A*, which processes each state individually, SA* processes states in sets, represented as Boolean functions. In our implementation we cluster those states together that have the same "$g + h$" cost. The graphs shown in Figure 5 (a) and (b) illustrate the effect of clustering.

In Figure 5(a) we see that in A* the length of the *Closed* list is linear in the explored state space, as expected. The reason for this is that A* handles the states explicitly. Note that the length of the curve is dependent on the informedness of the heuristic, and as described above, indicates the proportion of the search space that is examined.

In Figure 5(b) we see that for SA*, the *Closed* lists are more than 2 orders of magnitude shorter than those in Figure 5(a). This is direct consequence of clustering. We also see that the *Closed* lists for the strongest heuristic in SA* (Figure 5(b)) are longest, but also more erratic in behaviour. The weakest heuristic has the shortest *Closed* lists and most consistent behaviour.

The reason that the strong heuristic h_3 has longest *Closed* lists is that a well informed heuristic causes the search algorithm to quickly descend deep into the search space, and this results in a greater spread of costs (i.e. "$g + h$") in the visited nodes. As the states in the BDDs are clustered by the cost, this greater spread result in longer *Closed* lists. Conversely, when the heuristic is weak, the costs of states are similar, and hence more states can be clustered, and the *Closed* list is shorter.

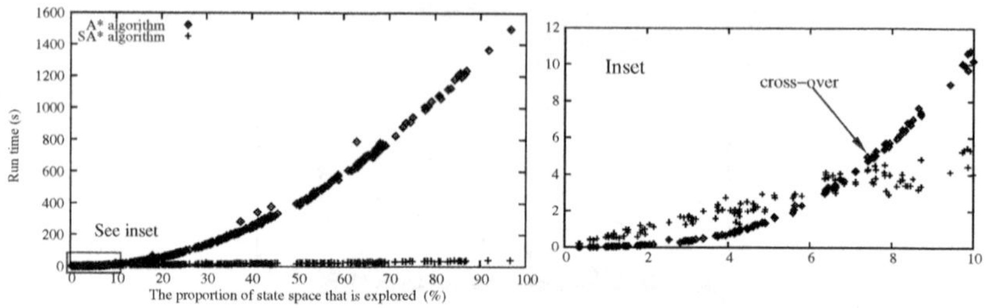

Fig. 6. Relative speed for the weakest heuristic, h_1, with an inset on the right

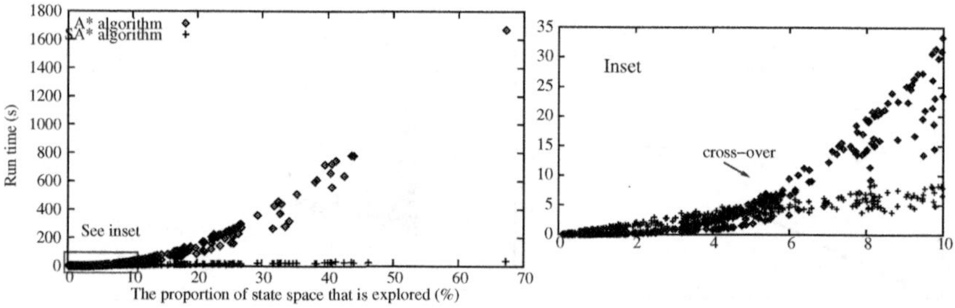

Fig. 7. Relative speed for heuristic h_2 with an inset on the right

Performance Comparison of A* and SA* To understand the performance difference between the two algorithms directed by different heuristics, we plot in Figure 6, 7 and 8 the time taken to find the solution for both algorithms for all 500 different start configurations. The graph in Figure 6 shows the time consumed by both algorithms for weakest heuristic h_1. We order the run times according to the proportion of the explored state space. A dramatic performance difference can be observed in the graph when the explored state space is greater than 20%. To further illustrate the behaviour, we magnify the left-lower corner of Figure 6. One can see that there is a cross-over at about 7%. A* takes less time to find the goal than SA* for all tests less than 7%, and vice versa above 7%. The graph shown in Figure 7 depicts the performance of A* and SA* using the next heuristic, h_2. We again magnify the left-lower corner area as we did above. The performance of both algorithms directed by h_2 is similar to the performance of h_1. There again is (or appears to be) a cross-over effect between A* and SA*, but this time at a slightly lower point close to 6%. The time taken to find the goal using the third and strongest heuristic is plotted in Figure 8. Note that both algorithms only need to explore no more than 6% of the state space to find the goal. This contrasts with 70% for h_2, and roughly 95% for h_1. Most of the data in Figure 8 lie in the area less than 3.5%. Though there are just a few data points above 3.5%, there is still a suggestion of a cross-over at above 5%.

One could expect the time taken to process the nodes in A* to be linear as each element in the *Closed* list represents one state. In Figures 6, 7 and 8 we

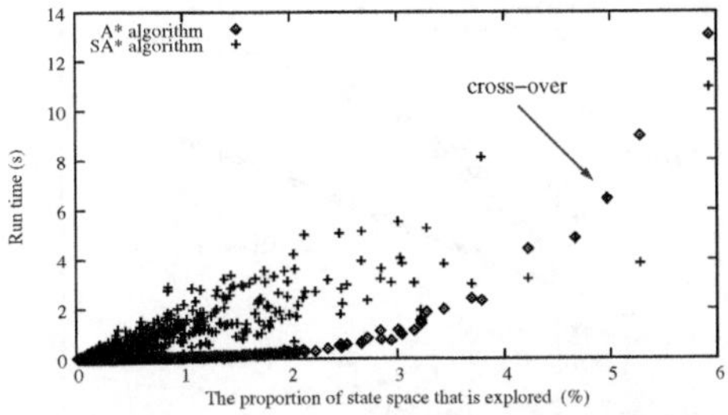

Fig. 8. Relative speed when the Manhattan distance h_3 is employed

see however that it is worse than linear for all 3 heuristics. The explanation for this behaviour lies in the computation of successor nodes: a node is considered *Closed* when we have expanded all its successors and added them to the *Open* list. The branching factor is the average number of successors for each node that is visited. As the branching factor in the sliding tile puzzle is approximately 2, it takes longer than linear time to process the *Closed* nodes.

In the case of SA*, we see the behaviour of the processing times for the *Closed* nodes in SA* *is* close to linear. Each *Closed* node in SA* is of course a BDD. While the branching factor in SA* and A* is of course the same, state space expansion based on BDD representation is different from the explicit state representation in A*. When expanding the new states for the search tree, SA* can compute all successors with just one BDD operation (see line 9 in Figure 1). As well, due to the clustering effect of SA*, the number of *Closed* nodes is often less than those in A*. Hence, the linear-like behaviour of SA* shows the advantage of combining sets of states and processing them together. This effect is most pronounced in the case of a weak heuristic when a large part of the state space must be searched.

For the weakest heuristics (h_1 and h_2), there can be a performance difference of 2 orders of magnitude. In the case where either a solution is very close to the start configuration or the heuristic employed is very strong, however, A* outperforms SA* as depicted by the cross-overs in Figure 6, 7 and 8. So, in the case of a weak heuristic, provided the goal is very close to the start configuration, A* outperforms SA*. The reason is that SA* needs to manage the operation on current states, transition relation and heuristic partitions [6], which all involve possibly complicated BDDs. If the heuristic employed is very strong, SA* has almost no advantage over A* although for a few tests SA* runs faster than A*. This behaviour is direct consequence of the fact that only a small part of the state space is searched.

Memory Consumption We have compared the memory consumption of A* and SA* for each of the 3 heuristics. The results are shown in Figure 9(a), (b) and (c). Note that we have opted to keep the scales the same for each of the

[6] The time taken to compute the transition relation BDD and heuristic partition BDDs has not been taken into account since they are computed *a priori*

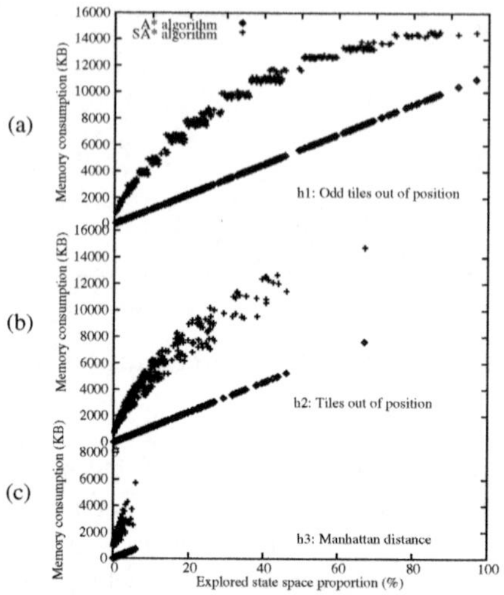

Fig. 9. Memory consumption of both the algorithms directed by the 3 heuristics

3 heuristics to illustrate the disparity in the proportion of state space that is explored. Each graph does however contains 500 data points. One could expect that SA* should consume less memory than A* due to the compact data structure (BDDs) employed. This is not the case however. As shown in the 3 graphs, A* always consumes less memory than SA*, by a factor of about one order of magnitude. The reason for this difference is not clear. While a BDD node in SA* can be represented by fewer bytes than an explicit node in A*, there are many more BDD nodes than explicit nodes. Furthermore, A* computes the successors and heuristic values on-the-fly whereas SA* needs to keep a transition-relation BDD and a number of heuristic-partition BDDs in memory for the purpose of expanding the search space. Note that it seems that there is no obvious way to avoid storing these BDDs in memory because once a state is encoded in a BDD, its relationship with the successors cannot be seen without the transition relationship.

The memory usage curves for A* in Figures 9(a), (b) and (c) show that there is a linear relationship between the memory consumption and the proportion of the explored state space. This result is expected because one state means one node in A*. In the case of SA*, however, we observe that for all the heuristics, the memory consumption increases rapidly at the start but tapers off as the search paths get longer. Of course, as we noted earlier, in the case of the strongest heuristics the search paths never get long.

In Figure 9 (a), we observe "steps", which indicate that the memory consumption remains constant as states are added. This occurs when the states are similar and have the same cost, and this is characteristic of a weak heuristic giving rise to more efficient clustering. If the heuristic is strong, new (successor) states are generally quite different from those in the existing *Closed* list, or have a different cost, so each new state states will impact on the BDD size. In Figure 9(b) there is still evidence of "steps" but it is less pronounced, and in Figure 9(c) the "steps" disappear.

4 Conclusions and Future Work

In this work, we have presented a symbolic, BDD-based A* algorithm called SA* and described its application to simple puzzles. We have also compared the performance of this algorithm to a conventional A* algorithm, using heuristics that have varying degrees of informedness.

The experimental results show the ability of SA* to cluster sets of states. This dramatically reduces the number of iterations needed by the algorithm to find an optimal path. We observe however that the benefit to be gained from clustering depends on the informedness of the heuristic: the more informed a heuristic is, the less the impact of clustering. In general, one can expect symmetries in state configurations will result in more clustering when the heuristic is poorly informed. We also note that the longer the search path the greater the benefit from clustering.

The performance comparison between A* and SA* revealed surprises:

- If only a weak heuristic is available, then SA* is faster than A*.
- If a strong heuristic is available, or the goal happens to be very close to the start configuration, A* is faster than SA*.

For the sliding tile puzzle, the performance cross-over happens at between 5%-7% of the explored state space, irrespective of heuristic.

A* uses substantially less memory than SA*, irrespective of the heuristic, although the proportion of the state space that is searched is much smaller when the heuristic is well informed of course. In the case of a very weak heuristic the memory usage of the two algorithms converge, and clustering leads to a step-like behaviour in the memory usage.

Note that optimisation techniques to condense data or speed up BDD operations are not studied in this research. A comparison of optimised algorithms is a next step. We have repeated the experimental work using another BDD package, called BuDDy[7]. The same behaviour was observed.

This work has only considered the sliding tile puzzle. We must be very careful extrapolating our results to other applications, but this is also an obvious next step. This research is a step toward applying AI techniques into formal verification.

---

[7] `http://www.it-c.dk/research/buddy/`

References

[1] Korf, R.: Iterative-deepening A*: An optimal admissible tree search. In: Ninth International Joint Conference on Artificial Intelligence(IJCAI-85), LA,California,USA, Morgan Kaufmann (1985) 1034–1036

[2] Korf, R.: Linear-space best-first search. Artificial Intelligence **62** (1993) 41–78

[3] Korf, R.: Space-efficient search algorithms. Computing Surveys **27** (1995) 337–339

[4] Korf, R.: Divide-and-conquer bidirectional search: First results. Proceedings of the Sixteenth International Joint Conference on Artificial Intelligence, IJCAI 99, Stockholm, Sweden **2** (1999) 1184–1189

[5] Ghosh, S., Mahanti, A., Nau, D.: Improving the efficiency of limited-memory heuristic search. Technical Report CS-TR-3420 (1995)

[6] Robnik-Sikonja, M.: Effective use of memory in linear space best first search (1996)

[7] Russell, S.: Efficient memory-bounded search methods. In: ECAI 92: 10th European Conference on Artificial Intelligence Proceedings, Vienna, Austria, Wiley (1992) 1–5

[8] Bryant, R.: Graph-based algorithms for boolean function manipulation. IEEE Transactions on Computers, C-35-8 (August 1986) 677–691

[9] Hu, A. J., Dill, D. L.: Efficient verification with BDDs using implicitly conjoined invariants. In: Computer Aided Verification. (1993) 3–14

[10] Hu, A. J., Dill, D. L., Drexler, A., Yang, C. H.: Higher-level specification and verification with BDDs. In: Computer Aided Verification. (1992) 82–95

[11] Janssen, G. L. J. M.: Application of BDDs in formal verification. In: the 22-nd Intl. School and Conf. on CAD, Yalta-Gurzuff, Ukraine. (May 1995) 49–53

[12] McMillan, K.: Symbolic model checking. Kluwer Academic Publishers, Boston,MA,USA (1994)

[13] Andersen, H. R.: An introduction to binary decision diagrams. Technical report, Department of Information Technology, Technical University of Denmark (1998)

[14] Bollig, B., Wegener, I.: Improving the variable ordering of OBDDs is NP-complete. IEEE transactions on computers **45** (1996) 993–1002

[15] Somenzi, F.: Binary decision diagrams (1999)

[16] Yang, B., Chen, Y. A., Bryant, R. E., O'Hallaron, D. R.: Space- and time-efficient BDD construction via working set control. In: Asia and South Pacific Design Automation Conference. (1998) 423–432

[17] Edelkamp, S.: Directed symcolic exploration in AI-planning. Proceeding of the 5th European conference on planning(ECP-99) (1999)

[18] Edelkamp, S., Reffel, F.: OBDDs in heuristic search. In Herzog, O., Günter, A., eds.: KI-98: Advances in Artificial Intelligence. Volume 1504 of Lecture Notes in Artificial Intelligence. Springer-Verlag, New York–Heidelberg–Berlin (1998) 81–92

[19] Hansen, E., Zhou, R., Feng, Z.: Symbolic heuristic search using decision diagrams. Symposium on Abstraction, Reformulation and Approximation (SARA2002), Kananaskis, Alberta, Canada (July 2002)

[20] Jensen, R., Bryant, R., Velso, M.: SetA*: An efficient BDD-based heuristic search algorithm. Proceedings of AAAI-2002, Edmonton, Canada (August 2002)

[21] Groce, A., Visser, W.: Model checking java programs using structural heuristics. In: International Symposium on Software Testing and Analysis. (2002) 12–21

[22] Groce, A., Visser, W.: Heuristic model checking for java programs. In: SPIN Workshop on Model Checking of Software. (2002) 242–245

[23] Santone, A.: Heuristic search + local model checking in selective mu-calculus. IEEE Transaction on Software Engineering **29** (2003) 510–523

BN+BN: Behavior Network with Bayesian Network for Intelligent Agent

Kyung-Joong Kim and Sung-Bae Cho

Dept. of Computer Science, Yonsei University
134 Shinchon-dong, Sudaemoon-ku, Seoul 120-749, Korea
uribyul@candy.yonsei.ac.kr
sbcho@cs.yonsei.ac.kr

Abstract. In the philosophy of behavior-based robotics, design of complex behavior needs the interaction of basic behaviors that are easily implemented. Action selection mechanism selects the most appropriate behavior among them to achieve goals of robot. Usually, robot might have one or more goals that conflict each other and needs a mechanism to coordinate them. Bayesian network represents the dependencies among variables with directed acyclic graph and infers posterior probability using prior knowledge. This paper proposes a method to improve behavior network, action selection mechanism that uses the graph of behaviors, goals and sensors with activation spreading, using goal inference mechanism of Bayesian network learned automatically. Experimental results on Khepera mobile robot show that the proposed method can generate more appropriate behaviors.

1 Introduction

There are many Action Selection Mechanisms (ASMs) to combine behaviors for generating high-level behaviors including spreading activation network, subsumption architecture and hierarchical ASM [1]. The ASM is essential in behavior-based robotics because it selects appropriate one among candidate behaviors and coordinates them. Usually, ASMs cannot insert goals into the model in explicit or implicit manner. Behavior network, one of ASM, can contain goals of robot in implicit manner and propagates activation of behaviors in two directions (forward and backward) through the network for dynamic selection [2].

Behavior network can have many goals that need to be active in different environments and status. User can insert prior knowledge of goal activation into behavior network in the design stage of behavior network but it is difficult to capture and represent the knowledge. There are some computational methodologies to represent knowledge into graph model with inference capability such as Bayesian network, fuzzy concept network and fuzzy cognitive map [3,4,5]. Above all, Bayesian network has been used practically to infer goals of software users in Microsoft Excel [6].

T.D. Gedeon and L.C.C. Fung (Eds.): AI 2003, LNAI 2903, pp. 979-991, 2003.

In previous work [7], we proposed a method to combine behavior modules evolved on CAM-Brain [8] using behavior network. In this paper, we attempt to apply Bayesian network to represent prior knowledge about goal activation in behavior network and infer posterior probability with observed variables. Bayesian network estimates the importance of goals with some observed sensor data. Structure of Bayesian network can be learned from the data that are collected with some different approaches such as conditional independence tests [9], scoring-based optimization [10] and hybrid of the two approaches [11]. In this paper, scoring-based method is adopted to construct Bayesian network because manual construction of network needs much effort and is impossible with observed raw sensor data.

In general, there are three kinds of agent architectures: deliberation, reactive agent and hybrid architectures. In imitation of the cognitive architecture of the human's brain, a new type of agent architecture is proposed and a robot soccer team underlying this framework is established [12]. To model complex systems, software agents need to combine cognitive abilities to reason about complex situations, and reactive abilities to meet hard deadlines. Guessoum proposed an operational hybrid agent model which mixes well known paradigms (objects, actors, production rules and ATN) and real-time performances [13]. There are some works related to hybrid agent architectures like combination of reactive control and deliberative planning [14,15,16].

Experiments are conducted on Khepera mobile robot simulator with four basic behaviors. Bayesian network structure is learned using the data that are collected from the experimental environments. Experimental results show that the proposed method of behavior network with Bayesian network is promising.

2 Behavior Network

Competition of behaviors is the basic characteristics of behavior network. Each behavior gets higher activation level than other behaviors from forward and backward activation spreading. Among candidate behaviors, one that has the highest activation level is selected and has control of robot. Activation level a of behavior is calculated as follows. Precondition is the sensor that is likely to be true when the behavior is executed. Add list is a set of conditions that are likely to be true by the execution of behavior and delete list is a set of conditions that are likely to be false by the execution of behavior. Figure 1 is a typical example of behavior network.

Forward propagation: Activation a is updated as the value added by environmental sensors that are precondition of the behavior. n means the number of sensors, and a_s is the activation level of the sensor.

$$\Delta a_1 = \sum_{i=1}^{n} f(a_{s_i}) \tag{1}$$

$$f(a_{s_i}) = \begin{cases} \phi \times a_{s_i} & s_i \in \text{precondition} \\ 0 & s_i \notin \text{precondition} \end{cases} \tag{2}$$

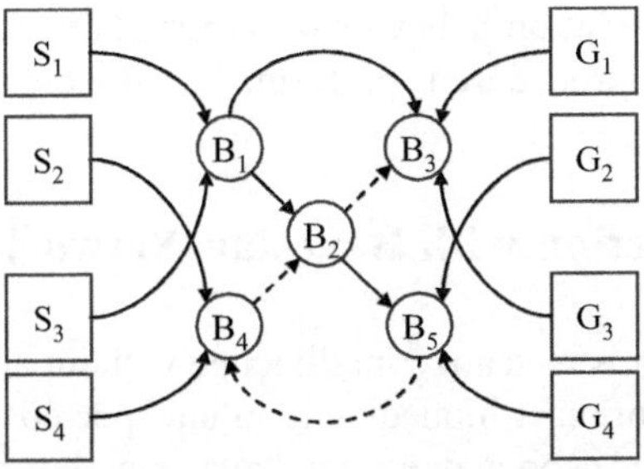

Fig. 1. An example of behavior network (S: sensors, B: behavior, G: goal). Solid line among behaviors represents predecessor link and dashed line represents successor link

Backward propagation: Activation a is updated as the value by goals that are directly connected to the behavior. If the execution of the behavior is desirable for the goal, positive goal-behavior link get active between them. Otherwise, negative goal-behavior link get active between them. n means the number of goals, and a_g is the activation level of the goal.

$$\Delta a_2 = \sum_{i=1}^{n} f(a_{g_i}) \tag{3}$$

$$f(a_{g_i}) = \begin{cases} \gamma \times a_{g_i} & g_i \in \text{positive link} \\ 0 & g_i \notin \text{negative link} \end{cases} \tag{4}$$

Internal spreading: Activation a is updated as the value added by other behaviors that are directly connected. If the execution of behavior B is desirable for behavior A, predecessor link from A to B and successor link from B to A get active. If the execution of behavior B is not desirable for behavior A, conflictor link from A to B is active. Here, n is the number of behaviors, and a_b is the activation level of the behavior.

$$\Delta a_3 = \sum_{i=1}^{n} f(a_{b_i}) \tag{5}$$

$$f(a_{b_i}) = \begin{cases} a_{b_i} & \text{predecessor link from } b_i \\ \phi / \gamma \times a_{b_i} & \text{successor link from } b_i \\ -\delta / \gamma \times a_{b_i} & \text{conflictor link from } b_i \\ 0 & \text{otherwise} \end{cases} \tag{6}$$

Finally, the activation of a is updated as follows.

$$a' = a + \Delta a_1 + \Delta a_2 + \Delta a_3 \tag{7}$$

If the activation level a' is larger than threshold θ and the precondition of the behavior is true, the behavior becomes a candidate to be selected. Among candidate behaviors, the highest activation behavior is chosen. Threshold θ is reduced by 10% and the activation update procedure is performed until there are candidate behaviors.

3 Behavior Selection with Bayesian Network

Behavior-based approach is to realize intelligence without any explicit representation. This property makes robot react immediately to unexpected situation such as "navigation on unknown planet." Robot does not have complex internal representation to process input signal. For higher behaviors, it is desirable to combine the reactive behaviors using some extra mechanisms like ASM.

Adding learning, inference and planning capabilities can improve ASM. Designing ASM is not an easy task because there are many variables to consider and knowledge about environment is not enough. Learning algorithm can determine the structure of ASM automatically and change a part of structure adaptively to the environments. Inference module uses computational model such as Bayesian network, fuzzy concept network and fuzzy cognitive map to represent prior knowledge and estimates unknown variables. ASM is not adequate to insert knowledge for inference and cannot select behaviors properly when the problem contains uncertainty. Planning optimizes the sequence of behaviors for solving the task.

In this paper, we focus on inference mechanism for ASM. Bayesian network is used to infer goals of behavior network that has one or more goals. Robot has several sensors that are represented as real value. Behavior-based robot uses only the sensor information to infer the status of environment. It is impossible to construct Bayesian network manually using the information of sensors. Therefore, Bayesian network structure is determined by learning from data that are collected from the environment.

3.1 Bayesian Network

Consider a domain U of n discrete variables $x_1,\ldots,x_n$, where each x_i has a finite number of states. A Bayesian network for U represents a joint probability distribution over U by encoding (1) assertions of conditional independence and (2) a collection of probability distributions. Specifically, a Bayesian network B can be selected as the pair (B_s, Θ), where B_s is the structure of the network, and Θ is a set of parameters that encode local probability distributions [17]. Figure 2 shows an example of Bayesian network that has six variables.

The joint probability for any desired assignment of values $<y_1, \ldots, y_n>$ to the tuple of network variables $<Y_1, \ldots, Y_n>$ can be computed by the following equation:

$$P(y_1,\ldots,y_n) = \prod_{i=1}^{n} P(y_i \mid Parents(Y_i)) \tag{8}$$

where $Parents(Y_i)$ denotes the set of immediate predecessors of Y_i in the network.

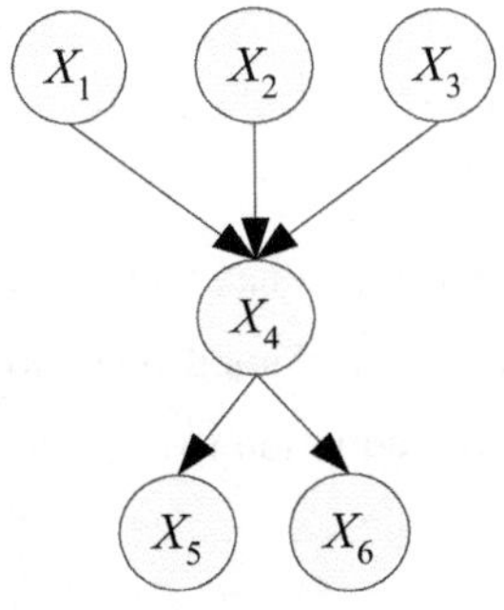

Fig. 2. An example of Bayesian network

3.2 Bayesian Network Learning

In this paper, we focus on learning structure of Bayesian network from data. The problem of learning a Bayesian network can be informally stated as: Given a training set $D=\{u_1,\ldots,u_N\}$ of instances of U, find a network B that best matches D. This optimization process is implemented in practice by using heuristic search techniques to find the best candidate over the space of possible networks. Scoring function assesses each Bayesian network using Bayesian formalism or minimum description length (MDL). Algorithm B based on greedy search is a representative method to find the structure of Bayesian network using scoring function [10]. Algorithm B is as follows.

Step 1: Initialize the network with no arc.
Step 2: Select arc($A\rightarrow B$) that has maximum increase of scoring function when the arc is inserted.
Step 3: Insert $A\rightarrow B$ into directed acyclic graph (DAG).
Step 4: Detect and remove cycles and remove it.
Step 5: Repeat 2-4 until there is no improvement or no arc inserted.

Usually, each Bayesian network is represented as a matrix and each element of the matrix represents link between nodes. Detection of cycle is easily conducted by checking links between ancestors and descendants of a node.

3.3 Behavior Network with Inference

Behavior network has one or more goals to be achieved in the environments. Coordination of their activation cannot be fixed in design stage of behavior network because there is uncertainty. Bayesian network is adopted to infer activation of goal from some observations from the environments. The structure of Bayesian network is automatically learned from the data that are collected from wandering. Observed variables are sensor information that can be collected from the robot sensors including distance, light and velocity. From these data, it is possible to estimate unknown variables including area information, emergency level and cooperation with other agents. Equation (3) is modified as follows:

$$\Delta a_2 = \sum_{i=1}^{n} \sum_{j=1}^{m} f(a_{g_i}) \times r_{i,j} \times P(v_j \mid observation) \tag{9}$$

where m is the number of unknown variables, and $r_{i,j}$ is the relevance of goal i with respect to variable j. This value is determined manually. $P(v_j \mid observation)$ is calculated using Bayesian network and equation (8).

4 Experimental Results

Khepera was originally designed for research and education in the framework of a Swiss Research Priority Program (see Figure 3). It allows confrontation to the real world of algorithms developed in simulation for trajectory execution, obstacle avoidance, pre-processing of sensory information, and hypothesis test on behavior processing. Khepera robot has two wheels. Eight infrared proximity sensors are placed around the robot.

Fig. 3. Mobile robot, Khepera

Two different experimental environments are used (Figure 4). Environment-I (E-I) is designed to test the proposed method with manually constructed Bayesian network. Environment-II (E-II) is for automatically constructed Bayesian network. In E-I, there are two points (A, B) where robot must pass. Light source is positioned in A. Robot can detect position A with the level of light sensor. Problem of E-I is to detect robot's unexpected stop and make appropriate change of goal. In the design stage of behavior network for E-I, it is difficult to incorporate of how to manage this situation. Problem of E-II is different with one of E-I. Robot must pass light sources in Area 2 and Area 3.

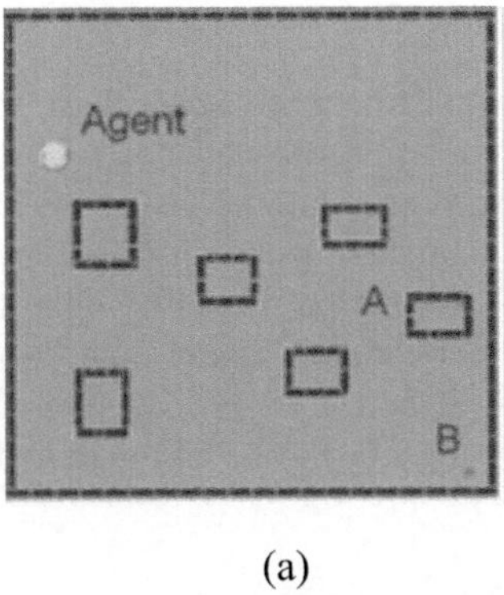

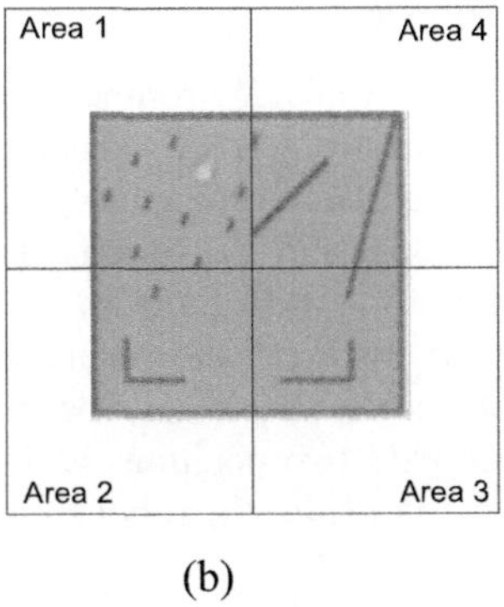

Fig. 4. Two different experimental environments. (a) Environment-I (b) Environment-II

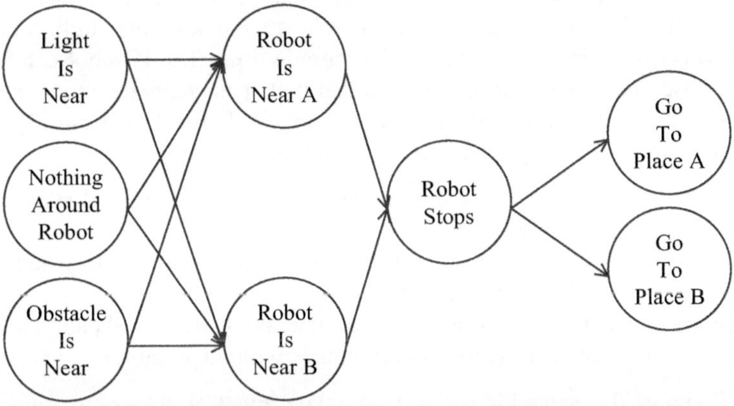

Fig. 5. Bayesian network that is manually constructed

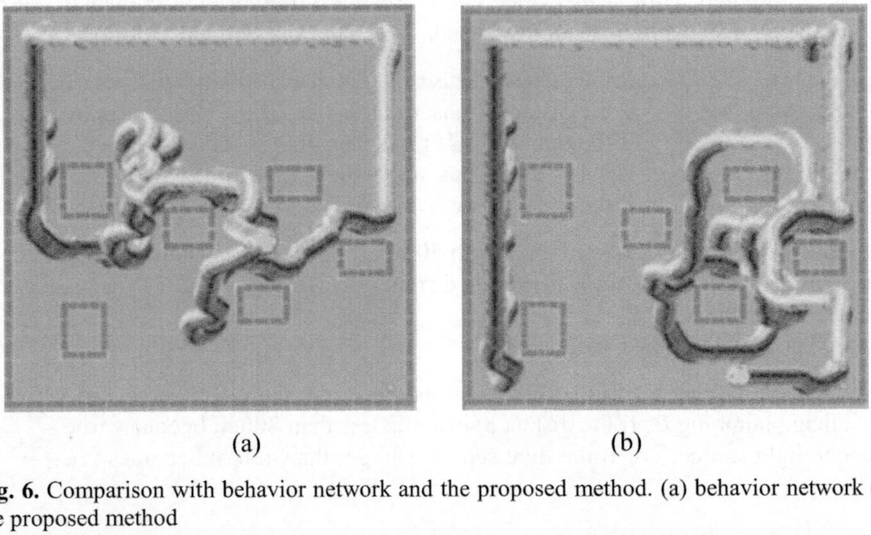

Fig. 6. Comparison with behavior network and the proposed method. (a) behavior network (b) the proposed method

4.1 Experiment I

Prior knowledge of this environment I is that robot frequently stops because it considers light source as obstacle. Also robot stops in position B by the small obstacle that is not easy to detect. Goal of agent is to go A and B sequentially (A and B are noted in Figure 4-(a)). Bayesian network is designed manually in a very simple form. The position of robot is estimated from the sensor information including light sensors and distance sensors. Figure 5 shows this Bayesian network and figure 6 shows a comparison with behavior network and the proposed method. The proposed method shows that robot can pass two positions without stopping but the behavior network without Bayesian network makes the robot stop at A.

4.2 Experiment II

There are four different areas in E-II (Area1, Area2, Area3, and Area4). Area1 is a start position that has many small obstacles. Area2 contains two light sources and Area3 contains one light source. Area 4 has simple obstacles. If robot can recognize area using observed information, behavior network can generate more appropriate behavior sequences. Robot uses three behaviors evolved on CAM-Brain [7]. They are "Avoiding obstacles," "Going straight," and "Following light."

Following light: The robot goes to stronger light. It must operate this module to go to the light source.
Avoiding obstacle: If there is an obstacle around the robot, it avoids obstacles without bumping against it.
Going straight: If there is nothing around the robot, it goes ahead. This module makes it to move continuously without stopping.

Figure 7 shows the behavior network for experiment II. There are two different links among behaviors. Predecessor link is represented with solid line and successor link is represented with dashed line. There are five different environmental sensors that can be inferred from original raw sensor data. They are defined as follows.

Obstacle is near: If the distance sensor is less than 1000, it becomes true.
Nothing around robot: If the distance sensor is less than 1000, it becomes true.
Light level I: If the light sensor is less than 400, it becomes true.
Light level II: If 400 < the light sensor < 450, it becomes true.
No light source: If the light sensor is larger than 450, it becomes true.

Light sensor has the value ranged from 50 to 500 and 50 means that light source is near. Distance sensor has the value ranged from 0 to 1024 where 1024 means that the obstacle is very near.

Three goals are defined as follows.

Minimizing bumping A: If the distance sensor is larger than 1020, it becomes true.
Minimizing bumping B: If the distance sensor is less than 800, it becomes true.
Going to light source: If the light sensor is larger than 450, it becomes true.

If the robot is in Area1, the value of distance sensor is large and needs to avoid bumping. If the robot is in Area4, the value of distance sensor is small and needs to go straight. If the light sensor has lower value, robot will propagate signals to "Following light."

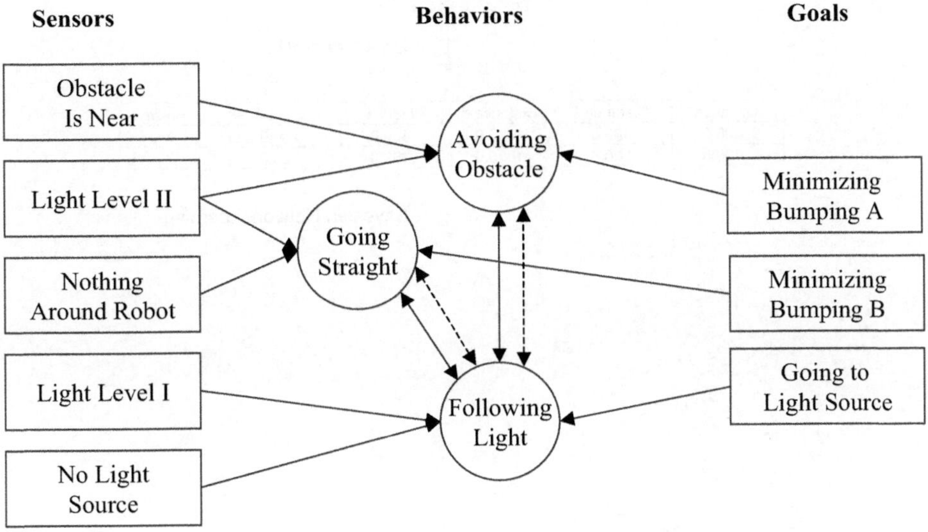

Fig. 7. Behavior network for experiment II

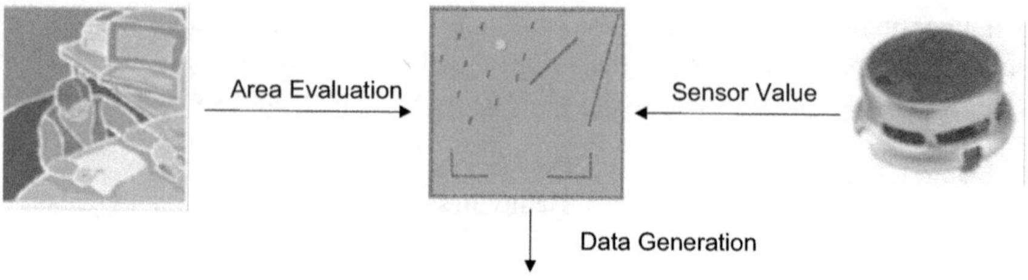

Distance1	Distance 2	Distance 3	Light 1	Light 2	Area
500.0	400.0	1022.0	499.0	200.0	1
498.0	300.0	1020.0	450.0	220.0	2

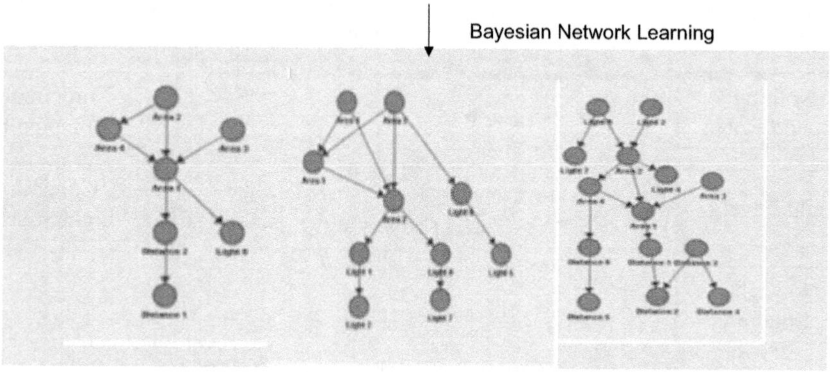

Fig. 8. Data collection for Bayesian network learning

Figure 8 shows the procedure of collecting data from the experimental environments. Robot moves randomly and user evaluates robot's position as one of "Area1," "Area2," "Area3," and "Area4." Khepera robot has 8 distance sensors and 8 light sensors of real value. Training data for Bayesian network have 20 variables (16 sensor values and 4 area information). If robot is in Area1, the variable for Area1 is 1.0 and other area variables are 0.0.

Bayesian network is learned using algorithm the *B* based on Tabu search [10]. Figure 9 shows the structure of Bayesian network that determines the position of robot. Bayesian network consists of 14 nodes (four nodes are area variables and ten nodes are related with distance and light sensors). We expected that light information was very important because light source can be used as criterion to classify bottom area with top area. Area2 and Area3 can be classified with the strength of light because Area2 has two light sources but area3 has one light source. The learned Bayesian network represents the information well. Two light sensor nodes directly link to Area2.

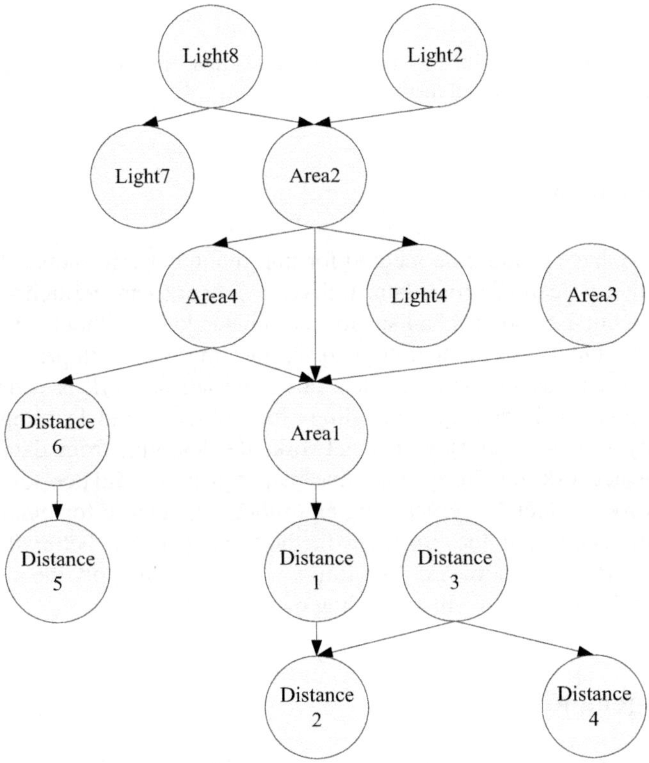

Fig. 9. Bayesian network that are learned from the collected data

Figure 10 shows experimental results in E-II. In (a,b), robot navigates the area with behavior sequence that is determined using the behavior network. The network selects one behavior at a time and executes it. In (c), robot navigates the area with the combination of behavior network and Bayesian network learned. Bayesian network determines the conditional probability of area1, area2, area3, and area4 with observed sensors. Unlike the robots in (a) and (b), robot passes light source in (c). Goal of robot is to visit four areas without exception and if there is a light source, it must go to pass the source.

(a) (b) (c)

Fig. 10. Experimental results with Bayesian network learned. (a,b) Behavior network without Bayesian network (c) The proposed method

5 Conclusions

In this paper, we have proposed a method for improving behavior network with Bayesian network that is learned from data. Behavior network can implicitly insert goals into the model but has no mechanism to coordinate goals without conflict. In this reason, some computational methodology to support inference with prior knowledge is needed. There are many inference models but Bayesian network is widely used and can provide sound mathematical foundations for inference and learning. To reduce user's difficulty in design of Bayesian network, the learning from data is adopted. Fuzzy concept network and fuzzy cognitive map can be candidates for inference of behavior network but there is no learning algorithm of structure for them. The different experiments show that the proposed method can perform better than behavior network without Bayesian network. As a future work, we will apply the model to more complex and realistic problems like avoiding movable obstacle.

Acknowledgement

This research was supported by Biometrics Engineering Research Center, and Brain Science and Engineering Research Program sponsored by Korean Ministry of Science and Technology.

References

[1] Pirjanian, P.: Behavior coordination mechanisms-state-of-the-art. Tech-report IRIS-99-375, Institute for Robotics and Intelligent Systems, School of University of Southern California, Oct (1999)

[2] Maes, P.: How to do the right thing. Connection Science, vol. 1, no. 3 (1989) 291-323

[3] Pearl, J.: Probabilistic Reasoning in Intelligent Systems. Morgan Kaufmann Publishers (1998)

[4] Chen, S.-M. and Wang, J.-Y.: Document retrieval using knowledge-based fuzzy information retrieval techniques. IEEE Transactions on Systems, Man and Cybernetics, vol. 25, no. 5 (1995) 793-803

[5] Stylios, C.D. and Groumpos, P.P.: Fuzzy cognitive maps: A model for intelligent supervisory control systems. Computers in Industry, vol. 39, no. 3 (1999) 229-238

[6] Horvitz, E., Breese, J., Heckerman, D., Hovel, D. and Rommelse, K.: The lumiere project: Bayesian user modeling for inferring the goals and needs of software users. Proceedings of the Fourteenth Conference on Uncertainty in Artificial Intelligence (1998) 256-265

[7] Kim, K.J. and Cho, S.B.: Dynamic selection of evolved neural controllers for higher behaviors of mobile robot. IEEE International Symposium on Computational Intelligence in Robotics Automation, Banff, Canada (2001) 467-472

[8] Garis, H. de and Korkin, M.: The CAM-Brain Machine (CBM): An FPGA based tool for evolving a 75 million neuron artificial brain to control a lifesized kitten robot. Journal of Autonomous Robots, vol. 10, no. 3 (2001) 236-249

[9] Pearl, J. and Verma, T. S.: Equivalence and synthesis of causal models. Proceedings of the 6th Conference on Uncertainty in Artificial Intelligence (1990) 220-227

[10] Campos, L. M. de, Fernandez-Luna, J.M., Gamez, J. A. and Puerta, J. M.: Ant colony optimization for learning Bayesian networks. International Journal of Approximate Reasoning, vol. 31 (2002) 291-311

[11] Singh, M. and Valtorta, M.: Construction of Bayesian network structures from data: A brief survey and an efficient algorithm. International Journal of Approximate Reasoning, vol. 12 (1995) 111-131

[12] Shi, L., Zhen, Y. and Zengqi-Sun.: A new agent architecture for RoboCup tournament: Cognitive architecture. Proceedings of the 3rd World Congress on Intelligent Control and Automation, vol. 1 (2000) 199-202

[13] Guessoum, Z.: A hybrid agent model: A reactive and cognitive behavior. The International Symposium on Autonomous Decentralized Systems (1997) 25-32

[14] Lyons, D.M., and Hendriks, A.J.: Planning as incremental adaptation of a reactive system. Robotics and Autonomous Systems, vol. 14, no. 4 (1995) 255-288

[15] Murphy, R. R., Hughes, K., Marzilli, A., and Noll, E.: Integrating explicit path planning with reactive control of mobile robots using Trulla. Robotics and Autonomous Systems, vol. 27, no. 4 (1999) 225-245

[16] Aguirre, E., and Gonzalez, A.: Fuzzy behaviors for mobile robot navigation: Design, coordination and fusion. International Journal of Approximate Reasoning, vol. 25, no. 3 (2000) 255-289

[17] Chickering, D. M., Heckerman, D. and Meek, C.: A Bayesian approach to learning Bayesian networks with local structure. Microsoft Technical Report MSR-TR-97-7, 1997.

Effectiveness of Syntactic Information for Document Classification

Kyongho Min[1] and William H. Wilson[2]

[1] School of Computer and Information Sciences
Auckland University of Technology
Private Bag 92006, Auckland, 1020 New Zealand
kyongho.min@aut.ac.nz
[2]School of Computer Science and Engineering, UNSW
Sydney, 2052 Australia
billw@cse.unsw.edu.au

Abstract. This paper compares effectiveness of document classification algorithms for a highly inflectional/derivational language that forms monolithic compound noun terms, like Korean. The system is composed of three phases: (1) a Korean morphological analyser called HAM [10], (2) compound noun phrase analysis and extraction of terms whose syntactic categories are noun, proper noun, verb, and adjective, and (3) various document classification algorithms based on *preferred class score* heuristics. We focus on the comparison of document classification methods including a simple voting method, and preferred class score heuristics employing two factors, namely *ICF* (inverse class frequency) and *IDF* (inverse document frequency) with/without term frequency weighting. In addition, this paper compares algorithms that use different class feature sets filtered by four syntactic categories. Compared to the results of algorithms that are not using syntactic information for class feature sets, the algorithms using syntactic information for class feature sets shows performance differences in this paper by -3.3% - 4.7%. Of the 20 algorithms that were tested, the algorithms, *PCSIDF FV* (i.e. Filtering Verb Terms) and *Weighted PCSIDF FV,* show the best performance (74.2% of F-measurement ratio). In the case of the Weighted PCSICF algorithm, the use of syntactic information for selection of class feature sets decreased the performance on document classification by 1.3 - 3.3%.

1 Introduction

A hard task for an information retrieval system is to understand meanings of terms in the documents and queries of users. Semantic keyword extraction is difficult to apply in information retrieval systems because of the complexity of semantic interpretation and parsing of enormous documents. Automatic document classification would im-

T.D. Gedeon and L.C.C. Fung (Eds.): AI 2003, LNAI 2903, pp. 992-1002, 2003.

prove the performance of information retrieval by applying a) an effective indexing scheme for task document contents and/or b) an efficient ranking system to pinpoint precise and necessary information for users, rather than to show large numbers of retrieved documents, many of them irrelevant [11].

Most automatic document classification systems have used class feature sets where the features are direct terms in the testing and training data, excluding stop words and low frequency words (e.g. words with less than 3 occurrences may be ignored). However, most systems have not considered the syntactic categories of extracted terms in document classification. This paper focuses on the hypothesis that knowing the syntactic categories of open class words in class features would affect the performance of classification algorithms based on ICF (Inverse Class Frequency) and IDF (Inverse Document Frequency).

Many researchers [4, 9, 12, 14] have studied comparisons of different classification algorithms rather than the influence of syntactic information (e.g. whether the word is a name, noun, verb, or adjective) in document classification. This paper compares the effectiveness of class features belonging to different syntactic categories.

We implemented a range of document classification algorithms for Korean newspaper articles, based on:

(a) HAM – Hangul (Korean) Analysis Module [10];
(b) a compound noun term analyser based on a longest substring algorithm [8] employing an agenda-based chart parsing technique [5];
(c) extraction of class features based on syntactic information; and
(d) a *preferred-class-score* document classification algorithm without a training process but with two classification factors, namely ICF (inverse class frequency) and IDF (inverse document frequency), and three filtering schemes based on syntactic categories of terms extracted from the testing documents.

The term, *preferred class score,* means that each term used in the classification process has a preferred score for a specific class on the basis of the classification algorithm. The algorithm proceeds by computing a number of scores, using these to decide the preferred class rather than complex classification methods and training processes. Thus we refer to the algorithm and the scores using the label "preferred class score" (PCS).

Previous researchers have studied the effects of the following factors for document classification: the number of features, the rareness of documents of a certain class, a term's frequency in a testing document, the rule-based Boolean combination of the terms for the class feature sets [1], different learning methods [7], and different similarity measurement methods based on an ICF and IDF [9]. In addition previous studies have shown that the performance of various classification methods has not been consistent because of different training techniques and testing environments employed.

However, some researchers have studied aspects of syntactic information in information retrieval and document classification. [6] studied the effectiveness of proper nouns in document processing. [2, 3] studied the usefulness of syntactic information and compound terms in a text classification task. Their systems showed a small improvement in performance (> 1%). The system employing Wordnet to extract class feature sets improved the performance of a rule-based Ripper classifier [13]. They found the use of linguistic information affected the performance of IR systems for filtering or document classification.

This paper focuses on a simple but effective classification algorithm that does not employ time-consuming and complex learning and document similarity measurement algorithms, the application of various numbers of feature sets filtered by syntactic categories (i.e. noun, name, adjective, and verb), and the comparisons of classification methods employing an ICF/IDF with filtered feature sets. Our method of classifying a document to a topic area, using a compound noun term analyser and document classification algorithms, (five basic algorithms and 15 algorithms filtering terms by three syntactic categories (i.e. proper noun, verb, and adjective)), will be described in the next section. In section 3, our experimental results will be discussed and section 4 describes possible further improvements of the current approaches, and conclusions.

2 Methods of Document Classification

This section describes methods of document classification employing syntactic information to extract terms for class features from the training and testing documents. The document classification system is composed of a) morphological analysis based on the HAM (Hangul Analysis Module [10]) system, b) a term extraction method and analysis of compound noun terms, and c) document classification algorithms based on *Preferred Class Score* (*PCS*).

2.1 Morphological Analysis and Compound Noun Analysis

Korean is a highly inflected language requiring very complex computation for Korean NLP. Thus, HAM for Linux [10] was employed for efficient and effective Korean morphological analysis; the HAM system has been continually enhanced since 1993. HAM reads a Korean text file and returns the multiple morphological analyses of each word in the text. However, HAM analyses words that are not in the lexicon as Nouns and their real syntactic categories are manually classified.

The morphological ambiguities of each word were manually removed to choose the best morphological analysis and extract terms, whose syntactic categories are noun, verb, adjective, and name, from documents. The number of ambiguities of a word ranged from 1 to 4, in our data set, which included 17470 words.

HAM focused on the analysis of inflectional and derivational morphemes [10]. However, HAM does not analyse compound noun terms. The identified compound noun term (e.g. "□□□□" – economic growth) is analysed to use each noun term (e.g. "□□" - economy, "□□" – growth) as a class feature, rather than use the compound noun term itself. The compound noun terms are processed by a longest substring algorithm [8] combined with an agenda-based chart parsing technique [5] to handle many ambiguities. A simple penalty scheme selects one analysis from the alternatives of the compound noun analysis. A penalty score for a found substring in the compound noun term is computed as the number of remaining bytes left behind after processing the substring.

After morphological and compound noun analysis, the class feature sets are extracted from the training data. The class feature sets are composed of the roots of nominals, proper names, verbal, and adjectival terms, rather than raw terms from a document, and excluding stop words. The words whose syntactic categories are adverbs, case-markers, and affixes are not considered for class features, and the remaining words are treated as unknown words, which might be numbers, alphanumeric words (e.g. "KBS2" – "Korea Broadcasting System 2") or compound words with Korean and English parts, in either order.

2.2 Classification Algorithms Using Syntactic Information

We tried five basic algorithms based on *preferred class score* (Table 1) and each algorithm employed four different filtering methods. The filtering methods used terms' syntactic categories to limit the number of class feature sets used in the classification process. When computing a class score for a term in the document, no probability or similarity measurement is used in the classification process. Only each term's class score is considered on the basis of either a simple voting score, ICF, or IDF algorithm to decide which class is to be preferred. The basic research task is to provide reliable class feature sets without semantic parsing and without a complex learning process requiring domain- and system-dependent parameters.

Five basic algorithms were implemented, for performance comparison, along with variants that filtered by syntactic categories. The five basic algorithms are as follows:

1. Preferred Class Score – One&Only (PCSO&O)

PCSO&O classifies documents by a simple voting scheme. When the system applies the voting scheme to each term T_{ik} in test document D_k, a preferred class score of 1 is only allocated to the class with maximum frequency for term T_{ik} in the class feature sets C_1, C_2, …, C_n, where n = 8 in this paper. For example, if the frequency of term T_{ik} in class feature set C_j, $Cf(T_{ij})$, is highest among class feature sets, then the preferred class score 1 is allocated to class C_j (i.e. $PCS(T_{ik}) = 1$) and other classes, C_1, C_2, …, C_{j-1}, C_{j+1}, …, C_n, have preferred class score of 0 for the term T_{ik}.

Table 1. Five Basic Algorithms

Basic Algorithm	Description
PCSO&O	Preferred Class Score based on a One&Only method
PCSICF	Preferred Class Score based on an Inverse Class Frequency
Weighted PCSICF (*WPCSICF*)	Preferred Class Score based on an Inverse Class Frequency weighted by term frequency in a test document
PCSIDF	Preferred Class Score based on an Inverse Document Frequency
Weighted PSCIDF (WPCSIDF)	Preferred Class Score based on an Inverse Document Frequency weighted by term frequency in a test document

After summation of each class's score over all terms in the document, the class with the maximum score is chosen for the class of the document. Let document D_k have terms $\{T_{1k}, \ldots, T_{ik}\}$ with frequencies $Cf(T_{ij})$ for each class feature set $\{C_1, \ldots, C_j\}$.

$$PCSO\&O(D_k) = C_j, \text{ if } C_j \text{ maximises } MAX_j\left\{\sum_{i=1}^{m} PCS(T_{ik})(C_j)\right\} \quad (1)$$

2. *Preferred Class Score – Based on Inverse Class Frequency (PCSICF)*

Two algorithms, *PCSICF* and *Weighted PCSICF*, are based on Inverse Class Frequency (ICF). [7], [9] compute similarity between a tested document and training data. The algorithm *PCSICF* computes the score by summing $Cf(T_{ij})*\log(N/icf(T_i))$ and the algorithm based on *Weighted PCSICF* computes the score by summing $Df(T_{ik})*Cf(T_{ij})*\log(N/icf(T_i))$ where $Cf(T_{ij})$ is the frequency of term T_i in class C_j in training documents and $Df(T_{ik})$ is the frequency of term T_i in document D_k in the test document. After computing and summing up scores of each class with *PCSICF* or *Weighted PCSICF* algorithm for all terms in the testing document, the class with the maximum score is allocated to the document.

Thus, the preferred class score is used in the classification algorithms rather than, for example, a cosine similarity computation. The *Weighted PCSICF* algorithm weighs positive local information in a testing document so that more frequently occurring terms are preferred for classification. The two algorithms are:

$$PCSICF(D_k) = C_j, \text{ if } C_j \text{ maximises } MAX_j\left\{\sum_{i=1}^{m} Cf(T_{ij})*\log(N/icf(T_i))\right\} \quad (2)$$

$$Weighted\ PCSICF(D_k) = C_j, \text{ if } C_j \text{ maximises}$$

$$MAX_j\left\{\sum_{i=1}^{m} Df(T_{ik})*Cf(T_{ij})*\log(N/icf(T_i))\right\} \quad (3)$$

3. *Preferred Class Score – Based on Inverse Document Frequency (PCSIDF)*

Two algorithms, *PCSIDF* and *Weighted PCSIDF*, are based on Inverse Document Frequency (IDF) rather than Inverse Class Frequency (ICF). Thus the algorithms compute the scores for each class for a testing document using $N/idf(T_i)$ where N is the total number of training documents (i.e. 720 in this paper). The *PCSIDF* algorithm computes the score for each class by $Cf(T_{ij})*log(N/idf(T_i))$, and the *Weighted PCSIDF* algorithm computes the score for each class as $Df(T_{ik})*Cf(T_{ij})*log(N/idf(T_i))$. The class with the maximum score is allocated to the testing document. The two algorithms are:

$$PCSIDF(D_k) = C_j, \text{ if } C_j \text{ maximises } MAX_j\left\{ \sum_{i=1}^{m} Cf(T_{ij}) * \log(N / idf(T_i)) \right\} \quad (4)$$

Weighted PCSIDF$(D_k) = C_j$, if C_j maximises

$$MAX_j\left\{ \sum_{i=1}^{m} Df(T_{ik}) * Cf(T_{ij}) * \log(N / idf(T_i)) \right\} \quad (5)$$

Algorithms Filtering Terms by Syntactic Information

Five basic algorithms based on either Inverse Document Frequency (IDF) or Inverse Class Frequency (ICF) [9, 7] were combined with three different filtering schemes. The schemes used syntactic categorical information rather than a complex learning algorithm to restrict the number of class features used in document classification. The filtering schemes are FP (Filtering Proper Noun terms), FV (Filtering Verb terms), and FPV (Filtering both Proper Noun and Verb terms). For example, *PCSICF FP* algorithm is based on ICF without using class feature sets whose syntactic categories are Proper Noun. Thus the algorithm is

$$PCSICF\ FP\ (D_k) = C_j, \text{ if } C_j \text{ maximises } MAX_j\left\{ \sum_{i=1}^{m} Cf(T_{ij}) * \log(N / icf(T_i)) \right\} \quad (6)$$

iff $(T_{ij}) \neq$ Proper Noun, where T_{ij} is term i in class feature set j.

The added condition for FV (Filtering Verb terms) is, iff $(T_{ij}) \neq$ Verb, where T_{ij} is term i in class feature set j and the condition for FPV (Filtering both Proper Noun and Verb terms) is, iff $(T_{ij}) \neq$ (Proper Noun or Verb), where T_{ij} is term i in class feature set j. Thus, there are 20 algorithms based on five basic algorithms × 4 different filtering schemes (FV, FP, FPV, and no filtering).

3 Experimental Results

The experimental results show that algorithms that do not use verb terms as their class feature sets perform better than those that do not use name terms as their class feature sets. The training data and testing data were collected from a Korean online newspaper. The system is implemented in Perl5.0 on a Linux platform. Surprisingly, the algorithm based on ICF performs better when it uses class feature sets after filtering out terms whose syntactic categories were Names.

3.1 Data Collection and Class Feature Set Extraction

From 8 topic classes, 100 news articles for each topic from the major Korean daily newspapers from 01/08/2000 to 31/08/2000 were collected for training (i.e. for extracting class feature sets) and testing the algorithms. The eight classes are domestic news (DSO), economy (DEC), world news (DIN), information and science (DIS), culture and life (DCL), political news (DPO), sports (DSP), and entertainment (DST). From each class, 10 articles were selected randomly for the performance tests with the various algorithms and the remainder were used for extraction of class feature sets. However, it would be possible for other human experts to classify some articles into different classes - the class we used for each article is based on the classification given by the newspaper.

In this paper, some words, whose syntactic categories are not Name, Noun, Verb, or Adjective, were not considered as either feature sets or terms for a document. Some terms in the testing documents do not occur in the class feature sets that are obtained from the training data. In this case, the words (8.2% (1259 terms) of the total terms (14670 terms) in testing document) from 80 testing documents are recognised as unclassified words by the training data. If these terms were used in the document classification process as class feature sets, then they would possibly affect the results of various algorithms in this paper.

Table 2 shows ratios of syntactic categories obtained from all the words occurring in 80 testing documents. In the case of a lexically ambiguous term, the syntactic category NAME is preferred to NOUN. The preference of a syntactic category for lexically ambiguous terms is necessary because the morphological system, HAM, does not disambiguate the lexical categories of terms between Name and Noun.

Table 2. Syntactic Categories Ratios of Testing Documents

Syntactic Category	DCL	DEC	DIN	DIS	DPO	DSO	DSP	DST	AVG
NOUN (%)	66	76	72	72	70	73	65	66	70.0
NAME (%)	7	3	6	7	8	7	15	9	7.8
VERB (%)	26	21	22	21	22	20	20	24	22.2

In Korean, there is no Noun/Verb ambiguity, in English words (e.g. "□□" – "study" as Noun, "□□□□" – "study" as Verb). A Korean word whose syntactic category is an adjective is categorised as VERB for simplicity because Korean adjectives function like English predicate BE+ADJP. For example, the word '□□□□' in the Korean sentence "□□□ (she) □□□□ (is beautiful)" is equivalent to a predicate "be beautiful" in English.

3.2 Results of Document Classification Based on Filtering Schemes

In terms of the precision ratio of document classification on the 80 testing texts (8 document classes × 10 sample texts) in Fig. 1, the algorithms filtering Verb terms (FV) performed better than those filtering Proper Nouns (FP), or Proper Nouns and Verbs (FPV). However, for the algorithm based on the factor ICF, the version filtering verb terms performed worse than that filtering terms with other syntactic categories. Thus, the choice of class feature sets and classification algorithms - ICF and IDF - greatly affected the document classification performance (Fig. 1). The average precision ratio of the 20 algorithms is 68.8%.

Among the 20 methods, the best performance was obtained by *PCS(IDF) FV* with/without weighting term frequency in documents, and filtering verb terms. This algorithm correctly classified 72.5% of testing documents. The simplest algorithm, *One&Only FV/FPV*, performs 2.5% worse than the best algorithm. Compared to algorithms not employing local document information (i.e. term frequency in the testing document), the algorithms weighted by term frequency performed classification better by less than 3.8%.

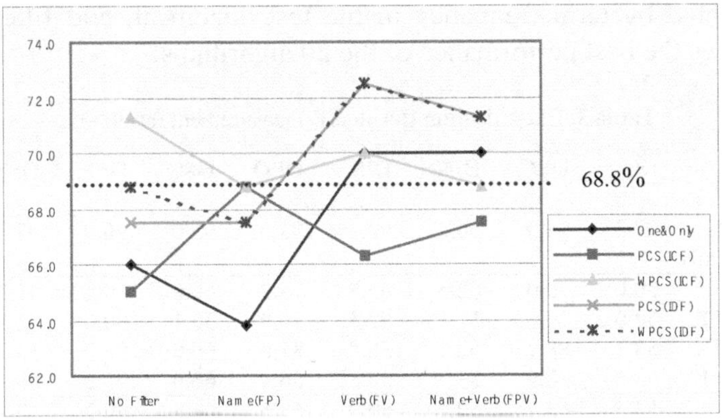

Fig. 1. Precision Ratio based on POS filtering

When the algorithm based on IDF was employed, the performances of methods were consistent: the performance difference between *PCS(IDF)* and *WPCS(IDF)* was fairly independent of the type of filtering (Fig. 1). However, the algorithms based on ICF showed that the methods weighted by term frequency performed better than those not weighted by term frequency.

In terms of the number of the terms used for document classification, that were extracted from testing documents, the largest number of terms used for classification algorithms were the algorithms without filtering schemes (100% terms used), filtering Proper Noun terms (92.2% terms used), filtering Verb terms (77.8% terms used), and filtering Proper Noun and Verb terms (70% terms used) in order (e.g. No filtering > FP > FV > FPV) (see Table 2). The results in Fig. 1 show that the number of the terms used for the classification process does not influence the performance of algorithms in this paper. Average performances of the five algorithms based on filtering schemes are 67.7% for no filtering algorithms, 67.3% for filtering Proper noun (i.e. Name) terms, 70.3% for filtering Verb terms, and 69.8% for filtering both Proper noun and Verb terms.

The classification methods based on PCS(ICF) algorithm depend greatly on syntactic categories, compared with those based on PCF(IDF). The class features whose syntactic categories are Verb do not affect document classification performance. However, algorithms based on the ICF factor were affected by higher value of log(N/ICF).

The F-measurement ratio in Table 3 is computed as (2 × precision ratio × recall ratio) / (precision ratio + recall ratio) (i.e. the harmonic mean of the precision and recall ratios). Across classes, DSP shows the best F-measurement ratio and DCL shows the worst F-measurement ratio. The simple algorithms, *One&Only and One&Only FX*, show comparable performance to some of the other 16 algorithms. The algorithm *WPCS(IDF) FV*, based on IDF, weighted by term frequency in the test document, and filtering Verb terms, shows the best performance of the 20 algorithms.

Table 3. Classification Results (F-measurement ratio(%))

Algorithm	DCL	DEC	DIN	DIS	DPO	DSO	DSP	DST	Average
Training data with One&Only	97.3	94.9	96.5	92.1	93.6	95.0	98.9	97.1	95.7
One&Only	50.0	75.0	62.5	62.5	78.2	60.8	100.0	42.9	70.3
One&Only FP	50.0	70.6	57.1	62.5	76.9	57.1	90.0	42.9	68.3
One&Only FV	53.3	88.9	62.5	58.8	81.8	66.6	95.2	53.3	72.6
One&Only FPV	53.3	88.9	53.3	58.8	86.9	66.6	95.2	53.3	72.6
PCS(ICF)	54.5	66.7	50.0	60.0	69.5	63.2	95.2	63.2	66.1
PCS(ICF) FP	56.0	70.6	57.1	60.0	80.0	63.2	95.2	63.2	70.8
PCS(ICF) FV	38.1	70.6	53.3	60.0	75.0	72.7	95.2	60.0	67.1
PCS(ICF) FPV	45.5	70.6	57.1	60.0	80.0	60.0	95.2	66.6	69.3
PCS(IDF)	52.2	75.0	57.1	58.8	76.9	60.8	95.2	60.0	70.2
PCS(IDF) FP	53.9	75.0	66.7	66.7	80.0	57.1	95.2	44.4	70.0
PCS(IDF) FV	57.1	82.4	66.7	73.7	80.0	69.5	95.2	52.7	74.2
PCS(IDF) FPV	54.5	82.4	66.7	73.7	80.0	63.6	95.2	52.7	73.1
WPCS(ICF)	52.2	82.4	75.0	63.6	81.8	70.0	95.2	52.7	72.9
WPCS(ICF) FP	56.0	77.8	62.5	57.1	81.8	63.2	95.2	55.6	69.8
WPCS(ICF) FV	45.5	82.4	75.0	63.6	81.8	66.6	95.2	52.7	71.5
WPCS(ICF) FPV	52.2	77.8	62.5	57.1	81.8	66.6	95.2	55.6	69.6
W PCS(IDF)	50.0	75.0	66.7	66.7	80.0	69.5	95.2	44.4	71.0
WPCS(IDF) FP	53.9	75.0	66.7	66.7	80.0	57.1	95.2	44.4	70.0
WPCS(IDF)FV	57.1	82.4	66.7	73.7	80.0	69.5	95.2	52.7	74.2
WPCS(IDF)FPV	54.5	82.4	66.7	73.7	80.0	63.6	95.2	52.7	73.1
Average	52.6	77.8	62.9	64.5	79.9	64.6	95.2	54.3	70.8

4 Conclusion

Document classification based on statistical methods will contribute to the efficiency and effectiveness of information search and information filtering. However, previous techniques have employed complex training and classification algorithms. This paper describes a parsimonious document classification algorithm without a complex learning process, pure comparisons of both ICF and IDF factors for document classification algorithms, and an efficient method of Korean compound noun analysis. When analysing Korean compound noun terms, an agenda-based chart parsing technique was applied to longest substring extraction, with a heuristic penalty score system.

The method described in this paper extracted the root forms of words of four categories (noun, verb, proper noun, and adjective) for class features and classification, after applying morphological analysis to the document rather than the use of direct terms excluding stop words and low frequency words.

The performance of the simple classification algorithm, One&Only, was satisfactory, compared to complex algorithms based on *ICF* and *IDF*, on the basis of complexity of classification and training processes. The One&Only algorithm correctly classifies 66% of the testing documents. *PCS(ICF)* weighted by term frequencies in the testing document resulted in 71.6% success rate of correct classification. When the algorithm is weighted by term frequency in documents, *ICF(PCS)* was greatly affected but not *IDF(PCS)*. The overlapping rate between class feature sets affects classification results and the class with the least overlapping class features showed the best classification result.

Acknowledgement

We greatly appreciate the permission of Prof. Kang, S. from Kookmin University, to use the Linux version of his Korean Morphology Analysis, HAM. His application contributes to the implementation of document classification system in this paper.

References

[1] Apte, C., Demerau, F., Weiss M.: Automated Learning of Decision Rules for Text Categorization. ACM Transactions on Information Systems. 12(3) (1994) 233-251

[2] Basili, R., Moschitti, A., and Pazienza, M.: NLP-driven IR: Evaluating Performance over a Text Classification Task. Proceedings of IJCAI'01. Seattle Washington (2001)

[3] Basili, R., Moschitti, A., and Pazienza, M.: Empirical Investigation of Fast Text Classification over Linguistic Features. Proceedings of ECAI'02. Lyon France (2002)

[4] Cohen, W., Singer, Y.: Context-Sensitive Learning Methods for Text Categorization, ACM Transactions on Information Systems, 7(2) (1999) 141-173
[5] Earley, J.: An Efficient Context-Free Parsing Algorithm. CACM. 13(2) (1970) 94-102
[6] Friburger, N., Maurel, D., Giacometti, A.: Textual Similarity Based on Proper Names. Proceddings of ACM SIGIR Conference on Mathmatical/Formal Methods in Information Retrieval, Tampere, Finland, (2002)
[7] Han, K., Sun, B., Han, S., Rim, K.: A Study on Development of Automatic Categorization System for Internet Documents. The Transactions of The Korean Information Processing Society. 7(9) (2000) 2867 – 2875
[8] Hirshberg, D.S.: Algorithms for the Longest Common Subsequence Problem. The Journal of ACM. 24(4) (1977) 664 - 675
[9] Joachims, T.: A Probabilistic Analysis of the Rocchio Algorithm with TFIDF for Text Categorization. Proceedings of International Conference of Machine Learning (CIML97). (1997) 143-151
[10] Kang, S.: Korean Morphological Analysis Program for Linux OS, http://nlp.kookmin.ac.kr. (2001)
[11] Lewis, D., Jones, K.S.: Natural Language Processing for Information Retrieval. Communication of the ACM. 39(1) (1996) 92-101
[12] Li, Y., Jain, A.: Classification of Text Documents. The Computer Journal. 41(8) (1998) 537 – 546
[13] Scott, S., and Matwin, S.: Using Lexical Knowledge in Text Classification. Technical Report (TR-98-03), Computer Science Department, University of Ottawa, Ottawa Canada, (1998)
[14] Yang, Y., Liu, X.: A Re-examination of Text Categorization Methods. Proceedings of ACM SIGIR Conference on Research and Development Retrieval. (1999) 42 – 49

Off-Line Signature Verification and Forgery Detection System Based on Fuzzy Modeling

Vamsi Krishna Madasu[1], Mohd. Hafizuddin Mohd. Yusof[2], Madasu Hanmandlu[3], and Kurt Kubik[1]

[1] School of Information Technology and Electrical Engineering
University of Queensland
QLD 4067, Australia
{madasu,kubik}@itee.uq.edu.au
[2] Faculty of Information Technology, Multimedia University, Cyberjaya
64100 Selangor D.E., Malaysia
hafizuddin.yusof@mmu.edu.my
[3] Dept. of Electrical Engineering, I.I.T. Delhi, Hauz Khas
New Delhi – 110016, India
mhmandlu@ee.iitd.ernet.in

Abstract: This paper presents an innovative approach for signature verification and forgery detection based on fuzzy modeling. The signature image is binarized and resized to a fixed size window and is then thinned. The thinned image is then partitioned into a fixed number of eight sub-images called boxes. This partition is done using the horizontal density approximation approach. Each sub-image is then further resized and again partitioned into twelve further sub-images using the uniform partitioning approach. The features of consideration are normalized vector angle (α) from each box. Each feature extracted from sample signatures gives rise to a fuzzy set. Since the choice of a proper fuzzification function is crucial for verification, we have devised a new fuzzification function with structural parameters, which is able to adapt to the variations in fuzzy sets. This function is employed to develop a complete forgery detection and verification system.

Keywords: Signature Verification, Forgery detection, Fuzzy Modeling, Box-Method.

1 Introduction

Signature verification and forgery detection relate to the process of verifying signatures automatically and instantly to determine whether the signature is genuine or forged. There are two main types of signature verification: static and dynamic. Static or off-line verification is the process of verifying an electronic or paper signature after it has been made, while dynamic or on-line verification takes place as a subject creates his signature on a digital tablet or a similar device. The signature in question is then compared with the previous samples of the signer's signature, which

T.D. Gedeon and L.C.C. Fung (Eds.): AI 2003, LNAI 2903, pp. 1003-1013, 2003.

constitute the database or knowledge base. In the case of an ink signature on paper, the computer requires the sample to be scanned for analysis, whereas a digital signature is already stored in a data format that signature verification can use.

The design of any signature verification system generally requires the solution of five sub-problems: data acquisition, pre-processing, feature extraction, matching, and performance evaluation [1]. Surveys of the state of the art off-line signature verification systems designed up to 1993 appear in [1, 2, and 3]. Another survey article [4] has summarized the approaches used for off-line signature verification from 1993-2000. We present here a review of a few papers in this field, which have not been covered in the survey articles. The emphasis of these papers is mainly on fuzzy based techniques for off-line signature verification.

An off-line signature system consisting of signature recognition and verification is proposed in [5]. In this, the recognition phase is based on the multi-stage classifier and a combination of global and local features whereas the verification is done using fuzzy concepts. HMM based approach in [6] derives the author dependent parameters dynamically and automatically to set up an optimal decision rule for off-line verification process. Here the cross validation principle is used to derive not only the best HMM models, but also an optimal acceptation/ rejection threshold for each author. This threshold leads to a high discrimination between the authors and impostors in the context of random forgeries.

Signature verification is also attempted using the Pseudo-Bacterial Genetic Algorithm (PBGA) [14], which was applied for the discovery of fuzzy rules. The rules are units themselves and they are constituted by several parameters to be optimized, however, the performance of a fuzzy system is obtained synergistically as a sum of the outputs of several rules. The PBGA was then applied for the extraction of personal features for signature verification. A pseudo-outer product based fuzzy neural network drives the signature verification system in [7]. This system is primarily used for verifying skilled forgeries. Signature verification using TS model is reported in [9] and features for this model are drawn from the box approach of [8]. In the present work, we follow the same features as in [8] but the TS model is modified to enhance its capability for the detection of forgeries.

Automatic examination of questioned signatures did not come into being until the advent of computers in the 1960s. As computer systems became more powerful and more affordable, designing an automatic forgery detection system became an active research subject. Most of the work in off-line forgery detection, however, has been on random or simple forgeries and less on skilled or simulated forgeries. Before looking into the landmark contributions in the area of forgery detection, we briefly explain the types of forgeries. The forgeries involved in handwritten signatures have been categorized based on their characteristic features [2]. This classification includes the following types:

- *Random Forgery* - The signer uses the name of the victim in his own style to create a forgery known as the simple forgery or random forgery.
- *Unskilled Forgery* - The signer imitates the signature in his own style without any knowledge of the spelling and does not have any prior experience.
- *Skilled Forgery* – undoubtedly the professional impostors or persons who have experience in copying the signature create the most difficult of all forgeries.

In the 1980's, Ammar et al. [10] published their work on the detection of skilled forgeries. They have calculated the statistics of dark pixels and used them to identify changes in the global flow of the writing. The later work of Ammar [11] is based on reference patterns, namely the horizontal and vertical positions of the signature image. The projection of the questioned signature and the reference are compared using Euclidean distance. Guo et al. [12] have presented an algorithm for the detection of skilled forgeries based on a local correspondence between a questioned signature and a model obtained a priori. Writer-dependent properties are measured at the sub-stroke level and a cost function is trained for each writer.

The paper is organized as follows: The pre-processing is presented in Section 2. Feature extraction is discussed in Section 3. The proposed signature verification and forgery detection system is explained in Section 4. The results of implementation of our system on sample signatures are given in Section 5 followed by conclusions in Section 6.

2 Pre-Processing of Signature Images

The original scanned signatures are pre-processed involving size normalization, binarization and thinning before features are extracted from each of them. These features constitute the knowledge base, which is then used for verifying the genuine signatures and detecting the forgeries. The components of pre-processing are discussed here.

The signatures were handwritten on a white sheet of paper with a black pen. The signature images were then scanned at 200 *dpi* resolution and resized by 50% using a B-Spline filter. Some typical signatures along with their forgeries are given in Fig. 1.

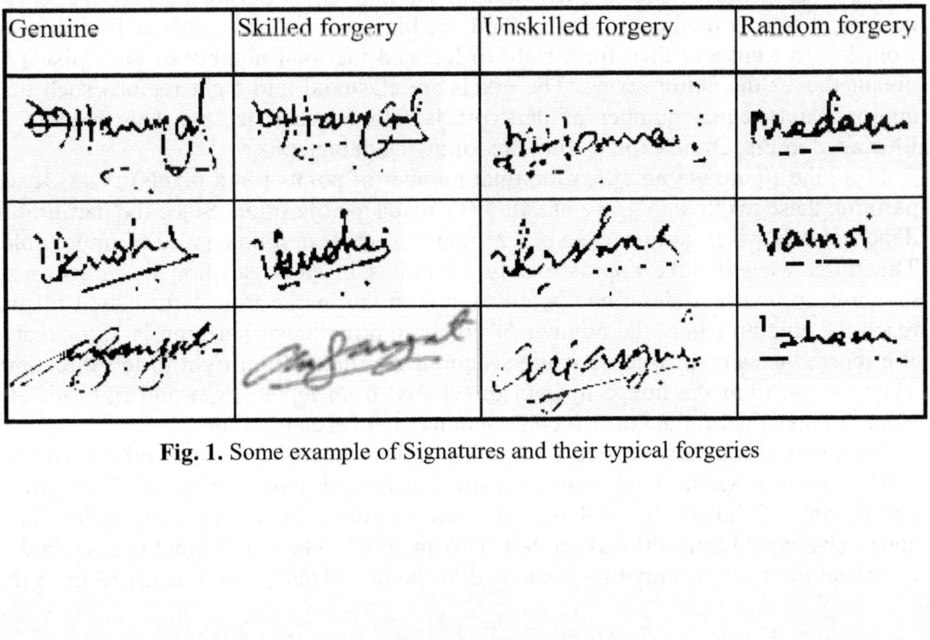

Fig. 1. Some example of Signatures and their typical forgeries

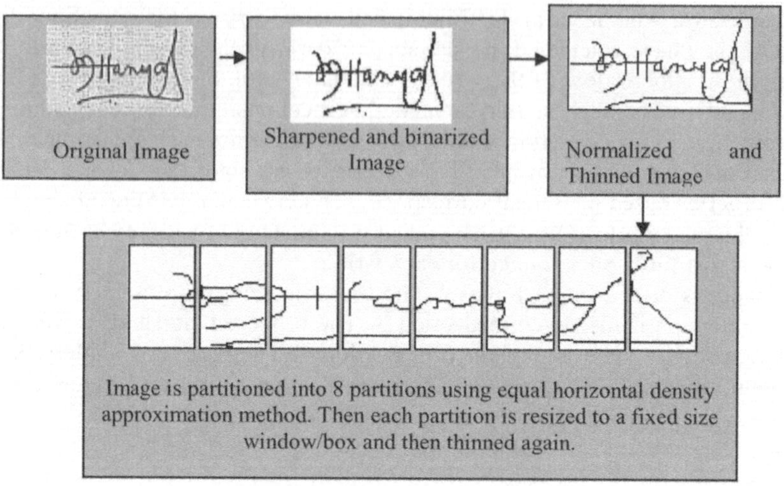

Fig. 2. Steps in Pre-Processing

As part of normalization, all the signature images are first resized to a fixed window of size (120 × 60 pixels), then binarized and thinned using the modified SPTA thinning algorithm [9]. Features are then extracted from this pre-processed signature image for the creation of the knowledge base.

3 Feature Extraction

The pre-processed image is then partitioned into eight portions using the equal horizontal density method. In this method, the binarized image is scanned horizontally from left to right and then from right to left and the total number of dark pixels is obtained over the entire image. The pixels are clustered into eight regions such that approximately equal number of dark pixels falls in each region. This process is illustrated in Fig. 2 and explained in the following paragraphs.

From the figure we note that the total number of points (dark pixels) is 48. If we partition these pixels into 4, we get 48/4=12 pixels per partition. Since the partition is done column wise, getting exactly 12 points in each partition is quite impossible. Therefore, we will take approximately 12 points in each partition using two-way scanning approach as described below. We scan the image from left to right till we reach the column where the number of points in a particular partition is 12 or more. We repeat the same procedure while scanning the image from right to left direction. Then we partition the image in both directions: from left to right and right to left. Next, we take the average of two column numbers in each partition.

Each partition is now resized to a fixed window (box of size 38 × 60 pixels) size and is thinned again. Each box is again partitioned into 4 rows × 3 columns, constituting 12 boxes. In total we have 96 partitions for a single signature. This approach adapted from [8] was termed "Box method". The idea behind this method is to collect the local information contained in the box. This method, evolved from the earlier ring and sector techniques [13, 8, 9], was fraught with the moving centroid. The centroid was treated as the reference point with respect to which the local information contained in a ring or a sector was collected. In the box method any corner point can be chosen as the reference point. The information can be gathered in the form of normalized distances and angles (See [8] for more details).

For the present problem of signature verification and forgery detection, we have experimented with both distance distributions and angle distributions. We have found that the angle distribution has an edge over distance distribution. Hence the choice fell on extracting angle information from the boxes. We now discuss the computational aspects. For this, we calculate the summation of the angles of all points in each box with respect to the bottom left corner of the box, which is taken as the reference. The summation of angles is normalized with the number of pixels in the box. These angles constitute the feature database for a given signature. Table 1 below gives the angle features of one of the signatures used for training. The eight rows stand for the eight partitions of the signature while the columns symbolize the further divisions within each partition.

Table 1. Knowledge base for a particular signature

	1	2	3	4	5	6	7	8	9	10	11	12
1	21.1	46.8	38.1	23.9	41.6	40.1	0	0	0	0	0	0
2	0	0	0	44.8	54.8	41.7	24.7	54.5	76.1	0	0	0
3	0	0	40.1	46.9	39.8	0	72.4	60.2	21.0	46.1	56.4	0
4	0	0	30.5	20.5	3.7	49.4	70.2	81.9	58.0	57.5	54.2	0
5	0	0	35.6	12.7	21.3	90	0	0	61.6	63.3	63.3	0
6	0	0	54.7	13.6	26.9	45.6	50.6	79.9	60.9	63.3	60.5	0
7	0	0	36.1	30.1	39.4	33.9	73.4	0	59.6	61.3	59.6	57.4
8	0	0	52.1	90	0	47.1	56.6	30.7	57.4	59.1	63.3	0

4 Proposed System

Since the main thrust here is to establish the genuineness of the signature thereby detecting the forgeries, we go in for fuzzy modeling of angle features. For the purpose of signature verification and detection of forgeries, we have employed the Takagi-Sugeno model. In this, we are following the same concept as outlined in [13]. According to this concept, each feature forms a fuzzy set when its values are gathered over large samples. This is because the same feature exhibits variation in different samples giving rise to a fuzzy set. So, our attempt is to model this variation/ uncertainty through a fuzzy model such as the TS model.

Let x_k be the kth feature in a fuzzy set A_k, so the kth fuzzy rule IF THEN rule in TS model has the following form

Rule k: IF x_k is A_k

$$\text{THEN } y_k = c_{k0} + c_{k1} x_k \tag{1}$$

Each feature will have a rule so we have as many rules as the number of features. The fuzzy set A_k is represented by the following exponential membership function that includes two structural parameters s_k and t_k :

$$\mu_k(x_k) = \exp -\left[\frac{(1-s_k) + s_k^2 \left| x_k - \overline{x_k} \right|}{(1+t_k) + t_k^2 \sigma_k^2} \right] \tag{2}$$

where $\overline{x_k}$ is the mean σ_k^2 is the variance of k th fuzzy set. This choice of exponential membership function was the result of extensive experimentation undertaken with various types of membership functions such as Gaussian and Generalized Gaussian etc., for representing characters in [8] and signatures in [9], where the use of structural parameters was not made. It can be observed from the sample signatures that their angle distributions do not follow the Gaussian function. Improved results have been obtained with exponential function rather than with Gaussian function. Selection of a proper membership function is an important step in the development of an effective system and this requires a few trails with widely used membership functions.

Note that the inclusion of these parameters will help track the variations in the handwriting of signatures. When $s_k = 1$ and $t_k = -1$, the membership function is devoid of structural parameters and hence it is solely governed by the mean and variance. The motivation for this modified membership function is two-fold: (i) The dynamics of mean and variance is reflected through s and t, and (ii) There is no need for any sophisticated learning techniques because the range of variation to track in s and t is less than that in mean and variance. Since the structural parameters account for the variations in mean and variance, these can be thought of embedded dynamics of mean and variance.

The numerator and denominator of the exponential function in (2) contain a constant term (i.e., 1) plus a function of parameter (s or t) and the known variation (i.e., either change over mean or variance). This choice is guided by the consideration of no role for parameters if the signatures of a person don't change. But this need not be the case for other applications.

The strength of the rule in (1) is obtained as

$$w_k = \mu_k(x_k) \tag{3}$$

The output is expressed as

$$Y = \sum_{l=1}^{L} w_l y_l \tag{4}$$

where L is the number of rules.

We define the performance function as

$$J = (Y_r - Y)^2 \tag{5}$$

where, Y and Y_r denote the output of the fuzzy model and the real system respectively. If Y_r is not available, it can be assumed to be unity. In the present system under consideration, its performance is judged by the above function. In addition, we will present an innovative approach that relies on the maximum value of the membership functions.

In order to learn the parameters (i.e., s_k and t_k), involved in the membership function and the consequent parameters c_{k0} and c_{k1}, (5) is partially differentiated with respect to each of these parameters.

Accordingly, we have

$$\frac{\partial J}{\partial c_{k1}} = \frac{\partial J}{\partial Y} \cdot \frac{\partial Y}{\partial y_k} \cdot \frac{\partial y_k}{\partial c_{k1}} = 2\left(Y - Y_r\right) w_k x_k \tag{6}$$

$$\frac{\partial J}{\partial c_{k0}} = \frac{\partial J}{\partial Y} \cdot \frac{\partial Y}{\partial y_k} \cdot \frac{\partial y_k}{\partial c_{k0}} = 2\left[Y - Y_r\right] w_k = 2\delta w_k \tag{7}$$

$$\frac{\partial J}{\partial s_k} = \frac{\partial J}{\partial Y} \cdot \frac{\partial Y}{\partial w_k} \cdot \frac{\partial w_k}{\partial t_k} = 2\delta y_k \mu_k \left[\left\{1 - 2s_k \left|x_k - \overline{x_k}\right|\right\} / T\right] \tag{8}$$

$$\frac{\partial J}{\partial t_k} = \frac{\partial J}{\partial Y} \cdot \frac{\partial Y}{\partial w_k} \cdot \frac{\partial w_k}{\partial t_k}$$

$$= 2\delta y_k \mu_k \left\{(1 - s_k) + s_k^2 \left|x_k - \overline{x_k}\right|\right\}\left\{1 + 2t_k \sigma_k^2\right\} / T^2 \tag{9}$$

where $\delta = Y - Y_r$, $T = (1 + t_k) + t_k^2 \sigma_k^2$ and k denotes the rule number.

We use the gradient descent learning for the parameters as follows

$$c_{ki}^{new} = c_{ki}^{old} - \in_1 \frac{\partial J}{\partial c_{ki}} \quad i = 0,1 \tag{10}$$

$$s_k^{new} = s_k^{old} - \in_2 \frac{\partial J}{\partial s_k} \tag{11}$$

$$t_k^{new} = t_k^{old} - \in_3 \frac{\partial J}{\partial t_k} \quad k = 0,1,\ldots,L \tag{12}$$

where $\in_1, \in_2, \in_3$ are the learning coefficients such that $\in_1, \in_2$ *and* $\in_3 > 0$. The choice of the initial values of these coefficients is not crucial in the learning process.

5 Implementation and Results

The proposed fuzzy modeling based technique discussed above has been applied on a signature database, developed in the Graphics Visualization & Games Development (GVGD) lab at the Multimedia University, Cyberjaya, Malaysia.

The signature database consists of a total of 510 handwritten signature images. Out of these, 255 were authentic signatures and others were forged ones. These signatures were obtained from 17 volunteers with each person contributing 15 signatures as shown in Table 2. The signatures were collected over a period of a few weeks to account for variations in the signature with time. The forgeries of these signatures were collected over a similar time frame. The random forgeries were obtained by supplying only the names of the individuals to the casual forgers who did not have any access to the actual genuine signatures. The unskilled forgeries, in turn, were obtained by providing sample genuine signatures to the forgers who were then allowed to practice for a while before imitating them to create the forged signatures. Each volunteer had to provide five imitations of any one of the genuine signatures, apart from his or her own signatures. These samples constituted the set of unskilled forged signatures for the set of genuine signatures. We then requisitioned the services of a few expert forgers who provided five forgeries of each genuine signature in the test set to create the skilled forged samples of all the persons.

Forged samples of a genuine signature are not readily available as it is difficult to imitate the various styles of signatures by amateurs for producing the unskilled forgeries and by professional impostors for producing the skilled forgeries. Keeping this point in mind and considering the real-world scenario, we have trained our system with only genuine signatures, i.e., none of the forgeries were used for training the system. We have set the number of training signatures for each individual at ten.

We now explain the process of signature recognition. If we take $Y_r = 1$ and also consider (15), then (5) becomes

$$J = (1 - \frac{1}{L}\sum_{i=1}^{L}\mu_i)^2 \tag{13}$$

With the above performance index, we compute $\frac{\partial J}{\partial s_i}$ and $\frac{\partial J}{\partial t_i}$ in order to update the structural parameters s_i and t_i; $i = 1,..,96$. Using these values, we compute the membership functions for all the features. This process is repeated for all training samples of a person. Here, we have devised an innovative approach for the classification of all signatures (i.e., test signatures and random, skilled and unskilled forgeries) of a person. In order to know the extent of variation in the genuine signatures, we determine the maximum and minimum membership functions for each feature over all signatures in the training set. The difference between these two gives the inherent variation in the signatures of a person.

We now use the inherent variation to judge the test signatures. We will also explain its utility in the testing phase. For a particular feature, if the membership value lies within the range of variation, which is given by the difference of minimum and maximum thresholds, it is counted as 'true'. The total number of 'true' cases for a particular signature divided by the total number of features (i.e., 96) gives the percentage. For example, in Fig. 3a, the test signature has 99% of its features lying well within the threshold as can be seen from the membership function (i.e., 95 out of 96 features are with the range of inherent variation). The skill-forged and unskilled forged signatures have corresponding figures of 88.5% (Fig. 3b) and 82.3% (Fig. 3c) respectively. We set the minimum limit or acceptable percentage for genuine signature at 91% referring to the output result of signature of a signer. Signatures that have percentage less than 91% are treated as forged.

Table 2. Signature Database

Type of Signature	Training set	Testing set	Total
Genuine	17 × 10	17 × 5	170 + 85 = 255
Skilled forgery	-	17 × 5	85
Unskilled forgery	-	17 × 5	85
Random forgery	-	17 × 5	85

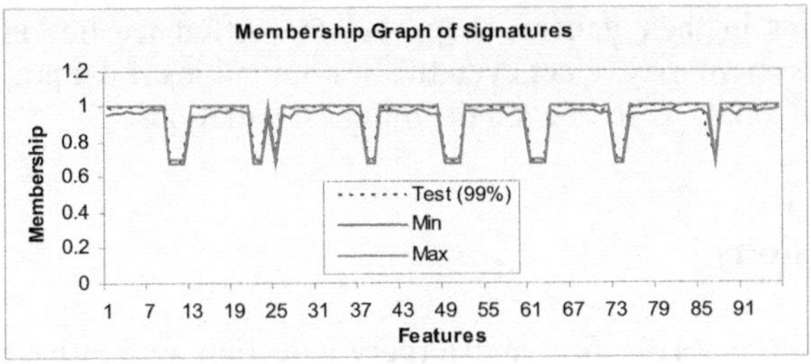

(a)

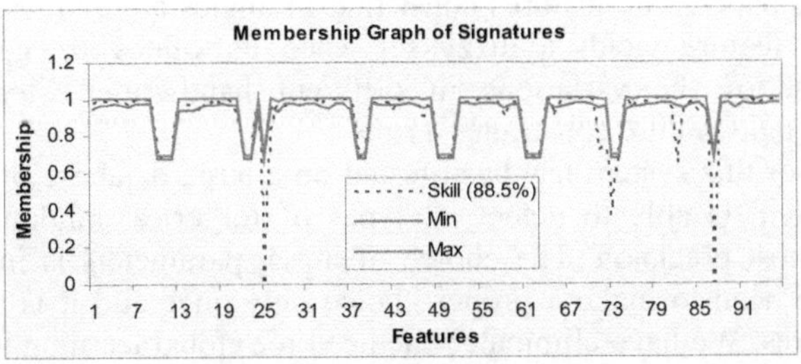

(b)

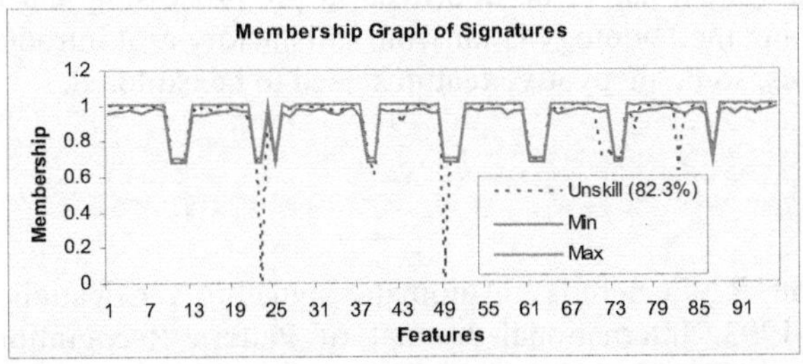

(c)

Fig. 3. Membership Graphs for a particular signature

Table 3. Results of Forgery Detection

Type of Signatures	Total number of samples	False Acceptance Ratio	False Rejection Ratio
Genuine Signatures	85	0%	0%
Skilled forgery	85	3.5%	0%
Unskilled forgery	85	0%	0%
Random forgery	85	0%	0%

The results of signature verification and forgery detection using this innovative approach are tabulated below. Although, all the test signatures were correctly identified as genuine and random and unskilled forgeries were rejected, a slight error was encountered in the case of skilled forgeries. The system accepted three skilled forgeries as genuine signatures thus achieving a False Acceptance Ratio (FAR) of just 3.5%. This may be due to the fact that the skilled forgeries almost resemble the genuine signatures in their general shape but the difference lies in their scale and orientation. The system may reject even the best imitations if the pre-processing stage is improved upon to take care of even the minutest variations.

6 Conclusions

An off-line signature verification and forgery detection system based on TS model, and involving structural parameters in its exponential membership function was presented in this paper. The features consisting of angles are extracted using the box approach. Each feature yields a fuzzy set when its values are gathered from all samples because of the variations in different handwritten signatures. In this formulation, a single feature constitutes a rule.

The efficacy of this system has been tested on a large database of signatures. The verification system is able to detect all types of forgeries: random, unskilled and skilled with utmost precision. The choice of initial parameters is important but not crucial. But, we need to make a proper choice only once and it is applicable to all types of signatures. We have eliminated the need for global learning techniques at the implementation stage by the choice of structural parameters.

This proposed system has to be improved in pre-processing and feature selection stages. Our thinning methodology is far from satisfactory as it introduces barbs. Other than angle features, some innovative features need to be explored.

References

[1] Plamondon, R., Leclerc, F., Automatic signature verification: the state of the art 1989-1993. International Journal of Pattern Recognition and Artificial Intelligence. 8 (1994) 643-660.

[2] Plamondon, R., Lorette, G., Automatic signature verification and writer identification: the state of the art. Pattern Recognition. 22 (1989) 107-131.

[3] Sabourin, R., Plamondon, R., Lorette, G., Off-line identification with handwritten signature images: Survey and Perspectives. In: Structured Image Analysis, Springer-Verlag, New York (1992) 219-234.

[4] Plamondon, R., Srihari, S.N., On-line and off-line Handwriting Recognition: A Comprehensive Survey. IEEE Transactions on Pattern Analysis and Machine Intelligence. 22 (2000) 63-84.

[5] Ismail, M.A., Gad, S., Off-line Arabic signature recognition and verification. Pattern Recognition. 33 (2000) 1727-1740.

[6] El-Yacoubi, Justino, E.J.R., Sabourin, R., Bortolozzi, F., Off-line signature verification using HMMS and cross-validation. In: Proceedings of the IEEE Workshop on Neural Networks for Signal Processing (2000) 859-868.

[7] Quek, C., Zhou, R.W., Antiforgery: a novel pseudo-outer product based fuzzy neural network driven signature verification system. Pattern Recognition Letters. 23 (2002) 1795-1816.

[8] Hanmandlu, M., Mohan, K.R.M., Chakraborty, S., Goel, S., Choudhury, D.R., Unconstrained handwritten character recognition based on fuzzy logic. Pattern Recognition. 36 (2003) 603-623.

[9] Hanmandlu, M., Mohan, K.R.M., Chakraborty, S., Garg, G., Fuzzy modeling based signature verification system. In: Proceedings of the sixth International Conference on Document Analysis and Recognition (2001) 110-114.

[10] Ammar, M., Yoshida, Y., Fukumura, T., A new effective approach for off-line verification of signatures by using pressure features. In: Proceedings of the International Conference on Pattern recognition (1986) 566-569.

[11] Ammar, M., Progress in verification of skillfully simulated handwritten signatures. International Journal of Pattern Recognition and Artificial Intelligence. 5 (1991) 337-351.

[12] Guo, J.K., Doermann, D., Rosenfeld, A., Off-line skilled forgery detection using stroke and sub-stroke properties. In: Proceedings of the International Conference on Pattern Recognition (2000) 355-358.

[13] Hanmandlu, M., Mohan, K.R.M., Gupta, V., Fuzzy logic based character recognition. In: Proceedings of the IEEE Conference on Image Processing (1997) 1714-717.

[14] Xuhua, Y., Furuhashi, T., Obata, K., Uchikawa, Y., Study on signature verification using a new approach to genetic based machine learning. In: Proceedings of the IEEE International Conference on Systems, Man and Cybernetics (1995) 4383-4386.

Case Study: A Course Advisor Expert System

Ovidiu Noran

Griffith University, Brisbane QLD (Australia)
noran@cit.gu.edu.au
http://www.cit.gu.edu.au/~noran

Abstract. This article presents the development of a knowledge-based expert system for a non-trivial application domain, i.e. selecting a customised postgraduate study program from existing- and targeted sets of skills. Following a customised spiral development paradigm, the paper describes the problem domain, followed by the concepts and requirements involved in the expert system design. The main design constraints are stated and the design decisions are justified in light of the specific task. Attention is paid to the knowledge representation / acquisition and rule base design processes. Implementation and deployment are explained, together with a sample run and result evaluation. Finally, further work is detailed in light of existing features and limitations of the prototype.

Keywords: Constraints, Expert Systems, Knowledge Acquisition, Knowledge Representation.

1 Introduction

The changes brought about by the Information and Communication Technology (ICT) revolution on the human society have also had an impact on the required set of skills of the workforce. While for the recently qualified professionals basic ICT knowledge has been built into the education process, the more mature workforce has to acquire this knowledge separately and in 'backwards compatibility' mode with their existing skills.

Particularly at higher levels of education, individuals may already possess some of the required ICT skills. These abilities may exist in the form of tacit knowledge or formal/informal awareness acquired during past undergraduate study[1] or current professional activity. Typically, this knowledge (although not structured in a consistent way) has the potential to be reused in the process of acquiring and formalizing the necessary ICT skills of the individual. Potential hurdles in the reuse of such existing knowledge are:

- unambiguously eliciting and self-assessing existing knowledge, which assumes the owner's awareness of his/her skills (not always trivial);
- devising a flexible learning process to incorporate the identified skills;
- matching a study program to a particular set of existing- and targeted skills.

[1] e.g. most engineers have undertaken at least one course of structured programming during their undergraduate studies.

T.D. Gedeon and L.C.C. Fung (Eds.): AI 2003, LNAI 2903, pp. 1014–1027, 2003.

This paper illustrates the problem through a case study describing the use of an expert system to customise the necessary study program for particular sets of existing- vs. targeted knowledge. The case study is limited to an expert system *prototype*, dealing with postgraduate students with a non-IT background who wish to acquire ICT formal education in an accredited University environment. For the purpose of this paper, the study program will be called a *Conversion Course*[2]. The system will offer guidance to potential students and to course designers about the necessary structure of a particular study program.

2 The Problem Domain

The existing Web-based University course description system presents a number of limitations. Occasionally, erroneous or stale information is provided, which may lead to misleading students in their enrolments / exams. In addition, the present system does not verify that a student may enrol in subjects in a way contradicting the University policies. Finally, the present subject and course descriptions may be confusing for prospective students[3]. Such limitations could adversely affect the image of the teaching institution.

The advisory expert system is intended to provide preliminary assistance and advice for prospective postgraduates, extracting and inferring the necessary information from a current knowledge base within a consultation session. This will provide a customized, realistic assessment of the alternatives, requirements and expectations for a particular student, and thus help prevent enrolment errors.

In the current teaching system, subjects are considered to be an indivisible entity, non-customizable, with a fixed number of points awarded on completion. Thus, prior knowledge covering part of a subject does not benefit a student, which still has to enrol in the *whole* subject and be awarded *all* the subject points if successful. To enable and support the proposed expert system, the teaching system should allow subject *modularity*, with points allocated to separately assessed course components.

3 Developing the Expert System

In developing the expert system it is desirable to start with a *prototype* of the complete expert system. A prototype can assess the feasibility of a project without full financial or resource commitment and may then be submitted for evaluation to users / stakeholders to obtain their feedback and commitment. User and host institution acceptance is a multidimensional aspect [15], which ultimately decides the usefulness of the entire system development effort. The development

[2] In this paper it is assumed that *subject* equals a single study unit (e.g. 'Programming 2') while *course* is a collection of *subjects* leading to the awarding of a degree (e.g. 'Master of Information Technology').

[3] e.g. compulsory/elective subjects, or a particular subject enrolment order for a course

of the prototype could not be left to the knowledge engineer alone[4], as it needs the knowledge elicitation from at least one domain expert [6] (in this case, a subject convener).

3.1 Life Cycle Models

Several life cycle paradigms have been considered, such as waterfall, incremental, linear, spiral, etc [6, 12, 13]. A modified *spiral* paradigm has been successfully adopted, observing some guidelines in adding new facts and rules, such as: maintain integrity (do not contradict the existing facts and rules), avoid redundancy (do not represent knowledge already existent in the rule base), prevent scattering the knowledge over an excessive amount of rules / facts, etc. Thus, a balance must be struck between the facts' and rules' complexity, expressive power and number[5].

3.2 Concepts of the Expert System

The expert system is based on several concepts aiming to improve the subject selection and education processes. These concepts are *modularity* of subjects[6], *prerequisites* and *outcomes* for subject modules and *credit* for previous studies.

Hence, the aim is to establish *modules* within the subjects, having their own prerequisites, outcomes and credit points awarded on completion. The granularity of (i.e. number of modules within-) a subject must maintain a balance between flexibility and processing / development time required[7].

3.3 User Requirements for the Expert System Prototype

The user will provide a preferred type of occupation (targeted set of skills) and previous knowledge (potentially usable to satisfy some of the prerequisites for the modules composing the study program) as requested by the system. The system will provide a study course to achieve the targeted occupation and may also suggest corrections if the user skills (as stated) are too limited or too high.

3.4 Design of the Expert System Prototype

System Requirements The system requirements represent a translation of the user requirements into the system domain[8]. For this particular case they may take the form:

---

[4] unless the knowledge engineer was also a domain expert, in which case some method of triangulation should be employed, e.g. member checking.

[5] i.e., a large number of simpler rules vs. fewer, more complex- and expressive rules.

[6] subject modules may be inferred e.g. from its syllabus (which may also assist subject decomposition)

[7] a very small number of modules defeats the purpose of decomposing the subjects, while a large number of modules may require too many resources.

[8] the accuracy of this translation should be subsequently *validated* with the end user.

- the expert system must rely on the user's tacit knowledge[9];
- modules are seen as objects, having interfaces (in fact, its prerequisites and outcomes) contained in special lists, further contained within module facts;
- a module prerequisite may be satisfied by at most one outcome, (either of another module, or declared as 'known' by the user);
- the consultation starts with a set of initial facts, asserted at run-time according to the user's answers to 'job domain' and 'job type' queries. These facts provide the initial list of unsatisfied prerequisites;
- the system attempts to match these unsatisfied prerequisites (first against outcomes declared 'known' by the user, and if unsuccessful, against the 'outcomes' lists of the other module facts in the knowledge base;
- when a matching outcome is found for a prerequisite, the module owning the outcome is marked as 'needed' and its prerequisites list is then in its turn scanned for matches with other outcomes. The prerequisite for which the match has been found is marked 'done';
- any prerequisites found not to be either declared 'known' by the user, or not already satisfied by other modules' outcomes will trigger additional questions for the user;
- according to the user's answers, prerequisites are either marked 'known' (user already has the required skill), or added to the unsatisfied prerequisites list;
- the process is repeated until all prerequisites are satisfied - either by other modules' outcomes ('done') or by previous skills of the user ('known');
- the list of all modules marked as 'needed' (that the user will need to enrol in) is printed out[10]. Note that some special modules (e.g. project) are automatically added, regardless of the user's prior knowledge.

System functionality is shown diagrammatically in Fig. 1.

Knowledge Representation and Method of Inference Design decisions had to be made regarding the formalism to be used in representing the knowledge contained in the teaching system. The type of knowledge to be represented (namely components of the various courses composing current study programs) matched the form of a collection of isolated facts, rather than a structured set of knowledge or a concise theory. Therefore, rules / assertions were preferred to frames or algorithms in a first decision round [16]. Production rules [21] have been chosen in preference to logic and semantic nets / frames [19] for being more suitable for solving design and planning problems [17].

Bottom-up reasoning / inference was decided upon, whereby a starting fact set (provided by the user) is matched against the conditions of the production rules in the rule base (constructed upon the policies governing the University study programs). Bottom-up inference has led to forward chaining as a preferred model of conflict resolution, with possible prioritisation by assigning production rules appropriate weights [17].

---

9 i.e. basic knowledge assumed by a prospective postgraduate student (e.g. Maths) - as in [5]. Not to be confused with the 'previous knowledge' stated to the system.

10 depending on the granularity of the subject decomposition, some 'needed' modules may have 'bonus' outcomes (i.e. which do not satisfy any prerequisites).

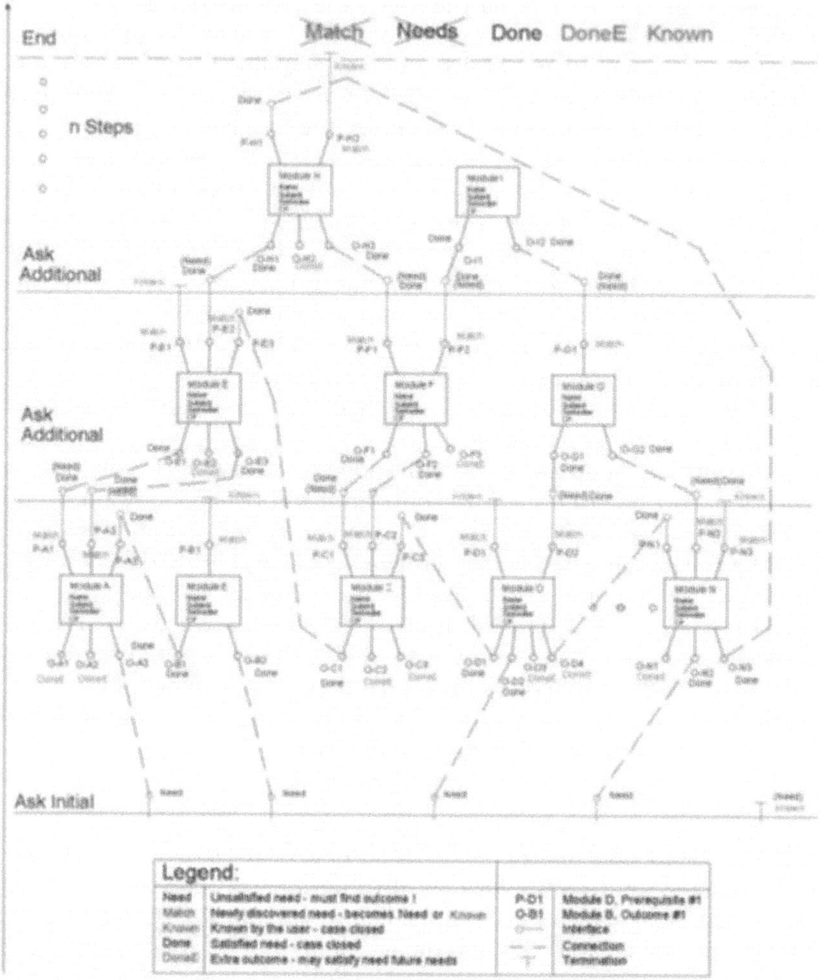

Fig. 1. Expert system prototype functionality

Knowledge Acquisition The knowledge acquisition methods chosen for the prototype have been the questionnaire and structured interview. This decision owes to several factors, such as prototype size, nature of the problem domain and availability of domain experts. Questionnaires are well suited to future automated processing, which benefit the knowledge acquisition process [2]. The questionnaire and interview designs have acknowledged the gap between the descriptions of domain specialists (subject conveners) and the resulting computational models [4, 18] and the social issues underlying knowledge creation [10][11]. The interview design has loosely followed the COMPASS procedure [23].

Design Constraints Constraints are necessary in order to enable a finite solution to be produced. Examples:
- the outcomes of a module must be distinct;
- two modules may not produce the same outcome: doing so would produce a redundancy in the teaching structure which needs to be resolved[12];
- all the module prerequisites contained in the knowledge base are satisfied by outcomes of other modules in the base[13].
- nil prerequisites for modules are allowed; however, they still require basic graduate knowledge, such as maths, physics, etc (the tacit knowledge previously mentioned);
- cyclic dependencies between any two modules are disallowed (e.g. if module A has a prerequisite satisfied by module B, module B must not have a prerequisite satisfied only by module A). Should this situation occur, the offending modules must be reassessed with regards to their prerequisites / outcomes[14];

Further constraints may be added in the developing the expert system, e.g.:
- maximum number of year **n** modules: may conflict with the concept of a Conversion Course and limit the flexibility of the expert system;
- maximum number of modules per Semester (the prototype violates this constraint). In real life, subjects tend to be equally distributed in all Semesters;
- balanced number of modules in each Semester: see previous.

The Expert System Conceptual Model The knowledge base should contain *facts* and *rules* referring to the prerequisites and outcomes of modules of the subjects offered in the University[15]. The facts are either 'fixed' (such as the modules information) or run-time asserted (e.g. the user's answers to the expert systems' questions). The inference engine must be chosen to match previous requirements and design decisions. The user interface, preferably graphical and integratable with currently and commonly used operating systems and enabling technology infrastructure (e.g. Internet), would preferably be implemented in the same language as the inference engine.

---

[11] the questionnaire and the interview structure must take into consideration the *culture*(e.g. shared values, beliefs) and policies of the institution hosting the domain experts and the system.

[12] thus this system may be used to identify and eliminate overlapping areas within different subjects

[13] this constraint does not include 'trivial' prerequisites, i.e. basic knowledge that a graduate is expected to have. Also, the constraint may be relaxed by accepting that students satisfy some prerequisites by enrolling outside the teaching institution.

[14] this should also be disallowed in real life, since there is no way for a student to enrol in either of the modules, *unless* he/she has previous skills covering the prerequisites of one of the offending modules.

[15] supplementary information is also provided, such as semester, credit points, parent subject, etc. Additional information may also be introduced (e.g. module convenor).

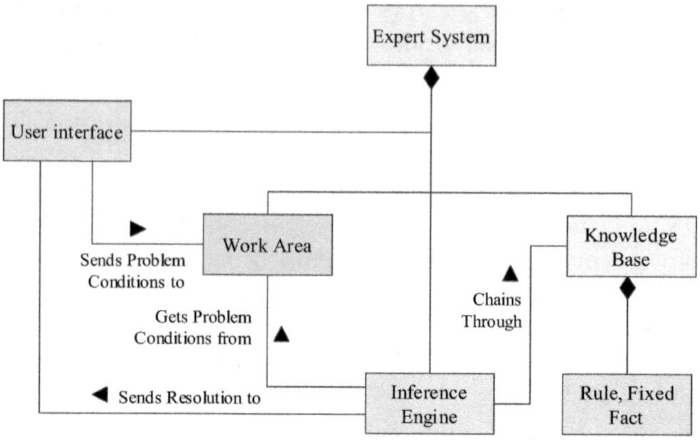

Fig. 2. Object diagram of the expert system (based on [8])

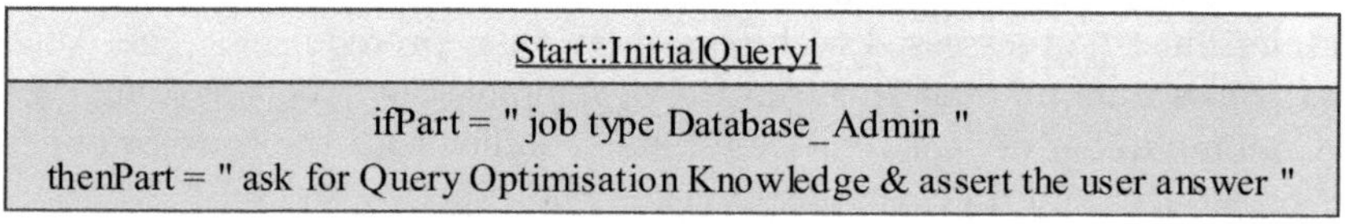

Fig. 3. UML rule representation

A Unified Modelling Language (UML, [25]) model of the expert system is presented in Fig. 2. In this figure, the user-expert system interaction occurs through the user interface, which sends the problem conditions (answers to questions) to the work area that holds all the temporary data. The inference engine uses the knowledge base for the rules and fixed facts, and the work area for the dynamic facts (asserted at run-time). The solution is delivered to the user interface. The classes shown in the figure will be further specified in the Implementation section.

The Knowledge Base In an UML representation, rules may be represented as objects, displaying *attributes* and *behaviour* as shown in Fig. 3.

The expressiveness of Fig. 3 may be improved by employing UML extension mechanisms[16], such as presented in Fig. 4. In this version, each rule may also represent the rules that call- and that are called by the rule in question.

3.5 Implementation

The virtual machine hierarchy as described in [10] provides a good guide towards the expert system prototype implementation. Several web-enabled expert system

[16] such custom representations (which in fact create new modelling languages) must be at least expressed by *glossaries* and a consistent *metamodel*, unambiguously describing the structure of the extended modelling language to the intended audience.

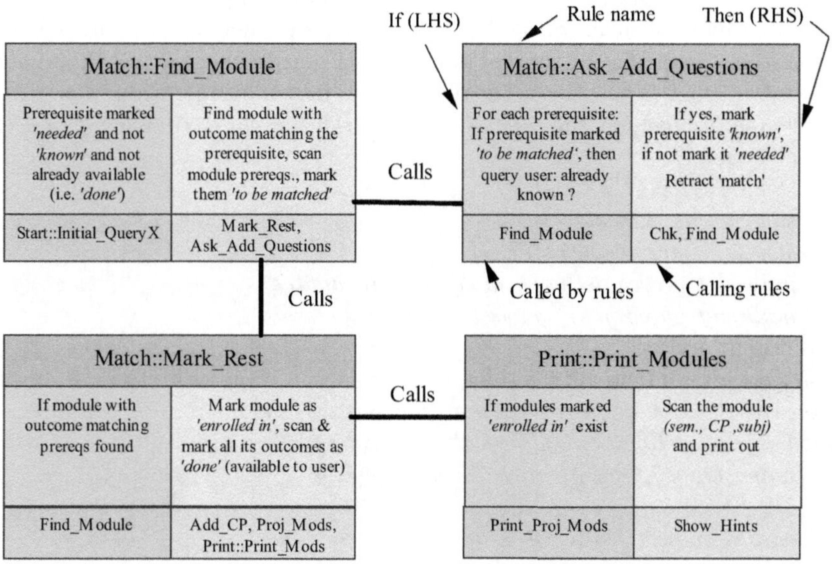

Fig. 4. Possible customised rule representation

shells have been considered for the specific problem domain. The shell has to be matched to the task [14] in order to assist in the knowledge representation exercise. Most shells impose a particular production rule formalism, chaining and structure[17] to the rule set.

Considering the size of the prototype, the resources available and the problem domain, the choice has been an expert system shell written in Java (JESS - The Java Expert System Shell [9]), emulating the CLIPS [11] language, with a simple pre-made graphical user interface. This solution provides the power of the CLIPS language and the flexibility of the Java cross-platform concept. Although the Java AWT (Abstract Window Toolkit) is available in JESS, JConsult [24] has been preferred as an off-the-shelf basic JESS graphical user interface. The inference engine is indirectly provided by CLIPS. The CLIPS language uses forward-chaining and the RETE fast pattern-matching algorithm [7].

Thus, in Fig. 2 the Work Area is the Java applet, the user interface is the applet's window, the inference engine is provided by the Java CLIPS implementation and the knowledge base resides in CLIPS file(s).

The Knowledge Base Implementation JESS, similar to CLIPS (and LISP), uses the *list* construct to hold the data. The facts may be asserted manually (i.e. hardcoded in the knowledge base) or at run-time. They are implemented as lists containing one or more other lists. Example:

(attribute (type job)(value "Database Administrator"))

[17] or lack thereof - refer e.g. EMYCIN [26]

Templates are similar to classes and contain models for the *facts*. Three types of templates have been used: *goal* (a special type of fact, used to initialize the expert system), *attribute* (a general purpose fact consisting of a *type* and a *value*) and *module* (a fact type describing the module information). Example:

```
(deftemplate module ; the module information template, equivalent of a class
    (slot name)
    (slot CP (type INTEGER))
    (slot parent_subject)
    (slot semester (type INTEGER))
    (multislot prerequisites) ; module prerequisites
    (multislot outcomes) ; module outcomes
); end deftemplate
```

A *slot* declares a field within the template - actually, a list in itself. A *multislot* declares a *multifield*, i.e. a list with more than two members. For example:

```
(module ; a particular module - the equivalent of an object
    (name Query_Optimisation_Evaluation)
    (CP 2)
    (parent_subject Database_Management_Systems)
    (semester 1)
    (prerequisites SQL_Data_Definition_Language Data_Storage)
    (outcomes Query_Optimisation Relational_Operators)
); end module
```

This is a *module* object example where the *prerequisites* and *outcomes* list both have more than two members. In this way, a variable number of members may be accomodated within a list without any special requirements.

The rules in the JESS environment take the form *if-part => then-part*:

```
(defrule RuleDB1a
 (attribute (type job)(value "Database Administrator"|"Database Designer"))
  =>
  (printout t "Databases Knowledge." crlf crlf
      "Do you have Database Architectures knowledge ?|explanatory|
      This type of job requires Database Architectures knowledge|yes|no|end")
  (assert (attribute (type Database_Architectures) (value (readline))))
); end RuleDB1a
```

An extensive coverage of the development environment and the user interface is beyond the scope of this document.

3.6 Testing / Verification: Running a Consultation

The system will initially require a target domain and specific employment opportunity (within the chosen domain); this will create the first set of modules needed. The system will then query the user on previous knowledge usable to satisfy the prerequisites of this first set of modules. In addition, the system will seek appropriate modules whose outcomes satisfy the prerequisites in question.

The majority of modules in the knowledge base have non-nil prerequisites. The system will seek to satisfy all the prerequisites of a module before enrolling

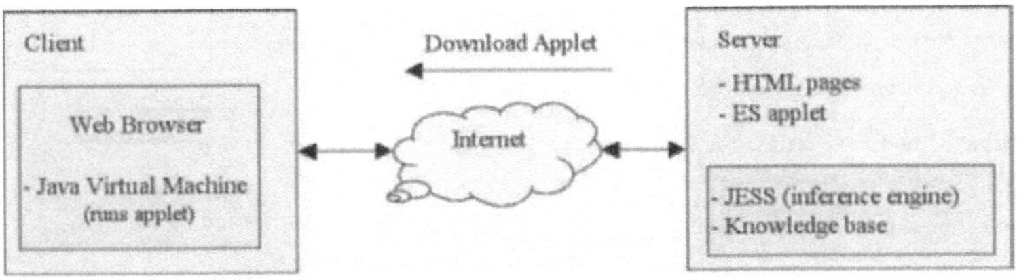

Fig. 5. The Web-based expert system

the user in that module. This process will occur recursively - i.e. a module with **n** prerequisites may require up to **(n-k)** modules (where **k** represents outcomes provided by the user) with outcomes that satisfy those prerequisites. These **(n-k)** module(s) may also have **p** prerequisites that need to be satisfied, and so on. Every time a new prerequisite is discovered, firstly the user is queried whether he/she has the knowledge to satisfy it. If not, a suitable outcome of another module is searched to satisfy the prerequisite. Although this would seem to create a larger pool of prerequisites each time, in fact many prerequisites are satisfied by one module, and per total there can be no unsatisfied prerequisites (meaning that the University caters for all the teaching needs of the student).

At the end of the consultation, the expert system will produce an output containing: Stated (existing) knowledge, Necessary modules with Parent Subject, Semester, Credit Points, Project modules (the Conversion Course must include a 40CP Project) and Total Credit Points necessary for the course[18]. Boundary values are as follows: needed CP $<$ 70 - existing knowledge may be overstated; 70 $<$ CP $<$ 105 - Ok.; 105 $<$ CP $<$ 150 - existing knowledge may have been understated; CP$>$ 150: the envisaged occupation / skills may be unsuitable for that person (previous knowledge too limited).

A sample run of the expert system for the job of *'Artificial Intelligence Researcher'* (with prior knowledge) has produced the output shown partially in Fig. 6. Note that at this stage there is no mechanism to ensure a *balanced* distribution of the modules within the Semesters[19].

3.7 Deployment

Figure 5 shows the current method of deployment of the prototype[20], which was deemed appropriate for the restricted initial scope of the problem domain.

[18] in a customised Course, the number of total credit points depends on the student's prior knowledge.

[19] the algorithm for that kind of function involves careful subject planning and further knowledge elicitation.

[20] at the time of writing, the expert system prototype is available for evaluation on `http://www.cit.gu.edu.au/~noran`

```
***************** Consultation Output Started **************
Job targeted: Artificial Intelligence Researcher .
---------------------------------
Prior (existing) Knowledge stated:
---------------------------------
Java_API .
Information_Systems_Concepts .
Entity_Relationship_Model .
Inference_Nets .
Artificial_Intelligence_Trends .
Knowledge_Representation_Principles .
=======================================================
The following Modules are needed:
--------------------------------|
Semester I:
-----------
Programming_I_2, subject Programming_II, Sem. 1, 5 CP.
Advanced_Information_Systems_Development, subject Introduction_to_Information_Systems_Development,
Sem. 1, 5 CP.
Programming_Language_Implementation_1, subject Programming_Language_Implementation, Sem. 1, 5 CP.
Programming_Language_Implementation_2, subject Programming_Language_Implementation, Sem. 1, 5 CP.
Introduction_to_Artificial_Intelligence_1, subject Introduction_to_Artificial_Intelligence, Sem. 1, 5 CP.
Programming_I_1, subject Programming_II, Sem. 1, 5 CP.
Natural_Language_Processing_Basics, subject Natural_Language_Processing, Sem.1, 5 CP.
Natural_Language_Processing_2, subject Natural_Language_Processing, Sem.1, 5 CP.
------------
Semester II:
------------
------------
------------

---------------------------------------
The following Project Modules are needed:
---------------------------------------
Artificial_Intelligence_Project_Part_1, subject Artificial_Intelligence_Project, Sem. 1, 20 CP.
Artificial_Intelligence_Project_Part_2, subject Artificial_Intelligence_Project, Sem. 3, 20 CP.
=======================================================
Total Credit Points = 113 .
****************************************
WARNING: you seem to require over (the required) 100 Credit Points.
----------
-> It is possible that you haven't stated ALL your existing knowledge.
-> You may also examine various courses of study by ASSUMING you had some of the knowledge.
----------
ADVICE: *Restart* and answer 'yes' to all existing / assumed knowledge.

***************** End of Consultation Output **************
```

Fig. 6. Extract from the Expert system output

3.8 Maintenance / Modification

Deployment is not the end of the spiral development paradigm. In fact, another spiral, comprising periodic maintenance and updating will be necessary to keep the knowledge base current. All additions / modifications must preserve the currently applicable knowledge base constraints and be properly validated [1]. Specialised maintenance constraints[21] and querying the reasoning of the expert system must also be available to the knowledge engineer.

[21] e.g. a mechanism to check for unsatisfied prerequisites for modules per total (included in the prototype).

4 Further Work on the Prototype and Beyond

Knowledge acquisition has proved to be the major bottleneck in this case study. Thus, the knowledge elicitation techniques will need to be improved, so as to shorten the knowledge acquisition turn-around time (e.g. via on-line questionnaires, and automated assessment and knowledge base input). Interview techniques should be selected to support this automation (e.g. [20]). Low-level automated knowledge acquisition may also be derived from CLIPS [11].

Substantial improvements may be made in the user interface. Also, an algorithm could be implemented to evenly distribute the modules by the semester they are offered in, together with a mechanism to limit the number of modules per semester (or else extend the study period).

The download time of the current expert system implementation could also be shortened by shifting data / input processing on the server side, using servlets (with user-program interaction occurring either via HTML pages containing e.g. FORM requests, or through applet / servlet interaction).

Although the presented prototype is reusable and scalable to a certain degree, a fully featured expert system may have different needs in terms of the set of development and knowledge acquisition tools [3].

5 Conclusions

This paper has presented the application of an expert system to a non-trivial problem domain that involves producing a customised study program for postgraduate students with previous knowledge. The design and implementation decisions have been highlighted and justified, and sample run results have been provided. The development and testing of the prototype have validated the concept of a knowledge-based advisory expert system, provided that a set of essential guidelines are followed and due attention is paid to the knowledge acquisition process design and implementation. Further details on the design and implementation of this system are presented in [22].

Similar to other rule-based expert systems, (notably EMYCIN [26]), the expert system prototype described in this paper may be reused by (1) establishing that the new problem domain suits the type of design decisions taken for the current system (e.g. pattern matching, forward chaining, Web enabling, etc) and (2) replacing the knowledge base with one specific to the new problem domain.

References

[1] Benbasat, I. and Dhaliwal, J. S. A Framework for the validation of knowledge acquisition. Knowledge Acquisition, 1(2), (1989) 215–233 1024

[2] Bergadano, F., Giordana, A. and Saitta, L. Automated vs. Manual Knowledge Acquisition: A Comparison in Real Domain. In H. Motoda and R. Mizoguchi and J. Boose and B. Gaines (Eds.), Knowledge Acquisition for Knowledge-based Systems. Amsterdam, The Netherlands: IOS Press.(1991) 1018

[3] Boose, J. H. A survey of knowledge acquisition techniques and tools. Knowledge Acquisition, 1(1) (1986) 3–37.

[4] Boose, J. H. and Gaines, B. R. (Eds.). Knowledge Acquisition Tools for Expert Systems: Academic Press.(1988)

[5] Collins, H. M., R. H., G. and Draper, R. C. Where's the expertise ? Expert systems as a medium of knowledge transfer. In M. Merry (Ed.), Expert Systems 85: Cambridge University Press.(1985)

[6] Diaper, D. An Organizational Context for Expert System Design. In D. Berry and A. Hart (Eds.), Expert Systems: Human Issues (pp. 214-236). London: Chapman and Hall.(1990)

[7] Forgy, C. RETE: A Fast Algorithm for the Many Pattern / Many Object Pattern Match Problem. Artificial Intelligence, 19 (1985) 17–37.

[8] Fowler, M. and Scott, K. UML Distilled (2nd ed.). Reading, MA: Addison-Wesley.(1999)

[9] Friedman-Hill, E. JESS - The Java Expert System Shell (Report# SAND98-8206). Livermore, CA: DCS, Sandia National Laboratories.(1998)

[10] Gaines, B. and Shaw, M. Foundations of Knowledge Acquisition. In H. Motoda, R. Mizoguchi, J. Boose and B. Gaines (Eds.), Knowledge Acquisition for Knowledge-based Systems. Amsterdam, The Netherlands: IOS Press.(1991)

[11] Giarratano, J. CLIPS User's Guide (Vol. 1: Rules): NASA.(1992)

[12] Giarratano, J. and Riley, G. Expert Systems: Principles and Programming. Boston, MA: PWS Publishing Company.(1998)

[13] ISO_JTC1/SC7. ISO/IS 12207: Software Engineering - Life cycle Processes. (2000)

[14] Jackson, P. Introduction to Expert Systems (3rd ed.). Harlow, England: Addison-Wesley.(2000)

[15] Jagodzinski, P., Holmes, S. and Dennis, I. User-Acceptance of a Knowledge-Based System for the Management of Child Abuse Cases. In D. Berry and A. Hart (Eds.), Expert Systems: Human Issues. London: Chapman and Hall.(1990)

[16] Kline, P. and Dolins, S. Designing Expert Systems. N. Y.: John Wiley & Sons.(1989)

[17] Lucas, P. and Van der Gaag, L. Principles of Expert Systems. Wokingham, England: Addison-Wesley.(1991)

[18] Marcus, S. (Ed.). Automating Knowledge Acquisition for Expert Systems. Boston: Kluwer Academic Publishers.(1988)

[19] Minsky, M. A framework for representing knowledge. In P. H. Winston (Ed.), Phychology of Computer Vision. New York: McGraw-Hill.(1975)

[20] Mizoguchi, R., Matsuda, K. and Yasushiro, N. ISAK: Interview System for Acquiring Design Knowledge. In H. Motoda and R. Mizoguchi and J. Boose and B. Gaines (Eds.), Knowledge Acquisition for Knowledge-based Systems. Amsterdam, The Netherlands: IOS Press.(1991)

[21] Newel, A. and Simon, H. A. Human Problem Solving. Englewood Cliffs, NJ: Prentice-Hall.(1972)

[22] Noran, O. A Course Advisor Expert System. School of CIT, Griffith University (`http://www.cit.gu.edu.au/~noran`) (2000)

[23] Prerau, D. S. Developing and Managing Expert Systems. Reading, MA: Addison-Wesley.(1990)

[24] Reichherzer, T. JConsult v1.3. Institute of Human and Machine Cognition, University of West Florida.(2000)

[25] Rumbaugh, J., Jacobson, I. and Booch, G. The Unified Modelling Language Reference Manual. Reading, MA: Addison-Wesley.(1999)

[26] van Melle, W., Scott, A. C., Bennett, J. S. and Peairs, M. The EMYCIN Manual (STAN-CS-81-16): Computer Science Department, Stanford University.(1981)

Applications of the Ecological Visualization System Using Artificial Neural Network and Mathematical Analysis

Bok-Suk Shin[1], Cheol-Ki Kim[2], and Eui-Young Cha[1]

[1] Department of Computer Science, Pusan National University
Busan 609-735 Korea
[2] Department of Computer Engineering, Miryang National University
Miryang, 627-702 Korea

Abstract. This paper presents a 3D visualization system with artificial neural network algorithm that tracks the motion of particles flowing in the water, where we get a great deal of variable information, and predicts the distribution of particles according to the flowing of water and the pattern of their precipitation. Various particles and their mutual collision influencing the force such as buoyancy force, gravitational force, and the pattern of precipitation are considered in this system and we control it by normalizing momentum equation and sequential equation. Flowing particles whose motion is changed with the environment can be visualized in the system presented here as they are in real water. We can track the motion of the particles efficiently and predict the pattern of the particles as well.

1 Introduction

Technique to control the motion of the particles in the water plays important role in applications such as computer applications, utilization of water, hydrodynamics, engineering, ecology. We track the motion of particles and analyze the pattern of their precipitation to control the flowing particles in the water. To predict the result, we need various pre-processing and operation because we extract a lot of variable informations. An efficient system is necessary for exact prediction to reduce the cost and operating time during the course from collecting data to extracting information.

In the fields such as utilization of water, hydrodynamics and engineering, we are studies the resistance of water in irrigation canal, the motion of particles in the water to flood control. In recent years, in order to disposal the deposits in the water and analyze the structure of substratum ecosystem occurred by the precipitation of containment, the studies on the motion and precipitation of particles are being made most of which are only in laboratory[1,2,3,4,5]. Most of such studies need a great deal of cost and time for the reason that we have to observe and collect the data and analyze it through the simulation in artificial waterway to observe the flowing particles in the

T.D. Gedeon and L.C.C. Fung (Eds.): AI 2003, LNAI 2903, pp. 1027-1037, 2003.

water and measure the pattern of precipitation[6]. The flowing particles and sediment transport are widely studied subjects. Extensive reviews of earlier works were provided by Chow [7], Vanoni [8], French [9], Chang [10], Yalin [11,12] and Nezu and Nakagawa [13], among others. Here, recent advances in numerical modeling of river hydraulics are summarized; and this is followed by a review of the recent computational modeling of sediment transport. Recent developments in related fields of particle resuspension, transport, and deposition in turbulent flow are also briefly described.

In the case of too much data with which we extract the necessary information, we do a pre-processing that reduces the dimensions with clustering technique.
Neural Network is a statistics analysis tool. It consists of a lot of components that work parallel and show the biological neural system. It is trained to operate the complex functions in various applications, such as pattern recognition, identification, classification, voice/vision control and so on. The purpose of the clustering is to set the similar samples in the same cluster. We obtain the attributes of the whole data by analyzing the cluster and apply it to decision making.

Clustering technique includes SOM, K-mean, ART and so on. K-mean clustering defines the similarity as the distance between the data in input space and minimizes the mean distance between the samples in the same cluster. It is easy to understand and realize. However, its result varies greatly according to the center of the cluster and it sensitive to the features which generate recognition error. SOM changes the weights of the neural network and performs the mapping of the data in high-dimension space into the two-dimension grid. The data that are close to each other in the input space are also close to each other in the two-dimension space. So the similarity between the data is easily visualized, and we reduce the influences of the model's initialization by adjusting the learning rate.

In this paper, in order to minimize the cost during analyzing the data and make the experiment possible in real time, we visualize the particles flowing in the water in the GUI-environment system presented here. As well, advanced and biological artificial neural network is used here to extract the motion of the particles more efficiently which have a lot of variable information.

2 System Structure

To make the experimental environment more real, we select the position, size and weight of the particles. And we can control the speed freely to create much flowing and the speed that decides the flowing. Due to the fluid energy, particles' feature, and the force between particles, there are many change in motion. We conduct the computation by inner step and visualize the result to satisfy the interaction. The visualization is more efficient than that in real environment. We denote a 3D-fluid space to analyze the particles and the direction of flowing, the depth of water and the width of water are indicated by X, Y, Z axis respectively. The particles flowing in the water have velocity V_{f_y}, which is being changed by buoyancy force(F_b) and gravitational force(F_g). The system presented here track the particles in various angles and shows the result in 3D-viewing space.

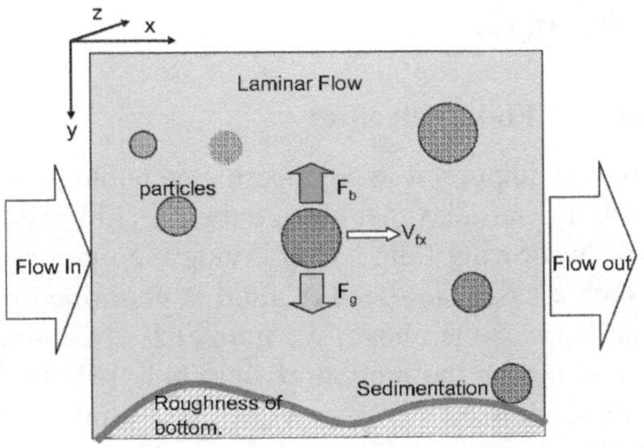

Fig. 1. The visualization model for the moving particles in water

2.1 Processing of Visualization Model

The visualization system is as follows: Fig 2. summarize the process

1. The system not only generates different weight particles, but controls the quantity of particles.
2. We can control the speed of flowing in real-time that influences the water flowing and precipitation.
3. Make the particles move in the conditions of water.
4. The particles that possess different momentum collide with each other frequently. These collisions become new parameter of particles' following.
5. We detect the particles' collisions while flowing efficiently with artificial neural network algorithm and it makes the operation fast.
6. Make the particles moving in various conditions accumulated if they get to bottom boundary condition.
7. The particles with space coordinates are visualized by the graphic library while moving.
8. Analyze the precipitation state that is changing with the motion direction and bottom structure by the visualization.

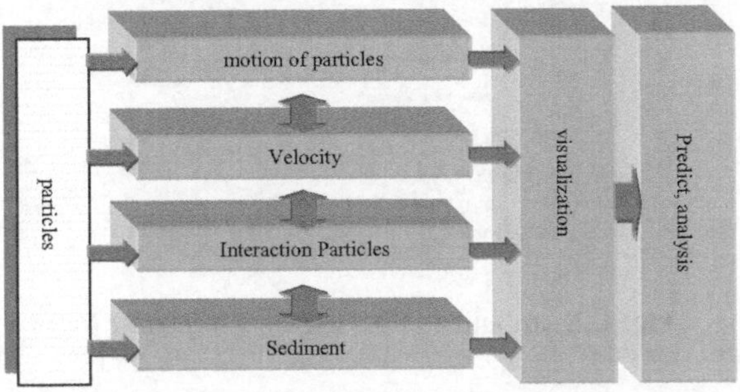

Fig. 2. Flow chart of visualization system

3 Motion of Particles

3.1 The Simulation of Fluid Velocities

The speed of fluid is very important in engineering technology. We should decide the speed of fluid to track the particles that change their loci by the flowing of fluid. We show the function of the flowing fluid by observing the motion of particles that pass the same point. In such a case, the speed of fluid is controlled by the space position and the time. On the base of the above, we normalize the momentum equation and sequential equation and search the motion of fluid [14, 15]. The flow of fluid keeps the general flow-laminar flow as shown in Fig 3, 4. In this case, the distribution of speed a parabola and the speed is regular to the time. The function of fluid speed is as following. Where , V_{fx} is velocity of fluid in XZ-plane, V_{fy} is is velocity of fluid in XY-plane

$$V_{fx} = -(y-b)^2 + V_{x_{\max}} \tag{1}$$

$$V_{fy} = K \cdot y \tag{2}$$

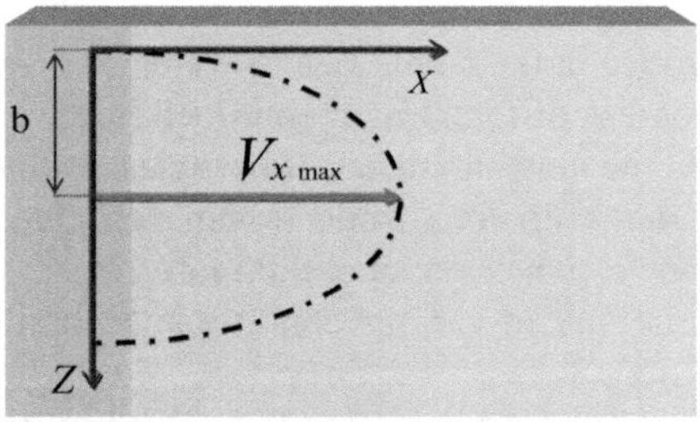

Fig. 3. Distribution of velocity in XZ-plane (V_{fx})

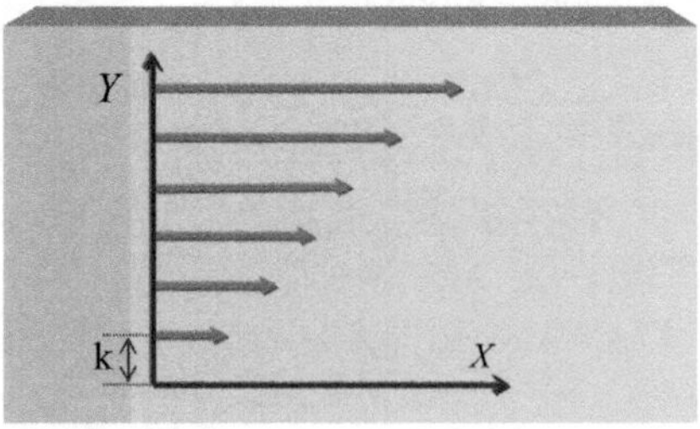

Fig. 4. Distribution of velocity in XY-plane (V_{fy})

3.2 Particle's Momentum

The momentum of the moving particles is calculated by the fluid speed. The motion of moving particles holds F_b (buoyancy force) and F_g (gravitational force), and they are decided by function (3), (4) and (5) as follows:

$$F = m \cdot a_z = F_b - F_g \tag{3}$$

$$F_b = g(\rho_p - \rho_w)v \tag{4}$$

$$F_g = \frac{1}{2} \cdot C_g \cdot A \cdot \rho_w \cdot V_{fx}^{\ 2} \tag{5}$$

In Eq. (4), F_b is buoyancy force. Where g is acceleration of gravity, ρ_p is density of particle, ρ_w is density of fluid and v is the volume of particle. In Eq. (5), F_g is gravitational force, And C_g is a coefficient of gravitational force which is related to Reynols number, according to Hind[16], The cross section area of particle A whose value is $\frac{\pi r^2}{2}$.

3.3 Particles Interaction

The particles that move in various directions collide with each other frequently. Due to the collision, there are some changes in the particles' motion. It needs much time to track the motion and detect the collision. To predict such collision in the visual system, the system presented here uses ART2 (Adaptive Resonance Theory), which is efficient in clustering, and detects the collision changes the motion[17,18]. ART2 generates the clusters dynamically, so it isn't influenced by the number of particles.

ART2 is a self organizing clustering neural network by unsupervised learning and fast on-line learning is possible in ART neural network. It creates a new cluster dynamically or integrates new patterns with the old clusters through the vigilance test of similarity measure between a given input pattern and ART cluster. It preserves the pre-learned knowledge stably and has an ability to accommodate the new patterns. The learning algorithm of ART2 neural network are followed.

(1) Let x_k be the k th input pattern vector, and w_i be the i th exemplar cluster pattern of neural network.

(2) Given a new input pattern, a MINNET is adopted to select the winner cluster j^*, which yields the minimum distance. Generally, the Euclidean distance is used.

$$\|x_{k-} w_{j*}\| = \min \|x_{k-} w_{i*}\| \tag{6}$$

(3) The vigilance test is carried out for the input pattern. If the distance between the input pattern and the winner cluster is smaller than vigilance parameter ρ which determines the radius of a cluster, it means that the input pattern is similar to the winner cluster. The input pattern is integrated into the winner cluster and the weight of the j^* is adjusted by

$$\text{If} \left\| x_{k-} w_{j*} \right\| < \sigma$$

$$\overset{new}{Cluster\ j*} = \frac{x_k + \overset{new}{w\,j*} \left\| \overset{old}{Cluster\ j*} \right\|}{\left\| \overset{old}{Cluster\ j*} \right\| + 1} \tag{7}$$

where, $\left\| Cluster_j \right\|$ denotes the number of members in cluster j. Otherwise, a new cluster is created with the input pattern.

(4) Repeat (1) ~ (3) for all of the input patterns.

(5) If the iteration exceeds the predefined number of iteration or the cluster exemplar pattern is converged, then stop learning.

4 Particles' Sedimentation Algorithm

To make the distributed particles accumulated until getting to bottom, we change the central positions of the particles as shown in Fig. 5. (here, $m = x - 1, n = y - 1$) To decide the particles' position, we detect if the central position of particles p get to the bottom, and if other particles are accumulated under it. As well, considering the change of position due to the falling of the particles, we search the right and left side to find new position. The precipitation algorithm of system is described as follows:

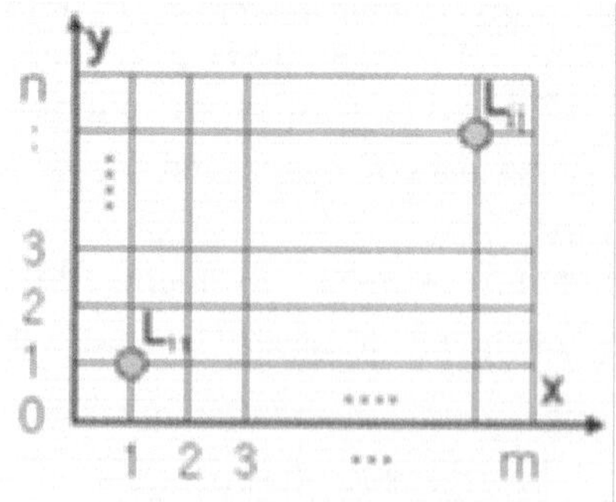

Fig. 5. Positional lattices for sedimentation ($m \times n$)

Algorithm : Sedimentation algorithm

```
Input : vertex data  P = {(x_c, y_c)}
Output :  Lattice L_ij  and  P' = {(x_c, y_c)}

All Let L_ij = 0
for  each point (x_c, y_c) in  P  do
Let  i = ⌊x_c⌋
Let j = ⌊y_c⌋
if (L_{i,j-2} ≠ 0) then
Let  j = j - 2  and  L_ij = 1
else
call  L_ij = process  (-1,-2,*i,*j)
if  L_ij ≠ 1  then    call  L_ij = process  (1,2,*i,*j)
endif
Let (x_c, y_c) = (i, j)
endfor


Process (m_1, m_2, i, j) {
if  L_{i+m_1, j-2} ≠ 0  then
if  L_{i-m_1, j} ≠ 1  and L_{i-m_2, j} ≠ 1 then
Let  i = i - m_1  and  j = j - 2  and    L_ij = 1
else
L_ij = 0
else if  L_{i+m_2, j-2} ≠ 0  then
if  L_{i-m_1, j-2} ≠ 1  then     Let j = j - 2
else
L_ij = 0
endif
return  L_ij
}
```

5 Results

The system presented here is implemented with Microsoft Visual C++ .Net, Silicon Graphics Library-OpenGL 1.1. We generate a 30cm□ 20cm□ 20cm cube space. In this space, the fluid flows from left to right and we generate ten thousand different weight virtual particles while the fluid speed is 10cm/s. Fig. 6(a),(b) shows the distribution of precipitation when the fluid speed is 10cm/s respectively. Fig. 6(b) shows result when substratum is added. These results make it possible to analyze the different distribution of precipitation due to the different substratum.

To analyze the distribution of the precipitated particles, we divide into five sites in direction of x-axis and twenty position in direction of z-axis as shown in Fig. 9 and table 1. We select three positions and analyze four different weight particles. The results are shown in Fig. 7. Comparing particle 1-th and particle 4-th, whose weight is the heaviest and the lightest respectively, we see the former is flowing forward but not far due to the fluid speed while the latter is flowing far away. Fig. 8. shows the result that the precipitation position varies according to the fluid speed. The particles don't move far away and sink where the fluid speed is slow, while move so far at the opposite condition.

(a)

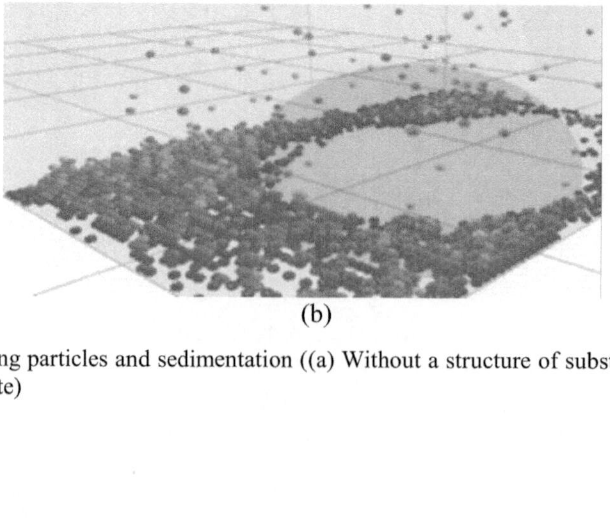

(b)

Fig 6. The moving particles and sedimentation ((a) Without a structure of substrate, (b) With a structure substrate)

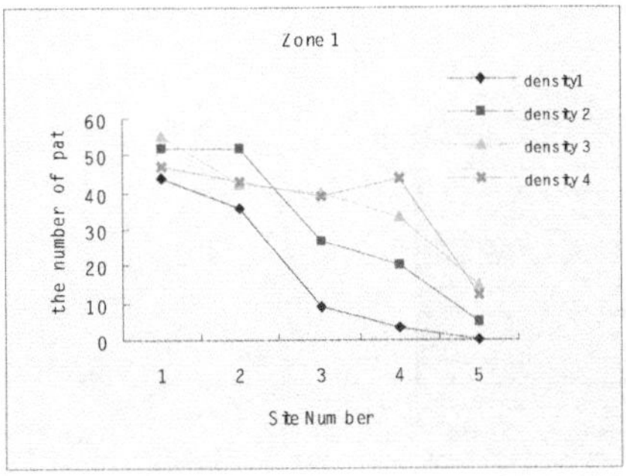

(a)

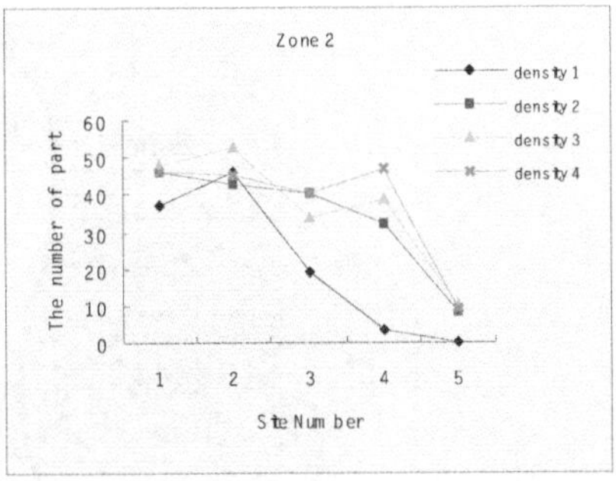

(b)

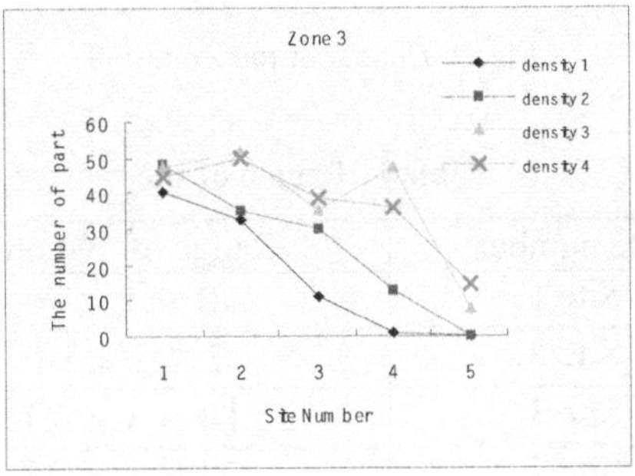

(c)

Fig. 7. The deposited particles ((a) Zone 1, (b)Zone 2, (c) Zone 3)

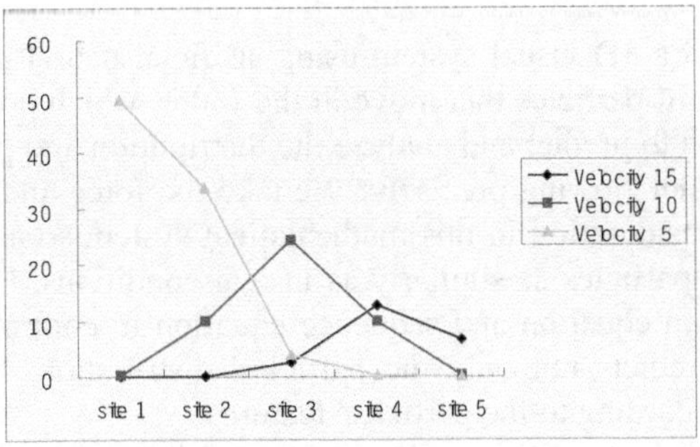

Fig. 8. Relationship between velocity and position of sedimentation

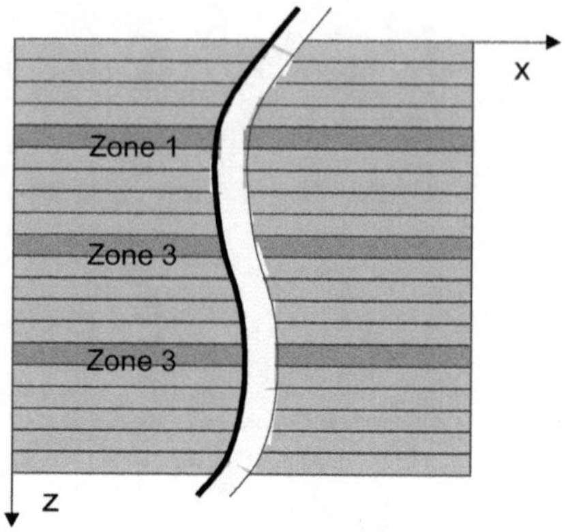

Fig. 9. Details of the Zone 1-3

Table 1. Region of site

Site number	Range of X-cord.
Site 1	$0 \le x < 7$
Site 2	$7 \le x < 14$
Site 3	$14 \le x < 21$
Site 4	$21 \le x < 28$
Site 5	$28 \le x < 30$

6 Conclusion and Future Work

This paper presented a 3D visual system using artificial neural network algorithm to visualize the motion of particles that move in the water, which possesses a lot of variable information, and to predict and analyze the distribution and precipitation of particles according to water flowing precisely. We take the force and interaction between the particles into consideration in this mathematical system so as to monitor the loci and precipitation of particles as similarly as in real conditions. Furthermore, we normalize the momentum equation and sequence equation to control as in actual condition. This system predicts not only the particles' distribution, but the precipitation pattern that varies according to the particles' features.

We need a boundary condition such as a curved line in the future work so as to predict the precipitation pattern that varies according to the structure of waterway.

References

[1] Angradi, T. R. Inter-habitat variation in benthic community structure, function, and organic matter storage in 3 Appalachian headwater streams. J. N. Am. Benthol. Soc. 15(1):42-63, 1996.

[2] Forman R.T.T. Land mosaics: the ecology of Landscapes and regions. Cambridge University Press, Great Britain, 1995.

[3] Johnson, P.D., Brown K.M., Covell, C.V 1994. Effects of heavy metals on a macroinvertebrate assemblage from a Rocky Mountain stream in experimental microcoms. J.N. Am. Benthol. Soc., 13, 4

[4] Stwvenson, R.J. Resource thresholds and stream ecosystem sustainability. J. N. Am. Benthol. Soc., 16, 2, 1997.

[5] Ramirez, A., Pringle, C.M. 1995. Structure and Production of a benthic insect assemblage in a neotropical stream. J. N. Am. Benthol. Soc., 14, 3

[6] M. Shamsa, G. Ahmadi, and DH. Smith, Computational modeling of flow and sediment transport and deposition in meandering rivers. Elsevier Science 25, 6, 689-699. 2002.

[7] Chow VT. Open channel hydraulics. New York: McGraw-Hill; 1959.

[8] Vanoni VA. ASCE sedimentation engineering, Manual and Report on Engineering Practice, No. 54, 1975.

[9] French RH. Open channel hydraulics. New York: McGraw-Hill; 1985.

[10] Chang HH. Fluvial processes in river engineering. New York: John Wiley; 1988.

[11] Yalin MS. Mechanics of sediment transport. Oxford: Pergamon Press; 1977.

[12] Yalin MS. River mechanics. Oxford: Pergamon Press; 1992.

[13] Nezu I, Nakagawa H. Turbulence in open-channel flow. Rotterdam: A A Balkema; 1993.

[14] John A.R, Clayton T.C. Engineering Fluid Mechanics, New York: John Wiley and Sons; 1997.

[15] Ahn SH, Hydraulics, Dongmyong Press, 2002.

[16] Hinds WC. Aerosol technology, properties behavior, and measurement of airborne particles. New York: John Wiley and Sons; 1982.

[17] G. A. Carpenter, S. Grossberg, The ART of Adaptive Pattern Recognition by a self-organizing Neurla Network, Computer, 21, 3, 77-88, 1988,

[18] G. A. Capenter, S. Grossverg, ART2: Self-Organization of Stable Category Recognition Codes for Analog Input patterns, Applied Optics, 26,23, 4919-4930, 1987.

Dynamic Games to Assess Network Value and Performance

Gregory Calbert, Peter Smet, Jason Scholz, and Hing-Wah Kwok

Command and Control Division, Defence Science and Technology Organisation
Edinburgh, South Australia, 5011
Greg.Calbert@dsto.defence.gov.au

Abstract: This paper looks at the analysis of network effectiveness and vulnerability through dynamic games. Each opposing side consists of a collection of vertices (pieces) connected in a network. Only vertices in the largest sub-graph exhibit mobility. We use the mobility and piece removal rules in the game checkers and modify the evaluation function to include sub-graph size balance. With this modified evaluation function, win-lose results of richly versus sparsely connected and centralised versus decentralized topologies are analysed. The results are compared with the current vulnerability studies in networks of varying topology. Finally we use temporal difference learning to calculate advisor weights for richly or sparsely connected vertices.

1 Introduction

The study of networks and their vulnerability to disruption or deliberate attack is a topic of increasing importance for society [2]. Many complex systems may be described in terms of networks where vertices represent the agents of the system and the edges represent interactions between agents [1]. With this view of networked agents, broad statistical properties of the system as a whole may be described. For example, the level of interaction within the system can be inferred from the edge richness within the network. Once a topologically dependent model of network function is found, critical agents within the system may be identified [2].

A wide ranging class of naturally occurring networks have similar statistical properties, in that the degree of such networks follows a power law [1]. Such networks include chemical interactions in animal metabolism, sexual partner networks, routing connections in the world-wide-web and power generation grids, to describe a few [1]. These networks, termed scale-free are characterised by a small proportion of highly connected vertices. In contrast, random networks, where each vertex has equal probability of being connected with any other vertex, do not have these highly connected vertices to any statistical significance [2].

Of importance is the performance of such scale free networks compared to randomly connected networks. Performance is defined either through the average diameter of the network or the size of largest connected sub-graph [2]. For the later

T.D. Gedeon and L.C.C. Fung (Eds.): AI 2003, LNAI 2903, pp. 1038-1050, 2003.

performance measure, as vertices are removed, the proportion of vertices in the largest connected subgraph drops to zero. For scale free networks, it has been shown that when random vertices, independent of their degree are removed or are no longer functional, the largest subgraph size decreases slowly [2]. However the largest subgraph size decreases drastically under network attack, when vertices with the highest degree are preferentially removed [6]. In contrast, random networks, having no central vertices of high degree show no difference in functionality between random vertex error or deliberate attack [2, 6].

Though these models of network performance are highly insightful, the results are generated assuming that under network attack, all vertices may be accessed with equal probability. There is no underlying model of the how an attacking system could manage to either sequentially or simultaneously disable vertices of high degree. In order to understand the true vulnerabilities of a networked system, one must model the dynamics of the attacking system at the outset. For example, if considering a computer virus disabling router nodes in a complex network, one must model the dynamics of transmission and replication through the network. When considering a police force attempting to disable a criminal network, one should simultaneously model the functionality of the criminal network as lynch-pin criminals are removed, as well as the communications/interactions of the legal/police network. When viewed this way, we have interactions between networks which we call a *network game*. Within each network, agents cooperate to achieve a non-cooperative goal against the opposing network.

In this paper we discuss non-cooperative dynamic games *between* two opposing networks. By this, we mean that within each network a cooperative game is played, in which a strategy is chosen by an agent controlling connected vertices to maximise some cost to the other network. In turn, an agent controlling the opposing network selects a strategy to minimize some cost function imposed by the original network. Vertices in these games are not only purely characterised by links with other vertices. In general, vertices will be characterised by a number of states, such as internal states (alive or removed from game) and possibly spatial states (coordinates of each vertex). Opposing network games are distinct from other *within* network non-cooperative games, in which each agent in the network adopts a strategy (such as choosing a packet routing path, or acceptance/rejection of market contract bids) to maximise its utility when competing against its direct neighbouring agents [11]. Such games have been extensively studied with communications or economics applications in mind.

Following this Introduction, we discuss formally modelling games between opposing networks in Section 2. We then discuss the network checkers game in Section 3. We next look at some results on the vulnerability of networks with differing topologies in Section 4 and look at valuing the network through reinforcement learning in Section 5. Discussion and Conclusion follow in Section 6.

2 Modelling Opposing Network Games

Formally, in order to specify the a game between two networks, we define at each time $t = 0,1,2,3,\cdots$ a sequence of graph pairs. Each graph pair corresponds to the new vertex-edge set of each side, following the application of some strategy by the opposing network, such as the removal of an edge or the removal of a vertex. Thus, as the game progresses, we define the sequence of graph pairs as

$$(G_1(0), G_2(0)), (G_1(1), G_2(1)),...., (G_1(t), G_2(t)),.... \quad (1)$$

where

$$G_i(t) = (V_i(t), E_i(t)) \quad (2)$$

specifies the number of vertices or elements and the topology of edges between them at the time t. The strategy used by a network will depend on the vertex states and the connectivity between them. In general, the strategy will be a map

$$S : G_1 \times G_2 \rightarrow G_j, j = 1,2. \quad (3)$$

In particular, let $S(v_j)$ be the strategy set for vertex $v_j \in V_i(t)$ at some time of the game t. Now suppose that vertices $\{w_1, w_2,...., w_k\} \in V_i(t)$ comprise a connected sub-graph of the network. Then at time t, the sub-graph has strategy set

$$S(t) = S(w_1) \oplus S(w_2) \oplus .... \oplus S(w_k). \quad (4)$$

When coding an agent to play such network games, strategies can be evaluated through the use of heuristics or the more sophisticated game playing techniques. In particular, our approach is to use a linear evaluation function of weighted game features in combination with either min-max decision tree search or the use of temporal difference learning to find appropriate weights for the features chosen. Furthermore, we assume that the strategy set is found from the maximal subgraph, which is the connected graph with the largest number of vertices. The largest sub-graph may be found by applying Djistraka's algorithm on the current network [9].

3 Networked Checkers

We chose the game of checkers as our template game, in which the pieces become vertices for an underlying network [7]. There are several principal reasons for this choice. First the checkers pieces exhibit spatial manoeuvre. Our aim is to model manoeuvre based conflict between opposing sides, as seen in human sporting games and warfare, thus the game of checkers is seen as an important start. Coupled with manoeuvre is the attrition of forces in the game, again a feature of warfare. Finally, checkers with a well defined end-set of states[1] has had considerable research into the agent based play. This means we are able to conduct Monte Carlo simulations across many games in order to assess the performance of a particular network topology. There are three end states in the game of checkers- a win, loss or draw. Our measure

[1] Here, we assume the game ends either with the loss of all pieces or no mobility from one side or no pieces taken after 40 moves.

of network performance is the observed frequency of wins plus half the draws across games. If we define, for each end-state graph pair (G_1, G_2) the function

$$\varphi(G_1, G_2) = \begin{cases} 1 \text{ for win,} \\ 1/2 \text{ for loss,} \\ 0 \text{ otherwise.} \end{cases} \quad (5)$$

then we estimate the expectation $E(\varphi(G_1, G_2))$. If $(G_1, G_2)^1, (G_1, G_2)^2, \ldots, (G_1, G_2)^n$ are independent and identically distributed random outcomes of the end-states of the network game, then our unbiased estimate of the expectation is

$$p = \frac{1}{n} \sum_{i=0}^{n} \varphi\left((G_1, G_2)^i\right). \quad (6)$$

Assuming we have a binomial model, then the unbiased estimate of the variance is

$$\operatorname{var}(p) = \frac{p(1-p)}{n}. \quad (7)$$

Network topologies are specified at the commencement of the game and only alter due to the removal of a vertex by the opposing network. Topologies are specified by the number of links between vertices at the instigation and the network growth rule. The maximum number of vertex to vertex connections is $\binom{12}{2}$ links.

Given a number of edges between vertices, the structural aspects of the topology are governed by a number of rules. In particular, the baseline topology is the *random network*. Here, two vertices are selected at random and an edge placed between them if not connected. This process continues till the link budget is exhausted. In this paper, we compare the performance of random networks with that of *hub-spoke networks*. As the name implies, the graph consists of one or number of centralised hubs, connecting to the spokes of the network. With a link budget of 11, the network is fully connected, with one central hub. Link budgets greater than 11 have additional hubs, with a new hub at link budgets of 12, 23, 33, 42, 50, 57 and 63 edges.

We have chosen the random and hub-spoke topologies to compare performance with communication network results [2]. In particular, we will compare the performance of centralised (hub-spoke) networks under network destruction, as the game is played, with that of decentralized random networks.

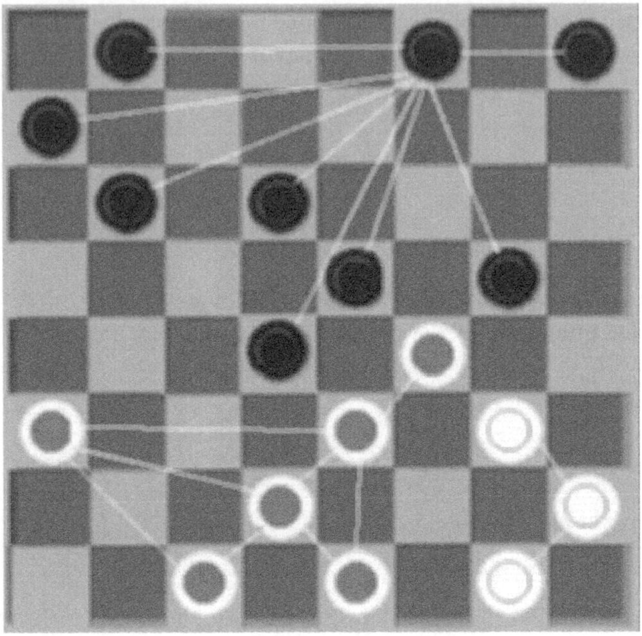

Fig. 1. Example of the network checkers game. Here the white side is connected with a random network topology, the red with a hub-spoke topology. Pieces on the white side which exhibit maneuver, the largest connected subgraph, are shown in shading

The results of Monte Carlo simulations will be dependent on the evaluation function used in the game. Indeed any performance results obtained must be coupled with the structure of the evaluation function [8]. At this stage of the research, we chose an simple as possible evaluation function that highlighted the features of materiel balance, value of more mobile kings as opposed to unkinged pieces, the relative rank balance[2] of pieces (to encourage king making) and the balance of the largest subgraph sizes, this being a measure of the strategy set and the network viability. Thus our evaluation function takes on the form

$$V = 100(V_1 - V_2) + 250(KV_1 - KV_2) + R_1^2 - R_2^2 + 100(N_1 - N_2), \quad (8)$$

where V_i, KV_i and N_i $i = 1,2$ are the number of unkinged, king and largest subgraph pieces respectively. R_i, $i = 1,2$ is the rank of each side.

It should be noted that the values for coefficients for the materiel terms in the evaluation have been taken from, a commonly played Java checkers web-site [7]. The value of 100 for the coefficient in the largest subgraph balance term was chosen as the minimal value congruent with materiel and mobility. There is however no a-priori experience for setting this value and machine learning techniques such as supervised and reinforcement learning are appropriate, as seen in Section 5.

[2] The rank is the sum of the distances of unkinged pieces from its own baseline.

4 Experiments with Network Checkers.

Monte Carlo simulations were conducted in the network checkers game. The focus of our research was on the frequency of performance success (wins plus half draws) as a function of different opposing network characteristics. In one set of trials, we considered the performance success as a function of the ratio of vertices to edges. As our starting number of edges is small, rather than considering this ratio, we only considered the absolute number of edges up to the maximum of 66. The commencing network topology in this set of experiments was random for each side.

Next we conducted a series of experiments to look at the performance of two different opposing topologies. In order to draw conclusions purely from the topological structure, the number of links was kept constant for each side during each set of simulations. The experiments focused on the random topology against the hub-spoke topology. The relative performance of the random topology was simulated, both as the level of edge connectivity increased and for differing strategy search depths of 0 (random legal moves played) to 3- ply search.

4.1 Results with Differing Edge Number

Simulations show that having an increased number of edges than the opponent does confer an advantage. This is not surprising as increased network connectivity implies both a larger maximum subgraph size and robustness over the game.

Of note is the performance of the white network against a red network either with 36 edges or 60 edges – nearly fully connected. Performance gains are not substantially greater, highlighting diminishing returns as edge numbers increase.

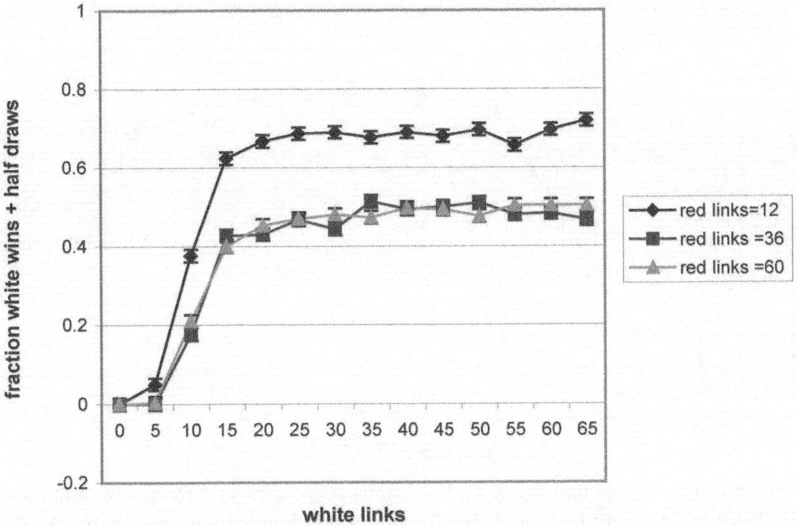

Fig. 2. Graph of the performance measure as a function the number of white edges

4.2 Results Comparing Different Topologies

Network performance and vulnerability models compare differing topologies to highlight the relative strengths and weaknesses of one structure over another. In these models, the network is eroded either by the removal of vertices or links and measures of performance, such as throughput or largest subgraph size are compared [1]. There are no assumptions on the network structure of the opposing agents. We call these models net versus no-net models. Here, we take the approach of comparing topologies by direct competition, a net versus net model. In this case, the removal of vertices is dependent on the performance of the adversary network, hence our model of performance may be seen as richer, as it does not make poor assumptions about the behaviour of the collection of agents attempting to destroy or disrupt a network.

In order to make a direct comparison between net versus no-net and net versus net results, suppose we considered the performance of a hub-spoke network and a random network. In a net versus no-net model, the hub-spoke architecture is more vulnerable either in measures of subgraph size or diameter [2]. If the hub vertex is removed both subgraph size drops to zero and diameter goes to infinity. In a random network, there is a non-zero probability that the sub-graph size will still be greater than one after removal of the vertex with highest degree. For example, in a random graph with 4 vertices with 3 edges, if the topology is not hub-spoke (which occurs with probability $4/6^3$), then removing the vertex with highest degree results in a sub-graph size of two. Hence the expected subgraph size after the removal of the vertex with highest degree is

$$E(\text{subgraph size}) = 2\left(1 - 4/6^3\right) \approx 1.96. \quad (9)$$

Hence the net versus no-net model considers the random network less vulnerable.

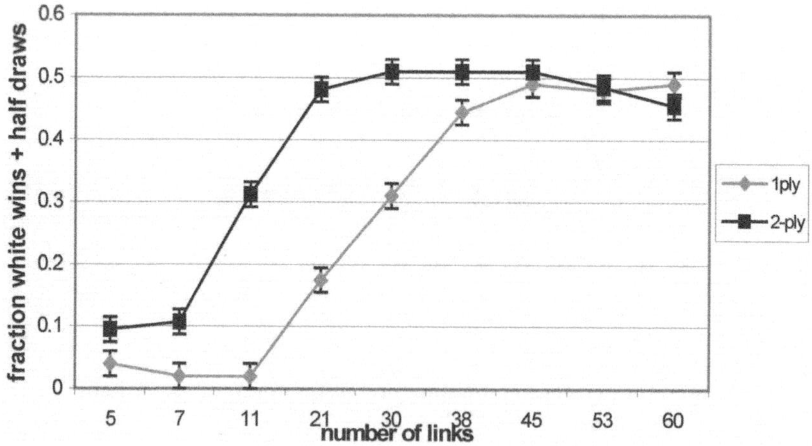

Fig. 3. Performance of a random against a hub-spoke topologies at different edge numbers and for search depth 1-2 ply. Our measure is the fraction of random net wins plus half draws

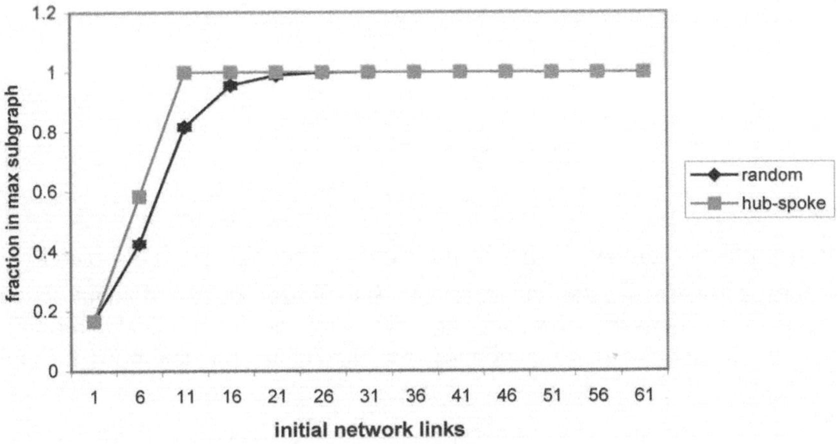

Fig. 4. Fraction of vertices in the largest subgraph, for random and hub-spoke topologies, assuming different edge budgets up to a maximum of 66

The figure on the previous page shows the performance of a random against a hub-spoke topology in the net versus net model. In contrast to the net versus no-net model, the hub-spoke topology shows higher performance than that of the random network, each with the same number of links. This is explained by considering the largest sub-graph size for the hub-spoke and the random networks given a fixed edge budget, prior to the removal of any vertices. Here, the hub-spoke network has the largest sub-graph size with the smallest number of edges. For a hub-spoke network, the addition of an edge always increases the sub-graph size by one if this sub-graph size is less than the number of vertices. This property is not assured in the random graph. The following simulation compares largest subgraph sizes in hub-spoke and random networks with 12 vertices at differing edge budgets.

The result that the hub-spoke topology performs better than the random topology shows that the dynamics of the interacting networks must taken into account. At the search depth of 2-ply, [8] the opponents response to a move is taken into account. Typically in observations of these games, the hub vertex is not placed in the position as one of the front pieces and is thus protected.

5 Learning the Value Function

We chose the value of 100 as our coefficient for the balance in sub-graph sizes arbitrarily. In general, we cannot assume that the coefficients for materiel balance and mobility will remain the same in the network game either. Finding the relative weights for these game features, termed advisors, can be done through reinforcement learning. Here, we briefly review the approach taken.

Given the board is in state s, we specify a action-value function $Q^{\pi}(s,a)$. This action-value function is the expected reward from winning the game given that the policy π determines what action to take at each particular state. At a terminal state s_T we define the action-value function (reward) to be

$$Q^{\pi}(s_{T,}\bullet) = \begin{cases} 1 \text{ for a win}, \\ 1/2 \text{ for a draw}, \\ 0 \text{ for a loss}. \end{cases} \tag{10}$$

For a particular policy π, this action-value function $Q^{\pi}(s,a)$ is therefore the probability that a win is achieved, given we start from state s and adopt action a. Furthermore, in network checkers, as there are no intermediate rewards or discounting, the action-value function satisfies the Bellman equation [3] for optimal return

$$Q^{*}(s_t, a_t) = \mathrm{E}^{\pi^*}\left\{ \max_a Q^*(s_{t+1}, a_{t+1}) \right\}, \tag{11}$$

where the expectation is taken over the optimal policy π^*. Indeed for the optimal policy π^* the expected difference between successive state action-value pairs will be zero [3],

$$\mathrm{E}^{\pi^*}(Q^*(s_t, a_t) - Q^*(s_{t+1}, a_{t+1}) \mid s_t, a_t) = 0. \tag{12}$$

It is the observation of equation (12) that is the basis of the temporal-difference learning approach, in the sense that the difference between successive action-value functions should be close to zero according to Bellman's equation and learning is achieved by "shifting" the action value function $Q(s_t, a_t)$ towards $Q(s_{t+1}, a_{t+1})$ [3].

In the network checkers game, we approximate the action-value function by some function $\tilde{Q}: S \times \Re^n \to \Re$. We adopt the non-linear approximation of mapping the linear evaluation function onto the standard sigmoidal function, meaning that

$$\tilde{Q}(s,a) = \left(1 + \exp(-\beta V_n(s, \vec{w})\right)^{-1} \tag{13}$$

where $V_n(s, \vec{w})$ is the principal value taken in a max-min search of depth n from state s and $\vec{w}$ is the vector of weights (advisors) we wish to calculate [5]. Adopting the sigmoidal function has the advantage that for the set of states and actions, $S \times A(s)$,

$$\tilde{Q}: S \times A(s) \to (0,1). \tag{14}$$

When the evaluation function is large, $\tilde{Q}$ is close to one, indicating a win, large and negative, $\tilde{Q} \approx 0$ indicating a loss. The constant β determines the sensitivity of $\tilde{Q}$ to adjustments of weights in the evaluation function.

Weight changes are determined through gradient descent [10] and are done on-line, meaning that weights of the currently applied action-value function are changed during the course of the game [3]. In particular, as a pair of states are visited during the game, an increment in the weight vector is given by

$$\Delta\vec{w} = \alpha\left(\tilde{Q}(s_{t+1}, a_{t+1}) - \tilde{Q}(s_t, a_t)\right)\nabla_{\vec{w}}\tilde{Q}(s_t, a_t), \tag{15}$$

where α is the learning rate and ∇ is the gradient operator. Here,

$$\nabla_{\vec{w}}\tilde{Q}(s_t, a_t) = \beta\tilde{Q}(1-\tilde{Q})\nabla_{\vec{w}}V_n(s, \vec{w}). \tag{16}$$

The learning rate is taken to satisfy stochastic convergence conditions given in [3]. To approximate the optimal policy, we determine the action for states s_t, s_{t+1} through an ε-greedy policy of choosing action $\arg\max_a \tilde{Q}(s,a)$ with probability $(1-\varepsilon)$ and a random action with probability ε. This approach directs the policy towards the optimal one, as specified by Bellman's equation, whilst assuring the exploration of novel states through the random action [3].

5.1 Results of Evaluation Function Learning

We applied the learning method described, referred to as SARSA(0) [10], to the network checkers game, in order to explore values for the subgraph balance advisor. In applying TD(0), the opponent's advisor weights were fixed at values of 1.0, 2.5, 1,0, 0.01 for the unkinged, kinged, subgraph and rank balance weights respectively. The initial weights chosen for learning were 1.0, 1.0, 1.0 and 0.01. Min-max search was conducted at a depth of 3-ply and the value of β was 0.5. We ran the SARSA(0) algorithm over 2500 games.

Our initial trials showed that altering the rank balance advisor value is generally troublesome to learning the advisor values of the other three variables. As games progressed, the rank balance fervently swung from positive to negative values. In our current game formulation, kings have a rank of zero and we believe that this is the cause of the rank balance instability. A side that is close to victory may have either a positive rank balance (many unkinged pieces versus few unkinged pieces) or negative rank balance (many kinged pieces against a few unkinged pieces). For this reason, we chose to set the rank balance advisor weight at a constant value of 0.01.

First, we ran the SARSA(0) method when the network was fully connected, with 66 edges. Next we ran the algorithm with a random topology connected 20 edges.

Our results show that a fully connected network, after 2500 games, the subgraph advisor weight is slightly lower than that of kinged piece (advisor values 0.29 and 0.31 respectively). Not surprisingly, the kinged pieces have the highest value, approximately double that of an unkinged piece (value 0.15). As kings have the highest mobility, in checkers they are considered to have the highest value [8].

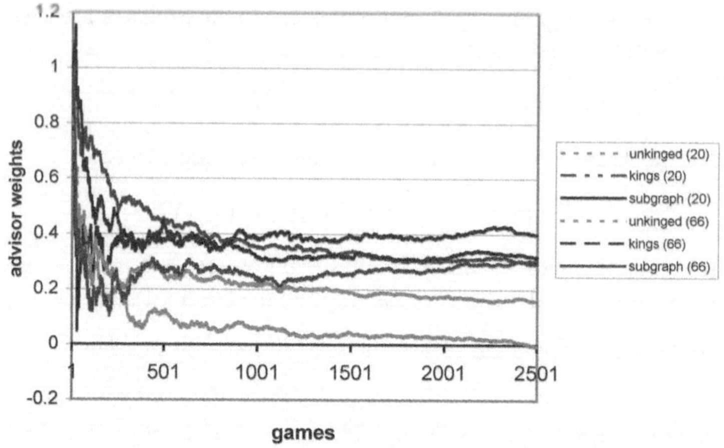

Fig. 5. Graph of the unkinged, kinged and subgraph advisor weights for 2500 games

With a sparse network, maintaining the subgraph size is crucial, as the loss of a single vertex with high degree may fragment the network into isolated small subgraphs or single vertices [2]. As we assume that only the largest subgraph exhibits mobility, a loss of a single piece may cause an incommensurate loss in mobility. The subgraph balance advisor weight reflects this, as it has the highest value in the learned advisor weights for a sparsely connected network (20 links, value 0.39).

At this stage of our research, the advisor weights are only taken as an indicator of the importance of unkinged, kinged and subgraph balance for networks either abundantly or sparsely connected. It is suggested that when attempting to find the exact optimal values for the advisor weights, one should apply the $SARSA(\lambda)$ algorithm for λ values close to one initially, then refine the values by choosing λ close to zero [4]. This is the next stage of our research.

6 Discussion and Conclusions

Our research goals are to understand the impact of element connectivity and communications in conflicts/games where two or more groups of agents cooperate with one another against an opposing group of elements. Several directions need to be taken for greater understanding of the role of networked gaming vertices. First, we assumed in our simulations that once an vertex is isolated from the main subgraph, it no longer exhibits any maneuver. This type of assumption may be termed *control by direction* in the sense that each vertex requires direction or guidance from peers to exhibit any mobility. Any truly autonomous act distinct from the main group is excluded.

The results presented in this paper may be contrast with agent models, where control is done by *veto*. Here, agents only recognize peers connected in their own subgraph and will act autonomously unless vetoed with other connected agents. Since control is by veto, this means that agents not connected will not recognize each other as peers, but will still act autonomously. Agents of the same group may eliminate one another. With this new assumption, the same set of simulations may be run, for example, comparing outcomes with differing connectivity to agent ratios (vertex to edge) and comparing outcomes with centralised versus de-centralised connectivity.

In real-world situations agents such as humans naturally establish or cut links between other agents because of either environmental or adaptive reasons [12]. This means a dynamic network is formed, depending on the state of each agent. In this sense, network formation may be seen as a strategic decision along with the maneuver of agents. Such adaptive dynamic network games are worth modelling, however this generates a cost in the problem domain. Consider a game with an approximate movement branching factor of 8 (as is checkers) and 20 edges to be placed amongst 12 agents. The total branching factor will be the product of the movement branching factor and the number of ways to place 20 edges amongst 12 agents, $8\binom{20}{12} \approx 800000$. This simple analysis implies that games, of dynamic network formation are of complexity vastly exceeding games such as go or backgammon.

In conclusion, we examined games in which mobility of vertices was determined by a network structure connecting vertices. The complexity of strategy was also dependent on the size of the largest subgraph connecting those vertices. Mobility was determined through the rules of checkers.

Through Monte-Carlo simulation, we examined game outcomes where one side had a differing edge to vertex ratio to the opponent. Larger edge to vertex ratios increased the probability of winning.

We also compared game outcomes opposing differing topologies with the same edge number on both sides. In this respect only the topological properties of the network were considered. We found through simulation that centralised networks with a hub vertex fare better against decentralized networks for two reasons. Firstly a centralized hub network connects the maximum number of vertices together with the minimum number of links. Second, once a hub vertex is protected from attack through manoeuvre, the network shows the greatest resilience to attack, as the loss of a vertex only decreases the subgraph size by one.

Finally, we conducted a series of machine learning experiments, implementing reinforcement/temporal difference methods in the play of the game. Through implementing gradient based $SARSA(0)$, we determined the values of the specific advisors for the importance of materiel and subgraph balance. For fully connected networks, the king balance advisor had the greatest value. With a sparse network, the subgraph balance advisor had the greatest value, due to the sensitivity of subgraph size as vertices as lost.

References

[1] Albert, R. and Barabasi, A. L.: The statistical mechanics of complex networks. Reviews of Modern Physics. Vol. 47, (2002), 47-94.

[2] Albert, R. *et al.*: Error and attack tolerance of complex networks. Nature. Vol. 206 (2000), 378-381.

[3] Barto, A.G. and Sutton R. S.: Reinforcement Learning: An Introduction. MIT Press, (1998).
[4] Baxter, J. *et. al.:* Experiments in Parameter Learning Using Temporal Differences. ICCA Journal, Vol. 211(2), (1998), 84-99.
[5] Beal, D. F. and Smith, M.C.: Learning Piece Values Using Temporal Differences. ICCA Journal, Vol. 20(3), (1997), 147-151.
[6] Holme, P. *et. al.:* Attack vulnerability of complex networks. Physical Review E. Vol. 65, (2002), Section on Complex Networks.
[7] Huang, V. and Huh, S.H. `http://www.cs.caltech.edu/~vhuang/cs20/c/applet/s11.html`.
[8] Levy, D.: Computer Gamesmanship: Elements of Intelligent Game Design. Simon and Schuster, New York, (1984).
[9] Lynch, N.: Distributed Algorithms. Morgan Kaufmann, 1997.
[10] Sutton, R. S.: Learning to Predict by the Methods of Temporal Differences. Machine Learning. Vol. 3, (1988), 9-44.
[11] Washburn, A. and Wood, K.: Two Person Zero Sum Games for Network Interdiction. Operations Research, Vol. 43(2), (1995), 243-251.
[12] Watts, A:. A Dynamic Model of Network Formation. Games and Economic Behaviour. Vol.34, (2001), 331-341.

Design and Implementation of an Intelligent Information Infrastructure

Henry C.W. Lau, Andrew Ning, and Peggy Fung

Department of Industrial and Systems Engineering,
The Hong Kong Polytechnic University, Hunghom, Hong Kong

Abstract. The lack of seamless data interchange and efficient data analysis hinders the formation of an effective information infrastructure that serves as the platform for data interchange across heterogeneous database systems. Information infrastructure has become an emerging platform to enable business partners, customers and employees of enterprises to access and interchange corporate data from dispersed locations all over the world. In general, information infrastructure is a browser-based gateway that allows users to gather, share, and disseminate data through Internet easily. This paper proposes the design and implementation of an information infrastructure embracing the emerging eXtensible Markup Language (XML) open standard, together with an intelligent data mining technique combining neural networks and On-Line Analysis Process (OLAP) and rule-based reasoning approaches to support knowledge discovery. To validate the feasibility of this approach, an information infrastructure prototype is developed and tested in a company with description of this case example covered in this paper.

1 Background

Recently, data mining technology, which aims at the conversion of clusters of complex data into useful information, has been under active research [1,15,23,24]. Data mining, in general, identifies and characterizes interrelationships among multivariable dimensions without requiring human effort to formulate specific questions. In other words, data mining is concerned with discovering new, meaningful information, so that decision-makers can learn as much as they can from their valuable data assets. Data mining tools require different data formats in relational and multi-dimensional database systems. The shared data access interface of data mining tools will enable easier exchange of information among different sources. There are many commercial products for data mining and a typical example of such product is Microsoft SQL server which has incorporated the On-Line Analytical Processing (OLAP) technology that provides a service for accessing, viewing and analyzing on large volume of data with high flexibility and performance [23,27].

T.D. Gedeon and L.C.C. Fung (Eds.): AI 2003, LNAI 2903, pp. 1051-1062, 2003.

While OLAP is able to provide numerical and statistical analysis of data in an efficient and timely way, it lacks the predictive capability, such as the projection of possible outcomes based on the past history of events so as to decide on the action to be taken. In this respect, it seems necessary that a certain “ingredient” of intelligence element needs to be added to OLAP to enable the self-learning capability of the whole system.

Neural network is a technology that has typically been used for prediction, clustering, classification and alerting of abnormal pattern [7,26]. They create predictive networks by considering a “training set” of actual records. In theory, the formation of neural network is similar to the formation of neural pathways in the brain as a task is practiced. Also a neural network refines its network with each new input it considers. To predict a future scenario, the neural network technology is able to work with a training set of data from the past history of records. It will use the training set to build a network, based on which the neural network is able to predict future scenario by supplying a set of attributes. Like the OLAP technology, there are many commercial products that have incorporated neural network technology. The inclusion of computational intelligence knowledge into a data mining technology can significantly enhance the “machine learning” capability of the system and is undoubtedly an issue that is justified to be addressed.

In creating decision support functionality, a mechanism, which is able to combine and coordinate many sets of diversified data into a unified and consistent body of useful information, is required. In larger organizations, many different types of users with varied needs must utilize the same massive data warehouse to retrieve the right information for the right purpose. Whilst data warehouse is referred as a very large repository of historical data pertaining to an organization, data mining is more concerned with the collection, management and distribution of organized data in an effective way. The nature of a data warehouse includes integrated data, detailed and summarized data, historical data and metadata. Integrated data enable the data miner to easily and quickly look across vistas of data. Detailed data is important when the data miner wishes to examine data in its most detailed manner while historical data is essential because important information nuggets are hidden in this type of data. OLAP is an example of architectural extension of the data warehouse.

Since after the setup of a data warehouse, the attention is usually switched to the area of data mining, which aims to extract new and meaningful information. In other words, a pool of ‘useful information' that has been stored in a company data warehouse becomes ‘intelligent information', thereby allowing decision-makers to learn as much as they can from their valuable data assets. In this respect, neural network can be deployed to enhance the intelligence level of the OLAP application.

Neural network searches for hidden relationships, patterns, correlation and interdependencies in large databases that traditional information gathering methods (such as report creation and user querying) may have overlooked. The responsibility of the neural network is to provide the desire change of parameters based on what the network has been trained on. Intrinsically, a sufficient amount of data sample is a key factor in order to obtain accurate feedback from the trained network. As neural network is meant to learn relationships between data sets by simply having sample data represented to their input and output layers [9], the training of the network with input and output layers mapped to relevant realistic values with the purpose to develop the correlation between these two groups of data will not, in principle, contradict the basic principle of neural network.

With a trained network available, it is possible that recommended action can be obtained with the purpose to rectify some hidden problems, should that occur at a later stage. Therefore, in the training process of the neural network, the nodes of the input layer of the neural network represent the data from the OLAP and those of the output layer represent the predictions and extrapolations. The data flow of the NOLAPS has been depicted in Fig. 1. It should be noted that the output information from the OLAP could be used to refine the OLAP data cube so as to continually update the database over time.

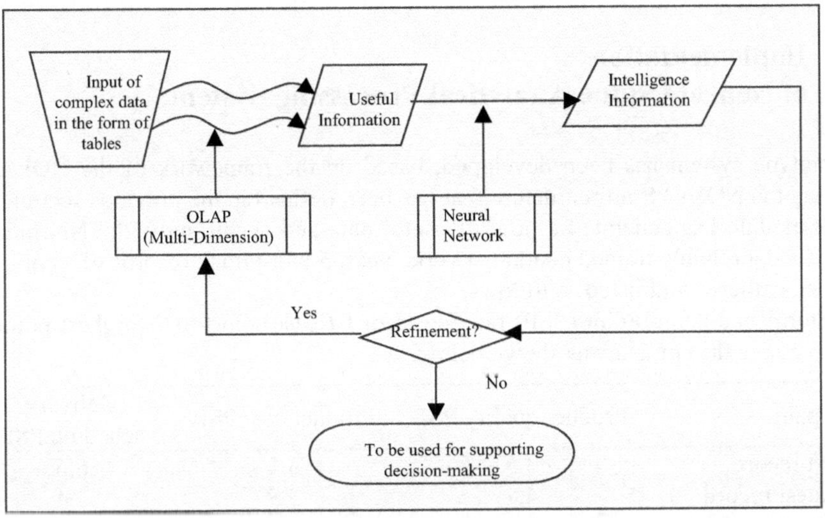

Fig. 1. Information flow of NOLAP

The data interchange within the NOLAPS encompasses three modules, namely OLAP module, Data Conversion (DC) module and Neural Network (NN) module (Fig. 2). The data repository, which aims to support efficient data interchange among the three modules, is essential for the coordination and updating of information from various sources. As for the OLAP module, it consists of descriptive data (dimensions) and quantitative value (measures), both of which generate the OLAP data cube by building up two elements, namely, fact table and dimension [6]. In the fact table, the required data and user-defined methods for analysis are specified clearly. In the descriptive data of OLAP, the different dimension levels are defined for further computational use on different views of OLAP data cube. Typical dimension includes location, company and time whereas typical measure includes price, sales and profit. With a multidimensional view of data, the OLAP module provides the foundation for analytical processing through flexible access to information. In particular, this distinct feature can be used to compute a complex query and analyze data on reports, thereby achieving the viewing of data in different dimensions in a more easy and efficient way.

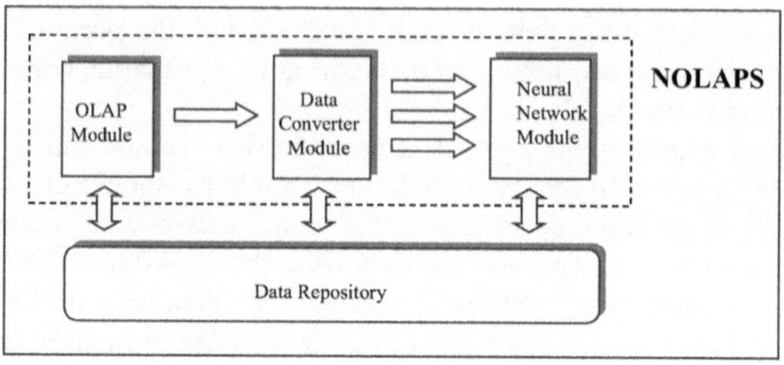

Fig. 2. Infrastructure of NOLAPS

2 Implementation of Neural Online Analytical Processing System

A prototype system has been developed, based on the framework of the NOLAPS. Pursuing the NOLAPS infrastructure that has been defined in the previous section, the OLAP module has generated a pool of useful data and accordingly, the NN module has created a reliably trained neural network. Next, 5 latest track records of a company has been gathered and listed as follows.

Performance Score Point (PSP) ranging from 1 (least point) to 7 (highest point) is used to assess the company as shown below.

Company A	Product quality PSP	Product cost PSP	Delivery schedule PSP
Latest record	3.5	6.5	6.6
2nd latest record	4.7	5.5	5.4
3rd latest record	5.0	5.1	4.8
4th latest record	5.6	4.1	4.4
5th latest record	4.0	4.0	3.0

After such information has been input, the NN module gives an assessment report back to user, thus supporting user to take action if deemed necessary. In the following table, "0" output from the NN node indicates a negative suggestion to the associated statement and "1" is the positive suggestion whereas "0.5" indicates that there is not enough data to justify a firm suggestion.

Company A	Output from NN module
Potentially competent	0.5
Suggested to be replaced	0
Service quality is compromised to meet the quoted price	1
Further assessment of company performance is required	1
Delivery time seems to be inconsistent due to certain company problems	1

Referring to Figures 3 & 4 and based on the NN output results as shown in the table, Company A seems to have problem in meeting the delivery schedules and it is suggested that further assessment regarding the company's performance is needed. Based on the suggestion of this assessment report, Company A has been approached in order to find out the reason behind the continual downgrade of performance in terms of product quality. After an organized investigation of the issue, it has been found that several of the senior staff of the product quality assurance group have left the company to start their own business. Because of this unexpected change, the company has suffered an unprecedented "brain-drain", resulting in the sudden decline of quality level of certain mainstream products.

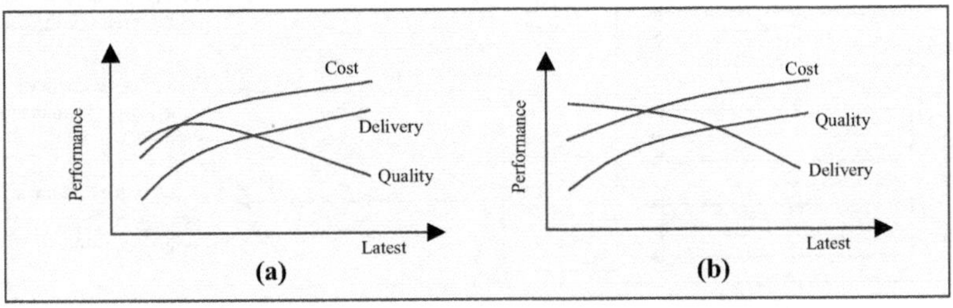

Fig. 3. Performance of a company based on past records

Because of the situation, Company A has been suggested to adopt some best practices related to quality assurance. In this case, the Total Quality Management (TQM) practice has been adopted and some necessary tools have also been acquired in order to implement such practice in the company. At this stage, it is still difficult to tell if the company could significantly reverse the downturn performance in terms of product quality. However, because of the signal generated from the NOLAPS, the problem of a business partner has been alerted and quick decision has been made with supporting assessment report, thus avoiding the loss of a trusted business partner, which can in turn weaken the overall performance of the VEN.

This case example indicates that the introduction of the neural network module to the OLAP module is able to significantly upgrade the decision support functionality. However, the results obtained, so far, is by no means perfect although it demonstrates that the suggested NOLAPS is basically viable and therefore it is justifiable to have further investigation along this line of research.

3 Neural Fuzzy Model

Fuzzy logic is a superset of conventional (Boolean) logic that has been extended to handle the concept of partial truth. According to Zadehi [29,30,31], rather than regarding fuzzy theory as a single theory, people should regard the process of "fuzzification" as a methodology to generalize any specific theory from a crisp (discrete) to a continuous (fuzzy) form. In particular, fuzzy logic has been deployed to replace the role of mathematical model with another that is built from a number of rules with fuzzy variables such as output temperature and fuzzy terms such as relatively high and reasonably low [2,3,12,22].

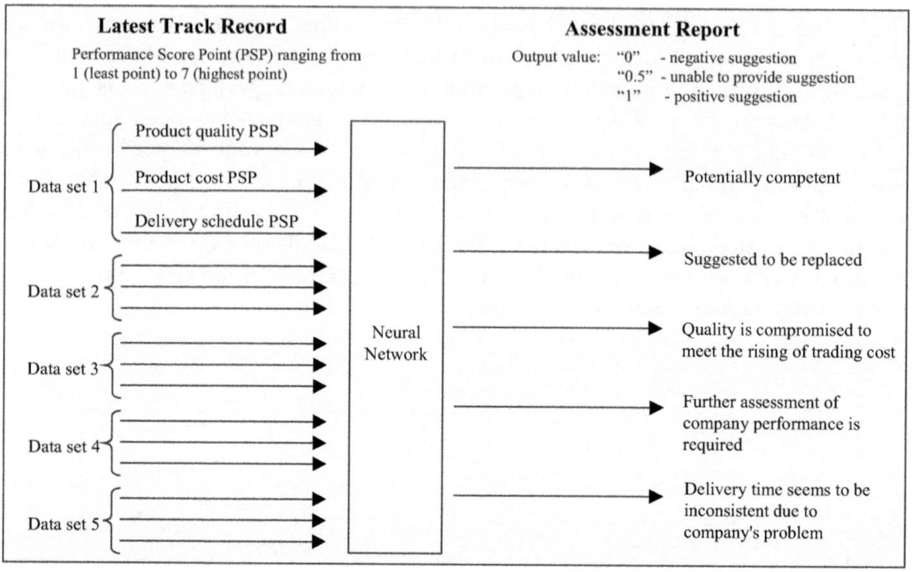

Fig. 4. Mapping of input and output nodes of Neural Network (NN) module of the NOLAPS

In an environment that there are multiple input and output parameters interacting with each others, many authors suggest that a fuzzy logic inference architecture can be employed to deal with the complex control issues [4,11,12,17,21,28]. Driankov *et al* [5] suggests the computational structure of a Fuzzy Knowledge base Controller (FKBC) to handle parameter-based control, which encompasses five computational steps including input scaling (normalization), fuzzification of inputs, inference or rule firing, defuzzification of outputs and output denormalization. It is not the intention of this paper to justify any detailed analysis, methodologies and techniques behind the fuzzy inference architectures covered in the above articles. However, most of the techniques described are very useful in understanding the design and operations of the fuzzy inference architecture that the development of a fuzzy expert system can be based on.

Whilst most of the fuzzy rules of these techniques are being set based on experience, past history and theoretical background, it is obvious that more can be done regarding the formulation of an algorithmic solution and the structure finding from existing data, which may lead to any generation of rules. In this respect, neural network has the ability to learn the relationship among input and output data sets through a training process, thus is able to "induce" output data if a new set of input data is made available. Although it can be said that neural network cannot do anything that cannot be done using traditional computing techniques, but they can handle some tasks that would otherwise be very difficult. For example, the rules generated by the neural network training process are more acceptable than those decided by expert knowledge that may not be independent and fair [31,32].

While fuzzy logic systems allow the use of linguistic terms representing data sets in the reasoning process, neural network is able to discover connections between data sets by simply having sample data represented to its input and output layers [8]. Neural network can be regarded as processing device, and it usually has some sort of "training" rule whereby the weights of connections are adjusted on the basis of presented patterns.

In order to enhance machine intelligence, an integrated neural-fuzzy model, which aims to make use of benefits generated by the synergy of neural network and fuzzy logic intelligence techniques. The main feature of the neural-fuzzy model is that it uses the trained neural network to generate 'If-Then' rules, which are then fuzzified prior to undergoing a fuzzy inferencing process, resulting in the generation of new process parameters for enhancing machine intelligence. The neural-fuzzy system described in this paper adopts inductive learning through the network, indicating that the system induces the information in its knowledge base by example. That induced information is then fuzzified and fed to the fuzzy logic system prior to fuzzy reasoning process. The significance of the neural-fuzzy model is that it attempts to formulate a technique for eliminating the knowledge acquisition bottleneck, thus enhancing the intelligent monitoring of a parameter-based control situation.

4 Demonstration of Neural-Fuzzy Model

The responsibility of the neural network model element is to provide the desire change of parameters based on what the network has been trained on. Intrinsically, a sufficient amount of data sample is a key factor in order to obtain accurate feedback from the trained network. In actual situations, recommended action about the required change of parameters to cope with the dimensional inconsistency is essential. In view of this situation, neural network can be regarded as a better option, if the dimensional values are mapped to the nodes of the input layer and heat transfer parameters are mapped to the output layer nodes, thus resulting in a control model that is the reverse of the heat transfer model. In the light of the fact that in an actual thermal system design, the required overall heat transfer is first determined from the system analysis. Then the rib geometry is chosen according to the nearest overall heat transfer performance determined from experimental investigations. Very often the difference between the designed overall heat transfer and the experimental performance data can be quite significant.

With a neural network, the correlation between the deviations of heat transfer parameters in response to the deviations of the occurring dimensional values can be trained based on a wide spectrum of actual sample data. As neural network is intended to learn relationships between data sets by simply having sample data represented to their input and output layers [9], the training of a network with input and output layers mapped to dimensional deviation values and heat transfer deviation values respectively with the purpose to develop the correlation between these two groups of data will not contradict the basic principle of neural network.

With a trained network available, it is possible that recommended action about the change of parameters can be obtained with the purpose to optimize the design of rib geometry, should that occur at a later stage. Therefore, in the training process of the neural network, the nodes of the input layer of the neural network represent the deviation of the dimensional values and those of the output layer represent the deviation of the heat transfer parameters.

If there is dimensional inconsistency on the heat transfer model, the values at the nodes from the neural network (representing the parameter deviations) may provide some hints for possible dimensional correction. With the availability of this information, a fuzzy logic approach can then be employed to provide a modified set of recommended parameter change based on the original output values from the neural network. The motive for using fuzzy logic reasoning in this model is to take advantage of its ability to deal with imprecision terms which fit ideally in the parameter-based control situations where terms such as "rib spacing could be increased slightly" are used. Furthermore, the vagueness and uncertainty of human expressions is well modeled in the fuzzy sets, and a pseudo-verbal representation, similar to an expert's formulation, can be achieved.

During fuzzy reasoning process, the input and output values of the neural network are generally fuzzified into linguistic terms so that fuzzy rules can be developed. The method of obtaining the corresponding output membership values from the "fired" fuzzy rule is called fuzzy logic reasoning. Many reasoning strategies have been developed, including Sup-bounded-product [18], Super-drastic-product [17,18,19], Sup-min [13], and Sup-product [11]. Since it is not the intention of this paper to present a review of fuzzy logic reasoning strategies, the mentioned reasoning strategies are not further explained in this paper. In this paper, the Sup-product strategy is adopted due to its simplicity and relatively less calculation time.

After the fuzzification process with the generation of fuzzy rules, it is necessary to have a defuzzification process. The defuzzification process is a process of mapping from a space of inferred fuzzy control results to a space of non-fuzzy control action in a crisp form. In fact, a defuzzification strategy is aimed at generating a non-fuzzy control action that best represents the possibility distribution of the inferred fuzzy control results. The Mean of Maximum (MOM) and Centre of Area (COA) are two common defuzzification methods in fuzzy control systems, and the latter method is selected in this neural-fuzzy model to defuzzify the reasoned fuzzy output (the parameters value). Proposed parameter change is carried out and the dimensional outcome, resulting from the change is checked against the expected dimension.

5 Cross Platform Intelligent Information Infrastructure

The real challenge for the implementation of information infrastructure in enterprises is to achieve *seamless* data interchange in the sense that data from various sources with dissimilar formats can integrate directly with the database system of any individual companies in a fully automatic way, i.e. without any human intervention. In this respect, data conversion and mapping of data-fields to match the formats of various enterprises are the pre-requisites of meeting this data integration criterion.

Since 1996, the new eXtensible Markup Language (XML) has been developed for overcoming the limitations of HTML, and has received worldwide acceptance in terms of data interchange in Internet-based information systems. In XML files, the content of data and the presentation of data are separated. Thus, various formats of data specified by XML are able to be transferred through the Internet and used on different platforms [16]. In brief, XML is a subset of Standard Generalized Markup Language (SGML) which is an international standard (ISO 8879) for defining and representing documents in an application-independent form [25].

In order to realize the potential of information infrastructure as an effective data processing gateway, data mining technology needs to be incorporated. In brief, data mining is seen as a technological approach that provides sophisticated analysis based on a set of complex data, enabling the management of different data formats in various database systems. The shared data access interface of data mining tools will enable exchange of data as well as results among various computer systems. The typical example of data mining tool is On-Line Analytical Processing (OLAP) that provides a service for accessing, viewing and analyzing on large volumes of data with high flexibility and performance. The essential characteristic of OLAP is that it performs a numerical and statistical analysis of data which are organized in the form of multi-dimensions (as opposed to the two-dimensional format of traditional relational data tables).

Cross platform intelligent information infrastructure allows users to access and retrieve data of any database format from distributed sources, followed by the automate assimilation of the data in the user's own database system. The information infrastructure also allows collaboration and data sharing and provides users with the ability to generate business transactions, reports and dynamic updates through the common interface. Based on the robust XML data interchange open standard, the information infrastructure can integrate virtually with all existing systems and database in real time and distrbute data automatically to the required destinations. In order to enable users to access timely information efficiently, an OLAP tool with intelligent functionality such as rule-based reasoning is required.

Fig. 5. shows the configuration of the information infrastructure which is a client-server system. The clients are users' (business partners, customers, employees) desktops who connect to the common interface using Internet browers. Thus, users can quickly access and retrieve all information and data needed through the information infrastructure, i.e. the Internet server with the XML and the intelligent OLAP facilities.

Currently in most client-server systems, the data requested from the client side is done by human users, such as requesting news on the Internet server or filling in the order forms. When the request originates from a database or a program on the client side, i.e. a database on the client side connects to the database on the server side, a standard data interchange format is needed. At this stage, a common data format such as the XML seems to be the right choice to meet the requirement. In order to create the XML data file, a XML translator is built in the information infrastructure so that different database systems can achieve data interchange among themselves.

Other than the data interchange through the information infrastructure, users may intend to perform data analysis such as statistical analysis. This task is normally referred as data mining. OLAP technology provides high flexibility and performance for the data mining and uses multidimensional data model for users to formulate complex queries and to arrange the query result on a report. The multidimensional data model organizes data into a hierarchy that represents levels of details on the data. Thus the data structure of multidimensional data model is clearer than the relational database which is the structure of most current databases. In the information infrastructure, a multidimensional database is implemented for the OLAP application.

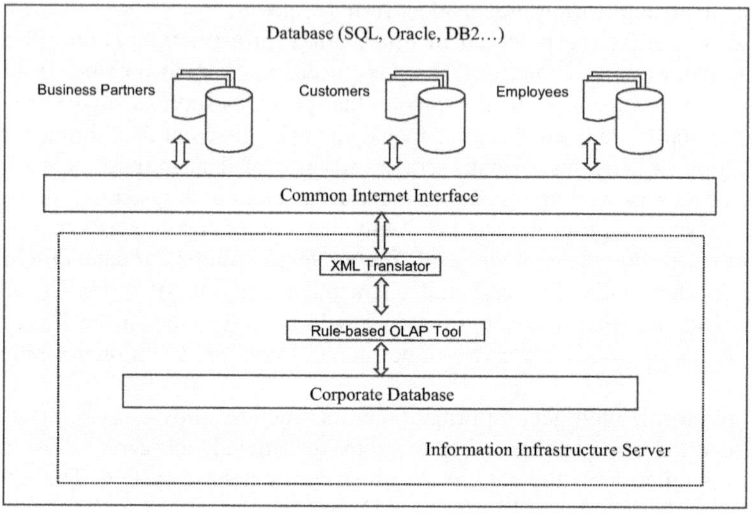

Fig. 5. Outline of Cross Platform Intelligent Informaton Infrastructure

Query on the large volume of data from the database by the OLAP tool is a complicated analysis and may not aggregate enough data. In the cross platform intelligent information infrastructure, the rule-based reasoning facility is incorporated to support the OLAP tool. The tasks of the rule-based reasoning are (i) to assist the setup of a corporate database, (ii) to help the extraction of useful information from the database, and (iii) to provide suggestions of efficient OLAP query methods. Thus, rule-based OLAP provides users with the access to the knowledge hidden in the databases, thereby discovering the trends and relationships in the data in order to make predictions using that information.

The rule-based reasoning facility includes knowledge base and inference engine. It retrieves knowledge (stored in the knowledge base) and uses the forward and backward inference logic to produce the actions (provide suggestions). In other words, it supports decision-makers (users) to get as much useful information as they can from their valuable knowledge base. In this respect, the rule-based reasoning can be deployed to enhance the intelligence level of the OLAP application.

The knowledge base of the rule-based reasoning facility consists of declarative knowledge and procedural knowledge [10]. The declarative knowledge contains what is currently known about the data interchange and process in the information infrastructure. The procedural knowledge is developed from repeated data interchange and process experience, and usually relates specific conditions with their likely outcomes, consequences, or indicated actions.

6 Conclusion

After a period of increasing professionalism in almost all areas of research and technology, the new era is to embrace the interaction of different approaches, to form multi-functional disciplines. The information infrastructures introduced above demonstrates the benefits of using combinations of technologies to form an integrated system, which capitalize on the merits and at the same time offset the pitfalls of the involved technologies. Further research on the infrastructure framework for knowledge discovery particularly the seamless integration of the technologies is needed in order to ensure and sustain the benefits of the infrastructures.

References

[1] Berson, A. and Smith, S.J. (1997), *Data Warehousing, Data Mining, & OLAP*, McGraw-Hill, New York.

[2] Buchanan, B. and Shortliffe, E.H. (1989). *Rule-based expert systems: the MYCIN experiments of the Stanford Heuristic Programming Project.* Addison-Wesley series in artificial intelligence. Reading, Mass.: Addison-Wesley.

[3] Burns R. (1997). Intelligent manufacturing. *Aircraft Engineering and Aerospace Technology*; 69(5): 440-446.

[4] Chiueh T. (1992) Optimization of Fuzzy Logic Inference Architecture, Computer, May; 67-71.

[5] Driankov D, Hellendoorn H, Reinfrank M. (1996) An Introduction to Fuzzy Control. Springer; 149-163.

[6] Erik, T., George, S. and Dick, C. (1999), *Microsoft OLAP Solutions*, John Wiley & Sons, New York.

[7] Haykin, S. (1994), *Neural networks, a comprehensive foundation*, Macmillan College Publishing Company.

[8] Haykin S. (1999). *Neural network, a comprehensive foundation.* 2nd edition. Upper Saddle River, N.J.: Prentice Hall.

[9] Herrmann, C.S. (1995), A hybrid fuzzy-neural expert system for diagnosis, *Proceedings of International Joint Conference on Artificial Intelligence*, pp. 494-500.

[10] Inference Corporation, 1992, *ART-IM 2.5 Reference Manuals* (Los Angeles).

[11] Kaufman A. (1975). *Introduction to theory of fuzzy subsets.* New York, Academic.

[12] Leung KS, and Lam W. (1988). Fuzzy Concepts in Expert Systems. *IEEE,* September, 43-56.

[13] Mamdani EH. (1974) Applications of fuzzy algorithms for control of a simple dynamic plant. *Proceedings of IEEE, 1974*; 121: 1585-1588.

[14] Merwe, J.v.d. and Solms, S.H.v. (1998), Electronic commerce with secure intelligent trade agents, *Computers & Security*, Vol. 17, pp. 435-446.

[15] Michael, L.G. and Bel, G.R. (1999), Data mining - a powerful information creating tool, *OCLC Systems & Services*, Vol.15, No. 2, pp81-90.

[16] Microsoft Corporation, 2000, *Microsoft BizTalk jumpstart kit*, Feb.
[17] Mizumoto M, Fukami S, Tanaka K. (1979). Some Methods of Fuzzy Reasoning. Advances in Fuzzy Set Theory and Applications. North-Holland, Amsterdam; 117-136.
[18] Mizumoto M. (1981) Note on the arithmetic rule by Zedeh for fuzzy reasoning methods. *Cyben System*; 12: 247-306.
[19] Mizumoto M. (1990) Fuzzy controls by product-sum-gravity method. In Advancement of fuzzy theory and systems in China and Japan, *Proceeding of Sino-Japan Joint Meeting on Fuzzy Sets and Systems Oct. 15-18*, Beijing, China, International Academic; 1-4.
[20] New Era of Networks, Inc., (2000), Powering the new economy.
[21] Nguyen HT. (2000) *A first course in fuzzy logic*. 2nd edition. Boca Raton, Fla: Chapman & Hall/CRC.
[22] Orchard A. (1994) *FuzzyCLIPS Version 6.02A User's Guide*. National Research Council. Canada.
[23] Peterson, T. (2000) *Microsoft OLAP unleashed*, 2nd edition, Sams Pubishing, Indianapolis.
[24] Robert S.C., Joseph A.V. and David B. (1999), *Microsoft Data Warehousing*, John Wiley & Sons.
[25] Salminen, A., Lyytikäinen, V. and Tiitinen, P., (2000), Putting documents into their work context in document analysis, *Information Processing & Management* 36, Issue 4, July 1, 623-641.
[26] Tandem Computers Incorporated (1997), Object Relational Data Mining Technology for a Competitive Advantage (White Paper), *Decision Support Solutions*.
[27] Thomsen, E. (1999), *Microsoft OLAP solutions*, J. Wiley, New York.
[28] Whalen T, Schott B. (1983) Issues in Fuzzy Production Systems. *International Journal of Man-Machine Studies*; 19:57.
[29] Zadeh, F. (1996) Fuzzy sets, fuzzy logic, and fuzzy systems: selected papers. Singapore, World Scientific.
[30] Zadeh, F. (1965) Fuzzy sets. *Information and Control* 8:338-53.
[31] Zadeh, F. (1993) The role of fuzzy logic and soft computing in the conception and design of intelligence systems. Klement, Slany.
[32] Zahedi F. (1991) An introduction to neural network and a comparison with artificial intelligence and expert systems. *Interfaces*; 21(2): 25-28.
[33] Zahedi F. (1993) Intelligent Systems for Business: Expert Systems with Neural Network, Wadsworth, Belmont, CA.

MML Classification of Music Genres

Adrian C. Bickerstaffe and Enes Makalic *

School of Computer Science and Software Engineering
Monash University (Clayton Campus)
Clayton, Victoria 3800, Australia

Abstract. Inference of musical genre, whilst seemingly innate to the human mind, remains a challenging task for the machine learning community. Online music retrieval and automatic music generation are just two of many interesting applications that could benefit from such research. This paper applies four different classification methods to the task of distinguishing between rock and classical music styles. Each method uses the Minimum Message Length (MML) principle of statistical inference. The first, an unsupervised learning tool called Snob, performed very poorly. Three supervised classification methods, namely decision trees, decision graphs and neural networks, performed significantly better. The defining attributes of the two musical genres were found to be pitch mean and standard deviation, duration mean and standard deviation, along with counts of distinct pitches and rhythms per piece. Future work includes testing more attributes for significance, extending the classification to include more genres (for example, jazz, blues etcetera) and using probabilistic (rather than absolute) genre class assignment. Our research shows that the distribution of note pitch and duration can indeed distinguish between significantly different types of music.

1 Introduction

The task of successfully identifying music genres, while trivial for humans, is difficult to achieve using machine learning techniques. However, applications of automated music genre recognition are numerous and significant. For example, a large database of music from unknown sources could be arranged to facilitate fast searching and retrieval. To illustrate, retrieval of different pieces from the same genre would become easily possible. Successful models of musical genres would also be of great interest to musicologists. Discovering the attributes that define a genre would provide insight to musicians and assist in automatically generating pieces of a particular style.

Research toward music classification is reasonably well-established. Soltau et al. developed a music style classifier using a three-layer feedforward neural network and temporal modelling [1]. The classifier was trained using raw audio samples from four genres of music: rock, pop, techno and classical. Cilibrasi et

* Author list order determined stochastically.

T.D. Gedeon and L.C.C. Fung (Eds.): AI 2003, LNAI 2903, pp. 1063–1071, 2003.

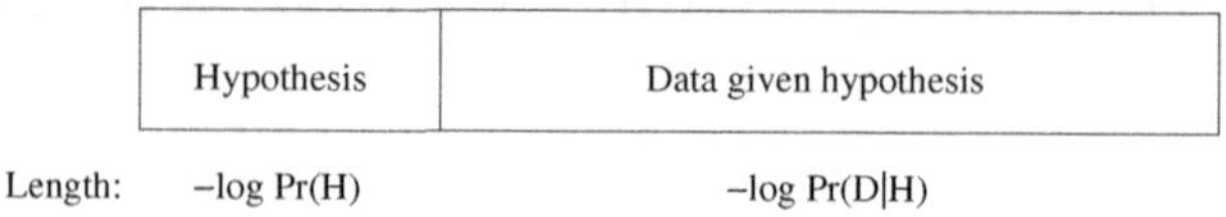

Fig. 1. Two-part message

al. developed a general similar measure to build phylogeny trees to cluster music [2]. Using a corpus of 118 pieces, their method was successful at distinguishing between classical, jazz and rock music.

This paper examines the task of distinguishing between two genres which are obviously different to the human ear - rock music and classical music. Melody, rather than performance styles of the pieces, was examined. We compared four Minimum Message Length (MML) based approaches to classification: mixture modelling, decision trees, decision graphs and neural networks. A wide variety of attributes were tested for significance and the set of attributes reduced to only those which appear to contribute to defining the genres.

An overview of MML is given in Section 2 followed by a description of the four classification tools used (Section 3). Results are discussed in Section 4 whilst limitations and future work are outlined in Section 5.

2 Overview of MML

Given a set of data, we are often interested in inferring a model responsible for generating that data. In the context of this paper, we have musical pieces and wish to infer their genre. MML [3, 4] is a Bayesian framework for statistical inference and provides an objective function that may be used to estimate the goodness of an inferred model.

Consider a situation in which a sender wishes to transmit some data to a receiver over a noiseless transmission channel. The message is transmitted in two parts (see Fig. 1):

1. an encoding of the model θ, and
2. an encoding of the data given the model, $x|\theta$.

Clearly, there may be many alternative models of a single dataset. Using MML, the optimal model is selected to be that which minimises the total message length (for example, in bits). In this way, MML is a quantitative form of Occam's Razor.

The most commonly used MML approximation is MML87 [4]. The approximation states that the total message length for a model Θ with parameters $\boldsymbol{\theta}$ is:

$$\text{msgLen}\,(\Theta) = -\log\left(\frac{h(\boldsymbol{\theta})}{\kappa_n^{n/2}\sqrt{F(\boldsymbol{\theta})}}\right) - \log f(x|\boldsymbol{\theta}) + \frac{n}{2} \quad (1)$$

where $h(\boldsymbol{\theta})$ is the prior probability, $f(x|\boldsymbol{\theta})$ is the likelihood function, n is the number of parameters, $\kappa_n^{n/2}$ is a dimension constant[1] and $F(\boldsymbol{\theta})$ is the determinant of the expected Fisher information matrix, whose entries (i,j) are:

$$\sum_{x \in \mathcal{X}} f(x|\boldsymbol{\theta}) \frac{\partial^2}{\partial \theta_i \partial \theta_j} \left(-\log f(x|\boldsymbol{\theta})\right) \ . \tag{2}$$

The expectation is taken over all data x in the dataspace $\mathcal{X}$. The optimal MML87 model is that which minimises the total message length (1).

3 MML Classifiers

Four different classification methods were used in our experiments. A brief summary of each tool is given below.

3.1 Mixture Modelling

Using mixture models, a single statistical distribution is modelled by a mixture of other distributions. That is, given S things and a set of attributes for each thing, a mixture model describes the things as originating from a mixture of T classes. Snob [3, 5] is a mixture modelling tool to perform unsupervised classification. Snob allows attributes from Gaussian, discrete multi-state, Poisson and Von Mises distributions. Some applications of Snob include:

- Circular clustering of protein dihedral angles [6]
- The classification of depression by numerical taxonomy [7]
- Perceptions of family functioning and cancer [8]

3.2 Decision Trees and Decision Graphs

A decision tree [9] is an example of a supervised classification method. Given a set of things, each comprising independent attributes, decision trees can be used to infer a dependent attribute (for example, the class to which a thing belongs). The attributes may be discrete or continuous. For example, the independent attributes may be gender, height and weight. The dependent attribute could then be one which poses a question such as 'plays sport?'. Referring to Fig. 2, the chance of a male person playing sport is $\frac{50}{75}$. A female person of height 170cm or more has a probability of $\frac{90}{102}$ of playing sport.

The leaves of a decision tree represent the probability distribution of a dependent attribute. Each fork consists of the independent attribute on which to split and, for continuous attributes, a cut point. Decision graphs [10, 11] are a generalisation of decision trees allowing multi-way joins in addition to splits (see Fig. 2). MML provides a method of growing decision trees/graphs that generalise well.

[1] Also called lattice constants. The first two are: $\kappa_1 = \frac{1}{12}$, $\kappa_2 = \frac{5}{36\sqrt{3}}$.

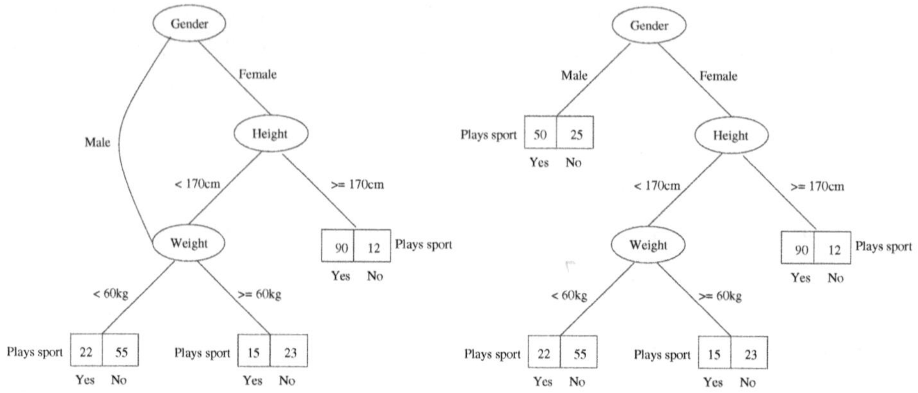

Fig. 2. An example decision graph (left) and decision tree (right)

3.3 Neural Networks

Neural networks, like decision trees/graphs, are often used for supervised classification problems. A neural network consists of many simple processing elements referred to as neurons. Neurons are grouped into logical groups or layers based on their functionality. A neural network comprises an input layer, zero or more hidden layers and one output layer.

Perhaps the most popular type of neural network in use today is the Multilayer Perceptron (MLP). A MLP is a feedforward, fully-connected neural network where each neuron in layer l is connected to each neuron in layer $l+1$. For the purposes of this paper, we are only concerned with single hidden layer neural networks (see Fig. 3). It has been shown that such networks can model any continuous function to arbitrary precision provided enough hidden neurons exist [12].

A neural network must be trained before it may be used for classification. A large number of training algorithms exist for neural networks of which second-order methods (for example, Levenberg-Marquardt [13]) are the most popular. The success of the training process largely depends on the chosen neural network architecture. We use an MML87 based method for inferring the optimal network architecture [14]. Previously, neural networks have been used in computer music research [1], but the experiments of Section 4 are the first to use MML based neural networks.

4 Results and Discussion

4.1 Attributes

Prior to classification experiments, a variety of musical attributes were tested for significance. Table 1 describes these attributes. Different sets of attributes were used in conjunction with decision trees to examine the influence of each attribute

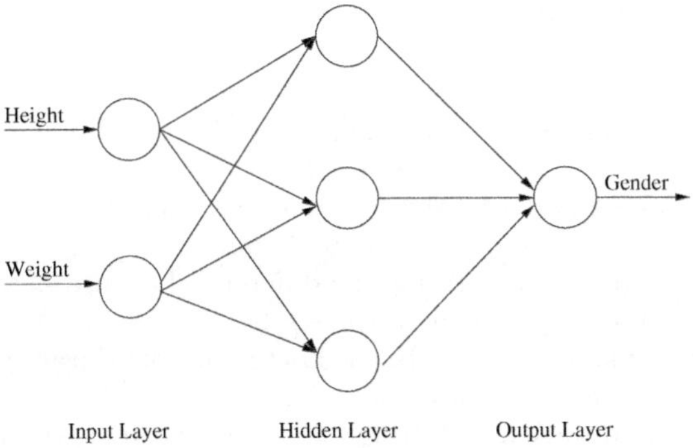

Fig. 3. An example single hidden layer neural network

on genre detection. Second order Markov models of pitch and duration contour were also implemented. Dissonance, syncopation and melodic direction stability were originally modelled as quantized five state multinomial distributions. Poor results using this approach prompted us to remodel these attributes as a single averaged value per piece.

Clearly, there are a large number of attribute combinations possible. Subsequently, we systematically eliminated all but the most significant attributes. These are:

- Pitch - mean and standard deviation
- Duration - mean and standard deviation
- Distinct pitch counts
- Distinct rhythm counts

4.2 Corpus

We chose 50 rock songs and 50 classical pieces for classification. All songs were encoded using the Musical Instrument Digital Interface (MIDI) format. Each piece was carefully chosen so as to achieve a true representation of the genres. That is, 'classic' rock and prominent classical pieces were selected. A complete list of all songs and their composers can be accessed on the web at `http://www.csse.monash.edu.au/~music-nnets/index.html`.

The corpus was divided into two sub-corpora - a training set and a validation set. The training set, consisting of 50 songs, was formed from 25 randomly selected songs from each genre. The remaining 50 songs were used for validation.

4.3 Mixture Modelling

Snob was the first classification method trialled. The classes found by Snob were examined to see whether the music genre could be inferred from knowledge of the class membership. Snob showed poor performance for *all* sets of test attributes. In each case, no indication of genre dependent classification was shown.

Table 1. Classification attributes

Attribute name	Attribute description
Pitch - mean and standard deviation	Gaussian offsets from middle C in semitones.
Duration - mean and standard deviation	Absolute note duration values taken as real numbers.
Pitch interval - mean and standard deviation	Semitone difference between note pitch $(p_i - p_{i-1})$.
Duration interval - mean and standard deviation	Difference between note duration $(d_i - d_{i-1})$.
Contour - pitch and duration	A trinomial distribution of whether pitch p_i is greater than, equal to or less than pitch p_{i-1}.
Tempo	Microseconds per quarter note.
Dissonance	Real value in range $[0, 1]$ representing the average dissonance.
Syncopation	Real value in range $[0, 1]$ representing the number of notes which start on the beat and have a rhythm value of a beat or more, compared to the total number of beats.
Melodic direction stability	Real value in range $[0, 1]$ representing the ratio between the number of consecutive pitch steps in the same direction and the total number of steps.
Note count	Modelled with a Poisson distribution where r is the average note count per part.
Distinct counts - pitch and rhythm	Modelled with a Poisson distribution where r is the average number of distinct pitches (and rhythms).
Consecutive identical pitches	Modelled with a Poisson distribution.
Big jump followed by step back count	Large semitone jump followed by a pitch. jump in the opposite direction. Modelled with a Poisson distribution.

Unexpectedly, six or more classes were common, often with very uneven class sizes. These findings indicate that unsupervised classification using Snob is not suitable for this task.

4.4 Decision Trees and Decision Graphs

With attributes of pitch mean and standard deviation, the decision tree identified two genres excellently. The leaves were 92% pure with only 4 misclassifications per genre (see Fig. 4). The misclassifications of rock occur when the average pitch falls close to middle C. Statistics show that the vast majority of rock songs center

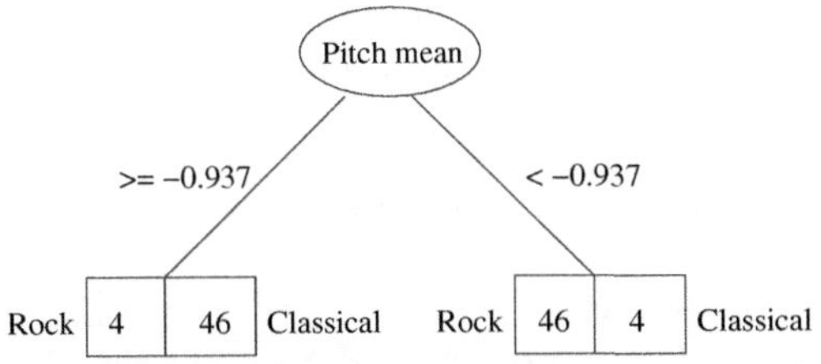

Fig. 4. Pitch mean decision tree

below middle C. Conversely, classical music appears to have the pitch mean above middle C. Classical songs with negative pitch mean were misclassified as rock songs.

Duration mean and standard deviation showed slightly worse results with a total of 12 misclassifications, six per genre. Here, classification was based upon the standard deviation rather than the mean itself. Classical pieces featured a relatively small standard deviation when compared to rock songs.

Distinct pitch and rhythm counts were again slightly worse with a total of 20 misclassifications. The results suggest that the number of distinct pitches for classical songs is greater than that for rock. This result is somewhat expected since classical pieces are generally regarded to be more complex than rock songs.

For each of the attribute pairs, the decision graph software produced decision graphs with no joins (i.e. decision trees). These decision graphs were structurally identical to those produced by the decision tree software. However, whilst the same attributes were split on, the cut points of these attributes were different. When using pitch mean and standard deviation, seven misclassifications resulted as compared to eight for decision trees. The duration mean and standard deviation attributes produced the same number of misclassifications for decision trees and decision graphs. Finally, the decision graph software performed better than the decision tree equivalent, with 18 misclassifications for distinct pitch and rhythm counts.

4.5 Neural Networks

As with decision trees/graphs, the attributes of pitch mean and standard deviation showed best results. Using MML87, we found that a two hidden neuron neural network was the optimal architecture for this problem. Only three misclassifications occurred during validation with no misclassifications in the training stage. This is equivalent to a 97% success rate.

Attributes of duration mean and standard deviation resulted in 12 misclassifications. Eight of these were classical music pieces. A single hidden neuron network was inferred to be optimal for this test.

Again, distinct pitch and rhythm attributes performed worse than pitch and duration. A total of 16 misclassifications occurred, equivalent to a success rate of 84%. This is four fewer misclassifications than the decision tree model and two fewer than the decision graph using these attributes.

5 Limitations and Future Work

Although the results presented in Section 4 are promising, several limitations remain to be addressed. Most prominently, the current classifiers are binary, and discriminate between two very different genres of music. More interesting experiments would involve at least four musical genres of varying similarity. For example, one may use classical, jazz, blues and techno pieces. Here, classical is likely to be very different to the remaining three genres, yet jazz and blues could sometimes appear similar. When classifying four or more genres, probabilistic rather than absolute class assignment is favourable. Genres may overlap, and the class assignment should reflect this.

A comparison with other non-MML classifiers, such as C5 [15], would be of interest. Furthermore, there are many more possible attributes to be examined for significance. The size of the attribute set described in Section 4.1 was simply limited by time constraints. Obviously, the number of combinations of attributes increases exponentially with the number of attributes modelled. Finally, a larger dataset is always desirable for such problems.

6 Conclusion

This paper has compared a variety of musical attributes and used a subset of these attributes to classify songs from two different genres. Unsupervised classification using mixture models was shown to be unsuited to the task at hand. Conversely, decision trees and graphs performed well, at best only misclassifying eight and seven pieces respectively. MML based neural networks performed better still. Using pitch mean and standard deviation, the inferred neural network exhibited a 97% success rate. Interestingly, the performance of the three attribute pairs tested ranked equally for decision trees, decision graphs and neural networks. Absolute pitch information was most influential, followed by duration information, and finally distinct pitch and rhythm counts. Various combinations of the aforementioned attributes did not improve performance. Our findings are in keeping with research showing that listeners are particularly sensitive to pitch distribution and frequency information within cognitive processes [16]. Decision graphs were seemingly of no advantage to decision trees for this particular application.

Acknowledgements

We thank Dr. Peter Tischer for his useful comments on drafts of this paper.

References

[1] Soltau, H., Schultz, T., Westphal M., Waibel, A.: Recognition Of Music Types. Proc. of the 1998 IEEE International Conference on Acoustics, Speech and Signal Processing (ICASSP) **2** (1998) 1137–1140 1063, 1066

[2] Cilibrasi, R., Vitanyi, P., de Wolf, R.: Algorithmic Clustering of Music. Submitted to the 2004 Special Issue of the IEEE Information Theory Transactions on "Problems on Sequences: Information Theory and Computer Science Interface" (2003) 1064

[3] Wallace, C. S. and Boulton, D. M.: An Information Measure for Classification. Computer Journal **11(2)** (1968) 195–209 1064, 1065

[4] Wallace, C. S. and Freeman, P. R.: Estimation and inference by compact encoding (with discussion). Journal of the Royal Statistical Society series B **49** (1987) 240–265 1064

[5] Wallace, C. S. and Dowe, D. L.: MML clustering of multi-state, Poisson, von Mises circular and Gaussian distributions. Statistics and Computing **10(1)** (2000) 73–83 1065

[6] Dowe, D. L., Allison, L., Dix, T. I., Hunter, L., Wallace, C. S., Edgoose, T.: Circular clustering of protein dihedral angles by Minimum Message Length. Proc. 1st Pacific Symposium on Biocomputing (PSB-1) (1996) 242–255 1065

[7] Pilowsky, I., Levine, S., Boulton, D. M. The classification of depression by numerical taxonomy. British Journal of Psychiatry **115** (1969) 937–945 1065

[8] Kissane, D. W., Bloch, S., Burns, W. I., Patrick, J. D., Wallace, C. S., McKenzie, D. P.: Perceptions of family functioning and cancer. Psycho-oncology **3** (1994) 259–269 1065

[9] Wallace, C. S., Patrick, J. D.: Coding decision trees. Machine Learning **11** (1993) 7 22 1065

[10] Tan, Peter J., Dowe, David L.: MML Inference of Decision Graphs with Multi-way Joins. Australian Joint Conference on Artificial Intelligence (2002) 131–142 1065

[11] Oliver, J. J.: Decision Graphs - An Extension of Decision Trees. Proc. of the Fourth International Workshop on Artificial Intelligence and Statistics (1993) 343–350 1065

[12] Hornik, K., Stinchcombe, M. and White, H.: Multilayer feedforward networks are universal approximators. Neural Networks **2** (1989) 359–366 1066

[13] Hagan, M. T. and Menhaj, M. B.: Training feedforward networks with the Marquardt algorithm. IEEE Transactions on Neural Networks **5(6)** (1994) 989–993 1066

[14] Makalic, E., Lloyd A., Dowe, David L.: MML Inference of Single Layer Neural Networks. Proc. of the Third IASTED International Conference on Artificial Intelligence and Applications (2003) 1066

[15] Quinlan, J. R.: C4.5: Programs for machine learning. Morgan Kaufmann (1993) 1070

[16] Saffran, J. R., Johnson, E. K., Aslin, R. N., Newport, E. L.: Statistical Learning of Tone Sequences by Human Infants and Adults. Cognition **70** (1999) 27–52 1070

Author Index

Lecture Notes in Artificial Intelligence (LNAI)

Vol. 2700: M.T. Pazienza (Ed.), Information Extraction in the Web Era. XIII, 163 pages. 2003.

Vol. 2702: P. Brusilovsky, A. Corbett, F. de Rosis (Eds.), User Modeling 2003. Proceedings, 2003. XIV, 436 pages. 2003.

Vol. 2703: O.R. Zaïane, J. Srivastava, M. Spiliopoulou, B. Masand (Eds.), WEBKDD 2002 – Mining Web Data for Discovering Usage Patterns and Profiles. Proceedings, 2003. IX, 181 pages. 2003.

Vol. 2705: S. Renals, G. Grefenstette (Eds.), Text- and Speech-Triggered Information Access. Proceedings, 2000. VII, 197 pages. 2003.

Vol. 2711: T.D. Nielsen, N.L. Zhang (Eds.), Symbolic and Quantitative Approaches to Reasoning with Uncertainty. Proceedings, 2003. XII, 608 pages. 2003.

Vol. 2715: T. Bilgiç, B. De Baets, O. Kaynak (Eds.), Fuzzy Sets and Systems – IFSA 2003. Proceedings, 2003. XV, 735 pages. 2003.

Vol. 2718: P. W. H. Chung, C. Hinde, M. Ali (Eds.), Developments in Applied Artificial Intelligence. Proceedings, 2003. XIV, 817 pages. 2003.

Vol. 2721: N.J. Mamede, J. Baptista, I. Trancoso, M. das Graças Volpe Nunes (Eds.), Computational Processing of the Portuguese Language. Proceedings, 2003. XIV, 268 pages. 2003.

Vol. 2734: P. Perner, A. Rosenfeld (Eds.), Machine Learning and Data Mining in Pattern Recognition. Proceedings, 2003. XII, 440 pages. 2003.

Vol. 2741: F. Baader (Ed.), Automated Deduction – CADE-19. Proceedings, 2003. XII, 503 pages. 2003.

Vol. 2744: V. Mařík, D. McFarlane, P. Valckenaers (Eds.), Holonic and Multi-Agent Systems for Manufacturing. Proceedings, 2003. XI, 322 pages. 2003.

Vol. 2746: A. de Moor, W. Lex, B. Ganter (Eds.), Conceptual Structures for Knowledge Creation and Communication. Proceedings, 2003. XI, 405 pages. 2003.

Vol. 2752: G.A. Kaminka, P.U. Lima, R. Rojas (Eds.), RoboCup 2002: Robot Soccer World Cup VI. XVI, 498 pages. 2003.

Vol. 2773: V. Palade, R.J. Howlett, L. Jain (Eds.), Knowledge-Based Intelligent Information and Engineering Systems. Proceedings, Part I, 2003. LI, 1473 pages. 2003.

Vol. 2774: V. Palade, R.J. Howlett, L. Jain (Eds.), Knowledge-Based Intelligent Information and Engineering Systems. Proceedings, Part II, 2003. LI, 1443 pages. 2003.

Vol. 2777: B. Schölkopf, M.K. Warmuth (Eds.), Learning Theory and Kernel Machines. Proceedings, 2003. XIV, 746 pages. 2003.

Vol. 2780: M. Dojat, E. Keravnou, P. Barahona (Eds.), Artificial Intelligence in Medicine. Proceedings, 2003. XIII, 388 pages. 2003.

Vol. 2782: M. Klusch, A. Omicini, S. Ossowski, H. Laamanen (Eds.), Cooperative Information Agents VII. Proceedings, 2003. XI, 345 pages. 2003.

Vol. 2792: T. Rist, R. Aylett, D. Ballin, J. Rickel (Eds.), Intelligent Virtual Agents. Proceedings, 2003. XV, 364 pages. 2003.

Vol. 2796: M. Cialdea Mayer, F. Pirri (Eds.), Automated Reasoning with Analytic Tableaux and Related Methods. Proceedings, 2003. X, 271 pages. 2003.

Vol. 2797: O.R. Zaïane, S.J. Simoff, C. Djeraba (Eds.), Mining Multimedia and Complex Data. Proceedings, 2002. XII, 281 pages. 2003.

Vol. 2801: W. Banzhaf, T. Christaller, P. Dittrich, J.T. Kim, J. Ziegler (Eds.), Advances in Artificial Life. Proceedings, 2003. XVI, 905 pages. 2003.

Vol. 2807: V. Matoušek, P. Mautner (Eds.), Text, Speech and Dialogue. Proceedings, 2003. XIII, 426 pages. 2003.

Vol. 2821: A. Günter, R. Kruse, B. Neumann (Eds.), KI 2003: Advances in Artificial Intelligence. Proceedings, 2003. XII, 662 pages. 2003.

Vol. 2829: A. Cappelli, F. Turini (Eds.), AI*IA 2003: Advances in Artificial Intelligence. Proceedings, 2003. XIV, 552 pages. 2003.

Vol. 2831: M. Schillo, M. Klusch, J. Müller, H. Tianfield (Eds.), Multiagent System Technologies. Proceedings, 2003. X, 229 pages. 2003.

Vol. 2835: T. Horváth, A. Yamamoto (Eds.), Inductive Logic Programming. Proceedings, 2003. X, 401 pages. 2003.

Vol. 2837: N. Lavrač, D. Gamberger, H. Blockeel, L. Todorovski (Eds.), Machine Learning: ECML 2003. Proceedings, 2003. XVI. 504 pages. 2003.

Vol. 2838: N. Lavrač, D. Gamberger, L. Todorovski, H. Blockeel (Eds.), Knowledge Discovery in Databases: PKDD 2003. Proceedings, 2003. XVI. 508 pages. 2003.

Vol. 2842: R. Gavaldà, K.P. Jantke, E. Takimoto (Eds.), Algorithmic Learning Theory. Proceedings, 2003. XI, 313 pages. 2003.

Vol. 2843: G. Grieser, Y. Tanaka, A. Yamamoto (Eds.), Discovery Science. Proceedings, 2003. XII, 504 pages. 2003.

Vol. 2850: M.Y. Vardi, A. Voronkov (Eds.), Logic for Programming, Artificial Intelligence, and Reasoning. Proceedings, 2003. XIII, 437 pages. 2003.

Vol. 2854: J. Hoffmann, Utilizing Problem Structure in Planning. XIII, 251 pages. 2003.

Vol. 2871: N. Zhong, Z.W. Raś, S. Tsumoto, E. Suzuki (Eds.), Foundations of Intelligent Systems. Proceedings, 2003. XV, 697 pages. 2003.

Vol. 2873: J. Lawry, J. Shanahan, A. Ralescu (Eds.), Modelling with Words. XIII, 229 pages. 2003.

Vol. 2891: J. Lee, M. Barley (Eds.), Intelligent Agents and Multi-Agent Systems. Proceedings, 2003. X, 215 pages. 2003.

Vol. 2902: F. Moura Pires, S. Abreu (Eds.), Progress in Artificial Intelligence. Proceedings, 2003. XV, 504 pages. 2003.

Vol. 2903: T.D. Gedeon, L.C.C. Fung (Eds.), AI 2003: Advances in Artificial Intelligence. Proceedings, 2003. XVI, 1075 pages. 2003.

Vol. 2863: P. Stevens, J. Whittle, G. Booch (Eds.), «UML» 2003 – The Unified Modeling Language. Proceedings, 2003. XIV, 415 pages. 2003.

Vol. 2864: A.K. Dey, A. Schmidt, J.F. McCarthy (Eds.), UbiComp 2003: Ubiquitous Computing. Proceedings, 2003. XVII, 368 pages. 2003.

Vol. 2865: S. Pierre, M. Barbeau, E. Kranakis (Eds.), Ad-Hoc, Mobile, and Wireless Networks. Proceedings, 2003. X, 293 pages. 2003.

Vol. 2867: M. Brunner, A. Keller (Eds.), Self-Managing Distributed Systems. Proceedings, 2003. XIII, 274 pages. 2003.

Vol. 2868: P. Perner, R. Brause, H.-G. Holzhütter (Eds.), Medical Data Analysis. Proceedings, 2003. VIII, 127 pages. 2003.

Vol. 2869: A. Yazici, C. Şener (Eds.), Computer and Information Sciences – ISCIS 2003. Proceedings, 2003. XIX, 1110 pages. 2003.

Vol. 2870: D. Fensel, K. Sycara, J. Mylopoulos (Eds.), The Semantic Web - ISWC 2003. Proceedings, 2003. XV, 931 pages. 2003.

Vol. 2871: N. Zhong, Z.W. Raś, S. Tsumoto, E. Suzuki (Eds.), Foundations of Intelligent Systems. Proceedings, 2003. XV, 697 pages. 2003. (Subseries LNAI)

Vol. 2873: J. Lawry, J. Shanahan, A. Ralescu (Eds.), Modelling with Words. XIII, 229 pages. 2003. (Subseries LNAI)

Vol. 2874: C. Priami (Ed.), Global Computing. Proceedings, 2003. XIX, 255 pages. 2003.

Vol. 2875: E. Aarts, R. Collier, E. van Loenen, B. de Ruyter (Eds.), Ambient Intelligence. Proceedings, 2003. XI, 432 pages. 2003.

Vol. 2876: M. Schroeder, G. Wagner (Eds.), Rules and Rule Markup Languages for the Semantic Web. Proceedings, 2003. VII, 173 pages. 2003.

Vol. 2877: T. Böhme, G. Heyer, H. Unger (Eds.), Innovative Internet Community Systems. Proceedings, 2003. VIII, 263 pages. 2003.

Vol. 2878: R.E. Ellis, T.M. Peters (Eds.), Medical Image Computing and Computer-Assisted Intervention - MICCAI 2003. Part I. Proceedings, 2003. XXXIII, 819 pages. 2003.

Vol. 2879: R.E. Ellis, T.M. Peters (Eds.), Medical Image Computing and Computer-Assisted Intervention - MICCAI 2003. Part II. Proceedings, 2003. XXXIV, 1003 pages. 2003.

Vol. 2880: H.L. Bodlaender (Ed.), Graph-Theoretic Concepts in Computer Science. Proceedings, 2003. XI, 386 pages. 2003.

Vol. 2881: E. Horlait, T. Magedanz, R.H. Glitho (Eds.), Mobile Agents for Telecommunication Applications. Proceedings, 2003. IX, 297 pages. 2003.

Vol. 2883: J. Schaeffer, M. Müller, Y. Björnsson (Eds.), Computers and Games. Proceedings, 2002. XI, 431 pages. 2003.

Vol. 2884: E. Najm, U. Nestmann, P. Stevens (Eds.), Formal Methods for Open Object-Based Distributed Systems. Proceedings, 2003. X, 293 pages. 2003.

Vol. 2885: J.S. Dong, J. Woodcock (Eds.), Formal Methods and Software Engineering. Proceedings, 2003. XI, 683 pages. 2003.

Vol. 2886: I. Nyström, G. Sanniti di Baja, S. Svensson (Eds.), Discrete Geometry for Computer Imagery. Proceedings, 2003. XII, 556 pages. 2003.

Vol. 2887: T. Johansson (Ed.), Fast Software Encryption. Proceedings, 2003. IX, 397 pages. 2003.

Vol. 2888: R. Meersman, Zahir Tari, D.C. Schmidt et al. (Eds.), On The Move to Meaningful Internet Systems 2003: CoopIS, DOA, and ODBASE. Proceedings, 2003. XXI, 1546 pages. 2003.

Vol. 2889: Robert Meersman, Zahir Tari et al. (Eds.), On The Move to Meaningful Internet Systems 2003: OTM 2003 Workshops. Proceedings, 2003. XXI, 1096 pages. 2003.

Vol. 2891: J. Lee, M. Barley (Eds.), Intelligent Agents and Multi-Agent Systems. Proceedings, 2003. X, 215 pages. 2003. (Subseries LNAI)

Vol. 2893: J.-B. Stefani, I. Demeure, D. Hagimont (Eds.), Distributed Applications and Interoperable Systems. Proceedings, 2003. XIII, 311 pages. 2003.

Vol. 2894: C.S. Laih (Ed.), Advances in Cryptology - ASIACRYPT 2003. Proceedings, 2003. XIII, 543 pages. 2003.

Vol. 2895: A. Ohori (Ed.), Programming Languages and Systems. Proceedings, 2003. XIII, 427 pages. 2003.

Vol. 2897: O. Balet, G. Subsol, P. Torguet (Eds.), Virtual Storytelling. Proceedings, 2003. XI, 240 pages. 2003.

Vol. 2898: K.G. Paterson (Ed.), Cryptography and Coding. Proceedings, 2003. IX, 385 pages. 2003.

Vol. 2899: G. Ventre, R. Canonico (Eds.), Interactive Multimedia on Next Generation Networks. Proceedings, 2003. XIV, 420 pages. 2003.

Vol. 2901: F. Bry, N. Henze, J. Maluszyński (Eds.), Principles and Practice of Semantic Web Reasoning. Proceedings, 2003. X, 209 pages. 2003.

Vol. 2902: F. Moura Pires, S. Abreu (Eds.), Progress in Artificial Intelligence. Proceedings, 2003. XV, 504 pages. 2003. (Subseries LNAI).

Vol. 2903: T.D. Gedeon, L.C.C. Fung (Eds.), AI 2003: Advances in Artificial Intelligence. Proceedings, 2003. XVI, 1075 pages. 2003. (Subseries LNAI).

Vol. 2904: T. Johansson, S. Maitra (Eds.), Progress in Cryptology – INDOCRYPT 2003. Proceedings, 2003. XI, 431 pages. 2003.

Vol. 2905: A. Sanfeliu, J. Ruiz-Shulcloper (Eds.), Progress in Pattern Recognition, Speech and Image Analysis. Proceedings, 2003. XVII, 693 pages. 2003.

Vol. 2916: C. Palamidessi (Ed.), Logic Programming. Proceedings, 2003. XII, 520 pages. 2003.